CONSTITUTIVE MODELS FOR RUBBER X

PROCEEDINGS OF THE 10TH EUROPEAN CONFERENCE ON CONSTITUTIVE MODELS FOR RUBBER (ECCMR X), MUNICH, GERMANY, 28–31 AUGUST 2017

Constitutive Models for Rubber X

Editors

Alexander Lion & Michael Johlitz

University of the Bundeswehr Munich, Neubiberg, Germany

CRC Press
Taylor & Francis Group
Boca Raton London New York Leiden

CRC Press is an imprint of the
Taylor & Francis Group, an **informa** business

A BALKEMA BOOK

CRC Press/Balkema is an imprint of the Taylor & Francis Group, an informa business

Typeset by V Publishing Solutions Pvt Ltd., Chennai, India
Printed and bound in Great Britain by CPI Group (UK) Ltd, Croydon, CR0 4 YY

Published by: CRC Press/Balkema
Schipholweg 107C, 2316 XC Leiden, The Netherlands
e-mail: Pub.NL@taylorandfrancis.com
www.crcpress.com – www.taylorandfrancis.com

ISBN: 978-1-138-03001-5 (Hbk)
ISBN: 978-1-315-22327-8 (eBook)

Constitutive Models for Rubber X – Lion & Johlitz (Eds)
© 2017 Taylor & Francis Group, London, ISBN 978-1-138-03001-5

Table of contents

Constitutive models and their implementation in FEM

Experimental characterisation

Constitutive Models for Rubber X – Lion & Johlitz (Eds)
© 2017 Taylor & Francis Group, London, ISBN 978-1-138-03001-5

Foreword

The 10th European Conference on Constitutive Models for Rubber takes place in Munich in Germany and is organised by Michael Johlitz, Alexander Lion and the International Conferences and Courses Ltd. (ICC). At first, the organisers are very thankful to Marcia Öchsner and her team for the professional management of this conference. Next, we thank all members of the scientific committee for their assistance and cooperation. We also express our gratitude to the organisers of the 9th ECCMR in Prague for their valuable support. Since such a conference cannot be realised without sponsoring, the organisers also thank the companies Boge Rubber & Plastics, Brabender, Endurica, Michelin, Netzsch, Springer and Synopt for their massive financial support. Finally, we express our gratitude to Manfred Mahlig and Michaela Lochbihler for their assistance concerning the organisation of the conference proceedings.

Elastomers are well-known from various applications in aerospace, shipping, civil or automotive engineering. Traditional applications are tyres, sealings, tubes, bridge bearings and suspensions or engine mounts. Essential advantages of elastomers are their broad availability and cheapness in combination with their mechanical flexibility and the unlimited number of possibilities to modify their material properties by adding fillers or other substances. A fundamental drawback of these materials is their unavoidable temporal degradation behaviour: elastomers are not stable but exhibit irreversible ageing which evolves faster under increasing temperature. During the first ECCMR conferences, the constitutive representation of the stress strain behaviour of rubber under uniaxial or multiaxial loads and isothermal conditions using models of nonlinear finite elasticity or viscoelasticity was in the centre of interest. In this context, a number of phenomenological and micromechanically-based models were developed, fitted to experimental data, compared and discussed. In order to represent experimentally observed curves both concepts are comparable. In order to understand the material behaviour micromechanical models possess advantages. Also during this period, constitutive approaches to model the Mullins and the Payne effect as well as the dynamic behaviour of elastomers were developed and presented on ECCMR conferences. In combination with finite element implementations, such models can be applied to compute stress, strain or temperature distributions in elastomer parts in dependence on external loads. During the last ten or twenty years, industrial and academic researchers are increasingly interested to model, understand and simulate the fatigue properties, the crystallisation behaviour under thermomechanical loads and the irreversible ageing behaviour of elastomers under realistic thermomechanical loading histories. Other researchers spare no effort to develop and simulate elastomers whose damping and stiffness behaviour is instantaneously adjustable by external electromagnetic fields. The latter topics developed more and more during the last ECCMR conferences.

The organisers are very delighted that three distinguished researchers, namely Julie Diani from the Ecole Polytechnique in France, Iva Petrikova from the University of Liberec in the Czech Republic and Jörn Ihlemann from the Technical University of Chemnitz in Germany, accepted their invitation for plenary lectures. Besides these lectures, the ECCMR conference contains a large number of very interesting oral presentations and posters on all topics of elastomers which are relevant for academic an industrial research and development. The topics of the conference cover experimental methods and processing, filler reinforcement, electromagnetically sensitive elastomers, dynamic properties, material modelling and FEM implementation, stress softening, ageing, fatigue and durability. The organisers are pleased that this conference appeals to a wide international audience.

Michael Johlitz and
Alexander Lion,
August 2017

Constitutive Models for Rubber X – Lion & Johlitz (Eds)
© 2017 Taylor & Francis Group, London, ISBN 978-1-138-03001-5

Foreword (Volume 1)

The extraordinary stress-strain behavior of rubber has presented an opportunity for inventive engineers and a challenge for scientists since the mid-ninteenth century, and continues to do so today. Major branches of theory, such as the statistical theory of rubber elasticity and finite strain elasticity theory, have been spawned by the properties of rubber. Until recently, however, the theoretical framework for large deformations found little application among rubber engineers because the mathematics rapidly becomes intractable for all but the simplest components. The advent of affordable and powerful computers has changed all this, and brought the chellenge of rubber new sets of people-software engineers and desk-top, as opposed to empirical, designers.

The development of the statistical theory of rubber elasticity in the 1940s, of finite strain elasticity theory in the 1950s, and of convenient forms for the strain energy function in the 1970s, all focused on modeling the elastic characteristics of rubber. Although much literature has appeared in recent years following this theme, the Physics of Rubber Elasticity by L.R.G. Treloar (3rd Edition, Clarendon Press, Oxford, 1975) and the proceedings of a Discussion on Rubber Elasticity (Proc.Roy.Soc.London, 1976, A351, No. 1666, 295–406) remain very valuable reviews.

The treatment of rubber as a 'hyperelastic' material—that is, a material modelled by a strain-energy function for finite strain—was implemented into finite strain element analysis in the 1980s and is now widely available in commercial software packages.

However, only a few engineering elastomers—such as unfilled natural rubber and some grades of polyurethane—really conform to the "hyperelastic" ideal. Most other engineering elastomers incorporate "reinforcing" fillers, needed to confer adequate strength properties and also to improve processing characteristics and to enable adjustment of hardness over a wide range. The stress-strain characteristics of such filled elastomers depart significantly from elasticity. While ways of thinking about these departures—such as "dynamic to static ratio" of rubber springs—may have satisfied a previous generation of design engineers, there is now an opportunity to apply more sophisticated models.

One major current challenge is thus to model those aspects of the inelastic behavior that are relevant to engineers, and to do this in such a way that the models are implementable in finite element analysis.

Although potentially the involvement of representatives of several disciplines should facilitate progress, this is only the case if they talk to each other. In practice, software engineers might rely on the literature and on desktop engineers as sources of information about rubber, and fail to achieve as good a balance of understanding as they could if they listened also to experimental rubber scientists and empirical designers. Applied mathematicians might develop phenomenological models which address issues of secondary interest to designers, or which misrepresent important aspects of the experimentally observed behavior. Experimentalists might develop models without reference to the existing framework of continuum mechanics, resulting in internal inconsistencies and difficulty in implementation in software packages. The First European Conference on Constitutive Models for Rubber sprang from the idea of providing a forum for multi-disciplinary discussion, seeking to bring the fragmented strands of recent research together.

Within the UK a start has been made in this direction—through a workshop on deformation Modelling for Solid Polymers (Oxford University, 1997) and a seminar on Finite Element Analysis of Elastomers (Institution of Mechanical Engineers, London, 1997). The proceedings of the latter are available as a publication of the same name (Professional Engineering Publications, London, 1999). Similarly, in Germany a workshop of Finite Element Analysis—Basics and Future Trends was organized by the Deutsche Institute für KautschukTechnology (Hannover, 1998). The interest in these essentially national meetings suggested that further cross-fertilisation should be stimulated by providing a European forum for discussion.

The contributions to this Proceedings cover a wide range of subjects. Consistent with the analysis given above, few authors chose to present hyperelastic models for rubber; readers interested in this topic will find ample references to earlier work. Several contributions address inelastic effects associated with filled FM.indd xv 5/8/2013 11:21:57 AM xvi elastomers—such as Mullins' effect and quasi-static hysteresis. For others—most obviously in processing uncured rubber—the interest is in modeling viscoelasticity. In addition to stress-strain behavior, work is presented on frictional contact and on mechanical failure. Looking at the applications side, computational techniques are addressed and applied to a diverse range of components, including tyres, earthquake isolation bearings and intervertebral discs. Overall, authors have achieved progress in a wide range of areas—including experimental results, theory and practical utility. They raise many questions as well, as one might expect from the first forum of this kind.

We would like to thank our colleagues on the Scientific Committee (R.W. Ogden, Chairman; D. Besdo, R. de Borst, K.N.G. Futler, H.A. Mang, H. Menderez, G. Meschke and H. Rothert) and all the authors who have worked with us to produce this book.

A. Dorfmann
A.H. Muhr
Vienna/Hertford, June 1999

Constitutive Models for Rubber X – Lion & Johlitz (Eds)

Sponsors

Springer

NETZSCH

MICHELIN
A better way forward

Endurica

Brabender®

BOGE
RUBBER & PLASTICS

SYN
OPT

Keynote lectures

Constitutive Models for Rubber X – Lion & Johlitz (Eds)
© 2017 Taylor & Francis Group, London, ISBN 978-1-138-03001-5

Effect of filler content and crosslink density on the mechanical properties of carbon-black filled SBRs

J. Diani
Laboratoire de Mécanique des Solides, CNRS, École Polytechnique, France

M. Brieu
Laboratoire de Mécanique de Lille, CNRS, École Centrale de Lille, France

P. Gilormini
Laboratoire de Procédés et Ingénierie en Mécanique et Matériaux, CNRS, École des Arts et Métiers, France

ABSTRACT: A set of non-crystallizing styrene butadiene rubbers with various crosslink densities, and amount and type of carbon-black fillers was manufactured to study the impact of the microstructure parameters on the mechanical response with an emphasis on material softening and resistance to mode I fracture. Cyclic uniaxial tests at small and large strains were applied to study the Payne and the Mullins effects, while single edge notch monotonic tension tests were considered for mode I fracture. Results are discussed, from a mechanical point of view, with two main goals, defining original modeling that provides quantitative impact of the microstructure parameters and providing general trends for material design.

1 INTRODUCTION

Since 2008, we have been benefiting from a large material plan provided by Michelin and consisting of a styrene butadiene rubber (SBR) matrix filled with various amounts and types of carbon-black and various crosslink densities. These materials have been submitted to mechanical testing in order to better characterize the impact of carbon-black content and crosslink density on their mechanical behavior. When possible, a model was proposed to quantify this impact. In this contribution, we will focus on three aspects of the mechanical behavior: the nonlinear behavior at small strain known as the Payne effect, the strain-softening recorded at large strain and known as the Mullins effect, and the properties at break measured during SENT tests. We will give an (non exhaustive) overview of the common results that can be found in the literature, present some original results, and open some discussions.

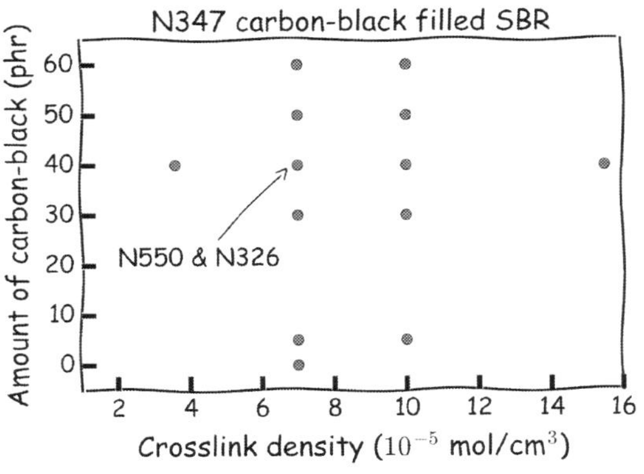

Figure 1. Material plan.

2 MATERIAL PLAN

The star-branched solution of SBR contains 15% of styrene and its molar mass is $M_n = 120\ 10^3$ g/mol. The glass transition measured by differential scanning calorimetry (DSC) shows at –48°C. The rubber matrix was filled with various amounts of N347 carbon-black ranging from 0 to 60 phr. The material crosslink density was varied by changing the amount of sulfur added before curing. Finally, in order to study the impact of carbon-black structure and fineness, two materials filled with 40 phr of either N326 or N550 were manufactured.

3 NONLINEAR BEHAVIOR AT SMALL STRAIN: PAYNE EFFECT

Unlike unfilled rubbers, the viscoelastic responses of filled rubbers during cyclic loadings depend on the applied strain amplitude. Materials exhibit nonlinear viscoelasticity despite sinusoidal stress responses to sinusoidal strain stimuli. This nonlinear behavior is known as the Payne effect due to the significant contribution of this author. The effect may be recorded during shear tests or uniaxial compression tests (Payne 1963), and presented as the decrease of the complex modulus with respect to the strain amplitude. The complex modulus being $X^* = X' + iX''$, with X' and X'' the storage and loss moduli measured in tension or shearing, Payne proposes to calculate its norm $|X^*| = \sqrt{X'^2 + X''^2}$. Then, the Payne effect is characterized by the ratio of the norm value at low strain over its value at high strain. Figure 2 presents an illustration of the change of complex shear modulus with respect to strain amplitude for a 50 phr filled SBR. According to the literature, here is how the microstructure affects the Payne effect:

- The degree of vulcanization does not affect the Payne effect (Payne and Whittaker 1971),
- The Payne effect increases with the amount of carbon-black (Payne 1963),
- The Payne effect increases with poor filler dispersion (Payne 1966),
- The Payne effect increases with the carbon-black fineness (Payne and Watson 1963).

Actually, a rather complete review on the experimental results for the Payne effect may be found in (Wang 1998) Among other results, it is shown that increasing the surface area of the carbon-black increases the Payne effect while no easy conclusion can be drawn with regard to the carbon-black structure.

The recent viscohyperelastic models for Payne effect (Rendek and Lion 2010, Delattre et al. 2016), are assuming that relaxation time or their associated moduli are dependent on the strain amplitude. The fact that the Payne effect is not affected by the crosslink density, which in return affects the relaxation time, may suggest that other paths remain to be considered for Payne effect modeling.

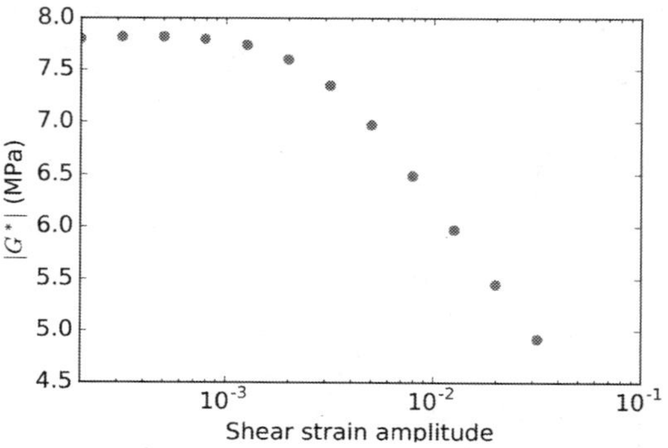

Figure 2. Payne effect recorded for 50 phr N347 filled SBR.

4 STRAIN SOFTENING: MULLINS EFFECT

Upon first stretch, filled rubbers undergo a substantial softening known as the Mullins softening due to Mullins' remarkable contribution to the phenomenon (Mullins 1969). In the latter review, one may find qualitative comparisons of the Mullins softening according to material parameters like the increase of Mullins softening with respect to the amount of carbon-black (Dorfmann and Ogden (2004) for instance). Recently, an objective parameter was proposed (Merckel et al. 2012) that allows to compare quantitatively the Mullins softening of different materials.

4.1 *An objective damage parameter for Mullins softening*

When carbon-black is added to a rubber matrix, two competing effects are witnessed, the material becomes stiffer (Figure 3) but also evidences substantial softening that increases with the applied stretching (Figure 4) and known as the Mullins effect. Using the concept of strain amplification factor introduced by Mullins and Tobin (1965), the fillers are assumed rigid, and the rubber gum supports the strain solely. As a consequence, the strain supported by the rubber matrix is equal to the applied strain amplified by a factor X depending on the amount of effective fillers. The Mullins softening is understood as a degradation of the amount of effective fillers. Therefore the strain undergone by the rubber gum, Λ_{gum} is evaluated as,

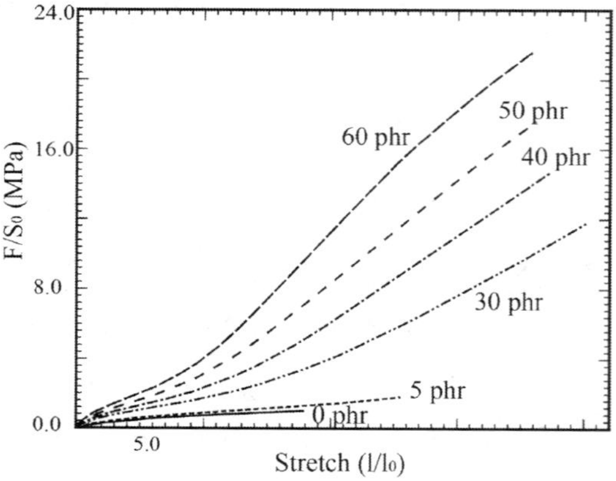

Figure 3. Stress-stretch response during monotonic uniaxial tension of a SBR matrix filled of different amounts of N347 carbon-black.

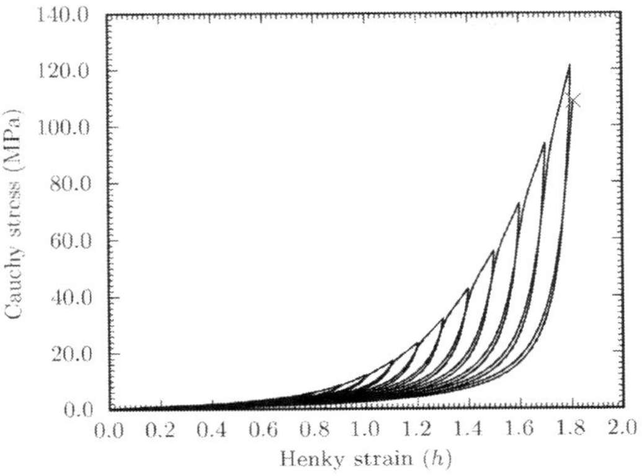

Figure 4. 40 phr filled SBR uniaxial stress-strain response during a cyclic loading with increasing maximum strain.

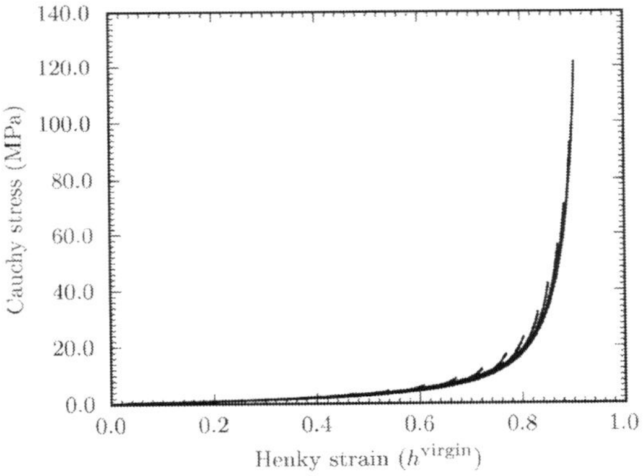

Figure 5. 40 phr filled SBR stress-strain response superimposition using intensity factors that depends on the maximum strain applied.

$$\Lambda_{gum} = X(1-D)\Lambda \tag{1}$$

with Λ the strain supported by the composite material and D the Mullins softening parameter depending on the applied maximum strain. For a given material, the values of the factor $X(1-D)$ according to the maximum applied strain are determined by optimization of the superimposition of the unloading stress-strain responses displayed in Figure 4. Figure 5 presents such a superimposition and for more details on this aspect see (Merckel et al. 2012). When the Hencky strain is chosen for Λ, D conveniently evolves linearly with the maximum applied strain Λ_{max} (Figure 6).

The damage parameter D obtained on large strain data is similar to the damage parameter defined at small strain by extracting the tangent modulus E

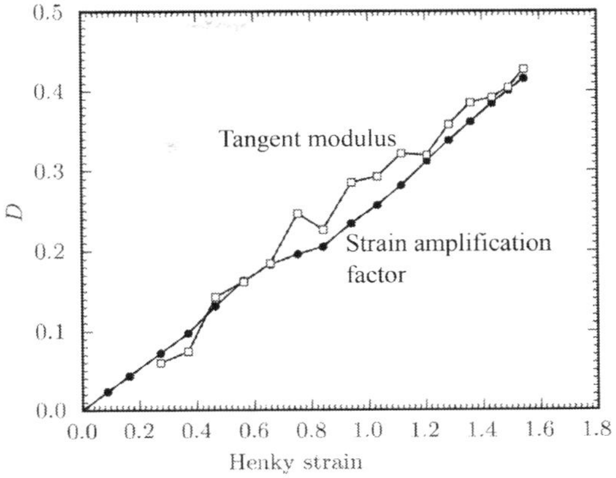

Figure 6. Mullins softening parameter with respect to maximum strain extracted from stress-strain responses shown in Figure 4.

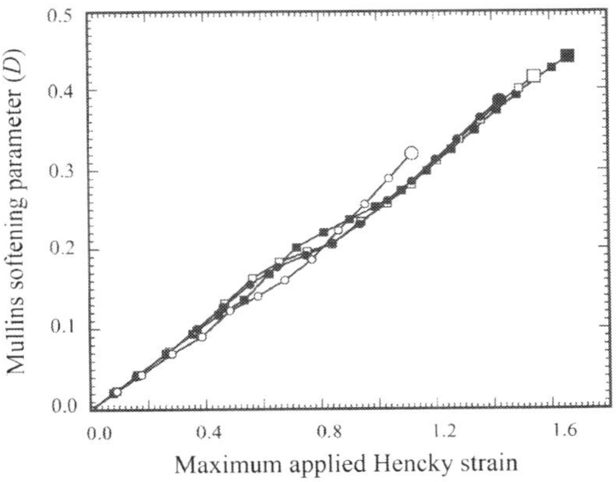

Figure 7. Mullins softening estimates for four SBRs filled with 40 phr of N347 carbon-black and characterized by various crosslink densities.

from the unloading responses according to the maximum applied strain and comparing it to the rubber gumtangent modulus (Merckel et al. 2011),

$$E = X(1-D)E_g \tag{2}$$

where denotes E_g the tangent modulus of the rubber gum. Figure 6 presents a comparison of the values of D estimated according to Eqs. (1) and (2).

4.2 *Effect of the crosslink-density*

The parameter D has been estimated for the SBRs filled with 40 phr of N347 carbon-black and various crosslink densities appearing in Figure 1. Figure 7 shows the same Mullins softening for the four tested materials. Consequently, the Mullins softening is not affected by the material crosslink density.

4.3 *Carbon-black impact on Mullins softening*

The impact of the amount of carbon-black on the Mullins softening is shown in Figure 8. As expected, the Mullins softening increases with the amount of fillers. More interestingly, since D evolves close to linearly with the Hencky strain, one may extract the slope of D and compare the Mullins softening of materials that are very different.

Last, the impact of the type of carbon-black was tested. Figure 9 presents the Mullins softening parameter D for SBR filled with 40 phr of N326, N347 and N550 carbon-black. Albeit these

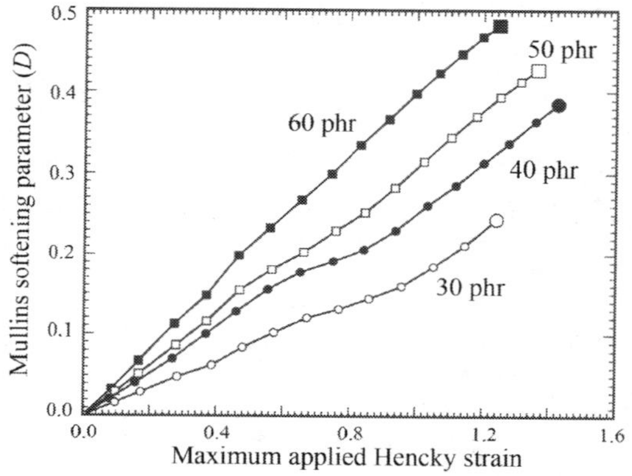

Figure 8. Effect of the amount of carbon-black on the Mullins softening of carbon-black filled SBR of similar crosslink density ($\simeq 7.10^{-5}$ mol/cm^3).

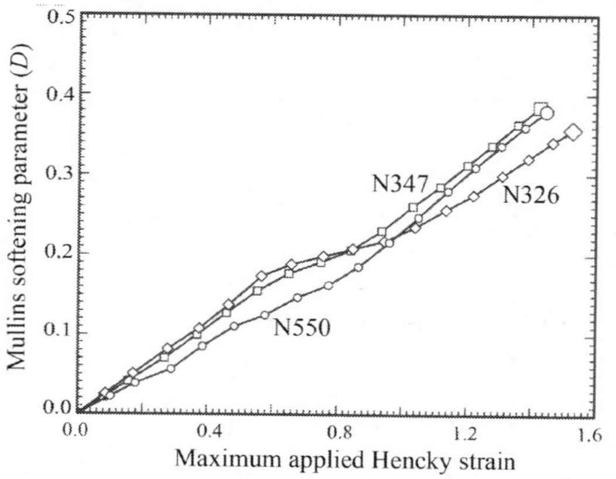

Figure 9. Mullins softening for three 40 phr SBR of similar crosslink densities filled with different types of carbon-black.

carbon-black types have different finenesses and structures the Mullins softening is similar.

5 PROPERTIES AT BREAK

The resistance to failure of carbon-black filled rubbers is a major issue for the rubber industry. When the resistance to mode I failure of these materials is tested with single edge notch tension samples, a steady failure is recorded and a Griffith (1921) analysis may be anticipated. Such an analysis was followed by Rivlin and Thomas (1953) and Greensmith (1963) and conducted to the following simple formula to calculate the critical energy release rate for rubbers:

$$G_c = \frac{2KcU_b}{\sqrt{\lambda_b}} \tag{3}$$

with $K \simeq 3$, U_b the elastic energy stored at crack propagation, c the initial length of the notch, and λ_b the stretch at break. While this formula was established for non-filled crystallizing natural rubber, assuming only localized crystallization near the crack tip, its extension to filled (crystallizing or non-crystallizing) rubbers remains open and we are currently working on it. Nonetheless, it is possible to run SENT test and compare the strain and stress at break changes according to the amount and type of carbon-black and to the crosslink densities.

5.1 *Experimental protocol*

Small samples of 40 mm length, 12 mm width and 2.5 mm thick were punched and a notch measuring between 0.5 and 1.6 millimeters was made with a razor blade. Ten tests were run at 5 mm/min and the applied stretch was characterized with the recorded crosshead displacement. The stretch at failure with respect to the notch initial length is plotted in Figure 10. As one can read, apart from the dispersion inherent to failure tests, no trend could be extracted. This is a little different from what Hamed and Park (1999) reported for a similar range of initial cut length, showing contradictory results with the tearing energy increasing and the stress at break decreasing with the increase of the initial cut length. Actually, when looking closely at Hamed and Park (1999) results, the trend is mostly determined by the measures obtained for even smaller initial notches that were not considered here. Albeit, it would be interesting to take a close look at the trend for smaller notches, we had decided to limit ourselves to the cited notch length range, test about ten samples for each materials and to extract the average engineering stress at break and the average stretch at break.

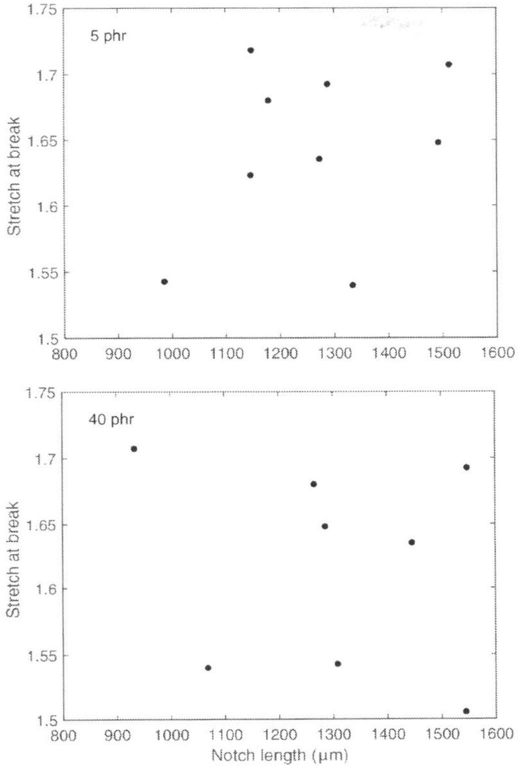

Figure 10. Strain at break recorded during SENT tests for two materials (filled with 5 phr and 40 phr) and plotted with respect to the initial notch length.

Finally, note that the fracture analysis may depend on the crosshead speed since filled rubbers are viscoelastic and quasi-static tests are targeted ideally. The moderate crosshead speed of 5 mm/min was chosen to cope with the large number of tests (more than 200). Nevertheless, the crosshead speed has been lowered to 0.5 mm/min on two materials and the results show that while the stretch (and consequently the stress) at break increases, the trends obtained when comparing materials presented in Figure 1 remain unchanged.

5.2 *Effect of average rubber chain length*

In order to study the impact of the crosslink density measured by swelling, not only the amount of sulphur added to the compound filled with 40 phr of N347 was varied but also some samples were submitted to three days at 100 C in vaccuo. The exposure to heat enhances the molecular mobility favoring the adhesion of the rubber matrix around the filler particles (Luo et al. 2004). As a result, the crosslink density measured by swelling increases appreciably. Figure 11 shows the stretch and stress at break obtained on the SENT samples with respect to the crosslink density. As shown by the error bars, the result scattering is significant and justifies the need of testing at least ten samples for each material. Figure 11 reveals that loose networks show better resistance to failure. This result is consistent with data from De and Gent (1996) and with Hamed (2000) scenario of failure based on more dissipative mechanisms in lightly crosslinked rubbers relieving local stress concentration contributing to a better distribution of the load among network chains.

Since exposing the materials to heat enhances the adhesion at the rubber/filler interface, it was legitimately expected to potentially increase the dissipative mechanism of chain desorption from the fillers. To the contrary, materials exposed to heat have shown lower resistance to failure due to the tighter network (Figure 11). This result allows

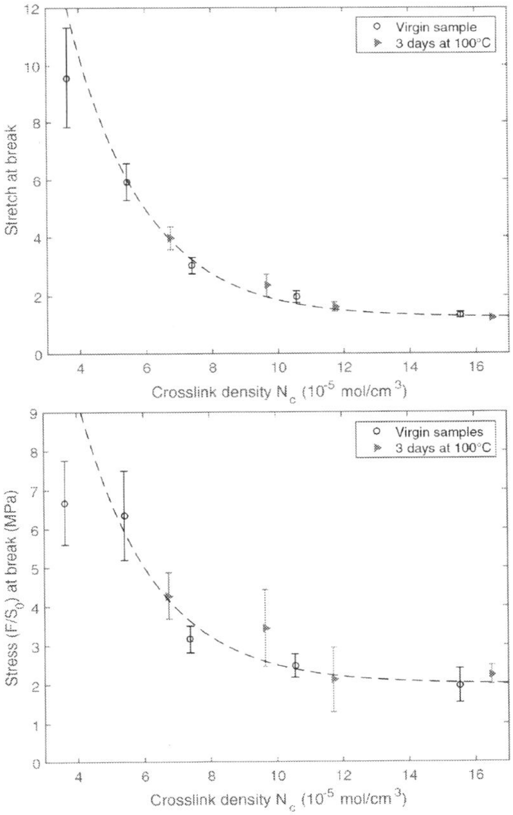

Figure 11. Mode-I failure properties of 40 phr N347 filled SBR with respect to the material crosslink density. Error bars indicate twice the standard deviation.

us to better understand the role of carbon-black fillers in the process of reinforcement of rubbers. It is probably not so much the fact that rubber chains may debond from the surface of fillers as a dissipative mechanism that improves the filled rubbers resistance to crack propagation. For networks with shorter chains, the material stress-strain response stiffens and the stretch at break reduces significantly and consequently the stress at break decreases. Therefore, the extensibility provided by longer chains is key in the resistance to failure. Actually, the role of long chains might be to prevent stress concentration by distributing the load as proposed by Hamed (2000).

5.3 *Effect of the amount of fillers*

The positive effect of carbon-black on SBR failure was already reported in the literature (Hamed and Park 1999, Medalia 1987, Gherib et al. 2010, among others). The interest of our contribution stands in the range of carbon-black added to the SBR (from 5 to 60 phr) allowing us to study how the material strength evolves with respect to the amount of carbon-black. Results are plotted in Figure 12. As expected, the stress and stretch at break increase when the amount of fillers increases. More interestingly, neither a threshold effect at the lower amount of fillers nor a plateau effect at the higher amount of fillers is noticed for the stress at break. In relative terms, the stress at break increases somewhat more than the stretch at break. One can even notice that except for the lowest amount of carbon-black (5 phr), the stretch at break evolves moderately with the amount of fillers and seems mainly dependent on the crosslink density, while the stress at break is very dependent on the amount of fillers. Sucha result is also observed when uniaxial tension tests are applied on unnotched samples (Merckel et al. 2013). Actually, it is worth noticing that, like shortening the polymer chains, adding fillers stiffens the material but without reducing the stretch at break. The latter is improved by adding fillers. Therefore, the addition of fillers combines two positive effects. One may consider that the carbon-black fillers act like stress supports, limiting stress concentration within the matrix.

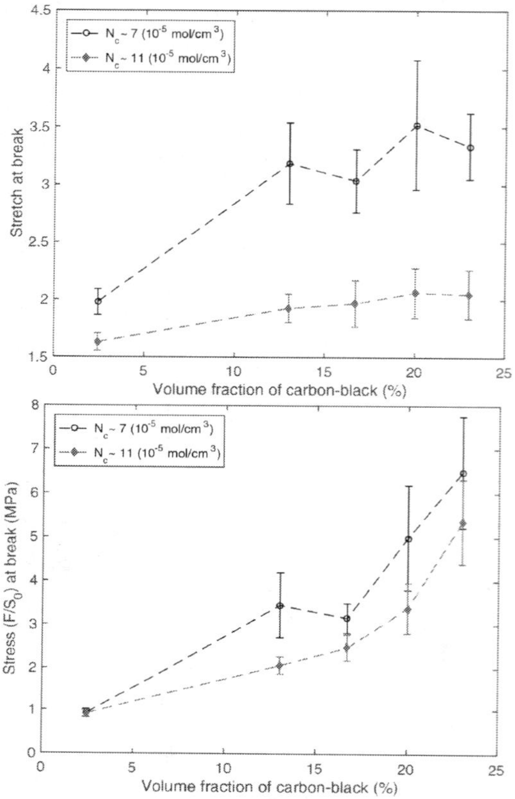

Figure 12. Mode-I failure properties of N347 carbon-black filled SBRs according to the volume fraction of fillers. Error bars indicate twice the standard deviation.

5.4 *Effect of the type of filler*

As presented in Figure 1, three types of carbon-black were used N326, N347 and N550 to manufacture three SBRs filled with 40 phr of carbon-black and presenting the same crosslink densities. N326 and N347 carbon-blacks have close finenesses characterized by their specific surfaces of 81 and 95 m^2/g respectively, while N550 shows a specific surface of 41 m^2/g (Vilgis et al. 2009). Therefore, the size of the primary particles is larger for N550. The structures of the aggregates are characterized by the amount of void volume that is measured by dibutylphtalate adsorption DBP. Typical measures of DBP recorded for N326, N550 and N347 are 68, 114 and 124 ml/100 g respectively (Kraus 1971). Therefore, compared to N347, N550 allows testing the carbon-black fineness and N326 the carbon-black structure. Figure 13 shows the properties at break measured during SENT tests. N550 compares unfavorably to N347 while N326 shows significant improvement of the resistance to failure. For similar finenesses (N326 and N347), the structure parameter has a significant impact on the resistance to fracture. N347 shows a stiffer stress-strain response but consequently a significantly lower stretch at break than N326. The stiffer response shown by N347 compared to N326 is due to more rubber gum trapped in the carbon-black void. This is known as the occluded

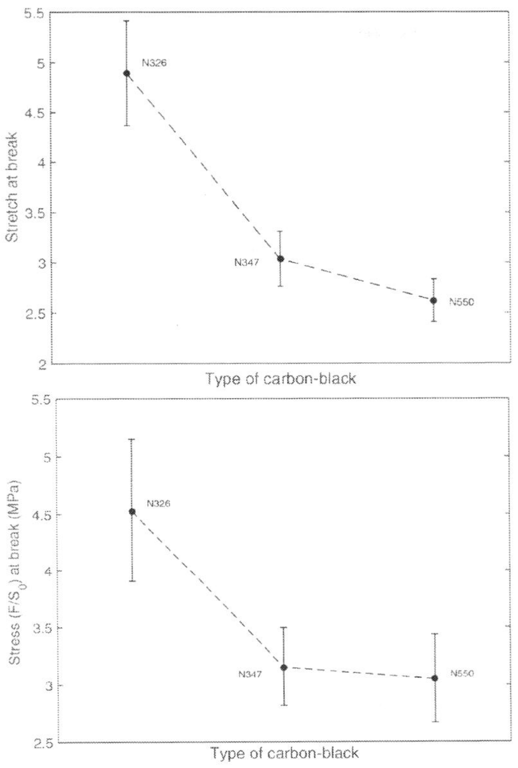

Figure 13. Effect of the fineness and structure of the carbon-black on the mode-I failure properties measured on SENT samples.

rubber gum and increases the effective amount of filler (Medalia 1972). It is observed to reduce the rubber extensibility leading to an unsatisfactory resistance to crack propagation. Similarly, N550 shows a stiffer stress-strain response and a lower stretch at break than N347. It is believed that in the case of fine carbon-black, fillers spread better in the matrix creating a filler network sustaining the stress and limiting stress concentration in the matrix. According to Figure 13, fineness is the first parameter to consider, since N550 compares poorly to N347 despite a favorable structure recognized by a smaller DBP value.

6 CONCLUSION

During this study, it was noted that the softening witnessed in carbon-black filled rubbers is not impacted by the material crosslink density, whereas it increases strongly with the amount of carbon-black added to the rubber gum. Moreover, while the type of carbon-black seemed to impact the Payne softening it was meaningless for the Mullins softening.

For the resistance to mode I failure, both the amount of carbon-black and the crosslink densities have a significant impact. On one side, the larger deformability offered by long chains extend the stretch at break significantly. On the other side, when increasing the amount of carbon black, the stretch at break first increases and quickly reach a plateau while the stress at break increases continuously.

ACKNOWLEDGEMENTS

The authors are thankful to Michelin for providing the set of materials, for the numerous open discussions and for its financial support.

REFERENCES

De, D. & A. Gent (1996). Tear strength of carbon-black filled compounds. *Rubber Chem. Technol. 69*, 834–850.

Delattre, A., Lejeunes, F. Lacroix, & S. Méo (2016). On the dynamical behavior of filled rubbers at different temperatures: Experimental characterization and constitutive modeling. *Int. J. Solids Struct. 90*, 178–193.

Dorfmann, A. & R. Ogden (2004). A constitutive model for the Mullins effect with permanent set in particle-reinforced rubber. *Int. J. Solids Struct. 41*, 1855–1878.

Gherib, S., L. Chazeau, J. Pelletier, & H. Satha (2010). The fluid dynamics of river dunes: a review and some future research directions. *J. Appl. Polym. Sci. 118*, 435–445.

Greensmith, H.N. (1963). Rupture of rubber. X. The change in stored energy on making a small cut in a test piece held in simple extension. *J. Appl. Polym. Sci. 7*, 993–1002.

Griffith, A.A. (1921). The phenomena of rupture and flow in solids. *Philos. Trans. R. Soc. London A221*, 163–198.

Hamed, G. (2000). Reinforcement of rubbers. *Rubber Chem. Technol. 73*, 524–533.

Hamed, G.R. & B. Park (1999). The mechanism of carbon black reinforcement of SBR and NR vulcanizates. *Rubber Chem. Technol. 72*, F04S05.

Kraus, G. (1971). A carbon-black structure-concentration equivalence principle. Application to stress-strain relationships of filled rubbers. *Rubber Chem. Technol. 59*, 199–213.

Luo, H., M. Klüppel, & H. Schneider (2004). Study of filled SBR elastomers using NMR and mechanical measurements. *Macromolecules 37*, 8000–8009.

Medalia, A.I. (1972). Effective degree of immobilization of rubber occluded within carbon black aggregates. *Rubber Chem. Technol. 60*, 1171–1194.

Medalia, A.I. (1987). Effect of carbon black on ultimate properties of rubber vulcanizates. *Rubber Chem. Technol. 60*, 45–61.

Merckel, Y., J. Diani, M. Brieu, & J. Caillard (2013). Effects of the amount of fillers and of the crosslink density on the mechanical behavior of carbon-black filled styrene butadiene rubbers. *J. Appl. Polym. Sci. 129*, 2086–2091.

Merckel, Y., J. Diani, M. Brieu, P. Gilormini, & J. Caillard (2011). Characterization of the Mullins effect of filled rubbers. *Rubber Chem. Technol. 84*, 402–414.

Merckel, Y., J. Diani, M. Brieu, P. Gilormini, & J. Caillard (2012). Effect of the microstructure parameters on the Mullins softening in carbon-black filled styrene butadiene rubbers. *J. Appl. Polym. Sci. 123*, 1153–1161.

Mullins, L. (1969). Softening of rubber by deformation. *Rubber Chem. Technol. 42*, 339–362.

Mullins, L. & N. Tobin (1965). Stress softening in rubber vulcanizates. part I. Use of a strain amplification factor to describe the elastic behavior of filler-reinforced vulcanized rubber. *J. Appl. Polym. Sci. 9*, 2993–3009.

Payne, A. (1963). The dynamic properties of carbon blackloaded natural rubber vuclanizates. Part I. *Rubber Chem. Technol. 36*, 432–443.

Payne, A. (1966). Effect of dispersion on dynamic properties of filler-loaded rubbers. *Rubber Chem. Technol. 39*, 365–374.

Payne, A. & W. Watson (1963). Carbon black structure in rubber. *Rubber Chem. Technol. 36*, 147–155.

Payne, A. & R. Whittaker (1971). Low strain dynamic properties of filled rubbers. *Rubber Chem. Technol. 44*, 440–478.

Rendek, M. & A. Lion (2010). Strain induced transient effects of filler reinforced elastomers with respect to the Payne-effect: experiments and constitutive modelling. *Z. Angew. Math. Mech. 90*, 436–458.

Rivlin, R.S. & A.G. Thomas (1953). Rupture of rubber. I. Characteristic energy for tearing. *J. Polym. Sci. 122*, 301–310.

Vilgis, T.A., G. Heinrich, & M. Klüppel (2009). *Reinforcement of Polymer nano-Composites. Theory, Experiments and Applications*. New York: Cambridge University Press.

Wang, M. (1998). Effect of polymer-filler and filler-filler interactions on dynamic properties of filled vulcanizates. *Rubber Chem. Technol. 71*, 520–589.

Constitutive Models for Rubber X – Lion & Johlitz (Eds)
© 2017 Taylor & Francis Group, London, ISBN 978-1-138-03001-5

Experimental research and numerical simulation of the damping properties of Magnetorheological Elastomers

I. Petríková & B. Marvalová
Technical University of Liberec, Liberec, Czech Republic

ABSTRACT: The dependence of dynamical moduli and loss factor of Magnetorheological Elastomers (MREs) on the external magnetic field intensity and on the frequency of applied cyclic shear deformation was studied. Isotropic samples of MRE were made of the silicon rubber matrix filled with carbonyl iron micro-particles. The magnetic field produced by an electromagnet was applied in course of cyclic loading on double-shear samples of MRE under controlled shear strain. Dynamical moduli and the loss angle were determined as the function of the magnetic field intensity and of the frequency and amplitude of cyclic deformation in shear. The dynamic stiffness of MRE depends on magnetic flux density and increases with increasing testing frequency. The loss factor of MRE samples is tunable by the magnetic flux density and it depends also on the testing frequency and amplitude.

1 INTRODUCTION

A variety of different rubber dampers are used in engineering applications to isolate structures from unwanted vibrations. Smart elastomeric composites, so-called Magnetorheological Elastomers (MREs), are increasingly being used as damping elements in the vibration absorbers (Carlson, 2000).

MREs, also referred to as magnetosensitive (MS) elastomers, are smart materials composed of micron-sized magnetically polarizable particles dispersed in an elastomeric matrix. The unique characteristic of MREs is that their shear modulus can be continuously controlled by the external magnetic field (Jolly, 1996a).

The MR effect is increased by choosing the material of the particles with high permeability. The particles have a typical size of 1 to 5 microns and they should be of a material with high magnetic saturation such as iron, Terfenol-D, carbonyl iron, or newly ferromagnetic shape-memory Ni-Mn-Ga (Faidley, 2006). The magnetorheological response of hybrid MREs consisting of two different magnetic filler particles was studied by Aloui and Klüppel. They were focused on an optimization of mechanical and magnetic properties of MREs by combining two fillers—Magsilica nanoparticles formed by 85% Fe_2O_3 coated with a 3 nm-thick silica layer and micro-sized carbonyl iron particles. The research concluded that the reinforcing potential is higher for nano-scale magnetic filler particles in comparison to micro-scale magnetic filler particles. (Aloui, 2014). The matrix materials commonly used for MREs are natural rubber, silicon rubbers, vulcanized rubbers filled with carbon black or with silica.

MRE composites inherit main properties of the elastomeric matrix such as large deformations, stress softening effect, amplitude and frequency dependency, reduction of stiffness at cyclic loading and viscoelastic time-dependent features (Lion, 1998).

MRE composites, however, have further interesting feature brought by the particles. Deformation of the MRE composite in the presence of the magnetic field causes field dependent elastic modulus which rises monotonically with applied magnetic field. The percentage of maximum increase in shear modulus in the presence of the magnetic field (MR effect) is reported to be between 30–60% of the zero-field modulus (Ginder, 2002). Calculations using finite element analysis show (Davis, 1999) that for typical elastomers the increase in shear modulus due to interparticle magnetic forces at saturation is about 50% of the zero-field modulus. For volume fraction of particles ϕ, the shear modulus G_{ran} of rubber filled with randomly dispersed, rigid particles is given with acceptable accuracy by (Guth, 1945)

$$G_{ran} = G_0\left(1 + 2.5\phi + 14.1\phi^2\right) \tag{1}$$

where G_0 is the shear modulus of the unfilled rubber. The optimum particle volume fraction for the largest change in modulus at saturation is predicted to be 27%. Calculations of the zero-field shear modulus perpendicular to the chain axis indicate that it does not exceed the modulus of a filled elastomer with randomly dispersed particles of the same concentration (Davis, 1999). It is stated that to obtain the maximal MR effect while retaining the mechanical properties of the composite the optimum filler fraction should not exceed 30 vol% (Lokander & Stenberg, 2003).

To maximize the magnetic permeability of the composite, the filler particles in the MRE should be aligned in the direction of the applied magnetic field (Carlson, 2000). The particles alignment is effected by an external magnetic field applied during the cross-linking of the MRE, the particles form columnar structures and become locked in place upon final cure. The MR effect can be controlled by the particles alignment to the magnetic field direction. It is possible to obtain higher MR effect not by increasing of the particles volume fraction, but by the formation of appropriate microstructure (Boczkowska, 2012). Therefore, the enhancement of MR effect can be achieved for lower particles volume fraction, what decreases weight of devices based on the MREs. The deformation of the MR elastomer changes the magnetic field distribution and therefore also magnitude of the magnetic field (Marvalova & Petrikova, 2015).

1.1 *Response of MREs to dynamic loading*

The MR effect implies not only an immediate and reversible increase in the modulus and stiffness of a MRE but also MREs exhibit a field dependent damping (Kallio, 2005). MREs can be used as mounts tunable by the magnetic field as springs for active control of the response of vibrating systems (Cantera, 2017). The loss factor of MREs in dynamic compression at low frequencies was found to increase by about 30% in magnetic field (Kallio et al. 2007). The damping and stiffness properties of aligned MREs depend on the mutual directions of load, magnetic field and the particle alignment in the composite (Kallio 2005, Ivaneyko 2015).

The response of MREs under dynamic compressive or shear loading has been studied experimentally by many investigators. Jolly et al. (1996b) presented a model of the magneto-viscoelastic effect of aligned anisotropic MREs in shear taking into account shear stress induced by the magnetic forces of particle dipoles. The quasistatic model presented is valid for strain rates up to 3.1 s^{-1} (1% at 50 Hz). The results of their experiments showed the material moduli increasing monotonically with applied field until magnetic saturation of the composite occurs between 0.6 T and 0.8 T. The dynamic modulus increases slowly with frequency but the change of the modulus induced by the magnetic field is relatively insensitive to frequency. The dynamic modulus depends on strain amplitude and it declines sharply in the vicinity of 1–2% strain due to the onset of magnetic yielding. Obviously, all MRE composites show more or less pronounced Payne effect i.e. the storage modulus decrease with increasing strain amplitude due to the breakdown of the filler network. The Payne effect increases with increasing concentration of filler material in the composite (Kallio, 2005).

Loss factor of the shear modulus is also strongly strain amplitude dependent even for very small amplitudes (Li, 2010). The loss factor exhibits the increasing trend with increasing of the dynamic strain amplitude. The damping capability of MREs decreases with increasing temperature (Zhang, 2011).

The effect of temperature on the MRE rheological behavior was studied (Molchanov et al. 2014). Temperature dependences of the dynamic moduli were measured at oscillation frequency of 1 Hz. Experimental results showed that G' increases while G'' decreases with temperature in the range from 20 to 100°C.

1.2 *Modeling of dynamic behaviour of MREs*

Main time-dependent features of MREs are derived from the viscoelastic properties of the elastomeric matrix. The dynamic properties depend obviously on the external magnetic field and on the content of particles. The model of the dynamic response of MREs should encompass all these factors. Several phenomenological models developed as the combinations of classical viscoelastic rheological models whose parameters depend on the intensity of magnetic field are reported (Li 2014, Cantera 2017).

Four-parameter viscoelastic model was developed (Li et al. 2010). The model is based on the classical standard solid model combined with one additional parallel spring which represents the dependence of modulus on the magnetic field. Their experimental study contains the harmonic strain controlled shear loading with various strain amplitudes and frequencies at various magnetic fields. The presented diagrams of stresses and strains have elliptical shapes, the areas of which increase steadily with the increment of the magnetic fields. Authors showed the good agreement between experimental data and the results predicted by the model.

A complex linear viscoelastic model for isotropic MREs was presented recently by Xin and collaborators (Xin et al. 2016). The viscoelasticity of MREs was divided into two parts: mechanical and magnetic. The mechanical shear storage and loss moduli are developed using the Kraus model. The magnetic shear storage and loss moduli corresponding to magnetic viscoelasticity are derived based on the magneto-elastic theory with consideration of the magnetic saturation. The proposed model was evaluated by the experimental data.

Fractional rheological models are frequently applied recently to describe fundamental properties of different materials, in particular, rheological behaviour of linear viscoelastic media. There have been numerous attempts to simulate damping properties of MREs with the use of a fractional rheological element. This leads to a significant reduction in the number of parameters that is necessary to determine from experiments.

The model of magneto-sensitive isotropic rubber confined to small strain was presented by Blom & Kari (Blom, 2011). Their constitutive model characterises the amplitude dependence of magnetic sensitivity, elasticity and viscoelasticity of isotropic MRE. The stress response is the sum of three time dependent components—the elastic stress $\tau_e(t)$ is linearly related to the instantaneous strain $\gamma(t)$, the viscoelastic stress $\tau_{ve}(t)$ is linearly related to the history of the strain rate and the friction stress $\tau_f(t)$ is nonlinearly related to the strain. The viscoelastic dependence is described by Abel type integral equation solved via fractional derivative calculus. The model includes the influence of changes in the magnetic field on the parameters.

New constitutive model for isotropic MREs was developed recently (Agirre-Olabide et al. 2017). The model is based on fractional derivatives, which are combined with particle-matrix interaction and the magneto-induced modulus model. The viscoelastic behaviour was modelled using a four-parameter fractional derivative model. The evolution of each parameter depends on particle content and magnetic flux density. Material parameters were determined from very carefully planned and evaluated experimental measurements. Storage and loss moduli dependence on frequency and magnetic field was modelled successfully. Previous experimental study of the authors (Agirre-Olabide et al. 2014) defined the limits of application of the linear viscoelasticity for the magneto-dynamic characterisation of MREs. The linear viscoelasticity region limits were examined depending on the particle content, frequency, external magnetic field, the inner structure of the samples and the temperature.

Poynting-Thomson model containing a fractional element has been proposed to describe dynamic properties of MREs in magnetic field (Nadzharyan et al. 2016). MRE samples containing different concentrations of iron particles have been tested and the frequency dependence of the storage and loss moduli has been determined for various magnetic field intensity. The strain amplitude has been kept constant in the limit of linear viscoelastic regime. The four model parameters were found by fitting the experimental data. The paper presents also a brief useful overview of theory of fractional calculus and its use in the modelling of viscoelastic material response.

The magneto-viscoelastic parametric model was developed by Zhu and collaborators (Zhu et al. 2012) by use of fractional derivative. This model includes four parallel rheological elements. The first two of them, the linear spring and the fractional derivative damper, characterize the viscoelastic properties of the matrix and the other two, the non-linear spring and the analogue damper, represent the damping characteristic dependence on the changing magnetic field.

Another parametric constitutive model representing MREs viscoelastic mechanical behaviour and magneto-induced characteristics was presented by Guo (Guo et al. 2014). The model connects the linear spring and Abel dashpot in series to describe matrix characteristic. And the nonlinear spring element is attached in parallel branch (Shen et al. 2004) to include the dependence of model on magnetic field intensity.

1.3 *Application of MREs*

Engineering applications of MREs are in general very recent and involve, for example, tuned vibration absorbers, stiffness-tunable mounts and suspensions and automotive bushings. In particular, vibration isolators made of MS rubber are shown to be more effective than traditional rubber isolators (Blom & Kari 2011, Alberdi-Muniain et al. 2013).

Aligned and isotropic MREs were tested in dynamic compression with a magnetic field (Kallio et al. 2007). This study investigated the possibility of using magnetorheological elastomers as tunable spring elements. The compressive dynamic stiffness of the aligned MREs was increased proportionally to increasing magnetic flux density. A coil device was developed for testing the MRE spring elements in compression. The study concluded on the basis of the results that aligned MREs can be used as active spring elements due to the substantial MR effect in compressive dynamic loading.

Three different configurations of MRE absorbers were examined (Lerner and Cunefare, 2008) for use as an alternative to tuned vibration absorbers in aircraft fuselages. MREs were placed into three vibration absorber configurations—shear, longitudinal, and squeeze modes. The squeeze mode device exhibited the largest frequency shift of 507%.

A phenomenological model which captures the nonlinear response of MRE seismic isolator was presented by Yang (Yang et al., 2013). This model incorporates the Bouc-Wen nonlinear element in parallel with the Kelvin-Voigt model. The model correctly reproduces the nonlinear character of hysteresis loops which differ from elliptical shape due to the application of magnetic field. MRE seismic isolator was designed, fabricated and tested. The Bouc-Wen element is widely used in structural and mechanical engineering. In contrast, Yu (Yu et al. 2015) used the improved Lu-Gre friction element, which has been included to the phenomenological model for structural control and identification of MRE base isolator.

Broad research of MRE based on waste tire rubber was undertaken by (Imaduddin et al. 2016) and field-dependent characterizations of the material under dynamic conditions were presented. Experimental examinations of microstructural magnetic and physico-chemical properties and thermal glass transition were evaluated. An electromagnetic device for compression test was developed and its function simulated by finite element analysis. The device was then integrated into the fatigue dynamic machine for MRE compression tests. It was demonstrated that the MREs based on reclaimed rubber possess a wide tunable capability of damping.

A broadband frequency-tunable MRE dynamic vibrations absorber was presented (Komatsuzaki et al. 2016) which works over a wide-frequency range. The frequency adjustability was investigated experimentally and the real-time vibration control performance was evaluated.

Sandwich structures formed by embedding magnetorheological elastomers between constrained layers of carbon fiber reinforced plastic laminates were prepared and examined by Kozlowska and collaborators (Kozlovska et al. 2016). The MREs were obtained by mechanical stirring of a mixture of elastomer with carbonyl-iron particles, followed by orienting the particles into chains under an external magnetic field. Sandwich structures were obtained by compressing MREs between two laminate strips. The free vibration responses of the three-layered beams with a MRE core were studied at different magnetic field levels. The free vibration tests revealed that the magnetic field causes a shift in natural frequency values and a reduction in the vibration amplitudes.

An extensive review on development of MREs and their application can be found in Li et al. (2014), Cantera et al. (2017) and Sutrisno et al. (2015).

2 EXPERIMENTAL

The magneto-rheological elastomer composites were prepared using RTV silicone rubber as the matrix material and carbonyl iron micro-particles have been used as the filler.

Globular carbonyl iron particles supplied by Sigma-Aldrich had diameter from 2–5 μm. The raw particle size of the filler and particle distribution were examined by a scanning electron microscope. SEM of MRE composite sample is in Figure 1. The carbonyl iron particle size distribution is in Figure 2. Particles are of the regular mainly spherical shape.

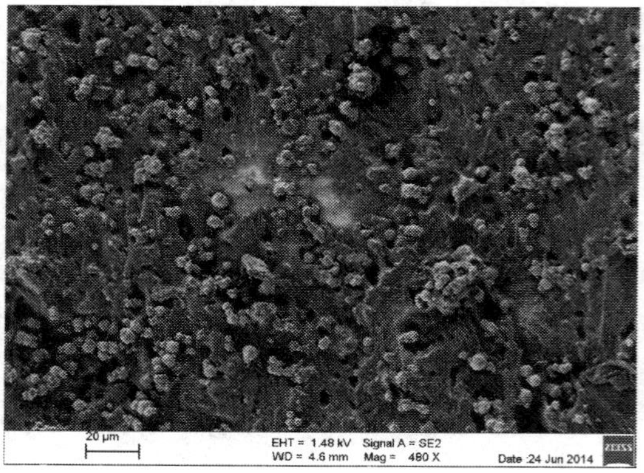

Figure 1. SEM picture of MRE with carbonyl iron particles (Petrikova et al., 2015).

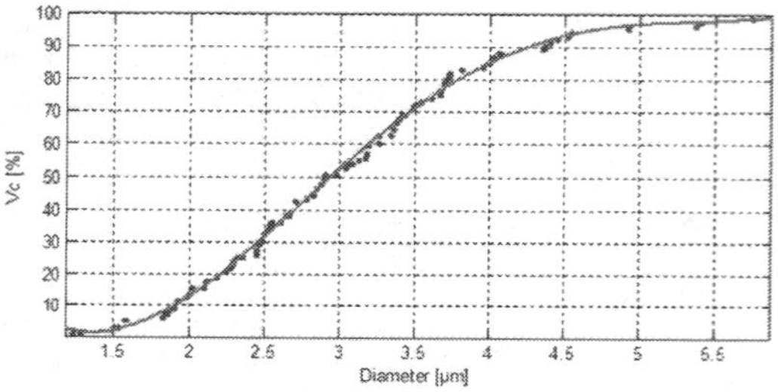

Figure 2. Carbonyl iron particle size distribution.

The matrix material is a two component silicone rubber compound mixed of ZA 22 base and RZA 22 curative in the ratio of 1:1. Carbonyl iron micro-particles of 30 vol% were interfused with silicone oil and then mixed to the silicon rubber mixture. The mixing and polymerization process were carried out at room temperature. The mixture was evacuated during the polymerization in order to remove bubbles.

Polymerization was carried out without the presence of magnetic field, MRE samples are therefore isotropic. Samples were cut into rectangular shape with dimension $20 \times 20 \times 5$ mm and glued together with aluminum strips to form the double lap shear specimen in Figure 3. Cyanoacrylate glue with activator were applied to ensure a perfect connection between the elastomer and the aluminum strips.

The electromagnet of U type in Figure 3 was designed and fabricated for the purpose of testing of MREs. Magnetic field in the central region of the gap in the core of electromagnet is homogeneous and the maximum value measured in the center of gap is around 0.75 T as seen from the characteristic curve in Figure 4.

Dynamic viscoelastic measurements were carried out on Instron Electropuls 3 kN. Double lap

Figure 3. U electromagnet with double lap shear specimen.

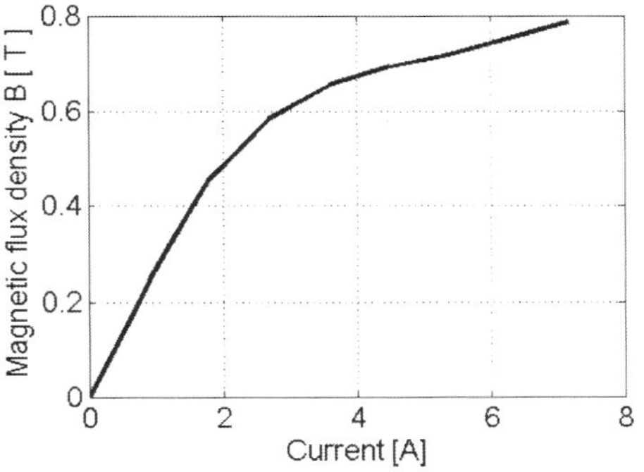

Figure 4. Magnetic field in the centre of gap of electromagnet.

shear specimens were loaded in cyclic shear under displacement control in order to study the changes in the dynamic moduli and loss angle depending on the intensity of the magnetic field and on the frequency of the load. The amplitude of the cyclic shear strain was 0.1 and the frequency was changed in steps from 1 to 10 Hz. The samples were first tested without the magnetic field to obtain the basic dynamical MRE properties, namely storage and loss moduli and the loss angle. Then the electromagnet current was switched on and the samples were cyclically loaded again under the magnetic field with the magnetic flux density increasing from 0.1 to 0.8 T.

3 RESULTS AND DISCUSSION

In order to determine the storage and loss moduli of the material, the static predeformation u_0, the amplitude Δu and the frequency of strain controlled loading, corresponding parameters of the force response and the phase angle δ, must be extracted from the recorded raw signals. We suppose that the raw signals, i.e. displacement $u(t)$ and force response $F(t)$, are harmonic functions approximately and we use the discrete Fourier transform (DFT) in order to determine the phase shift δ at the main excitation frequency (Petrikova et al. 2011).

Amplitudes of shear strain γ_a and shear stress τ_a were determined from recorded signals. Dynamic moduli were calculated as

$$G' = \frac{\tau_a}{\gamma_a}\cos(\delta), \quad G'' = \frac{\tau_a}{\gamma_a}\sin(\delta). \tag{2}$$

Resulting storage and loss moduli and the loss angle of the MRE specimens are in Figures 5–7. The graphs show that there has been a marked increase of dynamic moduli and damping capability of MRE due to the increase of magnetic field intensity. The expected slight rise in the values of these parameters with frequency is also apparent. The increments of storage and loss moduli between the zero-field and 0.78 T are 32% and 38% respectively, the loss angle increase was around 6% for the frequency 5 Hz as presented in Figure 8.

The cyclic test of silicone rubber specimens showed that the behaviour of the silicone rubber is

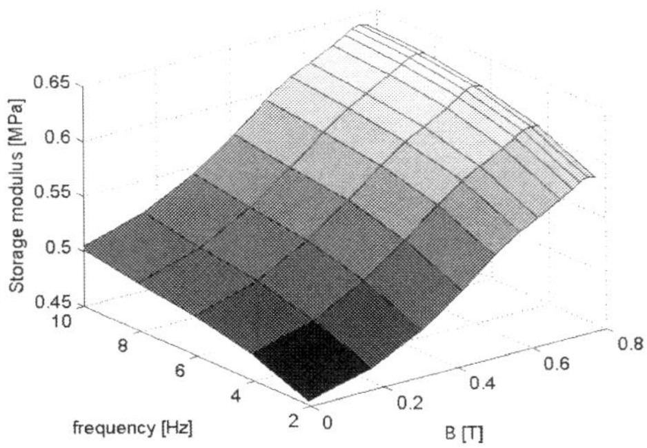

Figure 5. Storage modulus as a function of frequency and magnetic flux density.

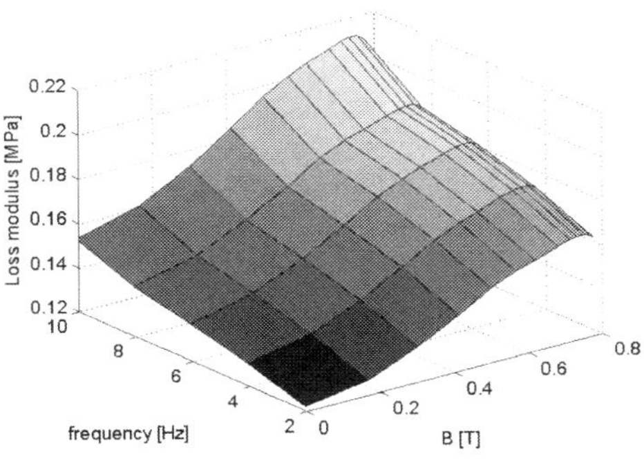

Figure 6. Loss modulus as a function of frequency and magnetic flux density.

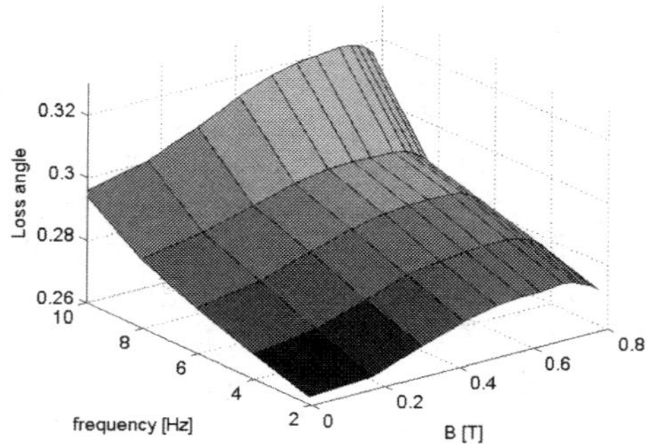

Figure 7. Loss angle as a function of frequency and magnetic flux density.

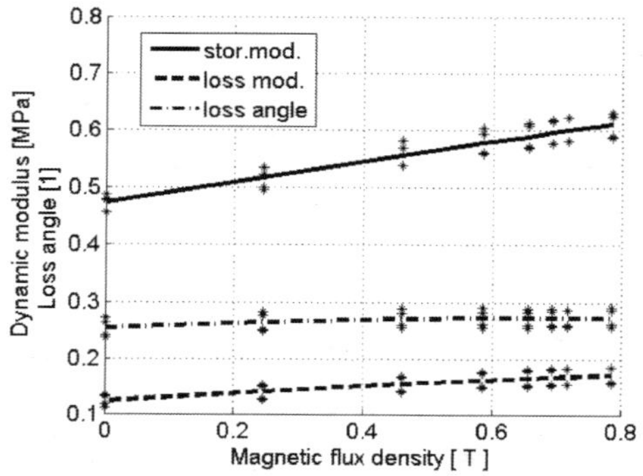

Figure 8. Response to rising magnetic field intensity, frequency of displacement loading 5 Hz.

very similar to natural rubber which exhibits only small hysteresis. The behaviour of samples with embedded particles is completely different—they showed a clear hysteresis and damping in cyclic loading even without the external magnetic field. The comparison of the dynamic response of filled and unfilled silicone rubber is in Figure 9.

Dynamic response of MRE in the zero magnetic field is approximately linearly viscoelastic. The character of hysteresis loop becomes nonlinear when the magnetic field is switched on, as is seen from Figure 10. The comparison of raw hysteresis loop and the loop obtained by approximation based on dynamic modules (2) is shown in Figure 11. The findings suggests that this method of experimental data processing has some spread as the amplitudes of signals cannot be determined by DFT because the energy of raw signal is smeared among the rest of signal frequencies.

Therefore we chose a different approach—we calculated the loss modulus from the dissipated energy density which corresponds to the area of the hysteresis loop (Liu & Qi, 2010). The dissipation of mechanical energy in the course one cycle can be expressed as

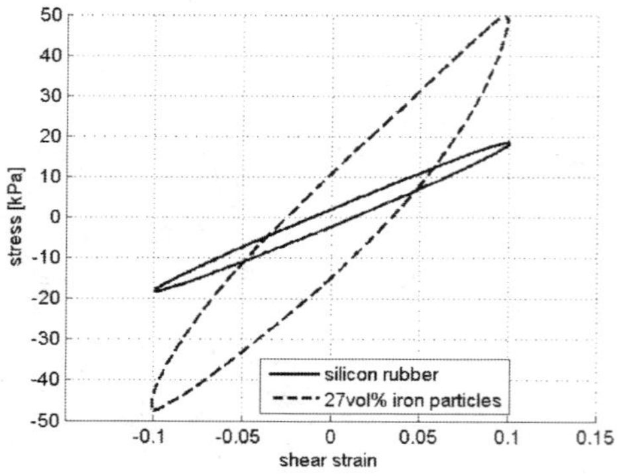

Figure 9. Hysteresis loops for unfiled silicon rubber and SR filed by carbonyl iron particles (without MG field).

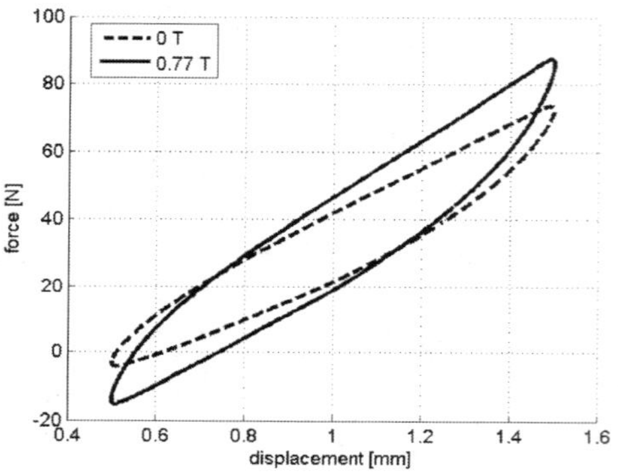

Figure 10. Hysteresis loops of MRE sample in zero field and in MG field, the frequency of displacement loading is 5 Hz.

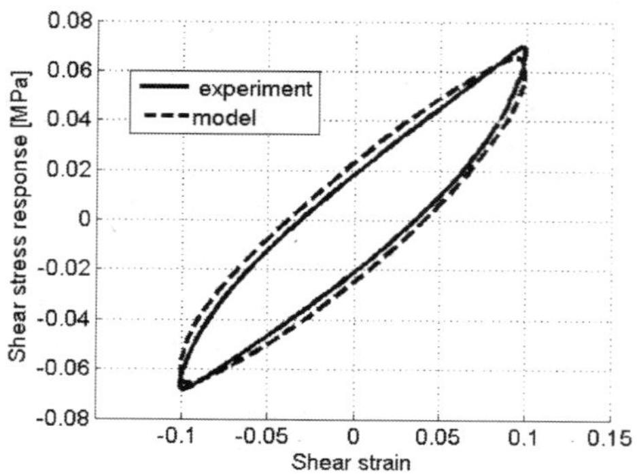

Figure 11. Comparison of experiment and linear viscoelastic model, frequency 5 Hz and mg flux density 0.77 T.

$$D = \int_0^T \tau(t)\,\dot{\gamma}(t)\,dt = \pi\gamma_a^2\, G\sin\delta = \pi\gamma_a^2\, G''. \qquad (3)$$

Density of dissipated energy was calculated numerically as the mean value of the areas of twenty consecutive steady hysteresis loops in each experiment.

The graph of dissipated energy for different frequencies and magnetic flux densities is outlined in Figure 12. Dynamic loss modulus values determined from the density of dissipated energy according to the relation (3) are displayed in Figure 13. As is apparent, the values are slightly different from those in Figure 6 which were determined by the first method using DFT.

Commercial rheometers are used in DMA measurements of MREs published by most authors. Evaluation of measured data is accomplished using apparatus' internal software. Therefore, the raw signal evaluation can be crucial for successful modeling of MREs response (Stolbov et al. 2010).

Material model parameters should be fitted to experimental hysteresis loop points rather than to dynamic moduli determined by an unknown internal software.

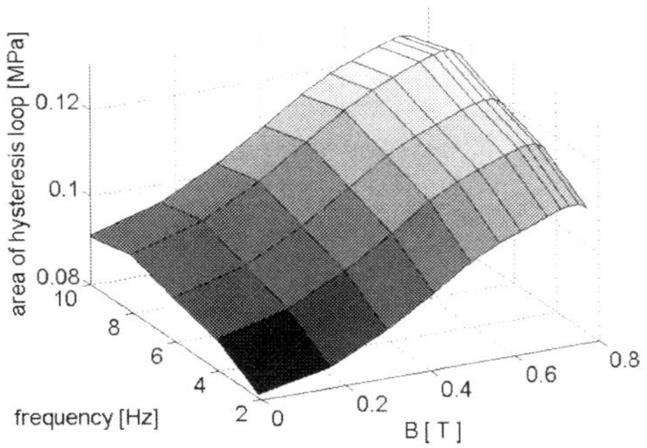

Figure 12. Area of hysteresis loops—dissipated energy density for different frequencies and magnetic flux densities.

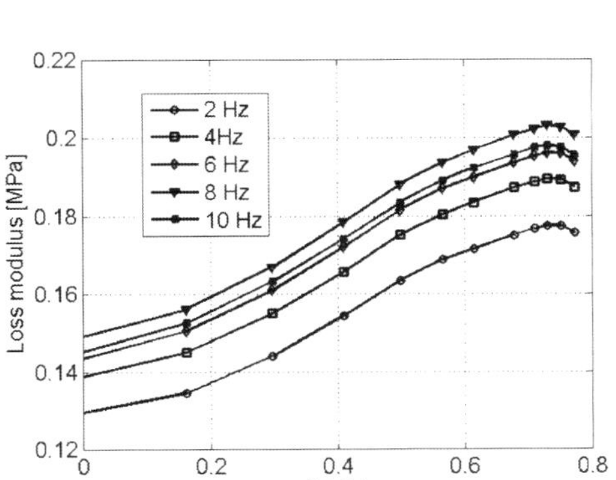

Figure 13. Loss modulus determined from hysteresis loops for different frequencies and magnetic flux densities.

4 CONCLUSION

This study investigated dynamical properties of magnetorheological elastomers with the isotropic distribution of ferrous particles. The paper presents results of the measurement of the response of double lap shear specimens to strain controlled cyclic loading under a uniform gradually changing external magnetic field which were perpendicular to the shear deformation of specimens. MRE samples containing 30 vol% concentrations of carbonyl iron particles have been synthesized and the frequency dependences of the storage and loss moduli have been measured in various magnetic fields. The shear storage and loss moduli and the loss factor were evaluated on the basis of linear viscoelastic model. The results confirmed that the dynamical properties namely the loss factor depend on the applied loading frequency and on the intensity of magnetic field. Results showed that MREs can be prospective as active damping elements in the shear mode.

ACKNOWLEDGEMENT

This work was supported by the Ministry of Education, Youth and Sports of the Czech Republic within the Institutional Endowment.

REFERENCES

Agirre-Olabide I., Berasategui J., Elejabarrieta, M.J. & Bou-Ali, M.M., 2014. Characterization of the linear viscoelastic region of magnetorheological elastomers. *Journal of Intelligent Material Systems and Structures*, 25(16), pp.2074–2081.

Agirre-Olabide I., Lion, A. & Elejabarrieta, M.J. 2017. A new three-dimensional magneto-viscoelastic model for isotropic magnetorheological elastomers. *Smart Materials and Structures*, 26(3), p.035021.

Alberdi-Muniain A., Gil-Negrete N. & Kari L. 2013. Modelling energy flow through magneto-sensitive vibration isolators. *International Journal of Engineering Science* 65: 22–39.

Aloui S. & Klüppel M. 2014. Magneto-rheological response of elastomer composites with hybrid-magnetic fillers. *Smart Materials and Structures*, 24(2), p.025016.

Blom P. & Kari L. 2011. A nonlinear constitutive audio frequency magneto-sensitive rubber model including amplitude, frequency and magnetic field dependence. *Journal of Sound and Vibration*, 330(5), pp.947–954.

Blom P. & Kari L. 2012. The frequency, amplitude and magnetic field dependent torsional stiffness of a magneto-sensitive rubber bushing. *International Journal of Mechanical Sciences* 60: 54–58.

Boczkowska A., Awietjan S.F., Pietrzko, S. & Kurzydłowski K.J. 2012. Mechanical properties of magnetorheological elastomers under shear deformation. *Composites Part B: Engineering*, 43(2), pp.636–640.

Cantera M.A., Behroo, M., Gibson R.F. & Gordaninejad F. 2017. Modeling of magneto-mechanical response of magnetorheological elastomers (MRE) and MRE-based systems. *Smart Materials and Structures*, 26(2), p. 023001.

Carlson J.D. & Jolly M.R. 2000. MR fluid, foam and elastomer device. *Mechatronics* 10: 555–569.

Davis L.C. 1999. Model of magnetorheological elastomers. *Journal of Applied Physics*, 85(6), pp.3348–3351.

Faidley L.E., Dapino M.J., Washington, G.N. & Lograsso, T.A. 2006. Modulus increase with magnetic field in ferromagnetic shape memory Ni–Mn–Ga. *Journal of intelligent material systems and structures*, 17(2), pp. 123–131.

Ginder J.M., Clark S.M., Schlotter W.F. & Nichols, M.E. 2002. Magnetostrictive phenomena in magnetorheological elastomers. *International Journal of Modern Physics B* 16: 2412–2418.

Guo F., Du C.B. & Li R.P., 2014. Viscoelastic parameter model of magnetorheological elastomers based on abel dashpot. *Advances in Mechanical Engineering*. Volume 2014, http://dx.doi.org/10.1155/2014/629386.

Guth E. 1945. Theory of filler reinforcement. *Journal of applied physics*, 16(1), pp. 20–25.

Imaduddin F. et al. 2016. A new class of magnetorheological elastomers based on waste tire rubber and the characterization of their properties. *Smart Materials and Structures*, 25(11), p. 115002.

Ivaneyko D., Toshchevikov V. & Saphiannikova M. 2015. Dynamic moduli of magneto-sensitive elastomers: a coarse-grained network model. *Soft matter*, 11(38), pp. 7627–7638.

Jolly M.R., Carlson J.D. & Munoz B.C. 1996a. A model of the behaviour of magnetorheological materials. *Smart Materials and Structures*, 5(5), p. 607.

Jolly M.R., Carlson J.D., Muñoz B.C. & Bullions T.A. 1996b. The magnetoviscoelastic response of elastomer composites consisting of ferrous particles embedded in a polymer matrix. *Journal of Intelligent Material Systems and Structures*, 7(6), pp. 613–622.

Kallio M. 2005. The elastic and damping properties of magnetorheological elastomers, *PhD thesis*, (p. 149). VTT.

Kallio M. et al. 2007. Dynamic compression testing of a tunable spring element consisting of a magnetorheological elastomer. *Smart Mater. Struct.* 16: 506–514.

Komatsuzaki T., Inoue T. & Terashima O. 2016. Broadband vibration control of a structure by using a magnetorheological elastomer-based tuned dynamic absorber. *Mechatronics*, 40, pp. 128–136.

Kozlowska, J. et al. 2016. Novel MRE/CFRP sandwich structures for adaptive vibration control. *Smart Materials and Structures*, 25(3), p. 035025.

Lerner A.A. & Cunefare K.A. 2008. Performance of MRE-based vibration absorbers. *Journal of Intelligent Material Systems and Structures*, 19(5), pp.551–563.

Li W.H., Zhou Y. & Tian T.F. 2010. Viscoelastic properties of MR elastomers under harmonic loading. *Rheologica acta*, 49(7), pp. 733–740.

Li Y. et al. 2014. A state-of-the-art review on magnetorheological elastomer devices. *Smart materials and structures*, 23(12), p.123001.

Lion, A. 1998. Thixotropic behaviour of rubber under dynamic loading histories: experiments and theory. *Journal of the Mechanics and Physics of Solids*, 46(5), pp. 895–930.

Liu, J. & Qi H. 2010. Dissipated energy function, hysteresis and precondition of a viscoelastic solid model. *Nonlinear Analysis: Real World Applications*, 11(2), pp. 907–912.

Lokander M. & Stenberg B. 2003. Improving the magnetorheological effect in isotropic magnetorheological rubber materials. *Polymer Testing* 22: 677–680.

Marvalova, B. & Petrikova, I. 2015. Modelling of Magneto-mechanical coupling in COMSOL Multiphysics. In Marvalova (ed.) *Proceedings of the 9th European conference on constitutive models for rubber (ECCMR); Prague, 30 August – 2 September* (pp. 657–662) Balkema.

Molchanov V.S. et al 2014. Viscoelastic properties of magnetorheological elastomers for damping applications. *Macromolecular Materials and Engineering*, 299(9), pp. 1116–1125.

Nadzharyan T.A. et al. 2016. A fractional calculus approach to modeling rheological behavior of soft magnetic elastomers. *Polymer*, 92, pp. 179–188.

Petrikova I. et al. 2015. Experimental research of the damping properties of magnetosensitive elastomers. In Marvalova (ed.) *Proceedings of the 9th European conference on constitutive models for rubber (ECCMR); Prague, 30 August – 2 September* (pp. 663–668) Balkema.

Petrikova I., Marvalova B. & Nhan P.T. 2011. Influence of thermal ageing on mechanical properties of styrene-butadiene rubber. In Jerrams S. (ed.) *Proceedings of the 7th European conference on constitutive models for rubber (ECCMR); Dublin, 20–23 September* (pp. 77–83) Balkema.

Shen Y. Golnaraghi M.F. & Heppler G.R. 2004. Experimental research and modeling of magnetorheological elastomers. *Journal of Intelligent Material Systems and Structures*, 15(1), pp. 27–35.

Stolbov O.V. et al. 2010. Low-frequency rheology of magnetically controlled elastomers with isotropic structure. *Polymer Science Series A*, 52(12), pp. 1344–1354.

Sutrisno J., Purwanto A. & Mazlan S.A., 2015. Recent progress on magnetorheological solids: materials, fabrication, testing, and applications. *Advanced engineering materials*, 17(5), pp. 563–597.

Xin F.L., Bai X.X. & Qian, L.J. 2016. Modeling and experimental verification of frequency-, amplitude-, and magneto-dependent viscoelasticity of magnetorheological elastomers. *Smart Materials and Structures*, 25(10), p. 105002.

Yang J. et al. 2013. Experimental study and modeling of a novel magnetorheological elastomer isolator. *Smart Materials and Structures*, 22(11), p. 117001.

Yu, Y., Li Y. & Li J. 2015. Parameter identification and sensitivity analysis of an improved LuGre friction model for magnetorheological elastomer base isolator. *Meccanica*, 50(11), pp. 2691–2707.

Zhang W., Gong X., Xuan S. & Jiang W. 2011. Temperature-dependent mechanical properties and model of magnetorheological elastomers. *Industrial & Engineering Chemistry Research*, 50(11), pp. 6704–6712.

Zhu J.T., Xu Z.D. & Guo Y.Q. 2012. Magnetoviscoelasticity parametric model of an MR elastomer vibration mitigation device. *Smart Materials and Structures*, 21(7), p. 075034.

Constitutive Models for Rubber X – Lion & Johlitz (Eds)
© 2017 Taylor & Francis Group, London, ISBN 978-1-138-03001-5

Efficiency of rubber material modelling and characterisation

H. Donner, L. Kanzenbach & J. Ihlemann
Professorship of Solid Mechanics, Chemnitz University of Technology, Chemnitz, Germany

C. Naumann
Freudenberg Technology Innovation SE & Co. KG, Weinheim, Germany

ABSTRACT: Efficiency requirements concerning rubber material modelling and characterisation methods are evolving because of high calculation efforts due to the complex behaviour of industrial rubber materials, large elastic as well as inelastic strains and increasing precision requirements for industrial components.

Two main contributions to the improvement of efficiency are presented in this paper: A new specimen setup for combined high precision tension-compression measurements with an extended strain range from a compression strain of –45% up to a tension strain of 400% and an approach for a new rheological element with time-rescaling invariance. This means a stress response invariance under uniform time-rescaling of the strain history. The element consists of two coupled differential equations. It is qualitatively compared to a generalized Maxwell model with constant loss modulus.

Moreover, a new, interesting and efficient approach for modelling complex inelastic material behaviour at large strains based on rheological models is presented. The basic kinematic assumption is the additive decomposition of the deformation rate. The framework is implemented in Abaqus via the user interface UMAT using an objective time stepping scheme. The approach is not at all limited to polymers.

1 INTRODUCTION

Numerous technical applications of rubberlike material lead to large deformations. Furthermore, inelastic effects are often of great significance to the function of rubber containing structures. Experimental investigation of filled rubber reveals several complex properties like hysteresis, permanent set and Mullins-effect. The latter is also associated with an induced anisotropy. In the time-domain relaxation and creep are observed. These properties require highly efficient methods concerning experimental characterisation, modelling, parameter identification, numerics and FE-implementation.

Furthermore, the efficiency requirements for industrial components are continuously increasing with time. Related to that, the precision requirements for simulation results and also the acceptance of higher simulation costs are increasing, too. In this context, also the efficiency requirements concerning rubber material modelling and rubber characterisation methods are evolving.

The properties of rubber materials emerge in a complex manner from simple interactions of the basic components. Thus, in many situations phenomenological descriptions of the material behaviour together with appropriate characterisation methods are a good choice. To this end, rheological models, i.e. nested serial and parallel connections of simple elements, are a flexible and therefore adaptive and powerful tool. However, to be successful, the basic elements have to cover all the basic properties of the material to be modelled. This is why a new rheological element (with a time-rescaling invariant behaviour) is proposed within this contribution.

2 SPECIMEN SETUP FOR HIGH PRECISION TENSION-COMPRESSION TESTS

For the phenomenological understanding of rubber materials, a new specimen setup, which enables high precision tension-compression-tests, is presented. A homogeneous deformation with compression strains up to –45% and tension strains up to 400% can be achieved by a special mounting design.

2.1 *Development of a special mounting geometry*

In Figure 1 the principal idea of the mounting algorithm is presented. With a prestressed modal

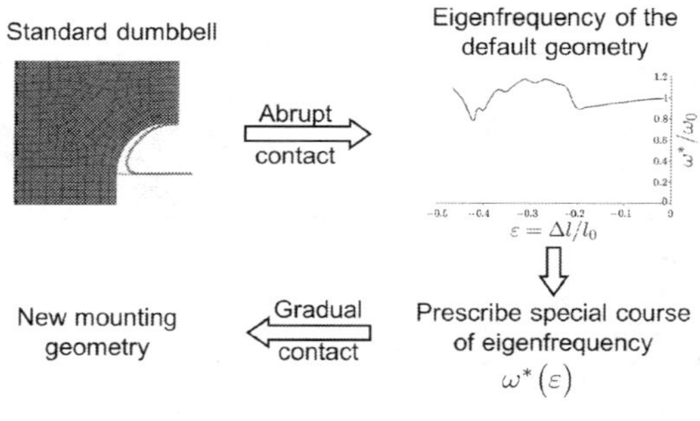

Figure 1. Principal idea of the mounting algorithm.

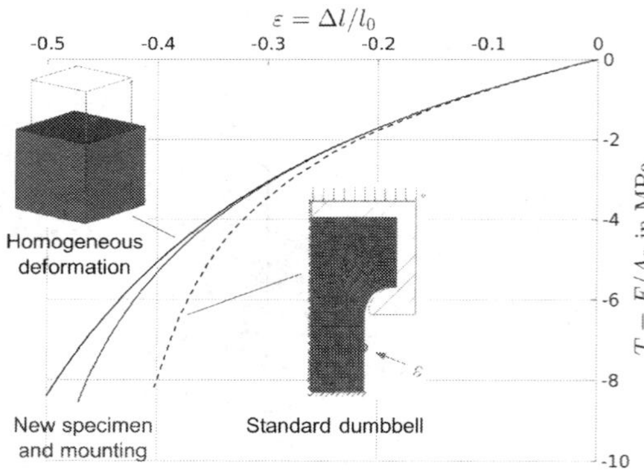

Figure 2. Stress-strain diagram for the new specimen (cf. Kanzenbach et al. 2016 Kanzenbach et al. 2016).

analysis the eigenfrequencies ω_j of a slender structure can be calculated, see equation (1).

$$([K_L^u]+[K_{NL}^u]-\omega_j^2(u)[M])[\phi_j]=[0] \quad (1)$$

If the value of ω_1 is zero, the stiffness matrices $[K_L^u]$ and $[K_{NL}^u]$ get singular and the structure is endangered by buckling. The value ω_1 describes a measure for the distance to instability, i.e. the first course of eigenfrequencies gives information about the buckling risk of a slender structure (Figure 1 right). Our approach to develop new mounting geometries is to prescribe a special course of eigenfrequencies and calculate the corresponding mounting. This method leads to an gradual contact with a homogeneous stress state between the specimen and the mounting geometry. In comparison to the standard dumbbell (Alshuth et al. 2007), no abrupt contact and no inhomogeneous stress states occur.

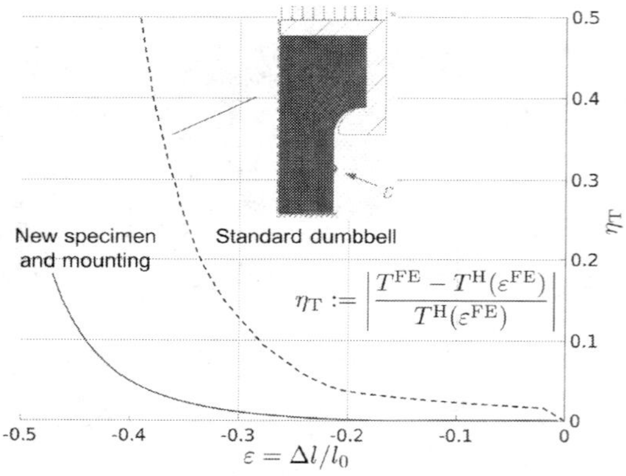

Figure 3. Homogeneity of the new specimen.

2.2 *FE-Simulation and homogeneity investigations*

With the developed mounting geometry, the homogeneity behaviour of the new specimen can be investigated in more detail. In Figure 2 the stress-strain curves for the new specimen (Kanzenbach et al. 2016) and the standard dumbbell (Alshuth et al. 2007) can be seen. There is only a small deviation between the stress state of the new specimen setup and the homogeneous deformation state. The benefit in homogeneity and maximum achievable level of compression can be seen in Figure 3. It demonstrates that the new method is a powerful tool to design an optimised specimen by taking into account its stability. The depicted measurement error η_T describes the relative difference between the stress state of the specimen T^{FE} and the homogeneous stress state $T^H(\varepsilon^{FE})$.

2.3 *Experimental investigations*

Additionally, the new specimen setup was tested experimentally. The investigations were performed with filled rubber and EPDM. First, the stability behaviour of the new specimen setup was proved. In Figure 4, the results of a simple buckling test, which was conducted with and without mounting geometry, can be seen. The value of stability κ is defined as the sum of the left and right deflection of the specimen, divided by the initial radius. Furthermore, the same tests were performed by FE-simulations with an inelastic material model (Gelke and Ihlemann 2016) and a small force disturbance. The good agreement between experiment and simulation is also shown in Figure 4. For the strain measurement an optical measurement system (greyscale correlation) was used. Figure 5 shows the new specimen at the maximum achievable compression. Even for such high compression no buckling

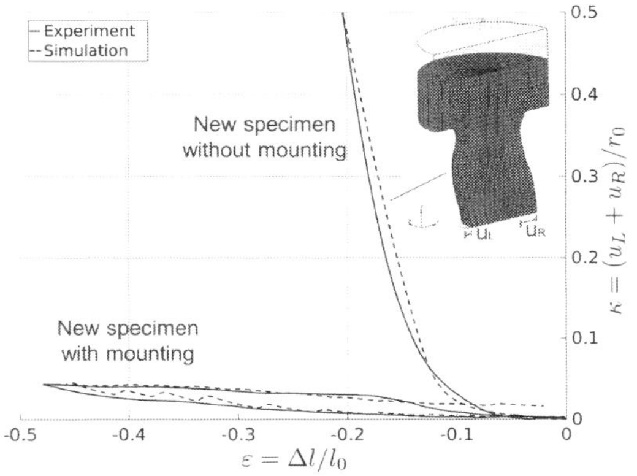

Figure 4. Stability test with and without mounting geometry.

Figure 5. Stability test with the new specimen setup.

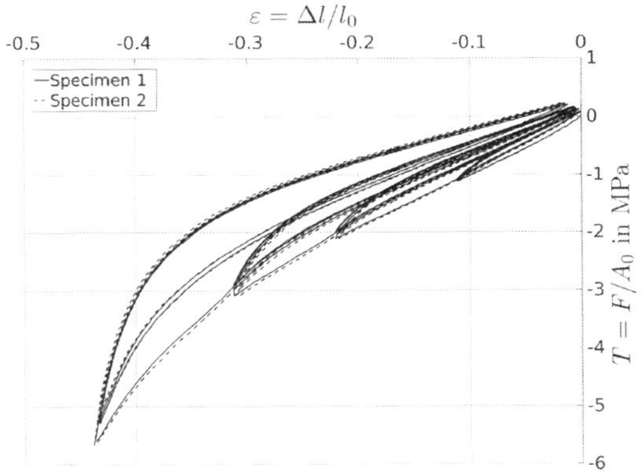

Figure 6. Stress-strain diagram in compression (filled EPDM).

occurs and the strain field is nearly homogeneous (Kanzenbach et al. 2016).

Additionally to the stability tests, a multi-hysteresis test in compression and tension was performed. Figures 6 and 7 show the corresponding

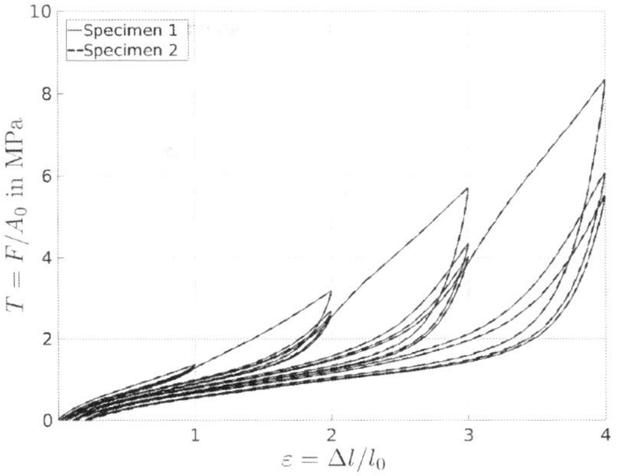

Figure 7. Stress-strain diagram in tension (filled NR) (cf. Kanzenbach et al. 2016).

stress-strain curves, with a high repeatability for different specimens. The test sequence was designed in such a way that all typical properties of rubber like hysteresis, permanent set, softening and Mullins-effect can be detected easily.

The new specimen setup has a large range of applications. Cyclic tension-compression tests as well as relaxation and recovery tests in compression and tension can also be done. Consequently, there is no longer a need for different types of specimens in tension and compression. In addition to rubber characterisation, the new specimen setup can be used for the identification of model parameters (Gelke and Ihlemann 2016).

3 OBJECTIVE EULERIAN FRAMEWORK FOR RHEOLOGICAL MODELS

Rheological models, i.e. nested serial and parallel connections of elements representing elasticity, viscosity and plasticity, are state of the art. They are often used to find phenomenological driven constitutive models for complicated material behaviour. A new approach combines advantages and avoids drawbacks of current approaches and is suitable for rubber-like materials, inelasticity and large strains.

Currently the multiplicative split of the deformation gradient is the dominant basis for rheological models. However, this approach is not unique concerning rotations of the incompatible intermediate configuration and it is not invariant under a permutation of the sequence of rheological elements within a serial connection. Moreover, the handling is complex due to the different stresses and strains on particular configurations.

Instead, we follow the idea published by Palmow (1984) to combine the advantages of the additive decomposition of the deformation rate $\underline{\underline{D}}$ with the concept of hyperelasticity. The basic equations for $\underline{\underline{D}}$ and the Kirchhoff stresses $\underline{\underline{T}}$ for parallel and serial connections are straightforward.

$$\text{parallel}: \underline{\underline{D}} = \underline{\underline{D}}_1 = \underline{\underline{D}}_2; \underline{\underline{T}} = \underline{\underline{T}}_1 + \underline{\underline{T}}_2 \tag{2}$$

$$\text{serial}: \underline{\underline{D}} = \underline{\underline{D}}_1 = \underline{\underline{D}}_2; \underline{\underline{T}} = \underline{\underline{T}}_1 = \underline{\underline{T}}_2 \tag{3}$$

Moreover, the balance of the stress power densities is very simple, too.

$$\underline{\underline{D}} \cdot\cdot \underline{\underline{T}} = \underline{\underline{D}}_1 \cdot\cdot \underline{\underline{T}}_1 + \underline{\underline{D}}_2 \cdot\cdot \underline{\underline{T}}_2 \tag{4}$$

Thus, for each basic element we need constitutive equations, which enables the computation of the elements stress tensor as a function of the history of its deformation rate. This simple context is very advantageous for the integration of new elements (e.g. the rescaling invariant element described in section 4).

The evolution of the deformation of element i is defined by an differential equation for the left Cauchy-Green tensor $\underline{\underline{b}}$, its isochoric part $\overline{\underline{\underline{b}}}$ and its Jaumann rate $\overset{*}{\overline{\underline{\underline{b}}}}$.

$$\overset{*}{\overline{\underline{\underline{b}}}}_i = \underline{\underline{D}}_i^{\mathrm{D}} \cdot \overset{*}{\overline{\underline{\underline{b}}}}_i + \overset{*}{\overline{\underline{\underline{b}}}}_i \cdot \underline{\underline{D}}_i^{\mathrm{D}} \tag{5}$$

The specific, simple and well-known example of a Maxwell element (cf. section 4) demonstrates the algorithm. For a given $\underline{\underline{D}}$ the split into $\underline{\underline{D}}_{el}$ and $\underline{\underline{D}}_{vis}$ has to be determined.

$$\text{Kinematics}: \underline{\underline{D}}^{\mathrm{D}} = \underline{\underline{D}}_{el} + \underline{\underline{D}}_{vis}. \tag{6}$$

$$\text{Viscosity}: \underline{\underline{T}}_{vis} = 2\eta \underline{\underline{D}}_{vis}^{\mathrm{D}}. \tag{7}$$

The elastic deformation results from the ODE. From this we conclude the elastic stresses (Neo-Hookean type).

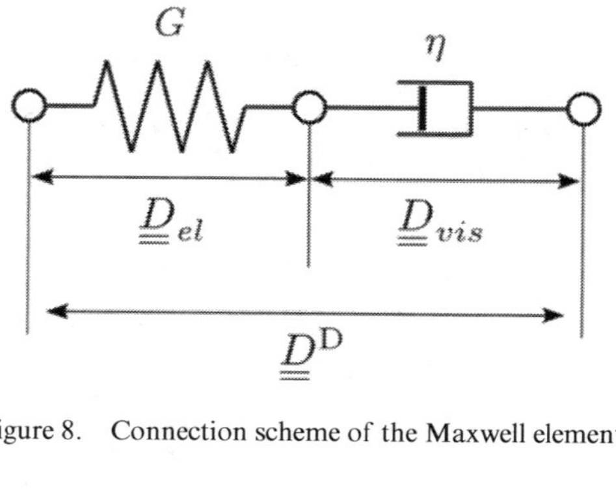

Figure 8. Connection scheme of the Maxwell element.

$$\overset{*}{\overline{\underline{\underline{b}}}}_{el} = \underline{\underline{D}}_{el}^{\mathrm{D}} \cdot \overline{\underline{\underline{b}}}_{el} + \overline{\underline{\underline{b}}}_{el} \cdot \underline{\underline{D}}_{el}^{\mathrm{D}}; \underline{\underline{T}}_{el} = G\overline{\underline{\underline{b}}}_{el}^{\mathrm{D}} \tag{8}$$

The (iterative) determination of $\underline{\underline{D}}_{el}$ and $\underline{\underline{D}}_{vis}$ results from the stress equivalence (cf. equation (3)).

$$\underline{\underline{T}}_{el} = \underline{\underline{T}}_{vis} \left(= \underline{\underline{T}}^{\mathrm{D}}\right) \tag{9}$$

Finally, volumetric stresses have to be added. The basis of the numerics are the incremental kinematics and the co-rotational framework proposed by Rashid (1993) with the relative deformation gradient $\Delta\underline{\underline{F}}$ and its polar decomposition.

$$\Delta\underline{\underline{F}} = \underline{\underline{F}}(t+\Delta t) \cdot \underline{\underline{F}}^{-1}(t) = \Delta\underline{\underline{V}} \cdot \Delta\underline{\underline{R}} \tag{10}$$

This results in a time integration scheme which allows an objective estimation of the deformation rate on the basis of FEM input data:

$$\underline{\underline{D}} = \frac{\ln\left(\left(\Delta\underline{\underline{V}}\right)\right)}{\Delta t} \tag{11}$$

and objective solutions for the differential equations of the elastic elements.

$$\overline{\underline{\underline{b}}}_i(t+\Delta t) = \exp\left(\left(\underline{\underline{D}}_i^{\mathrm{D}}\Delta t\right)\right) \cdot \Delta\underline{\underline{R}} \cdot \overline{\underline{\underline{b}}}_i(t) \cdot \Delta\underline{\underline{R}}^{\mathrm{T}} \cdot \exp\left(\left(\underline{\underline{D}}_i^{\mathrm{D}}\,\Delta t\right)\right) \tag{12}$$

The framework is implemented in Abaqus via the user subroutine UMAT (Donner and Ihlemann (2017)).

This approach combines thermodynamic consistency, strong objectivity, invariance under the change of reference configuration (Shutov and Ihlemann (2014)), exact inelastic incompressibility and invariance under permutations of serial-connected elements with an efficient and robust FEM-implementaion and simple connection results, also in the case of complex rheological models. It is suitable for large strain analysis and avoids artificial and incompatible intermediate configurations and hypoelasticity.

4 TIME-RESCALING INVARIANCE OF CONSTITUTIVE MODELS

Time-rescaling invariance means a new class of constitutive descriptions. This class is characterised by embracing characteristics which seemed to be exclusive in the past for elastoplasticity or for viscoelasticity. In the following, first results of ongoing research are presented.

4.1 *Experimental results and state of the art*

In the case of cyclic loading numerous industrial used rubber materials show very low rate dependency over a wide frequency range. Often the amount of this dependency is close to the experimental accuracy (see e.g. Ihlemann (2003)). Therefore, several material models have been designed partially or even in total rate independent (e.g. the model proposed by Rabkin (Freund et al. 2011) or the MORPH-model (Besdo and Ihlemann 2003)).

But cyclic tests, especially with different periodic variations of strain with time, give also contrary indications. In the case of triangular strain signals, the stress-strain-cycles always show sharp points at the extremities. But with sinusoidal signals the extremities are rounded in a small but clearly detectable area (see Figure 9). However, in the case of perfect rate independency the cycles would be identical. Even more pronounced deviations arise, if the sine function is cubed (see Figure 10).

$$\varepsilon(t) = \hat{\varepsilon}\sin^3(\omega t) \tag{13}$$

Beyond cyclic processes viscoelastic effects become obvious especially in creep and relaxation tests. It should be noted that within this paper, the terms creep and relaxation are used in a general manner. A relaxation/creep phase means a loading period with constant strain/stress only. Sometimes these terms are additionally associated with conditions for the foregoing loading process or with the kind or response of the material. This is not implied in this paper.

To model such a material behaviour, often generalized Maxwell models are used (for instance Landgraf (2016) provides a detailed overview and a very fresh literature survey). Some problems, especially with the considered frequency range, are discussed by Diercks (2016). Calculation times for those models are high because of the high number of differential equation which have to be solved. Another approach is viscoelasticity with process-dependent viscosities (Haupt and Lion (2002), Haupt (2002)).

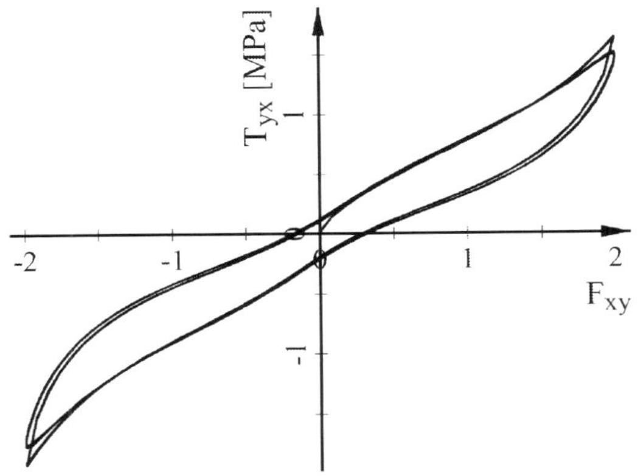

Figure 9. Simple shear test with carbon black filled SBR; performed by TARRC: Experimental steady state loops obtained with sinusoidal (the loop with rounded extremities) or triangular variation of strain with time (Besdo et al. (2003)).

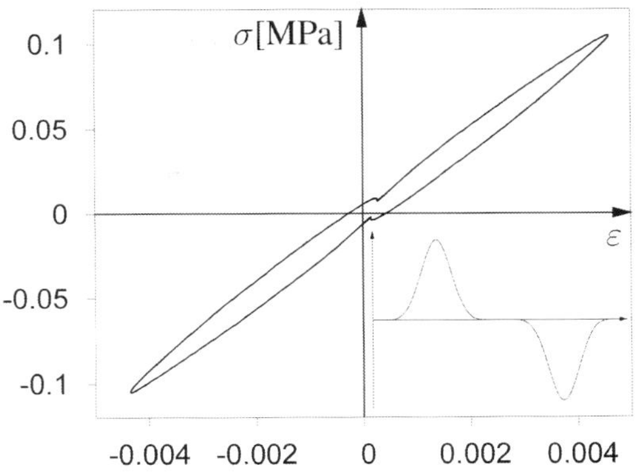

Figure 10. Tension-compression test with carbon black filled EPDM (cf. section 2); small deformations for approximate tension-compression symmetry: Experimental steady state loop obtained with an input strain signal following a cubed sine function (cf. the small diagram).

Ahmadi et al. (2007) showed some novel experiments (primary stages in Ahmadi et al. 2005) which allow new insights. These experiments consist of cyclic phases with changes of amplitude and frequency in between and additional phases of constant strain (relaxation phases). The amplitude and frequency changes are made in such a way that the strain rate is always continuous (cf. Figure 11). The experiments were made with two different materials and with input signals of triangular as well as sinusoidal variations of strain with time. Moreover and of particular interest in this context, the experiments were repeated under uniform time-rescaling of the prescribed strain history.

$$\varepsilon_1(t) = \varepsilon_2(\alpha t) \quad \forall\, t \quad \text{with:} \quad \alpha > 0 \tag{14}$$

Thus, all of the associated strain rates differ by the same factor α.

$$\dot{\varepsilon}_1(t) = \alpha\,\dot{\varepsilon}_2(\alpha t) \quad \forall\, t \tag{15}$$

Remarkably, the measured stress response shows the same time-rescaling in high accuracy (cf. Ahmadi et al. (2007)).

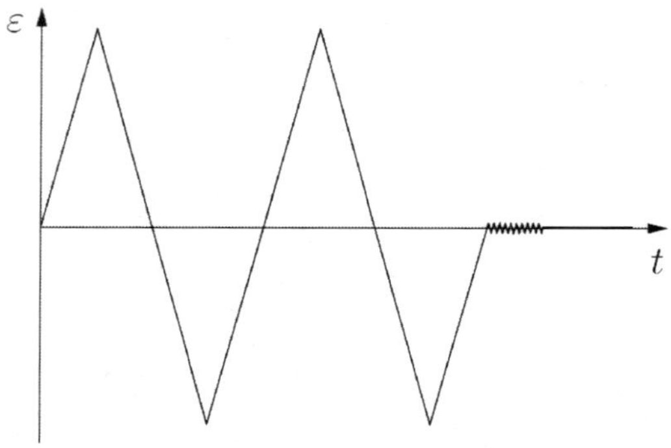

Figure 11. Schematic plot of the input signals for Figures 12 and 13. The signal is intended for investigating the influence of small perturbations like noise within an experimental setup. During the cyclic phases the amount of the strain rate is constant.

$$\sigma_1(t) = \sigma_2(\alpha t) \quad (16)$$

This means for instance, that a faster loading history results in a reduced relaxation time. This stress response invariance under uniform time-rescaling of the strain history is what we abbreviate with timerescaling invariance.

Somewhat similar formulations can be found in the literature to define rate independency. Mielke and Roubíček (2015) and already Truesdell and Noll (1965) in the context of hypoelasticty stated that rate independency means that any change of time-scale of the input leads to the same change of time-scale of the output. In contrast to this, in this paper we do not consider arbitrary changes of time-scale with a variable α (cf. equation (14)), but only a uniform rescaling with a constant α throughout the loading history. Thus, rate independency (elasticity, elastoplasticity) is a simple case of time-rescaling invariance. However, we are mainly interested in more complex timerescaling invariant models which show relaxation and creep, because this combination of characteristics is close to rubber materials. Up to now, those models could not be found in the literature.

4.2 *Modelling time-rescaling invariance*

In the case of a simple cyclic loading (for instance a triangular strain-signal with constant amplitude and constant frequency) of a time-rescaling invariant model which tends to a steady state in such a situation, storage and loss modulus could be calculated and would be frequency independent because of the invariance of the model. Thus, in that case, the phenomenology would be the same as in the case of a rate independent model.

A constant loss modulus over an arbitrary frequency range is realisable with an appropriate adjusted generalised Maxwell model (cf. for instance Landgraf (2016)). But those models necessarily show a significantly increasing storage modulus. Thus, time-rescaling invariance is not achievable with generalized Maxwell models. However, Scheffler (2009) showed that those models are capable to reproduce at least the time-rescaling invariant relaxation behavior within the measurements shown by Ahmadi et al. (2007). That is why only those generalized Maxwell models with constant loss modulus will be used within this paper as a reference to evaluate timerescaling invariant models (section 4.3).

The special experiments presented by Ahmadi et al. (2007) and described above as well as considerations of artificial intrinsic time scales (Haupt and Lion (2002), Haupt (2002)) inspired Naumann to search for a model with a rubber-like time-rescaling invariance (Naumann and Ihlemann (2014)). As an initial point he chose the differential equation of a single Maxwell element, a serial connection of a spring and a dashpot.

$$\dot{\sigma} = E\dot{\varepsilon} - \beta\sigma \quad (17)$$

Naumann's fundamental idea was to formulate an inner evolution equation for β with respect to the strain history. He integrated such an approach into a one-dimensional version of the MORPH-model. Finally, he used a 3-d generalisation technique proposed by Freund et al. (2011) to implement an advanced MORPH-model with a realistic relaxation behavior into the finite element method (Naumann and Ihlemann (2014)). Within this application the timerescaling invariance was only a special case. Slight deviations from this idealisation were also possible.

On that basis a rheological element with timerescaling invariance is the objective of current research in this field and of this paper. As a first step the theory is developed considering the one-dimensional case and geometrically linear theory.

First, the basic differential equation (17) with time dependent variable β is specified for the two input signals in equation (14).

$$\begin{aligned} \dot{\sigma}_1(t) &= E\dot{\varepsilon}_1(t) - \beta_1(t)\,\sigma_1(t) \\ \dot{\sigma}_2(\alpha t) &= E\dot{\varepsilon}_2(\alpha t) - \beta_2(\alpha t)\,\sigma_2(\alpha t) \end{aligned} \quad (18)$$

Let $\dot{\sigma}_1(t)$ be a solution of the first differential equation. The question now arises which relationship between $\beta_1(t)$ and $\beta_2(\alpha t)$ is needed to ensure that $\sigma_2(\alpha t) = \sigma_1(\alpha t)$ (cf. equation (16)) is a solution of the second differential equation?

Equation (16) implies:

$$\sigma_2(\alpha t) = \sigma_1(t) \Rightarrow \dot{\sigma}_1(t) = \alpha\dot{\sigma}_2(\alpha t). \quad (19)$$

The σ_1-version of equation (18) then reads:

$$\alpha\dot{\sigma}_2(\alpha t) = E\dot{\varepsilon}_1(t) - \beta_1(t)\,\sigma_1(t). \tag{20}$$

Inserting equations (15) and (16) leads to:

$$\alpha\dot{\sigma}_2(\alpha t) = E\alpha\dot{\varepsilon}_2(\alpha t) - \beta_1(t)\sigma_2(\alpha t). \tag{21}$$

This equation is identical with the σ_2-version of (18) and thus, time-rescaling invariance is achieved, if β fulfils:

$$\beta_1(t) = \alpha\beta_2(\alpha t). \tag{22}$$

Beside this important condition for time-rescaling invariance, thermodynamic consistency requires $\beta > 0$ (Haupt and Lion 2002). A simple and already known solution is the so-called endochronic element (Krawietz (1986), Kießling et al. (2016)) with k as a material constant.

$$\beta(t) = k|\dot{\varepsilon}(t)| \rightarrow \dot{\sigma} = E\dot{\varepsilon} - k|\dot{\varepsilon}|\sigma \tag{23}$$

This model is rate independent and therefore also time-rescaling invariant. Thus, equation (22) is validated. But the endochronic element is not the form which is searched for in this context, because stresses are constant during phases of constant strain. More interesting models are accessible with a differential equation for an evolving β. The time-rescaling properties of the rate of β follow from differentiating equation (22).

$$\beta_1(t) = \alpha\beta_2(\alpha t) \Rightarrow \dot{\beta}_1(t) = \alpha^2\dot{\beta}_2(\alpha t) \tag{24}$$

Therefore, for $\dot{\beta}$ terms like ε^2, β^2, or $\varepsilon\,\beta$ come into consideration.

Some early approaches of evolution equations for β had problems with the requirement that small perturbations of the input signal (for instance noise within an experimental setup) should cause likewise small deviations of the response signal. This requirement is also critical in the case of the endochronic element. To consider those effects, an input signal with three phases is used (see Figure 11). The first two phases have periodic triangular signals with different amplitudes, but with the same absolute value of the strain rate. The third phase is a phase of absolutely constant strain. Note that the second phase with the much smaller amplitude represents a phase of almost constant strain. Thus, the behaviour in the second and in the third phase should be similar, at least in the case of a very small amplitude during the second phase.

Figure 12 shows the desired behavior using the example of a Maxwell element.

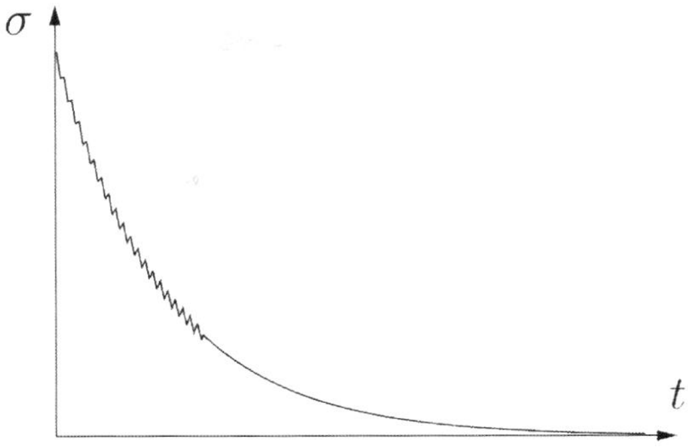

Figure 12. Stress response of a Maxwell element to a strain input signal like the one in Figure 11 with an amplitude ratio of 300. The stress signal during the second (small amplitude) and third (relaxation) phase is shown.

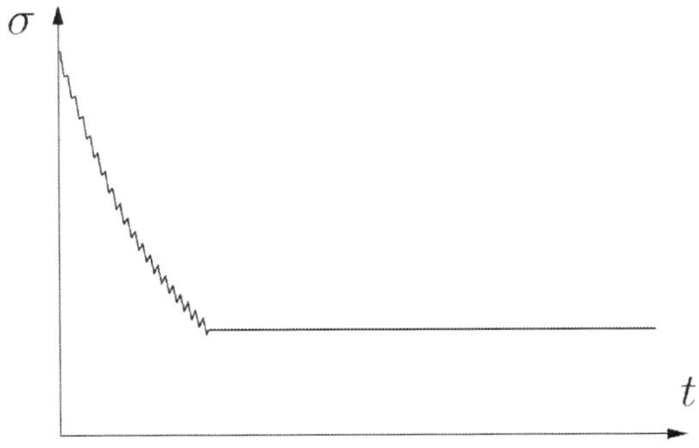

Figure 13. Stress response of an endochronic element analogue to Figure 12.

In contrast, the results for the endochronic element (Figure 13) show a kink between the second phase with small cycles and the phase of absolutely constant strain. This can be explained by the fact that $|\dot{\varepsilon}|$ is constant during the first two phases because of the triangular input signal and the special adjustment of amplitudes and frequencies. Thus, during these two phases equation (23) is equal to the evolution equation of a simple Maxwell element (cf. equation (17)). This is why Figures 12 and 13 are equal in the beginning. In contrast, throughout the third phase the endochronic element's characteristic (cf. equation (23)) is reduced to elasticity because of: $\dot{\varepsilon} = 0$. Thus, during this phase of constant strain the stresses are constant, too. The resulting kink would be just as striking, if the amplitude of the second phase is invisibly small. The stress response of such a material would be strongly influenced by low measurement noise. Such a material behaviour seems to be unthinkable.

Our search for a time-rescaling invariant, but not rate independent model without those kinks led so far to the following approach for an evolution equation of β according to equation (24) with two material constants p and k in combination with the evolution equation (17) for the stresses.

$$\dot{\sigma} = E\dot{\varepsilon} - \beta\sigma \quad \text{with:} \quad \dot{\beta} = p(k|\dot{\varepsilon}| - \beta)|\dot{\varepsilon}| \tag{25}$$

This model is investigated below. First of all timerescaling invariance should be proven. To this end, corresponding to equation (18), the just introduced evolution equation for β is specified for the two input signals in equation (14).

$$\begin{aligned} \dot{\beta}_1(t) &= p(k\,|\,\dot{\varepsilon}_1(t)\,| - \beta_1(t)\,|\dot{\varepsilon}_1(t)| \\ \dot{\beta}_2(\alpha t) &= p(k\,|\,\dot{\varepsilon}_2(\alpha t)\,| - \beta_2(\alpha t)\,|\dot{\varepsilon}_2(\alpha t)| \end{aligned} \tag{26}$$

Actually, these two variants coincide if equations (14), (22) and (24) are applied. Thus, if $\beta_2(t)$ is a solution of the second variant and if equations (14) is fulfilled, then the function $\beta_1(t)$ following equation (24) solves the first variant and therefore time-rescaling invariance is fulfilled.

However, this model expressed by the two coupled differential equations (25) is not rate independent. The model rather shows a distinct and at least qualitatively rubber-typical relaxation behaviour (cf. Figure 14).

As a reminder, because of the time-rescaling invariance of the model, the stress-strain curve given in Figure 14 would remain completely unchanged, even if the input signal as a whole would be multiplied or divided by a positive, but otherwise arbitrary constant α (cf. equation (16)).

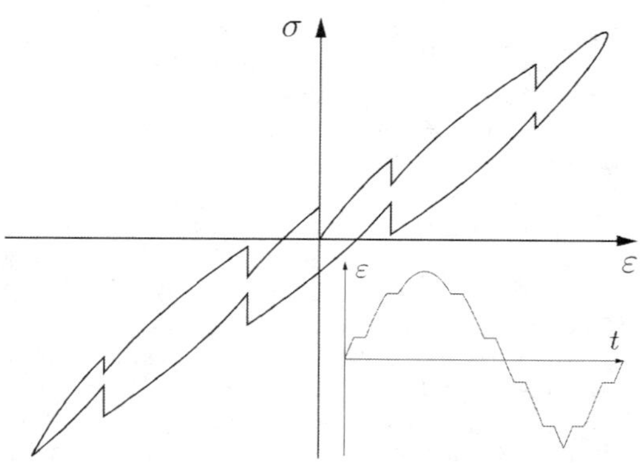

Figure 14. Rubber-like prediction of the new material model given in equation (25) obtained with a cyclic strain input signal with relaxation phases (small diagram). For better comparisons with rubber material behaviour and with other models the new model is supplemented by a parallel spring.

In the simple case of a perfect relaxation phase $(\dot{\varepsilon} = 0)\ \dot{\beta}$ is zero throughout the whole phase. Thus, β is constant and the model behaves exactly like a Maxwell element (cf. equation (17)) - with stress relaxation. In other situations $(\dot{\varepsilon} \neq 0)\ \beta$ trends to $k\,|\,\dot{\varepsilon}\,|$ (cf. equation (25)). The velocity of this evolution is controlled by parameter p.

Concerning input signals superimposed by noise, again the simple testing signal shown in Figure 11 is suitable to understand the model behaviour. Up to the relaxation phase with constant strain, the absolute value $|\dot{\varepsilon}|$ of the strain rate is constant throughout. Thus, after a transient phase during the big cycles, β is approximately equal to $k\,|\dot{\varepsilon}|$. From this moment on, the new model behaves like a Maxwell element with the corresponding material constants. From the moment of the start of the relaxation phase with $|\dot{\varepsilon} = 0|$ the value of β will keep its momentary value and thus, the model will continue behaving like the corresponding Maxwell element (cf. Figure 12) with stress relaxation, but without a kink.

4.3 *Rheological element of time-rescaling invariance*

The proposed model of two coupled differential equations (25) simulates a rate dependent material with rubber-typical relaxation behaviour, which is perfectly time-rescaling invariant. Thus, with these two differential equations the material behaviour is already defined for all frequencies in that way, that storage and loss modulus are constant.

In the following, this model is compared to a generalized Maxwell model with approximately constant loss modulus over a suitable frequency range (see also the beginning of section 4.2). The comparison is purely qualitative (therefore no axis scaling within simulation plots). No parameter identification has been done. The parameters of the two models have been adjusted to each other to be comparable in a good manner. Also for this purpose the new model is supplemented by a parallel spring, because of the high storage modulus of the generalised Maxwell model. The same parallel connection was already used to produce Figure 14.

First of all the stress responses to sinusoidal and triangular strain signals with equal amplitudes and frequencies are considered (Figures 15 and 16). Due to the adjustment of the parameter values of the two models, the stress responses to the triangular signals are similar. But the effects of the sinusoidal input signal differ significantly. In the case of the generalized Maxwell model the form of the stress-strain curve is almost elliptical and the hysteresis is increased (compared to the triangular input signal). In contrast, the relation of the two stress responses of the new model (Figure 16) is close to the relation within the measurements

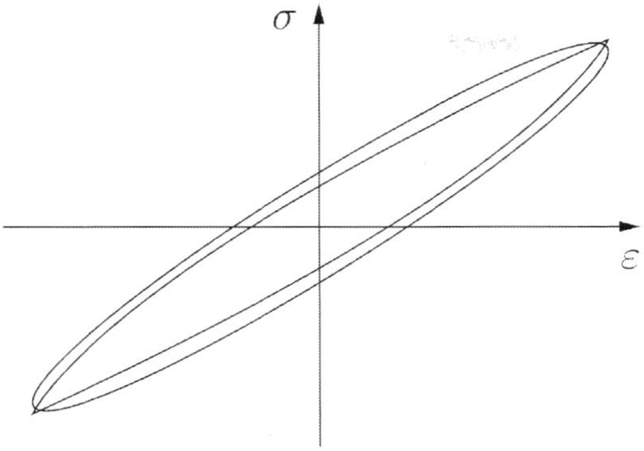

Figure 15. Steady state predictions of a generalized Maxwell model obtained with sinusoidal (the loop with rounded extremities) or triangular variation of strain with time.

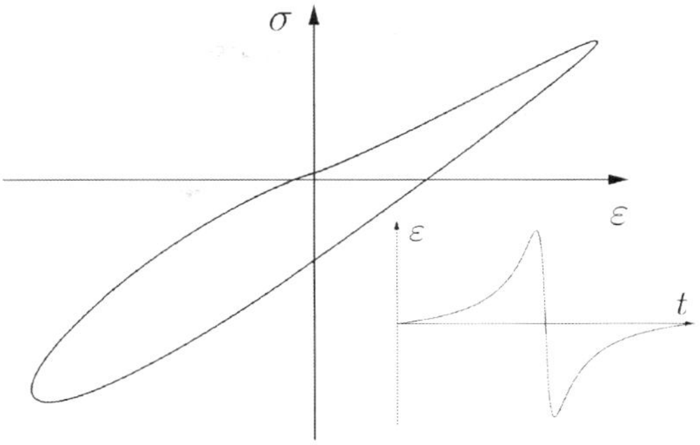

Figure 17. Steady state prediction of a generalized Maxwell model obtained with the input strain signal given in the small diagram (distorted harmonic function over time).

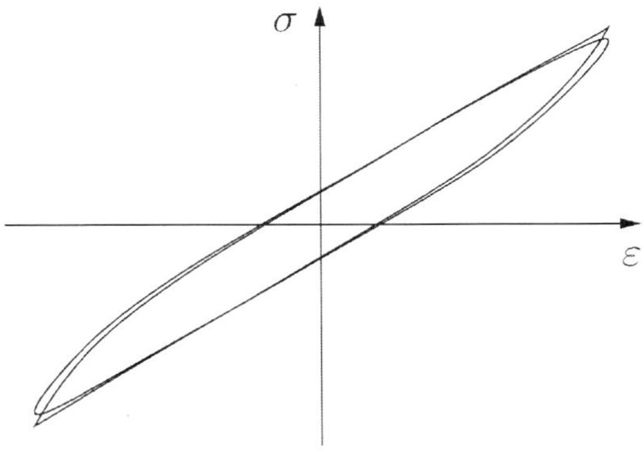

Figure 16. Steady state predictions of the new material model given in equation (25) obtained with sinusoidal (the loop with rounded extremities) or triangular variation of strain with time.

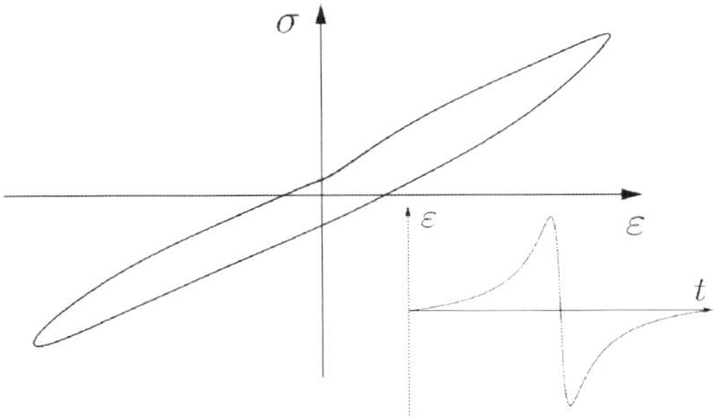

Figure 18. Simulation with the new material model given in equation (25) analogous to figure 17.

shown in Figure 9. The rounding of the extremities is restricted to a small area and the hysteresis is almost unchanged.

Note that in the case of a strain input signal in the form of a cubed sine signal (equation (13)) both models give stress-strain curves which are qualitatively close to each other and moreover close to the extraordinary form of the corresponding measured cycle shown in Figure 10. In general, there are a lot of periodic strain signals without qualitative differences in the stress responses of the two models.

A specific search for signals with remarkably different reactions led to two exceptional signals coming from distortions of harmonic functions. The first signal is the basis of Figures 17, 18 and 19 and it is displayed in each of these figures. In the

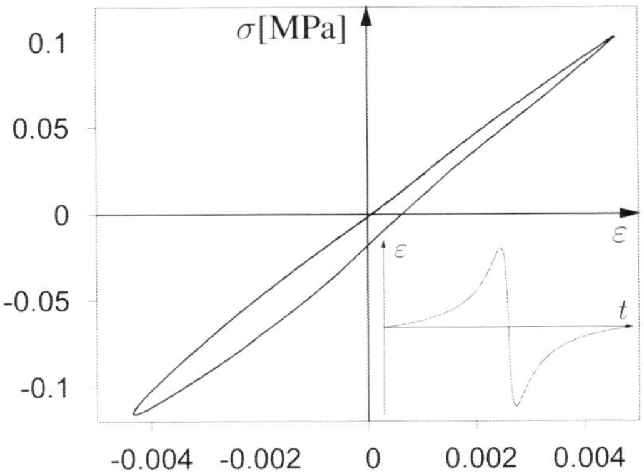

Figure 19. Tension-compression test with carbon black filled EPDM (cf. section 2); small deformations for approximate tension-compression symmetry: Experimental steady state loop obtained with the input strain signal given in the small diagram (distorted harmonic function over time).

same way the second signal is the strain input of Figures 20, 21 and 22. To be actually able to realize those signals even with high-end testing machines (e.g. Zwick/Roell) we developed an external target value control (Kanzenbach et al. 2016). This technique has also been used for the measurement shown in Figure 10.

With the first strain input signal, the generalized Maxwell model shows a remarkable asymmetry concerning the hysteresis area (Figure 17) whereas the new model shows another asymmetry concerning mainly the slope of the cycle (Figure 18). The experiment (Figure 19) shows its own effects, but similar to the generalised Maxwell model, an asymmetry of the hysteresis area is clearly visible.

The second strain signal combines small peaks in the one (the negative) direction and a spacious run of the curve in the opposite direction. To this periodic input signal the generalised Maxwell model responds with a pronounced transient phase. In the steady state (cf. Figure 20) the cycles are shifted so that the stresses during the small peaks are much higher than in the opposite direction. In contradiction to that, the stress strain curve of the new model is almost unchanged compared to the curve with a purely sinusoidal strain signal (Figure 21). With the response of the generalized Maxwell model in mind, however, it is noticeable, that also in the case of the new model the stresses differ with the same tendency. The measurements are somewhere between the two models (Figure 22). However, a vertical shift between the steady state cycles is clearly visible.

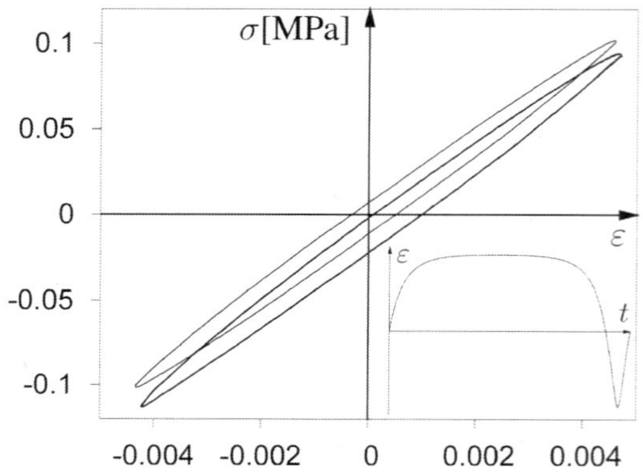

Figure 22. Tension-compression test with carbon black filled EPDM (cf. section 2); small deformations for approximate tension-compression symmetry: Experimental steady state loops obtained with (thin line) a sinusoidal strain signal and (thick line) the input strain signal given in the small diagram (distorted harmonic function over time).

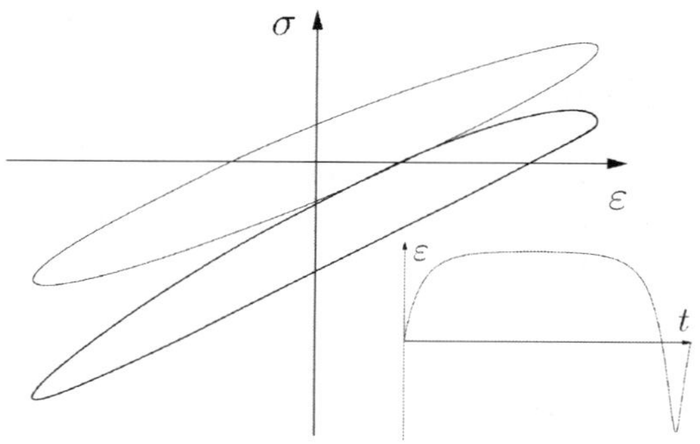

Figure 20. Steady state predictions of a generalized Maxwell model to two input strain variations with time. Thin line: sine function, thick line: distorted harmonic function given in the small diagram.

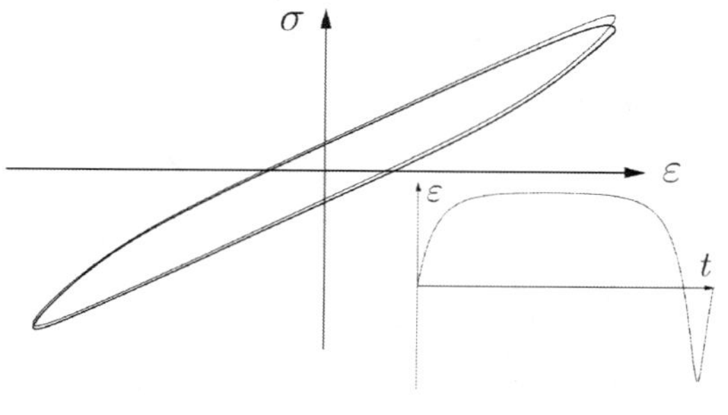

Figure 21. Simulation with the new material model given in equation (25) analogous to Figure 20.

5 CONCLUSIONS

Efficiency requirements concerning rubber material modelling and characterisation methods are evolving because of high calculation efforts due to the complex behaviour of industrial rubber materials, large elastic as well as inelastic strains and increasing precision requirements for industrial components.

Two main contributions to the improvement of efficiency are presented in this paper: A new specimen setup for combined high precision tension compression measurements and an approach for a new rheological element with time-rescaling invariance. Moreover, a draft of a new framework for rheological models is presented, which is ideally suited for the FE-application of the new rheological element within complex parallel and serial connections with other elements. However, the approach is not at all limited to polymers.

The new specimen setup provides new possibilities in the field of rubber characterisation. Several complex properties like hysteresis, permanent set, oftening, Mullins- and Payne-effect can be investigated in tension and compression. The range of strain amplitudes in compression is extended up to –45%, by a nearly homogeneous deformation field.

Time-rescaling invariance means a stress response invariance under uniform time-rescaling of the strain history. Thus, a stress-strain curve would remain completely unchanged, even if an input strain signal as a whole would be multiplied or divided by a positive, but otherwise arbitrary constant. Rate independency is a simple case of time-rescaling invariance. But the material model presented here is far from that and shows stress relaxation and creep.

ACKNOWLEDGEMENT

Parts of this work comes from industry-financed projects. The authors gratefully acknowledge financial support from Vibracoustic GmbH, Freudenberg Technology Innovation, ContiTech AG, Goodyear S.A. and Mehler Engineered Products GmbH. Moreover, we thank Ralf Landgraf for extensive support, fruitful discussions and proofreading the article.

REFERENCES

Ahmadi, H. R., D. Besdo, J. Ihlemann, J. G. R. Kingston, & A. Muhr (2005). Transient response of inelastic materials to changes of amplitude. In P.-E. Austrell and L. Kari (Eds.), *Constitutive Models for Rubber IV*, pp. 305–311. Leiden: Balkema.

Ahmadi, H. R., J. Ihlemann, & A. H. Muhr (2007). Timedependent effects in dynamic tests of filled rubber. In A. Boukamel, L. Laiarinandrasana, S. Méo, and E. Verron (Eds.), *Constitutive Models for Rubber V*, pp. 305–310. London: Taylor & Francis.

Alshuth, T., C. Hohl, & J. Ihlemann (2007). Vergleichende Messungen und Simulationen annähernd homogener Belastungsverteilungen. *PAMM 7*(1), 4060053–4060054.

Besdo, D. & J. Ihlemann (2003). A phenomenological constitutive model for rubberlike materials and its numerical applications. *International Journal of Plasticity 19*(7), 1019–1036.

Besdo, D., J. Ihlemann, J. G. R. Kingston, & A. H. Muhr (2003). Modelling inelastic stress-strain phenomena and a scheme for efficient experimental characterization. In J. J. C. Busfield and A. H. Muhr (Eds.), *Constitutive Models for Rubber III*, pp. 309–317. Lisse: Balkema.

Diercks, N. (2016). *The dynamic behaviour of rubber under consideration of the Mullins and the Payne effect*. München: Verlag Dr. Hut.

Donner, H. & J. Ihlemann (2017). An objective eulerian framework for rheological models at large strains. In A. S. Khan, H.-Y. Yu, and S. Habib (Eds.), *Plasticity in Conventional and Emerging Materials: Theory and Applications*, pp. 70–72. USA: Neat Press.

Freund, M., J. Ihlemann, & M. Rabkin (2011). Modelling of the Payne effect using a 3-d generalization technique for the finite element method. In S. Jerrams and N. Murphy (Eds.), *Constitutive Models for Rubber VII*, pp. 235–240. London: CRC Press.

Gelke, S. & J. Ihlemann (2016). Simulation of a chassis bushing with regard to strain induced softening of filled rubber. *PAMM 16*(1), 339–340.

Haupt, P. (2002). *Continuum Mechanics and Theory of Materials*. Berlin: Springer.

Haupt, P. & A. Lion (2002). On finite linear viscoelasticity of incompressible isotropic materials. *Acta Mechanica 159*(1), 87–124.

Ihlemann, J. (2003). *Kontinuumsmechanische Nachbildung hochbelasteter technischer Gummiwerkstoffe*. Düsseldorf: VDI.

Kanzenbach, L., C. Naumann, & J. Ihlemann (2016). Specimen design for high precision tension-compression tests. *PAMM 16*(1), 207–208.

Kanzenbach, L., M. Stockmann, & J. Ihlemann (2016). Extension of the Testing Machine Control for High-precision Uniaxial Tension Measurements. *Materials Today: Proceedings 3*(4), 993–996.

Kießling, R., R. Landgraf, R. Scherzer, & J. Ihlemann (2016). Introducing the concept of directly connected rheological elements by reviewing rheological models at large strains. *International Journal of Solids and Structures 97–98*, 650–667.

Krawietz, A. (1986). *Materialtheorie: Mathematische Beschreibung des phänomenologischen thermomechanischen Verhaltens*. Berlin: Springer.

Landgraf, R. (2016). *Modellierung und Simulation der Aushärtung polymerer Werkstoffe*. München: Verlag Dr. Hut.

Mielke, A. & T. Roubíček (2015). *Rate-Independent Systems—Theory and Application*, Volume 193 of *Applied Mathematical Sciences*. New York: Springer.

Naumann, C. & J. Ihlemann (2014). Extended morph model for application-oriented simulation of inelastic large strain rubber. In A. S. Khan and H.-Y. Yu (Eds.), *Multi-scale Modeling and Plasticity Characterization of Advanced Materials*, pp. 118–120. USA: Neat Press.

Palmow, W. A. (1984). Rheologische Modelle für Materialien bei endlichen Deformationen. *Technische Mechanik 5*(4), 20–31.

Rashid, M. M. (1993). Incremental kinematics for finite element applications. *International Journal for Numerical Methods in Engineering 36*(23), 3937–3956.

Scheffler, C. (2009). *Simulation gekoppelter Relaxations- und Erholungsprozesse bei technischen Gummiwerkstoffen mittels rheologischer Modelle*. Diploma thesis, Chemnitz University of Technology, http://nbnresolving.de/urn:nbn:de:bsz:ch1-200900574.

Shutov, A. V. & J. Ihlemann (2014). Analysis of some basic approaches to finite strain elasto-plasticity in view of reference change. *International Journal of Plasticity 63*, 183–197.

Truesdell, C. & W. Noll (1965). The Non-Linear Field Theories of Mechanics. In S. Flügge (Ed.), *Encyclopedia of Physics III/3*. Berlin: Springer.

Ageing

Constitutive Models for Rubber X – Lion & Johlitz (Eds)
© 2017 Taylor & Francis Group, London, ISBN 978-1-138-03001-5

Modelling of reaction-diffusion induced oxidation of elastomers in two spatial dimensions by means of ADI method

A. Herzig, M. Johlitz & A. Lion
University of the Bundeswehr Munich, Neubiberg, Germany

ABSTRACT: In the majority of applications ageing of elastomers is affected by heterogeneous effects. Probably the most important one is the influence of oxygen from the ambient environment. Oxygen molecules are absorbed by the surface layer and get transported into the interior by diffusion before reactions with the elastomer structure occur. Besides the mechanical properties, thermo-oxidative reactions affect the diffusion process itself and hence the prediction of oxidation is a current challenge when modelling ageing. This work provides a reaction-diffusion approach based on Fick's second law of diffusion. Diffusivity is described as a function of temperature and ageing to simulate the decreasing permeability by progressing oxidation. In order to describe the reaction rate of oxidation an Arrhenius approach is chosen which considers the available amount of oxygen. The progress of ageing is given by an evolution equation depending on temperature and reaction rate. Since heterogeneous oxidation is influenced by the shape and size of the elastomer sample, the model is solved and visualized in two spatial dimension using Alternating Direction Implicit (ADI) method. This numerical algorithm is able to solve the developed model with unconditional stability and sufficient accuracy. This work illustrates the effect of ageing on diffusivity and is able to simulate Diffusion-Limited-Oxidation (DLO) which is a key factor of heterogeneous oxidation.

1 INTRODUCTION AND STATE OF THE ART

In order to predict the mechanical properties and finally the lifetime of elastomers, a detailed knowledge of the ongoing ageing processes is essential. Although the ageing of polymers is driven by several processes running in parallel, oxidation is one of the primary mechanisms due to the omnipresence of oxygen in a lot of applications. Oxidation starts at the surface and continues more and more to the interior of the sample, since the oxygen molecules almost exclusively originate from the ambient environment. This can provoke an inhomogeneous distribution of oxygen and hence a heterogeneous occurrence of oxidation. The latter is of vital importance for a reliable lifetime prediction of elastomers, especially if mechanical properties are of interest.

Therefore, the process of oxygen getting from the surrounding air into the material and its subsequent reaction with the polymer should be investigated in detail. Generally, ageing of polymers caused by oxygen can be divided into three steps: absorption of oxygen molecules by the polymer, further diffusion of the oxygen into the interior and the oxidative reaction, which affects the molecular structure of the material. When trying to describe the oxidation behaviour by a model, these three parts should be considered. The driving force of solubility is the adsorption affinity of the polymer surface for O_2 molecules, which creates a gradient of oxygen concentration between the superficial layer and the interior bulk (Atkins & Ludwig 2008). The amount of oxygen being absorbed depends e.g. on the temperature, the oxygen partial pressure and the chemical composition of the polymer (Lin & Freeman 2005, Scheirs *et al.* 1995). Since the amount of absorbed oxygen provides no information about the oxygen contribution in the sample as well as about the amount of oxygen reacted with the polymer structure, no information about heterogeneous oxidation effects is available. Nevertheless, the amount of absorbed oxygen is a first indicator for oxidative ageing and is able to be determined by experimental methods (Herzig *et al.* 2014, 2015, 2016).

Solubility comes along with diffusion which is finally responsible for the heterogeneous allocation. The diffusion of oxygen molecules in polymers is usually described by Fick's second law (Van Amerongen 1950). Therefore, the most significant material parameter with respect to diffusion is the diffusion coefficient which strongly depends on the ageing temperature. On the one hand diffusion is accelerated by high temperatures but on the other hand the reaction rate of oxygen with the molecular structure is increased in parallel. These thermo-oxidative reactions provoke a decline of diffusivity so that the further diffusion is reduced or even suppressed. A dualism of both effects occurs and the influence of oxidation can be restricted to a small area near the surface or to a superficial layer. This phenomenon is known as Diffusion-Limited-Oxidation (DLO) (Wise *et al.* 1997, Gillen & Clough 1992). Severe oxidation of outer regions can decline the diffusion coefficient so much that no further oxygen can diffuse into the interior of the sample. This entails a high heterogeneous ageing behaviour and has huge influence on the mechanical properties of the entire elastomer component. DLO is a function of material composition, temperature, ageing time, ambient oxygen pressure, mechanical loads, sample dimensions and geometry as well as the presence of antioxidants (Herzig *et al.* 2016).

The reaction of oxygen molecules with the elastomer structure is an irreversible process. The formation of oxidised groups includes e.g. chain scission and the building of new covalent bonds between the chains (Mailhot *et al.* 2004). Several reaction schemes exist to describe the ongoing chemical reactions, one of which is, for instance, the "Basic Autoxidation Scheme" based on initial radical reactions (Steinke 2013, Verdu 2012).

This work offers an approach to model the heterogeneous oxidation behaviour of elastomer samples by means of a modified diffusion-reaction equation based on Fick's law of diffusion. Thereby the diffusivity is defined as a function of temperature and the previous ageing progress. This enables the model to describe the DLO effect by simulating a regional decline of the diffusion coefficient at the outer layer which prevents the diffusion of oxygen to the interior of the sample. The model is numerically solved by means of the Alternating Direction Implicit (ADI) method in two spatial dimensions which was implemented using MATLAB®.

2 MODELLING OF HETEROGENEOUS OXIDATION

After oxygen is absorbed from the ambient environment a concentration gradient at the outmost layer occurs which provokes diffusion to the interior of the sample. The diffusion of the oxygen continues until an equilibrium state is reached or the penetration is inhibited elsewhere. A constant oxygen concentration C_0 is assumed in the ambient environment around the material, that means Dirichlet boundary conditions are used. A common way to describe diffusion processes in materials is given by Fick's second law which describes the distribution of the oxygen concentration C (Equation 1).

$$\frac{\partial C}{\partial t} = D_{ox}(p,\theta) * \left(\frac{\partial^2 C}{\partial x^2} + \frac{\partial^2 C}{\partial y^2} \right) \tag{1}$$

D_{ox} is the diffusion coefficient which can alter during the process of ageing. The variables x and y stand for the depth of penetration in the two spatial directions. If oxygen molecules react with the elastomer or with antioxidants, they are no longer available to further diffusion. The influence of antioxidants on the heterogeneous ageing will not be considered in this work. In order to take oxygen consumption by oxidation into account, a reaction term r_{ox} is subtracted from the right-hand side of Equation 1.

$$\frac{\partial C}{\partial t} = D_{ox}(p,\theta) * \left(\frac{\partial^2 C}{\partial x^2} + \frac{\partial^2 C}{\partial y^2} \right) - r_{ox}(C,\theta) \tag{2}$$

The reaction rate r_{ox} describes the rate of oxygen consumption due to thermo-oxidative ageing. It is a function of the oxygen concentration since in case of no oxygen available no oxidative reaction can occur. Furthermore, the reaction rate depends on the temperature θ which is described by an Arrhenius function.

$$r_{ox} = \frac{C}{C_0} * r_o * e^{-\frac{E_r}{R\theta}} \tag{3}$$

The typical Arrhenius approach is complemented by the quotient C/C_0 which is zero in case there is no oxygen available. The maximum possible oxygen concentration is given by the boundary condition C_0. In this case the quotient reaches the value one. r_0 represents a constant factor and E_r the activation energy of the oxidative reaction. R is the universal molar gas constant.

As stated before, the diffusion coefficient is influenced by the ageing temperature and the progress of ageing. Therefore, an Arrhenius approach is used to describe the temperature dependency of the diffusion coefficient D_{ox}, which is modified by a term constituting the progress of ageing. The ageing expression p represents the progress of ageing between the values 0 and 1, thereby 0 stands for no ageing occurred and 1 stands for a completely aged condition (Herzig *et al.* 2016).

$$D_{ox} = D_o * e^{-\frac{E_D}{R\theta}} * (1-p) \tag{4}$$

Moreover, D_{ox} depends on the location within the sample since the progress of ageing is a function of the position. The process of the ageing parameter p is described by an evolution equation which is also used e.g. in (Johlitz 2012, 2015).

The value of p increases with ongoing oxidative reactions and the velocity is governed by the reaction rate. Hence, the evolution equation is complemented by the reaction rate r_{ox}. r_{ox0} is used to avoid inconsistences since the evolution equation uses no units. Since oxidation is an irreversible process, a decline of p is impossible.

$$\dot{p} = \nu_p * e^{-\frac{E_p}{R\theta}} * (1-p) * \frac{r_{ox}}{r_{ox0}} \tag{5}$$

The ongoing oxidation provokes a decline of the diffusion coefficient D_{ox} (Equation 4). Thus, the model is able to describe the effect of DLO, i.e. that a diffusion stop occurs caused by oxidation. The absorbed oxygen molecules react with the polymer in a certain distance to the surface. For instance, methyl or polar groups are formed which provoke a decrease of diffusivity and prevent other oxygen molecules from further diffusion to the interior (Van Amerongen 1950). Changes in morphology due to oxidation are also reasons which affect the diffusivity. Chain scission and reformation of the polymer network have effects on the permeation of oxygen molecules.

The model used in this contribution was introduced and implemented one-dimensionally in a previous work (Herzig *et al.* 2016). Therefore, the implicit Crank-Nicolson algorithm offered a reliable and stable numerical procedure. Since a one-dimensional approach is a massive simplification of physical circumstances, the model is formulated in two spatial dimensions here.

3 NUMERICAL IMPLEMENTATION

An extension of the Crank-Nicolson method to two-dimensional problems is possible but it is associated with some difficulties. The solution is quite easy for one-dimensional problems since a tridiagonal system has to be solved for every time step, which is a fast and straightforward way. In the case of two dimensions the implicit method requires the solution of a two-dimensional Poisson equation which is much more complicated and costly to solve. An alternative is provided by the Alternating Direction Implicit (ADI) method which splits every time step up in two single steps. This method is also known as implicit Peaceman-Rachford algorithm of alternating directions (Peaceman & Rachford 1955). The 2D problem is transformed in two 1D problems which can be solved by implicit schemes, that means the numerical solution can be simplified to two tridiagonal systems per time step (Li & Chen 2008).

Here, the ADI method is used to solve the reaction-diffusion equation (Equation 2) of the model shown above. The solution is achieved by separating the partial differential equation into two steps. The first step provides the oxygen concentration C in the interim time step $m^{+1/2}$ (Equation 6).

$$\begin{aligned} C_{i,j}^{m+\frac{1}{2}} = C_{i,j}^{m} + \frac{D_{ox\,i,j}^{\,m}\,\Delta t}{2h^2}\Big(& C_{i-1,j}^{m+\frac{1}{2}} + C_{i+1,j}^{m+\frac{1}{2}} - 2C_{i,j}^{m+\frac{1}{2}} \\ & + C_{i,j-1}^{m} + C_{i,j+1}^{m} - 2C_{i,j}^{m}\Big) \\ & - \frac{1}{2}\left(C_{i,j}^{m+\frac{1}{2}} + C_{i,j}^{m}\right)\frac{r_o}{C_0} e^{-\frac{E_r}{R\theta}}. \end{aligned} \tag{6}$$

$\frac{1}{2}\Delta t$ represent the half time step between m and $m+\frac{1}{2}$. The concentration $C_{i,j}^{m+1/2}$ is an interim solution calculated at every grid point of the area with $i,j = 1,\ldots,N$. N is the number of grid points in x and y direction and h represents the constant distance between two points in horizontal as well as in vertical direction. The diffusion coefficient is updated every full time step according to Equation 4 and 5. It is assumed that the diffusivity is constant over a single time step. The oxygen concentration used for the reaction term is chosen as a mixture of explicit and implicit calculation. The final oxygen concentration $C_{i,j}^{m+1}$ after a whole time step is calculated by a second step (Equation 7).

$$\begin{aligned} C_{i,j}^{m+1} = C_{i,j}^{m+\frac{1}{2}} + \frac{D_{ox\,i,j}^{\,m}\,\Delta t}{2h^2}\Big(& C_{i,j-1}^{m+1} + C_{i,j+1}^{m+1} \\ & - 2C_{i,j}^{m+1} + C_{i-1,j}^{m+\frac{1}{2}} + C_{i+1,j}^{m+\frac{1}{2}} - 2C_{i,j}^{m+\frac{1}{2}}\Big) \\ & - \frac{1}{2}\left(C_{i,j}^{m+1} + C_{i,j}^{m+\frac{1}{2}}\right)\frac{r_o}{C_0} e^{-\frac{E_r}{R\theta}}. \end{aligned} \tag{7}$$

The name of the method "alternating direction" comes from the fact that first the concentration is calculated in x-direction and then in y-direction. The computation of Equation 6 contains the solution of N tridiagonal equation systems of the following form (Equation 10) with

$$\alpha_{i,j} = \frac{D_{ox\,i,j}^{\,m}\,\Delta t}{h^2} \tag{8}$$

and

$$\beta = \frac{r_o}{C_0} e^{-\frac{E_r}{R\theta}} \tag{9}$$

$$\begin{aligned} & -\frac{\alpha_{i,j}}{2} C_{i-1,j}^{m+\frac{1}{2}} + \left(1 + \alpha_{i,j} + \frac{\beta}{2}\right) C_{i,j}^{m+\frac{1}{2}} - \frac{\alpha_{i,j}}{2} C_{i+1,j}^{m+\frac{1}{2}} \\ & = \frac{\alpha_{i,j}}{2} C_{i,j-1}^{m} + \left(1 - \alpha_{i,j} - \frac{\beta}{2}\right) C_{i,j}^{m} + \frac{\alpha_{i,j}}{2} C_{i,j+1}^{m}. \end{aligned} \tag{10}$$

Analogously to this, the solution of Equation 7 contains another N tridiagonal equation system (Equation 11).

$$-\frac{\alpha_{i,j}}{2}C_{i,j+1}^{m+1}+\left(1+\alpha_{i,j}+\frac{\beta}{2}\right)C_{i,j}^{m+1}-\frac{\alpha_{i,j}}{2}C_{i,j-1}^{m+1}$$
$$=\frac{\alpha_{i,j}}{2}C_{i-1,j}^{m+\frac{1}{2}}+\left(1-\alpha_{i,j}-\frac{\beta}{2}\right)C_{i,j}^{m+\frac{1}{2}}+\frac{\alpha_{i,j}}{2}C_{i+1,j}^{m+\frac{1}{2}}. \quad (11)$$

That means $2N$ tridiagonal systems of dimension N have to be solved at every time step to get from m to $m + 1$, which can be solved by Thomas algorithm. The coefficient matrix differs for each tridiagonal system since the $\alpha_{i,j}$ depend on $D_{oxi,j}{}^{m}$ which is a function of location and time. After finishing the calculation of a time step the oxygen concentration $C_{i,j}$ with $i,j = 1,\ldots,N$ is obtained which is used to update the ageing parameter p and subsequent the diffusion coefficient D_{ox} at every grid point. ADI method is unconditional stable and its accuracy is of second order in time and space (Li & Chen 2008).

4 RESULTS AND DISCUSSION

As an output of the model the ageing parameter p can be plotted as a function of ageing time over the entire cross section. The value represents the progress of ageing from virgin condition to a fully aged state. Here we assume a bar-shaped sample of infinite length and determine the ageing condition of the cross-sectional area. The edge length of the cross-section is divided in equidistant elements and the differential equations are solved for every grid point. In order to get a first impression of the outcome, Figure 1 shows the ageing parameter p at three different ageing times. The parameters of the model are adjusted so that ageing effects can be observed in all figures after short time of exposure.

At the beginning the initial condition is an unaged sample and hence p is zero over the whole cross-section. After oxygen is absorbed, the molecules start to diffuse into the sample and react with the polymers structure. Thus, the degree of ageing increases first in the outermost region whereby the interior stays almost unaffected. The heterogeneous effect of oxidation on the sample can be obviously seen in Figure 1. The corners of the sample are more affected by the oxidation than the rest of the cross section. This is an effect of diffusion since at these points oxygen is penetrating simultaneously from two surfaces and hence more oxygen is available for oxidation in these areas. Consequently, the corners are most vulnerable to ageing and therefore to embrittlement and crack growing. If the mechanical properties of the elastomer are

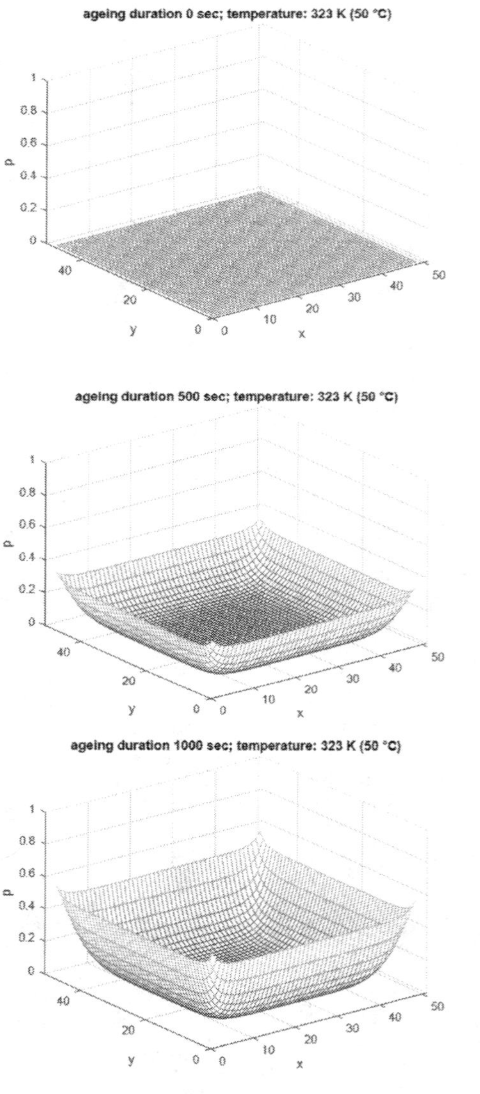

Figure 1. Ageing parameter p over the cross-section of an infinitely long and bar-shaped sample at different ageing durations (0, 500 and 1000 s) and constant temperature.

important at this points, oxidation can already become a problem although most of the sample shows a low state of ageing. An example are tyre treads since a lot of right-angled edges exist and surface properties are of vital importance. A counterexample would be a cylindrical rubber damper whose mechanical properties as a bulk are important and almost no corners exist. This emphasises the influence of the sample's geometry on its ageing behaviour. A large surface area provides a lot of contact to the oxygen containing environment and hence absorption is encouraged.

Another important phenomenon of heterogeneous oxidation is the effect of size which is illustrated in Figure 2. All three plots are calculated using equal material parameters and circumstances. The only difference is the size of the sample. The first plot (L = 1) has a normalized edge length of 1, the second graphic is scaled up by the factor 2 (L = 2) and the bottom graphic has edges of the length 4 (L = 4). The figure shows that the oxidized surface has an equal thickness for all three sample sizes but the ratio of aged and unaged material decreases with larger samples. An example of such a behaviour can be easily imagined when comparing the oxidation behaviour of very thin elastomer sealings and large scaled rubber dampers. After a certain time of exposure, the former is totally oxidized and reaches his lifetime whereat a large bulk of elastomer shows the same degree of ageing of the outmost layer but most of the material is almost not affected and operability is not endangered yet. That means DLO can produce a kind of protection layer with reduced diffusivity which protects the interior of the sample against the influence of ambient oxygen. Therefore, the sample has to have a certain dimension which depends on material properties and environmental circumstances.

The influence of ageing temperature and duration on the DLO effect can be illustrated if the model is solved for different temperatures and the solution is visualized for unified ageing durations. Figure 3 shows simulations of the ageing parameter p using the same set of parameters as for the model used in Figure 2. The temperatures of exposure are chosen to 20, 60, 100 and 150°C. The plots are made after 330, 660, 1000 and 5000 seconds of ageing. The initial condition at the beginning of exposure is not shown due to lack of space. Before ageing starts the samples are supposed to be unaged.

Figure 3 shows that at low temperatures the oxidative reaction rate is very low and ageing effects are not limited to the surface area but occur in a more homogeneous way (20°C/5000 s). Effects of oxidation can be clearly observed at a temperature of 60°C and although the surface is affected, oxygen reaches the core of the sample. Note that ageing at 60°C/5000 seconds exhibits the highest ratio

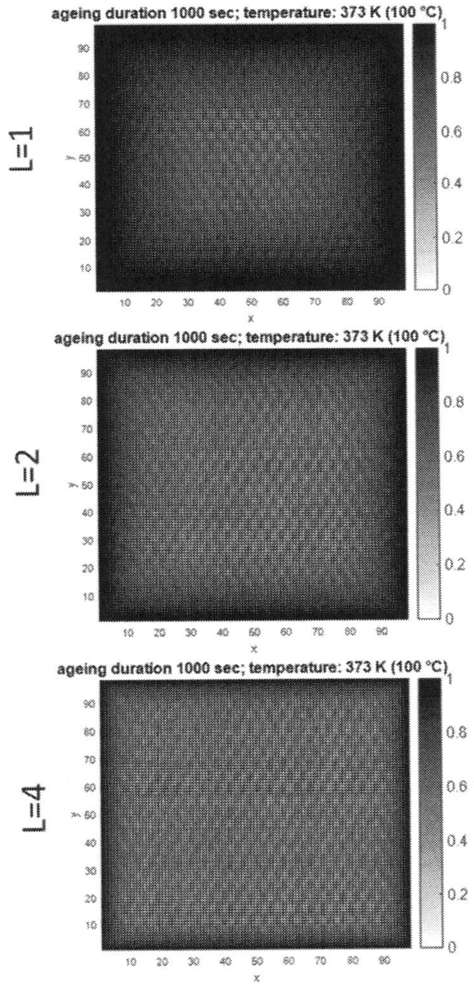

Figure 2. Effect of sample size on the oxidation behaviour p of elastomer samples at constant temperature and ageing duration (1000 seconds/100°C).

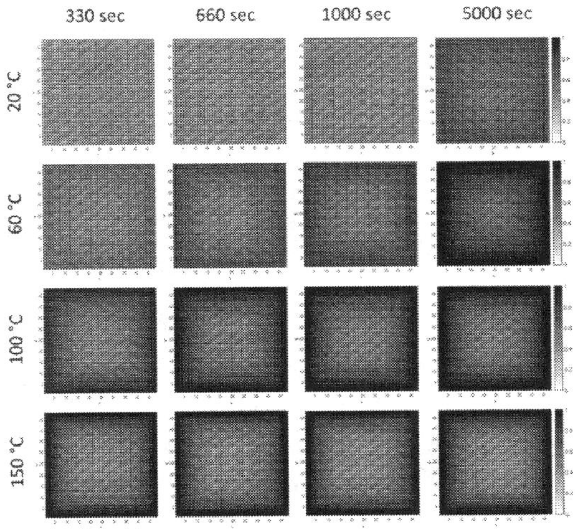

Figure 3. Heterogeneous ageing (ageing parameter p) as a function of temperature and ageing duration.

of oxidation in Figure 3. At higher temperatures, the oxidation of the very outside region processes very quickly but simultaneously the unaffected section in the interior is more pronounced. This can be obviously seen when comparing the plots 60°C/5000 s and 150°C/5000 s. Despite the higher ageing temperature the latter is not as much affected by oxidation than that at 60°C. The reason for this is DLO which reduces the diffusivity in the outmost region very much due to the high rate of oxidative reactions. Penetration of oxygen molecules to the interior is reduced or even prevented and thus ageing of the samples core fail to appear.

5 CONCLUSION AND OUTLOOK

The model introduced offers a possibility to simulate heterogeneous oxidation of elastomers as a function of temperature and ageing duration. The final output is given by the parameter p which describes the progress of ageing between a virgin and a fully aged condition. The model bases on a modified reaction-diffusion equation which considers the consumption of oxygen molecules by oxidative reactions with the elastomer. Furthermore, the penetration is influenced by chemical reactions which decline the diffusion coefficient. As a result, the model is able to consider the effect of DLO and visualizes the influence of shape and size of the elastomer component. Therefore, the model is solved in two spatial dimensions my means of ADI method which provides a stable and accurate numerical algorithm.

In order to consider samples of more complex geometry the model has to be extended to 3-dimensional simulation for instance by using the method of finite elements (FEM). Further, for most applications the mechanical properties play a more essential role than the ageing progress p which is difficult to grasp. Hence, future effort will be spent in the advancement of the presented model to provide a reliable tool for the prediction of elastomer ageing.

REFERENCES

Atkins, P.W. & Ludwig, R. 2008. *Kurzlehrbuch Physikalische Chemie: [mit 600 Übungen], Bachelor*, 4. Aufl. Weinheim: Wiley-VCH.

Gillen, K.T. & Clough, R.L. 1992. Rigorous experimental confirmation of a theoretical model for diffusion-limited oxidation, *Polymer*, Vol. 33 No. 20: pp. 4358–4365.

Herzig, A., Johlitz, M. & Lion, A. 2014. "An experimental set-up to analyse the oxygen consumption of elastomers during ageing by using a differential oxygen analyser", *Continuum Mechanics and Thermodynamics*.

Herzig, A., Johlitz, M. & Lion, A. 2015. "Experimental Investigation on the consumption of oxygen and its diffusion into elastomers during the process of ageing". *Constitutive Models for Rubbers: Proceedings of the 9th European conference on constitutive models for rubber (ECCMR)*: pp. 23–28.

Herzig, A., Johlitz, M. & Lion, A. 2016. "Consumption and diffusion of oxygen during the thermoxidative ageing process of elastomers". *Materialwissenschaft und Werkstofftechnik* Vol. 47 No. 5–6: pp. 376–387.

Herzig, A., Sekerakova, L., Johlitz, M. and Lion, A. (Submitted 2016). "A modelling approach for the heterogeneous oxidation of elastomers". *Continuum Mechanics and Thermodynamics*.

Johlitz, M. 2012. "On the Representation of Ageing Phenomena", *The Journal of Adhesion* Vol. 88 No. 7: pp. 620–648.

Johlitz, M. 2015. "Zum Alterungsverhalten von Polymeren: Experimentell gestützte, thermo-chemo-mechanische Modellbildung und numerische Simulation", *Habilitationsschrift*, Fakultät für Luft- und Raumfahrttechnik, Universität der Bundeswehr München, Neubiberg.

Mailhot, B., Bussière, P.-O., Rivaton, A., Morlat-Thérias, S. & Gardette, J.-L. 2004. "Depth Profiling by AFM Nanoindentations and Micro-FTIR Spectroscopy for the Study of Polymer Ageing". *Macromolecular Rapid Communications* Vol. 25 No. 2: pp. 436–440.

Peaceman, D.W. & Rachford, J.H.H. 1955. "The Numerical Solution of Parabolic and Elliptic Differential Equations". *Journal of the Society for Industrial and Applied Mathematics* Vol. 3 No. 1: pp. 28–41.

Steinke, L. 2013. *Ein Beitrag zur Simulation der thermo-oxidativen Alterung von Elastomeren, Fortschritt-Berichte VDI Reihe 5, Grund- und Werkstoffe, Kunststoffe*. Vol. 749, Düsseldorf: VDI-Verlag.

Van Amerongen, G.J. 1950. "Influence of structure of elastomers on their permeability to gases", *Journal of Polymer Science*, Vol. 5 No. 3: pp. 307–332.

Verdu, J. 2012. *Oxidative ageing of polymers*. London, Hoboken, NJ: Wiley.

Wise, J., Gillen, K.T. & Clough, R.L. 1997. "Quantitative model for the time development of diffusion-limited oxidation profiles". *Polymer*, Vol. 38 No. 8: pp. 1929–1944.

Constitutive Models for Rubber X – Lion & Johlitz (Eds)
© 2017 Taylor & Francis Group, London, ISBN 978-1-138-03001-5

A study on characterising ageing phenomena via the dynamic flocculation model

N.H. Kröger, R. Zahn & U. Giese
Simulation and Continuums Mechanics, Elastomer Chemistry, Deutsches Institut für Kautschuktechnologie e.V., Hanover, Germany

ABSTRACT: Thermal oxidative ageing is limiting the life time of rubber products. Beside environmental factors, the micro-structure of used polymers, the cross linking, fillers and additives are influencing components. Especially, the chemical degradation of the polymer network influences highly the mechanical behaviour. In this study we concentrate on characterising the ageing of an elastomer model compound (low- and high-cross linked EPDM) based on parameter identifications of the Dynamic Flocculation Model (DFM). The different compounds are tested in multi-hysteresis experiments in order to obtain an insight in the cyclic mechanical response.

The DFM is developed based on the extended non-affine tube model to capture the basic behaviour of the rubber matrix and also on the entropic driven material behaviour of the polymer network with the polymer-filler and filler-filler interaction. The latter is considering the breakage and re-agglomeration of filler clusters in the polymer matrix during repeated deformation cycles. Therefore, the DFM is capable to describe physically motivated effects in elastomers. The parameters of the model provide an insight of the structure of the elastomers, and are therefore investigated for different ageing states. In the evaluation of the model parameters for those different states, the DFM is examined if it is a suitable candidate for extension to describe the ageing phenomena.

1 INTRODUCTION

In the machine building industry as well as the automotive engineering elastomers are required for mechanically highly stressed, temperature— and chemical resistant component parts. Typical examples are tires, toothed and vee-belts, engine and aggregate bearings, air springs and chassis bushings as well as static and dynamic seals. The requirements regarding reliability and performance are particularly high in connection with the highly sensitive, safety-relevant functions of the components. In this context, it is particularly noteworthy that elastomers are relatively prone to ageing in comparison to other materials, such as metals or mineral-based materials, and are difficult to predict with respect to their lifetime.

While moderate mechanical and thermal stresses of filled elastomers do not cause an irreversible change in the material behaviour, oxidative processes, favoured by UV light, cause a progressive destruction of the material. The expiring processes are attributed to multi stage degradative mechanisms (Scott 1963, Hoff & Jacobsson 1982, Scott 1990, Coran 1994, Van Duin & Dikland 2003, Naskar & Noordermeer 2004). Recombination reactions of radicals formed during the ageing process lead to stable C-C linkages, resulting in higher cross linking densities, stiffness and embrittlement. Moreover, the formation of oxygen-containing side groups (hydroxyl, epoxide, carbonyl, carboxyl groups, etc.) on the polymer chain increases the glass transition temperature, which means a deterioration in the low-temperature properties. A further competing chemical mechanism is the irreversible damage of the polymer network by oxidative chain scission (Coran 1994). Experimental results indicate that, in addition to the reaction conditions (temperature, oxygen supply), the content of chain-centered double bonds and tertiary carbon atoms in the chain, as well as the nature of the substituents, determine the course of oxidative ageing (Norling, Lee, & Tobolsky 1965, Keller 1985, Bhattacharjee, Bhowmick, & Avasthi 1991, Bender & Campomizzi 2001, Van Duin & Dikland 2003).

Due to the degradative chemical reaction especially the mechanical response of the elastomers are influenced as well. A material model, which is to serve as the basis for mapping the influence of oxidative material ageing on the mechanical material behaviour, has to reproduce the softening phenomena (Mullins effect), hysteresis as well as moderate inelastic effects.

Especially it is necessary in the case of the typically inhomogeneous stress distributions within technical components to be able to predict the distortions in a realistic manner.

Such models were presented, for example, by Lion (1996), Pawelski (1998), Ogden & Roxburgh (1999), Klüppel & Schramm (1999) or Nasdala (2000) and by Besdo & Ihlemann (2003), Ihlemann (2003). Further developments allow to simulate the anisotropy induced by the softening into the material (Ihlemann 2005, Ihlemann 2007, Freund, Lorenz, Juhre, Ihlemann, & Klüppel 2011). The approach of Ogden & Roxburgh seems particularly suitable as a basis for an expansion around the influence of chemical ageing processes. This approach, described by Muhr, Gough, & Gregory (1999) and Holzapfel, Stadler, & Ogden (1999) has been further investigated and partially extended in (Naumann & Ihlemann 2014), is characterized by its compact structure, its numerical robustness and simple parameter identification. Despite the very low number of only three material constants (in the basic version), the softening can be reproduced. An Alternative approach is given by (Johlitz, Diercks, & Lion 2014). However, a stationary hysteresis and residual deformation can not be detected as well as the material parameters of the model give no further insight in the physics. A recent modelling approach by Syed, Stratmann, Hempel, Klüppel, & Saalwächter (2016) connects the cross linking agent concentration via the physically motivated extended non-affine tube model to mechanical properties.

In our work the aim is to apply the dynamic flocculation model based on the extended non-affine tube model (Klüppel & Schramm 1999, Klüppel, Meier, & Dämgen 2005) as an indicator of physical changes within the polymer network structure. The physical interpretable parameters of the model allow to obtain a hint to the underlying structure by the mechanical properties in multi-hysteresis tests, cf. (Plagge & Klüppel 2015).

2 EXPERIMENTAL PROCEDURE

The investigated material is a technical EPDM compound varied within the sulphur-accelerator system to obtain two different cross link networks as starting conditions. The recipe includes 100 phr Keltan 2450 as elastomeric basis, 50 phr low-active carbon black N550 for reinforcement, 5 phr zinc oxide, 3 phr stearic acid, 5 phr Sunpar 2280 as plasticizer, as well as TBBS, TBzTD-70 and sulphur with 1, 3.5 and 3 phr for the low cross linked system and 2, 7 and 6 phr for the high cross linked system. In order to enhance ageing effects an anti-ageing agent is not considered. The system is mixed within a 1.5 litre mixer and its test specimens vulcanized at 160°C temperature with t_{90}.

Real time ageing in components parts is taken place within the time span of years. To overcome this limiting factor in the experimental work, an accelerated ageing of the test specimens at 140°C in an oven is conducted. The test specimens (dumbbells) are aged for 0, 1, 4, 7, 14 and 28 days, before being investigated in uniaxial multi-hysteresis tests. The mechanical characterisation is done on a Zwick test machine for an initial strain rate ε_0 of 100% per minute. Overall upto six loading levels at −5% to 20%, −10% to 40%, −15% to 60%, −20% to 80% and −25% to 100%, dependent on maximal elongation at break after ageing, are applied. Unfortunately, at later ageing states (14 and 28 days) only the first two amplitudes could be recorded due to the low maximal elongation.

In comparison with (Plagge & Klüppel 2015), compression was also considered.

3 DYNAMIC FLOCCULATION MODEL (DFM)

The uniaxial version of the DFM is used for the parameter identification. A detail description of the model can be found in (Lorenz, Klüppel, & Heinrich 2012). The important eight parameters of the model are:

- G_c elastic modulus proportional to density of the network junctions,
- G_e tube constraint modulus proportional to the entanglement density of the rubber,
- n_e number of statistical chain segments between two successive entanglements ($n_e/T_e = n$, whereas

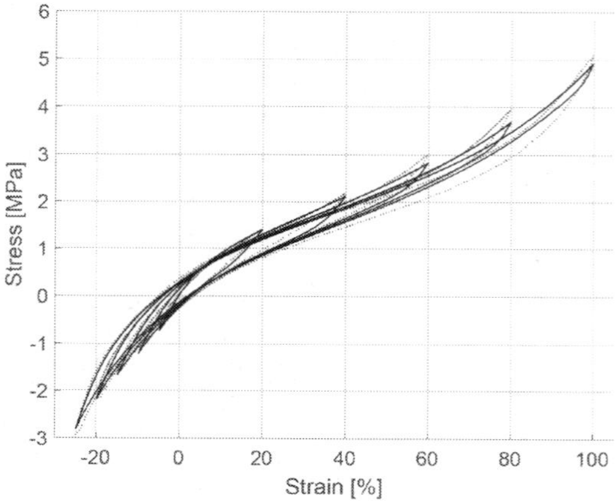

Figure 1. Exemplarily: Material model fit for unaged high cross linked EPDM (dotted: experiments, bold: fit).

n the finite extensibility parameter and T_e the trapping factor characterizing the portion of elastically active entanglements),

- s_v representing the strength of the virgin filler-filler bonds,
- s_d representing the strength of the damaged filler-filler bonds,
- x_0 average related cluster size,
- Φ_{eff} mechanical effective filler volume fraction,
- $\sigma_{set,0}$ phenomenological description of the inelastic behaviour.

4 PARAMETER IDENTIFICATION

As input for the parameter identification only the fifth cycle of an amplitude is used such that one can assume all virgin clusters broken at the corresponding stress of the amplitude. To ensure every cycle is weighted equally, for each cycle 400 data points are considered. The error function measures the sum of average mean absolute errors within each cycle. The optimization is realized in MATLAB with *fmincon* using a start parameter variation at first and different successively varied optimization algorithms, namely the interior-point (1st step including start parameter variation), the SQP- (2nd step—using results from 1st step) and the active-set-algorithms (3rd step—using results from 2nd step).

In order to obtain meaningful identified parameters some are assumed to be unaffected from ageing, namely the contribution of the filler system s_v, s_d, x_0, Φ_{eff}. This procedure is supported by results of a free parameter identification for the low cross linked system, where $s_v \in [14.740, 15.221]$ [MPa] with a mean of 14.862 [MPa], $s_d \in [38.225, 38.374]$ [MPa] with a mean of 38.262 [MPa], $x_0 \in [5.378, 5.971]$ with a mean of 5.818 and $\Phi_{eff} \in [0.242, 0.296]$ with a mean of 0.264. These parameter are identified for unaged samples and kept constant in further ageing states. The fixed values are for low cross linked EPDM $s_v = 14.763$ [MPa], $s_d = 38.225$ [MPa], $x_0 = 5.971$, $\Phi_{eff} = 0.248$ and for the high cross linked system $s_v = 13.966$ [MPa], $s_d = 38.227$ [MPa], $x_0 = 60.064$, $\Phi_{eff} = 0.282$. This also coincides with the constant filler content in both systems.

5 RESULTS AND CONCLUSION

The material model can describe the mechanical behaviour up to strains of 80% strain sufficiently well. For high strains the identification lags in quality throughout the experiments. The results of the parameter identification indicate certain expected insights in the ageing process. The increasing G_c correlates with a density of the network junction which means a post cross linking with respect to the ongoing oxidation. In case of the high cross linked EPDM the high temperature yields at first an predominant irreversible oxidative chain scission in opposite to post cross linking see Fig. 3. In contrast to (Plagge & Klüppel 2015) the competing nature of both chemical mechanism can be identified in the early stage ageing. Corresponding to the increasing network junction density the number of statistical chain segments between two successive entanglements decreases as expected, see Fig. 5.

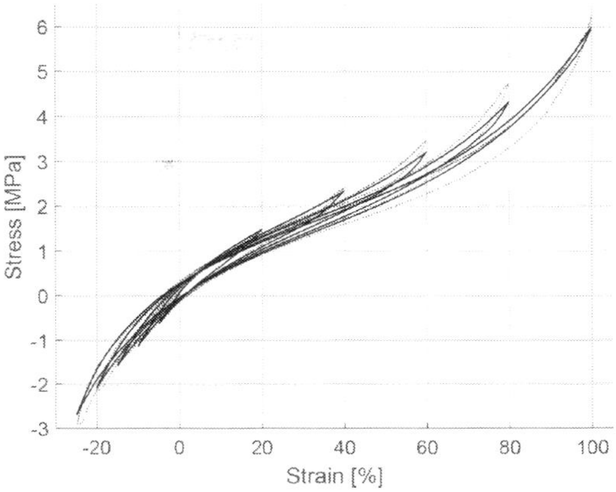

Figure 2. Exemplarily: Material model fit for four days aged low cross linked EPDM (dotted: experiments, bold: fit).

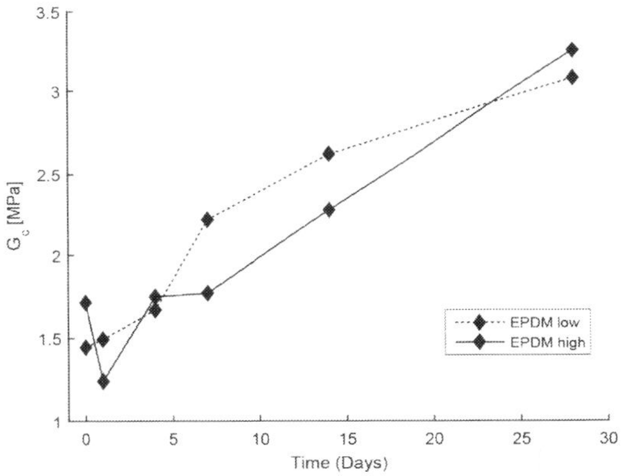

Figure 3. G_c elastic modulus for different ageing states.

Neglecting the results for 28 days of low cross linked EPDM, the increasing density entanglement of the rubber shows opposite trends for both systems. For the low system it yields hereby similar results as in (Plagge & Klüppel 2015). The effect of the high sulphur contents to the entanglement is to be examined in the future.

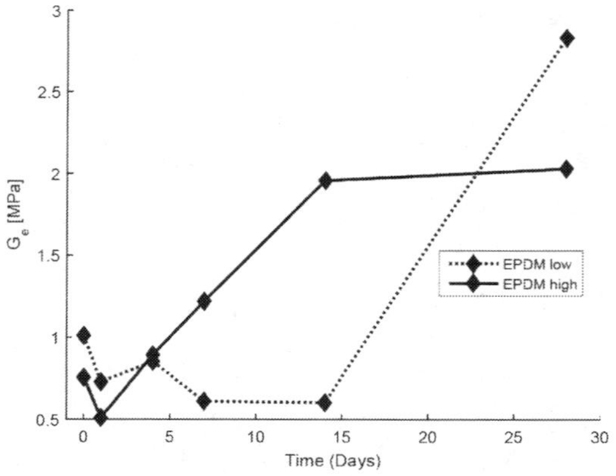

Figure 4. G_e tube constraint modulus for different ageing states.

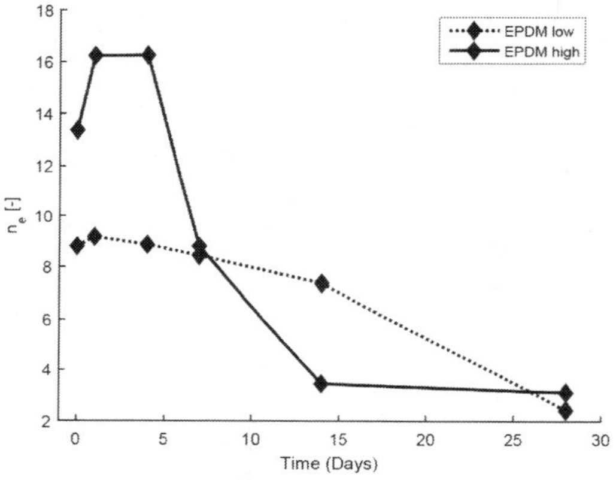

Figure 5. n_e number of statistical chain segments for different ageing states.

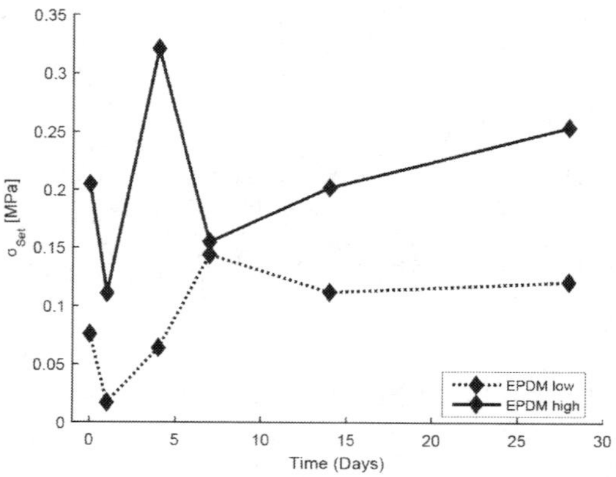

Figure 6. $\sigma_{set,0}$ for different ageing states.

Overall the DFM is a suitable tool to analyse the ageing phenomena with certain restrictions with respect to uniqueness of the material parameters. It is advisable to ensure sufficiently large number of stable cycles of the multi-hysteresis test within one ageing stage - 4 to 5. In case of large strains deviation of the experimental data and model have to be expected, such that a experimental plan should focus on lower loadings in order to obtain meaningful parameters.

ACKNOWLEDGEMENT

The experimental work was part of the AiF-project “Sensorintegration in Elastomerdichtungen für die Zustandsüberwachung”.

REFERENCES

Bender, H. & E. Campomizzi (2001). Improving the heat resistance of hydrogenated nitrile rubber compounds. *Kautschuk Gummi Kunststoffe 54*, 14–21.

Besdo, D. & J. Ihlemann (2003). A phenomenological constitutive model for rubberlike materials and its numerical applications. *International Journal of Plasticity 19*, 1019–1036.

Bhattacharjee, S., A.K. Bhowmick, & B. Avasthi (1991). Degradation of hydrogenated nitrile rubber. *Polymer degradation and stability 31*(1), 71–87.

Coran, A.Y. (1994). *Science and Technology of Rubber: Vulcanisation*, Chapter 7, pp. 372. Academic Press.

Freund, M., H. Lorenz, D. Juhre, J. Ihlemann, & M. Klüppel (2011). Finite element implementation of a microstructure-based model for filled elastomers. *International Journal of Plasticity 27*(6), 902–919.

Hoff, A. & S. Jacobsson (1982). Thermal oxidation of polypropylene close to industrial processing conditions. *Journal of Applied Polymer Science 27*(7), 2539–2551.

Holzapfel, G.A., M. Stadler, & R.W. Ogden (1999). Aspects of stress softening in filled rubbers incorporating residual strains. In *Constitutive Models for Rubber*, pp. 189–193. Balkema Rotterdam.

Ihlemann, J. (2003). *Kontinuumsmechanische Nachbildung hochbelasteter technischer Gummiwerkstoffe*. VDIVerlag Düsseldorf.

Ihlemann, J. (2005). Richtungsabhängigkeiten beim Mullins-Effekt. *Kautschuk Gummi Kunststoffe 58*(9), 438–447.

Ihlemann, J. (2007). Generalization of one-dimensional constitutive models with the concept of representative directions. In *Constitutive Models for Rubber V*, London, pp. 29–34. Taylor & Francis: A. Boukamel, L. Laiarinandrasana, S. Méo & E.V (Editoren).

Johlitz, M., N. Diercks, & A. Lion (2014). Thermo-oxidative ageing of elastomers: A modelling approach based on a finite strain theory. *International Journal of Plasticity 63*, 138–151.

Keller, R.W. (1985). Oxidation and ozonation of rubber. *Rubber chemistry and technology 58*(3), 637–652.

Klüppel, M., J. Meier, & M. Dämgen (2005). Modeling of stress softening and filler induced hysteresis of elastomer materials. In *Constitutive Models for Rubber*, Volume 4, pp. 171. Balkema.

Klüppel, M. & J. Schramm (1999). An advanced micromechanical model of hyperelasticity and stress softening of reinforced rubbers. In *Constitutive Models for Rubber*, pp. 211–218. Balkema Rotterdam.

Lion, A. (1996). A constitutive model for carbon black filled rubber: Experimental investigations and mathematical representation. *Continuum Mechanics and Thermodynamics 8*(3), 153–169.

Lorenz, H., M. Klüppel, & G. Heinrich (2012). Microstructure-based modelling and FE implementation of filler-induced stress softening and hysteresis of reinforced rubbers. *Zeitschrift für Angewandte Mathematik und Mechanik 92*(8), 608–631.

Muhr, A.H., J. Gough, & I.H. Gregory (1999). Experimental determination of model for liquid silicone rubber: Hyperelasticity and mullins' effect. In *Constitutive Models for Rubber*, pp. 181–187. Balkema Rotterdam.

Nasdala, L. (2000). *Ein viskoelastisches Schädigungsgesetz für den stationär rol lenden Reifen*. Ph.D. thesis, Universität Hannover.

Naskar, K. & J.W.M. Noordermeer (2004). Multifunctional peroxides as a means to improve properties of dynamically vulcanized PP/EPDM blends. *Kautschuk Gummi Kunststoffe 57* (5), 235–239.

Naumann, C. & J. Ihlemann (2014). Simulation von oxidativen Alterungsprozessen in Elastomerbauteilen. *Kautschuk Gummi Kunststoffe 10*, 68–75.

Norling, P.M., T. Lee, & A. Tobolsky (1965). Structure and reactivity in oxidation of elastomers. *Rubber Chemistry and Technology 38* (5), 1198–1213.

Ogden, R.W. & D.G. Roxburgh (1999). An energy-based model of the mullins effect. In *Constitutive Models for Rubbers*, pp. 23–28. Balkema Rotterdam.

Pawelski, H. (1998). *Erklärung einiger mechanischer Eigenschaften von Elastomerwerkstoffen mit Methoden der statistischen Physik*. Shaker.

Plagge, J. & M. Klüppel (2015). Application of a microstructure based model to thermally aged filler reinforced elastomer compounds. In *Constitutive Models for Rubber IX*, pp. 29. CRC Press.

Scott, G. (1963). Antioxidants. *Chemistry and Industry 16*, 271–281.

Scott, G. (1990). *Mechanisms of polymer degradation and stabilisation*. Springer Netherlands.

Syed, I.H., P. Stratmann, G. Hempel, M. Klüppel, & K. Saalwächter (2016). Entanglements, defects and inhomogeneities in nitrile butadiene rubbers: macroscopic vs microscopic properties. *Macromolecules 49*, 9004–9016.

Van Duin, M. & H.G. Dikland (2003). Effect of third monomer type and content on peroxide crosslinking efficiency of EPDM. *Rubber chemistry and technology 76*(1), 132–144.

Constitutive Models for Rubber X – Lion & Johlitz (Eds)
© 2017 Taylor & Francis Group, London, ISBN 978-1-138-03001-5

Comparison between thermo-oxidative aging and pure thermal aging of an industrial elastomer for anti-vibration automotive applications

M. Broudin
Vibracoustic, CAE and Durability Prediction Department, Carquefou Cedex, France

Y. Marco & V. Le Saux
ENSTA Bretagne—Institut de Recherche Dupuy de Lôme (IRDL), FRE CNRS 3744, Brest, France

P. Charrier & W. Hervouet
Vibracoustic, CAE and Durability Prediction Department, Carquefou Cedex, France

P.Y. Le Gac
Laboratoire Comportement des Structures en Mer, IFREMER (French Ocean Research Institute), Plouzané, France

ABSTRACT: Elastomeric anti-vibration parts found in automotive parts are usually massive, usually, a thickness of 10 mm or more is observed. Aging therefore leads to heterogeneous properties, possibly induced by several mechanisms due to the availability, or not, of oxygen in the part's bulk. To better understand the effects of oxygen in the degradation process, this paper aims at studying the aging of a rubber material (conventional vulcanization system) in aerobic (with oxygen) and anaerobic (without oxygen) conditions for a wide range of temperatures, relevant for under hood applications. A specific protocol to realize aging under anaerobic conditions was defined and validated. The consequences on the mechanical behavior are analyzed throughout monotonic tensile tests and fatigue campaigns results. The effects and kinetics of the two aging conditions are compared. In order to check the hypothesis of a physical aging related to desorption of low molecular weight ingredients, common to both aerobic and anaerobic conditions, a specific investigation is led on a material tested after swelling in toluene.

1 INTRODUCTION

1.1 *Industrial and collaboration backgrounds*

Rubber parts for anti-vibration systems are designed to meet several mechanical requirements: efficient damping on given ranges of frequencies, reliable static response over the service time, resistance to creep and fatigue. The automotive applications furthermore require the assessment of these features for thermal conditions inducing aging. This aging is related both to the effect of oxygen but also, for these massive parts, to the effect of temperature only. Several recent features require to go beyond the empirical rules. The first one is an ever reducing development time, which leads to assess compounds more quickly and/or to validate "carry over" strategies (checking that parts previously developped for one manufacturer specification will pass another's ones). A second point is the need to design parts meeting the requirements for any locations on the globe, increasing the severity profiles of users (roads and weather conditions). A last one is connected to the environmental standards (REACH and EURO6) that restrict the use of specific ingredients and lead to higher service temperatures. Integrating efficiently the aging effects in the design loop therefore becomes a crucial need.

This is the task of a research group bringing together ENSTA Bretagne, Chemnitz University of Technology, Freudenberg Technology Innovation and Vibracoustic. In a PhD thesis lead at ENSTA Bretagne, the focus is given on the understanding of the physico-chemical mechanisms and the description of the mechanical consequences of thermo-oxidative aging.

1.2 *Scientific background*

Thermo-oxidative degradation of rubber components is a topic widely investigated in the literature (Bolland & Gee, 1946, Broudin et al., 2015, Herzig et al., 2015, Kamarudding et al., 2011, Naumann & Ihlemann, 2013, Schlomka et al., 2017), because this aging induces severe modifications of the mechanical properties, both for constitutive response (Johlitz & Lion, 2013) and failure properties for monotonic tension (Gillen et al., 2005), crack propagation (South et al., 2003) or fatigue tests (Charrier et al., 2011). Nevertheless, relating the physico-chemical mechanisms to the evolution of the mechanical properties is not an easy task as many aging mechanisms could be at stake. A first range of mechanisms is of course related to oxidation, leading to strong modifications of the elastomer network (main-chain scissions, crosslink formation and/or crosslink breakage) (Verdu, 2012). Large-sized automotive parts age heterogeneously due to the availability, or not, of oxygen in the part's bulk. This leads to a classical analysis using a DLO approach based on the competition of the kinetics of oxygen diffusion and consumption (Celina et al., 2000). Nevertheless, the hardness profiles observed at the core of thick parts and samples (Le Saux, 2010) also illustrate that aging leads to mechanical changes even without oxygen. Numerous studies investigated the effect of temperature aging in the absence of oxygen on the mechanical properties of rubber materials. A first range of studies investigated the evolution of the tensile properties (Kohman, 1929, Shelton & Winn, 1944) for anaerobic conditions and found a limited increase of stiffness and little reduction of the ultimate tensile properties. Then, Kim et al. (1994) illustrated that the change of fatigue crack growth resistance of carbon-black filled vulcanizates for various heat aging environments leads to higher fatigue crack growth rate evolutions under air than under nitrogen because the oxidative degradation of the rubber causes a high reversion inducing a dominant effect on the mechanical fatigue resistance. This effect is also confirmed by Soma et al. (2010) to a lesser extent. Dealing with fatigue, a recent study showed that a first common drop of fatigue lifetimes can be observed for anaerobic and aerobic conditions, followed by stabilization in the case of anaerobic aging and further degradation for aerobic aging (Broudin et al., 2015). Nevertheless, the origins of these evolutions of the mechanical properties both for stiffness or failure remain unclear. Several explanations could be suggested: plasticizers extraction (Calvert & Billingham, 1979), sulfur network evolution (Kim & Lee, 1994), structural relaxation (Struik, 1978) and/or fillers network evolution (Roychoudhury & De, 1993).

The present paper aims at two objectives. The first one is to provide a more extensive comparison of the mechanical consequences of aging under aerobic and anaerobic conditions, based on the results of monotonic tensile tests and fatigue campaigns. The second one is to better understand the aging mechanisms under anaerobic conditions and to investigate the hypothesis of a physical aging induced by desorption of low molecular species.

2 EXPERIMENTAL SETTINGS

2.1 *Materials*

2.1.1 *Industrial compound*

The material studied here is a carbon black-filled NR/IR blend. Its recipe is given in Table 1.

All samples were made using the same material batch to limit properties scattering due to mixing, for example. This material will be called "C" in the Figures presented here.

2.1.2 *Material without soluble fraction*

To determine the effect of the extraction of the plasticizers during aging, it would be very convenient to consider the same compound but without plasticizers. Nevertheless, a filled material like the one investigated could not be mixed without plasticizers. It was therefore decided to extract the low molecular weight components (including plasticizers) thanks to swelling in toluene. The sample is finally dried in order to get back to a geometry close to the one before swelling. This material will be called "B" in the Figures hereafter.

2.2 *Aging conditions*

2.2.1 *With oxygen*

Aging with oxygen was carried out in ovens with air-flux control. To get reliable tests, several authors noticed the importance to specify the aging conditions, such as the rate of the air and air exchanges (Spetz, 1994). Therefore, to have quite similar aging conditions, all used ovens have been checked in terms of air speed and air renewal rates thanks to an anemometer and according to (ISO 188:2011). During aging, a thermocouple checks the temperature close to the samples and is connected to continuous strip-chart recorders. The Table 2 gives the tested temperatures and durations.

Table 1. Recipe of the industrial compound.

Formulation	
NR	75.00 phr
IR	25.00 phr
Zinc oxide	4.10 phr
Plasticizers	7.00 phr
Carbon black	29.00 phr
Stearic acid	2.00 phr
Antioxidant	3.00 phr
Antiozonant	3.00 phr
Accelerators	1.00 phr
Sulfur	3.38 phr

Table 2. Aging conditions with oxygen.

	Aging temperature [°C]						
	40	50	60	80	100	110	120
Aging duration [days]							0.25
						0.5	0.5
					1	1	1
				3	3	3	3
				7	7		
			14	14	14		
	21	21	21	21	21		
	42	42	42	42	42		
		90	90				
		180	180				
	360	360	360				
		720					

2.2.2 *Without oxygen*

An experimental protocol was developed to achieve aging without oxygen. To package the samples, a multilayer film is used (SIDEC CA40 NF H 00 310 Class IV). This film is impermeable to oxygen and temperature resistant. Wedges are placed such as no bending/torsion of the samples occurs when the vacuum is performed (Fig. 1, on the left). It is worth noting that the wedges material is chosen to present a thermal expansion coefficient close to the one of rubber, in order to avoid any preload due to differential thermal expansions.

Finally, to remove oxygen inside the package, a vacuum chamber machine is used (Audionvac VMS 133) (Fig. 1, on the right).

The Table 3 gives the tested temperatures and durations.

2.3 *Tensile tests campaign*

The samples used are S2 samples. The S2 samples are cut from plate and then placed in oven to carry out the various aging conditions. The tensile tests were achieved on testing machine equipped with a 1 kN load cell. The tests were displacement controlled (grip speed of 200 mm/min). An optical extensometer was used to measure the local elongation. For each aging condition, at least three samples were tested and the results averaged.

2.4 *Fatigue campaign*

Diabolo samples are obtained by injection molding in order to reproduce industrial manufacturing conditions of mass-produced automotive parts. The compound has been cured for 5 min, with a mold temperature set to 165°C. The fatigue tests were displacement controlled at a frequency of 5 Hz to keep a limited heat build-up. All tests were performed at

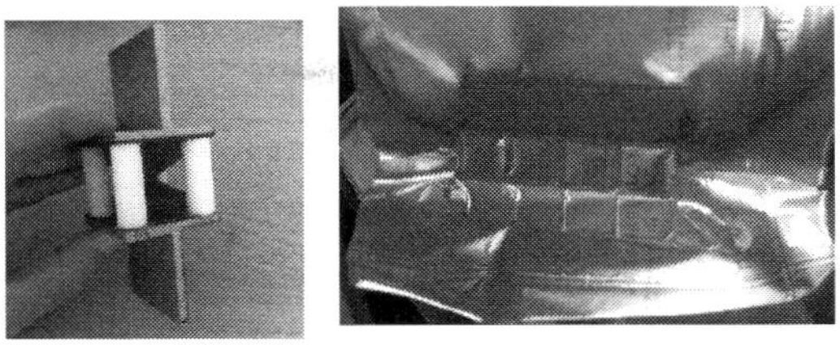

Figure 1. AE2 sample with PTFE wedges (left); sealed package of AE2 samples for aging without oxygen (right).

Table 3. Aging conditions without oxygen.

	Aging temperature [°C]						
	40	50	60	80	100	110	120
Aging duration [days]						0.5	0.5
				3	3	3	3
				14	14		
	21		21	21	21		
	42	42	42				
					90		
		180	180				

a controlled temperature of 23°C. The fatigue samples geometry and the fatigue protocol have been detailed in a former publication and will not be recalled here (Ostoja Kuczynski et al., 2003).

The first test campaign was targeted to investigate the evolution of the fatigue strength at room temperature. To keep a reasonable number of samples despite the numerous investigated aging conditions, for each of them, 5 AE2 samples are aged and then tested in fatigue for a given enforced displacement. In this campaign, 3 levels of displacement have been studied, 4.6 mm, 8.1 mm and 14.7 mm which correspond respectively to strains of 80%, 150% and 280%. These strains are evaluated by FE simulations from the geometry of the unaged sample.

3 RESULTS AND DISCUSSION

3.1 *Aging campaign with oxygen*

On the next page, Figure 2 to 4 (left column) illustrate the results obtained for thermo-oxidative aging. Figure 2 highlights typical results obtained for tensile tests on samples aged in presence of oxygen (here for 80°C and several durations). These observations are valid whatever the aging temperature. One can observe a drop of the strain at break and a stiffening of the mechanical response, which is quite a classical result and could be related to the changes of the vulcanized network. Nevertheless, it should be noted that the evolutions in stiffness

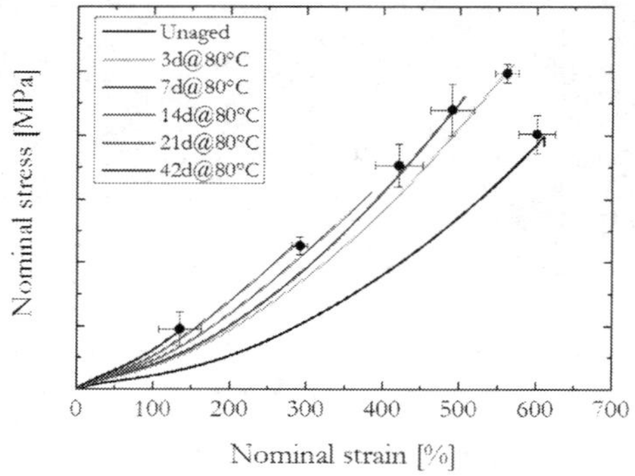

Figure 2. Material C. Aging with oxygen. Tensile test results after various preliminary heat aging (from 3 days to 42 days @ 80°C).

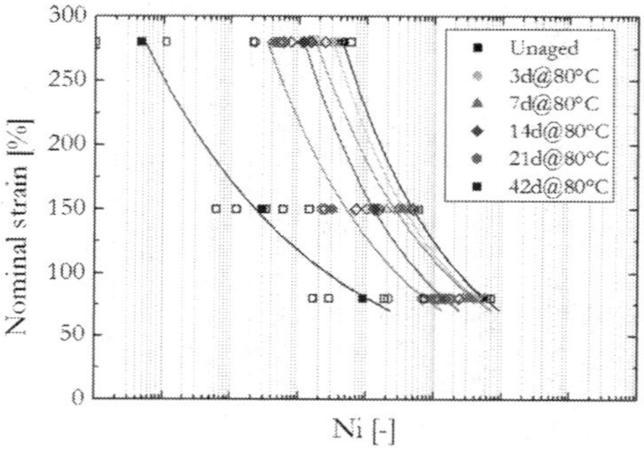

Figure 3. Material C. Aging with oxygen. Wöhler curves after various preliminary heat aging (from 3 days to 42 days @ 80°C).

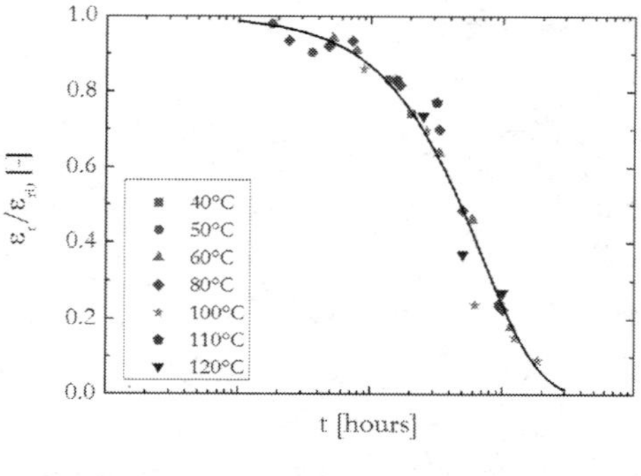

Figure 4. Material C. Aging with oxygen. Master curve obtained on the relative evolution of the strain at break.

and strain at break observed for short durations is already significant. Figure 3 presents the results of the fatigue campaigns for the same aging conditions. The drop of the fatigue lifetime with the aging durations is clearly visible. It is worth noting that the curves keep the same slope, except for the last aging condition, considered to be extremely severe.

A single aging mechanism is clearly not probable for this complex aging. Nevertheless, the identification of a time-temperature superposition seems possible, as illustrated by the master curve obtained on the relative evolution of the strain at break (Fig. 4). The deduced activation energy is of 94 kJ/mol. A comparable result can be obtained from the fatigue tests with an activation energy of 95 kJ/mol (Broudin et al., 2015). Both values are very close to other values found in the bibliography for various aging indicators (Celina et al., 2000, 2005).

3.2 *Aging campaign without oxygen*

On the next page, Figure 5 to 7 (right column) illustrate the results obtained for thermal aging without oxygen. Figure 5 highlights a drop of strain at break and a stiffening of the mechanical response. Both variations seems to occur early and then to stabilize, and are less severe than for oxidative aging. Figure 6 presents the results of the fatigue campaigns for the same aging conditions. Here again, the drop of the

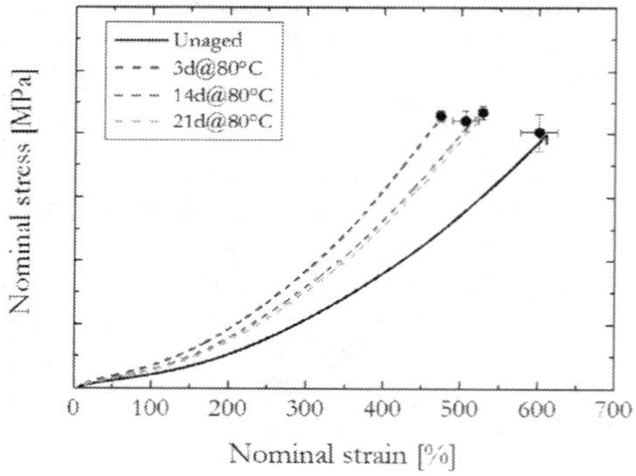

Figure 5. Material C. Aging without oxygen. Tensile test results after various preliminary heat aging (from 3 days to 21 days @ 80°C).

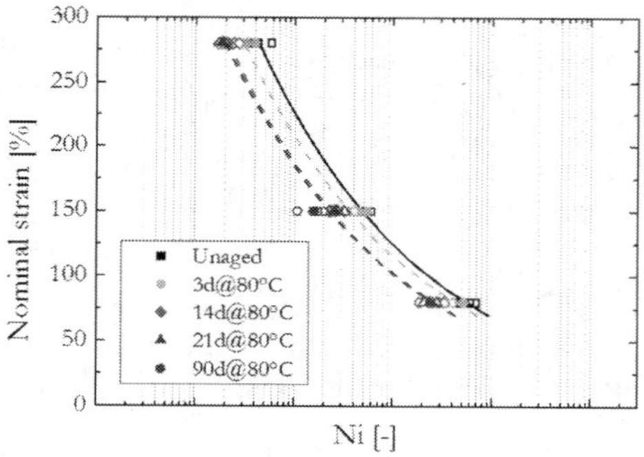

Figure 6. Material C. Aging without oxygen. Wöhler curves after various preliminary heat aging (from 3 days to 90 days @ 80°C).

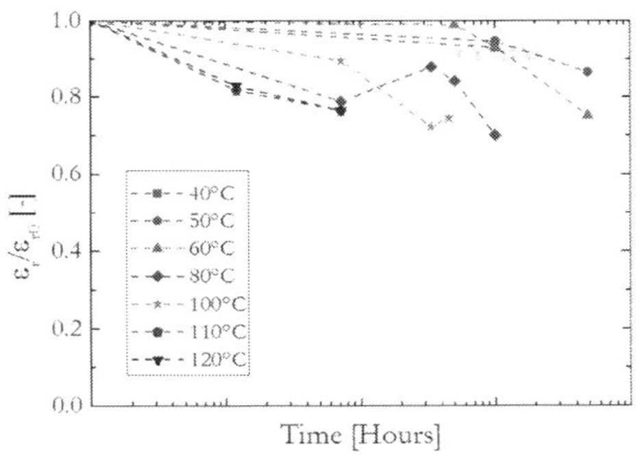

Figure 7. Material C. Aging without oxygen. Relative evolutions of the strain at break.

fatigue lifetime with the aging durations occurs early and then stabilizes to a level much less severe than for long oxidative conditions.

The identification of a time-temperature superposition seems in this case not possible, as illustrated on Figure 7.

3.3 *Comparison of the results obtained after both aging protocols and also to the results obtained for samples after swelling in toluene*

Figure 10 summarizes the evolution of the relative strain at break for tensile tests performed on samples after aerobic and anaerobic aging durations at 80°C. Figure 11 provides the same synthesis for the variations of the relative fatigue lifetime (for a given enforced displacement). One can observe on these curves that the drops of tensile and fatigue properties seem to present a common kinetic for the early aging durations. Then, the loss of the failure properties reaches a stabilized level for the aging conditions without oxygen, whereas the loss of failure properties continues for the aging conditions with oxygen.

In order to check if this common initial drop of tensile and fatigue properties, for conditions with or without oxygen, could be related to the extraction of plasticizers, tensile and fatigue tests have been performed on the material after swelling in toluene.

It should be underlined here that the results obtained are presented for 80°C only due to the restricted format of the paper, but similar conclusions can be drawn from the results obtained for lower and higher aging temperatures like 60°C and 100°C.

Figure 8 illustrates that, for tensile tests, the material after swelling in toluene presents a mechanical response and a strain at break consistent with the material aged without oxygen at 80°C after 14 days. The curve obtained for the same temperature and duration but under oxidative aging, presents a higher stiffening and a lower ultimate tensile strain.

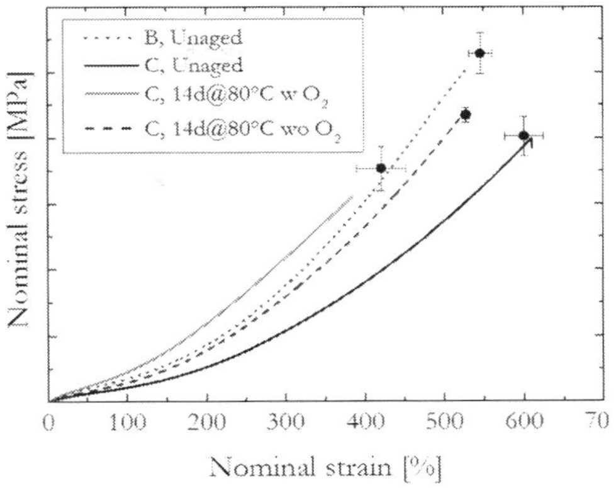

Figure 8. Tensile curves obtained after aging with and without oxygen at 80°C for 14 days and after swelling in toluene (mat. B).

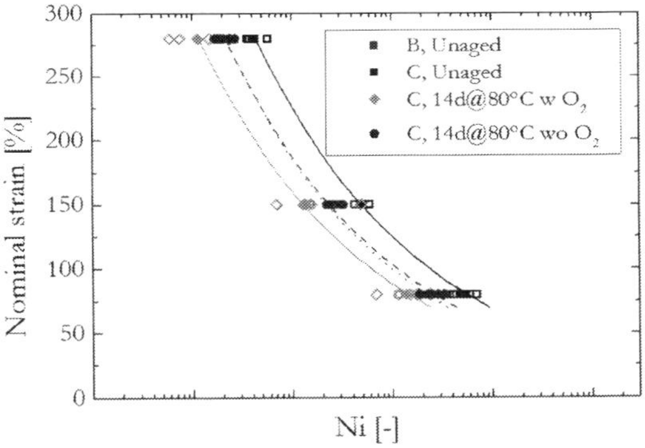

Figure 9. Fatigue curves after aging with and without oxygen at 80°C for 14 days and after swelling in toluene (mat. B).

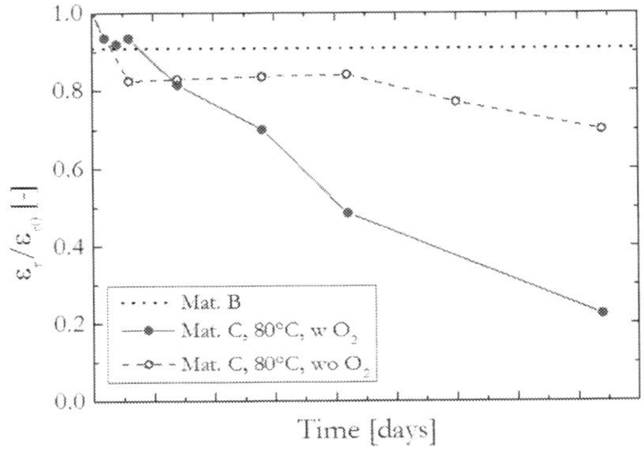

Figure 10. Evolution along time of the relative strain at break obtained after aging with and without oxygen at 80°C and after swelling in toluene (mat. B).

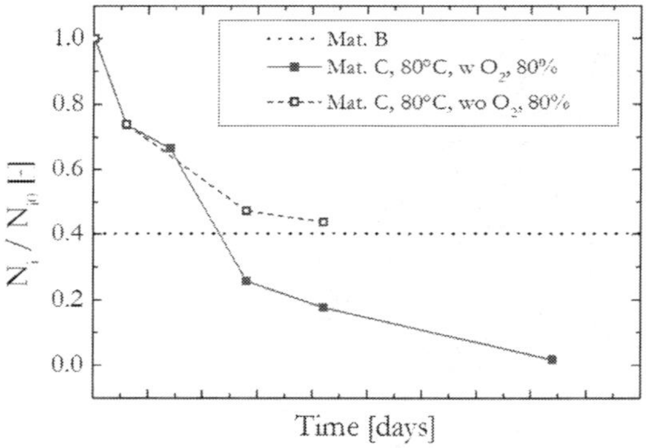

Figure 11. Evolution along time of the relative fatigue lifetime obtained after aging with and without oxygen at 80°C and after swelling in toluene (mat. B).

Figure 9 presents the results of the fatigue campaigns, for the same aging conditions. Here again the drop of the fatigue lifetime for the sample after swelling in toluene seems to correlate well with the one observed for aging without oxygen.

In order to check these first comparisons, the values obtained for the samples tested after swelling in toluene were compared with the evolutions of the samples aged with or without oxygen. These comparisons are illustrated in Figure 10 and 11 as the dotted line is the value obtained for the samples tested after swelling in toluene.

Figure 10 and 11 illustrate that the drop of strain at break and fatigue lifetime for anaerobic conditions stabilize for values correlated to the ones obtained for the material after swelling in toluene. It therefore seems that the initial drops of strain at break and fatigue properties observed both under aerobic and anaerobic conditions seem well correlated to the loss of low molecular weight components. The nature of these components is still to be determined even if it seems very probable that they are plasticizers. TGA and GC-MS measurements are currently done in order to clarify this point.

4 CONCLUSION AND PERSPECTIVES

In this paper, we focus on the aging of a rubber material (conventional vulcanization system) in aerobic (with oxygen) and anaerobic (without oxygen) conditions, over a wide range of temperatures relevant for under hood applications. Tensile and fatigue tests are analyzed. For both aging cases, progressive drops of the strain at break and the fatigue properties are observed as well as a progressive increase of the stiffness. Under anaerobic and aerobic conditions, these evolutions seem to exhibit the same kinetics for the short durations. Nevertheless, the evolution of the mechanical indicators reaches a plateau for anaerobic aging, whereas the evolutions go on for aerobic aging.

In order to investigate if these evolutions can be related to a physical aging induced by the extraction of low molecular weight components (plasticizers, for example), tensile and fatigue tests have been performed on samples after swelling in toluene. It appears that the values obtained on these samples are well correlated to the plateau observed for anaerobic aging. This hypothesis seems therefore reasonable and this mechanism could be common to aging conditions with or without oxygen.

Further investigations are running to identify the nature of the components extracted in toluene and to understand better the other possible aging mechanisms at stake, both for anaerobic and aerobic aging.

ACKNOWLEDGEMENTS

The authors would like to thank Vibracoustic for its financial support and all the partners, ENSTA Bretagne—IRDL, Freudenberg Technology Innovation, and Chemnitz University of Technology. A special thanks to Pierre-Yves Le Gac from IFREMER for fruitful scientific discussions and technical support.

REFERENCES

Bolland, J.L. & Gee, G. 1946. Kinetic studies in the chemistry of rubber and related materials. II. The kinetics of oxidation of unconjugated olefins. Trans. Faraday Soc. 42, 236–243.

Broudin, M., Le Saux, V., Marco, Y., Charrier, P., Hervouet, W. 2015. Investigation of thermal aging effects on the fatigue design of automotive anti-vibration parts. In ECCMR IX, 53–59. CRC Press.

Calvert, P.D. & Billingham, N.C. 1979. Loss of additives from polymers: A theoretical model. J. Appl. Polym. Sci. 24, 357–370.

Celina, M., Gillen, K.T., Assink, R.A. 2005. Accelerated aging and lifetime prediction: Review of non-Arrhenius behaviour due to two competing processes. Polym. Degrad. Stab. 90, 395–404.

Celina, M., Wise, J., Ottesen, D., Gillen, K.., Clough, R. 2000. Correlation of chemical and mechanical property changes during oxidative degradation of neoprene. Polym. Degrad. Stab. 68, 171–184.

Charrier, P., Marco, Y., Le Saux, V., Ranaweera, R.K.P. 2011. On the influence of heat ageing on filled NR for AVS automotive applications. In ECCMR VII, 381–387. CRC Press.

Gillen, K.T., Bernstein, R., Derzon, D.K. 2005. Evidence of non-Arrhenius behaviour from laboratory aging and 24-year field aging of polychloroprene rubber materials. Polym. Degrad. Stab. 87, 57–67.

Herzig, A., Johlitz, M., Lion, A., 2015. Experimental investigation on the consumption of oxygen and its diffusion into elastomers during the process of ageing. In ECCMR IX, 23–28. CRC Press.

ISO. Caoutchouc vulcanisé ou thermoplastique—Essais de résistance au vieillissement accéléré et à la chaleur. ISO 188:2011, 20.

Johlitz, M. & Lion, A. 2013. Chemo-thermomechanical ageing of elastomers based on multiphase continuum mechanics. Contin. Mech. Thermodyn. 25, 605–624.

Kamarudding, S., Le Gac, P.Y., Marco, Y., Muhr, A. 2011. Formation of crust on Natural Rubber after long periods of ageing. In ECCMR VII, 197–201. CRC Press.

Kim, S.G. & Lee, S.-H. 1994. Effect of crosslink structures on the fatigue crack growth behavior of NR vulcanizates with various aging conditions. Rubber Chem. Technol. 67, 649–661.

Kohman, G.T., 1929. The Absorption of oxygen by rubber. J. Phys. Chem. 33, 226–243.

Le Saux, V., 2010. Fatigue et vieillissement des élastomères en environnements marin et thermique: de la caractérisation accélérée au calcul de structure. Université de Bretagne Occidentale—Brest.

Naumann, C., Ihlemann, J. 2013. Chemomechanically coupled finite element simulations of oxidative aging in elastomeric components. In ECCMR VIII, 43–49. CRC Press.

Ostoja Kuczynski, E., Charrier, P., Verron, E., Marckmann, G., Gornet, L., Chagnon, L. 2003. Crack initiation in filled natural rubber: experimental database and macroscopic observations. In ECCMR III, 41–47. CRC Press.

Roychoudhury, A., De, P.P. 1993. Reinforcement of epoxidized natural rubber by carbon black: Effect of surface oxidation of carbon black particles. J. Appl. Polym. Sci. 50, 181–186.

Schlomka, C., Ihlemann, J., Naumann, C., 2017. Simulation of oxidative aging processes in elastomer components using a dynamic network model. In ECCMR X (in press).

Shelton, J.R., Winn, H. 1944. Aging of GR-S vulcanizates in air, oxygen, and nitrogen. Ind. Eng. Chem. 36, 728–730.

Soma, P., Tada, N., Uchida, M., Nakahara, K., Taga, Y. 2010. A Fracture mechanics approach for evaluating the effects of heat aging on fatigue crack growth of vulcanized natural rubber. J. Solid Mech. Mater. Eng. 4, 727–737.

South, J.T., Case, S.W., Reifsnider, K.L. 2003. Effects of thermal aging on the mechanical properties of natural rubber. Rubber Chem. Technol. 76, 785–802.

Spetz, G. 1994. Improving precision of rubber test methods: Part 2 — ageing. Polym. Test. 13, 239–270.

Struik, L.C.E. 1978. Physical aging in amorphous polymers and other materials. Elsevier, New York.

Verdu, J., 2012. Oxidative ageing of polymers. Wiley-ISTE, London.

Constitutive Models for Rubber X – Lion & Johlitz (Eds)
 ISBN 978-1-138-03001-5

Influence of antioxidant type on the thermo-oxidative aging of rubber vulcanizates

K. Reincke, K. Oßwald & S. Sökmen
Polymer Service GmbH Merseburg, Merseburg, Germany

B. Langer
Hochschule Merseburg, Merseburg, Germany
Polymer Service GmbH Merseburg, Merseburg, Germany

W. Grellmann
Martin-Luther-Universität Halle-Wittenberg, Germany
Polymer Service GmbH Merseburg, Merseburg, Germany

ABSTRACT: Former investigations (Reincke 2015) of a carbon-black filled Styrene–Butadiene and a Natural Rubber vulcanizate (SBR and NR) showed clear differences in the degradation behaviour of the used antioxidant during an artificial weathering. Now, in a further study, the behaviour during a thermo-oxidative aging was investigated for carbon-black filled NR vulcanizates with three different antioxidants. The duration of the thermo-oxidative aging at different temperatures was realized up to 1500 h. To characterize the material properties in dependence on the aging time and temperature as well as type of antioxidant, different methods of mechanical testing were applied. Focus was on the characterization of the viscoelastic material behaviour by using dynamic mechanical analysis. Here, the influence of aging and type of antioxidant on the Payne effect was studied.

1 INTRODUCTION

Aging of polymers is still a prevailing problem because it can result in sudden and unexpected component failure or generally in a shortened lifetime. Due to the demands for reduction in material input or for long-time resistance, it is necessary to ensure the required material properties over time. In the case of elastomers, this entails the design of a suitable recipe. A first aspect here is the selection of the polymer. Rubbers with double bonds within the main chain like NR, SBR or butadiene rubber (BR) are particularly susceptible to oxidative processes (Santoso 2007, Huneke 2006). Without protection, a direct reaction of the polymer chain with oxygen takes place and hydro peroxides will develop (Shelton 1957) and so, the degradation process is started. To prevent this degradation, synthetic polymers themselves can be stabilized or aging protection substances can be added to the mixture. These substances can react with media/oxygen to prevent the reaction with the polymer chain. In principle, aging protection substances can be classified in antioxidants, antiozonants and fatigue protection (Osen 2004).

The modulus of filled rubbers decreases with increasing amplitudes of the applied dynamic strain (see Fig. 1). This is known as Payne effect according to the detailed study of the low frequency dynamic properties of filled natural rubber carried out by Payne (Payne 1964). Usually, the reduction in modulus with increasing deformation amplitude is contributed to the breakdown of filler–filler bonds.

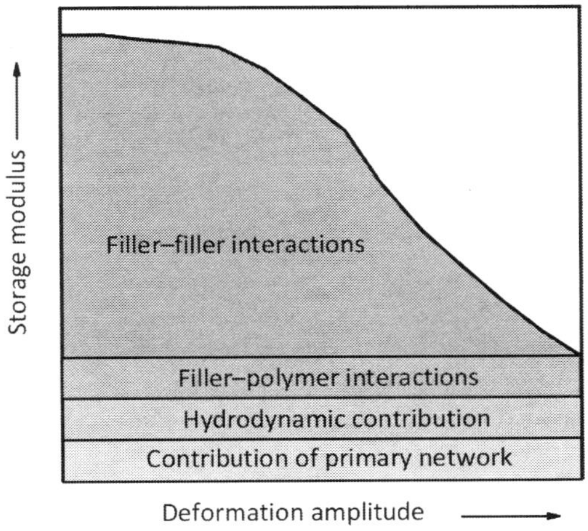

Figure 1. Schematic representation of the storage modulus of a filler-reinforced elastomer in dependence on the deformation amplitude according to Röthemeyer 2001.

The aim of this work was to investigate the influence of antioxidant type on dynamic and mechanical properties of elastomers after thermo-oxidative aging. The focus of this paper was to analyse the viscoelastic behaviour of carbon-black filled natural rubber. For this, the carbon black content and the type of antioxidant were varied. Besides, the effects of different aging temperatures and aging periods were investigated.

2 EXPERIMENTAL

2.1 *Materials*

The natural rubber compounds with different filler contents and different antioxidants have been manufactured and investigated. The recipe of the mixture is given in Table 1. The polymeric basis is natural rubber SVR10, which was masticated on the roll mill for 3 min as a first processing step. Highly active carbon black N234 having 119 ± 5 m^2/g BET surface area was used as filler. Zinc oxide and stearic acid were added as activators for the vulcanisation reaction realized via a sulfur–accelerator (CBS) system.

All rubber compounds were prepared in laboratory batch mixers, Haake 300p (Corp. Thermo Haake GmbH, Germany), and Brabender N50 (Corp. Brabender GmbH & co. KG, Germany). A starting temperature of 50°C, a roller rotation speed of 50 rpm and a fill factor of 70% were used.

The vulcanization temperature for pressing the sample plates was 160°C. The cure time t_{90} (indicating the time when 90% of the crosslinking reaction is over) was determined by using the vulcameter Göttfert Elastograph (Corp. Göttfert Werkstoff-Prüfmaschinen GmbH, Germany). From the sample plates, specimens for the different tests were obtained.

2.2 *Methods*

In order to determine the resistance of the different rubber vulcanizates to oxidation, accelerated aging test was used. Aging of the samples at three different temperatures (40°C, 60°C and 80°C) was studied besides room temperature at 23°C. The specimens/samples were stored in direct contact to oxygen in air-circulated ovens and in standard room climate, respectively. Tests were performed after time periods of 250 h, 500 h and 1500 h.

The mechanical properties were characterized by tensile and hardness tests. For the tensile test, dumb-bell shaped specimens of type S2 according to DIN 53504 were used. 5 specimens were tested for each series. The tests were done with a universal testing machine ZWICK Z020 at room temperature. The initial clamp distance was 50 mm and extension rate was selected 200 mm/min. The Shore A hardness was determined according to DIN ISO 7619-1 with a testing time of 3 s. For each mean value, 10 single measurements were performed.

Dynamic mechanical analysis (DMA) according to ISO 6721-4 was used to investigate the dynamic mechanical properties of rubbers before aging. From the sample plates, specimens of the size 56 × 13 × 2 mm^3 were cut. Dynamic properties were measured in tensile mode with TA Instruments (Corp. TA Instruments GmbH, Germany) under nitrogen atmosphere. Further DMA experiments were performed by using the rubber process analyzer (RPA). The RPA is a rotorless rotational rheometer to characterize raw polymers, raw rubber mixtures and also elastomers. It consists of a closed test chamber and a dynamic rotational shear system. RPA data allow the calculation of the tan (δ) as well as complex shear modulus (G*), shear storage modulus (G') and loss modulus (G") which are measured by rotational test mode. By using the RPA, the materials were tested before, during and after aging. In the beginning, samples were vulcanized within RPA. These samples were stored at the defined aging conditions. For the tests, the samples were put back into the RPA chamber. Due to the specific form, this is easily to realize. The amplitude sweep test was then carried out from 0.01 to 80% and a frequency of 10 Hz and a temperature of 60°C.

Table 1. Recipe of the investigated materials.

Ingredients	Content (phr)
NR or SBR	100
Carbon Black (CB) N234	0, 20, 40, 60
Sulfur	1.7
Zinc oxide	3.0
Stearic acid	1.0
CBS	2.5
Antioxidants (IPPD, TMQ, 6PPD)	1.5

3 RESULTS

In Fig. 2, the tensile strength σ_{max} and the tensile strain at break ε_R of NR vulcanizates with 3 different antioxidants are shown in dependence on the filler content. As can be seen, both material parameters have a maximum values at 20–40 phr carbon black content. The different antioxidant types have no significant influence on the strength, but on the deformability, characterized by ε_R. Mostly, latter has the lowest values for the NR compound with IPPD.

Figure 3 shows results of the DMA tests. The typical behaviour of the storage modulus E' is visible with a high level of E' in the energy-elastic

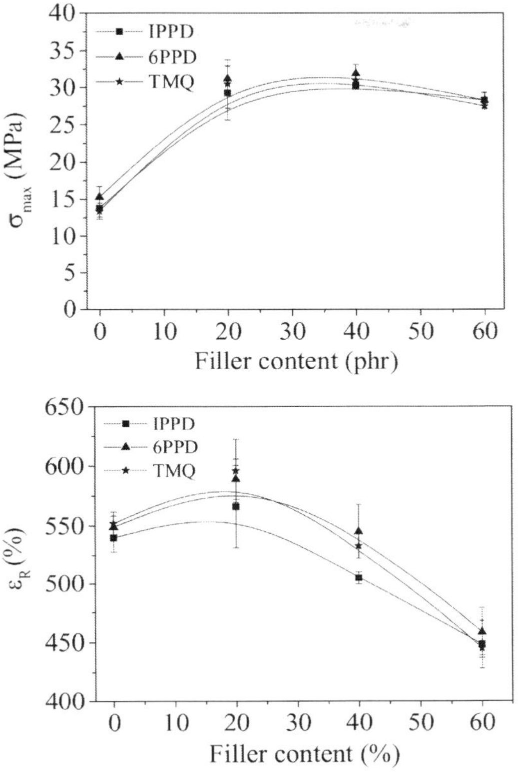

Figure 2. Tensile strength σ_{max} and tensile strain at break ε_R of NR vulcanizates with different antioxidants determined from tensile tests.

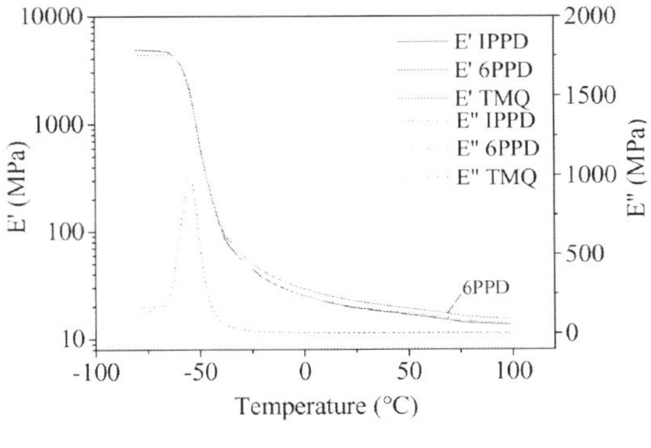

Figure 3. Storage and loss modulus E' and E'' of NR vulcanizates with 60 phr CB and different antioxidants as a function of the temperature.

range, the glass-transition area with a strong drop in E' and the application-related entropy-elastic range with a very low modulus. The glass transition (also indicated by the maximum of E'') is not influenced by the antioxidant, but the height of E' above T_g is slightly higher for the compound with 6PPD compared to the other two antioxidants. Differences in E'' could not be observed.

The Payne effect can be seen for the series NR materials with IPPD and higher filler contents of 40 and 60 phr, respectively (Fig. 4). As generally known, a pronounced decreasing in G' can be observed for vulcanizates with a filler content in the range or above the percolation threshold. Therefore, latter can be expected at a carbon-black content of 40 phr. The strongest loss in storage modulus was reached with 60 phr carbon black. This means, here, the filler–filler interactions play an important role for the mechanical behavior. Similar results were obtained for the other antioxidants. Partly, the height of the initial value of G' is different, depending on the type of antioxidant.

Not only had the filler content, but also the aging temperature had an influence on the course of the shear modulus G'. The following Fig. 5

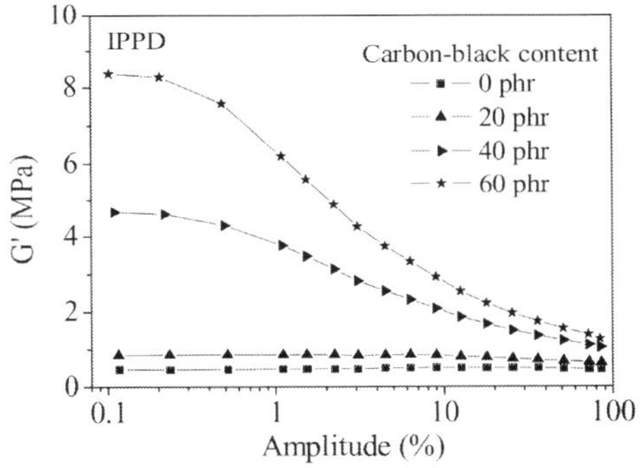

Figure 4. Shear modulus G' of NR vulcanizates with IPPD and different filler contents in the virgin state as a function of deformation amplitude.

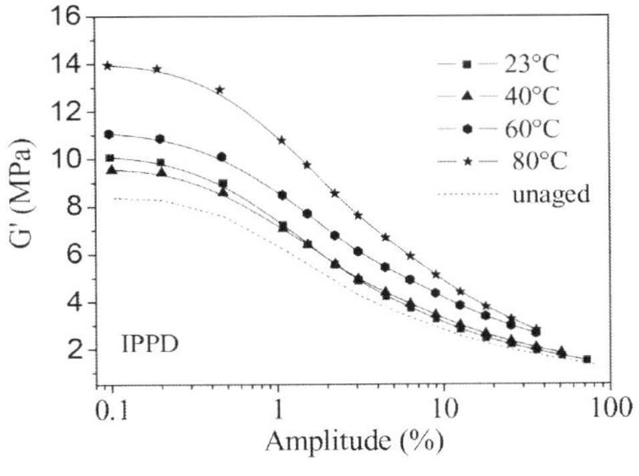

Figure 5. Shear modulus G' of the NR vulcanizate with 60 phr CB and IPPD after 1500 h thermo-oxidative aging at different temperatures as a function of deformation amplitude.

illustrates the increase especially in the initial values of G' at small deformation amplitudes. Interestingly, the NR material, which was aged at 40°C has a lower G' at small amplitudes than that from the room temperature storage. The highest increase in G' was found for the highest aging temperature of 80°C. Further, a decrease in the maximal reachable deformation amplitude was found with increasing aging temperature. This indicates an appearing stiffening of the material.

In Fig. 6, the courses of G' of the NR materials with different contents of carbon black and with TMQ and 6PPD after 1500 h aging are shown.

Again, no contribution of the filler network can be observed for the unfilled and the compound with 20 phr carbon black. The comparison of the initial values of G' indicates that there are differences in stiffness for the materials with the different antioxidants. Also the maximum reachable deformation amplitude depends on the type of antioxidant. To get a quantitative criteria for a comparison of the antioxidants, the difference of maximum and minimum of the complex modulus ΔG* was determined. In Fig. 7, the influence of the aging time on ΔG* of NR with 60 phr carbon black and with different antioxidants is shown. Depending on the antioxidant, the difference in G* becomes larger. This is due to an increase especially in the initial value of G* at small deformation amplitudes. This means, the Payne effect seems to be more pronounced with increasing aging time. According to the theory, this would be a sign of an influence of the thermo-oxidative aging not only on the polymer or low-molecular substances, but also on the filler–filler bonds.

To explain this result, more research work is necessary, for example with other fillers. Beside this relatively unexpected result regarding the Payne effect, it can be seen from the result in Fig. 6 that the kind of antioxidant has an influence also on ΔG*.

According to the literature (Kumar 2004, Chakraborty 2010), a stiffening during aging can be explained by a post-crosslinking. For the materials investigated here, the vulcanization was realized up to t_{90}, this means a post-crosslinking is quite a reason for the change in G' with increasing aging temperature (see Fig. 5).

An increase in hardness and modulus during a thermo-oxidative aging can also be explained by the conversion from polysulfidic to mono- and disulfidic bonds (Ismail 1997). The loss of low molecular-weight substances having a plasticizing effect like stearic acid is a third possible explanation of an increase in stiffness of an elastomeric material during aging. These possible effects are seemingly influenced by the type of antioxidant. For further clarification, investigations of the network density and the network characteristics are necessary. Also chemical analysis may be helpful to explain the observed aging behavior.

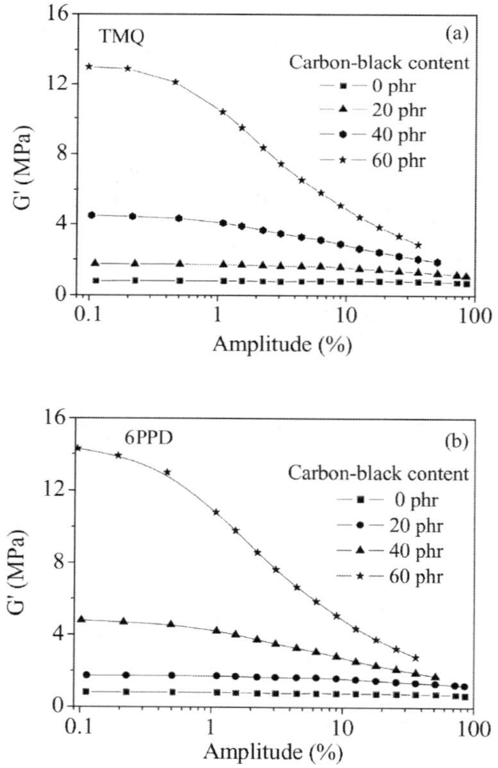

Figure 6. Shear modulus G' of NR vulcanizates with different amounts of carbon black and with TMQ (a) and 6PPD (b) after 1500 h thermo-oxidative aging @ 80°C as a function of deformation amplitude.

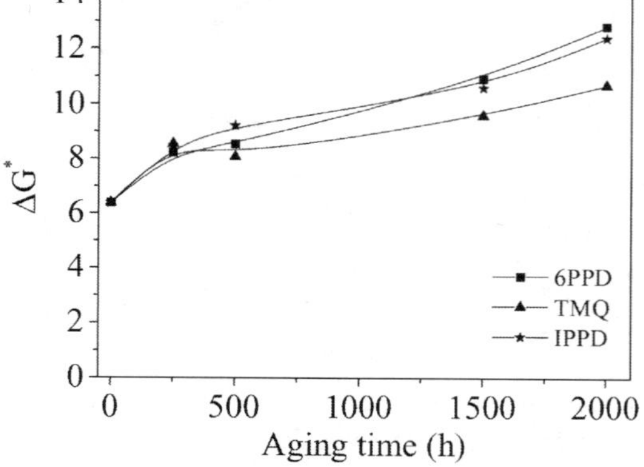

Figure 7. Difference in complex modulus ΔG* of NR vulcanizates with different antioxidants in dependence on the aging time.

4 CONCLUSIONS

NR vulcanizates with different contents of carbon black and different antioxidants were investigated. It was found that the antioxidant type influences the general mechanical property level, but also the dynamic-mechanical behaviour. With increasing aging time and temperature, the Payne effect becomes more pronounced and is influenced by the antioxidant type.

ACKNOWLEDGEMENTS

The authors acknowledge the Company Avokal GmbH, Wuppertal (Germany) for the antioxidants and the company Trinseo Deutschland GmbH, Schkopau (Germany) for the S-SBR polymer.

REFERENCES

Chakraborty, S., Kar, S., Dasgupta, S. & Mukhopdhyay, R. (2010). *Journal Elastomers and Plastics* 42: 443.

Huneke, B. & Klüppel, M. 2006. *Kautschuk Gummi Kunststoffe* 59: 242–250.

Ismail, H., Ishiaku, U.S., Azhar, A.A. & Mohd Ishak, Z.A. (1997). *Journal Elastomers and Plastics* 29: 270.

Kumar, A., Commereuc, S. & Verney, V. 2004. *Polymer Degradation and Stability* 85: 751.

Osen, E. 2004. *Das Alterungsverhalten von Gummi. VDI-Seminar* "Schadensanalyse an Elastomerbauteilen", 14/15. Juni 2004.

Payne, A.R. 1964. *Journal of Applied Polymer Science* 8: 1661–1667.

Reincke, K., Schoßig, M., Poser, S., Langer, B., Döhler, S., Heuert, U., Frank & W., Grellmann, W. 2015. *Experimental Characterization of the Ageing Resistance of Elastomers.* In: Maralova & B., Petrikova, I. (Eds): *Constitutive Models for Rubber IX. Proceedings of the European Conference on Constitutive Models for Rubbers IX (ECCMR IX)*. 2015, Prague, September 1st – 4th, CRC Press Taylor & Francis Group London, pp. 33–37.

Röthemeyer, F. & Sommer, F. 2001. *Kautschuktechnologie.* Carl Hanser Verlag München Wien.

Santoso, M., Giese, U. & Schuster, R.H. 2007: *Kautschuk Gummi Kunststoffe* 60:192–198.

Shelton, R.J. 1957. *Rubber Chemistry and Technology* 30: 1251.

Constitutive Models for Rubber X – Lion & Johlitz (Eds)
© 2017 Taylor & Francis Group, London, ISBN 978-1-138-03001-5

On the thermal aging of a filled butadiene rubber

K.D. Ahose, S. Lejeunes & D. Eyheramendy
LMA, Aix-Marseille University, CNRS UPR 7051, Centrale Marseille, Marseille Cedex 20, France

F. Sosson
SMAC, Toulon Cedex 9, France

ABSTRACT: In this study we investigate the influence of thermal aging on the mechanical properties of a butadiene rubber filled with carbon black. To emphasize the influence of both the crosslink density and the crosslink lengths, we consider three different materials based on the same formulation. Dynamic and quasi-static characterizations are realized periodically at room temperature to study of the impact of aging on various mechanical characteristic such as the equilibrium hysteresis, the Payne effect, etc. Crosslink density is followed by swelling tests.

1 INTRODUCTION

The impact of thermal aging phenomena on physical properties such as mechanical stiffness, tensile strength limit, loss angle, shore hardness, has been studied by many authors, see for instance Tomer, Delor-Jestin, Singh, & Lacoste 2007, Shabani 2013, Kumar, Commereuc, & Verney 2004, Kartout 2016, Ben Hassine 2013 and references therein. Obviously, aging phenomena are material dependent and different issues arrise depending on monomers or vulcanizate agents (see Choi, Kim, & C.-S. 2006 for a comparison of EPDM and NBR). These phenomena are also strongly coupled with the environment, chemical reactions can occur with air (oxidation), salt water (see for instance Gac, Saux, Paris, & Marco 2012, Rabanizada, Lupberger, Johlitz, & Lion 2015), etc. Furthermore, mechanical state during thermal aging may also have an influence on aging phenomena (see for instance Ciutacu, Budrugeac, Mare, & Boconcios 1990). The chemo-physical evolutions involved during aging may lead both to the formation of new crosslinks and to the dissociation of existing ones. For sulfur vulcanized system a commonly admitted mechanism is the degradation of poly-sulfur crosslinks into di-sulfur of mono-sulfur ones. Free sulfurs can eventually migrate and form new crosslinks. This phenomenon is of interest for mechanicians as it can occur during high-cycles fatigue in particular when the heat-build is significant or when the thermal environment plays an important role.

In this study we propose to investigate the consequence of thermal aging on a butadiene rubber reinforced with carbon black fillers and vulcanized with sulfur. To study the influence of the poly-sulfur crosslinks we consider a single rubber formulation cured under different conditions (temperature and time of cure). This lead to different crosslink networks and different crosslink densities. We proceed to mechanical characterizations with quasi-static and dynamic tests on unaged and aged specimens.

2 EXPERIMENTAL SETUP

2.1 *Materials*

The material is a polybutadiene rubber reinforced with carbon black and vulcanized with sulfur. Due to confidentiality reasons, the formulation of this rubber can not detailed here. The vulcanization system is efficient (sulfur to accelerator ratio is greater than one) and the filler mass fraction is about 45%. This system was cured at three different temperatures: 130°C, 150°C and 170°C. The time of cure for each temperature was previously determined with the help of a rheometer: time of cure was obtained when the measured torque reach 98% of the maximum observed one at each temperature. We obtained the following time of cure: 170°C–7 min, 150°C–17 min, 130°C–55 min. To simplify, in the following each couple time/temperature is refereed as a different material even if the initial composition is the same. Each material can be considered as fully vulcanized and differs from each others from its network structure: for high temperature, we have higher crosslink lengths (poly-sulfurs) and a smaller crosslink density than for smaller temperatures of cure.

2.2 *Thermal aging*

Thermal aging of H2 tensile specimens was done with the help of a oven. Each sample was previously put inside an individual plastic bag in which air was partially removed with a vacuum pump. The objective was to minimize thermo-oxidative phenomena and to limit non homogeneous aging. Some specimens, 2 by material, were periodically removed from the oven for mechanical, and chemical characterizations at room temperature.

2.3 *Mechanical characterizations*

Mechanical characterizations on tensile specimen consisted into the following sequence: a softening phase to eliminate the Mullins effect, relaxations by step (relaxations at different increasing and decreasing amplitudes), cyclic sinusoidal tests with two static amplitudes and three dynamics amplitudes (frequencies of 0.1 Hz, 3 Hz, 6 Hz, 10 Hz). These tests were all done at room temperature.

3 UNAGED BEHAVIOR

3.1 *Chemical and thermal characterizations*

We realized swelling tests on small specimens that were put in xylene. Figure 1, shows the evolution of the relative swelling ratio in mass upon the time spent in solvent. We define the relative swelling ratio in mass from:

$$q = \frac{m_s - m_0}{m_0} \tag{1}$$

where m_s is the mass of the specimen after swelling and m_0 is the initial mass. Swelling tests are good indicators of the crosslink density of the vulcanized network. The crosslink density could be estimated using the equation proposed by Kraus 1963 for filled rubber, however, we do not had access to the volume fraction of unfilled rubber in the swollen rubber phase. Nevertheless, it can be seen from

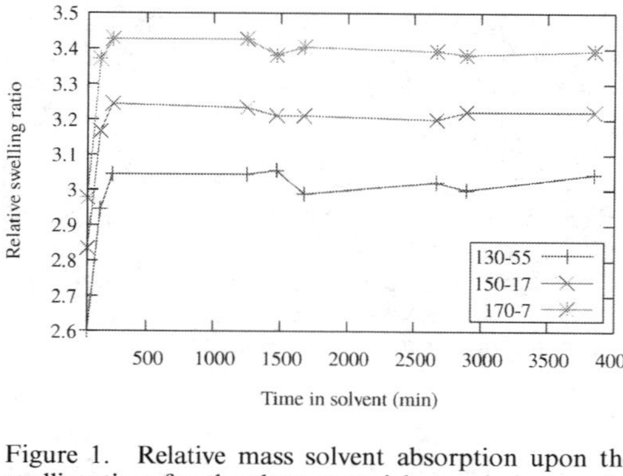

Figure 1. Relative mass solvent absorption upon the swelling time for the three materials on virgin samples (not submitted to thermal aging).

Kraus or Flory-Rhener works (Flory & Rehner 1943) that crosslink density is related to the inverse of the previously defined swelling ratio. We can therefore admit that crosslink density is higher for the material cured at 130°C–55 min than for those cured at 150°C–17 min and 170°C–7 min. We also realized thermal characterizations with a DSC and optical dilatometry with DIC. The glass-transition temperature was measured at −82°C for each materials and the thermal dilatation behavior was also very close for each materials.

3.2 *Mechanical behavior*

Figures 2 and 3 show some obtained results on unaged samples. It can be seen that, in accordance with the statistical theory of rubber network, the higher is the crosslink density the higher is the stiffness. Dynamic tests show that the dynamic softening effect (Payne effect) is nearly the same in the three unaged materials, hysteresis area are larger for the material that was cured at 170°C that for this at 150°C, which show also a larger hysteresis area than the one cured at 130°C, see for instance Figure 4.

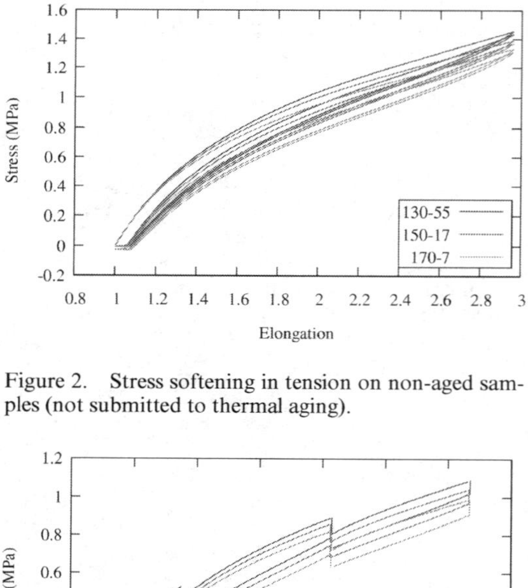

Figure 2. Stress softening in tension on non-aged samples (not submitted to thermal aging).

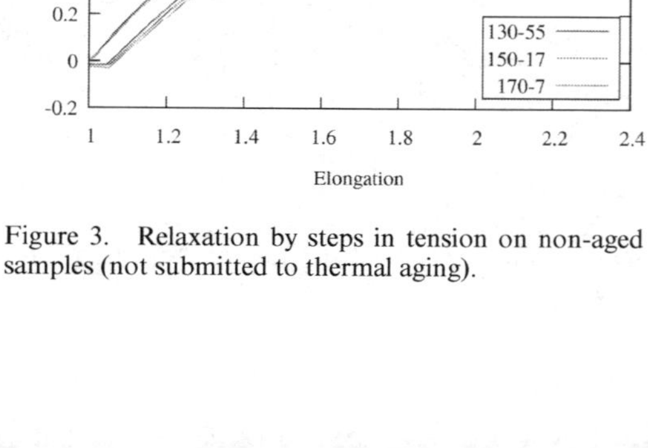

Figure 3. Relaxation by steps in tension on non-aged samples (not submitted to thermal aging).

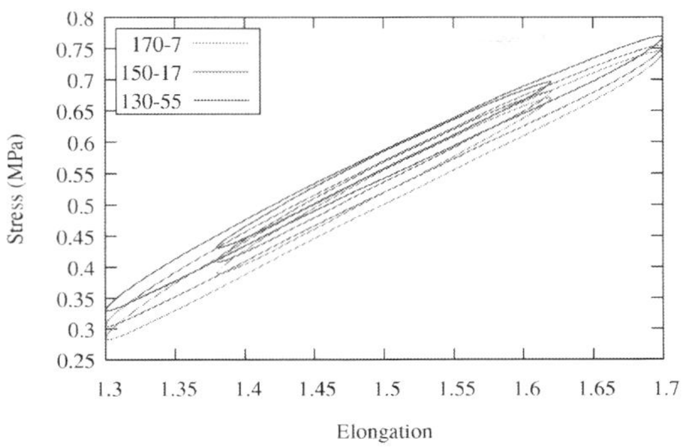

Figure 4. Stabilized hysteresis in tension on non-aged samples for cyclic tests at 3 Hz.

3.3 *Modeling of the unaged behavior*

We adopt a modeling framework that is very close to the one of Delattre et al. 2016. The equilibrium part of the stress is considered as hyperelastic and non-equilibrium one is modeled with a series of Maxwell elements (Generalized Maxwell model) with an accounting of the Payne effect as described in Delattre et al. 2016. The Cauchy stress, $\boldsymbol{\sigma}$, is therefore defined from:

$$\boldsymbol{\sigma} = \boldsymbol{\sigma}_{eq} + \boldsymbol{\sigma}_{neq} - p\mathbf{1} \qquad (2)$$

where p is the hydrostatic pressure. For the equilibrium part, we adopt a Mooney-Rivlin potential defined such as:

$$\rho_0 \psi_{eq} = sC_{10}(I_1 - 3) + sC_{01}(I_2 - 3) \qquad (3)$$

where ρ_0 is the reference density, I_1, I_2 the two first strain invariants. The material coefficients, C_{10}, C_{01} are classical Mooney parameters (which are identical for each materials) and s is a parameter that is fixed to 1 for the material 130–55 and that is identified for other materials. The identification of C_{10}, C_{01}, s_{150}, s_{170} is done by minimizing the least-square distance from the end of relaxation points (extracted from Figure 3) to the predicted ones for the three materials at the same time. We obtain the material parameters given at Table 1 (see also Figure 5). It can be remarked that the obtained values for s are nearly equal to the inverse of the relative swelling ratio such as we recover the hypothesis of proportionality of the rubber elasticity upon the crosslink density that is made in statistical theories. We can postulate:

$$s_x = \frac{q_{130}}{q_x} \qquad (4)$$

where x stands for the material 150–17 or 170–7.

Table 1. Material parameters of the equilibrium part.

C_{10}	C_{01}	s_{150}	s_{170}
0.185 MPa	0.127 MPa	0.948	0.9
q_{130}/q_{150}	q_{130}/q_{170}		
0.944	0.898		

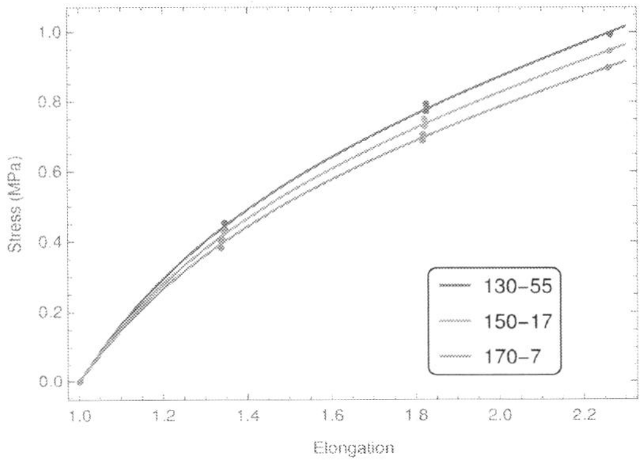

Figure 5. Identification of the equilibrium contribution with a Mooney-Rivilin potential.

For the non-equilibrium part, we adopt a Neo-Hooken potential and a Maxwell viscous flow:

$$\begin{cases} \rho_0 \psi_{neq} = \sum_{i=1}^{n} (G_i / s)\omega_i (I_1^{ei} - 3) \\ \dot{\overline{\mathbf{B}}}_{ei} = \overline{\mathbf{L}}\overline{\mathbf{B}}_{ei} + \overline{\mathbf{B}}_{ei}\overline{\mathbf{L}}^T - \frac{1}{s\tau_i}\overline{\mathbf{B}}_{ei}^D\overline{\mathbf{B}}_{ei} \\ \dot{\omega}_i = \frac{1}{h_i}\left\langle \omega_i - \left(\frac{3}{I_1(\overline{B})}\right)^{r_i} \right\rangle, \qquad \omega_i(t=0) = 1 \end{cases} \qquad (5)$$

where $\overline{\mathbf{B}}_{ei}$ is the *ith* isochoric left Cauchy Green tensor coming from the *ith* multiplicative split of the transformation gradient into viscous and elastic part, $\overline{\mathbf{L}}$ is the isochoric tensor of eulerian velocities, ω_i is the *ith* internal variable to account of the Payne effect, and G_i, τ_i, h_i, r_i are material parameters ($i = 1..n$).

The material parameters identification of the non equilibrium part is done only on the data of the material 130–55 (by setting $s = 1$). We have therefore a viscoelastic model that can account of a variation of the crosslink density, using the relation of eq. (4). A good agreement with the response of the two other materials is observed as can be seen from the Figures 6 and 7.

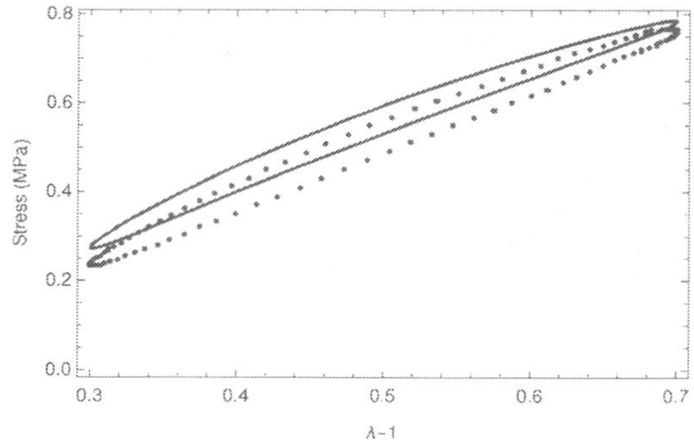

Figure 6. Validation of the identified parameters of the non-equilibrium contribution with the material 170–7 (f = 10 Hz).

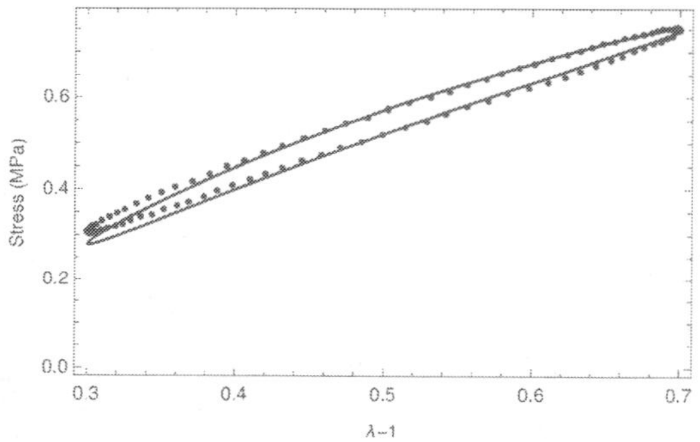

Figure 7. Validation of the identified parameters of the non-equilibrium contribution with the material 150–17 (f = 3 Hz).

4 AGED BEHAVIOR

We investigate, in this section, the behavior of aged tensile samples that were put in a oven at 90°C. No significant variation of mass or volume, of the samples, was observed after aging.

4.1 *Chemical characterizations*

The results of the swelling tests done at different aging time is synthesized with Figure 8. We plotted the inverse of the swelling ratio q relative to the initial (unaged) swelling ratio q_0 for each material. It can be seen that the material that has the larger intial mean crosslink length (the larger number of polysulfur crosslink) exhibits the stronger evolution of the crosslink density. After 11 days of aging, the swelling ratio, q, is very close for each material as the initial concentration of sulfur is the same in each material.

4.2 *Mechanical characterizations*

To investigate the results of relaxation tests, we have plotted the stiffness obtained from the end of relaxation tests done at the maximum amplitude. It can be seen from Figure 9 that the equilibrium behavior become stiffer. Furthermore, the material 170–7 seems to become more rigid than the two others. We can also notice that the dispersion of the results increases with aging.

For dynamic tests, we calculated the median line and the area of stabilized hysteresis curves. In the following, the term dynamic stiffness stands for the median line. Typical results are shown with Figures 10 and 11. As for the equilibrium stiffness, the dynamic stiffness increases with aging while the hysteresis area decreases. The dispersion is stronger than for equilibrium results and seems to increase with aging.

Figures 12, 13 and 14 show that the effect of the dynamic amplitude of loading (Payne effect) and the effect of the frequency of loading are not impacted by aging (at least after 15th days of aging): we only see a vertical translation of the curves.

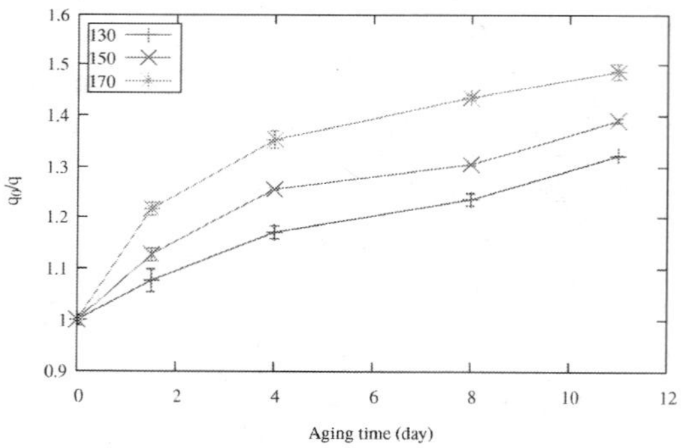

Figure 8. Swelling ratio evolution after aging.

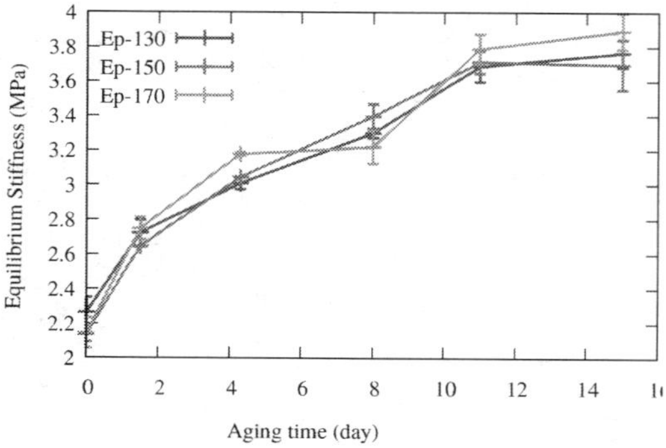

Figure 9. Equilibrium stiffness of aged samples.

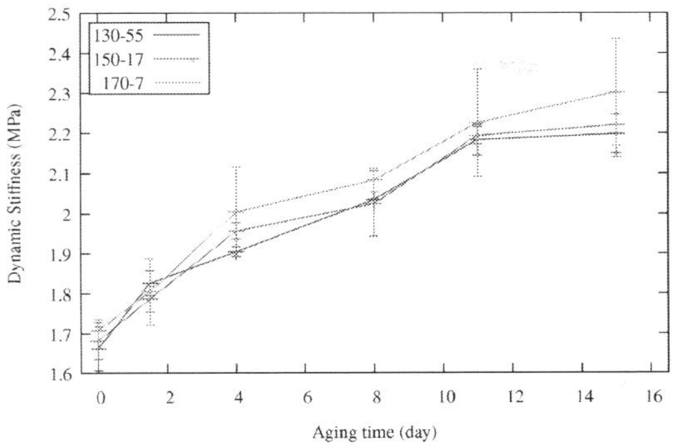

Figure 10. Dynamic stiffness of aged samples (for $f = 10$ Hz at 20% of dynamic amplitude with 30% of static preload).

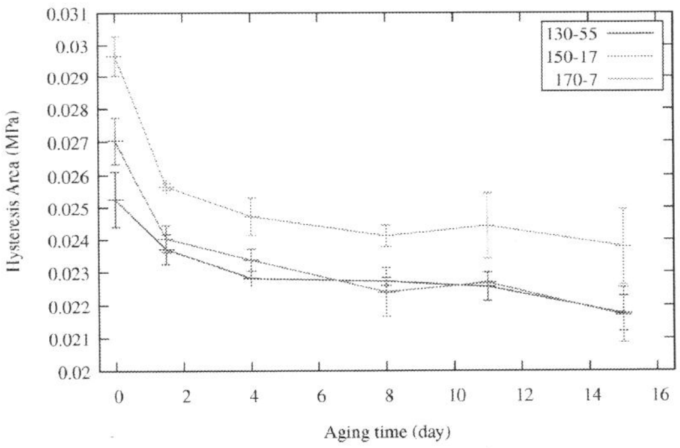

Figure 11. Hysteresis area of aged samples (for $f = 10$ Hz at 20% of dynamic amplitude with 30% of static preload).

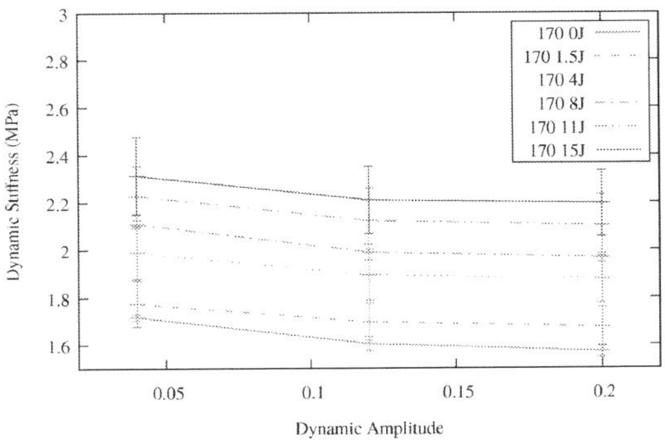

Figure 12. Dynamic stifness on aged samples upon the amplitude of the dynamic amplitude at 3 Hz with 50% of elongation preload.

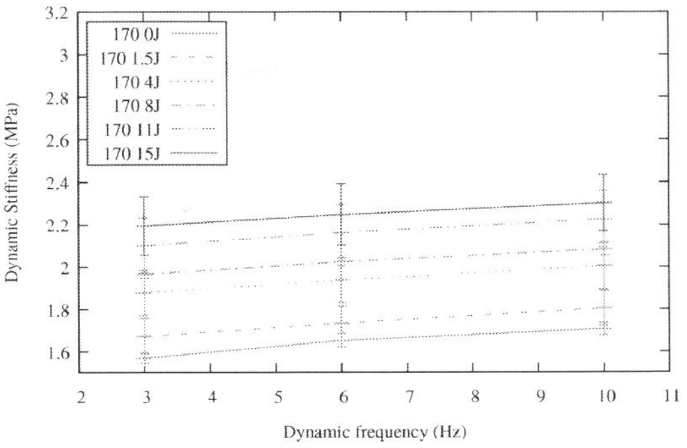

Figure 13. Dynamic stifness on aged samples upon the amplitude of the dynamic amplitude at 3 Hz with 50% of elongation preload.

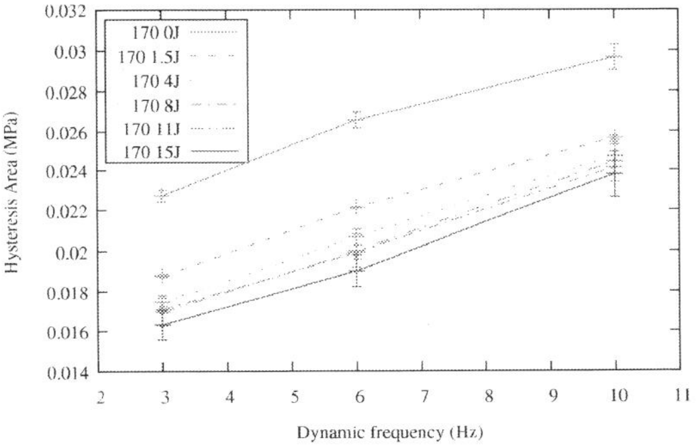

Figure 14. Hysteresis area on aged samples upon the amplitude of the dynamic amplitude at 3 Hz with 50% of elongation preload.

5 DISCUSSION

Based on the previous experimental results we can made the following remarks:

- *Crosslink density and mean crosslink length.* Under non-oxidative conditions, thermal aging leads to an increase of the crosslink density of a sulfur vulcanized filled rubber and this evolution is related to the initial (and current) mean crosslink length.
- *Fillers/rubber network interactions.* Aging mechanisms seem to not impact in a significant manner the fillers/network interactions (at least at 90°C for aging). Amplitude and frequency of loading effects are not modified by aging.
- *Rubber network.* The crosslink density may not be the only parameter to take into account in a modeling of non-oxidative aging. In this campaign, we show that the aged mechanical behavior can be slightly different between the three materials even if the results of swelling tests are

very close after aging. We guess that the aged rubber network is not exactly the same for the three materials after aging, even if the initial formulation is exactly the same.

- *Dispersion of experimental results.* We saw an increase of the dispersion of experimental results during aging. This effect can have many different origins, this can be due to damage that can occurs during mechanical test. As the limit of chains extensibility is reduced due to aging, damage can occur for smaller load amplitudes. This can also be a consequence of a non homogeneous aging that can have different origins: heterogeneity of temperature in the oven or oxidation (even if we tried to minimize it).

For the modeling part, taking into account of the previous remarks, we will have to introduce at least two supplementary variables that will be related to the mean crosslink length and the crosslink density. We already have introduced a relative crosslink density variable in the unaged behavior (eq. 4). This variable should eventually be complemented with another one.

6 CONCLUSION

We have investigated the consequence of thermal aging for a sulfur vulcanized filled rubber. From the results of the experimental campaign, we have shown that aging is strongly related to the initial rubber network (crosslink density and mean crosslink length). As already shown by previous authors, the increase of stiffness, the decrease of dissipated energy and the increase of crosslink density are the three main phenomena but the kinetic is dependent on the initial vulcanization system and on the initial crosslink lengths and crosslink densities.

To investigate further aging mechanisms and their kinetic, we need to do other experimental tests with different temperatures of aging. We will also need to study the impact of a permanent mechanical loads during aging, as done by Johlitz, Diercks, & Lion 2014.

From the modeling point of view the challenge is to describe the kinetic with few supplementary variables. We also need to have a good knowledge of the initial network structure and swelling tests may not be sufficient for that.

REFERENCES

Ben Hassine, M. (2013, October). *Modelling of thermal and mechanical ageings of an external protection of a cold shrinkable junction made of EPDM rubber*. Phd thesis, Ecole nationale supérieure d'arts et métiers - ENSAM.

Choi, S.-S., J.-C. Kim, & W.C.-S. (2006). Accelerated thermal aging behaviors of epdm and nbr vulcanizates. *Bulletin of the Korean Chemical Society 27*(6), 936–938.

Ciutacu, S., P. Budrugeac, G. Mare, & I. Boconcios (1990). Accelerated thermal ageing of ethylene-propylene rubber in pressurized oxygen. *Polymer Degradation and Stability 29*(3), 321–329.

Delattre, A., S. Lejeunes, F. Lacroix, & S. Méo (2016). On the dynamical behavior of filled rubbers at different temperatures: Experimental characterization and constitutive modeling. *International Journal of Solids and Structures 90*, 178–193.

Flory, P. & J. Rehner (1943). Statistical mechanics of crosslinked polymer networks ii. swelling. *The Journal of Chemical Physics 11*(11), 521–526.

Gac, P.L., V.L. Saux, M. Paris, & Y. Marco (2012). Ageing mechanism and mechanical degradation behaviour of polychloroprene rubber in a marine environment: Comparison of accelerated ageing and long term exposure. *Polymer Degradation and Stability 97*(3), 288–296.

Johlitz, M., N. Diercks, & A. Lion (2014). Thermo-oxidative ageing of elastomers: A modelling approach based on a finite strain theory. *International Journal of Plasticity 63*, 138–151. Deformation Tensors in Material Modeling in Honor of Prof. Otto T. Bruhns.

Kartout, C. (2016, March). *Thermo-oxidative ageing and fracture behavior of an EPDM*. Phd thesis, Université Pierre et Marie Curie - Paris VI.

Kraus, G. (1963). Swelling of filler-reinforced vulcanizates. *Journal of Applied Polymer Science 7*, 861–871.

Kumar, A., S. Commereuc, & V. Verney (2004). Ageing of elastomers: a molecular approach based on rheological characterization. *Polymer Degradation and Stability 85*(2), 751–757.

Rabanizada, N., F. Lupberger, M. Johlitz, & A. Lion (2015). Experimental investigation of the dynamic mechanical behaviour of chemically aged elastomers. *Arch Appl Mech 85*(85), 10111023.

Shabani, A. (2013, May). *Thermal and radiochemical of neat and ATH filled EPDM: establishment of structure/properties relationships*. Phd thesis, Ecole nationale supérieure d'arts et métiers - ENSAM.

Tomer, N., F. Delor-Jestin, R. Singh, & J. Lacoste (2007). Cross-linking assessment after accelerated ageing of ethylene propylene diene monomer rubber. *Polymer Degradation and Stability 92*(3), 457–463.

Constitutive Models for Rubber X – Lion & Johlitz (Eds)
© 2017 Taylor & Francis Group, London, ISBN 978-1-138-03001-5

Role of strain-induced crystallization on fatigue properties of natural rubber after realistic aerobic ageing

F. Grasland, J.M. Chenal & L. Chazeau
MATEIS, INSA-Lyon, CNRS UMR 5510, Université de Lyon, Lyon, France

J. Caillard & R. Schach
Manufacture Française des Pneumatiques Michelin, Centre des technologies, Clermont Ferrant Cedex 9, France

ABSTRACT: A scenario is proposed to describe the network evolution of a conventional crosslinked Natural Rubber (NR) during aerobic ageing. The effect of such ageing on Strain Induced Crystallization (SIC) of Natural Rubber is studied by in situ Wide Angle X-Rays Scattering (WAXS) experiments performed at crack tip. A relation is established between the network evolution caused by aerobic ageing, the fatigue properties and the ability to strain crystallize in such conditions and for such materials.

1 INTRODUCTION

Natural rubber is largely used in the tire industry due to its excellent mechanical properties, e.g. its very good resistance to fatigue crack growth at high strain (Trabelsi 2002, Rublon 2014). It is generally accepted that this outstanding behavior is related to its ability to crystallize under strain. Such phenomenon, so-called SIC, strongly depends on parameters like temperature (Rault et al. 2006), strain rate (Brüning 2013) as well as the architecture of the rubber network (Candau 2014). The microstructure of this network is formed during the crosslinking process and depends on the vulcanization system, i.e. "Efficient" or "Conventional" (Chapman et al. 1988). The former vulcanization recipe consists in the formation of short- or mono-sulfide bridges in the elastomer network whereas the latter (necessary to ensure a good adhesion between metallic and rubber parts in a tire) will mainly create longer polysulfide bridges. During its life, the tire will be submitted to a slow aerobic ageing which will cause structural modifications of the initial network and therefore an evolution of the rubber ability to crystallize under strain and to resist against crack propagation. In general, the structural modifications are caused by complex chemical mechanisms, highly sensitive to temperature, leading to chain scission and crosslinking (Colin et al. 2007). They can also involve sulfur bridge reorganization when NR is conventionally vulcanized. Nevertheless, most of the literature on NR ageing has been performed on efficiently crosslinked NR (Trabelsi 2002, Rublon 2014, Candau 2014), and in thermal conditions which are much too severe to be representative of the material ageing in tire applications (Colin et al. 2007). Within this frame, our objective is to study this material when it is aged at 77°C in air. Such parameters have been identified as capable of reproducing more realistically and over a reasonable duration, the ageing of rubber in some use conditions (Bauer et al. 2007). After characterization of the evolution of the aged materials microstructure, their crack propagation resistance will be studied at 0.01 Hz for different values of macroscopic deformations. Time resolved Wide-Angle X-ray scattering (WAXS) measurements, carried out at room temperature, will then provide information on the crystallization process around the crack tip. Based on these results, the relation between the network evolution during ageing, the fatigue properties and the ability to strain crystallize in such conditions will be established.

2 EXPERIMENTS

2.1 *Materials*

The materials are crosslinked unfilled NR, obtained by gum vulcanization according to the recipe given in Table 1. The Sulfur/CBS* ratios are different for both systems in order to obtain network with monosulfides bridges for "Efficient" system and polysulfides bridges for "Conventional" system. The material has been processed following the Rauline patent (Rauline 1993). First, the gum is introduced in an internal mixer and sheared for 2 min at 60°C. Then, the vulcanization

Table 1. Samples compositions and network elastically active chain density estimated from swelling measurements.

Sample code	NR1.5_Conv	NR2.5_Eff
Rubber	100	100
6PPD, phr	3	3
Stearic Acid, phr	2	2
ZnO, phr	5	1.5
CBS*, phr	0.5	2.52
Sulphur, phr	3	1.6
ν**	1.5	2.5

* N-Cyclohexyl-2-benzothiazole sulfenamide.
** Average density of elastically active chains from swelling measurement (in 10^{-4} mol.cm^{-3}).

ingredients are added and the so-formed mixture is sheared for 5 minutes. The material is subsequently sheared in an open mill for five minutes at 60°C. Sample sheets are finally obtained by hot pressing at optimal temperature. The cure time is estimated from the torque measurements performed with a Monsanto analyzer. "Pure Shear" test pieces, with a 6 mm gauge length, 35 mm long and 0.8 mm thickness, are machined.

2.2 *Ageing conditions*

NR1.5_Conv and NR2.5_Eff are submitted to aerobic and anaerobic ageing at 77°C from 0 to 21 days. To fulfill anaerobic conditions, the samples were first submitted to a degassing step in order to ensure the absence of residual oxygen. All the samples were stored at –20°C to avoid any supplementary degradation.

2.3 *Swelling measurements*

Swelling measurement provides the elastomer network chain density (from the swelling ratio) and the fraction of the chains which do not belong to the elastically active network (from the soluble fraction). The swelling procedure is as follows: samples with an initial mass m_i are introduced in two different solvents (toluene and cyclohexane) for 5 days in order to reach the swelling equilibrium at room temperature; then, the swollen material with the weight m_s is dried under vacuum at 50°C during one day and the final dried mass m_d is obtained. The volumic swelling ratio Q of the polymer and the soluble fraction F_s are then calculated through the equations 1 and 2:

$$Q = 1 + \frac{\rho_p}{\rho_s} * \left(\frac{m_s}{m_d} - 1 \right) \quad (1)$$

$$F_s = 100 * \left(\frac{m_i - m_d}{m_i} \right) \quad (2)$$

where ρ_p and ρ_s correspond to the density of the polymer and the solvent respectively. From swelling ratio, network chain density ν can be estimated using the Flory-Rehner equation (Flory et al. 1943). ν is the summation of the active chain density created by chemicals crosslinks and of the active chains created by entanglements trapped during the vulcanization process:

$$\nu = \frac{\ln(1-V) + V + \chi V^2}{V_1 \left(V^{\frac{1}{3}} - 2\frac{V}{f} \right)} \quad (3)$$

where V is the volume fraction of polymer in the swollen mass which corresponds to the inverse of swelling ratio Q. χ is the Flory-Huggins interaction parameter, V_1 is the molar volume of the solvent (107.9 cm^3.mol^{-1} for cyclohexane and 106.3 cm^3.mol^{-1} for toluene) and f is the crosslink functionality. Given the low sol fraction of the processed material suggesting that most chains belong to the rubber network, f can be taken equal to 4. As mentioned recently by Valentin et al (2008). χ has a strong impact on the calculated ν. A widely used equation to estimate this parameter is based on the Hildebrand solubility constants:

$$\chi = \chi_s + \frac{V_1}{RT} \left(\delta_p - \delta_s \right)^2 \quad (4)$$

where δs and δp are the solubility parameters for solvent and polymer respectively, R the ideal gas constant and T the absolute temperature. The first factor χ_s is an entropic contribution and was found empirically equal to 0.34 for non-polar systems (Blanks et al. 1964).

2.4 *Determination of tearing energy (G) and fatigue crack propagation*

The study of crack propagation resistance in fatigue tests basically consists in measuring the crack propagation velocity during a cyclic test as a function of the energy dissipated during fracture per unit of fracture surface area newly created by the crack propagation. Considering thin planar samples of uniform undeformed thickness t and denoting c the crack length, this "tearing energy" G is derived considering a constant displacement l as:

$$G = -\frac{1}{t} \frac{\partial W}{\partial c}\bigg|_l \quad (5)$$

The majority of crack growth rate measurements were carried out using a "pure shear" test piece as shown in Figure 1. In this case, tearing energy G can be simply calculated by:

$$G = Wh_0 \tag{6}$$

where h_0 is the initial sample length in the tensile direction. W can be simply estimated from the area under the stress-strain curve measured up to the maximum strain of the applied dynamic cycle on uncracked samples. The crack propagation velocity (dc/dn) is calculated as the crack length variation per strain dynamic cycle. After a first transition regime, in most of the rubbery materials the crack growth rate obeys a power law dependency of the tearing energy as follows:

$$\frac{dc}{dn} = AG^{\beta} \tag{7}$$

where A and β are material constants. The power exponent β is generally equal to 2 for natural rubber (NR) and 4 for styrene butadiene rubber (SBR) for G above 1000 kJ/m^2 (Lake 1972, Papadopoulos 2008, Kim 1994). This different fatigue behavior is generally explained by the ability of NR to crystallize under strain. Crack growth propagation experiments were performed on a home-made tensile device X-JICO allowing to run tensile cyclic tests at 0.01 Hz. In order to suppress the Mullins effect, cyclic tests were conducted on the samples before the creation of the initial crack, at different λ from 1.1 to 3. The pre-crack is created in the sample at one edge (10 mm-long). A short preliminary cyclic test (10 cycles at 0.01 Hz) is performed to transform the cutting incision into a fatigue crack. Afterwards, the procedure of fatigue cracks growths experiments consists in cycling the pre-crack sample for different global stretch ratio λ

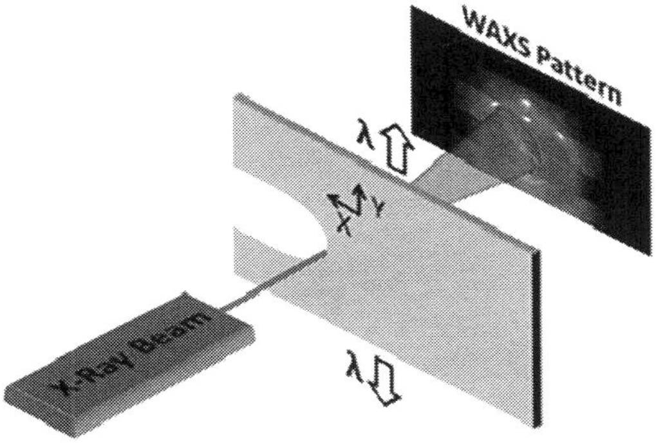

Figure 1. Schematic view of the experiment set-up for in-situ WAXS at the crack tip.

(corresponding to different values of G). The test is stopped after 100 cycles to measure the position of the crack tip with a camera from which we deduced the crack growth rate (dc/dn in nm per cycle). Several tests on various samples to estimate their repeatability and the error. A conventional way to represent the results is a log-log dc/dn vs. G curve as shown in Figure 4.

2.5 *In situ WAXS experiments*

The in-situ WAXS experiments are carried out on the BM2 beamline of the European Synchrotron Radiation Facility (ESRF). The X-Ray wavelengths is 1.54 Å and the beam size is approximately 300 μm × 200 μm. To identify the crack tip position, a specific procedure was performed. The pre-cracked sample was stretched, then, it was moved under the beam with respect to the beam axis (X & Y) and simultaneously the transmitted intensity was recorded as an indicator of the presence of the rubber in the beam (see Figure 1). Once the sample crack tip was positioned under the beam, the two-dimensional WAXS patterns were then recorded by an X-Ray detector with an exposure time of 10 s during the dynamic tests with a strain rate of 0.01 mm.s^{-1}. Finally, the evolution of the crystalline structure at the crack tip was measured until the catastrophic crack propagation. For the sake of simplicity, the calculation of the crystallinity index CI is not fully detailed is this manuscript.

3 RESULTS AND DISCUSSION

3.1 *Chemical mechanism*

The evolution of the active chain density as a function of the ageing time is presented on Figure 2. During the first 7 days of anaerobic ageing, NR1.5_Conv shows an increase of the active chain crosslinked density whereas no evolution is noticed for the material crosslinked by a so-called efficient vulcanization (NR2.5_Eff). This evolution, observed for the latter is consistent with literature which suggests that monosulfudic crosslinks are relatively stable thermally up to high temperature typically above 100°C (Morrison 1984).

The large increase of active chain density during anaerobic ageing in the case of conventional system can be ascribed to the shortening of polysulfidic linkages which cause the formation of additional crosslinks with a lower sulfur level as mentioned by Mathew (1983).

During aerobic ageing, polar groups can be created along the polymer chain (Colin et al. 2007) which consequently modifies the value of the Flory-Huggins parameter χ used in (3). By the use of two solvents (cyclohexane and toluene), the

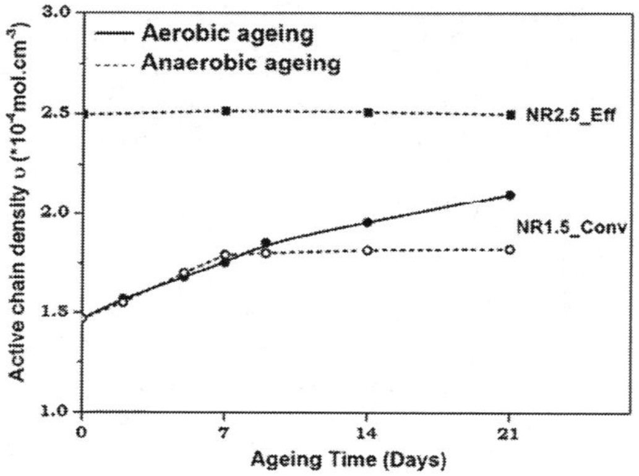

Figure 2. Evolution of the active chain density as a function of the ageing time at 77°C.

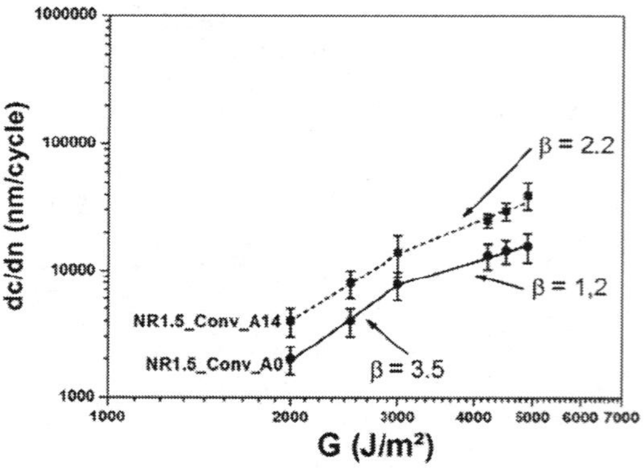

Figure 4. CI evolution at crack tip as a function of tearing energy G for non-aged and aged NR1.5_Conv samples.

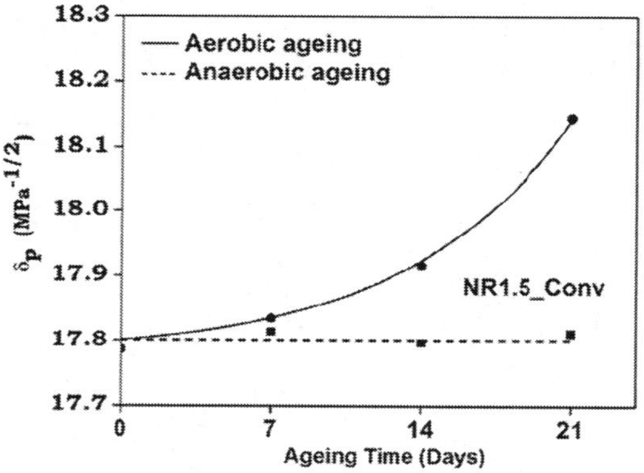

Figure 3. Evolution of solubility parameter of NR1.5_Conv as a function of the ageing time at 77°C.

solubility parameter of the polymer δ_p used in (4) can be deduced and provide a numerical estimate of the degree of polarity in a system. The evolution of δ_p as a function of ageing time is presented on Figure 3 for NR1.5_Conv. The solubility parameter increases during aerobic ageing and as expected remains constant for anaerobic ageing. This suggests the insertion of polar chemicals groups starting from 7 days.

The evolution of the active chain density during aerobic ageing is compared to the one obtained by anaerobic ageing for the conventional cross linked rubber in Figure 2. The curves superimpose between 0 and 7 ageing days. After 7 days, aerobic ageing leads to supplementary crosslinking. Besides, the soluble fraction deduced from (2) stays low (5%) which indicate that aerobic ageing mainly induces crosslinking. From these data, it can be proposed that during aerobic ageing at 77°C, in a conventionally crosslinked NR, we have first the completion of the rubber crosslinking by a shortening of polysulfudic bridges and then crosslinking by oxidative mechanisms.

3.2 *Fatigue crack growth properties*

All crack growth characterizations were performed at 0.01 Hz at ambient temperature. The fatigue crack properties are presented in Figure 4 for non-aged and aged NR1.5_Conv aged in air for 14 days. Ageing in the presence of oxygen strongly increases the crack growth rate, even a low G values. Each curve can be fitted with two regression lines. For both materials, the slope β is equal to 3.5 for G below 3500 J/m². Above this value, a new slope, lower, is found. A slope of 3.5 is characteristic of a rubber which cannot crystallize whereas a slope below 2 should be the result of strain induced crystallization mechanism (Lake 1972, Papadopoulos 2008). To confirm that strain-induced crystallization is the main factor responsible for the better crack growth resistance for G above 3500 J/m², WAXS measurements were performed at the crack tip for G below and above this value.

3.3 *Analysis of the SIC phenomenon at crack tip*

Figure 5 presents the *CI* evolution as a function of *G*. On the one hand, SIC appears for both materials after 4000 J/m² and tend to increase for higher tearing energy. On the other hand, the network evolution due to aerobic ageing reduces the ability of NR to strain crystallize at the crack tip for

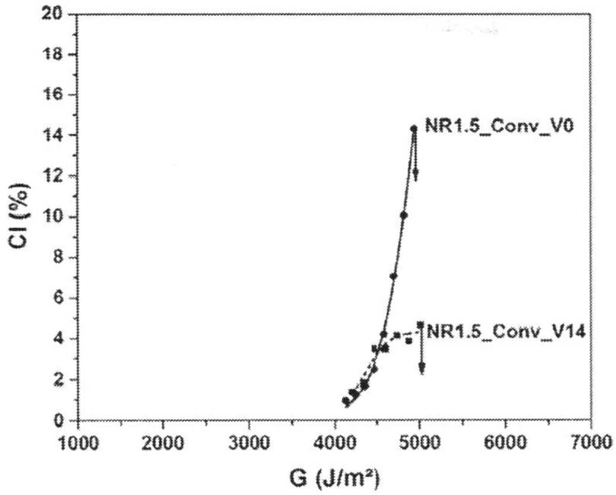

Figure 5. CI evolution at crack tip as a function of tearing energy G for non-aged and aged NR1.5_Conv samples.

G above 4500 J/m^2. Thus, Figure 4 and Figure 5 clearly show the correlation between the SIC at the crack tip and the slowing of the crack propagation. Moreover, in the G domain where undoubtedly SIC occurs at the crack tip, the crack propagation is related to the crystalline volume fraction in the crack tip vicinity.

4 CONCLUSION

In conventionally vulcanized natural rubber, ageing at 77°C in air causes shortening of the polysulfidic sulfur bridges and crosslinking by oxidative mechanisms. Both mechanisms are responsible for increasing the density of elastically active chains. This decreases the ability of the material to crystallize under strain. Thus, the volume fraction at the crack tip generated during fatigue test is decreased, which in turn reduces the fatigue crack growth resistance. Our results also suggest that the fatigue crack growth resistance is strongly correlated to the crystal volume fraction which appeared at the crack tip.

ACKNOWLEDGMENTS

The authors are indebted to the European Synchrotron radiation Facility (ESRF) and the local contacts Nathalie Boudet and Isabelle Morfin for providing the necessary beam-line time and technical assistance in the experiments on the BM2 beamline.

REFERENCES

Bauer D.R., Baldwin J.M., 2007, Rubber Oxidation and Tire Ageing—A review, *Rubber Chemistry and Technology*, 92, 110–117.

Blanks R.F., Prausnitz J.M., 1964, Thermodynamics of Polymer Solubility in Polar and Nonpolar, *Industrial & Engineering Chemistry Fundamentals,* 3, 1–8.

Brüning K., Schneider K., Roth S., 2013, Strain-induced crystallization around a crack tip in natural rubber under dynamic load, *Polymer*, 54, 6200–6205.

Candau N., Laghmach L., Chazeau L., Chenal J.M., 2014, Strain-Induced Crystallization of Natural Rubber and Crosslink Densities Heterogeneities, *Macromolecules*, 47, 5815–5824.

Chapman A.V., Porter M., 1988, Sulphur Vulcanization Chemistry, *Natural Rubber Science and Technology*, 12, 516.

Colin X., Audouin L., Verdu J., 2007, Kinetic modelling of the thermal oxidation of polyisoprene elastomers. Part 1: Unvulcanized unstabilized polyisoprene, *Polymer Degradation and Stability,* 92, 886–897.

Colin X., Audouin L., Verdu J., 2007, Kinetic modelling of the thermal oxidation of polyisoprene elastomers. Part 3: Oxidation induced changes of elastic properties, *Polymer Degradation and Stability,* 92, 906–914.

Flory P.J., Rehner J., 1943, Statistical Mechanics of CrossLinked Polymer Networks I. Rubberlike Elasticity, *The Journal of Chemical Physics,* 11, 512.

Kim S.G., 1994, Effect of Crosslink Structures on the Fatigue Crack Growth Behavior of NR Vulcanizates with Various Ageing Conditions, *Rubber Chemistry and Technology*, 67, 649–661.

Lake G.J., 1972, Mechanical fatigue of rubber, *Rubber Chemistry and Technology,* 45, 309–328.

Mathew N.M., De S.K., 1983, Thermo-oxidative ageing and its effect on the network structure and fracture mode of natural rubber vulcanizates, *Polymer,* 24, 1042–1054.

Morrison N.J., Porter M., 1984, Temperature effects on the stability of intermediates and crosslinks in sulfur vulcanization, *Rubber Chemistry and Technology*, 57, 63–85.

Papadopoulos I.C., Thomas A.G., Busfield J.J.C., 2008 Rate Transitions in the Fatigue Crack Growth of Elastomers I., *Journal of Applied Polymer Science*, 109, 1900–1910.

Rauline, R., 1993, *U.S. Patent No. 5,227,425*. Washington, DC: U.S. Patent and Trademark Office.

Rault J., Marchal J., Judenstein P., 2006, Chain orientation in natural rubber, Part II: 2H-NMR study, *Journal of Applied Polymer Science*, 109, 1900–1910.

Rublon P., Huneau B., Verron E., 2014, Multiaxial deformation and strain-induced crystallization around a fatigue crack in natural rubber, *Engineering Fracture Mechanics*, 123, 59–69.

Trabelsi S., Albouy P.A., Rault J., 2002, Stress-induced crystallization around a crack tip in natural rubber, *Macromolcules*, 35, 10054–10061.

Valentín J.L., Carretero-Gonzàles J., 2008, Uncertainties in the Determination of Cross-Link Density by Equilibrium Swelling Experiments in Natural Rubber, *Macromolecules,* 41, 4717–4729.

Constitutive Models for Rubber X – Lion & Johlitz (Eds)
© 2017 Taylor & Francis Group, London, ISBN 978-1-138-03001-5

Service life determination of rubber fuel hose used in aircraft applications

R.J. Pazur & C.G. Porter
Polymer and Textile Science, Quality Engineering Testing Establishment (QETE), National Defense of Canada, Ottawa, Ontario, Canada

ABSTRACT: The service life of rubber hoses exposed to jet fuel has been determined using accelerated immersion aging. The inner tube rubber has been identified as chlorinated poly(ethylene) or CPE. Immersion aging in jet fuel (JP8 +100) took place from 50° to 125°C in increments of 15°C, for various exposure times. Aged samples were tested for mechanical properties, volume swell resistance and network chain density by solvent swell in order to assess the level of rubber deterioration. Time-Temperature Superposition (T-TS) was applied to the testing results using 95°C as a reference temperature. Hose samples initially softened then progressively stiffened leading to a loss in ultimate properties (both tensile strength and elongation at break). Volume swell slowly increased during aging but remained under 80%. The network chain density slowly increased before declining under severer conditions. Diffusion-Limited Oxidation (DLO) was operating and influenced test results collected above 95°C. A linear Arrhenius behavior however, was observed using elongation at break as the material response and an activation energy of 50 ± 10 kJ/mol for the overall degradation process was calculated. Possible service lives in the 6 to 10 year range have been put forth including the predicted ultimate properties and volume swell changes. An 8 year service life corresponds to a loss of 80% and 40% respectively of the tensile strength and elongational properties of the rubber. Volume swell is expected to reach 65%. This is a reasonable upper limit for the rubber property losses and is in line with the expectations of a CPE rubber used in a fuel environment.

1 INTRODUCTION

Lightweight flexible metal braided rubber fuel hoses are commonly used in military aircraft. They are designed for use between 14 to 69 bar within the temperature limits of −54° to 121°C and must meet stringent specifications for use in aircraft fuel and/or oil systems (MIL-DTL 83797C 2010). Their service life after installation will have a large dependence on the operating environment including the compatibility of the inner tube rubber with the fuel as a function of time and temperature.

The rubber in the fuel hoses was identified as chlorinated poly(ethylene) elastomer (CPE). CPE possesses thermal stability up to 150°C due to a saturated main chain structure and its fuel resistance depends directly on the concentration of chlorine which can range from 25 to 45% (see Fig. 1). Cure systems such as amine-thiadiazole/magnesium oxide or organic peroxides are commonly used for vulcanization (Colbert 2001).

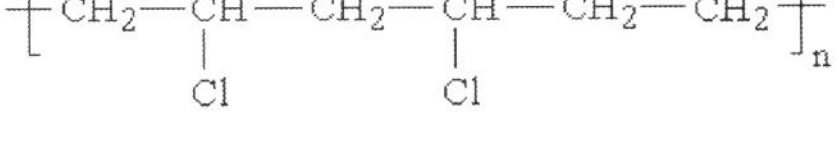

Figure 1. Simplified chemical structure of CPE.

Thermal analysis has identified two major transitions during CPE degradation: dehydrochlorination reactions with evolution of HCl followed by chain destruction at higher temperatures (Pan 1990). A minor transitional region in which crosslinking takes place was identified between the two major processes (Stoeva 2000). It is expected that the CPE degradation under thermo-oxidation possesses similarities to those of chlorosulfonated polyethylene (CSM). CSM cable jacketing materials stiffen as a function of aging time with both crosslinking and chain breaking reactions occurring in competition (Gillen 2006).

The purpose of this study is to estimate the service life of CPE rubber hoses using accelerated fuel immersion aging from 50 to 125°C. The jet fuel is JP-8 +100. Stress-strain, volume swell and network chain swelling measurements are used to follow the sample degradation.

2 EXPERIMENTAL

2.1 *Sample preparation and characteristics*

The metal braided fuel hose chosen for evaluation is rated for an operating pressure of 55.2 bar.

A photograph showing its cut cross-section is provided in Fig. 2. The hose contains two layers of stainless steel wire braid reinforcement.

The outer braided metal cover was carefully removed as well as the secondary layer attached to the rubber surface in order to isolate the inner rubber tube. Rectangular 1.5 cm wide strips were then cut in the longitudinal direction of the tube. Tensile dumbbells were then die cutt using a MET 4482 sized die. The rough outer surface of the dumbbell samples was smoothly ground in a buffing apparatus. Final sample thickness' were about 2.5 mm. Two 25 mm × 50 mm rectangular strips that were cut from the un-buffed hose sample were used for the volume swell determination according to ASTM D471.

From internal analytical analysis by thermogravimetric analysis, the CPE rubber is reinforced with both carbon black and inorganic fillers and contains 16% by weight of a trimellitate plasticizer. The chlorine content is within the range of 36–40 wt%.

JP-8 +100 test fuel is composed of 21.3% aromatics, 0.7% olefins and 78.0% saturates, according to ASTM D1319 testing. Jet fuels are normally formulated with antioxidants, metal deactivators, static dissipaters and corrosion and icing inhibitors. The +100 additive package contains proprietary chemicals that increase the thermal stability and heat capacity of the jet fuel.

2.2 *Test equipment*

Tensile testing was carried out on a Tensitech tensile tester (TechPro) fitted with a 500 N load cell. Stress-strain data were collected as per the testing methodology outlined in ASTM D412 and ASTM D471. Five dumbbell samples were tested for the control and each aging condition. Median values were used for calculations.

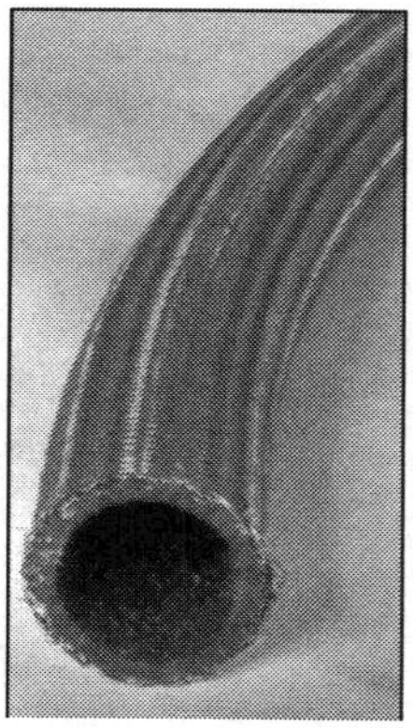

Figure 2. Picture showing the construction of the fuel hose including metal braiding and inner rubber tube. The ID of the inner tube is 2.54 cm.

2.3 *Testing procedure*

Tensile elongation dumbbells and volume swell specimens were suspended in a 300 mL tube containing the jet fuel and subjected to the testing regime shown in Table 1 using a heating mantle system. Water cooling condensers were necessary for the immersion aging above 80°C.

2.4 *Network swelling ratio (Q)*

The network chain density was followed by calculating the swelling ratio Q using chloroform as the swelling agent. Test specimens were allowed to expand under equilibrium conditions (72 h/23°C). The value of Q is calculated by using equation 1 where,

$$Q = \frac{\frac{m_d}{\rho_d} + \frac{m_s}{\rho_s}}{\frac{m_d}{\rho_d}} \tag{1}$$

m_d = mass of dried sample, ρ_d = density of dried sample, m_s = mass of solvent uptake and ρ_s = density of chloroform (1.489 g/cm^3). Results are presented in terms of the percent relative change (ΔQ) of the network swelling ratio which eliminates the need to correct for the filler volume fraction.

2.5 *Data interpretation*

Through Time-Temperature Superposition (T-TS), the aging data was shifted to a reference temperature of 95°C by estimating shift factors a_T. Assuming steady-state Arrhenius behavior of the degradation process, equation 2 was used to calculate the activation energy (E_a).

Table 1. Heat aging matrix used for accelerated fuel immersion aging of the hose samples. One additional set of samples was aged at ambient for 1680 h.

	Time, h					
T, °C	168	336	504	840	1680	2520
50				X	X	X
65				X	X	X
80			X		X	X
95		X		X	X	
110	X		X	X		
125	X	X	X			

$$a_T = \exp\left[\frac{E_a}{R}\left(\frac{1}{T_{ref}} - \frac{1}{T}\right)\right] \qquad (2)$$

R is the ideal gas constant (8.314 $JK^{-1}\ mol^{-1}$).

3 RESULTS AND DISCUSSION

3.1 *Initial properties*

The unaged samples possess tensile strength and elongation at break values of 17.8 MPa and 125% respectively. Apart from the reinforcement provided by the reinforcing fillers, the high tensile strength of CPE is aided by the favorable intrachain interactions through the polar C-Cl bonds. Positive interactions between the high loading of trimellitate plasticizer and polar C-Cl bond likely contribute to the high tensile value. Very low levels of crystallinity were detected by differential scanning calorimetry in the 40–45°C range (internal data) but were likely not sufficient to enhance the tensile strength. The elongation at break is low in value but the metal braiding prevents inner tube extension and/or expansion under use.

3.2 *Tensile strength*

The change in tensile strength upon jet fuel immersion is presented in Fig. 3. The loss of tensile strength is significant under mild aging conditions. Ten weeks of immersion aging at room temperature is sufficient to bring about a 45% loss in tensile strength (9.5 MPa nominal value). This can be largely attributed to the replacement of trimellitate plasticizer by the aromatic fraction of the jet fuel which exerts a strong softening effect on the rubber network causing the modulus to decrease. The severest aging and loss in tensile strength is taking place at 95°C and above. In fact, close to 100% tensile loss is observed after ten weeks at 95°C. Such large losses in tensile strength were not attainable at higher temperatures due to diffusion-limited oxidation (DLO) effects which operate in CPE as in other elastomers (Celina 2013). The low strain modulus data (not reported) attest to this effect due to their increase during higher temperature aging. The T-TS graph shown in Fig. 3b displays relatively good overlap of the data points onto the 95°C dashed master curve as a function of shifted time. Three levels of degradation (60, 70 and 80% tensile loss) with their corresponding shifted times have been selected and identified on this figure for service life calculations.

3.3 *Elongation at break*

The change in elongation at break properties is provided in Fig. 4a with the corresponding shifted data in Fig. 4b. Elongation loss varies from about 10% up to 60% depending on the test condition. The low temperature aging displays a plateau centred at −35%. This region is likely experiencing only small levels of degradation, but is dominated by the effect of plasticizer extraction and replacement by the jet fuel. The T-TS of Fig. 4b appears more suitable for the higher aging data where more pronounced elongation loss was observed. As with the tensile strength, three levels of degradation at 95°C based on elongation losses of 40, 50 and 60% have been marked on the graph.

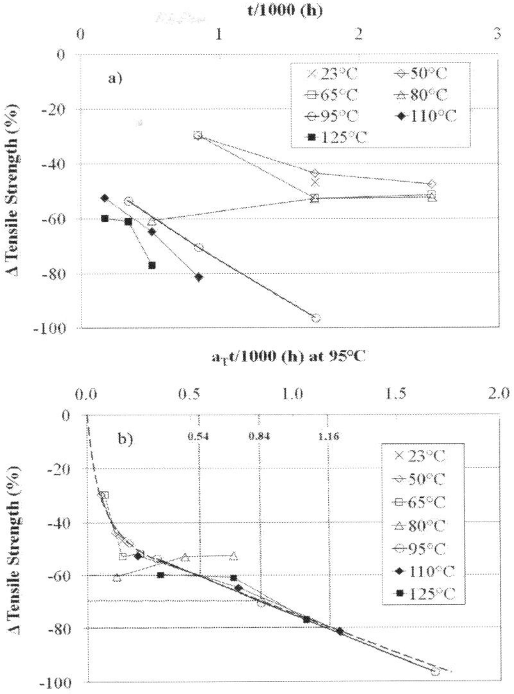

Figure 3. Plots displaying the a) change in tensile strength as a function of time and temperature and b) all data points shifted onto a dashed master curve drawn for T_{ref} = 95°C (a_T = 0.1 for 23°C data point).

3.4 *Volume swell*

The volume swell data are presented in Fig. 5. A positive swell is taking place from 10 to up to 80% during immersion aging. The aromatic fraction of the jet fuel is rapidly being absorbed by the rubber and the plasticizer is quickly removed and solubilized into the test jet fuel. For room temperature aging, a swelling level of about 25% was measured. Applying T-TS as seen in Fig. 5b, shows a general progression of increasing volume swell as a function of shifted time. The volume swell increase

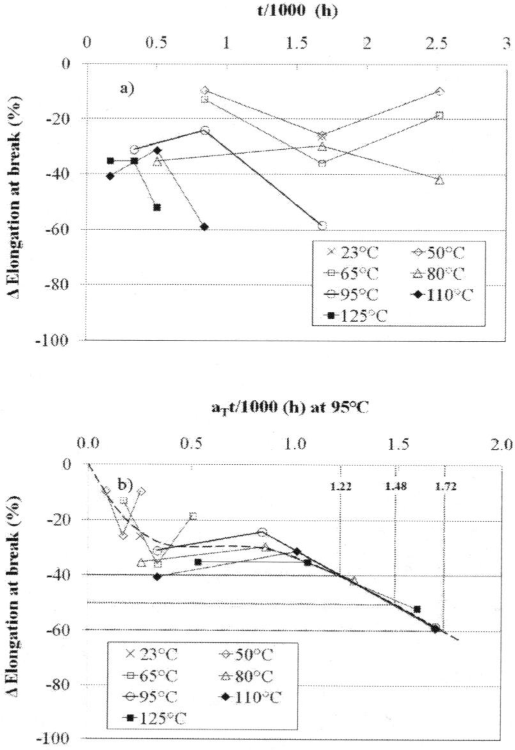

Figure 4. Graphs showing a) the change in elongation at break as a function of time and temperature and b) the shifted data using T-TS for T_{ref} = 95°C (a_T = 0.15 for 23°C data point).

is more prominent for temperatures higher than 80°C. After replacement of the plasticizer by the jet fuel, an increase in volume swell generally relates to a loss of the network chain density either through loss of chemical crosslinks or scission of the load bearing chains. The times to attain volume swells of 50, 60 and 70% at 95°C have been identified on Fig. 5b.

3.5 *Network chain density through swelling (Q)*

The effect of thermal fuel immersion aging on the swelling ratio is illustrated in Fig. 6. Results are presented in terms of the percent change of Q with respect to the unaged results. The network swelling is shown to initially fall 5–10% and then rise and become positive under severer aging conditions. DLO effects limit the network swelling at 100 and 115°C. The results indicate a small increase in network density under mild aging conditions followed by a decrease in its value for longer shifted times. Dehydrochlorination generates in-chain unsaturation which can undergo oxidative crosslink addi-

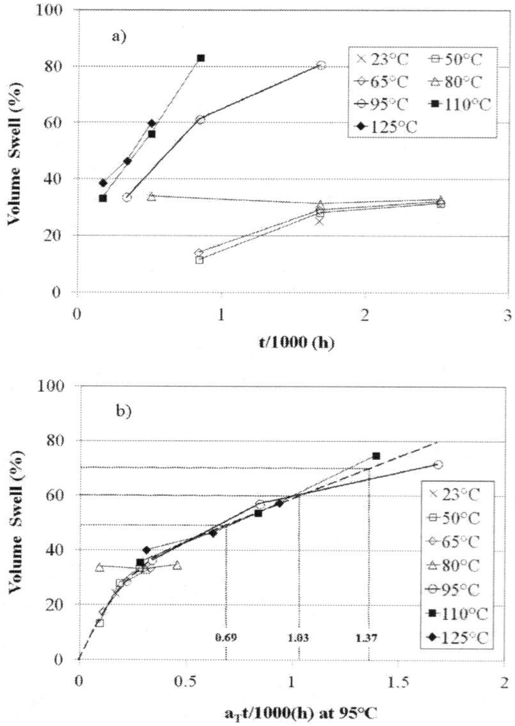

Figure 5. Plots illustrating a) the volume swell during immersion aging in JP-8 +100 as a function of time and temperature and b) its corresponding T-TS using T_{ref} = 95°C (a_T = 0.1 for 23°C data point).

tion reactions at the allylic position. The onset of destructive degradative effects is particularly evident at 95°C. A decline in network chain density is likely due to the advent of chain scission reactions. These testing results concord with the volume swell increase observed with the jet fuel.

3.6 *Arrhenius plot*

The Arrhenius behavior of the shifted aging results was plotted according to equation 2 on Fig. 7. A linear behavior of the shift factors over the tested temperature range was only noted with the elongation at break data. Significant deviation was observed for the tensile strength, volume swell and network chain density above and below 95°C. The deviant behavior above 95°C (shown by the dashed line) is due to DLO effects which are known to cause the appearance of lower than expected activation energies. The reasoning for the higher drop in shift factors at 80 and 65°C for tensile strength, volume swell and network chain swelling results is unknown. The exchange process between the plasticizer and jet fuel components is likely an issue.

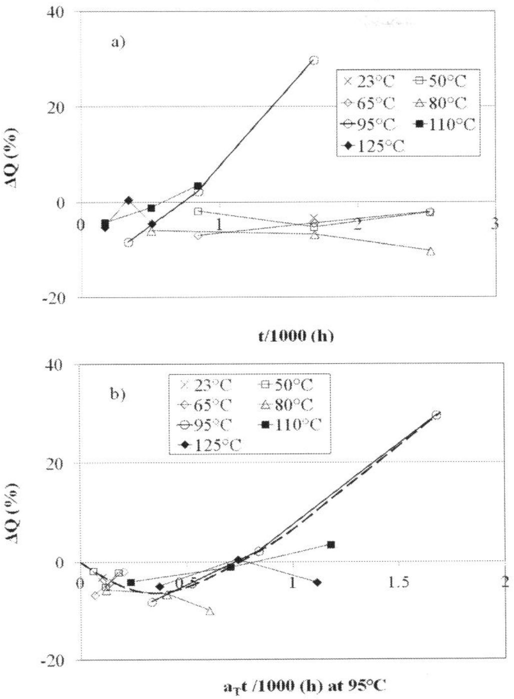

Figure 6. Graphs showing the a) change in network swelling as a function of time and temperature and b) its corresponding T-TS onto the dashed master curve for T_{ref} = 95°C (a_T = 0.06 for 23°C data point).

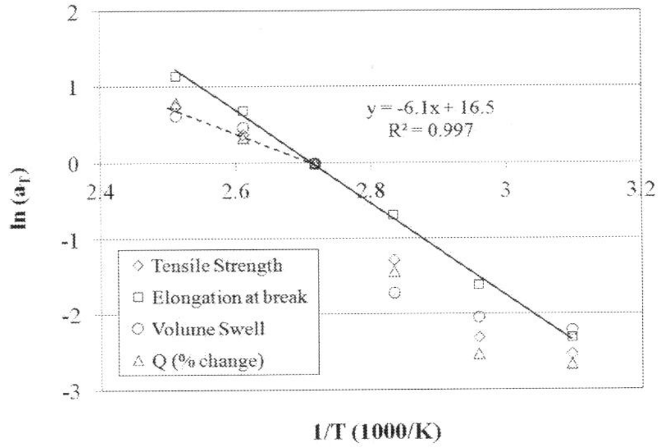

Figure 7. Arrhenius plots according to equation 2 using the shift factors (a_T) derived from the T-TS of the tensile strength, elongation at break, volume swell and network chain density data for the temperature range of 50 to 125°C.

Loss of residual crystallinity is possibly another factor. The mild increase in network chain density may restrict volume swell in this temperature range. Such network building reactions cause a mild decline in elongational characteristics. Tensile elongation has been successfully employed to follow the degradation behavior in CSM cable jacketing (Gillen 2006). Upon using elongation at break, the overall rubber aging process which includes both crosslink density increases and decreases, follows an activation energy which is approximately 50 ± 10 kJ/mol from 50 to 125°C. This value will be assumed to be representative of all test responses. The activation energy in liquid media testing is similar to that of other elastomers (Pazur 2015).

It is now possible to estimate the service life of the CPE fuel hoses as depicted in Fig. 8, at an operational temperature of 23°C using eq. 2. The amount of property loss is plotted on the y axis *versus* the estimated service life. Ultimate properties indicate a percentage loss whereas volume swell represents the actual nominal increase of the hose swelling. The vertically drawn lines indicate 6, 8 and 10 year service lives.

As an example, after 6 years of fuel hose use in contact with JP-8 +100, there is a predicted loss of about 70% tensile strength, 30% elongation at break with a corresponding nominal volume swell of roughly 55%. After 8 years, the tensile and elongation losses are now at 80 and 40% respectively with a volume swell of 65%. The volume swell is over 70% and the tensile and elongation at break losses are at 90 and 50% respectively at 10 years of service. All of these property changes are due to degradation of the rubber upon exposure to the jet fuel including its additives, dissolved oxygen and temperature (Pazur 2015).

No quantitative indication of acceptable rubber degradation was provided. It was stated that the rubber should be able to resist 48 hours of exposure to jet fuel at 121°C without any signs of

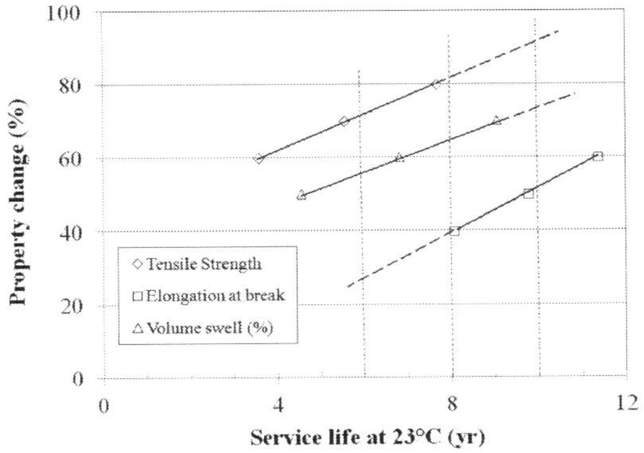

Figure 8. Actual property change depending on measurement technique as a function of service life for an operational temperature of 23°C. The long dashed lines indicate an extrapolation of the property behavior.

disintegration, solubility of component parts, porosity, blistering or collapse (MIL-DTL 83797C, 2010). Upon inspection of the rubber at 125°C, this extreme level of degradation was not seen. The initial 45% loss in tensile strength properties is due to the absorption of the aromatic component of the jet fuel and the replacement of the trimellitate plasticizer. The high loading of plasticizer (16 wt%) exacerbates this drop. Assuming that any degradation taking place at room temperature is negligeable and that the property loss is due only to the volume swell increase, the corrected tensile strength loss due to degradation effects becomes 25, 35 and 45% for 6, 8 and 10 years of service life. In the end, the final tensile strength values of 5.4, 3.6 and 1.8 MPa correspond to elongation at break values of 75, 63 and 50%. The hose metal braiding construction aids to restrict inner tube extension or collapse during use thus preventing tensile failure. Needless to say, extensive stiffening of the inner tube surface due to degradation can lead to rubber pitting and cracking and eventually to hose leaking and failure when under operating pressures. The onset of this extent of degradation may also bring about the release of rubber particles from the inner tube surface which could contaminate the jet fuel.

The main mechanism that dominates at long aging times is the chain length loss due to scission reactions which corresponds to the decline in tensile properties. According to the service life data of Fig. 8, eight years appears to be a maximum upper limit for hose use during service before replacement.

4 CONCLUSIONS

Upon accelerated immersion aging in JP-8 +100 (50° to 125°C), the hose rubber initially softened leading to a decrease in modulus and a progressive loss in ultimate properties (both tensile strength and elongation at break). The rubber stiffened with longer aging times and the volume swell increased and approached 80%. The network chain density slightly increased for mild aging conditions and then dropped dramatically due to degradation caused by chain scission reactions.

Diffusion-limited oxidation was taking place and influenced test results collected above 95°C. Elongation at break testing results provided a linear Arrhenius fit through the tested temperature range. An activation energy of 50 ± 10 kJ/mol was used for service life calculations.

An 8 year hose service life corresponds to a loss of 80% of the tensile strength and 40% of the elongational properties of the rubber inner tube. After plasticizer loss due to fuel replacement, the actual tensile loss caused by degradation is in reality 35%. Volume swell is expected to be 65%. This is a reasonable upper limit for the property losses seen in the rubber and is in line with the expectations of a chlorinated polyethylene rubber used in a fuel environment.

REFERENCES

Celina, M.C. 2013. Review of polymer oxidation and its relationship with materials performance and lifetime prediction. *Polym Degrad Stab.* *98*: 2419–2429.

Colbert, G.P. 2001. Solvent Resistant Elastomers – B. Neoprene, Hypalon and Chlorinated Polyethylene. In K.C. Baranwal, H.L. Stephens (eds), *Basic Elastomer Technology 1st Edition:* 382–417. Akron, OH: The Rubber Division American Chemical Society.

Gillen, K.T., Assink, R., Bernstein, R., Celina, M.C. 2006. Condition monitoring methods applied to degradation of chlorosulfonated polyethylene cable jacketing materials. *Polym Degrad Stab*. 91: 1273–1288.

MIL-DTL 83797C. 2010 Hose, Rubber, Lightweight, Medium Pressure, General Specification for.

Pan, W-P, Whiteley, MJ, Serageldin, MA, Cai G. 1990. *Thermochim Acta*. Decompostion of chlorinated polyethylene in air without ignition. 173:85–93.

Pazur, R.J., Cormier J.G. 2015. A Compatibility Study of Low-Temperature-Capable Fluoroelastomers in Jet Fuels. *RUBBER CHEM TECH.* *88*(4): 660–675.

Pazur, R.J., Kennedy, T.A.C 2015. Effect of plasticizer extraction by jet fuel on a nitrile hose compound. *RUBBER CHEM TECH.* *88*(2): 324–342.

Stoeva, St., Gjurova, K, Zagorcheva, M. 2000. Thermal analysis study on the degradation of the solid-state chlorinated poly(ethylene). *Polym Degrad Stab.* *67*: 117–128.

Constitutive Models for Rubber X – Lion & Johlitz (Eds)
© 2017 Taylor & Francis Group, London, ISBN 978-1-138-03001-5

Simulation of oxidative aging processes in elastomer components using a dynamic network model

C. Schlomka & J. Ihlemann
Professorship of Solid Mechanics, Technische Universität Chemnitz, Chemnitz, Germany

C. Naumann
Freudenberg Technology Innovation SE & Co. KG, Weinheim, Germany

ABSTRACT: The majority of rubbery components in industrial applications is subjected to environmental influences. Especially exposition to oxygen leads to a change in material behaviour due to reaction of the network structure with oxygen. Its availibility is limited by diffusion processes leading to a deceleration of the oxidative aging processes in the inner regions, also known as Diffusion-Limited Oxidation (DLO). Using a dynamic network model presented in previous works, this coupled problem can be solved efficiently using a staggered solution algorithm. While classical two-network models are limited to two-step processes, this model is capable of simulating oxidative ageing under varying mechanical loads by describing the evolution of a single network which continuously softens and stiffens due to oxidation processes. The solution algorithm is implemented in ABAQUS and numerical results are shown. Comparison to experimental data shows the valid prediction of the change in mechanical behaviour for inhomogeneously aged parts.

1 INTRODUCTION

Demands on the durability of elastomeric components in industrial applications continously increase. Therefore the understanding and prediction of aging phenomena gain in importance. In elastomeric components exposed to oxygen, a change in material properties can be observed which is linked to chemical oxidation processes. Two main causes of the change in material behaviour can be observed. Firstly, chemical reactions lead to scission of polymer chains and crosslinks and thus to softening of the material. Secondly, stiffening occurs due to the creation of new crosslinks. The speed of these reactions depends on the availability of oxygen, which is determined by a diffusion process. This leads to a deceleration of the oxidative aging processes in the inner regions of components, also known as Diffusion-Limited Oxidation (DLO). The here presented research deals with the mathematical description and numerical simulation of these aging phenomena in elastomeric components.

2 THE DYNAMIC NETWORK MODEL

Generalizations of the two-network model of Tobolsky (Andrews, Tobolsky, & Hanson 1946) are frequently used to simulate the effects of aging on the behaviour of elastomers. A primarily existing network degenerates over time due to chain and crosslink scission. Additionally, an independent secondary network forms due to further crosslinking. However, these models are limited to processes with a single load level as no further network is avaiable to adapt to another load level.

In contrast, elastomer components are often subjected to varying mechanical loads in industrial applications. Realistic simulations of the oxidative aging of such components can be realised using the dynamic network model (Naumann & Ihlemann 2015a) instead. For better understanding of the results shown here, essential considerations of the model will be outlined briefly.

In the dynamic network model, no distinction between separate networks is drawn. Instead the evolution of a single network is described, which continuously softens and stiffens while always considering the current deformation of the material. This is realised based on the idea of Tobolsky that new crosslinks are always inserted stress-free. As a consequence, oxidative ageing of a material point at non-zero deformation will lead to a chemical adaptation of the polymer network to this deformation. If all external loads are removed, a new stress-free state will become visible. For continuum mechanical modelling, a new configuration $\hat{\mathcal{K}}$ is introduced, which will describe this stress-free state after ageing. At the beginning of the aging process, $\hat{\mathcal{K}}$ is equal to the initially stress-free isochoric reference configuration $\bar{\bar{\mathcal{K}}}$. For the ease of reading, only the isochoric part of the deformation will be considered in the following.

The isochoric part of the deformation gradient $\bar{\underline{\underline{F}}}$ is decomposed multiplicatively into two parts. $\underline{\underline{F}}_1$ can be considered as the deformation from the isochoric reference configuration $\bar{\mathcal{K}}$ to the stress-free configuration $\hat{\mathcal{K}}$ while $\underline{\underline{F}}_2$ contains the stress-producing part of the deformation:

$$\bar{\underline{\underline{F}}} = \underline{\underline{F}}_2 \cdot \underline{\underline{F}}_1 \tag{1}$$

Hence, a free energy of Neo-Hookean type can be formulated with respect to the right CAUCHY-GREEN tensor $\underline{\underline{C}}_2$ of the second deformation:

$$\tilde{\rho}\psi = C_{10}\nu(t)\mu(t)\left(\underline{\underline{C}}_2 \cdot\cdot \underline{\underline{I}} - 3\right) \tag{2}$$

$$= C_{10}\nu(t)\mu(t)\left(\underline{\underline{C}}_1^{-1} \cdot\cdot \bar{\underline{\underline{C}}} - 3\right). \tag{3}$$

The factors $\nu(t)$ and $\mu(t)$ are introduced to describe the softening and stiffening of the material. Both start at initial values of 1 and evolve depending on the chemical reactions which is denoted by their dependence on time t. By evaluating the CLAUSIUS-DUHEM inequality, the following relation for the isochoric part of the second PIOLA-KIRCHHOFF stress tensor $\bar{\tilde{\underline{\underline{T}}}}$ is obtained:

$$\bar{\tilde{\underline{\underline{T}}}} = 2C_{10}\nu\mu \underline{\underline{\underline{\underline{P}}}}_C \cdot\cdot \underline{\underline{C}}_1^{-1}. \tag{4}$$

There, the projection tensor $\underline{\underline{\underline{\underline{P}}}}_C$ is used as $\bar{\tilde{\underline{\underline{T}}}}$ is the physical deviator:

$$\underline{\underline{\underline{\underline{P}}}}_C \cdot\cdot \underline{\underline{X}} = \left(\underline{\underline{X}} \cdot \bar{\underline{\underline{C}}}\right)' \cdot \bar{\underline{\underline{C}}}^{-1}. \tag{5}$$

Finally, the idea of Tobolsky, i.e. the stress-free insertion of new links, is implemented. Thus, in absence of scission processes and for constant deformation, the stress should remain unchanged even though crosslinking processes will lead to an increase of μ:

$$\overset{\Delta}{\bar{\tilde{\underline{\underline{T}}}}}\left(\overset{\Delta}{\bar{\underline{\underline{C}}}} = \underline{\underline{0}}, \dot{\nu} = 0\right) = \underline{\underline{0}}. \tag{6}$$

Under the assumption that the insertion of links is not dissipative, the following evolution equation for the stress-free configuration can be derived:

$$\overset{\Delta}{\underline{\underline{C}}}{}_1^{-1} = \frac{\dot{\mu}}{\mu}\left(\bar{\underline{\underline{C}}}^{-1} - \underline{\underline{C}}^{-1}\right). \tag{7}$$

Additionally, the Mullins-effect is modeled using the softening approach proposed in (Naumann & Ihlemann 2015b). It is an extension of the model of (Ogden & Roxburgh 1999) and enables arbitrary thermomechanically consistent material models to allow for the Mullins-effect.

3 SOLUTION OF THE COUPLED PROBLEM

To simulate the effect of aging on large sized components under time variable loads realistically, the DLO effect plays an important role and has to be taken into account by simulating the diffusion and reaction of oxygen (see e.g. (Steinke, Spreckels, Flamm, & Celina 2011, Charrier, Marco, Le Saux, & Ranaweera 2011, Naumann & Ihlemann 2013)). Combining the mechanical loading and the reaction/diffusion processes leads to a coupled problem. However, the time scales for the two sub-problems differ by several orders of magnitude. Thus, a fully coupled solution would require an unjustifiably high amount of computational resources. Instead, this problem can be solved efficiently for periodic loading using a staggered solution algorithm.

According to Haupt (1999), diffusion and reaction processes within deformed bodies can be described with the following equation:

$$\dot{c} = \underline{\tilde{\nabla}} \cdot D J_3 \underline{\underline{C}}^{-1} \cdot \underline{\tilde{\nabla}} c - r(c) \tag{8}$$

with the diffusion coefficient D and the third invariant of the deformation gradient J_3. For deformations varying in time, the solution depends on the variable tensor $J_3 \underline{\underline{C}}^{-1}$. However, fr periodic deformations, an approach from Naumann & Ihlemann (2013) can be used to calculate a coupling tensor $\underline{\underline{G}}$ for the periodic time T_L, assuming that a change of the concentration occurs too slowly to be significant during this timespan:

$$\underline{\underline{G}} := \frac{1}{T_L}\int_0^{T_L} D J_3 \underline{\underline{C}}^{-1} dt. \tag{9}$$

The resulting diffusion equation then reads:

$$\dot{c} := \underline{\tilde{\nabla}} \cdot \underline{\underline{G}} \cdot \underline{\tilde{\nabla}} c - r(c). \tag{10}$$

The solution of the evolution equation (7) for the stress-free configuration can be simplified by exploiting the fact of periodic loading. By introducing similar coupling tensors, an approximate analytical solution for arbitrary ageing durations can be found (Naumann 2017). By this, mechanical loading and reaction/diffusion can be simulated separately while passing coupling variables between them.

4 SIMULATION RESULTS

The dynamic network model and the staggered solution algorithm have been implemented into the finite element software ABAQUS. To demonstrate the applicability of the model, several simulations were performed and compared to experimental results. For that purpose, the parameters of the model were identified for an industrial natural rubber compound. All necessary experiments were conducted by Freudenberg Technology Innovation and Vibracoustic.

4.1 *Proof of concept*

The first example aims at proving the functionality of the model. It is a simulation of an AE42-buffer, being aged under dynamic loading. The geometry and the used boundary conditions are shown in Figure 1.

Top and bottom surface of the buffer are vulcanized to steel parts for mounting. Within the model they are omitted and replaced by the depicted displacement boundary conditions. Deflection of the top surface has been chosen to be force-driven, as a change in stiffness due to ageing will lead to a change in displacement response. This will in turn influence the geometry of the buffer, leading to a stronger coupling to the reaction and diffusion processes. The free surfaces are in direct contact with air, which is why a constant concentration is prescribed according to Henry's law.

The applied force load consists of a static preload and a harmonic oscillation, which is continued during the whole ageing process. Figure 2 shows the beginning of the signal.

As a first result, the stationary solution for the reaction/diffusion problem is obtained as shown in Figure 3. It is clearly visible that diffusion occurs

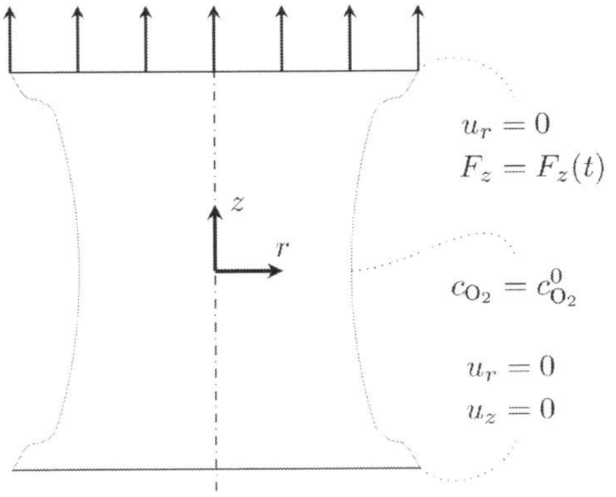

Figure 1. Geometry and boundary conditions of the simulated axisymmetric AE42-buffer: force F and displacement u on mounted surfaces and oxygen concentration c_{O_2} on free surfaces (HENRYS law).

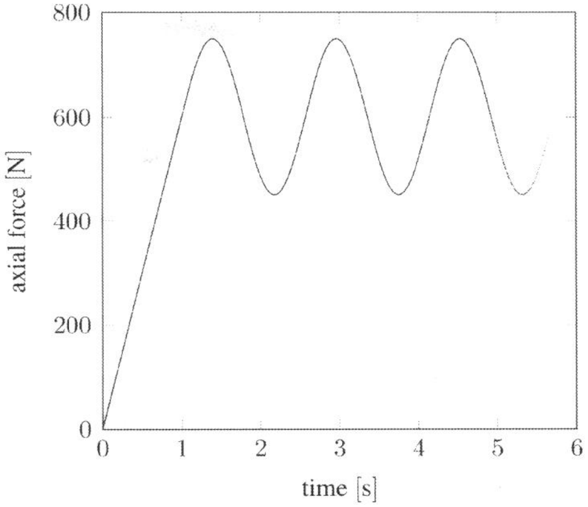

Figure 2. Load signal for the axial force applied on the AE42-buffer (beginning only).

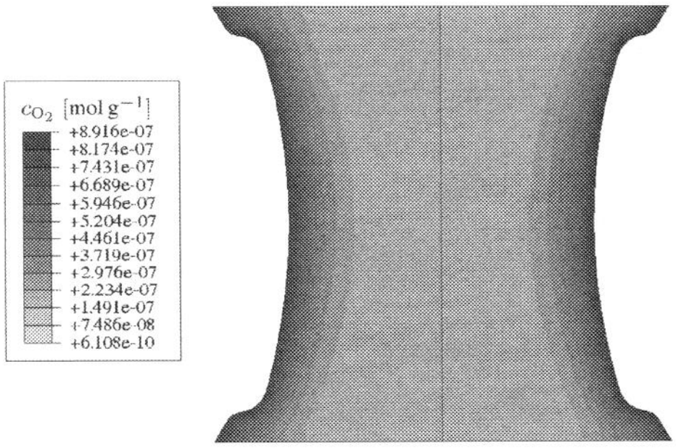

Figure 3. Stationary oxygen concentration in AE42-buffer.

too slowly to maintain a considerable oxygen concentration at the core of the buffer. Thus, the most significant changes in material properties are visible near the free surfaces and a very pronounced DLO-effect is observed.

Upon removing the exterior load after ageing, this inhomogeneity gets clearly visible. Because ageing occurred in a deformed state, the stress-free configuration of each material point evolved individually. Thus, the unloaded state of the buffer is clearly not stress-free, as shown in Figure 4. The permanent deformation is a result of balancing the different stress-free configurations in the part. While the regions close to the free surfaces push towards more deformation, the less aged core prevents this to some extent. It is important to notice that this results in considerable residual stresses. Especially near the surface, compression stresses can be observed.

Finally, the displacement response to the loading of the aged part can be compared to the original one, as shown in Figure 5. The permanent deformation is clearly visible in the offset of the curve.

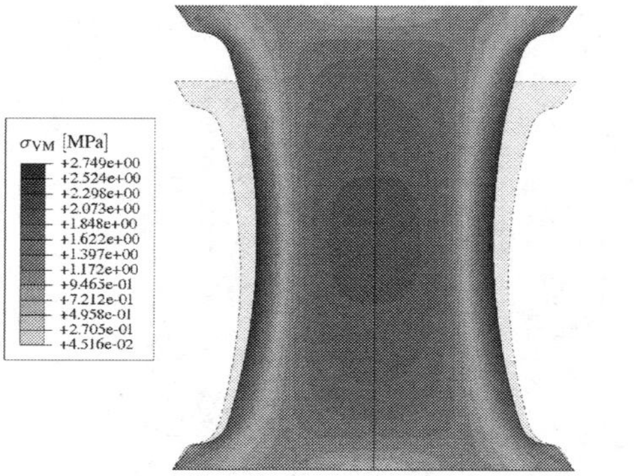

Figure 4. Permanent deformation of the AE42-buffer after ageing, compared to the original geometry (dashed) with residual von Mises stresses.

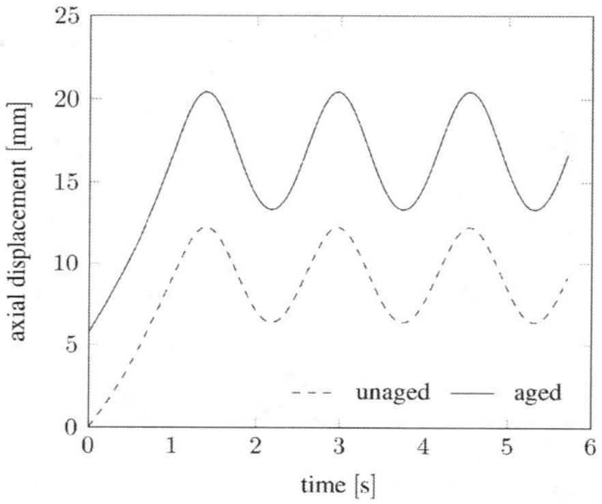

Figure 5. Axial displacement response of the AE42-buffer in unaged and aged state (beginning only).

Although one would intuitively expect the amplitude of the oscillation to be smaller because of stiffening, this is not the case here. The explanation for this result can be found in the stress distribution after applying the preload (see Figure 6). For the unaged buffer, the maximum stresses can be found near the skin of the buffer. The aforementioned residual compression stresses in this area of the aged part (see Figure 4) now almost compensate the stresses induced by the external loading. As a result, the load now concentrates on a smaller area in the lesser aged core of the buffer, leading to larger displacements.

For cross-checking purposes, an external compression force was simulated as well. There, the maximum stresses are located at the core for the unaged buffer as well. Thus, the expected decrease in displacement amplitude is observed in these simulations.

From these results it is derived that the consideration of the DLO-effect can be essential for realistic simulations of the oxidative ageing of elastomeric parts.

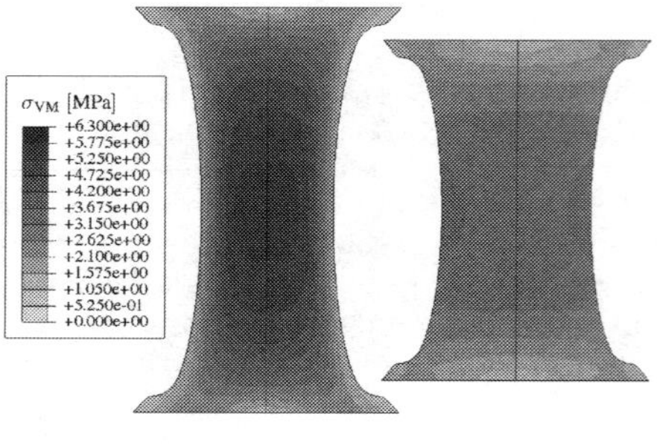

Figure 6. Von Mises stresses at preload of $F_z = 600$ in aged (left) and unaged (right) AE42-buffer.

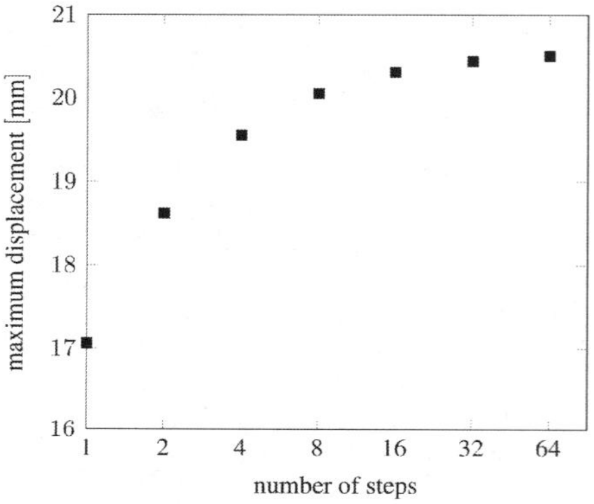

Figure 7. Evidence of convergence: maximum value of axial displacement response for the fully aged AE42-buffer for different numbers of staggered solution steps.

Based on the results outlined above and the strong coupling because of force-driven loading, the shown example can be regarded as challenging for the staggered solution algorithm and is further used to examine its convergence behaviour. For that purpose, the simulation was repeated with different numbers of staggered steps. The mechanical behaviour of the fully aged buffers was then compared using the maximum value of the axial displacement response as a criterion. The results can be seen in Figure 7.

For an increasing number of staggered solution steps, the maximum displacement increases, while the difference between the values decreases. This suggests that using more steps refines the approximation of the varying deformation, so the diffusion problem can be solved more accurately. As the relative difference of the displacement solutions between 32 and 64 steps amounts to only 0.3%, the algorithm can be assumed to be convergent. Adequate accuracy is reached for 8 steps already, with the difference to the final value ranging at 2%.

4.2 *Component stiffness*

A second example is presented for experimental validation of the numerical results. To isolate the prediction of the component stiffness from the influence of the permanent deformation, the AE42-buffer from above (see Figure 1) is aged without external loading ($F_z = 0$). Afterwards, displacement-controlled loading paths at 4 different load level tuples with 5 cycles each are applied once on the unaged buffer and on the aged one. The resulting force-displacement curves are shown in Figures 8 and 9 for experiment and simulation, respectively.

It has been discovered that the investigated rubber compound shows additional stiffening for a very short period of time directly after moulding. This is observed under nitrogen atmosphere in identical quantities, identifying the relevant processes as unrelated to oxidative ageing (Broudin et al. 2017). As such processes are out of the scope of the presented model, the samples regarded as "unaged" were stored before testing until the mentioned initial processes were negligible.

The mechanical parameters of the model (Neo-Hooke and Mullins effect) have been fitted to the unaged AE42-buffer, as a Neo-Hookean model is otherwise unlikely to match the buffers response nicely. In contrast, all of the remaining parameters (e.g. for softening and stiffening functions) were identified separately using homogeneous experiments.

Some mechanical features observed in the experiments, like permanent set, anisotropic softening and the slight upturn at higher displacements, cannot be resembled by the mechanical model because of its simplicity. However, capturing these effects is not the objective of the work presented here. Instead, the changes in mechanical behaviour of the buffer due to ageing are predicted realistically by matching the stiffening and the broadening of the Mullins hysteresis. Interestingly, the latter does not require a separate tuning of the parameters related to the Mullins effect. The used softening approach nicely integrates into the structure of the model and is covered by the oxidative softening and stiffening functions as well.

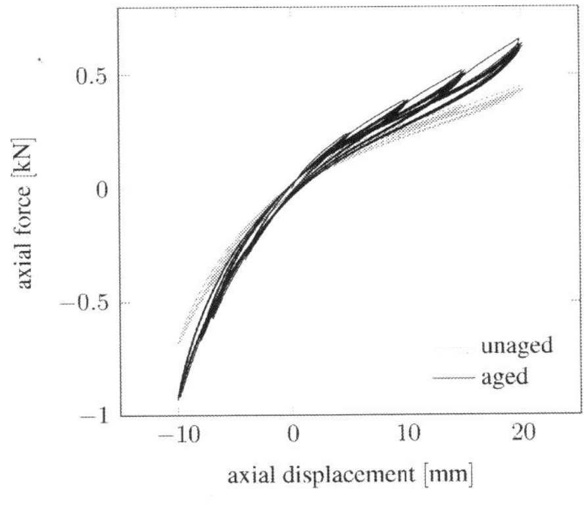

Figure 8. Experimental results for cyclic loading of AE42-buffers (4 load levels with 5 cycles each).

4.3 *Permanent deformation*

The third example focuses on the permanent deformation due to ageing, as is has been eliminated in the previous example. For this purpose, axisymmetric AE2 buffers have been used because their geometry includes a notch, as can be seen in Figure 10. In this region the stresses will localise, causing the most significant deformation to occur in the same, very limited area, which can be easily examined. The applied boundary conditions are similar to the ones used for the AE42-buffer (see Figure 1), but with a static axial displacement applied to the upper mounted surface instead of a force.

The parameters used for the numerical model are identical to the ones from the previous example. Mechanical parameters have not been re-identified for the AE2-buffer as the presented example is mainly a geometrical problem. Because only displacement boundary conditions are applied, the resultant deformation is mainly determined by the ratio of the bulk modulus to the shear modulus. Additionally, the applied external displacements are relatively small, as the experiments used for this example were

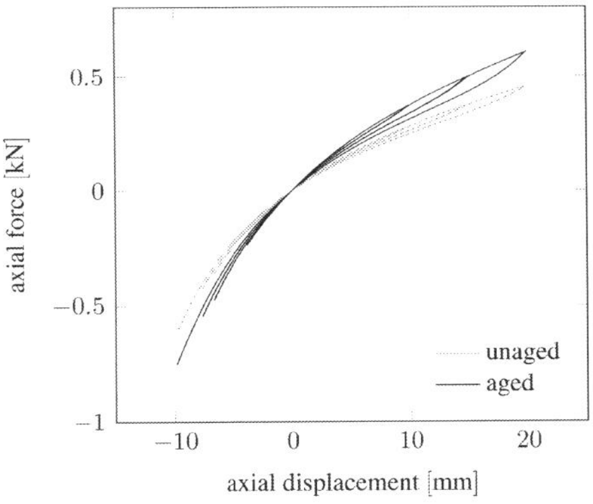

Figure 9. Simulation results for cyclic loading of AE42-buffers (4 load levels with 5 cycles each).

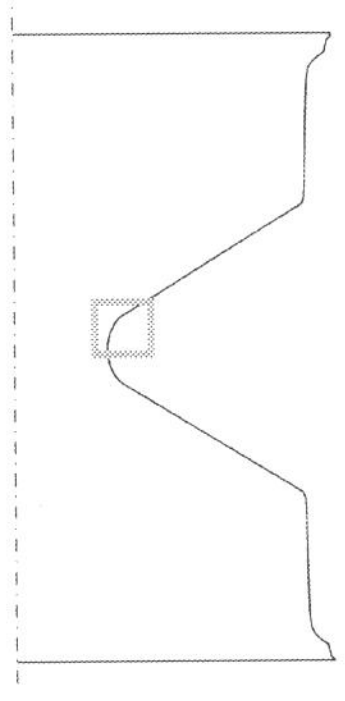

Figure 10. Geometry of the simulated axisymmetric AE2-buffer with area of interest highlighted.

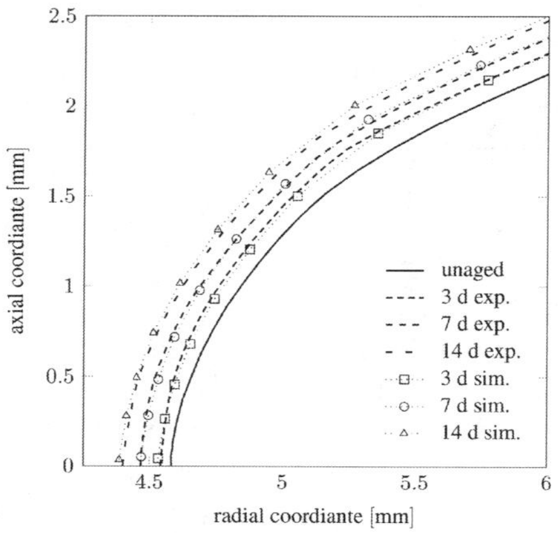

Figure 11. Permanent deformation of AE2-buffers for different ageing times: contour scans from experiments (exp.) and simulation results (sim.) with mesh nodes shown as symbols.

initially done for another purpose, but provided an interesting opportunity to test the model.

Like in the previous example, the "unaged" state does not refer to the virgin material. To overcome the initial processes overlaying the oxidative ageing effects of interest, the samples have been stored at testing conditions, but under a nitrogen atmosphere. Again, this was done for a limited amount of time only, until the mentioned initial processes were negligible.

To determine the geometry of the experimentally aged samples, an optical shape measurement was done, resulting in contour coordinates. The according contour shapes in the area of interest are shown in Figure 11 for various ageing times and in comparison to the permanent deformation obtained by numerical simulations. In all results, the symmetry planes of the buffers have been aligned.

It can be seen that the numerical results match the contour scans nicely, with a slight tendency for overestimation of the deformation for longer ageing durations. These findings complement the valid prediction of the shape of compression set specimen after ageing and the release of residual stresses by cutting them into two halves (Naumann 2017).

5 CONCLUSION

In this work, a dynamic network model is used for realistic simulations of the mechanical material behaviour during the ageing of elastomers. The model assumes that the elastomer consists of one network which continuously evolves due to chain and crosslink scission as well as crosslinking. Furthermore, the DLO-effect is captured by the model as well, enabling realistic simulations of whole elastomeric components. The presented results indicate the essentiality of this feature. It is shown that a staggered solution algorithm can be implemented for the model and its convergence has been demonstrated. Numerical predictions are shown to match with experimental results for component stiffness as well as permanent deformation.

ACKNOWLEDGEMENT

The authors gratefully acknowledge financial support and practical expertise as well as provision of experimental data from Freudenberg Technology Innovation and Vibracoustic. A special thank goes to Morgane Broudin from Vibracoustic for a fruitful scientific collaboration.

REFERENCES

Andrews, R.D., A.V. Tobolsky, & E.E. Hanson (1946). The Theory of Permanent Set at Elevated Temperatures in Natural and Synthetic Rubber Vulcanizates. *Journal of Applied Physics 17*(5), 352.

Broudin, M., Y. Marco, V. Le Saux, P. Charrier, & W. Hervouet (2017). Comparison between thermo-oxidative aging and pure thermal aging of an industrial elastomer for antivibration automotive applications. In *Constitutive Models for Rubber X*. CRC Press.

Charrier, P., Y. Marco, V. Le Saux, & R. Ranaweera (2011). On the influence of heat ageing on filled nr for avs automotive applications. In *ECCMR VII (European Conference on Constitutive Models for Rubber)*, pp. 381–388.

Haupt, P. (1999). *Continuum Mechanics and Theory of Materials*. Springer.

Naumann, C. & J. Ihlemann (2013). Chemomechanically coupled finite element simulations of oxidative ageing in elastomeric components. In N. Gil-Negrete and A. Alonso (Eds.), *Constitutive Models for Rubber VIII*, pp. 43–49. CRC Press.

Naumann, C. & J. Ihlemann (2015a). A dynamic network model to simulate chemical aging processes in elastomers. In B. Marvalová and I. Petríková (Eds.), *Constitutive Models for Rubber XI*, pp. 39–45. CRC Press.

Naumann, C. & J. Ihlemann (2015b). On the thermodynamics of pseudo-elastic material models which reproduce the mullins effect. *International Journal of Solids and Structures 69*, 360–369.

Naumann, C. (2017). *Chemisch-mechanisch gekoppelte Modellierung und Simulation oxidativer Alterungsvorgänge in Gummibauteilen*. Dissertation, Technische Universität Chemnitz. http://nbn-resolving.de/urn:nbn:de:bsz:ch1-qucosa-222075.

Ogden, R. & D. Roxburgh (1999). A pseudo-elastic model for the Mullins effect in filled rubber. *Proceedings of the Royal Society of London. Series A: Mathematical, Physical and Engineering Sciences 455*(1988), 2861–2877.

Steinke, L., J. Spreckels, M. Flamm, & M. Celina (2011). Model for heterogeneous aging of rubber products. *Plastics, Rubber and Composites 40*(4), 5.

Constitutive Models for Rubber X – Lion & Johlitz (Eds)
 ISBN 978-1-138-03001-5

Modeling and simulation of couplings between chemical aging and dissipative heating in dynamic processes on the example of an NBR elastomer

B. Musil, M. Johlitz & A. Lion
Institute of Mechanics, Univeristät der Bundeswehr München, Neubiberg, Germany

ABSTRACT: Typical elastomer components like tires, sealing rings, suspensions or engine mounts are exposed to different environmental influences, which can lead to changes in material properties. This is caused by irreversible modifications occurring in the elastomer network which is known as chemical ageing (Ehrenstein & Pongratz 2007). In this work, ageing behavior of Nitrile Butadiene Rubber (NBR) is investigated under the influence of different surrounding media as air and oil. Based on previous published works, in which chemical ageing was also modeled, a continuum mechanical approach is introduced here, whereby the rubber thermo-viscoelasticity is taken into account. Components produced from NBR are used in many industrial applications where dynamic loads also take place. For these reasons, a dissipative heating of such components occurs (cf. Johlitz et al. (2016)). The resulting temperature field can lead as an accelerator for chemical reactions, triggering the chemical ageing. Thus, the couplings between these two phenomena are also taken into account. Based on the example of NBR an illustrative boundary value problem is numerically solved using FEM and the results and properties of the model are discussed.

1 INTRODUCTION

Chemical ageing of elastomers has been already investigated in several studies by different experimental (chemical and mechanical) methods (see e.g. Tobolsky (1967), Celina et al. (1998) and Johlitz et al. (2014)). As it is widely known, chemical ageing of elastomers results in irreversible changes in the internal structure of the elastomer network. Influences of external media, especially oxygen, lead to scission of the primary network of the elastomer, thus in its degradation. In parallel, there is a creation of new network junctions, or so-called network reformation (cf. Tobolsky (1967), Shaw et al. (2005), Ehrenstein & Pongratz (2007)). The continuum mechanical modeling approach of chemical ageing is based on these modifications occurring in the elastomer network. An internal variable q_d represents the network degradation and q_r represents the network reformation. This type of modeling was also presented in the work of Johlitz et al. (2014) and it is based on the finite strain theory. In this current work the modeling approach of homogenous chemical ageing is extended by finite thermo-viscoelasticity.

To consider dissipative heating of elastomers, an approach motivated by the work of Johlitz et al. (2016) and Dippel (2015) is used. In these works viscoelasticity was also considered and a thermo-mechanical coupling was created. In many industrial applications dynamic loads occur, which lead to self-heating of such loaded elastomer components. The temperature increase in these elastomer components can finally trigger or accelerate chemical ageing. Thus, in this paper chemical ageing, finite thermo-viscoelasticity, dissipative heating and their interactions are taken into account and modeled accordingly.

2 MODELING APPROACH

2.1 *Kinematics*

Regarding the general motion of a deformable body, the concept of the multiplicative split of the deformation gradient $\mathbf{F}$ is considered. First a split into a thermal $\mathbf{F}_\theta$ and a mechanical part $\mathbf{F}_M$ is introduced, which motivates a thermo-mechanic intermediate configuration (TMIC). A further split of the mechanical part into a volumetric part $\bar{\mathbf{F}}$ and an isochoric part $\hat{\mathbf{F}}$ introduces a volumetric-isochoric intermediate configuration (IVIC). Splitting $\hat{\mathbf{F}}$ into the isochoric-elastic part $\hat{\mathbf{F}}_e$ and isochoric-inelastic part $\hat{\mathbf{F}}_i$, the viscoelasticity is considered, whereby an elastic-inelastic intermediate configuration is introduced (EIIC). The right Cauchy-Green deformation tensor $\mathbf{C}$ and the Green strain tensor $\mathbf{E}$ operate on the reference configuration.

Performing push forward operations of $\mathbf{E}$, following strain tensors are established

$$\begin{aligned} \mathbf{E}_{TM} &= \frac{1}{2}(\mathbf{C}_M - \mathbf{I}) + \frac{1}{2}(\mathbf{I} - \mathbf{B}_\theta^{-1}) = \mathbf{E}_M + \mathbf{E}_\theta \\ \mathbf{E}_{IV} &= \frac{1}{2}(\hat{\mathbf{C}} - \mathbf{I}) + \frac{1}{2}(\mathbf{I} - \bar{\mathbf{B}}^{-1}) = \hat{\mathbf{E}} + \bar{\mathbf{E}}, \end{aligned} \quad (1)$$

where

$$\begin{aligned} &\mathbf{C}_{\mathrm{M}} = \mathbf{F}_{\mathrm{M}}^{T} \cdot \mathbf{F}_{\mathrm{M}} \\ &\hat{\mathbf{C}} = \hat{\mathbf{F}}^{T} \cdot \hat{\mathbf{F}} = J_{\mathrm{M}}^{-\frac{2}{3}} \mathbf{C}_{\mathrm{M}} \\ &J_{\mathrm{M}} = \det \mathbf{F} \det \mathbf{F}_{\theta}^{-1} = J\, J_{\theta}^{-1} \\ &\mathbf{F}_{\theta} = \left(\alpha\left(\theta - \theta_0\right) + 1\right)\mathbf{I}. \end{aligned} \tag{2}$$

The strain tensor $\mathbf{E}_{\mathrm{TM}}$ operates on the TMIC and the strain tensor $\mathbf{E}_{\mathrm{IV}}$ operates on the IVIC.

2.2 *Material model*

Regarding the thermomechanical consistent formulation of the material model, the second law of thermodynamics in the form of Clausius-Duhem inequality

$$-\rho_0 \dot{\psi} + J_{\theta} \mathbf{S}_{\mathrm{M}} : \dot{\mathbf{E}}_{\mathrm{M}} - \rho_0 s \dot{\theta} - \frac{\mathbf{Q}}{\theta} \cdot \mathrm{Grad}\,\theta \geq 0 \tag{3}$$

needs to be fulfilled. This inequality is formulated in terms of the TMIC (cf. Dippel (2015)). The mechanical stress tensor $\mathbf{S}_{\mathrm{M}}$ acting on the TMIC is obtained using Piola push forward operation of the 2nd Piola-Kirchhoff stress tensor $\mathbf{S}$. The variable s respresents the specific entropy, $\mathbf{Q}$ is the heat flux vector, ρ_0 is the density and θ is the absolute temperature. According to the work of Lion et al. (2014) and Johlitz et al. (2016) the stress power $\mathbf{S}_{\mathrm{M}} : \dot{\mathbf{E}}_{\mathrm{M}}$ can be rewritten as the sum of volumetric and isochoric contributions

$$\mathbf{S}_{\mathrm{M}} : \dot{\mathbf{E}}_{\mathrm{M}} = -p \dot{J}_{\mathrm{M}} + J_{\mathrm{M}} \hat{\mathbf{S}} : \dot{\hat{\mathbf{E}}}, \tag{4}$$

which will be essential for the formulation of the material model. The stress tensor $\hat{\mathbf{S}}$ acting on the IVIC is introduced here, which can be calculated from the deviatoric part of the Cauchy stress tensor $\mathbf{T}^{\mathrm{D}}$ as

$$\hat{\mathbf{S}} = \hat{J} \hat{\mathbf{F}}^{-1} \cdot \mathbf{T}^{\mathrm{D}} \cdot \hat{\mathbf{F}}^{-T} = \hat{\mathbf{F}}^{-1} \cdot \mathbf{T}^{\mathrm{D}} \cdot \hat{\mathbf{F}}^{-T}. \tag{5}$$

The material model is motivated by a generalized Maxwell model with two springs of the equilibrium elasticity connected in parallel (see Fig. 1). The first spring element represents the primary network (degrading network) of the elastomer and the second spring element represents the secondary network (reforming network). A characteristic property of the secondary network is its stress-free reformation under temporally constant strain. For this reason a hypoelastic formulation of constitutive equation will be used, with the corresponding Helmholtz free energy function in the form of a history functional (see Eq. $(6)_3$). A similar model was used in the work of Hossain et al. (2009) and Lion and Johlitz (2012). The Maxwell elements

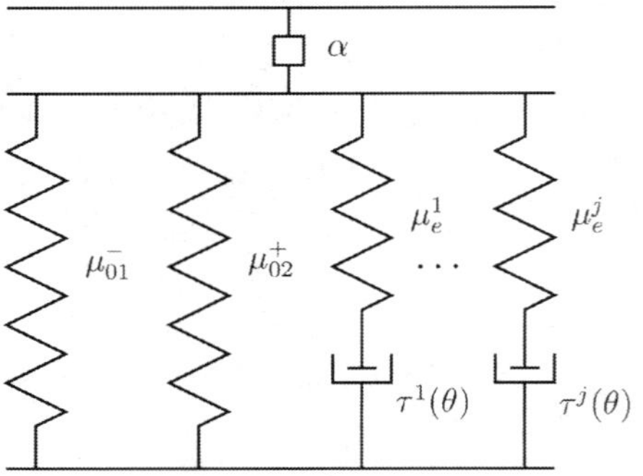

Figure 1. Reological model for the description of thermoviscoelasticity and ageing behavior.

represent the temperature-dependent viscoelasticity. In addition to the temperature-dependent Maxwell elements a thermal element with the parameter α is connected in series. This represents the thermal expansion of the material. From all these considerations an additive split of the Helmholtz free energy function ψ is motivated

$$\begin{aligned} &\psi = \psi_{\theta}(\theta) + \psi_{\mathrm{eq}}^{\mathrm{vol}}(J_{\mathrm{M}}) + \psi_{\mathrm{eq}}^{\mathrm{iso}}(\hat{\mathbf{C}}, q_{\mathrm{d}}, q_{\mathrm{r}}) + \psi_{\mathrm{neq}}^{\mathrm{iso}}(\hat{\mathbf{C}}_{\mathrm{e}}) \\ &\psi_{\mathrm{eq}}^{\mathrm{iso}} = \psi_{\mathrm{d}}(\hat{\mathbf{C}}, q_{\mathrm{d}}) + \psi_{\mathrm{r}}(\hat{\mathbf{C}}, q_{\mathrm{r}}) \\ &\rho_0 \psi_{\mathrm{r}} = \frac{1}{2} \int_0^t \left(\overset{4}{\mathbf{D}'}_{\mathrm{IV}}(s, q_{\mathrm{r}}) : [\hat{\mathbf{E}}(t) - \hat{\mathbf{E}}(s)] \right) \\ &\qquad : [\hat{\mathbf{E}}(t) - \hat{\mathbf{E}}(s)]\, ds. \end{aligned} \tag{6}$$

Calculating the time derivative of ψ and using the relation $\dot{\hat{\mathbf{E}}} = \frac{1}{2} \dot{\hat{\mathbf{C}}}$ Eq. (3) can be reformulated to

$$\begin{aligned} &-\left(\rho_0 s + \rho_0 \frac{\partial \psi_{\theta}}{\partial \theta}\right) \dot{\theta} - \left(p J_{\theta} + \rho_0 \frac{\partial \psi_{\mathrm{eq}}^{\mathrm{vol}}}{\partial J_{\mathrm{M}}}\right) \dot{J}_{\mathrm{M}} \\ &-\frac{\mathbf{Q}}{\theta} \cdot \mathrm{Grad}\,\theta + \left(\frac{1}{2} J \hat{\mathbf{S}} - \rho_0 \frac{\partial \psi_{\mathrm{d}}}{\partial \hat{\mathbf{C}}} - \frac{1}{2} \hat{\mathbf{S}}_{\mathrm{r}} \right. \\ &\left. -\rho_0 \hat{\mathbf{F}}_{\mathrm{i}}^{-1} \cdot \frac{\partial \psi_{\mathrm{neq}}^{\mathrm{iso}}}{\partial \hat{\mathbf{C}}_{\mathrm{e}}} \cdot \hat{\mathbf{F}}_{\mathrm{i}}^{-T} \right) : \dot{\hat{\mathbf{C}}} + \frac{\partial \psi_{\mathrm{d}}}{\partial q_{\mathrm{d}}} \dot{q}_{\mathrm{d}} \\ &+\rho_0 \frac{\partial \psi_{\mathrm{neq}}^{\mathrm{iso}}}{\partial \hat{\mathbf{C}}_{\mathrm{e}}} : \left(\hat{\mathbf{C}}_{\mathrm{e}} \cdot \hat{\mathbf{L}}_{\mathrm{i}} + \hat{\mathbf{L}}_{\mathrm{i}}^{T} \cdot \hat{\mathbf{C}}_{\mathrm{e}}\right) \geq 0, \end{aligned} \tag{7}$$

where

$$\hat{\mathbf{S}}_{\mathrm{r}} = \int_0^t \overset{4}{\mathbf{D}'}_{\mathrm{IV}}(s, q_{\mathrm{r}}) : [\hat{\mathbf{E}}(t) - \hat{\mathbf{E}}(s)]\, ds. \tag{8}$$

2.3 *Constitutive equations*

Based on the selected material functions constitutive equations of the material model are derived in this section. For the equilibrium elasticity and the viscoelasticity a Neo-Hookean model is used (cf. Mooney (1940))

$$\rho_0 \psi_{\mathrm{d}}(\hat{\mathbf{C}},q_{\mathrm{d}}) = \frac{1}{2}\mu_{01}^{-}\left(I_{\hat{\mathbf{C}}}-3\right)$$
$$\rho_0 \psi^{j}_{\mathrm{neq}}(\hat{\mathbf{C}}^{j}_{e}) = \frac{1}{2}\mu^{j}_{e}\left(\mathrm{I}_{\hat{\mathbf{C}}^{j}_{e}}-3\right). \tag{9}$$

The Helmholtz free energy function for the volumetric part is chosen as

$$\rho_0 \psi^{\mathrm{vol}}_{\mathrm{eq}}(J_{\mathrm{M}}) = \frac{1}{2}K\left(J_{\mathrm{M}}-1\right)^2. \tag{10}$$

In case of nearly mechanically incompressible material, bulk modulus K has to be chosen approximately three orders of magnitude higher than the shear modulus. According to ψ_θ an experimentally motivated approach is used (cf. Dippel (2015))

$$\rho_0 \frac{\partial^2 \psi_\theta(\theta)}{\partial \theta^2} = -\frac{1}{\theta}(A+B\theta), \tag{11}$$

where A and B are material parameters from caloric experiments.

Thus, keeping the $\frac{1}{2}\hat{J}_{\mathrm{M}}\hat{\mathbf{C}}^{-1}:\dot{\hat{\mathbf{C}}}=0$ incompressibility constraint in mind, the dissipation inequality (7) can be evaluated for selected material functions. Hence, following constitutive relationships are obtained:

$$\begin{aligned}
p &= -J_\theta^{-1}K\left(J_{\mathrm{M}}-1\right)\\
\mathbf{S} &= -pJ\mathbf{C}^{-1}\\
&+J_{\mathrm{M}}^{-\frac{2}{3}}\mathbf{F}_\theta^{-1}\left[\mu_{01}^{-}\left(\mathbf{I}-\frac{1}{3}\mathrm{tr}(\hat{\mathbf{C}})\hat{\mathbf{C}}^{-1}\right)\right.\\
&+\left(\hat{\mathbf{S}}_{\mathrm{r}}-\frac{1}{3}\left(\hat{\mathbf{S}}_{\mathrm{r}}:\hat{\mathbf{C}}\right)\hat{\mathbf{C}}^{-1}\right)\\
&\left.+\mu^{j}_{e}\left((\hat{\mathbf{C}}^{j}_{\mathrm{i}})^{-1}-\frac{1}{3}\mathrm{tr}\left((\hat{\mathbf{C}}^{j}_{\mathrm{i}})^{-1}\cdot\hat{\mathbf{C}}\right)\hat{\mathbf{C}}^{-1}\right)\right]\mathbf{F}_\theta^{-T}\\
s &= -\frac{1}{\rho_0}\left(\rho_0\frac{\partial\psi_\theta(\theta)}{\partial\theta}\right).
\end{aligned} \tag{12}$$

One can see that the stress tensor **S** has a modular structure. Thus, is additively splitted into the volumetric part $\mathbf{S}_{\mathrm{vol}}$, the degradative part $\mathbf{S}_{\mathrm{d}}$, the reformative part $\mathbf{S}_{\mathrm{r}}$ and the overstress parts $\mathbf{S}^{j}_{\mathrm{neq}}$.

Regarding the reformative part, an integral form of the stress tensor $\hat{\mathbf{S}}_{\mathrm{r}}$ (Eq. (8)) is reformulated by using the time differentiation to the hypoelastic formulation

$$\dot{\hat{\mathbf{S}}} = \overset{4}{\mathrm{D}}_{\mathrm{IV}}(t,q_{\mathrm{r}}):\dot{\hat{\mathbf{E}}} = \frac{1}{2}\overset{4}{\mathrm{D}}_{\mathrm{IV}}(t,q_{\mathrm{r}}):\dot{\hat{\mathbf{C}}}. \tag{13}$$

The fourth-order material tensor $\overset{4}{\mathrm{D}}_{\mathrm{M}}$ is valid for the TMIC and can be defined similar to Bonet and Wood (1997) or Hossain et al. (2009) as

$$\overset{4}{\mathrm{D}}_{\mathrm{M}}(t,q_{\mathrm{r}}) = 4\,q_{\mathrm{r}}(t)\frac{\partial^2 w}{\partial \mathbf{C}^2_{\mathrm{M}}}. \tag{14}$$

In this equation, the variable w represents the strain energy density of the network reformation process. Similar to Eq. (9) a Neo-Hookean approach is chosen

$$w = \frac{1}{2}\mu^{+}_{02}\left(I_{\hat{\mathbf{C}}}-3\right). \tag{15}$$

In order to transform the tensor $\overset{4}{\mathrm{D}}_{\mathrm{M}}$ to the IVIC, where Eq. (13) is valid, following transformation rule can be derived

$$\overset{4}{\mathrm{D}}_{\mathrm{IV}} = J_{\mathrm{M}}^{1/3}\overset{4}{\mathrm{D}}_{\mathrm{M}}. \tag{16}$$

2.4 *Temperature dependencies*

The remaining terms in the dissipation inequality (7) are still to be evaluated. The first term related to ψ_{neq} is evaluated similarly to Johlitz et al. (2016), so that a thermomechanical consistent evolution equation for the inelastic isochoric right Cauchy-Green tensor $\hat{\mathbf{C}}_{\mathrm{i}}$ of each Maxwell element can be formulated

$$\dot{\hat{\mathbf{C}}}^{j}_{\mathrm{i}} = \frac{2}{\tau^{j}(\theta)}\left(\hat{\mathbf{C}}-\frac{1}{3}\mathrm{tr}\left(\hat{\mathbf{C}}\cdot(\hat{\mathbf{C}}^{j}_{\mathrm{i}})^{-1}\right)\hat{\mathbf{C}}^{j}_{\mathrm{i}}\right). \tag{17}$$

In this equation $\tau^{j}(\theta)=\eta^{j}(\theta)/\mu^{j}_{e}$ represent the temperature-dependent relaxation times. Using e.g. the standard WLF-equation (cf. Williams et al. (1955)) for the formulation of the temperature-dependent viscosities $\eta^{j}(\theta)$, the temperature-dependent behavior of the Maxwell elements is included.

The second term with respect to ψ_{d} is treated similar to Lion and Johlitz (2012) or Johlitz et al. (2014). This results in some conditions for the dependence of the shear modulus of the primary network on the internal variable q_{d}

$$\mu_{01}^{-} = \mu_{01}(1-q_{\mathrm{d}}), \tag{18}$$

where the parameter μ_{01} is the shear modulus of the unaged elastomer. For this purpose an evolution equation for the internal variable q_{d} of the network degradation is needed, which takes into account the temperature dependence of the ageing process

$$\dot{q}_{\mathrm{d}} = \nu_{\mathrm{d}}\,e^{-\frac{E_{\mathrm{d}}}{R\theta}}(1-q_{\mathrm{d}}), \quad q_{\mathrm{d}}(0)=0. \tag{19}$$

In the same sense, an evolution equation for the internal variable q_{r} of the network reformation is formulated

$$\dot{q}_{\mathrm{r}} = \nu_{\mathrm{r}}\,e^{-\frac{E_{\mathrm{r}}}{R\theta}}(1-q_{\mathrm{r}}), \quad q_{\mathrm{r}}(0)=0. \tag{20}$$

Here, R represents the universal gas constant and the parameters v_d, v_r and E_d, E_r are pre-exponential factors and activation energies.

With the previously introduced equations a one-directional coupling between the mechanical properties and the temperature is reached. To describe the second coupling direction, which leads to deformation-induced temperature changes, the balance of internal energy ϵ

$$\rho_0 \dot{\epsilon} = J_\theta \mathbf{S_M} : \dot{\mathbf{E}}_\mathbf{M} - \mathrm{Div}\, \mathbf{Q} + \rho_0 r, \tag{21}$$

i.e. the first law of thermodynamics, is considered and evaluated. Eq. (21) can be reformulated with help of the Legendre transformation $\epsilon = \psi + \theta s$ and the calculated stress power and the temporal derivative of the free energy function ψ is inserted, whereby the heat radiation term $\rho_0 r$ is neglected. The derived heat balance finally takes the form

$$\rho_0 \theta \dot{s} + \mathrm{Div}\, \mathbf{Q} - \frac{1}{2}\mu_e^j (\hat{\mathbf{C}}_\mathrm{i}^j)^{-1} \cdot \hat{\mathbf{C}} \cdot (\hat{\mathbf{C}}_\mathrm{i}^j)^{-1} : \dot{\hat{\mathbf{C}}}_\mathrm{i}^j \\ -\frac{1}{2}\mu_{01}\left(I_{\hat{\mathbf{C}}} - 3\right)\dot{q}_\mathrm{d} = 0. \tag{22}$$

Thus, the model is coupled in both directions. On the one hand, the stress tensor and the internal variables of ageing (network degradation and reformation) depend on the temperature; on the other hand, dynamical deformations lead to the self-heating of the material. The last term in Eq. (22) shows that the process of network degradation also contributes to the dissipation of energy.

According to the heat flux vector $\mathbf{Q}$ the Fourier's law of heat conduction is taken into account

$$\mathbf{Q} = -\lambda_\theta \mathbf{Grad}\,\theta, \tag{23}$$

with the coefficient of thermal conductivity λ_θ. Thus, it is ensured that the term with $\mathbf{Q}$ always satisfies the Clausius-Duhem inequality (3).

Another option for modeling thermo-visco-elasticity can be done by the assumption that the viscosities $\eta^j(\theta)$ depend on the temperature in the same way as the shear moduli $\mu_e^j(\theta)$ of the Maxwell elements. This assumption leads automatically to the temperature-independent relaxation times, the shear moduli are considered to be temperature-dependent. In such a case the entropy

$$s = -\frac{1}{\rho_0}\left(\rho_0 \frac{\partial \psi_\theta(\theta)}{\partial \theta} + \frac{1}{2}\frac{\partial \mu_e^j(\theta)}{\partial \theta}\left(\mathrm{I}_{\hat{\mathbf{C}}_e^j} - 3\right)\right) \tag{24}$$

is directly affected by the mechanical properties and the deformation. The temperature-dependency of $\mu_e^j(\theta)$ can be modeled e.g. by using the approach proposed by Johlitz et al. (2009).

3 FE-IMPLEMENTATION

The presented constitutive model was implemented in the FE-software Comsol Multiphysics using the weak form of the balance of momentum based on the total Lagrangian formulation (see e.g. Bathe (1996))

$$\int_\Omega \tilde{\mathbf{S}} : \delta\mathbf{E} d\Omega = \int_\Gamma \mathbf{t} \cdot \delta\mathbf{u}\, d\Gamma, \tag{25}$$

whereby the displacement field $\mathbf{u}$ is the primary variable. A standard displacement-based method cannot be applied directly, because the shape functions are unable to properly describe the volumetric preserving deformation. In such a case volumetric locking or numerical instabilities will occur. Therefore a mixed u-p finite element formulation is used (cf. Sussman and Bathe (1987)). The stress tensor S is replaced in Eq. (25) by a modified version $\tilde{\mathbf{S}}$

$$\tilde{\mathbf{S}} = \mathbf{S} + (p - \tilde{p}) J C^{-1}. \tag{26}$$

In this equation p is the hydrostatic pressure computed according to Eq. $(12)_1$ and an auxiliary interpolated pressure $\tilde{p}$ is introduced. The interpolated pressure is obtained through an additional weak constraint

$$\int_\Omega \left(\frac{\tilde{p}}{K} - \frac{p}{K}\right)\delta\tilde{p}\, d\Omega = 0, \tag{27}$$

where it is forced to be equal to the hydrostatic pressure p. The order of the shape function for the auxiliary pressure variable should be one order less than that for the displacements.

The second primary variable represents the absolute temperature θ, which is calculated using the implemented weak form of the derived heat balance

$$\int_\Omega \rho_0 \theta \dot{s} \delta\theta\, d\Omega + \int_\Omega \lambda_\theta \mathrm{Grad}\,\theta \mathrm{Grad}(\delta\theta) d\Omega \\ -\int_\Omega \frac{1}{2}\Big(\mu_e^j (\hat{\mathbf{C}}_\mathrm{i}^j)^{-1} \cdot \hat{\mathbf{C}} \cdot (\hat{\mathbf{C}}_\mathrm{i}^j)^{-1} : \dot{\hat{\mathbf{C}}}_\mathrm{i}^j \\ + \mu_{01}\left(I_{\hat{\mathbf{C}}} - 3\right)\dot{q}_\mathrm{d}\Big)\delta\theta\, d\Omega = -\int_\Gamma \mathbf{n} \cdot \mathbf{Q} \delta\theta\, d\Gamma. \tag{28}$$

Implementation of a convective boundary condition is possible via Newtons law of cooling, where it applies for the heat flux on the surface q

$$q = -\mathbf{n} \cdot \mathbf{Q} = h\Delta\theta. \tag{29}$$

In this equation h represents the heat transfer coefficient and $\Delta\theta$ is the difference between bulk temperature of the adjacent fluid and the calculated temperature on the surface.

The inelastic strain tensor $\hat{\mathbf{C}}_\mathrm{i}^j$, the stress tensor $\hat{\mathbf{S}}_\mathrm{r}$ and the internal variables q_d, q_r are treated as additional degrees of freedom. The shape functions are chosen to be one order lower than those used for the displacements, because these variables add to the strains and stresses computed from displacement derivatives. These variables also do not require continuity between the mesh elements so discontinuous shape functions are used.

To solve the resulting coupled system of nonlinear differential equations a segregated (staggered) solution approach is used (cf. Holzapfel (2000) and the citations therein). The segregated approach treats each problem sequentially using the results of the previously solved field to evaluate the loads and material properties for the next field being solved. First a solution of the mechanical problem at isothermal conditions takes place, followed by a solution of the heat conduction problem at a fixed configuration. In the last step the ageing problem is solved, i.e. the corresponding evolution equations. This solution sequence is then repeated within a time step, as long as the convergence is not reached. When the segregated approach is applicable, it will converge to the same result as the fully coupled approach. In addition, the thermo-mechanical problem and the ageing problem can be solved using different time discretization. Within this concept the coupled system can be solved sequentially with lower computational cost.

4 RESULTS AND DISCUSSION

4.1 *Parameter identification*

As mentioned in the Sect. 1 two dominant processes take place in the elastomer network during ageing: chain scission (network degradation) and the creation of new network junctions (network reformation). In order to study these processes experimentally (phenomenologically), continuous and intermittent tests are required.

The continuous relaxation tests are used to determine the process of network degradation as a function of time and temperature. In this experiment the sample is brought to a constant strain and the stress response is observed. Network reformation has no influence here because the secondary network develops stress-free under constant strain in time. However, while in the classical relaxation test the stress drop is caused by viscoelastic effects, the stress drop in this experiment is based on chemical degradation of the primary network. In order to determine the influence of temperature on the degradation process, this experiment is usually performed at different elevated temperatures. This tests can be also performed in different surrounding medium, so that e.g. ageing in air and oil can be investigated. With subsequent parameter identification, the parameters of network degradation (v_d and E_d) can be identified.

The intermittent tests are carried out in addition to the aforementioned continuous relaxation tests. In these tests an isothermal load-free ageing of the samples takes place in different fixed ageing media (air, oil). At certain times ageing is intermitted and the samples are cooled down to room temperature. Relaxation tests or tensile tests can be then carried out with the aged samples. In this work, uniaxial tensile tests with some holding phase at the end were performed. The strain rate was set to 0,01%/s, such that possible influence of ageing on viscoelasticity was not investigated. Results of these tests are shown in Fig. 2 and Fig. 3. It can be seen that the NBR becomes stiffer as the ageing time increases, which is caused by the reformed network. In the case of oil ageing at 100°C, the degree of oxidation is much lower than that in the air at 80°C. This effect can be seen well at the level of the measured engineering stress P_{11}. In the following, an evaluation of the observed tensile curves at certain strains takes place. Combining these data with those from the continuous relaxation tests, so that the contribution of network degradation is subtracted, one obtains the evolution of the stress of the reformed network over

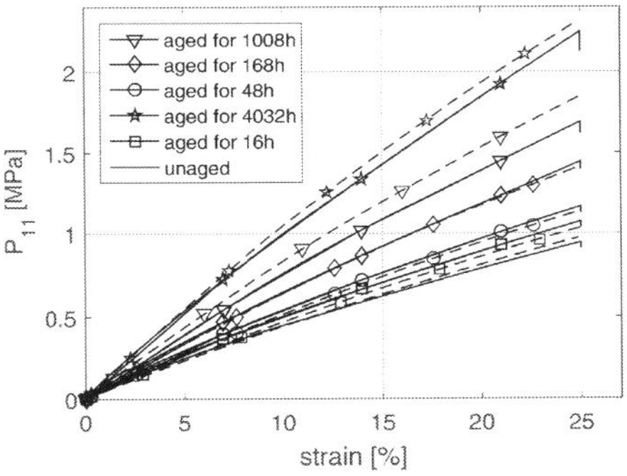

Figure 2. Intermittent tensile tests with NBR specimens aged in air at 80°C. Solid line denotes measurement, dashed line denotes FE-simulation.

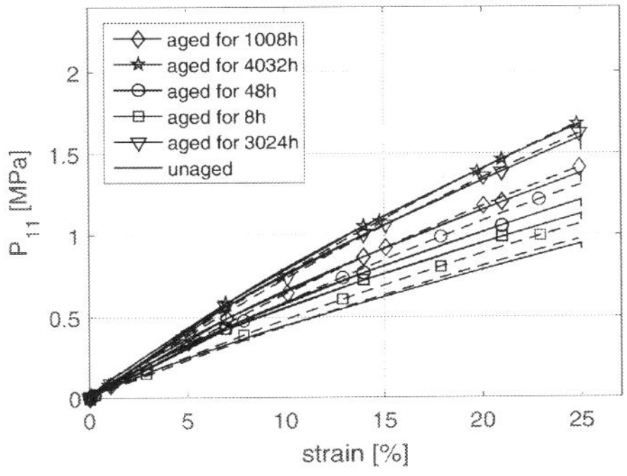

Figure 3. Intermittent tensile tests with NBR specimens aged in oil at 100°C. Solid line denotes measurement, dashed line denotes FE-simulation.

the ageing time. Followed by parameter identification, the parameters of network reformation (ν_r and E_r) and shear modulus μ_{02}^{+} of the reformed network can be identified. For details on parameter identification see Johlitz et al. (2014).

In order to identify the mechanical and thermomechanical properties of unaged NBR, various standardized tests for material characterization are also required (see e.g. Dippel (2015)).

4.2 *FE-simulations*

In the first FE-simulation the intermittent tensile tests are simulated. First a load-free isothermal ageing of the NBR tensile specimens is simulated in different media (air, oil). Then uniaxial tensile tests at room temperature are simulated with the aged specimens, the strain rate corresponds to 0,01%/s. The used material parameters, which resulted from the parameter identification, are listed in Table 1. Comparison of the simulation results with the measurements are shown in Fig. 2 and Fig. 3. It can be seen that the model can represent the mechanical and the ageing behavior of the material well. Thus, resulting stiffening of the material can be simulated in a good way with respect to different ageing media. It is important to note that only one specimen was tested for each ageing time, which can produce some statistical error and so affect the deviations between simulation and experiment. Some deviations between simulation and experiment can also result from the inaccuracies in the parameter identification.

In the second FE-simulation numerical experiments under dynamic loading are performed. So we would like to point to the simulation of coupling between dissipative heating and chemical ageing. A simple shear test in the plane strain mode is computed. Numerical values of the model parameters are listed in Table 2. On the left and right boundary of the square elastomer specimen adiabatic boundary conditions are applied. On the top and bottom of the specimen convective boundary conditions are modeled (see Eq. (29)). This is assumed to be a free convection between the elastomer surface and surrounding air, whereby the associated heat transfer coefficient is selected according to the work of Sae-Oui et al. (1999). The bulk temperature of the surrounding air and the initial temperature of the specimen is set to $\theta_0 = 353\,\mathrm{K}$. Sinusoidal shear loading conditions $u(t) = u_0 \sin(2\pi f t)$ are also applied, with a frequency of $f = 4\,\mathrm{Hz}$. A shear angle amplitude of 15° is chosen.

The resulting temperature field is inhomogeneous and is presented in Fig. 4. Similar results were also presented in the work of Johlitz et al. (2016).

Now, two different cases are considered: once an unaged elastomer specimen is exposed to dynamic

Table 1. Set of parameters used for the first FE-simulation.

Parameter	Medium	Value	Parameter	Medium	Value
μ_{01}	–	1,53 MPa			
μ_e^1	–	0,078 MPa	τ^1	–	4,2 s
μ_e^2	–	0,111 MPa	τ^2	–	60,4 s
μ_e^3	–	0,068 MPa	τ^3	–	$9{,}65\cdot 10^2$ s
μ_e^4	–	0,062 MPa	τ^4	–	$3{,}844\cdot 10^4$ s
μ_{02}^{+}	air	4,42 MPa	μ_{02}^{+}	oil	2,96 MPa
E_d^1	air	$1{,}14\cdot 10^5$ J mol^{-1}	E_d^1	oil	$8{,}36\cdot 10^4$ J mol^{-1}
E_d^2	air	$1{,}005\cdot 10^5$ J mol^{-1}	E_d^2	oil	$1{,}13\cdot 10^5$ J mol^{-1}
E_d^3	air	$1{,}005\cdot 10^5$ J mol^{-1}	E_d^3	oil	$6{,}49\cdot 10^4$ J mol^{-1}
ν_d^1	air	$4{,}83\cdot 10^{10}$ s^{-1}	ν_d^1	oil	$5{,}41\cdot 10^6$ s^{-1}
ν_d^2	air	$3{,}37\cdot 10^7$ s^{-1}	ν_d^2	oil	$2{,}94\cdot 10^9$ s^{-1}
ν_d^3	air	$3{,}37\cdot 10^7$ s^{-1}	ν_d^3	oil	95,3 s^{-1}
E_r^1	air	$3{,}93\cdot 10^4$ J mol^{-1}	E_r^1	oil	$5{,}67\cdot 10^4$ J mol^{-1}
E_r^2	air	$1{,}07\cdot 10^5$ J mol^{-1}	E_r^2	oil	$6{,}28\cdot 10^4$ J mol^{-1}
E_r^3	air	$1{,}07\cdot 10^5$ J mol^{-1}	E_r^3	oil	$3{,}44\cdot 10^4$ J mol^{-1}
ν_r^1	air	0,85 s^{-1}	ν_r^1	oil	$1{,}77\cdot 10^3$ s^{-1}
ν_r^2	air	$4{,}68\cdot 10^8$ s^{-1}	ν_r^2	oil	$1{,}59\cdot 10^2$ s^{-1}
ν_r^3	air	$4{,}77\cdot 10^8$ s^{-1}	ν_r^3	oil	$5{,}6\cdot 10^{-3}$ s^{-1}

Table 2. Set of parameters used for the second FE-simulation.

ρ_0 kg/m³	α 1/K	$A + B\theta$ J/kg·K	λ_θ W/m·K
1130	$2{,}155 \cdot 10^{-4}$	$540 + 3{,}6\theta$	0,225
μ_{01} MPa	μ_e^1 MPa	η^1 MPa s	μ_{02}^+ MPa
1,05	0,8	$5 \cdot 10^4$	3,15
E_d J/mol	v_d 1/s	E_r J/mol	v_r 1/s
$1{,}02 \cdot 10^5$	$1{,}4 \cdot 10^9$	$7{,}27 \cdot 10^4$	$1{,}16 \cdot 10^5$

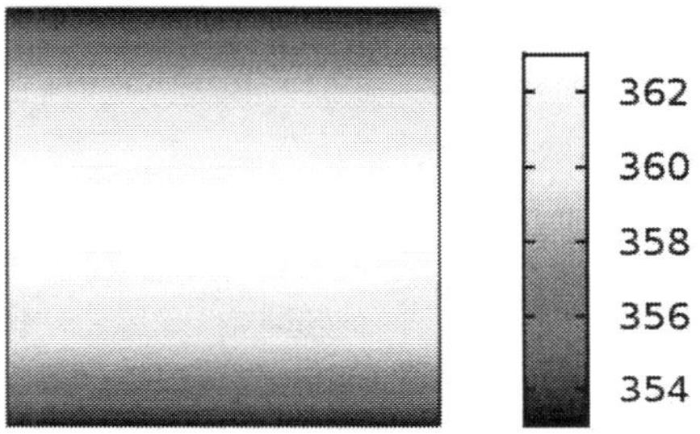

Figure 4. Resulting temperature field in K of the dynamically loaded elastomer specimen.

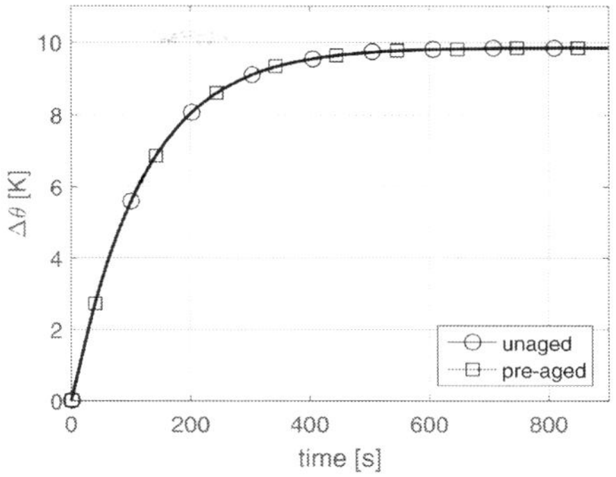

Figure 5. Temperature profile in the middle point of the dynamically loaded elastomer specimen.

loading; secondly the investigated specimen is first pre-aged load-free for 168 hours and then exposed to dynamic loading. Within this numerical experiment we would like to point out that ageing has no significant influence on the self-heating of the material. This effect is presented in Fig. 5. Here, resulting nonlinear temperature difference $\Delta\theta$ between the calculated and the initial temperature, from the middle point of the specimen, is plotted over time. Thus, no dependence on the pre-ageing is visible. A contribution of ageing is present in the last term of the heat balance (Eq. (22)); however, it is negligibly small according to the other terms. If an effect of ageing on the viscoelasticity would be proved experimentally, it can be then incorporated into the presented constitutive model. This would automatically increase the influence of ageing on self-heating. According to this, further investigations are necessary.

On the other hand, the influence of self-heating on ageing is evident. Therefore the unaged elastomer specimen, which has been subjected to dynamic loading, will be discussed now. The resulting temperature field can lead, as an accelerator of chemical reactions, to the initiation resp. acceleration of the ageing processes. This phenomenon can be very well represented by the $\mathbf{S}_r$ reformative part of the stress tensor. This-one develops in time and is caused by ageing, which is directly dependent on the temperature field. In Fig. 6 this reformed stress $S_{r,12}$ component is evaluated on the deformed

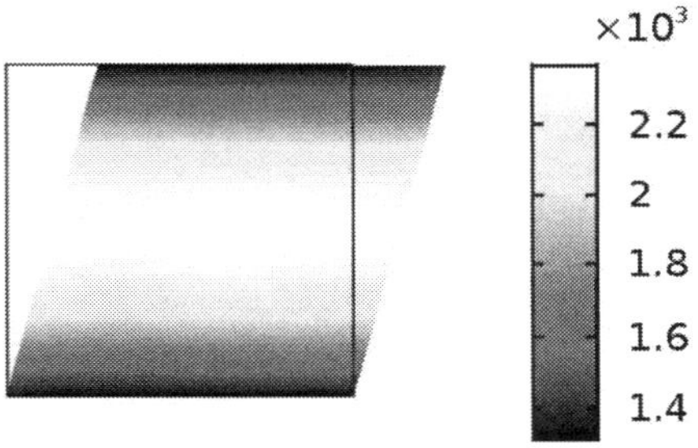

Figure 6. Resulting inhomogeneous distribution of the reformed stress. The $S_{r,12}$ component in Pa is plotted.

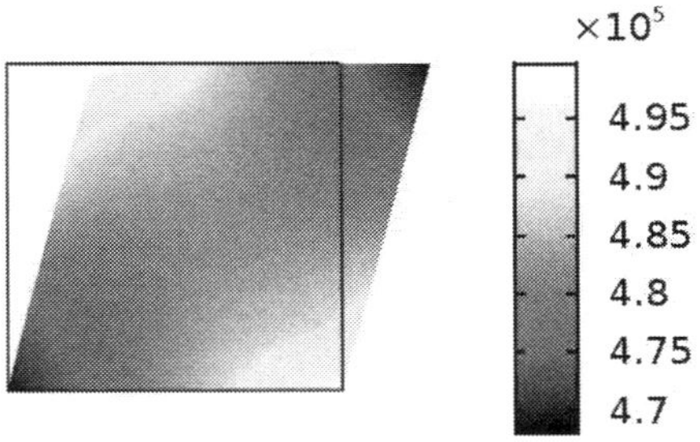

Figure 7. Distribution of the 2nd Piola-Kirchhoff stress. The S_{12} component in Pa is plotted.

specimen. It can be seen that due to the arising self-heating, the ageing process is accelerated locally and thus the reformed stress becomes inhomogeneous. Similar inhomogeneous distribution is also observed for the internal variables q_d and q_r.

It is important to note that the influence of $\mathbf{S}_r$ on the entire stress $\mathbf{S}$ is still barely recognizable (see Fig. 7). A time period of 900 seconds of loading was simulated in the time domain, which is of course not enough to provide such a pronounced ageing. Here, a time period in the range of weeks or months would be necessary to simulate in order to strengthening the presented effects.

5 CONCLUSION

In this work the modeling of the chemical ageing of NBR in different surrounding media was studied with regard to its thermo-viscoelasticity. The proposed continuum mechanical approach was based mainly on the works of Lion and Johlitz (2012), Johlitz (2012), Johlitz et al. (2014), Dippel (2015) and Johlitz et al. (2016). A thermodynamically consistent derivation of the constitutive equations was introduced. Selection of material functions, which are necessary for the formulation of the material model, was based on the experimental investigations of the aged or unaged NBR and subsequent parameter identification. The proposed constitutive model was implemented in a commercial FE-software. In general, the simulations with the implemented model showed a good agreement with the underlying experiments. Some illustrative boundary value problems were also presented within this paper in the context of the self-heating phenomena occurring in many elastomer components. Thus, the couplings between chemical ageing, finite thermo-viscoelasticity and dissipative heating were modeled and simulated. Here, the capability of the model to simulate this complex material behavior was shown and some possible improvements have been discussed. In the case of industrial applications this model can be used to simulate the chemical ageing of elastomers taking their thermo-viscoelasticity and self-heating into account and thus e.g. to depict the material behavior during dynamic loading in time, which may be necessary for a lifetime prediction of the rubber components.

ACKNOWLEDGEMENT

Financial support of the project by the Deutsche Forschungsgemeinschaft (DFG) under the grant number JO 818/3-1 is gratefully acknowledged.

REFERENCES

Bathe, K.J. (1996). *Finite Elemente Procedures* (2nd ed.). Prentice-Hall, Englewood Cliffs, New York.

Bonet, J. & R.D. Wood (1997). *Nonlinear continuum mechanics for finite element analysis*. Cambridge University Press, Cambridge.

Celina, M., J. Wise, D. Ottesen, K. Gillen, & R. Clough (1998). Oxidation profiles of thermally aged nitrile rubber. *Polymer degradation and stability 60*(2–3), 493–504.

Dippel, B. (2015). *Experimentelle Charakterisierung, Modellierung und FE-Berechnung thermomechanischer Kopplungen am Beispiel eines rußgefüllten Naturkautschuks*. Ph. D. thesis, Universität der Bundeswehr München.

Ehrenstein, G. & S. Pongratz (2007). *Beständigkeit von Kunststoffen*. Carl Hanser Verlag.

Holzapfel, G.A. (2000). *Nonlinear Solid Mechanics*. JohnWiley & Sons, Chichester.

Hossain, M., G. Possart, & P. Steinmann (2009). A finite strain framework for the simulation of polymer curing. Part I: elasticity. *Comput. Mech. 44*, 621–630.

Johlitz, M. (2012). On the representation of ageing phenomena. *J. Adhesion 88*(7), 620–648.

Johlitz, M., B. Dippel, & A. Lion (2016). Dissipative heating of elastomers: a new modelling approach based on finite and coupled thermomechanics. *Continuum Mechanics and Thermodynamics 28*(4), 1111–1125.

Johlitz, M., N. Diercks, & A. Lion (2014). Thermo-Oxidative Aging of Elastomers: A Modelling Approach Based on a Finite Strain Theory. *Int. J. Plasticity 63*, 138–151.

Johlitz, M., D. Scharding, S. Diebels, J. Retka, & A. Lion (2009). Modelling of the thermo-viscoelastic material behaviour of polyurethane close to the glass transition temperature. *Z. Angew. Math. Mech. 90*, 387–398.

Lion, A. & M. Johlitz (2012). On the representation of chemical ageing of rubber in continuum mechanics. *Int. J. Solids Struct. 49*(10), 1227–1240.

Lion, A., B. Dippel, & C. Liebl (2014). Thermomechanical material modelling based on a hybrid free energy density depending on pressure, isochoric deformation and temperature. *Int. J. Solids Struct. 51*, 729–739.

Mooney, M. (1940). A theory of large elastic deformation. *J. Appl. Phys. 11*, 582–592.

Sae-Oui, P., P. Freakley, & P. Oubridge (1999). Determination of heat transfer coefficient of rubber to air. *Plastics, rubber and composites 28*(2), 65–68.

Shaw, J., S. Jones, & A. Wineman (2005). Chemorheological response of elastomers at elevated temperatures: experiments and simulations. *J. Mech. Phys. Solids 53*, 2758–2793.

Sussman, T. & K.J. Bathe (1987). A finite element formulation for nonlinear incompressible elastic and inelastic analysis. *Comput. Struct. 26*, 357–409.

Tobolsky, A.V. (1967). *Mechanische Eigenschaften und Struktur von Polymeren*. Berliner Union Stuttgart.

Williams, M.L., R.F. Landel, & J.D. Ferry (1955). The temperature dependence of relaxation mechanisms in amorphous polymers and other glass-forming liquids. *J. Am. Chem. Soc. 77*(14), 3701–3707.

Constitutive Models for Rubber X – Lion & Johlitz (Eds)
© 2017 Taylor & Francis Group, London, ISBN 978-1-138-03001-5

Nitrile rubber—the influence of acrylonitrile content on the thermo-oxidative aging

L. Vozarova, M. Johlitz & A. Lion
Institute of Mechanics, Bundeswehr University Munich, Neubiberg, Germany

M. Köberl
Institute for Construction Materials, Bundeswehr University Munich, Neubiberg, Germany

ABSTRACT: Macromolecules with saturated bonds and unbranched chains are resistant to oxidation and their speed of oxidation is very small. The increasing content of double bonds in the macromolecule leads to lower resistance to oxidation, which may result in cleavage of the macromolecules, in cross-linking or in the formation of new functional groups. Due to its double bonds, rubber is very responsive to oxidation. Directly by ATR-FTIR measuring the absorbances after thermal degradation at high temperatures, it was possible to determine quantitatively each of the functional groups after the degradation of Nitrile-Butadiene Rubber (NBR). Ageing degradation leads to dramatic modifications of the chemical structure of the macromolecules. The objective of this contribution is studying the influence of the acrylonitrile content on the thermo-oxidative aging and the investigation of the spectral changes of functional groups as a function of the thermal aging time.

1 INTRODUCTION

Rubber is class of material that finds applications in a broad range of fields, especially where toughness, elasticity and impermeability are required. Practically, rubber is blended with filler to improve the toughness and the tensile strength, which provides added durability as well as reductions of the costs (He et al., 2016). Acrylonitrile butadiene rubber (NBR), or nitrile rubber, is commonly considered as the workhorse of many industrial and automotive rubber products. In recent years, high effort is given to evaluate the ageing behaviour of polymeric materials exposed to weathering in the presence of air or other environment.

The degradation of NBR and other diene rubbers can be roughly classified into 3 categories: thermal degradation (Celina et al., 1997, Ahlblad et al., 1996), photo degradation (Adam et al., 1990, Jouan et al., 1989), and ozone degradation (Hara et al., 1998). The polymers undergo primary chain cleavage due to peroxy radical reactions, leading to a decrease in the molecular weight and the generation of various new functional groups in the molecules through oxidation reactions (Kawashima and Ogawa, 2005).

In the process of polymer degradation, disruption of macromolecular chains by the impact of physical (heat, mechanical stress, light, or other radiation) and chemical (oxidation and hydrolysis) influences takes place. High temperatures favour oxidation reactions and significantly accelerate the formation of radicals, which further react with oxygen. Therefore, even a small degree of oxidation causes a significant change in the properties of elastomers.

The technique of Attenuated Total Reflection (ATR) is a special method of Fourier Transform Infrared Spectroscopy (FTIR) for surface investigation of non-transparent samples, e.g. filled elastomers. This non-destructive analytical method allows even smallest samples to be quickly examined. The crux of this method is an ATR crystal e.g. silicon, germanium or diamond. The sample is pressed using a compression rod onto a crystal surface, through which the IR radiation is directed (Smith, 2011).

This paper focuses on the quantitative analysis of several functional groups in NBR, such as a carbon-carbon double bond (butylene group, C=C), carbon-oxygen double bond (carbonyl group, C=O) and a carbon-nitrogen triple bond (nitrile group, C≡N) to investigate the thermo-oxidative sensitivity of NBR bearing different acrylonitrile (ACN) contents.

2 EXPERIMENTAL

2.1 *Materials*

Systematic investigations were performed on the aging stability of three types of vulcanized NBR with different ACN contents. The NBR samples consist of ACN (18 wt%, 28 wt% and 39 wt%), identical amounts of zinc oxide (ZnO), carbon black and other components. The preparation of the NBR was provided according to the following

Table 1. Components of the NBR mixture.

Component	Amount [phr*]		
	NBR 1846	NBR 2846	NBR 3946
ACN content (wt%)	18	28	39
Polymer	100	100	100
Filler (carbon black N 550)	60	60	60
Plasticizer (DEHP)	20	20	20
Antioxidant (6-PPD)	2	2	2
ZnO	5	5	5
Stearic acid	1	1	1
CBS	1,5	1,5	1,5
TMTM-80	0,5	0,5	0,5

* Parts per hundred NBR rubber.

formulation in Table 1. Long strips, approximately 5 × 2 × 70 mm, of each material were stored for a certain time and temperature. Afterwards, the FTIR measurements were made.

2.2 *Thermal aging*

Accelerated thermal aging of the strips cut from the rubber sheets (~2 mm thickness) was carried out in temperature controlled (with sensitivity of ±1°C), commercial air-circulating aging ovens at 80°C and 100°C. The duration of the thermal aging was up to 3024 hours.

2.3 *Attenuated Total Reflection-Fourier Transform Infrared spectroscopy (ATR-FTIR)*

ATR-FTIR spectroscopy is well suited to follow the chemical degradation process within the first microns of the surface as function of the aging time. The oxidation profile of the aged samples was monitored at room temperature using a Thermo Fisher Scientific Nicolet iS10 FTIR spectrometer in the wave number range of 600–4000 cm^{-1}. The spectra were measured with germanium as the ATR crystal using a series of 16 scans at a resolution of 4 cm^{-1}. The infrared radiation (IR) penetrates the surfaces of the samples for the first few microns. The average of six scans for each sample was taken for the peak identification. The collected spectra were analyzed using the Omnic software (Thermo Fisher Scientific). No baseline correction by the software was applied to the spectra.

3 RESULTS AND DISCUSSION

3.1 *Identification of qualitative analysis*

Figure 1 shows the infrared absorption spectra of raw NBR samples with different contents of ACN obtained by ATR technique. The broad peak

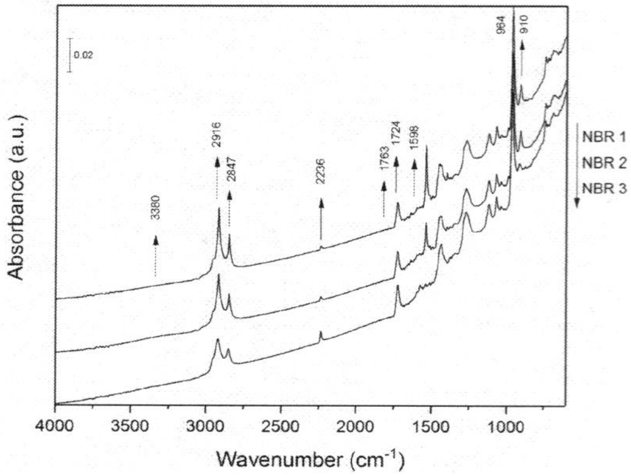

Figure 1. FTIR spectra of NBR with different contents of ACN (NBR 1 - 18%, NBR 2 - 28%, and NBR 3 - 39%) before heating.

at 3380 cm^{-1} belongs to hydroxyl groups (OH). A characteristic feature of NBR elastomer is its acrylonitrile group. Nitrile bands are usually in the range of 2215–2240 cm^{-1} in the IR spectrum. The peak at 2236 cm^{-1} corresponds to nitrile stretching (C≡N). Since NBR is prepared by the copolymerization of butadiene and acrylonitrile, the NBR molecule contains a second functional group in the form of a butylene double bond. The band at 964 cm^{-1} is assigned to a =CH out of plane deformation vibration and therefore represents a characteristic feature for the butylene part of the NBR.

The peaks in the range of 2800–3000 cm^{-1} are related to superimposed, symmetrical and asymmetrical –CH vibrations of components of the polymer structure. The band at 1460 cm^{-1} can be attributed either to symmetric or asymmetric CH_2 deformation vibrations. Furthermore, the rocking vibration at 723 cm^{-1}, which occurs in aliphatic hydrocarbons with a chain length of at least four CH_2 units, is due to the polymer structure.

The peaks at wavenumbers of 1763 cm^{-1}, 1724 cm^{-1}, 1598 cm^{-1} is attributed to carbonyl groups (C = O). The bis(2-ethylhexyl) phthalate (DEHP) added as a plasticizer can also be identified in the spectrum of the starting materials. The bands at about 1730 cm^{-1} are an indication of carbonyl bands of organic esters. If this occurs in connection with C–O–C vibrational stretching at about 1270 cm^{-1}, this serves as a proof for esters. Furthermore, the asymmetric C–O single bonds (1040 cm^{-1}, 1071 cm^{-1}, 1120 cm^{-1} and 1271 cm^{-1}) are further indicia for ester compounds.

3.2 *Spectral changes in the IR*

The FTIR spectra of NBR samples containing different weight contents of ACN before and after thermal aging at 100°C for 3024 hours were displayed in Figures 2a, b, c.

It can be seen that characteristic peaks in the spectrum compared to the unaged NBR are changed with increasing aging time. During the thermal oxidation of the NBR, a general broad increase in the hydroxyl (3380 cm^{-1}) and carbonyl (1763 cm^{-1} and 1598 cm^{-1}) regions of the spectrum is observed.

The considerable decrease in the range of methylene (2800–3000 cm^{-1}), symmetric and asymmetric groups, and butylene double bonds (965 cm^{-1}) is noticed. The growth and the loss of these bands were clearly illustrated in Figure 2. It can be seen that the characteristic nitrile band at 2236 cm^{-1} remains unchanged and stable, except for NBR1 containing 18% ACN (aging at 100°C and 3024 hours).

The decrease of the absorption band of the carbonyl group at 1723 cm^{-1} (the carbonyl group of plasticizer), is noticed. Due to high temperatures and aging times, the plasticizer diffuses from the interior of the sample to the surface to the extent that the plasticizer content of the sample is strongly reduced.

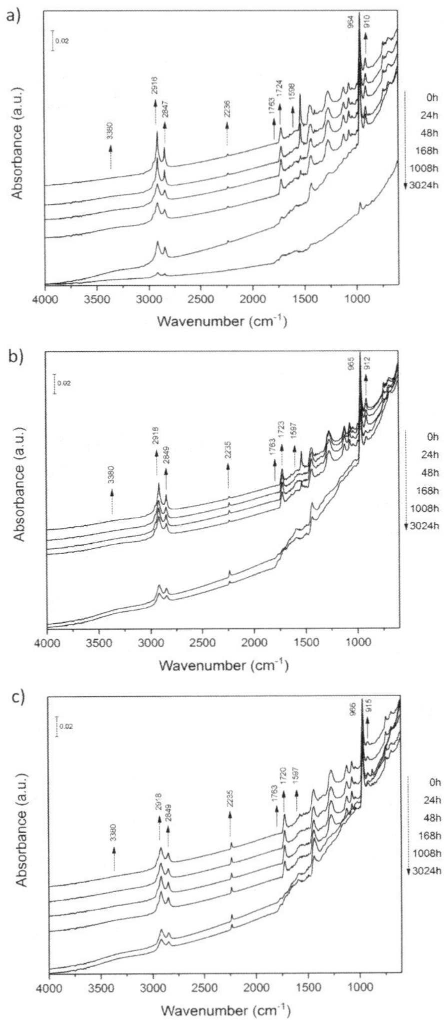

Figure 2. Series of ATR-FTIR spectra displaying vibrational changes of several functional groups during thermal aging at 100°C for NBR samples with a) 18%, b) 28%, and c) 39% ACN content.

3.3 *Relative oxidation*

Degradation- related spectral changes are observed primarily in the hydroxyl and carbonyl regions.

Since the nitrile group at 2235 cm^{-1} seems to be stable during aging, it is considered as the ideal internal reference band. The rate of the relative oxidation was calculated as the ratio of the peak areas (A) of the carbonyl bands ($A_{1500\text{–}1700\ cm^{-1}}/A_{2235\ cm^{-1}}$) and of the hydroxyl bands ($A_{3600\text{–}3000\ cm^{-1}}/A_{2235\ cm^{-1}}$) to the nitrile band.

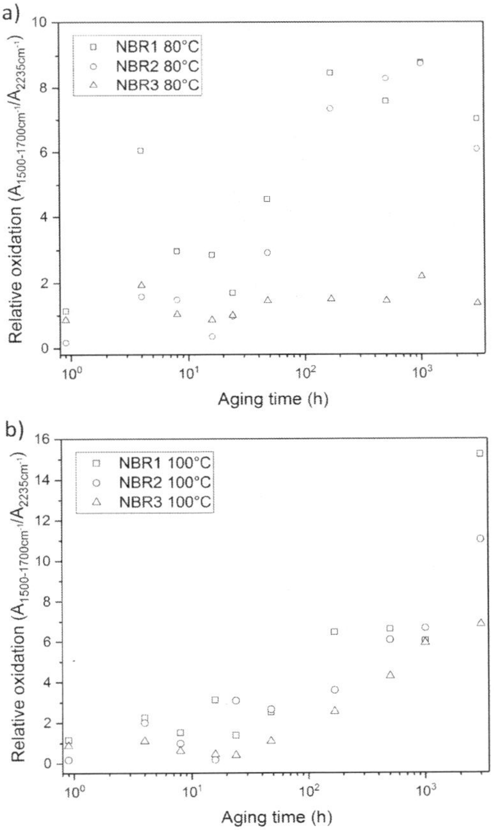

Figure 3. Growth of relative oxidation of NBR samples as ratio of peaks areas of carbonyl band to nitrile band during aging a) at 80°C, b) at 100°C in air.

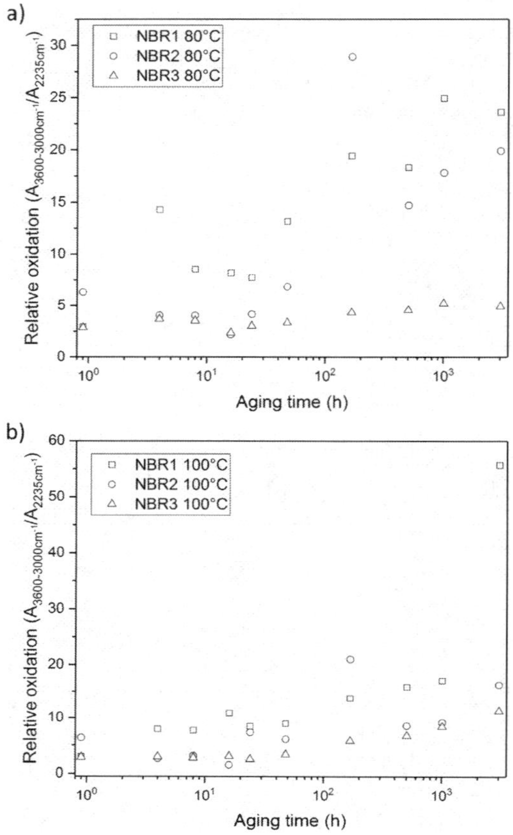

Figure 4. Growth of relative oxidation of NBR samples as ratio of peaks areas of hydroxyl band to nitrile band during aging a) at 80°C, b) at 100°C in air.

The internal normalization allows the comparison of the degree of relative oxidation with the unaged material. Figures 3 and 4 show the real values for the relative degree of oxidation of NBR samples during aging a) at 80°C, and b) at 100°C in air. It can be seen that the samples during aging in air at 100°C exhibit clear indications for oxidation of the surface. The increase of the aging time leads to a higher degree of surface oxidation. The degradation time (peak generation or loss) takes the least time for the lowest ACN level (highest unsaturation) and slows upon going toward higher ACN contents. It was shown that the carbonyl group growth correlates well with the stiffening of the rubber specimens and could be described using diffusion-limited oxidation model.

4 CONCLUSION

ATR-FTIR spectroscopy is convenient to monitor the chemical degradation process of elastomers as function of the aging time. Oxidation processes during aging effect degradation of the primary network and the build-up process of the secondary network. A similar degradation or production evolution is observed by ATR-FTIR, regardless of the ACN content in NBR, with the generation of hydroxyl- and carbonyl- based products. Loss of the butylene double bond takes place. The magnitude of the IR active group generation or loss is the most sensitive in the lowest content of ACN in NBR1. With increasing acrylonitrile content of the NBR and temperature during thermal aging, the content of relative oxidation decreases. Some peak absorption maxima are noticed to shift slowly towards higher wave numbers, such as the methylene peaks, because of the increased oxidative degradation experienced by the surface. Heat-induced thermal oxidation, which introduces additional polarity because of the formation of oxygenated functional groups, causes the vibrations to shift to higher wavenumbers.

AKNOWLEDGEMENTS

Financial support of the project by the Deutsche Forschungsgemeinschaft (DFG) under the grant number JO 818/3-1 is gratefully acknowledged.

REFERENCES

Adam, C., Lacoste, J., Lemaire, J. (1990). Photo-oxidation of elastomeric materials: Part 3—Photo-oxidation of acrylonitrile-butadiene copolymer. *Polymer degradation and stability* 27 (1), 85–97.

Ahlblad, G., Jacobson, K., Stenberg, B. (1996). Accelerated aging of nitrile-butadiene rubber studied by chemiluminescence. *Plastics rubber and composites processing and applications* 25, 464–468.

Celina, M., Wise, J., Ottesen, D. K., Gillen, K. T., Clough, R. L. (1997). Oxidation profiles of thermally aged nitrile rubber. *Polymer degradation and stability* 60, 493–504.

Hara, T., Ogawa, T., Osawa, S., Nakamura, S., Yoshida, S., Minegishi, K. (1998). Ozone deterioration of nitrile rubber surface. *Materiels Life* 10 (2), 93–101.

He, X., Li, T., Shi, Z., Wang, X., Xue, F., Wu, Z., Chen, Q. (2016). Thermal-oxidative aging behavior of nitrile-butadiene rubber/functional LDHs composites. *Polymer degradation and stability* 133, 219–226.

Jouan, X., Adam, C., Fromageot, D., Gardette, J., Lemaire, J. L. (1989) Microspectrophotometric determinations of product 'Profiles' in photooxidized matrices. *Polymer degradation and stability* 25 (2–4), 247–265.

Kawashima, T., Ogawa, T. (2005). Prediction of the Lifetime of Nitrile-Butadiene Rubber by FT-IR. *Analytical Sciences* 21, 1475–1478.

Smith, B. C. (2011). *Fundamentals of Fourier Transform Infrared Spectroscopy*, New York.

Constitutive models and their implementation in FEM

Constitutive Models for Rubber X – Lion & Johlitz (Eds)
© 2017 Taylor & Francis Group, London, ISBN 978-1-138-03001-5

A time-dependent hyperelastic approach for evaluation on rubber creep and stress relaxation

R.K. Luo, M. Easthope & W.J. Mortel
Department of Engineering and Technology, Trelleborg IAVS, Leicester, UK

ABSTRACT: This report presents an engineering approach to evaluate the time-dependent responses, i.e. creep and stress relaxation, for rubber anti-vibration components. A time-dependent damage function was introduced into hyperelastic models. This function can be expressed in three forms. A typical rubber product and a dumbbell specimen were selected to validate the proposed approach. It has been demonstrated that the predictions obtained from this method are consistent with the experimental data. It has also been established that the time-dependent response of industrial products can be predicted based on the responses from simple specimens, e.g., dumbbell specimen. In addition, it is possible to obtain a creep response based on a relaxation response and vice versa using the proposed approach, which has also been observed experimentally in the literature. The proposed function can also be easily incorporated into commercial finite element software (e.g., Abaqus). It has been shown that the proposed method may be used at an appropriate design stage.

1 INTRODUCTION

Rubber material is an excellent option for anti-vibration components in industry with a long term service. However, its time dependent behaviour is an undesirable issue in engineering applications. When a constant load is applied to a rubber spring, the deflection increases with time; this is known as creep. On the other hand, when a constant deformation is held on a rubber spring, the stresses set up gradually decrease with time. This phenomenon is known as stress relaxation. The time-dependent effect is one of the critical factors when considering engineering design and applications on anti-vibration systems. For example, an important requirement of rail vehicle design is to control rubber anti-vibration components not to exceed its structural limitation due to the creep effect and to avoid early failure over its service life required. Creep of rubbers easily exceeds structural limitations, and fracture due to creep often occurs due to long-term loading. Because long-term creep tests are very expensive and time-consuming, many accelerated methods for creep tests on polymers have been developed to predict long-term creep behaviour based on the short-term experiments. Based on the fact that higher stresses increase the creep or relaxation rate of viscoelastic materials, which is similar to the effect of higher temperatures, several time–temperature–stress superposition principles (TTSSP) have been proposed using Boltzmann's superposition principle and its modified forms; these principles have predicted creep for longer periods based on short-period tests. From the experimental point of view, the long term behaviour of rubber components can be obtained from the short-term tests. From the modelling point of view, it may also be possible to predict the long-term response of the rubber structures based on simulations on short-term reaction. The most short-term reactions of rubber springs are under mechanical loading conditions. The most simulation models used in industry are hyperelastic models based on strain energy density. Hence, investigations have been carried out for simulations on time-dependent responses of rubber anti-vibration systems using the hyperelastic approach with an added time function. The objective of this method is to predict creep/stress relaxation of rubber anti-vibration systems using available/easily modified models in industry so that an efficient approach can be used in an engineering design stage.

2 CONSTITUTIVE MODELS

The short-term behaviour of rubber material can be described using hyperelastic constitutive laws. There are several hyperelastic material models commonly used to describe rubber and other elastomeric materials based on strain energy potential or strain energy density. These hyperelastic models for rubber materials can be expressed in a general form:

$$W = W_I(\bar{I}) + W_J(J) \quad (1)$$

where $W_I(\bar{I})$ is the deviatoric part of the strain energy density of the primary material response and $W_J(J)$ is the volumetric part of the strain energy density. $\bar{I}$ can be further expanded into $\bar{I}_1$ and $\bar{I}_2$, which are the first and second deviatoric strain invariants, which are defined as:

$$\bar{I}_1 = \bar{\lambda}_1^2 + \bar{\lambda}_2^2 + \bar{\lambda}_3^2 \quad (2)$$

$$\bar{I}_2 = \bar{\lambda}_1^{-2} + \bar{\lambda}_2^{-2} + \bar{\lambda}_3^{-2} \quad (3)$$

where $\bar{\lambda}_1$ are the deviatoric stretches and J is the elastic volume ratio.

The corresponding stresses may be written as:

$$\sigma_{ij} = \frac{2}{J}\left[\frac{1}{J^{\frac{2}{3}}} + \left(\frac{\partial W}{\partial \bar{I}_1} + \bar{I}_1 \frac{\partial W}{\partial \bar{I}_2}\right) B_{ij} - \left(\bar{I}_1 \frac{\partial W}{\partial \bar{I}_2} + 2\bar{I}_2 \frac{\partial W}{\partial \bar{I}_2}\right)\frac{\delta_{ij}}{3} - \frac{1}{J^{\frac{4}{3}}} \frac{\partial W}{\partial \bar{I}_2} B_{ik} B_{kj}\right] + \frac{\partial W}{\partial J}\delta_{ij} \quad (4)$$

where B_{ij} is the component of the left Cauchy-Green deformation tensor $\boldsymbol{B}$, and δ_{ij} is the Kronecker delta.

A number of hyperelastic models have been derived from Equation (1). One of them is polynomial series which has been widely used in industry to predict rubber response under external loading. Equation (1) can be expressed by the following polynomial series and be used for static simulation:

$$W = W_I(\bar{I}) + W_J(J) = \sum\nolimits_{i+j=1}^{N} C_{ij}(\bar{I}_1 - 3)^i (\bar{I}_2 - 3)^j + = \sum\nolimits_{i=1}^{N} \frac{1}{D_i}(J-1)^{2i} \quad (5)$$

where C_{ij} and D_i are material constants for a given rubber compound.

Function (1) is a classic model limited to mechanical loadings without reference to time history. To determine the creep behaviour of the rubber materials under constant loading or to determine the stress relaxation under constant deformation, an additional function R_c is introduced into classic strain energy function (1). This concept is based on "damage models" used for Mullins effect simulations. This function R_c should relate the creep/stress relaxation behaviour with both the material "damage" under loading and the time elapsed. Hence, two dependent variables are required. The principle is to link the rubber creep/stress relaxation to the material loading condition and to the elapsed loading time. It is assumed that the creep/stress relaxation effect from loading on the material is related to rubber deformation, i.e., invariant $\bar{I}_1$ and $\bar{I}_2$, and the elapsed loading time T. Therefore, the extended hyperelastic model is not only related to the loading condition of the polymer material through the invariant $\bar{I}_1, \bar{I}_2$ but also to the elapsed time T:

$$W = W_I(\bar{I}) + W_J(J) + R_c(\bar{I}, T) \quad (6)$$

$$R_c(\bar{I}, T) = kT^{\zeta}(\bar{I}_1 + \bar{I}_2) \quad (7)$$

where T is the time elapsed from an end of mechanical loading, i.e., $T = 0$ when an initial loading ends, and k and ζ are parameters to be defined.

From a mathematical perspective, the proposed function (7) is continuous. From a scientific perspective, the above parameters have physically meaningful concepts. k is an amplitude parameter that describes the strength of the creep damage, ζ can be considered to be a hardening parameter that describes how quickly creep damage develops with the elapsed time T, $\bar{I}_1$ and $\bar{I}_2$ are the deformation state of rubber material. Thus, we can define k as an amplitude indicator for creep/stress relaxation damage and define ζ as a hardening index for creep/stress relaxation damage.

We have extended the capability of the hyperelastic models to include the time effect through a damage function $R_c(\bar{I}_1, \bar{I}_2, T)$. Based on engineering principles, there is a possibility to assume the function R_c related to deformation through either $\bar{I}_1$ or $\bar{I}_2$ only in addition to the time variable. Hence, two alternative functions are proposed which are similar to equation (6):

For invariant $\bar{I}_1$

$$W = W_I(\bar{I}) + W_J(J) + R_{c1}(\bar{I}_1, T) \quad (8)$$

where

$$R_{c1}(\bar{I}_1, T) = k_1 T^{\zeta_1} \bar{I}_1 \quad (9)$$

For invariant $\bar{I}_2$

$$W = W_I(\bar{I}) + W_J(J) + R_{c2}(\bar{I}_2, T) \quad (10)$$

where

$$R_{c2}(\bar{I}_2, T) = k_2 T^{\zeta_2} \bar{I}_2 \quad (11)$$

3 DUMBBELL SPECIMEN

Simple standard rubber specimens are usually utilized to obtain basic material properties. On the other hand, rubber components are complex parts and are usually loaded under compression and shear conditions in engineering applications. To cover the characteristics from both sides, two types of rubber specimens were selected. The first one was a type of simple standard dumbbell specimen (BS ISO 37:2011), shown in Figure 1 (a).

A three-dimensional finite element model was generated, as shown in Figure 1 (b). 1200 rubber elements with 8-node linear brick, hybrid with constant pressure were used. The total number of degrees of freedom was approximately 8400. For the static loading, a compressible polynomial form of the strain energy density (N = 3) was employed. For the creep and stress relaxation, we utilized the above polynomial series and the damage function (8) and (9) to perform the time-dependent simulation:

$$\begin{aligned} W = {} & C_{10}(\bar{I}_1-3)+C_{01}(\bar{I}_2-3)+C_{20}(\bar{I}_1-3)^2 \\ & +C_{11}(\bar{I}_1-3)(\bar{I}_2-3)+C_{02}(\bar{I}_2-3)^2+C_{30}(\bar{I}_1-3)^3 \\ & +C_{21}(\bar{I}_1-3)^2(\bar{I}_2-3)+C_{12}(\bar{I}_1-3)(\bar{I}_2-3)^2 \\ & +C_{03}(\bar{I}_2-3)^3+\frac{1}{D_1}(J-1)^2+\frac{1}{D_2}(J-1)^4 \\ & +\frac{1}{D_3}(J-1)^6+kT^{\zeta}(\bar{I}_1+\bar{I}_2) \end{aligned} \qquad (12)$$

For the static analysis, the load-deflection curves obtained from both the simulation and the experiment are compared in Figure 2(a). The simulated response agrees well with those observed in the experiment, which implies the suitability of the material parameters used in the model.

For the creep simulation, the material coefficients (C_{ij} and D_i), are unchanged. The two parameters of the damage function are specified as $k = -0.00395$ and $\zeta = 0.247$, which were determined from the creep experimental data. A FORTRAN code was used to transform equation (8) and (9) into a user subroutine and integrated with Abaqus. The creep comparison between the simulations and experiments is plotted in Figure 2 (b). It is demonstrated that the predicted creep curve is close to the measured data. We used the same model to

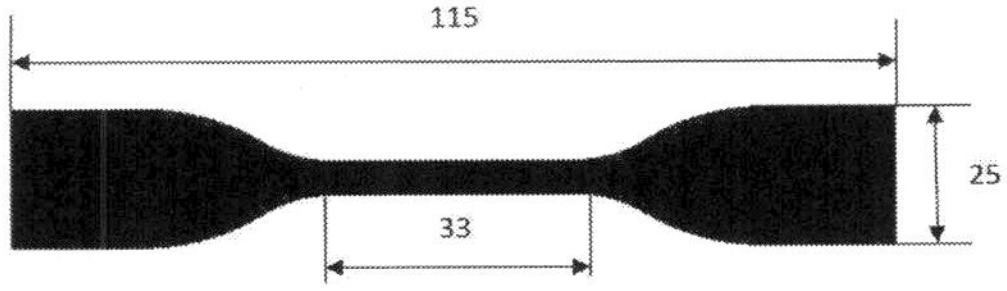

(a) Dumbbell specimen

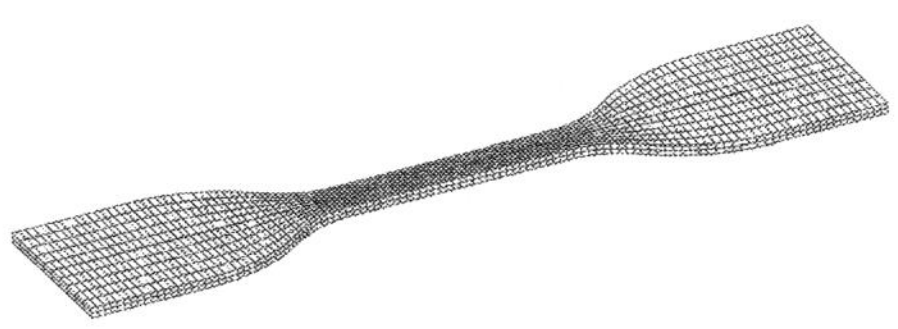

(b) Finite element model for Dumbbell specimen

Figure 1. The illustration of the dumbbell specimen.

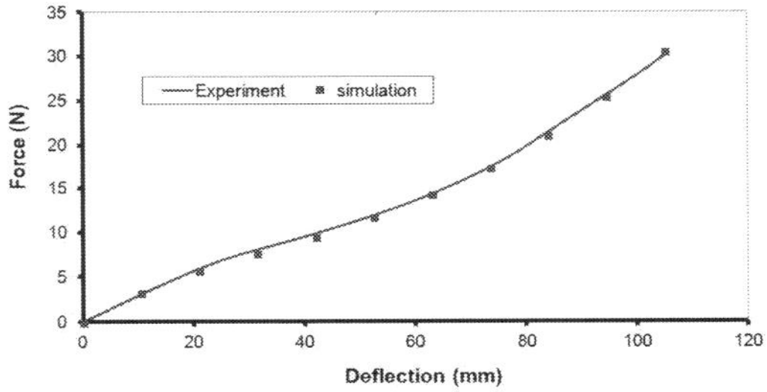

(a) Static loading comparison

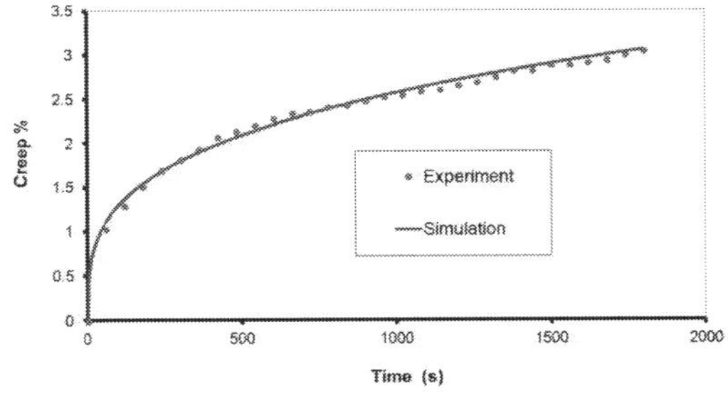

(b) Creep loading comparison

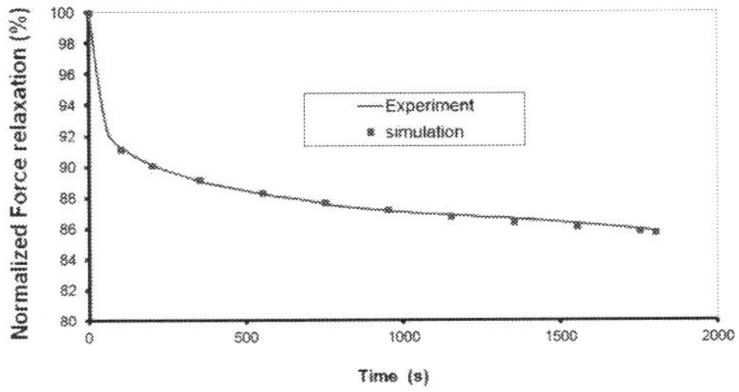

(c) Relaxation loading comparison

Figure 2. Comparisons between the simulation and experimental data for the dumbbell, see Luo et al. 2016.

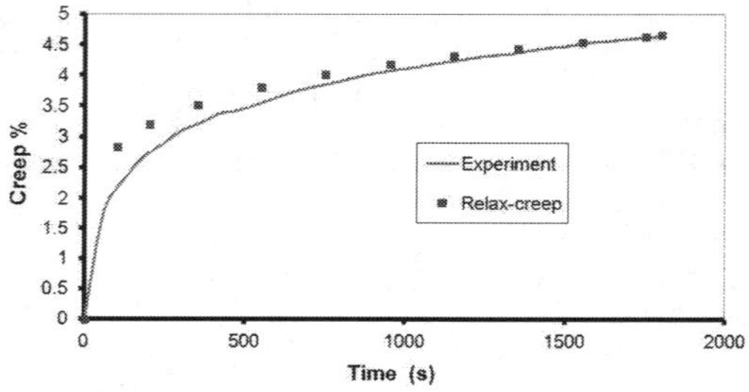

Figure 3. Creep comparison between the experimental data and the prediction using relaxation coefficient (k decreased by approximately 61%) for the dumbbell specimen, from Luo 2017.

perform the stress relaxation analysis with $k = -0.0113$ and $\zeta = 0.173$. Figure 2 (c) shows the comparison between the simulation result and the experiment data. The two results are consistent. All of the comparisons on the dumbbell specimen have indicated that the proposed approach can produce very good results for both the creep and the stress relaxation. Both creep and relaxation are time-dependent responses with shared characteristics between them. In order to observe the capability of this approach, we applied relaxation parameters to creep evaluations for the dumbbell specimen. Figure 3 illustrated the comparison between the simulated result (by decreasing approximately 60% of K) and the experimental data.

Hence, it is possible to obtain a creep response based on a relaxation response and vice versa (by changing K value only) using the proposed approach, which was also observed experimentally in the literature.

4 CIRCULAR MOUNT

This rubber spring in Figure 4, which measured maximum 260 mm in diameter and 170 mm in height, was used in vehicle mounts. Seven layers of rubber and six metal interleaves with top and bottom metal plates were used for this mount. The thickness of each rubber layer was 20 mm. The rubber compound was a filled synthetic high cis polyisoprene with shear modulus 0.8 MPa. Compression loading in the vertical direction was applied to the spring after the residual strain from Mullins effect had been removed. A force of approximately 60 kN was applied in the vertical direction. The maximum deflection was approximately 40 mm.

A creep experiment and a stress relaxation experiment were followed after the quasi-static experiment. For the creep experiment, a constant compression force (approximately 60 kN) was applied for 1800 seconds. The maximum creep strain was approximately 1.1% at the end of the test. The same procedure was also applied to the stress relaxation experiment. The maximum value of the force was approximately 4%.

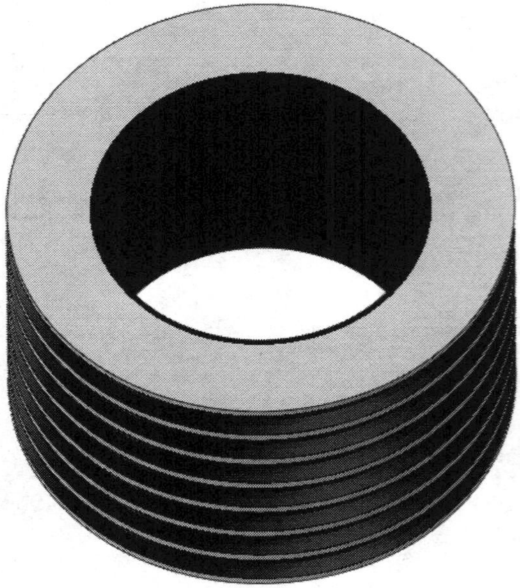

Figure 4. The seven-layer circular mount.

The finite element model of this circular mount is illustrated in Figure 5. For clarity, the entire component is displayed. 24,000 axisymmetric elements were used and the total number of degrees of freedom was approximately 67,000.

It can be further postulated that the time-dependent response of industrial rubber products can be predicted based on the responses from simple specimens, e.g. the dumbbell specimen. Figure 6 (a) and (b) illustrate the creep response and the stress response of the circular mount with the experimental data based on the calculation parameters from the dumbbell specimen. For both comparisons, the hardening index ζ was unchanged. For the creep comparison on the circular mount, the parameter k on the creep simulation from the dumbbell specimen was reduced by approximately 20%. The similar reduction (approximately 20%) was also applied to the relaxation comparison for the circular spring based on the parameter used on the relaxation calculation for the dumbbell specimen. The comparisons have indicated good agreements between the predicted results and the experimental data. These findings are very useful for engineering design and applications, which could make significant savings on both cost and time in industries.

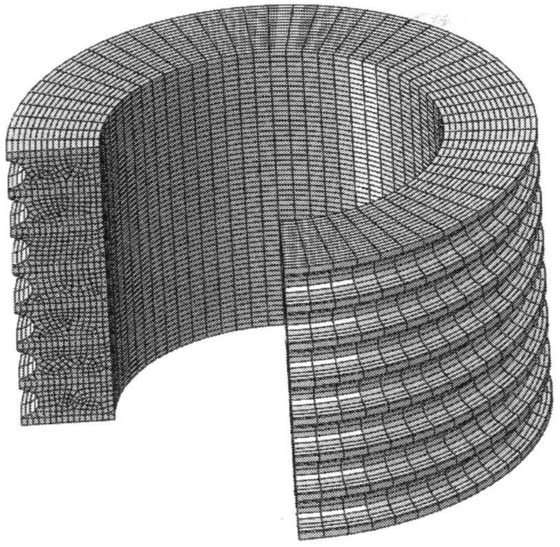

Figure 5. The finite element model of the 7-layer-circular rubber mount.

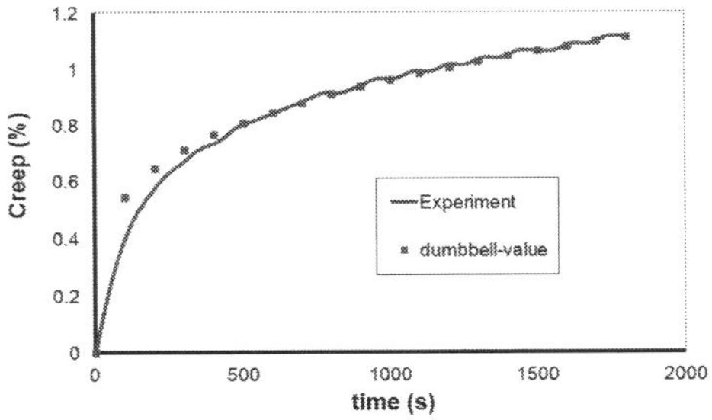

(a) Creep, dumbbell's k decreased by approximately 20%

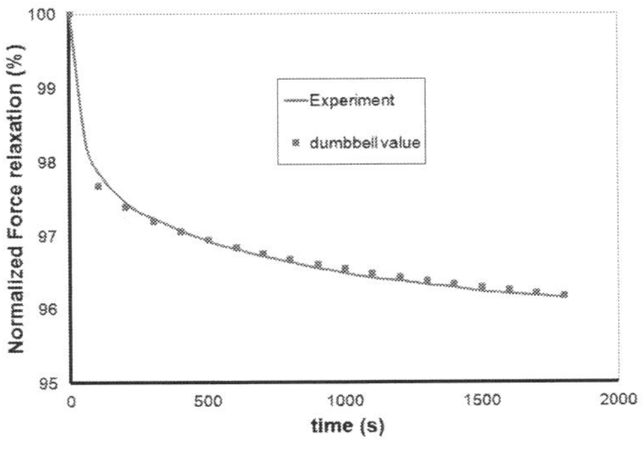

(b) Stress relaxation, dumbbell's k decreased by approximately 20%.

Figure 6. Comparison between the experimental data and the prediction using dumbbell parameter k for the 7 layers circular spring, see Luo 2017.

5 CONCLUSIONS

Time-dependent response of polymer materials is an important issue in engineering applications. In anti-vibration applications, many customers require these data, i.e. creep and stress relaxation of products, to meet the specifications. In industries, empirical methods are usually used to evaluate possible effects from creep and stress relaxation as no universal theoretical methods are available. On the other hand, traditional hyperelastic models, with no reference to time, have been used in industries over many years and approved to be a reliable method to predict mechanical loading responses. In order to meet industrial requirement on engineering applications, a time-dependent hyperelastic model has been developed and verified.

In engineering principle, both creep and stress relaxation of rubber materials are time-dependent responses. Hence, there is a relationship between these two phenomena. The similarities of these responses between the two loading cases have been observed (also established experimentally in the literature), on the two different experimental results. The two parameters, one is k for the amplitude of the response and the other is ζ for the shape of the response curves, can define the responses correctly. These definitions have been verified by these two different samples under different loading conditions. In addition, responses from the circular mount can be obtained from the parameters of the dumbbell sample by adjusting k only. It has been shown that shapes of the two response curves are similar but the amplitude of the responses is variable due to different components and different loading cases.

REFERENCES

BS ISO 37:2011. 2011. Rubber, vulcanized or thermoplastic. Determination of tensile stress-strain properties, BSI British Standards Limited.

Dassult Systems. 2016, Abaqus user manual, Dassult Systems, USA.

Du P. September/October 1989. Permanent set and creep in rubbers, Materials & Design, Vol. 10, No. 5, 264–265.

Flamm M, Grob E, Steinweger T and Weltin U. 2005. Creep prognosis for elastomeric parts through accelerated tests. In: P Austrell and L Kari (eds) Constitutive models for rubber. London, UK: Taylor & Francis, 445–447.

Gent A.N. 1962. Relaxation processes in vulcanized rubber. 1. Relation among stress relaxation, creep, recovery, and hysteresis, journal of applied polymer science, vol. VI, issue no. 22, 433–441.

Gent A.N. 1962. Relaxation Processes in Vulcanized Rubber. ii. Secondary Relaxation Due to Network Breakdown, journal of applied polymer science vol. VI, issue no. 22, 442–448.

Kolarik J. and Pegoretti J. 2006. Non-linear tensile creep of polypropylene: time-strain superposition and creep prediction. Polymer; 47: 346–356.
Kolarik J. and Pegoretti J. 2008. Proposal of the Boltzmann like superposition principle for nonlinear tensile creep of thermoplastics. Polym Test; 27: 596–606.
Lakes R.S. 1998. Viscoelastic solids. Boca Raton, FL: CRC Press.
Luo, Robert Keqi (罗克奇). 2017. Numerical prediction & case validation for rubber anti-vibration system, ISBN: 978-3-330-04435-7, Lambert Academic Publishing, Germany.
Luo Robert Keqi, Zhou Xiaolin & Tang Jinfeng. 2016. Numerical prediction and experiment on rubber creep and stress relaxation using time-dependent hyperelastic approach, Polymer Testing: 52, 246–253.
Luo R.K. 2016. Creep simulation and experiment for rubber springs, Proc IMechE Part L: J Materials: Design and Applications, Vol. 230, No. 2, 681–688.
Luo Robert Keqi. 2016. Creep modelling and unloading evaluation of the rubber suspensions of rail vehicles, Proc IMechE Part F: J Rail and Rapid Transit, Vol. 230(4), 1077–1087.
Luo Robert Keqi, Wu Xiaoping and Peng Limin. 2015. Creep loading response and complete loading-unloading investigation of industrial anti-vibration systems, Polymer Testing, Volume 46, 134–143.
Luo R.K., Gahagan P.G., Wu W.X. & Mortel W.J. 2001. Determination of rubber material properties for finite element analysis from actual products, In: D. Besdo, R.H. Schuster & J. Ihlemann (eds), Constitutive Models for Rubber II, Swets & Zeitlinger Publishers, The Netherlands, 257–260.
Luo R.K. & Mortel W. J. 2015. Rebound energy approach to evaluate rubber unloading-reloading process for industrial products with residual strain, In: Marvalova and Petrikova (eds): Constitutive Models for Rubbers IX, 473–478.
Luo Robert Keqi, Peng Li Min, Wu Xiaoping, Mortel William J. 2014. Mullins effect modelling and experiment for anti-vibration systems, Polymer Testing, 40, 304–312.
Luo R.K. 2015. Mullins damage effect on rubber products with residual strain, International Journal of Damage Mechanics, Vol. 24(2), 153–167.
Nagdi K. 1993. Rubber as an Engineering Material: Guideline for users. Hanser Publishers, Germany.
Ogden R.W. and Roxburgh D.G. 1999. An energy-based model of the Mullins effect. In: Dorfmann A., Muhr A.(eds), Constitutive Models for Rubber, Rotterdam, The Netherlands: Balkema, 23–28.
Ogden R.W. and Roxburgh D.G. 1999. A pseudo-elastic model for the Mullins effect in filled rubber, Proc. R. Soc. Lond. A 455, 2861–2878.
Schapery R.A. 1969. On the characterization of nonlinear viscoelastic materials, Polymer Engineering and Science 9(4), 295–310.

Constitutive Models for Rubber X – Lion & Johlitz (Eds)
© 2017 Taylor & Francis Group, London, ISBN 978-1-138-03001-5

Modeling the Payne effect with Marc in the frequency response of rubber

A.P. de Graaf
MSC Software, Munich, Germany

ABSTRACT: Many rubber parts used in industrial products are subject to harmonic loads superposed on static pre-loads. The accompanying deformation processes are in general highly nonlinear. The nonlinearity is not only present in the static deformation through geometric nonlinearity and hyperelasticity, but also in the frequency response through vibration amplitude dependent stiffness and damping behavior. The storage and loss modulus of a filled rubber may strongly depend on the size of the vibration amplitude, which is known as the Payne effect. A heuristic model for the Payne effect combining the thixotropic model and the Kraus model (both available in Marc) is presented and it has been implemented in Marc with the help of user subroutine UPAYNE for the analysis of a rubber bushing.

1 INTRODUCTION

Analysis of small amplitude vibrations in deformed viscoelastic elastomers are often carried out as linear perturbations around a static equilibrium state (Kim & Youn 2001, Morman & Nagtegaal 1983). The effects on stiffness and damping of the prestress and the vibration frequency are included in the formulation, but the effect of the vibration amplitude is often not included. However, filled rubbers show a pronounced dependence of the storage and loss moduli on the vibration amplitude, known as the Payne effect (Payne 1962a, b). It is generally observed that the storage modulus is high for very low amplitudes and decreases for increasing amplitudes to reach some asymptotic minimum at high amplitudes, whereas the loss modulus reaches a maximum at some intermediate amplitude, as illustrated in Figure 1.

The deformation processes accompanying the Payne effect are assumed to be reversible, so the material largely recovers when the dynamic loading ceases. In Section 2 we describe a model of viscoelasticity in the time domain including thixotropic effects. In Section 3 we apply this model in the frequency domain to small amplitude vibrations. In Section 4 we describe a heuristic model that combines this thixotropic model with the Kraus model to obtain a model with greater flexibility in simultaneously describing the frequency and amplitude dependence of the harmonic response and in Section 5 this heuristic model is used to analyze the frequency response of a rubber bushing.

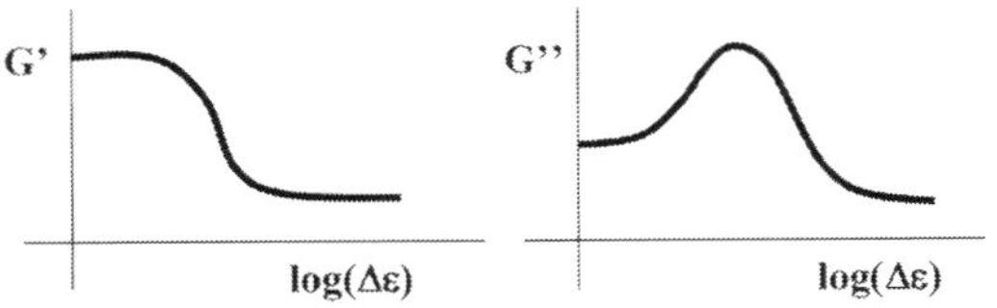

Figure 1. Storage modulus (left) and loss modulus (right) as a function of vibration amplitude.

2 A MODEL OF VISCOELASTICITY INCLUDING THIXOTROPIC EFFECTS

The model of Simo for finite strain viscoelasticity (Simo 1987) is a nonlinear extension of a linear generalized Maxwell model where a number of Maxwell elements (springs and dashpots in series) are combined in parallel. Following the ideas developed by Lion and coworkers (Höfer & Lion 2009, Rendek & Lion 2010) process dependent relaxation times can be introduced to this model to describe thixotropic effects leading to amplitude dependent behavior under harmonic loading histories. Simo introduces a Helmholtz free energy function of the form (repeated indices in tensor expressions imply summation)

$$\psi\left(E_{ij},Q_{ij}^{(k)}\right)=W^{0}\left(E_{ij}\right)-\sum_{k=1}^{N}Q_{ij}^{(k)}E_{ij}+\chi\left(Q_{ij}^{(k)}\right) \quad (1)$$

The material is characterized by the instantaneous (short term) strain energy function $W^0(E_{ij}(t))$, where $E_{ij}(t)$ is the Green-Lagrange strain at time t, from which the instantaneous 2nd Piola Kirchhoff stress is evaluated as

$$S_{ij}^{0}(t)=\frac{\partial W^{0}}{\partial E_{ij}} \tag{2}$$

The variables $Q_{ij}^{(k)}$ in Equation 1 are internal variables that describe the relaxation process and they are governed by the following evolution equations (superimposed dots denote differentiation with respect to time)

$$\dot{Q}_{ij}^{(k)}+\frac{\dot{z}_k}{\tau_k}Q_{ij}^{(k)}=g^{(k)}\frac{\dot{z}_k}{\tau_k}S_{ij}^{0}(t)\ \ (k=1,2,..,N) \tag{3}$$

$$Q_{ij}^{(k)}(0)=0$$

For each Maxwell element we have a time dependent relaxation time $\tau_k/\dot{z}_k$, where τ_k represents its initial relaxation time and $g^{(k)}$ its Prony factor. The equilibrium (long term) Prony factor $g^{(\infty)}$ is given as

$$g^{(\infty)}=1-\sum_{k=1}^{N}g^{(k)} \tag{4}$$

The relationship between the amplitude dependence and the time dependent relaxation effects is modeled by the dimensionless function $\dot{z}_k$ which after Lion is assumed to have the following form

$$\dot{z}_k(t)=1+d_k q_k(t) \tag{5}$$

Here d_k is a material time parameter and q_k is an internal variable whose evolution is driven by the deformation process. The internal variable q_k of each Maxwell element is assumed to be governed after Lion by the following evolution equation

$$\dot{q}_k=\frac{1}{\lambda_k}(\|\boldsymbol{d}\|-q_k) \tag{6}$$

$$q_k(0)=0$$

Here $\|\boldsymbol{d}\|$ is a norm of the rate of deformation tensor which provides a measure of amplitude when the problem is considered in the frequency domain

$$\|\boldsymbol{d}\|=\sqrt{\frac{2}{3}d_{ij}d_{ij}}=\|\boldsymbol{C}^{-1}\dot{\boldsymbol{E}}\| \tag{7}$$

with $\boldsymbol{C}^{-1}$ being the inverse of the right Cauchy strain tensor and $\dot{\boldsymbol{E}}$ the time derivative of the Green-Lagrange strain tensor. The constants λ_k are material parameters that are interpreted by Lion as relaxation times of the microstructure. The function $z_k(t)$ can thus be interpreted as defining an intrinsic process time of Maxwell element k. It must be a monotonically increasing function in time, which is guaranteed when the material constants λ_k and d_k are positive.

A standard argument employing the Clausius-Duhem inequality for isothermal processes (Simo 1987).

$$S_{ij}\dot{E}_{ij}-\dot{\psi}\geq 0 \tag{8}$$

finds the total stress at time t as

$$S_{ij}(t)=S_{ij}^{0}(t)-\sum_{k=1}^{N}Q_{ij}^{(k)}(t) \tag{9}$$

The solution of Equation 6 can be given as

$$q_k(t)=\frac{1}{\lambda_k}\int_0^t e^{-\frac{t-\tau}{\lambda_k}}\|\boldsymbol{d}(\tau)\|\,d\tau \tag{10}$$

which is used to integrate Equation 5 to find the intrinsic time z_k of Maxwell element k. The general stress response in terms of an arbitrary strain history $E_{ij}(t)$ can now be given as

$$S_{ij}(t)=g^{(\infty)}S_{ij}^{0}(t)+\sum_{k=1}^{N}\left[g^{(k)}S_{ij}^{0}(t)-\frac{g^{(k)}}{\tau_k}\int_0^t e^{-\frac{z_k(t)-z_k(\tau)}{\tau_k}}\dot{z}_k(\tau)S_{ij}^{0}(\tau)d\tau\right] \tag{11}$$

Each Maxwell element k is characterized by four material parameters: the Prony factor $g^{(k)}$, the initial relaxation time τ_k, the intrinsic time parameter d_k and the relaxation time of the microstructure λ_k. The equilibrium response is characterized by $g^{(\infty)}$ and the instantaneous strain energy function can be of any type capable of characterizing the static response of a rubber material.

In this section the strain energy of the rubber was represented by a general strain energy function $W^0(E_{ij}(t))$. In most cases, however, the material behavior is incompressible or near incompressible and the viscoelastic response is present in the deviatoric response only and not in the volumetric response. The theory outlined here must then be applied to the deviatoric response only to describe the deviatoric stress history, an appropriate uncoupling of the two responses must be made and the hydrostatic pressure contribution to the total stress response must be found from other considerations. (see e.g. Simo 1987, Simo & Hughes 1998). More details about the thixotropic model discussed in this section can be found in Marc 2016, Volume A.

3 SMALL AMPLITUDE VIBRATIONS AROUND STATIC EQUILIBRIUM STATES

When considering small sinusoidal variations of strain around a static equilibrium state E^*_{ij} (quantities related to this equilibrium state are superscripted with an asterisk "*") the total strain history can be written as the sum of an equilibrium part and a time dependent part

$$E_{ij}(t) = E^*_{ij} + \Delta\bar{E}_{ij}\sin\omega t \tag{12}$$

where ω is the angular frequency of the vibration. The total instantaneous stress then also becomes the sum of an equilibrium part and a time dependent part

$$S^0_{ij}(t) = S^{0*}_{ij} + \Delta S^0_{ij}(t) \tag{13}$$

For small amplitudes of the strain variations the instantaneous stress variations can be approximated as

$$\Delta S^0_{ij}(t) \approx \frac{\partial^2 W^0(E^*_{ab})}{\partial E_{ij}\partial E_{kl}}\Delta\bar{E}_{kl}\sin\omega t = \Delta\bar{S}_{ij}\sin\omega t \tag{14}$$

In a steady state the total stress response at time t can then be given as the sum of a static equilibrium stress response and a periodic stress response

$$S_{ij}(t) = g^{(\infty)}S^{*0}_{ij} + \Delta S_{ij}(t) \tag{15}$$

Equation 6 for the internal variable governing the intrinsic time scale becomes

$$\dot{q}_k + \frac{1}{\lambda_k}q_k = \frac{\omega}{\lambda_k}\|\Delta\bar{\boldsymbol{\varepsilon}}\||\cos(\omega t)| \tag{16}$$

where $\|\Delta\bar{\boldsymbol{\varepsilon}}\|$ is the norm of the true strain increment approximated from Equation 7 as

$$\|\Delta\bar{\boldsymbol{\varepsilon}}\| = \|\boldsymbol{C}^{*-1}\Delta\bar{\boldsymbol{E}}\| \tag{17}$$

which serves as the measure of vibration amplitude.

In the steady state the solution of q_k in Equation 16 is periodic and can be given by a Fourier series from which the intrinsic time z_k can be given as

$$z_k(t) = \left(1 + d_k q^{(0)}_k\right)t + \Delta_k(t) \tag{18}$$

with

$$\Delta_k(t) = \sum_{m=1}^{\infty}\left[\frac{d_k q^{(2m)}_k}{2m\omega}\sin(2m\omega t) - \frac{d_k r^{(2m)}_k}{2m\omega}\cos(2m\omega t)\right] \tag{19}$$

where only even terms appear in the Fourier series having coefficients $q^{(0)}_k$, $q^{(2m)}_k$ and $r^{(2m)}_k$ (m = 1, 2, 3, ...) found as

$$q^{(0)}_k = \frac{2}{\pi}\omega\|\Delta\bar{\boldsymbol{\varepsilon}}\|$$

$$q^{(2m)}_k = \frac{-\frac{4}{\pi}(-1)^m}{1+(2m\omega\lambda_k)^2}\omega\|\Delta\bar{\boldsymbol{\varepsilon}}\| \tag{20}$$

$$r^{(2m)}_k = \frac{-\frac{4}{\pi}(-1)^m 2m\omega\lambda_k}{1+(2m\omega\lambda_k)^2}\omega\|\Delta\bar{\boldsymbol{\varepsilon}}\|$$

The uneven Fourier terms vanish since the right hand side of Equation 16 is a periodic function with time period π/ω, which is half the time period of the excitation with angular frequency ω.

When employing partial integration and noting that any effects of nonzero initial conditions die out over time we can find the periodic stress response from Equation 11 as

$$\Delta S_{ij}(t) = g^{(\infty)}\Delta S^0_{ij}(t) + \sum_{k=1}^{N}\frac{g^{(k)}}{e^{\frac{T}{\tau^*_k}}-1}\int_0^T e^{\frac{\tau}{\tau^*_k}}e^{\frac{\Delta_k(t+\tau)-\Delta_k(t)}{\tau_k}}\frac{d\left[\Delta S^0_{ij}(t+\tau)\right]}{d\tau}d\tau \tag{21}$$

with

$$\tau^*_k = \frac{\tau_k}{1+d_k q^{(0)}_k} \tag{22}$$

Lion makes the assumption that the relaxation times of the microstructure λ_k are much larger than the time period $T = 2\pi/\omega$ of the excitation, i.e. $\lambda_k \gg T$ or $\omega\lambda_k \gg 2\pi$, which means that the $\Delta_k(t)$ in Equation 18 can be considered as a very small oscillation around a linearly increasing intrinsic time and can be ignored in Equation 21. With this assumption the periodic stress response can be approximated as

$$\Delta S_{ij}(t) \approx \left(g^{(\infty)} + g_V'(\omega, \| \Delta\bar{\varepsilon} \|)\right)\Delta\bar{S}_{ij}\sin(\omega t) + g_V''(\omega, \| \Delta\bar{\varepsilon} \|)\Delta\bar{S}_{ij}\cos(\omega t) \tag{23}$$

where $g^{(\infty)} + g_V'(\omega, \| \Delta\bar{\varepsilon} \|)$ is the storage factor of the in-phase response and $g_V''(\omega, \| \Delta\bar{\varepsilon} \|)$ is the loss factor of the out-of-phase response. For the non-constant term in the storage factor we find

$$g_V'(\omega, \| \Delta\bar{\varepsilon} \|) = \sum_{k=1}^{N} \frac{g^{(k)}\omega^2\tau_k^2}{\left(1 + d_k \frac{2}{\pi}\omega \| \Delta\bar{\boldsymbol{\varepsilon}} \|\right)^2 + \omega^2\tau_k^2} \tag{24}$$

and for the loss factor we find

$$g_V''(\omega, \| \Delta\bar{\varepsilon} \|) = \sum_{k=1}^{N} \frac{g^{(k)}\left(1 + d_k \frac{2}{\pi}\omega \| \Delta\bar{\boldsymbol{\varepsilon}} \|\right)\omega\tau_k}{\left(1 + d_k \frac{2}{\pi}\omega \| \Delta\bar{\boldsymbol{\varepsilon}} \|\right)^2 + \omega^2\tau_k^2} \tag{25}$$

We observe a frequency and amplitude dependence of the storage and loss factors. We also observe that the relaxation times of the microstructure λ_k do not appear in Equations 24, 25, they only appear if we include higher order harmonic terms. For the first order harmonic approximation we only need the material constants d_k to include the amplitude dependence for the Payne effect.

Figures 2, 3 show the simulated (sim) results of the thixotropic model with 8 Maxwell elements (N = 8) based on the experimental (exp) data reported in Höfer & Lion (2009) that was measured at three different frequencies (0.01 Hz, 1 Hz and 100 Hz). The representation of the storage data is quite satisfactory, but the representation of the loss data is poor, although the typical shape of the loss curves is captured quite well.

Tables 1, 2 give the material properties that were used for the simulated curves in Figures 2, 3. The long term shear modulus at zero amplitude

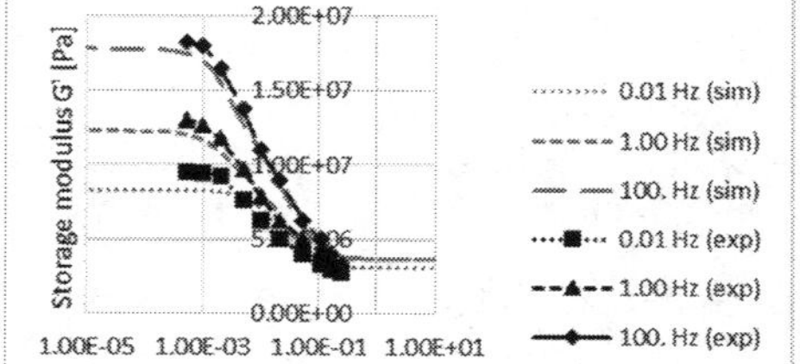

Figure 2. Simulated storage modulus of thixotropic model vs. experiment at three different frequencies.

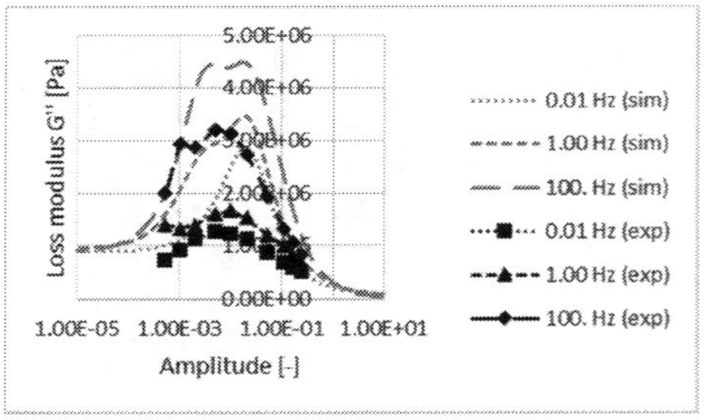

Figure 3. Simulated loss modulus of thixotropic model vs. experiment at three different frequencies.

Table 1. Shear stiffness of the material used in the simulated results (sim) of Figures 1, 2.

Long term shear stiffness $G^{(\infty)}$	9.0E+05 Pa
Short term shear stiffness $G^{(0)}$	1.794E+07 Pa
Long term Prony factor $g^{(\infty)}$	5.016E-02

Table 2. Thixotropic parameters of the material used in the simulated results (sim) of Figures 1, 2.

	τ_k	d_k	$g^{(k)}$
Term k	s	s	–
1	1.055E-03	2.924E+03	4.872E-05
2	1.441E-03	9.655E+03	1.662E-04
3	8.338E-03	3.873E+00	1.715E-01
4	1.269E-02	2.921E-01	1.412E-01
5	1.034E+00	5.601E+02	1.916E-01
6	1.427E+00	1.075E-04	3.009E-02
7	1.412E+02	1.000E+04	2.916E-01
8	1.000E+03	1.170E-03	1.238E-01

was estimated as 9×10^5 Pa, although this is not directly evident from the three measured curves (exp). There are 25 parameters to characterize the material: the short term shear stiffness and the $8 \times 3 = 24$ thixotropic parameters.

4 COMBINING THE THIXOTROPIC MODEL AND THE KRAUS MODEL

The thixotropic model of Section 3 is combined with the Kraus model (Kraus 1984) to obtain a better correlation with measured data. Based on considerations of breakage and recovery of bonds at the micro level Kraus derived the following relations for the storage and loss modulus as a function of amplitude

$$G_K'(\| \Delta\bar{\boldsymbol{\varepsilon}} \|) = G^{(\infty)'} + \frac{G^{(0)'} - G^{(\infty)'}}{1 + \left(\| \Delta\bar{\varepsilon} \| / \gamma_c\right)^{2m}} \tag{26}$$

and

$$G''_K(\|\Delta\bar{\boldsymbol{\varepsilon}}\|) = G^{(\infty)''} + \frac{2\left(G^{(m)''} - G^{(\infty)''}\right)\left(\|\Delta\bar{\boldsymbol{\varepsilon}}\|/\gamma_c\right)^m}{1+\left(\|\Delta\bar{\boldsymbol{\varepsilon}}\|/\gamma_c\right)^{2m}} \tag{27}$$

The material parameters in this model have following meaning:

$G^{(0)\prime}$ = storage modulus at very low amplitude
$G^{(\infty)\prime}$ = storage modulus at very high amplitude
$G^{(\infty)\prime\prime}$ = loss modulus at very high amplitude
$G^{(m)\prime\prime}$ = maximum loss modulus
γ_c = amplitude where loss modulus has maximum
m = model exponent

The Kraus model does not have a frequency dependence. This is introduced by combining it with the thixotropic model in a multiplicative sense writing the complex modulus of the combined model as (i being the imaginary unit, $i^2 = -1$)

$$\hat{G}(\omega,\|\Delta\bar{\varepsilon}\|) = \left[g^{(\infty)} + (g'_V + ig''_V)(g'_K + ig''_K)\right]G^0 \tag{28}$$

Here G^0 represents the instantaneous stiffness of the material at zero amplitude.

There is a non-uniqueness in Equation 28 since the two terms in the product are not unique. If one term is scaled with some factor and the other term is scaled with the inverse factor their product is unaffected. The g'_K and g''_K are the storage and loss factors of the Kraus moduli when normalizing them with the instantaneous stiffness. If we define the Prony factors in the thixotropic model in the usual way such that

$$g^{(\infty)} + \sum_{k=1}^{N} g^{(k)} = 1 \tag{29}$$

then it is apparent that the storage modulus at very low amplitude $G^{(0)\prime}$ of the Kraus model represents the instantaneous stiffness at zero amplitude. By normalizing the stiffness parameters in the model with respect to the instantaneous stiffness we obtain a unique representation.

Figures 4, 5 show the simulated (sim) results of the combined model with 6 Maxwell elements (N = 6) based on the same experimental data (exp) as was used in Section 3. The combined model fits the measured data better than the thixotropic model alone, especially the fits in Figure 5 for the loss data are much better than the ones in Figure 3.

Tables 3–5 give the material properties that were used for the simulated curves in Figures 4, 5. The long term shear modulus at zero amplitude is taken the same as in Section 3. Again there are 25 parameters to characterize the material: the short term shear stiffness, the 6 × 3 = 18 thixotropic parameters and the 6 Kraus parameters.

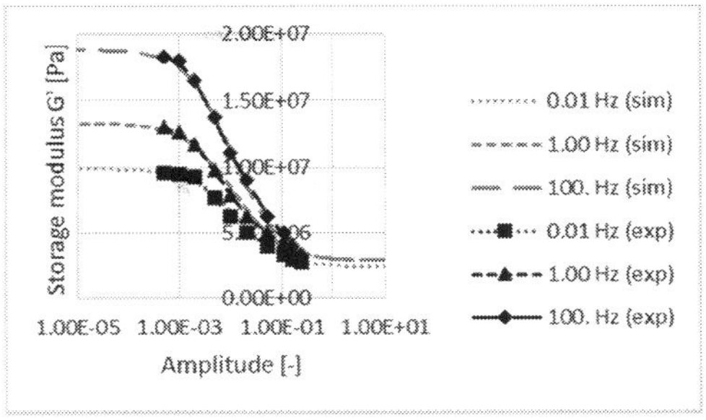

Figure 4. Simulated storage modulus of combined model vs. experiment at three different frequencies.

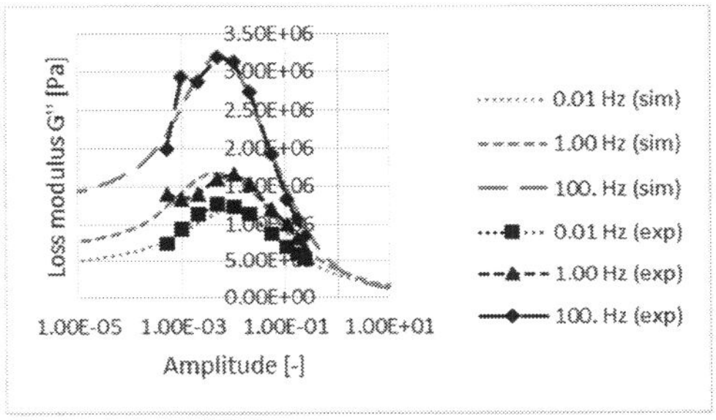

Figure 5. Simulated loss modulus of combined model vs. experiment at three different frequencies.

Table 3. Shear stiffness of the material used in the simulated results (sim) of Figures 4, 5.

Long term shear stiffness $G^{(\infty)}$	9.0E+05 Pa
Short term shear stiffness $G^{(0)}$	1.915E+07 Pa
Long term Prony factor $g^{(\infty)}$	4.7E-02

Table 4. Thixotropic parameters of the material used in the simulated results (sim) of Figures 4, 5.

	τ_k	d_k	$g^{(k)}$
Term k	s	s	–
1	6.456E-03	6.857E-02	1.970E-01
2	6.645E-03	7.584E-01	1.050E-01
3	6.024E-01	3.614E+02	3.744E-02
4	1.600E+00	1.466E-04	1.467E-01
5	1.000E+03	1.018E-04	3.624E-01
6	1.000E+03	3.130E-04	1.045E-01

Table 5. Normalized Kraus parameters of the material used in the simulated results (sim) of Figures 4, 5.

$g^{(0)\prime}$	1.000E+00	$g^{(\infty)\prime}$	1.693E-01
$g^{(\perp)\prime\prime}$	1.081E-07	$g^{(m)\prime\prime}$	1.014E-01
γ_c	1.380E-02	m	4.873E-01

5 RUBBER BUSHING EXAMPLE

As an example we analyze the frequency response of a rubber bushing. The bushing is fixed on a shaft and inside a housing which is assumed to be fixed to the environment, as schematically shown in Figure 6. The shaft carries a concentrated mass representing the primary inertia of the model. The bushing is subject to a static axial pre-load combined with axial frequency excitations between 2 Hz and 100 Hz at three different harmonic load levels, $1 \times F$, $1.5 \times F$ and $2 \times F$ (F being a nominal harmonic load level). For the instantaneous strain energy function W^0 we have taken a simple Neo-Hookean strain energy function

$$W^0 = C_{10}\left(I_1 - 3\right) \tag{30}$$

where I_1 is the first invariant of deformation defined as the trace of the right Cauchy strain tensor. The initial instantaneous shear stiffness of the material is then given as

$$G^0 = 2C_{10} \tag{31}$$

Figure 6 shows the harmonic displacement response of the pre-stressed rubber bushing using the combined thixotropic/Kraus model of Section 4 with the material data listed in Tables 3–5. We observe the nonlinearity (i.e. the Payne effect) in the frequency response, since with increasing harmonic load level the frequency with the peak response makes a left shift and this peak response is not linearly proportional to the applied load. Such phenomena cannot be analyzed by a linear frequency response that excludes the Payne effect. The shifts observed in Figure 6 should not be confused with shifts that can occur as a result of a varying static pre-stress. In a linear frequency response analysis the peak response is always at the same frequency, irrespective of the harmonic load level. The analysis was carried out with Marc implementing the combined model with the help of user subroutine UPAYNE.

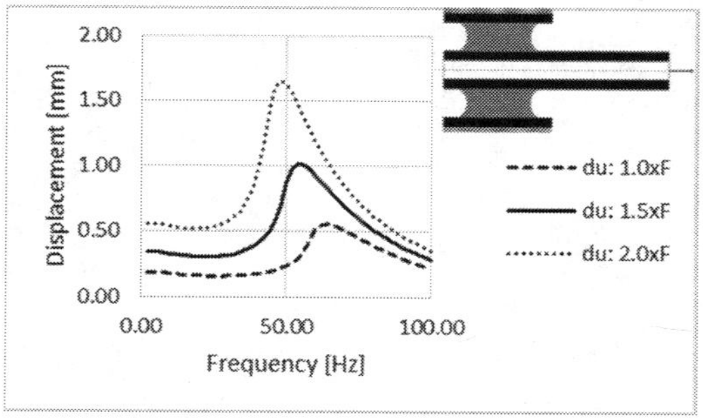

Figure 6. Frequency response of a rubber bushing at three different harmonic load levels.

6 CONCLUSIONS

In this paper material models available in Marc were presented that can be used in frequency response analyses of pre-stressed rubber components. The analyses are carried out as perturbation analyses around nonlinear static equilibrium states. The material models are capable of including effects of vibration frequency and vibration amplitude. A new combined thixotropic/Kraus model was presented, which fits the experimental data better than the thixotropic model alone. Including the vibration amplitude (Payne effect) leads to nonlinear frequency response analyses and this allows for capturing effects that cannot be captured with a purely linear frequency response approach as was demonstrated in the example.

REFERENCES

Höfer, P. & Lion, A. 2009. Modelling of frequency- and amplitude-dependent material properties of filler-reinforced rubber. *Journal of the Mechanics and Physics of Solids* 57: 500–520.

Kim, B.K. & Youn, S.K. 2001. A viscoelastic constitutive model of rubber under small oscillatory load superimposed on large static deformation. *Archive of Applied Mechanics* 71(11): 748–763.

Kraus, G. 1984. Mechanical Losses in Carbon-Black-Filled Rubbers. *Journal of Applied Polymer Science: Applied Polymer Symposium* 39: 75–92.

Marc 2016. *Volume A: Theory and User Information*. MSC Software.

Morman, K.N. (Jr.) & Nagtegaal, J.C. 1983. Finite element analysis of sinusoidal small-amplitude vibrations in deformed viscoelastic solids. Part I: Theoretical Development. *International Journal for Numerical Methods in Engineering* 19(7): 1079–1103.

Payne, A.R. 1962a. The Dynamic Properties of Carbon Black Loaded Natural Rubber Vulcanizates. Part I. *Journal of Applied Polymer Science* 6(19): 57–63.

Payne, A.R. 1962b. The Dynamic Properties of Carbon Black Loaded Natural Rubber Vulcanizates. Part II. *Journal of Applied Polymer Science* 6(21): 368–372.

Rendek, M. & Lion, A. 2010]. Amplitude dependence of filler-reinforced rubber: Experiments, constitutive modelling and FEM-Implementation. *International Journal of Solids and Structures* 47(21): 2918–2936.

Simo, J.C. 1987. On a fully three-dimensional finite-strain viscoelastic damage model: formulation and computational aspects. *Computer Methods in Applied Mechanics and Engineering* 60(2): 153–173.

Simo, J.C. & Hughes, T.J.R. 1998. *Computational Inelasticity*. Springer, New York.

Constitutive Models for Rubber X – Lion & Johlitz (Eds)

Eversion of tubes: Comparison of material models

H. Baaser
Mechanical Engineering, University of Applied Sciences, Bingen, Germany

B. Nedjar
University of Paris–Est IFSTTAR/MAST/EMMS, Marne–la–Vallée, France

R.J. Martin & P. Neff
Faculty of Mathematics, University of Duisburg-Essen, Essen, Germany

ABSTRACT: The basic open problem of nonlinear elasticity theory is to find a meaningful formulation for the basic elasticity describing the material behavior of elastomer components in order to obtain reasonable parameter calibrations in the range of large deformations. The exponentiated HENCKY energy in (Neff, Ghiba, & Lankeit 2015) based on the logarithmic strain tensor $\epsilon_{\log} = \ln \lambda$ combines several advantageous features for the description of rubber–like behaviour: an isochoric-volumetric split, invertible CAUCHY stress–stretch relation, extremely large ellipticity domain, strain hardening response and only four material parameters in the compressible setting with a clear physical meaning. We apply this formulation to the problem by eversion of cylindrical tubes. Finite element simulations show the possibility to capture the experimentally observed deformation response, see (Gent & Rivlin 1952) and (Liang, Tao, & Cai 2015). A comparison with neo–HOOKE– and MOONEY–RIVLIN–formulations is given.

1 INTRODUCTION

Already (Gent & Rivlin 1952) investigated rubber tubes which are turned inside out experimentally and compared the findings with theoretical results in the large strain regime for incompressibility and isotropy. They obtained quite good agreements for $\partial W / \partial I_1$ and $\partial W / \partial I_2$ dependencies based on a stored–energy function $W = c_1(I_1 - 3) + c_2(I_2 - 3)$ of RIVLIN–SAUNDERS type of the two deformation invariants I_1 and I_2.

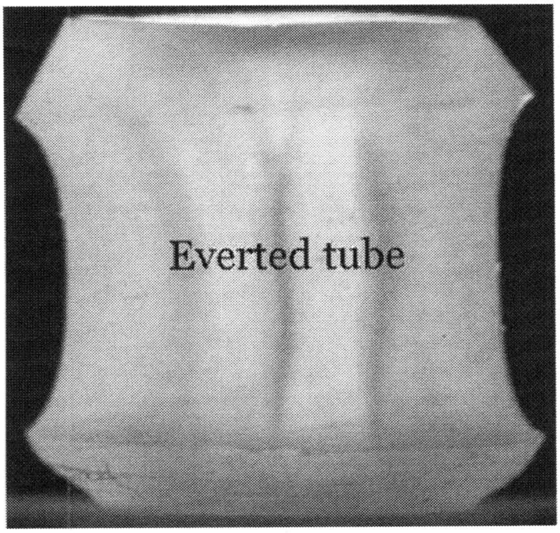

Figure 1. Photo of everted tube from literature.

Recently, one can find equivalent experimental settings in (Liang, Tao, & Cai 2015), where additionally some nice photos from rubber tubes turned inside out are given, see Figure 1.

We use such examples of everting tubes to demonstrate the capabilities of different hyperelastic material models implemented in the Finite Elemente framework ABAQUS.

2 THEORY

The *exponentiated HENCKY–logarithmic model* was recently introduced by (Neff, Ghiba, & Lankeit 2015). It is induced by the *exponentiated* HENCKY *strain energy*

$$\widehat{W}_{eH}(\boldsymbol{F}) = \frac{\mu}{k} \exp\left[k \,\|\operatorname{dev}\log \boldsymbol{V}\|^2\right] + \frac{\kappa}{2\hat{k}} \exp\left[\hat{k}\,\|(\ln\det \boldsymbol{V})^2\right], \tag{1}$$

where $\mu > 0$ is the (infinitesimal) *shear modulus*, $\kappa > 0$ is the (infinitesimal) *bulk modulus*, k and $\hat{k}$ are additional dimensionless parameters determining material hardening, $\boldsymbol{V} = \sqrt{\boldsymbol{F} \cdot \boldsymbol{F}^{\mathrm{T}}}$ is the *left stretch tensor* corresponding to the *deformation gradient* $\boldsymbol{F}$, $\log \boldsymbol{X}$ denotes the *principal matrix logarithm* on the set of positive definite symmetric matrices $\boldsymbol{X}$, $\operatorname{dev}\boldsymbol{X} = \boldsymbol{X} - \frac{\operatorname{tr}\boldsymbol{X}}{3}\boldsymbol{I}$ and $\|\boldsymbol{X}\| = \sqrt{\operatorname{tr}\boldsymbol{X}^T\boldsymbol{X}}$ are the *deviatoric part* and the *Frobenius matrix norm* of an 3 × 3-matrix $\boldsymbol{X}$, respectively, and tr denotes the *trace operator*. The exponentiated HENCKY energy is based on the so–called *volumetric* and *isochoric logarithmic strain measures*, see (Neff, Eidel, & Martin 2015), $\omega_{\text{iso}} = \|\operatorname{dev}\log\boldsymbol{V}\|$ and $\omega_{\text{vol}} = |\operatorname{tr}\log\boldsymbol{V}| = |\ln\det\boldsymbol{V}|$.

Since the formulation is hyperelastic, see (Holzapfel 2000), the first PIOLA–KIRCHHOFF stress tensor and the KIRCHHOFF stress tensor resulting from (1) are

$$\mathbf{P} = \frac{\partial \widehat{W}_{eH}}{\partial \mathbf{F}}; \quad \tau = 2\frac{\partial \widehat{W}_{eH}}{\partial \mathbf{V}} \cdot \mathbf{V}. \tag{2}$$

Likewise, the second PIOLA–KIRCHHOFF ($\boldsymbol{S}$) is given by the well known *pull back* and *push forward* operations, respectively, based on (2).

If we write the stress tensors

$$\boldsymbol{S} = \sum_{k=1}^{3} S^k\, \boldsymbol{N}^{(k)} \otimes \boldsymbol{N}^{(k)}; \boldsymbol{\tau} = \sum_{k=1}^{3} \tau_k\, \boldsymbol{n}^{(k)} \otimes \boldsymbol{n}^{(k)}, \tag{3}$$

in principal axis, one obtains the respective principal stresses

$$S^k = \frac{1}{\lambda_k} P^k \quad \text{and} \quad \tau_k = \lambda_k\, P^k \tag{4}$$

with $k = 1,2,3$ (no summation over k) scaled by the actual stretch λ_k.

Now using the push–forward procedure, the spatial tangent modulus is given by

$$\begin{aligned} \tilde{\mathrm{C}} = &\sum_{k=1}^{3}\sum_{l=1}^{3}\left[\frac{\partial^2 \widehat{W}_{\mathrm{eH}}}{\partial(\log\lambda_k)\partial(\log\lambda_l)} - 2\delta_{kl}\,\tau_l\right] \\ &\boldsymbol{n}^{(k)} \otimes \boldsymbol{n}^{(k)} \otimes \boldsymbol{n}^{(l)} \otimes \boldsymbol{n}^{(l)} \\ &+ \sum_{k=1}^{3}\sum_{l=1, l\neq k}^{3} \tau_k\lambda_l^2 - \tau_l\lambda_k^2\lambda_k^2 - \lambda_l^2 \\ &\boldsymbol{n}^{(k)} \otimes \boldsymbol{n}^{(l)}\left\{\boldsymbol{n}^{(k)} \otimes \boldsymbol{n}^{(l)} + \boldsymbol{n}^{(l)} \otimes \boldsymbol{n}^{(k)}\right\}, \end{aligned} \tag{5}$$

where we have to treat the situation of $\lambda_k \equiv \lambda_l$ in the second term of (5) as given e.g. in (Holzapfel 2000).

3 IMPLEMENTATION

The components of the stress tensor in (3) and especially the terms in (5) containing the derivatives of $\widehat{W}_{\mathrm{eH}}$ w.r.t. the stretch are obtained consequently by automatic generation using the MATLAB Symbolic Toolbox.

Exemplary, the first component of the KIRCHHOFF stress principal axes τ_1 results from

```
tau_1 = xlam1*diff(W_expH, xlam1)
```

in MATLAB and ends up as FORTRAN77 code for the first 47 columns in a form like

```
tau(1)
& = –xlam1*((AM*exp(xk1*(log(1.0D0/
XJ**(1.&g(1.0D0/XJ**(1.0D0/3.0D0)*xlam2)**2
+ log(1&am3)**2))*(log(1.0D0/XJ**(1.0D0/3.0D
0)*xl &J**(1.0D0/3.0D0)*xlam2)*2.0D0 + log(1.0
D0/X&.0D0)*(1.0D0/3.0D0))/xlam1-(BULK*
exp(xk2*&1)
```

Analogously, for the tangent modulus mainly the derivative of the 2. PIOLA–KIRCHHOFF stress components w.r.t. the stretches are obtained. Here, we end up for the first 47 columns in

```
GAMA(1,1)
&=xlam1**3*(1.0D0/xlam1**2*((AM*exp(xk1
&.0D0)*xlam1)**2+log(1.0D0/XJ**(1.0D0/3.0
&**(1.0D0/3.0D0)*xlam3)**2))*(log(1.0D0/X
&4.0D0)+log(1.0D0/XJ**(1.0D0/3.0D0)*xlam2
&D0/3.0D0)*xlam3)*2.0D0)*(1.0D0/3.0D0)
)/x&...
```

where just the first five lines of 19 lines of the second derivative in (5) are given.

Using that *tool* and procedure, the *automatic generation* and implementation of models of this type is straightforward and less prone to error as well from a continuum mechanical point of view as from coding.

Thus, the correct implementation of stress response and quadratic convergence behaviour of the global iteration scheme is ensured. Special aspects of the ABAQUS implementation are addressed later in Sect. 5.

4 PARAMETER CALIBRATION

In order to realize a suitable parameter fit, we formulate (2) as

$$P_i^* = 2\mu\exp\left(k\{\ln^2\bar{\lambda}_1 + \ln^2\bar{\lambda}_2 + \ln^2\bar{\lambda}_3\}\right)\frac{\ln\lambda_i}{\lambda_i} \tag{6}$$

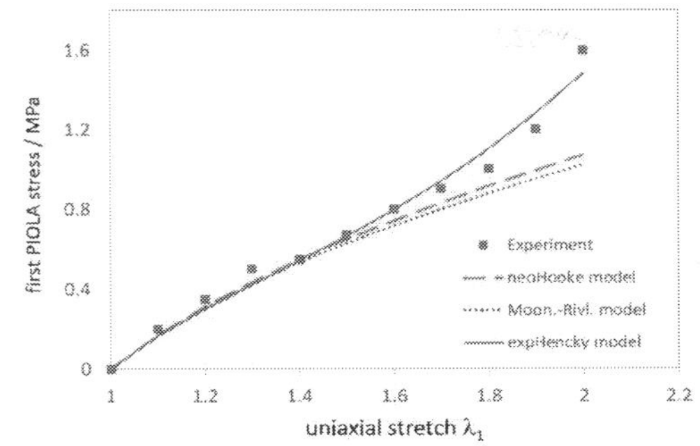

Figure 2. Uniaxial parameter fit for the exp. HENCKY model; in comparison resulting neo HOOKE model and MOONEY–RIVLIN model with $c_1 = \frac{5}{11}G$ and $c_2 = \frac{5}{110}G$, so that $c_2 = \frac{1}{10}c_1$ and $G = 2(c_1 + c_2)$.

in principal axis for (ideal) incompressibility with $\det \mathbf{F} = \lambda_1\lambda_2\lambda_3 = J \equiv 1$ and $\bar{\lambda}_i = J^{-1/3}\lambda_i = \lambda_i$ in that case; no summation over i.

With (6) the stress state is defined except for the hydrostatic pressure p, so that

$$P_i = -\frac{1}{\bar{\lambda}_i} p + P_i^* \tag{7}$$

for $i = 1,2,3$ ends up for the principal first PIOLA–KIRCHHOFF stresses.

For uniaxial test data with deformation state $\mathbf{F} = \mathrm{diag}\{\lambda_1, \frac{1}{\sqrt{\lambda_1}}, \frac{1}{\sqrt{\lambda_1}}\}$ from the uniaxial stretch λ_1 and the stress boundary conditions $P_2 = P_3 \equiv 0$ in perpendicular direction, we obtain

$$P_1 = 3\mu \exp\left(\frac{3}{2} k \ln^2 \bar{\lambda}_1\right) \frac{\ln \bar{\lambda}_1}{\bar{\lambda}_1} \tag{8}$$

after some calculations from (6).

For a silicone rubber as given in Fig. 2, we obtain by a simple least square fit the free model parameters $\mu = G = 0.612$ MPa and $k = 1.173$. In comparison, the equivalent uniaxial stress–stretch result for a neo–HOOKE model using the above calibrated $\mu = G$ is given.

5 FEM FORMULATION AND IMPLEMENTATION

The implementation within the FE system ABAQUS is realized by the umat user–subroutine in order to obtain the CAUCHY stress tensor $\sigma = \tau / \det \boldsymbol{F}$ in $(3)_2$ as given in (Holzapfel 2000).

The representation of the spatial tangent operator $\tilde{C}$ in (5) is modified as discussed in (Wriggers 2001) or more recently in (Ji, Waas, & Bazant 2013). In that case, we realize the first term in brackets in (5) in terms of

$$[\cdot] = \lambda_k^2 \lambda_l \frac{\partial S_k}{\partial \lambda_l} \tag{9}$$

as function of the second PIOLA–KIRCHHOFF stress tensor $\boldsymbol{S}$ in principal axis with $k,l = 1,2,3$.

Afterwards, the resulting modulus $\tilde{\boldsymbol{C}}$ is modified in each term $i\,j\,k\,l$ with $i,j,k,l = 1,2,3$ by

$$\left\{\tilde{C}^{\mathrm{ABAQUS}}\right\}_{ijkl} = \left\{\tilde{C}\right\}_{ijkl} + \frac{1}{2}\left(\tau_{ik}\delta_{jl} + \tau_{jk}\delta_{il} + \tau_{il}\delta_{jk} + \tau_{jl}\delta_{ik}\right) \tag{10}$$

in order to represent the JAUMANN derivatives as expected by ABAQUS environment for consistent linearization therein; here again, δ_{ab} indicates the KRONECKER symbol for $a,b = 1,2,3$. Note that the ABAQUS implementation is fully hyperelastic in this case.

6 SIMULATION AND RESULTS

We realize the numerical simulation of model experiments, see (Gent & Rivlin 1952) and recently (Liang, Tao, & Cai 2015), within the Finite Element Method as depicted in Fig. 3 exemplarily. Here, tubes with different inner radius r are everted

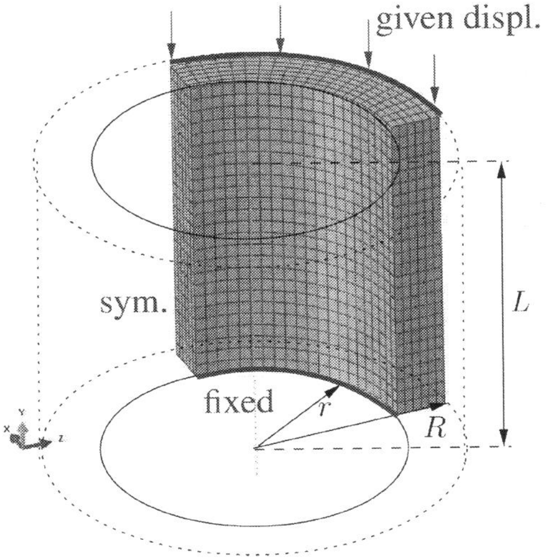

Figure 3. Model of elastomeric tube to be everted. Variation of inner radius r.

inside–out. The resulting deformation state with focus on the inner and the outer radius and on the axial length at the end of the process is observed for hyperelastic, time–independent material behaviour in the model. The eversion of the modeled tubes is realized by a given displacement of the double (axial) tube length at the outer circle signed in Fig. 3 in axial z–direction, whereas the nodes on the inner circle are fixed. Due to symmetry just a quarter of the tube is modeled—with symmetry conditions at both cutting planes at $x \equiv 0$ and $y \equiv 0$.

As result, four different deformation states at 20%, 50%, 75% and 100% of eversion are shown in Fig. 4 to Fig. 7; here, the shaded contours represent the maximal principal logarithmic strain within the bulk.

In order to compare different material models, we show in Fig. 8 the (global) reaction force everting the tube models: All three models (neo HOOKE, MOONEY–RIVLIN and exp. HENCKY) are applied with comparable infinitesimal shear modulus ($\mu = G = 0.612$ MPa) as mentioned in Sect. 4, previously. Fig. 8 shows the overall (axial) reaction force vs. the ratio of eversion of the tube models.

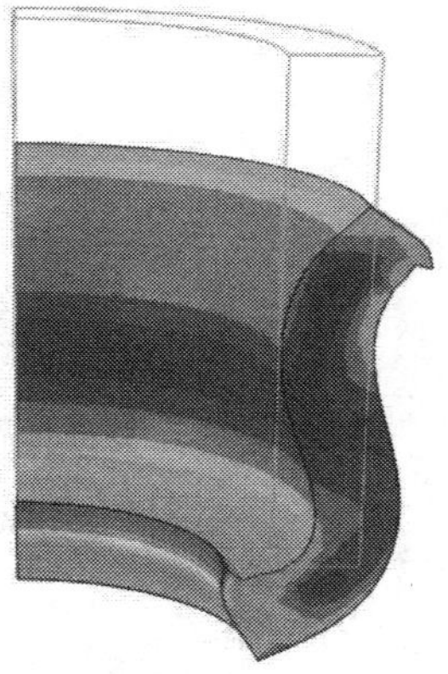

Figure 4. Eversion of tube: State @ 20% deformation.

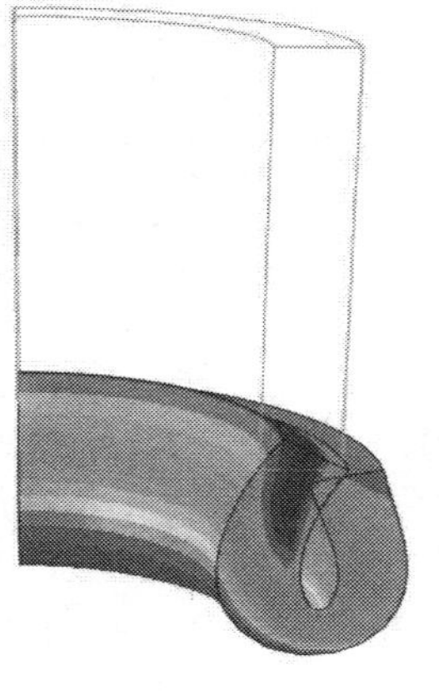

Figure 5. Eversion of tube: State @ 50% deformation.

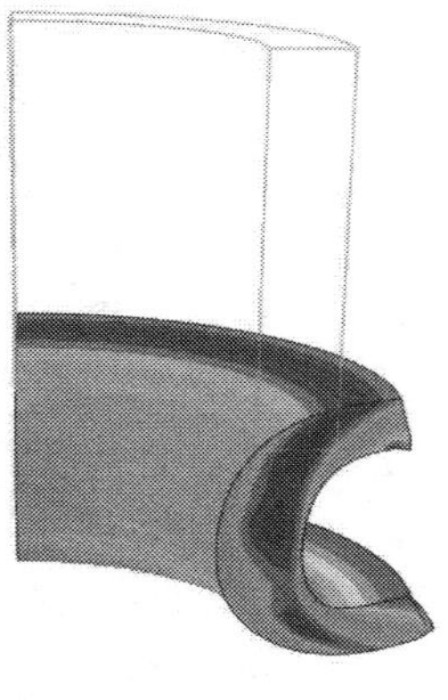

Figure 6. Eversion of tube: State @ 75% deformation.

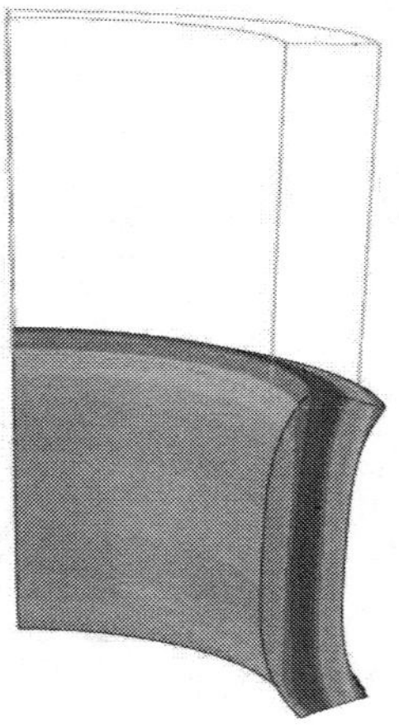

Figure 7. Eversion of tube: State @ 100% deformation.

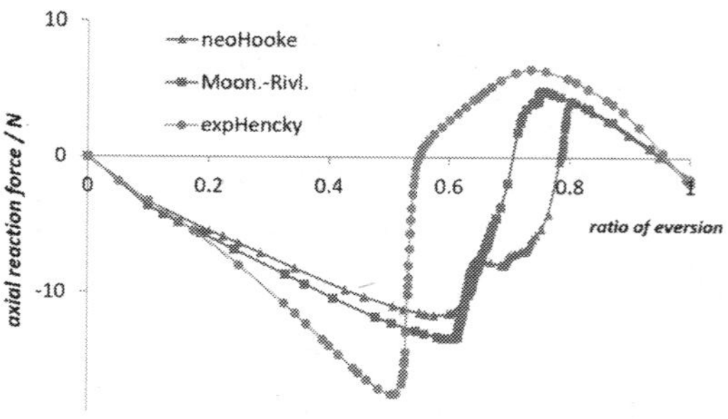

Figure 8. Global reaction force to evert the tubes. (Blue) triangles—neo HOOKE; (red) squares – MOONEY–RIVLIN; (green) dots—exp. HENCKY model.

The applied models result in a typical course of compressing the tube in axial direction to more than half of axial deformation, and then turning the sign into a tension characteristics. Here, the exp. HENCKY model shows this characteristics much earlier than the MOONEY–RIVLIN model, whereas the neo HOOKEian model seems to run through an instability point in that configuration of $r = 4.5$ mm and $R = 6.0$ mm. Further investigation have to follow by varying the inner and the outer radius using different material models with comparable material parameters.

7 SUMMARY AND CONCLUSION

The numerical simulation of model experiments everting rubber tubes inside–out is realized using different hyperelastic models. The treatment of such a deformation process is a crucial task within the FE regime itself; nevertheless, we use this (i) to prove the numerical implementation of enhanced and highly nonlinear material models, and (ii) to check for realistic results in comparison to experimental observations from the literature.

In conclusion, the realistic modeling of highly nonlinear elastic material behavior remains a challenging task. It can be seen that the uniaxial calibration ends up in reasonable results, but does not lead to comparable predictions for a three–dimensional deformation process in different but still hyperelastic material models.

REFERENCES

Gent, A. & R. Rivlin (1952). Experiments on the mechanics of rubber I: Eversion of a tube. *Proc. Phys. Soc.*.

Holzapfel, G. (2000). *Nonlinear Solid Mechanics*. Wiley.

Ji, W., A. Waas, & Z. Bazant (2013, July). On the importance of work–conjugacy and objective stress rates in finite deformation incremental finite element analysis. *J. Applied Mechanics 80*. DOI: 10.1115/1.4007828.

Liang, X., F. Tao, & S. Cai (2015). Creasing of an everted elastomer tube. *Soft Matter 12*(37). DOI: 10.1039/C6SM01381C.

Neff, P., B. Eidel, & R. Martin (2015). Geometry of logarithmic strain measures in solid mechanics. *Archive for Rational Mechanics and Analysis DOI: 10.1007/s00205-016-1007-x*.

Neff, P., I.-D. Ghiba, & J. Lankeit (2015). The exponentiated Hencky–logarithmic strain energy. Part I: Constitutive issues and rank–one convexity. *Journal of Elasticity 121*(2), 143–234.

Wriggers, P. (2001). Nichtlineare Finite–Element– Methoden. Springer.

Constitutive Models for Rubber X – Lion & Johlitz (Eds)
© 2017 Taylor & Francis Group, London, ISBN 978-1-138-03001-5

A new constitutive model for carbon-black reinforced rubber in medium dynamic strains and medium strain rates

F. Carleo
Division of Engineering Science, Soft Matter Group, School of Engineering and Materials Science, Materials Research Institute, Queen Mary University of London, London, UK

J.J.C. Busfield
Soft Matter Group, School of Engineering and Materials Science, Materials Research Institute, Queen Mary University of London, London, UK

R. Whear
Jaguar Land Rover Ltd., Gaydon, Warwick, UK

E. Barbieri
Division of Engineering Science, School of Engineering and Materials Science, Materials Research Institute, Queen Mary University of London, London, UK

ABSTRACT: Modelling the viscoelastic behaviour of rubber for use in component design remains a challenge. Previous reviews (Diani, Fayolle, & Gilormini 2009) and our studies presented in this paper highlight the issues of using of the most common viscoelastic non-linear constitutive models (Besdo & Ihlemann 2003; Bergström & Boyce 1998; Ogden & Roxburgh 1999). In detail, such models cannot reproduce or predict the experimental stress data for filled natural rubber loaded under the typical operating conditions. Examples of such conditions include cyclic strain history with constant strain rates and variable amplitude. This paper examines the behaviour of natural rubber elastomers filled with different percentages of carbon black. The elastomers chosen are typical of the materials used in vibration damping or automotive suspensions. We show that a constitutive model based on the fractional calculus can provide a good agreement for cyclic uniaxial tensile tests at a constant amplitude. The proposed model can capture, for example, the hysteresis and cyclic stress softening observed in the experimental data.

1 INTRODUCTION

The research on the viscoelastic properties of rubberlike materials has attracted much attention particularly in the automotive sector. Two different philosophies exist in formulating constitutive models: the phenomenological and the mechanistic approach. The first is based on direct observation, on the measure and curve-fitting of the experimental data, the second consists in a derivation of the strain energy function from statistical considerations on the molecular motion of polymeric chains. Notwithstanding a significant number of the non-linear viscoelastic models proposed in the literature, previous reviews (Diani, Fayolle, & Gilormini 2009) and our studies show that none of them can fit the real behaviour of filled rubber.

The existing models are each developed under specific loading conditions. For quasi-static applications, numerous and well-established hyperelastic models exist in the literature. An hyperelastic constitutive law consists of an equation relating the strain energy density to the three invariants of the strain tensor. These models can predict with few material parameters the rubber behaviour in tension, shear, compression or biaxial load. Also, these models are the basis for numerous models that try to predict the viscoelastic response. For dynamic applications, the distinction between linear and non-linear viscoelastic theories is important. Filled rubber has a highly non-linear viscoelastic behaviour. Different models for different loading condition exist that account for the high non-linearity of the filled vulcanised rubber, but no current constitutive law can model all the non-linear viscoelastic effects that the filled rubbers exhibit in the regime of interest.

The phenomenological models can roughly be classified into two large groups: the damage models and the rheological models with serial and parallel combination of elastic and viscous elements.

The damage models (Ogden & Roxburgh 1999, Dorfmann & Ogden 2003) are not able to model the different loading and reloading stress-strain path with the consequence that they fail to predict the cyclic stress softening phenomenon.

Rheological models (Bergström & Boyce 1998, Hurtado, Lapczyk, & Govindarajan 2013) with elastic and viscous component require too many calibration parameters that change with the experimental input, hence these models are not predictive.

This paper presents a new constitutive model able to represent the mechanical response of a filled rubber loaded with cyclic strain history at medium strain amplitude. Section 2 presents the experimental campaign. Cyclic uniaxial tensile tests were carried to take a close look at the nonlinear behavior of the vulcanized Natural Rubber (NR) filled with different amount of carbon black. Section 3 shows the fractional calculus method and its use in formulating the proposed constitutive model. In particular, section 3.1 reviews fractional calculus in the context of nonlinear viscoelasticity. The new model is presented in section 3.2. Sections 4 and 5 present the preliminary results, the limitations of the actual model and future works.

2 EXPERIMENTAL TESTS

Cyclic uniaxial tensile tests were carried out to examine the nonlinear behavior of the vulcanized Natural Rubber (NR) filled with different amounts of carbon black. In addition, the two separate strain history inputs adopted (Figure 1 and Figure 2) allow analysis of phenomenon such as hysteresis, cyclic stress softening, recovery, permanent set and pre-strain effect.

In the range where automotive components are expected to operate, the filled vulcanised elastomers display high nonlinearity, large hysteresis, Mullins effect and permanent set.

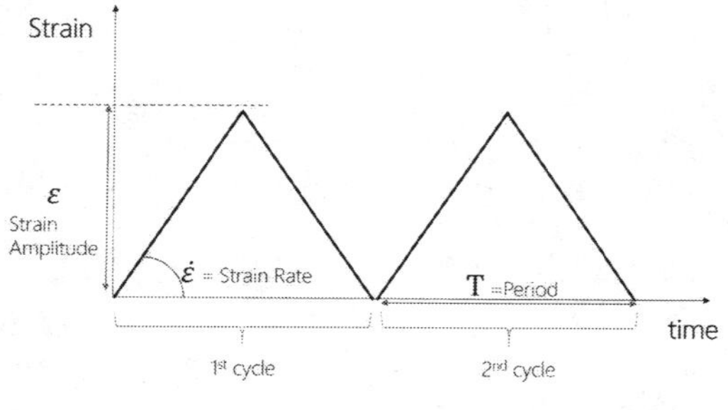

Figure 1. Schematic layout: Cyclic strain history with constant strain amplitude and strain rate.

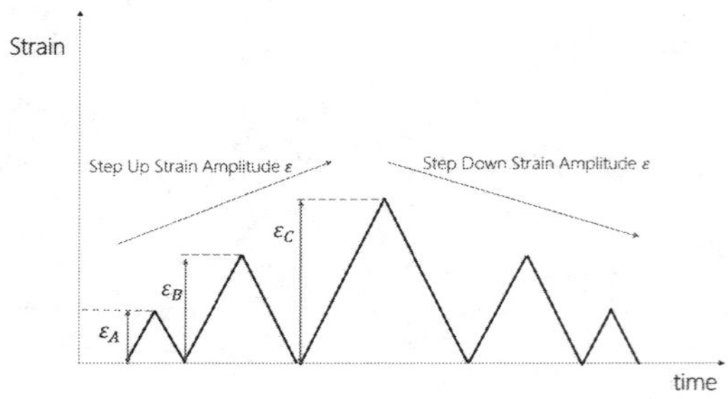

Figure 2. Schematic layout: Cyclic strain history with constant strain rate and various strain amplitude organised in step up and step down.

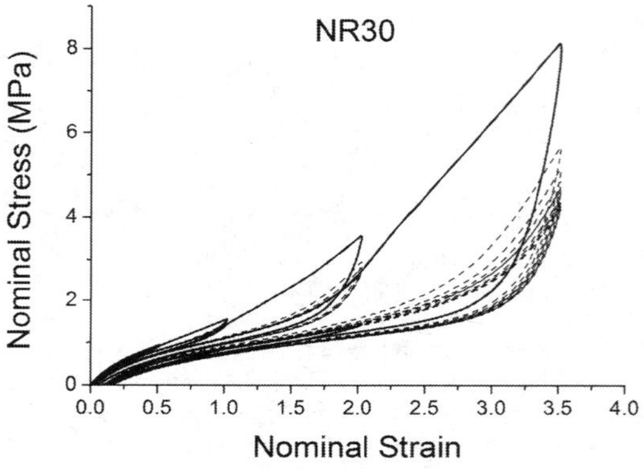

Figure 3. Stress-Strain response of Natural Rubber filled with 30 phr of carbon black (NR30) submitted to cyclic uniaxial tension with a strain history input as in Figure 2.

The hysteresis increases with the amount of carbon black in the compound.

There is a reduction in the stress on each successive loading at the same strain amplitude. The decline is largest in the first and second loading-unloading cycles and becomes rather small in the following cycles (Figure 4). The effects of stress softening are not a major issue for a compound with low content of CB.

The behaviour of rubber is rate dependent, and there is an enhancement of stress when the deformation rate is increased.

The unloading and reloading responses differ.

The material tries to return to the virgin loading path whenever the load increases beyond its previous maximum value, after a transient behaviour. The stress-strain path is highly dependent on the loading history (Figure 3).

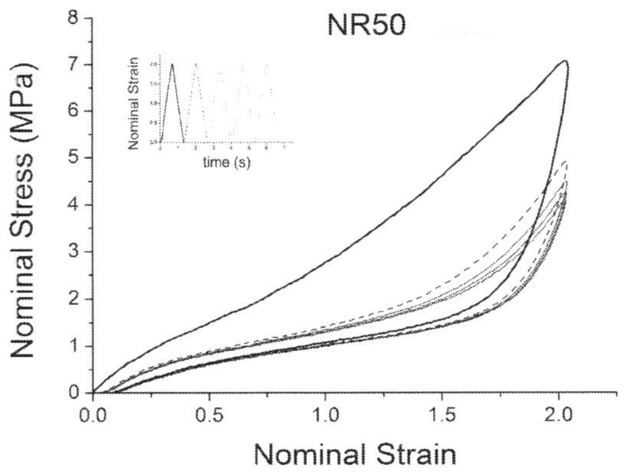

Figure 4. Stress-Strain response of Natural Rubber filled with 50 phr of carbon black (NR50) submitted to cyclic uniaxial tension with a strain history input as in Figure 1.

3 FRACTIONAL CALCULUS (FC) METHOD

3.1 *Literature review*

Fractional Calculus (FC) is a branch of mathematics that generalizes the operators of differentiation and integration from integer to fractional order. Some important properties are:

- the operator is linear;
- the zero order derivative of a function returns to the function itself;
- when the order is an integer, FC gives the same results of the ordinary operation.

In the literature, different definitions are available for the fractional derivative (FD), such as the Riemann-Liouville, Caputo or Grunwald- Letnikow (Eq. 1).

$$D_x^\alpha f(x) = \lim_{h \longrightarrow 0} \frac{1}{h^\alpha} \sum_{j=0}^{\frac{x}{h}} (-1)^j \binom{\alpha}{j} f(x - jh) \quad (1)$$

These operators are not a merely mathematical curiosities but have numerous applications in different fields. One example is the study of materials containing elastic and viscous components.

The FD method is an interpolation between elastic and viscous behaviour. The elastic material (spring) connects the stress with the strain (zero order derivative), while in the viscous material the stress is proportional to the first time derivative of the strain. In the rheological models, this component is called a spring-pot.

A justification for the use of the empirically developed FC models exists. Bagley & Torvik (1983) proved a link between molecular theories and FC approach, showing that Rouses molecular theories describing viscoelasticity produced a fractional derivative relationship between stress and strain.

In the literature, the applications of fractional derivative models are numerous in linear viscoelastic problems such as creep and relaxation (Schmidt & Gaul 2001; Pritz 2003; Sasso, Palmieri, & Amodio 2011; Zhou, Wang, Han, & Duan 2011). Linear fractional derivative models, such as Fractional Kelvin or Fractional Maxwell, are popular because they describe the behaviour of viscoelastic dampers with a small number of parameters.

Several attempts to model the nonlinearity have also been made (Sjöberg & Kari 2003; Ramrakhyani, Lesieutre, & Smith 2004), particularly in the case of cyclic loading condition with small strain amplitude, where the filled rubber shows a dependence of the viscoelastic storage modulus on the magnitude of the applied strain. This effect is known as the Fletcher-Gent effect or the Payne effect (Lion & Kardelky 2004).

The main advantage is that fractional models describe the behaviour of viscoelastic materials with a small number of parameters. The drawback is that it is difficult to estimate the parameters of the model, which are highly dependent on the choice of the first attempt solutions (Lewandowski & Chorażyczewski 2010).

3.2 *A model for non linear viscoelasticity*

The idea to use the fractional derivative to describe phenomena at medium strain amplitude derives from the ability of this operator to reproduce

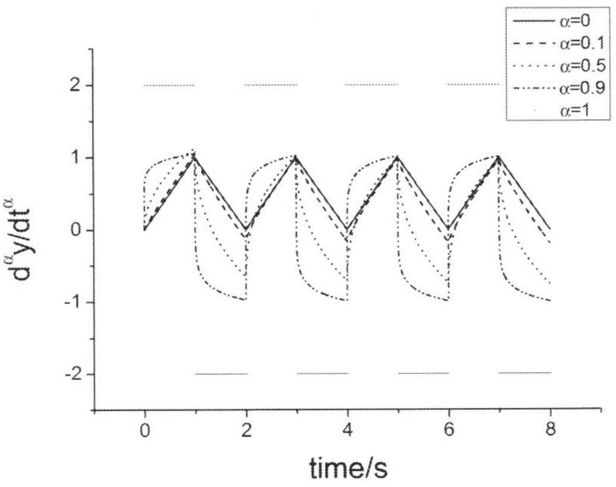

Figure 5. Fractional derivative of a triangular function.

hysteresis loops for a cyclic strain history. Figure 5 shows various orders of fractional derivatives of a triangular function.

The total stress in the present paper is obtained by adding to a viscous damaging part an hyperelastic stress (Eq. 2).

The formulation of this new model is a modification of the Nonlinear Fractional Voigt model. The basic behaviour is well represented by an hyperelastic model as the Yeoh theory (Eq. 3, for uniaxial tension test). The remaining part of the total stress is obtained multiplying the fractional derivative of the strain with a linear function expressed in equation 4 as the strain input function.

$$\sigma = \sigma_H + \sigma_{FC} \tag{2}$$

$$\sigma_H = 2\left(C_1 + 2C_2(I_1 - 3) + 3C_3(I_1 - 3)^2\right)\left(\lambda^2 - \tfrac{1}{\lambda}\right) \tag{3}$$

$$\sigma_{FC} = \varepsilon(t) \cdot D_t^{\alpha}\varepsilon(t) \tag{4}$$

This formulation has three parameters for the hyperelastic part (C_1, C_2, C_3) and one parameter (α) for the order of the fractional derivation. The model uses two different orders of derivation. One order of derivation needs to define the primary loading path. A second value of α is used to fit the first unloading curve and the following cycles. The model in total needs just 5 parameters.

4 PRELIMINARY RESULTS AND LIMITATIONS

The parameters are determined during an optimisation algorithm that minimises the mean square error. Figures 6, 7 and 8 show the comparison of experimental and calculated data.

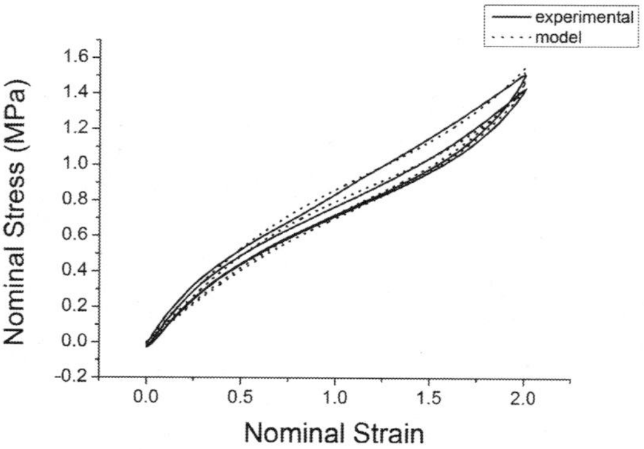

Figure 6. Best-fit curves: Comparison of calculated and experimental data for NR10.

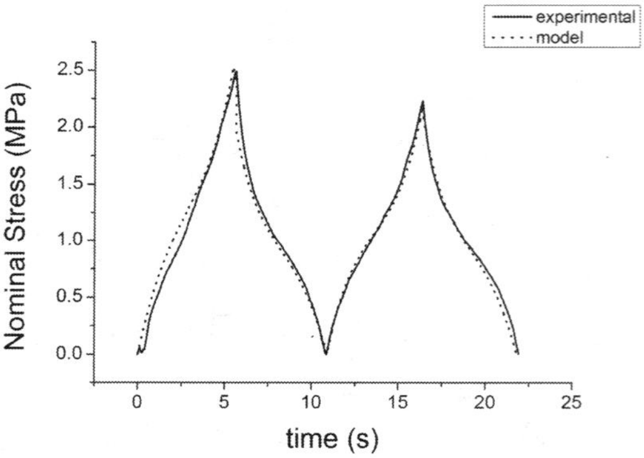

Figure 7. Best-fit curves: Comparison of calculated and experimental data for NR20.

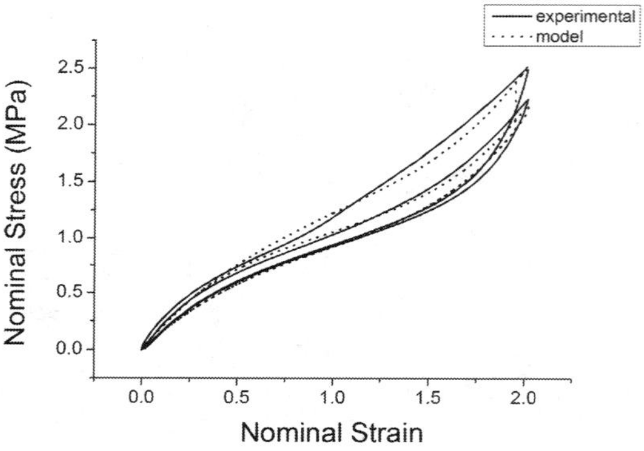

Figure 8. Best-fit curves: Comparison of calculated and experimental data for NR20.

For compounds with a small amount of carbon black, the model reproduces qualitatively the first loading and the relaxed cycle. However, the theoretical first unloading curve is unable to reproduce the high tangent modulus that the experimental data show at high strain amplitude.

5 CONCLUSIONS AND FUTURE WORKS

The model described with use of FC provides a good fitting, requires only few parameters, and the approach is physically justified (Bagley & Torvik 1983). Future improvements on the model are expected to define the best fitting in the first unloading path at high strain amplitude. Also, the model has to be able to predict the behaviour of other compounds with a higher amount of carbon black and a higher nonlinear viscoelastic behaviour. Furthermore, the model will be tested on more complex strain history.

REFERENCES

Bagley, R.L. & P. Torvik (1983). A theoretical basis for the application of fractional calculus to viscoelasticity. *Journal of Rheology 27*(3), 201–210.

Bergström, J. & M. Boyce (1998). Constitutive modeling of the large strain time-dependent behavior of elastomers. *Journal of the Mechanics and Physics of Solids 46*(5), 931–954.

Besdo, D. & J. Ihlemann (2003). A phenomenological constitutive model for rubberlike materials and its numerical applications. *International Journal of plasticity 19*(7), 1019–1036.

Diani, J., B. Fayolle, & P. Gilormini (2009). A review on the mullins effect. *European Polymer Journal 45*(3), 601–612.

Dorfmann, A. & R. Ogden (2003). A pseudo-elastic model for loading, partial unloading and reloading of particlereinforced rubber. *International Journal of Solids and Structures 40*(11), 2699–2714.

Hurtado, J., I. Lapczyk, & S. Govindarajan (2013). Parallel rheological framework to model non-linear viscoelasticity, permanent set, and mullins effect in elastomers. *Constitutive Models for Rubber VIII 95*.

Lewandowski, R. & B. Chorażyczewski (2010). Identification of the parameters of the kelvin–voigt and the maxwell fractional models, used to modeling of viscoelastic dampers. *Computers & structures 88*(1), 1–17.

Lion, A. & C. Kardelky (2004). The payne effect in finite viscoelasticity: constitutive modelling based on fractional derivatives and intrinsic time scales. *International Journal of Plasticity 20*(7), 1313–1345.

Ogden, R. & D. Roxburgh (1999). A pseudo–elastic model for the mullins effect in filled rubber. In *Proceedings of the Royal Society of London A: Mathematical, Physical and Engineering Sciences*, Volume 455, pp. 2861–2877. The Royal Society.

Pritz, T. (2003). Five-parameter fractional derivative model for polymeric damping materials. *Journal of Sound and Vibration 265*(5), 935–952.

Ramrakhyani, D.S., G.A. Lesieutre, & E.C. Smith (2004). Modeling of elastomeric materials using nonlinear fractional derivative and continuously yielding friction elements. *International journal of solids and structures 41*(14), 3929–3948.

Sasso, M., G. Palmieri, & D. Amodio (2011). Application of fractional derivative models in linear viscoelastic problems. *Mechanics of Time-Dependent Materials 15*(4), 367–387.

Schmidt, A. & L. Gaul (2001). Fe implementation of viscoelastic constitutive stress-strain relations involving fractional time derivatives. *Constitutive models for rubber 2*, 79–92.

Sjöberg, M. & L. Kari (2003). Nonlinear isolator dynamics at finite deformations: an effective hyperelastic, fractional derivative, generalized friction model. *Nonlinear Dynamics 33*(3), 323–336.

Zhou, H., C. Wang, B. Han, & Z. Duan (2011). A creep constitutive model for salt rock based on fractional derivatives. *International Journal of Rock Mechanics and Mining Sciences 48*(1), 116–121.

Constitutive Models for Rubber X – Lion & Johlitz (Eds)
© 2017 Taylor & Francis Group, London, ISBN 978-1-138-03001-5

An affine full network model for strain-induced crystallization in rubbers

A. Nateghi, M.A. Keip & C. Miehe
Institute of Applied Mechanics (Civil Engineering), Chair I, University of Stuttgart, Stuttgart, Germany

ABSTRACT: Upon stretching a natural rubber sample, polymer chains orient themselves in the direction of the applied load and form crystalline regions. If the sample is retracted, the original amorphous rubber network is restored. Due to crystallization, properties of rubber change considerably. Reinforcing effect of the crystallites stiffens the rubber and increases the crack growth resistance in it. Hence, it is of great importance to understand the mechanism leading to strain-induced crystallization and to be able to model it. However, limited theoretical work has been done on investigation of the kinetics of strain-induced crystallization. A key characteristic observed in the stress-stretch diagram of strain-crystallizing rubber is the hysteresis. This hysteresis is entirely attributed to strain-induced crystallization. In this work, we propose a micro-mechanically motivated material model for strain-induced crystallization in rubbers. To this end, we construct a model for a single crystallizing polymer chain based on *non-Gaussian chain statistics*. The modified chain model is then incorporated into the *affine full network model*. The proposed model is numerically implemented and its performance is compared with experimental data.

1 INTRODUCTION

Natural rubber crystallizes under stretch and its amorphous structure becomes semi-crystalline. Crystallization changes the mechanical response of rubber dramatically. The tensile strength, the fatigue life and the crack growth resistance of rubber increases with crystallization. In industrial applications of rubbers, predicting their lifespan is an important part of the design process. Thus, it becomes important to acquire a quantitative understanding of strain-induced crystallization. In this work we will present a multi-scale model for strain-induced crystallization in rubbers.

Micromechanical treatment of rubber elasticity was first addressed within the framework of *Gaussian statistical mechanics* of polymer chains, see Kuhn (1934) and Kuhn (1936). Extensions to higher stretches were made with the *non-Gaussian statistical mechanics* of polymer chains, see Kuhn & Grün (1942) and James & Guth (1943). Micromechanically-based network models for rubber elasticity are categorized into two groups: I-*affine models*, see for example Treloar (1954), Treloar & Riding (1979) and Wu & van der Giessen (1993) II-*non-affine models* such as Arruda & Boyce (1993) and Miehe et al. (2004). We will base our model on the *affine full network model* originally proposed by Treloar (1954).

Strain-induced crystallization in rubbers was first discovered by Katz (1925) using X-ray diffraction. Since then, this technique has been widely used to study strain-induced crystallization, see for example Gehman & Field (1939), Mitchell (1984) and Trabelsi et al. (2002). Recent progress in X-ray diffraction techniques has made it possible to evaluate the real time crystalline mass fraction during loading and unloading of the rubber specimen, see Toki et al. (2003).

Strain-induced crystallization was first theoretically treated by Flory (1947). Recently, computational modeling of strain-induced crystallization has attracted the attention of researchers. A few recent contributions can be found in Kroon (2010), Mistry & Govindjee (2014), Dargazany et al. (2014) and Guilié et al. (2015). Among the above-mentioned contributions, the latter is constructed on the affine full network model and serves as a basis for our work.

The paper is organized as follows. In Section 2 we discuss the macroscopic structure of the model. In Section 3 a modification of the non-Gaussian chain model for crystallizing rubbers is proposed. We then construct a *thermodynamically-consistent* framework for the evolution of crystallization in the chain level. In Section 4 the modified chain model is incorporated into the affine full network model. Finally, in Section 5 the performance of the model is evaluated and the results are compared with the experimental data.

2 MACROMECHANICS OF CRYSTALLIZATION

This section outlines the macroscopic constitutive equations of the finite inelasticity problem of crystallization. We begin with the introduction of the basic fields and the governing equation. Considering nearly incompressible materials, the stress response is decoupled into volumetric and isochoric contributions and the decoupled stress contributions are presented.

2.1 *Basic fields and governing equation*

Let the one-to-one map φ define the deformation of a body $\mathcal{B}$ undergoing finite strains

$$\varphi:\begin{cases}\mathcal{B}\times\mathbb{R}_+\to\mathcal{S}\\(\boldsymbol{X},t)\mapsto x=\varphi(\boldsymbol{X},t).\end{cases} \tag{1}$$

φ maps at time $t\in\mathbb{R}_+$ the material points $\boldsymbol{X}\in\mathcal{B}$ of the reference configuration $\mathcal{B}\subset\mathbb{R}^3$ to spatial points $x\in\mathcal{S}$ in the current configuration $\mathcal{S}\subset\mathbb{R}^3$. In order to exclude interpenetration of matter, the deformation gradient $\boldsymbol{F}:=\nabla\varphi(\boldsymbol{X},t)$ has to fulfill the constraint $J:=\det[F]>0$. The spatial velocity gradient is denoted by l and can be expressed by $\boldsymbol{l}=\dot{\boldsymbol{F}}\boldsymbol{F}^{-1}$. Furthermore, let $\boldsymbol{g}=\delta_{ab}\boldsymbol{e}^a\otimes\boldsymbol{e}^b$ and $\boldsymbol{G}=\delta_{AB}\boldsymbol{E}^A\otimes\boldsymbol{E}^B$ be the covariant spatial and material metric tensors, respectively. The boundary $\partial\mathcal{B}$ of the body is split into two parts, $\partial\mathcal{B}_\varphi$ and $\partial\mathcal{B}_T$, such that $\partial\mathcal{B}=\partial\mathcal{B}_\varphi\cup\partial\mathcal{B}_T$ and $\partial\mathcal{B}_\varphi\cap\partial\mathcal{B}_T=\varnothing$. The body is subjected to Dirichlet boundary condition $\varphi=\bar{\varphi}$ on $\partial\mathcal{B}_\varphi$ and Neumann boundary condition $[\boldsymbol{\tau}\boldsymbol{F}^{-T}]\cdot\boldsymbol{N}=\bar{\boldsymbol{T}}$ on $\partial\mathcal{B}_T$. Here, $\bar{\varphi}(\boldsymbol{X},t)$ is the prescribed deformation on $\partial\mathcal{B}_\varphi$, $\bar{\boldsymbol{T}}(\boldsymbol{X},t)$ is the prescribed nominal traction on $\partial\mathcal{B}_T$, $\boldsymbol{N}(\boldsymbol{X})$ is the unit outward reference normal to the boundary $\partial\mathcal{B}$ of the body and τ is the Kirchhoff stress tensor. The boundary value problem under consideration is to find the deformation field $\varphi(\boldsymbol{X},t)$ such that the quasi-static equilibrium equation is locally satisfied

$$\mathrm{Div}[\boldsymbol{\tau}\boldsymbol{F}^{-T}]+\bar{\gamma}=0\quad\text{in }\mathcal{B}, \tag{2}$$

together with the Dirichlet and Neumann boundary conditions. In equation (2), $\bar{\gamma}(\boldsymbol{X},t)$ is the prescribed body force per unit reference volume.

2.2 *Macroscopic free energy, stresses and moduli*

We choose the macroscopic free energy Ψ per unit reference volume of crystallizing rubber to be a function of the current metric g and a vector of internal variables $\mathcal{I}$ characterizing strain induced crystallization

$$\Psi=\Psi(\boldsymbol{g},\mathcal{I};\boldsymbol{F}). \tag{3}$$

Following a standard Coleman-Noll procedure, the relation between the Kirchhoff stress and the free energy is given by the so-called Doyle-Ericksen formula

$$\boldsymbol{\tau}=2\partial_g\Psi(\boldsymbol{g},\mathcal{I};\boldsymbol{F}), \tag{4}$$

together with the reduced form of the Clausius-Planck inequality $\mathcal{D}_{mac}=-\partial_{\mathcal{I}}\Psi\cdot\dot{\mathcal{I}}\geq 0$. This inequality demands that the local macroscopic dissipation be non-negative in order to have the second law of thermodynamics satisfied. We will, however, take care of thermodynamical consistency on the micro level.

2.3 *Decoupled volumetric-isochoric response*

A multiplicative decomposition of the deformation gradient into volumetric and isochoric parts is considered

$$\boldsymbol{F}=\boldsymbol{F}_{vol}\bar{\boldsymbol{F}}, \tag{5}$$

where $\boldsymbol{F}_{vol}:=J^{1/3}\boldsymbol{1}$ and $\bar{\boldsymbol{F}}:=J^{-1/3}\boldsymbol{F}$. Furthemore, the stored energy is additively split into volumetric and isochoric parts

$$\Psi=U(J)+\bar{\Psi}(\boldsymbol{g},\mathcal{I};\bar{\boldsymbol{F}}). \tag{6}$$

Following (6), the stresses in (4) additively decompose into volumetric and deviatoric contributions

$$\boldsymbol{\tau}=p\boldsymbol{g}^{-1}+\mathbb{P}:\bar{\boldsymbol{\tau}}, \tag{7}$$

where $\mathbb{P}$ is the fourth order deviatoric projection tensor. The volumetric variable p and the isochoric Kirchhoff stresses $\bar{\boldsymbol{\tau}}$ are defined by

$$p:=JU'(J)\qquad\text{and}\qquad\bar{\boldsymbol{\tau}}:=2\partial_g\bar{\Psi}(\boldsymbol{g},\mathcal{I};\bar{\boldsymbol{F}}). \tag{8}$$

3 MICROMECHANICS OF CRYSTALLIZATION

Polymers are made of long chain molecules which can assume a large number of configurations due to the rotational freedom about the chemical bonds. In statistical treatment of chain molecules, polymer molecules are commonly approximated by idealized chains composed of rigid links of equal length, the so-called *Kuhn segments*. These segments are assumed to have full rotational freedom at their ends. Therefore, they do not necessarily represent the structural unit of the polymer molecule. The distribution of the configurations of these idealized *freely-jointed chains* has been treated in the classical literature of rubber elasticity under the titles *Gaussian* and *non-Gaussian* distributions. It is well-known that Gaussian chain statistics gives unacceptable results at high extensions. In what follows, we modify the non-Gaussian chain model and construct a model that can describe the response of a single chain undergoing crystallization.

3.1 *Statistical mechanics of a single crystallizing chain*

Let us consider a semi-crystalline chain molecule, i.e., a chain composed of a crystalline portion and an amorphous portion. Let N_c denote the number of the segments present in the crystalline part of the chain and N_a the number of the segments in the amorphous part of the chain. Thus, $N_c + N_a = N$. The degree of crystallization of the chain is defined as the fraction of the segments which occur in the crystalline part

$$\xi := \frac{N_c}{N}. \tag{9}$$

N_c and N_a can then be written in terms of the total number of segments and the crystallization degree

$$N_c = \xi N \quad \textbf{and} \quad N_a = (1-\xi)N. \tag{10}$$

For simplification of our treatment we assume that the crystalline portion of the chain evolves parallel to the vector $\boldsymbol{r}$ connecting the ends of the chain. A graphical sketch of the sample chain is given in Figure 1. The total displacement length r of the chain is equal to the sum of the length of the crystalline portion r_c and the length of the amorphous portion r_a, i.e., $r = r_c + r_a$. The length of the crystalline portion of the chain is $r_c = N_c l = \xi N l$, where l is the length of Kuhn segments. The relative stretch in the amorphous part of the chain is defined as ratio of the displacement length r_a traversed by the amorphous portion of the chain to the totally extended length of the amorphous part

$$\lambda_a^r := \frac{r_a}{N_a l} = \frac{r - \xi N l}{(1-\xi)N l} = \frac{\lambda/\sqrt{N} - \xi}{1-\xi}, \tag{11}$$

where the definition of the total stretch $\lambda := r/(\sqrt{N} l)$ is used. Following the same logic as in Flory (1947) the *absolute* configurational entropy of the semi-crystalline chain is computed with respect to a hypothetical *fully-crystalline* chain. To do so, two imaginary steps are considered: 1) melting of N_a segments of the fully-crystalline chain to form the amorphous part of the semi-crystalline chain and 2) assignment of the ends of the semi-crystalline chain to the corresponding cross-linkage position in space such that the displacement length r is traversed. The total entropy change is obtained by adding the entropy changes of the steps mentioned above. The entropy change corresponding to the first step is

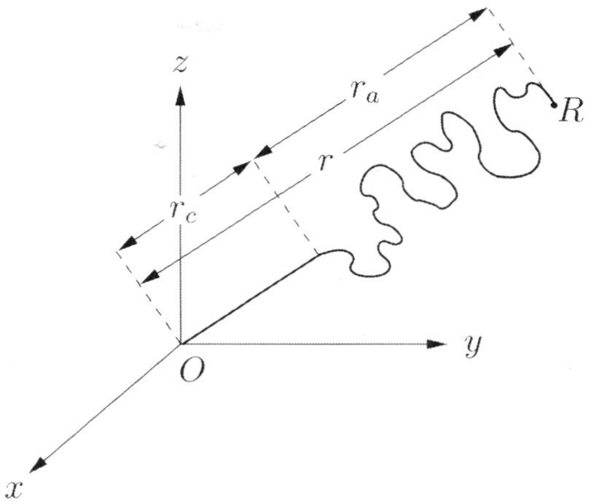

Figure 1. A prototype semi-crystalline chain has been considered. The degree of crystallization ξ is defined as the number of segments in the crystalline part of the chain divided by the total number of segments.

$$s_1 = N_a s_f. \tag{12}$$

Here, s_f is the entropy of fusion per segment. The entropy change corresponding to the second step is equal to the configurational entropy of the amorphous portion of the chain. This can be written in analogy to the classical expression for the configurational entropy of an amorphous chain

$$s_2 = -N_a k\left(\lambda_a^r \mathcal{L}^{-1}(\lambda_a^r) + \ln\frac{\mathcal{L}^{-1}(\lambda_a^r)}{\sinh\mathcal{L}^{-1}(\lambda_a^r)}\right), \tag{13}$$

where k is the Boltzmann constant and $\mathcal{L}^{-1}$ is the inverse Langevin function. Thus, the total configurational entropy of the semi-crystalline chain with respect to the fully-crystalline chain is

$$s = s_1 + s_2. \tag{14}$$

If we plug $\xi = 0$ and $\lambda = 1$ in (14) we will end up with the entropy of the chain in the initial most probable configuration

$$s_0 = s(\lambda_a^r, \xi)\,|_{\lambda_a^r = \frac{1}{\sqrt{N}}, \xi = 0}. \tag{15}$$

Thus, the entropy change corresponding to a deformation from the initial state of the chain to a stretched and crystallized state is the subtraction of (15) from (14)

$$\Delta s = s - s_0. \tag{16}$$

Crystallization in rubbers is accompanied by release of latent heat, see Treloar (1975) (p.16). Taking h_f as the heat of fusion of a segment, the heat change ΔQ due to crystallization of the sample unstrained amorphous chain is $\Delta Q = -\xi N h_f$. The stored energy in a single chain due to stretch and crystallization is then given by

$$\psi_{sc} = \Delta Q - T\Delta s. \tag{17}$$

So far we have only considered individual crystallizing chains and the interaction between a crystallizing chain and the rubber network has not been taken into account. In order to include the reinforcing effect of the crystallites in the model a *phenomenological* term is added to the free energy of a single semi-crystalline chain (17). We use the exponential function of Guilié et al. (2015) for this purpose

$$\psi_h(\lambda) = \begin{cases} 0 & \text{if} \quad \lambda < \lambda^* \\ \dfrac{h}{1+p}(\lambda - \lambda^*)^{1+p} & \text{if} \quad \lambda \geq \lambda^*, \end{cases} \tag{18}$$

where h, p and λ^* are constants. The total free energy of the chain can then be written as

$$\psi(\lambda, \xi) = \psi_{sc}(\lambda, \xi) + \psi_h(\lambda). \tag{19}$$

Applying the standard Coleman-Noll procedure, we can define the chain axial force and the crystallization driving force by

$$\mathfrak{f} := \partial_\lambda \psi(\lambda, \xi) \text{ and } \mathfrak{F} := -\partial_\xi \psi(\lambda, \xi). \tag{20}$$

The reduced microscopic dissipation has to be non-negative to ensure thermodynamical consistency

$$\mathcal{D}_{mic} := \mathfrak{F}\dot{\xi} \geq 0. \tag{21}$$

3.2 *Evolution of crystallization*

Equation (21) imposes a restriction on the constitutive equation for evolution $\dot{\xi}$ of the chain crystallization degree. In order to ensure the satisfaction of (21) we use the concept of *dissipation potential*. We define a convex dissipation potential ϕ in terms of the chain crystallization driving force $\mathfrak{F}$. Thus, ϕ satisfies the condition

$$\beta\phi(\mathfrak{F}_1) + (1-\beta)\phi(\mathfrak{F}_2) \geq \phi(\beta\mathfrak{F}_1 + (1-\beta)\mathfrak{F}_2), \tag{22}$$

for all $\beta \in [0,1]$ and for all $\mathfrak{F}_1$ and $\mathfrak{F}_2$. In addition we demand that ϕ be positive $\phi(\mathfrak{F}) \geq 0$ and normalized $\phi(0) = 0$. For differentiable dissipation potentials $\phi(\mathfrak{F})$, thermodynamical consistency is ensured by defining the evolution in the form $\dot{\xi} = \partial_{\mathfrak{F}}\phi(\mathfrak{F})$. The partial derivative $\partial_{\mathfrak{F}}\phi(\mathfrak{F})$ can be generalized to the subdifferential for non-smooth dissipation potentials. The real value g is a *subgradient* of a convex function f at point x_0 if

$$f(x) \geq f(x_0) + g(x - x_0) \qquad \forall x. \tag{23}$$

The *subdifferential* of f at x is denoted by $\partial_x f(x)$ and is the set of all subgradeints of f at x, see Rockafellar (1997) (p.214). We define the evolution $\dot{\xi}$ of the crystallization degree to be a subgradient of ϕ. Alternatively, we can say that $\dot{\xi}$ is an element of the subdifferential $\partial_{\mathfrak{F}}\phi(\mathfrak{F})$

$$\dot{\xi} \in \partial_{\mathfrak{F}}\phi(\mathfrak{F}). \tag{24}$$

It can be proved that for an evolution equation of the form (24), with the dissipation potential ϕ satisfying the conditions mentioned above, the dissipation inequality (21) is a priori satisfied. In order to complete a thermodynamically-consistent evolution framework for crystallization in the chain level, we need to specify a dissipation potential ϕ that satisfies the above-mentioned conditions. To do so, we assume that crystallization is governed by a *rate-independent evolution law with threshold*. The dissipation potential $\phi(\mathfrak{F})$. is defined by

$$\phi(\mathfrak{F}) = \begin{cases} 0 & \text{if} \quad \mathfrak{F} \in [\mathfrak{F}_0, \mathfrak{F}_t] \\ +\infty & \text{if} \quad \mathfrak{F} \notin [\mathfrak{F}_0, \mathfrak{F}_t], \end{cases} \tag{25}$$

see Figure 2 a). Here, the upper threshold $\mathfrak{F}_t$ is a constant and the lower threshold $\mathfrak{F}_0$ is the crystallization driving force in the initial state of the chain. Note that in order to satisfy the normality condition for the dissipation potential ϕ a non-negative value should be selected for $\mathfrak{F}_t$, i.e., $\mathfrak{F}_t \geq 0$. Furthermore, the material parameters should be selected such that the initial crystallization driving force $\mathfrak{F}_0$ be non-positive

$$\mathfrak{F}_0 = -\partial_\xi \psi(\lambda, \xi)|_{\lambda=1, \xi=0} \overset{!}{\leq} 0. \tag{26}$$

Using (24) and (25) the evolution equation can be written

$$\dot{\xi} \in \partial_{\mathfrak{F}}\phi(\mathfrak{F}) = \begin{cases} [0, +\infty] & \text{if} \quad \mathfrak{F} = \mathfrak{F}_t \\ 0 & \text{if} \quad \mathfrak{F} \in (\mathfrak{F}_0, \mathfrak{F}_t) \\ [-\infty, 0] & \text{if} \quad \mathfrak{F} = \mathfrak{F}_0. \end{cases} \tag{27}$$

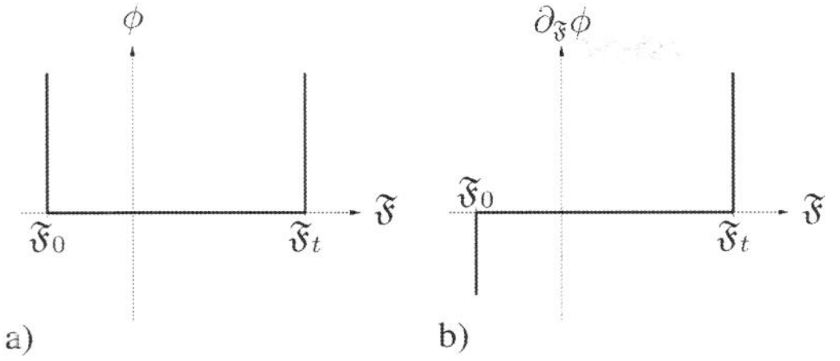

Figure 2. a) Dissipation potential function $\phi(\mathfrak{F})$. b) Subdifferential $\partial_{\mathfrak{F}}\phi(\mathfrak{F})$ of the dissipation potentail function. The evolution $\dot{\xi}$ is an element of the subdifferential, i.e., $\dot{\xi} \in \partial_{\mathfrak{F}}\phi(\mathfrak{F})$.

See Figure 2 b) for a graphical representation of the subdifferential $\partial_{\mathfrak{F}}\phi(\mathfrak{F})$. The chain crystallizes when $\mathfrak{F} = \mathfrak{F}_t$ and it melts when $\mathfrak{F} = \mathfrak{F}_0$. From (27) it can inferred that for an evolution governed by the dissipation potential (25), the driving force $\mathfrak{F}$ is bounded by a lower bound $\mathfrak{F}_0$ and an upper bound $\mathfrak{F}_t$, i.e., $\mathfrak{F}_0 \leq \mathfrak{F} \leq \mathfrak{F}_t$.

4 AFFINE FULL NETWORK MODEL FOR CRYSTALLIZATION

In this section, we aim to link the micro-variables of the chains in rubber network to mechanical macro-variable. We assume that deformation is affine, see Treloar & Riding (1979) and Wu & van der Giessen (1993), that is, we assume that the stretch in a chain with a specific direction is equal to the stretch in the bulk rubber in that direction. Microscopic free energies of polymer chains are homogenized over the three-dimensional orientation space to obtain the macroscopic free energy of the rubbery material. This is then used to derive the macroscopic stresses in terms of the micro forces.

4.1 *The affine stretch assumption*

Consider a unit Lagrangian orientation vector $\boldsymbol{r}$ and its corresponding co-vector $\boldsymbol{r}_\flat := \boldsymbol{G}\boldsymbol{r}$ obtained by mapping $\boldsymbol{r}$ with the reference metric $\boldsymbol{G}$. Thus, we have $|\boldsymbol{r}|_{\boldsymbol{G}} = \sqrt{\boldsymbol{r}_\flat \cdot \boldsymbol{r}} = 1$. The spatial isochoric orientation vector $\boldsymbol{t}$ is obtained by mapping $\boldsymbol{r}$ with the isochoric part $\bar{\boldsymbol{F}}$ of the deformation gradient defined in (5), i.e.,

$$\boldsymbol{t} = \bar{\boldsymbol{F}}\boldsymbol{r}. \tag{28}$$

The length of vector $\boldsymbol{t}$ is the stretch of a line element in the direction of $\boldsymbol{r}$ in the reference configuration

$$\bar{\lambda} = |\boldsymbol{t}|_{\boldsymbol{g}} = \sqrt{\boldsymbol{t}_\flat \cdot \boldsymbol{t}} \quad \textbf{with} \quad \boldsymbol{t}_\flat := \boldsymbol{g}\boldsymbol{t}, \tag{29}$$

where $\boldsymbol{g}$ is the current metric. The assumption of affine deformation gives the following relationship between the macro-stretch $\bar{\lambda}$ in the continuum and micro-stretch λ in the chain level

$$\lambda = \bar{\lambda}. \tag{30}$$

4.2 *Macroscopic isochoric free energy*

Let us consider an idealized network consisting of n randomly-jointed chains per unit volume. The chains are composed of equal number of Kuhn segments. Therefore, the network is *isotropic* and *homogeneous* in the undeformed state. The sum of the n energies ψ^i stored in the chains is equal to the macroscopic isochoric energy $\bar{\Psi}$ stored in the network

$$\bar{\Psi}(\boldsymbol{g}, \mathcal{I}; \bar{\boldsymbol{F}}) = \sum_{i=1}^{n} \psi^i(\bar{\lambda}^i, \xi^i), \tag{31}$$

where the assumption of affine stretches (30) has been used. Macro-stretches $\bar{\lambda}^i$ can be computed using (28) and (29). At each point, the above discrete sum can be replaced by a continuous average value of the stored energies in chains with evenly distributed orientation vectors $\boldsymbol{r}$ in space

$$\bar{\Psi}((\boldsymbol{g}, \mathcal{I}; \boldsymbol{F})) = n\langle \psi(\bar{\lambda}, \xi)\rangle = \langle n\psi(\bar{\lambda}, \xi)\rangle, \tag{32}$$

where, the operator $\langle\Diamond\rangle$ denotes the continuous average of the argument $\Diamond$ over evenly-distributed orientations in the space. Following Miehe et al. (2004), we define the average operator in (32) on a *micro-sphere* $\mathcal{S}$ with unit radius via

$$\langle v\rangle = \frac{1}{|\mathcal{S}|\}} \int_{\mathcal{S}} v(\boldsymbol{r})dA, \tag{33}$$

where $|\mathcal{S}| = 4\pi$ is the surface area of the micro-sphere and $\boldsymbol{r}$ is the unit orientation vector. It becomes relevant to introduce a macroscopic measure of crystallization using the average operator of (33). The average degree of crystallization $\bar{\xi}$ is defined as the average of the chain degrees of crystallization over the micro-sphere

$$\bar{\xi} = \langle \xi\rangle. \tag{34}$$

4.3 *Isochoric stresses and moduli*

Having the isochoric macro-energy (32) at hand, we are able to derive the isochoric contributions to the stresses in $(8)_2$. To do so, the derivative of the macro-stretch (29) with respect to the current metric is required

$$\partial_g \bar{\lambda} = \frac{1}{2}\bar{\lambda}^{-1}\, \boldsymbol{t} \otimes \boldsymbol{t}. \qquad (35)$$

Using $(8)_2$, (32), (35) and $(20)_1$, the isochoric part of the Kirchhoff stress can be written in terms of the chain forces

$$\bar{\tau} \equiv \left\langle \bar{\lambda}^{-1} n \mathfrak{f}(\bar{\lambda}, \xi)\, \boldsymbol{t} \otimes \boldsymbol{t} \right\rangle. \qquad (36)$$

5 EVALUATION OF THE MODEL

The affine full network model for crystallization in rubbers, can be readily implemented into finite element codes and used for solving representative boundary value problems. In this section we will compare the results predicted by our model with the experimental results of Toki et al. (2003). In their experimental setup, three types of rubber, namely, sulfur-vulcanized natural rubber (NR-S), peroxide vulcanized natural rubber (NR-P) and sulfur vulcanized synthetic polyisoprene rubber (IR) were stretched up to a total stretch of 6 at 0°C and then were retracted to the initial length. They measured the mass fraction of the strain-induced crystallites at different stages of loading using synchrotron wide-angle X-ray measurements. They concluded that even at high strains, the majority of the molecules remain amorphous.

We design a boundary value problem which simulates the experimental setup in a straightfroward manner. A cube of rubber is stretched up to a total stretch of $\lambda_t = 6$ in vertical e_3 direction and then is unloaded to the stretch of $\lambda_t = 1$. For the discretization of this simple boundary value problem we use only one Q1-P0 element. The nominal stress at the top of the element and the degree of crystallization have been compared with the experimental data for the three above-mentioned rubber types in Figure 3. In the crystallization plots, the averaged crystallization degree $\bar{\xi}$ is considered to be the macroscopic measure of crystallization. It can be computed locally at Gauss points. Note that in this simple problem the choice of the Gauss point does affect the results.

The model is able to reproduce the experimental crystallization data very well. However, the hysteresis in the stress-stretch is underestimated. Note that unlike the results of Kroon (2010) and Mistry & Govindjee (2014) in our treatment after full retraction of the material no remnant crystallization is present. This is in line with the experimental observations.

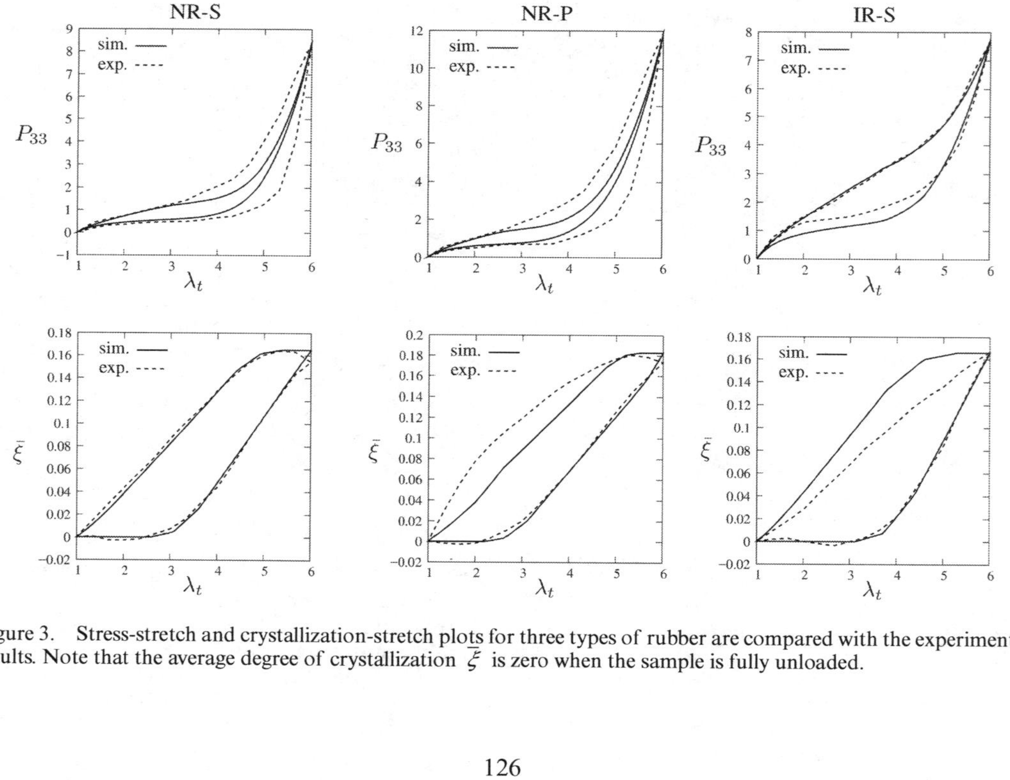

Figure 3. Stress-stretch and crystallization-stretch plots for three types of rubber are compared with the experimental results. Note that the average degree of crystallization $\bar{\xi}$ is zero when the sample is fully unloaded.

6 CONCLUSION

We proposed a micro-mechanically motivated material model for strain-induced crystallization in rubbers. Our point of departure is constructing a micro-mechanical model for a single crystallizing polymer chain. A thermodynamically consistent framework for the evolution of crystallization in the chain level is proposed. This modified chain model is then incorporated into the affine full network model. Finally, the model is numerically implemented and its performance is compared with the experimental results.

REFERENCES

Arruda, E.M. & M.C. Boyce (1993). A three–dimensional constitutive model for the large stretch behavior of rubber elastic materials. *Journal of theMechanics and Physics of Solids 41*, 389–412.

Dargazany, R., V.-U. Khiêm, E.-A. Poshtan, & M. Itskov (2014). Constitutive modeling of strain-induced crystallization in filled rubbers. *Physical Review E 89*, 022604.

Flory, P.J. (1947). Thermodynamics of crystallization in high polymers. I. Crystallization induced by stretching. *Journal of Chemical Physics 15*, 397–408.

Gehman, S.D. & J.E. Field (1939). An X-ray investigation of crystallinity in rubber. *Journal of Applied Physics 10*, 567–572.

Guilié, J., T.-N. Le, & P. Le Tallec (2015). Micro-sphere model for strain-induced crystallisation and three-dimensional applications. *Journal of the Mechanics and Physics of Solids 81*, 58–74.

James, H.M. & E. Guth (1943). Theory of elastic properties of rubber. *The Journal of Chemical Physics 11*, 455–481.

Katz, J. (1925). Was sind die Ursachen der eigentümlichen Dehnbarkeit des Kautschuks? *Kolloid-Zeitschrift 36*, 300–307.

Kroon, M. (2010). A constitutive model for strain-crystallising rubber-like materials. *Mechanics of Materials 42*, 873–885.

Kuhn, W. (1934). über die Gestalt fadenförmiger Moleküle in Lösungen. *Kolloid-Zeitschrift 68*, 2–15.

Kuhn, W. (1936). Beziehungen zwischen Molekülgröße, statistischer Molekülgestalt und elastischen Eigenschaften hochpolymerer Stoffe. *Kolloid-Zeitschrift 76*, 258–271.

Kuhn, W. & F. Grün (1942). Beziehungen zwischen elastischen Konstanten und Dehnungsdoppelbrechung hochelastischer Stoffe. *Kolloid-Zeitschrift 101*, 248–271.

Miehe, C., S. Göktepe, & F. Lulei (2004). A micro-macro approach to rubber-like materials-Part i: The non-affine microsphere model of rubber elasticity. *Journal of the Mechanics and Physics of Solids 52*, 2617–2660.

Mistry, S.J. & S. Govindjee (2014). A micro-mechanically based continuum model for strain-induced crystallization in natural rubber. *International Journal of Solids and Structures 51*, 530–539.

Mitchell, G.R. (1984). A wide-angle X-ray study of the development of molecular orientation in crosslinked natural rubber. *Polymer 25*, 1562–1572.

Rockafellar, R.T. (1997). Convex analysis. princeton landmarks in mathematics.

Toki, S., I. Sics, S. Ran, L. Liu, & B.S. Hsiao (2003). Molecular orientation and structural development in vulcanized polyisoprene rubbers during uniaxial deformation by in situ synchrotron X-ray diffraction. *Polymer 44*, 6003–6011.

Trabelsi, S., P.A. Albouy, & J. Rault (2002). Stress-induced crystallization around a crack tip in natural rubber. *Macro-molecules 35*, 10054–10061.

Treloar, L. (1954). The photoelastic properties of short-chain molecular networks. *Transactions of the Faraday Society 50*, 881–896.

Treloar, L.R.G. (1975). *The Physics of Rubber Elasticity* (3rd ed.). Clarendon Press, Oxford.

Treloar, L.R.G. & G. Riding (1979). A non–gaussian theory of rubber in biaxial strain. i. mechanical properties. *Proceedings of the Royal Society London A 369*, 261–280.

Wu, P.D. & E. van der Giessen (1993). On improved network models for rubber elasticity and their applications to orientation hardening in glassy polymers. *Journal of the Mechanics and Physics of Solids 41*, 427–456.

Constitutive Models for Rubber X – Lion & Johlitz (Eds)
© 2017 Taylor & Francis Group, London, ISBN 978-1-138-03001-5

Constitutive modelling of the amplitude and rate dependency of carbon black-filled SBR vulcanizate and its implementation into Abaqus

M. Fujikawa
Department of Mechanical Systems Engineering, University of the Ryukyus, Okinawa, Japan

N. Maeda
Graduate School of Engineering and Science, University of the Ryukyus, Okinawa, Japan

J. Yamabe
International Research Center for Hydrogen Energy, Kyushu University, Fukuoka, Japan
International Institute for Carbon-Neutral Energy Research, Kyushu University, Fukuoka, Japan

M. Koishi
The Yokohama Rubber Co., Ltd., Kanagawa, Japan

ABSTRACT: This study developed a new constitutive model for capturing nonlinear viscoelastic stress-strain responses of a carbon black-filled styrene-butadiene rubber vulcanizate (SBR-CB) under a wide range of deformations. In the assumed micro-mechanical network structure, the isochoric free energy was decomposed into elastic equilibrium and viscoelastic overstress response. Based on our previous performance evaluations of hyperelastic material models, an eight-chain equilibrium network model was employed. For the nonlinear viscoelastic network, a micro-sphere model proposed by Miehe et al was used. A simple phenomenological form was assumed for the micro stress and strain equation. For the micro-inelastic strain evolution equation based on the Rendek and Lion model, a relaxation time depending on strain amplitude was developed in order to reproduce the stress-strain relationship at strain amplitudes of ≤ 50%. The proposed model could practically reproduce dynamic responses of SBR-CB under a wide range of applied strains and strain rates. In addition, the model was thermodynamically consistent and could be implemented in a commercial FEM software.

1 INTRODUCTION

Industrial rubbers are commonly used in tires and engine mounts and show nonlinear viscoelastic behavior. When designing such rubber components, it is important to capture its stress–strain response under arbitrary loading conditions. In general, the rubber materials exhibit linear viscoelastic behavior under small deformation, whereas they show nonlinear viscoelastic behavior under large deformation. Figure 1 shows the storage modulus, E', loss modulus, E'', and loss tangential, tan δ, of a carbon black-filled styrene-butadiene rubber (SBR-CB) measured by a dynamic measurement tester (DMA). Because of the well-known Payne effect, these properties show remarkable nonlinearity, depending on the strain amplitude and strain rate. In addition, these tendencies also depend on the strain history and loading condition. From the practical importance, various nonlinear viscoelastic constitutive laws have been proposed so for; however, there are few models with a capability to reproduce both dynamic characteristics and various loading conditions (uniaxial tension (UT), pure shear (PS), and equibiaxial tension (BT)) under large deformations.

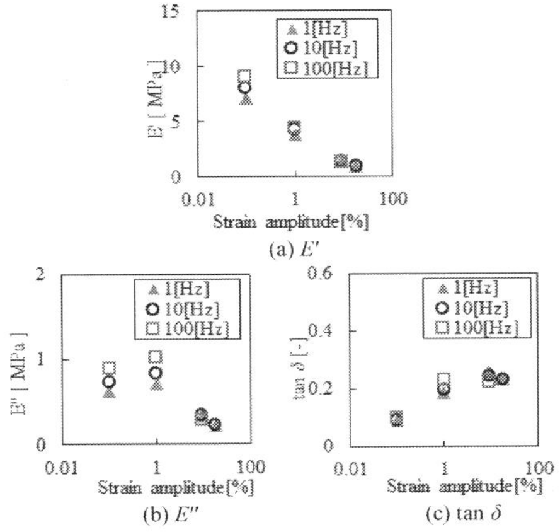

Figure 1. An example of the Payne effect for an SBR-CB.

The main purpose of this study is to develop a new constitutive model for capturing linear and nonlinear viscoelastic stress-strain responses of a SBR-CB. Under various experimental conditions, performances of the proposed model were verified and discussed in terms of reproducibility and predictability.

2 NONLINEAR VISCOELASTIC MODEL

2.1 *Outline of the proposed model*

The proposed constitutive law for viscoelastic materials is described in this section. As shown in Figure 2, a network structure for the rubber materials was assumed to consist of a free superimposed molecular chains (solid line) and chemically bonded ground-state network (dotted line). This is based on the viscoelastic model proposed by Bergström & Boyce (1998) and Miehe & Göktepe (2005), showing how the free molecular chain relaxes with time for the instantaneous response. We proposed a viscoelastic dynamics model as shown in Figure 3, based on the network structure shown in Figure 2. Following our previous study (Maeda, et al. 2015), the proposed model employed the eight-chain model (Arruda & Boyce 1993) for the chemically bound molecular chain motion. In contrast, based on the micro-sphere model (Miehe et al. 2004, Miehe & Göktepe 2005) and the evolution equation for the inelastic strain (Rendek & Lion 2010), a new phenomenological nonlinear viscoelastic constitutive equation was proposed for the free molecular chain motion.

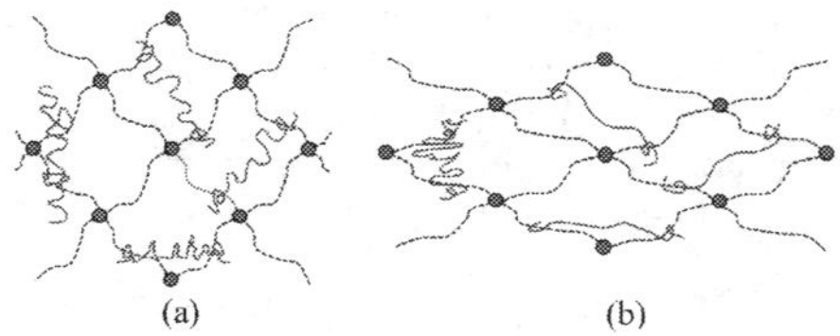

Figure 2. Illustration of the micromechanical mechanism of the assumed network structure: (a) undeformed and (b) deformed state immediately after instantaneous loading.

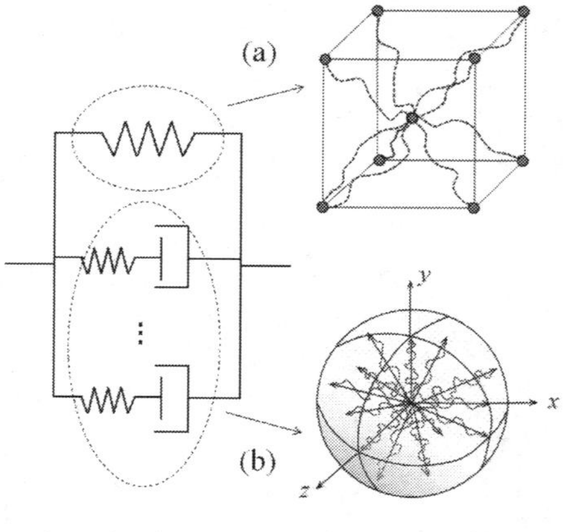

Figure 3. Rheology model: Superimposed free energy $\Psi = \Psi_{eq} + \Psi_{in}$ represents (a) an eight-chain model for the equilibrium response and (b) Micro-sphere model for the nonlinear time-dependent overstress response.

2.2 *Equilibrium response part*

The strain energy density of the eight-chain model is expressed by the following equation (Arruda & Boyce 1993):

$$\Psi_{eq} = \mu N\left\{\lambda_r \mathcal{L}^{-1}(\lambda_r) + \ln\frac{\mathcal{L}^{-1}(\lambda_r)}{\sinh \mathcal{L}^{-1}(\lambda_r)}\right\} + \frac{1}{d}(J-1)^2 + \Psi_0 \tag{1}$$

where μ, d and N are material parameters, J is the volume change rate expressed by det**F** using the deformation gradient tensor **F**, and

$$\lambda_r = \sqrt{\frac{\bar{I}_1}{3N}}. \tag{2}$$

where $\bar{I}_1$ is the first invariant of the isochoric right Cauchy green tensor and $\mathcal{L}^{-1}(\bullet)$ is an inverse Langevin function. The following approximate expression was used to calculate the inverse Langevin function (Cohen, 1991):

$$\mathcal{L}^{-1}(x) \approx x\frac{3-x^2}{1-x^2} \tag{3}$$

2.3 *Overstress response parts*

This section summarized calculation methods for a single molecular chain with respect to the nonlinear viscoelastic part. Details of the calculation methods for micro-to-macro transition were provided in the existing literatures (Miehe et al. 2004; Miehe & Göktepe 2005). In addition, our calculation assumed incompressibility.

2.3.1 *Molecular chain strain in Micro-sphere model*

An operation used to compute the strain for a single molecular chain in the micro-sphere model was summarized here. When the deformation of the single chain is assumed to be an affine stretch, the micro-stretch, λ, of the single chain is obtained by the following equation:

$$\lambda = |\mathbf{Fr}| \tag{4}$$

where **r** is orientation vector. From Equation 4, the Hencky strain, ε, of the single chain is expressed by the following equation:

$$\varepsilon = \ln(\lambda) \tag{5}$$

Then, ε was assumed to be decomposed into elastic strain, ε_c, and inelastic strain, ε_{in}, of the single chain as follows:

$$\varepsilon = \varepsilon_c + \varepsilon_{in} \tag{6}$$

2.3.2 *Consistency with thermodynamics*

According to the existing literature (Miehe & Göktepe 2005), the macroscopic free energy per unit volume in the reference configuration can be expressed as follows:

$$\Psi_{in} = \Psi_{in}(\mathbf{g}, \hbar; \mathbf{F}) \tag{7}$$

where $\hbar$ is an internal variable and **g** is the Kronecker's symbol. Kirchhoff stress is written as follows:

$$\boldsymbol{\tau} = 2\partial_g \Psi_{in}(\mathbf{g}, \hbar; \mathbf{F}) \tag{8}$$

Here, $\hbar$ must be consistent with the second axiom of thermodynamics, which demands a positive macroscopic dissipation:

$$\ell_{mac} := \boldsymbol{\kappa} \cdot \dot{\hbar} \geq 0 \quad \text{with } \boldsymbol{\kappa} = -2\partial_{\hbar} \Psi_{in}. \tag{9}$$

Based on the literature (Miehe & Göktepe 2005), the following simple phenomenological equation for a free molecular chain was assumed:

$$\Psi^{in} = \frac{1}{2} \sum_{a=1}^{s} \mu^a (\ln\lambda - \varepsilon_{in}^a)^2 \tag{10}$$

where μ^a is a material constant representing rigidity and s is the number of Maxwell models. From Equation 9, it is necessary to satisfy the following equation:

$$\ell_{mic} := \sum_{a=1}^{s} [\beta^a \dot{\varepsilon}_{in}^a] \geq 0 \tag{11}$$

where

$$\beta^a := -\partial_{\varepsilon_{in}^a} \Psi_{in} = \mu^a (\ln\lambda - \varepsilon_{in}^a). \tag{12}$$

2.3.3 *Evolution equation for inelastic strain*

From Rendek & Lion (2010), the evolution equation for ε_{in}^a was assumed to be as follows:

$$\dot{\varepsilon}_{in}^a = \frac{1}{Z^a}\left(1 + d^a q^a(t)\right)\left(\ln\lambda - \varepsilon_{in}^a\right). \tag{13}$$

where

$$\dot{q}^a(t) = \frac{1}{\gamma^a}\left(\left|\frac{d(\ln\lambda)}{dt}\right| - q^a\right), \tag{14}$$

and Z^a, d^a, γ^a are material constants. In order to develop a model having a capability to reproduce stress-strain relationship under a wide range of strains, Equation 14 was modified as follows:

$$\dot{\varepsilon}_{in}^a = \frac{1}{Z^a}\left(1 + \left\langle \left|\ln\lambda - \varepsilon_{in}^a\right| - \varepsilon_m \right\rangle^0 d^a q^a(t)\right)\left(\ln\lambda - \varepsilon_{in}^a\right) \tag{15}$$

where < > are the Macaulay's brackets defined by $\langle s \rangle = (s+|s|)/2$. This equation leads to the linear viscoelastic model with the relaxation time of Z^a when $|\varepsilon_c^a| < \varepsilon_m$ and the evolution equation proposed by Rendek & Lion (2010) when $|\varepsilon_c^a| > \varepsilon_m$. The proposed model satisfies the second law of thermodynamics when all the material constants are positive according to Equations 11 and 12.

2.4 *Implementation in Abaqus*

The proposed constitutive model was implemented in Abaqus via the user subroutine function UMAT. In Abaqus, the deformation gradient tensor, $\mathbf{F}_{n+1}$, at the current increment, the internal variable, $\boldsymbol{\alpha}_n$, at the previous increment, and the time increment, Δt, are inputted. Users can implement their own constitutive equations into Abaqus by programming the Cauchy stress, **T**, and consistent tangential stiffness matrix, $\mathbb{C}^{MJ}$, at the current increment. Here, $\mathbb{C}^{MJ}$ is expressed by the following equation using the tangent stiffness, $\mathbb{C}^{\tau J}$, of the Jaumann rate type of Kirchhoff stress:

$$\mathbb{C}^{MJ} = \frac{1}{J}\mathbb{C}^{\tau J} \tag{16}$$

where

$$\dot{\boldsymbol{\tau}}_{(J)} = \dot{\boldsymbol{\tau}} - \mathbf{W}\boldsymbol{\tau} - \boldsymbol{\tau}\mathbf{W}^T = \mathbb{C}^{\tau J} : \mathbf{D}. \tag{17}$$

where $\dot{\boldsymbol{\tau}}_{(J)}$ is the Jaumann rate of the Kirchhoff stress, **D** is the rate-of-deformation tensor, and **W** is the spin tensor. Details of the calculation process and formula were provided in some literatures (Miehe 1995, Sun et al. 2008).

2.4.1 *Computation of the Cauchy stress*

The Cauchy stress in the proposed model was programmed based on the backward Euler method.

The detailed formulation scheme is derived in the literatures (Miehe & Göktepe 2005, Rendek & Lion 2010). Note that the scheme does not require convergence judgment in the stress calculation.

2.4.2 *Computation of the consistent tangent matrix*

The tangential stiffness matrix was calculated by numerical differentiation. According to the literature (Sun et al. 2008), $\Delta\mathbf{F}^{(ij)}_{n+1}$ is defined as follows:

$$\Delta\mathbf{F}^{(ij)}_{n+1} = \frac{h}{2}\left(\mathbf{e}_i \otimes \mathbf{e}_j + \mathbf{e}_j \otimes \mathbf{e}_i\right)\mathbf{F}_{n+1} \tag{18}$$

where h is a small perturbation parameter and e_i (i = 1, 2, 3) represents the basis vectors in the spatial description. Using Equation 18, tangent stiffness matrices were computed by:

$$C^{MJ}_{ij} = \frac{1}{Jh}\left(\boldsymbol{\tau}\left(\mathbf{F}_{n+1} + \Delta\mathbf{F}^{(ij)}_{n+1}, \boldsymbol{\alpha}_n\right) - \boldsymbol{\tau}\left(\mathbf{F}_{n+1}, \boldsymbol{\alpha}_n\right)\right). \tag{19}$$

It is known that the result of the numerical differentiation includes roundoff and truncation error depending on the value of h; thus, this study empirically set h to 1×10^{-7}.

3 VERIFICATION OF EFFECTIVENESS OF THE PROPOSED METHOD

Section 3.1 outlines the method of determining material constants in this model for a carbon black-filled SBR with a volume fraction of 25% (SBR-CB25). In Sections 3.2 to 3.4, FEM analysis is conducted under the same conditions as the experimental results used to identify the material constants. The main purpose of these sections is to verify the reproducibility of the determined material constants. In section 3.5, the predicted performance of the model for a new material test was discussed. All numerical experiments from Sections 3.2 to 3.5 were performed with Abaqus using the material constants determined in Section 3.1.

3.1 *Identification of material constants*

The procedure of determining material constants is summarized below. Details of each material test are summarized in Sections 3.2 to 3.4 together with the results of experiment and simulation.

3.1.1 *Curve fitting for small deformation*

The material constants (μ, μ^a, Z^a) were determined from the master curve obtained by DMA testing.

3.1.2 *Curve fitting for large deformation*

The material constants were determined from UT, PS, and BT tests under large deformations. The real-coded GA (Higuchi et al. 2001) was used to determine the material constants under $\varepsilon_m = 0$.

3.1.3 *Superposition of material constants*

Using the material constants determined in subsections 3.1.1 and 3.1.2, appropriate values of ε_m were determined based on the measured E' and E'' depending on the strain amplitude. To eliminate the influence of the material parameters determined at small deformation on the stress-strain relationship at large deformation, we set d^a and λ^a to 10^{20} and 1.0, respectively. Additionally, if necessary, we finely adjusted other material constants. The material constants determined from the above procedure are summarized in Tables 1 and 2.

3.2 *Reproducibility for DMA*

This section confirmed whether the proposed model can reproduce the master curve at small deformation or not. In the experiment, E' and E'' were measured at a strain of 0.1% under temperature sweep and frequency sweep conditions; then, the master curve was estimated by means of the time-temperature correspondence principle. In the simulation, the stress-strain curves at the fifth cycle at test frequencies of 0.1, 10, 10^3, 10^5, 10^7, and 10^9 Hz were calculated. E' and E'' were obtained from the following equations:

Table 1. Equilibrium response material properties.

μ	N	d
1.143	1.957	1.00E-03

Table 2. Overstress response material properties.

a	μ^a	Z^a	γ^a	d^a	ε_m
1	1.61E+03	1.00E-11	1	1.00E+20	0.002
2	2.25E+03	1.00E-10	1	1.00E+20	0.002
3	9.50E+02	1.00E-09	1	1.00E+20	0.002
4	2.32E+02	1.00E-08	1	1.00E+20	0.002
5	9.85E+01	1.00E-07	1	1.00E+20	0.002
6	5.34E+01	1.00E-06	1	1.00E+20	0.002
7	3.38E+01	1.00E-05	1	1.00E+20	0.002
8	2.47E+01	1.00E-04	1	1.00E+20	0.002
9	1.99E+01	1.00E-03	1	1.00E+20	0.002
10	1.62E+01	1.00E-02	1	1.00E+20	0.002
11	1.31E+01	1.00E+01	1	1.00E+20	0.002
12	1.04E+01	1.00E+00	1	1.00E+20	0.002
13	7.83E+00	1.00E+01	1	1.00E+20	0.002
14	6.75E+00	1.00E+02	1	1.00E+20	0.002
15	4.95E+00	1.00E+03	1	1.00E+20	0.002
16	3.87E-01	1.00E+04	1	1.00E+20	0.002
17	3.17E-03	1.00E+05	1	1.00E+20	0.002
18	1.37E+01	4.77E+06	1.45E+09	1.01E+08	0.002

$$E' = (\sigma_{max} - \sigma_{min}) / (\varepsilon_{max} - \varepsilon_{min}) \tag{20}$$

$$E'' = (\sigma_{0,max} - \sigma_{0,min}) / (\varepsilon_{max} - \varepsilon_{min}) \tag{21}$$

where the definitions of σ_{max}, σ_{min}, $\sigma_{0,max}$, and $\sigma_{0,min}$ are shown in Figure 4(a). Results of experiment and simulation are shown in Figure 4(b), showing the proposed model successfully reproduced the experimental results of the DMA testing.

3.3 *Reproducibility for UT*

The following uniaxial tensile tests (Tests 1–6) were performed by using the dumbbell-shaped specimen shown in Figure 5. Tests 1 and 2 were one-cycle UT tests, Test 3 was a multi-cycle UT test, and Tests 4 to 6 were multi-relaxation tests. The experimental conditions of these tests are as follows:

- Test 1: $|d\lambda/dt| = 2.8 \times 10^{-2}$/s, $(\varepsilon_{min}, \varepsilon_{max}) = (0, 50\%)$
- Test 2: $|d\lambda/dt| = 5.7 \times 10^{-5}$/s, $(\varepsilon_{min}, \varepsilon_{max}) = (0, 50\%)$
- Test 3: $|d\lambda/dt| = 1.6 \times 10^{-2}$/s, $(\varepsilon_{min}, \varepsilon_{max}) = (0, 40\%)$ at ten cycles, followed by $(\varepsilon_{min}, \varepsilon_{max}) = (20, 40\%)$ at ten cycles
- Test 4: $|d\lambda/dt| = 0.1$/s with five-minute relaxation intervals
- Test 5: $|d\lambda/dt| = 2.8 \times 10^{-2}$/s with five-minute relaxation intervals
- Test 6: $|d\lambda/dt| = 5.7 \times 10^{-4}$/s with five-minute relaxation intervals

Figures 6 and 7 show the results of experiment and FEM simulation using Abaqus with the material constants shown in Section 3.1. The proposed model reproduced the uniaxial tensile-test results.

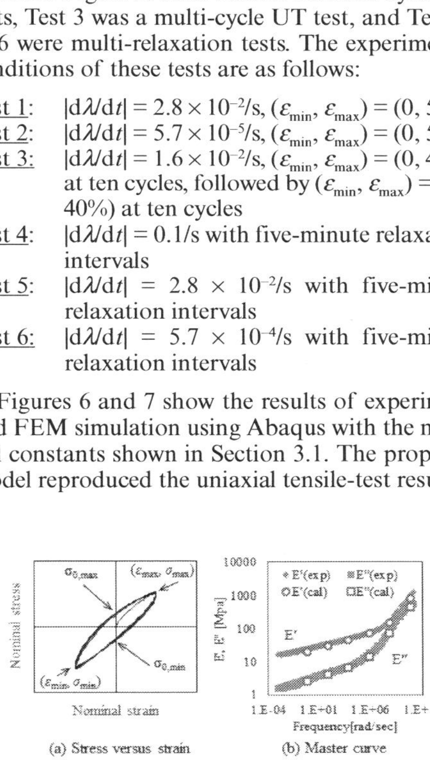

Figure 4. Experiments and simulations of DMA testing: (a) illustration of the stress–strain relationship and (b) master curve.

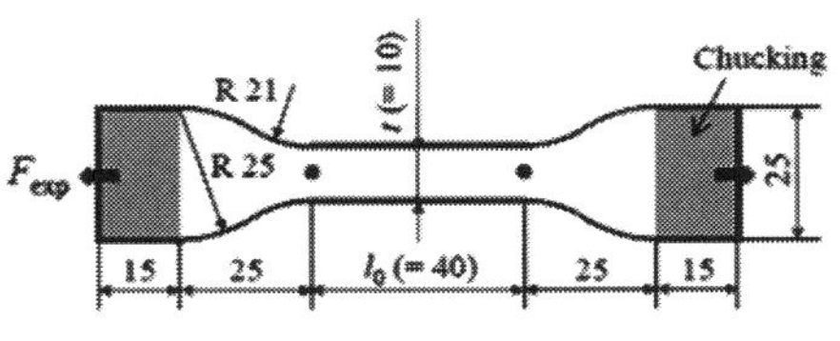

Figure 5. Geometry of the specimen.

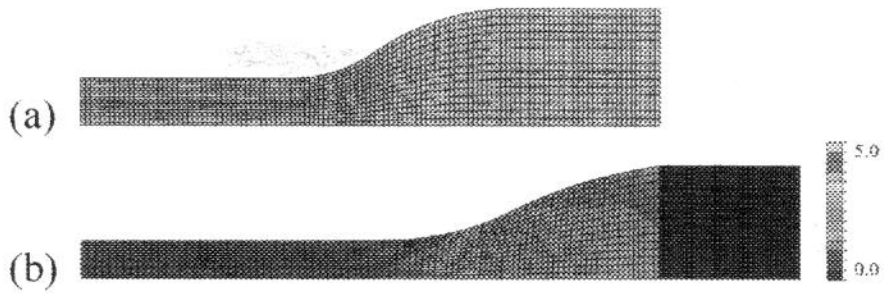

Figure 6. σ_{xx} distribution of the quarter-shape model using the proposed constitutive equation under UT: (a) before deformation and (b) after deformation (50% strain).

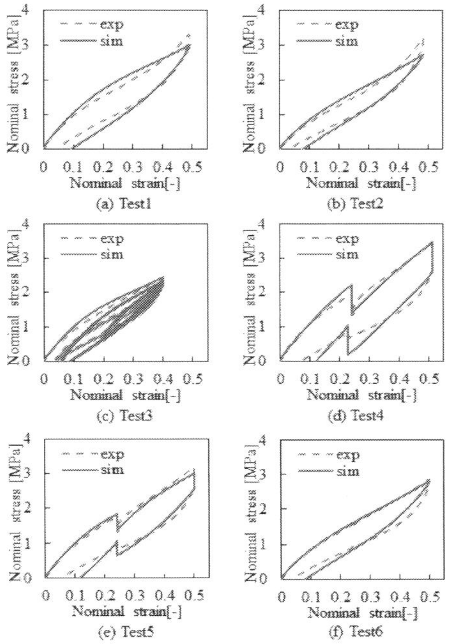

Figure 7. Experiments and simulations: (a), (b) 1cycle, (c) multi-cycle, and (d)-(f) multi-step relaxation UT tests.

3.4 *Reproducibility for PS and BT*

PS and BT tests (Tests 7–10) were conducted by using the pre-cut rectangular specimen shown in Figure 8 under the following conditions of strain rate and strain amplitude:

- Test 7: $|d\lambda/dt| = 2.8 \times 10^{-2}$/s, $(\varepsilon_{min}, \varepsilon_{max}) = (0, 50\%)$ for PS
- Test 8: $|d\lambda/dt| = 5.7 \times 10^{-4}$/s, $(\varepsilon_{min}, \varepsilon_{max}) = (0, 50\%)$ for PS
- Test 9: $|d\lambda/dt| = 2.8 \times 10^{-2}$/s, $(\varepsilon_{min}, \varepsilon_{max}) = (0, 50\%)$ for BT
- Test 10: $|d\lambda/dt| = 5.7 \times 10^{-4}$/s, $(\varepsilon_{min}, \varepsilon_{max}) = (0, 50\%)$ for BT

In these simulations, the nominal stress-strain relations were calculated from the measured tensile load and the stretch in the reference area (Fujikawa

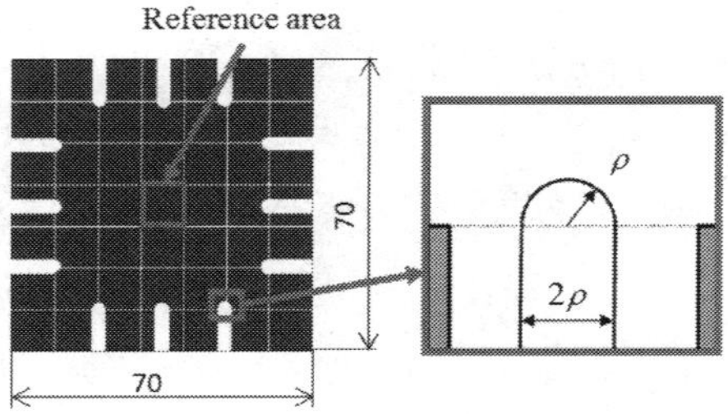

Figure 8. Geometry of the specimen of biaxial tensile tester for PS and BT.

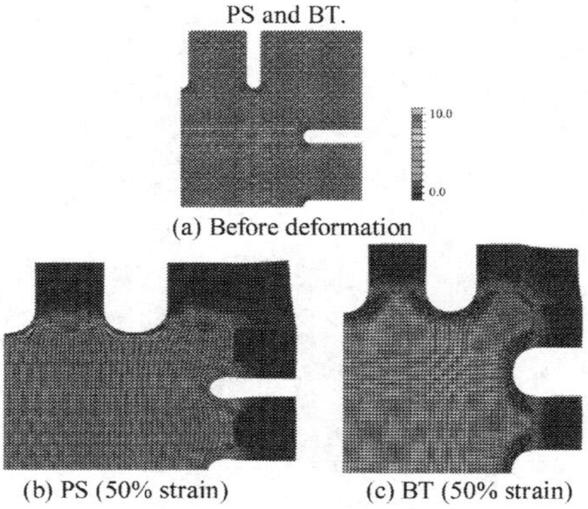

Figure 9. σ_{xx} distribution of the quarter-shape model using the proposed model (a) before deformation, (b) in PS and (c) BT.

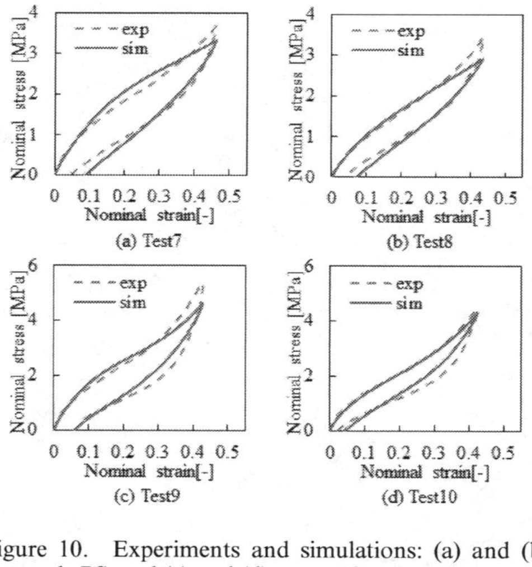

Figure 10. Experiments and simulations: (a) and (b) one-cycle PS, and (c) and (d) one-cycle BT tests.

et al. 2014). Figures 9 and 10 show the results of actual and numerical experiments and the proposed model also reproduced the PS- and BT-test results.

3.5 *Payne effect reproducibility*

Experiments and simulations associated with strain-amplitude dependence were performed. The FEM simulation was performed under strain amplitudes ranging from 0.1 to 20% at a test frequency of 0.313 Hz. E' and E'' were calculated by the same calculation procedures as those in Section 3.2. Results of experiment and simulation are shown in Figures 11 and 12. The proposed model successfully reproduced the experiment results related to the Payne effect.

3.6 *Prediction performance of the proposed model*

To evaluate the predictive performance of the proposed method, four UT tests were conducted under combinations of mean strain and strain amplitude: (5%, 5%), (30%, 5%), (10%, 10%), and

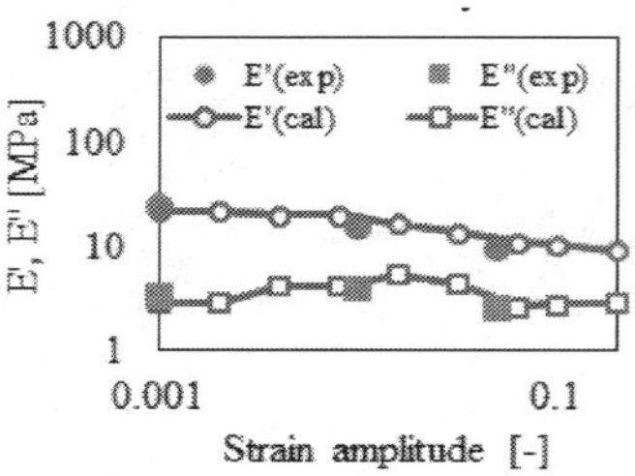

Figure 11. E' and E'' with various strain amplitudes in experiments and simulations at 0.313 Hz.

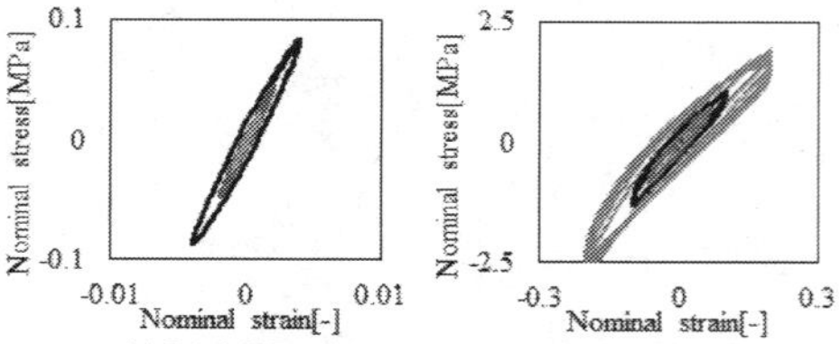

Figure 12. Stress-strain relationships at various strain amplitudes at 0.313 Hz using the proposed model.

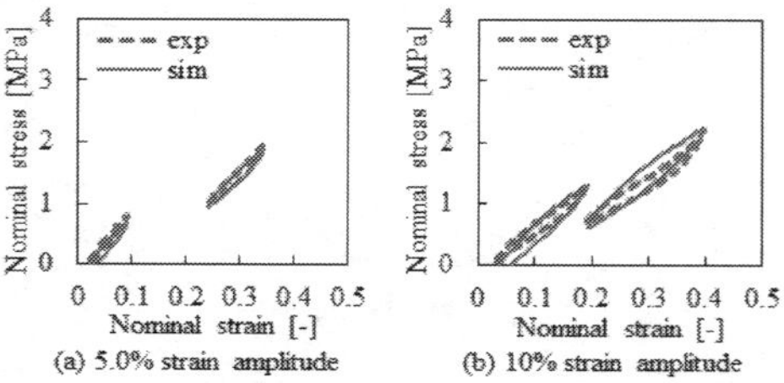

Figure 13. Prediction performance for the proposed model.

(30%, 10%). Results are summarized in Figure 13, representing the proposed model had a good predictive performance.

4 CONCLUSIONS

This paper proposed a new nonlinear viscoelastic model, based on the micro-sphere model proposed by Miehe et al. (2004) and the inelastic strain evolution law proposed by Rendek & Lion (2010). The key aspects of the proposed model can be summarized as follows:

- The proposed model satisfied thermodynamics laws, when all material constants were positive.
- The proposed model could be implemented into Abaqus via the user subroutine UMAT.
- Material constants were identified by material tests under small and large deformations, respectively. The superposition of the respective material constants enables to determine all the material constants.
- The proposed model successfully reproduced the test results of UT tests under a wide range of deformations and PS, and BT tests under large deformation. The predictive performance was also good within the range of the experimental results.

REFERENCES

Arruda, E.M. & Boyce, M.C. 1993. A three-dimensional constitutive model for the large stretch behavior of rubber elastic materials. *J. Mech. Phys. Solids* 41: 389–412.

Bergström, J.S. & Boyce, M.C. 1998. Constitutive modeling of the large strain time–dependent behavior of elastomers, *J. Mech. Phys. Solids*. 46: 931–954.

Cohen, A. 1991. A padé approximation to the inverse Langevin function, Rheological acta, 30–3: 270–273.

Fujikawa, M., Maeda, N., Yamabe, J., Kodama, Y. & Koishi, M. 2014. Determining stress-strain in rubber with in-plane biaxial tensile tester. *Exp. Mech.* 54: 1639–1649.

Higuchi, T., Tsutsui, S. & Yamamura, M. 2001. Simplex crossover for real-coded genetic algorithms. *Trans. Jpn. Soc. Artif. Intell.* 16: 147–155.

Holzapfel, G. 2000. Nonlinear solid mechanics: a continuum approach for engineering. Wiley, New York.

Maeda, N., Fujikawa, M., Makabe, C., Yamabe, J., Kodama, Y. & Koishi, M. 2015. Performance evaluation of various hyperelastic constitutive models of rubbers, *Constitutive Models for Rubber IX*: 271–277.

Miehe, C. & Göktepe, S. 2005. A micro-macro approach to rubber-like materials – Part II: The micro-sphere model of finite rubber viscoelasticity, *Journal of the Mechanics and Physics of Solids* 53: 2231–2258.

Miehe, C. 1995. Numerical computation of algorithmic (consistent) tangent moduli in large-strain computational inelasticity, Comput. Methods Appl. Mech. Engrg. 134: 223–240.

Miehe, C., Göktepe, S. & Lulei, F. 2004. A micro-macro approach to rubber-like materials – Part I: the non-affine micro-sphere model of rubber elasticity, *Journal of the Mechanics and Physics of Solids* 52: 2617–2660.

Rendek, M. & Lion, A. 2010. Amplitude dependence of filler-reinforced rubber: Experiments, constitutive modelling and FEM – Implementation, *International Journal of Solids and Structures*, 47: 2918–2936.

Sun, W., Chaikof, E.L. & Levenston, M.E. 2008. Numerical Approximation of Tangent Moduli for Finite Element Implementations of Nonlinear Hyperelastic Material Models, *Journal of Biomechanical Engineering*, 130: 061003.

Constitutive Models for Rubber X – Lion & Johlitz (Eds)
 ISBN 978-1-138-03001-5

Application and extension of the MORPH model to represent curing phenomena in a PU based adhesive

R. Landgraf & J. Ihlemann
Faculty of Mechanical Engineering, Professorship of Solid Mechanics, Chemnitz University of Technology, Chemnitz, Germany

ABSTRACT: In this contribution, the phenomenological modelling of the curing process and the corresponding mechanical behaviour of a PU based adhesive is considered. To characterise the underlying chemical reaction of the curing process, differential scanning calorimetry tests were carried out. Moreover, uniaxial tension tests on the fully cured material were conducted and revealed typical behaviour of elastomers, like hysteresis, Mullins effect and permanent set. Based on the experimental results, a constitutive model has been formulated to capture the mechanical behaviour of the material in the fully cured state as well as during the curing process. The rubber-like behaviour of the fully cured state is described by the help of the MORPH constitutive model, which is able to capture typical phenomena of elastomers. Additionally, the chemical curing process itself is regarded by a so called degree of cure. Based on this variable, it is shown how the MORPH model may be extended to capture the mechanical behaviour during the curing process.

1 INTRODUCTION

Lightweight constructions and smart structures are two crucial research aspects in engineering. The combination of both research fields is addressed within the Collaborative Research Center/Transregio 39 “PT-PIESA”, which deals with the research on series production of lightweight constructions with included actuator and sensor functionalities. One of the pursued research lines is the forming technology, where sheet metal compounds with embedded Macro-Fibre-Composites (MFCs) are developed (see Nestler et al. 2014). The specific approach is to place the MFCs within an adhesive layer between two planar metal sheets (see Fig. 1) and to form this so called Piezo-Metal-Compound (PMC) with already included MFC to a specific final shape.

Due to the fact that MFCs can only sustain small strains, particular protections of the MFCs during the forming process need to be regarded. To this end, the production steps depicted in Fig. 2 are applied. First the PMC is prepared in such a

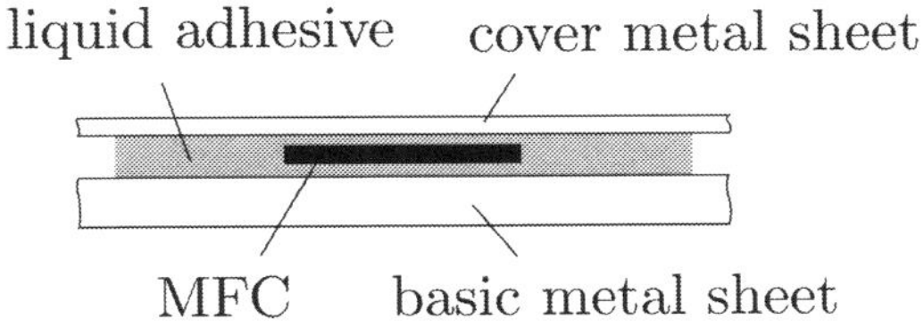

Figure 1. Piezo-Metal-Compound consisting of two metal sheets and an adhesive layer with embedded MFC.

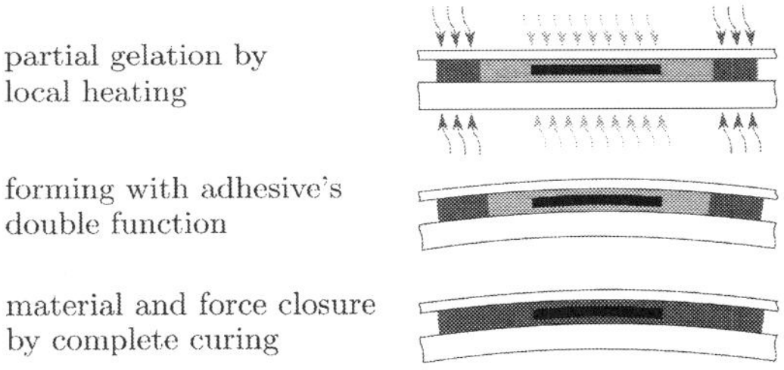

Figure 2. Schematic illustration of the basic steps for the production of Piezo-metal-compounds.

way that the MFC is embedded within an uncured adhesive layer. Next, the outer region of the adhesive layer is cured by local heating such that a basic connection between the metal layers is created but still a past-like adhesive core remains. Therewith, the MCF is protected from excessive loading during the subsequent sheet metal forming process. Only after forming, the adhesive layer is fully cured, which yields a complete material closure between the MFC and the metal sheets.

For the realisation of the described production approach, the heat curing polyurethane (PU) based adhesive *Teroson 1510* (Henkel 2016) has been chosen. The initial paste-like material includes encapsulated isocyanate, which gets released only at temperatures above 85°C. Once the isocyanate is released, the curing reaction starts and the process to complete solidification only takes few minutes. The final material exhibits rubber-like mechanical behaviour.

This contribution concentrates on the phenomenological modelling of the material behaviour of the PU based adhesive. Here, the mechanical behaviour of the fully cured material as well as the process of curing are taken into account. Thus, the phenomenological description of a so called degree of cure is included. The modelling approach is based on a general curing model (see for example Lion and Höfer 2007, Landgraf et al. 2014, Landgraf 2016) as well as the MORPH constitutive model by Ihlemann (2003) for the representation of rubber like materials. First, the experimental characterisation and the phenomenological modelling of the underlying cure kinetics are presented in Sec. 2. Next, Sec. 3 deals with the MORPH constitutive model and its adaptation to uniaxial tension test on the fully cured PU adhesive. Finally, in Sec. 4 the MORPH constitutive model is extended in such a way that changes of the mechanical behaviour during the curing process can be simulated.

2 CURE KINETICS

2.1 *Experimental observations by DSC*

In order to characterise the curing process, differential scanning calorimetry (DSC) experiments were performed. In particular, different temperature profiles including heating and cooling phases as well as an isothermal phase at different temperatures were prescribed. The temperature profiles and the corresponding measured heat flow data are depicted in Fig. 3. Therein, the varying signs at heating and cooling phases can be observed. Moreover, the non-zero heat flow during the isothermal phases (only at $T = 90°C$ and $T = 100°C$) reveals the exothermic behaviour of the underlying cross-linking reaction.

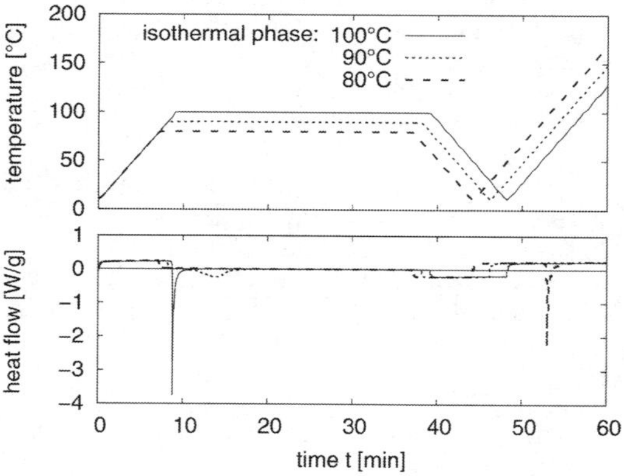

Figure 3. DSC-experiments: Prescribed temperature profiles (top) and measured heat flow (bottom).

The isothermal parts of the DSC-measurements have been used to define a degree of cure q, which is a measure for the progress of the curing process. The corresponding heat flow data of the isothermal phases is denoted by $\dot{h}_{iso}(t)$. It is employed to define the curing rate $\dot{q}$, which represents the rate of the curing process:

$$\dot{q} = \frac{\dot{h}_{iso}(t)}{h_{max}}, \quad q(t) = \int_0^t \frac{\dot{h}_{iso}(\bar{t})}{h_{max}} d\bar{t}. \tag{1}$$

Here, h_{max} is the specific ultimate heat, i.e. the maximum releasable curing heat per unit mass. A time integration of the curing rate $\dot{q}$ furthermore yields the degree of cure q, which is depicted in Fig. 4 for the case of isothermal curing at constant temperatures.

Here, it can be observed that curing is not initiated at 80°C curing temperature, i.e. the degree of cure does not evolve. Moreover, curing is accelerated with increasing curing temperatures.

2.2 *Phenomenological model*

The simulation of the reaction kinetics and the corresponding thermodynamic behaviour includes formulations for the degree of cure $q(t)$ and the specific enthalpy $h(t)$. First, the progress of the curing process is described by an autocatalytical reaction model of Kamal and Sourour (1973) and a diffusion factor $f_D(q, \theta)$ as introduced by Fournier et al. (1996):

$$\begin{aligned} \dot{q} &= \left(K_1(\theta) + K_2(\theta) q^{\alpha}\right)(1-q)^{\beta} f_D(q,\theta), \\ K_i(\theta) &= K_{i0} \exp\left[-\frac{E_i}{R\theta}\right], \quad i = 1,2, \\ f_D(q,\theta) &= \frac{2}{1+\exp\left[\frac{q-q_{end}(\theta)}{b}\right]} - 1. \end{aligned} \tag{2}$$

The factors $K_i(\theta)$ are temperature-dependent functions that show Arrhenius behaviour, $R = 8.3144$ J/(mol K) is the universal gas

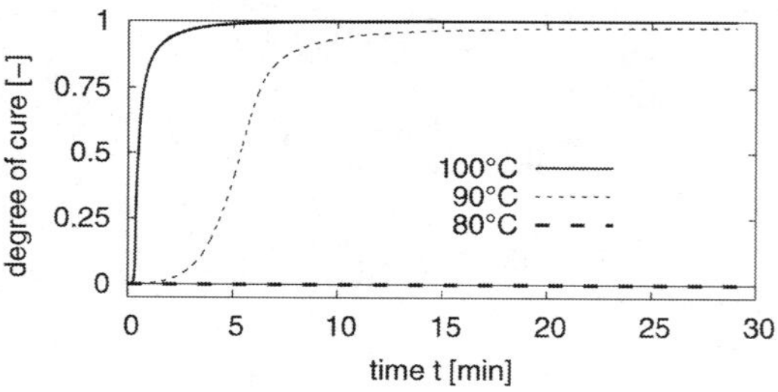

Figure 4. Degree of cure q calculated from isothermal phases in DSC-experiments.

constant, and K_{10}, K_{20}, E_1, E_2, α, β and b are material parameters. Moreover, the temperature-dependent function $q_{end}(\theta)$ is a measure for the maximum attainable degree of cure at certain temperatures. Specific mathematical approaches for this function are often related to the evolution of glass transition temperature during the curing process. However, since the glass transition temperature of the fully cured PU based adhesive is approximately −50°C, an alternative ansatz has been chosen:

$$q_{end}(\theta) = \frac{k_1}{1+\exp\left[-k_2\left(\theta-\theta_{ref}\right)\right]} - k_3. \quad (3)$$

Here, k_1–k_3 and θ_{ref} are further material parameters.

In order to simulate the DSC-experiments and the heat flow corresponding to the temperature profiles given in Fig. 3, a model for the specific enthalpy has to be formulated. Here, the basic approach of Kolmeder et al. (2011) has been applied. It includes a mixture rule of specific enthalpies of the initially uncured (Fluid) and the fully cured (Solid) material.

$$h(\theta,q) = h_F(\theta)(1-q) + \{\Delta h_{FS} + h_S(\theta)\}q \quad (4)$$

Moreover, the temperature dependent contributions $h_F(\theta)$ and $h_S(\theta)$ are modelled as follows:

$$h_i(\theta) = a_i\theta + \frac{b_i}{2}\theta^2 + \frac{c_i}{4}\theta^4, i = F,S. \quad (5)$$

In Eqs. (4) and (5), Δh_{FS}, a_i and b_i and c_i are further material parameters. The identification of the parameters in Eqs. (2) – (5) was conducted using MATLAB's optimization tool *lsqnonlin*. In a first step, Eq. (2) was adapted to the experimental data given in Fig. 4. Next, the identified model for the degree of cure was fixed and Eqs. (4) and (5) for the specific enthalpy model were adapted to the heat flow data given in Fig. 3. In a final optimization step, the parameters of the degree of cure and the specific enthalpy models were refined using both corresponding experimental data in parallel. Thereby, the identification results given in Figs. 5 and 6 as well as the corresponding model parameters listed in Eqs. (6) and (7) were achieved.

$$\begin{array}{|ll|}\hline K_{10} = 1.9469\cdot 10^9\ \mathrm{s}^{-1} & K_{20} = 1.5950\cdot 10^6\ \mathrm{s}^{-1} \\ E_1 = 8.4257\cdot 10^4\ \dfrac{\mathrm{J}}{\mathrm{mol}} & E_2 = 4.7094\cdot 10^4\ \dfrac{\mathrm{J}}{\mathrm{mol}} \\ \alpha = 1.2889 & \beta = 3-\alpha \\ b = 125.14 & k_1 = 288.85 \\ k_2 = 0.7423\ \mathrm{K}^{-1} & k_3 = 102.12 \\ \theta_{ref} = 363.4\ \mathrm{K} & \\ \hline\end{array} \quad (6)$$

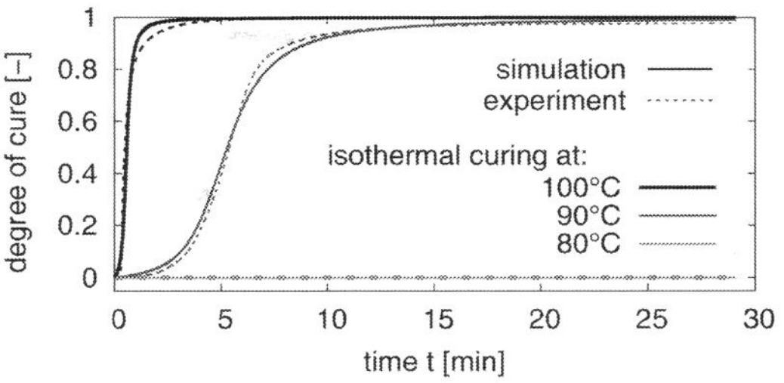

Figure 5. Degree of cure—comparison of experiments and simulation for isothermal curing at different temperatures.

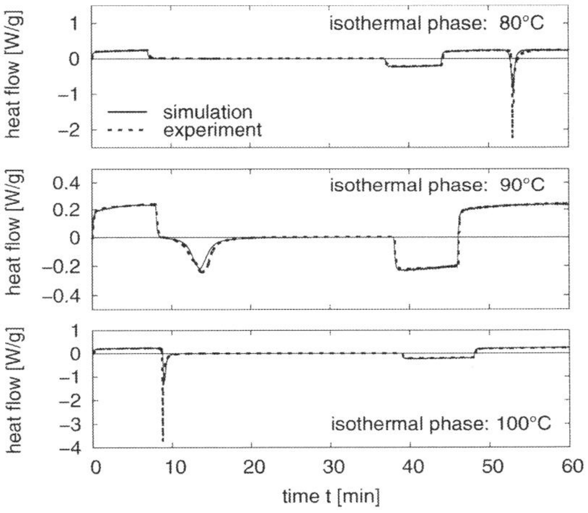

Figure 6. Heat flow—comparison of experiments and simulation for prescribed temperature profiles given in Fig. 3.

$$\begin{array}{|ll|}\hline \Delta h_{FS} = -38.164\ J/g & \\ a_F = 1.074\ \dfrac{\mathrm{J}}{\mathrm{gK}} & a_S = 1.149\ \dfrac{\mathrm{J}}{\mathrm{gK}} \\ b_F = 6.806\cdot 10^{-3}\ \dfrac{\mathrm{J}}{\mathrm{gK}^2} & b_S = 3.439\cdot 10^{-3}\ \dfrac{\mathrm{J}}{\mathrm{gK}^2} \\ c_F = -4.495\cdot 10^{-7}\ \dfrac{\mathrm{J}}{\mathrm{gK}^4} & c_S = -7.822\cdot 10^{-8}\ \dfrac{\mathrm{J}}{\mathrm{gK}^4} \\ \hline\end{array} \quad (7)$$

3 MECHANICAL BEHAVIOUR (FULLY CURED STATE)

In this section, the mechanical behaviour of the fully cured material is examined. First, the MORPH constitutive model for the simulation of rubber like materials is briefly recalled. Next, the results of uniaxial tension tests, which revealed rubber like material behaviour, as well as the model adaptation to these data are presented.

3.1 *Model of Rubber Phenomenology (MORPH)*

A suitable model to describe typical phenomena of rubber like materials is the Model of Rubber Phenomenology (MORPH) by Ihlemann (2003). It is able to capture the strong non-linear material behaviour including phenomena like the Mullins effect, static hysteresis and permanent set. In the following, the material model is introduced in an Eulerian description and by the application of the tensor notation used in Ihlemann (2003). Here, the order of tensors is denoted by the number of underlines. For example, the deformation gradient and the left Cauchy-Green tensor are given by $\underline{\underline{F}}$ and $\underline{\underline{b}} = \underline{\underline{F}} \cdot \underline{\underline{F}}^T$, respectively, and the Tresca invariant of the latter tensor can be calculated by its eigenvalues b_1, b_2 and b_3 as follows

$$b_T = \max\left[|b_I - b_J|\right]; \quad I,J = 1,2,3. \tag{8}$$

Moreover, the corresponding maximum Tresca invariant within the time span [0, t] is given by

$$b_T^S = \max\left[b_T(\tau); \quad 0 \le \tau \le t\right]. \tag{9}$$

It represents a measure of the maximum attained strain within the regarded deformation history. The stress response of the MORPH constitutive model is formulated by the following decomposition of the Cauchy stress tensor $\underline{\underline{\sigma}}$:

$$\underline{\underline{\sigma}} = 2\alpha \underline{\underline{b}}' + \left(\underline{\underline{\sigma}}^A\right)' - p\underline{\underline{I}}. \tag{10}$$

It includes a Neo-Hookean like contribution with an material function α, an additional stress $\underline{\underline{\sigma}}^A$, and the hydrostatic part containing the hydrostatic pressure p and the second order unit tensor $\underline{\underline{I}}$. Moreover, ()′ denotes the deviatoric part of a tensor. The additional stress $\underline{\underline{\sigma}}^A$ is defined by

$$\overset{*}{\underline{\underline{\sigma}}}{}^A = \beta \overset{*}{b}_T \left(\underline{\underline{\sigma}}^L - \underline{\underline{\sigma}}^A\right), \tag{11}$$

where $\overset{*}{\underline{\underline{\sigma}}}{}^A$ is the Jaumann rate of the additional stress, β is another material function and $\underline{\underline{\sigma}}^L$ is the limiting stress

$$\underline{\underline{\sigma}}^L = \gamma \exp\left(\left(p_7 \frac{\overset{*}{\underline{\underline{b}}}}{\overset{*}{b}_T} \frac{b_T}{b_T^S}\right)\right) + p_8 \frac{\overset{*}{\underline{\underline{b}}}}{\overset{*}{b}_T}, \tag{12}$$

which includes a third material function γ. Moreover, the factor $\overset{*}{b}_T$ occurring in Eqs. (11) and (12) is the Tresca invariant of $\overset{*}{\underline{\underline{b}}}$, i.e. the Jaumann rate of the left Cauchy-Green tensor. The material functions used in Eqs. (10) – (12) are defined by

$$\begin{aligned} \alpha &= p_1 + p_2 \Big/ \sqrt{1 + \left(p_3 b_T^S\right)^2} \\ \beta &= p_4 \Big/ \sqrt{1 + \left(p_3 b_T^S\right)^2} \\ \gamma &= p_5 b_T^S \left\{1 - 1 \Big/ \sqrt{1 + \left(b_T^S / p_6\right)^2}\right\}. \end{aligned} \tag{13}$$

Altogether, the MORPH constitutive model includes eight material parameters $p_1 - p_8$, which have to be identified by the help of experimental data.

3.2 *Uniaxial tension test and parameter identification*

Next, a reference material model representing the fully cured state of the PU based adhesive is defined. To this end, uniaxial tension tests with standard S2-dumbbell specimens and cyclic loading-unloading at constant velocity $\upsilon \approx \pm 50$ mm/min were performed. The exemplary loading program given in Fig. 7 reveals the different definitions of turning points. While the upper turning points are determined by prescribed strain amplitudes, the lower turning points were approached in such a way that the reaction force remains non-negative, in order to avoid buckling.

The depicted loading program and the resulting stresses were used as input for a parameter identification process, where the material parameters of the MORPH constitutive model (see Sec. 3.1) were identified. The experimentally obtained stress-strain curves as well as the corresponding simulation results with identified parameters are given in Fig. 8. The identified parameter set is listed in Eq. (14).

$p_1 = 10^{-5}$ **MPa**	$p_2 = 9.0228$ **MPa**	
$p_3 = 2.8566$	$p_4 = 17.658$	
$p_5 = 0.1006$ **MPa**	$p_6 = 10^{-5}$	(14)
$p_7 = 4.0020$	$p_8 = 1.0350$ **MPa**	

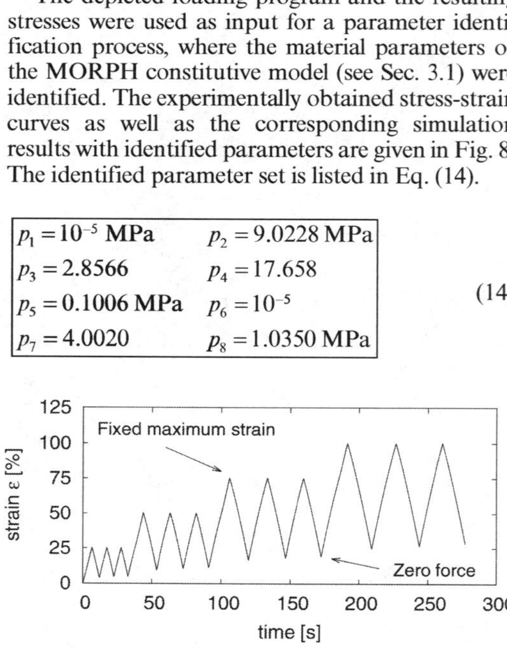

Figure 7. Exemplary loading program of the uniaxial tension tests.

From Eq. (14) it can be seen that both p_1 and p_6 tend to zero. Indeed, the value 10^{-5} was assigned as lower limit during the identification process. From this, it was concluded that parameter p_1 can be set to zero. However, the same procedure would not be possible for parameter p_6 (see parameter function γ in Eq. (13)). Instead, the corresponding material function was slightly modified as follows:

$$\gamma = p_5 b_T^S \left\{1 - p_6 \Big/ \sqrt{p_6^2 + \left(b_T^S\right)^2}\right\}. \qquad (15)$$

For the case $p_6 > 0$, both Eqs. (15) and $(13)_3$ are equal. However, the reformulated parameter function γ additionally allows for parameter values $p_6 \leq 0$. A repetition of the parameter identification process yielded the results depicted in Fig. 9. The corresponding parameters are listed in Eq. (16).

$p_1 = 0.0$ **MPa**	$p_2 = 8.8043$ **MPa**
$p_3 = 2.9503$	$p_4 = 17.807$
$p_5 = 0.1221$ **MPa**	$p_6 = -1.1601$
$p_7 = 3.5020$	$p_8 = 0.7580$ **MPa**

(16)

A comparison of Figs. 8 and 9 reveals the slightly better identification results with the modified MORPH model.

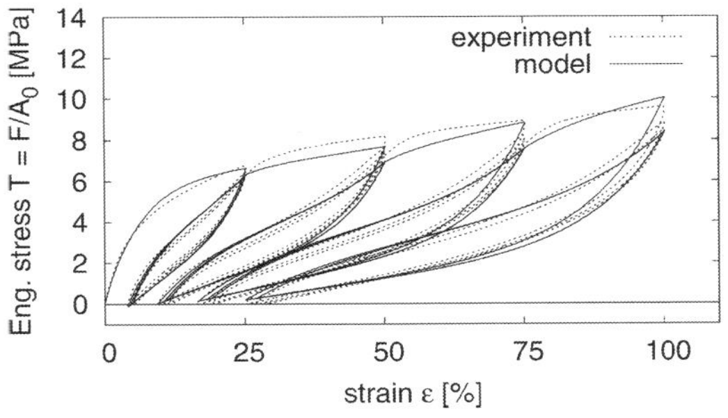

Figure 8. Comparison experiments and simulation results (original MORPH model).

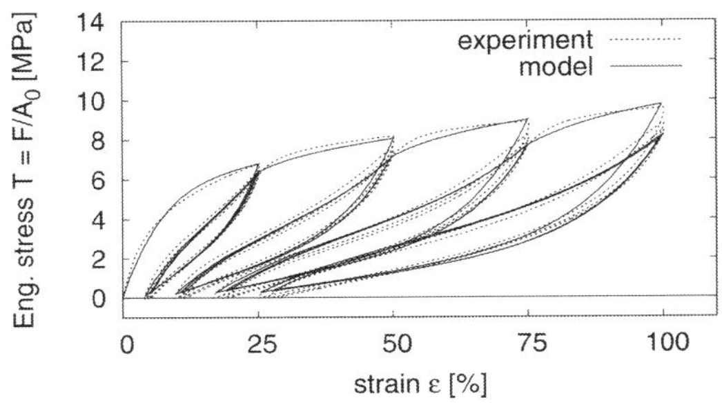

Figure 9. Comparison experiments and simulation results (modified MORPH model).

4 EXTENSION OF MORPH BY CURING PHENOMENA

4.1 *Analysis of MORPH*

Finally, the extension of the MORPH constitutive model for the representation of changing mechanical behaviour during the curing process is presented. To this end, first the model parameters $p_1 - p_8$ were analysed within a loading program including loading/unloading cycles with varied maximum strain values (see Fig. 10). The corresponding stress-strain curve of the modified MORPH model with the reference material parameters listed in Eq. (16) is given in Fig. 11.

Next, the model parameters were varied in order to evaluate their impact on the stress response. Thereby, two different sets of parameters with two distinct influences could be identified. The first set, which includes the parameters p_1, p_2, p_5 and p_8, is responsible for the overall stiffness while the basic curve shape remains the same. In the case, where these parameters are reduced to 1% of their original values, one obtains the stress-strain diagram given in Fig. 12. Compared to Fig. 11 it can

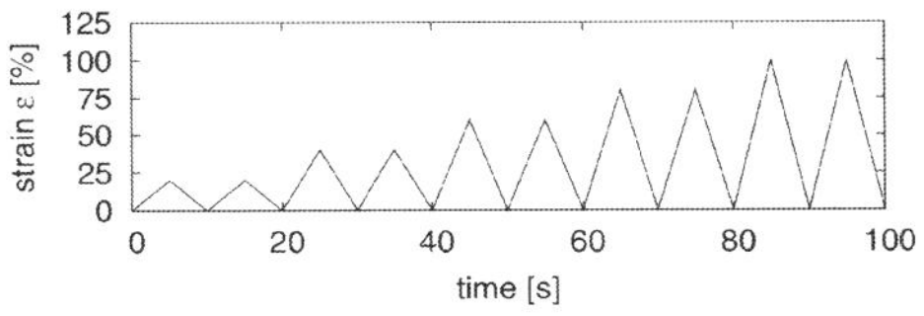

Figure 10. Loading program for the investigation of MORPH's parameters.

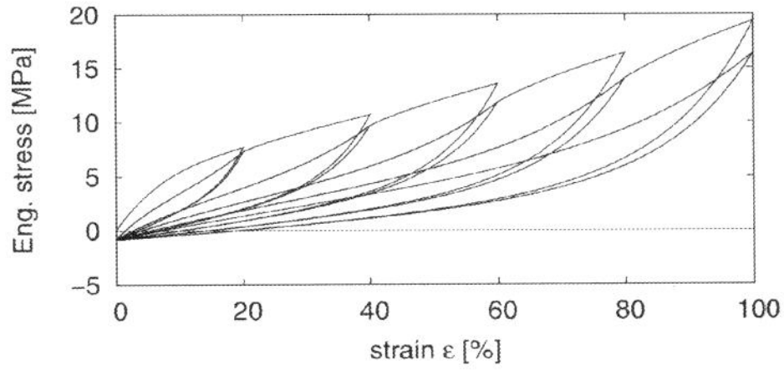

Figure 11. Stress-strain curve for the modified MORPH model.

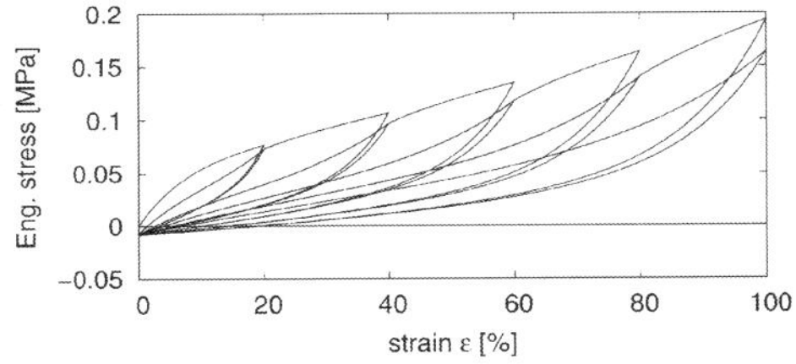

Figure 12. Stress-strain curve of the MORPH model with reduced parameters p_1, p_2, p_5 and p_8.

be seen that the basic curvature remains while the stress values are reduced to 1%. This behaviour of stiffness reduction is also indicated by the fact that each of the regarded parameters has the unit of stiffness (i.e. MPa).

The second set includes p_1, p_2, p_3, p_6 and p_7 and is mainly responsible for the basic shape of stress response. A reduction of p_1, p_2 to 0.01% and p_3, p_6 and p_7 to 1% of their original values yields the stressstrain curve given in Fig. 13. Note that the parameters p_1 and p_2 appear in both sets. This is due to the fact that basic changes in the shape of stress-strain curves are only visible if the stiffness of the Neo-Hookean part of Eq. (10) is reduced too. Otherwise, only pure elastic behaviour would be observed.

4.2 *Formulation of curing dependent material parameters*

The preliminary investigations described Sec. 4.1 motivated for the definition of curing dependent material parameters in a similar way as applied in Landgraf et al. (2014). Thereby, the fully cured material with the parameter set listed in Eq. (16) is used as reference material and the curing dependency is included by normalized function $k_s(q) \in [0, 1]$ and $k_f(q) \in [0, 1]$ as follows:

$$\begin{aligned} p_i(q) &= p_{i_0} \cdot k_s(q) \cdot k_f^2(q) \quad & i = 1,2 \\ p_j(q) &= p_{j_0} \cdot k_s(q) \quad & j = 5,8 \\ p_k(q) &= p_{k_0} \cdot k_f(q) \quad & k = 3,6,7 \\ p_4(q) &= p_{4_0} & \end{aligned} \tag{17}$$

Here, $k_s(q)$ is responsible for the basic stiffness whilst the curvature shape remains. In contrast, the factor $k_f(q)$ controls the curvature shape of the stress response. In a first approach, the two normalized function were defined by

$$k_i(q) = \Delta k + (1 - \Delta k) \cdot q^{n_i}, \quad i = s, f. \tag{18}$$

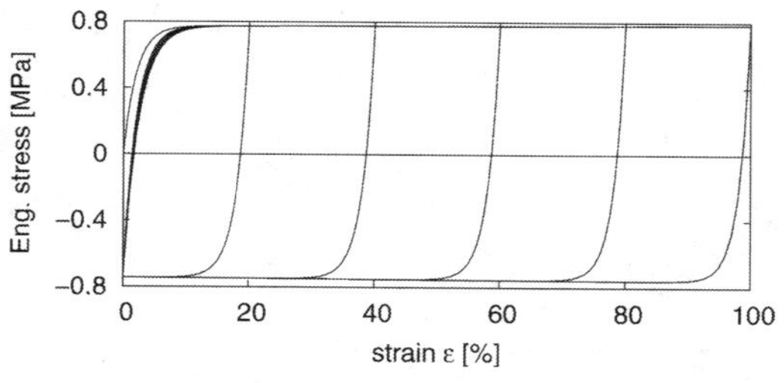

Figure 13. Stress-strain curve of the MORPH model with reduced parameters p_1, p_2, p_3, p_6 and p_7.

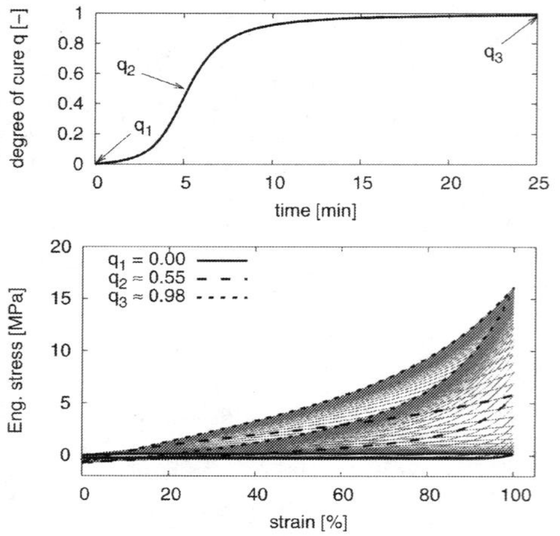

Figure 14. Degree of cure (top) and stress strain curve (bottom) for exemplary curing simulation. Specific curing states are highlighted.

Thereby, Δk, n_s and n_f are material parameters. Note that parameter Δk needs to be greater than zero in order to avoid zero valued parameters p_1–p_8.

Finally, an exemplary simulation of the curing process reveals the stress response at different curing states. To this end, a sinusoidal strain loading with frequency $f = 0.05$ Hz and maximum strain of 100% is applied to the modified MORPH model with the curing dependent material parameters of Eq. (17). For this simulation, the parameters of Eq. (18) are set to $\Delta k = 10^{-3}$, $n_s = 0.2$ and $n_f = 1$. Moreover, the underlying chemical reaction is simulated by the degree of cure model of Sec. 2.2. Here, a constant temperature of 90°C is prescribed. The complete evolutions of the degree of cure and the stress-strain response are given in Fig. 14. Moreover, specific curing states and the corresponding mechanical behaviour are highlighted.

5 CONCLUSIONS

In this paper, the phenomenological modelling of curing phenomena in a PU based adhesive was considered. To this end, the reactions kinetics was examined by DSC measurements and a models for the degree of cure and the specific enthalpy were formulated. Moreover, uniaxial tension tests revealed the elastomeric behaviour of the fully cured material and served for an adaptation of the MORPH constitutive model. Finally, the MORPH model was analyses and extended to simulate curing dependent mechanical behaviour. Further steps of the research include experimental investigations of the mechanical behaviour during the curing process and the parameter identification of the extended MORPH model with respect to these curing dependent data. Moreover, the MORPH model will be further extended by the representation of chemical shrinkage and heat expansion phenomena. The finite element implementation of the complete model will finally facilitate simulations and analyses of PMC production processes.

ACKNOWLEDGEMENT

The financial support by the German Research Foundation (DFG) within the Collaborative Research Center/ Transregio 39 is greatly acknowledged. Moreover, we thank Lars Kanzenbach and Rocco Sickel for conducting experiments.

REFERENCES

Fournier, J., G. Williams, C. Duch, & G.A. Aldridge (1996). Changes in molecular dynamics during bulk polymerization of an epoxide-amine system as studied by dielectric relaxation spectroscopy. *Macromolecules 29*(22), 7097–7107.

Henkel (2016). Technical data sheet Teroson PU 1510.

Ihlemann, J. (2003). *Kontinuumsmechanische Nachbildung hochbelasteter technischer Gummiwerkstoffe.* VDI, Düsseldorf.

Kamal, M.R. & S. Sourour (1973). Kinetics and thermal characterization of thermoset cure. *Polymer Engineering & Science 13*(1), 59–64.

Kolmeder, S., A. Lion, R. Landgraf, & J. Ihlemann (2011). Thermophysical properties and material modelling of acrylic bone cements used in vertebroplasty. *Journal of Thermal Analysis and Calorimetry 105*(2), 705–718.

Landgraf, R. (2016). *Modellierung und Simulation der Aushärtung polymerer Werkstoffe.* Verlag Dr. Hut, München.

Landgraf, R., M. Rudolph, R. Scherzer, & J. Ihlemann (2014). Modelling and simulation of adhesive curing processes in bonded piezo metal composites. *Computational Mechanics 54*(2), 547–565.

Lion, A. & P. Höfer (2007). On the phenomenological representation of curing phenomena in continuum mechanics. *Archives of Mechanics 59*(1), 59–89.

Nestler, M., W.-G. Drossel, S. Hensel, & R. Müller (2014). Fabrication method for series production of sheet metal parts with integrated piezoelectric transducers. *Procedia Technology 15*, 494–502.

Constitutive Models for Rubber X – Lion & Johlitz (Eds)
© 2017 Taylor & Francis Group, London, ISBN 978-1-138-03001-5

A RVE procedure to estimate the J-Integral for rubber like materials

M. Welsch
Institute for Mechanical Engineering and Computer-Assisted Product Development,
Helmut Schmidt University/Bundeswehr University Hamburg, Germany

ABSTRACT: The J-Integral is sensitive for nonlinear and multiaxial stress states. It can be used as a physical based failure predictor. This work presents the development of a RVE procedure and a FEM cell with an optimized mesh for high deformations to calculate the J-Integral for the whole range of multiaxial stress states and crack orientations. Based on this data, an empirical fitting equation is provided to estimate and to map the J-Integral in the post processing step of a FEM simulation. The equation is implemented as a simple subroutine in Abaqus to spot critical locations in different test specimens. Theses examples show that the J-Integral gives an adequate failure predictor and combines strain and hydrostatic stress load cases for rubber like materials.

1 INTRODUCTION

1.1 *Validity for nonlinear material*

Following the work of H. Buggisch (1981), a general balance of continuum mechanics in material notation can be given as:

$$\int_V \rho_0 dV \varphi = \int_A \psi_j n_j dA + \int_V \mathcal{X} dV, \tag{1}$$

The integrals in this equation are defined over the volume V and the corresponding surface A, given by the normal vector n. $\varphi(X_k,t)$, $\psi(X_t,t)$ and $\mathcal{X}(X_t,t)$ are functions of the material coordinate X_k and the time t. Substitution of these functions lead to the classical mass, energy and momentum balances of continuum mechanics. Furthermore, it is shown that a variation of the control volume by an infinitesimal translation or rotation leads to an alternative balance formulation:

$$J = \int_A H_{jk} dA_j = 0. \tag{2}$$

It contains the energy-momentum tensor,

$$H_{jk} = (\rho_0 \dot{\varphi} - \mathcal{X})\delta_{jk} - \psi_{j,k} \tag{3}$$

which can be calculated by neglecting volume forces and using the strain energy W as

$$H_{jk} = W\delta_{jk} - \sigma_{ij} u_{i,k}. \tag{4}$$

It reveals, that J is a conserved quantity if the condition $\mathrm{div}\,\sigma = 0$ is satisfied. That means, the J-Integral is valid for every kind of nonlinear deformation, like material and geometrical nonlinearity, unsteadiness, anisotropy and plasticity. Or in other words, the J-Integral is valid for every kind of static FEM simulation (in contrast to other fracture mechanic quantities like the stress intensity factors). J gives a scalar value with the dimension N/mm, representing an inner material load, that drives the crack propagation and generates free surfaces.

Unfortunately, there is no direct solution method to obtain the J-Integral for nonlinear problems (Le 1996). The boundary value problem of the crack has to be solved for each part and load case as well as crack shape, size and orientation.

1.2 *FEM implementation of the J-Integral*

The FEM implementation to estimate the J-Integral is classically formulated by an integral over the energy, stress and displacement gradient, that is collapsed to the crack tip:

$$J = \lim_{\Gamma\to\infty} \int_\Gamma \mathbf{n}\left(W\mathbf{I} - \sigma\frac{\partial \mathbf{u}}{\partial \mathbf{x}}\right)\mathbf{q}\,d\Gamma. \tag{5}$$

This approach goes back to Rice (1968) and C. Atkinson (1968). They worked out, that the J-Integral is a path independent quantity. However, eq. (5) is not suitable for FEM. Therefor C.F. Shih (1986) proposed a transformation to a closed-loop integral and afterwards to an area integral by using the Gauss theorem.

$$J = -\int_{C+C_+ +\Gamma+C_-} \mathbf{mH\bar{q}}\,d\Gamma - \int_{C_+ +C_-} \mathbf{t}\frac{\partial \mathbf{u}}{\partial \mathbf{x}}\mathbf{\bar{q}}\,d\Gamma \tag{6}$$

$$J = -\int_A \left(\frac{\partial}{\partial \mathbf{x}}\right)(\mathbf{H\bar{q}})\,d\Gamma - \int_{C_+ +C_-} \mathbf{t}\frac{\partial \mathbf{u}}{\partial \mathbf{x}}\mathbf{\bar{q}}\,d\Gamma \tag{7}$$

$$\mathbf{H} = W\mathbf{I} - \sigma\frac{\partial \mathbf{u}}{\partial \mathbf{x}}, \bar{\mathbf{q}}(\Gamma) = \mathbf{q}, \bar{\mathbf{q}}(C) = 0 \tag{8}$$

Eq. (7) is convenient for FEM. It is still path independent and the stress field around the crack tip must not be estimated directly, because the area adjacent to the crack tip must not be considered. Figure 1 shows both, the integration area and the corresponding representation in a finite element mesh.

However, it is allowed to set the inner path to the crack tip. In FEM solvers like Abaqus, Ansys and Marc/Mentat this implementation is known as VCE method (Helen 1975) or De Lorenzi Integral (de Lorenzi 1982). In theory this means, that the singularity must be approximated in high quality by a special element formulation (Kuna 2010). But in practice the influence of an underestimated stress field at the crack tip decreases very fast by taking more elements in the surrounding area.

The J-Integral is a 2D quantity. To get a 3D reference, the J-Integrals must be estimated all around the crack to get a mean or max value, that represents the whole crack.

1.3 *The J-Integral as failure predictor*

Working with the J-Integral in fracture mechanics means particulary: Anticipating where a crack could be critical, modelling a crack at this location and estimate the J-Integral with FEM. The estimated value is than compared to a reference value from experiments. This procedure is well accepted and a common proof concept in engineering for safeness.

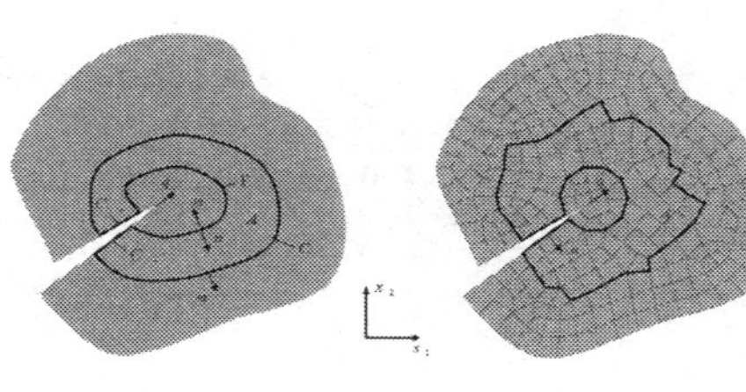

Figure 1. The J-Integral can be defined as area integral around a crack tip.

Another use case for the J-Integral in fracture mechanics is the simulation of the crack propagation.

Usually, the J-Integral is not used as general failure predictor for structural mechanic issues, like it is the case for the Mises stress.

This work deals with the idea to use the J-Integral as general failure predictor. To do so, a function like $J = f(\sigma,\in)$ must be provided, which explicitly defines the J-Integral for a given stress-strain state in the critical crack orientation. As mentioned above, there is no known direct way to get such general function for nonlinearity. The goal of this work is to present a solution of this problem.

2 J-INTEGRAL AS FUNCTION OF THE STRESS-STRAIN STATE

2.1 *The RVE model*

Figure 2 shows the developed representing volume element (RVE). The mesh is optimized for an accurate J-Integral estimate and for high deformation. It has 500.000 degrees of freedom and 125.000 elements. The crack is modelled ideally penny shaped, with 76 elements along the crack and 46 elements orthogonal to the tip. The length of the cube edges are six times the length of the crack radius. That is adequate to capture the singularity. The inner core is rotatable and tied to the outer cell by the concept of hanging nodes.

2.2 *FEM results*

The cell is loaded with two displacement and one stress condition. The conditions must cover all possible deformation and stress states. This requires some tricks. First, the most driving parameter for the J-Integral in the nonlinear domain is the hydrostatic pressure, which must be negative for crack opening,

$$p = -\frac{1}{3}(\sigma_{\max} + \sigma_{\text{mid}} + \sigma_{\min}). \tag{9}$$

All load cases with a positive hydrostatic pressure can be neglected. In all these cases the J-Integral is

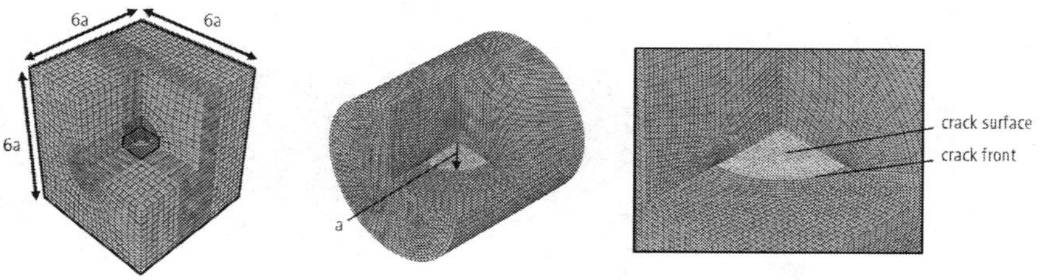

Figure 2. The RVE model with a rotatable inner core including a penny shaped crack.

zero by definition and nearly zero in simulation using a contact formulation between the crack faces. The hydrostatic pressure is used to iteratively set the stress boundary condition. The logarithmic invariants, developed by J.C. Criscione (2000), are used to set the other two displacement conditions:

$$I_A = \sqrt{\ln \lambda_{\max}^2 + \ln \lambda_{\mathrm{mid}}^2 + \ln \lambda_{\min}^2} \tag{10}$$

$$I_B = \frac{3\sqrt{6}}{I_A^3} \cdot \ln \lambda_{\max} \cdot \ln \lambda_{\mathrm{mid}} \cdot \ln \lambda_{\min} \tag{11}$$

I_A represents an equivalent strain measure, that gives a resulting amount of strain and I_B represents a load case equivalent, which goes continuously from equibiaxial $I_B = -1$ over planar (sometimes known as pure shear) $I_B = 0$ to uniaxial $I_B = 1$.

The necessary sets of boundary conditions for the FEM simulations are deduced from sets of p, I_A and I_B.

2.3 *An empirical regression model*

The proposed model to fit the numerical data is based on a set of exponential functions.

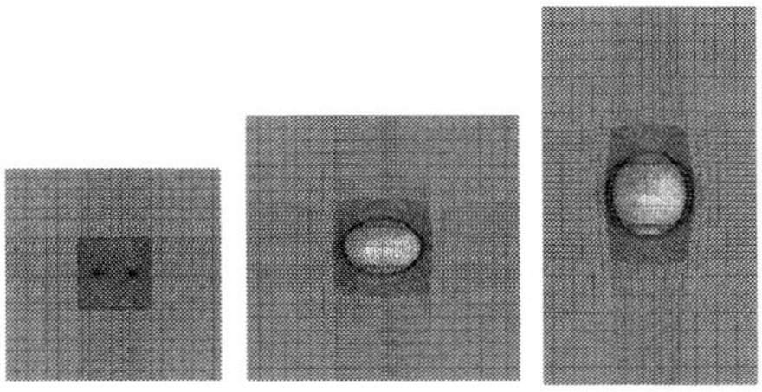

Figure 3. Crack opening for equibiaxial and uniaxial deformation close to the convergence limit.

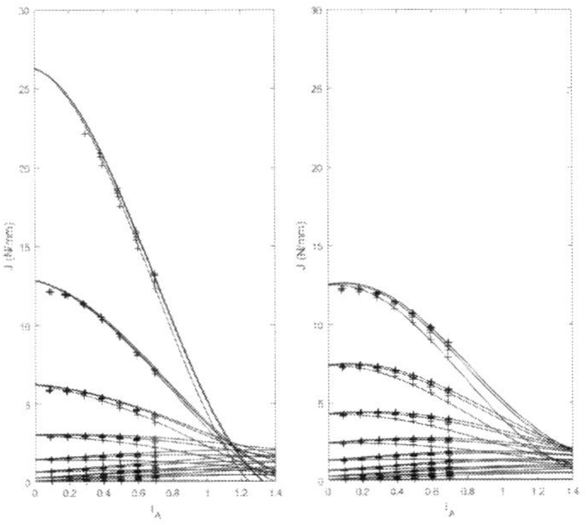

Figure 4. FEM results with Mooney-Rivlin 0.4/0.1 (left) and 0.3/0.2 (right) in comparison to eq. (12). The curves show iso-curves for hydrostatic pressure, that are spread by the load case equivalent I_B.

$$\begin{aligned}
f_1 &= (c_{11}e^{-I_A} + c_{12}e^{I_A})I_A + c_{13} \\
f_2 &= (c_{21}e^{-I_A} + c_{22}e^{I_A})I_A + c_{23} \\
f_3 &= (c_{31}e^{-I_A} + c_{32}e^{I_A})I_A + c_{33} \\
f_4 &= (c_{41}e^{-I_B} + c_{42}e^{I_B})I_B + c_{43} \\
f_5 &= (c_{51}e^{-p} + c_{52}e^{p})p + c_{53} \\
J(p, I_A, I_B) &= f_1 e^{f_2 p} p + f_3 p I_B + f_4 p I_A + f_5 I_A I_B
\end{aligned} \tag{12}$$

There are other ways for interpolation or fitting (for examples a KNN algorithm or a neuronal net, that both lead to very good regression models). However, the most important reason (quite apart from programming in Fortran77) for choosing this set of e-functions is due to the handling of the extrapolation behaviour. The cell is optimized for high deformations, but 65% macroscopic strain on the cell means up to 800–1000% strain at the crack tip. It is necessary to provide a smart assuming for

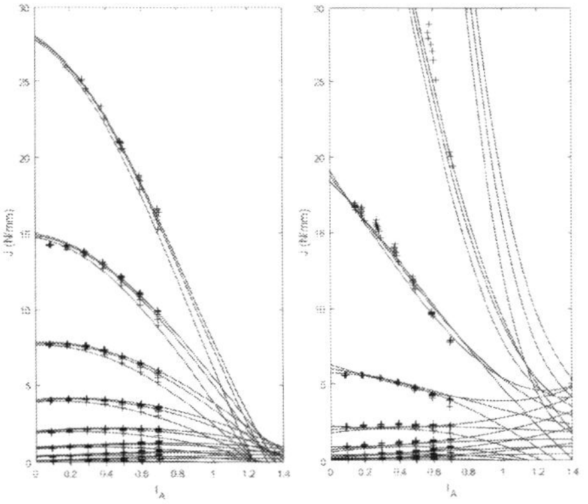

Figure 5. FEM results with Mooney-Rivlin 0.268/0.169 (left) and an advanced model, that takes the upturn into account (right).

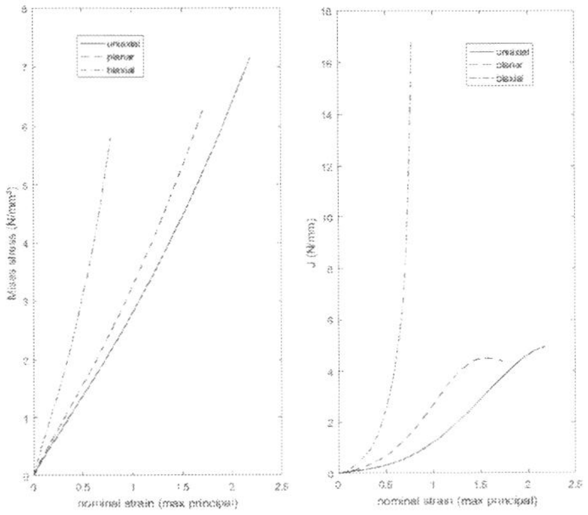

Figure 6. Comparison of Mises stress and J-Integral (eq. (12)) for MR 0.3/0.2 depending on multiaxial load case.

deformations around and beyond 100% of macroscopic strain, where no FEM results are available. There are no extra parameters to set the extrapolation tail in eq. (12). The tail is a direct result of the parameter fitting and how the tails of exponential functions work.

Figure 3 shows the crack opening close to the convergence maximum.

However, more than 100% strain always seems to be critical for durability and there is no big practical need for an estimate far beyond this region. Eq. (12) gives a reasonable extension from ca. 65% nominal strain up to 200% strain.

Figure 4 shows a comparison of the Mooney-Rivlin parameters 0.4/0.1 and 0.3/0.2 (both resulting in a shear modulus of G = 1 N/mm^2) to demonstrate the multiaxial sensitivity of the J-Integral. Additionally, an advanced material model, presented in (Welsch 2015), is investigated. The result is shown in Figure 5. The model fits the whole upturning shape for different multiaxial conditions of a measured specimen. It is compared with a corresponding Mooney-Rivlin parameter set, which gives the same result, but only for the first stretch region until the upturn occurs.

Figure 6 shows the comparison between the Mises stress and the J-Integral as functions of the nominal strain to get a better impression, how the J-Integral works in contrast to the Mises stress. The exponential extrapolation leads to a decrease of the J-Integral beyond 200% strain, that leads to an interesting effect, presented in the next section.

Figure 7. P0 specimen in compression. Left: Mises stress, middle: maximal principal logarithmic strain, right: J-Integral (eq. (12)) for MR 0.3/0.2.

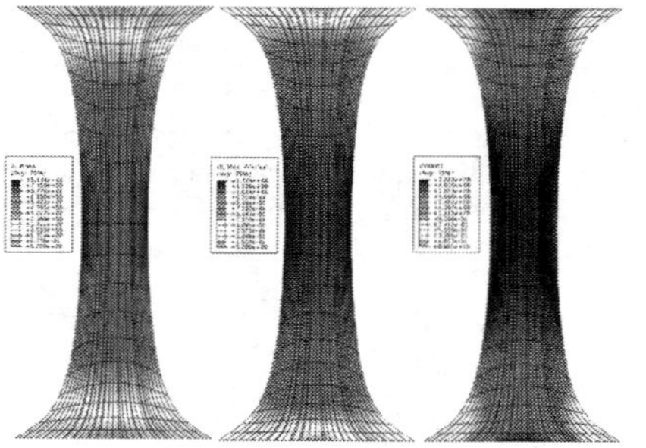

Figure 8. P0 specimen stretch. Left: Mises stress, middle: maximal principal logarithmic strain, right: J-Integral (eq. (12)) simulated for MR 0.3/0.2.

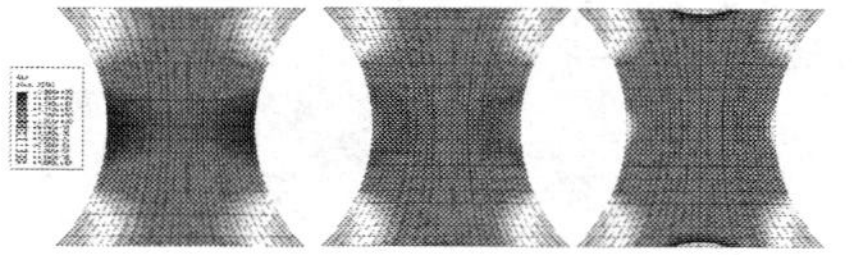

Figure 9. P0 specimen stretch with different pre-strain. The amplitude of the J-integral (eq. (12)) is mapped to the undeformed configuration.

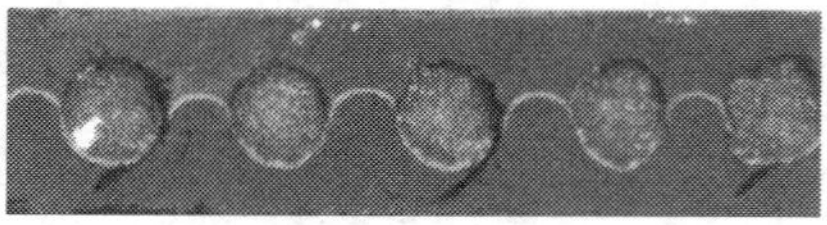

Figure 11. A cut through a durability test sample shows the typical failure propagation for an air spring sleeve. Single cracks grow from both fibre layers to the middle layer and merge to a bigger crack later on.

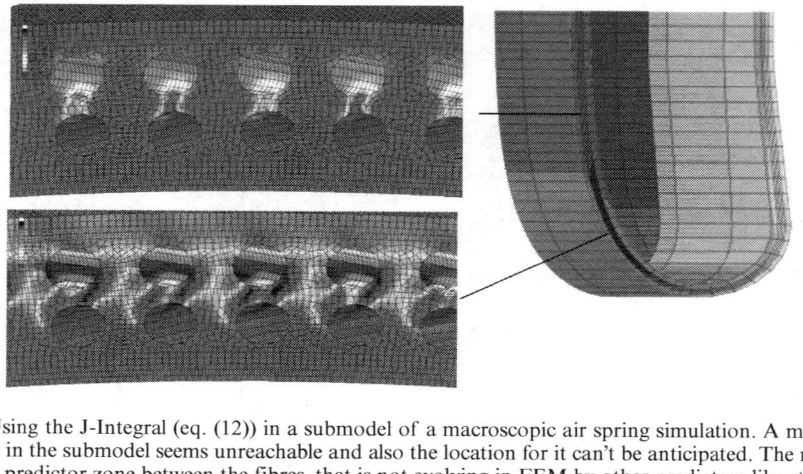

Figure 10. Using the J-Integral (eq. (12)) in a submodel of a macroscopic air spring simulation. A modeling of an accurate crack in the submodel seems unreachable and also the location for it can't be anticipated. The result shows a sharp crescent predictor zone between the fibres, that is not evolving in FEM by other predictors like stress, strain or energy.

Table 1. Coefficients for the regression model eq. (12).

	Standard model (Mooney-Rivlin)		Measurement fit with advanced model (Welsch 2015)	
Setting	MR 0.4/0.1	MR 0.3/0.2	MR equivalent	Ogden equivalent
c_{11}	−0.097783619	−0.067867754	0.109849078	0.034776199
c_{12}	−0.037126985	−0.037870567	−0.065825189	−0.047638242
c_{13}	0.155053296	0.165011137	0.228950989	0.169342445
c_{21}	0.197050801	0.220411633	−0.065309406	−0.129273742
c_{22}	−0.06229769	−0.064404719	−0.053764206	−0.093441481
c_{23}	1.185193622	0.990595503	1.111307215	1.527391216
c_{31}	0.030999577	−0.154087193	−0.363930084	0.824395638
c_{32}	0.033418541	−0.058956458	0.012107138	0.409288112
c_{33}	0.010461327	0.010830693	0.043025235	−0.038607931
c_{41}	0.119573215	0.142177787	0.145025359	−0.030823258
c_{42}	0.034442483	0.045872324	0.046534691	−0.116329205
c_{43}	0.976738549	0.9037564	0.539562579	0.772552111
c_{51}	−1.394634187	−0.868525446	−0.917194723	−2.287932685
c_{52}	−0.024575117	0.000033663	−0.001634637	−0.106979965
c_{53}	0.528326162	0.457854204	0.371863815	0.456047792

3 EXAMPLES

3.1 *P0-specimen*

The shape of the P0 specimen gives the opportunity to test characteristics and durability for compression, decompression, bending and torsional load cases in one specimen. For compression the J-Integral is zero (Figure 7). In experiments the durability for compression is 100 to 300 times higher than an equal strain level in stretching (M. Flamm 2003). The torsional load case it not shown. Here the J-Integral fails, because there is no crack opening without negative hydrostatic pressure. The shear case is adequate again. Obviously, the J-Integral does not cover every failure mode.

The uniaxial stretch simulation of the P0 specimen gives some interesting insights. In some load cases, the J-Integral spots a failure on the bottom of the plates and not in the middle of the specimen as assumed. The plate is a location where neither strain, Mises nor strain energy spot anything. Usually, this triaxial stress state at the spot is known as special failure mode with self-contained predictors.

The J-Integral handles this critical hydrostatic states in combination with strain failures by nature.

If the amplitude of the displacement is constant and the middle is shifted, the amplitude of the J-Integral shows a relieve at the most recent cross section before the J-Integral increases at the plate interface (Figure 7). This might be an effect of the extrapolation tail. The extrapolation drives all hydrostatic iso curves to zero with increasing strain. None of the effects are really unrealistic in comparison to experiments at all (J.-B. Le Cam 2008). On the contrary, the J-Integral may arise the key to these effects.

3.2 *Reinforced rubber sleeve*

The simulation of the P0-specimen demonstrates that the J-Integral as predictor combines a strain and hydrostatic stress predictor.

Therefore it can be used in complex stress states, like reinforcement analysis, where the inner load is not obvious and a crack can not be either modelled or anticipated. Figure 10 shows an analysis example. The J-Integral gives an accurate failure prediction (see Figure 11).

4 CONCLUSIONS

The J-Integral seems to give a valid predictor for failure analysis of rubber materials, because it describes the physics of crack propagation, that is reasonable most durability failures. The J-Integral represents a single value, that merges the influences of the load case and the material behaviour. Unfortunately, the use of the J-Integral as failure predictor forces the modelling of a crack at an anticipated location, since there is no direct function to calculate the J-Integral. So, there is no general method to estimate the J-Integral for uncracked deformations, that are usual results in FEM analysis. This work provides an approach to obtain such a function by using a RVE method and a fitting of the numeric data with a regression model. The model is highly influenced by the chosen material model and validated up to 65% nominal strain. Furthermore, the extrapolation solution opens the opportunity to use the model up to 200% strain. Implementing the model as subroutine in Abaqus (see appendix) a simple test specimen and a more complex reinforced air spring sleeve are investigated. The results are far different from classical approaches like strain, stress or energy predictors. It could be a first step for a unified theory of pre strained durability effects and the interaction of strain and stress at interfaces.

REFERENCES

Atkinson, C., J.E. (1968). The flow of energy into the tip of a moving crack. *International Journal of Fracture Mechanics*, 1–8.

Buggisch, H., D. Gross, K.K. (1981). Einige erhaltungssaetze der kontinuumsmechanik vom j-integral-typ. *Ingenieur- Archive 50*, 103–111.

Criscione, J.C., J.D. Humphrey, A.D.W.H. (2000). An invariant basis for natural strain which yields orthogonal stress response terms in isotropic hyperelasticity. *Journal of the Mechanics and Physics of Solids 48*, 2445–2465.

de Lorenzi, H. (1982). On the energy release rate and the jintegral for 3-d crack configurations. *International Journal of Fracture 19*, 183–193.

Flamm, M., T. Steinweger, U.W. (2003). Festigkeitshypothesen in der rechnerischen lebensdauervorhersage von elastomeren. *KGK Kautschuk Gummi Kunststoffe*, 582–585.

Helen, T. (1975). On the method of virtual crack extensions. *International Journal for Numerical Methods in Engineering 9*, 187–207.

Kuna, M. (2010). *Numerische Beanspruchungsanalyse von Rissen*. Wiesbaden: Vieweg + Teubner.

Le, K. (1996). *Kontinuumsmechanisches Modellieren von Medien mit veraenderlicher Mikrostruktur*.

Le Cam, J.-B., B. Huneau, B.E.V. (2008). Description of fatigue damage in carbon black filled natural rubber. *Fatigue Fracture of Engineering Materials 31*, 10311038.

Rice, J. (1968). A path independ integral and the approximate analysis of strain concentration by notches and cracks. *International Journal of Applied Mechanics*, 379–386.

Shih, C.F., B. Moran, T.N. (1986). Energy release rate along a three-dimensional crack front in a thermally stressed body. *International Journal of Fracture 30*, 79–102.

Welsch, M. (2015). A colloidal elastomer model. *Constitutive Models for Rubbers IX*, 197–204.

Abaqus Subroutine

```
  SUBROUTINE UVARM(UVAR, DIRECT, T,
  TIME, DTIME, CMNAME, ORNAME,
  1 NUVARM, NOEL, NPT, LAYER, KSPT,
    KSTEP, KINC, NDI, NSHR, COORD,
  2 JMAC, JMATYP, MATLAYO, LACCFLA)
C
  INCLUDE 'ABA PARAM.INC'
C
  CHARACTER*80 CMNAME, ORNAME
  CHARACTER*3 FLGRAY(15)
  DIMENSION UVAR(NUVARM), DIRECT
  (3, 3), T(3, 3), TIME(2)
  DIMENSION ARRAY(15), JARRAY(15), JM
  AC(*), JMATYP(*), COORD(*)
  REAL STRAIN1, STRAIN2, STRAIN3, STR
  ESS1, STRESS2, STRESS3, P,
  1 IA, IB, J , C12, C13, C31, C32, C33, C41, C42,
  C43, C51, C52, C53,
  2 F1, F2, F3, F4, F5
C
C Error counter :
  JERROR = 0
C Stress tensor :
  CALL GETVRM('SP', ARRAY, JARRAY,
  FLGRAY, JRCD, JMAC, JMATYP,
  1 MATLAYO, LACCFLA)
    JERROR = JERROR + JRCD
    STRESS1 = ARRAY(1)
    STRESS2 = ARRAY(2)
    STRESS3 = ARRAY(3)
C Strain tensor :
  CALL GETVRM('LEP', ARRAY, JARRAY,
  FLGRAY, JRCD, JMAC, JMATYP,
  1 MATLAYO, LACCFLA)
    JERROR = JERROR + JRCD
    STRAIN1 = ARRAY(1)
    STRAIN2 = ARRAY(2)
    STRAIN3 = ARRAY(3)
C Regression :
  C11=0.109849078397087
  C12 = -0.065825189166343
  C13 = 0.228950988894945
  C21 = -0.065309406176328
  C22 = -0.053764206037744
  C23 = 1.111307214851051
  C31 = -0.363930084322742
  C32 = 0.012107138393695
  C33 = 0.043025235383638
  C41 = 0.145025359121497
  C42 = 0.046534690756570
  C43 = 0.539562578961777
  C51 = -0.917194723194916
  C52 = -0.001634636963877
  C53 = 0.371863815234936
  P = -1*(-1.0/3.0*(STRESS1+STRESS2+STR
  ESS3))
  IA = SQRT(STRAIN1**2+STRAIN2**2+STR
  AIN3**2)
  IF(IA.GT.0.0)THEN
  IB = 3.0*SQRT(6.0)/(IA**3)*STRAIN1* STR
  AIN2*STRAIN3
  ELSE
  IB = 0.0
  ENDIF
C
  IF(P.GT.0.0)THEN
  F1 = C11*exp(-IA)*IA + C12*exp(IA)*IA+C13
  F2 = C21*exp(-IA)*IA + C22*exp(IA)*IA +
  C23
  F3 = C31*exp(-IA)*IA + C32*exp(IA)*IA+C33
```

```
F4 = C41*exp(–IB)*IB + C42*exp(IB)*IB+C43
F5 = C51*exp(–P)*P + C52*exp(P)*P+C53
J = F1*P*exp(P*F2) + F3*P*IB+F4*P*IA +
F5*IA*IB
ELSE
UVAR(1) = 0.0
ENDIF
IF(J .GT.0.0)THEN
UVAR(1) = J
ELSE
UVAR(1) = 0.0
ENDIF
UVAR(2) = IA
IF(IB.GT.1.0)THEN
UVAR(3) = 1.0
ELSEIF(IB.LT.–1.0)THEN
UVAR(3) = –1.0
ELSE
UVAR(3) = IB
ENDIF
UVAR(4) = –P
C If error, write comment to .DAT file :
IF(JERROR.NE.0)THEN
WRITE(6, *) 'REQUEST ERROR IN UVARM
FOR ELEMENT NUMBER ',
1 NOEL, 'INTEGRATION POINT NUMBER ',
NPT
ENDIF
RETURN
END
```

Constitutive Models for Rubber X – Lion & Johlitz (Eds)
© 2017 Taylor & Francis Group, London, ISBN 978-1-138-03001-5

Lateral stiffness of rubber mounts under finite axial deformation

A.H. Muhr
TARRC, UK

ABSTRACT: Theory for the combined bending and shear behaviour of multi-layer rubber-steel bonded blocks, used for estimating the lateral stiffness and critical axial load, was developed by Gent (1964) and is widely disseminated in the literature; the key equations from it also appear in most Standards for structural bearings. This theory is usually restricted to nominally small strain linear behaviour, and moderately high shape factor multilayer blocks, and the effect of axial compliance is not usually addressed. The material model used for the rubber is very simple, requiring only linear behaviour in simple shear and the approximation of incompressibility.

This paper describes the behaviour of rubber blocks that are unlike such customary laminated bearings in that (1) the number of layers is small, including the possibility of only one (2) the shape factor, and hence axial stiffness, is low (3) effects of kinematic nonlinearity and axial compliance are included. The work builds on an earlier experimental and FEA study, which addressed effects of non-linear lateral load-deflection behaviour and instability of single layer rubber blocks. More comprehensive experiments are reported and interpreted using an approximate analytical beam-column theory.

1 INTRODUCTION

In this paper the applicability of a recent theory for the lateral stiffness of a strip of rubber under a finite strain (Goodchild et al, 2017) to bonded blocks of rubber is investigated.

2 BEAM COLUMN THEORY FOR RUBBER MOUNTS

2.1 *The basic theory*

To derive the expression for lateral stiffness we first need a differential equation to describe the lateral static equilibrium deflection $v(x)$ of a beam-column as a function of the axial coordinate x, under the influence of the moment $M(x)$, arising from the axial load P, the shear load Q and the couple $M(0)$ applied to the end. It is assumed that the beam-column has uniform, continuous constitutive properties along its length. By elimination of the moment M the condition for static equilibrium can be expressed as (Gent 1964, Derham & Thomas 1981):

$$\left(\left(\frac{BR}{P+R}\right)\frac{d^3}{dx^3}+P\frac{d}{dx}\right)v+Q=0 \tag{1}$$

where the bending stiffness parameter, B, is the reciprocal of the bending compliance per unit length of the beam, and the shear stiffness parameter R is the reciprocal of the shear compliance per unit length of the beam.

Equation 1 may be cast into the form:

$$\left(\frac{d^3}{dx^3}+q^2\frac{d}{dx}\right)v+\frac{Q}{P}q^2=0 \tag{2}$$

where

$$q^2=\frac{P(P+R)}{BR} \tag{3}$$

Equation 2 has the solution:

$$y=a_1\cos(qx)+a_2\sin(qx)-\frac{Q}{P}x+a_3 \tag{4}$$

The constants of integration a_1, a_2, a_3 are found from the boundary conditions. For a beam-column, with ends constrained to remain parallel and normal to the original axis of the undeformed beam, but free to move in one lateral direction (that of the y axis) at the top, we have

$$\begin{aligned} &x=0, \quad y=0\Rightarrow a_3=-a_1 \\ &x=0, \quad \theta=0\Rightarrow\frac{dy}{dx}=\frac{Q}{R}\Rightarrow a_2=\frac{1}{q}\left(\frac{Q}{R}+\frac{Q}{P}\right) \qquad \text{(5a)} \\ &x=L, \quad \theta=0\Rightarrow\frac{dy}{dx}=\frac{Q}{R}\Rightarrow a_1=-\frac{1}{q}\left(\frac{Q}{R}+\frac{Q}{P}\right)\tau \end{aligned}$$

where

$$\tau \equiv \tan\left(\frac{qL}{2}\right) = \frac{1-\cos(qL)}{\sin(ql)} = \frac{\sin(qL)}{1+\cos(qL)} \tag{5b}$$

where the latter two identities are used in the algebraic manipulations.

The shear stiffness $k_s = Q/v(L)$ may be found by substituting equations (5a) into equation (4); using (3) for q^2 and (5b) for τ leads to

$$k_s = \left(\frac{2\tau}{q}\left(\frac{P+R}{RP}\right) - \frac{L}{P}\right)^{-1} = \frac{P^2}{2qB\tau - PL} \tag{6}$$

Equation (6) will be valid at small lateral deflections such that changes in P, B, R, and L are of second order in the lateral deflection. However, we may regard P, B, and R all to be functions of L, and thus subject to change in the axial deformation associated with application of the axial load P prior to the lateral deformation.

The unstrained beam-column is considered to be a prism with a number of uniformly spaced parallel planar cross-sections, normal to the axis, on which lateral strain of the rubber is constrained to be zero. Such planes represent inextensible reinforcing shims to which the rubber is bonded, so will be termed bonded shims. In this paper, the number of bonded shims is zero or two (one at each end), but the theory could readily be generalized for any higher integer. It will be convenient in this section to consider domains of applicability of the equations according to the Shape Factor S of the geometries of rubber block being considered, defined by

$$S \equiv \frac{\text{area of bonded shim}}{\text{stress-free surface area of the rubber between adjacent bonded shims}}$$

The concept of shape factor is especially useful for cylindrical rubber blocks, but is also useful for estimating stiffness for other shapes of block (eg cuboids) by analogy with the equations for cylindrical blocks, provided the ratio of maximum and minimum radii from the centroid of a cross section is not too extreme. Throughout this section the beam-column is considered to have axial compliance, in contrast to some forms of the theory presented for laminated bearings (eg Gent 1964, Derham & Thomas 1981), and other variables are expressed as functions of the axial extension ratio of the beam:

$$\lambda = \frac{L}{L_0} \tag{7}$$

where L is the length of the beam column and the subscript 0 indicates the unstrained (reference) state.

2.2 *Thomas theory for a rubber prism ($S = 0$)*

Equation (6) has been adapted by Thomas (Goodchild et al., 2017) for the case that an axial load—tensile or compressive—is applied to a simple homogeneous rubber block or strip. To this end, he developed appropriate expressions for the parameters P, B and R as functions of the axial elongation λ. Thus in contrast to Gent's original analysis of laminated bearings the axial compliance was explicitly included. This is the key first step in the attempt here to develop a unified beam-column theory for all geometries of rubber beams.

The equation for the axial force P is obtained for a homogeneous uniaxial strain using an appropriate strain-energy function; Thomas chose the Mooney-Rivlin equation which gives the tensile force N as:

$$N(\lambda) = 2A_0(\lambda - \lambda^{-2})\left(C_1 + \frac{C_2}{\lambda}\right) \tag{8}$$

where A_0 is the cross-sectional area of the prism in the unstrained state, as indicated by the subscript 0. Equation (8) complies with the convention that N is positive for tension ($\lambda > 1$); P as defined in Figure 1 is positive for compression ($\lambda < 1$) thus $P = -N$. Goodchild et al. (2017) pointed out that the Mooney-Rivlin strain energy function is only appropriate for the case of tension. In compression, the neo-Hookean model is better, which is a special case of the Mooney-Rivlin function with $C_2 = 0$. For our present, exploratory, purposes, the neo-Hookean model will be used for simplicity, though better models could be considered (eg Thomas 1955; Gent & Thomas, 1958). The bending stiffness B for the prism at an axial elongation λ is obtained from

$$B(\lambda) = E_B I \tag{9}$$

In Equation (9), E_B is the incremental Young's modulus, referred to the strained state:

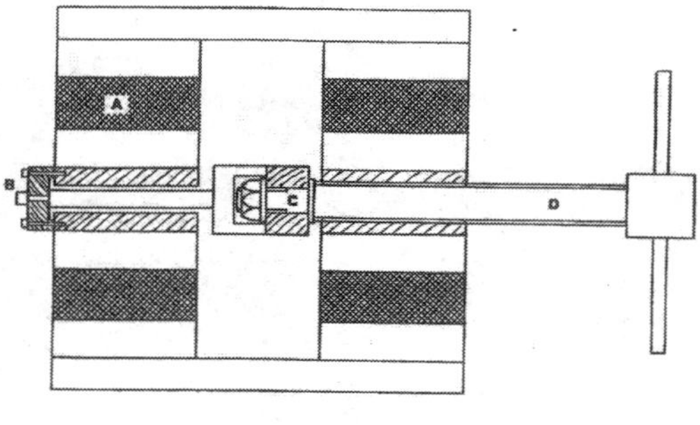

Figure 1. Quadruple shear apparatus. A—rubber block; B—load cell; C—spigot of lead screw with yoke etc enabling tensile or compressive force to be applied to load cell; D—lead screw.

$$E_B = \lambda \frac{d}{d\lambda}\left(\frac{P(\lambda)}{A}\right) = 2\left\{\left(2\lambda^2 + \frac{1}{\lambda}\right)C_1 + \left(\lambda + \frac{2}{\lambda^2}\right)C_2\right\} \quad (10)$$

and I is the second moment of area of a cross section through the deformed rubber prism normal to its axis:

$$I = \pi\frac{a^4}{4} = \pi\frac{a_0{}^4}{4\lambda^2} \text{ - cylinder of radius } a_0 \quad (11)$$

$$I = \frac{a^3 b}{12} = \frac{a_0{}^3 b_0}{12\lambda^2} \text{ - cuboid of cross section } a_0.b_0 \quad (12)$$

For the prism of rectangular cross-section the bending is about an axis parallel to the side of unstrained width b_0; for the purposes of this paper $a_0 < b_0$.

The equation for R may be derived for the limiting case of a small lateral shear strain superimposed on a finite axial strain for a hyperelastic material:

$$R = \frac{N}{1-\lambda^{-3}} = \frac{P}{\lambda^{-3}-1} \quad (13a)$$

Equation (13a) holds true whatever the strain energy function. In the case of $C_2 = 0$ (ie a neo-Hookean material), substituting (8) into (13a) gives

$$R = \lambda G A_0 \quad (13b)$$

Equation (13b), for a neo-Hookean material, is true not just in the limit of a small shear strain, but also for a finite shear strain imposed on a block under finite axial strain.

2.3 *Instability to lateral deflection*

From equation (6) it is evident that the lateral stiffness would fall to zero if τ were to become infinite, giving the criterion for lateral instability

$$\text{if } qL \to \pi \text{ then } t \to \infty \text{ so } k_s = \frac{P^2}{2qBt - PL} \to 0 \quad (14)$$

Using equations (7) (with $C_2 = 0$) to (13) to calculate qL, we find, for a neo-Hookean material

$$qL = \lambda L_0\sqrt{\frac{P(P+R)}{BR}} = L_0\sqrt{\frac{A_0(1-\lambda^3)}{I_0(1+2\lambda^3)}} \quad (15)$$

So the criterion for instability is

$$\frac{1-\lambda^3}{1+2\lambda^3} = \left(\frac{\pi}{L_0}\right)^2 \frac{I_0}{A_0} \equiv \chi \quad \Rightarrow \quad \lambda = \left(\frac{1-\chi}{1+2\chi}\right)^{\frac{1}{3}} \quad (16)$$

An interesting feature of equation (16) is that a beam with χ equal to or greater than 1 cannot go unstable. For example, for a cylindrical beam

$$\chi = \left(\frac{\pi a_0}{2L_0}\right)^2 \quad (17)$$

Thus the theory predicts that a neo Hookean cylinder of rubber will not go unstable if $a_0 > 2L_0/\pi$. A single rubber layer with bonded shims meeting this geometrical criterion would be considered to have a shape factor of $S_0 > 1/\pi = 0.3183$. This condition would be met by an individual layer in a typical laminated bearing.

3 COMPARISON OF THOMAS THEORY WITH EXPERIMENTAL RESULTS FOR BONDED BLOCKS

3.1 *Rationale for the comparison*

The Thomas theory, outlined above, is intended for application to small lateral deflections of rubber prisms in a prior state of homogeneous uniaxial finite strain. The boundary conditions, during the finite axial strain, should thus be of zero lateral stress at the planar ends. This poses a difficulty as to how to apply the lateral force Q to the prism to drive the subsequent small lateral deflection. This was resolved in the original paper, in which the experiments were restricted to tensile axial strain applied to a long rectangular strip, and Q was applied by means of fixtures lightly clamped a distance λL_0 apart to a portion of the prestretched strip. This arrangement overcame the perturbation to the homogeneous strain field caused by the terminal clamps of the strip. Our purpose here is to find out how significant the departures are for blocks of rubber bonded to endplates, of a range of shape factors, all low in comparison with those typical for rubber-steel laminated bearings.

3.2 *Testpieces and techniques*

Seven sets of unfilled natural rubber blocks were moulded with bonded steel endpieces; the dimensions for each set are given in Table 1. It is apparent that three of the rectangular blocks have a shape factor exceeding the critical value $1/\pi$, so that the theory predicts that they would not go unstable in compression, had they lubricated rather than bonded end conditions.

With the exception of the cylinders E and F, the steel endplates had the same lateral dimensions as the rubber mouldings. E and F had end plates of 50 mm diameter, and the rubber was moulded with a filet taking the rubber radius from a_0 up to the

Table 1. Geometry of testpieces.

	a_0	b_0	L_0	S	χ
Cuboids	mm	mm	mm	–	–
A	34	64	77.5	0.14	0.1583
B	54	66.5	10	1.49	23.983
C	53.5	66.5	20	0.74	5.8853
D	53.5	66	30	0.49	2.6157
Cylinders	mm		mm	–	–
E	15	–	80	0.094	0.0867
F	20	–	80	0.125	0.1542
G	25	–	80	0.156	0.2410

edge of the 50 mm diameter endplate over a length of approximately 10 mm.

The rectangular blocks (cuboids) are the same mouldings as featured in the experimental work of Fan et al. (1992), whereas the cylinders were moulded using a different unfilled NR material about ten years later, as recentering springs for the Rolling-ball rubber-layer seismic isolators studied in the PSTRBIS ECOEST 2 Project in 1999 (Guereiro et al, 2007; Dona et al, 2017). Their dimensions and shape factors are given in Table 1.

Combined compression and shear was carried out using two methods. The first, used for testpieces A, B, C and D, used a quadruple shear method (see Figure 1), with which the axial deformation was imposed by a uniaxial 100 kN Instron screw-driven machine, while the relative shear deflection was applied by hand using a lead screw. Some of the results using this technique were reported by Fan et al (1991). The technique has the advantage that the axial load can be monitored using the Instron load cell, but the disadvantage that only the relative shear deflection is controlled, and it was soon revealed that the shear deflections of the two sets of blocks are not in general equal and opposite. This is because, as instability is approached, the most stable configuration is not the symmetric one.

The second method used a double shear jig (see Figure 2). In this case, the axial deformation was applied by hand, either with a manual hydraulic ram (for compression) or using the nuts on the clamping screws (for tension). After locking the axial displacement with the nuts, the shear deflection was applied with an Instron servohydraulic machine. This approach had the advantage that the shear deflection was completely controlled by the servohydraulic actuator, but the disadvantage that the axial force could not be monitored during the shear.

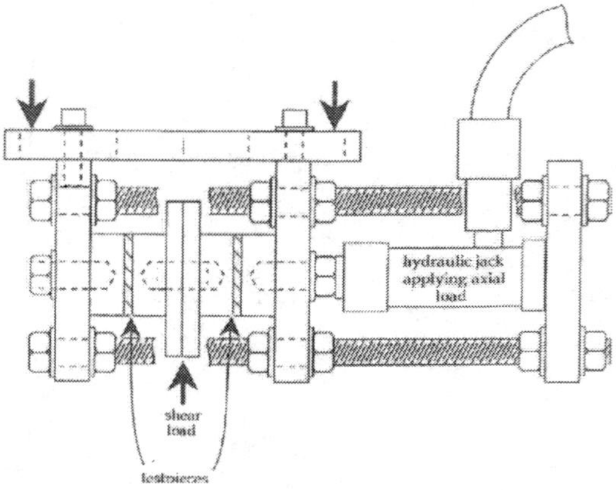

Figure 2. Double shear jig. The testpieces discussed in the paper are of much lower shape factor than those depicted, but could be accommodated by using four studs of appropriate length (two are concealed by those in view).

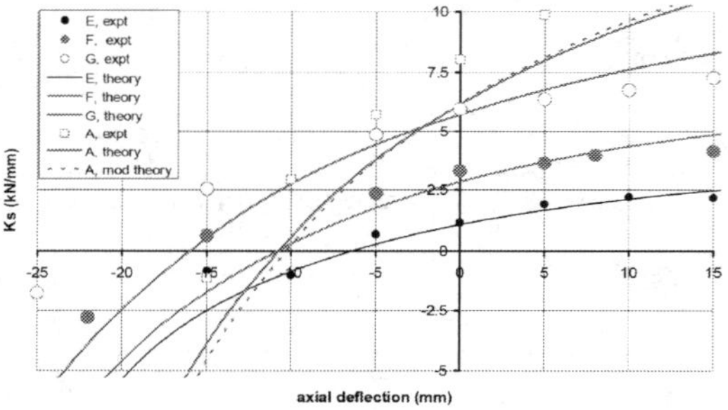

Figure 3. Small strain lateral stiffness of the cylindrical blocks and of the softest rectangular block (A) as functions of axial deflection.

3.3 *Comparison of the results with the theory*

Figures 3 and 4 compare the predicted values of the small-strain lateral stiffnesses of the blocks as functions of the axial deflection with the measurements. While the rectangular block of lowest shape factor A appears in both figures, to facilitate comparison, the other rectangular blocks are all much stiffer than the cylindrical blocks and results for them are best displayed—in Figure 4 - separately.

It is apparent that all the blocks show lateral instability, despite the fact that blocks B, C and D meet the criterion $\chi > 1$ given below equation (16). It might be thought that this has arisen because the bonding to the endplates is not in accord with the boundary conditions of the theory. To get more insight into this behaviour, and the reason for the experimentally observed instabilities for blocks with high χ values, the relative variation

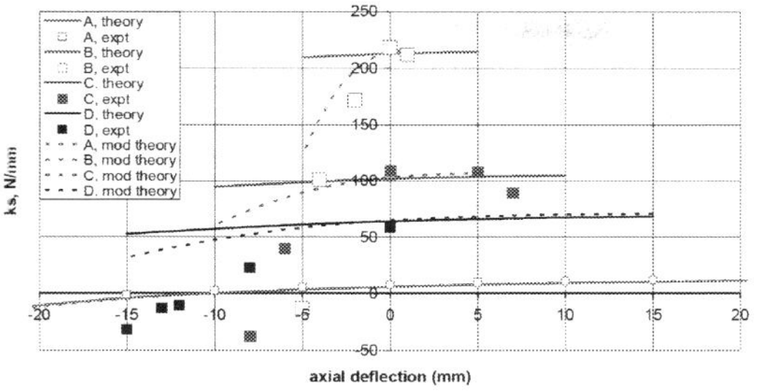

Figure 4. Small strain lateral stiffness of the rectangular blocks as functions of axial deflection. Solid lines, theory of Section 2.2 ($S = 0$), dashed lines, modified theory of Section 4 ($S > 0$).

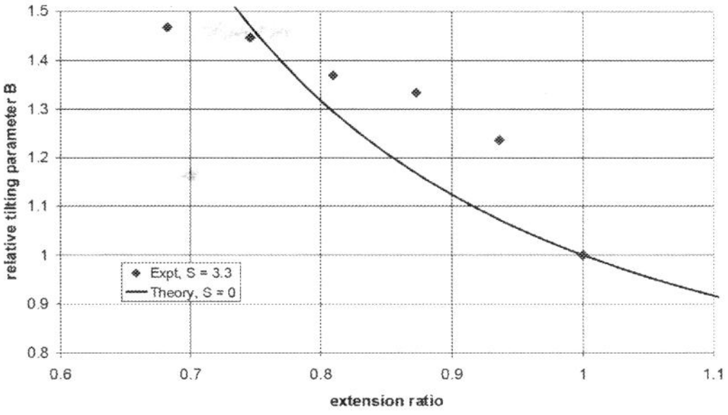

Figure 6. Dependence of relative tilting parameter B on extension ratio; full line—theory, equations (9) and (10).

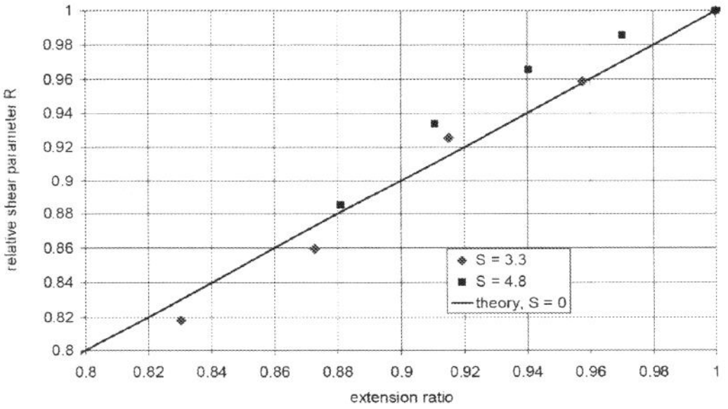

Figure 5. Dependence of relative shear parameter R on extension ratio; full line—theory, equation (13b).

of the parameters B and R with λ are plotted in Figures 5 and 6. The test results reported in these figures were obtained for pairs of bonded unfilled natural rubber rubber discs of 50 mm diameter; their shape factors are reported in the figures. To the author's knowledge, there are no other experimental data on the effect of finite axial strain on these parameters for thin bonded discs. The figures suggest that, to a fair approximation, the relative effect of finite axial strain is about the same as that given by the theory derived for the case of $S = 0$.

4 A REVISED THEORY FOR RUBBER BLOCKS WITH A PLATE BONDED AT EACH END

The aim here is to see if an appropriately modified beam-column theory might capture the essence of the behaviour shown in the previous section for single bonded blocks of rubber of low, rather than zero, shape factor. Only two of the equations will be modified. In particular, equation (8) will be replaced by

$$N \approx GA_0\left[\left(\lambda - \frac{1}{\lambda^2}\right) + 3S^2\left(1 - \frac{1}{\lambda^2}\right)\right] \tag{18}$$

Note that, for a neo-Hookean material and $S = 0$, this equation reduces to (8). The second term, in S^2, was proposed by Lindley (1966). He also showed that its magnitude and relative dependence on axial strain is in reasonable agreement with experiment. It has a somewhat more strongly rising rate with strain than the first term.

Equation 10 will be replaced by

$$E_B \approx G\left(2\lambda^2 + \frac{1}{\lambda}\right)\left(1 + \frac{2S^2}{3}\right) \tag{19}$$

The "shape factor" term $2S^2/3$ in Equation (18) is, according to Gent & Meinecke (1970), correct for a circular cross-section but an underestimate by about 14% for a square cross-section and an overestimate by about 20% for the rectangular cross-section of Block A (see Table 1). By good chance, it agrees with the more complicated expression given by Gent & Meinecke for rectangular blocks B, C and D to within 1% or so, justifying the use here of the simplified expression for the cases that the effect of shape factor is significant. Thus, equation (19) is devised to extrapolate to a useful approximate expression as $\lambda \to 1$, and to equation (10) as $S \to 0$, and takes into account the evidence from Figure 6 that the magnitude of the effect of λ on B is not wildly wrong for experimental results for a higher shape factor.

The predictions of the theory of Section 2.2, with equations (18) and (19) substituted for equations (8) and (10) respectively, are shown as dashed lines in Figure 4. It is evident that while the modified theory has some qualitative capability to capture the experimentally observed effect of compression on lateral stiffness, a large quantitative discrepancy remains.

5 DISCUSSION AND CONCLUSIONS

The objective has been to assess the applicability of the theory of Thomas (Goodchild et al., 2017), for the lateral stiffness of prisms of rubber under states of finite axial strain, to that of blocks of rubber bonded at each end to rigid plates. A summary of this theory has been presented, and it is shown that it predicts that circular blocks with nominal "shape factors" below $1/\pi$ will exhibit lateral instability for large enough compressive strain, whereas those with higher shape factors should remain stable at all axial compressions.

Although the theory works moderately well for the lateral stiffness of compressed rubber blocks with bonded endplates having shape factors S up to about 0.15, experiments show that bonded blocks such that $1/\pi < S < 1.49$ do show lateral instability at sufficiently high compression, in conflict with the theory.

As a working hypothesis, supported by what little experimental data is available, it is assumed that there is no effect of the constraint of bonding at the ends on the relative effect of axial strain on the tilting and shear parameters. However, it is noted that the "shape factor effect" due to suppression of lateral strain at the bonded ends will increase both the axial force and the tilting parameter B above the values appropriate for unbonded blocks, and approximate equations, based on the literature, quantifying these effects are presented. A modified form of the Thomas theory, using these equations, is given and found to have some qualitative capability to capture the experimentally observed effect of compression on lateral stiffness, but a large quantitative discrepancy remains.

Since FEA with a neo-Hookean material model captures the experimentally observed behaviour moderately well (Fan et al, 1992), it is not thought that the discrepancy between analytical theory and experiment is due to internal rupture (eg cavitation) reducing the tilting parameter B of the blocks. This leaves only two possibilities for the shortcoming of the analytical theory:

1. the basic uniform beam-column assumption, that B and R are constant along the axis of the blocks, may not be justifiable.
2. the proposed equations for the effect of axial strain on P and B (equations (18) and (19) respectively) may not be sufficiently accurate, in particular the ratio P/B would have to increase more strongly with compression than suggested.

It is of course possible that the discrepancy arises from a combined effect of both possibilities. A further refinement may also be necessary to get good quantitative fits, applicable also to the FEA modelling: to replace the neo-Hookean material model by a more specific and accurate material model.

ACKNOWLEDGMENTS

Thanks are due for experiments done by Lu Guang & Chen Yue Rui (data in Figures 5 and 6); by John Cook (original data for rectangular blocks), and by Fabrizia Cilento & Regina Vitale (cylindrical blocks and new data for rectangular blocks).

REFERENCES

Derham C.J. & Thomas A.G., The design of seismic bearings Report no. UCB/EERC-81/10 ed Kelly J.M. publ. University of California, 1981.

Donà M., Muhr A.H., Tecchio G., da Porto F., Experimental characterization, design and modelling of the RBRL seismic-isolation system for lightweight structures, *J Earthquake Engineering & Structural Dynamics,* 2016, published online in Wiley Online Library, DOI: 10.1002/eqe.2833.

Fan, L.J., Muhr, A.H., Parsons B. & Thomas A.G., 1992. Shear load-deflection behaviour of compressed rubber blocks. Rubbercon 92, Brighton, 14–19 June. TARRC Reprint 1420.

Gent A.N. & Meinecke E.A., 1970. Compression, bending and shear of bonded rubber blocks. *Polymer Engineering Science*, **10**, 48–53.

Gent A.N. & Thomas A.G., 1958. Forms for stored energy function for vulcanized rubber *J Polym Sci*, **28**, 625–628.

Gent A.N. 1964 Elastic stability of rubber compression springs. *J. Mech. Eng. Sci*, 6, 318. TARRC reprint 505.

Goodchild I.R., Muhr A.H. & Thomas A.G. The lateral stiffness and damping of a stretched rubber beam, submitted to *Plastics Rubber & Composites*, 2017.

Guerreiro L., Azevedo. J., Muhr A.H., Seismic Tests and Numerical Modeling of a Rolling-ball Isolation System. *J Earthquake Engineering*, **11**(1), 49–66, 2007.

Lindley P.B., 1966. Load-compression relationships of rubber units. *J Strain Analysis*, **1**, 190–195.

Thomas A.G., 1955. The departures from the statistical theory of rubber elasticity *Trans Farad Soc*, **51**, 569–582.

Constitutive Models for Rubber X – Lion & Johlitz (Eds)
© 2017 Taylor & Francis Group, London, ISBN 978-1-138-03001-5

Finite element implementation of a constitutive model of rubber ageing

J. Heczko & R. Kottner
NTIS—New Technologies for the Information Society, Faculty of Applied Sciences, University of West Bohemia, Pilsen, Czech Republic

ABSTRACT: Rubber ageing is nowadays a much discussed topic with a connection to other important phenomena, such as fracture or fatigue.

In this work, a phenomenological constitutive model of chemical ageing of rubber (based on the Lion-Johlitz model (Lion & Johlitz 2012)) was implemented using the finite element method. The main concern was to investigate the possible effects of coupling between thermomechanics and ageing and the effect of oxygen diffusion on the evolution of mechanical properties of aged material and their distribution in the computational domain. Numerical tests were performed to assess the applicability and limits of the model in predicting different types of observed behavior.

Computational tools that can simulate different types of material behavior may be useful not only to real-world simulations, but to meta-research as well, which may enable the development or revision of methods for structural optimization or parameter identification. In particular, numerical simulations are important in cases of inhomogeneous physical fields, which are common when studying elastomers.

1 INTRODUCTION

Polymers may undergo significant changes in their mechanical properies during the service life. Among the causes is usually mechanical or thermal loading or exposition to environmental effects, the latter being known as ageing. Chemical ageing changes the composition and homogeneity of a part, which may lead to stress redistribution and therefore it is needed to simulate such behavior and eventually consider ageing in the design process. A finite element implementation of an already existing phenomenological model of chemical ageing is presented in this contribution together with an example of its use on a real-world rubber part.

2 MATERIAL MODEL

The material model including diffusion is described in (Johlitz & Lion 2013) in great detail and this section only sums up the relations that were used in the presented implementation. In accordance with numerical examples in (Johlitz & Lion 2013), we assume isothermal conditions and therefore can drop the terms that depend on temperature gradient. Moreover, we neglect the viscoelastic terms as well.

The model assumes the existence of two polymer networks arranged in parallel and therefore an additive split of the stress tensor is used

$$\sigma = -p\mathbf{I} + \sigma_{CS} + \sigma_{R}, \tag{1}$$

where σ_{CS} is stress in the network that undergoes chain scission and σ_R is stress in the network that is built up by newly created bonds. The process of chain scission is captured by an internal variable q_{CS} and the creation of new bonds by q_R. Stress in the scission-affected network is given by

$$\sigma_{CS} = 2\mu_1(q_{CS})\varepsilon^D, \tag{2}$$

where ε^D is the strain deviator and μ_1 is the ageing-dependent shear modulus. Stress in the new network is given by a hypoelastic relation:

$$\sigma_R = \int_0^t 2\mu_2(q_R(s))\dot{\varepsilon}^D ds \tag{3}$$

with t being time, $\dot{\varepsilon}^D$ the strain deviator rate and μ_2 the ageing-dependent material parameter.

The dependence of μ_1 and μ_2 on internal variables is given by

$$\mu_1 = \mu_{10}(1 - q_{CS}), \tag{4}$$

$$\mu_2 = \mu_{20}\, q_R\,, \tag{5}$$

with μ_{10} and μ_{20} being material parameters.

The evolution equations for the internal variables are chosen as

$$q_{CS} = \nu_{CS}\, c \exp\left(-\frac{E_{CS}}{R\,\theta}\right), \tag{6}$$

$$q_R = \nu_R\, c \exp\left(-\frac{E_R}{R\theta}\right), \tag{7}$$

where ν_{CS}, ν_R, E_{CS}, and E_R are material parameters, c is the concentration of oxygen, θ is the absolute temperature, and $\sqrt{R = 8.314\ \text{J/mol}}$ is the universal gas constant.

The diffusion of oxygen and its consumption by the chemical reactions is described by the following balance equation:

$$\rho\frac{\partial c}{\partial t} + \nabla\cdot\mathbf{j} - \hat{c} = 0. \tag{8}$$

The oxygen flux $\mathbf{j}$ is given by

$$\mathbf{j} = -\lambda(\delta + \xi\theta)\nabla c \tag{9}$$

with λ, δ, and ξ being material parameters. The reaction term $\hat{c}$ is prescribed as

$$\hat{c} = -k(\theta, q_{CS}, q_R)(\delta + \xi\theta)c \tag{10}$$

$$k(\theta, q_{CS}, q_R) = \big(\alpha_{CS}(1 - q_{CS}) + \alpha_R(1 - q_R)\big)\exp\left(-\frac{\Delta E}{R\theta}\right) \tag{11}$$

with α_{CS}, α_R, and ΔE being additional material parameters.

3 BOUNDARY-VALUE PROBLEM

The boundary-value problem is defined on the computational domain Ω whose boundary Γ is divided into several parts $\Gamma_D^{\mathbf{u}}$, Γ_D^c, $\Gamma_N^{\mathbf{u}}$, and Γ_N^c, each of which corresponds to a boundary condition. Following relations hold: $\Gamma = \Gamma_D^{\mathbf{u}} \cup \Gamma_N^{\mathbf{u}} = \Gamma_D^c \cup \Gamma_N^c$ (see e.g. Sec. 2.7 in (Ehlers & Bluhm 2013)).

The weak formulation of the problem is: Find displacement $\mathbf{u}$ and oxygen concentration c such that fulfill the following equations for any testing functions $\delta\mathbf{u}$, δc

$$\int_\Omega \nabla\delta\mathbf{u} : \sigma\, \mathrm{d}V = \int_{\Gamma_N^{\mathbf{u}}} \delta\mathbf{u}\cdot\bar{\mathbf{t}}\,\mathrm{d}S\,, \tag{12}$$

$$-\int_\Omega \delta c(\rho\dot{c} + k c)\,\mathrm{d}V + \int_\Omega \nabla\delta c\cdot\mathbf{j}\,\mathrm{d}V = \int_{\Gamma_N^c} \delta c\,\bar{j}\,\mathrm{d}S\,, \tag{13}$$

with the Dirichlet boundary conditions

$$\mathbf{u} = \bar{\mathbf{u}} \qquad \text{on } \Gamma_D^{\mathbf{u}}\,, \tag{14}$$

$$c = \bar{c} \qquad \text{on } \Gamma_D^c\,, \tag{15}$$

and the initial conditions

$$\mathbf{u} = \mathbf{u}_0 \qquad \text{at } t = 0 \text{ in } \Omega\,, \tag{16}$$

$$c = c_0 \qquad \text{at } t = 0 \text{ in } \Omega\,. \tag{17}$$

3.1 *Reformulation of the balance equation*

In order to implement the boundary-value problem, the balance equation (12) was reformulated by differentiating it with respect to time:

$$\int_\Omega \nabla\delta\mathbf{u} : \dot{\sigma}\,\mathrm{d}V = \int_{\Gamma_N^{\mathbf{u}}} \delta\mathbf{u}\cdot\dot{\bar{\mathbf{t}}}\,\mathrm{d}S\,. \tag{18}$$

Taking eqs. (2) and (3) into account, the stress rate becomes

$$\dot{\sigma} = -\dot{p}\mathbf{I} + \underbrace{2\left(\dot{\mu}_{01}\varepsilon^D + \mu_{01}\dot{\varepsilon}^D\right)}_{\dot{\sigma}_{CS}} + \underbrace{2\mu_{02}\,\dot{\varepsilon}^D}_{\dot{\sigma}_R}\,. \tag{19}$$

4 NUMERICAL EXAMPLES

This chapter describes numerical examples that were used to test the implemented model. The examples have been implemented using the SfePy finite element framework (Cimrman 2014), which uses Python as its main coding language.

Following a study of a vulcanizing chamber (Keckstein, Jirasko, & Kottner 2016) (see Fig. 1), a simplified model of a rubber seal treated therein was studied. The seal is a critical part of the vulcanizing chamber and its tightness determines successful proces of tire manufacture. Therefore, it is subject to periodic replacement due to changes in mechanical properties, which are caused by mechanical loading at elevated temperatures.

The vulcanizing chamber itself is axisymmetric with the seal at its perimeter and the diameter of the chamber being much larger than the cross-section of the seal. Therefore, the state of plane strain was assumed in the model presented here, although the script can also handle three-dimensional problems.

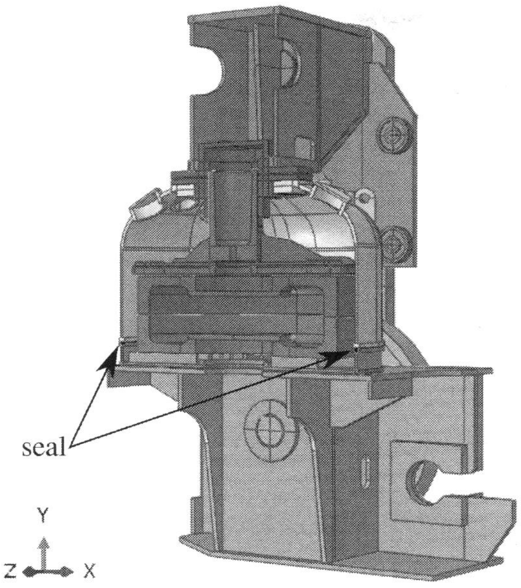

Figure 1. Curing press in section with the location of the vulcanizing chamber seal.

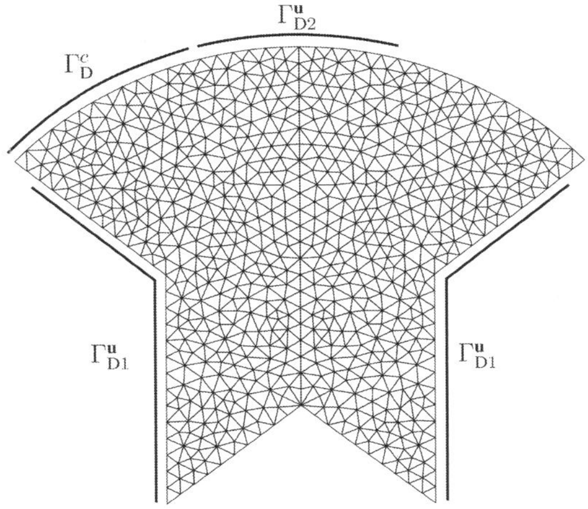

Figure 2. Undeformed mesh of the vulcanizing chamber seal.

The finite-element mesh in its undeformed state is depicted in Fig. 2 together with the regions where Dirichlet boundary conditions were prescribed. Bilinear quadrilateral elements were used.

The boundary conditions were:

$$\mathbf{u} = 0 \qquad \text{on } \Gamma^{\mathbf{u}}_{D1}, \tag{20}$$

$$\mathbf{u} = \bar{\mathbf{u}}(t) = \left[0, u_y(t)\right]^{T} \qquad \text{on } \Gamma^{\mathbf{u}}_{D2}, \tag{21}$$

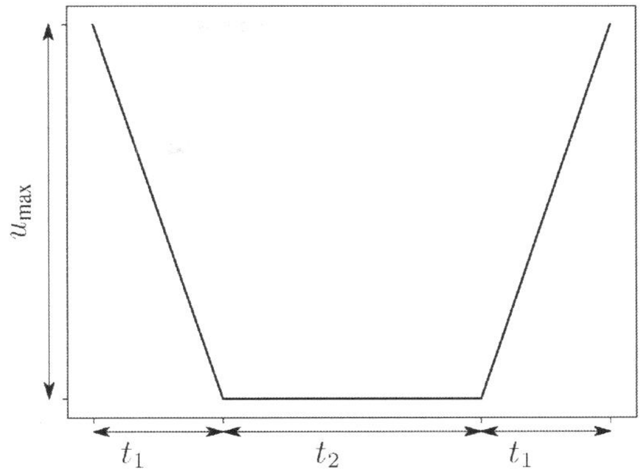

Figure 3. Prescribed vertical compressive displacement of the seal.

$$c = \bar{c} = 1 \qquad \text{on } \Gamma^{c}_{D}, \tag{22}$$

$$\bar{\mathbf{t}} = 0 \qquad \text{on } \Gamma^{\mathbf{u}}_{N}, \tag{23}$$

$$\bar{j} = 0 \qquad \text{on } \Gamma^{c}_{N}. \tag{24}$$

The function $u_y(t)$ is depicted in Fig. 3. Values of its parameters were $u_{max} = 0.8$ mm, $t_1 = 0.1$ h, $t_2 = 19.8$ h. The initial conditions were set to

$$c = 1 \qquad \text{in } \Omega, \tag{25}$$

$$q_{CS} = q_R = 0 \qquad \text{in } \Omega. \tag{26}$$

An implicit finite-difference solver was used for time-stepping and the Newton method (from the Scipy package) as the nonlinear system solver.

Following values of material parameters were used:

$$\begin{aligned} &\lambda = 10^{-6}\,\mathbf{kgs/m^3}, && \rho = 920\,\mathrm{kg/m^3}, \\ &E_{CS} = 5000, && E_R = 5000, \\ &\Delta E = 1.0, && \\ &\alpha_1 = 10.0, && \alpha_2 = 10.0, \\ &\nu_{CS} = 10^{-4}, && \nu_R = 10^{-4}, \\ &\mu_{10} = 1.0\,\mathrm{Pa}, && \mu_{20} = 1.0\,\mathrm{Pa}, \\ &\delta = 1.0, && \xi = 10^{-2}. \end{aligned} \tag{27}$$

5 RESULTS

Since the evolution equations for the internal variables q_{CS} and q_R are formally the same and since the same values of the corresponding material parameters were used, the resulting values of q_{CS} and q_R equal at all time-steps. The value of these internal variables is here assumed to quantify the degree of ageing and change in mechanical properties (by Eqs. (4) and (5)).

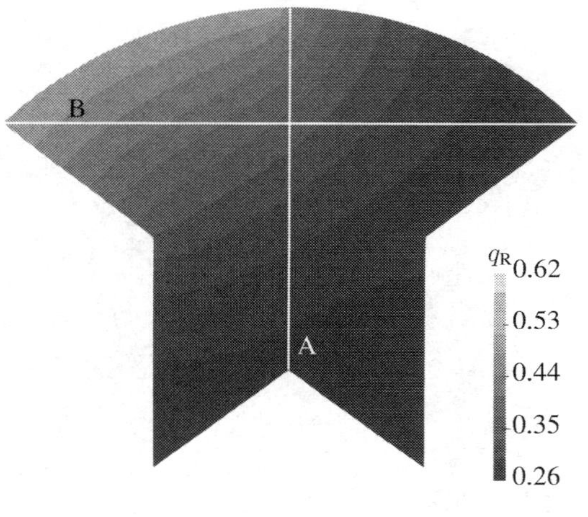

Figure 4. Resulting values of the internal variable q_R in the vulcanizing chamber seal after 20 hours at 25°C.

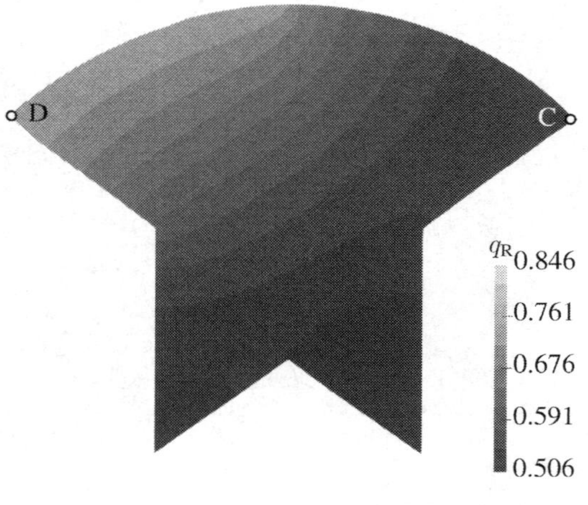

Figure 5. Resulting values of the internal variable q_R in the vulcanizing chamber seal after 20 hours at 170°C.

The spatial distribution of q_R at the end of the simulations is depicted in Figs. 4 and 5, each figure corresponding to different temperature (25°C or 170°C). In Figs. 6 and 7 are the results plotted along two lines in different directions and the concentration of oxygen c is plotted in addition to q_R. It can be clearly seen that the higher temperature results in higher degree of ageing, which was expected. The spatial distribution of oxygen concentration conforms to the boundary condition (22) in the sense that access of oxygen accelerates ageing.

Time dependence of the examined quantities at two different locations is shown in Fig. 8. At point C, the concentration c is held constant at all times, which results in rapid ageing. At point D, oxygen

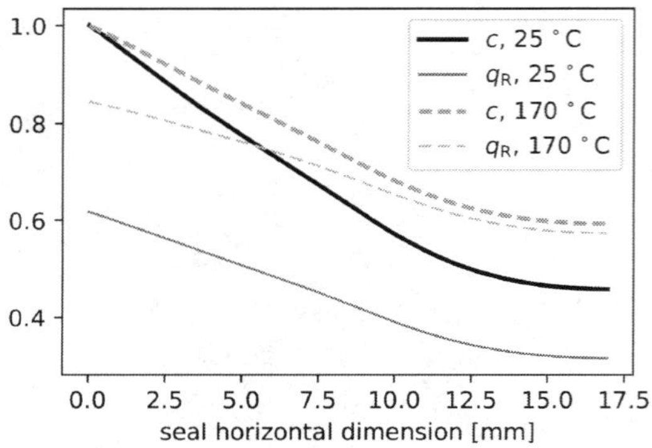

Figure 6. Resulting values of concentration c and internal variable q_R along the vertical axis of the seal (line A in Fig. 4).

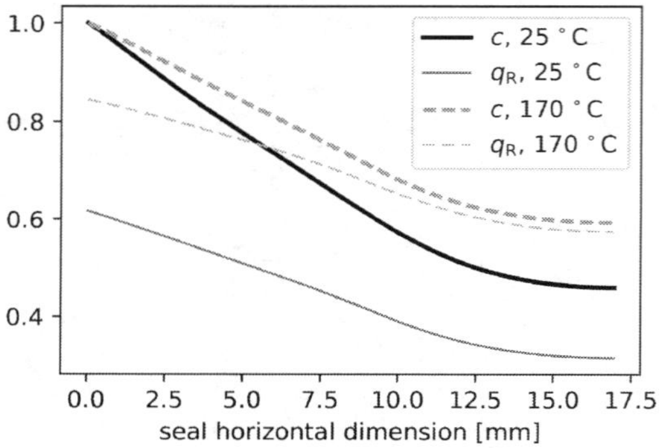

Figure 7. Resulting values of concentration c and internal variable q_R along the horizontal dimension of the seal (line B in Fig. 4).

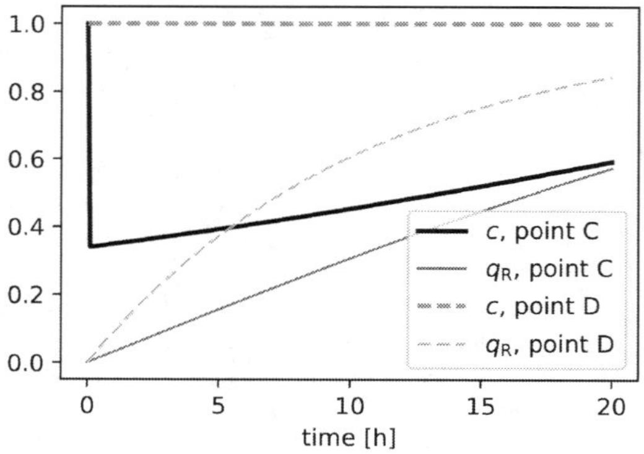

Figure 8. Oxygen concentration and internal variable q_R as functions of time at two different locations (see Fig. 5 for points C and D).

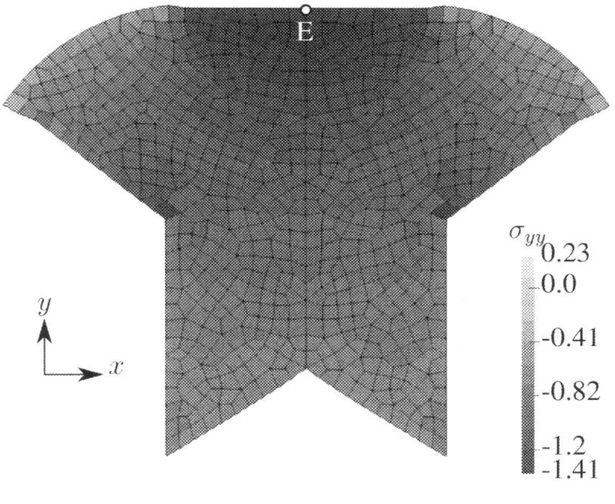

Figure 9. Stress distribution (σ_{yy}) at the beginning of the pressure-holding period ($t = 0.1$ h), temperature 25°C.

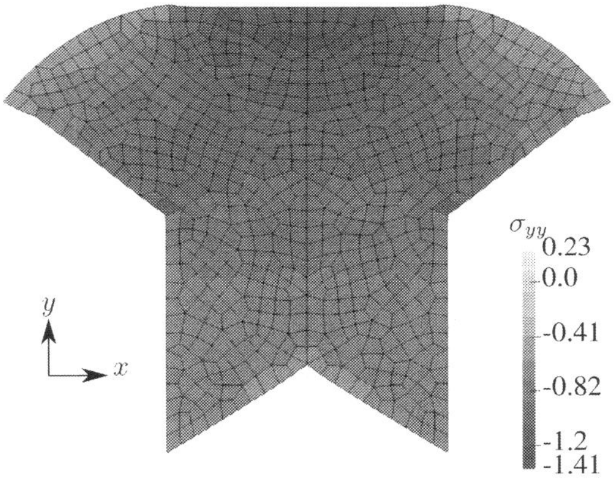

Figure 10. Stress distribution (σ_{yy}) at the end of the pressure-holding period ($t = 19.9$ h), temperature 25°C.

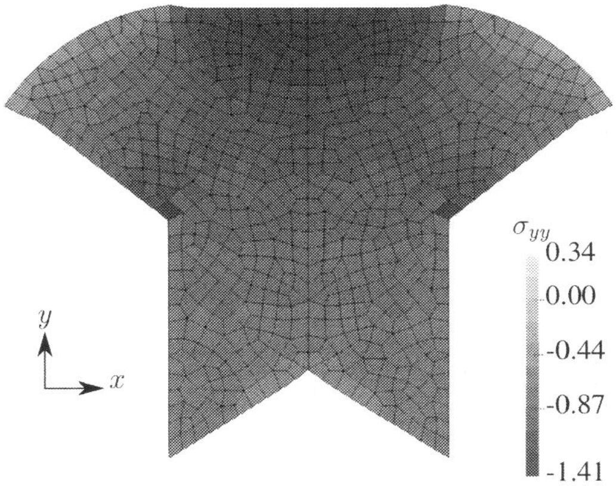

Figure 11. Stress distribution (σ_{yy}) at the beginning of the pressure-holding period ($t = 0.1$ h), temperature 170°C.

is consumed shortly after beginning of the simulation and diffusion is insufficient to supply enough oxygen resulting in significantly slower ageing.

The spatial distribution of stress in the deformed configuration is shown in Figs. 9 and 10 for the temperature 25°C and in Figs. 11 and 12 for the temperature 170°C. Only the normal component in the y-direction is shown here. A decrease in stress due to chain scission is clearly visible in the most loaded areas.

The evolution of individual stress parts (as in Eq. (1)) in time is depicted in Figs. 13 and 14. Again, the effect of ageing is more pronounced at the higher temperature. The reformation-related stress part, σ_R, is constant during the holding period (up to strain variations due to stress redistribution) and the effect of the newly created network becomes apparent after unloading. The effect

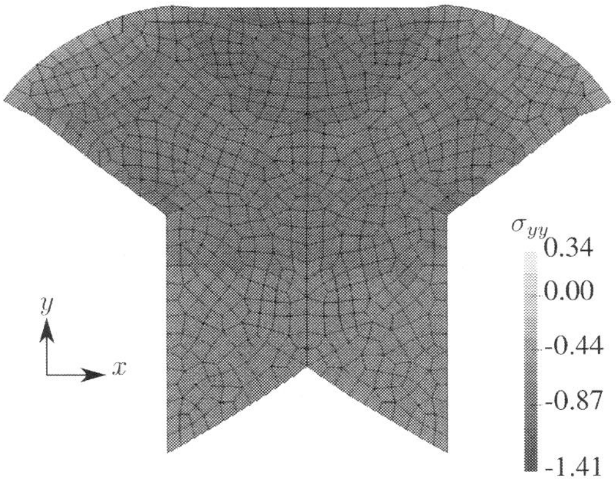

Figure 12. Stress distribution (σ_{yy}) at the end of the pressure-holding period ($t = 19.9$ h), temperature 170°C.

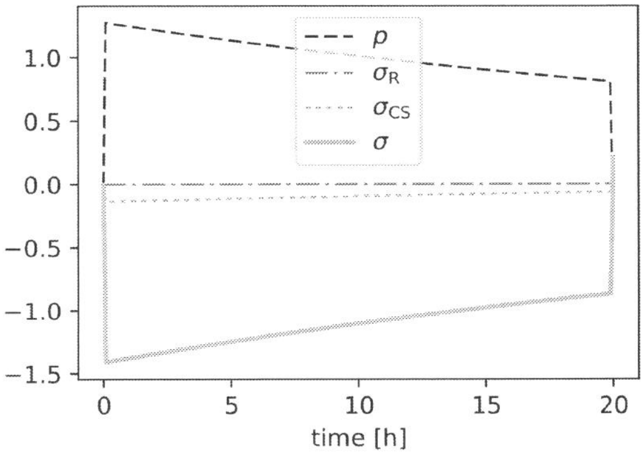

Figure 13. Stress evolution (σ_{yy}) in time at the location E (see Fig. 9), temperature 25°C.

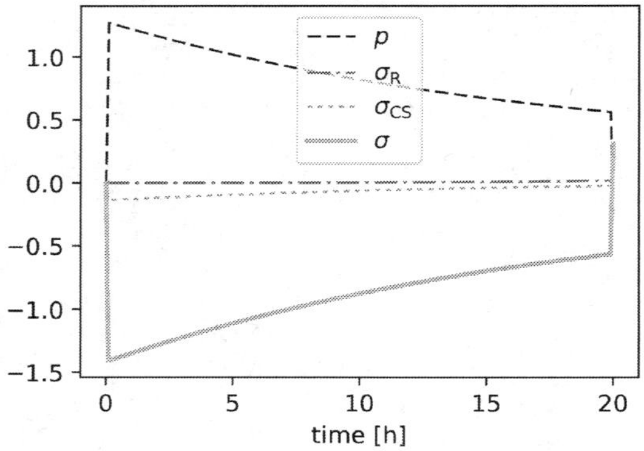

Figure 14. Stress evolution (σ_{yy}) in time at the location E (see Fig. 9), temperature 170°C.

of chain scission is apparent through the decrease of σ_{CS}.

6 SUMMARY

Inhomogeneous distribution of mechanical properties evolves as a consequence of the complex behavior in both space and time. The model may be used to predict the distribution after a period of combined mechanical and thermal loading. The present state of the implementation allows for simulating isothermal processes under small strains. While the extension by a thermo-mechanical coupling might be straightforward, thanks to the capabilities of the used SfePy finite element framework, simulating diffusion under large strain deformations will probably need additional theoretical treatment.

ACKNOWLEDGEMENT

This publication was supported by the project LO1506 of the Czech Ministry of Education, Youth and Sports under the program NPU I.

REFERENCES

Cimrman, R. (2014). SfePy—write your own FE application. In P. de Buyl and N. Varoquaux (Eds.), *Proceedings of the 6th European Conference on Python in Science (EuroSciPy 2013)*, pp. 65–70. http://arxiv.org/abs/1404.6391.

Ehlers, W. & J. Bluhm (2013). *Porous media: theory, experiments and numerical applications*. Springer Science & Business Media.

Johlitz, M. & A. Lion (2013). Chemo-thermomechanical ageing of elastomers based on multiphase continuum mechanics. *Continuum Mechanics and Thermodynamics 25*(5), 605–624.

Keckstein, T., J. Jirasko, & R. Kottner (2016). Finite element analysis of a curing press with focus on tightness of the vulcanizing chamber. In *ASME 2016 International Mechanical Engineering Congress and Exposition*, pp. V009T12A086–V009T12A086. American Society of Mechanical Engineers.

Lion, A. & M. Johlitz (2012). On the representation of chemical ageing of rubber in continuum mechanics. *International Journal of Solids and Structures 49*, 1227–1240.

Experimental characterisation

Constitutive Models for Rubber X – Lion & Johlitz (Eds)

Internal failure behavior of rubber vulcanizates under constraint conditions

E. Euchler, K. Schneider & G. Heinrich
Leibniz-Institut fuer Polymerforschung Dresden e. V., Dresden, Germany

T. Tada
Sumitomo Rubber Industries Ltd., Kobe, Japan

H.R. Padmanathan
Indian Institute of Technology, Kharagpur, India

ABSTRACT: The internal failure behavior of unfilled and silica filled Solution-Styrene-Butadiene-Rubber (SSBR) has been investigated with respect to cavity formation and growth during tensile deformation. The internal damage of rubber pancake specimen during tensile loading is caused by hydrostatic load due to constraint strain and incompressibility of rubber. Using a home-made dilatometric cell, stress as well as volume strain vs. apparent strain were determined simultaneously. Due to restricted shrinkage of pancake specimen in radial direction, the dilatometrically estimated volume change is related to cavitation. The present study focuses on investigations of the cavitation phenomena using mechanical tests and fracture surface analysis. According to results of former studies, network constitution (e.g. rubber-filler interaction) influences the onset of cavity formation and the rate of cavity growth for filled SSBR. It was found that size and size distribution of cavities determine the internal failure behavior of silica filled non-crystallizing rubber vulcanizates, like SSBR.

1 INTRODUCTION

Durability of rubber products, like tires, is strongly associated to their failure behavior. Especially, internal failure processes of rubber materials are still not fully understood. Internal damage in rubber vulcanizates occurs due to constraint strain conditions, where the typical shrinkage perpendicular to the deformation direction due to incompressibility of rubber is suppressed. The constraint conditions lead to increasing stress concentrations within the rubber vulcanizates and a large amount of volume strain is generated. This increase in volume under deformation is attributed to formation and growth of cavities.

Several research groups have investigated and discussed cavitation in rubber using experimental and analytical approaches. First experimental studies on cavitation in rubber were performed by Holt & McPherson (1937), Busse (1938), Yerzley (1939) and Jones & Yiengst (1940) using a dilatometry equipment to estimate the volume change of rubber samples during deformation. For example, they found out that the content of filler influences the amount of volume strain and that a characteristic "yield point" describes the onset of internal damage.

Following the methods and techniques of the earlier studies Bekkedahl (1949), Gee et al. (1950) and Gent & Lindley (1958) intensified investigations in this field. They used so-called pancake specimen to realize constraint strain conditions in rubber vulcanizates. Thereby, they observed an influence of the dimensions of the disk-shaped specimen and, therefore, introduced the "shape factor" in this context to describe their diameter-to-length ratio.

Especially Gent performed extensively experimental as well as analytical attempts to describe cavitation and introduced a mechanical criterion for the onset based on the Young's modulus of rubber vulcanizates (Gent & Lindley 1961). Thus, analytical determination of the onset of volume strain attributed to onset of internal failure due to cavity formation was possible. This approach was taken up for further analytical investigations on this topic to improve the validity of this mechanical criterion. For example Williams & Schapery (1965), Lindsey (1967), Ball (1982), Chou-Wang & Horgan (1989), Horgan & Polignone (1995), Sivaloganathan (1999), Li et al. (2008) and Kakavas & Perig (2015) extended the existing model for example by considering the contribution of surface energy of particles, which influences the amount of stress concentration around small cavities, or by addition of damage parameters depending on initial size of inclusions and material constants.

The extended and improved models describing the internal failure process were used for numerical investigations on cavitation in rubber by Blatz & Kakavas (1993), Dollhofer et al. (2004), Lopez-Pamies (2009), Hocine et al. (2011) and Lefèvre et al. (2015).

Nowadays, for experimental investigations on cavitation during deformation X-ray scattering as well as diffraction methods were applied. For example Shinohara et al. (2007) and Zhang et al. (2015) detected cavity formation at nanoscale by using small-angel-X-ray-scattering. Le Gorju Jago (2012) and Tada et al. (2015) performed computer tomography (CT) to evaluate cavity formation online during deformation.

Due to the long period of time in which volume strain effects in rubber vulcanizates attributed to cavitation were investigated, several reviews were published highlighting the most important findings, e.g. Stringfellow & Abeyaratne (1989), Gent (1990), Fond (2001), Dorfmann et al. (2002), Diani et al. (2009), Lopez-Pamies (2009) and Le Cam (2010).

However, there are still open questions regarding the mechanisms and criteria of cavity formation and cavity growth. Only comprehensive investigations will close the gap of understanding the origin and the development of internal damage under deformation. In this case, the evaluation of the internal failure process in silica filled rubber vulcanizates has rarely been published jet. For that reason, our study is focused on experimental investigations on internal failure behavior of non-crystallizing rubber materials filled with silica.

2 EXPERIMENTAL

2.1 *Materials*

Unfilled and silica filled rubber compounds were prepared using solution-styrene-butadiene-rubber (SSBR) BUNA VSL 2525–0M (Lanxess, Germany). The recipes of investigated rubber materials are shown in Table 1. The amounts of all ingredients are expressed in parts per hundred rubber (phr).

The mixing process consists of two steps: First, SSBR was mixed with stearic acid and zinc oxide in an internal mixer (HAAKE™ Rheomix 600p) at 110°C. For filled samples precipitated silica Ultrasil 7000GR (Evonik Industries, Germany) and disulfidic coupling agent Si266 (Evonik Industries, Germany) were added in this step. The total

Table 1. Investigated rubber materials.

Ingredients				
phr	SSBR-0	SSBR-30	SSBR-30 s	SSBR-50 s
VSL 2525-0M	100	100	100	100
Stearic acid	2	2	2	2
Zinc oxide	3	3	3	3
Ultrasil 7000GR	0	30	30	50
Si266	0	0	3	5
CBS	1.4	1.4	1.4	1.4
DPG	1.7	1.7	1.7	1.7
Sulfur	1.5	1.5	1.5	1.5
Rheometry				
min				
T_{90} (at 160°C)	4.99	11.51	7.65	11.42

mixing time for unfilled and filled samples was 5 min and 15 min, respectively. In the second step, the masterbatches were mixed with the curatives CBS, DPG and sulfur in a two-roll mill (Polymix 110 L) at 50°C for 10 min.

The obtained green rubber compounds were vulcanized in a compression molding press at 160°C to sample sheets of 1 mm thickness. The vulcanization times (T_{90}) were determined from rheometric measurements using a Rubber Process Analyzer (RPA) at 160°C.

The denotation of investigated rubber materials in Table 1 represents the following: The number behind "SSBR" refers to the amount of the silica and "s" refers to the presence of the coupling agent.

2.2 *Specimen geometry*

To investigate the internal failure behavior of rubber vulcanizates by mechanical tests, disk-shaped pancake specimens were applied. In this study, the vulcanized rubber samples were fixed between cylindrical polycarbonate sample holders with super-glue LOCTITE 406 (Henkel, Germany). Thus, constraint strain on top and bottom surfaces the rubber sample was generated (Fig. 1).

In the case of pancake specimen the length (*l*) in tensile direction is small compared to the dimension in radial direction (diameter, *d*), thus, certain hydrostatic tension arises because the rubber sample cannot shrink in radial direction. Therefore, the volume of the rubber sample increases as a function of applied tensile deformation. The amount of volume change depends on the degree of constraint strain, this means it depends on the diameter-to length ratio of the pancake specimen.

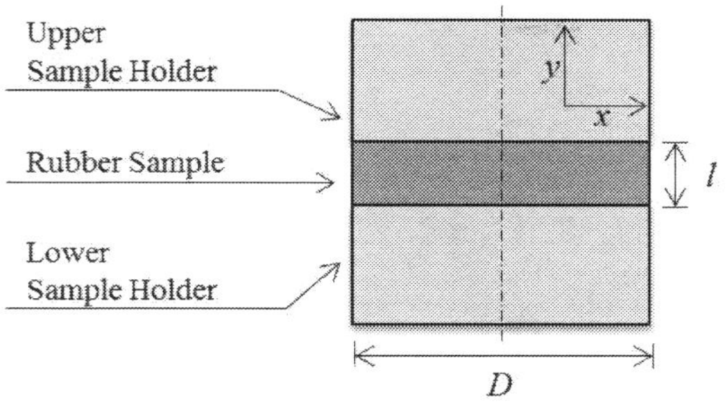

Figure 1. Principle of applied pancake specimen. D and l represent diameter and length of the sample, respectively. Upper sample holder is loaded in longitudinal direction (y), lower sample holder is fixed.

The diameter-to-length ratio is defined as shape factor (Gent and Lindley, 1959):

$$S = D / (4 \cdot l) \tag{1}$$

In this study, the pancake specimen was defined by a length l = 1 mm, a diameter D = 20 mm and a resulting shape factor S = 5.0.

2.3 *Testing equipment*

Using a home-made equipment, the volume change of a pancake specimen during tensile loading can be estimated. The pancake specimen was mounted in a glass vessel filled with a liquid. The volume change of the pancake specimen during deformation was estimated by measuring the displacement of the liquid column in a calibrated glass capillary connected to the glass vessel, using a video camera system. The results were expressed as volume strain of the pancake specimen, i.e. volume change divided by original volume. The authors described the testing equipment and its procedure detailed elsewhere (Heinrich et al. 2016).

Additionally, analysis of fracture surfaces was performed using digital light microscopy (Keyence VHX 5000, Japan) to evaluate the surface topology of rubber pancake specimen after fracture.

2.4 *Testing setup*

To investigate the internal failure behavior of SSBR under constraint strain monotonic loading was applied. The tensile tests were performed at room temperature with a crosshead speed of 1 mm/min.

3 RESULTS

3.1 *Internal failure behavior of SSBR pancake specimen with respect to the filler systems*

Figure 2 shows stress and volume strain as a function of apparent strain for the investigated rubber materials. Although tests were performed until fracture, for better visualization the results are presented in an apparent strain range from 0 to 1.

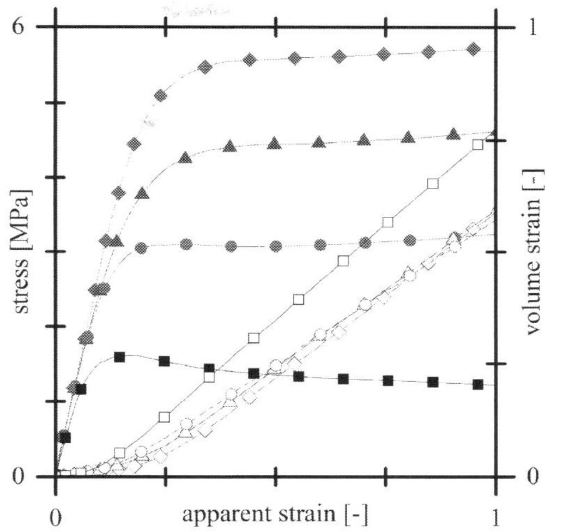

Figure 2. Stress- (full elements) and volume strain (blank elements) vs. apparent strain of SSBR pancake specimen: SSBR-0 (!), SSBR-30 (,), SSBR-30 s (7), SSBR-50 s (∧).

At low apparent strain, stress increases rapidly with increasing strain due to the high stiffness caused by constraint strain. Once a certain threshold is exceeded, stress ceases to increase with strain, as though the rubber has reached a stress value comparable to the yield point of thermoplastics. As the apparent strain increases beyond this threshold, stable deformation occurs under almost constant stress. Furthermore, when apparent strain is lower than the threshold, nearly no volume strain is detectable. But volume strain increases when apparent strain passes the threshold, referred to the transition from elastic to inelastic volume deformation. After passing this transition range, volume strain increases with apparent strain in a linear fashion.

The profile curves in Figure 3 illustrate the topologies of cross sections through the fracture surfaces (x ≈ 2000 μm) recorded by digital light microscopy. For better visualization the curves are shifted vertically. On the one hand, for unfilled SSBR, a fracture surface with distinct roughness can be observed, but no characteristic features can be detected. On the other hand, for silica filled SSBR significantly deep regions are visible, which characterize the dimensions of formed cavities. Representatively, the fracture surface of SSBR-30 s is shown and typical regions of cavities with a

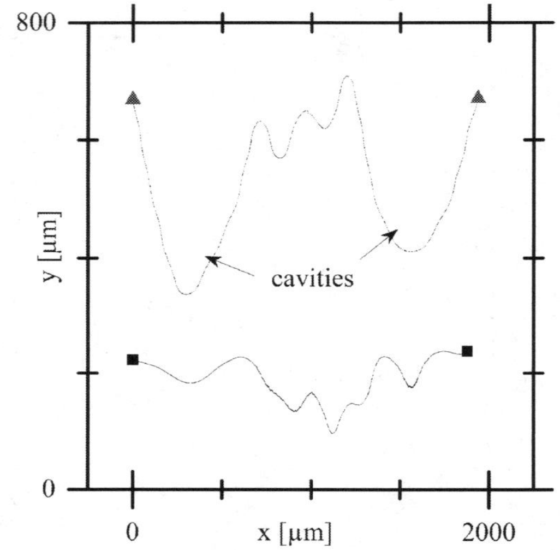

Figure 3. Profile curves of cross sections from fracture surfaces: SSBR-0 (!) and SSBR-30 s (7).

diameter about 700 µm and a depth of 300–350 µm are marked.

Furthermore, from surface analysis of silica filled SSBR the size and the size distribution of cavities depending on the rubber composition can be estimated. Using the digital light microscopy the detected characteristic length scale of cavity diameters is in the range of 60–850 µm. With respect to the rubber composition, the histogram (Fig. 4) quantifies the cavity size distribution within a representative sample area of 36 mm^2. In fact, the number of small and medium-size cavities increases with increasing filler content. By adding silane as coupling agent the number of larger cavities decreases.

3.2 *Internal failure behavior of SSBR pancake specimen characterized by volume strain deformation*

The characteristic transition range from elastic to inelastic deformation behavior is observed for both stress- and volume strain vs. apparent strain curves. The onset of this transition range can be determined by applying a tangent at the curves in the range where apparent strain is low. Thus, the starting point of deviation between the experimental data and the tangents can be understood as onset of the transition range. The critical stress and apparent strain values corresponding to this threshold can be determined. The results are summarized in Table 2.

Particularly, the transition range is shifted towards higher stress and apparent strain values

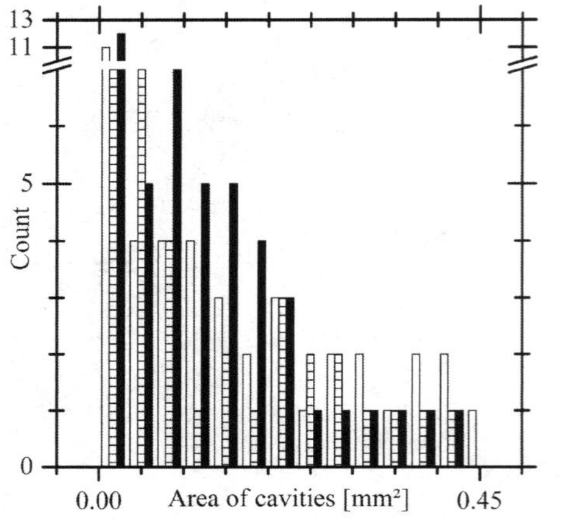

Figure 4. Histogram of cavity size distribution: SSBR-30 (∀), SSBR-30 s (&), SSBR-50 s (!).

Table 2. Critical stress and critical apparent strain values related to the onset of the transition range between elastic and inelastic volume deformation.

Critical stress MPa	SSBR-0	SSBR-30	SSBR-30 s	SSBR-50 s
estimated from stress	1.09	2.15	2.38	3.58
estimated from volume strain	1.19	2.32	2.41	4.05
Critical apparent strain %				
estimated from stress	5.1	7.9	9.6	13.2
estimated from volume strain	6.2	9.8	9.9	15.6

Table 3. Volume strain rate at inelastic volume deformation.

Property –	SSBR-0	SSBR-30	SSBR-30 s	SSBR-50 s
Volume strain rate	0.854	0.649	0.709	0.737

when the amount of filler increases. No significant influence of silanization can be detected regarding the onset of transition range. The critical values estimated from volume strain curves are consistently above those which were estimated using the stress vs. apparent strain data, although they are in the same ranking.

Table 3 shows the rates of volume strain in the stable but inelastic volume deformation range. The rate of volume strain is defined as the slope in volume strain vs. apparent strain curves and expresses the speed of volume increase with apparent strain. The results state that for unfilled SSBR the volume strain rate has a higher value compared to silica filled SSBR. In this case of silica filled SSBR the value of volume strain per apparent strain is approximately 0.70. The rate decreases slightly without silanization and increases with higher filler content.

4 DISCUSSIONS

Increasing volume strain of rubber vulcanizates under deformation with constraint strain conditions is attributed to formation and growth of cavities. This type of failure behavior occurs due to high hydrostatic stress appearing because of suppressed radial shrinkage of the rubber in pancake specimen. The phenomenon is more predominant in filled rubber materials. Cavity formation during tensile deformation under constraint strain conditions can be explained by following elementary processes occurring within rubber vulcanizates:

1. Disentanglement of molecular chains
2. Breakage of molecular chains
3. Breakage of crosslinks
4. Separation of filler agglomerates
5. Fragmentation between rubber and filler
6. Expanding process related inhomogeneities

The cavity formation and growth in silica filled SSBR was investigated in this study. It is revealed, that in highly filled rubber materials the higher number of filler particles and agglomerates support the formation of cavities which is attributed to predominant cohesive failure in filler agglomerates. Furthermore, the findings of this study confirm, that poor rubber-filler interactions lead to an acute internal failure processes. Particularly, in silica filled SSBR vulcanizates without addition of silane coupling agent relatively weak rubber-filler bonds break under deformation and get fragmented, leading to the initiation of cavity formation.

The onset of inelastic volume deformation for unfilled SSBR appears at lower apparent strain values compared to silica filled SSBR. Furthermore, the results of filled SSBR suggest, that the content of silica does not affect significantly the critical stress value of the onset of inelastic volume deformation. However, contrary to unfilled SSBR, the strength increases due to reinforcing effects of the filler and due to improved rubber-filler interaction by means of silanization. But, higher silica content and improved rubber-filler interaction enhances the stiffness of SSBR vulcanizates, which leads to increasing cavity growth (rate of volume strain). This effect is comparable to observations from crack propagation behavior of rubber vulcanizates during tear fatigue tests (Lake & Lindley 1965).

5 CONCLUSIONS

The internal failure behavior of silica SSBR vulcanizates under constraint strain conditions was investigated in this study. The filler content influences the mechanical properties and internal failure behavior of silica filled SSBR vulcanizates significantly. Particularly, the formation and growth of cavities is determined by the distribution of filler and their interactions with the rubber matrix, which can be improved by means of silane coupling agent. Future work will capture comprehensive investigations on silica filled SSBR by using X-ray scattering and diffraction methods to characterize the transition range from elastic to inelastic volume deformation, which might be forced by the formation of nanoscale cavities.

ACKNOWLEDGEMENT

Sumitomo Rubber Industries, Ltd. (Kobe, Japan) is acknowledged for their generous support of this project.

REFERENCES

Ball, J.M. 1982. Discontinuous Equilibrium Solutions and Cavitation in Nonlinear Elasticity. *Philosophical Transactions of the Royal Society of London A: Mathematical, Physical and Engineering Sciences* 306: 557–611.

Bekkedahl, N. 1949. Volume dilatometry. *J Res Natl Bur Stand (1934)* 43: 145–156.

Blatz, P.J. & Kakavas, P. 1993. A geometric determination of void production in an elastic pancake. *Journal of Applied Polymer Science* 49: 2197–2205.

Busse, W.F. 1938. Physics of Rubber as Related to the Automobile. *Journal of Applied Physics* 9: 438–451.

Chou-Wang, M.S. & Horgan, C.O. 1989. Void nucleation and growth for a class of incompressible nonlinearly elastic materials. *International Journal of Solids and Structures* 25: 1239–1254.

Diani, J., Fayolle, B. & Gilormini, P. 2009. A review on the Mullins effect. *European Polymer Journal* 45: 601–612.

Dollhofer, J., Chiche, A., Muralidharan, V., Creton, C. & Hui, C.Y. 2004. Surface energy effects for cavity growth and nucleation in an incompressible neo-Hookean material–modeling and experiment. *International Journal of Solids and Structures* 41: 6111–6127.

Dorfmann, A., Fuller, K.N.G. & Ogden, R.W. 2002. Shear, compressive and dilatational response of rubberlike solids subject to cavitation damage. *International Journal of Solids and Structures* 39: 1845–1861.

Fond, C. 2001. Cavitation criterion for rubber materials: A review of void-growth models. *Journal of Polymer Science Part B: Polymer Physics* 39: 2081–2096.

Gee, G., Stern, J. & Treloar, L.R.G. 1950. Volume changes in the stretching of vulcanized natural rubber. *Transactions of the Faraday Society* 46: 1101–1106.

Gent, A.N. 1990. Cavitation in Rubber: A Cautionary Tale. *Rubber Chemistry and Technology* 63: 49–53.

Gent, A.N. & Lindley, P.B. 1958. Tension Flaws in Bonded Cylinders of Soft Rubber. *Rubber Chemistry and Technology* 31: 393–394.

Gent, A.N. & Lindley, P.B. 1961. Internal Rupture of Bonded Rubber Cylinders in Tension. *Rubber Chemistry and Technology* 34: 925–936.

Gent, A.N. & Lindley, P.B. 1959. Internal Rupture of Bonded Rubber Cylinders in Tension. *Proceedings of the Royal Society of London. Series A. Mathematical and Physical Sciences,* 249: 195–205.

Heinrich, G., Schneider, K., Euchler, E., Tada, T. & Ishikawa, M. 2016. Internal rubber failure—Characterization and evaluation of cavitation in rubber vulcanizates under constrained loading conditions. *Tire Tech International*: 74–76.

Hocine, N.A., Hamdi, A., Naït Abdelaziz, M., Heuillet, P. & Zaïri, F. 2011. Experimental and finite element investigation of void nucleation in rubber-like materials. *International Journal of Solids and Structures* 48: 1248–1254.

Holt, W.L. & Mcpherson, A.T. 1937. Change of Volume of Rubber on Stretching: Effects of Time, Elongation, and Temperature. *Rubber Chemistry and Technology* 10: 412–431.

Horgan, C.O. & Polignone, D.A. 1995. Cavitation in Nonlinearly Elastic Solids: A Review. *Applied Mechanics Reviews* 48: 471–485.

Jones, H.C. & Yiengst, H.A. 1940. Dilatometer Studies of Pigment-Rubber Systems. *Industrial & Engineering Chemistry* 32: 1354–1359.

Kakavas, P.A. & Perig, A.V. 2015. Fracture initiation from initial spherical flaws in incompressible propellant materials. *Matéria (Rio de Janeiro)* 20: 407–419.

Lake, G.J. & Lindley, P.B. 1965. The mechanical fatigue limit for rubber. *Journal of Applied Polymer Science* 9: 1233–1251.

Le Cam, J.B. 2010. A review of volume changes in rubbers: The effect of stretching. *Rubber Chemistry and Technology* 83: 247–269.

Le Gorju Jago, K. 2012. X-ray Computed Microtomography of Rubber. *Rubber Chemistry and Technology* 85: 387–407.

Lefèvre, V., Ravi-Chandar, K. & Lopez-Pamies, O. 2015. Cavitation in rubber: an elastic instability or a fracture phenomenon? *International Journal of Fracture* 192: 1–23.

Li, J., Mayau, D. & Lagarrigue, V. 2008. A constitutive model dealing with damage due to cavity growth and the Mullins effect in rubber-like materials under triaxial loading. *Journal of the Mechanics and Physics of Solids* 56: 953–973.

Lindsey, G.H. 1967. Triaxial Fracture Studies. *Journal of Applied Physics* 38: 4843–4852.

Lopez-Pamies, O. 2009. Onset of Cavitation in Compressible, Isotropic, Hyperelastic Solids. *Journal of Elasticity* 94: 115–145.

Shinohara, Y., Kishimoto, H., Inoue, K., Suzuki, Y., Takeuchi, A., Uesugi, K., Yagi, N., Muraoka, K., Mizoguchi, T. & Amemiya, Y. 2007. Characterization of two-dimensional ultra-small-angle X-ray scattering apparatus for application to rubber filled with spherical silica under elongation. *Journal of Applied Crystallography* 40: s397-s401.

Sivaloganathan, J. 1999. On cavitation and degenerate cavitation under internal hydrostatic pressure. *Proceedings of the Royal Society of London. Series A: Mathematical, Physical and Engineering Sciences* 455: 3645–3664.

Stringfellow, R. & Abeyaratne, R. 1989. Cavitation in an elastomer: Comparison of theory with experiment. *Materials Science and Engineering: A* 112: 127–131.

Tada, T., Schneider, K., Heinrich, G. & Ishikawa, M. 2015. Effect of constrained strain on failure behavior of rubber vulcanizates. *Constitutive Models for Rubbers IX*. CRC Press.

Williams, M.L. & Schapery, R.A. 1965. Spherical flaw instability in hydrostatic tension. *International Journal of Fracture Mechanics* 1: 64–72.

Yerzley, F.L. 1939. Adhesion of Neoprene to Metal. *Industrial & Engineering Chemistry* 31: 950–956.

Zhang, H., Scholz, A.K., De Crevoisier, J., Berghezan, D., Narayanan, T., Kramer, E.J. & Creton, C. 2015. Nanocavitation around a crack tip in a soft nanocomposite: A scanning microbeam small angle X-ray scattering study. *Journal of Polymer Science Part B: Polymer Physics* 53: 422–429.

Constitutive Models for Rubber X – Lion & Johlitz (Eds)
© 2017 Taylor & Francis Group, London, ISBN 978-1-138-03001-5

Investigation of time dependence of dissipation and strain induced crystallization in natural rubber under cyclic and impact loading

K. Schneider, L. Zybell, J. Domurath & G. Heinrich
Leibniz-Institut fuer Polymerforschung Dresden e.V., Dresden, Germany

S.V. Roth, A. Rothkirch & W. Ohm
Photon Science at DESY, Hamburg, Germany

ABSTRACT: Strain Induced Crystallization (SIC) is one of the special features of Natural Rubber (NR), responsible for its outstanding mechanical performance. Especially, for the short term behavior of NR, e.g. within a rolling tire, the kinetics of SIC plays an important role. SIC can be observed to some extend by thermography, but it is not directly possible to separate the heating effects accompanying SIC from dissipative heating of the rubber material (LeCam et al., 2015, Spratte et al., 2017). By means of synchrotron x-ray diffraction, SIC can be monitored during deformation including cyclic deformations and impact loading with a time resolution of less than 10 ms (Brüning et al., 2013). Combining this method with thermography enables the separation of processes of structure formation and dissipation. This can serve as a basis for a more realistic modelling of the dynamic and thermomechanical behavior of natural rubber.

Within the presentation we report about investigations of cyclic loadings of differently Carbon Black (CB) filled NR samples with respect to SIC, melting of the crystallites as well as the dissipative heating and convective cooling of the rubber materials.

1 INTRODUCTION

Strain induced crystallization (SIC) is one of the reasons for the outstanding properties of natural rubber. Due to this crystallization the stiffness of the material is increased mainly on the positions of highest strain leading to a self-reinforcing effect. But due to the kinetics of SIC this reinforcing effect has a certain time dependence.

SIC can be followed directly by wide angle x-ray scattering (WAXS). Because of the increase in brilliance of contemporary synchrotron radiation sources and the effort in detector sensitivity and read out time, nowadays the exposure time for a single pattern can be reduced to the millisecond regime.

Thermography enables a space-resolved measurement of surface temperatures. By suitable thermodynamic models it is possible to derive from those temperatures the thermal behavior of samples caused by thermal active processes like energy dissipation or crystallization/melting of crystallites in a good approximation. The high time resolution of contemporary infrared cameras enables to follow time dependent thermal processes in detail.

By the combination of deformation experiments with simultaneous synchrotron x-ray scattering as well as thermography, it becomes now experimentally possible to follow SIC and simultaneously to separate the crystallization dependent heat production or adsorption from dissipative processes in the material and to follow the time dependence of these processes.

Due to the high time resolution as a consequence of the high brilliance of the synchrotron sources, meanwhile it becomes possible to investigate practically interesting loading regimes, like cyclic impact loading with frequencies of e.g. 10 ... 100 Hz, which is representative for the dynamic loading of a rolling tire.

2 EXPERIMENTAL

2.1 *Materials*

Unfilled and carbon black (CB) filled natural rubber compounds were investigated. The filler content was 0, 20 and 50 (weight) parts CB per hundred rubber (phr), they are named NR, NR20 and NR50 respectively.

2.2 *Sample geometry*

To guaranty a well-defined clamping, specimen with a cylindrical bulge were used. The parallel length of the specimen was 10 mm, the thickness 1.5 mm and the width 2.0 mm.

2.3 *Experimental arrangement*

The investigations were carried out at beamline P03, DESY, Hamburg (Schneider, G.J., 2010, Buffet, 2012). The general outline of the experiments is shown in Figure 1, it was described elsewhere (Schneider, K., 2010).

A home-made tensile machine enables quasi-static as well as dynamic loading of the specimen. Optionally the loading can be performed pneumatically driven impact-like.

2.4 *Estimation of relative crystallinity*

The WAXS pattern of NR is mainly characterized by an amorphous halo. During stretching a pre-orientation of the chains takes place which results in a change in the azimuthal intensity within the amorphous halo. Reaching the onset strain of SIC, the sharp reflexes of the crystallites appear mainly within the region of the amorphous halo.

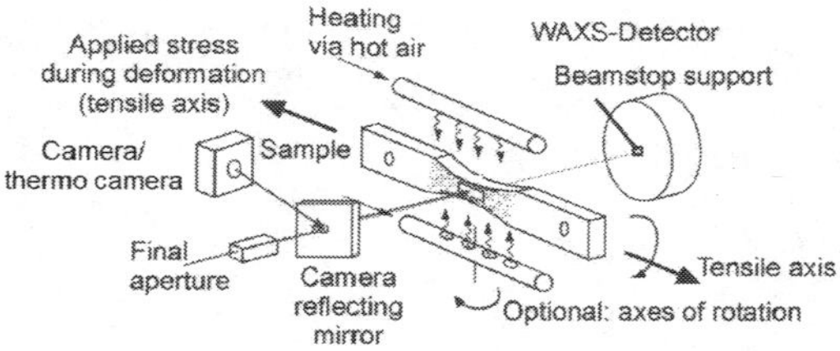

Figure 1. Experimental setup for the simultaneous estimation of SIC and temperature changes during deformation.

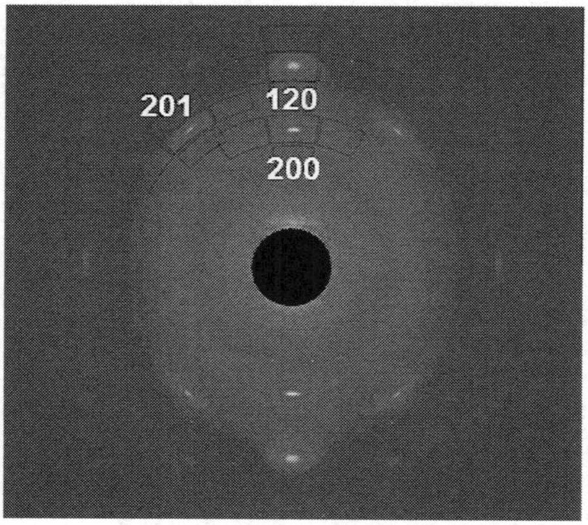

Figure 2. WAXS pattern of NR with the reflexes of SIC (stretching in horizontal direction) and their crystallographic indices as well as the regions which were used for crystallinity estimation.

The intensity of the peaks refers to the degree of crystallinity. With ongoing stretching also the thickness of the sample reduces, roughly proportional to $\lambda^{-1/2}$, where λ is the stretching ratio with $\lambda = 1+\varepsilon$ with the strain ε. Because the scattering signal is proportional to the scattering volume, one has to account for decreasing sample thickness when comparing intensities of crystalline reflexes.

For a first qualitative discussion the intensity of the crystalline peaks was estimated as the cumulative intensity within the region of the crystalline reflexes subtracted by the intensity of an equivalent region of the pattern in direct vicinity to the relevant peaks at same scattering angle.

The relevant crystallite peaks of SIC and the regions, which were used for the estimation of relative crystallinity, are shown in Figure 2.

3 RESULTS

3.1 *Free cooling behavior of the specimen*

After an impact loading or unloading of a sample, where spontaneous heating or cooling is induced, there follows a thermal equilibration to the surrounding. The heat exchange is mainly driven by the temperature difference between the surface and the surrounding, the surface area, heat transmission coefficient and the heat capacity of the sample. Under the preposition of thin samples with lateral dimensions which are large compared to the thickness, the time dependent temperature of the free cooling or heating sample can be approximated by

$$T(t) = T_\infty + (T_0 - T_\infty)\cdot \exp(-t/\tau), \tag{1}$$

with the characteristic time constant

$$\tau = \frac{m\cdot c}{\alpha\cdot A} = \frac{\rho\cdot c\cdot d_0}{\alpha\cdot\sqrt{\lambda}} = \frac{\tau_0}{\sqrt{\lambda}}, \tag{2}$$

where m is the mass of the sample, c the specific heat capacity, A the actual surface transferring the heat, α is the heat transmission coefficient, ρ the density of the material, λ the stretch ratio and d_0 the initial thickness of the sample. T_0 and T_∞ are the initial and the equilibrium (ambient) temperature.

For a NR sample with an initial thickness of 1.5 mm it follows $\tau_0 = 13.6$ s.

If the impact load crosses the threshold strain at which SIC starts, the time dependent heat production caused by the SIC overlays the heating of the sample caused by entropy elasticity and dissipation. Therefore the tip in the temperature-time curve is not so sharp like in the case of unloading, where the crystallites were dissolved immediately.

3.2 *Time dependence of SIC*

The typical time dependence of SIC in NR20 after impact loading is shown in Figure 3.

The kinetics of strain induced crystallization can be described by

$$\frac{d\Phi}{dt} = \left(k \cdot \left(\Phi_f - \Phi\right)\right)^n, \tag{3}$$

with the crystallinity Φ, the equilibrium crystallinity Φ_f, a constant k and the exponent n. The dissolution of the crystallites appears suddenly during unloading below a certain strain threshold; see also Brüning 2015. In the present case the exponent n for $\lambda > 4$ is about 0.2.

3.3 *SIC and temperature evolution during cyclic loading*

Below the onset of SIC there is a certain amount of energy dissipated within each loading cycle, where the energy loss within the mechanical cycle is identical to the heat input into the specimen. After a certain number of cycles, a final equilibrium temperature is reached where the heat loss to the surrounding is equal to the dissipated mechanical energy.

For triangular loading-unloading cycles with step-wise increased strain, the evolution of stress, SIC and temperature as a function of stretching ratio are shown in Figures 4 to 6. For each strain level the loading-unloading cycles are repeated 3 times.

The stress-stretch curves show the commonly known behavior with hysteresis. The first loading cycle to a new strain level is always slightly higher than the following cycles.

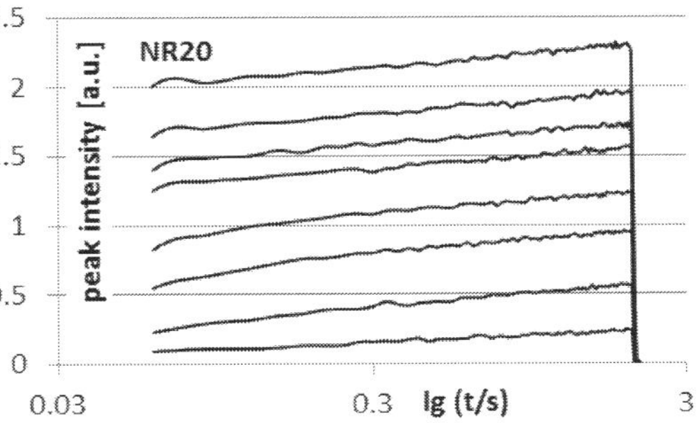

Figure 3. Time dependent increase of the intensity of the (1 2 0)-peak which is proportional to the strain induced crystallinity after impact loading of NR 20 for 2 s, stretching ratios are (from bottom curve up) 3.5, 4.5, 5.0, 5.5, 6.0, 6.5, 7.0 and 7.5.

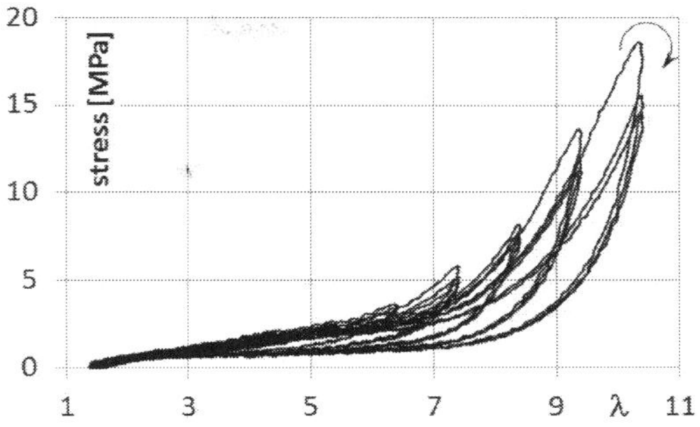

Figure 4. Stress vs. stretching ratio for NR during repeated cyclic loading with triangular strain profile to $\lambda = 4.4, 5.4, 6.4, 7.4, 8.4, 9.4$ and 10.4.

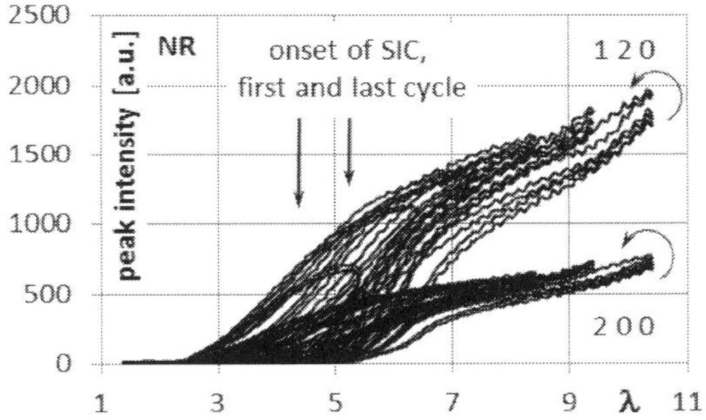

Figure 5. Peak intensity which is proportional to the strain induced crystallinity during cyclic loading of NR with triangular strain profile, evaluation of the two indicated different crystalline reflexes.

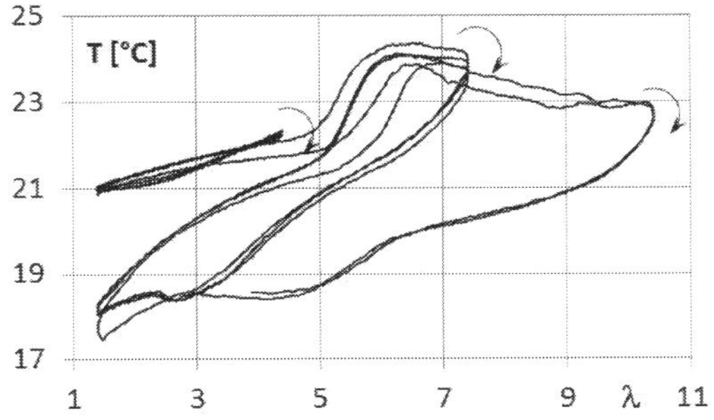

Figure 6. Cyclic loading of NR with triangular strain profile, temperature evaluation in dependence of the stretching ratio, $\lambda = 4.4, 7.4$ and 10.4.

The curves of peak intensity vs. stretch, which have the same form like the crystallinity-stretch curves show the onset of SIC at a stretch of about 3 to 4 and a hysteresis during unloading, see Fig. 5.

After some cycles to higher strains the onset value for SIC shifts to higher values.

Up to now it was not reported, that the crystallization process shows a different behavior at lower and higher strains. After an initial nearly linear increase of crystallinity with stretch the crystallization process seems to reaches a saturation. That means the crystallinity remains nearly constant during further stretching or increases only much slower than before. During unloading, nearly no hysteresis is observable in this range.

The temperature-stretch curves show a behavior which correlates strongly with the crystallization-stretch curves.

In the range before the onset of SIC, only a very small heating of the sample can be observed during a loading-unloading cycle. The final temperature is slightly higher than the initial one, the whole cycle is superposed by a steady heat release to the surrounding.

During the classical SIC, that is the first region of the crystallization curve, see Fig. 6, the system releases the heat of crystallization resulting in a strong increase of temperature associated with increased heat flux to the surrounding. There the crystallinity Φ is coupled with the specific crystallization heat Δh_{cr} and the increase of temperature ΔT by

$$\Delta h_{cr} \cdot \Phi = c_p \cdot \Delta T. \quad (4)$$

Because Δh_{cr} is not really known, the estimated crystallinity is shown here always only in arbitrary units.

Above a stretch of about 6 the temperature falls approximately exponential. This means that there is nearly no further heat production or at least a heat production which is much less than heat flux according to the natural cooling down.

3.4 *SIC and temperature evolution during cyclic impact loading*

The kinetics of SIC during cyclic impact loading is rarely independent on the holding times at low and high strain under the condition that the low strain stage is below the onset of SIC.

During stretching the thickness of the sample reduces and therefore also the absorption changes. This enables an accurate synchronization of the impact device with the scattering equipment and with the resulting vibrations of the sample. Interestingly, it seems here that the SIC requires a certain nucleation time, which impedes a strong reproducibility of the crystallinity vs. time curves.

So, at very short impacts of less than 10 ms the reproducibility of the crystallinity and also the reinforcement effect becomes less reproducible.

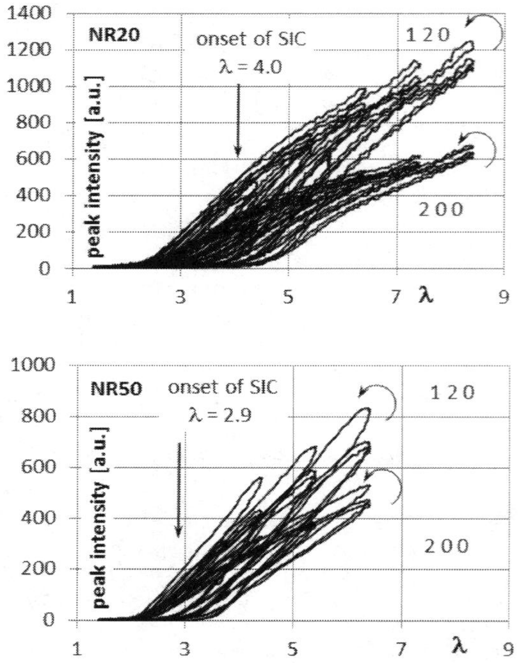

Figure 7. Peak intensity which is proportional to the strain induced crystallinity during cyclic loading of NR20 (top) and NR50 (bottom). The intensities of the (1 2 0) as well as the (2 0 0) reflexes are shown.

3.5 *Influence of filler content on SIC and temperature evolution during cyclic stretching*

The filler content has a characteristic impact on the SIC vs. strain curves. On the one hand, the total amount of crystallizable rubber becomes less, on the other hand, due to the distribution of the rubber between the filler, strain amplification takes place locally causing a shift of the onset of SIC to lower strains, see Fig. 7.

4 DISCUSSION

The stress-strain behavior of NR must be divided into 3 regions resp. regimes. In the first region, before the onset of SIC, the material behaves hyperelastically with superimposed dissipative behavior. This dissipative behavior is responsible for continuous heating of the samples during cyclic loading. According to the heat transfer to the surrounding, the sample finally reaches an asymptotic temperature by this process, normally above the ambient temperature.

Despite this dissipation the material behaves hyperelastically, the energy stored by entropy-elasticity is responsible for a reversible heating and cooling of the sample.

During SIC the heat of crystallization is responsible for a strong heating of the sample in the second region, during subsequent unloading the same amount of energy is released for the dissolution of the crystallites causing a cooling of the sample.

After reaching a certain degree of crystallinity, which is dependent on the individual rubber network and the filler content, in a third region a further deformation of the sample obviously behaves like an energy-elastic material without further heating of the sample.

In filled rubber materials the distinction of the stages of the strain crystallizing materials is less apparently like in the unfilled NR.

For highly dynamic loading of rubber the dynamics of SIC must be taken into account. Here, mainly the question of nucleation of SIC must be further investigated.

5 CONCLUSIONS

Coupling of thermography and synchrotron x-ray scattering during cyclic tensile tests is a powerful tool to separate dissipative processes within unfilled as well as filled rubber materials from structural changes like strain induced crystallization. For a quantitative evaluation of the energy balance, it is important to take into account the natural heat transfer to the surrounding. For a quantitative evaluation of the thermography measurements, not only the geometry change of the samples has to be taken into account, but also the changes in the emission coefficient of the samples due to deformation, in particular in the case of unfilled natural rubber.

The hysteresis of the stress-strain curves is similar to the hysteresis in the SIC. But at a certain strain the crystallization slows down or stops, while the sample temperature is mainly estimated by free cooling.

ACKNOWLEDGEMENT

DESY, Hamburg (Germany) is acknowledged for beamtime within the long term proposal II-2015 0042.

REFERENCES

Brüning, K., Schneider, K. & Heinrich, G., In-situ Characterization of Rubber during Deformation and Fracture, chap. 2 in Grellmann, W. et al. (eds.) 2013. *Fracture Mechanics and Statistical Mechanics of Reinforced Elastomeric Blends.* LNACM 70, Springer-Verlag Berlin, Heidelberg.

Brüning, K. Schneider, K., Roth, S., Heinrich, G. 2015. Kinetics of strain-induced crystallization in natural rubber: A diffusion-controlled rate law. *Polymer* 72: 52–58.

Buffet, A., Rothkirch, A., Döhrmann, R., Körstgens, V., Abul Kashem, M.M., Perlich, J., Herzog, G., Schwartzkopf, M., Gehrke, R., Müller-Buschbaum, P. & Roth S.V. 2012. P03, the Microfocus and Nanofocus X-ray Scattering (MiNaXS) beamline of the PETRA III storage ring: The microfocus endstation, *J. Synchr. Radiation* 19: 647.

LeCam, J.B., Samanca Martinez, J.R., Balandraud, X., Toussaint, E. & Caillard, J. 2015. Revisiting the mechanisms involved in rubber deformation using experimental thermomechanics. In B. Marvalová I. Petríková (eds), *Constitutive Models for Rubbers IX: 3–11.* Boca Raton et al.: CRC Press.

Schneider, G.J., Vollnhals, V., Brandt, K., Roth, S.V. & Göritz, D. 2010. Correlation of mass fractal dimension and cluster size of silica in styrene butadiene rubber composites, *J. Chem. Phys.* 133: 094902.

Schneider, K. 2010. Investigation of Structural Changes in Semi-Crystalline Polymers During Deformation by Synchrotron X-Ray Scattering. *J. Polymer Sci. B* 48: 1574–1586.

Spratte, T., Plagge, J., Wunde & M., Klüppel, M. 2017. Investigation of strain-induced crystallization of carbon black and silica filled natural rubber composites based on mechanical and temperature measurements. *Polymer* 115: 12–20. DOI: 10.1016/j.polymer.2017.03.019

Constitutive Models for Rubber X – Lion & Johlitz (Eds)
© 2017 Taylor & Francis Group, London, ISBN 978-1-138-03001-5

The study of local deformations of stretched filled rubber surface

I.A. Morozov & R.I. Izyumov
Institute of Continuous Media Mechanics UB RAS, Perm, Russia
Perm State University, Perm, Russia

O.K. Garishin
Institute of Continuous Media Mechanics UB RAS, Perm, Russia

ABSTRACT: An approach to determining local elongations in stretched filled rubbers is proposed. The structural-mechanical properties of stretched filled Styrene-Butadiene Rubbers (SBR) surface are investigated by means of Atomic Force Microscopy (AFM). Finite Element Modelling (FEM) is used to simulate indentation of an AFM-probe into an unfilled elastomer stretched to a certain length. Comparison of the FEM and AFM results makes it possible to calculate the local elongation ratio of the elastomer in the gaps between filler inclusions of macroscopically stretched sample. The obtained results show that the local elongation ratio can be larger than the macroscopic one. Analysis also revealed the areas where the matrix is surrounded by filler inclusions and remains almost undeformed.

1 INTRODUCTION

The properties of elastomer nanocomposites strongly depend on filler type, filler particle distribution in the matrix and polymer-filler interactions.

The application of an external load makes the local stress-strain state at the length scale of filler inclusions substantially non-uniform. Investigation of the distance between filler particles in stretched elastomers by AFM methods (Maas & Gronski 1995, Diagon et al. 2007) has shown that the filler inclusions are oriented along the axis of elongation; some of the inclusions maintain their initial relative positions (Morozov et al. 2012, Morozov 2016).

In this paper we propose the approach that allows one to determine the local elongation between filler inclusions in stretched rubber. The given method is based on a combination of the experimental data (maps of structural-mechanical properties of the surfaces of stretched samples) and the results of finite element modeling (the indentation of the stretched elastomer by the AFM probe).

2 MATERIALS AND METHODS

SBR-vulcanizates filled with 30 or 50 wt. parts of silica Seosil 115 Gr were studied. They will be labeled as SBR30 and SBR50 in further discussion. Silane Si 266 (2.4 wt. parts) was used to activate the filler surface. In the experiment, rubber strips were fixed in the stretching device, stretched to a certain length and placed under the AFM scanner.

The experiments were performed on the AFM Dimension Icon in the nanomechanical mapping regime (PeakForce Capture). In this mode the probe indents the surface with a frequency of 2 kHz. As a result, each point of the surface has its own load-displacement curve: $F(z) = kd$, where F is the applied load; z is the displacement of the cantilever base; k and d are the spring constant (bending stiffness) and the cantilever deflection.

The scan area was 1.5 μm × 1.5 μm with an image resolution of 256 × 256 pixels (each point has its own indentation curve). An example for $F(z)$ is given in Figure 1a, where one can see the loading (the probe moves toward the surface) and unloading parts of the indentation curve. The onset of indentation (the point (0;0) in Fig. 1a) is where the probe jumps to the surface. The indentation depth is calculated as follows: $u = (z–z_0) – (d–d_0)$. Tip-sample adhesion, F_{adh}, is the absolute value of the pull-off force.

The surface stiffness S is calculated from the loading part of indentation curve (see Fig. 1b) at $u = 10$ nm (see Fig. 1b): $S = \partial F/\partial u|_{u=10}$.

AFM probes—ScanAsyst-Air with calibrated tip radius ~5 nm and cantilever stiffness ~0.45 N/m were used in the experiments.

The FEM-model was used to calculate mechanical interaction between the AFM probe and the elastomer surface. In this case, the 3D contact

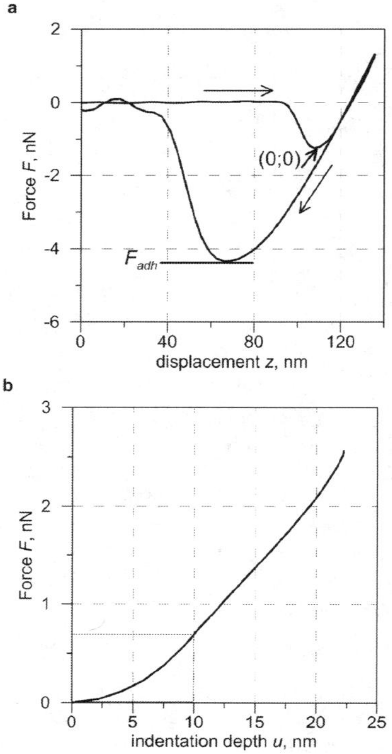

Figure 1. Interaction of the AFM-probe with the SBR surface: (a) force vs. displacement of the cantilever base; (b) force vs. indentation depth for the loading part of the indentation curve.

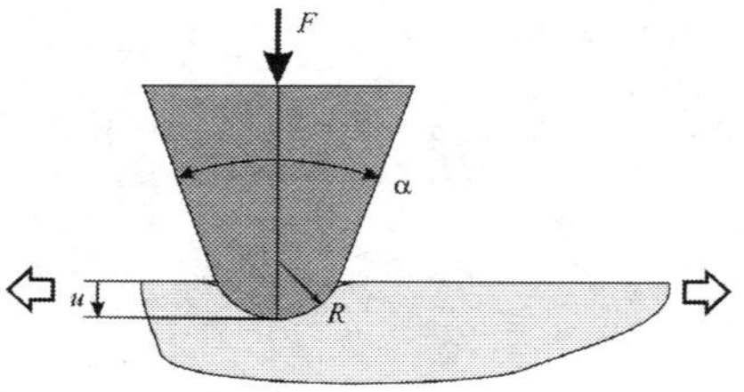

Figure 2. Scheme of the FEM model of indentation of the rigid probe into stretched elastomer.

boundary problem of indentation of a rigid cone with a rounded tip into a soft flat stretched surface was solved numerically. The mechanical properties of the elastomer were described by the elastic Ogden potential of the second order. Material constants were determined by approximation of the experimental uniaxial stress-strain curve of the corresponding unfilled SBR.

The scheme of the model is depicted in Figure 2.

The result of the FEM solution is the dependence of the force F on the indentation depth u for different elongation ratios λ of the material: $\lambda = l/l_0$, where l and l_0 are the given and initial length of the "sample".

3 RESULTS AND DISCUSSION

The results of FEM modeling are shown in Figure 3: force curves $F(u)$ for different elongation ratios (a) and the stiffness of the stretched material divided by the stiffness of the undeformed elastomer (calculated on the basis of these curves) (b).

The FEM simulation showed that the indentation stiffness of the elastomer increases with increasing tension and at a sevenfold elongation is almost two orders higher than the stiffness of the undeformed material.

The results of AFM mapping of the undeformed SBR30 are shown in Figure 4. The elevations on the height map correspond to filler inclusions. The filler is clearly visible on the adhesion (and indentation depth, which is not

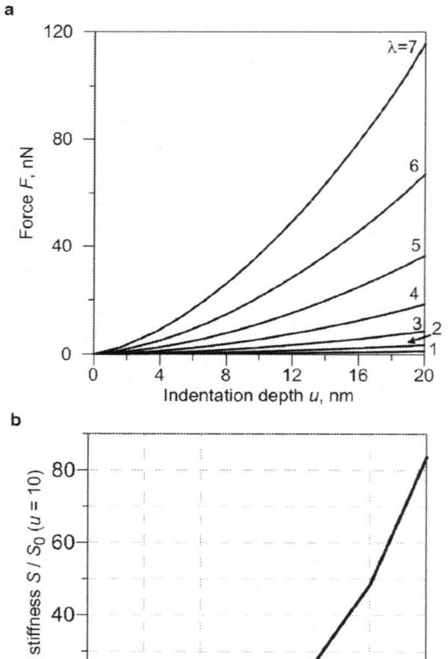

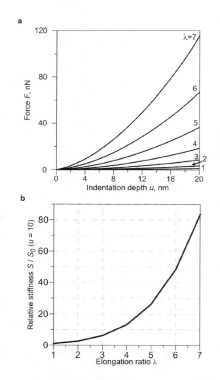

Figure 3. FEM simulation results: (a) applied force vs. indentation depth for different elongation ratios; (b) relative stiffness of the stretched elastomer.

shown here) map as the dark areas on the bright elastomer field. Therefore, the areas corresponding to filler were defined by setting a certain gray threshold of the adhesion map and marked by contours in Figure 4. The distribution of stiffness of the matrix (excluding the areas with filler) is shown in Figure 4, the average value is $<S_0> = 0.07$ N/m.

Local elongation ratio λ_{loc} between filler inclusions was evaluated by the next way: the stiffness map of the stretched rubber, which was calculated from AFM experiment, was divided by $<S_0>$ (obtained from the undeformed rubber,

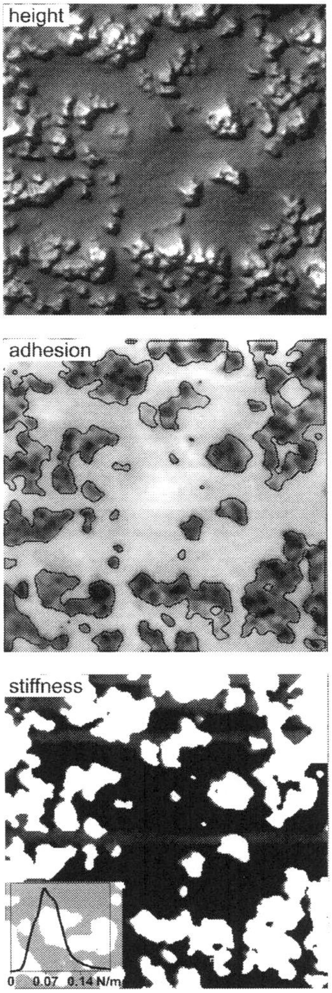

Figure 4. AFM images 1.5 × 1.5 μm of the undeformed SBR30: height, probe-surface adhesion and stiffness at $u = 10$ nm. The dark areas correspond to low values, and the bright ones to high values.

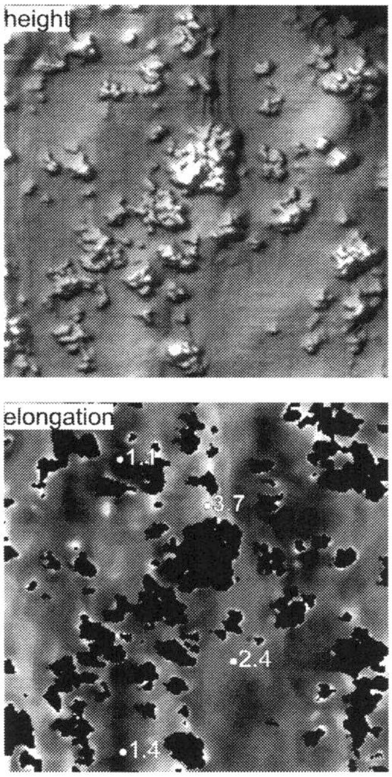

Figure 5. AFM images 1.5 × 1.5 μm of 2 times stretched filled SBR30: height and elongation map.

see Figure 4). The comparison of experimental stiffness value at the certain point with the relative stiffness of numerical solution (see Fig. 3b) gives the elongation ratio. By this way, the map of local elongation of the filled rubber matrix was constructed.

Figures 5 and 6 show the AFM-images of SBR30. The samples were macroscopically stretched in 2 and 3 times. The values of local elongation ratio in certain points are marked.

In SBR30 stretched 2 times (Fig. 5), the areas with local elongation that is higher (almost twice) or lower than macroscopic extension were observed. The latter case corresponds to so-called occluded rubber—part of the matrix which is shielded from the external load by filler. However, as the macroscopic strain increases, such low-loaded areas disappear and stiff elastomer strands start to form in the matrix between inclusions (see Fig. 6).

An increase in the filler fraction leads to the formation of a large number of stiff oriented polymer strands between filler inclusions. The local elongations in this case are higher than in the rubber with lower filler content.

Figure 7 shows height and elongation maps obtained for 3 times stretched SBR50. The dense

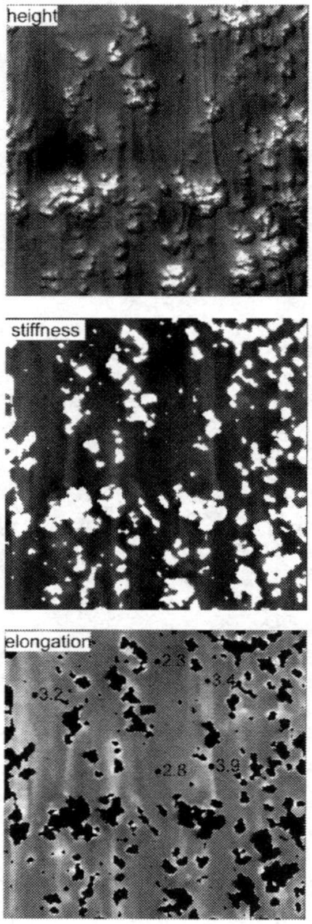

Figure 6. AFM images 1.5 × 1.5 μm of 3 times stretched SBR30: height, stiffness (at u = 10 nm) and elongation ratio map.

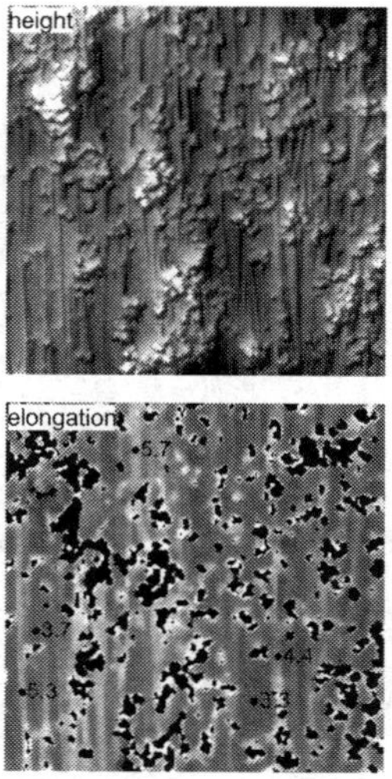

Figure 7. AFM images 1.5 × 1.5 μm of 3 times stretched SBR50: height and elongation ratio map.

mesh of oriented polymer strands between filler particles is observed. The elongation ratio of the matrix in the strands is almost twice higher than the macroscopic deformation of the sample. The elongation of the matrix between the strands also exceeds the macroscopic value.

4 CONCLUSIONS

The approach to evaluating local elongations between filler nanoparticles of the stretched filled rubber surface was suggested. The proposed method includes numerical FEM simulations and AFM-experiments.

The study of macroscopically stretched filled SBR samples revealed the areas, where the matrix elongation is 1.5…2 times higher than the applied macroscopic strain. The formation of stiff polymer strands between filler inclusions is observed at large applied strains (the more filler fraction the higher amount of these strands). The elongation ratio of the matrix in these strands is especially high.

The areas with weakly stretched matrix are also observed. This rubber is shielded from the external load by filler. However, the increase of strain (and filler content) diminishes their appearance.

ACKNOWLEDGEMENT

This work was supported by RFBR Grant 15-08-03881.

REFERENCES

Diagon, Le Y., Mallarino, S., Fretigny, C. 2007. Particle structuring under the effect of an uniaxial deformation in soft/hard nanocomposites. *The European Physical Journal E*. 22: 77–83.

Maas, S., Gronski W. 1995. Deformation of filler morphology in strained carbon black loaded rubbers. A study by atomic force microscopy. *Rubber chemistry and technology*. 68: 652–659.

Morozov, I.A., Lauke, B., Heinrich, G. 2012. Quantitative microstructural investigation of carbon-black-filled rubbers by AFM. *Rubber chemistry and technology*. 85: 244–263.

Morozov, I.A. 2016. Structural–Mechanical AFM Study of Surface Defects in Natural Rubber Vulcanizates. *Macromolecules*. 49: 5985–5992.

Constitutive Models for Rubber X – Lion & Johlitz (Eds)
© 2017 Taylor & Francis Group, London, ISBN 978-1-138-03001-5

Experimental characterisation and modelling of the thermomechanical behaviour of foamed rubber

H. Seibert & S. Diebels
Chair of Applied Mechanics, Saarland University, Saarbruecken, Germany

ABSTRACT: The present contribution deals with the thermomechanical modelling of a foamed rubber material including the experimental characterisation. In our case, a mixed open and closed cell porous material, consisting of an EPDM-bulk matrix with air filled pores, is investigated. This kind of material is mainly used as a sealing component in automotive applications, e.g. door seals or gaskets in the engine compartment. The sealing effect is mainly provided by multiaxial deformations due to local compressibility in contact with the sealing surface. Consequently, the experimental methods in this work are given by general biaxial tensile tests and hydrostatic compression tests. Furthermore, investigations regarding the temperature dependence of the material are executed with respect to applications in thermally loaded fields like the engine compartment. Besides the evaluation of the described experiments, the modelling of the viscoelastic porous material is challenging. Due to the structure of the material, a model based on the theory of porous media and the theory of mixtures is used. The material's point of compaction is modelled by an additive compression-term. A multiplicative split of the deformation gradient allows a separate identification of elastic and time dependent material parameters. A similar approach finds application to identify volumetric and isochoric deformations. The parameters identified at room temperature are modified afterwards in order to take temperature dependence into account.

1 INTRODUCTION

Nowadays, the computational simulation of materials has increasing significance within production processes. In a historical context, mainly load bearing parts have been of great interest. Based on the general trend towards multifunctional materials and the reduction of costs, non-load bearing parts get into focus, too.

In this context, the modelling of materially and structurally compressible materials gains importance. Especially porous rubber materials are of high interest, since they are applied as sealing components. Early works regarding this topic have been published by Blatz and Ko (1962) or Gent and Thomas (1959), followed by more recent works, e.g. Danielsson et al. (2004) or Raghunath and Juhre (2013).

The challenge in the experimental characterisation of porous rubber materials lies in the necessity of three-dimensional strain measurement. In contrast to single phase elastomers, there is no constraint concerning incompressibility of the material, which can be used for calculating the strain in thickness-direction. The three-dimensional strains are measured by an optical field method in the present work. The software ISTRA4D® by Dantec Dynamics based on Digital Image Correlation (*DIC*) is used. For more details about the general method see Sutton et al. (2009). The reader is referred to Seibert et al. (2014) and Seibert et al. (2016) for a detailed description of the particular procedure.

Special care has to be taken concerning the structural compressibility of porous rubber materials. It consists of a materially incompressible bulk phase and the compressible pore gas. The resulting structural compressibility shows remarkable behaviour under compressive load. After compressing the pore gas to a singular point, the mixture behaves like an incompressible rubber material due to the absence of the materially compressible gas phase. In order to describe such a biphasic material, a model based on the theory of porous media is applied. For general details concerning the theory of porous media, the reader is referred to the comprehensive historical review given by De Boer (2000). In this context, the point of compaction is described as proposed by Eipper (1998) using an additional compression term. The asymptotic behaviour is described by comparing the volume strain with the initial volume fraction of the solid phase. Koprowski-Theiß (2011) already presented a viscoelastic model with this approach but based on different experiments.

The thermomechanical coupling is realised following the method given by Johlitz et al. (2012). Therein, the viscous parameters identified at room temperature are changed in order to include temperature dependence. The necessary experimental data is generated in uniaxial tensile tests executed at several temperatures.

2 DETAILS CONCERNING THE MATERIAL

The material used in this contribution is a commercially available foamed EPDM rubber. The mixed open and closed cell structure has a porosity between 40–60% according to the manufacturer. Fig. 1 shows an optical microscopy image of the investigated material including a clearly detectable connection between two pores, illustrating the partially open cell character, see Fig. 1-Ⓐ. The material is available in several geometries, e.g. cords and membranes. In the present work, only the membrane material is used in order to characterise the foamed EPDM's behaviour.

An important property of the investigated material is the nearly ideal isotropic behaviour, which can be shown in a uniaxial tensile test with three-dimensional *DIC*, see Fig. 2. By stretching

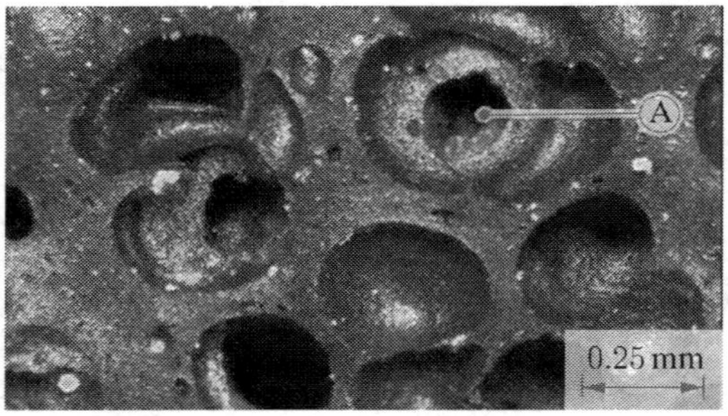

Figure 1. Optical microscopy image of cellular EPDM-cord including an open pore (A).

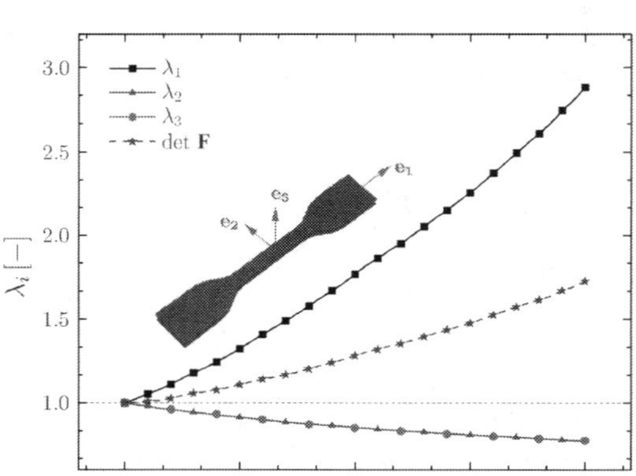

Figure 2. Isotropic material behaviour in uniaxial tensile test, stretch λ_i over machine displacement u_M.

the specimen in $\mathbf{e}_1$-direction, the equality of both transversal contractions in $\mathbf{e}_2$- and $\mathbf{e}_3$-direction can be observed. A further detail in Fig. 2 is the measurable material's compressibility, indicated by the plot of the Jacobian determinant det **F**, which is different of 1.

3 EXPERIMENTAL CHARACTERISATION

This section gives an outline of the experimental procedure of the characterisation process. Details about the method of the model generation can be found in Seibert et al. (2016). Typical applications for the investigated cellular rubber material define requirements for a useful material model. Especially in case of sealing components, the functionality of the material is guaranteed by multiaxial deformations combined with local compression. Regarding sealing applications in engine compartments, an additional load is given by increased temperatures. Moreover, viscous or viscoelastic effects occur due to pressure fluctuation in the separated medium. The mentioned effects have to be imitated in the experimental data base in order to predict the material's behaviour.

3.1 *The test procedure*

Summarising the requirements given above, the experimental procedure has to give information about four specific material effects:

- Multiaxial hyperelasticity,
- (Rate dependent) compressibility,
- Finite viscoelasticity,
- Temperature dependence.

Up to this point, stress softening and other inelastic effects are neglected, which is realised by a suitable preconditioning of the specimens, cf. Seibert et al. (2016).

A staggered experimental procedure is applied as displayed in Fig. 3 due to the multiple requirements for an appropriate material model described in the section above.

Since it has been shown in earlier works that the experimental data set has a high influence on the validity of a model in case of incompressibility, a general biaxial tensile test is used to characterise the material parameters for the hyperelastic case. This argumentation can be transferred to the isochoric deformation of a compressible material, as it will be described in a later section, by using an additional compression term. The material's behaviour under multiaxial deformations can be predicted correctly in combination with an appropriate free energy function, e.g. a Biderman approach, see Biderman (1958). The custom built test device is

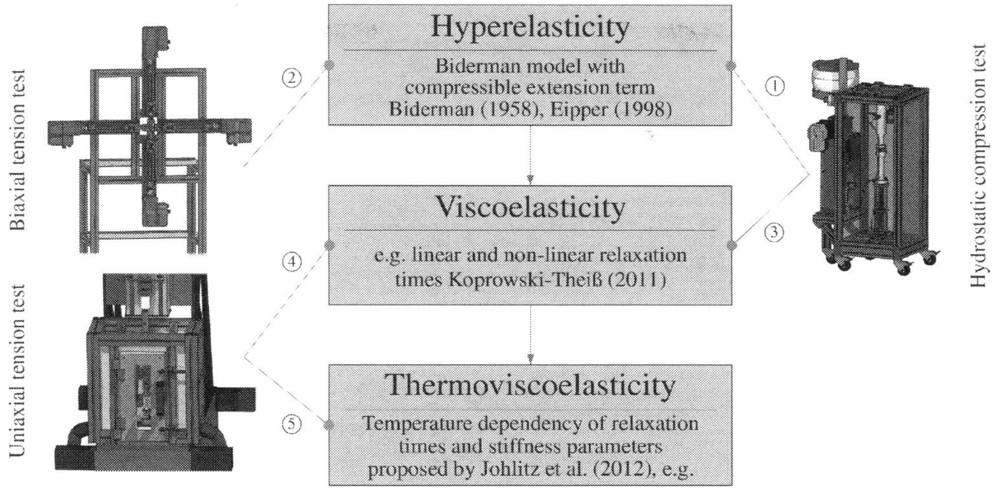

Figure 3. Experimental test procedure and corresponding modelling approaches.

presented in Johlitz and Diebels (2011) and Seibert et al. (2014), for example.

The compression term is characterised using a hydrostatic compression test. The testing device for this application is an in-house development, first time mentioned in Koprowski-Theiß et al. (2012). The main component of the device is a test chamber containing a hydraulic medium that mainly consists of water and the compressible specimen. The reduction of the chamber volume by an intruded piston rod induces a volume change in the specimen. The measurement of the corresponding chamber pressure yields a suitable stress-volume change relation for the material. For details about the testing setup and the evaluation procedure see Koprowski-Theiß et al. (2012) and Seibert et al. (2016).

The parameters of the hyperelastic isochoric and compressible model part can be identified based on the experiments mentioned above. The influence of deformation rates and cyclic loading is not taken into account, yet. In case of the hydrostatic compression test, the evaluation of the dynamic experiments is very similar to the quasi-static ones. Hence, the viscous parameters of the compressible model extension can be computed by means of the same experiments considering time dependence.

Since the time dependent evaluation of the biaxial tensile test is still in development, the rate dependent isochoric model is identified in a uniaxial tensile test with adapted thermal chamber. As a consequence of this procedure, the inelastic isochoric model part only has to provide the ability to predict uniaxial deformation states. The material parameters are identified at room temperature.

The final step in the staggered experimental characterisation is the adaptation of the viscous parameters regarding their temperature dependence. Therefore, the experiments executed for the identification of the parameters at room temperature are performed at several lower and higher temperatures. The method itself was already proposed by Johlitz et al. (2012).

3.2 *Analysis of the biaxial tensile test*

The benefit of a general biaxial tensile test is shown by the concept of the plane of invariants, cf. Treloar (1975) or Johlitz and Diebels (2011), among others. Originally, this concept was mentioned with regard to incompressible materials. When the hydrostatic parts of the deformation are removed, the remaining isochoric strains behave in the same manner. Following special paths concerning the machine movement, it is possible to enter arbitrary deformation states as displayed in Fig. 4. More details about the machine movement resulting in this deformation path over time parameter τ can be found in Seibert et al. (2014) or Seibert et al. (2016), respectively.

As mentioned before, the compressibility of the foamed rubber material is a great challenge with regard to the strain-measurement in the tensile tests. Since the constraint of incompressibility is missing in the evaluation process, a full field three-dimensional measurement is necessary. In this context, the software ISTRA4D® by Dantec Dynamics enables the observation of the specimen by several sets of stereoscopic camera systems.

A further challenge of the biaxial tensile test is the spatial discrepancy between the locations of

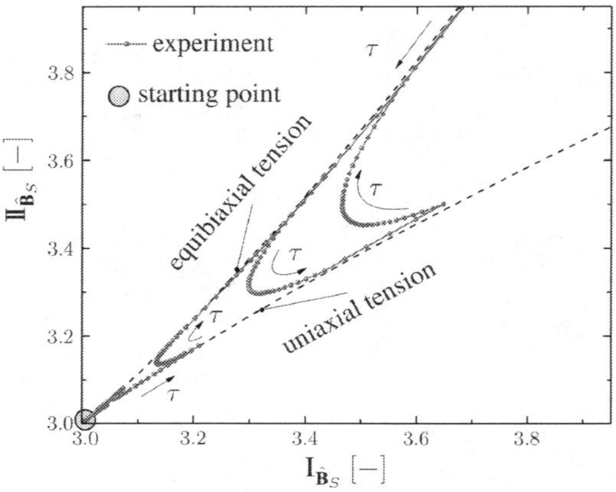

Figure 4. Biaxial tensile test: path in plane of invariants for isochoric deformations ($\hat{\mathbf{B}}_S$) in the specimen's centre.

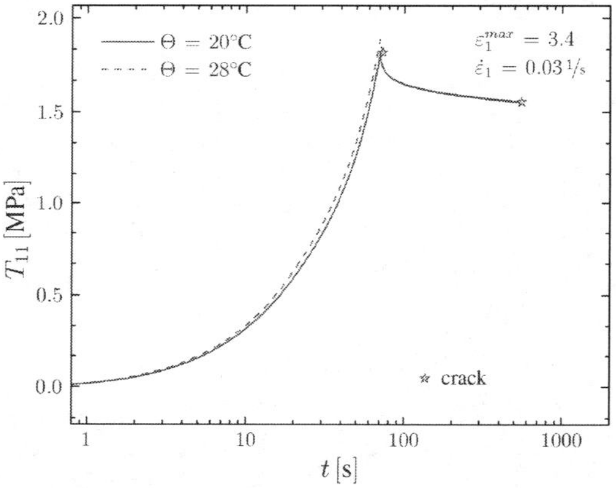

Figure 5. Relaxation experiments with damage effect at different temperatures, stress over time.

the strain measurement and the force measurement. Hence, the analysis of the experiments has to be performed as an inverse calculation. The measured forces are used as boundary conditions of the finite element simulation. The resulting deformations are compared to the experimentally measured deformations. The variation of the parameters is executed by evolution strategies like Genetic Algorithms or the Nelder-Mead-method, cf. Rechenberg (1994).

3.3 *Remarks concerning damage effects*

Execution of relaxation tests with the material under uniaxial tension shows an unusual damage effect. The effect occurs during the relaxation process after reaching the maximum stress. Although the load on the material is steadily decreasing, the material tends to break spontaneously during the relaxation. The time between reaching the maximum deformation and the destruction of the specimen can be influenced by changing the environmental temperature, the strain rate, and the maximum strain level during the relaxation experiment. Fig. 5 exemplarily displays the influence of the temperature. Since there is no further knowledge about this effect, relaxation experiments are excluded from the mechanical characterisation method presented in this contribution.

4 MODELLING APPROACH

In order to benefit from the staggered experimental procedure the mathematical material model is divided in different model parts corresponding to the occurring effects. The method given in this contribution is based on the theory of porous media, which deals with the theory of mixtures (see exemplarily Truesdell (1957) or Truesdell and Toupin (1960)) and the concept of volume fractions, cf. Bowen (1980). A comprehensive overview about the historical background can be found in De Boer (2000) and Ehlers (1996).

The theory applied in this approach assumes the parallel existence of multiple phases of a material. The special case investigated in this article is a hybrid model of first type consisting of an incompressible solid phase and the compressible pore gas, see De Boer (2000). Fig. 6 displays the concept of the coexistence of two phases. The index $(\cdot)_S$ indicates the solid phase, $(\cdot)_G$ the gas phase. Diffusion of pore-gas is neglected due to the small amount of open cells in the membrane material, which results in a kinematic coupling between the constituents. By means of the concept of volume fractions, a statistically homogenised model can be generated, see Fig. 6. The mixture can be interpreted as

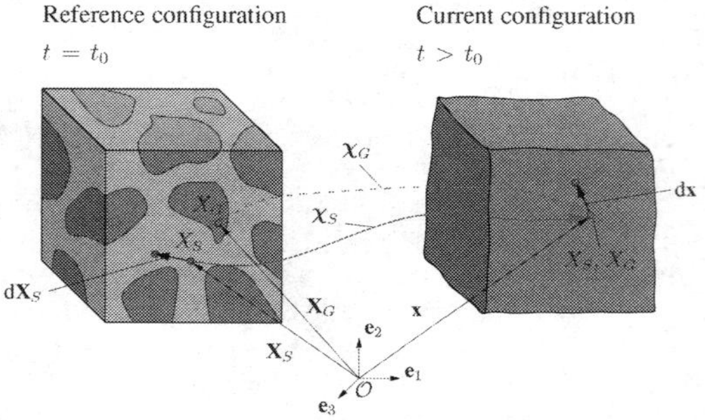

Figure 6. Kinematics of the mixture based on the theory of porous media.

a superimposed continuum. Each of the particular constituents has its own unique motion function. Therefore, every spatial point **x** is filled with parts of every constituent at each time step

$$\mathbf{x}=\chi_\alpha(\mathbf{X}_\alpha,t), \tag{1}$$

with χ_α the unique motion of phase φ^α and a material point in the reference configuration $\mathbf{X}_\alpha$.

Starting with the kinematics for each constituent as given in (1), the formalism of continuum mechanics can be carried out. The procedure used in this contribution is mainly inspired by the modelling approach proposed in Koprowski-Theiß (2011) and Koprowski-Theiß et al. (2012). The model parameters in tensile states and those in compressive states can be identified separately, when a multiplicative decomposition of the deformation gradient of the solid phase $\mathbf{F}_S$ into a volumetric part, indicated by $(\bar{\cdot})$ and an isochoric part, indicated by $(\hat{\cdot})$ is executed

$$\mathbf{F}_S=\bar{\mathbf{F}}_S\cdot\hat{\mathbf{F}}_S=\left(\mathrm{J}_S^{\frac{1}{3}}\,\mathbf{I}\right)\cdot\left(\mathrm{J}_S^{-\frac{1}{3}}\,\mathbf{F}_S\right), \tag{2}$$

with the Jacobian determinant of the solid phase J_S.

The advantage of the procedure is rooted in the possibility of separate parameter identifications. Parameters of the volumetric part are identified using hydrostatic pressure experiments without any isochoric deformations. The isochoric model part is determined afterwards by tensile tests.

This procedure is repeated with respect to elastic and inelastic material behaviour. Since inelasticity can be recognised in compression tests as well as in tensile tests, both the isochoric and volumetric deformation gradient have to be split in two parts. These are an inelastic part indicated by $(\cdot_i)$ and an elastic part, indicated by $(\cdot_e)$. The resulting decomposed deformation gradients can be written as

$$\bar{\mathbf{F}}_S=\bar{\mathbf{F}}_{S_e}\cdot\bar{\mathbf{F}}_{S_i}, \qquad \hat{\mathbf{F}}_S=\hat{\mathbf{F}}_{S_e}\cdot\hat{\mathbf{F}}_{S_i}. \tag{3}$$

The multiplicative split of the deformation leads to an additive decomposition of equilibrium and non-equilibrium stresses. The presented method has already been executed successfully in Koprowski-Theiß (2011) with a different free energy function for the isochoric part, i.e. a second order approach in the first principal invariant $I_{\hat{\mathbf{B}}_S}$. Replacing the hyperelastic part by a Biderman-model, see Biderman (1958), in the form

$$\begin{aligned}\rho_0^S\,\hat{\Psi}_{eq}^S &= c_{10}\left(\mathrm{I}_{\hat{\mathbf{B}}_S}-3\right)+c_{20}\left(\mathrm{I}_{\hat{\mathbf{B}}_S}-3\right)^2\\ &\quad+c_{30}\left(\mathrm{I}_{\hat{\mathbf{B}}_S}-3\right)^3+c_{01}\left(\mathrm{II}_{\hat{\mathbf{B}}_S}-3\right),\end{aligned} \tag{4}$$

offers the possibility to predict the multiaxially deformed behaviour under tensile load by the additional influence of the second principal invariant $\mathrm{II}_{\hat{\mathbf{B}}_S}$.

For the purpose of modelling the material under hydrostatic load, an additive compression term as proposed in Eipper (1998) is used in the form

$$\begin{aligned}\rho_0^S\,\bar{\Psi}_{eq}^S &= \frac{\Lambda^S}{\gamma\left(\gamma-1+\frac{1}{(1-n_0^S)^2}\right)}\left(\mathrm{J}_S^\gamma-1\right)\\ &\quad-\frac{\Lambda^S}{\left(\gamma-1+\frac{1}{(1-n_0^S)^2}\right)}\ln\left(\frac{\mathrm{J}_S-n_0^S}{1-n_0^S}\right)\\ &\quad+\frac{\Lambda^S}{\left(\gamma-1+\frac{1}{(1-n_0^S)^2}\right)}n_0^S\,\frac{\mathrm{J}_S-n_0^S}{1-n_0^S}.\end{aligned} \tag{5}$$

Therein Λ^S is the Lamé-parameter of the solid phase, γ is a dimensionless positive material parameter and n_0^S is the initial volume fraction of the solid phase. Using (5) ensures that the pressure tends to infinity if the point of compaction is reached, i.e. if the pores are closed.

The viscoelastic model part is exactly the same as already mentioned in Koprowski-Theiß (2011) and Koprowski-Theiß et al. (2011), since the viscoelastic parameters are identified by uniaxial tensile tests in addition to the multiaxially generated hyperelastic equilibrium part. Without giving any details in this article, the free energy functions for Maxwell element number j can be formulated as

$$\rho_0^S\,\hat{\Psi}_{neq}^{Sj}=\frac{1}{2}\hat{\mu}_{0e}^j\left(I_{\hat{\mathbf{B}}_{Se}^j}-3\right), \tag{6}$$

$$\begin{aligned}\rho_0^S\,\bar{\Psi}_{neq}^{Sj} &= \Lambda_e^j\left(1-n_0^S\right)\left[\mathrm{J}_{Se}^j-1\right]\\ &\quad-\Lambda_e^j\left(1-n_0^S\right)^2\left[\ln\left(\frac{\mathrm{J}_{Se}^j-n_0^S}{1-n_0^S}\right)\right],\end{aligned} \tag{7}$$

with the shear-modulus $\hat{\mu}_{0e}^j$, the Lamé-parameter Λ_e^j, the first principal invariant of the elastic isochoric left Cauchy-Green deformation tensor $I_{\hat{\mathbf{B}}_{Se}^j}$ and the elastic Jacobian determinant J_{Se}^j. The resulting stress-strain relations of the inelastic model parts are given by

$$\hat{\mathbf{T}}_{neq}^S=\sum_{j=1}^{n}\mathrm{J}_S^{-1}\left[\hat{\mu}_{0e}^j\,\mathrm{J}_{Se}^{j\,-\frac{2}{3}}\left(\mathbf{B}_{Se}^j-\frac{1}{3}\mathrm{I}_{\mathbf{B}_{Se}^j}\,\mathbf{I}\right)\right] \tag{8}$$

$$\begin{aligned}\bar{\mathbf{T}}_{neq}^S &= \sum_{j=1}^{n}\mathrm{J}_S^{-1}\left[\Lambda_e^j\left(1-n_0^S\right)\mathrm{J}_{Se}^j\right]\mathbf{I}\\ &\quad-\sum_{j=1}^{n}\mathrm{J}_S^{-1}\left[\Lambda_e^j\left(\frac{\mathrm{J}_{Se}^j\left(1-n_0^S\right)^2}{\mathrm{J}_{Se}^j-n_0^S}\right)\right]\mathbf{I}\end{aligned} \tag{9}$$

for n Maxwell elements.

The internal variables in the eqs. (8) and (9) have to be determined by evaluation of the evolution equation

$$\left(\mathbf{C}_{Si}^{j}\right)'_{S} = \frac{2\mathrm{J}_{S}^{-1}\hat{\mu}_{0e}^{j}\mathrm{J}_{Se}^{j\,-\frac{2}{3}}}{\hat{\eta}_{0}^{j}}\left(\mathbf{C}_{S} - \frac{1}{3}I_{\mathbf{B}_{Se}^{j}}\mathbf{C}_{Si}^{j}\right) + \frac{2\Lambda_{e}^{j}}{\overline{\eta}_{0}^{j}\mathrm{J}_{Si}^{j}}\left(1 - n_{0}^{S} - \frac{\left(1-n_{0}^{S}\right)}{\mathrm{J}_{Se}^{j} - n_{0}^{S}}\right)\mathbf{C}_{Si}^{j}. \tag{10}$$

for Maxwell element number j. For more details see Koprowski-Theiß (2011).

The last step in the staggered identification procedure is the modification of the viscosities regarding their temperature dependence. Since the thermal chamber is only applicable for a uniaxial tensile test so far, the previously identified parameters have to be modified in a post-process. A possible strategy for modifying identified material parameters can be found in Johlitz et al. (2012). In this contribution, isothermal experiments, e.g. cyclic loading tests, can be used to detect the variation of the shear-moduli and the viscosities after identifying them at room temperature. By modifying both parameters in the same way, the resulting relaxation times given by

$$\hat{r}^{j} = \frac{\hat{\eta}^{j}(\Theta)}{\hat{\mu}_{e}^{j}(\Theta)} = const. \tag{11}$$

stay constant. In eq. (16), Θ represents the current temperature. In contrast to a physically motivated approach, where the relaxation times are temperature dependent, the presented method allows to adjust the relaxation behaviour to a certain range of deformation velocities and to a certain temperature interval using a fixed number of relaxators. The modification proposed by Ferry (1980) and Johlitz et al. (2012) is an exponential Arrhenius approach based on the values identified at room temperature Θ_0,

$$\hat{\mu}_{e}^{j}(\Theta) = \hat{\mu}_{0e}^{j}\exp\left(b\left(1 - \frac{\Theta}{\Theta_{0}}\right)\right), \tag{12}$$

$$\hat{\eta}^{j}(\Theta) = \hat{\eta}_{0}^{j}\exp\left(b\left(1 - \frac{\Theta}{\Theta_{0}}\right)\right). \tag{13}$$

The weak temperature dependence of the hyperelastic partial model is neglected due to the limited temperature variations in the entropy elastic region. Thus, the temperature dependence is purely modelled as an effect influencing the viscous parameters without changing the hyperelastic model base.

5 CONCLUSION AND OUTLOOK

This contribution gives a small overview of effects mainly influencing the material's behaviour in real applications. Therein, particular characteristics of the material class are observed separately, which supports benefits of the staggered material model proposed in this study. In detail, hydrostatic compression tests and biaxial tensile tests are executed to develop a basic hyperelastic material model at room temperature. For the biaxial tensile tests, a three-dimensional optical strain-measurement is used. The resulting model is complemented by an inelastic model part identified by the use of cyclic experiments in hydrostatic compression and uniaxial tension. Relaxation effects are neglected due to an unusual damage mechanism in relaxation tests. In a final step, the parameters of the inelastic model part are adapted depending on the temperature in uniaxial tensile tests.

Future work will include the adaption of the thermal chamber to the biaxial tensile test in order to investigate the viscoelastic material behaviour under biaxial deformation and the temperature dependence of the multiaxially identified viscous parameters.

The observed damage effect during material's relaxation has to be investigated and included as well.

REFERENCES

Biderman, V. (1958). Calculations of rubber parts. *Rascheti na prochnost*, 40.

Blatz, P.J. & W.L. Ko (1962). Application of finite elastic theory to the deformation of rubbery materials. *Trans. Soc. Rheol. 6*(1), 223–251.

Bowen, R.M. (1980). Incompressible porous media models by use of the theory of mixtures. *Int. J. Eng. Sci. 18*(9), 1129–1148.

Danielsson, M., D. Parks, & M. Boyce (2004). Constitutive modeling of porous hyperelastic materials. *Mech. Mater. 36*(4), 347–358.

De Boer, R. (2000). *Theory of porous media: highlights in historical development and current state*. Springer.

Ehlers, W. (1996). Grundlegende Konzepte in der Theorie poröserMedien. *Techn. Mech. 16*, 63–76.

Eipper, G. (1998). *Theorie und Numerik finiter elastischer Deformationen in fluidgesättigten porösen Festkörpern*. Inst. f. Mechanik (Bauwesen) d. Univ. Stuttgart.

Ferry, J.D. (1980). *Viscoelastic properties of polymers*. JohnWiley & Sons.

Gent, A. & A.G. Thomas (1959). The deformation of foamed elastic materials. *J. Appl. Polym. Sci. 1*(1), 107–113.

Johlitz, M. & S. Diebels (2011). Characterisation of a polymer using biaxial tension tests. Part I: Hyperelasticity. *Arch. Appl. Mech. 81*(10), 1333–1349.

Johlitz, M., S. Diebels, & W. Possart (2012). Investigation of the thermoviscoelastic material behaviour of adhesive bonds close to the glass transition temperature. *Arch. Appl. Mech. 82*(8), 1089–1102.

Koprowski-Theiß, N. (2011). *Kompressible, viskoelastische Werkstoffe: Experimente, Modellierung und FE-Umsetzung*. Universität des Saarlandes.

Koprowski-Theiß, N., M. Johlitz, & S. Diebels (2011). Modelling of a cellular rubber with nonlinear viscosity functions. *Exp. Mech. 51*(5), 749–765.

Koprowski-Theiß, N., M. Johlitz, & S. Diebels (2012). Compressible rubber materials: experiments and simulations. *Arch. Appl. Mech.*, 1–16.

Raghunath, R. & D. Juhre (2013). Finite element simulation of deformation behaviour of cellular rubber components. *Mech. Res. Commun. 47*, 32–38.

Rechenberg, I. (1994). Evolutionsstrategie 94, volume 1 of Werkstatt Bionik und Evolutionstechnik. *Frommann Holzboog, Stuttgart.*

Seibert, H., T. Scheffer, & S. Diebels (2014). Biaxial testing of elastomers—Experimental setup, measurement and experimental optimisation of specimens shape. *Techn. Mech. 34*(2), 72–89.

Seibert, H., T. Scheffer, & S. Diebels (2016). Thermomechanical characterisation of cellular rubber. *Continuum Mech. Thermodyn.*, 1–15.

Sutton, M., J. Orteu, & H. Schreier (2009). *Image Correlation for Shape, Motion and Deformation Measurements: Basic Concepts, Theory and Applications*. Springer.

Treloar, L. (1975). *The physics of rubber elasticity, 3rd edn. Clarendon*. Oxford.

Truesdell, C. (1957). Sulle basi della termomeccanica. *Rend. Lincei 22*(8), 33–38.

Truesdell, C. & R. Toupin (1960). Principles of classical mechanics and field theory. *Handbuch der Physik 3*(1).

Constitutive Models for Rubber X – Lion & Johlitz (Eds)
© 2017 Taylor & Francis Group, London, ISBN 978-1-138-03001-5

Some cautions when applying nanoindentation tests on a fluoroelastomer: Experimental researches and application

C. Fradet, F. Lacroix, G. Berton & S. Méo
Centre d'Etude et de Recherche sur les Matériaux Elastomères, Laboratoire de Mécanique et Rhéologie, Université de Tours, France

E. Le Bourhis
Institut Pprime, Université de Poitiers, France

ABSTRACT: The local mechanical response of fluoroelastomer rubber is presented through nanoindentation results. The investigated property is the elastic reduced modulus also called indentation modulus. It is well known that this modulus (as the indentation hardness) is dependent on the applied experimental protocol with variables such as the maximum load and loading rates. This dependence is expected to be extensively observed on elastomers whose mechanical behavior is viscoelastic and time-dependent. This paper describes the methodology of nanoindentation on such materials and discusses the results which clearly present behavior variations depending on the protocol being used and on the sample morphology. These results are then used for a practical application in which vulcanization gradients are observed.

1 INTRODUCTION AND LITERATURE REVIEW

Nanoindentation is very powerful to probe the mechanical properties of materials at the local scale because of its high sensitivity. However, this sensitivity is also responsible of measurement distortions (Fischer-Cripps 2011, Němeček 2012). Indeed, a lot of factors are known to distort nanoindentation results. In literature, one can find three main kinds of error factors: instrumental, environmental and material factors. In more details, one can cite: the detection of the initial point of contact, the frame compliance, the hardness, geometry and tip default of the indenter, the thermal drift, the sample's roughness, adhesion and behavior (elastic, elastic-plastic or viscoelastic), the nature of the surrounding deformation (sinking-in or piling-up), the sample's positioning and preparation (residual stresses in the case of polished metals for example) and the presence or absence of a substrate and its nature. Of course the time scale (i.e. loading rates) has to be added to those influence parameters as well as the analysis model used to convert raw data into numeral mechanical properties.

In the present study, we worked on two of these factors. First, the influence of the samples' roughness on nanoindentation results is given. Variations in local (nanoindentation) and global (tensile and creep tests) responses as a function of tests' kinetics are then studied. Finally, a practical application providing vulcanization moduli gradient along the thickness of a rubber sample is presented.

The historical original model of the theory of contact elasticity from Boussinesq (1885) and Hertz (1881) is valid only in the ideal case involving perfectly smooth surfaces. The natural roughness of indented materials is also neglected by Love (1939) and Sneddon (1965) and so by any model resulting of these works.

The ISO standard 14577 specifies that the mean deviation roughness R_a must be less than 5% of the indentation contact depth so as to consider the measurement uncertainty due to surface defects negligible. This means that for small indentations (i.e. local measurements) the roughness has to be very low. On materials such as metals, one can proceed to a mirror finish polishing so that this "5% condition" can be easily fulfilled. However, such sample preparation is not worth considering for rubbers because of their high compliance and the potentially resulting damages and changes of properties due to friction, self-heating or induced residual stresses.

Some studies can be found in the literature about nanoindentation performed on rough surfaces. Bobji & Biswas (1998) developed a model to deconvolute the effects of the roughness from the measurements of the hardness through a numerical simulation of spherical nanoindentation on a fractal surface and experiments on copper. Jiang et al. (2008) compared by numerical simulation the results of the indentation hardness H and the indentation modulus E_r between a smooth thin film of copper and a rough one with two sizes of width of defects. Both the indentation modulus and hardness appeared to decrease with an increasing width of surface defects. Chen et al. (2013) studied the influence of different roughnesses (R_a) and different Berkovich radii on measured values of the reduced modulus and the hardness by finite element simulation of a reference stainless steel. They showed that the roughness causes important measurement deviations and overestimations of the calculated mechanical properties but that the consequences are less severe for large indenter radii and high indentation depths.

To summarize, the influence of the roughness on nanoindentation measurements is a function of the shape of the asperities, their spatial distribution and density, the distribution of their heights, their widths but also of the shape and radius on the indenter as well as the elastic-plastic properties of the indented material (Fischer-Cripps 2011, Němeček 2012).

Another reason of variations in nanoindentation results is the choice of test's rates especially for rubbers. Unfaithfully, the influence of such variables is almost not referenced in literature. Some suggestions can still be found. Jin et al. (2015) propose long loading and holding times but high-speed force removal on a poly(methyl methacrylate) so as to avoid inadequate shapes of unloading curves. These suggestions are also made by Tang & Ngan (2003) and Fu et al. (2014), according to studies performed on a polypropylene and an epoxy resin respectively.

2 EXPERIMENTAL

2.1 *Material*

Experimental characterizations of this study involve peroxide vulcanizates of a fluoroelastomer (FKM). This kind of rubber is used in blends that must withstand severe chemical and thermal stresses. This rubber is reinforced with carbon black and silica. Raw rubber blend was obtained by mixing in an internal mixer and was then pressed and vulcanized so as to obtain 2-millimeters-thick sheets.

2.2 *Nanoindentation experiments and analysis*

Local measurements were performed using a NanoTest from MicroMaterials Limited, allowing loads in the range 0.5 mN – 500 mN and measured depths up to 20 μm. This system is composed of a pendulum with an electromagnetic load actuator (coil and magnet) allowing the material's surface loading. The resulting displacement is measured with a capacitive displacement transducer. A three-sided Berkovich diamond tip was chosen due to its wide use. A load-control method was used employing a trapezoidal load history composed of a loading, dwelling (hold at maximum load) and unloading period. Several experimental protocols were explored so as to understand the mechanical response of the studied elastomer under different kinetics of loading (from few seconds to about a thousand seconds) at a maximum force of 1 mN. The curves analyses were made according to the Oliver & Pharr (2004) method. They give a mathematical description of the unloading curve:

$$P = A\left(h - h_f\right)^m \tag{1}$$

With P, the force and h, the displacement. A and m are two material constants and h_f is the final depth after complete unloading, estimated by the least square method.

By derivating this equation one can define the unloading slope S = dP/dh at P = P_{max} and h = h_{max}:

$$S = Am\left(h_{max} - h_f\right)^{m-1} \tag{2}$$

It is then possible to obtain the contact depth h_c between the sample and the tip:

$$h_c = h_{max} - \varepsilon \frac{P_{max}}{S} \tag{3}$$

With ε a correction factor equals to 0.75 for a Berkovich indenter.

The projected contact area between the material and the indenter is thus determined:

$$A_p = 24.49\, h_c^2 + C_1\, h_c^1 + C_2\, h_c^{1/2} + \cdots + C_8\, h_c^{1/128} \tag{4}$$

Finally, the indentation modulus also called reduced modulus, which is an elastic modulus, is described as:

$$E_r = \frac{S\sqrt{\pi}}{2\beta\sqrt{A_p}} \tag{5}$$

With β, a coefficient equals to 1.034 for a Berkovich indenter.

Each test was repeated at least 9 times and results graphs represent averages and standard deviations.

2.3 *Macroscopic tensile and creep tests*

Tensile and creep tests were carried out on a Zwick Z010 dynamometer with a 10 kN load cell. Standard H2 dumbbell samples were cut from the fluroelastomer 2-mm-thick sheet. The experimental protocol consisted in: a tensile phase up to 40 N with four different applied rates between 20 N/s and 0.04 N/s followed by a creep stage during 300 seconds. Under those conditions typical stress and strain undergone by the material are 4.5 MPa and 160% respectively. Each test was repeated three times and averages and standard deviations are represented on the graphs.

3 RESULTS AND DISCUSSION

3.1 *Importance of the sample preparation*

One of the critical factors in the case of elastomer samples, whether laboratory samples or real industrial parts, is the surface roughness. One can see on Figure 1 real differences of the surface quality between the molded surface and a cross-section prepared with a new (i.e. never used) razor blade. Observations made with the optical microscope indicates different morphologies at the two surfaces associated with a difference in roughness. The optical profiling system allows to confirm it. Indeed, Figure 1a,c presents several rough patches on the whole molded surface, generally with dimensions of many tens of micrometers while the prepared surface seems to be more regular and smooth (Fig. 1b,d). Typical values of the arithmetic mean deviation roughness R_a were calculated from the surface statistics provided by the profiling system after an average of at least 3 measurements, leading to 562 ± 76 and 137 ± 6 nanometers for the molded surface and the cross-section respectively. These results attest to a significant improvement of the surface quality by revealing a roughness twice as low as in the case of a sample preparation with an unused razor blade.

Regarding nanoindentation method and the typical size of indents made in this study, working on the second type of surface quality (smoother) is obviously much more relevant than working on the first one (rougher). In the present case, the main consequence of a high sample roughness is a lack of reproducibility (Chen et al. 2013) on raw data and so on extracted mechanical properties. An example is given in Figure 2, where it can be seen that, under the same experimental conditions, a rougher surface leads to higher measurement scatters. The direct impact of such a variability in hystereses is an important standard deviation with respect to indentation moduli.

Characteristic moduli's standard deviations represent at least 15%, even 20% of the mean value on the rough surface but only 3 to 7% on

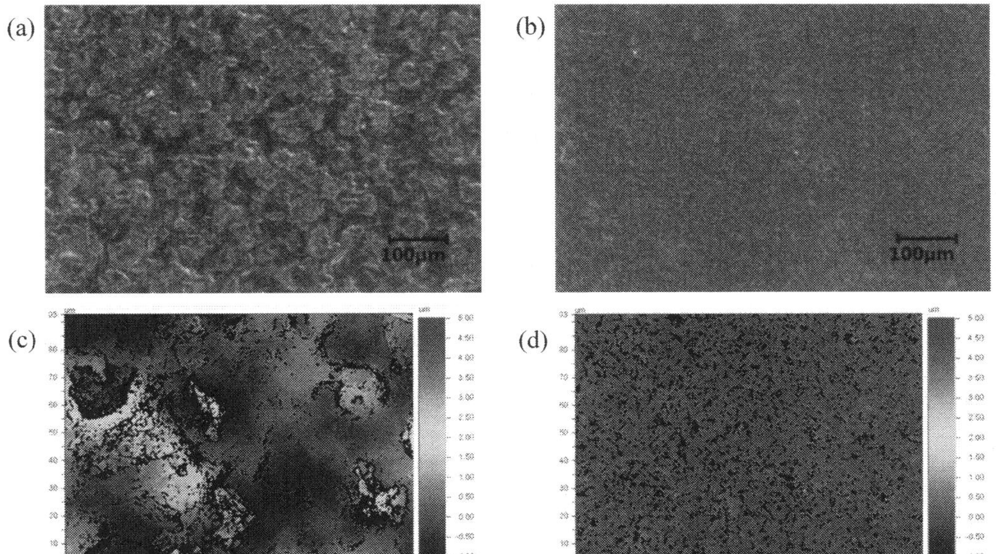

Figure 1. Optical micrographs of the molded surface (a) and the razor-blade-prepared cross-section (b) and corresponding topography pictures by scanning interferometry (c and d).

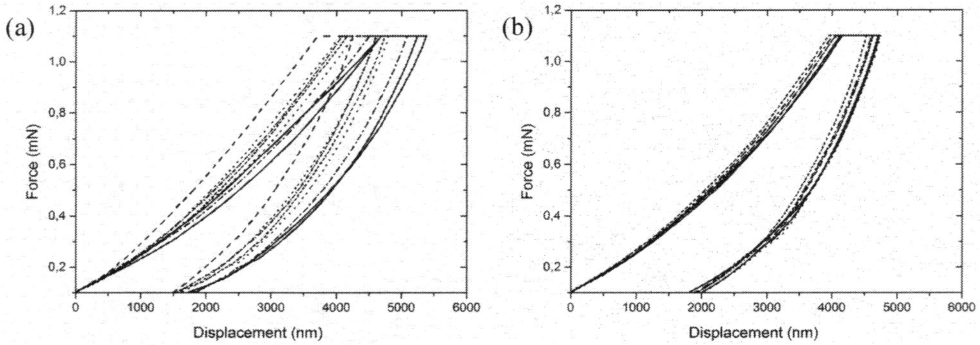

Figure 2. Reproducibility of nanoindentation force-displacement curves with 9 tests, (a) on the molded surface and (b) on the razor-blade-prepared cross-section.

the smoother one. Finally, considering that contact depths of indents performed for this paper are typically 2.7 or 3 microns, the condition given by the "5% rule" can be considered as fulfilled for tests performed in the cross-section.

3.2 *Importance of tests' parameters*

3.2.1 *Nanoindentation results*

Because of their time-dependent response, elastomers are known to be very sensitive to loading or displacement rates used in experimental characterizations. It has been decided to study this phenomenon at the local scale through various nanoindentation protocol kinetics. Figure 3 presents displacements during the loading $h_{loading}$, during the holding stage $h_{holding}$ and the total displacement $h_{max} = h_{loading} + h_{holding}$ as a function of the loading time with a constant holding time of 300 seconds. It can be seen that an increase in the loading time yields an increase of the displacement induced at this stage. This mechanical response is typical of the manifestation of the elastomer's viscosity. Moreover, one can notice that high displacement during the loading stage leads to a reduced creep response. Combined with the time-independence of h_{max}, this observation shows that loading and holding displacements balance one another out for a given maximal force. These results show that low times of loading i.e. high rates can be chosen without risk of an overestimation of the properties due to the viscosity contribution, thanks to the holding stage. This conclusion is essential because of the need to have well-estimated local mechanical properties as well as reasonable experiment durations.

Unloading rates influence has also been studied. As a reminder, it is the unloading curve which is used for the calculation of the elastic modulus (see Section 2.2). It is then important to apply a relevant unloading rate. According to Figure 4 the displacement recovery during unloading is dependent of the removal time of the force. Indeed, considering the viscous nature of the material, the

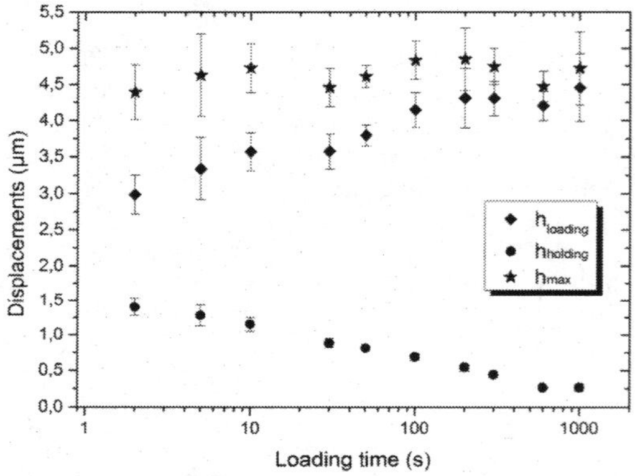

Figure 3. Nanoindentation loading, holding and maximum displacements as a function of the loading time.

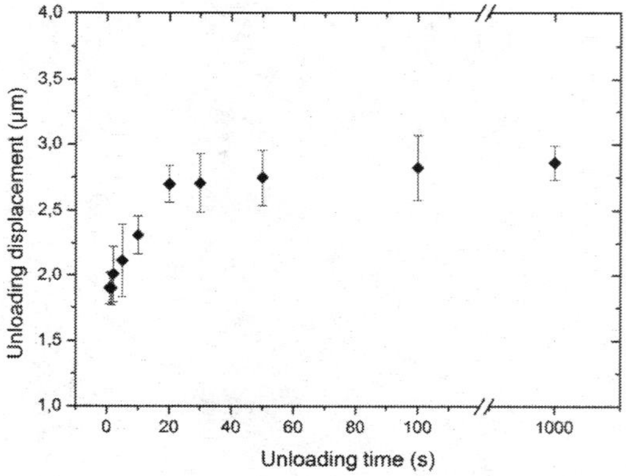

Figure 4. Nanoindentation unloading displacements as a function of the unloading time.

higher the unloading time, the higher is the tip displacement, as expected. Thus, to be in accordance with the literature hypothesis of an elastic contact during unloading, long times have to be chosen so as to remove as much as possible viscous effects.

3.2.2 *Tensile and creep tests results*

To confirm results presented on Figure 3, it has been decided to carry out some global tests, similar to nanoindentation ones i.e. with a loading stage and a holding one. Displacements during the tensile phase, during the holding of the force were measured and so were obtained the total displacements. Figure 5 summarizes the results. Similarly to nanoindentation observations, it can be seen that the total displacement is the same whatever the loading time. In other words, if the applied loading rate is too important, then the material cannot accommodate the high mechanical stress and the strain during the loading remains low; but thanks to the holding stage the missing strain will be recovered into creep displacement. Furthermore, the comparison between local (Fig. 3) and global (Fig. 5) results seems to show similar trends regarding the studied displacements but with higher standard deviations for nanoindentation whose measurements are very sensitive to any instrumental, environmental or material factor. Finally, these results obtained at two very different scales can be correlated in terms of creep quantity compared to the total displacement: $h_{holding}/h_{max} \times 100$, particularly for long loading times. Indeed, under both local and global conditions, the creep quantity represents 15% of the total displacement for a loading time of 100 seconds and 5% for 1000 seconds. Differences in creep quantity appear between macro—and microscopic results with shorter loading times, probably due to viscous effects which do not impact global and local measurements in the same way.

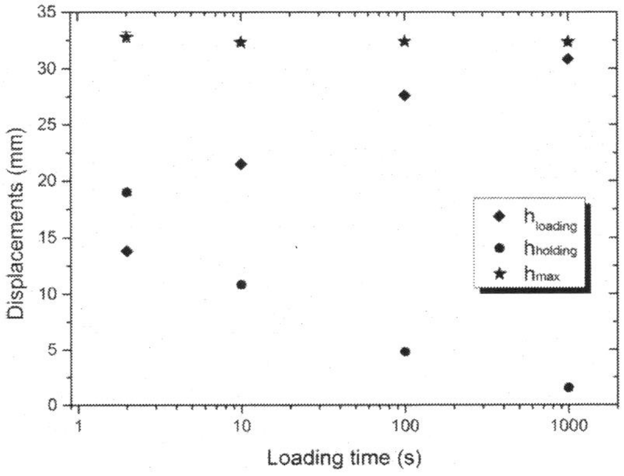

Figure 5. Macroscopic loading, holding and maximum displacements as a function of the loading time.

4 APPLICATION

4.1 *Influence of the indentation spacing*

In the literature about nanoindentation, one can find some values of the minimum distance required so as to avoid interactions between two side by side indentations: Hay & Pharr (2000) propose for example a spacing of 20 to 30 times the maximum depth. However, these values can be considered as questionable. First, their origin is confusing and the way of obtaining them is not clearly explicated. Then, they are given for elastic-plastic materials which present a residual imprint and whose response can change after removal of the stress (residual stresses, work hardening). However, deformation mechanisms in elastomers are much different: one can expect a larger elastically deformed area under complete loading but reversible induced strains. Figure 6 gives the modulus evolution as a function of the left space between two successive indents. It is worth reminding that typical contact depths are near 2.7 or 3 μm (Section 3.1) and so that contact diameters are around 30 microns. Beneath this value, one can suppose measurement interactions, thus different moduli. However, Figure 6 shows one single point with a higher modulus when indentations are very close to each other i.e. 3 μm spacing while with higher distances, moduli are nearly stable. Indeed, if indentation displacements are reversible, if the characteristic time of the material allows the displacement recovery during the time separating two indentations and if there is no damage induced by the test, very short indentation distances can be chosen. A priori, two indentations can be carried out as close as 5 μm from each other, but larger spacings (i.e. 10 or 20 μm) can be prefered so as to

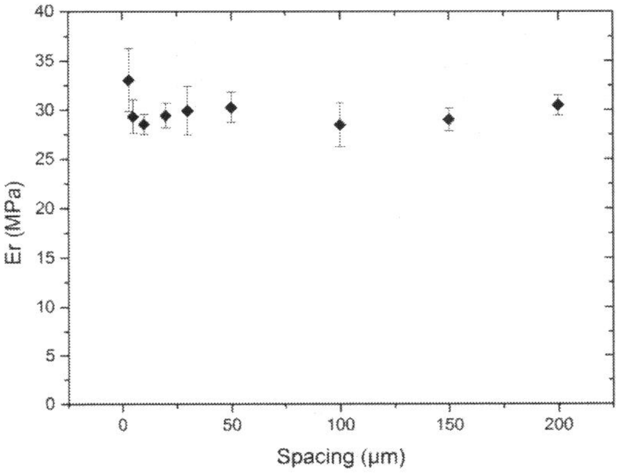

Figure 6. Indentation modulus as a function of the distance between two successive indentations.

obtain a good ratio between the mapping precision and its total time.

Future works will include repeating these tests at different maximum forces i.e. with different deformation volumes and finite element analysis so as to obtain an estimation of these volumes.

4.2 *Application to probe vulcanization gradients along the cross-section*

Previous results were the subject of a practical application so as to get information about differences in moduli along the thickness of the rubber sheet due to vulcanization gradients. The following experimental parameters were chosen: 100 seconds for the loading and unloading phases and 300 seconds of holding at the maximum force 1 mN. The first indentation was made at 20 micrometers from the edge i.e. far enough to prevent the tip from being, even partially, in the emptiness; and the last one at 1 millimeter in the bulk i.e. up to half of the sheet's thickness. The spacing between two successive indentations was 20 microns. Four profiles were carried out and then averaged (Fig. 7):

Results reveal moduli variations along the cross-section thickness. Indeed, one can see that moduli near the edge are lower than bulk moduli. This is the opposite of the generally accepted presence of a skin due to curing of bulk elastomer parts. Indeed, one can suppose that thermal gradients induced by the vulcanization process cause a high-modulus external surface and an important decreasing of this material's property as the part's core is approached. One cannot rule out edge effect that is a measurement artefact due to the fact that indentations were made too close to the external surface. However, this phenomenon is to be considered only at distances of the same

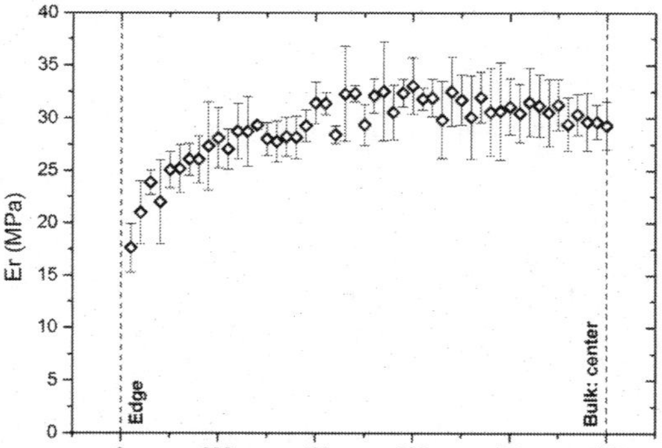

Figure 7. Evolution of the indentation modulus within the cross-section thickness.

magnitude as determined in former section (Fig. 6) while low moduli are observed quite far from the edge i.e. up to 400 µm. This loss in properties near the surface could also be due to a thermal or chemical damage even though the studied fluoroelastomer does not show any reversibility on its vulcanization curve in case of overcuring. One can also notice quite stable moduli in the bulk, starting from around 400 microns, so that sample properties are homogeneous on around 60% of its thickness. This non-evolution could be explained by the curing process of the elastomer sample, which includes, in addition to the vulcanization, a post-vulcanization and also by the thickness of the sample which, in comparison with thicker industrial products, allows a fairly uniform temperature gradient.

Moreover, it can be said that if indentations have to be carried out on the external surface, the previously described moduli evolution (Fig. 6) has to be taken into consideration as it can cause a depth-dependence of the results, in particular for small depths.

5 CONCLUSION

In this study, two factors known for influencing nanoindentation results have been studied: the sample roughness and the experimental variables during a test. Optical observations and reproducibility tests tend to prove a non-negligible influence of the surface quality on results. A satisfactory roughness could be obtained thanks to an unused razor blade and results with relatively low deviations were collected. On the other hand, viscous effects revealed to have a great impact on nanoindentation tests. Great care must be taken in choosing tests' rates so as to get relevant quantitative mechanical properties as well as reasonable experimental times.

Using nanoindentation as a local probe is unlimited in terms of applications. We chose here to apply it in order to see internal vulcanization gradients. The tests showing low indentation interactions (probably due to the elastic recovery of indents), even with short spaces between them, profiles within the rubber sheet thickness have been carried out. Those profiles seem to show homogeneous moduli along the cross-section that is an adequate curing process. However, what could be softening has been noticed in the area close to the external surface. Complementary tests should be repeated and performed with lower forces so as to gain measurement precision and to get moduli closer to the external surface and hence rule out edge effects. Thus, the whole potential of the nanoindentation technique would be used.

REFERENCES

Bobji, M.S. & S.K. Biswas (1998). Estimation of hardness by nanoindentation of rough surfaces. *J. Mat. Res. 13*, 3227–3233.

Boussinesq, M.J. (1885). *Application des potentiels à l'étude de l'équilibre et du mouvement des solides élastiques*. Gauthier-Villars.

Chen, L., A. Ahadi, J. Zhou & J.-E. Ståhl (2013). Modeling effect of surface roughness on nanoindentation tests. *Procedia CIRP 8*, 334–339.

Fischer-Cripps, A.C. (2011). *Nanoindentation* (3 ed.). Springer.

Fu, K., Y. Chang, Y. Tang & B. Zheng (2014). Effect of loading rate on the creep behaviour of epoxy resin insulators by nanoindentation. *J. Mater. Sci.: Mater. Electron. 25*, 3552–3558.

Hay, J.L. & G.M. Pharr (2000). Instrumented Indentation Testing. *ASM Handbook 8*, 232–243.

Hertz, H. (1881). On the contact of elastic solids. *J. Reine Angew. Math. 92*, 156–171.

Jiang, W.-G., J.-J. Su & X.-Q. Feng (2008). Effect of surface roughness on nanoindentation test of thin films. *Eng. Fract. Mech. 75*, 4965–4972.

Jin, T., X. Niu, G. Xiao, Z. Wang, Z. Zhou, G. Yuan & X. Shu (2015). Effects of experimental variables on PMMA nano-indentation measurements. *Polym. Test. 41*, 1–6.

Love, A.E.H. (1939). Boussinesq's problem for a rigid cone. *Q.J. Math. 10*, 161–175.

Němeček, J. (2012), *Nanoindentation in materials science* (2 ed.). InTech.

Oliver, W.C. & G.M. Pharr (2004). Measurement of hardness and elastic modulus by instrumented indentation: Advances in understanding and refinements to methodology. *J. Mat. Res. 19*, 3–20.

Sneddon, I.N. (1965). The relation between load and penetration in the axisymmetric Boussinesq problem for a punch of arbitrary profile. *Int. J. Engng. Sci. 3*, 47–57.

Tang, B. & A.H.W. Ngan (2003). Accurate measurement of tip–sample contact size during nanoindentation of viscoelastic materials. *J. Mat. Res. 18*, 1141–1148.

Constitutive Models for Rubber X – Lion & Johlitz (Eds)
 ISBN 978-1-138-03001-5

New ideas to represent strain induced crystallisation in elastomers

K. Loos, M. Johlitz & A. Lion
Faculty for Aerospace Engineering, Institute of Mechanics, Universität der Bundeswehr Munich, Germany

L. Palgen & J. Calipel
MICHELIN Technology Center, Clermont-Ferrand, France

ABSTRACT: Natural rubber experiences a natural reinforcement caused by high deformation, called strain induced crystallisation. This phenomenon was primarily observed by Katz (1925). SIC usually starts at strains of 200–400% in natural rubber. In industrial applications the usage of natural rubber is advantageous because of SIC and its superior qualities such as increasing crack growth resistance.

In the last years, SIC has been a quite active topic with numerous publications, recent developments and new paradigms. Although it is a complicated subject for mechanicians because of the thermodynamics, polymer physics and sophisticated kinetics it is are warding subject for industrial as well as academic applications.

In this study, the overall objective is to find constitutive equations which describe the material behaviour of elastomers focused on the phenomenon of strain induced crystallisation. The following paper specifies a thermomechanical approach to model strain induced crystallisation of elastomers. The model is based on the concept of representative directions and considers thermoelasticity and crystallisation.

1 INTRODUCTION

1.1 *Overview strain induced crystallisation*

The phenomenon of Strain Induced Crystallisation (SIC) in natural rubber is the focus of modelling and experimental investigations in several research studies during the last years. Since its discovery by Katz (1925) SIC became an active topic due to its advantageous impact such as the improvement of mechanical and ultimate material characteristics. Natural rubber's superior crack growth resistance and outstanding tensile properties are just two industrial benefits, see Albouy et al. 2012, Rublon 2013. The kinetics of strain induced crystallisation is a highly investigated topic. First studies have been conducted by Acken et al. 1932 and Long et al. 1934. Since these first analyses, SIC has been the subject of several theoretical (e.g. Yeh and Hong 1979, Gaylord and Lohse 1976, Flory 1947, Alfrey and Mark 1942), as well as experimental investigations where several studies have been conducted such as thermal measurements, birefringence and stress relaxation ((Tosaka et al. 2012)). Filled natural rubber was observed by Chenal et al. (2007) and different parameters governing SIC were detected. Synchrotron radiation permits realtime measurement (e.g. Trabelsi et al. 2003, Toki et al. 2005).

Especially the recent reviews and attempts (e.g. Albouy et al. 2014, Durin et al. 2017 and Candau et al. 2014) show the actuality of SIC's topic, whereas the modelling of SIC remains incomplete (Huneau 2011 and Tosaka et al. 2004). Further research in crack growth resistance emphasizes its importance for industrial applications (Tosaka et al. 2004, Beurrot et al. 2010). Experiments to fatigue crack observation have been conducted at the synchrotron Soleil by Rublon (2013). Tosaka et al. (2004) and Beurrot et al. (2010) found out that crystallites slow down, deviate and stop the crack growth inside the material due to strain induced crystallisation.

Lion & Johlitz (2015) applied a directional approach in constitutive modelling which is capable to represent strain induced anisotropies. This approach has similarities to the directional approach by Freund and Ihlemann (2010) and Shutov et al. (2013).

The present article is structured into an introducing part with fundamentals and the general idea of the model. The second part analyses and specifies the model in the contribution of strain induced crystallisation. The third part gives a first evaluation of the model by usage of fictitious material parameters.

1.2 *General constitutive model*

The present paper formulates some ideas to represent strain and temperature induced crystallisation of elastomers. The constitutive model consists of two energy contributions. The energy contribution of the Gibbs-type represents the volumetric part of the material behaviour. It depends on pressure, temperature and possibly on the total crystallinity. The second part is the energy contribution of Helmholtz-type which describes the isochoric part of the material behaviour.

$$\varphi = g + \hat{\psi} \tag{1}$$

The advantage of this decomposition of the energy is the consideration of both caloric and mechanical effects. Caloric experiments are usually carried out under constant pressure and prescribed temperature histories. If the material is isotropic, the resulting deformations are purely volumetric and the isochoric strain tensor vanishes. Since the volumetric strains are small for many polymers, the pressure usually depends linearly on them. The isochoric part of the energy depends on the isochoric strain tensor.

In order to take process-dependent anisotropies caused by strain induced crystallisation into account, the model uses the concept of representative directions and depends on directional crystallinities, directional stretches and temperature.

2 SIC MODELLING IDEAS

2.1 *Decomposition of volumetric and deviatoric parts*

The following concept is based on the general thermomechanical framework by Lion & Johlitz (2015) and especially on the directional approach in constitutive modelling by Lion et al. (2013). It includes an extension focused on strain induced crystallisation modelling.

According to Flory (1961), conformation (changes in shape) and volume changes are considered independently. The approach is to decompose the deformation gradient $\mathbf{F}$ multiplicatively into an isochoric part $\hat{\mathbf{F}}$ and a volumetric part $\bar{\mathbf{F}}$, therefore $\mathbf{F} = \bar{\mathbf{F}}\hat{\mathbf{F}}$. With the determinant $J = \det(\mathbf{F})$ the volumetric deformation forms to

$$\bar{\mathbf{F}} = J^{\frac{1}{3}}\mathbf{1} \tag{2}$$

and the isochoric conformation to

$$\hat{\mathbf{F}} = \bar{\mathbf{F}}^{-1}\mathbf{F} = J^{-\frac{1}{3}}\mathbf{F}. \tag{3}$$

With the basics of continuum mechanics (Truesdell and Noll 1992) and (Haupt 2002), the definition of the Green-Lagrange strain tensor $\mathbf{E} = \frac{1}{2}(\mathbf{C} - \mathbf{1})$ where $\mathbf{C} = \mathbf{F}^T\mathbf{F}$ is the right Cauchy-Green deformation tensor, has an isochoric analogon $\hat{\mathbf{E}} = \frac{1}{2}(\hat{\mathbf{C}} - \mathbf{1})$ with $\hat{\mathbf{C}} = \hat{\mathbf{F}}^T\hat{\mathbf{F}}$. Consequently, the Green-Lagrange strain tensor can be written as

$$\mathbf{E} = \frac{1}{2}(J^{\frac{2}{3}}\hat{\mathbf{C}} - \mathbf{1}) = J^{\frac{2}{3}}\hat{\mathbf{E}} + \frac{1}{2}(J^{\frac{2}{3}} - 1)\mathbf{1} \tag{4}$$

The Cauchy stress tensor, the 'true stress', is split into volumetric and deviatoric parts:

$$\mathbf{T} = -p\mathbf{1} + \mathbf{T}^D \tag{5}$$

Here p describes the hydrostatic pressure $p = -\frac{1}{3}tr(\mathbf{T})$ and $\mathbf{T}^D$ describes the deviatoric part: $\mathbf{T}^D = \mathbf{T} - \frac{1}{3}tr(\mathbf{T})\mathbf{1}$.

The second Piola-Kirchhoff stress is defined as $\tilde{\mathbf{T}} = J\mathbf{F}^{-1}\mathbf{T}\mathbf{F}^{-T}$ and analogous the deviatoric second Piola-Kirchhoff-type tensor is $\hat{\tilde{\mathbf{T}}} = J\hat{\mathbf{F}}^{-1}\mathbf{T}^D\hat{\mathbf{F}}^{-T}$. Thus, the second Piola-Kirchhoff stress can be written as an additive decomposition of a volumetric and a deviatoric part: $\tilde{\mathbf{T}} = -pJ^{\frac{1}{3}}\hat{\mathbf{C}}^{-1} + J^{-\frac{2}{3}}\hat{\tilde{\mathbf{T}}}$.

With the volume strain $\varepsilon_{vol} = J - 1$, a further consequence is the additive decomposition of the stress power into volumetric and deviatoric parts:

$$\begin{aligned} &\tilde{\mathbf{T}} : \dot{\mathbf{E}} \\ &= (-pJ^{\frac{1}{3}}\hat{\mathbf{C}}^{-1} + J^{-\frac{2}{3}}\hat{\tilde{\mathbf{T}}}) : \frac{d}{dt}\left(J^{\frac{2}{3}}\hat{\mathbf{E}} + \frac{1}{2}(J^{\frac{2}{3}} - 1)\mathbf{1}\right) \\ &= -p\dot{\varepsilon}_{vol} + \hat{\tilde{\mathbf{T}}} : \dot{\hat{\mathbf{E}}} \end{aligned} \tag{6}$$

The advantage is that this form separates the volumetric strain rate $\dot{\varepsilon}_{vol}$ and the hydrostatic pressure p.

2.2 *Thermodynamical consistency*

The Clausius-Duhem inequality ensures the thermodynamical consistency of constitutive equations such that the second law of thermodynamics is satisfied:

$$-\rho_R\dot{\psi} + \tilde{\mathbf{T}} : \dot{\mathbf{E}} - \rho_R s\dot{\theta} - \frac{\vec{q}_R \cdot \vec{g}_R}{\theta} \geq 0 \tag{7}$$

where the scalars ρ_R, θ, ψ, s denote the density, the thermodynamic temperature, the specific Helmholtz free energy and the entropy per unit mass related to the reference configuration.

By insertion of (6), the Clausius-Duhem inequality (7) forms to

$$-\rho_R\dot{\psi} - p\dot{\varepsilon}_{vol} + \hat{\tilde{\mathbf{T}}} : \dot{\hat{\mathbf{E}}} - \rho_R s\dot{\theta} - \frac{\vec{q}_R \cdot \vec{g}_R}{\theta} \geq 0 \tag{8}$$

The Legendre transformation

$$\rho_R\psi = \rho_R\varphi - p\varepsilon_{vol} \tag{9}$$

with its derivative

$$\rho_R\dot{\psi} = \rho_R\dot{\varphi} - \dot{p}\varepsilon_{vol} - p\dot{\varepsilon}_{vol} \tag{10}$$

inserted into the Clausius-Duhem inequality (8) leads to

$$-\rho_R\dot{\varphi} + \hat{\tilde{\mathbf{T}}} : \dot{\hat{\mathbf{E}}} + \dot{p}\varepsilon_{vol} - \rho_R s\dot{\theta} - \frac{\vec{q}_R \cdot \vec{g}_R}{\theta} \geq 0 \tag{11}$$

The Legendre transformation replaces hydrostatic pressure p and volumetric strain rate $\dot{\varepsilon}_{vol}$ with the hydrostatic pressure rate $\dot{p}$ and volumetric strain ε_{vol}. The dependencies for the specific hybrid free energy (1) are separated in

$$\varphi(\hat{\mathbf{E}}, p, \theta, ...) = g(p, \theta, ...) + \hat{\psi}(\hat{\mathbf{E}}, \theta, ...) \tag{12}$$

2.3 *Representative directions*

The concept of representative directions is a generalisation technique in order to transfer one-dimensional material models to three-dimensional states of stress and deformation (Freund and Ihlemann 2010, Freund et al. 2011). The general idea behind this approach is the projection of arbitrary states of deformation to one-dimensional elongations along certain directions. With the help of the directional stretches the material response as directional stresses is calculated. The directional stresses are then combined to generate the three-dimensional-stress tensor. The major advantages of this approach is that the formulation of constitutive equations becomes easier since the

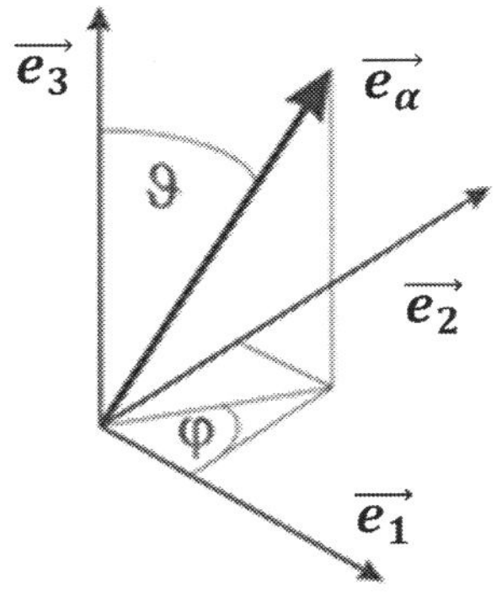

Figure 1. Unit vector in the orientation space of the reference configuration.

formulation is done for one-dimensional states of deformation. Moreover it includes direction-dependent effects like anisotropic behaviour naturally. The calculation is performed on the directional level and therefore specified to each direction $\vec{e}_\alpha$.

$$\vec{e}_\alpha = \sin\vartheta\cos\varphi\vec{e}_1 + \sin\vartheta\sin\varphi\vec{e}_2 + \cos\vartheta\vec{e}_3 \tag{13}$$

The averaging operator A is defined as

$$f(t) = \frac{1}{4\pi}\int_0^{2\pi}\int_0^{\pi} f_\alpha(\vartheta, \varphi, t)\sin\vartheta d\vartheta d\varphi \tag{14}$$

$$f = A[f_\alpha] \tag{15}$$

where f_α is a physical quantity such as stress or stretch related to the direction of $\vec{e}_\alpha$. Consequently, its first derivative is $\dot{f} = A[\dot{f}_\alpha]$.

With reference to this operator A, we find i.a. the physical quantities related to the direction $\vec{e}_\alpha$:

$\hat{\lambda}_\alpha = \sqrt{\hat{\mathbf{F}}\vec{e}_\alpha \cdot \hat{\mathbf{F}}\vec{e}_\alpha} = \sqrt{\vec{e}_\alpha \cdot \hat{\mathbf{C}}\vec{e}_\alpha}$ (directional stretch)

$\hat{s} = A[\hat{s}_\alpha]$ (isochoric part of the specific entropy)

$\hat{\tilde{\mathbf{T}}} = A[\hat{\tilde{\mathbf{T}}}_\alpha]$ (deviatoric 2nd Piola-Kirchhoff stress)

$x = A[x_\alpha]$ (total crystallinity)

With these definitions and the definition of the directional stress tensor $\hat{\tilde{\mathbf{T}}}_\alpha = \frac{\hat{\sigma}_\alpha}{\hat{\lambda}_\alpha}\vec{e}_\alpha \otimes \vec{e}_\alpha$, the isochoric part of the stress power can be formulated to

$$A\left[\hat{\sigma}_\alpha\dot{\hat{\lambda}}_\alpha\right] = A\left[\hat{\tilde{\mathbf{T}}}_\alpha : \dot{\hat{\mathbf{E}}}\right] = A\left[\hat{\tilde{\mathbf{T}}}_\alpha\right] : \dot{\hat{\mathbf{E}}} = \hat{\tilde{\mathbf{T}}} : \dot{\hat{\mathbf{E}}}. \tag{16}$$

2.4 *Structure of a constitutive model*

The thermoelasticity with crystallisation is modelled by the hybrid free energy where the volumetric and isochoric parts are split additively, see (1). The Gibbs-type energy for the volumetric material behaviour depends on the total degree of crystallinity

$$g = g(p, \theta, x) \tag{17}$$

whereas the Helmholtz-type energy for the directional material behaviour depends on the directional degree of crystallinity

$$\hat{\psi} = A[\hat{\psi}_\alpha(\hat{\lambda}_\alpha, \theta, x_\alpha)] \tag{18}$$

Thus, the first time-rate of the hybrid free energy is

$$\dot{\varphi} = \frac{\partial g}{\partial p}\dot{p} + \frac{\partial g}{\partial \theta}\dot{\theta} + \frac{\partial g}{\partial x}\dot{x} + A\left[\frac{\partial \hat{\psi}_\alpha}{\partial \hat{\lambda}_\alpha}\dot{\hat{\lambda}}_\alpha + \frac{\partial \hat{\psi}_\alpha}{\partial \theta}\dot{\theta} + \frac{\partial \hat{\psi}_\alpha}{\partial x_\alpha}\dot{x}_\alpha\right] \tag{19}$$

With $x = A[x_\alpha]$ this forms to

$$\dot{\varphi} = \frac{\partial g}{\partial p}\dot{p} + \frac{\partial g}{\partial \theta}\dot{\theta} + A\left[\frac{\partial \hat{\psi}_\alpha}{\partial \hat{\lambda}_\alpha}\dot{\hat{\lambda}}_\alpha + \frac{\partial \hat{\psi}_\alpha}{\partial \theta}\dot{\theta} + \left(\frac{\partial g}{\partial x} + \frac{\partial \hat{\psi}_\alpha}{\partial x_\alpha}\right)\dot{x}_\alpha\right] \tag{20}$$

The Clausius-Duhem inequality (11) with (16) and (20) forms to

$$-\rho_R A\left[\left(\frac{\partial g}{\partial x} + \frac{\partial \hat{\psi}_\alpha}{\partial x_\alpha}\right)\dot{x}_\alpha\right] + A\left[\left(\hat{\sigma}_\alpha - \rho_R\frac{\partial \hat{\psi}_\alpha}{\partial \hat{\lambda}_\alpha}\right)\dot{\hat{\lambda}}_\alpha\right] + \left(\varepsilon_{vol} - \rho_R\frac{\partial g}{\partial p}\right)\dot{p} - \rho_R\left(s + A\left[\frac{\partial \hat{\psi}_\alpha}{\partial \theta} + \frac{\partial g}{\partial \theta}\right]\right)\dot{\theta} - \frac{\vec{q}_R \cdot \vec{g}_R}{\theta} \geq 0 \tag{21}$$

With the following quantities

$$\varepsilon_{vol} = \rho_R\frac{\partial g}{\partial p} \quad \text{(volume strain)}$$

$$s = -\frac{\partial g}{\partial \theta} - A\left[\frac{\hat{\psi}_\alpha}{\partial \theta}\right] \quad \text{(specific entropy)}$$

$$\hat{\sigma}_\alpha = \rho_R\frac{\partial \hat{\psi}_\alpha}{\partial \hat{\lambda}_\alpha} \quad \text{(direct. stress)}$$

$$\dot{x}_\alpha = -\beta(\theta, \ldots)\left(\frac{\partial g}{\partial x} + \frac{\partial \hat{\psi}_\alpha}{\partial x_\alpha}\right) \quad \text{(direct. cryst. rate)}$$

$$\vec{q}_R = -\kappa \vec{g}_R \quad \text{(heat flux vector)}$$

the *2nd* Piola-Kirchhoff stress tensor is composed to

$$\tilde{\mathbf{T}} = -p(1+\varepsilon_{vol})^{\frac{1}{3}}\hat{\mathbf{C}}^{-1} + (1+\varepsilon_{vol})^{-\frac{2}{3}}\hat{\tilde{\mathbf{T}}} \tag{22}$$

where the volumetric part depends on the total crystallinity $p(x)$ but its isochoric part is a superposition of directional stresses which depend on directional crystallinities $\hat{\sigma}_\alpha(x_\alpha)$.

2.5 *Directional crystallinity*

The basic idea of introducing the degree of crystallinity into the constitutive model is to distinguish two phases of the total mass of the material m_α: crystalline $m_{\alpha c}$ and amorphous phase, where the amorphous phase again is subdivided into a crystallisable fraction $m_{\alpha a}$ and a non-crystallisable fraction $m_{\alpha n}$,

$$m_\alpha = m_{\alpha c} + m_{\alpha a} + m_{\alpha n} \tag{23}$$

Thus, the directional crystallinity is assumed as a mass fraction

$$x_\alpha = \frac{m_{\alpha c}}{m_\alpha} \tag{24}$$

An essential simplifying assumption is a linear dependence between the non-crystallisable fraction and the crystalline phase, i.e. to each crystalline part exists a non-crystallisable part: $m_{\alpha n} = \zeta m_{\alpha c}$. With the following calculation starting from (23) the crystallisable amorphous fraction can be expressed as

$$\begin{aligned} & 1 = \frac{m_{\alpha c}}{m_\alpha} + \frac{m_{\alpha a}}{m_\alpha} + \frac{m_{\alpha n}}{m_{\alpha c}}\frac{m_{\alpha c}}{m_\alpha} \\ \Leftrightarrow \quad & 1 = x_\alpha + \frac{m_{\alpha a}}{m_\alpha} + \zeta x_\alpha \\ \Leftrightarrow \quad & \frac{m_{\alpha a}}{m_\alpha} = 1 - (1+\zeta)x_\alpha \end{aligned} \tag{25}$$

The maximum value of the crystallisable amorphous phase is limited to 1, consequently

$$0 \leq x_\alpha \leq \frac{1}{1+\zeta} =: x_0 \tag{26}$$

Therefore, each fraction simplifies to

$$\frac{m_{\alpha c}}{m_\alpha} = x_\alpha \quad \text{(directional crystallinity)}$$

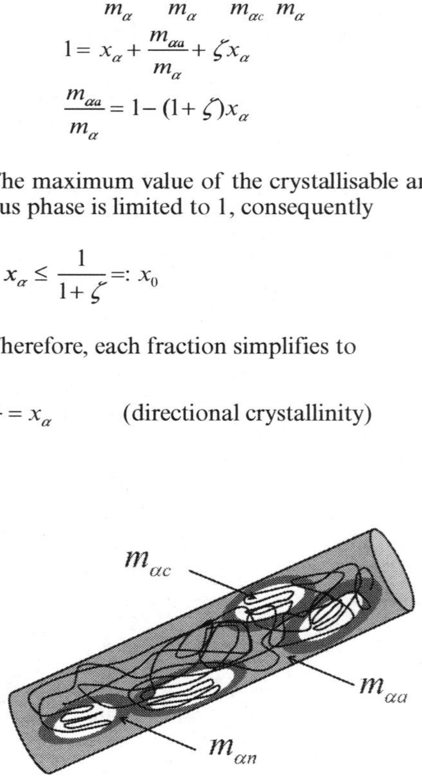

Figure 2. Fractions of specimen: crystallisable amorphous $m_{\alpha a}$, non-crystallisable amorphous $m_{\alpha n}$ and crystalline phase $m_{\alpha c}$.

$$\frac{m_{\alpha a}}{m_\alpha} = 1 - \frac{x_\alpha}{x_0} \quad \text{(cryst. amorphous frac.)}$$

$$\frac{m_{\alpha n}}{m_\alpha} = \left(\frac{1}{x_0} - 1\right) x_\alpha \quad \text{(non-cryst. amorphous frac.)}$$

The specific directional Helmholtz free energy of the single phases are stated as

$\hat{\psi}_{\alpha a}(\hat{\lambda}_\alpha, \theta)$ (cryst. amorphous phase)

$\hat{\psi}_{\alpha n}(\hat{\lambda}_\alpha, \theta)$ (non-cryst. amorphous phase)

$\hat{\psi}_{\alpha c}(\hat{\lambda}_\alpha, \theta)$ (crystalline phase)

The total Helmholtz free energy $\hat{\Psi}_\alpha$ includes the free energy of the individual phases

$$\hat{\Psi}_{\alpha 0} = m_{\alpha a}\hat{\psi}_{\alpha a} + m_{\alpha c}\hat{\psi}_{\alpha c} + m_{\alpha n}\hat{\psi}_{\alpha n} \tag{27}$$

and the entropy of mixing

$$S_{mix} = -m_\alpha \frac{R}{M}\gamma(x_\alpha), \tag{28}$$

thus results to

$$\hat{\Psi}_\alpha = \hat{\Psi}_{\alpha 0} - \theta S_{mix} \tag{29}$$

where R is the universal gas constant and M is the molar mass. The directional Helmholtz free energy per unit mass results to

$$\hat{\psi}_\alpha = \left(1 - \frac{x_\alpha}{x_0}\right)\hat{\psi}_{\alpha a} + x_\alpha \hat{\psi}_{\alpha c} + \left(\frac{1}{x_0} - 1\right) x_\alpha \hat{\psi}_{\alpha n} + \frac{R\theta}{M}\gamma(x_\alpha) \tag{30}$$

2.6 *Entropy of mixing*

The total entropy S of a gas differs after a change in volume $V_1 \rightarrow V_2$ or a change in temperature $\theta_1 \rightarrow \theta_2$.

$$S = nR\ln\left(\frac{V_2}{V_1}\right) + nc_V \ln\left(\frac{\theta_2}{\theta_1}\right) \tag{31}$$

where n is the amount of substance and c_V is its heat capacity.

After mixing two ideal materials with volumes V_A and V_B under constant pressure p and temperature θ, where both were initially separate systems of different composition, each in a thermodynamic state of equilibrium, an increase in total entropy is observed:

$$S_{mix} = m_A \frac{R}{M_A}\ln\left(\frac{N_A + N_B}{N_A}\right) + m_B \frac{R}{M_B}\ln\left(\frac{N_A + N_B}{N_B}\right) \tag{32}$$

where N is the number of particles, M is the molar mass and m is the mass of gas. The mixing is entirely accounted by the diffusion of each material into the final volume, which was not accessible before due to a partition between both volumes.

Next assumption is that a crystalline particle is assumed to consist of the mass of the crystallite and that of the sourrounding non-crystallisable amorphous material. All phases possess similar mean molar masses: $M = M_a = M_{cryst}$. Then the specific mixing entropy is

$$\frac{S_{mix}}{m_\alpha} = -\frac{R}{M}\gamma(x_\alpha) \tag{33}$$

where

$$\gamma(x_\alpha) = \frac{x_0 - x_\alpha}{x_0}\ln\left(\frac{x_0 - x_\alpha}{x_0}\right) + \frac{x_\alpha}{x_0}\ln\left(\frac{x_\alpha}{x_0}\right) \tag{34}$$

Next, the idea is to distinguish between entropy-elastic amorphous phase and energy-elastic behaviour of the crystalline phase. For the before divided crystallisable and non-crystallisable amorphous phase it is assumed to exhibit nearly the same thermomechanical behaviour $\hat{\psi}_{\alpha a} = \hat{\psi}_{\alpha n}$.

With another simplifying assumption it is possessed that crystallinity does not influence the volumetric behaviour of the elastomer $\frac{\partial g}{\partial x} = 0$. Since elastomers are incompressible in a good approximation, the pressure and the temperature dependence of the volume strain are assumed to be linear.

2.7 *Evolution equation for the directional crystallinity*

From the general form of the evolution equation

$$\dot{x}_\alpha = -\beta(\theta, \ldots)\left(\frac{\partial g}{\partial x} + \frac{\partial \hat{\psi}_\alpha}{\partial x_\alpha}\right) \tag{35}$$

and with the assumption $\hat{\psi}_{\alpha a} = \hat{\psi}_{\alpha n}$, the directional Helmholtz free energy per unit mass (30) reduces to

$$\hat{\psi}_\alpha = (1 - x_\alpha)\hat{\psi}_{\alpha a} + x_\alpha \hat{\psi}_{\alpha c} + \frac{R\theta}{M}\gamma(x_\alpha) \tag{36}$$

Thus, the evolution equation with the assumption $\frac{\partial g}{\partial x} = 0$, and the derivative of (34)

$$\frac{d\gamma(x_\alpha)}{dx_\alpha} = \frac{1}{x_0}\ln\left(\frac{x_\alpha}{x_0 - x_\alpha}\right) \tag{37}$$

results to

$$\dot{x}_\alpha = \beta(\theta,...)\left(\hat{\psi}_{\alpha a} - \hat{\psi}_{\alpha c} - \frac{R\theta}{Mx_0}\ln\left(\frac{x_\alpha}{x_0 - x_\alpha}\right)\right) \tag{38}$$

3 MODEL EVALUATION

The nondimensionalisation with the help of the variable $s = \frac{t}{T}$ leads to

$$\tilde{x}'_\alpha = T\beta(\theta,...)\left(\hat{\psi}_{\alpha a}(\tilde{\hat{\lambda}}_\alpha,\tilde{\theta}) - \hat{\psi}_{\alpha c}(\tilde{\hat{\lambda}}_\alpha,\tilde{\theta}) - \frac{R\tilde{\theta}}{Mx_0}\ln\left(\frac{\tilde{x}_\alpha}{x_0 - \tilde{x}_\alpha}\right)\right) \tag{39}$$

where $x_\alpha(t) = x_\alpha(Ts) = \tilde{x}_\alpha(s)$, $\hat{\lambda}_\alpha(t) = \hat{\lambda}_\alpha(Ts) = \tilde{\hat{\lambda}}_\alpha(s)$ and $\theta(t) = \theta(Ts) = \tilde{\theta}(s)$.

In case of infinitely fast thermal or mechanical excitations $T \to 0$, (39) tends to zero i.e. the directional crystallinity remains constant.

In case of infinitely slow thermomechanical excitations $T \to \infty$, the directional crystallinity evolves infinite slowly. Thus, the differential equation (39) can be used to compute the equilibrium values of both the directional crystallinity and the melting temperature.

$$0 = \hat{\psi}_{\alpha a} - \hat{\psi}_{\alpha c} - \frac{R\theta}{Mx_0}\ln\left(\frac{x_\alpha^{eq}}{x_0 - x_\alpha^{eq}}\right)$$
$$\Leftrightarrow x_\alpha^{eq}(\hat{\lambda}_\alpha,\theta) = x_0\left(1 + \exp\left(\frac{-x_0 M(\hat{\psi}_{\alpha a} - \hat{\psi}_{\alpha c})}{R\theta}\right)\right)^{-1} \tag{40}$$

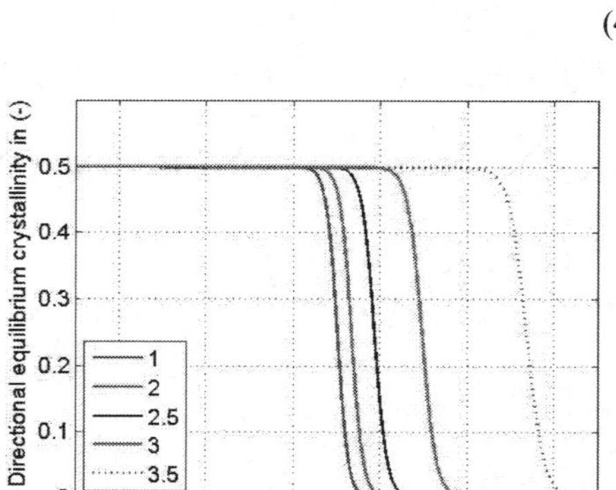

Figure 3. Equilibrium melting behaviour at various directional stretches $\lambda_\alpha = 1, \ldots, 3.5$.

The equilibrium melting temperature, shown in Figure 3, increases nonlinearly with the directional stretch. The transition region of the equilibrium crystallinity shifts to higher temperatures with increasing stretch. The explanation is that increasing stretch increases the crystallinity. Crystals need more energy to melt, which explains the increase in melting temperature.

Equation (40) shows that the equilibrium value of the directional crystallinity is a function of temperature θ and directional stretch $\hat{\lambda}_\alpha$ and is permanent in the interval $0 \leq x_\alpha^{eq} \leq x_0$.

4 CONCLUSION

A thermomechanical approach to model strain induced crystallisation was presented, which uses the separation of volumetric and isochoric effects. The presented model is based on the concept of representative directions.

The model distinguishes three different phases in the specimen: crystallisable amorphous phase, the non-crystallisable amorphous phase and the crystalline phase.

In the first step, thermoelasticity and crystallisation have been considered. In addition, the consideration of viscous phenomena and the Mullins effect is possible. In order to validate the model, the explicit relation for the equilibrium degree of crystallinity has been calculated. Furthermore, the stretch-dependent equilibrium melting temperature has been analysed. The validation of the model's properties show that transient effects due to process-dependent crystallinity can be simulated.

REFERENCES

Acken, M. F., W. E. Singer, & W. P. Davey (1932). X-Ray Study of Rubber Structure. *Rubber Chemistry and Technology 5*(1), 30–38.

Albouy, P. A., G. Guillier, D. Petermann, A. Vieyres, O. Sanseau, & P. Sotta (2012). A stroboscopic X-ray apparatus for the study of the kinetics of strain-induced crystallization in natural rubber. *Polymer (United Kingdom) 53*(15), 3313–3324.

Albouy, P. A., A. Vieyres, R. Pérez-Aparicio, O. Sanséau, & P. Sotta (2014). The impact of strain-induced crystallization on strain during mechanical cycling of cross-linked natural rubber. *Polymer (United Kingdom) 55*(16), 4022–4031.

Alfrey, T. & H. Mark (1942). A statistical treatment of crystallization phenomena in high polymers. *Rubber Chemistry and Technology 15*(3), 462–467.

Beurrot, S., B. Huneau, & E. Verron (2010). In situ sem study of fatigue crack growth mechanism in carbon black-filled natural rubber. *Journal of applied polymer science 117*(3), 1260–1269.

Candau, N., R. Laghmach, L. Chazeau, J.-M. Chenal, C. Gauthier, T. Biben, & E. Munch (2014). Strain-induced crystallization of natural rubber and cross-link densities heterogeneities. *Macromolecules 47*(16), 5815–5824.

Chenal, J. M., C. Gauthier, L. Chazeau, L. Guy, & Y. Bomal (2007). Parameters governing strain induced crystallization in filled natural rubber. *Polymer 48*(23), 6893–6901.

Durin, A., N. Boyard, J. L. Bailleul, N. Billon, J. L. Chenot, & J. M. Haudin (2017). Semianalytical models to predict the crystallization kinetics of thermoplastic fibrous composites. *Journal of Applied Polymer Science 134*(8).

Flory, P. (1961). Thermodynamic relations for hight elastic materials. *57*, 829–838.

Flory, P. J. (1947). Thermodynamics of crystallization in high polymers. i. crystallization induced by stretching. *The Journal of Chemical Physics 15*(6), 397–408.

Freund, M. & J. Ihlemann (2010). Generalization of onedimensional material models for the finite element method. *ZAMM-Journal of Applied Mathematics and Mechanics/Zeitschrift für Angewandte Mathematik und Mechanik 90*(5), 399–417.

Freund, M., H. Lorenz, D. Juhre, J. Ihlemann, & M. Klüppel (2011). Finite element implementation of a microstructurebased model for filled elastomers. *International Journal of Plasticity 27*(6), 902–919.

Gaylord, R. & D. Lohse (1976). Morphological changes during oriented polymer crystallization. *Polymer Engineering & Science 16*(3), 163–167.

Haupt, P. (2002). *Continuum mechanics and theory of materials*. Springer Science & Business Media.

Huneau, B. (2011). Strain-Induced Crystallization of Natural Rubber: a Review of X-Ray Diffraction Investigations. *Rubber Chemistry and Technology 84*(3), 425–452.

Katz, J. (1925). Röntgenspektrogramme von kautschuk bei verschiedenen dehnungsgraden. eine neue untersuchungsmethode für kautschuk und seine dehnbarkeit. *Chem. ztg 49*, 353.

Lion, A., N. Diercks, & J. Caillard (2013). On the directional approach in constitutive modelling: A general thermomechanical framework and exact solutions for Mooney-Rivlin type elasticity in each direction. *International Journal of Solids and Structures 50*(14–15), 2518–2526.

Lion, A. & M. Johlitz (2015). A thermodynamic approach to model the caloric properties of semicrystalline polymers. *Continuum Mechanics and Thermodynamics 28*(3), 799–819.

Long, J. D., W. E. Singer, & W. P. Davey (1934). Fibering of Rubber. Time Lag and Its Relation to Rubber Structure. *Rubber Chemistry and Technology 7*(3), 505–515.

Rublon, P. (2013). *Etude expérimentale multi-échelle de la propagation de fissure de fatigue dans le caoutchouc naturel*. Ph. D. thesis.

Shutov, A. V., R. Landgraf, & J. Ihlemann (2013). An explicit solution for implicit time stepping in multiplicative finite strain viscoelasticity. *Comp. Meth. Appl. Mech. Engrg. 265*(April 2013), 213–225.

Toki, S., I. Sics, B. S. Hsiao, M. Tosaka, S. Poompradub, Y. Ikeda, & S. Kohjiya (2005). Probing the nature of straininduced crystallization in polyisoprene rubber by combined thermomechanical and in situ x-ray diffraction techniques. *Macromolecules 38*(16), 7064–7073.

Tosaka, M., S. Kohjiya, S. Murakami, S. Poompradub, Y. Ikeda, S. Toki, I. Sics, & B. S. Hsiao (2004). Effect of Network-Chain Length on Strain-Induced Crystallization of NR and IR Vulcanizates. *Rubber Chemistry and Technology 77*(4), 711–723.

Tosaka, M., K. Senoo, K. Sato, M. Noda, & N. Ohta (2012). Detection of fast and slow crystallization processes in instantaneously-strained samples of cis-1, 4-polyisoprene. *Polymer 53*(3), 864–872.

Trabelsi, S., P. A. Albouy, & J. Rault (2003). Crystallization and melting processes in vulcanized stretched natural rubber. *Macromolecules 36*(20), 7624–7639.

Truesdell, C. & W. Noll (1992). *The non-linear field theories of mechanics*. Springer.

Yeh, G. S. & K. Hong (1979). Strain-induced crystallization, part iii: Theory. *Polymer Engineering & Science 19*(6), 395–400.

Constitutive Models for Rubber X – Lion & Johlitz (Eds)
© 2017 Taylor & Francis Group, London, ISBN 978-1-138-03001-5

Experimental investigation of the compression modulus at a technical EPDM, exposed to cyclic compressive hydrostatic loadings

O. Gehrmann, N.H. Kröger & P. Erren
Department Simulation and Continuum Mechanics, The German Institute of Rubber Technology, Hanover, Germany

D. Juhre
Institute of Mechanics, Otto von Guericke University Magdeburg, Madgeburg, Germany

ABSTRACT: In many finite element simulations rubber materials are commonly assumed to be nearly incompressible by applying a high ratio between compression modulus and shear modulus. The compression modulus is commonly given as a constant value. In reality this assumption is not fully correct. Influencing factors like the compressibility of the included filler or trapped air can lead to a notably change of the compression modulus during mechanical loading. The focus of this work is on the estimation of the cyclic evolution of the compression modulus for a technical EPDM by using cyclic volumetric compression tests.

1 INTRODUCTION

Although, in most material modeling approaches, rubber is said to be incompressible, many rubber compounds are at least slightly compressible. Zimmermann and Stommel showed in finite element (FE) analyses that the results for hydrostatic loads superimposed with non-hydrostatic loads can significantly vary with a moderate change of the compression modulus (Zimmermann & Stommel 2013). In former works, new modeling approaches for the compression modulus are suggested by e.g. Horgan and Murphy (Horgan & Murphy 2009a, Horgan & Murphy 2009b, Horgan & Murphy 2009c). The underlying material is assumed to be ideal hyperelastic and the validation with experimental data is based on investigations of Rivlin and Saunders from 1951. Further experiments regarding the characterization of the compressive behavior of FKM and HNBR rubber are documented in (Ilseng, Skallerud, & Clausen 2015). In (Rajagopal & Saccomandi 2009) and references within, the modulus of shear is given as a function of the hydrostatic pressure, considering the framework of small deformations. Important and well known effects like material softening and inelastic behavior for cyclic loading in case of rubber materials are not investigated in the cited literature. In this work the cyclic behavior of a technical EPDM in volumetric compression is investigated. The insights shall be applied for future modeling approaches of the compression modulus.

2 SPECIMEN PRODUCTION AND TEST SETUP

The volumetric compression test is conducted on a *MTS 831 Elastomer Test System*. The specimen is a cylinder of EPDM rubber with a diameter of 7:4 mm and a height of approximately 9:9 mm. The test setup is shown in Figure 1. The arrow symbolizes the motion of the stamp.

The EPDM mixture consists out of 50% polymer (Keltan 4450) and 50% other ingredients. It is prepared on a 1.5 liter mixer and admixed with the crosslinking agent and catalysts at the roller. The samples are vulcanized at 160°C for 19 min. Due to shrinkage of the rubber several specimens with varying diameter are molded. The specimens with the tightest fit are chosen. The experiment is displacement driven with a velocity of 0.05 mm/s, see Figure 1 (right). The repeated cyclic loading is realized between 0.1 mm and 0.4 mm.

3 ANALYTIC EXAMPLE OF A COMPRESSED CYLINDER

At first view, the applied test procedure (see Figure 1) is not applicable to provide any statements about the material behavior of rubber under hydrostatic loadings. The deformation is not fully hydrostatic (see Figure 2). However, it can be shown, that the globally measured reaction force is mainly a cause of the hydrostatic fraction of the specimen's total deformation.

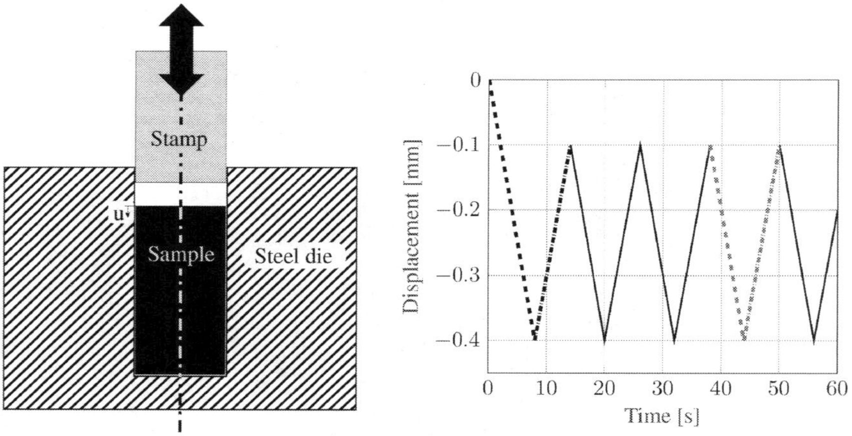

Figure 1. Scheme of the test setup (left) and applied deformation (right; *1st* cycle in red, *4th* cycle in blue, cf. Figure 8).

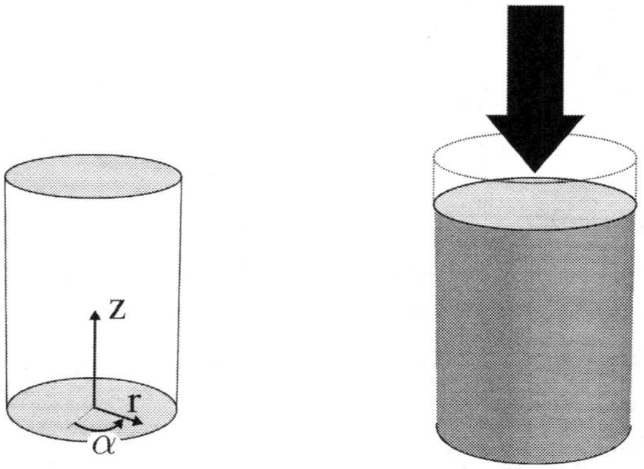

Figure 2. Cylindrical coordinates and exaggerated deformation of the cylindrical specimen during the test.

The displacement vector in cylindrical coordinates is given by:

$$u = (u_r, u_\alpha, u_z) \tag{1}$$

For the compression test shown in Figure 1 the assumption of small deformations is reasonable. For a cylinder with rotational symmetry the strain components can be written as e.g. shown in (Zienkiewicz, Taylor, & Zhu 2013):

$$\begin{aligned} &\varepsilon_{rr} = \frac{\partial u_r}{\partial r} \quad \varepsilon_{\alpha\alpha} = \frac{u_r}{r} \quad \varepsilon_{zz} = \frac{\partial u_z}{\partial z} \\ &\varepsilon_{zr} = \frac{1}{2}\left(\frac{\partial u_r}{\partial r} + \frac{\partial u_z}{\partial z}\right) \quad \varepsilon_{r\alpha} = \varepsilon_{\alpha z} = 0 \end{aligned} \tag{2}$$

For the demonstration a linear elastic material is assumed. In this case, the stress components in cylindrical coordinates are defined as

$$\begin{aligned} \sigma_{rr} &= \frac{(1-\nu)E}{(1+\nu)(1-2\nu)}\left(\varepsilon_{rr} + \frac{\nu}{1-\nu}(\varepsilon_{\alpha\alpha} + \varepsilon_{zz})\right) \\ \sigma_{\alpha\alpha} &= \frac{(1-\nu)E}{(1+\nu)(1-2\nu)}\left(\varepsilon_{\alpha\alpha} + \frac{\nu}{1-\nu}(\varepsilon_{rr} + \varepsilon_{zz})\right) \\ \sigma_{zz} &= \frac{(1-\nu)E}{(1+\nu)(1-2\nu)}\left(\varepsilon_{zz} + \frac{\nu}{1-\nu}(\varepsilon_{rr} + \varepsilon_{\alpha\alpha})\right) \\ \sigma_{zr} &= \frac{E}{(1+\nu)}\varepsilon_{zr} \end{aligned} \tag{3}$$

with ν as the Poisson's ratio and E as the Young's modulus. The boundary condition in radial direction is given as a hindered deformation $u_r = 0$. Only a deformation in z-direction takes place. By applying the equation

$$K = \frac{E}{3(1-2\nu)} \tag{4}$$

one obtains for the stress components written in matrix shape

$$\sigma = \begin{bmatrix} \frac{3K\nu\varepsilon_{zz}}{(1+\nu)} & & \\ & \frac{3K\nu\varepsilon_{zz}}{(1+\nu)} & \\ & & \frac{3K(1-\nu)\varepsilon_{zz}}{(1+\nu)} \end{bmatrix} \tag{5}$$

with K as the compression modulus. The decomposition into a non-hydrostatic and a hydrostatic part, yields:

$$\frac{\sigma}{K\varepsilon_{zz}} = \begin{bmatrix} \frac{3\nu}{(1+\nu)} - 1 & & \\ & \frac{3\nu}{(1+\nu)} - 1 & \\ & & \frac{3(1-\nu)}{(1+\nu)} - 1 \end{bmatrix} + \begin{bmatrix} 1 & & \\ & 1 & \\ & & 1 \end{bmatrix} \tag{6}$$

Exemplarily, but typically, values for the compression modulus of 1000 MPa and of 3 MPa for the Young's modulus are assumed. The Poisson's ratio can be calculated by

$$\nu = \frac{3K - E}{6K} \tag{7}$$

and equals to $\nu = 0{,}4995$ in this case. The application of this value on equation 6 shows that the globally measured behavior can almost be completely referred to the hydrostatic stress response since the nonhydrostatic share is neglectable (see equation 8).

$$\frac{\sigma}{K\varepsilon_{zz}} = \begin{bmatrix} -6.7\cdot 10^{-4} & & \\ & -6.7\cdot 10^{-4} & \\ & & 1.3\cdot 10^{-3} \end{bmatrix} + \begin{bmatrix} 1 & & \\ & 1 & \\ & & 1 \end{bmatrix} \tag{8}$$

4 NUMERICAL SIMULATION OF THE COMPRESSED CYLINDER

A numerical simulation of the compression test is performed. For the simulation, a two-dimensional axisymmetric finite element (FE) model is used. The FE model for the compressive deformation is schematically shown in Figure 3. The symmetry line is symbolized by the dot-dash line and the contact lines are symbolized by the dashed lines. A gap of 5 μm is given between the rubber specimen and the outer contact line.

Two different material models are compared. First, two uniaxial tensional loading cycles are simulated (see Figure 4). The Neo-Hooke material model (Mooney 1940) has a pure hyperelastic behavior since the loading path equals to the unloading path. The MORPH (Model of Rubber Phenomenology) is able to describe (next to the non-linear behavior) material softening, permanent set and hysteresis loop of a filled rubber material under large deformations (Besdo & Ihlemann 2003, Juhre, Ihlemann, Alshuth, & Klauke 2011).

In case of simulating the compression test from Figure 1, no significant difference between these material models is visible (see Figure 5). The definitions of both models is identical in case of pure hydrostatic loadings. For both models a constant compression modulus of 2000 MPa is applied. The gap of 5 μm between the rubber specimen and the

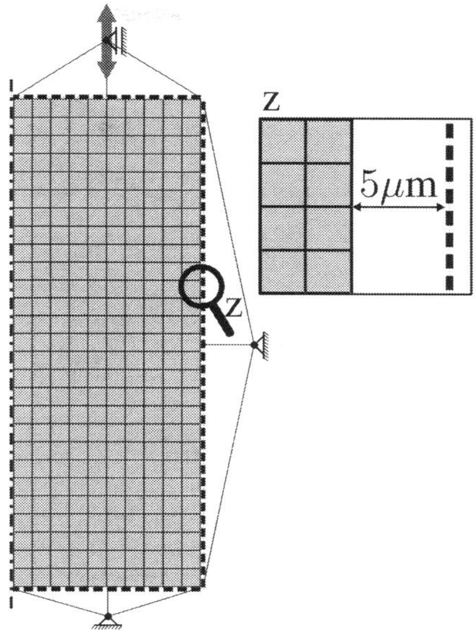

Figure 3. Sketch of the FE model for the compressive deformation with dashed lines as contact lines.

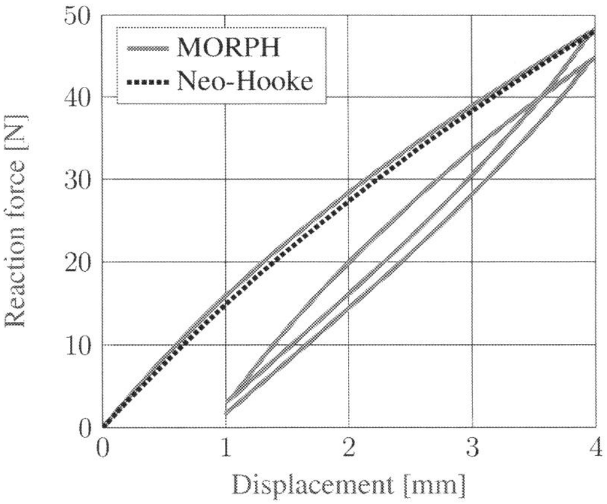

Figure 4. Comparison of the hyperelastic Neo-Hooke material model with the MORPH by using the global reaction force over the global uniaxial deformation of the specimen.

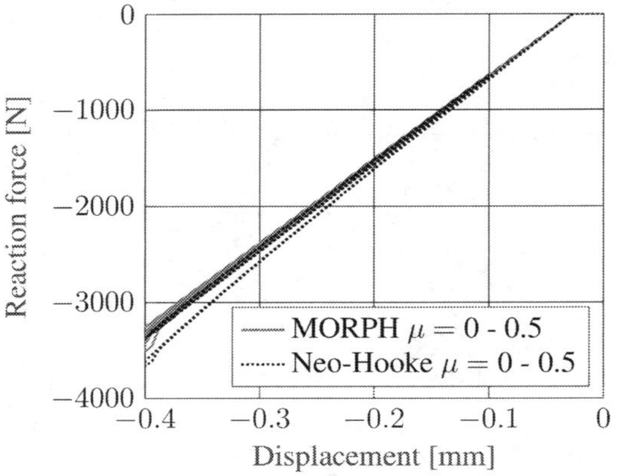

Figure 5. Comparison of the hyperelastic Neo-Hooke material model with the MORPH at the compression test.

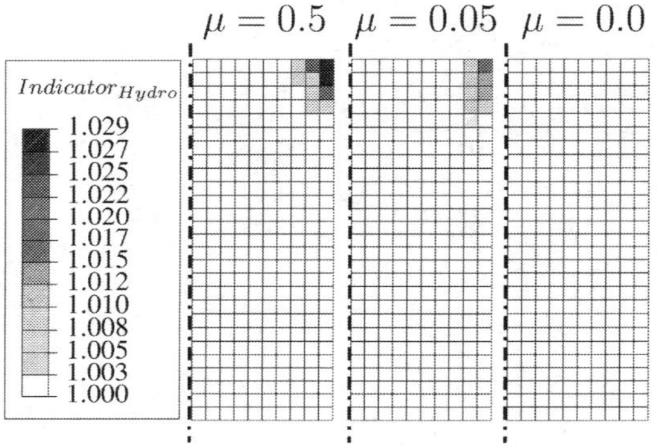

Figure 6. The $Indicator_{Hydro}$ mapped on the undeformed FE mesh with a varied friction coefficient.

outer contact line (see Figure 3) creates a plateau at the beginning of the global deformation. A similar behavior can be observed in the experimental data and is the motivation for that gap in the FE model.

An indicator is defined to measure the level of hydrostatic stress state within the specimen. The indicator is defined as

$$Indicator_{Hydro} = \frac{\sigma_1 + \sigma_2 + \sigma_3}{3\sigma_1} \tag{9}$$

with $\sigma_i(i = 1,2,3)$ as the principal Cauchy stresses. The indicator equals to 1.0 for a pure hydrostatic load. Figure 6 shows the specimen exposed to the maximum global deformation of 0.4 mm and with a varied friction coefficient. The numerical values are mapped on the undeformed FE mesh. In general it can be stated that the indicator is close to 1.0 for different friction coefficients and different regions in the rubber cylinder. The deviations to 1.0, especially for the highest friction coefficient of 0.5, do not result in a relevant amount of dissipation of energy within the MORPH based simulations (see Figure 5). The maximum distortion, caused by friction to the contact lines, takes place in the area of the max ($Indicator_{Hydro}$). The findings from the simulations support the statements from the analytical investigations (see section 3). It shows that the behavior of the globally measured reaction force over the globally measured displacement is mainly driven by the hydrostatic load of the specimen. Non-hydrostatic loadings can be neglected in any further investigations.

5 EXPERIMENTAL RESULTS AND EVALUATION

Figure 7 shows the smoothed force-displacement curve of the cyclic hydrostatic compression test. A clear dependency on the loading path is observable. The material properties of the investigated EPDM shows similar behavior like in non-hydrostatic multi hysteresis tests for rubbery materials. The dissipated energy appears as a hysteresis loop in the force displacement diagram. Additionally, material softening is observable.

To investigate, if only the possible friction between the specimen and the surrounding fixture is responsible for the hysteresis loop, a second test with the use of a lubricant is conducted. The lubricant causes a decrease of the friction coefficient. A comparison of both tests, results in the conclusion that the dissipated energy due to friction is neglectable compared to the dissipated energy of the rubber material itself, see Figure 7. The simulation behind Figure 5 (see section 4) shows the same results regarding the influence of the friction coefficient on the global dissipated energy. The experimental compression modulus is derived by:

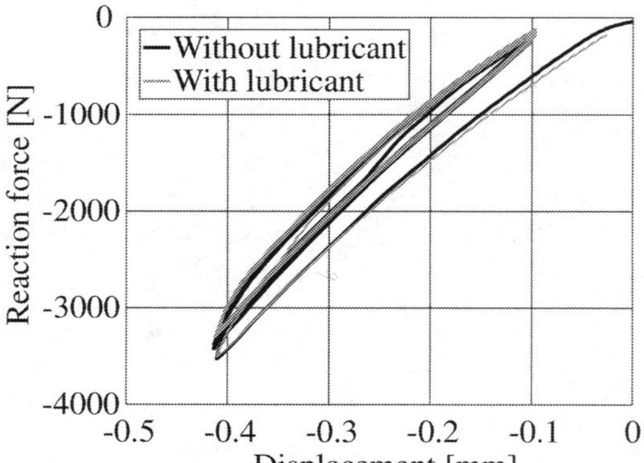

Figure 7. Comparison of resulting smoothed data with or without using soap as lubricant.

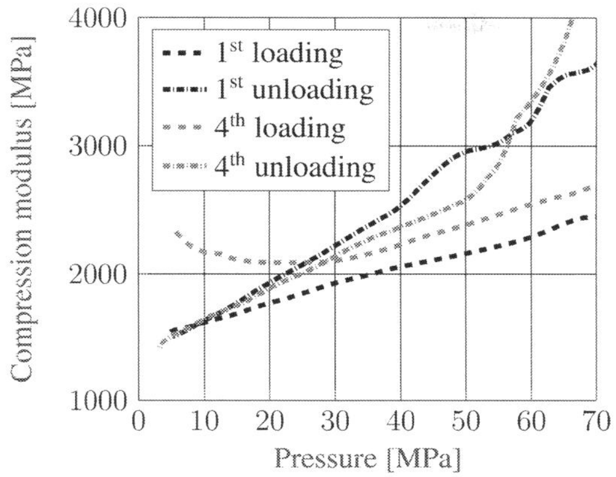

Figure 8. Comparison of the resulting compression modulus (experiment conducted with lubricant) for different stages of loading (cf. Figure 1 (right)).

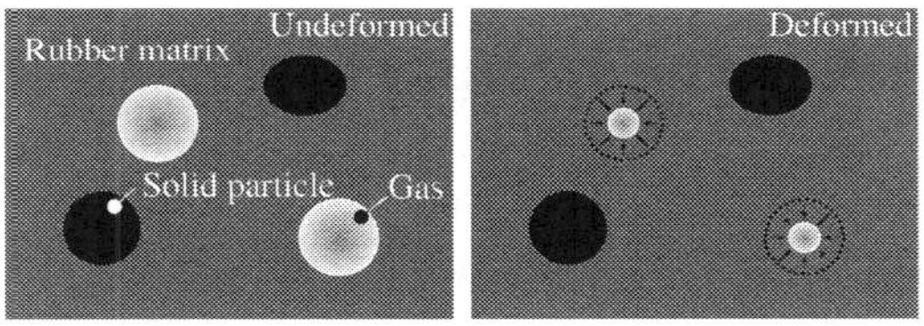

Figure 9. A possible explanation for the observed material behaviour under hydrostatic loading.

$$K = \frac{\partial F_s}{\partial u} \cdot \frac{h}{A} \tag{10}$$

where u is the smoothed displacement signal, F_s is the smoothed force signal and h is the height of the specimen. Due to the smoothing the derived data near the turning points of the loading is ambiguous and not considered in the results shown in Figure 8. However, the trends for the dependency of the compression modulus on load path and cyclic repetition are visible. For the first loading the compression modulus depends nearly linear on the applied pressure. That is less true for further stages of loading and unloading.

6 CONCLUSION AND OUTLOOK

Experimental investigations, focusing on the mechanical behavior of a rubber material under hydrostatic loading, are performed. For the investigations, a repeated loading of a cylindrical shaped specimen in a tight-fitting steel die is chosen. Furthermore, two different conditions are applied. These two conditions differ in terms of the friction coefficient between the specimen and the solid surrounding. The influence of the dissipated energy by friction is numerically and experimentally proven to be small in comparison with the dissipated energy by the rubber material. Analytical and numerical investigations are applied to reconstruct the experimental investigations. The analytical investigations demonstrates that the applied test procedure can be used to measure the material behaviour under hydrostatic loading. The numerical investigations exhibits that an advanced material model, like MORPH, can describe the inelastic material behaviour under non-hydrostatic loadings, but behaves like the Neo-Hooke model (by definition) under hydrostatic loadings. By evaluating the test results under hydrostatic loadings, the following can be extracted.

- The compression modulus is a function of the loading intensity and the loading history.
- The dependency of the loading history is significantly more distinct for the loading compared to the unloading.
- A general increase of the compression modulus with repeated loading is visible. For pressure levels, lower than 10 MPa this increase amounts to more than 55%.
- The compression modulus is relatively independent on the load history for the unloading case.
- A nearly steady material behavior is reached after the first loading cycle.

A possible explanation for the observed material behaviour under hydrostatic loadings is shown in Figure 9. Filled rubber contains inhomogeneities with different compression behavior. As an example, solid inhomogeneities (e.g. accumulated filler) or gaseous inhomogeneities (e.g. trapped air) are considered. The compression modulus of the solid inhomogeneities (e.g. carbon black particles) and the rubber matrix is serval magnitudes greater than the compression modulus of gaseous inhomogeneities. On a small scale, the globally applied small hydrostatic deformations can therefore lead to non-hydrostatic large deformations. As it is well known, strain energy is dissipated under these conditions.

Thus, working with a constant compression modulus is only a rough description of the reality for strong hydrostatic loadings. For these cases, a modeling of the compression modulus as a function of the pressure level, load history and loading/unloading distinction could lead to more realistic results in finite element applications. To verify these statements, additional investigations with a test device designed to be more sensitive to small deformations is advisable.

REFERENCES

Besdo, D. & J. Ihlemann (2003). A phenomenological constitutive model for rubberlike materials and its

numerical applications. *International Journal of Plasticity, 19, pp. 1019–1036.*
Horgan, C.O. & J.G. Murphy (2009a). Compression tests and constitutive models for the slight compressibility of elastic rubber-like materials. *Int J Eng Sci 47 (2009), 1232–1239.*
Horgan, C.O. & J.G. Murphy (2009b). A generalization of hencky's strain energy density to model the large deformations of slightly compressible rubbers. *Mechanics of Materials 41 (2009), 943–950.*
Horgan, C.O. & J.G. Murphy (2009c). On the volumetric part of strain-energy functions used in the constitutive modeling of slightly compressible solid rubbers. *Int J Solids Struct 46 (2009), 3078–3085.*
Ilseng, A., B.H. Skallerud, & A.H. Clausen (2015). Volumetric compression of HNBR and FKM elastomers. *Constitutive Models for Rubber IX (2015), 235–241.*
Juhre, D., J. Ihlemann, T. Alshuth, & R. Klauke (2011). Some remarks on influence of inelasticity on fatigue life of filled elastomers. *Plastics, rubber and composites. - London: Taylor and Francis, Bd. 40.2011, 4, pp. 180–184.*
Mooney, M. (1940). A theory of large elastic deformation. *Journal of Applied Polymer Science 11, pp. 582–592.*
Rajagopal, K.R. & K.R. Saccomandi (2009). The mechanics and mathematics of the effect of pressure on the shear modulus of elastomers. *Proc R Soc A 465 (2009), 3859–3874.*
Zienkiewicz, O., R. Taylor, & J. Zhu (2013). *The Finite Element Method: Its Basis and Fundamentals.*
Zimmermann, J. & M. Stommel (2013). The mechanical behavior of rubber under hydrostatic compression and the effect on the results of finite element analyses. *Arch Appl Mech 83 (2013), 293–302.*

Constitutive Models for Rubber X – Lion & Johlitz (Eds)
© 2017 Taylor & Francis Group, London, ISBN 978-1-138-03001-5

A novel algorithm: Tool to quantifying rubber blends from infrared spectrum

Sanjoy Datta, Jan Antoš & Radek Stoček
Centre of Polymer Systems, Tomas Bata University in Zlín, Zlín, The Czech Republic

ABSTRACT: Infrared (IR) spectroscopy can be used to detect the polymers present in a blend, provided the unique characteristic peaks of the polymers are known. However, quantification of such a blend is not possible without modifying the spectrum because of vertical baseline shifting, attributed to various reasons. The present work deals with the explanation of a uniquely derived algorithm for baseline creation on and subsequent subtraction from the originally obtained spectrum. Using the algorithm, a scientific baseline fit was achieved which carved the way to find the exact relative peak heights of the polymers constituting a blend. The logical approach to frame the algorithm is the subject matter of the work. An introduction to using the algorithm for experimental quantification of a binary rubber blend is also provided. Of course, it is a general approach and can be used with equal effectiveness for any other binary compositions.

1 INTRODUCTION

Compounded and cured rubber products are composites that are composed of either a single rubber or blends of rubbers, and various compounding ingredients (Amari et al. 1999, Cui et al. 1999). Many such rubber products contain reinforcing carbon black filler. Such products are sometimes needed to be reverse engineered to understand their constituting components. Attenuated total reflection (ATR) Fourier transform infrared (FT-IR) studies can qualitatively detect the rubbers. However, it is not possible to quantify a carbon black compounded rubber product from an infrared (IR) spectrum, due to an up sloping baseline towards decreasing wavenumber, attributed to absorption of IR radiation by carbon black. A proper baseline fitting to and subsequent subtraction from the original IR spectrum is thus needed for such quantification.

Some earlier works related to IR baseline corrections can be found in the literature. However, these discoveries do not eliminate the possibility of finding yet another method to perform baseline correction.

In a relevant work, Lee et al. (Lee et al. 2007) first corrected the FT-IR baselines and then calibrated the binary blends of natural rubber (NR)-styrene butadiene rubber (SBR), NR-polybutadiene rubber (BR), and SBR-BR, aimed towards using these calibrations to find the compositions of unknown rubber blends. They used exponential fitting of curves which were greatly generalized.

The present work deals with a mathematical equation relating the characteristic IR absorption peak height ratios of two known rubbers to their concentrations, through an IR binary blend parameter (P_{IR}), completely defining a binary rubber blend.

Cured binary blends of NR and SBR at three different blend ratios, including 50 phr (parts per hundred rubber by weight) carbon black, were subjected to ATR FT-IR studies. The fingerprint wavenumber peak height ratios of NR at 1375 cm^{-1} and SBR at 699 cm^{-1} for the three blends were calculated. The magnitudes of these peaks were precisely read from a modified spectrum, which was generated by fitting each of the initially obtained spectra with a created baseline that was then subtracted from the spectrum in question.

The dataset of intensity versus wavenumber, obtained from the IR spectrometer was processed using a completely new algorithm to create and subtract the baseline.

The logic behind the design of such an algorithm and a brief discussion of the algorithm is the subject matter of the present paper. An introduction to the use of the algorithm for experimental quantification of a binary rubber blend is also discussed in brief.

2 THEORITICAL BACKGROUND

2.1 *Explanation of the new algorithm*

Instead of a smooth baseline fitting over the entire range of data, the algorithm focuses on finding a baseline separately for each peak. It was assumed that at the base of each peak, mathematically consisting of two local minima, the baseline was almost a straight line. Therefore, an array of paired points was separately fitted at the base of all peaks and these points were then joined by straight lines to obtain the new baseline.

The algorithm accepts some initial parameters set by the user, based on which the spectrum is modified accordingly. These parameters are as follows:

1. x1, x2 - range of wavenumber selected from the data
2. k1, k2 - cut-off values for differentiation on the initial dataset and on the regenerated dataset in the negative output respectively.
3. ky - cut-off value for negative area in output.

During an intermediate stage, a dataset is generated in the negative output which demands for a cut-off parameter in the negative output also.

The cut-off parameters k1, k2 and ky, on the y-axis, are required to eliminate small and unwanted peaks. Based on the requirement of smaller peaks, these parameters should be adjusted to lower values.

In the process of creating the modified spectrum, the dataset generated by the IR spectrometer is imported and processed using the derived algorithm, which is implemented using Wolfram Mathematica. It executes the following main steps to obtain the final modified dataset in (x,y) (wavenumber, absorbance) from the initial source:

1. Creates sign vectors of derivatives according to the set parameter k1
2. Converts sign vector to string
3. Searches regular expression "NO*P" resulting in substrings within the string
4. Determines regular expressions from the positions of the minima
5. Calculates equations of lines from each of two consecutive minima
6. Assigns the basepoint y co-ordinates corresponding to x co-ordinate values in each line
7. Calculates the output as $\{\mathbf{out}_x, \mathbf{out}_y\}: = \{\mathbf{dr}_x, \mathbf{dr}_y - \mathbf{bP}\}$, where $\mathbf{out}_x$ is x value of output data, $\mathbf{out}_y$ is y value of output data, $\mathbf{dr}_x$ is data ranged and **bP** is basepoint
8. Also generates some data in negative output
9. Calculates other minima in the negative output using k2 as the cutoff value of differentiation and ky as the cutoff value for the negative area output
10. Adds the minima in the negative output to the previous set of minima and calculates new equations of lines
11. Calculates new set of basepoints based on new equations of lines
12. Subtracts new basepoints to generate new output dataset as the result

The calculation of processing in the application named baseline creator is now explained. The processing of data consists of three important steps:

1. Import data
2. Calculate baseline
3. Export results

Suppose that the data is imported and parsed. This work focuses only on the explanation of calculating baselines based on imported data. Other less important parts of the algorithm are omitted for the sake of simplicity.

In the calculation, data is found, loaded, parsed and processed into the form of List of xy coordinates. Data is then restricted only to a ranged area according to the entered range <x1,y1>, thus resulting in the ranged data (Figure 1). The reason of using range is to focus only on the specific part of data where specific setting can be applied.

Minima are then found within the input (ranged) data. Based on the list of minima, the list of basepoints is calculated, so that every x (input) value gets its corresponding y value. A more detailed explanation of calculating minima and basepoint list is described later.

After the first process of finding of minima, the output list is calculated as the subtraction between the input (ranged) data and the baseline points. Even though the process of finding minima was performed, there may be some areas within the output where the negative magnitudes of y values are significant. This effect may be caused by several factors such as the shape of the input data, the setting of cut-off parameters resulting in specific amount and places of minima. This can be solved by searching of some other minima in the negative output and recalculating the baseline points. However, some small deviations in the negative area may be

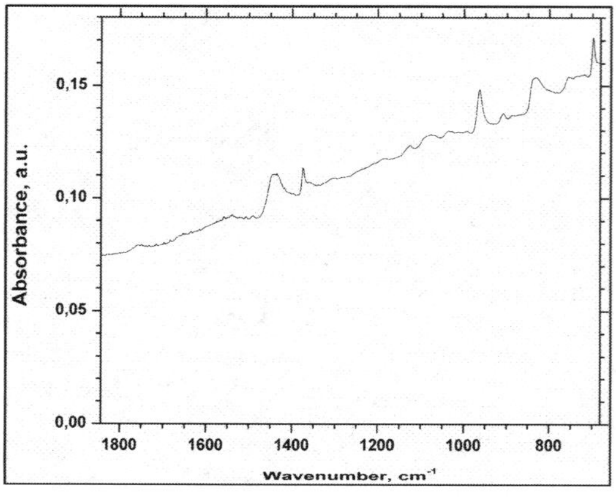

Figure 1. The ranged data.

allowed; therefore, the forbidden area in the negative output can be determined by the ky parameter. Below this area, within the rest of the negative output, other minima can be found and added to the previous list of minima (united and ordered by x). According to the new list of minima, new baseline points have to be calculated. Figure 2 depicts the united minima whereas Figure 3 shows the input with united minima and corrected baseline.

The output is calculated in the same way as described before and the corrected output is shown in Figure 4.

The process of finding the minima is now elaborated. In order to find the minima in the input data, it has to be analyzed somehow. There are many ways to do this; here the data was analyzed in the differentiation way. The vector of differentiation is first calculated.

$$\mathbf{d} = \Delta \mathbf{y} / \Delta \mathbf{x} \quad (1)$$

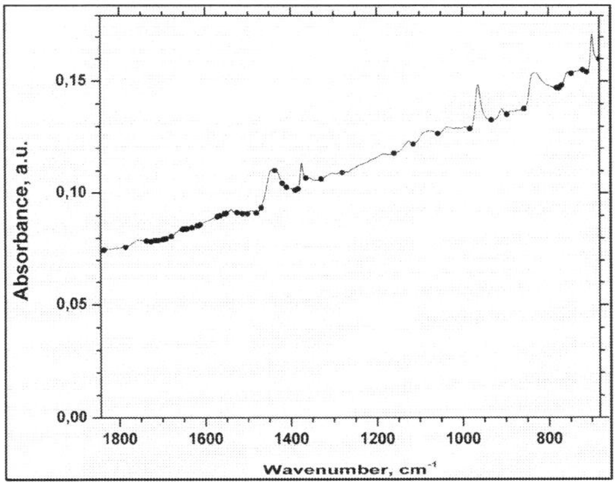

Figure 2. United minima.

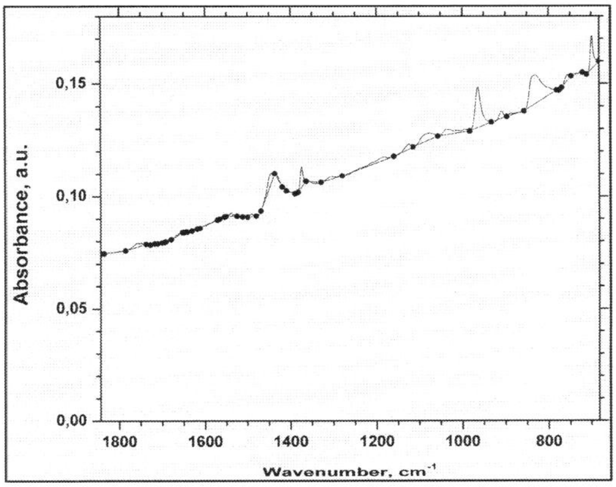

Figure 3. Input, united minima and corrected baseline.

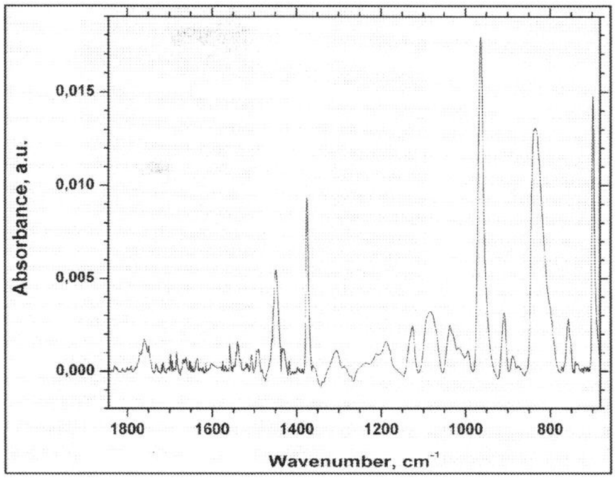

Figure 4. Corrected output.

This vector could be normally used to find the minima; however this may lead to large amount of minima wrong location of places (for instance in the middle of logical peak). Due to the small deviations of the input signal, the way of the processing of the data should be different. This can be done by the determination of the cut-off parameter k1.

The differentiation vector is converted to sign vector **s** as follows:

$$\begin{matrix} \mathbf{d}_i \le -k1 & -k1 < \mathbf{d}_i < k1 & \mathbf{d}_i \ge k1 \\ \Rightarrow \mathbf{s}_i = -1 & \Rightarrow \mathbf{s}_i = 0 & \Rightarrow \mathbf{s}_i = 1 \end{matrix} \quad (2)$$

Sign vector then defines the direction of differentiation. The places of minima can be then easily found in the middle of this series −1, 0, ..., 0, 1. The task is to then find some specific pattern within the sign vector. One easy way is to convert this vector to the text characters and then to utilize the functions of finding subtext within the text. Hence, the sign vector is converted to the string vector **ss**, which is then concatenated to only one string. Graphically it may be depicted as follows:

$$\begin{matrix} \mathbf{s}_i = -1 & \mathbf{s}_i = 0 & \mathbf{s}_i = 0 & \mathbf{s}_i = 0 & \mathbf{s}_i = 1 \\ \mathbf{ss}_i = "N" & \mathbf{ss}_i = "O" & \mathbf{ss}_i = "O" & \mathbf{ss}_i = "O" & \mathbf{ss}_i = "P" \end{matrix} \quad (3)$$

where, “N”—negative, “O”—Zero, “P” positive differentiation.

A convenient way of finding the subtext within the text is to use regular expression and then apply some searching algorithm. Therefore the pattern “NO*P” is defined, which means that the object of finding is defined as "N" followed by any numbers of "O" (even zero times) followed by "P". Using the searching algorithm, the positions of subtexts are obtained. Minima are then located in the middle of these subtexts and the vector positions can be calculated as follows:

$$\mathbf{im}_i = Ceiling(\mathbf{iss}_e - \mathbf{iss}_b) \qquad (4)$$

where, $\mathbf{im}_i$—position of minimum, $\mathbf{iss}_b$—beginning of subtext, $\mathbf{iss}_e$—end of subtext.

From the positions of minima, it is very easy task to assign their corresponding x and y values, thus finally resulting in the vector of minima **m**.

Once the minima points are determined, the baseline points have to be calculated. There may be several logical ways to connect minima to each other in order to create the baseline curve. In the present case, the baseline is supposed to be constructed of straight line segments between two adjacent minima. In order to calculate points between two minima, the vector of line equations needs to be obtained. The equation of lines can be easily calculated if two points lying on it are available. The resultant vector of equations **eL** then consists of pairs of slopes and absolute coefficients.

Once the vector is determined, the algorithm has to find x values lying between two adjacent minima and calculate the baseline points. There is no need of considering the period of measurement since the samples may not be uniformly distributed over the x axis, which may otherwise corrupt the matching between x values of baseline points and those of the input data.

Finally, the output can be calculated as the subtraction of the baseline points from the input data.

2.2 *Calculation of P_{IR} from Beer Lambert law*

From Beer lambert law (Lee et al. 2007), the intensity of vibration (A) in infrared spectroscopy is proportional to the concentration of the sample in some appropriate unit (c), and the thickness of the sample traversed by the radiation (l) and is equated through a proportionality constant (ε), called the molar extinction coefficient as follows:

$$A = \varepsilon \cdot c \cdot l \qquad (5)$$

If a binary blend of two rubbers designated as *1* and *2* are considered, then Equation 5 may be written as

$$\frac{A_1}{A_2} = P_{IR}\frac{(1-c_2)}{c_2} \qquad (6)$$

where, $c_1 + c_2 = 1$ and $P_{IR} = \varepsilon_1/\varepsilon_2$

3 EXPERIMENTAL

Three batches representing different blend ratios of NR and SBR at 0.3:0.7, 0.5:0.5 and 0.7:0.3 were mixed, cured and finally analyzed by the derived algorithm modified FT-IR spectroscopic study to find the fingerprint wavenumber peak height of NR at 1375 cm^{-1} and SBR at 699 cm^{-1}.

4 RESULTS AND DISCUSSION

Three algorithm modified, superimposed spectra of NR and SBR in the blend ratio 0.3:0.7 are shown in Figure 5. The figures for the other two blends are not shown to avoid unnecessary lengthening of the text.

The results obtained after feeding the initially obtained dataset with the cut off parameters and running the algorithm are shown in Table 1.

In all the three cases, the calculated values of P_{IR} were very close and the median value was 0.26.

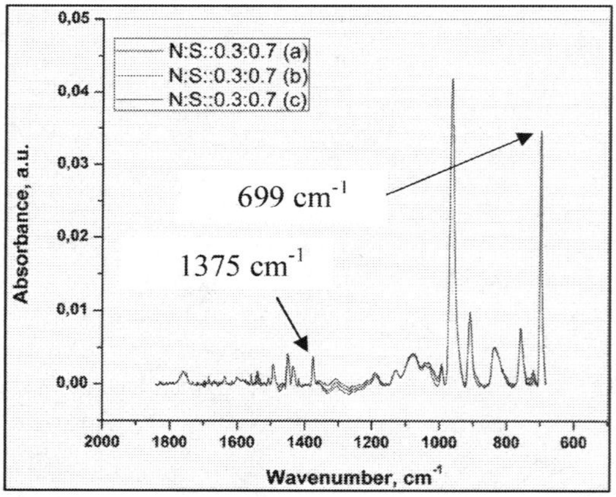

Figure 5. Algorithm modified superimposed spectra for the blend NR:SBR::0.3:0.7.

Table 1. Calculated values of P_{IR}.

$c_1 = (1 - c_2)$	c_2	A_1/A_2	$P_{IR}\,(\varepsilon 1/\varepsilon 2)$	P_{IR}
0.3	0.7	0.11	0.26	**0.26**
0.5	0.5	0.26	0.26	
0.7	0.3	0.63	0.27	

5 CONCLUSIONS

A derived numerical algorithm for baseline fitting and subtraction aimed to modify an IR spectrum, required to find exact IR peak heights was devised. These heights were used to calculate a binary infrared blend parameter (P_{IR}), which was found to be a constant for a set of known binary blend of rubbers and was independent of the blend ratio.

ACKNOWLEDGEMENTS

This article was written with the support of Operational Program Research and Development for Innovations co-funded by the European Regional Development Fund (ERDF) and national budget of the Czech Republic, within the framework of the project CPS—strengthening research capacity (reg. number: CZ.1.05/2.1.00/19.0409) as well supported by the Ministry of Education, Youth and Sports of the Czech Republic—Program NPU I (LO1504).

REFERENCES

[1] Amari, T., Themelis, N.J., & Wernick, I.K. 1999. Resource recovery from used rubber tires. *Resources Policy* 25: 179–188.

[2] Cui, H., Yang, J., & Liu, Z. 1999. Thermogravimetric analysis of two Chinese used tires. *Thermochimica Acta* 333: 173–175.

[3] Lee, Y.S., Lee, W.K., Cho, S.G., Kim, I., & Ha, C.S. 2007. Quantitative analysis of unknown compositions in ternary polymer blends: A model study on NR/SBR/BR system. *J. Anal. Appl. Pyrol* 78: 85–94.

Constitutive Models for Rubber X – Lion & Johlitz (Eds)
© 2017 Taylor & Francis Group, London, ISBN 978-1-138-03001-5

Influence of dissipative specimen heating on the tearing energy of elastomers estimated by global and local characterization methods

S. Dedova
Leibniz-Institut für Polymerforschung Dresden e. V., Germany
Institut für Werkstoffwissenschaft, TU Dresden, Germany

K. Schneider & G. Heinrich
Leibniz-Institut für Polymerforschung Dresden e. V., Germany

ABSTRACT: The methods for the analysis of fatigue and fracture behavior of elastomers are based on two different approaches: the global method, called as conventional TFA method, developed by Rivlin and Thomas (Rivlin & Thomas 1952) and the local method, known as "J-Integral", developed by Cherepanov and Rice (Cherepanov 1967, Rice 1968). The both methods were developed for an elastic materials. However, the mechanical behavior of rubber is far from being elastic, making the simplified approaches questionable (Dedova 2015). In this work the dissipative energy was analysed with respect to thermal effects during dynamic loading. It has been provided, that most of the mechanical produced energy is converted into thermal energy during the loadung process. Thereby only a certain amount of the heat can be transferred again in mechanical energy during unloading, while the other amount is lost as dissipative heat. On the other side, one part of the energy, which is transferred to the sample is saved in the specimen as heat, while another part goes by heat-exchange into the surrounding. These contributions were discussed and analysed separately in this work. The understanding of the dissipative effects during cyclic loading and the energy transport processes enables to apply the methods for analysing of tearing energy of rubber.

1 INTRODUCTION

Nowadays, the methods for the analysis of fatigue and fracture behaviour of elastomers are based on two different approaches: the global method, called as conventional TFA method, developed by Rivlin and Thomas (Rivlin & Thomas 1952) and the local method, known as "J-Integral", developed by Cherepanov and Rice (Cherepanov 1967, Rice 1968). The conventional method with "pure shear" specimen is based on the energy approach of Griffith (Griffith 1920). The local method is based on an energy balance and was verified for the special case of a straight crack in an elastic material under static conditions. Under these conditions the J-integral corresponds to the tearing energy. However, the mechanical behaviour of rubber is far from being perfectly elastic, making the simplified approaches inapplicable (Dedova 2015). During loading and unloading of rubber an inherent dissipation occurs which has to be taken into account by calculating of the tearing energy.

1.1 *Global approach*

The conventional method for analysing of tearing energy was developed by Rivlin and Thomas (Rivlin & Thomas 1952) and is based on the energy approach for elastic materials of (Griffith 1920). Due to relatively simple application the method becomes famous in industry. Here the tearing energy or energy release rate is defined as energy, which leads to creation of a new crack surface, and can be written in the form as

$$T = -\frac{dW_{el}}{bdc}\Big|_{l=const}, \qquad (1)$$

here T – tearing energy; W_{el} – elastic energy of the specimen; b – thickness of specimen; dc – crack growth step; l – deformed specimen length. According to this approach, the tearing energy is equivalent to the change of elastic energy during crack propagation. But this method has some restrictions: the specimen has to have a rectangular shape ("pure shear" specimen), the crack has to grow without deviation in the direction normal to applied force, and the strain field around the crack tip shifts conform in crack propagation direction. These ideal conditions are hardly fulfilled by material under real loading conditions. Also the dissipative effects within the material are not be taken into account by using of this method.

1.2 *Local approach*

For more local applications an energy based approach was developed by Cherepanov (Cherepanov 1967) and Rice (Rice 1968). This method, the so-called J-integral, is based on the rule of conservation of energy and is written in a form of path independent integral

$$J = \int_{\Gamma} (w_{el}\bar{n} - \bar{\bar{\sigma}}\bar{n}\nabla\bar{u})ds, \tag{2}$$

here Γ – path of integration; w_{el} – elastic energy density; $\bar{n}$ – path normal; $\bar{\bar{\sigma}}$ – stress tensor; $\nabla\bar{u}$ - displacement gradient; ds – path element.

The integration path is located in the elastically deformed region of the material. The method is path independent and can be used to analyse deviating cracks. It offers an opportunity to analyse a tearing energy in the complicate load condition. The method was confirmed, but it works correct only for material, which behaves elastic along the integration path.

The known concepts were developed for elastic materials and include the determination of an elastic energy density, but rubber materials are visco-elastic. The dissipation effects in rubber are relatively big compared with other materials.

In this work by means of thermography the dissipated energy, which is lost during cyclic loading and causes an increase of temperature of the samples, will be separated from the reversible heating and cooling of the sample due to the conversion of mechanical energy into internal energy and vica versa according to the entropy elasticity of rubber. A part of the recoverable stored energy will be consumed for crack propagation.

By calculating the tearing energy this processes have to be taken into account. The separation of different effects will be discussed in details.

2 THEORETICAL BACKGROUND OF THE ENERGY BASED CONCEPT

The local approach for analysing of tearing energy (by the J-Integral) is based on a law of conservation of energy. It can be written in a form as (Cherepanov 1967, Cherepanov 1974).

$$dW + dQ = dW_k + dU + d\Pi, \tag{3}$$

here is dW – on the system performed mechanical work, dQ – thermal energy supplied to the system; dW_k – change in the kinetic energy of the system; dU – change in the internal energy of the system; Π – loss of energy, due to the creation of new crack surface.

The time dependent description of the above mentioned processes can be described by equation 4, where the dot on the quantities means the derivative with respect to time:

$$\dot{W} + \dot{Q} = \dot{W}_k + \dot{U} + \dot{\Pi}. \tag{4}$$

In the case of the J-Integral approach for crack propagation the change of kinetic energy is 0 and during the infinitesimal time step of crack propagation no exchange of thermal energy to the surrounding takes place.

By external loading the energy

$$dW = \sigma_{ij} d\varepsilon_{ij}, \tag{5}$$

is transferred to the sample. Here is σ – applied stress, ε – applied strain.

As long as there is no heat transfer backwards to the surrounding, the whole energy increases the internal energy of the sample

$$dU = dW. \tag{6}$$

As long as the material shows pure entropy elasticity, the whole increase of internal energy will be found in a temperature increase according to

$$dU = \delta Q = c_p \rho V dT, \tag{7}$$

Here is ρ – density, c_p – specific heat capacity, dT step of temperature changing, V – volume of body.

With respect to the internal energy this energy contribution can be split into elastically stored (recoverable) energy and dissipated energy.

During unloading, only the recoverable part of the internal energy can be converted back into mechanical energy, practically due to further dissipation the finally recovered energy will be less than this part. After a whole hysteresis cycle the mechanicallost energy, that is the energy represented by the hysteresis loop, will be converted to heat.

Additional to the processes described above, steadily heat will be transferred to the surrounding, as long as there is a temperature difference between the sample and the surrounding according to

$$\frac{dQ}{dt} = \dot{Q} = -\alpha A(T_1 - T_2), \tag{8}$$

here is α – coefficient of heat transfer, A – surface in a contact with environment, T_1 – temperature of the body, T_2 – temperature of surrounding, t – time.

For a whole description of the deformation processes with respect to the energy balance the energy conversion must be considered together with dissipation and energy exchange to the surrounding.

3 EXPERIMENTAL PROCEDURE

3.1 *Heating conditions during cyclic loading*

The aim of the presented experiments was to measure and analyse the contribution of thermal energy on the total energy during cyclic loading process. The experiments were performed with natural rubber with 20 phr carbon black, crosslinked by sulphur. For the biaxial experiments square specimen (77 mm × 77 mm × 2 mm) were used. The experiments were carried out on a biaxial tester at IPF (for more information about the machine possibilities see (Schneider 2014)).

Additional to the stress-strain measurement the temperature of the samples was measured continuously by a thermo camera VarioCam from InfraTec, Germany, and after that analysed using IRBIS software.

Generally the samples were biaxial cyclical loaded (sinusoidal with 1 Hz) with different maximum strains equal to 30%, 60% and 90%. Simultaneous the sample temperature was recorded with high time resolution and the temperature of the environment was measured. Selected cycles were evaluated. Table 1 gives an overview about the performed experiments.

To estimate the parameters for the heat transfer to the environment according to eq. 8, at the end of the cyclic loading at certain strain the temperature of the samples was measured over time until it reached the temperature of surrounding.

For the sample temperature during free cooling of an equibiaxial stretched incompressible sample follows from eq. 8

$$T = T_{env} + (T_{start} - T_{env})\exp(-t/\tau), \tag{9}$$

where

$$\tau = \tau_0\lambda_3 = \frac{c_p\rho d_0\lambda_3}{2\alpha} = \frac{c_p\rho d_0}{2\alpha\sqrt{\lambda}}, \tag{10}$$

here is T_{env} – temperature of environment, T_{Start} – temperature of the specimen at the beginning of the cooling process, λ – strain ($\lambda = \varepsilon + 1$), λ_3 – is the change of the sample thickness d/d_0, τ – time constant of the cooling process.

Table 1. Experemental data.

	Environment	Loading (sinus)		Cooling (hold on)	
Test	T [°C]	$\varepsilon_x = \varepsilon_y$ [%]	t [s]	$\varepsilon_x = \varepsilon_y$ [%]	t [s]
No 1a	23	30	300	0	600
No 1b	23	30	300	30	600
No 2a	22.5	60	300	0	600
No 2b	22.5	60	300	60	600
No 3a	23	90	300	0	600
No 3b	22.2	90	300	78	600

In a second step, the dissipated mechanical energy during different cycles ΔW_{diss} was estimated from the hysteresis loop in the stress-strain-curves (sum of both loading directions). The heat loss of the whole sample to the environment within the hysteresis cycle can be estimated by

$$\frac{\dot{Q}}{d_0 A_0} = \frac{\dot{Q}}{V_0} = -\frac{c_p\rho}{\tau_0\lambda_3}(T - T_{env}). \tag{11}$$

For the whole hysteresis cycle the total heat loss can be estimated by integration as

$$\int\frac{\dot{Q}}{V_0}dt = -\frac{\delta Q_{loss}}{V_0} = -\frac{c_p\rho}{\tau_0}\int\frac{(T - T_{env})}{\lambda_3}dt. \tag{12}$$

For the increment in internal heat $\delta Q_{residual}$ of the sample during the whole cycle follows

$$\frac{\delta Q_{residual}}{V_0} = -c_p\rho(T_{end} - T_{start}). \tag{13}$$

From the principle of conservation of energy it follows

$$\Delta W_{diss} = -\frac{\delta Q_{loss}}{V_0} - \frac{\delta Q_{residual}}{V_0} = c_p\rho\left(\frac{1}{\tau_0}\int\frac{(T - T_{env})}{\lambda_3}dt + (T_{end} - T_{start})\right). \tag{14}$$

By this equation the heat capacity multiplied by the density of the sample can be estimated independent from a calorimetrical measurement.

With the separately estimated parameters τ and c_p the integration over the whole loading cycle can be repeated to estimate, whether the whole mechanical energy, which is stored as internal energy, is stored in the form of thermal visible energy. This means, it will be possible to check, whether all the elastically stored energy is stored on the basis of entropy elasticity.

4 RESULTS AND DISCUSSION

4.1 *Analysis of mechanical properties*

The stress-strain curves of an analysed material are shown in the Figure 1. The curve progression stay in the same range for every load state.

The temperature profile during the cyclic biaxial loading to 30, 60 and 90% strain and the

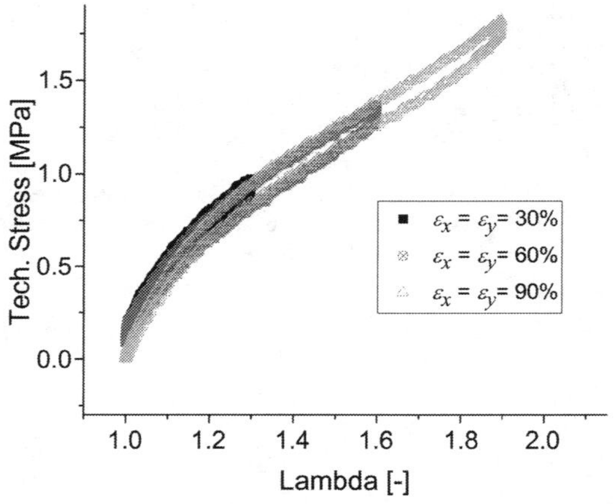

Figure 1. Stress-strain curves of equibiaxial sinus loading $\varepsilon_x = \varepsilon_y$= 30, 60 and 90%.

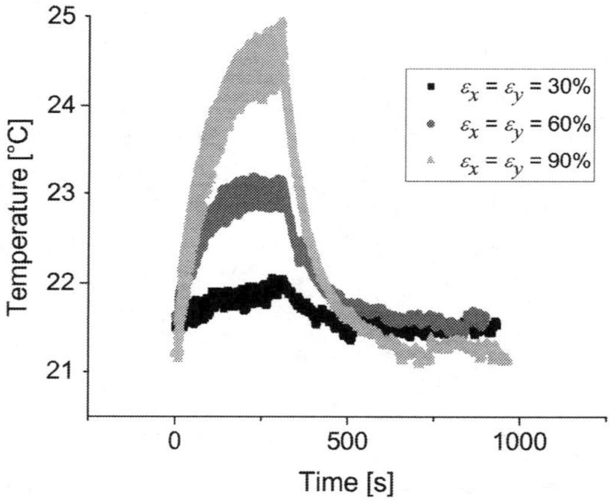

Figure 2. Heating of specimen during equibiaxial sinus loading. $\varepsilon_x = \varepsilon_y$ = 30, 60 and 90%. Cooling condition with $\varepsilon_x = \varepsilon_y$ = max.

subsequent cooling is shown in Figure 2. During each loading cycles (equibiaxial sinus loading, 1 Hz) the specimen is heated, due to internal dissipative processes in the material and conversion of mechanical energy into internal heat. With increasing overall temperature, the heat exchange to the environment increases resulting finally in an asymptotic mean temperature of the samples.

This exponential trend is shown in Figure 3. for the first 300 cycles at the different loading amplitudes.

The dissipated mechanical energy during a single cycle was estimated from the stress-strain curves by integration of eq. 5 and shown for the strain of 90% in the Figures 4 and 5.

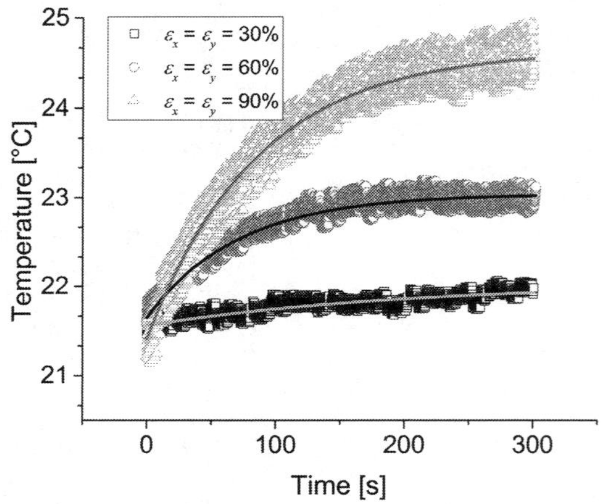

Figure 3. Heating curves for equibiaxial sinus loading $\varepsilon_x = \varepsilon_y$ = 30, 60 and 90% and an expotential trend during 300 cycles.

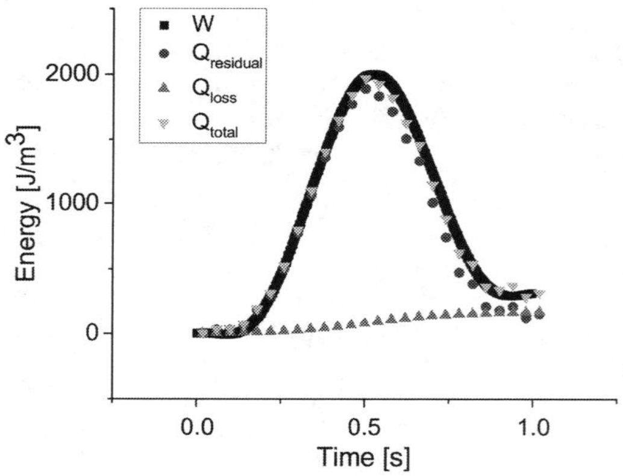

Figure 4. Mechanical energy and thermal energies of the 1st cycle for sinus loading $\varepsilon_x = \varepsilon_y$ = 90%.

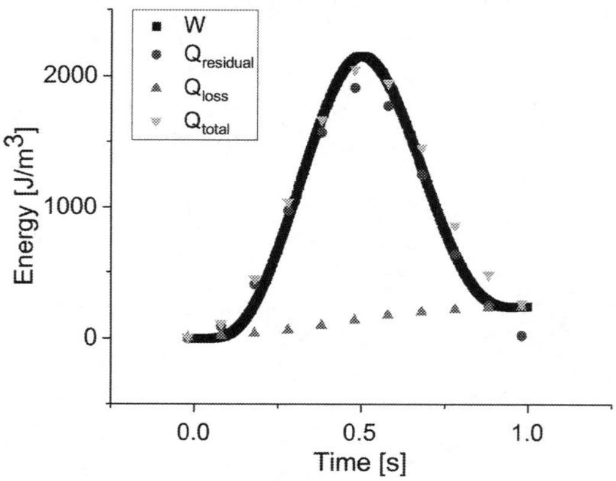

Figure 5. Mechanical energy and thermal energies of the 300th cycle for sinus loading $\varepsilon_x = \varepsilon_y$ = 90%.

Table 2. Summary of the results. Relationship of contributions of mechanical and thermal energies for analysed cycles [J/m^3]

	$\varepsilon_x = \varepsilon_y = 30\%$				$\varepsilon_x = \varepsilon_y = 60\%$				$\varepsilon_x = \varepsilon_y = 90\%$			
Cycle	W_{diss}	Q_{total}	$Q_{residual}$	Q_{loss}	W_{diss}	Q_{total}	$Q_{residual}$	Q_{loss}	W_{diss}	Q_{total}	$Q_{residual}$	Q_{loss}
1	79.2	56.7	35.4	21.3	198.2	193.5	106.1	87.4	310.8	314.0	147.6	166.3
50	76.7	73.3	39.3	33.9	178.3	153.2	74.7	78.4	260.1	243.5	94.6	148.8
100	76.8	73.5	35.4	38.1	172.2	175.9	94.5	81.3	248.7	243.6	47.5	196.1
150	72.2	76.8	35.4	41.4	170.2	123.3	31.6	91.7	247.4	246.6	47.4	199.2
200	74.4	75.0	35.4	39.6	167.6	199.8	88.7	111.1	241.3	306.3	71.4	234.9
250	72.1	78.2	31.5	46.6	163.4	130.6	0.3	130.3	238.2	279.7	25.6	254.0

4.2 *Analysis of heating conditions of the whole process*

First, the time constant for the heat transfer τ_0 to the surrounding was calculated from the cooling part of experiment. This coefficients, estimated by the different experiments, is material constant and have the same value.

By the procedure described above (see eq. 9 – 14) the heat capacity of the sample, multiplied by the density, was estimated. This value fits quite well with the externally by DSC estimated heat capacity (c_p) in the relevant temperature range of c_p = 1705 ... 1710 [J/(kg·K)].

4.3 *Analysis of heating conditions of separate cycles*

The curves of temperature changing during individual cycles of sinus load experiment were analysed according to the procedure described above—estimating the total lost and the dissipated heat in comparison to the mechanical energy.

The results for the first and the 300th cycle with the applied strain of 90% are shown in the Figures 4 and 5.

The figures show the correlation between mechanical and thermal energies. The mechanical energy stays in the same range for all the cycles with the same amplitude. The contributions of the saved and loss thermal energies per cycle change due to the increase of heat loss according to the increase of the mean temperature of the specimen.

In the first loading cycle (Figure 4) the $Q_{residual}$ shows the maximum values for every analysed amplitude, but the Q_{loss} stay at there minimum. The temperature of the specimen and environment are at the same level.For the time per cycle of 1 sec most of produced heat energy stays dissipated in the specimen due to the small heat flow to the environment due to the small temperature differences. The saved thermal energy leads to a temperature increase at the next cycle. This process repeats every step. So it causes the general heating of the specimen (see Figure 3).

The last cycles (Figure 5) show another relationship between $Q_{residual}$ and Q_{loss}. Here the temperature of the mean specimen is already near to thermal equilibrium with the environment (see Figure 3) in so far, that the energy input during the cycle is comparable to the energy transfer to the surrounding.

The total thermal energy shows the same trend for all cycles and is in a line with the applied respective recovered mechanical work. The ratio between the absolute values will be discussed in detail for the different loading ranges. The summary of the different energy contributions—mechanical dissipated energy (W_{diss}), thermal saved energy ($Q_{residual}$), thermal energy lost during interchange with the surrounding (Q_{loss}) – is shown in the Table 2. The results show, that most of the saved mechanical energy appear as thermal contribution to the internal energy. This means, that entropy elasticity is the mean contribution to the hyperelastic behavior of the material.

5 CONCLUSIONS

In this work the cyclic loading process of rubber material was analysed with the respect to dissipative specimen heating. The mechanical work and thermal energies were separated from each other. It was established, that most of the dissipated energy, which is produced by mechanical work, goes into heating of the specimen, a lower part is transferred to the surrounding. With increasing number of the loading cycles the temperature increases with an exponential trend and stops to growth, when the energy exchange of loaded body and environment attains equilibrium conditions. Most of the stored mechanical energy within the loading cycle is stored in the frame of a change of the entropy of the system. This work shows, that the thermal effects have a strong influence on the energy balance of the loaded specimen and that they have to be taken into account during the calculation of tearing energy for rubber like materials.

ACKNOWLEDGEMENTS

The first author would like to thank the Fa. Goodyear for the financial support. The author thanks Roberto Lombardi, Arnaud Vieyres, Karsten Brüning, Jan Domurath for the valuable discussion and Lutz Zybell for helping with the experimental procedure too. The acknowledgements also go to Kerstin Arnhold for the DSC measurement and Peter Dowidat (Fa. Coesfeld) for the support with the biaxial tester software.

REFERENCES

Cherepanov, G.P. (1967). Crack propagation in Continuous Media. *Journal of Applied Mathematics and Mechanics 31*(3), 10.

Cherepanov, G.P. (1974). *Mechanics of Brittle Fracture*. Moscow: Publishing office "Nauka", Mathematical, The Main Editors Office of Physical and Mathematical Literature (russ.).

Dedova, S., e. a. (2015). Analysis of crack propagation in elastomers based on global and local characterization methods. *Constitutive Models for Rubber IX* (September), 367–371.

Griffith, A.A. (1920). The Phenomena of Rupture and Flow in Solids.

Rice, J.R. (1968). A Path Independent Integral and the Approximate Analysis of Strain Concentration by Notches and Cracks. *Journmal of Applied Mechanics 35*, 379–386.

Rivlin, R. & A. Thomas (1952). Rupture of Rubber. I. Characteristic Energy for Tearing. *Journal of Polymer Science Vol. X, No. 3*, 291–318.

Schneider, K., e. a. (2014). Charakterisierung und Versagensverhalten von Elastomeren bei dynamischer biaxialer Belastung. *KGK 4*, 48–52.

Constitutive Models for Rubber X – Lion & Johlitz (Eds)
© 2017 Taylor & Francis Group, London, ISBN 978-1-138-03001-5

Crack growth under long-term static loads: Characterizing creep crack growth behavior in hydrogenated nitrile

W.V. Mars
Endurica LLC, Findlay, Ohio, USA

K. Miller
Axel Products, Inc., Ann Arbor, Michigan, USA

S. Ba & A. Kolyshkin
Schlumberger, Houston, Texas, USA

ABSTRACT: When a load is carried over an extended period, a crack in a viscoelastic material might grow, even if the load is less than the static tearing strength T_c for unstable rupture, and even in the absence of dynamic cycles. The rate of crack growth under these conditions is governed by the instantaneous value of the energy release rate. In the oil and gas industry, a seal or packer may well be required to tolerate high pressures encountered for an extended time without developing a crack in the material, even in the absence of dynamic cyclic loading. In this case, understanding the development of time-dependent cracking is essential to the successful material selection and design. A finitely scoped, strain-controlled procedure was used in which static strain is slowly increased on a planar tension test piece over a prespecified total test duration while a camera records images of crack development. The procedure was applied on a carbon black-filled hydrogenated nitrile. The information obtained in the experiment can be used to evaluate creep-crack growth, or in combination with measurements of the fatigue crack growth rate law, to analyze cases where crack growth might occur under mixed conditions.

1 INTRODUCTION

Crack growth sometimes occurs under loads less than those required to produce an unstable rupture. In dynamic applications, for example, cyclic loads can induce significant crack growth (Thomas 1958; Gent & Mars 2012). In static applications, given sufficient time, crack growth might also occur due to a creep mechanism (Greensmith & Thomas 1955; Lake & Lindley 1964a, Kadir & Thomas 1981, Bhowmick 1986).

There are many applications involving long-term, subcritical static loading that exceeds the intrinsic strength. The intrinsic strength represents conditions under which an elastomer may be expected to operate indefinitely without crack growth. The project being presented in this paper was motivated by the need to characterize creep-crack growth rate behavior for high-pressure, high-temperature (HPHT) applications in the energy industry (Zhong 2016, Mody et al. 2013, Shell 1980). Thermoplastic elastomers are another application where creep effects are sometimes significant (Mars and Ellul 2017).

The characterization of creep-crack growth is founded upon the energetic framework of Rivlin and Thomas (1953). In their framework, crack tip conditions are parameterized by means of the energy release rate T, which relates prospective changes in the total stored energy U to prospective changes in the projected surface area A of a given crack, given by the following equation:

$$T = -\frac{dU}{dA} \tag{1}$$

Equation (1) represents the energy released during fracture per unit of newly created fracture surface area. The energy release rate has been applied extensively to strength and fatigue problems (Rivlin & Thomas 1953, Gent & Mars 2012, Mars 2007), where its ability to objectively relate results across a wide range of test pieces and crack configurations has no rival.

Subject to certain limitations, this energy release rate might also be applied to comprehending static applications involving creep, particularly when the creep process is predominantly localized at the crack tip (Netzker et al. 2013), as it often will be in problems involving the development of small cracks (Aït-Bachir et al. 2012). The work presented in this paper applies an experimental procedure using the system shown in Figure 1 for observing

Figure 1. Testing system with environmental chamber, back-lit planar tension specimen, and camera.

creep-crack growth, and to derive the characteristic creep-crack growth rate curve.

In general, the total rate of crack growth is the sum of the cycle-dependent and time-dependent components according to Lake & Lindley (1964a) and Busfield et al. (2002), and as given by:

$$\frac{dc}{dN} = \frac{\partial c}{\partial N} + \frac{\partial c}{\partial t}\frac{dt}{dN} \tag{2}$$

Here, c = crack length
T = time and
N = number of applied cycles.

The first term is the cyclic fatigue crack growth rate, which is a function of the peak energy release rate T_{max}, and the ratio $R = T_{min}/T_{max}$. T_{min} is the minimum energy release rate of the applied cycle, given by the following:

$$\frac{\partial c}{\partial N} = f(T_{max}, R) \tag{3}$$

The factor $\partial c/\partial t$ in the second term is the creep-crack growth rate. This growth rate depends on the instantaneous energy release rate $T(t)$, and is important in non-crystallizing rubbers. It follows this rule:

$$\frac{\partial c}{\partial t} = g(T(t)) \tag{4}$$

The factor dt/dN in the second term of equation (2) is the time duration of each cycle.

Commercial fatigue codes like fe-safe/Rubber™ and Endurica CL™ enable the user to independently specify equations (3) and (4), so that they may be considered in the critical plane analysis (Barbash and Mars 2016) of elastomer part durability. The purpose of the work being presented in this paper is to measure the creep component of the relationship $g(T)$.

2 EXPERIMENTAL

The material being tested in this work is a proprietary filled, hydrogenated nitrile (HNBR) compound used in the oil and gas industry.

The experiment used a single-edge-cut planar tension specimen (see Figure 2). After cutting the specimen with a razor, the specimen was loaded under a slowly and linearly increasing strain during the test period. As the strain increased, the crack length was imaged continuously. Simultaneously, measurements of the load and displacement were collected, enabling calculation of the energy release rate on the crack at any given time during the test.

The dimensions of the planar tension specimen were 150 mm × 10 mm × 2 mm. The width was large relative to the height so that homogeneous straining was assumed (Castellucci et al. 2008). The initial cut, made via razor blade to a depth of 25 mm, was at least 2.5 times larger than the gauge height of the specimen to avoid undesired edge effects.

From prior measurements made on this material, it was known that the intrinsic strength (Lake & Thomas 1967, Lake & Yeoh 1980, Bhowmick 1983, Mars et al. 2016) is T_0 = 93 J/m^2, and that the critical tearing energy at unstable rupture is T_c = 5030 J/m^2.

2.1 *Control*

The strain ε applied to the specimen was initially 5%, and was increased linearly at a rate of 4.5%/hour until the test was completed (see Figure 3). The testing system features a temperature chamber. In the present work, the chamber temperature was maintained at 100°C.

Figure 2. Edge crack in a planar tension specimen.

2.2 *Measurement*

Strain in the specimen was determined by dividing the crosshead displacement by the gauge height h. The engineering stress σ was determined from the force recorded by a 4-kN load cell. The non-cracked cross-sectional area was computed based on the unstressed specimen dimensions, and taking into account the instantaneous length of the developing crack, which was recorded with a camera. One can spot the camera position in front of the specimen in Figure 1. Figure 4 shows typical images taken by the camera with the time each picture was taken recorded.

The strain energy density W is estimated by integrating the stress-strain curve. This integration can be done numerically by computing the area in between the stress-strain curve.

$$W = \int \sigma d\varepsilon \qquad (5)$$

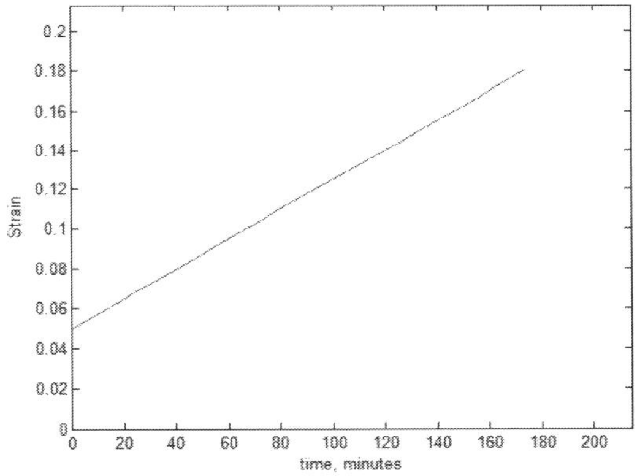

Figure 3. Strain ramp used to drive creep crack growth.

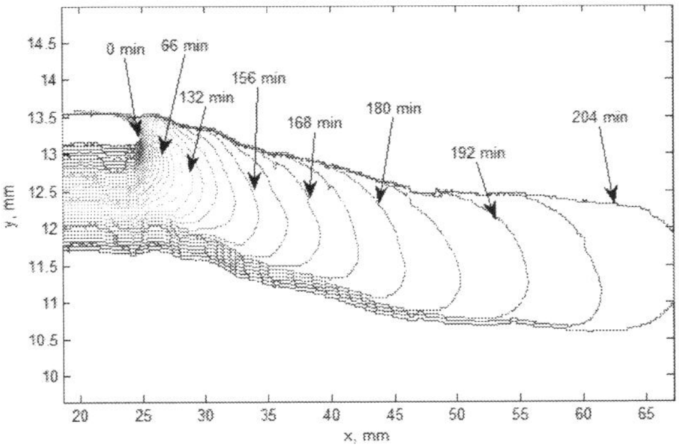

Figure 4. Crack images recorded and overlaid to show the crack tip progression during the experiment. Note that the y—axis is exaggerated relative to the x—axis.

3 RESULTS AND ANALYSIS

The energy release rate, T, at each instant was computed using equation (6), where h, the specimen gauge height, was 10 mm.

$$T = Wh \qquad (6)$$

If it is assumed that the creep-crack growth rate law follows a power law relation between the energy release rate and the crack growth rate, it then follows that the length history $c(t)$, for a given history of strain energy density W, is given by:

$$c = c_0 + r_q \left(\frac{h}{T_q} \right)^F \int_0^t W^{F_q} dt = c_0 + A \int_0^t W^{F_q} dt \qquad (7)$$

where c = the crack length, mm
C_0 = the initial crack length, prior to application of any cycles, mm
T = time, s

A is a parameter derived from the curve fitting process that reflects the combined influence of the material parameters r_c, T_c, and F, and the specimen gauge height h.

Crack images collected during the experiment were digitized and plotted as shown in Figure 4. In this figure, the colors represent time, with blue indicating early times and red corresponding to late times. From this information, the crack length history can be derived. Figure 5 shows typical crack length vs. time history, and the curve fit of equation (7) to the observations.

An error sum E was minimized to obtain the fit parameters c_0, A, and F_q. The integral was evaluated numerically during minimization to reflect the actual observed history $W(t)$ as follows:

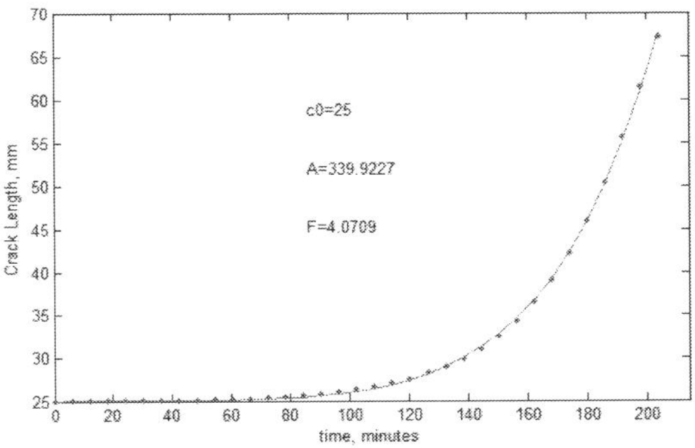

Figure 5. Crack length history, observed (dots) and fitted (solid line).

$$E = \left(c(t) - \left[c_0 + A\int W^{F_q} dt\right]\right)^2 \tag{8}$$

Crack growth rates *dc/dt* were determined by differentiating equation (7) with respect to time at each of the times for which crack length and stress/strain were recorded by the testing system. The crack growth rate was then plotted as a function of the energy release rate *T*, Figure 6, with the curve-fit parameters given in Table 1. The following power law was then fit to the results:

$$\frac{dc}{dt} = BT^{F_q} = r_q\left(\frac{T}{T_q}\right)^{F_q}. \tag{9}$$

Due to the differentiation, it might be a challenge to have robust numerical fitting of the data. It has been found that using the method above for determining the slop F_q allows to have good precision on the value from about three replicates only. This is a great advantage as it reduces the testing time required to get good estimate of the slope.

To avoid the strange units on fit parameter *B*, the parameters r_q and T_q are chosen to normalize the results relative to the uppermost point of the curve.

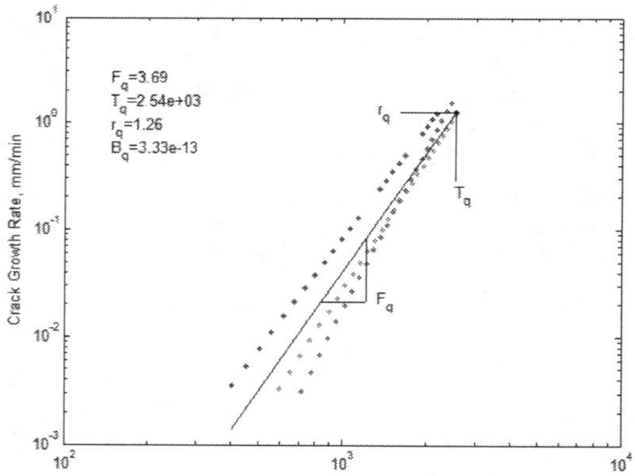

Figure 6. Creep-crack growth rate curves (3 replicates).

Table 1. Fitted power law creep crack growth parameters. (mm/min vs. J/m²).

Material	HNBR
Fq	3.69
Tq, kJ/m²	2.54
rq, mm/min	1.26

4 DISCUSSION

It is interesting to compare the fatigue (cyclic) crack growth with the creep crack growth per cycle. For computing the creep crack growth per cycle, one needs to integrate it over one cycle similar to what was done by Busfield et al. (2002). For example, the incremental crack length can be expressed as:

$$\Delta c = \int_0^{t_0} B_q (T(t))^{F_q} dt \tag{10}$$

where t_0 represents the time period of one fatigue cycle.

For the subject material, the fatigue crack growth rate law exhibited a powerlaw slope of F0 = 2.55, a critical tearing energy Tc = 3.05 kJ/m², and rc = 0.40 mm/cyc. Figure 7 plots combined creep-fatigue rate curves, via equation (2), for a series of different values of the cycle period dt/dN ranging from 0.1 seconds to 10 seconds. Longer times are seen to produce larger per-cycle rates, due to the creep effect, as expected. Shorter times are seen to eliminate the creep effect, with the rate curve approaching the cyclic fatigue crack growth rate curve.

Also, the slop $F_q = 3.69$ is higher than the value $F_c = 2.55$ obtained for the same material in pure fatigue crack growth test. The steeper slope of the creep crack growth rate law seems to suggest that the crack tip dissipation occurring during the creep process is perhaps weaker than that occurring during the cyclic process, since low slope is known to be associated with the strength of crack tip dissipation processes (Lake and Thomas 1967).

Note also that this result implies that at long times (or low frequency), and at high loads, it becomes increasingly important to characterize both fatigue and creep crack growth contributions,

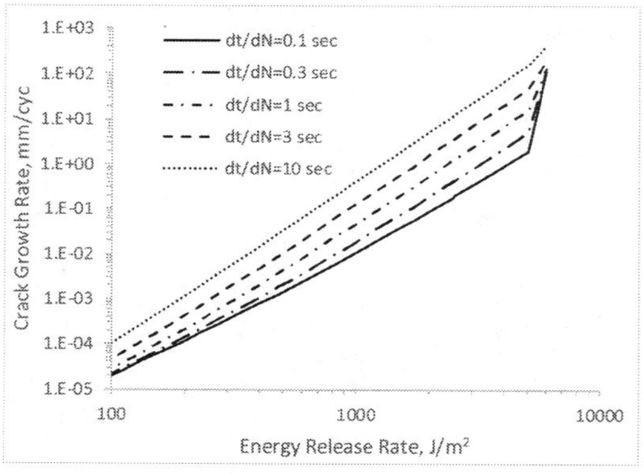

Figure 7. Total cyclic crack growth rate for a series of different cycle periods, computed via equation (2).

and to use the full form of equation (2) to estimate the service life.

5 CONCLUSION

An experimental procedure for observing the time-dependent creep-crack growth rate law has been implemented. The measurement is useful when designing for loads that must be supported over a long period, and when analyzing fatigue performance under conditions involving mixed cyclic and time-dependent crack growth. The slope of the creep-crack growth rate law is not necessarily the same as the slope of the fatigue crack growth rate law. Indeed, in the case of filled HNBR, the creep slope was significantly higher than the fatigue slope.

REFERENCES

Aït-Bachir, M., Mars, W.V., Verron E, 2012. "Energy release rate of small cracks in hyperelastic materials." *International Journal of Non-Linear Mechanics* 47, no. 4: 22–29.

Barbash, K.P., & Mars, W.V. (2016). Critical Plane Analysis of Rubber Bushing Durability under Road Loads (No. 2016-01-0393). SAE Technical Paper.

Bhowmick, A.K., A.N. Gent, and C.T.R. Pulford. 1983. "Tear strength of elastomers under threshold conditions." *Rubber Chemistry and Technology* 56, no. 1: 226–232.

Bhowmick, A.K. 1986. Tear strength of elastomers over a range of rates, temperatures, and crosslinking: tearing energy spectra. *Journal of Materials Science*, 21(11), 3927–3932.

Busfield, J.J.C., K. Tsunoda, C.K.L. Davies, and A.G. Thomas. 2002. "Contributions of time dependent and cyclic crack growth to the crack growth behavior of non-strain-crystallizing elastomers." *Rubber Chemistry and Technology* 75, no. 4: 643–656.

Castellucci, M.A., A.T. Hughes, and W.V. Mars. 2008. "Comparison of test specimens for characterizing the dynamic properties of rubber." *Experimental Mechanics* 48, no. 1: 1–8.

Gent, A.N., and W.V. Mars. 2012. "Strength of elastomers." *Science and Technology of Rubber*: 419–454.

Greensmith, H.W., and A.G. Thomas. 1955. "Rupture of rubber. III. Determination of tear properties." *Journal of Polymer Science* 18, no. 88: 189–200.

Kadir, A., & Thomas, A.G. 1981. Tear behavior of rubbers over a wide range of rates. *Rubber Chemistry and Technology*, 54(1), 15–23.

Lake, G.J. Lindley P.B., 1964a. "Cut Growth and Fatigue of Rubbers. II. Experiments on a Noncrystallizing Rubber", *Journal of Applied Polymer Science*, Vol. 8, pp. 455–466.

Lake, G.J., and A.G. Thomas. 1967. "The strength of highly elastic materials." *Proceedings of the Royal Society of London A: Mathematical, Physical and Engineering Sciences*, vol. 300, no. 1460, pp. 108–119. The Royal Society.

Lake, G.J., and O.H. Yeoh. 1980. "Measurement of rubber cutting resistance in the absence of friction." *Rubber Chemistry and Technology* 53, no. 1: 210–227.

Lake G.J., Lindley P.B., 1964b. Ozone Cracking, flex cracking and fatigue of rubber, *Rubber Journal*, Vol 146, No. 11, pp. 30–39, 1964b.

Mars, W.V. 2007. "Fatigue life prediction for elastomeric structures." *Rubber Chemistry and Technology* 80, no. 3: 481–503.

Mars, WV, Ellul MD, 2017. Fatigue Characterization of a Thermoplastic Elastomer, *Rubber Chemistry and Technology*, 90, no. 2.

Mars W.V., R. Stocek, C. Kipscholl, October 10–13, 2016, Intrinsic Strength Analyzer Based on the Cutting Method, Paper #79, Fall 190th Technical Meeting of the Rubber Division of the American Chemical Society, Inc., Pittsburgh, PA, ISSN: 1547–1977.

Mody, R., D. Gerrard, and J. Goodson. 2013. "Elastomers in the Oil Field." *Rubber Chemistry and Technology* 86, no. 3: 449–469.

Netzker, C., T. Horst, K. Reincke, R. Behnke, M. Kaliske, G. Heinrich, W. Grellmann. 2013. "Analysis of stable crack propagation in filled rubber based on a global energy balance." *International Journal of Fracture* 181, no. 1: 13–23.

Rivlin, R.S., and A.G. Thomas. 1953. "Rupture of rubber. I. Characteristic energy for tearing." *Journal of Polymer Science* 10, no. 3: 291–318.

Shell, Robert L. 1980. "Test for Evaluating Extrusion Resistance for Oil Well Packer Applications." *Rubber Chemistry and Technology* 53, no. 5: 1239–1260.

Thomas, A.G. 1958. "Rupture of rubber. V. Cut growth in natural rubber vulcanizates." *Journal of Polymer Science* 31, no. 123: 467–480.

Zhong, Allan. 2016. "Challenges for High-Pressure High-Temperature Applications of Rubber Materials in the Oil and Gas Industry." In Residual Stress, Thermomechanics & Infrared Imaging, *Hybrid Techniques and Inverse Problems*, Volume 9, pp. 65–79. Springer International Publishing.

Constitutive Models for Rubber X – Lion & Johlitz (Eds)

ISBN 978-1-138-03001-5

Mechanical characterization under CO_2 of HNBR and FKM grade elastomers for oilfield applications—effects of 10GE reinforcements

E. Lainé, J.C. Grandidier & G. Benoit
Institut Pprime, CNRS, ISAE-ENSMA, Université de Poitiers, Futuroscope, France

F. Destaing & B. Omnès
Centre Technique des Industries Mécaniques, Nantes Cedex 3, France

ABSTRACT: Despite its importance in sealing, the fundamental understanding of the behavior of elastomeric seals in a gas environment is still incomplete. Further experimental investigations are required to obtain accurate predictive coupled model able to estimate the combining effects of gases and mechanical loading. The effect of CO_2 pressure (2 and 6 MPa) on the mechanical behavior of HNBR and FKM seals at two temperatures (60 and 130°C) is described in the paper. Seals of these two 10GE reinforced matrices are also tested to assess the contribution or not of these reinforcements. The results allow to classify the four materials according to their properties and indicates that HNBR is the most appropriate in this context. Moreover, they show that the reinforced materials suffer damage after the return to zero pressure at 60 and 130°C (6 MPa exposure), inducing a loss of rigidity and therefore, such materials are not suitable for the gaseous environment.

1 INTRODUCTION

Carbon dioxide is a natural gas which is frequently encountered in hydrocarbon environments. Relatively low concentrations of CO_2 in hydrocarbon mixtures can cause significant swelling of the seal. More significantly, the effect of absorbed CO_2 upon Rapid Gas Decompression (RGD) can be catastrophic if consideration is not given to the choice of polymer, cure, and particle reinforcement. Hydrogenated Nitrile Butadiene Rubber (HNBR) and fluorocarbon rubber (FKM) are well known for their resistance to chemical and thermal degradation (Alcock et al. 2015, Grelle et al. 2015, Qamar et al. 2013). Nonetheless, performance and functional life of polymer materials can be significantly affected when submitted to gas environment. Indeed, saturation conditions, especially in CO_2, are known to alter the gas permeability due to plasticizing effects in almost all polymers and this effect is more pronounced for elastomer components (Briscoe et al. 1994, 1996, Major et al. 1996, Davies et al. 1999). In addition, they are specially designed for resistance to RGD. However, the fundamental understanding of the behavior of elastomeric seals in a gas environment is necessary because it is still incomplete and more particularly on coupling effects (diffuso-mechanical). In this work, we are interested in HNBR and FKM, non-reinforced and reinforced with 10GE, in order to evaluate the influence of these reinforcements on mechanical characteristics. Mechanical compression tests were carried out on O-rings under pressure and after desorption of CO_2. By comparing the different mechanical responses in the different configurations (temperature and CO_2 pressure), this procedure will evaluate the behavior and durability of each material.

2 EXPERIMENTAL

2.1 *Materials and sample*

Two elastomer types, a hydrogenated nitrile rubber (HNBR) and a fluorelastomer type (FKM) were selected for these investigations.

The compounds were produced with a nominal hardness of 80 and 90 Shore A respectively for the HNBR and the FKM. HNBR and FKM were compounded as summarized in Table 1, respectively, using a base polymer which is 96% saturated with 36% acrylonitrile content and a 70% fluorine content.

These two matrices reinforced by nanofilled (10GE) were produced.

The dimensional characteristics of the O-rings are 50.17 mm × 5.33 mm.

Table 1. Composition of the HNBR and FKM assessed in the paper.

Materials	HNBR phr*	FKM phr*
NBR/FKM	100	100
N-330 HAF carbon black	70	20
Antioxydant agent	1.5	1.5
Vulcanisation agent		8
Vulcanizing agent	8	2.5
Vulcanizing accelerator	2	3
	+ Nanofilled	+ Nanofilled

*Parts per hundred rubber parts in weight.

2.2 *Experimental device*

An Instron 8802 tensile machine was fitted with a pressure and temperature regulated chamber which allows mechanical testing in gaseous nitrogen, hydrogen or carbon dioxide. For safety reason related to hydrogen, the volume of the chamber is small (1.77l with a diameter of 150 mm and length of 100 mm). More information on this device can be found in previous papers (Sun 2011, Castagnet 2012).

To observe the tests, the enclosure has an optical access through a central cylindrical sapphire window of 25 mm diameter. Depending on the pressure conditions, the enclosure can be closed by two cylindrical doors, one allowing testing up to 4 MPa and the other up to 40 MPa. These two doors are equipped with a window of respective diameter 40 mm and 25 mm. The door window resistant to 4 MPa is centered, unlike the window of the second door which is offset.

The traction machine, with a maximum capacity of 20 kN, can operate up to a frequency of 20 Hz. Its maximum stroke is 25 mm (limitation to the dimensions of the lower cooled line). However, the height of the chamber limits the dimensions of the various assemblies and test pieces. The traction machine is provided with an external load cell and a pressurized column containing a pressure compensated internal load cell. The presence of the internal load cell allows direct measurement of the applied load without any sealing friction. The whole of this device is temperature controlled.

The HYCOMAT device couples a purely mechanical load to a load of gas, under nitrogen (N_2), hydrogen (H_2), or carbon dioxide (CO_2). Pressure and temperature regulation are assured by PID regulators.

2.3 *Specific assembly*

One specific assembly to test the 0-rings were dimensioned and manufactured, for the compression tests (Figure 1). A load reversal system is used to perform the mechanical compression tests. This inverter assembly consists of two rigid U-shaped structures making it possible to place the sample on the upper U connected to the fixed jaw and to compress it by means of the lower U connected to the movable jaw (Figure 1).

The compression tests were limited to a 1.5 mm crushing of the seal since this is equivalent to a local deformation of the order of 25% in the section thereof. In practice, this type of seal in operation is compressed by a maximum of 15 to 25%.

2.4 *Protocol*

The compression tests for each of the materials were carried out according to an identical protocol. This one consists in several following steps:

Step A: Positioning of the seal for compression test.
Step B: Temperature stabilization at 60 or 130°C (at least six hours).
Step C: Compression tests at constant temperature (60 or 130°C).
Step D: Pressurization under CO_2 (2 or 6 MPa) at a speed between 2 and 4 MPa/min, then stabilization for at least six hours (Figure 2).

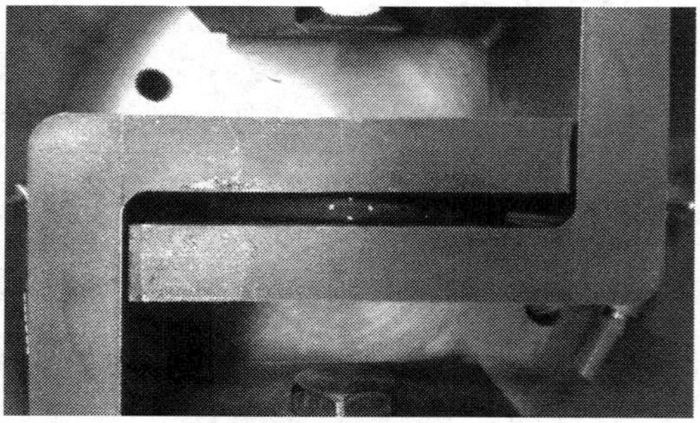

Figure 1. Specific U-shaped mounting for seal compression.

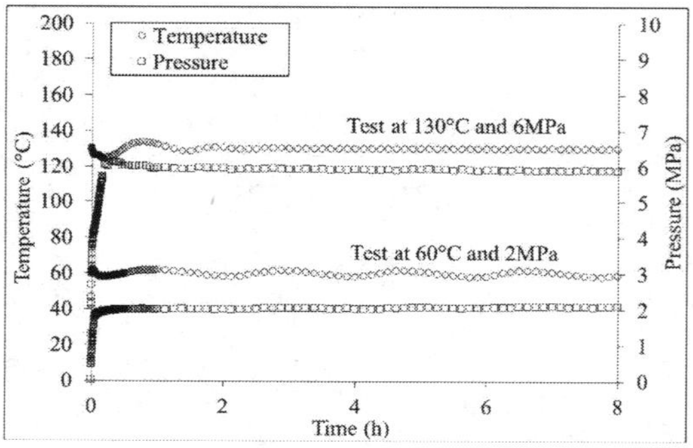

Figure 2. Time of stabilization.

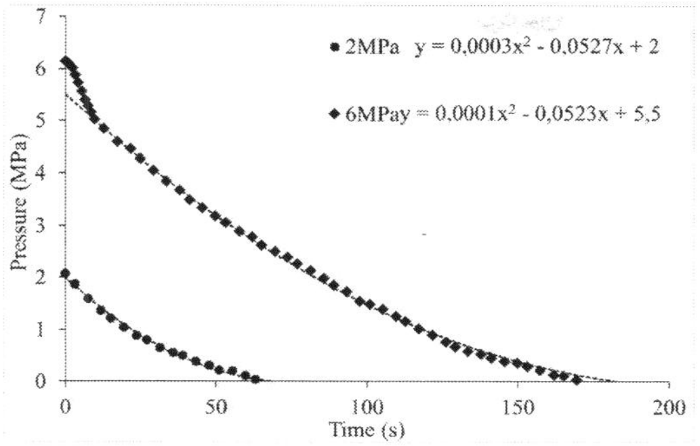

Figure 3. Decompression speed.

Step E: Compression tests at constant temperature (60 or 130°C) and constant CO_2.

Step F: Back to a zero CO_2 pressure at a speed between 2 and 4 MPa/min (Figure 3), then stabilization for at least six hours.

Step G: Compression tests at constant temperature (60 or 130°C).

3 RESULTS AND DISCUSSIONS

3.1 *HNBR (Compression tests at 60° and 130°C)*

Compression tests on an HNBR seal at different CO_2 pressures (2 and 6 MPa) and at two temperatures (60° and 130°C) were carried out on four different O-rings. The load - displacement curves are shown in Figure 4 and Figure 5 respectively at 60° and 130°C. It can be seen that if the CO_2 absorption at 60°C decreases the stiffness, this is less significant at 130°C because the curves are almost superimposed.

3.2 *FKM (Compression tests at 60° and 130°C)*

Compression tests on an FKM seal at different CO_2 pressures (2 and 6 MPa) and at two temperatures (60 and 130°C) were carried out on four different O-rings. The load - displacement curves are shown in Figure 6 and Figure 7 respectively at 60° and 130°C. As for HNBR, the absorption of CO_2 at 60°C decreases the stiffness is in accordance with the CO_2 pressure. At 130°C, the FKM behavior is different in comparison to HNBR. Indeed, the FKM has a notable influence on the stiffness, and particularly on the response at 6 MPa.

3.3 *Compression tests after desorption of CO_2 (HNBR & FKM)*

The last step of the test protocol (see 2.4) consisted in carrying out compression tests after total desorption of the CO_2 seal. The objective being to evaluate the impact of the pressure level on the final

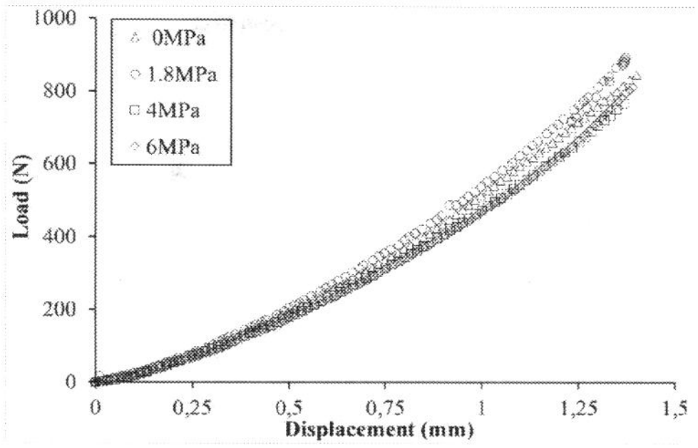

Figure 4. HNBR – Compression tests at 60°C.

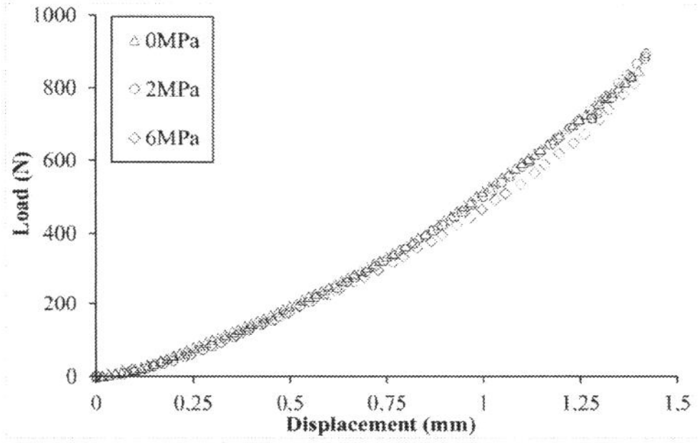

Figure 5. HNBR – Compression tests at 130°C.

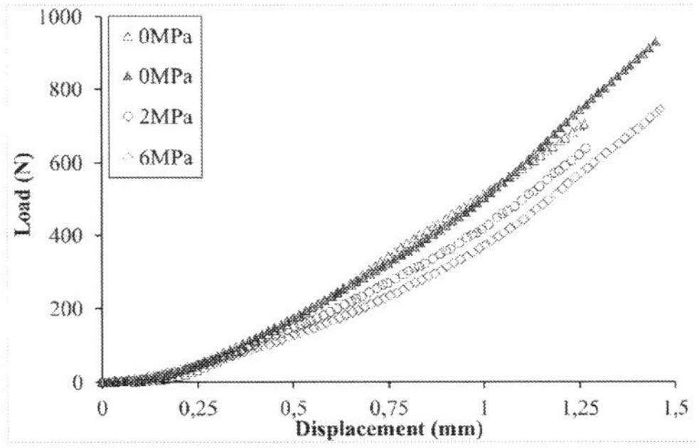

Figure 6. FKM – Compression tests at 60°C.

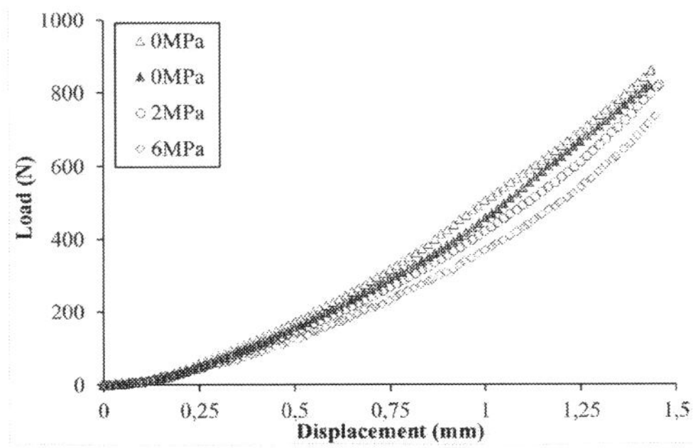

Figure 7. FKM – Compression tests at 130°C.

stiffness of the material. Thus, it was observed visually and mechanically that for the HNBR material, the O-ring suffered no damage and perfectly recovered its initial stiffness (Figure 8).

For the O-ring in FKM it is the same at 60°C (Figure 9). On the other hand, at 130°C, we observed visually during decompression, local swelling of the seal which led to damage (Figure 10). It results a loss of rigidity after a return of 2 and 6 MPa under CO_2. Although this loss of stiffness is small, one can imagine that a succession of sorption/desorption at these pressures would lead to the rapid degradation of the FKM seal.

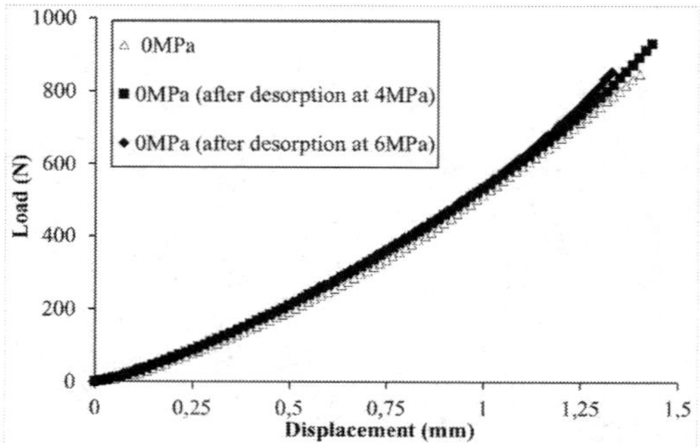

Figure 8. HNBR – Compression tests at 130°C.

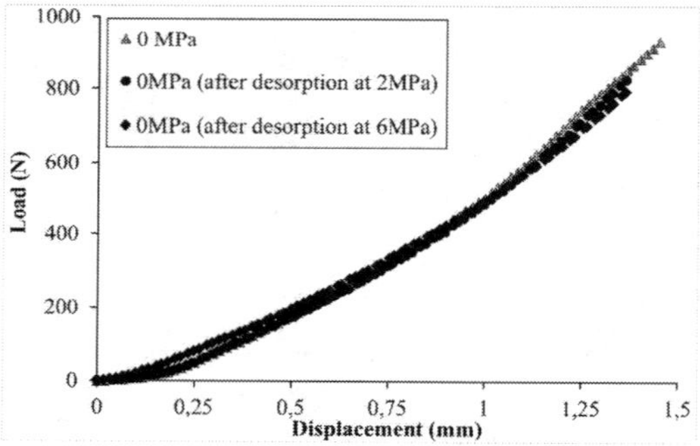

Figure 9. FKM – Compression tests at 60°C and zero pressure.

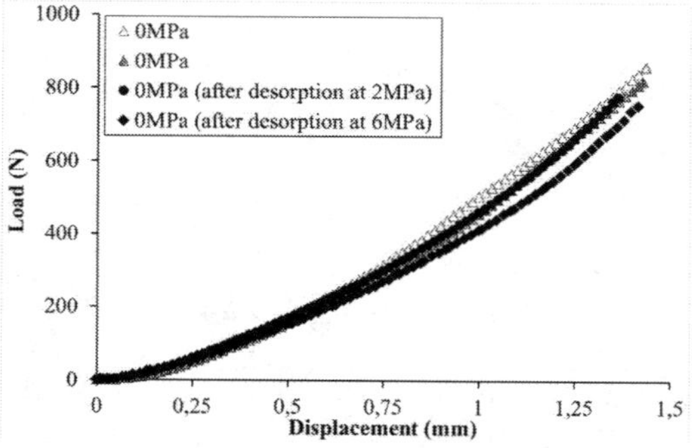

Figure 10. FKM – Compression tests at 130°C and zero pressure.

3.4 *Effects of 10GE reinforcements on the matrix HNBR*

Compression tests on an HNBR-10GE seal at different CO_2 pressures (2 and 6 MPa) and at two temperatures (60 and 130°C) were carried out on four different O-rings. The load - displacement curves are shown in Figure 11 and Figure 12 respectively at 60 and 130°C. The results of compression tests after decompression are presented on Figure 13 and Figure 14.

For HNBR-10GE, it is noticed in Figure 11 and Figure 12 that the latter is sensitive to 60°C, since its rigidity decreases progressively with the increase in pressure. On the other hand, at 130°C, although this is considerably less marked, it seems that the stiffness remains dependent on the CO_2 pressure.

It may be thought that loads intensify this trend as this was not the case for unfilled HNBR material (Figure 5). In compression tests after CO_2 desorption, it is clear that this reinforced material (HNBR-10GE) is modified since its residual stiffness is greatly reduced at 6 MPa. In addition, this is observed both at 60 and 130°C (Figure 13 and Figure 14). It is therefore obvious that,

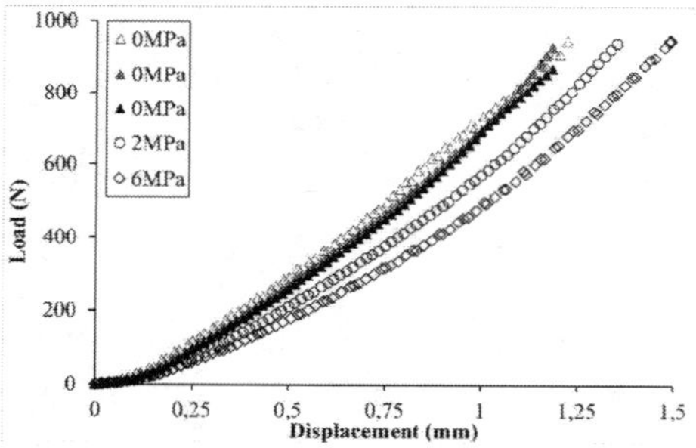

Figure 11. HNBR-10GE – Compression tests at 60°C.

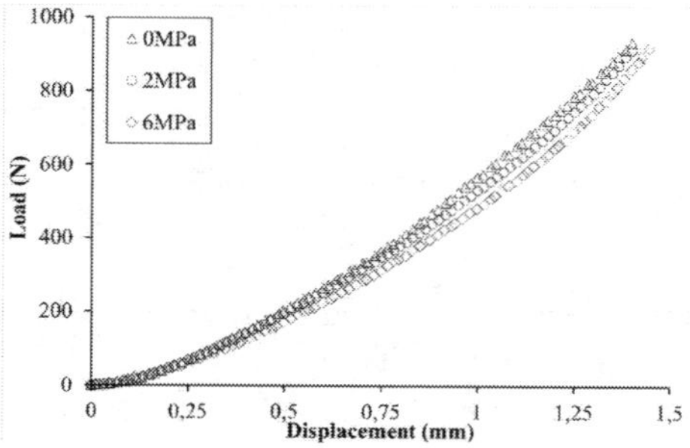

Figure 12. HNBR-10GE – Compression tests at 130°C.

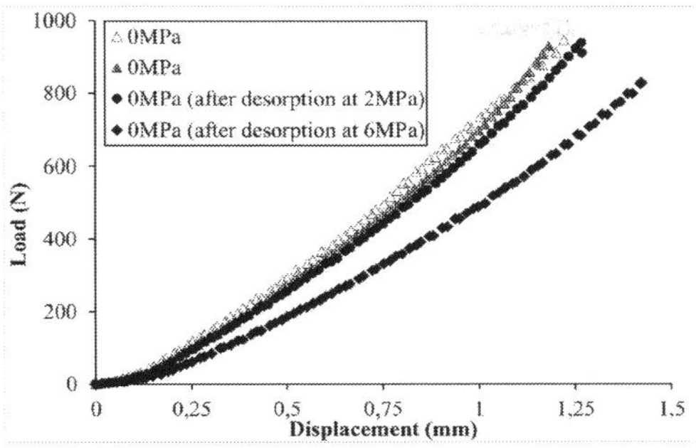

Figure 13. HNBR-10GE – Compression tests at 60°C and zero pressure.

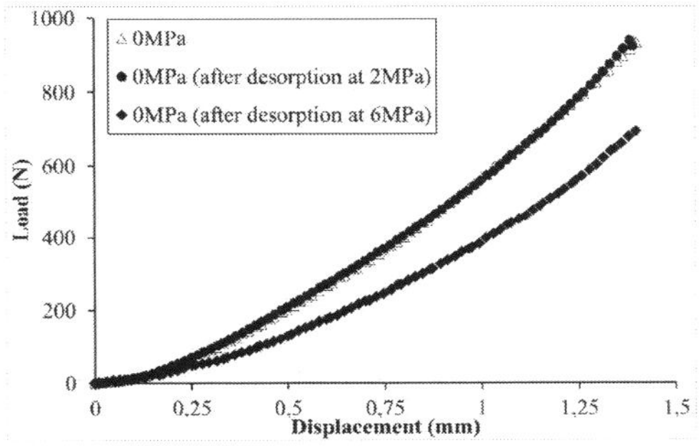

Figure 14. HNBR-10GE – Compression tests at 130°C and zero pressure.

compared to the responses under the same conditions (temperature and pressure) with the HNBR matrix, this reinforced material degrades under the effect of CO_2 (pressure and/or desorption).

Thus, the 10GE reinforcements weak the matrix subjected to a certain pressure of CO_2 (between 2 and 6 MPa) and a CO_2 desorption following the imposed speed (Figure 3).

3.5 *Effects of 10GE reinforcements on the matrix FKM*

Compression tests on an HNBR-10GE seal at different CO_2 pressures (2 and 6 MPa) and at two temperatures (60° and 130°C) were carried out on four different O-rings. The load - displacement curves are shown in Figure 15 and Figure 16 respectively at 60° and 130°C. The results of compression tests after desorption of CO_2 are presented on Figure 17 and Figure 18.

For the FKM-10GE, it can be seen on Figure 15 and Figure 16 that it is sensitive to 60°C, since its rigidity decreases progressively with increasing the hold pressure value. On the other hand, at 130°C, although this is less significant, the

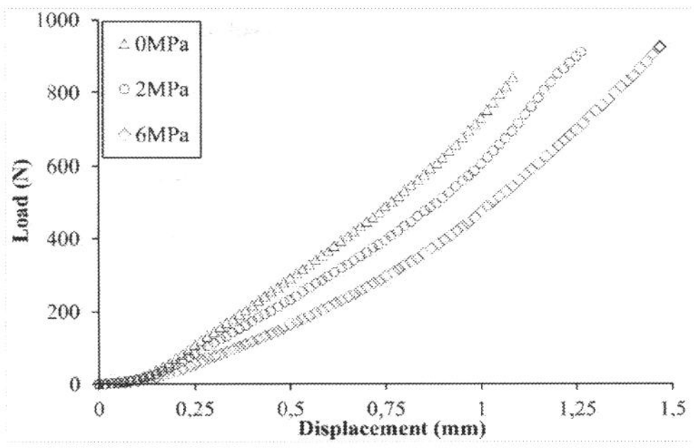

Figure 15. FKM-10GE – Compression tests at 60°C.

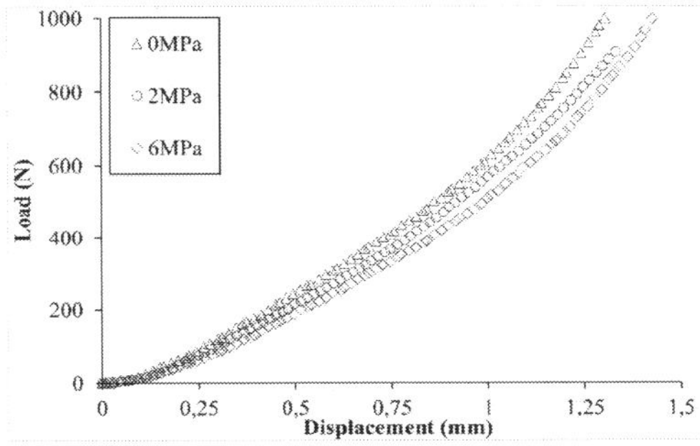

Figure 16. FKM-10GE – Compression tests at 130°C.

stiffness remains dependent on the CO_2 pressure. It can be assumed that the loads intensify this trend as it already appeared for the reinforced material (FKM, Figure 10).

During compression tests after desorption of CO_2, it clearly appears that this reinforced material is modified since its residual rigidity is reduced to 6 MPa. Moreover, this is true for the two study temperatures (Figure 17 and Figure 18).

As with HBNR-10GE, a material with 10GE reinforcements in an FKM matrix suffers a loss of stiffness. This appears from a certain pressure under CO_2 of between 2 and 6 MPa.

3.6 *Observations*

The acquisition of the images carried out during the tests makes it possible to observe the in-situ damage during CO_2 desorption.

Most damages observed during the CO_2 desorption step are located on the molding junction of the manufacturing process. They were only found on the HNBR-10GE (Figs. 19a,b) and FKM-10GE (Figs. 20a,b) seals. It may be thought that the nature of the filler and/or the process of implementation is at the origin of this weakness.

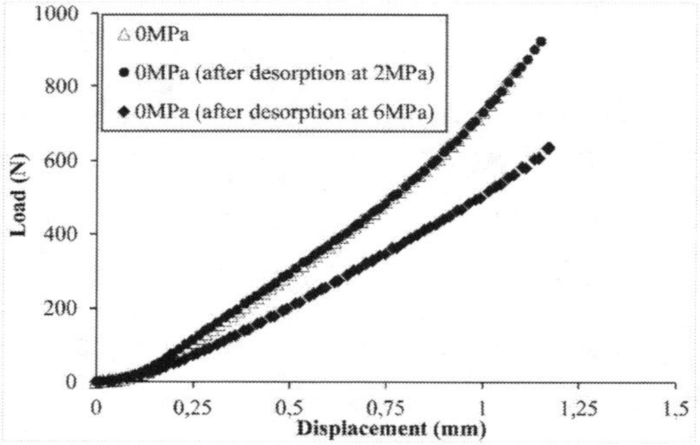

Figure 17. FKM-10GE – Compression tests at 60°C and zero pressure.

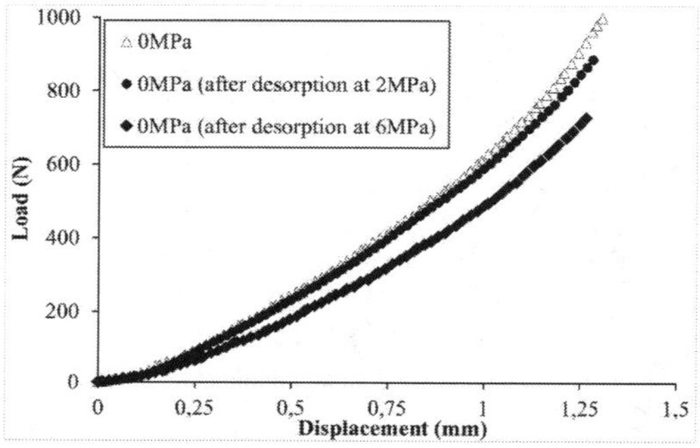

Figure 18. FKM-10GE – Compression tests at 130°C and zero pressure.

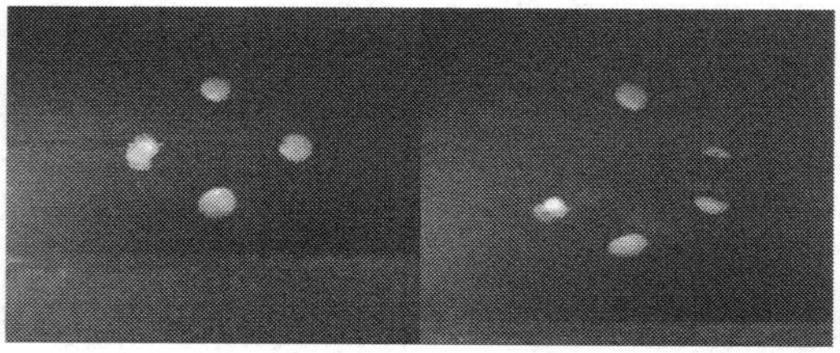

Figure 19. HNBR-10GE seal (compression test, 130° and 6 MPa) – (a) Photo before decompression t = 0 s – (b) Photo during décompression t = 7 min.

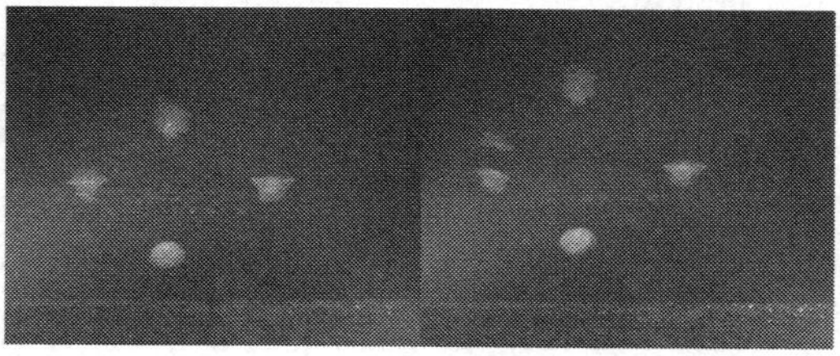

Figure 20. FKM-10GE seal (compression test, 130° and 6 MPa) – (a) Photo before decompression t = 0 s – (b) Photo during décompression t = 3 min.

4 CONCLUSIONS

HNBR seems unaffected by CO_2. Its behavior after decompression is not modified, whatever the initial condition (temperature and CO_2 pressure).

The behavior of the FKM under pressure of CO_2 is affected in compression (loss of rigidity). On the other hand, the effect of CO_2 desorption on compression stiffness seems to have little effect. Only after having been under 6 MPa of CO_2 and at 130°C, the material slightly sees its rigidity decreased.

The behavior of HNBR-10GE is affected by CO_2 in compression particularly at 60°C. On the other hand, the pressure at 6 MPa associated with the decompression cause damage to the joint. After returning to zero pressure, the O-ring lost in rigidity.

The behavior of the FKM-10GE is affected by CO_2 in compression (loss of rigidity). Thus, the effect of the pressure under 6 MPa of CO_2 and the desorption lead to damage to the FKM-10GE which results in a loss of compression stiffness after return at zero pressure.

Finally, HBNR is the most suitable material for these critical environments (temperature, CO_2 pressure). On the other hand, reinforcements seem to weaken the matrix (HNBR and FKM), and lead to materials damage with a loss rigidity according to the imposed conditions. (6 MPa at 60° and 130°C).

REFERENCES

Alcock, B. & Peters; T.A. & Gaarder, R.H. & Jørgensen, J.K. 2015. *Polymer Testing*, **47**, 22–29.

Briscoe, B.J. & Kelly, C.T. 1996. *Polymer*, **15**, 3405–3410.

Briscoe, B.J. & Savvas, T. & Kelly, C.T. 1994. Explosive decompression failure of rubbers—A review of the origins of pneumatic stress-induced rupture in elastomers, *Rubber Chem. Technol.* 67 (3), 384–416.

Briscoe, B.J. & Zakaria, S. 1991. *Journal of Polymer Science B. Polymer Physics*, **29**, 989–999.

Castagnet, S. & Grandidier, J.C. & Comyn, M. & Benoit, G. 2012. Mechanical testing of polymers into pressurized hydrogen: tension, creep and ductile fracture. *Exp. Mech.*, 52 (3), 229.

Davies, O.M. & Arnold, J.C. & Sulley, S. 1999. *Journal of Materials Science*, **34**, 417–422.

Grandidier, J.C. & Baudet, C. & Boyer, S.A.E.; & all 2015. *Oil & Gas Sci. and Tech.—Rev. IFPEN*, **70**, 251–266.

Grelle, T. & Wolff, D. & Jaunich, M. 2015. *Polymer Testing*, **48**, 44–49.

Major, Z. & Lang, R.W. 2010. *Engineering Failure Analysis*, **17**, 701–711.

Qamar, S.Z. & Akhtar, M. & Pervez, T. & Al-Kharusi, M.S.M. 2013. *Materials and Design*, **43**, 487–496.

Sun, Z. & Benoit, G. & all 2011. Fatigue crack propagation under gaseous hydrogen in a precipitation-hardened martensitic stainless steel, *Int. J. Hydrogen Energy*, 36 (14), 8641.

Constitutive Models for Rubber X – Lion & Johlitz (Eds)
ISBN 978-1-138-03001-5

Sequential automated time-temperature algorithm for dynamic mechanical analysis

C. Costes
Institut de Recherche en Genie Civil et Mecanique, UMR CNRS 6183, École Centrale de Nantes, France
DCNS Research, Technocampus Ocean, Bouguenais, France

F. Le Lay
DCNS Research, Technocampus Ocean, Bouguenais, France

E. Verron & M. Coret
Institut de Recherche en Genie Civil et Mecanique, UMR CNRS 6183, École Centrale de Nantes, France

J.F. Sigrist
DCNS Research, Technocampus Ocean, Bouguenais, France

ABSTRACT: Dynamic Mechanical Analysis (DMA) is a widely used experimental technique to provide information about frequency and temperature-dependent mechanical response of polymer materials. Classically, it is used to produce master curves assuming the Time Temperature Superposition principle relating measures to a reference temperature behavior; it leads to the determination of both vertical and horizontal shift factors which respectively affect stiffness and reduced frequency. In this paper, we propose a robust automated generic algorithm that evaluates sequentially the vertical and horizontal shift factors avoiding over-fitting and polynomial instabilities. The derivation of this algorithm is detailed. Then its validation and comparison with other methods are presented considering synthetic data; the influence of noise is investigated. The method is finally applied and compared to other methods on real experimental data.

1 INTRODUCTION

Viscoelastic properties of soft materials are more and more used in industrial applications. Experiments dedicated to measure these properties can be separated in two main categories: time dependent experiments and harmonic experiments such as Dynamic Mechanical Analysis (DMA). DMA consists in prescribing harmonic (angular frequency ω) strain $\gamma(t)$ (or stress $\sigma(t)$) while measuring stress (or strain). The mechanical response is expected to be a combination of multiple fundamental modes. In small strain,

$$\gamma(t) = \gamma_s + \gamma_d \sin(\omega t), \quad (1)$$

and only the fundamental mode is measured and used to identify the dynamic amplitude of stress σ_d and the phase φ (Wineman and Rajagopal 2000)

$$\sigma(t) = \sigma_s + \sigma_d \sin(\omega t + \Phi). \quad (2)$$

The complex modulus G^* that relates dynamic stress and strain

$$G^* = \frac{\sigma_d}{\gamma_d}\left(\cos(\varphi) + i\sin(\varphi)\right) \quad (3)$$

can be separated into a conservative G' and a loss G'' parts:

$$G^* = G' + iG''. \quad (4)$$

One application of DMA is the determination of the Time Temperature Superposition (TTS) law of materials. TTS provides an equivalence between experimental data measured at a given temperature T_i and the mechanical response of the material at a reference temperature T_0. The efficiency of this method has been proved for amorphous polymers (Ferry, Child, Zand, Stern, Williams, and Landel 1957, Rubinstein and Colby 2003). TTS involves two independent parameters: the horizontal shift factor a_{T_0} which relates the frequency f at a temperature T and the reduced frequency f_r at the reference temperature T_0, and the vertical shift factor b_{T_0} relates the measured complex modulus at temperature T and the reference one at temperature T_0, such that

$$G^*(\omega,T) = b_{T_0}(T)G^*(\omega\, a_{T_0}(T),T_0) \quad (5)$$

In order to use the TTS principle for a given material, and then to establish master curves, the determination of shift factors is mandatory.

Different procedures for the establishment of master curves have been recently proposed (Duperray 1994, Sihn and Tsai 1999, Dealy and Plazek 2009, Rouleau 2013). Duperray (1994) developed an algorithm which fits $G^*(\omega,T)$ with a bi-polynomial function:

$$G^*(\omega,T) = \sum_{\substack{k_\omega \in [0,7] \\ k_T \in [0,3]}} A^*_{k_\omega k_T}\left(\log(\omega)\right)^{k_\omega} T^{k_T} \quad (6)$$

which involves 32 complex parameters ($A^*_{k_\omega k_T}$) to be identified. Assuming $b_{T_0} \approx 1$, the problem reduces to the identification of a_{T_0} for each tested frequency. The dependency in frequency of a_{T_0} is faded by averaging results with each tested frequency. A few years later, Sihn and Tsai (1999) proposed another automated method for TTS, in which isotherms are sequentially shifted and fitted one after another. Recently, Rouleau (2013) proposed a new method with interesting features. First it considers as many parameters for $a_{T_0}(T_i)$ and $b_{T_0}(T_i)$ as the number of isotherms used for computation; initial values of $a_{T_0}(T_i)$ are guessed by fitting a polynomial on shifted data. Second, the method extends, smoothes and linearises shifted curves for phase (named Φ_{xtd}) and complex modulus amplitude (named $|G^*|_{xtd}$). Third, it uses Kramers-Kronig relationships to fit a phase (Φ_{fit}) with $|G^*|_{xtd}$ and a complex modulus amplitude $|G^*|_{fit}$ with Φ_{xtd}. Finally, a cost function which compares both extended functions ($|G^*|_{xtd}$,Φ_{xtd}) with their computed fits ($|G^*|_{fit}$,Φ_{fit}) is minimized.

The present paper proposes the derivation of a new method that evaluates in two sequential steps both vertical and horizontal shift factors. Details are given in the next section; then our method is validated with noisy synthetic data, and finally compared to two other methods (Duperray 1994, Rouleau 2013).

2 METHODS

2.1 *Overview of the sequential method*

The present method consists in evaluating sequentially the vertical and horizontal shift factors. It involves two different programs referred to as b_T *Program* and a_T *Program* in the following. Both programs have very similar architectures as depicted in Figure 1 and involve the same functions, e.g. F_{Shift}, F_{Comp}, and F_{L3SPL} in the figure. (ω_m,T_m,G'_m,G''_m) represent the measured data for (ω,T,G',G''). The choice of variables (x_m, y_m) gives the generic (abscissa, ordinate) representation that is used further in both programs. For b_T *program*, the chosen plan is a Wicket one (with $x_m = log(|G^*_m|)$ and $y_m = \Phi_m$). A third representation axis (z) is needed for a_T *program*. This third axis helps to consider two plans in which disparities are to be reduced simultaneously: (x_m,y_m) and (x_m,z_m). As detailed in Figure 1 for a_T *program*, the common x axis represents logarithmic angular frequencies while y and z stand for conservation and loss corrected logarithmic modulus.

The method can be separated into five steps.

1. Transposing the problem in the (x_m, y_m) plan, in which translations (on x axis) are applied in step 4.
2. Separating (x_m, y_m) data into isotherms (x_{iT}, y_{iT}).
3. Converting each isotherm into the best Leave One Out Cross Validation (LOOCV) cubic spline $(SPL0)_{iT}$ using the function F_{L3SPL}. $(SPL0)_{iT}$ and T_{iT} represent a data reduction and approximation of (x_m, y_m) for each temperature T_i.

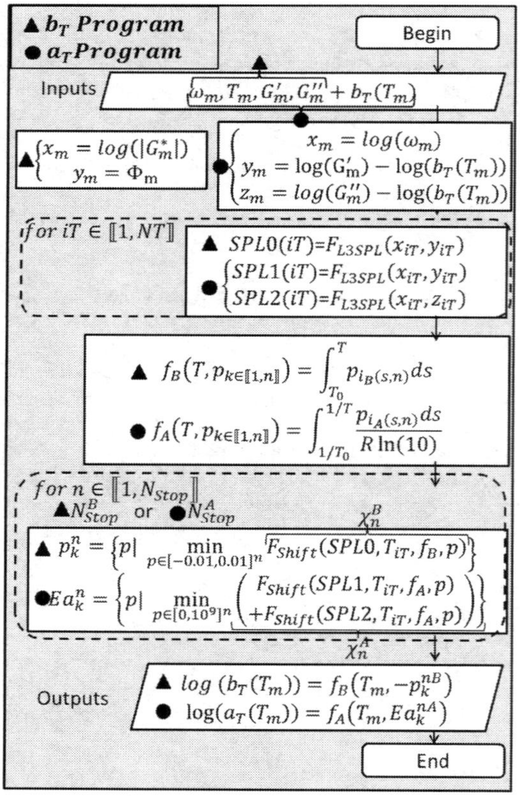

Figure 1. Generic representation of b_T *Program* and a_T *Program*.

4. Applying TTS to horizontally shift each spline $(SPL0)_{iT}$. Resulting splines are denoted $(SPLShT)_{iT}$.
5. Evaluating the RMS standard deviation of shifted splines in both overlap and empty regions between them, with the help of function F_{Shift}. Its minimization gives final shift factors.

2.2. *Useful functions*

2.2.1 *Data reduction*

The function F_{L3SPL} considers data (x_{iT}, y_{iT}) as input and calculates (xBP, yBP, dBP), the coordinates of cubic spline breakpoints (BP). (dBP) is the spline derivative value at each BP. A LOOCV method is used to determine the spline complexity, starting with a simple straight line up to multiple BP cubic spline. The LOOCV least square (LSQ) deviation takes into account the number of parameters of each regression:

$$\text{LOOCV LSQ} = \sqrt{\frac{\sum_{k=1}^{N_{iT}} (y_k - \hat{y}_k)^2}{N_{iT} - n_p + 1}} \quad (7)$$

where N_{iT} is the number of datapoints (x_{iT}, y_{iT}) of isotherm iT, n_p is the number of parameters of the regression, y_k the ordinate of the k-th point, and $\hat{y}_k$ the ordinate of the k-th LOOCV regression evaluation for $x = xk$.

2.2.2 *Standard deviation for a spline set*

The function F_{Comp}, presented in Figure 2, considers a group of splines $(SPL)_{i_T \in [1,NT]}$ that represents (x_m, y_m) data and evaluates a global deviation on some given intervals. In this figure, $U(SPL)$ is the set of overlapping intervals (between at least two splines) and of those not covered by the splines. For each abscissa, the function $I(x)$ furnishes the index of affected splines:

$$\forall x \in [\min((x_{BP})iT), \max((x_{BP})iT)]$$
$$I(x) = \begin{cases} if\ iT\ Find(x) \neq \varnothing \text{ then } iT\ Find(x) \\ else\ \{iT\ Inf(x), iT\ Sup(x)\} \end{cases} \quad (8)$$

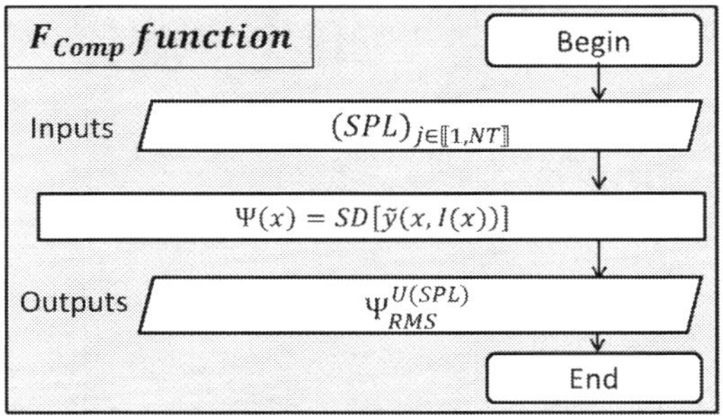

Figure 2. Function F_{Comp}.

where $iT\ Find(x)$ stands for the list of splines which contain x in their breakpoints, $iT\ Inf(x)$ is the spline which first breakpoint is the closest one to x and $iT\ Sup(x)$ the spline which last breakpoint is the closest one to x. The function $\tilde{y}((x, I(x))$ represents the vector of values given by all $I(x)$ splines for abscissa x, and the function $\Psi(x)$ gives the standard deviation of $\tilde{y}$. $\Psi_{RMS}^{U(SPL)}$ is the RMS value of Ψ over the $U(SPL)$ set, and is the output of F_{Comp}; it gives the global evaluation of splines deviation.

2.2.3 *Cost function minimization*

In both bT *Program* and aT *Program*, after data reduction, a cost function F_{Shift} performs minimization; it is presented in Figure 3. Its input arguments are: initial splines $SPL0$ with temperature of isotherms T_{iT}, a function *func*, and parameters p. The function *func* computes the relation between p, T_{iT} and horizontal shift (in (x, y) plan) for each isotherm. With this method, F_{Shift} can be used with both b_T *Program* and a_T *Program*. Average standard RMS deviation evaluated by F_{Comp} is finally weighted with the number of parameters np in vector p.

2.3 *bT Program*

First, data are represented by curves $y_m = \Phi_m$. vs. $x_m = log(|\ G_m^*)$. In this Wicket plot, $log(b_T)$ acts as a horizontal shift. In this case the temperature function *func* for F_{Shift} (see Fig. 3) is f_B, defined as an integral function from T_0 to T as shown in Figure 1. This method defines a piecewise linear regression over the temperature range. The parameter p defines the slopes of f_B along the T_m range. The concerned slopes indexes with variable temperature s are given by the function $i_B(s, n)$:

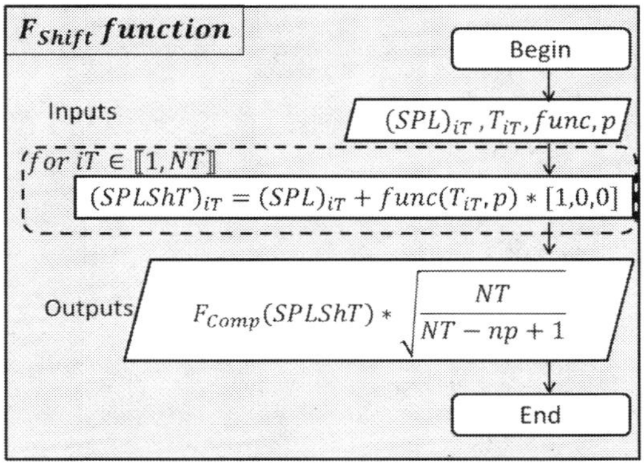

Figure 3. Function F_{Shift}.

$$i_B(s,n) = \left\lfloor \frac{(n-1)[s - min(T_m)]}{max(T_m) - min(T_m)} + \frac{3}{2} \right\rfloor \quad (9)$$

Then, starting with a parameter $p_0 = 0$ the cost function F_{Shift} is minimized with a limited memory Broyden-Fletcher-Goldfarb-Shanno (LM-BFGS) algorithm. Then, the optimized parameter is used to generate a new starting parameter p_k, length of which is increased by 1. Value of the cost function for each optimised parameter p_k^n is memorized with χ_n^B. Algorithm stops after attaining the maximum complexity N_{Stop}^B which keeps at least two isotherm temperatures for each slope of f_B. The final parameter value is p_k^{nB} with $nB\, nB$ the index of minimal values in χ_n^B.

2.4 *aT Program*

The structure of the a_T *Program* is similar to the one of b_T *Program* except few differences: first there are two splines steps for both storage and loss moduli, and second the temperature function f_A differs. As storage and loss moduli are simultaneously considered, the cost function to be minimized is the sum of F_{Shift} values computed with storage and loss splines. Function f_A is chosen as a piecewise description of an Arrhenius law in a decimal logarithmic base. In Fig. 1 for a_T *program*, parameter p is a n vector which values represent activation energies. The right parameter index to be considered along the temperature inverse variable s is given by $i_A(s, n)$:

$$i_A(s,n) = \left\lfloor \frac{(n-1)(s - 1/\max(T_m))}{1/\min(T_m) - 1/\mathbf{max}(T_m)} + \frac{3}{2} \right\rfloor. \quad (10)$$

This equation represents the equal distribution of activation energy indexes between $\min(1/T_m)$ and $\max(1/T_m)$. For the horizontal shift, the quantity kept into memory for each iteration of complexity is χ_n^A. In the same time, the associated parameter value (activation energy) Ea_k^n is kept constant. The stop condition of the algorithm is the same as in b_T *program*, i.e it keeps at least two isotherms for each slope of f_A. The corresponding maximum complexity number is named N_{Stop}^A. Final parameter to be taken is the activation energy vector Ea_k^{nA} with nA the minimal value of χ_n^A.

2.5 *Remak: Modification of Duperray's method*

Finally, we also propose a slight modification of Duperray (1994) approach that can substantially improve the results. A logarithmic form for G^* would be easier to fit with a polynomial function involving both $\log\omega$ and T.

3 RESULTS AND DISCUSSION

3.1 *Generation of synthetic data*

First synthetic data are considered to evaluate accuracy and robustness of our method, as compared to other published approaches.

Here, synthetic data are generated with a four parameter Zener Fractional (ZF) derivative model (Wineman and Rajagopal 2000). The corresponding rheological scheme is shown in Figure 4. In this figure, G_∞ is the asymptotic low frequency shear modulus and G_Δ is the difference with the high frequency modulus, τ is the relaxation time, and α the slope of modulus fading from low to high frequencies. Numerical values of these parameters are given in Table 1.

For the TTS principle, the horizontal shift $a_{T_0}(T)$ is defined with the Williams-Landel-Ferry (WLF) model (Ferry, Child, Zand, Stern, Williams, and Landel 1957) between the glass transition temperature T_g and $T_g + 100$ with two parameters C_1^0 and C_2^0. Outside of the WLF definition domain, a_{T_0} satisfies an Arrhenius law with activation energies E_{LT} and E_{HT} for low and high temperatures, respectively. Corresponding values are given in Table 2. The vertical shift factor b_{T_0} satisfies a Rouse model with thermal expansion ($\alpha_T = 2.10^{-4}\ K^{-1}s$) and entropic rigidification (Rouleau 2013).

$$b_{T_0}(T) = \left(\frac{T}{T_0(1 + \alpha_T * (T - T_0))^3} \right). \quad (11)$$

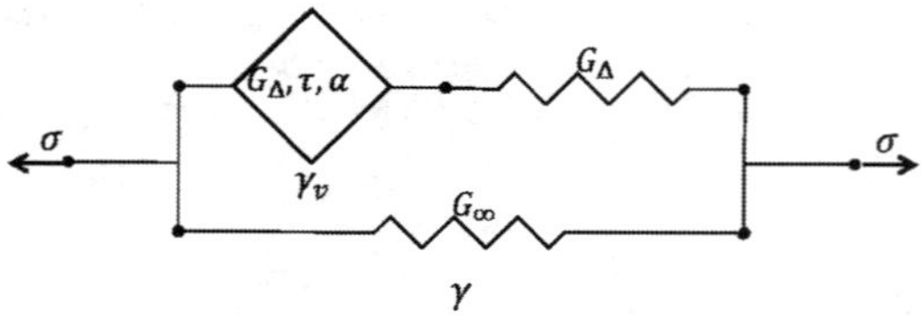

Figure 4. Fractional Zener model (ZF).

Table 1. ZF parameters at T_0 for data generation.

G_∞	G_Δ	τ	α
1 MPa	1 GPa	10^{-7} s	0.5

Table 2. Values for $a_{T_0}(T)$.

T_g	E_{LT}	C_1^0	C_2^0	E_{HT}
–60°C	50 kJ/mol	9.23	130°K	25 kJ/mol

Table 3. Measurement resolution.

δT	$\delta_{\gamma d}$	δG	$\delta\phi$
0.1	10^{-6}	0.01 MPa	10^{-6} rad

Next, realistic data are generated by considering the example of the Metravib DMA +300 industrial machine. The frequency range is set to $[1\,\text{Hz}, 120\,\text{Hz}]$, and temperature setpoints are distributed from −70 to −20 every 5, and then from −10 to +70 every 10. Concerning the resolution, a standard normal noise level for DMA is applied to measures of T, γ_d, G, and Φ. This standard noise deviation is set equal to the resolution of each measure given in Table 3. In order to artificially increase or reduce the noise level, a multiplicative coefficient is applied to the standard noise level.

3.2 *Comparison of methods on synthetic data*

25 synthetic DMA raw results are generated for each noise level (multiplicative coefficient from 1 to 10) following the above-mentioned model. Methods are investigated considering the standard noise level. Resulting shift factors vs. temperature curves are given in Figures 5 and 6.

As shown in Fig. 5, the only method able to estimate the vertical shift factor is the one presented here. Nevertheless, it seems that Rouleau's method gives the right tendency but with high dispersion. All methods are revealed able to estimate the horizontal shift factor (see Fig. 6). The method of Duperray (1994) overestimates a_T especially in low variability intervals. This can be explained by Eq. (6) used to fit data: a 3rd order polynomial with logarithmic reduced pulsation is unable to fit low variability on pulsations' boundaries and high variability on midrange ones. The modified method of Duperray (1994) suffers the same difficulty, even if fitting logarithmic modulus helps to reduce the center variability and then leads to better results for flat boundary regions. The Rouleau (2013) method shows a good fitting of the horizontal shift factor; and our sequential method gives more accurate results.

The efficiency of methods is quantitatively estimated by computing the RMS deviation to original shift factors as a function of noise level; the corresponding results are shown in Figures 7 and 8 for b_T and a_T, respectively. For b_T, Rouleau's method leads to dispersive results but also to more deviation than methods without vertical shift factor. It is also noticeable that deviation does not increase with noise level, then errors induced by this method might be larger or of the same order of magnitude than the noisiest level. The present method exhibits accurate results for the vertical

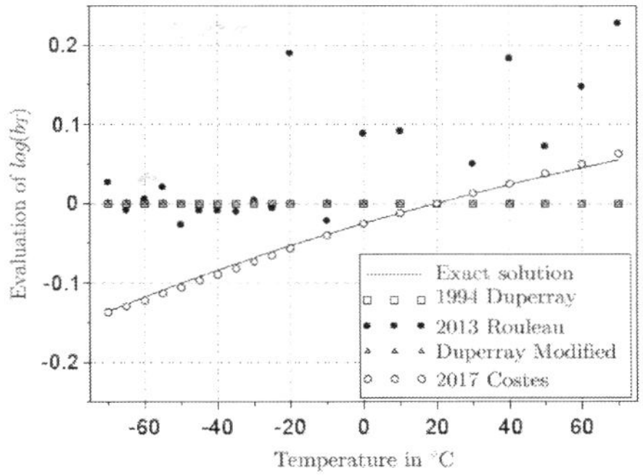

Figure 5. Vertical shift factor b_T obtained with the standard noise level.

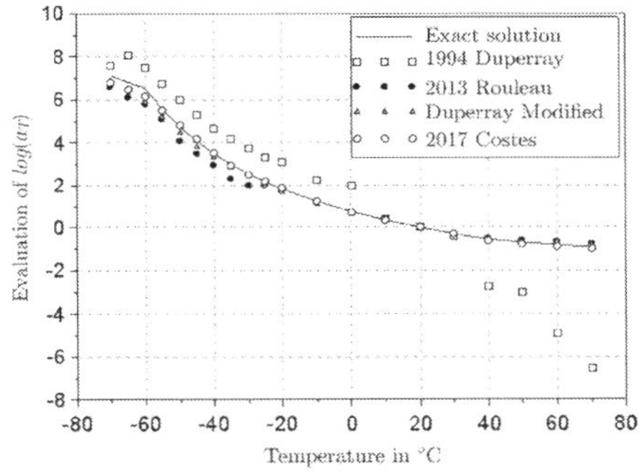

Figure 6. Horizontal shift factor a_T obtained with the standard noise level.

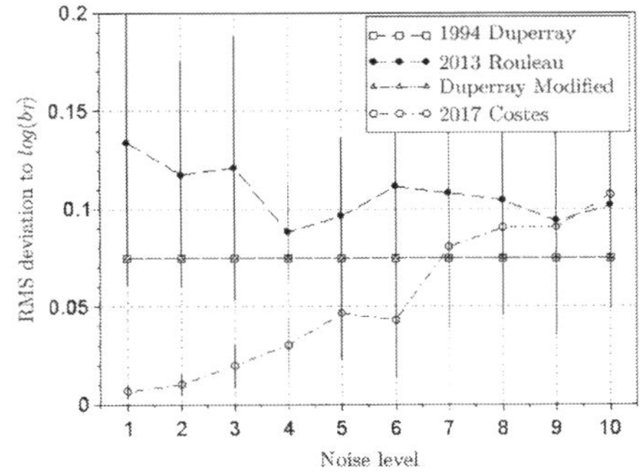

Figure 7. RMS deviation vs. noise level for b_T.

shift. Moreover, the RMS deviation increases with the noise level.

For the horizontal shift factor, RMS deviation issued from the method of Duperray (1994) does not increase with noise level. It is probably because there are more errors due to the polynomial fit than errors related to noise level. RMS deviation for the Rouleau's method remains

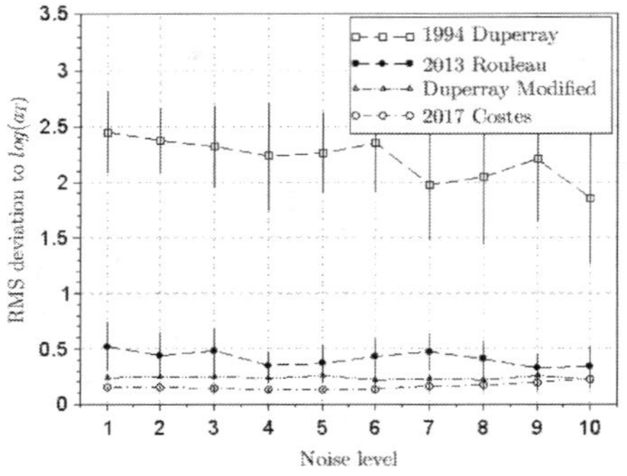

Figure 8. RMS deviation vs. noise level for a_T.

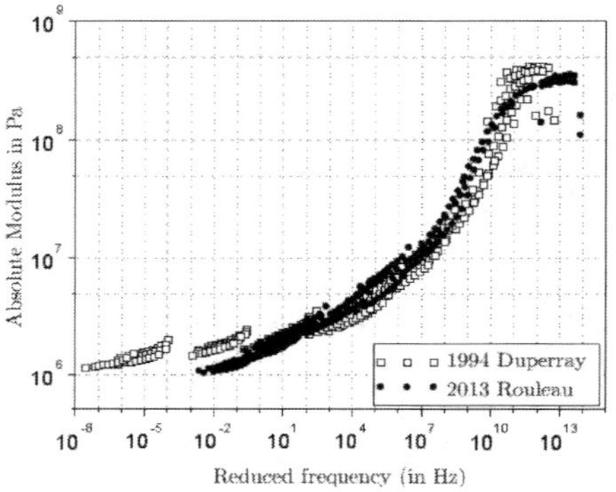

Figure 9. Master curves issued from Duperray and Rouleau methods.

constant (≈0.4 decades) with noise level, maybe because extensions performed in the method do not fit ZFM model and lead to a larger error than the noise itself. Duperray's modified method gives good results for horizontal shift (≈0.25 decades). Similarly than for the Duperray's original method, noise level does not affect the RMS deviation. Finally, our sequenced method gives accurate results. They are better (RMS deviation ≈0.15 decades) than all other methods for low noise levels, and equivalent to the modified Duperray's method for very noisy data.

3.3 *Application to real data*

Finally, methods are applied to experimental data obtained on double lap shear samples. The material is a soft polymer, glass temperature of which being –63 as measured by differential scanning calorimetry.

Figures 9 and 10 present the master curves issued Duperray and Rouleau methods on the one hand, and Duperray modified and our sequential methods on the other hand. All these results con-

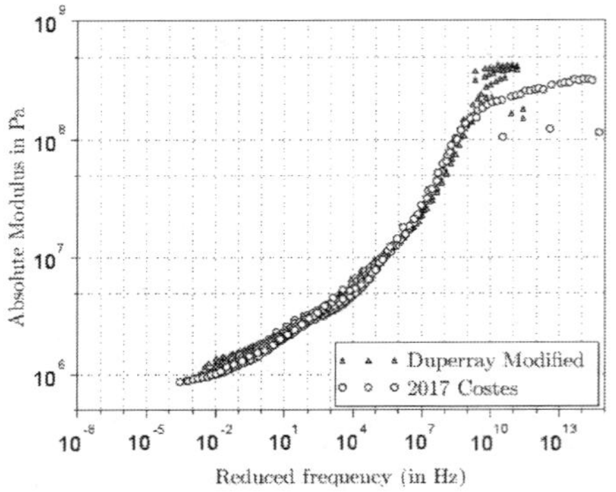

Figure 10. Master curves issued from Duperray modified and our sequential methods.

duct to similar conclusions than those obtained with artificial data.

4 CONCLUDING REMARKS

In this paper, a new sequential method dedicated TTS master curves has been detailed, assessed, and compared with other methods considering both synthetic and real DMA data. It has been shown that our sequential method leads to better results than all other methods examined here. Nevertheless, it presents some limitations. First, it is difficult to predict the noise level involved during a real test. Second, it is difficult to determine whether a material does not fulfill TTS or whether it admits a large vertical shift factor. Indeed, it has been shown that the vertical shift factor cannot be identified with sufficient accuracy in cases with high noise levels.

REFERENCES

Dealy, J. and D. Plazek (2009). Time-temperature superposition—a users guide. *Rheology bulletin*.

Duperray, B. (1994). Automated matser curves generation. *Belgian Plasic and Rubber Institues 25th anniversary International Conference*, 1–18.

Ferry, J. D., W. C. Child, R. Zand, D. M. Stern, M. L. Williams, and R. F. Landel (1957). Dynamic mechanical properties of polyethyl methacrylate. *Journal of colloid science*.

Rouleau, L. (2013). Application of kramerskronig relations to timetemperature superposition for viscoelastic materials. *Mechanics of Materials 65*, 66–75.

Rubinstein, M. and R. H. Colby (2003). *Polymers Physics*. Oxford university press.

Sihn, S. and S. W. Tsai (1999). Automated shift for time-temperature superposition. *Department of Aeronautics and Astronautics, Stanford University*.

Wineman, A. and K. R. Rajagopal (2000). *Mechanical response of polymers*. Press syndicate of the univsersity of Cambridge.

Constitutive Models for Rubber X – Lion & Johlitz (Eds)
© 2017 Taylor & Francis Group, London, ISBN 978-1-138-03001-5

New biaxial test method for the characterization of hyperelastic rubber-like materials

D.C. Pamplona, H.I. Weber & G.R. Sampaio
Department of Mechanical Engineering, PUC-Rio, Rio de Janeiro, Brazil

R. Velloso
Department of Civil Engineering, PUC-Rio, Rio de Janeiro, Brazil

ABSTRACT: The aim of this research is to discuss the characterization of hyperelastic materials by determining suitable parameters to the Yeoh Strain Energy Function (SEF), neglecting the viscoelastic and creep effects. The experimental analysis was conducted using a traction test machine Instron® model 3343. The experimental data from the uniaxial and nonhomogenous biaxial tests performed in membranes were used as an input for Abaqus® FE to obtain the parameters of the SEF in consideration. The curve obtained with the parameters of the uniaxial experiment presented smaller loads for the same vertical displacement biaxial experimental results. Only when the extensions were small was there a small difference between the numerical analysis with the parameters obtained by the biaxial adjustments and biaxial experimental results. We are confident that the proposed test for characterizing biaxially hyperelastic materials is an improvement because it is low-cost and can be performed using almost any existing uniaxial test machine.

1 INTRODUCTION

Hyperelastic materials are used in industries from aerospace to biomedicine. A correct characterization is crucial for the consistent design of structures. One of the most significant characteristics of the material is its ability to undergo large deformations with very small loads and return to its original configuration when the load is removed. Isotropic, hyperelastic materials are described by a Strain Energy Function (SEF), W, usually based on three strain invariants of deformation, I_1, I_2 and I_3, (Dolwichai et al. 2006). SEF represents the energy stored in a material per unit of reference volume (volume in the initial configuration) as a function of the strain in the material (Ali et al. 2010). Usually, the transformation is isochoric, because hyperelastic materials are considered incompressible, i.e., $I_3 = 1$.

In this research, the studies are conducted with hyperplastic membranes. However, there have been a limited number of experimental works, because most of the works to date have been theoretical. The works by Ogden (1972, 1997), Alexander (1971), Pamplona & Bevilacqua (1992) and Pamplona et al. (2001, 2006, 2012, 2014) are among the numerical and experimental investigations in this field. The present research compares the parameters of Yeoh SEF obtained with uniaxial and nonhomogenous biaxial experiments. The uniaxial tests are easy to perform and are well understood, but this can be insufficient; in fact, a second test using multiple directions is necessary to characterize the material. FE software packages such as Abaqus® offer several SEFs to describe the nonlinear behaviour of hyperplastic materials. Yeoh (Renaud et.al. 2009) constitutive model considered here is represented in the equation (1), and has good fit over a large strain range and can simulate various modes of deformation with limited data. This leads to reduced requirements for material testing.

$$W = C_{10}(I_1 - 3) + C_{20}(I_1 - 3)^2 + C_{30}(I_1 - 3)^3 \qquad (1)$$

where Cij are the parameters. However, it is known that the material should present stress softening during the first loading and that a cyclic load should be applied before stress strain measurements (preconditioning). To test the membrane that we were using, five cyclic indentation tests were performed, and the curves load-displacement with cyclic loads showed to be very similar, (Bianchi, 2015); thus, the membranes in this research were not pre-conditioned.

The present article shows a new low cost non homogenous biaxial test that can be done in any uniaxial tension test machine very easily. This study shows that the parameters obtained with the biaxial experimental data to describe the membrane, showed a good match with experimental data. However, the parameters obtained with the uniaxial experimental analysis presented reasonable agreement with the biaxial experimental analysis only for small deformations.

2 EXPERIMENTAL ANALYSIS

This section describes the standard tests performed to input stress-strain data into Abaqus® FE code and evaluate the material parameters. All rubber specimens were chosen from the same batch. Both experimental tests, uniaxial and biaxial, were performed using the traction test machine Instron® model 3343. This equipment has an extension capacity of 150 cm with a load cell of 50 N. The tests were performed at ambient conditions with a displacement rate of 150 mm/min.

2.1 *Material*

The membranes were chosen looking for membranes with dimensional homogeneity of mechanical isotropy coming from a rigorous fabrication process, dentistry membrane, having a thickness of 0.180 mm, which was successfully used previously in our laboratory.

2.2 *Uniaxial test*

To perform the uniaxial tests, rectangular samples of 40 mm of width (L) by 30 mm of height (H) were prepared. The samples were glued to wood sticks with cyanoacrylate glue. This assured a perfect attachment to the grip of the test machine, avoiding any slip. The test was performed at a constant elongation speed until rupture of the membrane. Figure 1 shows the test being performed. The data, relating the load and displacement, are exported to the software Excel, and the strain vs stress relation is obtained. These data are the input to the software Abaqus® and used for the characterization.

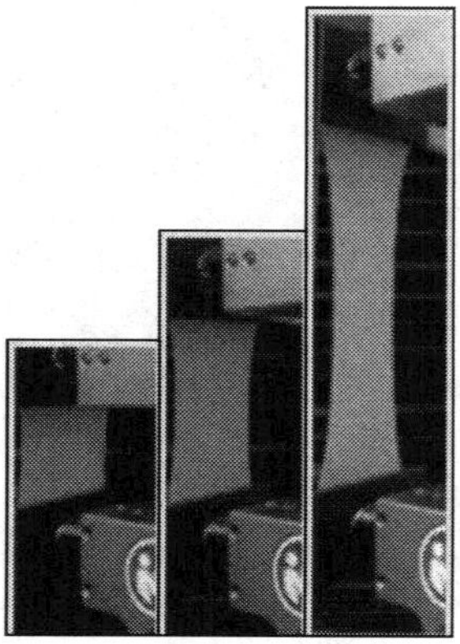

Figure 1. Test to characterize the membrane uniaxially.

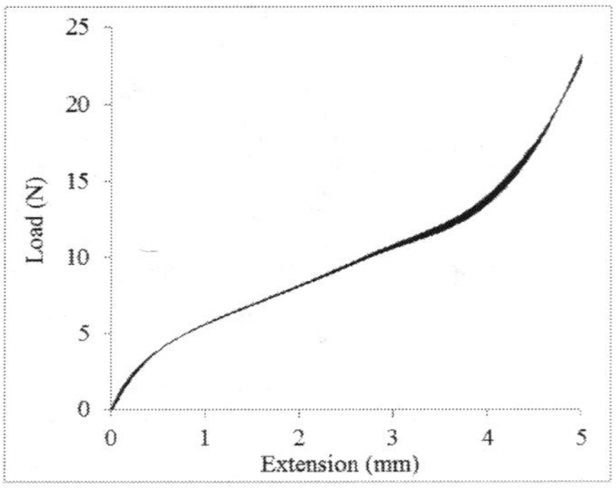

Figure 2. Results of the uniaxial test.

2.2.1 *Results of the uniaxial test*

Four tests were performed to obtain a curve that fits in a trustful way. All the tests were considered to an extension of 150 mm, with a nominal strain of 5.0. Figure 2 shows the obtained stress vs strain curve. The curve showing the behaviour of the membrane to the traction test presented 2 inflection points, one with a nominal strain of 1.0 and the other showing the beginning of stiffening at a nominal strain equal 4.0. Between those two points, the stress x strain curve is almost linear.

2.3 *Biaxial experiment*

The tests were performed using plane annular membranes with internal and external radiuses of 13 mm and 60 mm, respectively. The extension test was performed using the Instron® machine. To conduct the tests, an acrylic structure with height of 250 mm, base of 167 mm and thickness of 12 mm was built and attached to the base of the machine (Fig. 3(a)).

The rubber membranes were glued to the annular acrylic ring with internal and external radiuses of 60 mm and 75 mm, respectively. Eight screws and bolds attach the ring to the structure. In the centre of both sides of the membrane, two circular metal plates of radius 7.5 mm are glued to the membrane (Fig. 3(b)). The test consists of displacing the centre of the membrane vertically using an aluminum cylindrical rod (Fig. 4).

2.3.1 *Results of the biaxial experiment*

Four tests were performed to obtain a curve that fits in a trustful way the results obtained. The results of the tests are shown in Figure 5. All the tests were considered to an extension of 120 mm. The beginning of the curve is smooth, showing large deformation to small loads until a vertical displacement of 30 mm; afterwards the curve turns to a more linear relation between load and displacement.

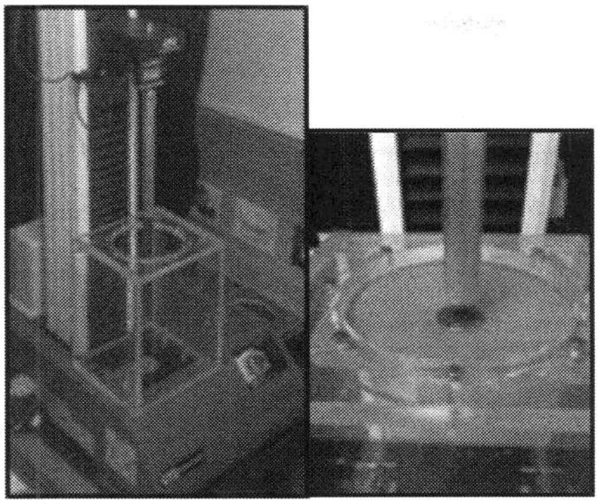

(a) (b)

Figure 3. (a) Test machine with the acrylic structure attached (b) annular membrane prepared to begin the test.

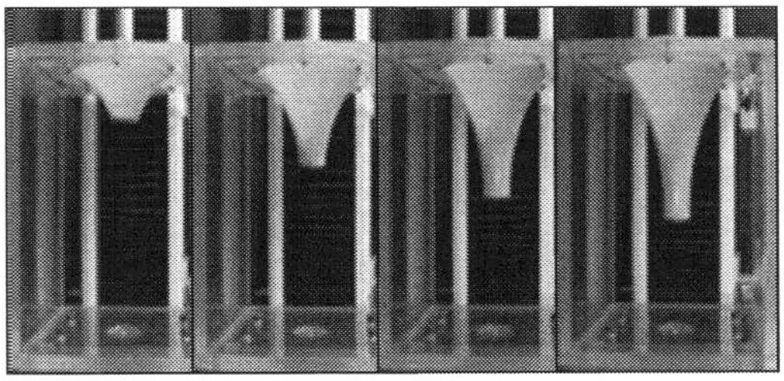

Figure 4. Biaxial test: vertical displacement of the annular membrane.

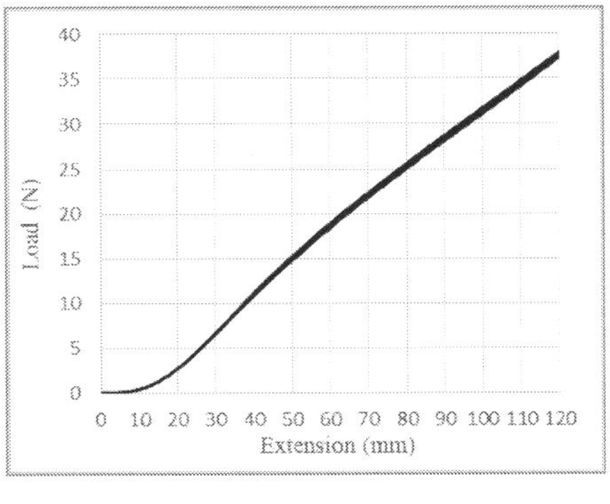

Figure 5. Results of the biaxial test.

3 NUMERICAL ANALYSIS

The numerical analysis uses the finite element programme Abaqus® to characterize the material of the membrane used in the experimental tests, the test data are specified as nominal stress–nominal strain data pairs for uniaxial and biaxial tests to determine the parameters.

3.1 *Uniaxial characterization*

The data obtained in the experimental uniaxial test (Fig. 2) were used as input stress-strain data in Abaqus® FE. Figure 6 show the adjustments with the experimental data. The parameters of the material model from the uniaxial test, $C_{10} = 0.25062$ (MPa), $C_{20} = -3.62088\text{e-}3$ (MPa) and $C_{30} = 8.21\text{e-}5$ (MPa).

3.2 *Biaxial characterization*

To obtain the parameters compatible with the biaxial experimental data, the data obtained experimentally (Fig. 5) were used as input load-vertical displacement data in Abaqus®. Simulations of the biaxial test were carried out using 56 quadratic axisymmetric membrane elements (MAX2), adjusted with the pattern search method. The parameters of the material model from the uniaxial test, $C_{10} = 0.26645$ (MPa), $C_{20} = -3.94674\text{e-}3$ (MPa) and $C_{30} = 7.41889\text{e-}5$ (MPa).

3.3 *Comparison of the two characterizations*

Figure 7 shows the simulation of the experimental biaxial test with the parameters adjusted by the uniaxial and biaxial tests. The curves obtained with the parameters of the uniaxial test are far from the experimental curve presenting smaller load for the same vertical displacement. For the curve obtained with the parameters of the biaxial test, FE simulations show that the behaviour is in good agreement with the experimental data. Only when the extensions are between 15 and 50 mm there is a small difference between the numerical analysis with the parameters obtained by the biaxial tests and the experimental results.

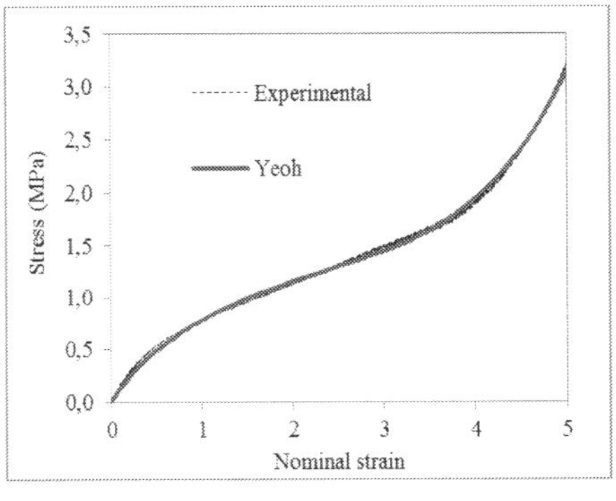

Figure 6. Adjustments of the uniaxial experimental data.

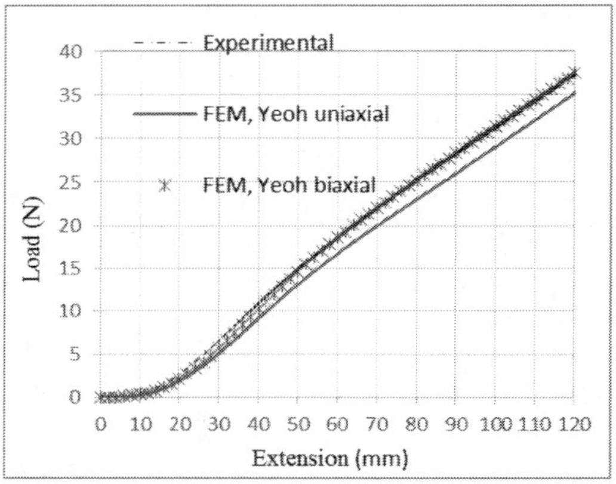

Figure 7. Adjustments of the biaxial experimental data.

4 CONCLUSIONS

The biaxial test is performed using plane annular membranes and consists of displacing the rigid centre of a membrane vertically using an aluminum cylindrical rod. Uniaxial tests were also performed with the membranes of the same sample of materials, and the results from both tests were compared. The experimental data from the tests were used as an input for Abaqus® FE to obtain the parameters of the Yeoh SEF. Analyzing the stress x strain curves obtained by the experimental uniaxial tests, one may observe three different behaviours. In the beginning of the uniaxial test, one can observe a material stiffening region. After this point, the curves become almost linear until reaching a second inflection point and strong stiffening. For the curve with the biaxial test, the beginning of the deformation presents a smooth increase, which becomes almost linear afterwards. The FEM modelling for the biaxial test was done using axisymmetric membrane elements MAX1, adjusted with the pattern search method; the data obtained experimentally were used as input load-vertical displacement data in Abaqus® FE. A comparison of the characterizations, uniaxial and biaxial, shows that the biaxial parameters fit very well the biaxial experiment. The results with the uniaxial parameters used to model the biaxial test show that the curve presented smaller loads for the same vertical displacement. When the extensions were between 15 and 50 mm there was a small difference between the numerical analysis with the parameters obtained by the biaxial tests and biaxial experimental results. Thus, we are confident that this analysis comparing the two characterizations and the different results obtained when the membrane is characterized uniaxially will give the researcher a good idea of the inaccuracy in his results.

ACKNOWLEDGEMENTS

The financial support from the Brazilian Council for Scientific and Technological Research (CNPq; grant number 301832/2009-9) were essential for this research.

REFERENCES

Alexander H, Tensile instability of initially spherical balloons, International Journal of Engineering Sciences 1971;9:151–162.

Ali A, Hosseini M and Sahari B. A review of constitutive models for rubber-like materials. American Journal of Engineering and Applied Sciences. 2010; 3(1):232–239. http://dx.doi.org/10.3844/ ajeassp.2010.232.239.

Bianchi R.S. Análise experimental da endentação de membranas hiperelásticas com e sem atrito 2015, Relatório de Estudo Orientado. Department of Mechanical Engineering PUC-Rio.

Dolwichai P, Limtragool J, Inban S and Piyasin S. In: Hyperelastic Material Models for Finite Element Analysis with Commercial Rubber, Technology and Innovation for Sustainable Development Conference (TISD 2006); 2006.

Khon Kaen, Thailand. Khon Kaen: Khon Kaen University; 2006. p.769–774.

Ogden RW, Non-Linear Elastic Deformations, fourth ed., Dover Publications, Mineola, NY, 1997.

Ogden RW, Large deformation isotropic elasticity-on the correlation of theory and experiment for incompressible rubberlike solids, Proc R Soc London A326 (1972) 565–584.

Pamplona D, Bevilacqua L, Large deformations under axial force and moment load of initially flat membranes. International Journal of Non-Linear Mechanics 1992;27: 639–650.

Pamplona D, Gonçalves PB, Davidovich M, Weber HI., Finite axisymmetric deformations of an initially stressed fluid-filled cylindrical membrane., International Journal of Solids and Structures 2001;38:2033–2047.

Pamplona D, Gonçalves PB, Lopes SRX. Finite deformations of cylindrical membrane under internal pressure. International Journal of Mechanical Sciences 2006;48(6):683–696.

Pamplona DC, Mota DES. Numerical and experimental analysis of inflating a circular hyperelastic membrane over a rigid and elastic foundation. International Journal of Mechanical Sciences 2012;65:18–23.

Pamplona DC, Weber HI, Sampaio GR. Analytical, numerical and experimental analysis of continuous indentation of a flat hyperelastic circular membrane by a rigid cylindrical indenter. International Journal of Mechanical Sciences 2014; 87:18–25.

Renaud C, Cros JM, Feng ZQ and Yang B. The Yeoh model applied to the modeling of large deformation contact/impact problems. International Journal of Impact Engineering. 2009; 36(5):659–666.

Constitutive Models for Rubber X – Lion & Johlitz (Eds)
© 2017 Taylor & Francis Group, London, ISBN 978-1-138-03001-5

Thermomechanical analysis of energy dissipation in natural rubber

J.-B. Le Cam
Institute of Physics, UMR 6251 - CNRS/University of Rennes 1, Campus de Beaulieu, Rennes, France
LC-DRIME: Imaging, Mechanics and Elastomers Laboratory—Cooper Standard/Institute of Physics, Campus de Beaulieu, Rennes Cedex, France

ABSTRACT: This paper presents an energetic analysis of the mechanical properties of crystallizable Natural Rubber (NR). For that purpose, mechanical tests have been performed with an unfilled NR at increasing maximum stretches. Temperature variations due to both the material deformation and the effects of heat diffusion were measured by using an infrared camera at the surface of specimens under quasi-static and uni-axial loading at ambient temperature. The heat sources produced and absorbed by the material during the deformation cycles are deduced from the temperature variations by using the heat diffusion equation. Energetic balance carried out has shown that no intrinsic dissipation is produced while energy dissipation is observed in the strain-stress relationship [1]. The question is therefore where does mechanical energy go? This paper answers this question. The experiments carried out, the theoretical framework used to perform energy balance and the identification of heat source are first presented in details. Then, energy balances accounting for both the non-entropic and entropic contributions to NR elasticity enable us to clearly identify the physical origin of the mechanical energy dissipated.

1 INTRODUCTION

Mechanical properties of Natural Rubber (NR), cis-1,4-polyisoprene, are mainly related to Strain-Induced Crystallization (SIC) (Yijing, Zhouzhou, Ya, Tongfan, and Zhiping 2017, Huneau 2011). Especially, SIC enables NR to better resist the crack growth (Lee and Donovan 1987, Lake 1995) and the fatigue (Cadwell, Merril, Sloman, and Yost 1940, Beatty 1964, Saintier 2000, Le Cam, Huneau, and Verron 2013). SIC in NR has been widely investigated by means of X-ray diffraction technique since the pioneer work by Katz (1925). A number of studies were dedicated to the crystallographic structure of NR (see for instance (Bunn 1942) and (Takahashi and Kumano 2004) for the monoclinic structure of crystal, and (Immirzi, Tedesco, Monaco, and Tonelli 2005) and (Rajkumar, Squire, and Arnott 2006) for the debate on the existence of orthorhombic cell). Several studies investigated the relationship between crystallinity and strain (Goppel and Arlman 1949, Toki, Fujimaki, and Okuyama 2000, Trabelsi, Albouy, and Rault 2003a). They have highlighted that during a deformation cycle, at a given strain, the crystallinity is higher during unloading than during loading. This was attributed to a supercooling effect by Trabelsi, Albouy, and Rault (2003a). But numerous other phenomena accompanying SIC are still misunderstood, for instance the opposite effects of SIC on stress, described by Flory (1947): on the one hand, SIC acts as network points, which increases the network chains density and leads to stress hardening; on the other hand, SIC induces stress relaxation of the amorphous phase (a plateau is observed in the strain-stress curve once SIC starts (Trabelsi, Albouy, and Rault 2003a, Toki, Fujimaki, and Okuyama 2000)). Moreover, as reported by Albouy and coworkers, SIC develops at constant strain of the amorphous chains (Albouy, Guillier, Petermann, Vieyres, Sanseau, and Sotta 2012, Albouy, Vieyres, Perez-Aparicio, Sanseau, and Sotta 2014). Clark, Kabler, Blaker, and Ball (1940) were the first who suggested experimentally that the mechanical hysteresis is closely related to SIC. Up to date, the idea that mechanical hysteresis is due to the kinetics difference between crystallization and crystallite melting has been widely disseminated, but no further information on the origin of this mechanical dissipation has been provided. The formation of a mechanical hysteresis is a complicated process and better understand the energetic nature of the hysteresis loop could provide information of importance on the reinforcement of NR. This is the aim of the present paper.

2 INTRODUCING ENERGIES AND CORRESPONDING POWERS

2.1 *Strain energy and mechanical energy dissipated*

The strain energy is the energy brought mechanically to deform the material. During the loading (unloading), the strain energy is denoted W_{def}^{load} (W_{def}^{unload}), it corresponds to the area under the load (unload) curve. It is calculated as follows:

$$W_{def}^{load} = \int_{\lambda_{\min}}^{\lambda_{\max}} \pi d\lambda \quad and \quad W_{def}^{unload} = \int_{\lambda_{\max}}^{\lambda_{\min}} \pi d\lambda \tag{1}$$

λ is the stretch defined as the ratio between current and initial lengths. π is the nominal stress tensor, defined as the force per unit of initial (undeformed) surface. The corresponding power is the rate of the strain energy or the strain power P_{strain} and is defined as:

$$P_{strain} = \pi\dot{\lambda} \tag{2}$$

For certain materials, a part of the mechanical energy brought is dissipated. The load and unload curves do no longer superimposed and a hysteresis loop forms. The corresponding energy dissipated is W_{hyst} over one cycle is determined as follows:

$$W_{hyst} = W_{def}^{load} - W_{def}^{unload} \tag{3}$$

The corresponding power P_{hyst} is obtained by dividing W_{hyst} by the cycle duration.

2.2 *Heat power density or heat source*

During the mechanical cycle, the material produces and absorbs heat. Under non adiabatic test conditions, corresponding temperature variations are influenced by heat diffusion effects. Determining the heat source from temperature measurements requires therefore to take into account these effects. For that purpose, the framework of the thermodynamics of irreversible processes and the heat diffusion equation is used. In the case of homogeneous heat power density, the heat diffusion equation writes (Chrysochoos 1995):

$$\rho C\left(\dot{\theta} + \frac{\theta}{\tau}\right) = s = s_{tmc} + d_1 \tag{4}$$

where ρ is the density, C is the specific heat. θ is the temperature variation, τ is a time constant that is identified from a natural return to ambient temperature. The right-hand side of Equation 4 represents the heat power density s or (heat source) produced or absorbed by the material itself. It can be divided into two terms that differ in nature:

- the intrinsic dissipation d_1: this positive quantity corresponds to the heat production due to mechanical irreversibilities during deformation process, for instance viscosity or damage;
- the thermomechanical couplings s_{tmc}: they correspond to the couplings between the temperature and the other state variables. In rubbers, the coupling between temperature and strain is due to entropic coupling s_{ent} (Samaca Martinez, Le Cam, Balandraud, Toussaint, and Caillard 2013b) and/or isentropic-type thermoelastic coupling s_{isent}, which is driven by the change in the internal energy.

In unfilled natural rubber, volume changes are small (Le Cam and Toussaint 2008, Chenal, Gauthier, Chazeau, Guy, and Bomal 2007) and the specific heat does not vary significantly, even when crystallization occurs (Vogt 1937, Boissonnas 1939, Mayor and Boissonnas 1948), the product ρC can be therefore assumed to be constant and the heat sources can be calculated in $W.m^{-3}$. The heat source due to thermomechanical couplings s_{tmc} being reversible, the temporal integration of this equation over one mechanical cycle provides the energy due to intrinsic dissipation $W_{intrinsic}$.

$$W_{intrinsic} = \int_{cycle} (s_{tmc} + d_1)dt = \int_{cycle} d_1 dt \tag{5}$$

The corresponding intrinsic dissipation (d_1) over one cycle is obtained by dividing $W_{intrinsic}$ by the cycle duration. Finally, the energy corresponding to heat diffusion effects $W_{thermal}$ is defined by:

$$W_{thermal} = \int_{cycle} \rho C\left(\frac{\theta}{\tau}\right)dt \tag{6}$$

These energies and corresponding powers were determined from mechanical tests and thermal measurements.

2.3 *Energy balances*

Energy balances require to be carried out at two different scales, from one mechanical cycle to another, and within each mechanical cycle. This is more precisely detailed in the oral presentation.

3 EXPERIMENTAL SECTION

The material studied is an unfilled natural rubber (NR) supplied by the "Manufacture Française des pneumatiques Michelin". Its chemical composition is given in Table 1. The compound was

Table 1. Chemical composition in parts per hundred rubber (phr).

Ingredient	Quantity
Natural rubber NR	100
Carbon black	0
Antioxidant 6PPD	1.9
Stearic acid	2
Zinc oxide ZnO	2.5
Accelerator CBS	1.6
Sulfur solution 2H	1.6

cured for 22 *min* at 150°*C*. The degree of crosslinking is 6.5 10^{-5} mol.cm^{-3}. In such NR formulation, SIC is observed in uniaxial tension starting from a stretch λ_c of about 4. During unloading, crystallite melting is complete at a lower stretch λ_m of about 3 (Toki, Fujimaki, and Okuyama 2000, Trabelsi, Albouy, and Rault 2003b, Le Cam and Toussaint 2008). The specimen geometry is 5 *mm* wide, 10 *mm* high and 1.4 *mm* thick.[1] Mechanical loading was applied using a 50*N* Instron 5543 testing machine. It corresponds to four sets of three cycles, for four increasing maximum stretch ratios: $\lambda_1 = 2$, $\lambda_2 = 5$, $\lambda_3 = 6$ and $\lambda_4 = 7.5$. λ_1 was chosen inferior to λ_c, λ_2 was superior but close to λ_c, λ_3 and λ_4 are superior to λ_c (λ_4 is close to the stretch at failure). The signal shape chosen was triangular, to ensure a constant strain rate during loading and unloading. Tests were performed at a loading rate equal to ± 100 *mm/min*. The temperature was measured using a Cedip Jade III-MWIR infrared camera (320 × 240 *pixels*, 3.5–5 *μm*).

4 RESULTS AND DISCUSSIONS

4.1 *Mechanical and thermal responses*

Figure 1 presents the mechanical response obtained in terms of the nominal stress in relation to the stretch. This figures shows that whatever the maximum stretch applied, the mechanical cycles have no significant effect on the mechanical response, in the sense that no stress softening is observed. For cycles at $\lambda_1 = 2 < \lambda_c$, a small hysteresis loop

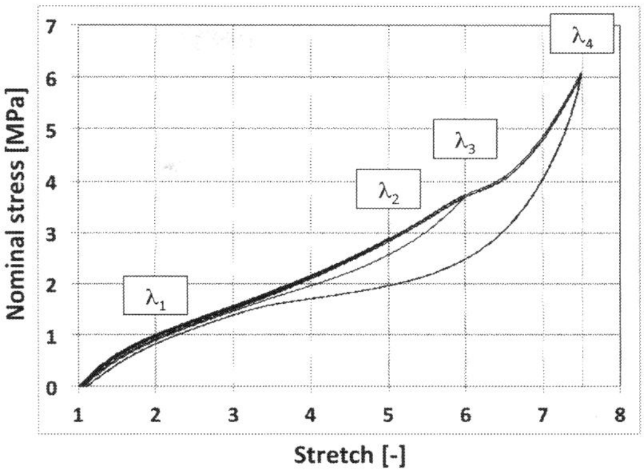

Figure 1. Mechanical response.

is observed. The corresponding power density P_{hyst} is equal to 1.6 10^3 W/m^3. For cycles at $\lambda_2 = 5 > \lambda_c$, the area of the hysteresis loop increases ($P_{hyst} = 4.8$ 10^3 W/m^3). It should be recalled that the maximum stretch applied exceeds that at which crystallization starts ($\lambda_c = 4$). These results are in good agreement with those reported in the literature for unfilled natural rubber studied with X-ray diffraction: a hysteresis loop forms if SIC takes place in the material. Classically, the hysteresis loop is assumed to be due to the difference in the kinetics of crystallization and crystallite melting (Toki, Fujimaki, and Okuyama 2000, Trabelsi, Albouy, and Rault 2003b, Le Cam and Toussaint 2008), without further explanations on the mechanism leading to the mechanical energy lost. For cycles at $\lambda_3 = 6 > \lambda_c$, the hysteresis loop is higher than for the previous stretches. For cycles at $\lambda_4 = 7.5 > \lambda_c$, a plateau is observed from $\lambda = 6$ on, followed by a high stress increase. The hysteresis loop is much higher than for the previous maximum stretch applied (4.5 10^4 W/m^3 versus 1.3 10^4 W/m^3). For higher stretches, a high increase in the nominal stress is observed: crystallites act as fillers and strongly reinforce the material stiffness. As assumed by (Flory 1947) and highlighted by (Toki, Fujimaki, and Okuyama 2000) and (Trabelsi, Albouy, and Rault 2003b), once crystallization occurs, relaxation is induced in the amorphous phase. The plateau observed is a manifestation of this relaxation.

The thermal responses obtained during the mechanical cycles are depicted in Figure 2. Temperature variations are mainly due to entropic elasticity. Thermal accommodation is reached at the third cycle. When SIC starts, a significant additional heat production is observed. It is highlighted by the dotted line (in red) plotted for the third (stabilized) cycles for $\lambda > \lambda_c$. The effect of the thermal dissipation on the hysteresis loop area was found negligible and will be discussed in the oral presentation.

[1] In this paper, the mechanical tests performed were those reported reported in Samaca Martinez's PhD, which investigated the thermal and caloric signature of the main phenomena involved in rubber deformation (Samaca Martinez, Le Cam, Balandraud, Toussaint, and Caillard 2013a, Samaca Martinez, Le Cam, Balandraud, Toussaint, and Caillard 2013b, Samaca Martinez, Le Cam, Balandraud, Toussaint, and Caillard 2013c, Samaca Martinez, Le Cam, Balandraud, Toussaint, and Caillard 2014, Le Cam, Samaca Martinez, Balandraud, Toussaint, and Caillard 2015).

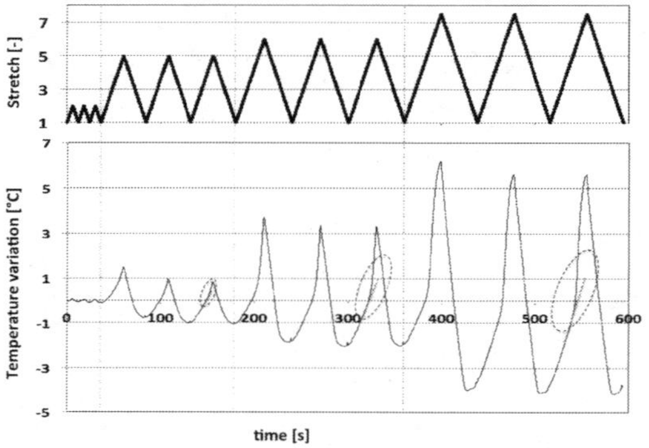

Figure 2. Thermal response.

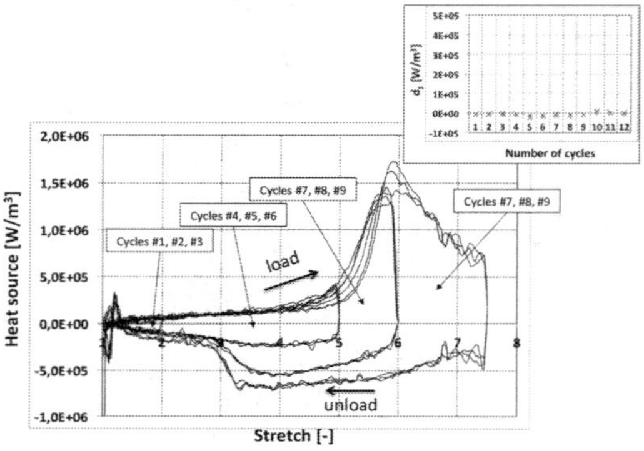

Figure 3. Calorific response.

4.1.1 *Calorimetric response*

Figure 3 depicts the calorimetric response of NR in relation to the stretch for a loading rate equal to ± 100 *mm/min*. Several comments can be drawn from this figure. First, the heat source is positive during loading and increases with the stretch. During unloading, the heat source is negative. This is due to preponderant entropic effects. Before SIC starts, the load-unload curves are symmetrical, meaning that the heat produced during the loading phase is equal to the heat source absorbed during the unloading phase. Thus, no intrinsic dissipation is detected.

Second, once SIC starts, the heat source evolution for loading and unloading are no longer symmetrical. This cannot be explained by entropic elasticity. During loading, the heat source evolves in a quasi-linear manner until a stretch close to 4 is reached. Dissymmetry occurs for stretches higher than 4, the stretch level at which SIC starts. During unloading, the heat source rate first increases in absolute value until reaching a stretch equal to 4, is constant until a stretch equal to 3.5 is reached, and then decreases. Moreover, corresponding thermal measurements highlight that temperature is higher during unloading than during loading (see Figure 3 in the present paper or Fig. 12(c) in (Samaca Martinez, Le Cam, Balandraud, Toussaint, and Caillard 2013b) where temperature variation is given in relation to stretch). This demonstrates that crystallization, which is exothermal, continues during unloading. For each cycle, the energy corresponding to the intrinsic dissipation $W_{intrinsic}$ has been calculated by integrating the heat source. Intrinsic dissipation d_1 is obtained by dividing $W_{intrinsic}$ by the duration of the cycle considered. The values obtained, which are reported in the diagram in Figure 3, show that no intrinsic dissipation was detected whatever the mechanical cycle considered. This is consistent with the fact that no self-heating was observed. Consequently, the only explanation for the dissymmetry is the difference in the kinetics of crystallization and crystallite melting. This is in good agreement with studies reported in the literature (Toki, Fujimaki, and Okuyama 2000, Trabelsi, Albouy, and Rault 2003b, Le Cam and Toussaint 2008). Third, it is observed that the area of the hysteresis loop increases with the maximum stretch reached, meaning that the higher the crystallinity, the higher the area (mechanical energy) of the hysteresis loop. For cycles at $\lambda > 6$, the evolution of the heat source differs from that obtained for lower stretches. Indeed, during the loading phase, instead of increasing continuously, the heat source decreases from $\lambda = 6$ on. This means that either heat due to crystallization continues to be produced (it remains positive), but at a lower rate, and/or that larger energetic effects take place (negative heat sources for positive stretch rates). The fact that the hysteresis loop area is larger than before (4.5 10^4 W/m^3 versus 1.3 10^4 W/m^3) pleads in favor of SIC.

4.1.2 *Where does mechanical energy go?*

The oral presentation will explain where does energy goes by linking cyclic internal energy changes and the mechanical energy dissipation. The physical origin of the internal energy changes will be discussed.

5 CONCLUSION

In this paper, calorimetric analyses shown that cyclic deformation of unfilled natural rubber does not produce heat, *i.e.* the heat absorbed during unloading is the same as that produced during loading. This is true whatever the material crystallizes or not. The crystallite melting absorbs the entire heat produced by SIC, meaning that natural rubber is not viscous even though its mechanical response exhibits a hysteresis loop. Energy balances performed at the scale on the mechanical cycles and within each mechanical cycle highlight the link between internal energy changes and the apparent mechanical energy dissipation. Finally, the demonstration that NR is able to dissipate energy brought mechanically without converting it into heat is a realistic way to explain its extraordinary resistance to fatigue and to crack growth.

ACKNOWLEDGEMENTS

Author acknowledges the "Manufacture Française des pneumatiques Michelin" for supporting this study. Author thanks J.R. Samaca Martinez, most of the results presented in the paper were obtained during his PhD Thesis. X. Balandraud, E. Toussaint, J. Caillard and D. Berghezan are acknowledged for the fruitful discussions.

REFERENCES

Albouy, P.-A., G. Guillier, D. Petermann, A. Vieyres, O. Sanseau, and P. Sotta (2012). A stroboscopic X-ray apparatus for the study of the kinetics of strain-induced crystallization in natural rubber. *Polymer 53*(15), 3313–3324.

Albouy, P.-A., A. Vieyres, R. Perez-Aparicio, O. Sanseau, and P. Sotta (2014). The impact of strain-induced crystallization on strain during mechanical cycling of cross-linked natural rubber. *Polymer 55*(16, SI), 4022–4031.

Beatty, J.R. (1964). Fatigue of rubber. *Rubber Chemistry and Technology 37*, 1341–1364.

Boissonnas, C. (1939). *Ind. Eng. Chem. 31*, 761.

Bunn, C. (1942). *Proc. R. Soc. London, Ser. 1 180*, 40.

Cadwell, S.M., R.A. Merril, C.M. Sloman, and F.L. Yost (1940). Dynamic fatigue life of rubber. *Industrial and Engineering Chemistry (reprinted in Rubber Chem. and Tech. 1940;13:304–315) 12*, 19–23.

Chenal, J.-M., C. Gauthier, L. Chazeau, L. Guy, and Y. Bomal (2007). Parameters governing strain induced crystallization in filled natural rubber. *Polymer 48*, 6893–6901.

Chrysochoos, A. (1995). Analyse du comportement des matériaux par thermographie infra rouge. In *Colloque Photomécanique*, Volume 95, pp. 201–211.

Clark, G.L., M. Kabler, E. Blaker, and J.M. Ball (1940). Hysteresis in crystallization of stretched vulcanized rubber from x-ray data. *Industrial and Engineering Chemistry 32*, 1474–1477.

Flory, P.J. (1947). Thermodynamics of crystallization in high polymers. i. crystallization induced by stretching. *The Journal of Chemical Physics 15*, 397–408.

Goppel, J. and J. Arlman (1949). *Appl. Sci. Res. Sect. A 1*, 462.

Huneau, B. (2011). Strain-induced crystallization of natural rubber: a review of X-ray diffraction investigations. *Rubber Chemistry and Technology 84*(3), 425–452.

Immirzi, A., C. Tedesco, G. Monaco, and A. Tonelli (2005). *Macromolecules 38*, 1223.

Katz, J. (1925). *Naturw 4*, 169.

Lake, G.J. (1995). Fatigue and fracture of elastomers. *Rubber Chemistry and Technology 68*, 435–460.

Le Cam, J.-B., B. Huneau, and E. Verron (2013). Fatigue damage in carbon black filled natural rubber under uni- and multiaxial loading conditions. *International Journal of Fatigue 52*, 82–94.

Le Cam, J.B., J.R. Samaca Martinez, X. Balandraud, E. Toussaint, and J. Caillard (2015). Thermomechanical Analysis of the Singular Behavior of Rubber: Entropic Elasticity, Reinforcement by Fillers, Strain-Induced Crystallization and the Mullins Effect. *Experimental Mechanics 55*, 771–782.

Le Cam, J.-B. and E. Toussaint (2008). Volume variation in stretched natural rubber: competition between cavitation and stress-induced crystallization. *Macromolecules 41*, 7579–7583.

Lee, D.J. and J.A. Donovan (1987). Microstructural changes in the crack tip region of carbon-black-filled natural rubber. *Rubber Chemistry and Technology 60*, 910–923.

Mayor, A.R. and C.-G. Boissonnas (1948). Variation de la chaleur spécifique du caoutchouc en fonction de l'allongement. *Helvetica Chimica Acta 31* (6), 1514–1532.

Rajkumar, G., J. Squire, and S. Arnott (2006). *Macromolecules 39*, 7004.

Saintier, N. (2000). *Prévisions de la durée de vie en fatigue du NR, sous chargement multiaxial*. Thèse de doctorat, École Nationale Supérieure des Mines de Paris.

Samaca Martinez, J.R., J.-B. Le Cam, X. Balandraud, E. Toussaint, and J. Caillard (2013a). Filler effects on the thermomechanical response of stretched rubbers. *Polymer testing 32*, 835–841.

Samaca Martinez, J.R., J.-B. Le Cam, X. Balandraud, E. Toussaint, and J. Caillard (2013b). Thermal and calorimetric effects accompanying the deformation of natural rubber. part 1: Thermal characterization. *Polymer 54*, 2717–2726.

Samaca Martinez, J.R., J.-B. Le Cam, X. Balandraud, E. Toussaint, and J. Caillard (2013c). Thermal and calorimetric effects accompanying the deformation of natural rubber. part 2: quantitative calorimetric analysis. *Polymer 54*, 2727–2736.

Samaca Martinez, J.R., J.-B. Le Cam, X. Balandraud, E. Toussaint, and J. Caillard (2014). New elements concerning the Mullins effect: A thermomechanical analysis. *European Polymer Journal 55*, 98–107.

Takahashi, Y. and T. Kumano (2004). *Macromolecules 37*, 4860.

Toki, S., T. Fujimaki, and M. Okuyama (2000). Strain-induced crystallization of natural rubber as detected real-time by wide-angle x-ray diffraction technique. *Polymer 41*, 5423–5429.

Trabelsi, S., P. Albouy, and J. Rault (2003a). Crystallization and melting processes in vulcanized stretched natural rubber. *Macromolecules 36* (20), 7624–7639.

Trabelsi, S., P.-A. Albouy, and J. Rault (2003b). Effective local deformation in stretched filled rubber. *Macromolecules 36*, 9093–9099.

Vogt, W. (1937). *The Chemistry and Technology of Rubber*. Reinhold Publ. N.-Y.

Yijing, N., G. Zhouzhou, W. Ya, H. Tongfan, and Z. Zhiping (2017). Features of strain-induced crystallization of natural rubber revealed by experiments and simulations. *Polymer Journal 49*(3), 309–317.

Constitutive Models for Rubber X – Lion & Johlitz (Eds)
© 2017 Taylor & Francis Group, London, ISBN 978-1-138-03001-5

Investigation of crosslinking kinetics of silicone rubber/POSS nanocomposites

İ. Karaağaç
İzocam San. ve Tic. A.Ş., Kocaeli, Turkey

G. Özkoç & B. Karaağaç
Department of Chemical Engineering, Kocaeli University, Kocaeli, Turkey

ABSTRACT: This study focus on investigation peroxide crosslinking kinetics of silicone rubber/Poly Oligomeric Silsesquioxane (POSS) nanocomposites by using a nonlinear phenomenological model and also evaluating performance of the model on peroxide vulcanization of silicone. In composite materials, silicone is needed to adhere reinforcing fibers in order to maintain internal pressures. POSSs can be used to improve the adhesion between silicone and common reinforcing fibers such as Kevlar and Rayon through the polar and/or reactive groups which are able to join the chemical structure during crosslinking process. In this study, Octavinyl-POSS (O-POSS), Methacryl-POSS (M-POSS), and octamaleamic acid-POSS (OM-POSS) are incorporated to a reference silicone compound. Effects of POSS type and ratio on the vulcanization kinetics are studied. It is revealed that this model can be successfully used for the nanocomposites. The kinetic parameters are found to be consistent with proposed reaction mechanisms and with rheological variations due to POSS incorporation.

1 INTRODUCTION

Rubber is an essential engineering material in our daily life with its excellent durability and satisfactory properties for both static and dynamic applications. Increasing demand for rubber based materials providing various high performance requirements such as high temperature resistance, low gas permeability, resistance to harch solvents and other chemicals imposed academic and industrial domain to produce different kinds of synthetic rubber. Silicone rubber falls into hybrid synthetic rubbers, which have both organic and inorganic chain structure. It is famous with its high temperature performance as well as good dielectric properties and biologically inert pattern. However, silicone exhibits poor adhesion characteristics with the other components in composite materials. For instance, in turbo charger hose construction, it is not possible to have a strong interface between silicone reinforcing fibers for obtaining satisfactory mechanical properties. Polyhedral oligomeric silsesquioxanes (POSSs), which have unsaturated sites to join the crosslinks or polar and/or reactive groups to physically and/or chemically bond to common reinforcing fibers (Kevlar and Rayon), can be used for silicone-fiber adhesion. POSSs are new generation nanofillers, providing flexible physical and chemical properties and relatively low cost; they are distinguished from other common nanofillers such as organoclays, nanosilica, and nanotubes. They are used for improving mechanical, thermal, and flame retardancy properties of polymers by copolymerization, grafting, and also by blending.

There are lots of mathematical models for evaluating vulcanization kinetics. These can be classified in two major topics; mechanistic and empirical (phenomenological). In mechanistic approach, amount of consumed and produced species during vulcanization reaction is a tool for mathematical models whereas vulcanization experimental data is fitted to nonlinear kinetic models in empirical approach. nth order vulcanization kinetics approach (Isayev and Deng model) is the most common kinetic model for vulcanization and given with Equation 1. It correlates total vulcanization reaction rate with isothermal reaction temperature and instant vulcanization time, in the curing region of the cure curve (between minimum and maximum torques) (Choi & Kim, 2008).

$$\alpha = \frac{\left[k\left(t-t_0\right)\right]^n}{1+\left[k\left(t-t_0\right)\right]^n} \tag{1}$$

where α = state of cure, k = reaction rate constant (min^{-1}), t = instant time (min), n = reaction order, and t_0 = induction time in minutes. In this model, it is assumed that vulcanization rate is maximum at the beginning of the reaction (Wang et al., 2007; Liu et al., 2007). Calculated k values for various reaction temperatures are used to determine activation energy of the reaction (E_a) by means of Arrhenius equation. Reaction order does not depend on vulcanization temperature; it is varied only by compound formulation.

In this study, O-POSS, M-POSS, and OM-POSS were incorporated to a model silicone rubber formulation to investigate effect of POSS type and loading ratio on vulcanization kinetics of the nanocomposites. Compounding was performed in a laboratory type internal mixer (Banbury). Cure curve data was processed for studying vulcanization kinetics through a common nonlinear empirical model. Performance of the model, which is usually used for modeling accelerated sulfur vulcanization, was evaluated for peroxide vulcanization.

2 EXPERIMENTAL

2.1 *Materials*

High temperature vulcanized (HTV) type VMQ silicone rubber (ELASTOSIL R401/70) and heat stabilizer (ELASTOSIL AUX-66), which is a metal oxide combination, were obtained from Wackers (Turkey). Pre-silanized silica (VP Coupsil 6508) was obtained from Evonik (Germany). POSSs were purchased from Hybrid Plastics Inc. (USA). Analytical grade dicumyl peroxide (Pergan, Turkey) was used as crosslinking agent. Molecular structures of the POSSs are given in Figure 1.

O-POSS can directly accompany with vulcanization reactions thanks to its multiple vinyl groups. There is no other functional group on O-POSS, enabling the comparative investigation to determine the effects of reactive groups carried by the other POSS types. M-POSS contains vinyl and polar acrylic groups; it can form polar interactions with common reinforcing fibers (through carboxylic acid and amine groups in Kevlar, through hydroxyl groups in Rayon). OM-POSS can accompany with vulcanization reaction through double bonds on maleamic acid. It can also form polar interactions with the reinforcing filler, as M-POSS do.

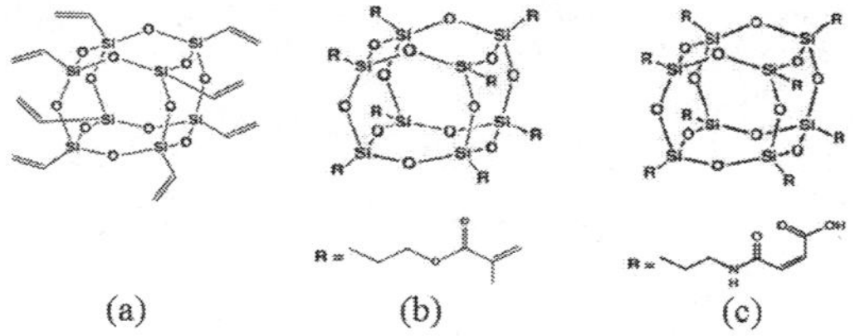

Figure 1. Molecular structure of POSSs (a) O-POSS (b) M-POSS (c) OM-POSS.

2.2 *Methods*

Thirteen different formulations were prepared in 300 cm^3 a Banbury mixer. Reference formulation contains 100 phr silicone rubber, 5 phr silica, 2 phr heat stabilizer, and 0.5 phr peroxide. Three different POSS types were incorporated into reference formulation in four different amounts (1, 3, 7, and 10 phr). Working temperature, mixing speed, and mixing cycle time was 50°C, 50 rpm, and 15 min, respectively.

Cure curves and rheological properties were obtained by using a moving die rheometer (Pinoeer MDR, Alpha Technologies). Test temperatures were selected as 160°C, 170°C, 180°C, and 190°C for calculating reaction kinetic parameters (k, n, E_a). Data points of the cure curves (torque versus time) were converted to state of cure (α, between 0 and 1) versus time (t) table. OriginPro 8.0® software was used to evaluate t-α data through Isayev and Deng kinetic model.

3 RESULTS AND DISCUSSIONS

3.1 *Rheological properties of silicone/POSS nanocomposites*

Rheological properties were investigated for the selected curing temperature (180°C) and important parameters are given in Table 1 (Formulation codes are given as short form that represents only initial of POSS type and amount in phr). As it can be seen from the table, all POSS types have distinguished

Table 1. Rheological properties of the compounds.

Code	Cure extent (dNm)	CRI
Ref	23.59	54.35
1O	24.95	55.25
3O	41.98	34.96
7O	75.90	27.62
10O	117.23	21.88
1M	17.56	19.88
3M	11.91	7.00
7M	4.68	1.24
10M	2.78	1.20
1OM	17.23	40.5
3OM	15.30	38.61
7OM	11.58	35.21
10OM	9.87	31.45

effects on compound rheology. The most distinct changes are in cure extent and cure rate index (CRI) values. Cure extent increases in parallel with POSS loading ratio in O-POSS containing nanocomposites; cure rate index exhibits a trend, vice versa. Compared with reference formulation, these results refer to higher activation energy for the beginning of the reaction and higher total crosslink density at the final stage. Sirin (2016) performed swelling measurements for silicone/O-POSS nanocomposites and revealed that higher O-POSS incorporation gave lower swelling ratios confirming the present results. Also, researchers correlated swelling ratio with mechanical properties of the nanocomposites. However, in case of both M-POSS and OM-POSS incorporation, cure extent decreases significantly. Furthermore, even for lower loading ratios (upward 3 phr), M-POSS deteriorates cure extent sharply. Both POSS types slow down the vulcanization reaction. Once again, effect of M-POSS is more pronounced. Briefly, both POSS types either decelerate vulcanization or decrease the amount of crosslinks. Both are expected to lead higher calculated activation energy in kinetic study.

3.2 *Vulcanization kinetics of silicone/POSS nanocomposites*

Cure curve data of the reference compound and POSS containing nanocomposites were processed for the investigation of vulcanization kinetics through nonlinear Isayev and Deng model (Equation 1). In modeling step, t-α data referring 160°C, 170°C, 180°C, and 190°C curing temperatures was evaluated by using OriginPro 8.0® software to calculate reaction kinetic parameters. Nonlinear regression results are given in Table 2 (k,n) and Figure 2 (E_a).

Table 2. Reaction kinetic parameters k(min^{-1}) and n through Isayev and Deng model.

	160°C		170°C		180°C		190°C	
Code	n	k	n	k	n	k	n	k
Ref	2.42	0.02	1.88	0.47	2.69	1.12	2.45	8.27
1O	1.86	0.04	1.76	1.50	1.87	2.15	2.03	4.86
3O	1.64	0.07	1.78	0.28	1.86	1.09	2.22	1.36
7O	1.52	0.06	1.64	0.15	1.66	0.96	2.10	3.50
10O	1.63	0.04	1.59	0.17	1.46	0.96	1.70	2.06
1M	–	–	–	–	2.11	0.25	1.71	1.01
3M	–	–	–	–	1.38	0.25	1.11	0.55
7M	–	–	–	–	–	–	1.10	0.16
10M	–	–	–	–	–	–	1.10	0.11
1OM	–	–	–	–	1.69	0.70	1.55	3.09
3OM	–	–	–	–	2.02	0.94	1.68	6.90
7OM	–	–	–	–	1.93	1.05	1.98	5.20
10OM	–	–	–	–	1.81	0.85	1.86	4.35

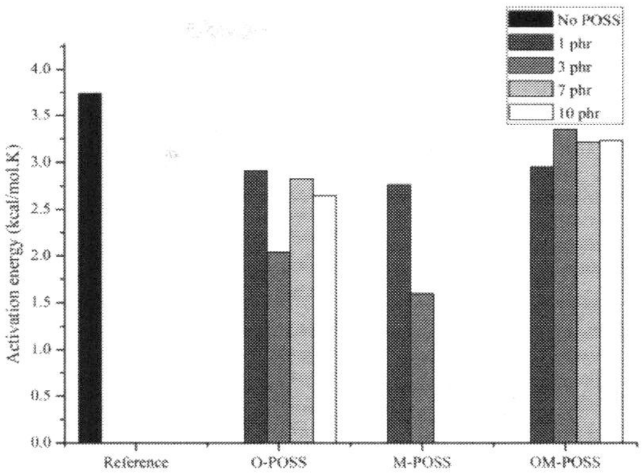

Figure 2. Reaction activation energies through Isayev and Deng model.

Isayev and Deng model exhibited that reaction order may possess various values different than 1 (represents the first order reaction kinetics) and expressing that rubber main chain is not the only species reacting in vulcanization. When we compare the order of reactions, it is seen that silicone/POSS nanocomposites have usually the order lower than 2 whereas reference compound has the order between 2.4 and 2.7. This is referred to the altered reaction stoichiometry stemming from altered reaction kinetics in POSS containing compounds as a result of its accompanying to the reaction mechanism.

Appropriate cure curves for M-POSS and OM-POSS could not be achieved at 160°C and 170°C due to poor reaction rate. In case of higher loading ratios such as 7 and 10 phr, appropriate cure curves for M-POSS could not be obtained even at 180°C. Ultimately, reaction kinetics could not be evaluated for a wide temperature range for these cases. Furthermore, for 7M and 10M nanocomposites, activation energy could not be calculated by using Arrhenius equation with a single vulcanization curve. In case of two available cure curves for two different temperatures, regression coefficient (R) always equals to 1; thus this value was not taken into consideration. R values were quite closed to 1 (0.85–0.99) representing good correlation through Isayev and Deng model with both the reference compound and silicone/O-POSS nanocomposites.

It is clear from Table 2 that vulcanization follows lower order reaction mechanism for silicone/POSS nanocomposites relative to the reference compound. Reaction rate constant also tends to decrease in these nanocomposites, especially for higher curing temperatures. This trend corresponds to lower cure rate index value given in Table 1. For silicone/O-POSS nanocomposites, activation energy is lower than the reference compound. Lower activation energy means easier crosslinking; thus cure extent values are higher for the same nanocomposites as seen in Table 1. Whern both findings are evaluated together, it can be concluded that O-POSS makes the vulcanization slower. Besides, it improves crosslinking with respect to enhanced reaction efficiency between silicone and peroxide. It is well kown that, crosslinking efficiency is higher in the presence of vinyl end groups in silicone rubber (Donnet & Custedero, 2005). Thus, the higher reaction efficiency in silicone/O-POSS nanocomposites is attributed to high vinyl end group level in O-POSS.

In peroxide crosslinking of saturated rubbers, if there are methyl side groups on the rubber main chain, reaction may undergo chain scission through hydrogen replacement from methyl group on the allylic radical. In Figure 3, this phenomenon is shown through an example, which is given for a saturated hydrocarbon based rubber, schematically (Kruzelak et al., unpubl.). In this case, multiple vinyl co-agents are used for suppressing chain scission in vulcanization media. Here, co-agents are assumed to immediately combine with the radical on the methyl group to create a new radical, and to prevent termination through chain scission (Figure 4). By this way, it is possible to have sufficient mechanical strength of vulcanizates without

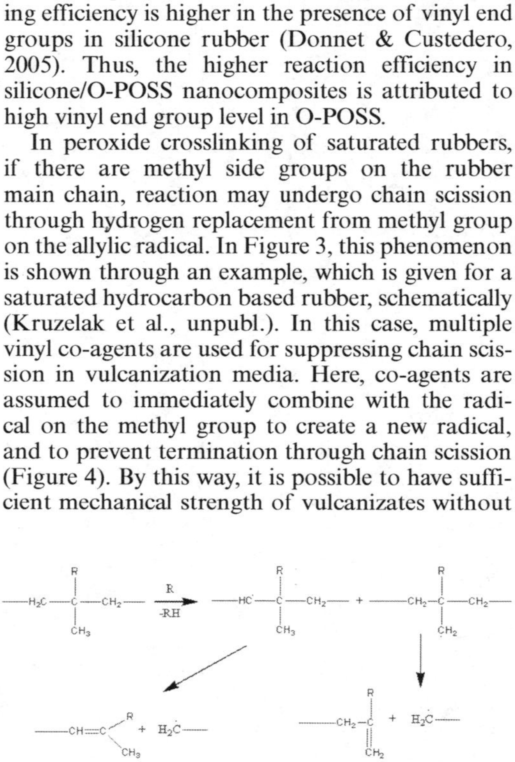

Figure 3. Chain scission on methyl group in peroxide crosslinking.

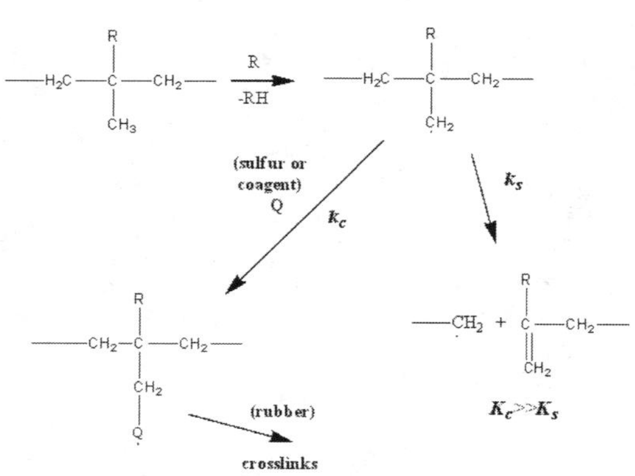

Figure 4. Co-agent role in peroxide crosslinking of saturated rubbers.

deterioration in good thermal properties. In this study, O-POSS intuitively serves as an appropriate co-agent in silicone crosslinking thanks to its multiple vinyl groups.

In case of OM-POSS incorporation, reaction order exhibits also a decrease; and this is more evident for higher vulcanization temperatures (190°C) related to O-POSS nanocomposites. It is clear from Table 2 that, higher OM-POSS loadings do not affect the reaction order as much as lower loadings do. Reaction rate drop is at lower level related to O-POSS nanocomposites. This also coincide with CRI values of OM-POSS nanocomposites, which are given in Table 1, and means that in case of OM-POSS incorporation, crosslinking efficiency is less affected than OM-POSS nanocomposites are. However, OM-POSS incorporation induces extent of cure in contrary to previous case (Table 1). Lower cure extent in spite of lower activation energy related to the reference compound indicates inhibition of silicone-peroxide reaction. Besides, this finding also indicates OM-POSS to combine to potential reactive sites on silicone rubber through side reactions. In our previous study, it was concluded that OM-POSS incorporation revealed improved adhesion between silicone and reinforcing fibers (Kevlar and Rayon) that can chemically interact with OM-POSS (Sirin et al., 2016). Improved adhesion was attributed to possible covalent bonds between silicone and OM-POSS as it has been demonstrated by means of reaction kinetic parameters, in this study.

In peroxide crosslinking of rubbers, chain scission, disproportional termination, oxygenation, dehydrohalogenation, radical transfer, elastomer branching, and acid-catalyzed decomposition of peroxide are the main side reactions (Kruzelak et al., unpubl.). Oxygenation takes place through combination of peroxy radicals with oxygen to give less stable hydroperoxy radicals that can readily degrade elastomeric chain. This causes both reduction in the amount of peroxy radicals, which can act as active crosslinking agent, and deterioration in mechanical properties of the vulcanizates by means of chain scission. Acidic species cause anionic and heterolytic decomposition of peroxide resulting a reduction in the required amount of peroxide in the reaction media.

In case of OM-POSS incorporation, it is believed that both oxygenation and acid-catalyzed peroxide decomposition may occur. Reduction in crosslinking efficiency (lower cure extent values) is attributed to reduced amount od peroxide through acid-catalyzed peroxide decomposition. Besides, oxygenation coincides with our previous study, in which adhesion between silicone rubber and reinforcing fillers was improved (Sirin et al., 2016). Both studies coincide with our present kinetic findings on OM-POSS effects in silicone rubber based nanocomposites.

M-POSS incorporation has an analogous effect on reaction order with the other POSS types. Reduction in the reaction order is more evident particularly at high vulcanization temperatures. This indicates that M-POSS significantly affect the reaction stoichiometry and it causes considerable drop in the number and/or amount of converted reactive species. In case of 1 phr M-POSS incorporation, reaction rate is one-tenth of reference compound's (Table 1). Moreover, reaction rate is hundred times lower for 10M nanocomposite. Reduction in cure extent and reaction activation energy is also remarkable. Lower cure extent despite lower activation energy particularly at higher curing temperatures indicates decomposition of M-POSS resulting lower activity at these temperatures and also M-POSS to consume a significant part of peroxide resulting lower crosslinking efficiency. This finding coincides with a previous study, which was completed by our working group (Sirin, 2016). In that study, it was shown that decomposition exotherm of M-POSS started at 105°C and 160°C consuming peroxide; hence, it causes a considerable drop in peroxide amount resulting low cure extent. Completed vulcanization reaction could not be obtained at relatively lower curing temperatures in the presence of M-POSS nanocomposites. This is also attributed to aforementioned decomposition of M-POSS.

4 CONCLUSIONS

In this study, O-POSS, M-POSS, and OM-POSS were incorporated into a reference silicone rubber formulation to investigate effects of POSS type and loading ratio on vulcanization kinetics. Cure curve data, which was obtained at various temperatures, was processed to fit nonlinear Isayev and Deng model. Important reaction kinetic parameters (k,n, E_a) were calculated and correlated with rheological and mechanical properties of the nanocomposites.

POSS types have several effects on rheological properties of the compounds. O-POSS increases activation energy of peroxide vulcanization; total state of cure values are higher than that of reference silicone compound. This finding is in a good correlation with swelling measurements performed in our previous study (Sirin, 2016). Both M-POSS and OM-POSS incorporation decelerate vulcanization and lower the total crosslink density. O-POSS is concluded to perform as a powerful co-agent in peroxide vulcanization of silicone rubber. Reaction efficiency between silicone and peroxide is significantly reduced by OM-POSS. This is attributed to side reactions such as oxygenation and acid-catalyzed peroxide decomposition that take place between OM-POSS and peroxide. In case of M-POSS incorporation crosslinking is inhibited due to considerable amount of peroxide consumption by M-POSS decomposition at even moderate temperatures that vulcanization has not start yet.

POSS incorporation reduces vulcanization reaction order in all cases due to accompanying peroxide reaction mechanism. Calculated regression coefficients in Arrhenius relation indicate that Isayev and Deng model can be used as a successful tool for modeling vulcanization kinetics of silicone/POSS nanocomposites with a sufficient approximation.

ACKNOWLEDGEMENTS

Authors gratefully thank to TÜBİTAK for financial support to the study (within project no: 113M258) and to attend the conference (within project no: 115Z811).

REFERENCES

Choi, D., Kader, M.A., Cho, B.H., Huh, Y., Nah, C., 2005. Vulcanization kinetics of nitrile rubber/layered clay nanocomposites. J. Appl. Polym. Sci. 98(4), 1688–1696. doi: 10.1002/app.22341.

Choi, S.S., Kim, J.C., 2008. Influence of the cure systems on long time thermal aging behaviors of NR composites. Macromol. Res. 16(6), 561–566. doi: 10.1007/BF03218560.

Donnet, J.B., Custodero, E., 2005. Reinforcement of elastomers by particulate fillers. In. The science and technology of rubber. Eds: Mark, J.E., Erman, B., Eirich, F.R. 3rd ed., Elsevier Academic Press, London, 367–400.

Kruzelak, J., Sykora, R., Hudec, I., unpubl. Vulcanization of rubber compounds with peroxide curing systems. Rubber Chem. Technol. doi: 10.5254/rct.16.83758.

Sirin, H., 2016. Improvement of interfacial adhesion and matrix properties in silicone elastomer/continues fiber composites with functional nanoparticles. Ph.D. Thesis. Kocaeli University Institute of Science, 438577.

Sirin, H., Kodal, M., Karaagac, B., Ozkoc, G., 2016. Effects of octamaleamic acid-POSS used as the adhesion enhancer on the properties of silicone rubber/silica nanocomposites. Compos. B. 98, 370–381. Doi: 10.1016/j.compositesb.2016.05.024.

Wang, P.Y., Chen, Y., Qian, H.L., 2007. Vulcanization kinetics of low-protein natural rubber with use of a vulcameter. J. Appl. Polym. Sci. 105(6), 3255–3259. doi: 10.1002/app.26488.

Constitutive Models for Rubber X – Lion & Johlitz (Eds)
© 2017 Taylor & Francis Group, London, ISBN 978-1-138-03001-5

Diffusion of oils in elastomers—determination of concentration profiles

T. Förster
Wehrwissenschaftliches Institut für Werk- und Betriebsstoffe (WIWeB), Erding, Germany

ABSTRACT: Elastomers exposed to liquids during application, frequently undergo various interactions. Amongst the most critical phenomena sorption of certain components of surrounding media and loss of elastomer additives has to be mentioned because of possibly negative impact on the elastomer material. Therefore, correlation of detailed material composition with mechanical properties is worth exploring as general conclusions on the long term stability can be derived. As the sorption of liquids follows the mechanisms of diffusion, knowledge on diffusion behavior of components in elastomers is very valuable. To quantitatively describe diffusion of mineral oils, a storage procedure in which elastomer specimens vertically dip in a test liquid only at the bottom side is described in this work. Analyses via thermal desorption of volatile components followed by gas chromatography and mass spectrometry of consecutive elastomer pieces along the diffusion axis allows the generation of concentration profiles of elastomer additives and absorbed mineral oil. It can clearly be shown that the swelling potential of different mineral oils is reflected by the diffusion behavior of these oils. Therefore, the procedure provides detailed information on diffusion processes and thus facilitates the prediction of lifetime.

1 INTRODUCTION

Elastomer materials can be exposed to various conditions during application, i.e. temperature or contact to liquids. As acrylonitrile butadiene rubber (NBR) exhibits very good resistance against nonpolar liquids, it is typically used for applications where elastomers have intense contact to fuels or lubricants (Röthemeyer 2006). Despite its principal resistance certain components of surrounding media can be absorbed by elastomers during application (Harogoppad 1991, Mathai 1996).

Considering the long term stability of elastomers, aging via thermal or oxidative routes and the uptake of liquids have to be taken into account (Förster 2017, Ogorodnikova 1982, Starmer 1993). Underlying processes follow the principles of diffusion described by the Fickian laws. Although the idea of storing elastomers in liquids is a common procedure to gain information on the material compatibility, no detailed information on processes on a molecular base can be derived. Usually, the determination of physical and mechanical properties, e.g. mass and volume change, hardness change, elongation at break, merely allow interpretations of the influence of the tested liquid. To quantitatively describe diffusion processes inside elastomers, locally resolved data are essential.

In this work a method to follow diffusion processes inside elastomer materials is described by storing NBR specimens in mineral oil and only allowing the diffusion from one side. Via analyses of slices along the diffusion axis by gas chromatography/mass spectrometry local resolution is determined.

2 EXPERIMENTAL

The elastomer used in this work was a carbon black filled acrylonitrile butadiene rubber (NBR) provided from Deutsches Institut für Kautschuktechnologie (DIK). The formulation is summarized in Table 1. Mineral oil based reference oils IRM 901, IRM 902 and IRM 903 were purchased from Fuchs Schmierstoffe.

Table 1. Elastomer composition.

Component	Content/phr
Polymer	100
Bis(2-ethylhexyl) phthalate (DEHP)	20
6-PPD	2
Carbon black, N550	60
Zinc oxide	5
Stearic acid	1
Sulfur	2
CBS	1.5
TMTM-80	0.5

Elastomers were stored in reference oils according to DIN ISO 1817 in the temperature range of 60 to 120°C and durations from 16 to 168 h. To determine concentration profiles elastomer specimens ($45 \times 5 \times 2$ mm^3) were stored in flasks while dipping in oil at the bottom side (several hundred micrometers) at various temperatures from 60 to 120°C. Typically the storage duration was 24 h. After cleaning of specimens with petroleum benzine, concentrations of mineral oils and elastomer additives were determined in thin slices (ca. 300–500 μm) of each specimen. Pieces were cut with a knife and thickness was measured afterwards.

Thermal desorption of adsorbed mineral oils and elastomer additives at 300°C was accomplished with a Frontier Lab Pyrolyzer Py2020iD. GC/MS analyses were performed on an Agilent 7890 A gas chromatograph coupled to an Agilent 5975 MSD mass spectrometer. Concentrations were calculated by normalizing the determined mass for each component on the mass of solid elastomer constituents, i.e. polymer and filler. Volume and mass change of the NBR was determined according to DIN ISO 1817 and Shore A hardness with a micro Shore A unit (Bareiss digi test II hardness tester) according to DIN ISO 7619-1.

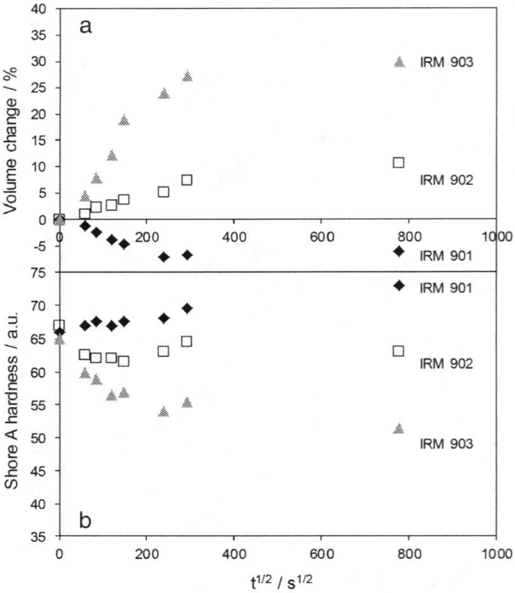

Figure 1. Volume change (a) and Shore A hardness (b) of investigated elastomer in dependence on the storage duration in reference oils IRM 901, 902 and 903 at 100°C.

3 RESULTS

Classical storage experiments allow the observation of general trends and can be used to make principal statements to compatibility issues. In Figure 1a the volume change of the tested elastomer after storage in reference oils IRM 901, 902 and 903 is displayed.

For each series the volume change is time dependent due to diffusion processes. After ca. one week the storage equilibrium is reached and no further changes of volume can be detected. Storage in IRM 902 and 903 leads to an increase of volume. IRM 903 causes the most pronounced volume increase because of its high swelling potential. It can be seen that a volume decrease after contact to IRM 901 (low swelling potential) is observed corresponding to a net reduction of volume and mass (not displayed here). However, no information can be derived whether the mass decrease only results from the reduction of elastomer additive concentration or whether oil is simultaneously absorbed. Thus, a net increase of mass and volume can correspond to a complete loss of elastomer additives while this decline is overcompensated by mineral oil uptake.

An important parameter for interpretation of interactions between liquid and elastomer is the effect of media on the Shore A hardness (Figure 1b). It strongly correlates to changes in the material's composition. Thus, a net uptake of components results in a decrease of hardness (IRM 902 and 903) whereas the net loss of constituents leads to a hardness increase. The hardening of elastomers could be explained by the extraction of plasticizer. However, uptake of mineral oil counteracts the loss of elastomer additive because it can also act as plasticizer. Therefore, explanations only based on physical mechanical investigations lack fundamental understanding on a molecular base and, thus, further investigations are required.

The experimental approach performed in this study is the storage of elastomer specimens vertically dipping in mineral oil (Figure 2a). Using this setup assures that diffusion only occurs from one side, namely upwards. This procedure also allows the determination of concentrations of mineral oil and elastomer additives simultaneously and locally resolved by analyzing slices of stored specimens, starting from the bottom side where elastomer and mineral oil stay in contact (i.e. penetration depth = zero). The determination of the layer thickness is the fundament for concentration profiles. Varying the slice thickness offers a straight forward parameter to optimize the local resolution in dependence on the diffusion progress. Currently a resolution of ca. 300 μm can be achieved.

Figure 2b gives an overview of chromatograms received after thermal desorption of all volatile components, i.e. mineral oil and elastomer additives. It can be seen that the mineral oil concentration (corresponds to the broad peak in the chromatograms) is highest at the elastomer edge dipping in mineral oil

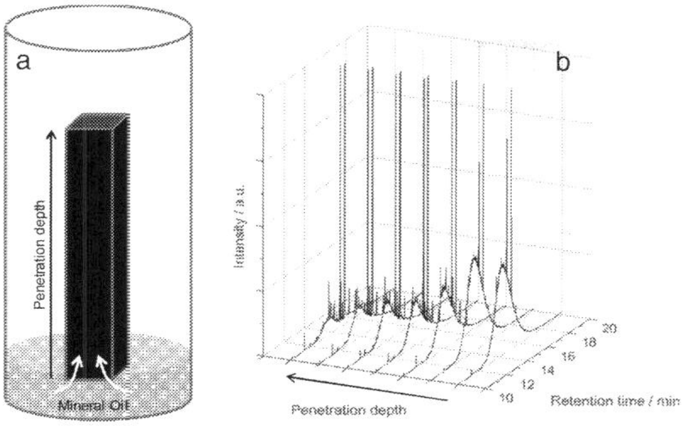

Figure 2. a) Storage conditions (schematic) for diffusion experiments and b) Chromatograms of thermally desorbed components from elastomer pieces in dependence of penetration depth (increasing from right to left) after storage in reference oil IRM 903 at 100°C for 24 h.

during storage (right chromatogram in Figure 2b). With increasing distance from the mineral oil reservoir lower concentrations of mineral oil are detected. In case of elastomer additives an inverse tendency is observed (peaks at ca. 14.7 and 15.1 min cut on top due to reasons of clarity). This clearly indicates the pronounced uptake of mineral oil by edge near regions of the elastomer and a parallel loss of elastomer additives by diffusion processes.

As the maximum uptake of mineral oil is observed on the elastomer surface, i.e. the area where elastomer and oil stay in contact, the surface concentration c_S is a very good indicator for the compatibility of the elastomer towards the tested oil.

In Table 2 an overview of surface concentrations of mineral oil is given. It can be seen that IRM 903 uptake is more pronounced compared to IRM 902. Additionally, it can be observed that the temperature has a stronger effect on the uptake of IRM 902 than on IRM 903. The surface concentration of the elastomer plasticizer is at the same time significantly reduced compared to its original concentration of ca. 15%, i.e. it nearly approaches zero. In both cases the oil uptake exceeds the loss of plasticizer. This equals a net uptake and is in very good agreement with classical storage experiments (Figure 1a).

However, the presented procedure allows a more detailed interpretation of data. To quantitatively describe diffusion processes, concentrations are plotted in dependence of the penetration depth, i.e. the specimen height. In Figure 3 the concentration of mineral oil IRM 903 (normalized to polymer and filler mass) and elastomer plasticizer are depicted exemplarily for the sample stored at 120°C for 24 h. During storage mineral oil is soaked into the elastomer whereas plasticizer is lost in the surface near regions of the specimen. This can qualitatively be derived from the shape of the curves. Mineral oil concentration decreases

Table 2. Surface concentrations c_S of mineral oil after storage of elastomer in mineral oils IRM 902 and IRM 903 for 24 h.

Temperature/°C	c_S/% IRM 903	c_S/% IRM 902
60	30.3	11.0
80	37.7	14.0
100	40.9	33.0
120	39.8	33.5

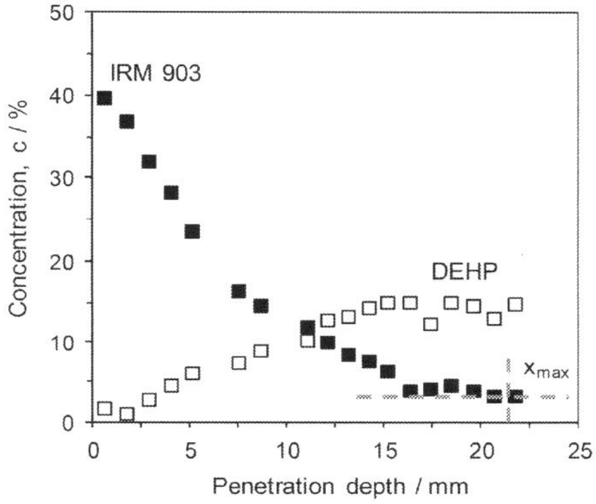

Figure 3. Concentrations of mineral oil IRM 903 and plasticizer DEHP in dependence of penetration depth after storage in reference oil IRM 903 at 120°C for 24 h.

with increasing penetration depth whereas plasticizer content increases up to the original concentration in the specimen bulk.

A further parameter describing the degree of interactions is the maximal penetration depth x_{max}. It is defined as the penetration depth up to which a significant difference to the observed baseline can be detected. In Table 3 an overview of x_{max} is displayed. Similar to the trends observed for the surface concentration, the maximum penetration depth increases with increasing storage temperature. Additionally, values determined for storage in IRM 903 are above those determined for storage in IRM 902.

By iterative fitting of equation (1) according to Hauffe (1966) the diffusion coefficient D for mineral oil can be calculated.

$$c(x,t) = c_S - (c_S - c_0)\, erf\left(\frac{x}{2\sqrt{Dt}}\right) \quad (1)$$

where $c(x, t)$ = concentration; t = time; x = penetration depth; c_S = surface concentration; and c_0 = starting concentration.

Similar to the trends observed for the surface concentration and maximal penetration depth, diffusion coefficients increase with increasing storage

Table 3. Maximum penetration depth x_{max} of mineral oil after storage of elastomer in mineral oils IRM 902 and IRM 903 for 24 h.

Temperature/°C	x_{max}/mm IRM 903	x_{max}/mm IRM 902
60	6	4
80	13	7
100	17	10
120	23	13

temperature. Diffusion coefficients for storage in IRM 903 were determined to 0.3 (60°C), 1.5 (80°C), 2.3 (100°C) and 3.9 10^{-6} cm^2 s^{-1} (120°C), respectively.

Hardness of elastomers is a relatively surface sensitive technique. The trend of Shore A hardness along the diffusion pathway is displayed in Figure 4 for example for the storage of elastomer in IRM 903 at 80°C for 24 h. From this plot it can be seen that the Shore A hardness is significantly reduced in the surface near region of the specimen after storage to mineral oil.

Additionally, the concentrations of mineral oil and plasticizer are displayed in Figure 4. At positions far away from the specimen surface, i.e. penetration depth > 15 mm, the elastomer is nearly in the original state. That means that in this region nearly no oil and the original plasticizer concentration are present. In consequence, the hardness equals the value of the untreated material. Approaching the elastomer surface the plasticizer concentration decreases whereas the one of mineral oil increases. At the same time the Shore A hardness decreases from 68 to 58. Therefore, a clear correlation of the material composition and mechanical properties can be derived. The mineral oil acts as plasticizer, but, its plasticizing capability is not as pronounced as the one of DEHP. This can be concluded from the observation that in the region of 5 to 13 mm the plasticizer concentration is not changed, whereas the oil concentration lays in the range from 2 to 15% and only a minor hardness reduction is observed. The most obvious reduction of hardness occurs when very high mineral oil concentrations over ca. 20% are detected.

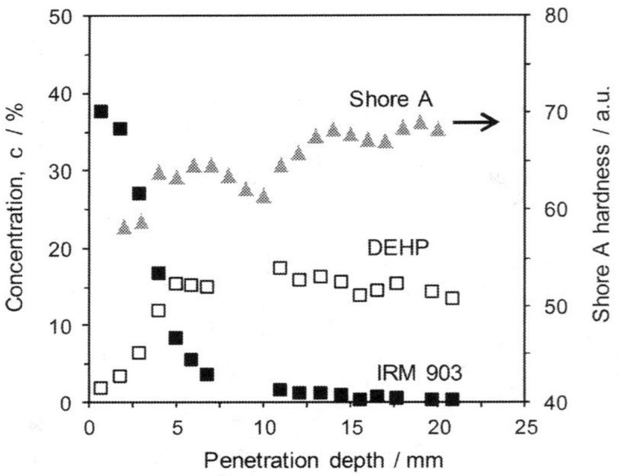

Figure 4. Concentrations of mineral oil IRM 903 and plasticizer DEHP and Shore A hardness in dependence of penetration depth after storage in reference oil IRM 903 at 80°C for 24 h.

4 CONCLUSION AND OUTLOOK

This paper describes a method to quantify different components in elastomers with high local resolution. From the determined concentration profiles valuable information on the compatibility of elastomers with mineral oils can be derived allowing the prediction of lifetime. Parameters determined are surface concentration, maximal penetration depth and diffusion coefficients.

Future work will focus on tracing concentrations of diffusing model substances for an even more precise prediction of the material's stability and compatibility properties.

ACKNOWLEDGEMENT

The author gratefully wants to acknowledge Nadine Weber, Artur Blivernitz and Johannes Bibinger for assisting with experimental efforts.

REFERENCES

Förster, T. 2017 Prediction of In-Service NBR Properties by TG-IR after storage in mineral oil. In A. Öchsner, L. da Silva & H. Altenbach (eds) *Properties and Characterization of Modern Materials Volume 33 of the series Advanced Structured Materials*: 399–411. Singapore: Springer-Verlag.

Harogoppad, S.B., Aminabhavi, T.M. 1991. Diffusion and Sorption of Organic Liquids through Polymer Membranes. II. Neoprene, SBR, EPDM, NBR and Natural Rubber versus n-Alkanes. *J Appl Polym Sci* 42: 2329–2336.

Hauffe, K. (2. Ed.) 1966. *Reaktionen in und an festen Stoffen*. Berlin: Springer-Verlag.

Mathai, A.E., Thomas, S.J. 1996. Transport of Aromatic Hydrocarbons Through Crosslinked Nitrile Rubber Membranes. *J Macromol Sci Phys B* 35: 229–253.

Ogorodnikova, G.F., Sinitsyn, V.V. 1982. Mechanism of Rubber Swelling in Oils and Grease. *Chem Tech Fuels Oils* 18: 306–308.

Röthemeyer, F., Sommer, F. (2. Ed.) 2006. *Kautschuk Technologie – Werkstoffe – Verarbeitung – Produkte*. Munich: Carl Hanser Verlag.

Starmer, P.H. 1993. Swelling of Nitrile Vulcanizates by Polar and Non-Polar Liquids - Part 1: Review of Parameters. *J Elastom Plast* 25: 59–73.

Constitutive Models for Rubber X – Lion & Johlitz (Eds)
 ISBN 978-1-138-03001-5

Mechanical characterization of highly aligned polyurethane microfibers

C.J. Tan & A. Andriyana
Centre of Advanced Materials, Department of Mechanical Engineering, Faculty of Engineering, University of Malaya, Kuala Lumpur, Malaysia

B.C. Ang
Centre of Advanced Materials, Department of Chemical Engineering, Faculty of Engineering, University of Malaya, Kuala Lumpur, Malaysia

G. Chagnon
Université Grenoble-Alpes, TIMC-IMAG, Grenoble, France
CNRS, TIMC-IMAG, Grenoble, France

ABSTRACT: In this study, highly aligned polyurethane fibers with potential application as scaffold in tissue engineering were fabricated using a simple dry spinning technique assisted by an electrical field. Morphology study shows that the fibers have an average diameter of 2.80 ± 0.06 µm without any observable bead. Subsequently, mechanical characterizations were conducted to study the mechanical responses of these fibers. Three uniaxial mechanical tests were conducted: monotonic, cyclic and relaxation tensile tests at the strain rate of 0.001 s^{-1}. Complex mechanical responses such as anisotropy, toe region, hysteresis, stress softening, residual deformation, and stress relaxation are found in these polyurethane elastomeric microfibers. Furthermore, it is highlighted that amount of hysteresis increases with the maximum strain, whereas stress relaxation is faster at a higher strain.

1 INTRODUCTION

Thousands of surgeries are performed yearly to save lives from tendon tissue degeneration in the world. This tissue degeneration may cause by various factors, such as illness, injury, trauma, genetic factor and so on. The damage to the tissue level of our body may pose serious health risk to us and may even cause fatality.

Tissue engineering is a remedy to these tissues diseases. It is aimed to regenerate the damaged tissues, instead of replacing the tissues from foreign source which may cause infection and rejection to the patient (O'Brien, 2011). In this tissue engineering, a biological construct or also known as scaffold is important to act as a support for tissue regeneration (Hutmacher, 2001, Ratcliffe et al., 2015).

Tendon tissues in our body are always in service since we are always moving. Therefore, the tendon scaffold needs to be strong or resilient. It is also found that tendon tissues exhibit viscoelastic mechanical behavior (Hooley et al., 1980, Johnson et al., 1994) like typical elastomers. Elastomeric biomaterials are biocompatible and resilient (Chen et al., 2013), thus suitable for the fabrication of tendon scaffolds.

Hence, polyurethane elastomer with biocompatibility (Hu et al., 2012, Pavlova and Draganova, 1993) is used in this study for the potential application of tendon scaffold. It is processed into highly aligned fibers to mimic the microstructure of the tissue tissues in order to enhance tissue regeneration (Orr et al., 2015, Xie et al., 2010). Then, the mechanical responses of the fibers under uniaxial monotonic, cyclic and relaxation tensile test are studied.

2 METHODOLOGY

2.1 *Materials*

MDI-polyester/polyether polyurethane was the raw material in fabricating the polymeric fibers. High purity N,N-Dimethylformamide (DMF) and Tetrahydrofuran (THF) were used to dissolve the MDI-polyester/polyether polyurethane beads. All chemicals were used as received.

2.2 *Dry spinning of polyurethane fibers*

Polyurethane fibers were drawn from the polymer solution in a syringe with 27G size needle to be collected on a rotating drum below 100 rpm. The distance between the tip of the needle and rotating drum was 15 cm and the polymer solution flow rate was kept at 0.05 ml/hr. 1 kV voltage was applied between the tip of the needle and rotating drum to further stretch the fibers.

After the fabrication, the fibers morphology was observed under Phenom ProX desktop scanning electron microscope. The diameter of 100 fibers in the micrographs was measured to get the average diameter.

2.3 *Mechanical characterizations*

2.3.1 *Sample geometry*

The specimen for tensile test was 6 cm in total length. 4 cm was the gauge length and 1 cm at each end of the specimen was to be clamped during the tensile test. The cross-sectional area of the specimens was estimated by multiplying the mass of the specimens with the cross-sectional area to mass ratio since the irregular cross-sectional area was hard to measure.

In order to obtain the ratio mentioned, 12 pairs of perpendicular diameters, D_1 and D_2 of the specimen were measured and the equivalent circular diameter was calculated using this equation, $D = \sqrt{D_1 D_2}$ (Sun et al., 2008). Average value of D was calculated and the cross section area to mass was computed. The cross-sectional area to mass for 8 specimens was computed and the average value, 93.4 $mm^2\ g^{-1}$ was obtained.

2.3.2 *Uniaxial tensile tests*

Shimadzu AGS-X Series Universal Tensile Testing Machine equipped with a 5 kN load cell was used to conduct three mechanical tests, uniaxial monotonic, cyclic and relaxation tensile tests under the strain rate of 0.001 s^{-1} on the polyurethane fibers. Figure 1 shows the loading profile for uniaxial cyclic and relaxation tensile tests. There are two maximum strain, 20% and 40% in the cyclic test and the unloading was set to stop when the stress is 0 MPa as shown in Figure 1(a). The fibers are held at 20% and 40% strain consecutively for 3600 s as shown in Figure 1(b).

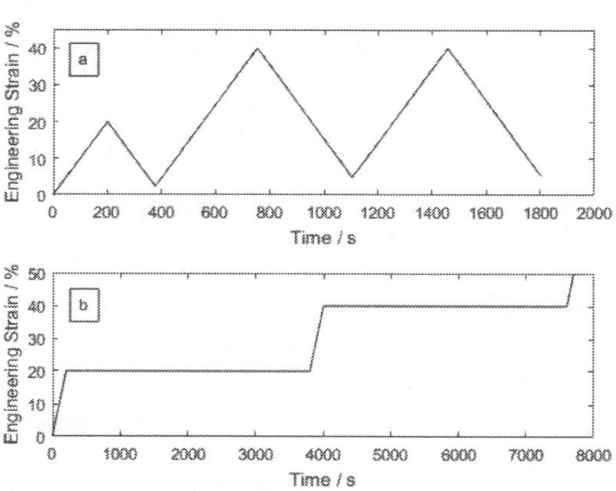

Figure 1. Loading profile for uniaxial (a) cyclic and (b) relaxation tensile test.

3 RESULTS AND DISCUSSION

3.1 *Morphology of highly polyurethane fibers*

The micrographs of the polyurethane fibers are shown in Figure 2. Based on this figure, the fibers are packed closely and aligned in one direction. The average diameter measured from 100 fibers in the micrographs is 2.80 ± 0.06 μm.

3.2 *Mechanical characterizations*

The average engineering stress-strain curve from 8 uniaxial monotonic tensile tests is plotted in Figure 3. The curve starts with a nearly linear trend until it reaches the ultimate tensile strength. A gradual failure is observed after reaching the ultimate tensile strength. Besides that, a small distinct

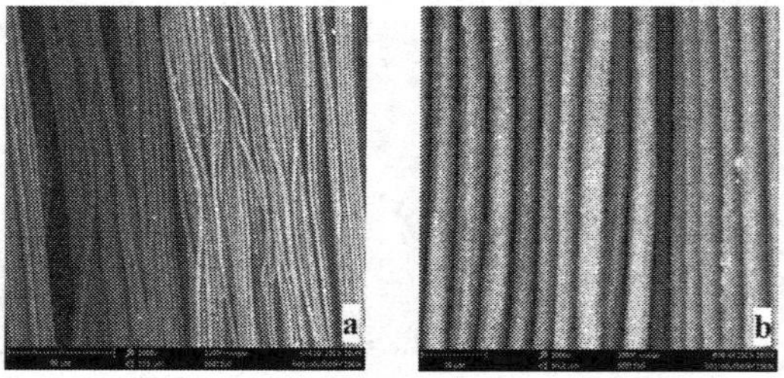

Figure 2. Micrographs of polyurethane fibers under (a) 1000 × magnification and (b) 5000 × magnification.

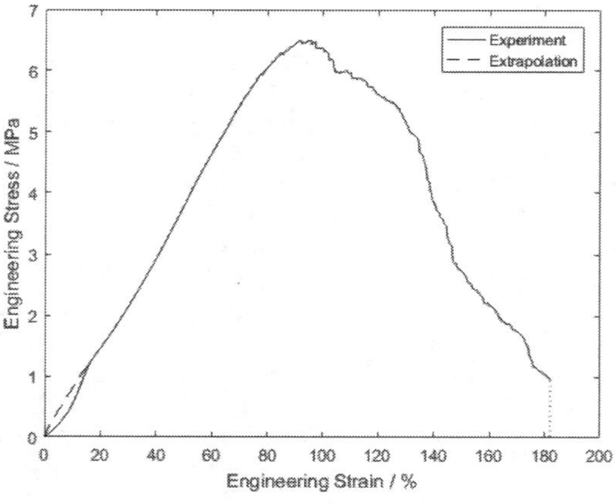

Figure 3. Average engineering stress-strain curve for uniaxial monotonic tensile test.

curvature is also seen at below 20% strain, similar to the toe region found in the stress-strain behavior of tissues (Tamura et al., 2015). It is the stress-softening effect prior to the excessive drawing force exerted on the fibers during the fabrication process. In order to make up for this stress-softening, an extrapolation is done as shown in Figure 3. The initial modulus of these fibers is 7.79 MPa with the ultimate tensile strength of 6.51 MPa.

Figure 4 shows the average engineering stress-strain curves from 2 uniaxial cyclic tensile tests. Typical mechanical behaviors of elastomers such as the stress-softening, hysteresis and residual deformation (Diani et al., 2009) can be observed in this curve.

The three cycles in Figure 4 are separated and shown in Figure 5 for better illustration. It appears that the amount of hysteresis increases as the

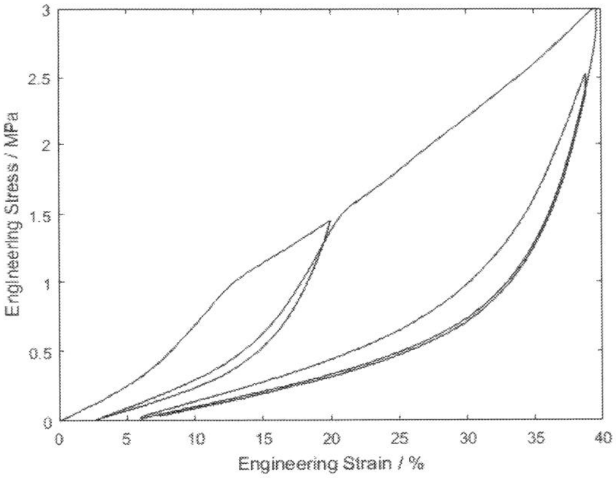

Figure 4. Average engineering stress-strain curves of uniaxial cyclic tensile test.

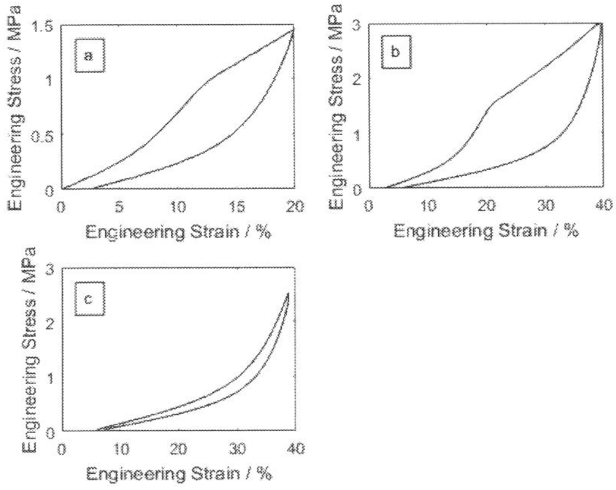

Figure 5. Hysteresis behavior to the maximum strain of (a) 20%, (b) 40% (1st cycle), and (c) 40% (2nd cycle).

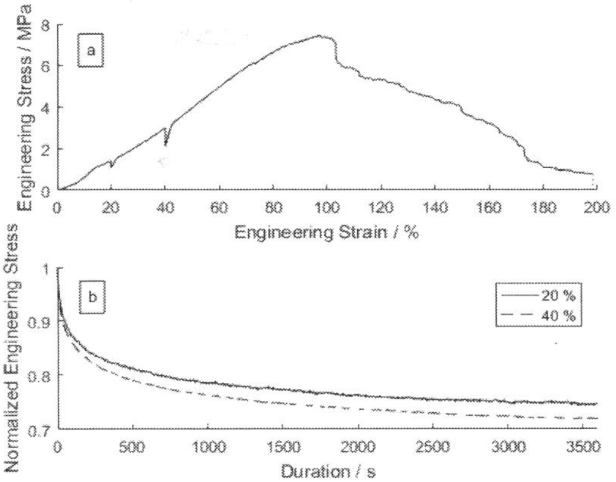

Figure 6. (a) Average engineering stress-strain curve of uniaxial relaxation tensile test; (b) Normalized engineering stress against relaxation duration.

loading to maximum strain increases from 20% to 40%. However, the hysteresis is very small if the maximum strain does not exceed the previous maximum strain.

Viscoelasticity or time dependent mechanical response of the fibers are manifested in the uniaxial relaxation tensile test as shown in Figure 6. The fibers were held at 20% and 40% engineering strain respectively as shown in Figure 1(b) for 3600s. A drop in engineering stress is observed during the holding period as shown in Figure 6(a).

Nevertheless, the rate of stress relaxation is high during the first 500 s as shown in Figure 6(b). Then, the rate of stress relaxation decreases as the holding period increases.

4 CONCLUSION

Polyurethane fibers with high alignment for the potential application of tendon scaffolds were successfully fabricated using a simple dry spinning technique assisted by an electrical field.

The morphology of the fibers was observed and studied. It is found that the fibers are having 2.80 ± 0.06 μm average diameter with no visible beads. All the fibers are observed to be oriented in one direction, thus having high alignment, similar to the fibrous microstructure of tendon tissues.

Uniaxial tensile tests were also conducted on the fibers. The initial modulus and ultimate tensile strength of the fibers are 7.79 MPa and 6.51 MPa respectively retrieved from the monotonic tensile test. Elastomeric mechanical behaviors such as hysteresis, stress softening, and residual deformation are clearly seemed in the cyclic tensile test whereas stress relaxation is seemed in the relaxation tensile test. All these mechanical behaviors are identical with the tendon tissues.

In short, the fibers fabricated in this study is similar to the tendon tissues in the aspect of aligned fibrous microstructure and mechanical behaviors. Nevertheless, biological characteristics such as the biostability, biocompatibility and biodegradability need to be studied further in details to justify the suitability of these fibers for scaffold application.

REFERENCES

Chen, Q., Liang, S. & Thouas, G.A. 2013. Elastomeric biomaterials for tissue engineering. *Progress in Polymer Science,* 38, 584–671.

Diani, J., Fayolle, B. & Gilormini, P. 2009. A review on the Mullins effect. *European Polymer Journal,* 45, 601–612.

Hooley, C.J., McCrum, N.G. & Cohen, R.E. 1980. The viscoelastic deformation of tendon. *Journal of Biomechanics,* 13, 521–528.

Hu, Z.-J., Li, Z.-L., Hu, L.-Y., He, W., Liu, R.-M., Qin, Y.-S. & Wang, S.-M. 2012. The in vivo performance of small-caliber nanofibrous polyurethane vascular grafts. *BMC Cardiovascular Disorders,* 12, 1–11.

Hutmacher, D.W. 2001. Scaffold design and fabrication technologies for engineering tissues—state of the art and future perspectives. *Journal of Biomaterials Science, Polymer Edition,* 12, 107–124.

Johnson, G.A., Tramaglini, D.M., Levine, R.E., Ohno, K., Choi, N.-Y. & Woo, S.L.-Y. 1994. Tensile and viscoelastic properties of human patellar tendon. *Journal of Orthopadic Research,* 12, 796–803.

O'Brien, F.J. 2011. Biomaterials & scaffolds for tissue engineering *materials today* 14 88–95.

Orr, S.B., Chainani, A., Hippensteel, K.J., Kishan, A., Gilchrist, C., Garrigues, N.W., Ruch, D.S., Guilak, F. & Little, D. 2015. Aligned multilayered electrospun scaffolds for rotator cuff tendon tissue engineering. *Acta Biomaterialia.*

Pavlova, M. & Draganova, M. 1993. Biocompatible and biodegradable polyurethane polymers. *Biomaterials,* 14, 1024–1029.

Ratcliffe, A., Butler, D.L., Dyment, N.A., Jr, P.J.C., Proctor, C.S., Ratcliffe, S.S. & Flatow, E.L. 2015. Scaffolds for tendon and ligament repair and regeneration. *Ann Biomed Eng.,* 43, 819–831.

Sun, L., Han, R.P.S., Wang, J. & Lim, C.T. 2008. Modeling the size-dependent elastic properties of polymeric nanofibers. *Nanotechnology,* 19, 1–8.

Tamura, A., Murakami, J., Sone, Y. & Koide, T. 2015. Strain rate effects on the tensile behavior of fiber bundles isolated from nerve root. *Journal of Sustainable Research in Engineering,* 2, 63–69.

Xie, J., Li, X., Lipner, J., Manning, C.N., Schwartz, A.G., Thomopoulos, S. & Xia, Y. 2010. "Aligned-to-random" nanofiber scaffolds for mimicking the structure of the tendon-to-bone insertion site. *Nanoscale.,* 2, 923–926.

Constitutive Models for Rubber X – Lion & Johlitz (Eds)
© 2017 Taylor & Francis Group, London, ISBN 978-1-138-03001-5

Sorption experiments on elastomers assisted by Gas Chromatography/Mass Spectrometry (GC/MS)

A. Blivernitz, T. Förster & S. Eibl
Wehrwissenschaftliches Institut für Werk- und Betriebsstoffe, Erding, Germany

A. Lion & M. Johlitz
Institute of Mechanics, Universität der Bundeswehr München, Neubiberg, Germany

ABSTRACT: Fuels are complex mixtures containing different compounds of various substance classes. Depending on the composition, particularly on the content of aromatic hydrocarbons, fuels show different swelling behavior towards elastomers. Therefore, understanding and quantitatively describing diffusion processes are essential requirements to prevent malfunctions as swelling can change material properties.

To get better insight in diffusion processes, model liquids containing different amounts of dodecane and m–xylene, which represent aliphatic and aromatic fuel fractions, were used to perform sorption experiments on Acrylonitrile Butadiene Rubber (NBR). The mass uptake of each compound was traced simultaneously and time resolved by extraction and subsequent Gas Chromatography/Mass Spectrometry (GC/MS).

Swelling potential of mixtures containing m–xylene and dodecane increases with higher aromatics content, which is indicated by higher equilibrium mass uptake and diffusion rate. This work provides an experimental approach to investigate interactions of fuel components and elastomers on a molecular level by quantifying diffusion processes of individual mixtures' components by GC/MS. The results allow understanding the diffusion behavior of single components in mixtures, and may be applied to compatibilityand lifetime predictions of sealing materials and in–service aviation fuels.

1 INTRODUCTION

Elastomers tend to swell during contact with fuels due to diffusion. The extent of swelling depends on the type of rubber and on the composition of the penetrating liquid. Fuels, for example, contain various different chemical compounds (Dooley et al. 2010). Particularly high contents of aromatic hydrocarbons lead to a pronounced substance uptake (Graham et al. 2006). As a consequence material properties can be affected negatively. Therefore, understanding and quantitatively describing diffusion processes are essential requirements to prevent malfunctions of e.g. gaskets during application.

A classical method to investigate diffusion processes is to carry out sorption experiments. Here, thin polymer specimens are immersed in the desired fluid and the mass uptake of the sample is traced gravimetrically over time until equilibrium swelling is reached (George et al. 2000). As in this case diffusion can be considered one–dimensional, the process can be quantified by an analytical solution of the Fickian law (Crank 1975). Diffusion coefficients and other specific parameters can be calculated from obtained data after proving that the investigated processes follow the assumption of Fickian diffusion. However, common sorption experiments are not capable of providing information for individual components in a diffusing mixture as the total mass uptake is measured. To investigate the impact of individual chemical compounds in complex mixtures on the swelling behavior it is necessary to trace the diffusion of single substances simultaneously.

Within this work model fuels containing different amounts of m–xylene and dodecane are prepared to carry out sorption experiments. An analytical method based on a gas chromatographic technique with mass spectrometric detection is established to measure concentrations of individual substances in elastomer specimens simultaneously and time resolved during the swelling process. This procedure allows the calculation of diffusion coefficients and equilibrium concentrations for every mixture constituent.

2 EXPERIMENTAL

2.1 *Materials and reagents*

The elastomer used in this study was a carbon black filled acrylonitrile butadiene rubber (NBR) with an acrylonitrile content of 18%, supplied by the Deutsches Institut für Kautschuktechnologie (DIK). Table 1 gives the composition of the material.

Analytical grade reagents dodecane, m–xylene and n–heptane (>99%, Sigma Aldrich) were used for preparation of the immersion mixtures and GC/MS–standard solutions.

2.2 *Instrumentation*

GC/MS analyses were performed on an Agilent 7890A gas chromatograph coupled with an Agilent 5975 MSD mass spectrometer. The used column was a 30 m DB–5MS (0.25 mm i.d.; 0.25 μm film). All masses were determined with an analytical balance AG 285 (Mettler Toledo).

2.3 *Procedure*

First thin elastomer sheets (30 × 10 × 1 mm^3) were extracted with acetone in a soxhlet apparatus for 5 days to remove additives. Two extracted specimens were stored in screw–tight glass vessels containing ca. 50 mL of each m–xylene/dodecane mixture. Table 2 gives the composition of the prepared mixtures.

Table 1. Elastomer composition.

Component	Content/phr
Perbunan 1846	100
Bis(2-ethylhexyl) phthalate	20
6–PPD*	2
Carbon black (N 550)	60
Zinc oxide	5
Stearic acid	1
Sulfur	2
CBS	1.5
TMTM–80**	0.5

*N-(1,3-dimethylbutyl)-N'-phenyl-p-phenylene diamine.

**80% tetramethylthiuram monosulfide, 20% elastomer binder and dispersing agents.

Table 2. Composition of model liquids.

	Concentration/mol–%	
Mixture	m–Xylene	Dodecane
X10D90	9	91
X20D80	20	80
X33D67	33	67
X50D50	50	50
X80D20	77	23

The weight of one of two specimens was measured after periodic intervals until equilibrium swelling was reached (gravimetric approach). Additionally, a small piece (3 to 6 mg) was cut out of the second elastomer sheet and placed in a GC–vial (GC/MS procedure). 1 mL n–heptane was added and the vial was sealed. To ensure, that diffusion was only considered in one dimension, the pieces were removed in adequate distances from sample edges and already generated holes. For specimens, which were soaked in the neat solvents dodecane and m–xylene only the mass uptake was traced, as GC/MS–analyses would not provide any additional information. All sorption experiments were carried out at room temperature (21°C). The prepared GC–vials were treated in an ultrasonic bath to accelerate the extraction of the sorbed compounds from the elastomer pieces by n–heptane. Subsequently the solutions were analyzed by GC/MS. A quantification of dodecane and m–xylene based on an external calibration was performed.

3 RESULTS AND DISCUSSION

3.1 *Comparison of classical and GC/MS assisted sorption experiments*

Results from classical (gravimetric) and GC/MS assisted sorption experiments were compared to evaluate the consistency of both methods. In Figure 1 sorption curves, i.e. the relative mass

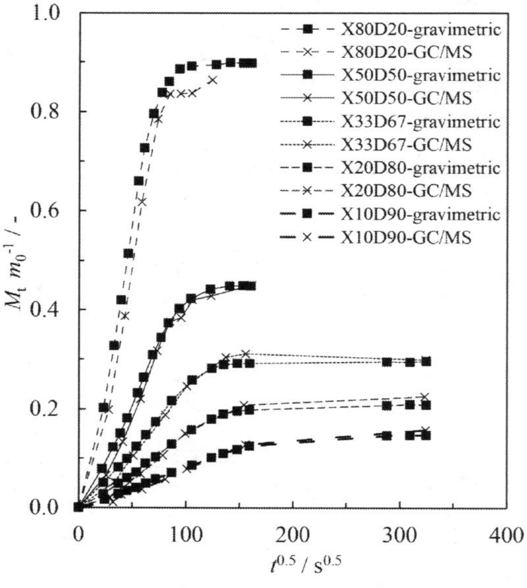

Figure 1. Comparison of sorption curves by gravimetric and GC/MS analysis.

uptake ($M_t\ m_0^{-1}$) versus the square root of time ($t^{0.5}$) are depicted for model liquids. The relative mass uptake is derived from GC/MS and gravimetric analyses.

M_t is the mass uptake at time t that can be calculated for the gravimetric method as

$$M_t = m_t - m_0 \quad (1)$$

Here m_t is the elastomer sheet mass at time t. m_0 is the initial mass of the elastomer sheet after extraction with acetone. In GC/MS assisted sorption experiments M_t is calculated as the sum of the masses of the sorbed m–xylene (m_{xylene}) and dodecane ($m_{dodecane}$) in the analyzed elastomer pieces according to following equation

$$M_t = m_{xylene} + m_{dodecane} \quad (2)$$

Accordingly, m_0 is calculated for each piece by subtracting the mass of the sorbed substances (M_t) from the total sample mass (m_t).

The obtained graphs show very good agreement between the two evaluated methods. However, minor deviations can be attributed to measurement uncertainty caused by weighing errors, as small samples are used for the GC/MS analysis.

From gravimetric and GC/MS based data diffusion coefficients D were calculated via iterative fitting of the theoretical equation for one–dimensional diffusion in a plane sheet according to Fickian law (Crank 1975).

$$\frac{M_t}{m_0} = \frac{M_\infty}{m_0}\left[1 - \frac{8}{\pi^2}\sum_{n=0}^{\infty}\left(\frac{1}{(2n+1)^2} e^{\frac{-(2n+1)^2\pi^2 Dt}{h^2}}\right)\right] \quad (3)$$

The iteration procedure considered the first 11 terms (n = 0 to 11) of equation (3). M_∞ is the mass uptake at equilibrium swelling and h is the sheet thickness (Aminabhavi & Munnoli 1994). In Figure 2 calculated diffusion coefficients and equilibrium concentrations from sorption experiments based on GC/MS and gravimetric measurement are compared.

The determined diffusion coefficients and equilibrium mass uptakes increase with growing m–xylene concentration due to more pronounced interactions between polar elastomer and the polar m–xylene.

For both methods similar diffusion coefficients and equilibrium concentrations are determined. Again deviations probably occur due to weighing errors and evaporation of sorbed substances from the elastomer samples during weighing. Nevertheless, the total agreement is fairly good for both parameters.

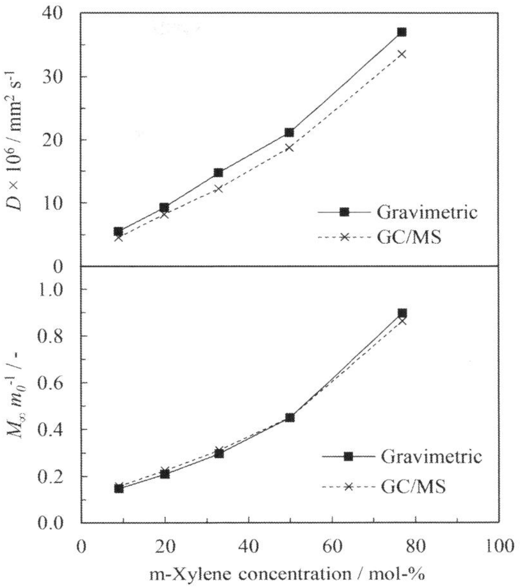

Figure 2. Comparison of diffusion coefficients and equilibrium concentrations for classical (gravimetric) sorption experiments and GC/MS assisted sorption experiments.

3.2 *Investigation of the diffusion behavior of individual substances in a mixture*

Individual sorption curves were generated for dodecane and m–xylene by using chromatographic data of each sorption experiment allowing a simultaneous tracing of their diffusion processes. Exemplarily in Figure 3 chromatograms of the sorption of mixture X50D50 after different storage durations are displayed. The intensity of peaks at ca. 6 min corresponds to the m–xylene concentration whereas the peaks at ca. 11 min represent dodecane, respectively.

Peak intensities of dodecane and m–xylene rise with increasing storage duration. Therefore, concentrations of both substances increase as the sorption proceeds.

Sorption curves of dodecane and m–xylene are depicted in Figure 4 for the model liquids X20D80 and X50D50. To receive sorption curves of the mixtures, the relative mass uptakes of every single component, derived from chromatographic data, was summed up.

The curves in Figure 4 indicate a distinct dependency of diffusion processes on the detailed composition of the mixture. Both, diffusion rate and equilibrium mass uptake, increase for either substance with increasing m–xylene concentration. In consequence, the overall diffusion coefficient and the equilibrium mass uptake increase when m–xylene concentration is higher. By iterative fitting of equation (3), D can be calculated separately for each constituent as well as the whole mixture. Calculated D and $M_\infty\ m_0^{-1}$ values are summarized in Table 3. As reference diffusion coefficients and

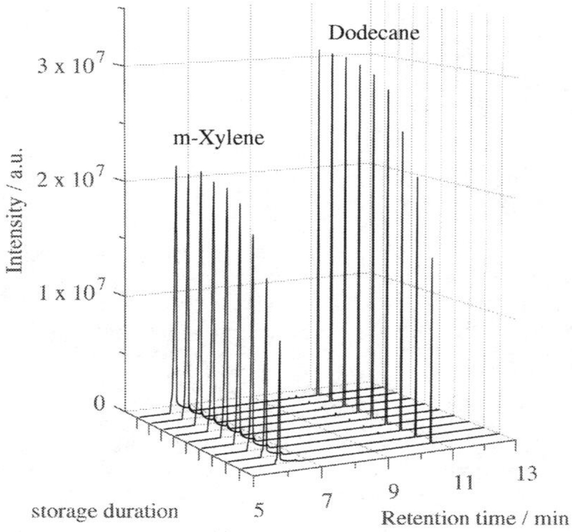

Figure 3. Chromatograms of GC/MS assisted sorption experiment with X50D50 after different storage durations.

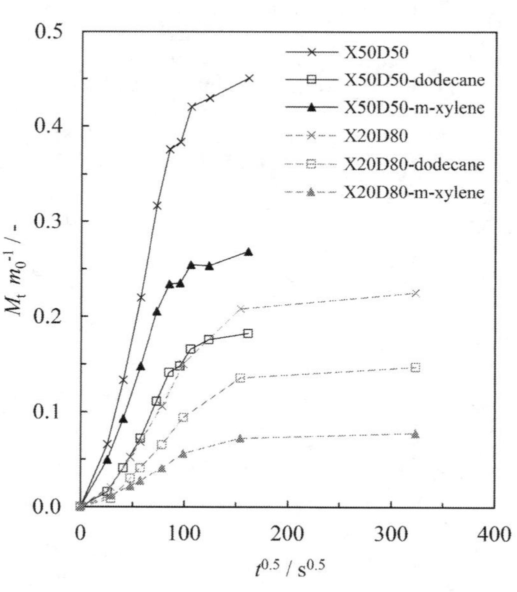

Figure 4. Sorption curves determined by GC/MS for mixtures X20D80 and X50D50 as well as their individual components.

Table 3. Calculated diffusion coefficients and equilibrium concentrations.

	$D \times 10^6$/mm^2 S^{-1}	$M_\infty\ m_0^{-1}$/–
Dodecane	4.1	0.109
X20D80	9.3	0.225
X20D80–**dodecane**	7.3	0.147
X20D80–**m–xylene**	9.4	0.078
X50D50	18.8	0.450
X50D50–**dodecane**	16.2	0.182
X50D50–**m–xylene**	20.9	0.268
m–Xylene	58.0	1.503

equilibrium mass uptake of neat dodecane and m–xylene, determined by gravimetric sorption experiments, are included.

The neat substances show in case of dodecane the lowest and in case of m–xylene the highest D and $M_\infty\ m_0^{-1}$ values, respectively. They represent the lower and upper limits of diffusion coefficients and equilibrium mass uptakes when a mixture of these two chemicals is given. Diffusion coefficients and equilibrium mass uptakes of model liquid mixtures and their individual components increase with growing m–xylene concentration. It is noteworthy, that D values of mixtures lie between the values of the corresponding individual constituents whereby diffusion coefficients of m–xylene are higher than those of dodecane. Generally, the diffusion behavior is dominated by the component with the higher diffusion coefficient and equilibrium mass uptake.

4 CONCLUSION

This work presents an approach to receive detailed information on the behavior of single components during sorption by using gas chromatography/mass spectrometry. In contrast to classical swelling experiments which are merely gravimetrical procedures, GC/MS assisted sorption experiments allow time resolved tracing of diffusion processes of mixtures' individual components in elastomers.

It was found, that the introduced and the classical method show very good agreement when diffusion coefficients and equilibrium concentrations are compared. Slight deviations can be probably attributed to weighing errors and evaporating substances from the elastomer during sample preparation. From GC/MS data, diffusion coefficients and equilibrium mass uptake were calculated for mixtures containing different amounts of dodecane and m–xylene as well as their individual components in an acrylonitrile butadiene rubber. D and $M_\infty\ m_0^{-1}$ of model liquids increase with a higher m–xylene concentration. Mixtures show D values which lie between those of the individual components. m–Xylene exhibits the highest diffusion coefficients in each case. In general, diffusion of a mixture is predominately governed by the concentration of the component with the higher individual D and $M_\infty \; m_0^{-1}$ values.

The results found, contribute to a better understanding of diffusion processes in elastomers. They may be applied to further compatibility considerations of sealing materials in fuels of various compositions.

REFERENCES

Aminabhavi, T. & R. Munnoli (1994). An assessment of chemical compatibility of bromobutyl rubber, chlorosulfonated polyethylene and epichlorohydrin membranes in the presence of some hazardous organic liquids. *Journal of Hazardous Materials 38*, 223–242.

Crank, J. (1975). *The Mathematics of Diffusion*. Oxford: Clarendon Press.

Dooley, S., S. Won, M. Chaos, J. Heyne, Y. Ju, F. Dryer, K. Kumar, C. Sung, H. Wang, M. Oehlschlaeger, R. Santoro, & T. Litzinger (2010). A jet fuel surrogate formulated by real fuel properties. *Combustion and Flame 157*, 2333–2339.

George, S., V.K.T., & S. Thomas (2000). Molecular transport of aromatic solvents in iso–tactic polypropylene/acrylonitrile–co–butadiene rubber blends. *Polymer 41*, 579–594.

Graham, J., R. Striebich, K. Myers, D. Minus, & W. Harrison (2006). Swelling of nitrile rubber by selected aromatics blended in a synthetic jet fuel. *Energy Fuels 20*, 759–765.

Constitutive Models for Rubber X – Lion & Johlitz (Eds)
© 2017 Taylor & Francis Group, London, ISBN 978-1-138-03001-5

Multi-objective optimization of hyperelastic material constants: A feasibility study

S. Connolly, D. Mackenzie & T. Comlekci
Department of Mechanical and Aerospace Engineering, University of Strathclyde, Glasgow, UK

ABSTRACT: A preliminary study of the use of multi-objective optimization methods in determining the material constants of a hyperelastic material model is presented. Classical experimental data is fitted to the third-order Ogden and the third-order Yeoh model hyperelastic material models through finite element analysis and multi-objective optimization, using the Simulia products Abaqus and Isight, respectively. The resulting hyperelastic material constants for both material models provide an improved curve-fit over the integrated fitting tool within Abaqus, in terms of average absolute difference (error). This indicates that this approach is suitable for development of an efficient and effective method for experimental characterization of rubber-like materials using both homogeneous and inhomogeneous experimental results.

1 INTRODUCTION

The ability to accurately simulate a component's behavior allows the engineer to gain a greater understanding of its expected service life. To do this successfully, the conditions the component is exposed to (structural and/or thermal) must be known, along with the corresponding material properties. In the case of rubber components, the description of a given material requires independent characterization due to the variation in properties resulting from chemical composition, processing and manufacturing methods. In addition, the stress-strain response of rubber is highly nonlinear and dependent on the applied mode of deformation, requiring measurement of its multiaxial response.

Hyperelastic material models are widely used in simulation or mathematical descriptions of the static response of rubber. These may have a micromechanical or phenomenological bases. Although the former is physical in its formulation, both approaches typically require the same degree of testing to calibrate the response. Compressibility effects may be taken into account by partitioning the deformation gradient into isochoric and deviatoric components, which requires an additional confined compression experiment. Models of more complex isothermal behaviors experienced by rubbers, such as the Mullins effect with permanent set and induced anisotropy (Diani, et al., 2009), hysteresis and viscoelasticity (Bergstrom & Boyce, 1999), are usually based on hyperelastic models. The remainder of this paper's content will study the isothermal and quasi-static material response of rubber and a novel method to obtain hyperelastic constants from experimental data.

The most common method of characterizing the static and incompressible response of rubber is through performing three homogeneous tests: uniaxial tension (UT), planar tension (PT) (equivalent to pure shear) and equibiaxial tension (ET). These tests were popularized by Treloar on 8% Sulphur rubber (Treloar, 1944) who collected the equibiaxial tension, or 2-dimensional extension, data with the use of an inflated rubber sheet. Uniaxial and planar tension data can both be collected using cut rubber sheets of simple geometry and a commonplace uniaxial testing machine. However, equibiaxial testing requires some form of bespoke testing equipment usually in the form of simultaneous loading of perpendicular axes within a 2D plane or perpendicularly applied inflation and punch tests.

Due to the difficulty of applying equibiaxial loading and gaining accurate results, several different methods have been developed to obtain this data. One method is to adapt a uniaxial testing machine with a device that translates a controlled ratio of the applied vertical force horizontally (Brieu, et al., 2006). The difficulty in this method is in achieving a homogeneous equibiaxial response, as the chosen geometry of the sample will affect the degree of biaxiality (Seibert, et al., 2014). Alternatively, a load can be applied radially to a circular sample using a complex pulley system. Other methods use inhomogeneous deformations and advanced imaging techniques to extract the required equibiaxial material response (Sasso, et al., 2008).

In this study, a multi-objective optimization procedure was developed to determine whether hyperelastic material constants can be obtained by an alternative method. Standard homogeneous tests are represented by simple finite element representations for use in the optimization process to obtain the optimal hyperelastic constants. If successful, this method will be extended to efficiently gain optimized material constants using simple inhomogeneous test data and equivalent finite element simulations.

2 OPTIMIZATION METHODOLOGY

2.1 *Material data and modelling*

Treloar's data for UT, PT & ET tests from (Steinmann, et al., 2012) were adopted for use in this optimization procedure. This data set is commonly used as a benchmark in the development and validation of hyperelastic material models.

Comparison studies investigating several material models in terms of the best fit to the entire data set (Marckmann & Verron, 2006) and the best fit to all data sets using data from a single test (Steinmann, et al., 2012, Hossain & Steinmann, 2013) indicate that the micro-mechanical extended-tube model (Kaliske & Heinrich, 1999) performed best overall. The best phenomenological model was the third-order Ogden model (Ogden, 1972) with six material constants, alhough this model is incapable of predicting other data sets when only one test's data was used.

Ideally, the extended-tube and third-order Ogden material models would therefore be used in this study, however, the extended-tube model was not available at the time of study and required a UHYPER or UMAT subroutine for its implementation. Therefore, the third-order Ogden model was used along side a third-order Yeoh model (Yeoh, 1993), which is a hybrid micro-mechanical model with a phenomenological adjustment. Since the number of coefficients is proportional to the time required for the optimization process, with only three material constants, the third-order Yeoh model provided an initial insight into the feasibility of the chosen optimization methods and a suitable comparison.

2.2 *Finite element modelling*

For each mode of deformation, an Abaqus input script was generated with only a single millimeter cubed element. Within this script, the hyperelastic constants were parameterized such that they could be easily identified and controlled by the optimizer. Displacement boundary conditions were applied for each variation to the appropriate nodes, shown for the case of equibiaxial tension in Figure 1. With displacements applied in millimeters, the nodal force reaction was extracted along with the nodal displacement to give stress and strain results in MPa and mm/mm respectively. Though the extracted force from a single node gives only a quarter of the actual stress value, the matched data was consequently quartered for comparison.

2.3 *Optimization method*

The optimization procedure is performed automatically using a combination of two Simulia products: Abaqus and Isight. Abaqus is used for the simulation of the input set of hyperelastic constants using the simple finite element models, discussed in section 2.2. Isight is responsible for assessing and directing the input of the hyperelastic constants based on the results of the comparison of the error between the finite element results and the experimental data sets. As there are multiple data sets, one for each mode of deformation, the optimization process is therefore multi-objective and a weight function is used to ensure that the error of each data set's magnitude is normalized. For all optimizations, the absolute difference (error) function is used.

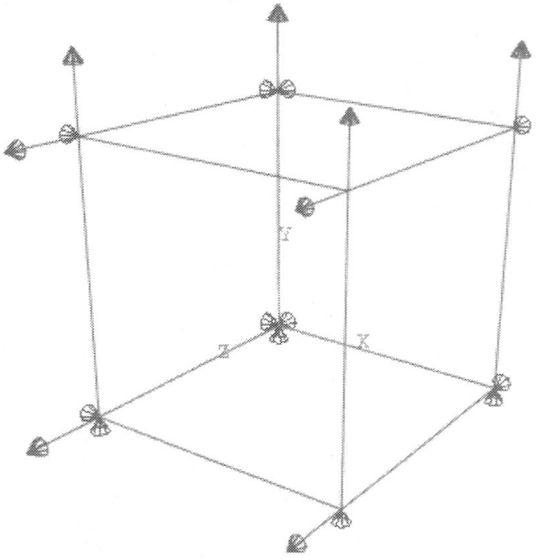

Figure 1. Equibiaxial Tension boundary conditions for the single element simulated model.

Prior to optimization, Abaqus' evaluation tool was used for comparison to gain material constants for the selected hyperelastic model. This tool requires only the input of the experimental data sets and the selection of the material models to be evaluated. Abaqus then generates the 'optimal' hyperelastic constants using linear least squares method or a Levenberg-Marquardt algorithm for the Yeoh and Ogden curve fits respectively. The stability is then checked for uniaxial, planar and equibiaxial deformation in tension and compression within the nominal strain range: $-0.9 \leq \varepsilon \leq 9.0$, using the Drucker stability criterion. (Abaqus, 2016)

In the selection of the optimization method, two distinct variations were used: the first used initial guesses based on the values obtained by the Abaqus evaluation and employed an optimization algorithm; the second method used trivial starting points and wide-ranging bounds with a hybrid optimization-exploratory algorithm. In the former, the Hooke-Jeeves algorithm (Hooke & Jeeves, 1961) was used due to its ability to find local minima. As for the latter, the parallel Pointer-2 algorithm (Van der Velden & Koch, 2010) was selected. Pointer-2 is an algorithm that controls four optimisation methods in serial or in parallel, ensuring that the design space is explored and most, if not all, minima are found within it. The Isight optimisation process for the Hooke-Jeeves method is shown in Figure 2.

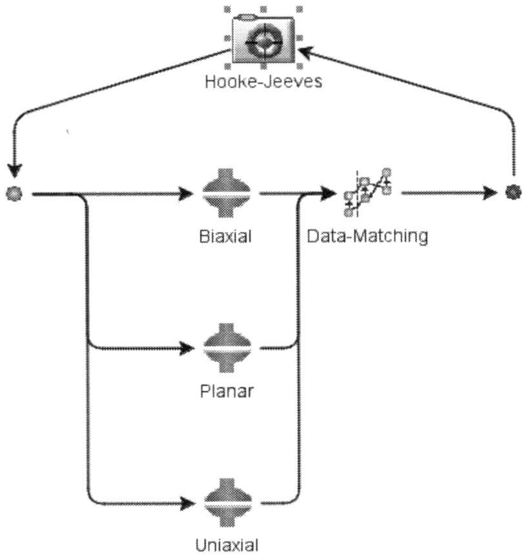

Figure 2. Hooke-Jeeves multi-objective optimization within Isight.

3 OPTIMIZATION RESULTS

For both material models, an optimization using each method was attempted. In the case of the Hooke-Jeeves optimization, which is somewhat dependent on the chosen starting point, the initial material constants were chosen to within one decimal point of the evaluated constant from Abaqus. As for the upper and lower bounds, the nearest whole integer was used rounding upwards and downwards respectively. In the Pointer-2 optimization, the starting points were chosen arbitrarily and the constants were bound by a wide range.

The solutions found to be optimal for the fit of all data sets were compared in terms of the absolute difference (error) function used for the optimization. Stress-strain graphs were plotted to visually assess their fit to the material data. In the presentation of the results, the following abbreviations are used: Error: absolute difference (error), Abq: Abaqus evaluated results, H-J: Hooke-Jeeves and P-2: Pointer-2. Other symbols used are the common notations for the hyperelastic material constants or have been previously defined.

3.1 *3rd-order Yeoh*

Given that the third-order Yeoh model was not ranked as a highly accurate model in the referenced comparison studies, it was not expected to achieve as close a fit as the Ogden model. This was observed to be the case in the initial Abaqus evaluation of the material models. The Hooke-Jeeves model was capable of gaining a better fit to the data, in terms of average error and the Pointer-2 optimization gained solutions with a significantly smaller error value. Solutions for the Pointer-2 and Hooke-Jeeves optimizations are shown in Table 1, along with the Abaqus evaluated results. The Drucker stability check revealed that all sets were stable for the assessed nominal strain range.

Table 1. Hyperelastic constants and absolute difference (error) for third-order Yeoh optimizations.

Yeoh:	Abq	H-J	P-2
c10	0.1852	0.1740	0.2032
c20 (E-3)	−1.449	−3.242E-3	−2.786
c30 (E-5)	3.973	2.460	6.847
Average Error	1.407	1.298	1.097
ET Error	2.367	0.8590	1.767
PT Error	0.4686	1.582	0.6573
UT Error	1.386	1.453	0.8663

The test data alongside the results of the fitted curves are shown in Figure 3. Sixth-order polynomial functions have been used to plot the Abaqus and optimization results for clarity. It can be seen that the Abaqus evaluated Yeoh constants seem to produce a visually better fit than the Pointer-2 optimization, regarding their uniaxial behavior. However, the Pointer-2 data was found to be mathematically the better fit in terms of the chosen error function for uniaxial tension and the overall average error. Different error functions will be investigated in future studies.

3.2 *Third-order Ogden*

The third-Ogden model was expected to capably achieve a solution for the simpler Hooke-Jeeves method. However, the Pointer-2 method was less likely to achieve the desired result due to the vast number of numerical combinations with six coefficients. After running the Pointer-2 optimization for a 24 hour period, the program was stopped and the best result was taken. As can be seen in Table 2, the result is significantly worse than the Abaqus and Hooke-Jeeves results, in terms of absolute difference (error), for all data sets. This set was also found to be unstable in uniaxial tension greater than $\varepsilon = 3.08$ and, the equivalent deformation mode, in biaxial compression less than $\varepsilon = -0.505$. The other sets were completely stable over the assessed range.

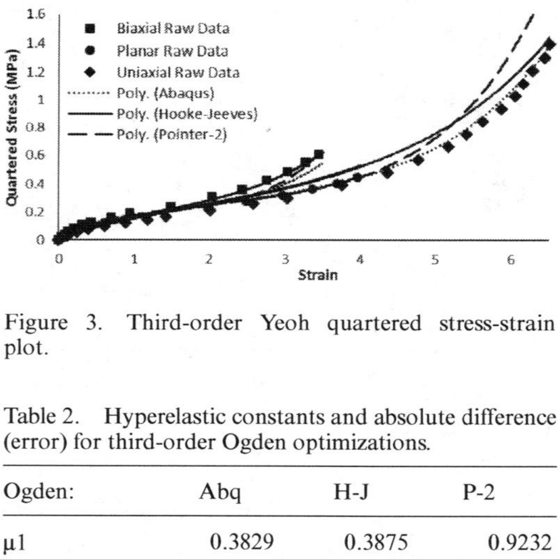

Figure 3. Third-order Yeoh quartered stress-strain plot.

Table 2. Hyperelastic constants and absolute difference (error) for third-order Ogden optimizations.

Ogden:	Abq	H-J	P-2
$\mu1$	0.3829	0.3875	0.9232
$\alpha1$	1.452	1.463	1.382
$\mu2$(E-3)	1.309	1.006	25.92
$\alpha2$	5.489	5.587	1.490
$\mu3$(E-2)	1.545	1.258	−54.33
$\alpha3$	−1.875	−1.963	0.591
Average Error	0.1954	0.1239	2.042
ET Error	0.1583	0.1205	3.174
PT Error	0.1213	0.1531	1.466
UT Error	0.3066	0.0982	1.488

The stress-strain results of the third-Ogden optimization are shown in Figure 4. As previously, the optimization results are plotted using sixth-order polynomials. Both the Abaqus and Hooke-Jeeves results are a good fit of the entire data set. However, as suggested by the numerical error, the Pointer-2 set did not manage to gain a suitable set of coefficients within the prescribed time. If run for long enough, a comparable or better result would be obtained. However, the required time for a solution would not be a feasible alternative means for material characterization.

3.3 *Observations*

The third-order Ogden model produced a significantly better fit of the data than the third-order Yeoh model. However, with six coefficients, the Ogden model did not gain a suitable solution with

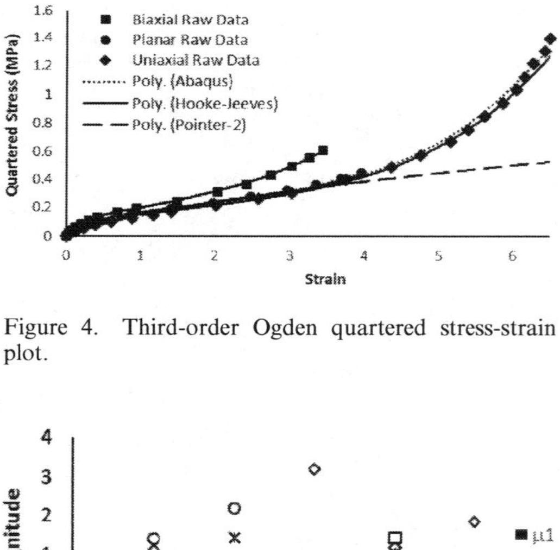

Figure 4. Third-order Ogden quartered stress-strain plot.

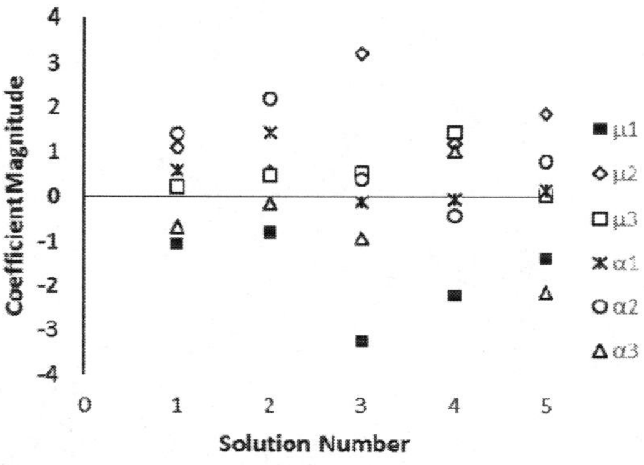

Figure 5. Third-order Ogden coefficients plotted for five optimization results with similar error.

the exploratory method. When viewing the sets of coefficients produced within the Pointer-2 optimization, the constants of the Ogden model were found to fluctuate significantly for solutions with similar magnitudes of error. This demonstrates that a unique set of constants may not exist for the third-order Ogden model for this data set due to it being phenomenological in nature, which is in agreement with Ogden et al. (Ogden, et al., 2004). This is shown in Figure 5, where the magnitude of the third-order Ogden coefficients are plotted for five solutions with similar error. This suggests that, where an approximate initial guess is not known, mechanically based models with fewer coefficients are preferable.

4 NOVEL MATERIAL CHARACTERISATION METHOD

Using the results of this feasibility study, it is proposed that this method could be utilized in an alternative means of characterizing rubber-like materials. This alternative method would take the results of homogeneous and inhomogeneous experiments and, using multi-objective optimization, find the optimal set of material constants to fit all experimental data sets. Simple uniaxial compression tests can achieve equivalent results to equibiaxial tension, with the assumption that rubber is incompressible, but these tests require negligible friction. By using the results from a bonded compression test and an equivalent simulated experiment within a multi-objective optimization, it is hypothesized that the equivalent material constants would be discovered.

This method, if successful, could provide an alternative to bespoke equibiaxial testing equipment when gathering the multi-axial response for rubber-like materials. The data to capably characterize the incompressible, static response is hypothesized to use three tests: uniaxial tension, planar tension and bonded uniaxial compression tests. The benefit of these tests is that they can all be performed using the same uniaxial testing machine, provided it is capable of producing the required loads in both tension and compression. Additionally, the required specimens are of simple geometry and easy to manufacture.

The accuracy of the method is largely dependent on both the experimental results and finite element results. The finite element results will be of approximately comparable accuracy to the material model's limitations, provided that finite element phenomena are appropriately considered, notably mesh convergence and volumetric-locking in this instance. However, the experimental error may be increased due to the inclusion of inhomogeneous tests. These may require significantly more cycling before a consistent response is attained, due to the propagation of stress-softening through the specimen, owing to the Mullin's effect. Also, the strain-rate will be somewhat variant throughout the material and will require consideration for the different tests to be consistent in this regard.

In order to validate this method, it will be important to use material models to generate simulated equibiaxial data for comparison to equibiaxial data gathered in a more conventional form. Also, it will be necessary to use compression specimens of different diameter to provide further validation and demonstrate repeatability.

5 CONCLUSIONS

Using Treloar's data and two multi-objective optimization techniques, the feasibility of integrating these techniques in an alternative method for material characterization has been investigated. The results of this study have found that a suitable initial value and approximate bounds for the hyperelastic coefficients is of significant importance. Also, higher-order phenomenological models are expected to be less appropriate unless an optimization process that exploits local minima is used.

In gaining optimized constants, the Hooke-Jeeves method was more efficient than the Pointer-2 method, as would be expected. However, the time taken to gain a solution using either multi-objective optimization method is substantially longer than the Abaqus evaluation tool. A significant portion of the time required is spent in 'housekeeping' tasks within Abaqus, some examples are: accessing the license server, writing the results files and reading output database files. For this type of optimization, the simple homogeneous deformations of the unit-cube models could be more efficiently simulated in a purely mathematical form.

ACKNOWLEDGEMENTS

This project was supported in full by an EPSRC Studentship grant related to (EP/N509760/1).

REFERENCES

Abaqus 2016. v2016 User's Manual, Dassault Systèmes, Providence, RI, USA

Bergstrom, J.S. & Boyce, M.C., 1999. Mechanical behavior of particle filled elastomers. Rubber chemistry and technology, 72(4), pp. 633–656.

Brieu, M., Diani, J. & Bhatnagar, N., 2006. A New Biaxial Tension Test Fixture for Uniaxial Testing Machine - A Validation for Hyperelastic Behavior of Rubber-like materials. Journal of Testing and Evaluation, 35(4), pp. 1–9.

Diani, J., Fayolle, B. & Gilormini, P., 2009. A review on the Mullins effect. European Polymer Journal, 45(3), pp. 601–612.

Hooke, R. & Jeeves, T., 1961. Direct Search Solution of Numerical and Statistical Problems. Journal of the ACM (JACM), 8(2), pp. 212–229.

Hossain, M. & Steinmann, P., 2013. More hyperelastic models for rubber-like materials: consistent tangent operators and comparative study. Journal of the Mechanical Behavior of Materials, 22(1–2), pp. 27–50.

Kaliske, M. & Heinrich, G., 1999. An extended tube-model for rubber elasticity: statistical-mechanical theory and finite element implementation. Rubber Chemistry and Technology, 72(4), pp. 602–632.

Marckmann, G. & Verron, E., 2006. Comparison of hyperelastic models for rubber-like materials. Rubber Chemistry and Technology, 79(5), pp. 835–858.

Ogden, R.W., 1972. Large Deformation Isotropic Elasticity—On the Correlation of Theory and Experiment for Incompressible Rubberlike Solids. Proceedings of the Royal Society of London A: Mathematical, Physical and Engineering Sciences, 326(1567), pp. 565–584.

Ogden, R.W., Saccomandi, G. & Sgura, I., 2004. Fitting hyperelastic models to experimental data. Computational Mechanics, 32(6), pp. 484–502.

Sasso, M., Palmieri, G., Chiappini, G. & Amodio, D., 2008. Characterization of hyperelastic rubber-like materials by biaxial and uniaxial stretching tests based on optical methods. Polymer Testing, 27(8), pp. 995–1004.

Seibert, H., Scheffer, T. & Diebels, S., 2014. Biaxial testing of elastomers-Experimental setup, measurement and experimental optimization of specimen's shape. Technische Mechanik, 34(2), pp. 72–89.

Steinmann, P., Hossain, M. & Possart, G., 2012. Hyperelastic models for rubber-like materials: consistent tangent operators and suitability for Treloar's data. Archive of Applied Mechanics, 82(9), pp. 1183–1217.

Treloar, L.R.G., 1944. Stress-strain data for vulcanised rubber under various types of deformation. Transactions of the Faraday Society, Volume 40, pp. 59–70.

Van der Velden, A. & Koch, P., 2010. Isight design optimization methodologies. ASM Handbook, Volume 22.

Yeoh, O.H., 1993. Some Forms of the Strain Energy Function for Rubber. *Rubber Chemistry and Technology,* 66(5), pp. 754–771.

Constitutive Models for Rubber X – Lion & Johlitz (Eds)
© 2017 Taylor & Francis Group, London, ISBN 978-1-138-03001-5

Strain-induced crystallization ability of hydrogenated nitrile butadiene rubber

K. Narynbek Ulu
Institut de Recherche en Génie Civil et Mécanique, UMR CNRS 6183, Ecole Centrale de Nantes, France
LRCCP, Vitry-sur-Seine, France

M. Dragičević & P.-A. Albouy
Laboratoire de Physique des Solides, UMR CNRS 8502, Université Paris-Sud, Orsay, France

B. Huneau
Institut de Recherche en Génie Civil et Mécanique, UMR CNRS 6183, Ecole Centrale de Nantes, France

A.-S. Béranger & P. Heuillet
LRCCP, Vitry-sur-Seine, France

ABSTRACT: Strain-Induced Crystallization (SIC) phenomenon in elastomers is widely investigated, especially for natural rubber. It is considered as a major reinforcement mechanism explaining the very good mechanical properties of natural rubber, particularly its resistance to crack growth. Despite the fact that most of the studies are focused on natural rubber, other elastomers also exhibit SIC. Here, we study the ability of Hydrogenated Acrylonitrile Butadiene Rubber (HNBR), a synthetic rubber with high resistance to heat and petroleum products, to crystallize when it is stretched. Few studies have reported results on the SIC of HNBR, but none present a quantitative evolution of the phenomenon for different elongations. In the present study, X-ray diffraction patterns are obtained at three temperatures (20°C, 30°C and 40°C) for HNBR cross-linked with peroxide. These are recorded during uniaxial tensile tests at low strain rate in order to establish the relationship between the index of crystallinity and the stretch ratio; the orientation of the amorphous part of the elastomer is also presented.

1 INTRODUCTION

Strain-induced crystallization (SIC) of rubbers, especially of natural rubber (NR), has been extensively studied. At room temperature and without deformation, NR is an amorphous elastomer. However, as first discovered by Katz in 1925 using wide angle X-ray diffraction (WAXD), the NR polymer chains arrange in a more ordered state under large deformations (Katz 1925). A lot of studies have been devoted to this phenomenon: the reader can refer to the reviews of Tosaka (2007) or Huneau (2011) for the works published until 2010; in the recent years, one can also cite the works of Albouy et al. (2012), Brüning et al. (2012), Rublon et al. (2013) and Candau et al. (2012, 2014) among others.

SIC is considered as a major reinforcement mechanism that explains the very good mechanical properties of NR, especially its resistance to crack growth either under monotonous or cyclic loading (Hamed et al. 1996). If most of the studies have focused on NR, one should note that other elastomers can also exhibit SIC. Among them, we focus our attention on hydrogenated acrylonitrile butadiene rubber (HNBR), which is a synthetic rubber used in the industry specifically for its high heat and oil resistances.

In the past, few studies revealed its ability to crystallize under strain. HNBR tends to undergo strain-induced crystallization only for high acrylonitrile (ACN) content (approximately greater than 35 wt.%) and with increasing hydrogenation (Kubo et al. 1987, Braun et al. 1992, Sawada 1994, Severe & White 2000). NBR, which is not hydrogenated, does not crystallize in comparison. Therefore, the tensile strength of HNBR is significantly higher than the one of NBR (Sawada 1994). Strain-induced crystallization is possible due to the polymeric chains having a fairly regularly alternating comonomer sequence (butylene and acrylonitrile units in almost equal amount) (Braun et al. 1992). Moreover, Braun et al. detected strain-induced crystallization of HNBR (48 wt.% ACN, hydrogenation percent not specified, not crosslinked) at elongation of 100% ($\lambda = 2$) for experiments at room temperature. Crystallization occurs at relatively low elongations for temperatures between the glass transition, T_g, and room temperatures. Below T_g, strain-induced crystallization does not occur and the structure is amorphous due to restricted mobility of the polymer chains; above the melting temperature, T_m, the structure tends to stay amorphous due to excessive flow during deformation and, therefore, it becomes significantly difficult to achieve SIC; for temperatures greater than 50°C, SIC does not occur (Braun et al. 1992).

In stretched HNBR samples, X-ray diffraction patterns have been observed only by few authors (Obrecht et al. 1986, Braun et al. 1992, Severe & White 2000, Osaka et al. 2013). Braun et al. first proposed a unit cell for HNBR containing 48% wt. ACN with $a = 0.5$ nm, $b = 0.36$ nm and $c = 0.77$ nm. Those values have been later confirmed, with small discrepancies, by Severe et al. (2000).

Overall, it can be said that a complete study has not been conducted on the SIC phenomenon of HNBR. At the best, some authors give the X-ray diffraction pattern for one (Obrecht et al. 1986, Severe & White 2000) or two (Braun et al. 1992, Osaka et al. 2013) stretch ratios. The first objective of this study, consequently, is to determine precisely the relationship between the crystallinity and the stretch ratio during *in-situ* tensile tests at three temperatures (20°C, 30°C and 40°C). The curves of the stress evolution as well of the orientation of the amorphous phase of the elastomer are also presented as a function of the stretch ratio. The discussion of these results is done with the few above-mentioned studies on HNBR and with more extended literature on NR.

2 MATERIAL AND EXPERIMENTAL PROCEDURE

2.1 *Material*

HNBR is a copolymer initially made of acrylonitrile (C_3H_5N) and butadiene (C_4H_6) as the well-known NBR (nitrile butadiene rubber), but having most of the double bonds of the butadiene saturated with hydrogen, inducing the transformation of butadiene to butylene (C_4H_8).

The considered material is a highly saturated (96% hydrogenation) HNBR containing 44 wt.% of acrylonitrile (ACN). It is crosslinked with peroxide and is denoted HNBR-44 in the following sections.

2.2 *In situ tensile DRX*

The tensile machine is mounted on a rotating anode X-ray generator (40 kV, 40 mA, focus size: 0.2 × 0.2 mm²) equipped with a copper anode whose K_α radiation (0.1542 nm) is selected by a doubly curved graphite monochromator. The sample is located at the focalization point, which ensures maximum diffracted intensity. Diffraction patterns are recorded via a pixel camera; depending upon the elastomer nature, exposure times as low as 0.3 s can be used. Sample holders are fixed onto two opposing actuators that are symmetrically translated. One of the actuators is further equipped with a load sensor. The stretching amplitude is 500 mm at a speed that can be varied between 1 and 800 mm/min. A polycarbonate enclosure, equipped with a Kapton window for X-ray exit, allows keeping the temperature constant between 0°C and 90°C.

In the present study, experiments were performed with an exposure time of 4 s, at 20 mm/min and for three temperatures: 20°C, 30°C and 40°C.

2.3 *Data processing*

Figure 1 gives a typical diffraction pattern of HNBR and the scattering geometry. The data treatment is based on angular φ scans centered on the so-called amorphous ring (that mainly originates from inter-chain correlations) as indicated by the two black dotted circles; it includes the most intense crystallization peak located in the equatorial position. This allows for an easier separation of the amorphous and crystalline contributions.

From the X-ray pattern, the amorphous contribution to the scattered intensity can be fitted by the following function $C_0(1+\sum_{n=1}^{\infty} P_{2n}^{RX} P_{2n}(\varphi))$ where φ is the azimuthal angle. For HNBR, first two coefficients $\langle P_2^{RX} \rangle$ and $\langle P_4^{RX} \rangle$ are needed to account for the experimental intensity modulation.

The crystallization peak is adjusted by a Pearson VII function, commonly used in crystallography for its ability to fit various peak profiles. We define the crystallinity index as a ratio of the intensity associated to crystalline spots to the total scattered intensity within the delimited zone.

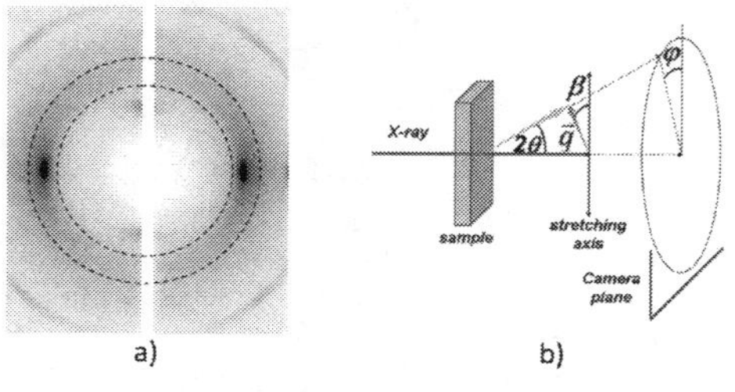

Figure 1. a) Representative diffraction pattern of HNBR-44 obtained at the maximum of mechanical cycling (acquisition time 4 s, stretch ratio 5.5 and vertical stretching direction). b) Schematic representation of the scattering geometry.

3 RESULTS AND DISCUSSION

3.1 *Index of crystallinity*

Figure 2 shows the evolution of the index of crystallinity with the stretch ratio for the three considered temperatures. Similar to NR, the crystallinity curve exhibits a hysteresis: the stretch ratio necessary for crystallization (i.e. SIC onset), denoted by λ_C, is greater than the stretch ratio corresponding to the melting of crystallites, λ_M. Both λ_C and λ_M increase with the temperature. Table 1 gives the values of λ_C and λ_M determined by the intersection of the dotted line (extrapolation of the linear part) with the abscissa axis.

Moreover, for a given stretch ratio, the incidence of SIC decreases with increased temperature as expected from thermodynamics. This has already been suggested by Braun et al. (1992), who noticed that SIC just about disappears at 50°C. The evolution of crystallinity with respect to both the stretch ratio and the temperature is finally very similar to the one observed for NR (Albouy et al. 2005, Candau et al. 2015).

3.2 *Stress-strain curve*

Figure 3 gives the engineering stress evolution with respect to the stretch ratio.

As expected, a stress upturn (stiffening of the material) is observed. It is shifted toward higher values of stretch ratio and becomes less pronounced with increasing temperature. This seems to be strongly related to the crystallinity evolution previously described in Figure 2. Indeed, the stress upturn occurs at the onset of SIC and is more distinct when the crystallinity is elevated. For HNBR, Sawada et al. (1994) already noticed that the stress upturn increases with the hydrogenation level, suggesting an effect of SIC since the capability of the elastomer to crystallize increases with the hydrogenation level. However, to the authors' knowledge, this effect of SIC on the mechanical properties has never been demonstrated so clearly for the HNBR by means of simultaneous measurements. Finally, it is important to note that this observation is consistent with the literature for NR (Albouy et al. 2005, Albouy et al. 2014) and seems to confirm the stiffening effect of SIC, irrespective of the considered elastomer.

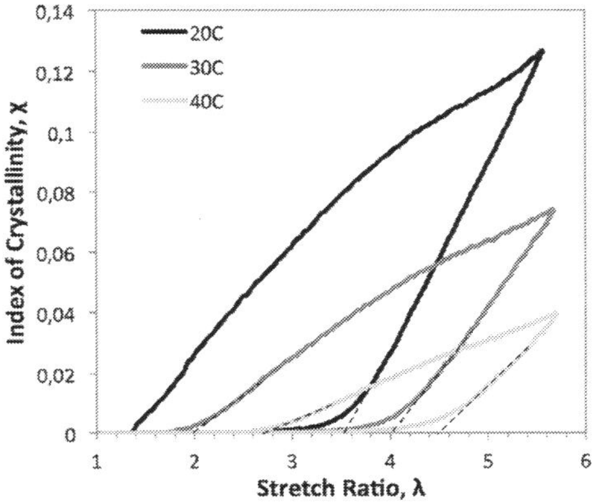

Figure 2. Index of crystallinity of HNBR-44 with respect to the stretch ratio at 20°C, 30°C and 40°C.

Table 1. Crystallization onset and melting stretch ratios of HNBR-44 at different temperatures.

	20°C	30°C	40°C
λ_C	3.5	4.0	4.5
λ_M	1.4	2.0	2.7

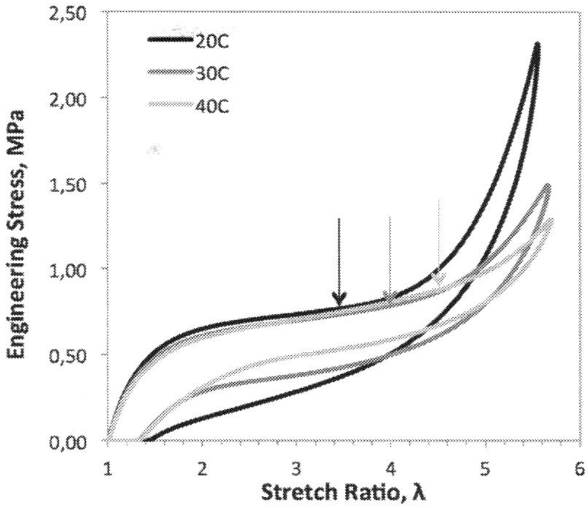

Figure 3. Engineering stress—stretch ratio curve of HNBR-44.

3.3 *Orientation of the amorphous phase*

Similar to the azimuthal intensity, the segmental orientation of the amorphous phase can be developed by a Legendre polynomial $\sum_{n=1}^{\infty}\langle P_{2n}\rangle P_{2n}(\vartheta)$ where ϑ is the angle between the stretching axis and a segment. A relationship $\langle P_{2n}\rangle \propto \langle P_{2n}^{RX}\rangle$ may be rigorously demonstrated for the scattered intensity related to intra-chain correlations. Arguments exist to extend it to the intensity associated to inter-chain correlations (Albouy et al. 2014, Zaghdoudi et al. 2015). With this assumption, $\langle P_2^{RX}\rangle$ is indicative of the evolution of the segmental orientation within the amorphous fraction.

Figure 4 presents the evolution of $\langle P_2^{RX}\rangle$ with the stretch ratio during the tensile test (the sign minus is added to avoid negative values).

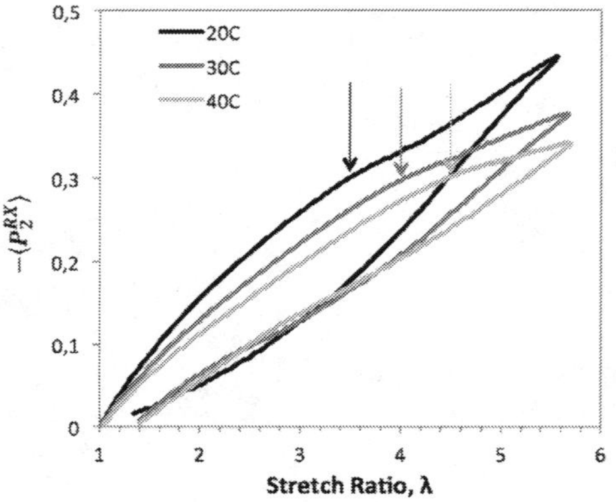

Figure 4. Evolution of the orientation parameter $\langle P_2^{RX} \rangle$ with the stretch ratio for HNRB-44.

It evolves quasi-linearly before the onset of crystallization. The first noticeable decrease of the slope starts approximately when the onset of SIC is reached for the three temperatures; afterwards, the slope is almost constant, except for 20°C where it slightly increases. In NR, this evolution is interpreted to represent a partial relaxation of the amorphous phase as SIC develops in the rubber. However, this phenomenon is not so marked for HNBR as compared to NR, where $\langle P_2^{RX} \rangle$ remains nearly constant after SIC onset. This lower effect of SIC on the orientation of the chains may be tentatively related to the intrinsic lesser stereoregularity of the HNBR chains and, hence, to the reduced ability to crystallize.

4 CONCLUSION

The present study reports some results on SIC of HNBR (containing 44% wt. ACN) by using X-ray diffraction during *in-situ* uniaxial tensile tests. The main results are listed below:

- crystallization of HNBR-44 starts at high elongations (SIC onset λ_C varies from 3.5 to 4.5 depending on the temperature);
- melting of crystallites occurs at a stretch ratio λ_M that is smaller than λ_C, thereby inducing hysteresis;
- the overall crystallinity decreases with increased temperature;
- SIC appears to have a strong influence on the mechanical properties and is related to the stiffening of the elastomer;
- the orientation of the amorphous phase, evaluated trough the parameter $\langle P_2^{RX} \rangle$, is only slightly affected by the SIC, contrary to what is observed in NR. This point requires further investigations.

REFERENCES

Albouy, P.-A., G. Guillier, D. Petermann, A. Vieyres, O. Sanseau & P. Sotta 2012. A stroboscopic X-ray apparatus for the study of the kinetics of strain-induced crystallization in natural rubber. *Polymer* 53(15): 3313–3324.

Albouy, P.-A., J. Marchal & J. Rault 2005. Chain orientation in natural rubber, Part I: The inverse yielding effect. *European Physical Journal E: Soft Matter and Biological Physics* 17(3): 247–259.

Albouy, P.-A., A. Vieyres, R. Pérez-Aparicio, O. Sanseau & P. Sotta 2014. The impact of strain-induced crystallization on strain during mechanical cycling of cross-linked natural rubber. *Polymer* 55(16): 4022–4031.

Braun, D., A. Haufe, D. Leiss & G. P. Hellmann 1992. Strain-induced crystallisation and miscibility behaviour of hydrogenated nitrile rubbers. *Angewandte Makromolekulare Chemie* 202: 143–158.

Brüning, K., K. Schneider, S.V. Roth & G. Heinrich 2012. Kinetics of Strain-Induced Crystallization in Natural Rubber Studied by WAXD: Dynamic and Impact Tensile Experiments. *Macromolecules* 45(19): 7914–7919.

Candau, N., L. Chazeau, J.-M. Chenal, C. Gauthier, J. Ferreira, E. Munch & C. Rochas 2012. Characteristic time of strain induced crystallization of crosslinked natural rubber. *Polymer* 53(13): 2540–2543.

Candau, N., R. Laghmach, L. Chazeau, J.M. Chenal, C. Gauthier, T. Biben & E. Munch 2014. Strain-Induced Crystallization of Natural Rubber and Cross-Link Densities Heterogeneities. *Macromolecules* 47(16): 5815–5824.

Candau, N., R. Laghmach, L. Chazeau, J.M. Chenal, C. Gauthier, T. Biben & E. Munch 2015. Temperature dependence of strain-induced crystallization in natural rubber: On the presence of different crystallite populations. *Polymer* 60: 115–124.

Hamed, G.R., H.J. Kim & A.N. Gent 1996. Cut growth in vulcanizates of natural rubber, cis-polybutadiene, and a 50/50 blend during single and repeated extension. *Rubber Chemistry and Technology* 69(5): 807–818.

Huneau, B. 2011. Strain-induced crystallization of natural rubber: a review of X-ray diffraction investigations. *Rubber Chemistry and Technology* 84(3): 425–452.

Katz, J.R. 1925. Rontgen spectrographic testings on expanded rubber and its possible relevance for the problem of the extension characteristics of this substance. *Naturwissenschaften* 13: 410–416.

Kubo, Y., K. Hashimoto & N. Watanabe 1987. Structure and properties of highly saturated nitrile elastomers. *Kautschuk Gummi Kunststoffe* 40(2): 118–121.

Obrecht, W., H. Buding, U. Eisele, Z. Szentivanyi & J. Thormer 1986. Hydrated Nitrile Rubber—a material with new properties. *Angewandte Makromolekulare Chemie* 145: 161–179.

Osaka, N., M. Kato & H. Saito 2013. Mechanical properties and network structure of phenol resin crosslinked hydrogenated acrylonitrile-butadiene rubber. *Journal of Applied Polymer Science* 129(6): 3396–3403.

Rublon, P., B. Huneau, N. Saintier, S. Beurrot, A. Leygue, E. Verron, C. Mocuta, D. Thiaudiere & D. Berghezan 2013. In situ synchrotron wide-angle X-ray diffraction investigation of fatigue cracks in natural rubber. *Journal of Synchrotron Radiation* 20: 105–109.

Sawada, H. 1994. Hydrogenated Nitrile Rubber. *International Polymer Science and Technology* 21(2): 64–71.

Severe, G. & J.L. White 2000. Physical properties and blend miscibility of hydrogenated acrylonitrile-butadiene rubber. *Journal of Applied Polymer Science* 78(8): 1521–1529.

Tosaka, M. 2007. Strain-induced crystallization of crosslinked natural rubber as revealed by X-ray diffraction using synchrotron radiation. *Polymer Journal* 39(12): 1207–1220.

Zaghdoudi, M., P.A. Albouy, Z. Tourki, A. Vieyres & P. Sotta 2015. Relation between stress and segmental orientation during mechanical cycling of a natural rubber-based compound. *Journal of Polymer Science Part B-Polymer Physics* 53(13): 943–950.

Fracture, fatigue and lifetime prediction of rubber

Constitutive Models for Rubber X – Lion & Johlitz (Eds)
© 2017 Taylor & Francis Group, London, ISBN 978-1-138-03001-5

Rubber reinforcing carbon fibre cord under tension and bending Part 1: Stress analysis

R. Tashiro & S. Yonezawa
NSG Technical Research Laboratory, Itami, Japan

C.A. Stevens
NGF Europe Limited, St Helens, UK

ABSTRACT: For the application of a rubber power transmission belt, the belt is reinforced with load bearing members. These operate under stress fields of tension and bending (around pulleys). This study is part of a project to help predict the fatigue life of carbon fibre cords reinforcing belts under local conditions of tension plus bending. There are two parts to the study: Part 1: FEA stress analysis of a cord in rubber under tension plus bending and Part 2: The fatigue to failure study of the same. The FE stress analysis has been completed for a 6 K carbon fibre rubber impregnated cord encased within rubber.

1 INTRODUCTION

1.1 *Mechanics of tension and bending*

If a uniform body is put under tension the simple analysis is that every point within the body will be subject to the same stress. For a linear elastic body, the internal stress σ will be proportional to the applied force, according to the modulus of the material, E. The strain in tension is given by $\varepsilon_T = \sigma/E$.

If a body is put into bending, the outside edge of the body is subject to an extension strain, the inside edge of the body is subject to a compression strain. In the centre of the body is a neutral axis that is not strained during bending. The strain ε_B is given by the height r, in the body above the neutral axis and the bending radius R: $\varepsilon_B = r/2R$.

If the body is subject to both tension and bending, then the strains are assumed to be independent and additive: $\varepsilon = \varepsilon_T + \varepsilon_B$.

1.2 *Rubber impregnated carbon fibre cord*

For the reinforcement of synchronous toothed belts, rubber impregnated cords are used. The majority of these rubber impregnated cords use glass filaments (E glass, or more recently high strength U and K glass). The impregnation used is a rubber latex, crosslinked with resorcinol formaldehyde resin (RFL). Glass is a high modulus material (~80 GPa) able to deliver reliable timing belt reinforcement for about 50% of the world's automotive engines.

There is a requirement to increase the power rating of timing belts for industrial use. One approach is to increase the modulus of the reinforcing material – and replace the glass with carbon fibres (300 GPa). For this project standard modulus carbon fibre with an adhesive size was used. The bundle of 6000 (6 K) filaments were impregnated with a latex system and then twisted to produce the reinforcing cord.

1.3 *Background to the study*

The body of literature on fatigue studies is very full of fatigue to failure studies of materials in pure tension, and of fatigue to failure studies of materials under three point bending. However, fatigue to failure studies examining the interaction of tension and bending together are not reported. For instance, the effect of tension upon the fatigue life of three point bend testing is not well understood. For the application of a rubber power transmission belt, the belt is reinforced with load bearing members. These operate under stress fields of tension and bending (around pulleys). The very newest belts (such as Gates Poly Chain®, ContiTech Synchrochain® and Optibelt Delta Chain®) have multi-filament carbon fibre cords for stiffness and very high power transmission. This study is part of a project to help predict the fatigue life of carbon fibre cords reinforcing belts under local conditions of tension plus bending.

2 STRESS ANALYSIS STUDY

2.1 *Aim*

The aim of the study was to study the internal stresses within the cord (as a non-uniform body), first under tension, and then under tension plus bending. The effect of frequency upon fatigue was incorporated into the modelling.

2.2 *FEA model*

The 0.7 mm diameter cord of 6000 filaments was modelled (Fig. 1) with a cross section of 300 element areas. This was located centrally within a 2 × 2 mm block of rubber of 416 element areas (Fig. 2). The length of the model was at least 12 mm, and increased as required to match the bending.

Each element was treated as a beam, with a composite tensile modulus derived from the carbon modulus mixed with the impregnated rubber modulus. The beam was angled according to the twisted fibre direction within the cord. The transverse stiffness of the beam was that of the rubber modulus. Several values of rubber modulus were studied, as rubber has a frequency dependent modulus. It was assumed that the carbon modulus was independent of frequency.

The linear tension was modelled by applying a force to the ends of the mesh (Fig. 3). The tension plus bending was modelled by applying a force to the ends of the mesh in a downwards direction, while the centre of the mesh was in contact with an immoveable incompressible frictionless pulley.

2.3 *FEA model conditions and outputs*

The linear tension applied force was normalised to a reference force. 10 levels of force from 0.006 to 0.835 were applied. Rubber moduli of 10, 20, 40 and 80 MPa were used for each force condition.

The tension plus bending was modelled for all combinations of 7 levels of forces from 0.06 to 0.695, pulley diameters 100, 50, 20 and 10 mm, and rubber moduli of 10, 20, 40 and 80 MPa. Unfortunately not all of the combinations could converge to solutions. At the highest condition, only the largest pulley diameter could be modelled. Overall, of the 112 combinations, 86 gave results.

The stress results were obtained for every element within the 3D mesh. Only elements across the mid-point of the mesh were analysed, to avoid any end constraint effects. For the tension and bending conditions, this meant analysing the data at the top of the pulley.

The fatigue of the cord will be mainly tensile along the length of the filaments, so the stresses along the (twisted, angled) directions of the beams are reported.

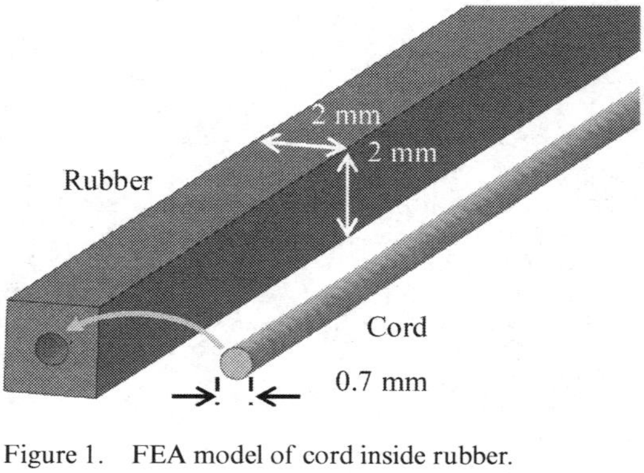

Figure 1. FEA model of cord inside rubber.

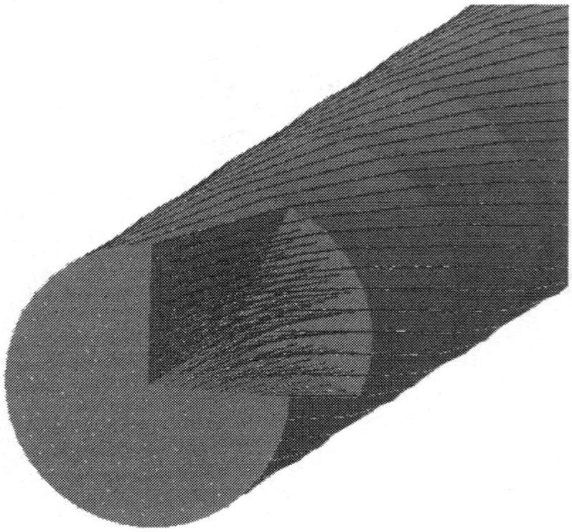
Figure 2. FEA section of cord showing beam directions.

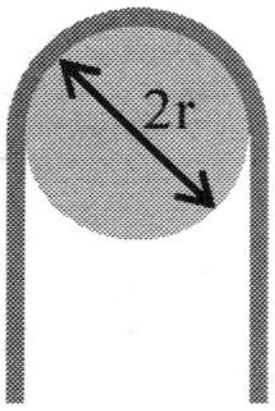

Figure 3. Modelling of tension plus bending by downward force upon the mesh in contact with a pulley of diameter 2r.

3 STRESS ANALYSIS RESULTS

3.1 *Linear tension*

The linear tension stress results for the carbon fibre cord in rubber showed that there was a stress concentration at the centre of the cord. This reflects the non-uniformity of the cord. The cen-

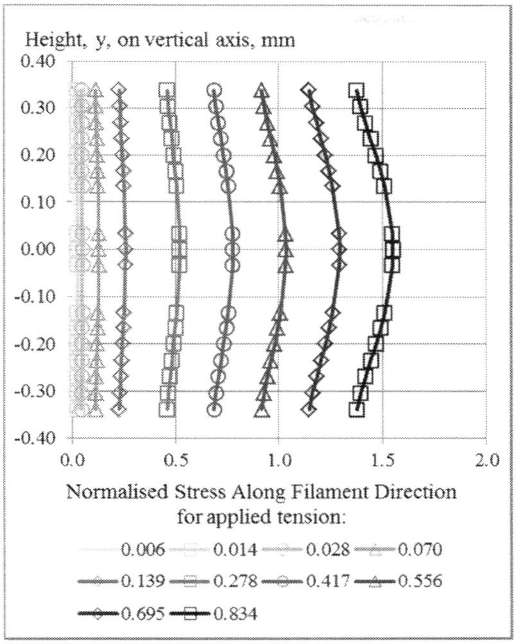

Figure 4. Normalised Stress (x) within the cord for different levels of linear tension for the location on the central vertical axis (y).

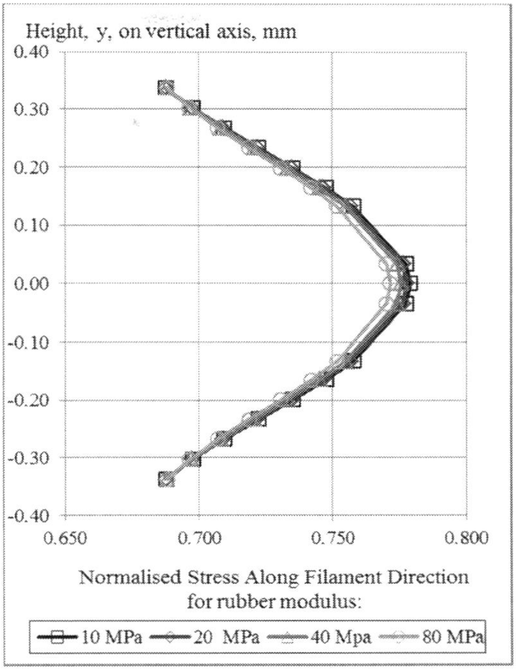

Figure 5. Normalised Stress (x) within the cord for linear tension for the location on the central vertical axis (y) as the rubber modulus is changed.

tral high stresses occurred because of the twisting of the fibres. A fibre at the centre of the cord is the same length as the cord, and is aligned in the direction of the applied tension. A fibre at the outside of the cord is longer than the length of the cord, and will therefore have a lower local strain for the macro applied strain. A fibre at the outside of the cord is also angled across the line of the cord, and therefore angled across the line of tension. Rather than present the 2D stress image of the stresses in the cross-section, only the vertical axis data are presented. These are presented (Fig. 4) as it looking at the cord, with the top of the cord position at the top pf the y-axis, the bottom of the cord at the bottom of the y-axis. The y-axis co-ordinates are referenced to the 0 mm starting height of the centre of the cord.

The stresses at low applied force (0.006) were very close to zero. As the applied force increased, the stress concentration at the centre of the cord increased in magnitude. The ratio of the filament stress at the outside of the cord to the centre of the cord was constant (91%).

If all of the applied highest force were applied to the cord (ignoring the rubber), then the uniform stress would have been about 0.84. The cord (in rubber) has much higher stress than the perfect uniform body, and the centre of the cord had almost double the nominal applied stress.

The effect of rubber modulus was small (Fig. 5). As the rubber modulus increased, the stress distribution within the cord was reduced. The drop in stress was about 1% for the four-fold increase in rubber modulus. This was because the stiffer rubber carried a little bit more of the load relative to the cord.

3.2 *Tension with bending: Low tension*

For a uniform body under bending, the stress versus height (radius) should be a straight line. The results at low load (0.028) were not linear (Figure 6). The stress results for the linear tension without bending are also shown. The tension plus bending curves were displaced from zero stress by the tension.

The tension was subtracted from the tension plus bending results to give the effect of bending (Figure 7). The y axis was changed from the absolute height to the relative height – relative to the central of the cord – which is the neutral axis for bending. This centred the bending on zero stress, and the stress distributions were not quite symmetrical in terms of tension versus compres-

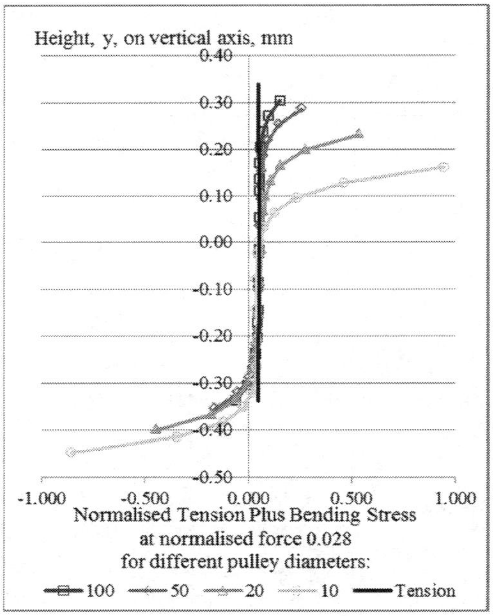

Figure 6. Normalised Stress (x) within the cord for low tension of 0.028 for linear tension and increasing bending curvature.

sion. The tension results were more spread out than the compression curves. This was the effect of the rubber underneath the cord applying a compression to the bottom of the cord.

That the bending data centred on zero stress indicated that the two effects were independent and additive at low load. Figure 8 compares the stress predicted for bending a uniform body in comparison with the FEA stress results for the carbon fibre cord in rubber.

The uniform body bent around a 52 mm pulley (50 mm plus 2 mm for the thickness of the rubber) with the same modulus as the cord would have a peak tension of 0.47, with a peak compression of the same magnitude (as the strain is the same). The stress is linear with height. The bending data for the carbon cord gave a peak tension of 0.21 (and the same in compression).

The cord did not develop the stresses expected. This was due to the fibres (elements) being twisted at an angle across the line of bending strain. Also, the fibres "stress shared" by transferring stresses along their length to positions of lower bending strain. The middle 0.4 mm of the 0.7 mm cord was not experiencing stress due to bending at this condition.

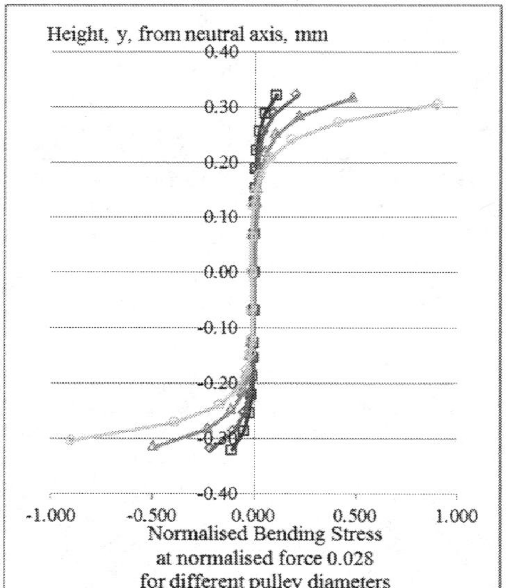

Figure 7. Normalised Stress (x) within the cord for low tension of 0.028 and bending, after subtraction of the linear tension. Y axis is height from horizontal mid-point (neutral axis) of cord.

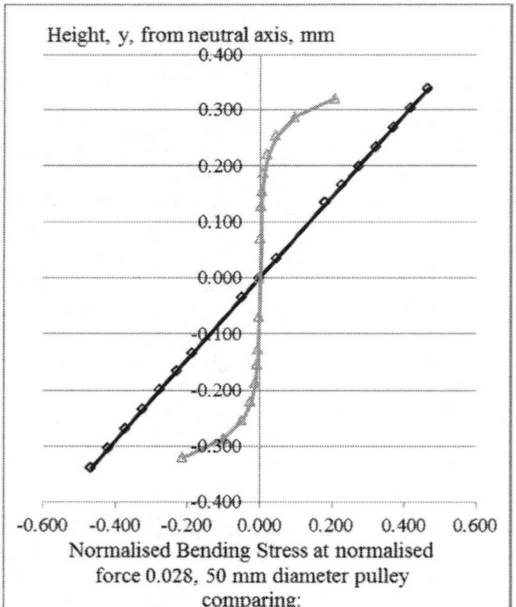

Figure 8. Normalised Stress (x) within the cord for low tension of 0.028, 50 mm diameter pulley, 20 MPa rubber, in comparison with the uniform body bending on a 52 mm pulley.

3.3 *Tension with bending: High tension*

The effect of increasing the applied tension to the 50 mm pulley stress field is shown in Figure 9. At low load, the middles of the cord stresses were close to zero, with tension and compression at high bending. At high loads (0.56, 0.70) the stress field had changed to an "S" shape.

The "bulge" in stress in the centre of the cord was the characteristic of the high stresses due to high tension already seen in Figure 4. As the tension was increased, the cord stress distribution dropped to lower y co-ordinates as the supporting rubber between the cord and pulley was compressed. The cord also was compressed, with the 0.7 mm high cord compressing to about 0.6 mm difference in top and bottom (y co-ordinate).

At low stress field went from tension into compression. When the tension reached 0.21 the stress field changed to tension-tension. In conjunction with this the difference between maximum and minimum stress also narrowed. This has significant consequences if the cord fatigue life is affected by R ratio. The R ratio for low tension was –0.95 (close to –1). At the mid tension of 0.21, the R ratio had passed through zero to a value of 0.17, and then increased to 0.70 for the highest tension applied.

3.4 *Tension with bending: Rubber modulus*

The changes in cord stress distribution with rubber modulus are shown in Figure 10. At high modulus the y co-ordinates of the cord were those of low tension. For the lower values of rubber modulus, increasing rubber compression was observed, with the stress distribution lowering in y co-ordinate. The cord height (change in y) was constant 0.63 mm, invariant of rubber modulus.

Increasing rubber modulus had the effect of increasing the maximum tension and lowering the minimum tensions ("compression" offset by the applied to tension). From 10 MPa to 80 MPa rubber modulus, the maximum tension stress increased from 0.54 to 0.60. The minimum tension decreased, from 0.17 to 0.10. The R-ratio decreased from 0.32 to 0.17. Although the rubber modulus had only a small effect upon the cord stresses in tension, there may be a large effect upon the fatigue life of the cord as the rubber modulus changes with frequency, and changes the R ratio.

3.5 *Tension with bending: High bending*

The cord was found to be subject to high stress fields for conditions of high tension and high

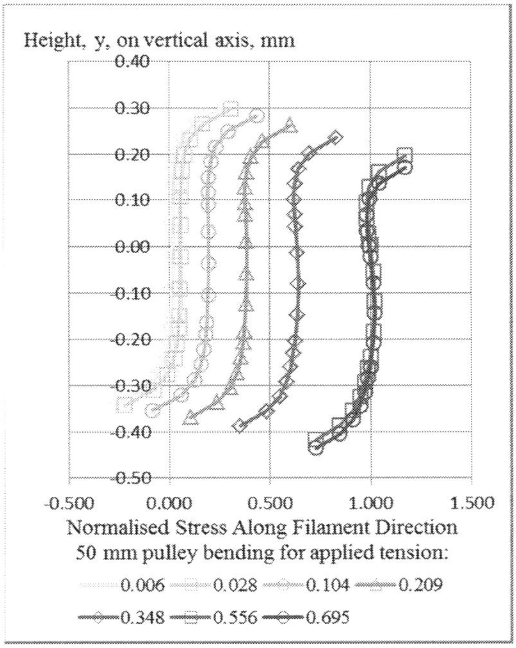

Figure 9. Normalised Stress (x) within the cord bent around a 50 mm pulley for increasing tensions with 80 MPa rubber.

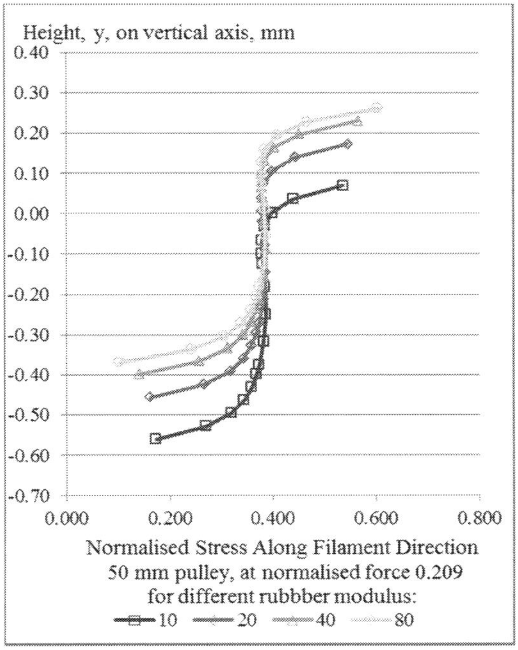

Figure 10. Normalised Stress (x) within the cord bent around a 50 mm pulley at 0.209 tension for different rubber modulus.

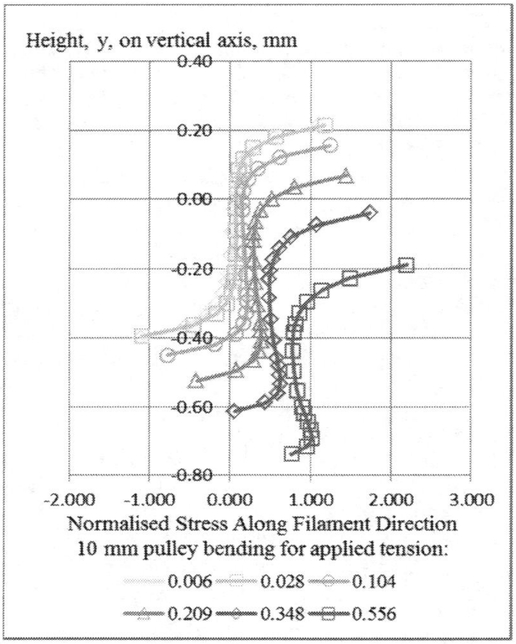

Figure 11. Normalised Stress (x) within the cord bent around a 10 mm pulley at increasing tension for 80 MPa rubber modulus.

bending curvature (small pulley diameter). The highest tension could not be modelled for the 10 mm pulley. At 0.56 tension, the internal stresses reached 2.2 for the 80 MPa rubber, the highest level of stress observed in this study. The increasing forces induced high compression into the rubber, with the cord descending in y co-ordinate.

The stresses for the 10 mm pulley versus applied tension are shown in Figure 11 for the 80 MPa rubber. At low strain the stresses were centred close to zero stress with about equal tension and compression. For the 50 mm pulley a tension of 0.21 was required to bring the stress distribution to tension-tension. The 10 mm pulley had higher compressive strains, and required 0.35 tension to bring the stress distribution to tension-tension. The R-ratio continually increased with applied tension.

The high bending with high load had significantly distorted the stress field close to the bottom of the cord. The stress field was very asymmetric about the cord centre. This may have been due to an interaction with the very highly compressed rubber between the cord and pulley at high bending and high load.

4 CONCLUSIONS

The finite element study of carbon fibre in rubber in tension and bending has been completed. The stress distribution for the vertical axis has been presented for changes in applied tension, rubber modulus and curvature of bending. The maximum stresses, minimum stresses and R-ratio data can be studied for conditions that will provide long fatigue life. These can then be applied in the real world to enhance the performance of synchronous toothed timing belts reinforced with rubber impregnated carbon fibre cords.

Constitutive Models for Rubber X – Lion & Johlitz (Eds)
© 2017 Taylor & Francis Group, London, ISBN 978-1-138-03001-5

Fatigue behaviour of unidirectional carbon-cord reinforced composites and parametric models for life prediction

Y. Tao, E. Bilotti & J.J.C. Busfield
School of Engineering and Materials Science and the Materials Research Institute, Queen Mary University of London, London, UK

C.A. Stevens
NGF Europe Limited, Lea Green, St Helens, UK

ABSTRACT: Unidirectional Carbon Cord reinforced HNBR composites (CF-HNBR) were prepared and fatigue tests under stress control were performed under non-relaxed tension-tension conditions. In this paper, various Constant Life Diagrams (CLD) that are based on different theoretical formulations have been applied to the measured fatigue data of the CF-HNBR composites. The results show that the predictions made by piecewise CLD and modified Harries CLD produce the most accurate results. In addition, a novel experimental set-up is described that replicates in a simplified way the real-pulley situation encountered under typical service conditions to investigate the effect of the bending curvature on the lifetime of the composite subject to coupled tension and bending conditions.

1 INTRODUCTION

Fibre reinforced composite materials are widely used in engineering applications. Hence, an accurate prediction of the service life of a composite structure is crucial. However, it is difficult to predict the lifetime because of material anisotropy and inhomogeneity compared to other isotropic material such as metal materials and other pure rubber materials (Busfield, 2011). A good understanding of the micromechanics of the composite failure process is required for life-prediction purposes. This has been studied extensively by Talreja (2015). For cord reinforced rubber material, the crack growth behaviour of the rubber matrix and peeling behaviour at the interface were investigated extensively by Busfield (2000, 2008, 2012, 2013). A number of cumulative damage models were proposed based on the fatigue damage induced by repeated loading which was reviewed by Socie and Morrow (1980) and Hwang and Han (1986). In the damage models, the damage function is defined as a certain material property that varies with the number of applied cycles and the applied stress level during fatigue, such as the residual modulus, resultant strain and residual strength. However, such methods require a detailed understanding of the way in which the fatigue damage affects the material's residual properties. In practice, the designer usually needs to assess the service life before this type of information is available. Therefore, another methodology based on the phenomenological modelling of the material is required. Various models such as the Linear CLD, Piecewise Linear CLD (Philippidis et al., 2004), Harris's CLD (2003), Kawai's CLD (2015), Boerstra's CLD (2007) and Kassapoglou's CLD (2007) have all been proposed to address this. This approach has no identified physical mechanism of degradation as it is only based on a mathematical analysis of the fatigue data over a wide range of stress ratios. Care must be exercised as a singularity occurs from all compressive region to tensile-compressive regions where R jumps from $-\infty$ to $+\infty$. In addition, at R = 1 (alternating stress equals zero), this type of loading could be considered as a constant creep load rather than a fatigue load (Vassilopoulos et al., 2010).

Moreover, the cord-rubber composites found in components such as a pneumatic tyre or a timing belt often undergo a combination of complex tension and bending deformations rather than in simple extension alone. And there appears a lack of scientific literature examining this type of fatigue loading.

In this paper, the applicability of five different CLD models on the CF-HNBR composites were evaluated. To apply this type of fatigue modelling to a real case incorporating both tension and bending will require closer interpretation of the stress states in the fibres that will use the results of the finite element analysis stress distributions presented in (Tashiro et al., 2017).

Figure 1. Testing setup for tension-bending coupling fatigue loading.

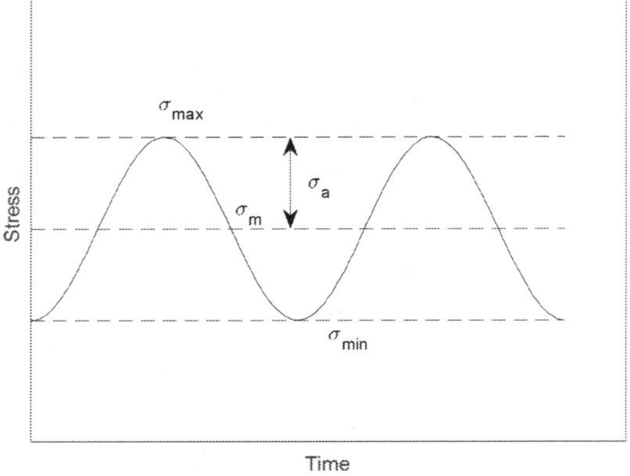

Figure 2. The definition of the commonly encountered test parameters used in this fatigue investigation.

2 EXPERIMENTAL PROCEDURES

2.1 *Materials*

Previously compounded HNBR was re-milled immediately prior to moulding using a 2-roll mill for 5 minutes and it was shaped into a flat sheet for compression moulding. This mixing also ensured an even dispersion of the additives after a relatively long storage time. The vulcanization behaviour was recorded using a rheometer (MDR2000, Alpha Technology, USA) at 180°C according to ASTM D–3182. The cords were placed in between two pieces of rubber with a 5 N pretension attached to either end of the cord and then moulded using a hydraulic hot press at 180°C for 20 min.

2.2 *Test procedures*

Tension-tension fatigue loading at 10 Hz: Carbon cord-HNBR specimens were subjected to a sinusoidal waveform under stress-controlled mode with each test being conducted to a constant maximum load for ranges of various positive R ratios and test frequencies. Five repeats were carried out under each condition.

Tension-bending coupling fatigue loading at $R = 0.4$: a novel experimental set-up was developed that replicates in a simplified way the real-pulley situation encountered under typical service conditions, as shown in Figure 1. The extent of bending strain in the test was achieved by using pulleys with different diameters (20 mm, 30 mm, 40 mm and 50 mm). The life-time was recorded as a function of the frequency and the stress level.

3 RESULTS AND DISCUSSIONS

3.1 *Tension-tension results and CLDs*

The commonly used parameters encountered in a fatigue test are all shown in Figure 2. The main parameters used in the constant life diagram in this paper were normalised stress amplitude, $a = \sigma_a / \sigma_u$, and normalised mean stress, $m = \sigma_m / \sigma_u$, where σ_u was a reference stress level.

The CLDs reveal the combined effect of the mean stress and alternating stress during fatigue testing. Five models based on different approaches were applied on the fatigue testing data of CF-HNBR composites. For the linear model, the fatigue data at $R = 0.3$ is used while that at $R = 0.3$, $R = 0.5$ and $R = 0.7$ was used to create the Piecewise Linear Model and the Harris Model. The main feature of Kawai Model is that it can be derived by the S-N data at the critical R ratio ψ_χ, defined as the ratio of the ultimate compressive strength to the tensile strength of the examined material. For CF-HNBR composite in this paper, $\psi_\chi = -0.01$. However, no fatigue data is currently available for this critical R ratio. Therefore the fatigue data at $R = 0.3$ which is the closest value to ψ_χ was used to construct the CLD. The Kassapoglou Model was generated by assuming the scale and shape parameters were the same as those of carbon cord in a two parameter Weibull distribution.

From Figure 3, it can be seen that the Piecewise Linear and Harris CLD gave accurate predictions compared to Linear CLD. However, Kawai's CLD failed to predict the lifetime and this might because the improper usage of fatigue data at $R = 0.3$ instead of data at the critical R ratio. The Kassapoglou Model provided very poor results for these materials as it is based on parameter taken from the static behaviour which were not related directly to fatigue data.

3.2 *Tension-bending results*

The S-N curve of the coupled tension-bending test using the smallest 20 mm wheel (therefore with the greatest bending contribution) at 5 Hz are shown in Figure 4. The y axis is the normalised peak stress. Figure 5 shows the preliminary results of the influence of the bending curvature and the lifetime under the combination of tension and bending loading. The finite element study of the carbon

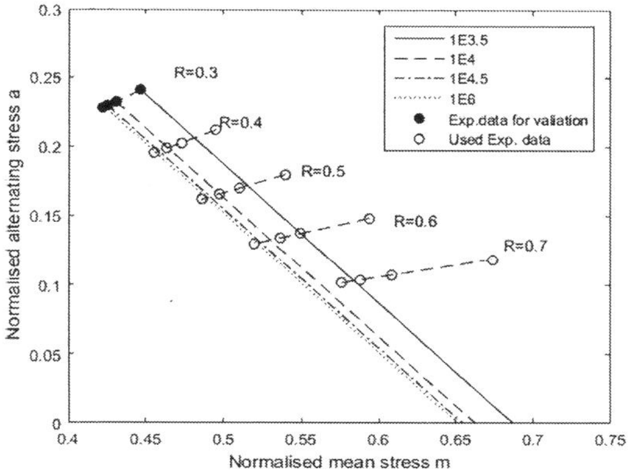

(a)

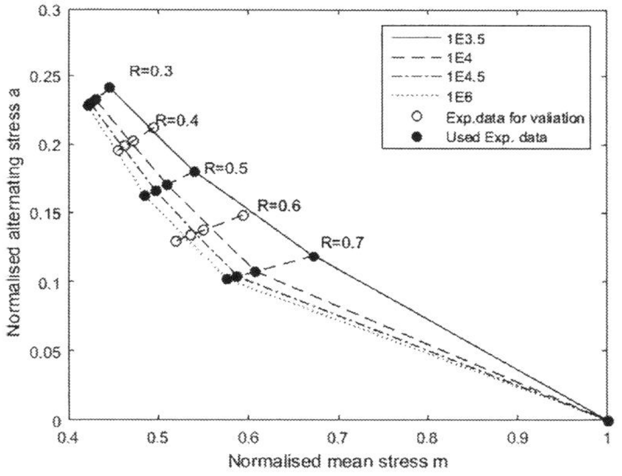

(b)

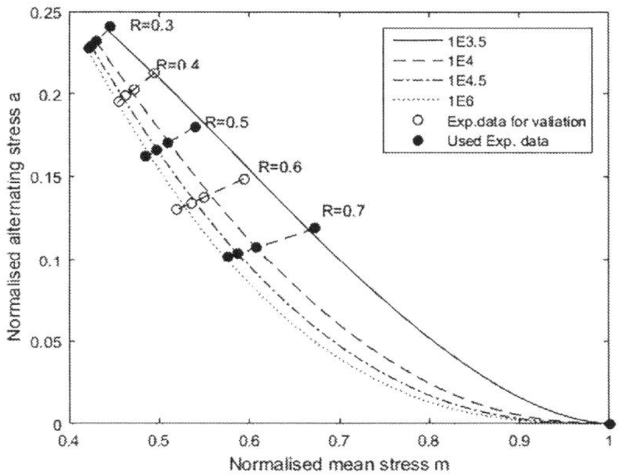

(c)

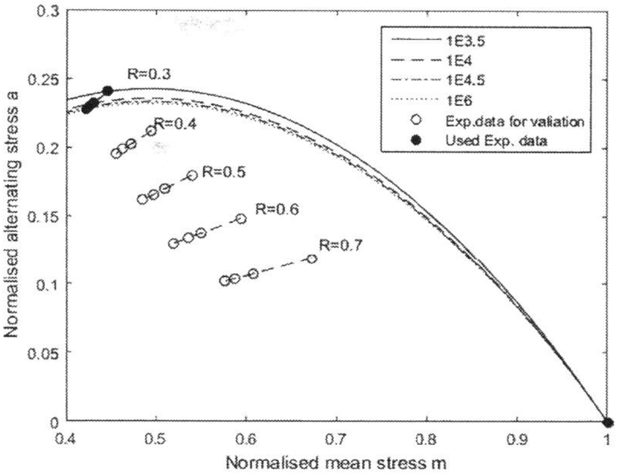

(d)

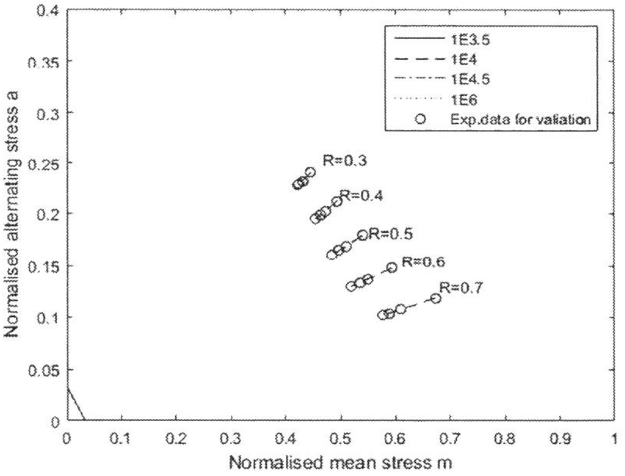

(e)

Figure 3. Different CLDs based on different approaches, (a) Linear, (b) Piecewise Linear, (c) Harris, (d) Kawai, (e) Kassapoglou.

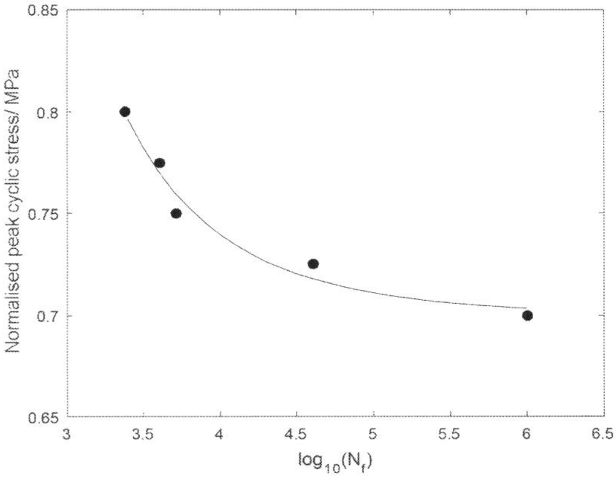

Figure 4. S-N curve of tension-bending coupling test on 20 mm wheel at 5 Hz.

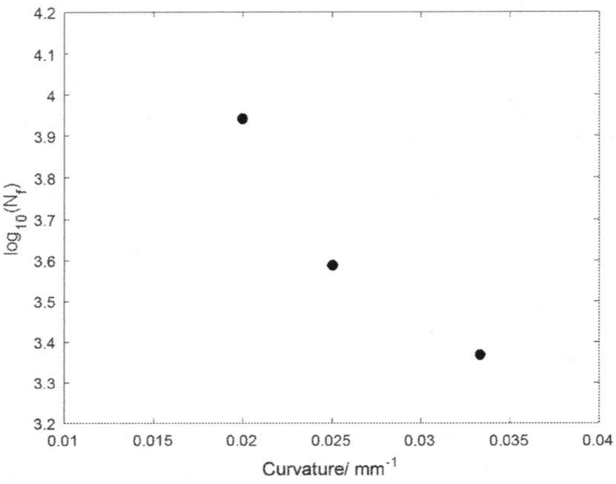

Figure 5. Preliminary results of the influence of the bending curvature and the lifetime.

cord in rubber under tension and bending conditions has also been studied (Tashiro et al., 2017).

It can be found that the bending fatigue life decreased with an increase in the bending curvature at certain percentage of the ultimate strength. A more comprehensive picture will be provided in the presentation when the full set of data becomes available.

4 CONCLUSION

Different CLDs were examined on the tension-tension behaviour of the CF-HNBR composites at various R ratios and the Harris's CLD and Piecewise Linear CLD turned out to be the most accurate ones for making a good life prediction.

Simple coupled tension and bending fatigue tests are currently still ongoing and data will be presented at the conference.

ACKNOWLEDGEMENTS

One of the authors, Yinping Tao would like to thank NGF for their financial support and for the supply of the various materials. She would also like to thank the China Scholarship Council for their financial support.

REFERENCES

Asare, S. & Busfield, J.J.C. 2011. Fatigue life prediction of bonded rubber components at elevated temperature. Plastics, *Rubber and Composites*, 40, 194–200.

Baumard, T., Thomas, A. & Busfield, J. 2012. Fatigue peeling at rubber interfaces. *Plastics, Rubber and Composites*, 41, 296–300.

Boerstra, G. 2007. The Multislope model: A new description for the fatigue strength of glass fibre reinforced plastic. *International journal of fatigue*, 29, 1571–1576.

Harris, B. 2003. A parametric constant-life model for prediction of the fatigue lives of fibre rein-forced plastics. *Fatigue in Composites*, 546–568.

Hwang, W. & Han, K. 1986. Cumulative damage models and multi-stress fatigue life prediction. *Journal of Composite Materials*, 20, 125–153.

Kassapoglou, C. 2007. Fatigue life prediction of composite structures under constant amplitude loading. *Journal of Composite Materials*, 41, 2737–2754.

Kawai, M. 2015. 8 – 2D woven fabric composites under fatigue loading of different types and in different environmental conditions. *In:* Car-Velli, V. & Lomov, S.V. (eds.) *Fatigue of Textile Composites*. Woodhead Publishing.

Papadopoulos, I., Thomas, A. & Bus-Field, J. 2008. Rate transitions in the fatigue crack growth of elastomers. *Journal of applied* polymer science, 109, 1900–1910.

Philippidis, T.P. & Vassilopoulos, A.P. 2004. Life prediction methodology for GFRP lam-inates under spectrum loading. *Composites Part A: applied science and manufacturing*, 35, 657–666.

Sakulkaew, K., Thomas, A.G. & Bus-Field, J.J. 2013. The effect of temperature on the tearing of rubber. *Polymer Testing*, 32, 86–93.

Socie, D. & Morrow, J. 1980. Review of con-temporary approaches to fatigue damage analysis. *Risk and Failure Analysis for Improved Performance and Reliability*. Springer.

Talreja, R. 2015. 1 – A conceptual framework for studies of durability in composite materials. *In:* CarVelli, V. & Lomov, S.V. (eds.) *Fatigue of Textile Composites*. Woodhead Publishing.

Tashiro R., Yonezawa S., Stevens C.A. Rubber reinforcing carbon fibre cord under tension and bending Part 1: Stress analysis. *EC-CMR2017*.

Tsunoda, K., Busfield, J., Davies, C. & Thomas, A. 2000. Effect of materials variables on the tear behaviour of a non-crystallising elastomer. *Journal of materials science*, 35, 5187–5198.

Vassilopoulos, A.P. 2010. Influence of the constant life diagram formulation on the fatigue life prediction of composite materials. *International journal of fatigue*, 32, 659–669.

Constitutive Models for Rubber X – Lion & Johlitz (Eds)
 ISBN 978-1-138-03001-5

Service life prediction under combined cyclic and steady state tearing

R.J. Windslow & J.J.C. Busfield
Soft Matter Research Group, Queen Mary University of London, UK
Materials Research Institute, Queen Mary University of London, UK

ABSTRACT: In this paper a novel approach for modelling failure under a combination of cyclic and static loading is presented. The cyclic fatigue and steady state tearing performance of two non-strain crystallising elastomers, NBR and EPDM, are initially characterised. A simplified component was then tested to failure using a combined cyclic/steady-state loading regime and its service life is compared against predictions from the developed analytical model.

1 INTRODUCTION

1.1 *Problem background*

For non-strain crystallising elastomers crack growth can either occur during cyclical loading conditions or under steady state strained conditions providing a threshold tearing energy has been exceeded, Sakulkaew *et al.*, (2011). Most conventional applications fail either under cyclic crack growth or as a result of steady state crack growth and hence the service life can simply be predicted by applying the relevant fracture principal to either one or other of these conditions. Nevertheless, there are some engineering applications, in which the cracks grow through a combination of cyclic and steady state tear conditions. This is much more complicated to model due to the inter-dependence between these two fracture phenomena, which prevents direct integration.

1.2 *Fracture and fatigue in elastomers*

As first demonstrated by Rivlin and Thomas (1953), fracture within elastomers is an energy based phenomena. They extended Griffith's (1921) fracture criterion to polymers such as elastomers by taking into consideration the additional energy dissipation effects. They proposed using the tearing energy, T. They quantified this as the strain energy, U, released as the fracture surface grew by area, a:

$$T = \left(\frac{\mathrm{d}U}{\mathrm{d}a}\right)_l \tag{1}$$

As demonstrated by Kadir and Thomas (1981), cracks grow at different rates, depending on the energy at the crack tip. This allows characteristic tear profiles to be built, with typical schematics for this data being shown, Figure 1.

When plotted on a logarithmic scale typical static tear profiles split into three linear regions, characteristic with how the crack grows over that range of tear rates, Fukahori *et al.*, (2013). Each of the three regions follow a power law relationship. This relates the change in crack length, c, with time, t, to the tearing energy, T, through two material constants, B and β:

$$\frac{\mathrm{d}c}{\mathrm{d}t} = B(T)^{\beta} \tag{2}$$

Similarly, cyclic fatigue profiles for elastomers break down into four regions when plotted using logarithmic scales, Figure 1b. Most engineering cyclic fatigue problems lie within Region III. Similar to the steady tearing data the data in this region for the rate at which the crack length, c, changes with cycle number, N, can be also be represented using a power law relationship between the tearing energy, T, and two material constants, A and F:

$$\frac{\mathrm{d}c}{\mathrm{d}N} = A(T)^{F} \tag{3}$$

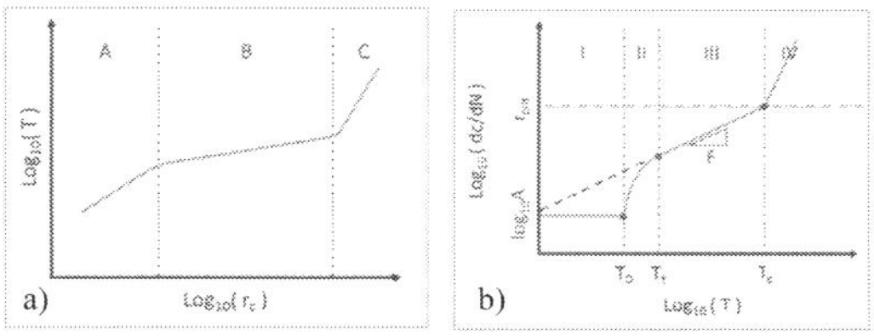

Figure 1. a) Static and b) Cyclic, tear profiles.

If one has a purely static or cyclic problem it is relatively simple to integrate the above equations to determine either the time or the number of cycles to failure. By finding the tearing energy as a function of crack length one can simply integrate the function to determine the time or number of cycles to failure.

$$N_f = \int_0^{N_f} \mathrm{d}N = \int_{c_0}^{c_f} \frac{\mathrm{d}c}{A\left[T = f(c)\right]^F} \tag{4}$$

2 MATHEMATICAL MODEL

2.1 *Complex fatigue system*

Some components exhibit a more complex fatigue profile, such as actuated seals. In this case, there are both cyclic and static aspects to the crack growth. During a sealing cycle the seal is deformed into contact with a counter surface before it is then held for a period.

Previously Busfield *et al.,* (2002) have demonstrated that crack growth per cycle can be split into the cyclic, loading phase, and a time dependent, hold phase, Eqn. 5. To differentiate between crack length after a single complete sealing cycle versus during the separate contributions capitals have been used for the full cycle.

$$\frac{\mathrm{d}C}{\mathrm{d}N} = \left(\frac{\mathrm{d}c}{\mathrm{d}N}\right)_{Cycle} + \left(\frac{\mathrm{d}c}{\mathrm{d}N}\right)_{Time} \tag{5}$$

In service these seals generally only undergo 500–1000 cycles, nevertheless the hold period can last up to 8 hrs. Although the number of sealing cycles is relatively low it is impractical to physically replicate this loading regime. This means mathematical models and F.E.A. are the only viable option for predicting and/or determining the component's lifetime.

By attempting to directly integrate Eqn. 5 to determine the number of cycles to failure the following equation is achieved, where c_o is the initial crack length and c_f is the crack length at failure.

$$N_f = \sum_{c_i = C_0}^{c_f - 2} \left[\int_{c_i}^{c_{i+1}} \frac{\mathrm{d}c}{A(T)^F} + \int_{c_{i+1}}^{c_{i+2}} \frac{\mathrm{d}c}{B(T)^\beta} \right] \tag{6}$$

This is nigh on impossible to solve due to the interdependence between the two contributions, Figure 2. A different approach is therefore required.

2.2 *Proposed approach*

The authors propose using an Explicit Euler scheme to model the problem. Taking Cycle 1, the cyclic crack growth contribution can be expressed as:

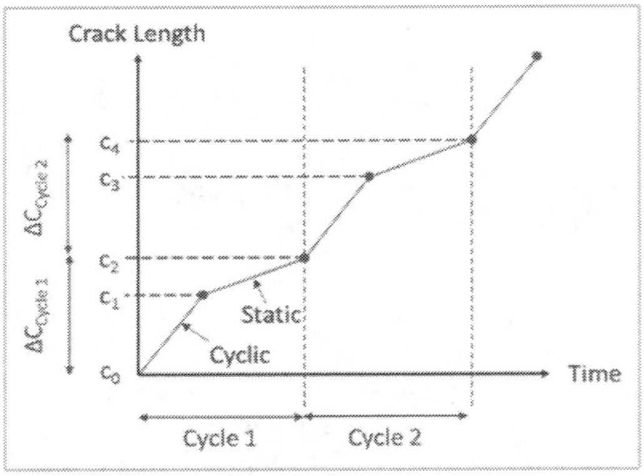

Figure 2. Complex crack growth due to mixed fatigue.

$$\left(\frac{\mathrm{d}c}{\mathrm{d}N}\right)_{Cycle} = \frac{c_1 - c_0}{\Delta N} = A\left[T_{Cycle}\right]^F \tag{7}$$

where c_0 is the initial crack length, c_1 is the crack length at the end of the loading stage, ΔN is the change in cycle number, which of cause for this case $\Delta N = 1$, A and F are material constants determined from the material fatigue profile, and T_{Cycle} is the peak tearing energy during loading. It is also a function of crack length:

$$T_{Cycle} = f(c_0) \tag{8}$$

T_{Cycle} as a function of crack length can simply be found by carrying out an iterative modelling scheme in F.E.A. As the initial crack length, c_0, should be known the only unknown in the expression is the crack length at the transition between the loading and holding stage, c_1, see Figure 2. Eqn. 7 can be adapted to find this unknown.

$$c_1 = \left(A\left[T_{Cycle}\right]^F\right) + c_0 \tag{9}$$

Similarly, the crack growth during the hold stage can be expressed as:

$$\left(\frac{\mathrm{d}c}{\mathrm{d}N}\right)_{Time} = \frac{c_2 - c_1}{\Delta t} = B\left[T_{Time}\right]^\beta \tag{10}$$

where c_2 is the crack length at the end of the hold stage, hence the crack length after the first sealing cycle, $c_2 = C_1$. Δt is the time of the hold period, B and β are the material constants from the static tear profile, and T_{Time} is the tearing energy during the hold stage. The only unknown in this expression is c_2 but it can be determined by rearranging the previous equations as before:

$$c_2 = \left(\Delta t * B\left[T_{Time}\right]^{\beta}\right) + c_1 \qquad (11)$$

Combining Eqn. 9 with Eqn. 11 forms a single expression for the crack length at the end of the sealing cycle, C_i, in terms of the initial crack length and the crack growth during the loading and hold stages:

$$C_1 = \left(\Delta t \cdot B\left[T_{Time}\right]^{\beta}\right) + \left(A\left[T_{Cycle}\right]^{F}\right) + c_0 \qquad (12)$$

Iterating to the next step the expression for the crack length at cycle 2, C_2, would be:

$$C_2 = \left(\Delta t \cdot B\left[T_{Time}\right]^{\beta}\right) + \left(A\left[T_{Cycle}\right]^{F}\right) + C_1 \qquad (13)$$

This scheme could be advanced iteratively and hence presents a viable option for modelling fatigue in these complex fatigue problems. However, a difficulty arises as T_{Time} is a function of both crack length, c_1, and time, t, due to viscoelastic relaxation:

$$T_{Time} = f\left(c_1, t\right) \qquad (14)$$

In essence, during the hold stage the tearing energy is gradually decaying with time causing the rate of crack growth to decrease.

3 MATERIAL CHARACTERISATION

To better understand this problem two elastomers were characterised before undergoing mixed mode fatigue tests, allowing the mathematical model to be compared against real test data. these tests were undertaken using the pure shear crack growth (also known as the planar tension crack growth) test piece geometry. By using this planar geometry, the tearing energies dependence on crack length was removed, Eqn. 15. This meant the cyclic crack growth should be constant and the static crack growth would only depend on time due to relaxation effects. The characterisation tests were therefore also carried out using this planar crack growth test piece geometry.

3.1 *Material selection*

The elastomers used in this study were an NBR and an EPDM typically used for sealing applications. Both elastomers were provided as uncured masterbatch by Clwyd Compounders.

3.2 *Cyclic crack growth*

The cyclic crack growth was characterised on an INSTRON 8801servo-hydraulic test machine using the planar crack growth test piece geometry. Samples were cut into rectangular strips of 175 mm × 45 mm × 2 mm, before being installed between the clamps, Figure 3. Once installed this left a test region of 175 mm × 15 mm × 2 mm.

A 30 mm flaw was cut into the sample such that the crack tip was a suitably distance from the free edge zones. For the planar crack growth test geometry, the tearing energy is given by:

$$T = wh \qquad (15)$$

where the tearing energy, T, is the product of the sample's unstrained height, h, and the strain energy density, w, in the pure shear test region. Nevertheless due to the complex loading regimes through the test piece it is difficult to determine the extent of the pure shear region. Using F.E.A. Busfield (1997) developed an expression for the tearing in a pure shear test piece in terms of the force displacement data, with a correction to account for edge effects. This approach has been successfully implemented in previous works, Baumard *et al.*, (2013), and Asare and Busfield, (2011) and was again adopted here:

$$T = \frac{\int_0^x F \mathrm{d}x}{t\left(L - c_X - \left(0.28\,h\right)\right)} \qquad (16)$$

In this equation, the force, F, displacement, x, curve during loading is integrated to give the strain energy. This is divided by a function of the sample's thickness, t, total length, L, unstrained height, h, and the horizontal component of the crack length, c_X.

The fatigue tests were carried out at 0.5 Hz at amplitudes ranging from 20% strain to 40% strain. A webcam was set up in front of the samples with a picture quality of 1240 × 1024 pixels at 10 fps. Prior to testing a picture was taken of a ruler on the surface of the samples. This image quantified

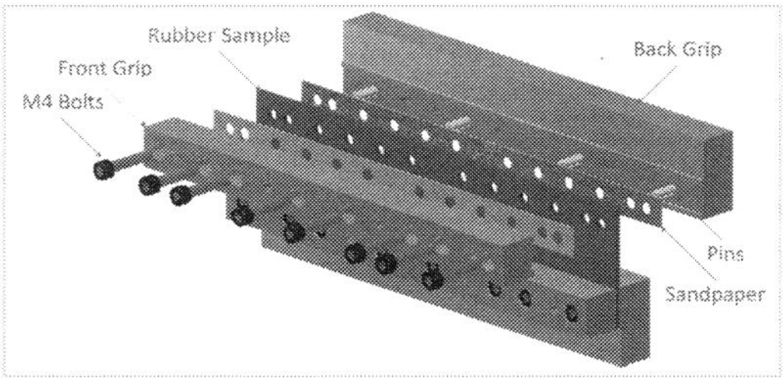

Figure 3. Exploded drawing of the planar test set-up.

the number of pixels per mm in the pictures. Time-lapse photo logging software was then used to take pictures every 50 seconds, Figure 4. As the test frequency was 0.5 Hz this equated to one picture every twenty five cycles.

Once the data had been collected it was post-processed to determine the crack length at different cycles, the plot of which provided the crack growth rate. As shown in Figure 5, the crack growth rate is initially overstated, which is due to two spurious phenomena. Firstly the crack tip is unrealistically sharp and secondly the modulus is higher as only a limited amount of cyclic stress softening has taken place. By ignoring the early data a linear region, more characteristic of the material's steady state behaviour is observed, forms.

The tearing energy was determined as an averaged value taken over several points in the linear region. Once the tearing energy and crack growth rate were determined the data was plotted to form fatigue profiles for both elastomers, Figure 6.

3.3 *Static crack growth*

In addition, the static crack growth behaviour was characterised. Due to viscoelastic relaxation the planar tension tear test geometry is not ideal for static tear tests. It was however important that the sample was tested after being exposed to a similar load history to the samples cycled in pure shear.

To achieve this planar samples were first pre-cycled to 40% strain for 1,000 cycles at 0.5 Hz. The test region was then cut into trouser tear specimens of $58 \times 15 \times 2$ mm, with the leg cut orientated to match the direction of flaws cut in to the original planar geometry. These were then characterised at tear rates from 0.05 mm/s to 0.0005 mm/s to form static tear profiles, shown in Figure 7. For trouser tear specimens:

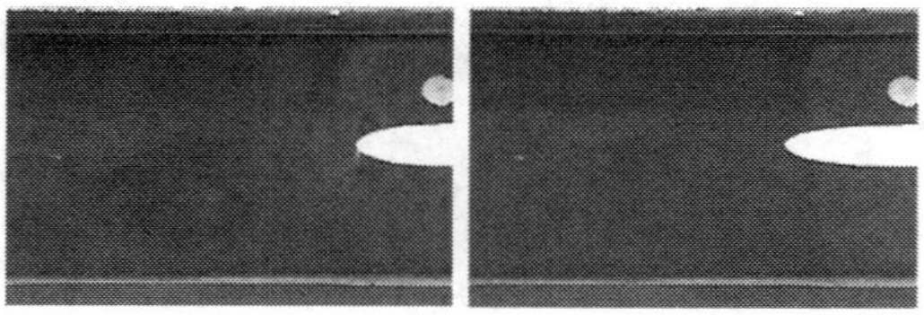

Figure 4. Crack tip image, N = 0 and N = 10000 cycles.

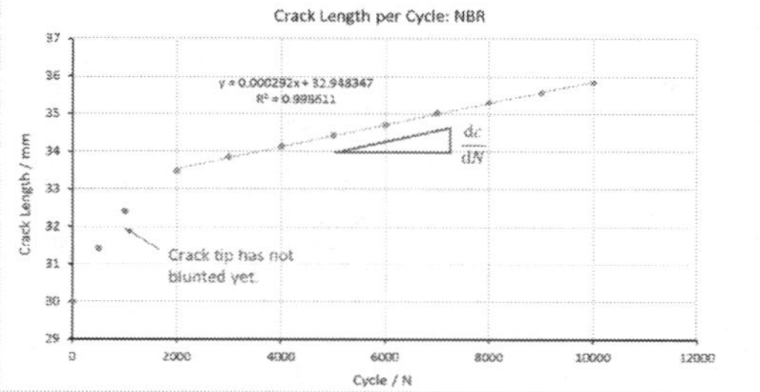

Figure 5. Variation in crack length with number of cycles.

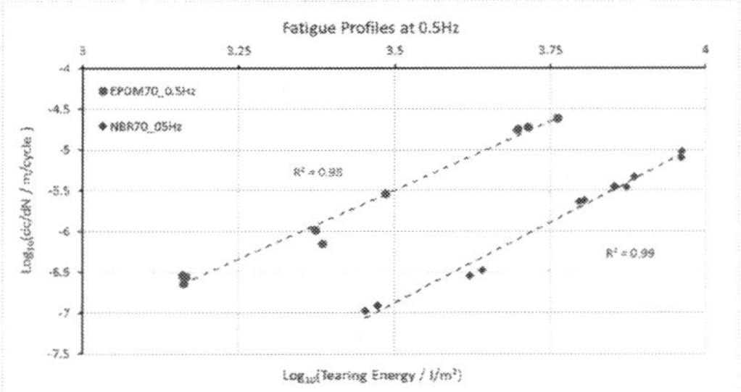

Figure 6. Fatigue profiles for the EPDM and NBR.

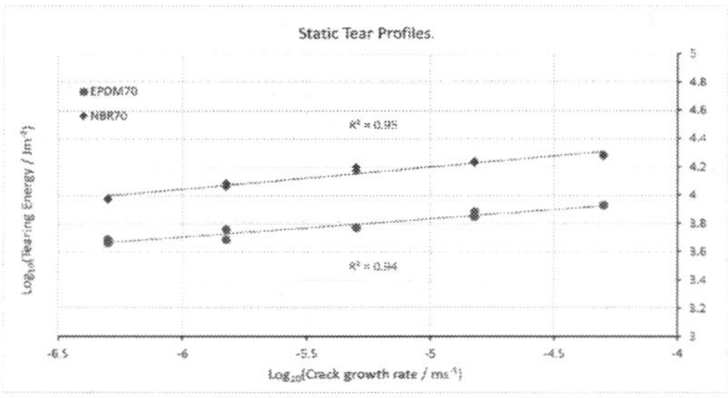

Figure 7. Static tear profiles for the EPDM and NBR.

$$T = \frac{2F\lambda}{h} - bw \tag{17}$$

Here the tearing energy is given by the force, F, tearing the sample, the extension of the trouser legs, λ, and the resulting strain energy density, w, in them and the sample's thickness, h, and total width, b.

3.4 *Cyclic stress relaxation*

Finally the cyclic stress relaxation behaviour was studied in this work. Viscoelastic relaxation in elastomers is split into two contributions, a physical contribution due to viscoelastic and filler effects, and a chemical contribution due to chemical degradation, Yamaguchi *et al.*, (2014). Over short time scales, the physical contribution dominates.

Cyclically loading under planar conditions was expected to induce some directionality to the polymer chains and filler degradation. This was

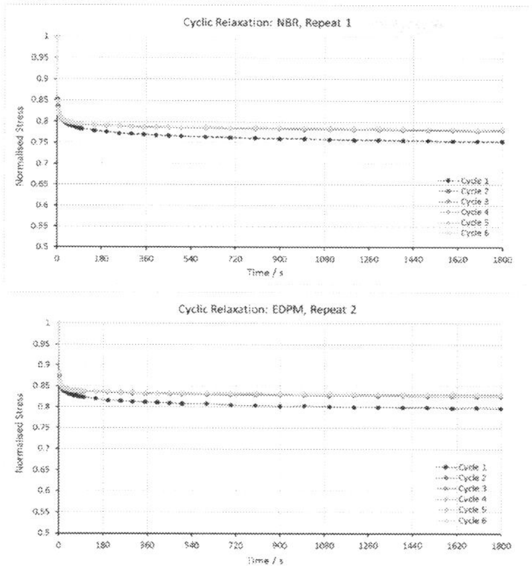

Figure 8. Cyclic relaxation of the NBR and EPDM.

likely to cause the relaxation to vary from cycle to cycle. To study this effect, planar samples were first pre-cycled for 1,000 cycles to 40% strain at 0.5 Hz, removing stress softening effects.

They then underwent trapezoid loading cycles, in which the samples were strained to 40% at a rate equivalent to 0.5 Hz before being held strained for 30 minutes, causing short term viscoelastic relaxation. They were then unloaded to complete the cycle. The data is shown in Figure 8.

As shown from the figures, an equilibrium profile is reached after only two cycles. The averaged relaxation data for each elastomer was used to characterise the tearing energy during the hold stage.

3.5 *Mixed mode fatigue*

Mixed mode fatigue tests were run using planar tension crack growth test pieces. To further simplify the problem the samples were first pre-cycled for 1,000 cycles to 40% strain at 0.5 Hz, such that a steady state in the material's stress softening behaviour was reached. The tearing energy per cycle should therefore be relatively constant.

A 30 mm flaw was cut into the specimens before they were put through a trapezoidal strain cycle. In the first second of the cycle the uncut section of the sample was loaded to 40% strain at a rate that was equivalent to loading at 0.5 Hz. The sample was then held for 1797 seconds before being unloaded in one second followed by a one second hold stage. Each test consisted of 120, 30 minute cycles. Using a similar set up as for the cyclic tear tests, pictures were taken of the crack tip every 30 minutes to determine the crack growth rate, Figure 9. The results are shown in Tables 1 and 2.

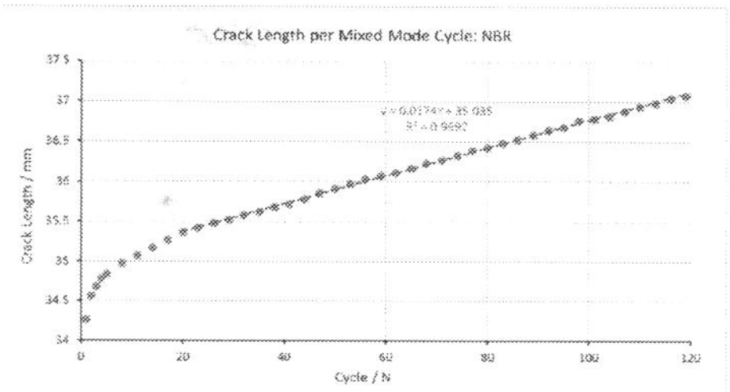

Figure 9. Variation in crack length with number of cycles.

Table 1. Predication vs test data for mixed cycle, EPDM.

Variable	Source	EPDM, R1	EPDM, R2
T_{peak}/Jm^{-2}	Test	6091.8	5408.8
dC/dN_{Test} /m·cycle^{-1}	Test	2.83E-04	2.36E-04
dc/dN_{Cycle} /m·cycle^{-1}	Model	2.93E-05	1.96E-05
dc/dN_{Time} /m·cycle^{-1}	Model	2.08E-03	8.76E-04
dC/dN_{Model} /m·cycle^{-1}	Model	2.11E-03	8.95E-04
$(dc/dN)_{Model}/(dc/dN)_{Test}$		7.44	3.80

Table 2. Predication vs test data for mixed cycle, NBR.

Variable	Source	NBR, R1	NBR, R2
T_{peak} /Jm^{-2}	Test	6884.6	7561.4
dC/dN_{Test} /m·cycle^{-1}	Test	1.74E-05	3.27E-05
dc/dN_{Cycle} /m·cycle^{-1}	Model	2.84E-06	4.11E-06
dc/dN_{Time} /m·cycle^{-1}	Model	2.38E-05	4.23E-05
dC/dN_{Model} /m·cycle^{-1}	Model	2.66E-05	4.65E-05
$(dc/dN)_{Model}/(dc/dN)_{Test}$		1.53	1.42

4 RESULTS

The peak tearing energy was defined as the averaged value taken at every 10th cycle. Once this was known the cyclic component of the crack growth was readily determined from the fatigue profiles, Figure 6.

Finding the growth during the hold stage was a little more difficult. The tearing energy should decrease in line with relaxation, hence the tearing energy as a function of time was known. The static tear profiles

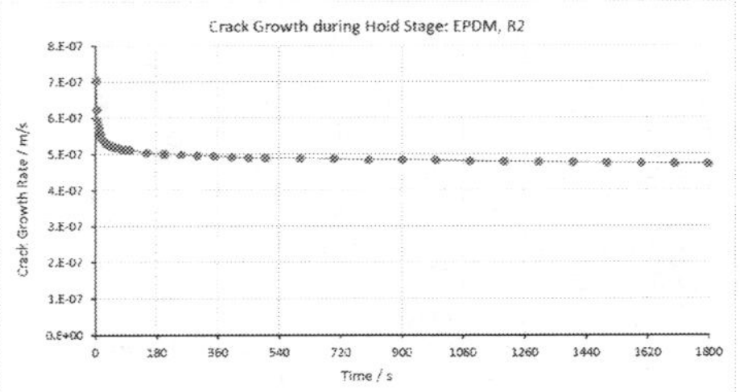

Figure 10. Crack growth during hold stage.

relate the crack growth rate to the tearing energy. By combining these, one can plot the crack growth rate with time, Figure 10, the integral of which gives the crack growth during the hold stage.

5 DISCUSSION

Considering the complications of the model, for the predications to be within the correct order of magnitude for both elastomers is a good result. The NBR data in particular is a very good prediction all things considered. Slight variances between the two were expected due to a culmination of errors from the three characterisation methods. Furthermore, predications for very simple, single contribution, fatigue problems generally do not achieve perfect results either.

The cyclic crack growth contribution is likely to be over predicted as stress relaxation will cause the tearing energy during the unloading phase to be much lower, nevertheless the cyclic contribution appears to be minor. The main error arises in the calculation of the crack growth during the hold stage.

As the hold time is 1797seconds, slight miscalculations in the static tear behaviour are amplified through the model. This is not helped by the logarithmic nature of the static tear plots which also magnifies small errors. Taking EPDM, R2 as an example, the correct predication requires the static tearing energy values to be 20% greater than the laboratory tests predicted. This demonstrates an impractical sensitivity to the static tear data which is notoriously difficult to characterise. The model provides a better predication for the NBR sample probably because the static term is less dominant and a better fit is achieved to for static tear data.

The NBR was more tear resistant than the EPDM, as shown in Figures 6 and 7. Analysing the mixed mode cycles this was benefitted by its relaxation behaviour. As the NBR relaxed further the tearing energy in it reduced faster, slowing the crack growth. This means that under fixed displacement tear conditions that viscoelastic relaxation actually enhances the component's fatigue life.

6 CONCLUSION

A mathematical model for predicting crack growth during mixed mode fatigue regimes was presented. The analytical model was compared against test data, and showed promising results, fitting within an order of magnitude for both elastomers tested. The difficulties encountered in matching the data was attributed to its sensitivity to errors in the static tear data. Of note, viscoelastic relaxation was shown to have a beneficial effect on the elastomer's fatigue resistance.

ACKNOWLEDGEMENTS

The authors would like to thank Cameron, a Schlumberger company, for their sponsorship and support in this research. They would also like to thank Clwyd Compounders who kindly supplied the elastomers.

REFERENCES

Asare, S., & Busfield, J.J.C., (2011). Fatigue life prediction of bonded rubber components at elevated temperature. *Plastics, Rubber and Composites*, *40*(4), 194–200.

Baumard, T.L.M., Thomas, A.G., & Busfield, J.J.C., (2013). Evaluation of the tearing energy in a radial tyre. In Alonso (Ed.), *Constitutive Models for Rubber VIII*. San Sebastian: CRC Press.

Busfield, J.J.C., Ratsimba. C.H., Thomas. A.G., (1997). Crack growth and strain induced anisotropy in carbon black filled natural rubber. *Journal of Natural Rubber Research*, 12, 131–141.

Busfield, J.J.C., Tsunoda, K., Davies, C.K.L., & Thomas, A.G., (2002). Contributions of time dependent and cyclic crack growth to the crack growth behavior of non strain-crystallizing elastomers. *Rubber Chem. Tech.*, *75*, 643–656.

Fukahori, Y., Sakulkaew, K., & Busfield, J.J.C., (2013). Elastic-viscous transition in tear fracture of rubbers. *Polymer*, *54*(7), 1905–1915.

Griffith, A.A., (1921). The Phenomena of Rupture and Flow in Solids. *Philosophical Transactions of the Royal Society A: Mathematical, Physical and Engineering Sciences*, *221*, 163–198.

Kadir, A., & Thomas, A G., (1981). Tear Behaviour of Rubbers over a Wide Range of Rates. *Rubber Chemistry and Technology*, *54*(1), 15–23.

Rivlin, R.S., & Thomas, A.G., (1953). Rupture of Rubber. I. Characteristic energy for tearing. *Journal of Polymer Science*, *10*(3), 291–318.

Sakulkaew, K., Thomas, A.G., & Busfield, J.J.C., (2011). The effect of the rate of strain on tearing in rubber. *Polymer Testing*, *30*(2), 163–172.

Yamaguchi, K., Thomas, A.G., & Busfield, J.J.C., (2015). Stress relaxation, creep and set recovery of elastomers. *International Journal of Non-Linear Mechanics*, *68*, 66–70.

Constitutive Models for Rubber X – Lion & Johlitz (Eds)
© 2017 Taylor & Francis Group, London, ISBN 978-1-138-03001-5

Impact of stress softening on tearing energy of filled rubbers as evaluated by the J-Integral

M. Wunde, J. Plagge & M. Klüppel
Deutsches Institut für Kautschuktechnologie e.V. (DIK), Hannover, Germany

ABSTRACT: Whereas unfilled elastomers show nearly ideally hyperelastic behavior, the differences in the stress response by adding filler are connected to the interaction between the polymer and the filler. Filling of the polymer leads not only to a reinforcement but increases also the hysteresis and stress-softening. By evaluating the displacements of an airbrushed pattern on notched Pure-Shear samples with an ARAMIS system, the displacement fields around the crack tip are obtained. Using a physically motivated model of stress softening and hysteresis of filled rubber the energy density and stress distribution can be calculated. We are able to determine the J-Integral J for closed contours around the crack tip. For purely elastic materials the value of J is path-independent but due to energy dissipation the J-Integral depends on the integration path. By variation of the strain amplitude and the integration path the impact of stress softening on the J-Integral is evaluated.

1 INTRODUCTION

Rice and Cherepanov introduced the J-Integral into the field of fracture mechanics (Rice, 1968; Cherepanov, 1967). The J-Integral is a line integral evaluating the stress and strain fields along a contour Γ. This path integral indicates the energy flow to the crack tip under deformation, which is not stored but dissipated by crack extension (Zehnder, 2012; Grellmann et al., 2001). For purely elastic material behavior the integral is independent of the selected contour. The value of J coincides with the elastic energy release rate $T = -dW/dS$. This so called tearing energy states the energy release W per change of surface area S due to an increasing crack (Rivlin & Thomas, 1953). For Pure shear samples the tearing energy is given by $T = wh$, where w is the energy density and h is the height of the sample.

Stress softening denotes the stress decrease in elastomers due to previous deformation (Mullins & Tobin, 1965). This so called Mullins effect is explained by interactions and restructuring of the filler and polymer network as reviewed by Vilgis et al. (Vilgis et al., 2009). In repeated straining after several cycles a steady-state cycle is reached.

2 EXPERIMENTAL

2.1 *Material*

A natural rubber (NR) compound filled with 50 phr carbon black (CB) is investigated. The composite is cross-linked semi-efficiently by sulfur in combination with the vulcanization accelerator N-cyclohexyl-2-benzothiazole sulfonamide (CBS). The sample was compounded with the processing and vulcanization additives stearic acid and ZnO and protected against aging by N-isopropyl-N-phenyl-P-phenylenediamine (IPPD). The full recipe is shown in Table 1.

2.2 *Sample preparation*

Mixing of the compound was done at Continental AG in industrial scale. At our institute vulcanization is performed at 150°C in a heat press up to 90% of the vulcameter torque maximum (t_{90} time). The vulcanizates are produced as Pure Shear samples with a height of 27.5 mm and width of 200 mm. Into the short edge a 50 mm long cut is inserted.

2.3 *Measurement of displacement fields*

For measuring the displacement fields an airbrushed pattern is sprayed onto the samples. The notched Pure Shear samples are strained in a Zwick Universal Testing machine 1445 five times at a speed of 20 mm/minute to the same maximum strain value varying between 10% and 50%.

Table 1. Recipe of the compound [phr].

NR	CB	CBS	Sulfur	IPPD	ZnO	Stearic acid
100	50	1.1	1.8	1.8	3	1

The deformation of the composites in the last cycle is followed with a camera with a spot size of approximately 20 mm × 20 mm, recording 1 image per second.

In the consecutive images of a deformation cycle the evolution of the strain distribution is received from the movements of the airbrushed black and white pattern. This way the displacement fields around the crack tip are calculated using the ARAMIS software from GOM.

2.4 *Evaluation of displacement fields*

For a deformation cycle the image with the highest deformations is evaluated. The deformed coordinates of this image are interpolated with a spline function. By comparing the deformated coordinates with the undeformed coordinates from the first image of the series the distribution of the deformation gradient $(\boldsymbol{F})_{ij} = dx_i/dX_j$ is obtained. The deformation gradient measures the change of the original coordinates X_j to the new coordinates x_i due to deformation (Holzapfel, 2000). The displacement vector is given by $\boldsymbol{u} = \boldsymbol{x}-\boldsymbol{X}$.

From the deformation gradient the eigenvalues and eigenvectors $\boldsymbol{b}^{(n)}$ of the left Cauchy-Green Tensor $\boldsymbol{B} = \boldsymbol{F}^T \boldsymbol{F}$ can be calculated. The principal strains λ_n are received as square root of the eigenvalues. The third principal strain is obtained from incompressibility condition.

2.5 *Model to calculate the energy density and stress distribution*

As the J-Integral represents the dissipated energy, the value of J for a closed contour, not involving the crack tip, corresponds to the energy dissipated inside that contour. The dissipated energy density is higher for higher strained regions close to the crack tip. Therefore the value dJ/dA, where A is the area of the contour around the crack tip, is increasing toward the crack tip. The stress softening is correlated to the dissipated energy.

For calculating the energy density and stress distribution a recently proposed model of stress softening and hysteresis (Plagge & Klüppel, 2017a) is used. This model is built on physical principles and can model any deformations. The model is presented also in this book (Plagge & Klüppel, 2017b).

To describe the hyper-elastic rubber matrix the model relies on the extended non-affine tube model (Heinrich & Kaliske, 1997; Klüppel & Schramm, 2000). Stress softening is regarded by decreasing hydrodynamic amplification of the rubber matrix. This is done by a power law distribution $P_X(X)$ of amplification factors X being shifted to smaller amplification factors:

$$P_X(X) = X^{-\chi} \frac{\chi - 1}{1 - X_{\max}^{1-\chi}} \tag{1}$$

where χ is the exponent of the power law and X_{max} is the maximum amplification factor. When the maximal strain in the deformation history, and thereby the maximum value of the first invariant $I_1 = \Sigma_i \lambda_i^2 - 3$, is increasing, the value of X_{max} is decreasing and the normalized distribution $P_X(X)$ is shifted to lower amplification factors X.

The amplified energy density w_X is given by

$$w_X(I_1, I^*, X_{\max}) = \int_1^{X_{\max}} dX\, P_X(X) w(X I_1, X_I^*) \tag{2}$$

where $I^* = \Sigma_i \lambda_i^{-1} - 3$ is a generalized invariant and $w(I_1, I^*)$ is the energy density of the rubber matrix given by the non-affine tube model.

The elastic Cauchy stress is calculated from the amplified energy density w_X via:

$$(\sigma_{el})_{ij} = \sum_{n=1}^{3} \frac{\lambda_n}{\lambda_1 \lambda_2 \lambda_3} \frac{\partial w_X}{\partial \lambda_n} \mathbf{b}_i^{(n)} \otimes \mathbf{b}_j^{(n)} \tag{3}$$

So far hysteretic contributions, although being described by the model, have not been implemented into the further calculations.

2.6 *Integration to obtain the J-Integral*

For a given path Γ around the crack tip the J-Integral can be calculated according to

$$J = \oint_\Gamma \left[w dy - \sigma_{jk} n_k u_{j,1} ds \right] \tag{4}$$

where Γ is the integration path; w is energy density; $\boldsymbol{\sigma}$ is Cauchy stress tensor; $u_{j,1}$ is du_j/dx_1; and $\boldsymbol{n}$ is the normal vector. By variation of the integration path different values for the J-Integral can be calculated. For an infinite long Pure Shear sample with a contour surrounding the edge of the sample, already Rice showed, that $J = wh$ (Rice, 1968). This equals the value of the tearing energy T for Pure Shear samples.

3 RESULTS AND DISCUSSION

3.1 *Strain fields*

The evaluation of the displacements by the ARAMIS software results in reasonable strain fields approaching up to approximately 1–1.5 mm to the crack tip. Closer to the crack tip the deformations could not be recorded reasonably. Figure 1 shows that $\Delta\varepsilon_{yy}=\varepsilon_{yy}-\varepsilon_{\infty}$ has a power law behavior in front of the crack tip. ε_{yy} is du_2/dx_2 and ε_{∞} is the value of ε_{yy} far away from the crack tip. The value of ε_{∞} is set appropriately, somewhat below the lowest value of ε_{yy}. The exponent of $\Delta\varepsilon_{yy}$, $\alpha \approx -1.3$ differs much from $\alpha=-0.5$ as expected from linear elasticity (Anderson, 1995).

3.2 *Multihysterese measurements and parameter identification*

In order to get the parameters in the model a multihysterese measurement was done on a dumbbell specimen. The dumbbell was stretched successively from 10% up to 100%, five times for each strain value. In Figure 2 is shown the measurement and the fit with the model. As so far hysteretic contributions have been neglected, the model is lying in between the loading and unloading curves. The parameters of the model are listed in Table 2.

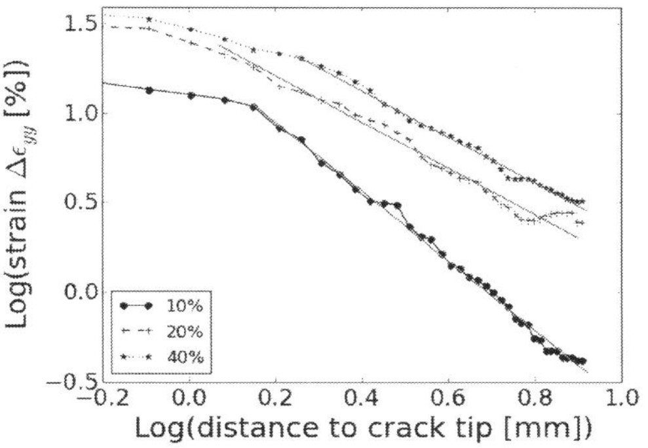

Figure 1. Log-log plot of the strain increase $\Delta\varepsilon_{yy} = \varepsilon_{yy}-\varepsilon_{\infty}$ due to the presence of the crack at $y = y_{crack\ tip}$ for varying distance from the crack tip. The straight lines indicate power law behavior. For clarity only stretching to 10%, 20% and 40% is shown.

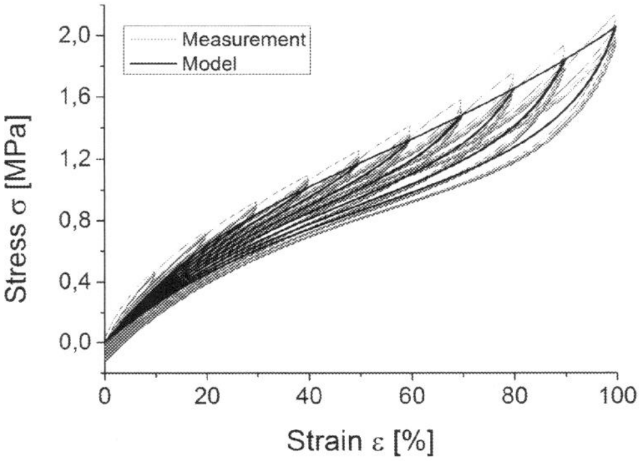

Figure 2. Multihysterese measurement on a dumbbell sample fitted by the model.

Table 2. Parameters of the compound.

G_c [MPa]	G_e [MPa]	n	ϕ_{eff}	χ	γ
0.118	0.326	40	0.2	2.10	0.408

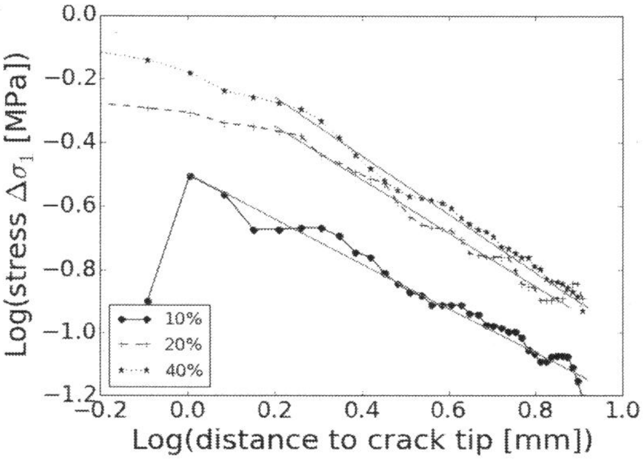

Figure 3. Log-log plot of the stress increase $\Delta\sigma_1 = \sigma_1-\sigma_{\infty}$ due to the presence of the crack at $y = y_{crack\ tip}$ for varying distance from the crack tip. The straight lines indicate power law behavior. For clarity only stretching to 10%, 20% and 40% is shown.

3.3 *Stress fields*

Modelling of the stress fields shows reasonable results approaching up to around 1–1.5 mm to the crack tip. Figure 3 shows that $\Delta\sigma_1 = \sigma_1-\sigma_{\infty}$ has a power law behavior in front of the crack tip. σ_1 is the first principal value of stress and σ_{∞} is the value of σ_1 far away from the crack tip. The value of σ_{∞} is set appropriately, somewhat below the lowest value of σ_1. The exponent of $\Delta\sigma_1$, $\alpha \approx -0.8$ is larger than $\alpha = -0.5$ as expected from linear elasticity (Anderson, 1995):

$$\sigma_{ij} = \frac{K}{\sqrt{2\pi r}} f_{ij}(\theta) \tag{5}$$

where K is the stress intensity factor, r is the distance from the crack tip and $f_{ij}(\theta)$ is a function of the angle θ. Accordingly stress softening modifies the stress and strain fields around the crack tip significantly. Nevertheless it is not clear whether the strain fields obtained reflect the real material behavior.

3.4 *Energy density*

The energy density distribution is exemplified in Figure 4. As the strain fields are underrated closer than 1–1.5 mm from the crack tip, the same is true for the energy density w.

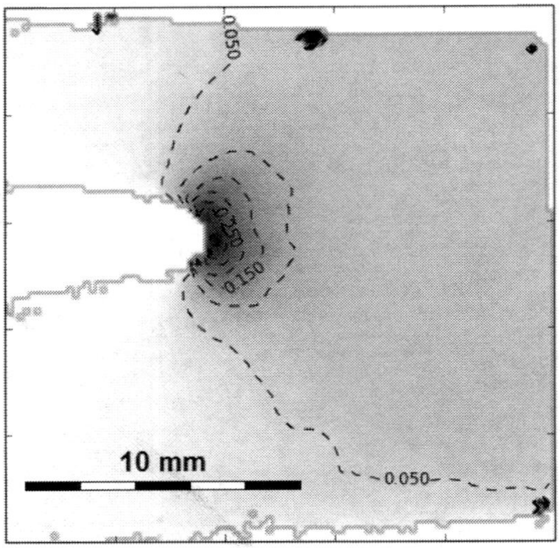

Figure 4. Energy density w [MPa] for the filled NR stretched to 20%.

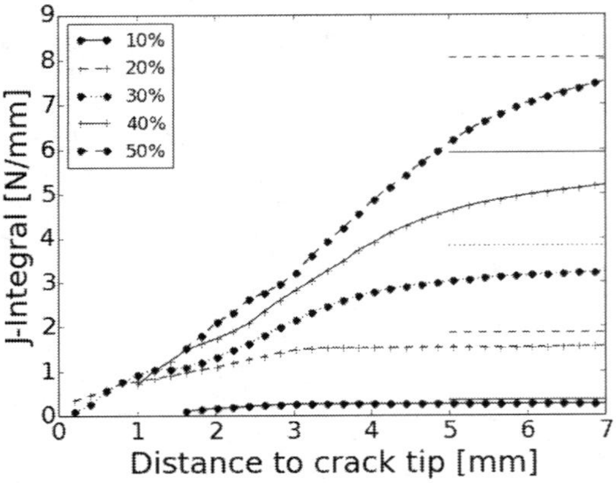

Figure 5. Value for the J-Integral for strain amplitudes between 10% and 60%. The J-Integral is evaluated on half spheres with radius r. The global tearing energy $T = wh$ for different strain amplitudes is shown as horizontal lines for comparison.

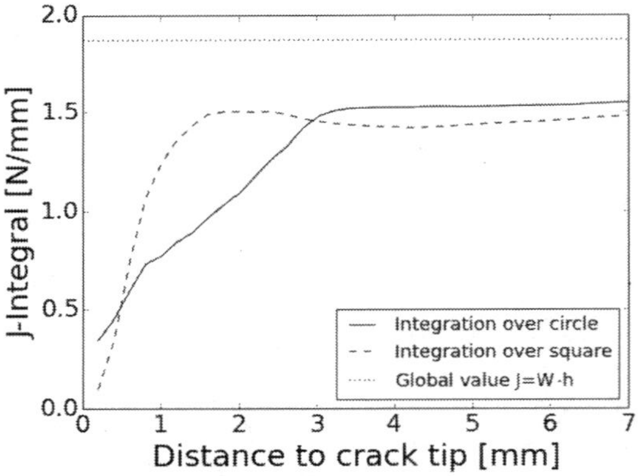

Figure 6. Comparison of the J-Integral of circle and square contours for the filled NR stretched to 20%.

3.5 *Integration over circles around the crack tip*

Although by integration over a rectangular path Equation 1 simplifies, the most intuitive path is given by a circle with radius r around the crack tip. The values of the J-Integral for circles with different radii are measured on the same sample. The J-Integral shows to be not path-independent but increases for small radii until a plateau value is reached (Figure 5). The J-Integral values below around 1–1.5 mm contour radius are underrated due to insufficient resolution as discussed in the previous chapters. From only finite strains being recorded, it follows that $J \to 0$ is obtained when approaching the crack tip. Nevertheless an increase of the J-Integral with rising distance to the crack tip is recognizable. The difference of J for two different radii corresponds to the energy dissipated in the area between the two integration paths. With rising strain amplitude the values for the J-Integral as well as the global tearing energy $T = wh$ is increasing (Figure 5). For calculating the global value T the energy density w in a region far away from the crack tip is taken.

3.6 *Comparison of square and circle integration paths*

A square surrounds a bigger area than a circle with the same diameter. Therefore the J-Integral of a square centered at the crack tip should surpass the value of the circle. The difference is especially pronounced at small distances from the crack tip (see Figure 6). This leads to a strong increase of J with increasing distance at small distances. In some cases our results show the increase followed by a decrease. This would mean that in the outwards areas energy is created and is not physical meaningful.

4 CONCLUSIONS

The displacement fields of a carbon-black filled NR compound could be obtained using an ARAMIS system on notched Pure Shear samples. Reasonable resolution could be achieved up to around 1–1.5 mm from the crack tip. A novel model (Plagge & Klüppel, 2016) is used to calculate the energy density and stress distribution surrounding the crack tip. The J-Integral, being path-independent for purely elastic materials, is decreasing for contours closer to the crack tip.

ACKNOWLEDGEMENT

We would like to thank the Deutsche Forschungsgemeinschaft (DFG) for financial support (grant KL1409/9-1) and our project partners IPF Dresden and Continental AG for very good cooperation. Continental is appreciated for mixing the compound in industrial scale.

REFERENCES

Anderson, T.L. 1995. *Fracture mechanics*. Boca Raton, USA: CRC Press.

Cherepanov, G.P. 1967. Crack propagation in continuous media. *Journal of Applied Mathematics and Mechanics* 31: 503–512.

Grellmann, W., Reincke, K., Lach, R. & Heinrich, G. 2001. Characterization of crack toughness behaviour of unfilled and filled elastomers. *Kautschuk Gummi Kunststoffe* 54: 387–393.

Heinrich, G. & Kaliske, M. 1997. Theoretical and numerical formulation of a molecular based constitutive tube-model of rubber elasticity. *Computational and Theoretical Polymer Science* 7 (3/4): 227–241.

Holzapfel, G.A. 2000. *Nonlinear solid mechanics*. Hoboken, USA: Wiley.

Klüppel, M. & Schramm, J. 2000. A generalized tube model of rubber elasticity and stress softening of filler reinforced elastomer systems. *Macromolecular Theory and Simulations* 9 (9): 742–754.

Plagge, J. & Klüppel, M. 2017a. A physically based model of stress softening and hysteresis of filled rubber including rate- and temperature dependency. *International Journal of Plasticity* 89: 173–196.

Plagge, J. & Klüppel, M. 2017b. A hyperelastic physically based model for filled elastomeres including continuous damage effects and viscoelasticity. *This book*.

Rice, J.R. 1968. A path independent integral and the approximate analysis of strain concentration by notches and cracks. *Journal of Applied Mathematics and Mechanics* 35: 379–386.

Rivlin, R.S. & Thomas, A.G. 1953. Rupture of Rubber. I. Characteristic energy for tearing. *Journal of Polymer Science* 10: 291–318.

Vilgis, T.A., Heinrich, G. & Klüppel, M. 2009. *Reinforcement of polymer nano-composites: Theory, experiments and applications*. Cambridge: University Press.

Zehnder, A.T. 2012. Fracture mechanics. *Lecture Notes in Applied and Computational Mechanics* 62: 33–54.

Constitutive Models for Rubber X – Lion & Johlitz (Eds)
© 2017 Taylor & Francis Group, London, ISBN 978-1-138-03001-5

Influence of discontinuous thermo-oxidative ageing on the fatigue life of a NR-compound used for engine-mount application

C. Neuhaus, A. Lion & M. Johlitz
Institute of Mechanics, Faculty of Aerospace Engineering, University of the Bundeswehr Munich, Neubiberg, Germany

P. Heuler
AUDI AG Ingolstadt, Germany

M. Barkhoff & F. Duisen
BOGE Elastmetall GmbH, Damme, Germany

ABSTRACT: Elastomeric parts used in the automotive industry can experience irreversible changes of the materials' properties when exposed to elevated temperatures. This phenomenon is known as thermo-oxidative ageing and has a large influence on the fatigue behavior of natural rubber. To determine the change in the fatigue life due to pre-ageing, an ageing campaign in which the samples are aged with a constant temperature over a certain time, is usually performed. However, in practice, the ageing process must be considered as a discontinuous process as, for instance, pauses may occur that interrupt or decelerate the ageing process. To date, only a few studies are available that investigate the influence of discontinuous ageing on the mechanical fatigue behavior, e.g. (Charrier et al. 2012). For a reliable lifetime prediction of rubber parts, it is however crucial to estimate the error that occurs when continuous pre-ageing is used during the ageing campaign. The goal of this experimental study is to examine the differences between continuous and discontinuous thermo-oxidative ageing with regard to the fatigue life of natural rubber. Three types of discontinuous pre-ageing are performed, i.e. intermittent ageing, multiple succeeding ageing-temperatures and -resting periods after continuous ageing. In comparison with conventionally pre-aged samples, considerable differences regarding the fatigue life become apparent.

1 INTRODUCTION

1.1 *Thermo-oxidative ageing*

Elastomeric parts in automotive application not only undergo mechanical loads but also thermal loads that can affect the mechanical behavior in a complex manner. Especially natural rubber (NR) is very vulnerable to elevated temperatures. Because of the chemical structure of NR elevated temperatures can cause an irreversible degradation of the polymer network, known as thermo-oxidative ageing. Since the fatigue life of NR strongly depends on the materials ageing state, the influence of pre-ageing has been widely investigated in the literature e.g. (Broudin et al. 2015), (Ngolemasango et al. 2008), (Woo et al. 2009).

The normal procedure for investigating the influence of ageing is to age samples artificially at a constant temperature for a certain time. After this procedure, the mechanical behavior, e.g. the fatigue life, is evaluated. When performing a lifetime prediction, it has to be kept in mind that artificial ageing does not necessarily correspond to the temperature loads elastomeric parts experience in real application.

1.2 *Discontinuous thermo-oxidative ageing*

The major difference between ageing in service and artificial ageing in the laboratory is the time-temperature profile. For simplicity, a constant ageing temperature is usually used for artificial ageing in the laboratory. However, the time-temperature profile elastomeric parts undergo in automotive application may differ highly from this simple profile. For instance, it is well known that the temperature profile is highly variable over the time and contains various different ageing temperatures (Schmid et al. 2010).

Moreover, elastomeric parts in automotive application are usually only exposed to elevated temperatures during the driving process (or shortly after the driving process during postheating) which means that the ageing process is interrupted very frequently (Charrier et al. 2012). Thus, while artificial ageing in laboratory is a continuous process, ageing in practice must be considered as a highly discontinuous process. It is therefore necessary to investigate the differences between continuous and discontinuous thermo-oxidative ageing concerning the fatigue life.

2 EXPERIMENTAL INVESTIGATION

2.1 *Materials and experimental set-up*

The material used in this paper is a carbon-black filled NR/BR-blend for automotive applications. The fatigue tests are performed load controlled (load ratio R = −1) at a testing frequency of 4 Hz. For evaluation of the fatigue tests, the tensile part of the displacement cycle in the steady state is used. As failure criterion, a drop of stiffness of 20% is chosen. All test are performed with dumbbell samples, Figure 1.

To investigate the influence of discontinuous ageing, three different types of thermo-oxidative pre-ageing are performed:

- Intermittent ageing
- Multiple ageing temperatures
- Rest periods

The experimental execution of these experiments is explained in detail in the following sections. To guarantee comparability of the results, all experiments in this paper are performed with samples from the same production batch. Moreover, the samples are aged in identical ovens with air-flux control to ensure consistent atmospheric conditions during the ageing process.

2.2 *Intermittent ageing*

In order to investigate the influence of intermittent ageing, samples are not aged with one single ageing period but with multiple ageing periods of the same length, see Figure 2. After one period is finished, the samples are taken out of the oven and cooled down to room temperature ($RT = 21°C$). It is assumed that the ageing process is interrupted when the samples reach room temperature. This

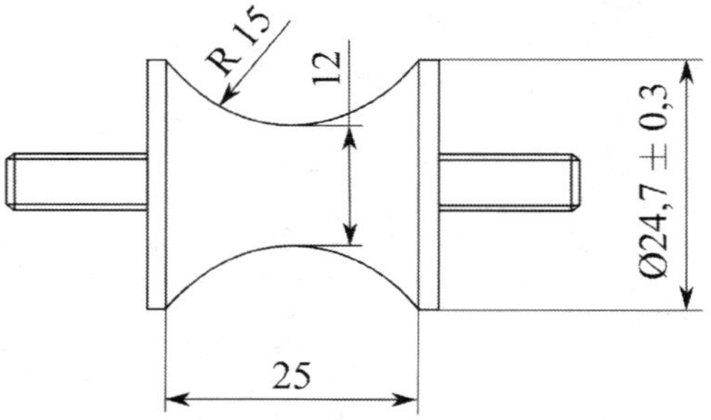

Figure 1. Dumbbell-specimen for uniaxial fatigue tests.

process of cyclic ageing is repeated until the desired cumulative ageing time t_a is reached.

The parameters for the investigation of intermittent ageing are specified in Table 1 for the different ageing temperatures. The S-N curves after intermittent ageing are presented in Figure 4. For comparison, also the S-N curves after continuous pre-ageing are depicted. Obviously, the fatigue life after intermittent ageing is not identical with the fatigue life after continuous ageing. For the ageing temperatures $T_a = 100°C$ and $T_a = 120°C$ the fatigue life is clearly higher after intermittent ageing for the same time. For $T_a = 85°C$ only a small difference can be recognized.

Of course, during the time in the oven the temperature is not constant inside the samples. Figure 3

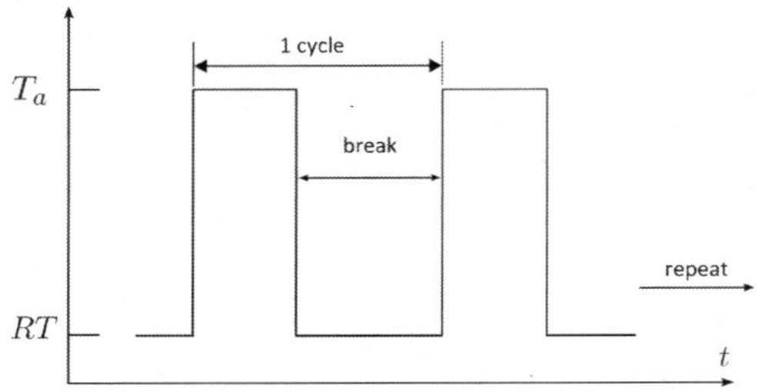

Figure 2. Ageing procedure for intermittent ageing.

Table 1. Experimental matrix for intermittent ageing.

	Ageing temperature T_a		
	85°C	100°C	120°C
ageing cycles	48 × 8 h	82 × 1 h	48 × 1/2 h
cooling time	>16 h	>22 h	>22 h
t_a (cumulative)	384 h	82 h	24 h

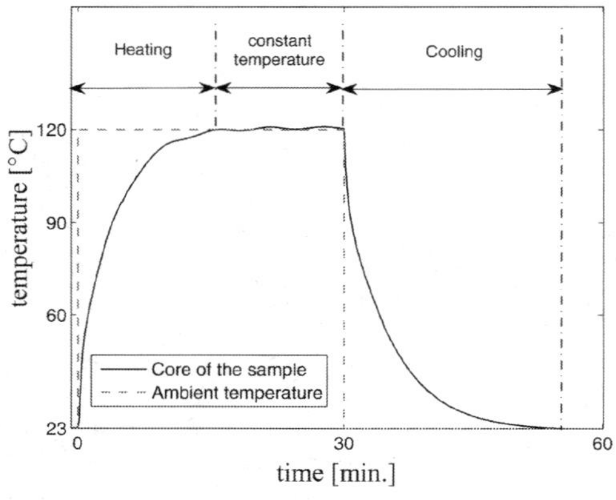

Figure 3. Core-temperature of dumbbell-sample during one ageing cycle.

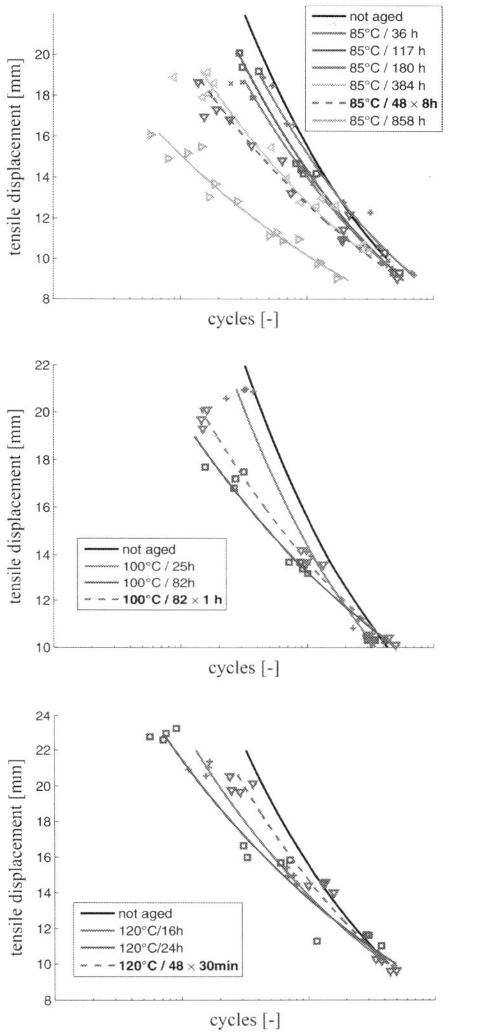

Figure 4. S-N curves after intermittent pre-ageing (dashed lines) compared to S-N curves after continuous pre-ageing (solid lines). All tests are performed at room temperature.

presents the temperature profile in the core of a sample aged at an ambient temperature of 120°C during one ageing cycle. The temperature is measured using a thermocouple. Although the ambient temperature is always maintained constant, it takes half of the time (15 minutes) to reach the target temperature. If the sample is taken out of the oven, it takes 27 minutes to reach room temperature. For this reason, the cumulative ageing time t_a must not be considered as the effective ageing time. Instead, an effective ageing time t_e must be calculated that takes into account the heating and the cooling of the samples.

The calculation is carried out with the Vogel-Fulcher-Tammann equation:

$$a = e^{\frac{E}{R}\left(\frac{1}{T_{\mathrm{ref}}} - \frac{1}{T_i}\right)} \tag{1}$$

In this equation, a is the time shift-factor that denotes the relation between the two ageing-times t_{ref} and t_i that are expected to cause the same progress of ageing. The activation energy E is a constant parameter and R is the gas constant. With the given temperature T_i over the time t_i, the effective ageing time t_e at the temperature T_{ref} is calculated by

$$t_e = a \cdot t_i. \tag{2}$$

Equation (2) is only valid if the temperature T_i is constant over the time. In the case of varying temperatures $T(t)$, the effective ageing time is calculated by integration:

$$t_e = \int_t^{t+\Delta t} e^{\frac{E}{R}\left(\frac{1}{T_{\mathrm{ref}}} - \frac{1}{T(t)}\right)} dt \tag{3}$$

The integral in this equation is solved numerically by the trapezoidal rule. The activation energy E in equation (3) was determined previously based on fatigue tests after preageing and varies from 77 kJ /mol to 93 kJ/mol.

The equivalent ageing times for the different stages of the ageing cycles that were calculated with equation (3) are presented in Table 2. For example, the heating phase of 15 minutes at an ambient temperature of 120°C is expected to cause the same progress of ageing concerning the fatigue life as an ageing time of 6.9 minutes at a constant temperature of 120°C. Also note that the cooling phase at this temperature is equivalent to only 24 seconds at a constant temperature of 120°C although the cooling phase is 27 minutes long. Using this approach, the total effective ageing times for the three different intermittent ageing procedures are calculated and presented in the last column of Table 2. For example, the intermittent ageing procedure at 120°C with the cumulative ageing time t_a = 24 h is assumed to have the same ageing effect as continuous ageing at 120°C for 18.4 h.

Figure 5 presents the evolution of the fatigue life at a tensile displacement of 19 mm over the preageing time. The dashed lines represent the fatigue life after continuous pre-ageing at the different preageing temperatures (The data is obtained from the S-N curves in Figure 4). The fatigue lifes after intermittend preageing are also plotted taking into account the cumulative ageing time t_a as well as the calculated total effective ageing times t_e.

Table 2. Calculation of the effective ageing times for intermittent ageing tests.

T_a	Phase	Time	Equivalen t_e @ T_a	$\Sigma\, t_e$ (one cycle)	Total t_e @ T_a
120°C	Heating	15 min.	6.9 min.		
	constant temperature	15 min.	15 min.	22.8 min.	× 48 = **18.4 h**
	Cooling	27 min.	24 sec.		
100°C	Heating	7 min.	2.5 min.		
	constant temperature	53 min.	53 min.	55.5 min.	× 82 = **76 h**
	Cooling	7.1 min.	8 sec.		
85°C	Heating	15 min.	7.7 min.		
	constant temperature	7.75 h	7.75 h	7.89 h	× 48 = **379 h**
	Cooling	28 min.	38 sec.		

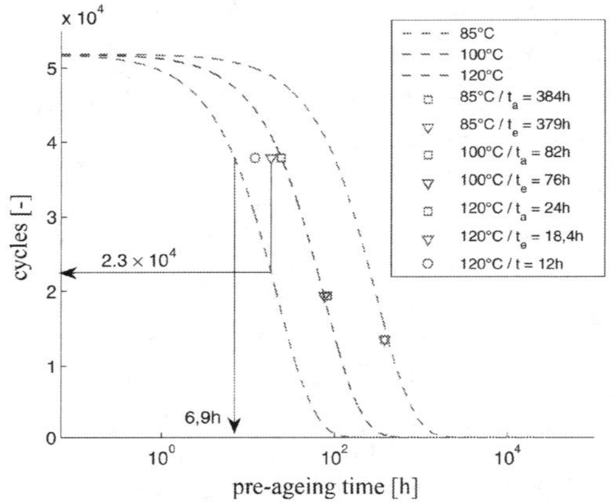

Figure 5. Fatigue lifes for intermittent ageing compared to the fatigue lifes after continuous pre-ageing (dashed lines) at a tensile displacement of 19 mm.

The fatigue life for intermittent ageing at 85°C is in good agreement with the curve for continuous pre-ageing for both t_e and t_a. For intermittent ageing at T_a = 100°C the fatigue life is higher compared to continuous ageing. However, considering the calculated effective ageing time t_e = 76 h the fatigue life is in perfect agreement with the curve obtained from continuous pre-ageing. In these cases, it is concluded that intermittent ageing has the same effect on the fatigue life as continuous ageing if the real ageing time is calculated by taking into account the heating and cooling phases.

A different result can be seen for intermittent ageing at 120°C. For a calculated effective ageing time of T_e = 18.4 h a fatigue life of approximately 2.3 × 10^4 cycles is expected. In fact, the samples have a fatigue life of approximately 3.8 × 10^4 cycles. However, this fatigue life would be expected after 6.9 h of continuous ageing, see Figure 5. Even under the assumption that the heating and cooling periods in a single cycle do not cause any ageing at all, the effective ageing time (15 minutes × 48 cycles = 12 h) is longer than the value of 6.9 h. This observation leads to the conclusion that, at least for ageing at 120°C, intermittent pre-ageing has much less impact on the fatigue life compared to continuous pre-ageing. One possible explanation is based on the fact that thermo-oxidative ageing is an autocatalytic chain reaction that contains several partial reactions (Bolland 1949), (Streit 2011). This would mean that the actual degradation of the polymer network requires an incubation phase (or induction period (Verdu 2012)) that appears in every single ageing block and therefore can not be neglected.

2.3 *Different pre-ageing temperatures*

In these experiments, samples are not aged with a single ageing temperature but with two blocks of different ageing temperatures, see Figure 7. The corresponding durations of the ageing blocks are chosen to have a similar effect on the fatigue life. Figure 6 illustrates the fatigue lifes at a tensile displacement of 16 mm that are obtained from the S-N curves after preageing with two different ageing temperatures. The fatigue tests are performed at room temperature. The single S-N curves are not presented in this work.

Obviously, the order of the ageing periods can have an important influence on the resulting fatigue life. If, for example, the samples are first aged at T_a = 70°C for 1000 h and at T_a = 85°C for 384 h afterwards the fatigue life is only half of what it is when the order of the ageing periods is reversed. The lower part of Figure 6 shows the steady state reaction force during the fatigue tests at a tensile displacement of 16 mm. In the majority of cases, samples have a considerably higher stiffness if they are first aged with the higher ageing temperature. Additionally, samples were aged with multiple mixed ageing periods instead of only two periods. The procedure containing the ageing periods 100°C/82 h and 70°CC/1500 h was divided into four periods of 20.5 h/100°C and three periods of

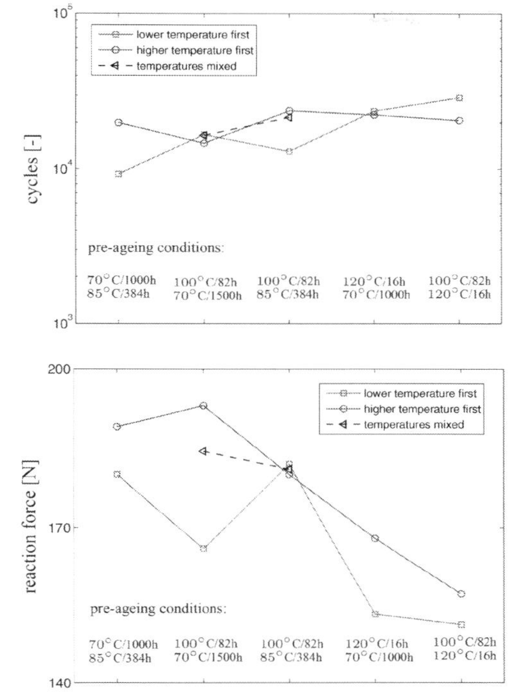

Figure 6. Fatigue life at a tension displacement of 16 mm (upper figure) and reaction force at this displacement (lower picture). All tests are performed at room temperature.

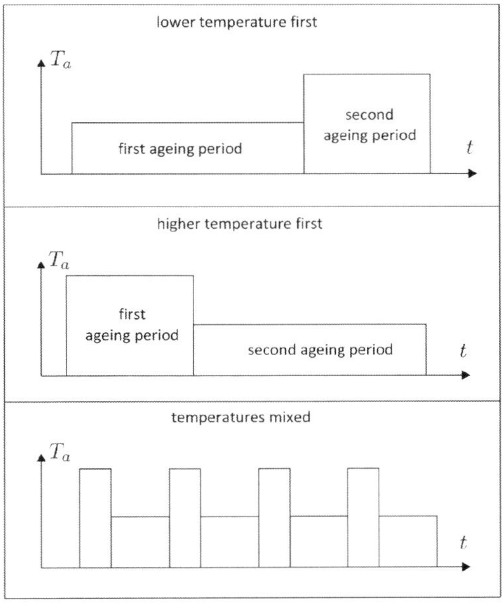

Figure 7. Illustration of the ageing procedures for pre-ageing with different temperatures.

360 h/70°C respectively. The last ageing period at 70°C had a length of 420 h.

The procedure containing the ageing periods 100°C/82 h and 85°C/384 h is devided into four periods of 20.5 h/100°C and four periods of 96 h/85°C. In these cases of mixed preageing temperatures, both the fatigue lifes and the reaction forces range between the values after ageing with two ageing periods.

2.4 *Rest periods after continuous ageing*

For the last experimental setup, samples are aged continuously at constant temperature without interruptions. After the ageing, half of the samples are tested within 24 hours. The rest of the

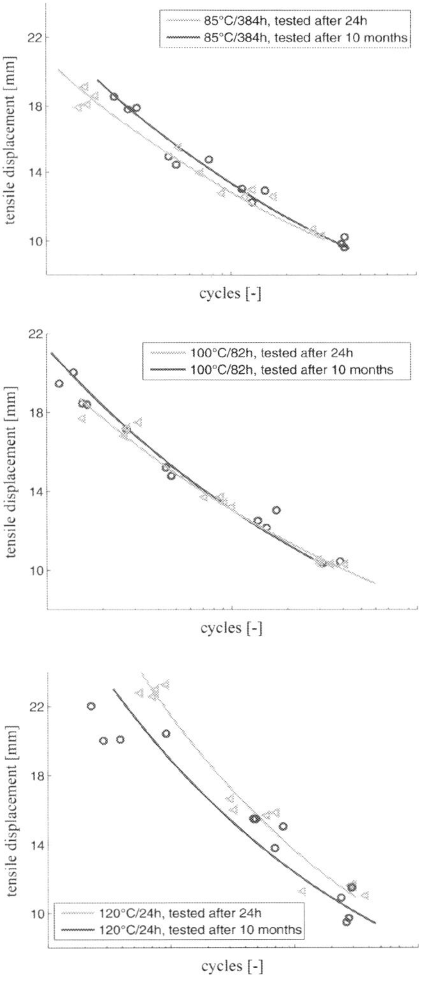

Figure 8. Effect of recovery at room temperature after pre-ageing. All tests are performed at room temperature.

samples is stored at controlled room temperature (RT = 20°C) and normal humidity for exactly 10 months. During this time, the samples are not exposed to any light. After this process that is called recovery in this paper, the samples are also tested. This process is performed with three different ageing procedures. The results are shown in Figure 8. After the recovery time, the samples aged at 85°C/384 h have a slightly higher fatigue strength compared to the samples that where tested directly after the ageing process. After ageing at T_a = 100°C for 82 h the resting period seems to have no influence on the fatigue life. Another result is achieved after ageing at 120°C/24 h. In this case the resting period leads to a considerable reduction of the fatigue life. It is assumed that during the ageing at 120°C the majority of the antioxidant was consumed which makes the material vulnerable to ageing processes even at room temperature. These results lead to the conclusion that the ageing state is not constant after artificial ageing but also evolves at room temperature. This is a very important finding since elastomeric parts are usually in use for several years. Moreover, the resting period may lead to both an increase or a reduction of the fatigue life depending on the pre-ageing procedure.

3 CONCLUSIONS

In this paper the influence of discontinuous ageing on fatigue life was investigated based on three different experimental setups and compared to the results after continuous preageing. Due to the wide range of possible parameters, it is not possible to give a complete overview on the effects of discontinuous ageing. Nevertheless, it was shown that the time sequence of the thermo-oxidative pre-ageing has a high influence on the fatigue life of NR.

It is concluded that the process of intermittent ageing should be examined more closely in the future. On the one hand, it seems to have the highest degree of relevance because the ageing process is frequently interrupted in real application. On the other hand, very high temperatures may occur as short temperature peaks (Zwittian 2009). The findings of this work lead to the assumption that these peaks could be much less harmful for natural rubber. In this case, a lifetime prediction based on continuous pre-ageing at very high temperatures could lead to a massive underestimation of the fatigue life.

REFERENCES

Bolland, J. (1949). Kinetics of olefin oxidation. *Quarterly Reviews, Chemical Society*, 1–21.

Broudin, M., V.L. Saux, Y. Marco, P. Charrier, & W. Hervouet (2015). Investigation of thermal aging effects on the fatigue design of automotive anti-vibration parts. *Constitutive Models for Rubber IX*, 53–59.

Charrier, P., Y. Marco, V.L. Saux, & R. Ranaweera (2012). On the influence of heat ageing on filled nr for automotive avs applications. *Constitutive Models for Rubber Vii*, 381–388.

Ngolemasango, F., E. Bennett, & J. Clarke (2008). Degradation and life prediction of a natural rubber engine mount compound. *Journal of Applied Polymer Science 110*, 348–355.

Schmid, A., B. Seufert, Z. Pezelj, U. Weltin, & J. Spreckels (2010). Kundennaher Betriebsfestigkeitsnachweis von Elastomerbauteilen im Fahrwerksbereich unter Berücksichtigung von Alterung und Ermüdung. *MP – Materials Testing 07–08*, 543–548.

Streit, G. (2011). *Elastomere Dichtungssysteme Werkstoffe-Anwendungen-Konstruktion-Normen.* Expert verlag.

Verdu, J. (2012). *Oxidative Ageing of Polymers.* John Wiley & Sons, Inc.

Woo, C., W. Kim, S. Lee, B. Choi, & H. Park (2009). Fatigue life prediction of vulcanized natural rubber subjected to heat-aging. *Procedia Engineering 1*, 9–12.

Zwittian, T. (2009). Versuchsoptimierung Kunde-Teststrecke-Prüfstand. *DVM-Bericht 676*, 181–189.

Constitutive Models for Rubber X – Lion & Johlitz (Eds)
© 2017 Taylor & Francis Group, London, ISBN 978-1-138-03001-5

Effect of filler-polymer interfacial phenomena on fracture of SSBR-silica composites

Mohammad Alimardani & Mehdi Razzaghi-Kashani
Department of Polymer Engineering, Faculty of Chemical Engineering, Tarbiat Modares University, Tehran, I.R. Iran

ABSTRACT: The idea of partitioning the equilibrium fracture strength (G_0) of adhesive joints into the contributions of cohesive (R), Interfacial (I) and Substrate (S) failures in is extended here for filled polymer composites. Using the silane spacer length and grafting density as tuning parameters, composites of silane treated silica-Solution Styrene Butadiene Rubber (SSBR) are prepared with controlled filler-polymer interfacial phenomena, and the mechanisms through which silane length modifies the fracture of composites is extracted. It was shown that chain length of silane greatly modifies the filler-filler and the filler-polymer energetic interactions. Tearing energy was assessed in trousers geometry and the share of matrix and the interface on the crack growth resistance of the composite is determined. The highest value of fracture strength was measured for composites having a balanced range of both dissipation characteristics and filler-polymer chemical compatibility.

1 INTRODUCTION

In unfilled rubbers when cohesive failures are the sole mechanism of fracture, tearing energy has been found to be a function of equilibrium fracture energy (G_0) multiplied by the rate (v) and temperature (T) dependent viscoelastic dissipations (i.e. $G_c = G_0(1+f(T,\nu))$ (Andrews 1974, Saulnier et al. 2004). The parameter G_0 is determined by the attributes of rubber network chains, most notably by the molecular weight of rubber (M), the molecular weight between entanglements (M_e) and the molecular weights between crosslinks (M_c) (Ahagon and Gent 1975). When a multi-component system of polymer-filler-interphase is concerned, contribution of each constituent in preventing crack growth should also be taken into consideration. In adhesive joints, the suggested formula by Andrews and Kinloch for G_0 accounts the resistance to crack growth in the bulk polymer, through the interface and within the substrate, as below (Andrews and Kinloch 1974):

$$G_0 = rP + iW_p + sF \qquad (1)$$

where r, i and s are the area fractions of polymer, interface and substrate and P, W_P and F refer to the energy per unit area required for a crack to propagate in each of these three media, respectively.

This idea, after modifications, may be extended to polymer nanocomposites where surface features of particles matters much to the rubber reinforcement and crack advancement.

Literature is well-documented on the use of long chains silanes in modifying the properties of particulate-filler polymer composites (Jenkins, Dauskardt and Bravman 2004). It has always been speculated that silane chain length dictates the properties of silane-treated silica filled polymers through two possible reinforcing mechanisms, namely energetic interaction and/or mechanical engagement. Energetic interaction is regarded as the energy variations resulting from the interaction of non-bonded constituents such as van der Waals and electrostatic interactions. Mechanical engagement refers to either possible entangling of grafted chains on the filler surface with surrounding polymer chains or simple modification of polymer-filler interfacial friction through mechanical interlocking. Even for the long chain silane, the existence of an entangling interaction is not reasonable, instead, mechanical interaction of frictional type makes more sense. Jenkins et al (Jenkins et al. 2004) reported that the presence of long chain silanes grafted on the silica surface affects fracture strength of polymer composites, probably through mechanical engagement of the silane and polymer chains. An augmented shear interfacial adhesion of composites having treated silica has also been attributed to such a mechanical interlocking between long chain silane and polymer chains (Jiang et al. 2007). Similar studies (Ladouce-Stelandre et al. 2003, Suzuki, Ito and Yatsuyanagi 2005) have been conducted to realize the significance of silane spacer length in modification of mechanical properties of the resulting composites, however, it is not yet obvious which of the energetic interactions or mechanical engagements should be regarded as the mechanism of effectiveness of silane spacer length.

Here, it is hypothesized that severity of energetic interactions can be truly reflected in surface energy of fillers. Silica is treated with both short and long chain silanes and the surface energy of the resulting powders are adjusted by controlling the grafting density to prepare systems of similar and diverse surface energy. This methodology would pave the way of realizing the true mechanism (i.e. energetic interaction, mechanical engagement) underlying the reinforcement. Having realized the reinforcing mechanisms, the fracture properties of the composites is assessed in trouser geometry and discussions over the prediction of tearing energy using the affecting parameters will be given.

2 EXPERIMENTAL SECTION

2.1 *Materials*

Solution-polymerized SBR (SSBR-Buna VSL 4526 2HM) having styrene content of 26 wt% and vinyl content of 44.5 wt% was purchased from LANXESS (Leverkusen, Germany). Untreated precipitated silica of Ultrasil VN3 having a specific surface area of 180 m²/gr was provided by Evonik-Degussa AG (Essen, Germany). Two kinds of silane namely propyltrimethoxysilane (C3-silane) and hexadecyl-trimethoxysilane (C16-silane) were purchased from Sigma-Aldrich, Germany. Surface of Ultrasil was treated with these two silanes in our laboratory. The commercial grade of Coupsil 8113 was also purchased from Evonik-Degussa AG (Essen, Germany). It has a total content of 11.3 wt% of bifunctional silane known as 3-triethoxysilylpropyl) tetra sulfide (TESPT).

2.2 *Characterization methods*

Surface energy values of silica powders were estimated using a sessile drop–based method and the two-component theory of Fawkes. Infrared spectrometer of Perkin–Elmer was used for evaluating the surface modification by silane in the wavenumber range of 400 to 4000 cm^{-1}. Differential Scanning Calorimetry (DSC) of NETZSCH type was used for measuring heat of vulcanisation. A rubber process analyzer (RPA 2000, Alpha Company) was also employed to analyze non-linear viscoelastic properties. This strain sweep experiment was conducted in a range of 0.27–75% at 60°C and frequency of 1 Hz. Tearing energy was measured in the trouser geometry where legs of the specimen are pulled apart in opposite directions to provide a state of mode III crack loading. Providing that the leg extension is adequately prohibited during the test, tearing energy (Gc) is calculated using the Equation 1:

$$G = 2F / t \tag{2}$$

where, F refers to the mean tearing force in Newton and t to the thickness of sample at the central part.

2.3 *Compound preparation*

Rubber compounds were prepared using an internal mixer and a two roll mill. After drying silica in 80°C for two days, a master batch of rubber and silica was prepared using an internal mixer (Brabender-W50ETH). The content of filler was chosen to be 60 phr (parts per hundred of rubber). The antiozonant 6PPD (Table 1) and cure activators of zinc oxide and stearic acid were also added to the compound during this state of mixing. The resulting master batch was compounded with sulfur and accelerators (CBS N-cyclohexylbenzothiazole-2-sulfenamide & DPG diphenylguanidine) on a two roll-mill (Brabender-PM2000) operating at the friction ratio of 1:1.5 for another 15 min. Vulcanizates were finally compression molded at 160°C.

More details on the grafting process of silica as well as preparation of the resulting composites can be found elsewhere (Alimardani et al. 2016).

The terms Ultrasil and Coupsil will be used to indicate the unmodified silica and the silica treated commercially by TESPT, respectively. Four other powders with definite surface energies and grafting densities were prepared using short and long chain silane. They will be designated by *C3-silica, *C16-silica, C3-silica and C16-silica. The characteristics of these homemade silica will be given in the *results and discussions* section. The designations will be used to indicate both the silica powders and the resulting compound.

Table 1. Surface energy of pristine and modified silica powders; polar (γ_s^p), dispersive (γ_s^d) and total (γ_s^t).

Silica type	γ_s^p	γ_s^d	γ_s^t
Ultrasil	30.4	37.27	67.67
Coupsil	12	39.75	51.75
C3-silica	18.8	40.64	59.48
C16-silica	0	39.45	39.45
*C3-silica	11.3	41.76	53.06
*C16-silica	11.3	39.4	50.7

3 RESULTS AND DISCUSSIONS

3.1 *Silica surface characterization*

Infrared (IR) spectra of unmodified and treated silica by two silanes of short and long chain length are depicted in Figure 1. Two remarkable peaks including an intense broad band in the 1100–1000 cm^{-1} range, and a less-intense peak at wave numbers around 3400 cm^{-1} are the silica characteristics in IR. The asymmetric Si–O–Si stretching is behind the former peak, and the latter peak is related to stretching of hydroxyl bonds (O-H). Treatment of silica by silane leads to appearance of new peaks around 2800–2950 cm^{-1}, being assigned to methyl symmetric and asymmetric C–H stretching and also methylene asymmetric and symmetric C–H stretching. Number of methylene groups in C16 silane is larger than C3-silica and it would result in a sharper peak compared to C3 silane. When comparing two spectra, peak intensity depends on not only the quantity of the material taken for testing but also the relative number of the bonds being available in each system. Therefore, state of surface modification can be tracked by pursuing any variations in bond numbers. Plotting two different spectra from a single baseline and comparing the relative intensity of two peaks in each of the spectrum is a reliable indication of the relative number of bonds in each system. During the treatment of silica, number of hydroxyl groups is reduced and at the same time siloxane groups are added. Increasing the peak height of siloxane is another reliable sign of surface modification particularly when the hydroxyl intensity is lowered.

3.2 *Silica surface engineering*

To understand the severity and the type of the operative mechanisms, first, degree of grafting is adjusted to provide two filler systems with diverse

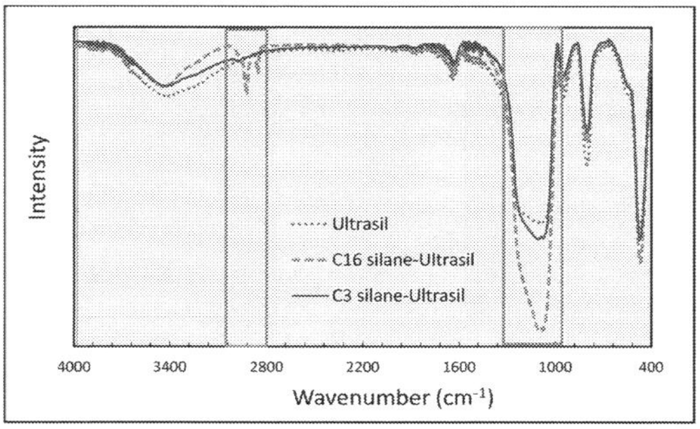

Figure 1. Infrared spectrums for treated and untreated silica; only two of the lab-treated silica powders are shown.

silane spacer length (3 & 16 carbon-length), but with almost identical surface energy (*C3-silica and *C16-silica). Then, by using the stoichiometric amounts of short and long chain silane during the surface modification, two new types of silica with different silane spacer length as before, but with diverse surface energies are obtained (C3-silica and C16-silica).

Considering surface energy values of filler in Table 1, it is seen that the dispersive part of unmodified silica remains almost constant during surface modification. Comparing surface energy of treated silica powders having equi-molar silanes of C3 and C16 (i.e. C3-silica and C16-silica), it is revealed that silica treated with long-chain silane (C16-silica) exhibited an exceptional hydrophobic character $(\gamma_s^p = 0)$ due to severe shielding of hydroxyl groups existing on the surface. By increasing the grafting density of C3-silane and lowering the grafting density of C16-silane, two new filler systems having similar polar components of surface energy were created (i.e. *C3-silica and *C16-silica). These two samples would have similar energetic interactions with the matrix; therefore, any variation in response of the resulting composites would be purely resulted from the mechanical engagement, with no interference from the energetic interactions. This is while any variation in mechanical response of the composites prepared by C3-silica and C16-silica would be expected both from energetic interactions and mechanical engagements.

3.3 *Evaluating the role of matrix vulcanisation*

In order to track the net result of silane spacer length on the composite mechanical properties, the influence of any other affecting parameter has to be considered. Among these, filler quality of dispersion and matrix state of vulcanization can be named. In reference (Alimardani et al. 2016), efforts were made to control filler dispersion using the thermodynamic-based theories. Here, to have an understanding of the degree to which matrix state of vulcanisation varies among composites, uncured samples were analyzed using DSC in a non-isothermal manner (Figure 2). The area under each peak from 160°C to 240°C represents the heat of vulcanisation.

Comparing the values, it is clear that there exists a connection between the total value of surface energy of fillers and the values of the peak area. Ultrasil and C16-silica form the lower and upper bound, and the other compounds can be ranked between these two values, based on their surface energy. The results can be well correlated to the theory that accounts the filler-filler interaction and bound rubber as the prime factor controlling degree of vulcanisation (Hosseini and Razzaghi-Kashani

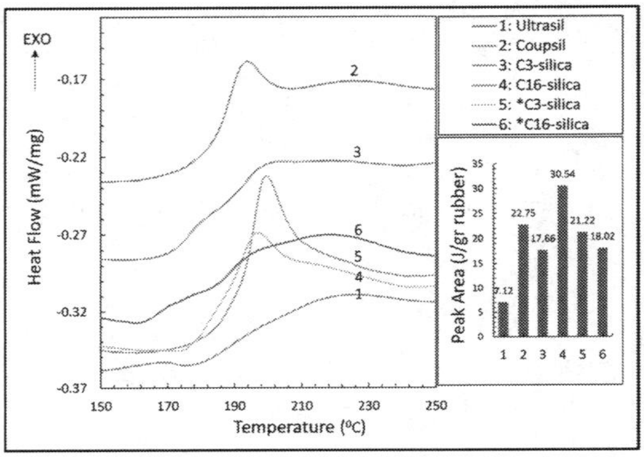

Figure 2. DSC graphs showing temperature sweep of uncured samples; the calculated area under each peak has also been denoted in the inset.

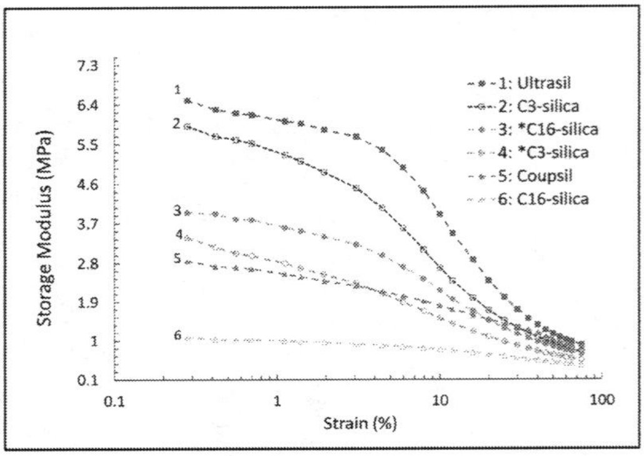

Figure 3. Strain sweep of vulcanizates at 60°C and frequency 1 Hz.

2014). Given the variance observed for the heat of vulcanisation, further investigation is required as discussed below to guarantee that vulcanisation is not mixed up with the role of silane spacer length.

3.4 *Non-linear viscoelastic properties*

As seen in Figure 3, effectiveness of silane spacer length on the reinforcement is studied using the strain-induced softening phenomena (the so-called Payne effect). The value of storage modulus at low strains for each composite is proportional to the polarity of the filler surface; being lowest for C16-silica; moderate for *C3-silica, *C16-silica, relatively higher for C3-silica and highest for Ultrasil. This is a direct role of filler-filler energetic interaction on rubber chain mobility and composite reinforcement. Interestingly, minor difference between the response of *C3-silica and *C16-silica vulcanizates can be detected which is attributed to the mechanical engagement resulting from the frictional interlocking of long chain silane and the surrounding polymer chains.

Given the similar trend observed for uncured samples (Mahtabani, Alimardani and Razzaghi-Kashani 2016), it is indicated that at such highly filled compounds (60 phr silica), the role of filler-filler interaction is so significant that variations in the matrix state of vulcanisation cannot greatly affect the mechanical properties of composites.

3.5 *Fracture and tearing energy*

In Table 2, the tearing energy measured using the trouser geometry is listed for the composites. The biggest value is observed for *C16-silica and *C3-silica, being followed by the Coupsil, C3-silica and Ultrasil. The lowest value is for C16-silica. The highest value of fracture strength was not observed

Table 2. Tearing energy (Gc) of the composites.

Compound	Tearing energy (J/m^2)
Ultrasil	21880 ± 1500
Coupsil	30000 ± 10000
C3-silica	29228 ± 2000
C16-silica	11338 ± 200
*C3-silica	33711 ± 1500
*C16-silica	34434 ± 1500

for composites with the maximum energy dissipation (i.e. Ultrasil). Instead, the sample with a mediocre range of dissipation characteristics holds the maximum. It implies that a simple connection between the tearing energy and viscoelastic dissipation does not exist.

Due to the existence of covalent bond at the interface of Coupsil, the tearing force is increased when facing a bond at the interface and sharply reduced when the bond is broken. This is led to a large standard deviation for the composite Coupsil.

Considering the formula suggested for adhesive joints, it seems logical that tearing energy may be scaled with Equation 3.

$$G_c \propto \frac{G''}{\gamma_{12} \times CD^{0.5}} \tag{3}$$

In this Equation, γ_{12} is the interfacial tension, G'' is the nonlinear viscoelastic dissipation and CD is a symbol for crosslink density that is scaled to the DSC peak area. Precisely, it implies that not only the energy per area required for bond breakage in the rubber matrix, but also the role of interface is considered with the energy to cope with the viscoelastic dissipations. The combination

Table 3. Prediction based on the Equation 3.

Compound	$G''/(y12*CD^{0.5})$
Ultrasil	9116
Coupsil	6026
C3-silica	9636
C16-silica	4691
*C3-silica	9826
*C16-silica	12758

of these parameters based on the Equation 3 and including the corresponding values for each composite yields the values listed in Table 3. Again, the biggest value is expected for *C16-silica and the least for the C16-silica. The others can roughly be put between these two boundaries.

The relation however only considers the energetic compatibility of polymer and filler at the interface and the role of covalent bonding as physical barrier is neglected. This aspect will be extensively pursued in a future publication.

4 CONCLUSION

The present study revealed that the long chain of silane affects composite mechanical properties through energetic filler-filler interaction (represented by surface energy of filler). Also, minor roles were shown to be played by mechanical engagement of long chain silane and polymer chains. Efforts were made to prevent filler quality of dispersion and matrix state of vulcanisation disguise the role of silane chain length. Considering the share of matrix, interface and also the contributions of viscoelastic properties, the trend of tearing energy among several number of rubber composites can be acceptably ranked.

REFERENCES

Ahagon, A. & Gent, A. (1975). Threshold fracture energies for elastomers. *Journal of Polymer Science: Polymer Physics Edition* 13: 1903–1911.

Alimardani, M., Razzaghi-Kashani, M., Karimi, R. & Mahtabani, A. (2016). Contribution of Mechanical Engagement and Energetic Interaction in Reinforcement of SBR-Silane-Treated Silica Composites. *Rubber Chemistry and Technology* 89: 292–305.

Andrews, E. (1974). A generalized theory of fracture mechanics. *Journal of Materials Science* 9: 750–757.

Andrews, E. & Kinloch, A. (1974). Mechanics of elastomeric adhesion. *Journal of Polymer Science: Polymer Symposia* 46: 1–14.

Hosseini, S. M. & Razzaghi-Kashani, M. (2014). Vulcanization kinetics of nano-silica filled styrene butadiene rubber. *Polymer* 55: 6426–6434.

Jenkins, M., Dauskardt, R. & Bravman, J. (2004). Important factors for silane adhesion promoter efficacy: surface coverage, functionality and chain length. *Journal of adhesion science and technology* 18: 1497–1516.

Jiang, Z. X., Meng, L. H., Huang, Y. D., Liu, L. & Lu, C. (2007). Influence of coupling agent chain lengths on interfacial performances of polyarylacetylene resin and silica glass composites. *Applied Surface Science* 253: 438–445.

Ladouce-Stelandre, L., Bomal, Y., Flandin, L. & Labarre, D. (2003). Dynamic mechanical properties of precipitated silica filled rubber: influence of morphology and coupling agent. *Rubber Chem. Technol.* 76: 145–159.

Mahtabani, A., Alimardani, M. & Razzaghi-Kashani, M. (2016). Further evidence of filler–filler mechanical engagement in rubber compounds filled with silica treated by long-chain silane. *Rubber Chemistry and Technology* In Press.

Saulnier, F., Ondarçuhu, T., Aradian, A. & Raphaël, E. (2004). Adhesion between a viscoelastic material and a solid surface. *Macromolecules* 37: 1067–1075.

Suzuki, N., Ito, M. & Yatsuyanagi, F. (2005). Effects of rubber/filler interactions on deformation behavior of silica filled SBR systems. *Polymer* 46: 193–201.

Constitutive Models for Rubber X – Lion & Johlitz (Eds)
© 2017 Taylor & Francis Group, London, ISBN 978-1-138-03001-5

True stress control for fatigue life experiments of inelastic elastomers

K. Narynbek Ulu
LRCCP, Vitry-sur-Seine, France
Institut de Recherche en Génie Civil et Mécanique (GeM), UMR CNRS 6183, Ecole Centrale de Nantes, Nantes, France

B. Huneau & E. Verron
Institut de Recherche en Génie Civil et Mécanique (GeM), UMR CNRS 6183, Ecole Centrale de Nantes, Nantes, France

A.-S. Béranger & P. Heuillet
LRCCP, Vitry-sur-Seine, France

ABSTRACT: Fatigue of industrial elastomers is an important direction of research. In this context, the most widely used approach in both academic and industrial research is fatigue life testing. In most studies, the experimental procedure considers a prescribed constant cyclic displacement. Currently, the relation between the prescribed displacement and stress/strain is derived assuming mostly elastic behavior. However, limitations arise when the elastomer exhibits strong inelastic behavior; the mechanical state parameters of the material, i.e. stress and/or strain, do not remain constant throughout the duration of the displacement-controlled experiment. The present work, firstly, states the limitations of conventional approaches. Then, a new experimental procedure based on true stress control is presented for fatigue testing, which alleviates some of the issues associated with conventional methods and leads to creation of a Wöhler curve for elastomers. Additionally, the procedure is extended for simultaneous parallel testing of several specimens for time efficiency and greater statistical significance.

1 INTRODUCTION

Elastomers have proven to be useful in a wide range of application and can be found in many manufactured products. Hence, lifetime prediction of elastomers is an important field of study at present. There are many damage processes that affect the lifetime of an elastomeric part in service conditions, and damage due to fatigue is one of the major ones.

The theoretical background of fatigue life testing comes from continuum mechanics, where one assumes that a history of loading can be used to predict lifetime. Fatigue life is defined itself as a number of cycles until initiation of a crack of a specific size or, in some cases, complete failure of a tested specimen.

For metals, where applied stresses are large and strain measurements are small, the experiments are carried out in force control for practical purposes. On the other hand for elastomers, the experiments are carried out mostly in displacement control. Usually, within industry and in academic research, the next step consists in calculating the relationship between displacement and strain, as determination of the stress state is complicated in elastomers (Mars & Fatemi 2003). This step is not straightforward and several major assumptions are considered to overcome the complications. The main assumption in use of displacement control in fatigue life testing is the elasticity of the material. Moreover, it becomes apparent that normalization of results, as it is done for metals for example, is problematic mainly due to complex response of elastomers.

It is important to note that true stress, not displacement or force, is what is experienced by the material and what drives the fatigue damage. In the present work, the drawbacks of testing in displacement (or force) control are discussed with presentation of simple examples. And in order to alleviate these issues, a procedure in true stress control is developed; constant true stress amplitude is maintained at constant R-ratio $\left(R = \sigma^{min} / \sigma^{max}\right)$. The procedure is straightforward and can be extended to testing of specimens in parallel, which is important in fatigue life testing of elastomers due to the large scattering observed in the results (Svensson 1981).

2 DRAWBACKS OF DISPLACEMENT AND FORCE CONTROL

2.1 *Background*

Before discussing the limitations of displacement (and force) control in fatigue life testing of elastomers, the difference between mechanical and experimental testing parameters should be discussed. The three main experimental parameters are displacement, force, and cross-section surface area; these are dependent on both material response and specimen geometry. On the other hand, strain and true (Cauchy) stress are the mechanical parameters and are intrinsic to the material itself. In practice, the mechanical parameters are calculated from the experimental; hence for example, strain is derived from displacement, and true stress can be calculated from actual cross-section area and force.

As mentioned above, fatigue life testing of elastomers is usually carried out in displacement control, starting from first studies by Cadwell et al. (1940), who calculated principal strain as a function of applied displacement. Another mechanical parameter that is utilized in displacement control testing, especially for multiaxial loading, is strain energy density. In both cases *S-N* (stress—number of cycles) curves, first introduced by Wöhler for metals (Schütz 1996), are replaced by λ-*N* (strain) or *W-N* (energy) for elastomers if tested in displacement control.

2.2 *Effects of inelastic response of materials on testing*

Theoretically, when performing tests in displacement control, one assumes mostly elastic behavior of the material; that is, it is possible to derive a relationship between experimental and mechanical parameters.

In reality, several issues arise because the response of elastomers is complex. Determination of mechanical parameters from experimental is much more complicated (Mars & Fatemi 2003). There are complex phenomena, especially in filled elastomers, which consist of many dissipative mechanical responses that materialize as hysteresis on stress-strain curves (Bergström & Boyce 1998).

The first phenomenon that is observed is referred to as the Mullins effect and it occurs when a filled elastomeric material is subjected to a new maximum loading (Diani et al. 2009): one can observe significant softening with a decrease in measured force with successive cycles. Thus, the Mullins effect is usually observed within the initial cycles of fatigue life testing. However, it can reappear after the initial cycles, if the material is subjected to a higher and new maximum loading. The consequence of the softening is presence of inelastic strain; when the material is fully unloaded, a specimen elongates and one can measure compression in bulky specimens or observe buckling in thin specimens.

Afterwards, with constant displacement loading conditions, one can observe long-term softening that is classically attributed to viscoelastic response of elastomers. Several authors (Kingston & Muhr 2007; Mars 2007; Ayoub et al. 2014) refer to it as 'cyclic stress relaxation' (and as 'cyclic creep' or 'ratcheting' in force controlled experiments).

The influence of these two relaxation phenomena on fatigue life testing depends on the degree of inelasticity that an elastomer in question exhibits. We can separate the influence into two simple cases: where the Mullins effect is present, but long-term cyclic stress relaxation can be considered insignificant; or, where both relaxation phenomena cannot be disregarded.

2.2.1 *Mullins effect with insignificant long-term stress relaxation*

In many studies on natural rubber (NR), displacement control in fatigue life testing is deemed to be sufficient. The Mullins effect is taken into account and the long-term stress relaxation is considered to be negligible; after a certain amount of cycles, it is assumed that there is stabilization of the stress-strain behavior and the relationship between the experimental and the mechanical parameters can be established (ASTM International 1999; Mars & Fatemi 2002; Mars 2007). However, there is no consensus on what is sufficient to 'accommodate' for the Mullins effect. For example in one study (Andre et al. 1999), 100 cycles are deemed to be satisfactory to have a stabilized stress-strain behavior with no long-term relaxation; but in others, this 'accommodation phase' is absent (Ismail & Jaffri 1999; Kim & Jeong 2005; Legorju-Jago & Bathias 2002).

Moreover, there is an influence on the *R*-ratio; due to the presence of inelastic strain and compression/buckling at zero displacement, the actual *R*-ratio in terms of stress $\left(R = \sigma^{min} / \sigma^{max}\right)$ or strain $\left(R = (\lambda^{min} - 1)/(\lambda^{max} - 1)\right)$ is lower as compared to displacement *R*-ratio $\left(R = u^{min} / u^{max}\right)$.

Overall, the main problem with this approach is that many fatigue life testing protocols exist, but these are not equivalent; interpretation of different studies and comparison of different elastomers is complex at best and erroneous at worst.

2.2.2 *Mullins effect with significant long-term cyclic stress relaxation*

Displacement control testing is inappropriate for fatigue life testing of elastomers that exhibit both the Mullins effect and significant long-term cyclic stress relaxation. With a simple uniaxial example, one can show that the mechanical parameters, stress and strain, are not constant with displacement control.

Figure 1 shows a simplified representation of loading path start- and end-points: where *A* represents the first cycle; *B* – some cycles after *A* that represents the end of softening due to the Mullins

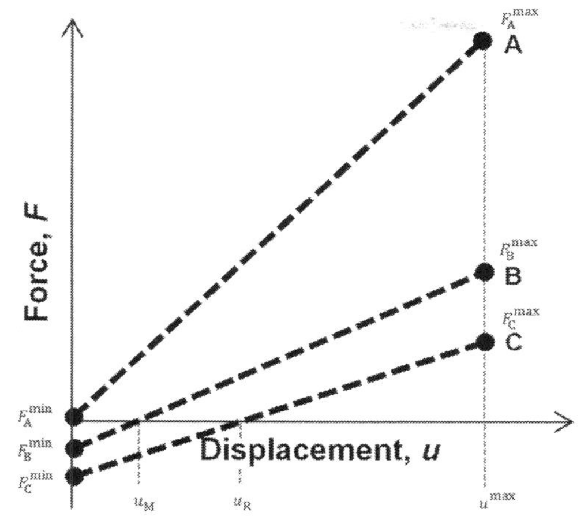

Figure 1. Simplified figure of the Mullins effect with long-term cyclic stress relaxation.

effect; and finally, cycle C occurs some cycles after B that represents the long-term cyclic stress relaxation. The strain at point A is:

$$\lambda_A^{max} = \frac{l_0 + u^{max}}{l_0} = \frac{l_{max}}{l_0} \tag{1}$$

where l_0 is the original reference length of the specimen. However, at B and C the reference lengths are not l_0 but $l_0 + u_M$ and $l_0 + u_R$ respectively, because the length of the specimen increases when the specimen is unloaded ($F = 0$). Thus the strain are the following:

$$\lambda_B^{max} = \frac{l_{max}}{l_0 + u_M} = \frac{\lambda_A^{max}}{\lambda_M} \tag{2}$$

$$\lambda_B^{max} = \frac{l_{max}}{l_0 + u_R} = \frac{\lambda_A^{max}}{\lambda_R} \tag{3}$$

where λ_M and λ_R are the inelastic strain with respect to the original reference length. Thus:

$$\lambda_A^{max} > \lambda_B^{max} > \lambda_C^{max} \tag{4}$$

To predict stress at points A, B, and C, the cross-section area needs to be calculated. Assuming incompressibility, $S = S_0 / \lambda$, and the cross-section area at A equals to:

$$S_A^{max} = \frac{S_0}{\lambda_A^{max}} \tag{5}$$

However, again at B and C, the reference cross-section areas are not S_0, but S_0 / λ_M and S_0 / λ_R respectively. Thus, the cross-section areas at B and C are:

$$S_B^{max} = \frac{S_0 / \lambda_M}{\lambda_B^{max}} = \frac{S_0}{\lambda_A^{max}} \tag{6}$$

$$S_C^{max} = \frac{S_0 / \lambda_R}{\lambda_C^{max}} = \frac{S_0}{\lambda_A^{max}} \tag{7}$$

The cross-section area remains constant in constant displacement control. Maximum true stress from A to C then decreases:

$$\sigma_A^{max} > \sigma_B^{max} > \sigma_C^{max} \tag{8}$$

as $F_A^{max} > F_B^{max} > F_C^{max}$. Similar deduction can be made for the minimum true stress, where:

$$\sigma_A^{min} > \sigma_B^{min} > \sigma_C^{min} \tag{9}$$

as cross-section area is S_0 and $F_A^{min} > F_B^{min} > F_C^{min}$.

Thus in general, there is continuous stress relaxation and diminution of strain as cycling loading progresses; both mechanical parameters are not constant under displacement control.

Furthermore, similar predictions can be made for experiments carried out in force control. Instead of long-term cyclic stress relaxation, there is long-term cyclic creep. It can be shown that true stress will increase under constant loading:

$$\sigma_A^{max} < \sigma_B^{max} < \sigma_C^{max} \tag{10}$$

Moreover, the relationship between the applied force and calculated strain is complex and difficult to predict because of the cyclic creep.

To summarize, neither displacement control (nor force) control can be used to carry out proper fatigue life testing on elastomers which exhibit strong inelastic behavior. The mechanical parameters, stress and strain, which drive the fatigue damage phenomenon in materials, are never constant over the duration of the tests. As such, the analysis performed on the obtained results cannot be valid. Therefore, the need for a testing procedure in true stress control is identified.

3 TRUE STRESS CONTROL PROCEDURE

3.1 *Testing of individual specimens*

True stress control has been reported for biaxial loading (Johnson et al. 2013). However, the complexity of achieving it comprises one of the reasons that true stress control is not widely utilized. The advantages of true stress control are apparent, since by controlling the mechanical parameter, the issues related to the Mullins effect, long-term stress relaxation, and other visco—elastic and—plastic behaviors are naturally taken into account.

The goal of the procedure is to attain specific true stress amplitude in cyclic loading for a given R-ratio:

$$\Delta\sigma = \sigma^{max} - \sigma^{min} \tag{11}$$

where $\sigma^{max} = \text{Max}(F/S)$ and $\sigma^{min} = \text{Min}(F/S)$; F is the actual force measured by a testing machine and S is the actual cross-section area of the specimen.

The cross-section area depends on the prescribed displacement as discussed earlier. For the present procedure, it is calculated *a priori*. Digital image correlation (DIC) measurements are made to predict the cross-section area as a function of applied displacement. It is assumed that the tested elastomer is incompressible. Additionally, it is useful to carry out finite element (FE) analysis to supplement the DIC results.

In order to implement the procedure on a testing machine, it is necessary to calculate true stress in real time. An algorithm is developed to achieve the goal set by Eq. 11. Because elastomers are subjected to relatively large displacements and relatively small forces, the algorithm is run in displacement control. At each cycle, the algorithm calculates the real-time stress and compares maximum and minimum values to the set target. Maximum and minimum stress are calculated at each cycle by:

$$\sigma_i^{max} = \text{Max}\left(\frac{F(t)}{S(u(t))}\right) \tag{12}$$

$$\sigma_i^{min} = \text{Min}\left(\frac{F(t)}{S(u(t))}\right) \tag{13}$$

where t is the elapsed time during cycle i, F is the force, and S depends on displacement. Depending on whether the maximum and/or the minimum true stress are higher or lower than the desired target, the algorithm increases or decreases the prescribed displacement in small increments for the next cycle. If the target(s) is (are) reached, the displacement is not adjusted. The procedure iterates until the specimen breaks.

3.2 *Testing of multiple specimens in parallel*

Application of the above procedure is not possible in parallel testing on a single machine. As one of the specimens breaks and/or significant cracks occur in one or more specimens, the effective cross-section surface area as well as overall force change cannot be easily taken into account in an experimental setting.

This limitation is resolved by first realizing the tests on individual specimens and then exporting the maximum and minimum displacements coordinates over the duration of the experiment for use in parallel testing. This ensures that in the two cases mentioned above, the remaining specimens will continue to experience a constant true stress load. For the sake of repeatability, individual tests are carried out on at least 2 specimens to ensure that the displacement master curves are consistent and are not affected by macro-defects such as internal voids and other heterogeneities.

4 VALIDATION OF THE PROCEDURE

4.1 *Material, specimen geometry and experimental apparatus*

Carbon black filled, peroxide cross-linked hydrogenated butadiene rubber (HNBR) is used to validate the procedure as it shows a "worst-case" scenario with significant Mullins effect and long-terms cyclic stress relaxation. The geometry of specimens is that of flat dumbbells with a reference length of 10 mm, where the stress state is in uniaxial tension. The specimens are cut using a die from compression molded sheets.

Testing is performed on an electro-mechanic machine, Instron E10000. A special grip setup is designed and manufactured for testing of up to 8 specimens in parallel; each specimen grip is equipped with a force load-cell (HBM S-type) to monitor fatigue life and calculate true stress.

4.2 *Calculation of cross-section area*

DIC measurement are made for cyclic loading of 30 mm, 45 mm, and 55 mm at $R = 0$ and a frequency of 3 Hz. Strain measurements are made at 10, 1000, and 10,000 cycles for each amplitude. To supplement DIC measurements, finite element analysis is carried out in commercial software Abaqus. The results of both approaches show good correlation; a polynomial of degree 6 is best-fit to represent the relationship between the applied displacement and the cross-section area.

4.3 *Implementation of the procedure*

4.3.1 *Single specimens*

The procedure is tested at three true stress loading levels – 4, 6, and 8 MPa—for $R = 0$. A constant average stress rate, $\dot{\sigma}$, of 20 MPa/s is utilized; thus, the average frequencies for 4, 6, and 8 MPa are 5, 3.33, and 2.5 Hz respectively. Finally, all specimens are tested at 120°C since this is the average operational temperature of this particular grade of HNBR.

4.3.2 *Testing in parallel*

For each loading level a master displacement curve is calculated as an average of 2 or more maximum and minimum displacements. The last 10% of cycles $\left(i \in \left[90\% * N_f, N_f\right]\right)$ with the individual test displacements are omitted as crack formation is observed, which changes the effective cross-section area of a specimen and gives incorrect values of true stress. Moreover, extrapolation of the master curves is performed past the original fatigue lives of the 2 or more individually tested specimens to 2 million cycles $\left(i \in \left[N_f, 2*10^6\right]\right)$. Due to the scattering of fatigue life results, one cannot guarantee that the maximum fatigue life has been attained in those experiments. A curve of best-fit is extrapolated from the last 50% of the cycles minus the last 10% $\left(i \in \left[50\% * \left(90\% * N_f\right), 90\% * N_f\right]\right)$.

4.4 *Results*

4.4.1 *Single specimens*

As a general example, Figure 2 shows minimum and maximum displacement of three experiments carried out on individual specimens at true stress amplitude of 6 MPa and R-ratio of zero. Figure 3 shows the true stress amplitude of the same three experiments. Primarily, the target of constant true stress amplitude is reached after about 100 cycles; the initial 100 cycles correspond to the machine envelope and the algorithm trying to reach the set amplitude. Secondly, there is good repeatability for the maximum and minimum displacements. It should be noted that the fatigue lives of three experiments are different (Test 1: 7800, Test 2: 122,400, and Test 3: 11,400 cycles), but the applied displacements are similar. Similar results are obtained for other loading levels; the results are not shown for brevity. Finally, master displacement curves can be created for tests in parallel.

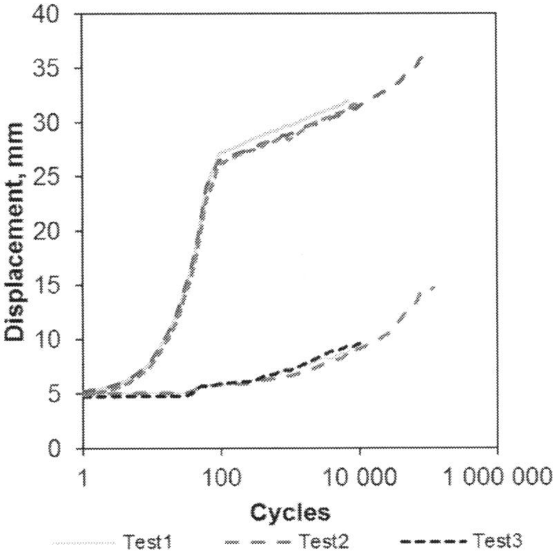

Figure 2. Evolution of maximum and minimum displacements for 3 individual tests; target stress – 6 MPa, $R = 0$.

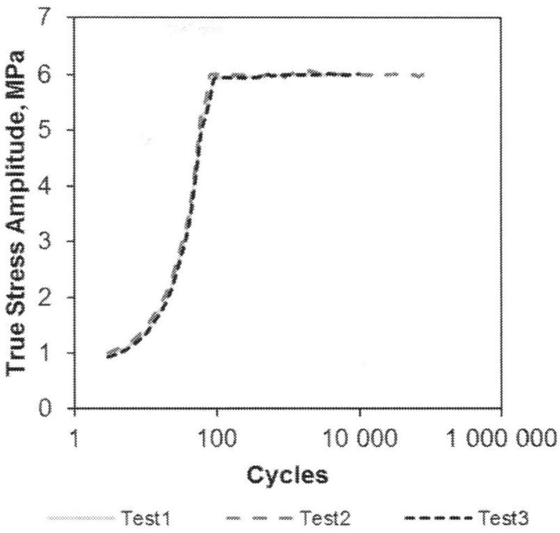

Figure 3. Evolution of true stress amplitude for 3 individual tests; target of 6 MPa, $R = 0$.

4.4.2 *Testing in parallel*

The evolution of true stress amplitude for specimens tested in parallel is shown in Figure 4. For all loading levels, it appears that the actual stress amplitude experienced by each specimen is not exactly equivalent to the target stress; however, calculated percent difference from the target amplitude does not exceed 10% in the 24 tested specimens. This phenomenon could be explained by the fact that during specimen installation (although extra precaution has been taken) each specimen is slightly in compression or in tension with respect to each other. Moreover, the stress appears to be almost constant for the majority of test duration.

4.5 *Wöhler curve*

Finally, a Wöhler curve can be plotted as shown in Figure 5. The present procedure allows plotting the actual applied stress instead of the target stress. The experiments were stopped after 2 million cycles for target amplitude of 4 MPa as indicated

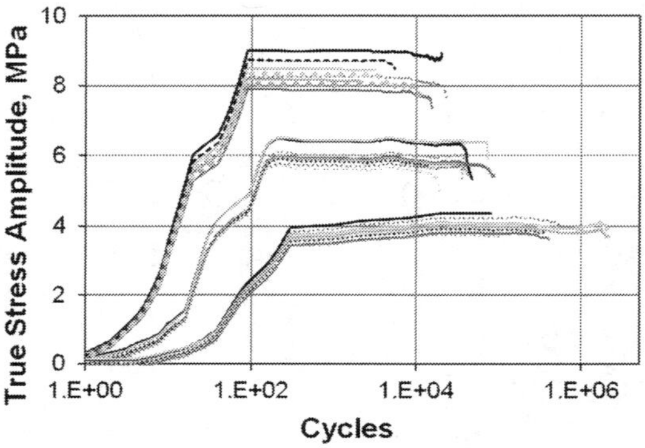

Figure 4. True stress amplitude for parallel tested specimens; target stress – 4, 6, and 8 MPa, $R = 0$.

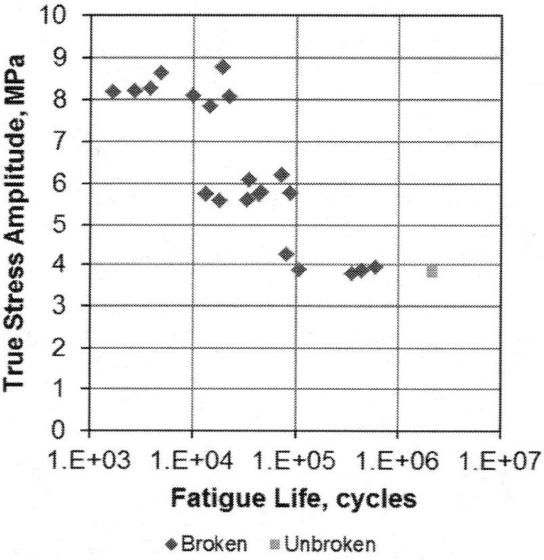

Figure 5. Wöhler curve in true stress amplitude, $R = 0$ for HNBR.

by squared points. As expected, there is scattering in terms of both, the fatigue life and the true stress amplitude.

5 CONCLUSION

It has been shown that conventional fatigue life testing in displacement (or force) control is not appropriate for elastomers showing significant inelastic behavior, because neither of the mechanical parameters is constant (stress nor strain). Thus, a true stress control procedure has been developed. The first results show that the procedure can maintain desired target true stress amplitude at a target *R*-ratio for individual tests. The procedure has been extended to test multiple specimens in parallel.

REFERENCES

André, N., Cailletaud, G. & Piques, R., 1999. Haigh diagram for fatigue crack initiation prediction of natural rubber components. *KGK-Kautschuk und Gummi Kunststoffe*, 52(2):120–123.

ASTM International, 1999. ASTM D4482 - 99: Standard Test Method for Rubber Property-Extension Cycling Fatigue.

Ayoub, G. et al., 2014. A visco-hyperelastic damage model for cyclic stress-softening, hysteresis and permanent set in rubber using the network alteration theory. *International Journal of Plasticity*, 54:19–33.

Bergström, J.S. & Boyce, M.C., 1998. Constitutive modeling of the large strain time-dependent behavior of elastomers. *Journal of the Mechanics and Physics of Solids*, 46(5):931–954.

Cadwell, S.M. et al., 1940. Dynamic Fatigue Life of Rubber. *Industrial & Engineering Chemistry Analytical Edition*, 12(1):19–23.

Diani, J., Fayolle, B. & Gilormini, P., 2009. A review on the Mullins effect. *European Polymer Journal*, 45(3):601–612.

Ismail, H. & Jaffri, R.., 1999. Physico-mechanical properties of oil palm wood flour filled natural rubber composites. *Polymer Testing*, 18(5):381–388.

Johnson, M. et al., 2013. The Equi-Biaxial Fatigue Characteristics of EPDM under True (Cauchy) Stress Control Conditions. *ECCMR (VIII)*.

Kim, J.-H. & Jeong, H.-Y., 2005. A study on the material properties and fatigue life of natural rubber with different carbon blacks. *International Journal of Fatigue*, 27(3):263–272.

Kingston, J.G.R. & Muhr, A.H., 2007. Effects of Strain Crystallisation on Cyclic Fatigue of Rubber. *6th Engineering Integrity Society International Conference on Durability and Fatigue*.

Legorju-Jago, K. & Bathias, C., 2002. Fatigue initiation and propagation in natural and synthetic rubbers. *International Journal of Fatigue*, 24(2–4):85–92.

Mars, W. & Fatemi, A, 2002. A literature survey on fatigue analysis approaches for rubber. *International Journal of Fatigue*, 24(9):949–961.

Mars, W.V. & Fatemi, A., 2003. Fatigue crack nucleation and growth in filled natural rubber. *Fatigue and Fracture of Engineering Materials and Structures*, 26(9):779–789.

Mars, W.V., 2007. Fatigue Life Prediction for Elastomeric Structures. *Rubber Chemistry and Technology*, 80(3):481–503.

Schütz, W., 1996. A history of fatigue. *Engineering Fracture Mechanics*, 54(2):263–300.

Svensson, S., 1981. Testing methods for fatigue properties of rubber materials and vibration isolators. *Polymer Testing*, 2(3):161–174.

Constitutive Models for Rubber X – Lion & Johlitz (Eds)
© 2017 Taylor & Francis Group, London, ISBN 978-1-138-03001-5

Experimental study of dynamic crack growth in elastomers

T. Corre
Institut de Recherche en Génie Civil et Mécanique, UMR CNRS 6183, École Centrale de Nantes, France
DCNS Research, Technocampus Ocean, Bouguenais, France

M. Coret & E. Verron
Institut de Recherche en Génie Civil et Mécanique, UMR CNRS 6183, École Centrale de Nantes, France

B. Leblé & F. Le Lay
DCNS Research, Technocampus Ocean, Bouguenais, France

ABSTRACT: An experimental method to investigate dynamic crack growth in elastomers is developed. It considers the propagation of a crack in a "pure shear" sample. Both pre-cutting extension step and fracture step are recorded with high resolution and high speed cameras, respectively. The evolution of crack speed along its travel throughout the sample is precisely measured. Using high precision digital image correlation measurement, the method allows to accurately compute kinematic fields, e.g. velocity and acceleration, during crack growth.

1 INTRODUCTION

Crack growth in elastomers is mostly studied in two different contexts: fatigue on the one hand, with a crack growing slowly under cyclic service loading conditions; and dynamic propagation on the other hand, under quasi-static loading conditions. Contrary to fatigue case, only few experimental studies tackle the problem of dynamic crack growth. Nevertheless, this approach is developing along with observation techniques since the 2000s, and is considered here.

The pioneering work of Rivlin & Thomas (1953) provided both theoretical results (tearing energy) and sample geometries, such as "pure shear" or "trousers" specimens, to handle the problem of crack growth in elastomers. Following these ideas, studies aimed at characterizing the tearing properties of elastomers (Thomas 1960) or analyzing the relationship between tearing energy and crack growth rate (Lake et al. 2000, Morishita et al. 2016). In most cases, studies consider specific samples that enable the computation of tearing energy with analytical formulas. Another approach focuses on the understanding of fracture kinematics, using for example popping balloons (Stevenson and Thomas 1979, Moulinet and Adda-Bedia 2015) or highly biaxially strained sheets (Gent and Marteny 1982, Petersan et al. 2004). In this case, the speed of the crack (Petersan et al. 2004), its path (Deegan et al. 2001) or the shape of its tip (Marder 2006, Morishita et al. 2016) are investigated.

Along with these experimental works, theoretical analyses have tried to extend the results of Linear Elastic Fracture Mechanics (LEFM) to the case of elastomers. Indeed, LEFM, the well-established fracture theory of brittle materials (Irwin 1957), has revealed its limits in the case of elastomers (Livne et al. 2008). To shed some light on the observations, expressions of the J-integral have been proposed for finite strain (Chang 1972). Similarly, asymptotic analyses of strain and stress fields in the neighbourhood of a crack tip have also been derived for various hyperelastic models (Geubelle and Knauss 1994, Long et al. 2011). Finally, the influence of dynamic effects on fracture properties has also been investigated (Freund 1998), although mainly considering the small strain assumption.

In the present study, we aim at providing experimental tools to fill the gap between experimental observations and theoretical studies of dynamic crack growth under large strain. An experimental set-up, based on crack growth in "pure shear" specimens, is proposed. Crack propagation is recorded by a set of complementary cameras, and the displacement fields is measured with Digital Image Correlation (DIC) technique. Finally, we show that all kinematic fields can be computed from the displacement field during crack propagation.

2 METHODS

2.1 *Experimental set-up*

Experiments are carried out with “pure shear”, or “strip geometry”, specimens. Dimensions of the specimens are $200 \times 40 \times 3$ mm^3. Samples are molded with two cylindrical bulges (diameter 15 mm) at the edge of their long side in order to fit into specially designed grips avoiding slippage during loading. The width/height ratio of the specimens working area is 5, the smallest ratio acceptable to obtain a strain field close to pure shear conditions in specimen center (Yeoh 2001). Note that this ratio is chosen to have the largest height and therefore the largest values of both tearing energy and crack growth speed (Lake et al. 2000).

To reach these high values of crack growth speed, the method consists in two steps: first a pure shear specimen without crack is pre-stretched to the prescribed stretch ratio, then it is cut at the center of one edge with a razor blade and the crack grows.

- During the first step, the specimen is stretched in a tensile machine Instron 5584 at a crosshead speed of 20 mm.min^{-1}. The specimen is extended until the displacement corresponds to the prescribed stretch ratio. This part of the experiment is recorded with a high resolution camera (29 Mpix), referred to as HR in the following. The spatial resolution is about 0.04 mm.pix^{-1}.
- The second step begins with the cut of a crack at one edge of the sample as shown in Figure 1. Crack propagation is recorded with a high speed camera Photon SAZ, referred to as HS throughout the rest of the paper. The frame rate is chosen for each test, and its maximum value is 40000 Hz for full resolution (1024×1024). The spatial resolution is about 0.2 mm.pix^{-1}. It is to note that the cut is made while the crosshead is still in motion; nevertheless due to the short duration of this second step, i.e. from 16 to 300 ms, we consider it fixed.

2.2 *Field measurements*

First, a black speckle patten is painted on the specimen, then the displacement field is measured by DIC during the two steps; the commercial software Vic-2D is used. As mentioned above, each step is recorded with a different camera. Practically, measurements are reset after the first step, and data issued from the HS camera must be completed with the ones of the HR camera in order to compute the displacement field along the whole experiment: the final HR displacement field is projected onto HS images, as explained hereunder.

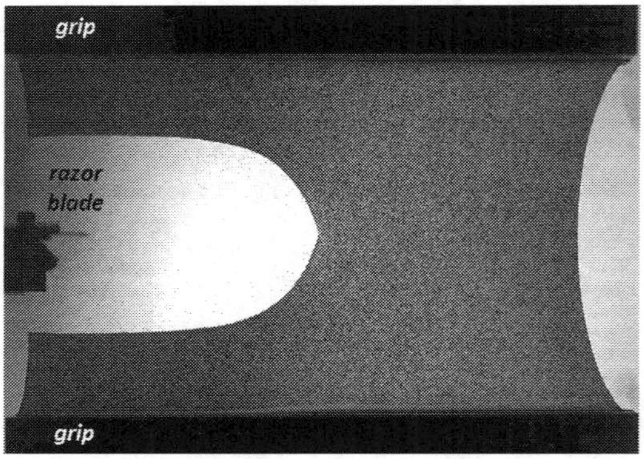

Figure 1. Crack growth in pure shear specimen. Prescribed stretch ratio: 3. HS camera (20000 fps, shutter speed 1/266666 s).

The DIC software provides the displacement field at nodes of a grid covering the useful area of the specimen. Each node is characterized by its coordinates in the first image of each film. Let denote X^{HR} and X^{HS} nodes coordinates of the HR and HS camera, respectively. The first image of the HR camera corresponds to the undeformed configuration of the specimen considered as the reference configuration $X^{\mathrm{HR}}_{\mathrm{ref}}$. Thus

$$X^{\mathbf{HR}} = X^{\mathrm{HR}}_{\mathrm{ref}}. \tag{1}$$

Then, for every image i, coordinates x_i^{HR} of a node are measured, and the displacement field U_i^{HR} is

$$U_i^{\mathbf{HR}} = x_i^{\mathrm{HR}} - X^{\mathrm{HR}}. \tag{2}$$

Consider now the last image li of the loading step film. Recalling both the low crosshead speed and the HR camera frame rate (0.5 Hz), it is assumed that this image shows the same mechanical state of the sample than the one in the first image of the HS camera (just before cutting the sample). Then, the HR displacement field U_{li}^{HR} is projected onto the HS grid, i.e.

$$U_1^{\mathbf{HS}}\left(X^{\mathrm{HS}}\right) = U_{li}^{\mathbf{HR}}\left(X^{\mathrm{HS}}\right), \tag{3}$$

which states that grids x_{li}^{HR} and X^{HS} are superimposed and U_{li}^{HR} is interpolated. Then, the HS displacement field at image i, u_i^{HS}, is corrected to take into account the loading step displacement field:

$$U_i^{\mathbf{HS}}\left(X^{\mathrm{HS}}\right) = u_i^{\mathbf{HS}}\left(X^{\mathrm{HS}}\right) + U_1^{\mathbf{HS}}\left(X^{\mathrm{HS}}\right) \tag{4}$$

However, X^{HS} does not correspond to the reference configuration. In fact, the HS camera has never "seen" the undeformed configuration; both deformation gradient, and then strain fields are computed from the position of HS grid nodes in the reference configuration, given by the expression:

$$x_{\text{ref}}^{\text{HS}} = X^{\text{HS}} - U_1^{\text{HS}}\left(X^{\text{HS}}\right). \tag{5}$$

Then, the usual relation, similar to equation (2), for image i of the HS camera is

$$x_i^{\text{HS}} = X_{\text{ref}}^{\text{HS}} + U_i^{\text{HS}}\left(X_{\text{ref}}^{\text{HS}}\right). \tag{6}$$

2.3 *Space and time derivatives*

Once the displacement field determined, the deformation gradient can be computed for any image i of the second step (crack growth):

$$\mathbf{F}_i^{\text{HS}} = \frac{\partial x_i^{\text{HS}}}{\partial X_{\text{ref}}^{\text{HS}}}. \tag{7}$$

Equation (7) is computed for each image using a finite element discretization. The mesh in made of triangular elements whose summits are the positions of nodes in the reference configuration for each camera. The deformation gradient at a node is then the mean of all the values in its surrounding elements. Afterwards, any strain field can be computed for any frame.

Moreover, the high frame rate of the HS camera leads to quite accurate computation of Lagrangian time derivative of any mechanical field during crack growth. Indeed, For each specific node of the grid, the value of any field evolves with frames and therefore with time. Its derivative with respect to time defines the Lagrangian time derivative. In practice, to reduce the noise, the evolution of the field to be derived at a node is fitted by a polynomial of degree 2 over 5 time steps, which is used to compute the value of the time derivative.

3 RESULTS

3.1 *Materials*

The method is applied for two different thermoset polyurethanes, referred to as PUB and PUJ in the following. Their chemical compositions differ: PUJ is made of toluene diisocyanate and polyether while PUB is made of diphenylmethane diisocyanate and polyester. Both materials are first subjected to standard uniaxial tension experiments; the corresponding nominal stress vs. stretch ratio curves are shown in Figure 2. Both materials exhibit a standard non-linear S-shape response; PUJ is initially stiffer than PUB. They are considered homogeneous, isotropic and incompressible, and the influence of their composition on their behaviour is not investigated here.

3.2 *Crack growth speed*

For all tests, fracture is initiated by a cut with a razorblade at the centrer of one edge, its displacement is controlled. The crack is considered to be growing freely when it grows faster than the blade (approximately 0.6 m.s^{-1}). The crack speed is measured from the time change in the remaining length of the sample in the deformed configuration. Results are shown in Figure 3; for comparability, the speed is plotted over the distance travelled by the crack and normalized by the total distance travelled. During a typical test ("λ_y = 3" in Fig. 3 for instance), crack propagation goes through three stages: the crack speed

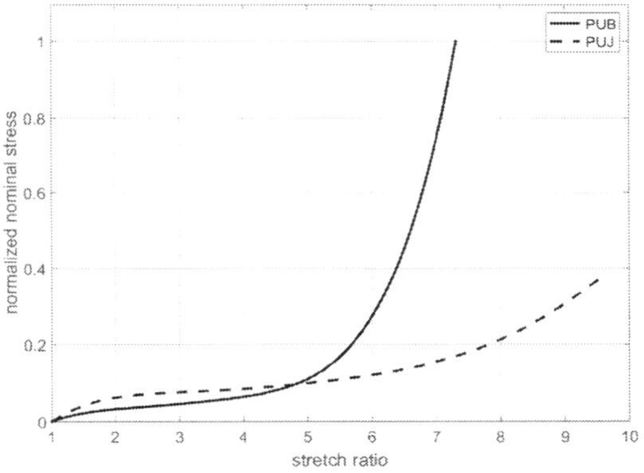

Figure 2. Stress-strain curves of materials. Stress is adimensionalized with respect to the ultimate tensile stress of PUB.

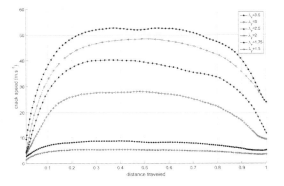

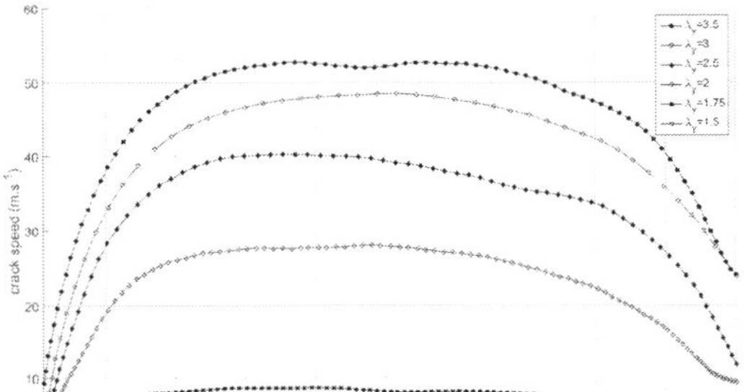

Figure 3. Crack speed along the travel distance in PUJ samples for different prescribed stretch ratios.

rises, reaches a plateau, and then decreases. As expected, the pure shear sample leads to a steady-state propagation of the crack, which can be related to the wide homogeneously extended area in the middle of the sample. The corresponding crack speeds are given in Table 1. It is to note that DIC leads to the true stretch ratio when fracture is initiated, in practice it varies a little from the prescribed one, due to some flow of the material out of the grips. Corresponding values are also given in Table 1.

3.3 *Kinematic fields*

In order to illustrate the accuracy of computed kinematic fields, displacement, velocity, and acceleration fields are presented respectively in Figure 4, 5, and 6.

They corresponds to a steady-state crack growth at about 20 m.s^{-1} in a PUB sample

Table 1. Prescribed stretch ratio, true (measured) stretch ratio in the middle of the specimen, and pure shear crack speed for PUJ.

prescribed stretch ratio	measured stretch ratio	pure shear crack speed (m.s^{-1})
1.5	1.457	5.1
1.75	1.696	8.2
2	1.953	26.9
2.5	2.343	38.3
3	2.899	46.5
3.5	3.375	50.7

stretched at 200%. Fields are represented by a coloured patch at each node of the HS grid (about 31000 nodes).

The kinematic fields can be represented either in the current configuration x_i^{HS} or in the reference one $X_{\mathrm{ref}}^{\mathrm{HS}}$. For instance, Fig. 4 shows the displacement field in the reference configuration. In this case, we retrieve the initial rectangular shape of the sample and its undeformed dimensions. The crack is then represented by a straight line, which corresponds to the nodes where DIC fails. In particular, Fig. 4 highlights the side effects: the horizontal displacement is close to zero in the middle of the sample and increases along the centre line toward the edge. Fig. 5 shows the velocity field in the current configuration, it exhibits the crack geometry during its growth. The actual position of the lips (measured in the picture and not with DIC) is represented by black dotted line. The horizontal velocity field (top image in Fig. 5) shows that there is motion in the x-direction at the maximum speed of 4 m. s^{-1}. As expected, upward and downward velocities of the material close to the lips are higher: it reaches 40 m.s^{-1} in this case, i.e. twice the crack growth speed! Such high velocities in the short time of the experiments lead to high values of acceleration: the vertical acceleration close to crack lips is about 6×10^4 m.s^{-2} (Fig. 6 bottom). These values of acceleration are difficult to discuss but they can give some clues to quantify the role of dynamic effects during crack growth.

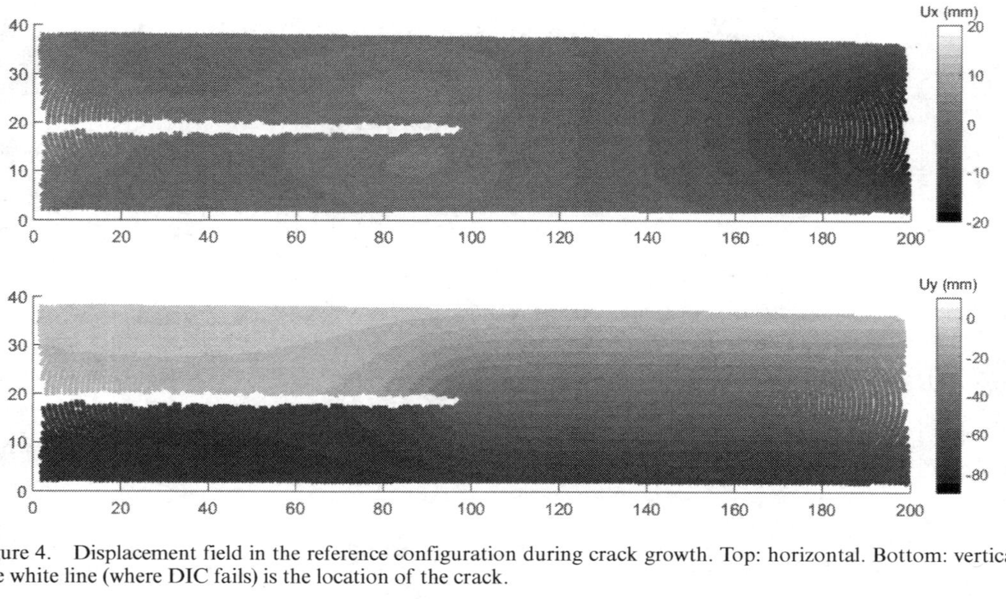

Figure 4. Displacement field in the reference configuration during crack growth. Top: horizontal. Bottom: vertical. The white line (where DIC fails) is the location of the crack.

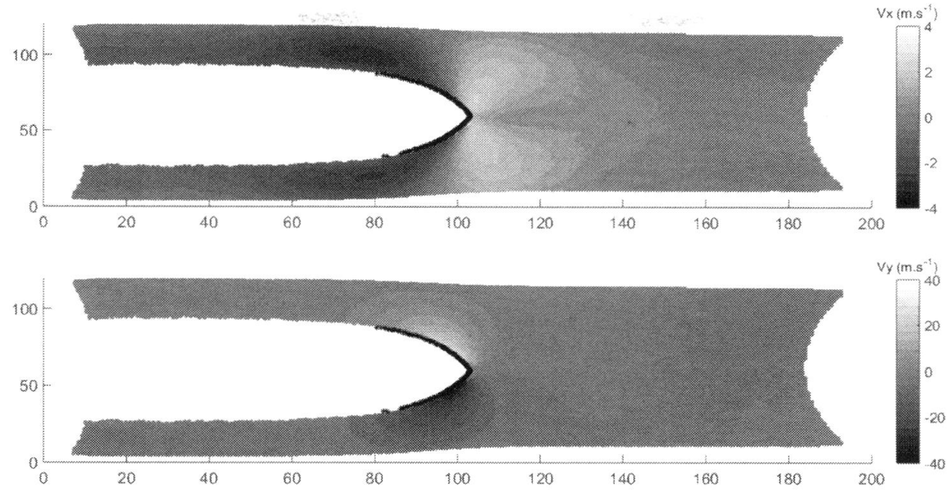

Figure 5. Velocity field in the deformed configuration during crack growth. Top: horizontal. Bottom: vertical. Prescribed stretch ratio: 3, crack speed: 20 m.s^{-1}. The black dotted line represents the crack lips.

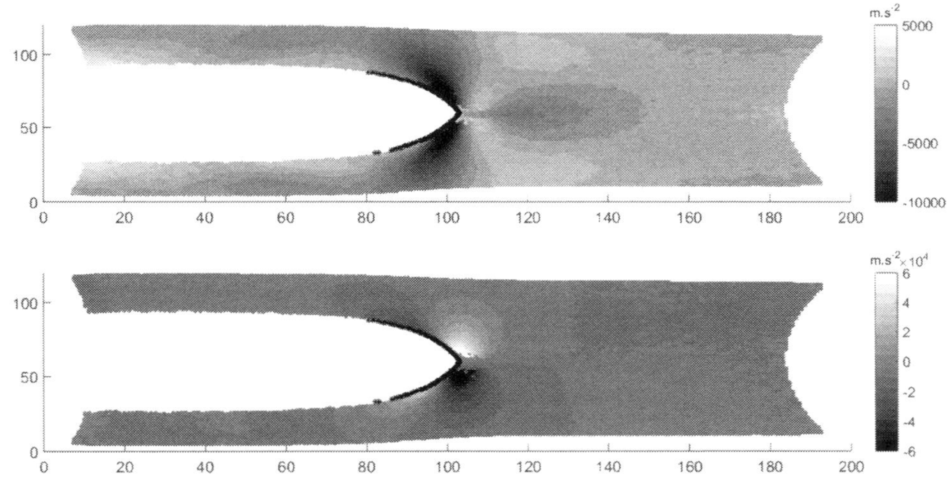

Figure 6. Displacement field in the reference configuration during crack growth. Top: horizontal. Bottom: vertical. Prescribed stretch ratio: 3, crack speed: 20 m.s^{-1}. The black dotted line represents the crack lips.

4 CONCLUSION

The two camera method developed in this study has proved its relevance to investigate dynamic crack growth in elastomers. The crack speeds reached are of the same order of magnitude of the fastest ones obtained recently in similar experiments (Morishita et al. 2016). Then, the steady state crack growth allows to consider theoretical results derived under this assumption (Freund 1998). Finally, accurate kinematic fields can help to characterize crack growth. For example, velocity and acceleration fields allows to "quantify" the dynamic effects: from which crack growth speed do we have to take into account dynamic effects, such as kinetic energy? Moreover, the careful definition of the reference configuration and the computation of the deformation gradient allows a mechanical analysis of the fracture process, with both Lagrangian and Eulerian points of view. Kinematic fields can also be completed with energetic and stress fields through the use of an appropriate constitutive model, and then can lead to the computation of fracture mechanics quantities, such as the J-integral, during crack growth.

ACKNOWLEDGMENTS

The authors thank Mr Pierre Corre for his help to build the cutting device.

REFERENCES

Chang, S.-J. (1972). Path-independent integral for rupture of perfectly elastic materials. *Zeitschrift für angewandte Mathematik und Physik ZAMP 23*(1), 149–152.

Deegan, R.D., P.J. Petersan, M. Marder, & H.L. Swinney (2001). Oscillating fracture paths in rubber. *Physical review letters 88*(1), 014304.

Freund, L.B. (1998). *Dynamic fracture mechanics*. Cambridge university press.

Gent, A.N. & P. Marteny (1982). Crack velocities in natural rubber. *Journal of Materials Science 17*(10), 2955–2960.

Geubelle, P.H. & W.G. Knauss (1994). Finite strains at the tip of a crack in a sheet of hyperelastic material: I. homogeneous case. *Journal of Elasticity 35*(1–3), 61–98.

Irwin, G.R. (1957). Analysis of ststress ans strains near the en of a crack traversing a plate. *Journal of applied mechanics 24*, 361.

Lake, G.J., C.C. Lawrence, & A.G. Thomas (2000). High-speed fracture of elastomers: Part i. *Rubber Chemistry and Technology 73*(5), 801–817.

Livne, A., E. Bouchbinder, & J. Fineberg (2008). Breakdown of linear elastic fracture mechanics near the tip of a rapid crack. *Physical review letters 101*(26), 264301.

Long, R., V.R. Krishnan, & C.-Y. Hui (2011). Finite strain analysis of crack tip fields in incompressible hyperelastic solids loaded in plane stress. *Journal of the Mechanics and Physics of Solids 59*(3), 672–695.

Marder, M. (2006). Supersonic rupture of rubber. *Journal of the Mechanics and Physics of Solids 54*(3), 491–532.

Morishita, Y., K. Tsunoda, & K. Urayama (2016). Velocity transition in the crack growth dynamics of filled elastomers: Contributions of nonlinear viscoelasticity. *Physical Review E 93*(4), 043001.

Moulinet, S. & M. Adda-Bedia (2015). Popping balloons: A case study of dynamical fragmentation. *Physical review letters 115*(18), 184301.

Petersan, P.J., R.D. Deegan, M. Marder, & H.L. Swinney (2004). Cracks in rubber under tension exceed the shear wave speed. *Physical review letters 93*(1), 015504.

Rivlin, R. & A.G. Thomas (1953). Rupture of rubber. i. characteristic energy for tearing. *Journal of Polymer Science 10*(10), 291.

Stevenson, A. & A. Thomas (1979). On the bursting of a balloon. *Journal of Physics D: Applied Physics 12*(12), 2101.

Thomas, A. (1960). Rupture of rubber. vi. further experiments on the tear criterion. *Journal of Applied Polymer Science 3*(8), 168–174.

Yeoh, O. (2001). Analysis of deformation and fracture of pure shear rubber testpiece. *Plastics, Rubber and Composites 30*(8), 389–397.

Constitutive Models for Rubber X – Lion & Johlitz (Eds)
© 2017 Taylor & Francis Group, London, ISBN 978-1-138-03001-5

Characterising the cyclic fatigue performance of HNBR after aging in high temperatures and organic solvents for dynamic rubber seals

B.H.K. Shaw & J.J.C. Busfield
Soft Matter Group, School of Engineering and Materials Science & The Materials Research Institute, Queen Mary University of London, London, UK

J. Jerabek & J. Ramier
Schlumberger Gould Research Centre, Cambridge, UK

ABSTRACT: Many industries work extensively with NBR and HNBR elastomers because they work well at high temperatures and in hostile chemical environments such as organic solvents. Both elastomers are widely used in applications such as dynamic sealing, where they can be exposed to high strain rates, high frequencies, high temperatures and in contact with oil based fluids. Therefore, characterising their fatigue properties under these extreme conditions is essential when trying to predict failure in service. Using a fracture mechanics based approach enables a prediction of the lifetime of a component. This requires the material's tear resistance to be measured first. This behaviour is expressed as the relationship between the strain energy release rate (sometimes known as the tearing energy) and the fatigue crack growth rate under different conditions. The experimental process required to measure cyclic pure shear fatigue crack growth test of two HNBR samples before and after ageing at 100°C for 168 hours is described and the results are discussed.

1 INTRODUCTION

HNBR exhibits enhanced ageing and heat resistance when compared with NBR. It is commonly used for extreme applications, including those found in the oil and gas industry such as dynamic seals, due to its excellent chemical and swelling resistance.

Fracture based fatigue crack growth testing provides a potential framework to help predict the service life of an elastomeric part. This requires cyclic crack propagation tests to be undertaken which measure the relationship between the strain energy release rate (T) (also known as the tearing energy) and the average rate of crack propagation rate per loading cycle (dc/dn). This enables predictions of the lifetime to be made for real elastomeric parts (Busfield et al., 2002).

The oil and gas industry experiences extreme conditions down well that include seal temperatures in excess of 200°C, pressures of hundreds of MPa and exposure to various different drilling fluids that are combinations of both oil and water. All of these various conditions can have an adverse effect on the performance of elastomers. Conditions such as being in contact with organic solvents involve several different physical and chemical interactions, some of which impart opposing effects. As the industry explores even more demanding oil well conditions, it is important that fatigue behaviour of these elastomers are evaluated in fatigue before and after ageing to check they are suitable for specific environmental conditions.

Elastomeric parts are used extensively for sealing and flow control applications in oil and gas field applications and they are used in both static and dynamic loading conditions. Dynamic loading conditions add additional viscoelastic complications as strain rate and cyclic softening effects also significantly change the properties of the elastomer.

To ensure both safe operation and to avoid environmental disasters it is important to have an understanding of how long a part will last prior to failure. For dynamic elastomeric seals a fatigue analysis that adopts a fracture mechanics approach is the preferred method of doing this. In cyclic fatigue analysis, it is possible to examine how many cycles it will take for a crack to propagate through a part and cause failure. This is done in a test lab by inserting a crack of a known size into a rubber test specimen and cyclically loading it whilst measuring the crack length change as the test progresses. By choosing a suitable geometry such as the pure shear crack growth test piece it is possible to measure the tearing energy versus crack propagation rate and from this generate a prediction of the lifetime of a real product.

The fracture mechanics approach that is currently used is based on work originally carried out by Griffith (1920) which investigated how the strength of a material was effected by cracks. He quantified the failure using an energetics approach. He assumed that if the elastically stored energy released was greater than the free energy required to create a new surface, then the crack would propagate through the sample. The results he made were found to be in good agreement for glass.

It was Rivlin and Thomas (1953) that adopted this approach for rubbery materials by modifying it to take into account irreversible viscoelastic energy dissipation effects around the tip of the crack. This lead to the expression for tearing energy T or strain energy release rate:

$$T = -\frac{dU}{da} \tag{1}$$

where U is the elastically stored energy and a is the area of a one of the fracture surfaces of the crack.

Rivlin and Thomas (1953) then established for rubber materials that this energy required to propagate the crack was a material characteristic. Therefore, in cyclic crack growth behaviour the crack growth per cycle is determined by the maximum strain energy release rate but not influenced much by the way it is reached. Therefore, the crack growth per cycle is a function of maximum strain energy release rate per cycle:

$$\frac{dc}{dn} = f(T) \tag{2}$$

Figure 1, Lake G.J., (1983), shows the cyclic crack growth rates for two rubber materials, natural rubber and SBR on logarithmic scales plotting tearing energy against crack growth rate per cycle. The behaviour of this plot can be split into three regions labelled I, II and III.

Region I shows the tearing energy to be less than the threshold tearing energy for the material. In this region, crack growth occurs only due to chemical effects such as ozone degradation where environmental ozone attacks the carbon-carbon double bonds. This process can be rapid in natural rubber and in various other olefin rubbers.

Crack growth in region II occurs as a result of a combination of mechanical degradation and chemical degradation such as ozone attack.

Region III has a power law dependency between crack growth rate and the tearing energy and is experienced with most rubber materials:

$$\frac{dc}{dn} = B\left(\frac{T}{T_u}\right)^{\beta} \tag{3}$$

where $T_u = 1$ Jm^{-2} is included to make T/T_u dimensionless, B and β are therefore measured material constants.

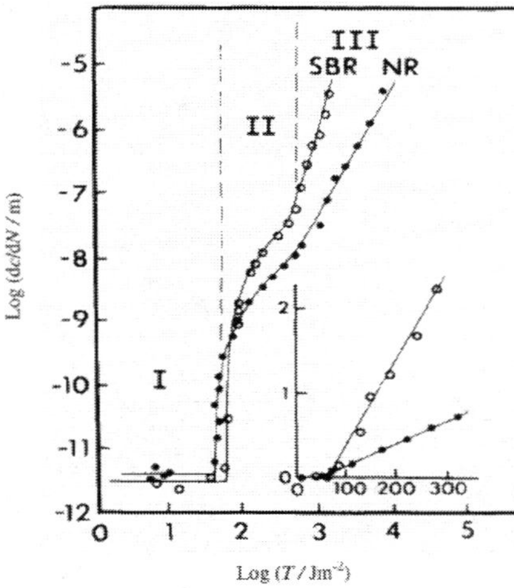

Figure 1. Fatigue crack growth rate versus tearing energy for two different elastomers. Lake G.J., (1983).

As this behaviour is a material characteristic, the results obtained should therefore be geometrically independent of the detailed test piece that is used to make the test. However, different test pieces have specific characteristics. In this work a pure shear geometry was used as the crack propagation rate is constant throughout the test and complexities due to the crack tip geometry can be avoided. This is done by testing a specimen whose length (w) is much greater than its height (l_0). The grips along the long edge ensure that there is no significant strain in the direction of crack extension.

The test specimen is split into four different loading regions shown in Figure 2.

Region A is the unstrained region beyond the crack tip and which has zero elastic stored energy, region B has a complex stress field around the crack tip, region D has a complex stress field due to edge effects and region C is deformed in uniform homogeneous pure shear. This means the elastic energy can be calculated as:

$$U_C = WV_C \tag{4}$$

where W is the strain energy density, U_C is the elastically stored energy in region C and V_C is the volume of this region. As the crack grows, region C gets smaller and region A gets larger. The tearing energy can then be calculated as:

$$T = Wl_0 \tag{5}$$

By using equation (5) and by measuring the crack propagation rate per cycle using a camera, equation (3) can be used to explore the relationship between tearing energy and crack propagation

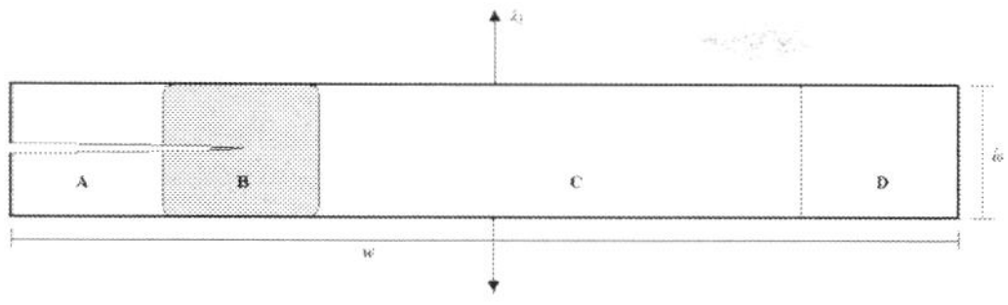

Figure 2. A pure shear crack growth test piece.

rate. This can be found to characterise the fatigue crack growth behaviour of the material.

This report describes research into cyclic fatigue tests carried out using pure shear fatigue crack growth specimens for dynamic sealing applications. It includes a description of the experimental procedure followed by the results and a short discussion and conclusion section examining the results.

2 EXPERIMENTAL PROCEDURE

2.1 *Method*

A pure shear specimen geometry was used, as it enabled the crack to propagate through the test piece at a constant rate and the complexities of the detailed behaviour around the crack tip can be neglected.

This experiment investigated the difference in crack growth fatigue resistance of unaged samples of HNBR rubber contrasted with samples that were aged in an air oven, under zero strain at 100°C for a week.

The test pieces were strained to 10%, 13.3%, 16.6% and 20% cyclically at a rate of 1 Hz. This meant the average strain rates were 20%/s, 26.6%/s, 33.2%/s and 40%/s.

The tests were conducted over at least 10,000 cycles with the force displacement data being collected every 100 cycles up to 1,000 cycles and then above 1,000 cycles it was collected every 1,000 cycles

Elastic stored energy, W, was calculated by integrating the force displacement data. From 1,000 cycles onwards the cyclic stress softening properties of the elastomer approached equilibrium, so data was used from 1000 cycles onwards.

2.2 *Test specimen*

The specimen were all made from HNBR filled with 60 phr of N550 carbon black, the formulation of which can be found in Table 2 in the appendix. Test samples were cut from a sheet of approximate thickness 2.2 mm. The sheet was then cut into strips of length $w = 175$ mm and height $h = 45$ mm. A horizontal cut of length 30 mm was introduced into one end of the specimen. When placed into the pure shear grips the two grips clamped approximately 15 mm at both the top and the bottom of the specimen, therefore the unstrained height between the grips, $l_0 = 15$ mm.

2.3 *Experimental set up*

Before any samples were placed within the grips, the cut cracks were propagated a very small distance by hand. The initial crack is placed in the sample using a razor blade which leads to a very sharp crack tip. The detailed shape of the crack tip can significantly affect the crack propagation rate until is naturally blunts and reaches the steady state conditions later on during the test (Papadopoulos, I.C., Thomas, A.G. and Busfield. J.J.C., 2008). Therefore, to ensure a more realistic test regime from the first cycle, the crack was hand propagated a small distance to blunt the initially cut crack tip. The specimen's dimensions were then measured and it was then placed within the grips and the load cell was reset to zero.

To measure the crack propagation rate, an HD camera was set to periodically take pictures of the crack. As the sample was cyclically loaded at a frequency of 1 Hz, with the camera being set to take pictures of the crack after every minute, pictures will be taken after every 60 cycles. The crack length was measured using image analysis software and this was linked to the cycle number it was taken at and the crack propagation rate was calculated.

3 RESULTS

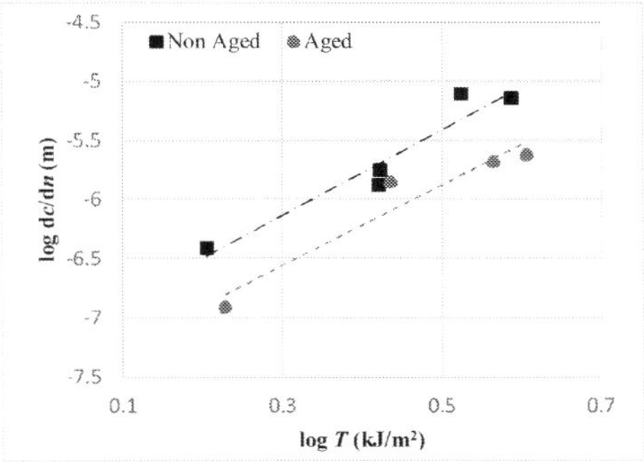

Figure 3. The fatigue crack growth rate plotted against tearing energy for both the unaged and the aged HNBR materials.

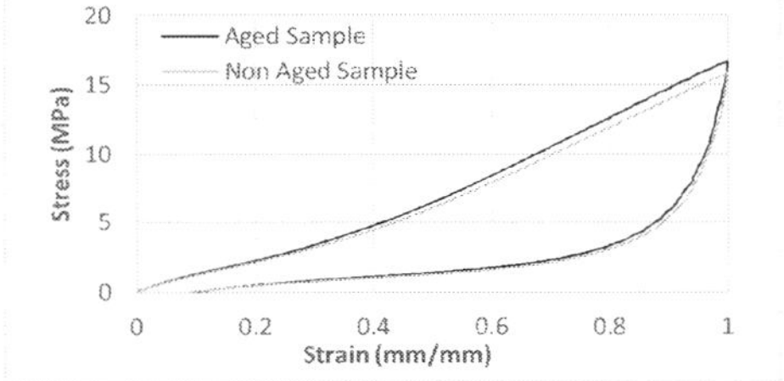

Figure 4. The stress strain behavior for both aged and un aged materials.

Table 1. The fitting parameters from Equation 3 used to fit the experimental behavior.

Non Aged	β	3.6
	B(m)	6.1×10^{-8}
Aged	β	3.4
	B(m)	2.6×10^{-08}

4 DISCUSSION AND CONCLUSIONS

From the results shown in Figure 3, it was clear that there is a significant difference between the fatigue crack growth behaviour of the aged samples and unaged samples. It was observed that after ageing that the samples, once removed from the oven and left to cool down, were tougher to cut and prepare for testing.

This observation was also born out in this measured data in that at similar tearing energies, the cracks propagate through the unaged specimen faster than the aged samples.

There are several potential reasons for this unexpected behaviour. Research such as Cadambi and Ghassemieh (2012) has shown similar behaviour when air ageing HNBR at temperatures of 150°C. They attribute this to a change in crosslink density as aging in high temperatures can cause the vulcanisation process to start again. A characterisation before and after ageing was made using equilibrium swelling and tensile stress strain measurements. The cross link density increased slightly from 1.736×10^{-3} mol/g for the non aged samples to 1.763×10^{-3} mol/g for the aged samples. The modulus, as shown in Figure 4, also slightly increased. This evidence of a modest increase in the cross link density was thought unlikely, on its own, to be able to account for the increase in the fatigue resistance of the aged materials. The origin of this increase is still under investigation.

The material constant β values for both sets of test data, found in Table 1, were very similar, at approximately 4. This is similar to other materials such as SBR rubber that exhibit no strain induced crystallisation, however the values for B are significantly different.

Further work that will be completed prior to the presentation to examine, the influence of ageing the specimens at elevated temperatures in deionised water and drilling fluids, on the fatigue crack growth behaviour.

ACKNOWLEDGEMENTS

One of the authors, Barnabas Shaw would like to thank Schlumberger Gould Research and EPSRC for providing the funding for this research and Queen Mary University of London for the use of laboratory facilities.

REFERENCES

Boochathum, P., Prajudtake, W., 2001, Vulcanization of cis- and trans-polyisoprene and their blends: cure characteristics and crosslink distribution, European Polymer Journal, 37(3), p. 417–427.

Busfield, J.J.C. et al, 2002, Contributions of Time Dependent and Cyclic Crack Growth to the Crack Growth Behaviour of Non Strain-Crystalizing Elastomers, Rubber Chemistry and Technology, Volume 75, No. 4, p. 643–656.

Busfield, J.J.C. et al, 1997, Crack Growth and Strain Induced Anisotropy in Carbon Black filled Natural Rubber, Journal of Natural Rubber Research, Volume 12(3), p. 131–141.

Cadambi, R.M. and Ghassemieh, E., 2012, The ageing behaviour of hydrogenated nitrile butadiene rubber/ nanoclay nanocomposites in various mediums, Journal of Elastomers & Plastics 44(4), p. 353–367.

Griffith, A.A., 1920, The phenomena of rupture and flow in solids, Philosophical Transactions of the Royal Society A, Vol. 221, p.163.

Lake G.J., 1983, Aspects of Fatigue and Fracture of Rubber, Progress of Rubber Technology, Vol. 45, pp. 89.

Papadopoulos, I.C., Thomas, A.G. and Busfield. J.J.C., 2008, Rate transitions in fatigue crack growth of elastomers, Journal of Applied Polymer Science, Vol. 109, p. 1900–1910.

Rivlin, R.S. and Thomas, A.G., 1953, Rupture of Rubber. Part 1. Characteristic Energy for Tearing, Journal of Polymer Science Vol. 10, p. 291.

Tunnicliffe, L.B. et al, 2013, The effect of fillers on crosslinking, swelling and mechanical properties of peroxide-cured rubbers, Constitutive Models for Rubbers VIII—Gill-Negrete & Alonso (eds).

APPENDIX

Table 2. The details of the compound formulation used in this investigation.

		HNBR (phr)
Elastomer	Therban 4364	100
Filler	N550	60
Antidegradant	Antidegradant	1.5
Plasticisers	Ester plasticiser	10
Crosslink	Peroxide	8
Co-Agent	Co-Agent	5
Coagent ionic X	Coagent ZDA	5

Constitutive Models for Rubber X – Lion & Johlitz (Eds)
© 2017 Taylor & Francis Group, London, ISBN 978-1-138-03001-5

Gradient damage models in large deformation

B. Crabbé & J.-J. Marigo
Laboratoire de Mécanique des Solides, Ecole Polytechnique, Palaiseau, France

E. Chamberland
Groupe Interdisciplinaire de Recherche en Eléments Finis, Université Laval, Québec (Québec), Canada

J. Guilié
Centre de technologies de Ladoux, Michelin, Clermont-Ferrand, France

ABSTRACT: In 1998, Francfort and Marigo proposed a variational approach to fracture (Francfort & Marigo 1998) in which the whole damage or crack evolution is governed by a least energy principle. In the framework of brittle materials, with softening damage behaviour, they constructed damage gradient models which rely on three principles: *irreversibility* of the damage evolution, *stability* and *energy balance*. They introduced regularized damage models: the total energy contains terms with spatial damage gradient weighted by a parameter called "characteristic length". These models were thoroughly developed in the past years (Marigo, Maurini, & Pham 2016) and enriched with couplings with plasticity (Alessi, Marigo, & Vidoli 2014), dynamics, temperature (Sicsic & Marigo 2013). Yet they stayed in the framework of small strains. The aim of this work is therefore to study the relevance of theses models in finite strains. To this end, both an analytical and a numerical study were conducted. The homogeneous and localized analytical response of a damaging hyperelastic 1D bar submitted to a quasi-static traction test were first established, then compared to the ones obtained *via* a 1D numerical implementation using the FEniCS library. Once the 1D analytical and numerical study has been conducted, numerical simulations were performed in higher dimensions with the finite element academic code MEF++. Compressible or quasi-incompressible neo-Hookean and Mooney-Rivlin solid laws were used on different geometries.

1 INTRODUCTION

1.1 *Gradient damage models*

In the past years, the gradient damage models have proven to be very efficient in modelling the crack initiation and propagation for various behaviours of materials in small deformation. To construct a damage model, one need several items:

- A scalar growing damage variable, called α, whose value is zero if the material is sound, and one if a crack has appeared.
- The dependence of the stiffness or elastic potential to the damage variable, a function $a(\alpha)$ which multiplies either the Young's modulus or the elastic potential, according to the chosen model. The material's stiffness decreases when α increases.
- An evolution law of α, which is based on thermodynamic principles.

The variational approach to fracture makes it possible to write the damage evolution problem under a variational formulation, which relies on three principles: irreversibility, stability and energy balance.

1.2 *Notations*

In this document, all second order tensors will be written with a double underline: $\underline{\underline{\boldsymbol{E}}}$ the Green-Lagrange deformation tensor, $\underline{\underline{\boldsymbol{F}}}$ the gradient deformation tensor, $\underline{\underline{\boldsymbol{\Pi}}}$ the first Piola-Kirchhoff stress tensor, $\underline{\underline{\boldsymbol{S}}}$ the second Piola-Kirchhoff stress tensor and $\underline{\underline{\boldsymbol{\sigma}}}$ the Cauchy stress tensor. In one dimension, their respective scalar values will be designated by E, F, Π, S, σ.

The damage scalar value will be called α, the displacement u. C_t is the set of kinematically admissible displacements at time t, and $D(\alpha_t)=\{\beta:\alpha_t\leq\beta\leq 1 \text{ in } \Omega\}$ is the set of accessible damage fields from α at time t, *i.e.* those who take into account the irreversibility condition.

2 FORMULATION OF THE PROBLEM

Let us write the damage evolution problem of a 3D structure. Let ψ be the hyperelastic potential of the considered material. The total energy of the structure ε is the integral over the domain Ω of the sum of the elastic potential and the dissipated energy due to the damage process

$$\mathcal{E}(u,\alpha)=\int_{\Omega}\left(a(\alpha)\psi(\underline{\underline{F}})+w(\alpha)\right.$$
$$\left.+12w_1\ell^2\underline{\underline{\nabla\alpha}}\cdot\underline{\underline{\nabla\alpha}}\right)dx. \quad (1)$$

The dissipated energy during a homogeneous damage process $w(\alpha)$ is chosen as a linear function of α

$$w(\alpha)=w_1\alpha, \quad (2)$$

W_1 a constant. With such a model, a purely elastic phase is guaranteed during the traction test before a critical stress value is reached and the stiffness softening begins.

We seek to minimize the total energy (1) with respect to the variables u and α. The variational form of the evolution problem can be written with the three following principles:

- Irreversibility
  $\dot{\alpha}\geq 0, \quad (3)$
- that is, α can only grow or stay constant, which can be interpreted as a "non healing" condition, preventing a damaged material to heal.
- First order stability, which states that the Gâteaux derivative of the total energy (1) in a admissible direction $(v,\beta)\in C_t\times D(\alpha_t)$ is always positive
  $\mathcal{E}'(u,\alpha)(v,\beta)\geq 0 \quad \forall(v,\beta)\in C_t\times D(\alpha_t). \quad (4)$
- Applying the first order stability condition (4) to the total energy (1) gives the equilibrium equation and the local damage criterion after an integration by part
  $a'(\alpha)\psi(\underline{\underline{F}})+w'(\alpha)-w_1\ell^2\Delta\alpha\geq 0. \quad (5)$
- Energy balance: it gives the consistency condition on the damage criterion, but will not be studied in this paper.

3 ONE DIMENSIONAL ANALYTICAL STUDY

In this section we will expose the analytical and numerical results obtained for a 1D study of a damaging bar under traction. The problem consists of a 1D bar whose displacement is imposed on both extremities. After having written the total energy with anadequate hyperelastic potential and having developed the first order stability condition, the evolution of the damage during the homogeneous phase and its localization is set.

3.1 *Model*

First of all, an adequate 1D hyperelastic potential has to be established. Starting from a 3D Ciarlet-Geymonat potential (Ciarlet & Geymonat 1982), it is reduced to a 1D potential.

To ensure that there exists at least one solution to a boundary value problem in finite elasticity (Ball 1976), such a potential has to follow some criteria: it needs to be polyconvex, and to verify conditions (6), (7) and (8)

$$\psi(\underline{\underline{F}})=\frac{\lambda}{2}(tr\underline{\underline{E}})^2+\mu\mathrm{tr}\left(\underline{\underline{E}}^2\right)+O\left(\left\|\underline{\underline{E}}\right\|^3\right), \quad (6)$$
$$\text{with } \underline{\underline{E}}=\frac{1}{2}\left(\underline{\underline{F}}\cdot\underline{\underline{F}}^T-1\right),$$

$$\psi(\underline{\underline{F}})\rightarrow+\infty \quad \text{when} \quad \det(\underline{\underline{F}})\rightarrow 0^+, \quad (7)$$

and

$$\psi(\underline{\underline{F}})\geq\left(\left\|\underline{\underline{F}}\right\|+\left\|\mathrm{cof}\,\underline{\underline{F}}\right\|+\det\underline{\underline{F}}\right) \quad (8)$$

where tr is the trace of a tensor, $\|\cdot\|$ is the euclidean norm and $\|\mathrm{cof}\,\underline{\underline{\mathbf{F}}}\|=\frac{1}{2}\left[(\mathrm{tr}\underline{\underline{F}}^T\underline{\underline{F}})^2-\mathrm{tr}(\underline{\underline{F}}^T\underline{\underline{F}})^2\right]$. Condition (7) prevents a volume from being reduced to zero with a finite deformation energy. Condition (8) is the coercivity of ψ.

Such a general 3D law takes the form

$$\psi(\underline{\underline{F}})=a\left\|\underline{\underline{F}}\right\|^2+b\left\|\mathrm{cof}\,\underline{\underline{F}}\right\|^2+c(\det\underline{\underline{F}})^2$$
$$-d\ln(\det\underline{\underline{F}})^2+e, \quad (9)$$

where a, b, c, d and e are real constants, and this gives the expression of a 3D Ciarlet-Geymonat potential. In 1D, (9) is written as

$$\psi(F)=(a+c)F^2-d\ln F^2+e. \quad (10)$$

Developing $\ln F^2$ up to the second order

$$\ln F^2=2E-\frac{(2E)^2}{2}+O(\|E\|^3), \quad (11)$$

and using condition (6) gives

$$(a+c)(1+2E)-d(2E-2E^2)+e=\frac{A}{2}E^2, \quad (12)$$

and we obtain

$$a+c=d=-e=\frac{A}{4}. \quad (13)$$

Thus we have a 1D potential which respects the conditions (6), (7), (8)

$$\psi(F)=\frac{A}{2}\left(E-\frac{1}{2}\ln(1+2E)\right), \quad (14)$$

and where the 1D Green-Lagrange deformation is

$$E = u' + \frac{u'^2}{2}. \tag{15}$$

Note that for small deformation, the development around zero of $\ln(1+2E) = 2\ln(1+u')$ gives the Hook's law

$$\psi(F) = \frac{A}{2}u'^2. \tag{16}$$

For large deformation, $u' \ll u'^2$ and $ln(1+u') \ll u'^2$ and we have

$$\psi(F) = \frac{A}{2}\frac{u'^2}{2}. \tag{17}$$

The second Piola-Kirchhoff stress is

$$S = \frac{A}{2}\frac{2E}{1+2E}, \tag{18}$$

and in 1D, the Cauchy stress and the first Piola-Kirchhoff stress are equal

$$\sigma = \Pi = \frac{A}{2}\frac{2E}{\sqrt{1+2E}}, \tag{19}$$

or

$$\sigma = \frac{A}{2}\left(1 + u' - \frac{1}{1+u'}\right). \tag{20}$$

3.2 *Total energy and first order stability condition*

Let (u, α) be a pair of admissible displacement/damage fields of $C_t \times D(\alpha_t)$. Without external forces, the total energy of the bar is given by

$$\begin{aligned}\mathcal{E}(u,\alpha) = \int_\Omega \Bigg[& \frac{A_0 a(\alpha)}{2}\left((u' + \frac{u'^2}{2}) - \ln(1+u')\right) \\ & + w_1\alpha + \frac{1}{2}w_1\ell^2(\alpha')^2 \Bigg] dx\end{aligned} \tag{21}$$

with

$$a(\alpha) = (1-\alpha)^2. \tag{22}$$

A_0 is the non-damaged value of the Young's modulus, so that when the material is sound ($\alpha = 0$), the stiffness is equal to the inital stiffness, and when the material is broken ($\alpha = 1$), the stiffness is null. The first order stability condition gives on one hand

$$\int_\Omega \frac{A_0 a(\alpha)}{2}\left(1 + u' - \frac{1}{1+u'}\right)v' = 0 \quad \forall v \in C_t \tag{23}$$

which after integrating by parts, gives the 1D equilibrium equation

$$\Pi'(x) = 0 \quad \forall x \in \Omega, \tag{24}$$

and on the other hand

$$\frac{A_0 a'(\alpha)}{2}\left(u' + \frac{u'^2}{2} - \ln(1+u')\right) + w'(\alpha) - w_1\ell^2\alpha'' \geq 0 \tag{25}$$

which is the local damage criterion for a hyperelastic material with a 1D Ciarlet-Geymonat potential, also written as in (5) for a 1D setting

$$a'(\alpha)\psi_0(F) + w'(\alpha) - w_1\ell^2\alpha'' \geq 0 \tag{26}$$

with ψ_0 the non damaged elastic potential.

3.3 *Homogeneous damage*

Let a bar of length L with a homogeneous section be submitted to traction with an imposed displacement depending on time t on $x = L$ such that $u_t(L) = tL$. On $x = 0$, we have $u_t(0) = 0$. The displacement field in the bar is $u_t(x) = tx$.

The damage criterion (25) is written with respect to time for $\alpha = 0$, with $u' = t$

$$f(t) = -A_0\left(t + \frac{t^2}{2} - \ln(1+t)\right) + w_1. \tag{27}$$

Since $f(0) = w_1 > 0$ and $f(t) \to -\infty$ when $t \to \infty$, there exists a time t_e for which the damage criterion is reached.

Writing the equality in (26) gives the expression of α with respect to the Green-Lagrange deformation E

$$\alpha = 1 - \frac{w_1}{A_0\left(E - \frac{1}{2}\ln(1+2E)\right)}. \tag{28}$$

Inserting (28) in (19) makes it possible to plot the Cauchy stress evolution with respect to E (Figure 1).

3.4 *Damage localization*

Let $(X_0 - d, X_0 + d)$ be an interval on the bar where the damage localizes. The damage profile can be found by solving the equality in the expression of the damage criterion (26), with a spatial derivative of α which is no longer zero

$$w_1\ell^2\alpha'' = w'(\alpha) - A_0(1-\alpha)\left(E - \frac{1}{2}\ln(1+2E)\right). \quad (29)$$

Multiplying (29) by α', using (22) and keeping in mind that $S = \partial\psi / \partial E$ where ψ is given by (14), we get

$$\left(\frac{1}{a(\alpha)} - 1\right)\left(\frac{\partial\psi(E)}{\partial E}E - \psi(E)\right) + w(\alpha) - \frac{w_1\ell^2}{2}(\alpha')^2 = 0. \quad (30)$$

ψ is a convex function of E, and its dual potential is $\psi^*(S) = SE - \psi(E)$. When S = 0, a crack appears on the bar in X_0, and the damage profile is given by

$$\frac{\ell}{\sqrt{2}}\frac{d\alpha}{\sqrt{\alpha}} = \pm dx, \quad (31)$$

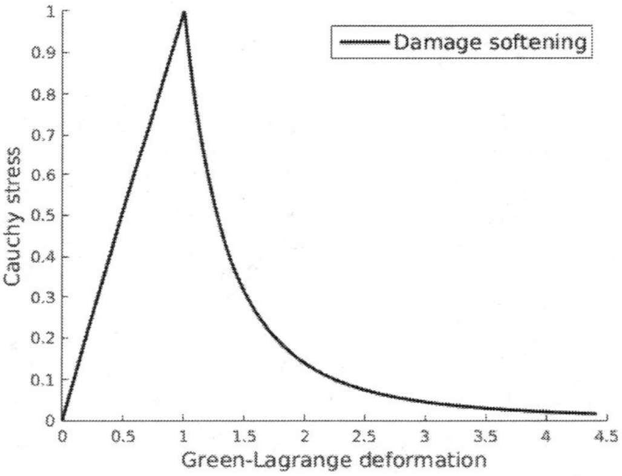

Figure 1. Homogeneous damage softening for a 1D bar.

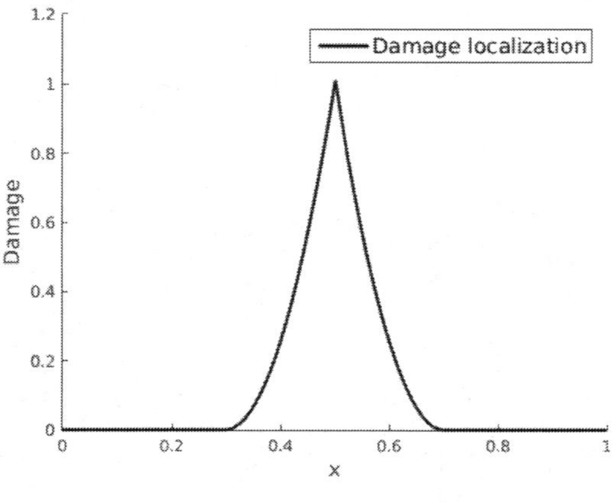

Figure 2. Damage localization on a 1D bar.

with the boundary conditions $\alpha(X_0 \pm d) = 0$ and $\alpha'(X_0 \pm d) = 0$ which enable to determine d, and to show that the size of damaged band is directly proportional to the characteristic length ℓ

$$2d = 2\sqrt{2}\ell. \quad (32)$$

This is the same profile as for an elastic linear material, plot on Figure 2.

4 NUMERICAL SIMULATIONS

4.1 *Numerical implementation*

The numerical problem consists of finding the displacement and the damage such that the potential energy of the system is minimized at each time step. For this purpose, a strategy of alternate minimization is used: at a given time step, each one of the variables is successively found while the other one is fixed, and this operation is repeated until the damage variable reaches convergence, *i.e.* the difference between the new value of α and its old value is smaller than a given tolerance *tol*.

Both the displacement and the damage problems are non linear and require the use of non linear variational solvers. Furthermore, the irreversibility of the damage evolution has to be taken into account, and to ensure that the damage at a given time step is never lower than the damage at the previous time step, a bound constrained solver is used.

Finally, to preserve the ellipticity of the problem, it is important to add a residual stiffness, that is a small parameter k_{ell} in the expression of (22) so that $a(\alpha)$ is never perfectly equal to zero

$$a(\alpha) = (1-\alpha)^2 + k_{ell}. \quad (33)$$

4.2 *1D simulations*

The first step is to validate the analytical results obtained in part 3. The FEniCS library (Alnæs, Blechta, Hake, Johansson, Kehlet, Logg, Richardson, Ring, Rognes, & Wells 2015) is used. A 1D bar of length L is discretized in N parts, its left end is blocked and a displacement is imposed on its right at each time step. To ensure damage localization on the bar, Dirichlet conditions $\alpha(0) = 0$ and $\alpha(L) = 0$ are imposed. The results are in agreement with the analytical ones.

4.3 *2D and 3D simulations*

Once the one dimensional study has been performed, and both analytical and numerical results are the same, simulations in higher dimensions can be carried out. An academic finite element code is used, MEF++ (GIREF 2016), which enables to perform simulations with neo-Hookean or Mooney-Rivlin hyperelastic potentials.

Depending on the value given to the bulk modulus κ, the material's behaviour can be either quasi-compressible or quasi-incompressible. We usually chose it between 0.1 and 100.

4.3.1 *2D plate*

A test that generalises the 1D experiment is a 2D plate submitted to traction on one end, and blocked on the other end. Dirichlet conditions are used for the damage. The damage profile in the reference configuration is the same the one in 1D. The main difference is in the fact that in 2D, a striction zone appears in the deformed configuration, due to the Poisson ratio and the large deformation zone in the middle of the sample. Figure 3 shows the displacement and the damage in the reference configuration when the sample is broken: in this case, the sample broke for a deformation of around 300%. This simulation was made with a neo-Hookean material, and the geometrical, material and numerical parameters are summarized in Table 1.

4.3.2 *2D plate with a hole*

A second example is a 2D plate with a hole in the middle. The same boundary conditions are applied, but due to the presence of the hole, the stress concentrates on the edges of the hole, making the damage appear first on this region and the plate break progressively. The Figure 4 shows the displacement in the deformed configuration (top), and the damage localization (bottom) for three different values of characteristic length: $\ell = 0.1, \ell = 0.15$ and $\ell = 0.2$. The size of the characteristic length changes the size of the damage band: the smaller it is, the smaller is the damage band.

The Table 2 shows the parameters used for this simulation. The bulk modulus is chosen relatively high in order to model a quasi-incompressible behaviour.

4.3.3 *3D hourglass specimen*

The last example presented here is a hourglass specimen following a compressible neo-Hookean law submitted to traction. The results of the traction test are shown on Figure 5 and Figure 6: the

Figure 3. 2D plate submitted to traction. Top: displacement in the reference configuration. Bottom: damage in the reference configuration. Both images show that the plate is broken in its middle part, where the damage localizes and reaches the value 1.

Table 1. Geometrical, material and numerical parameters for a 2D plate submitted to traction.

L	H	h	k_{ell}	*tol*
1.0	0.2	0.0033	1e-3	1e-4
κ	C_{10}	C_{01}	G_c	ℓ
100	10	5	150	0.2

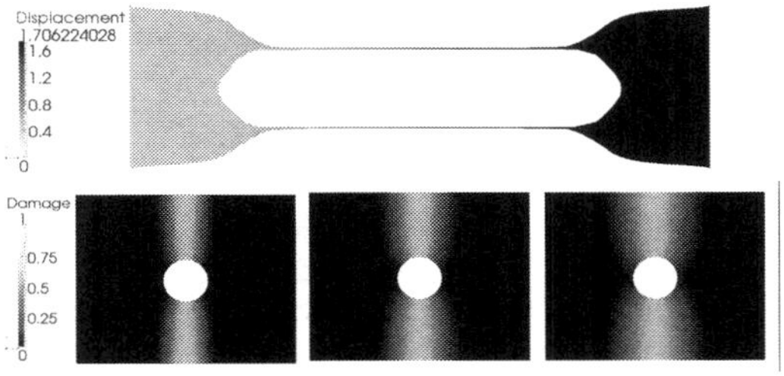

Figure 4. Rupture of a 2D plate with a hole. Top: displacement in the deformed configuration. Bottom: Damage in the reference configuration for $\ell = 0.1$, $\ell = 0.15$ and $\ell = 0.2$.

Table 2. Geometrical, material and numerical parameters for a 2D plate with a hole submitted to traction.

L	H	h	k_{ell}	*tol*
1.0	0.8	0.01	1e-3	1e-4
κ	C_{10}	C_{01}	G_c	ℓ
100	10	5	80	0.1, 0.15, 0.2

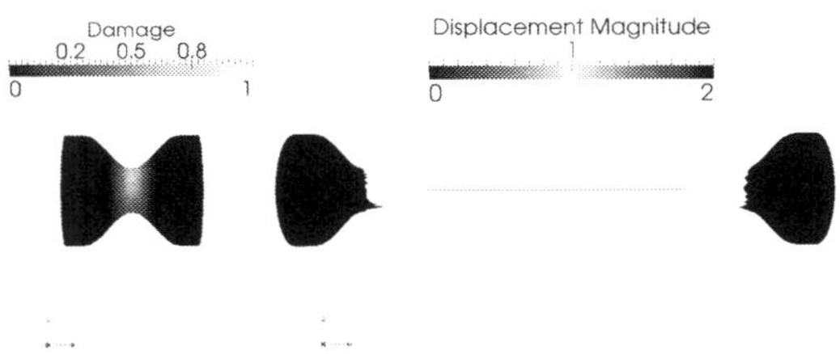

Figure 5. Rupture of a 3D hourglass specimen under traction. Left: damage in the reference configuration, localized in the constricted part of the sample. Right: displacement in the deformed configuration, the specimen is broken in two parts in its middle.

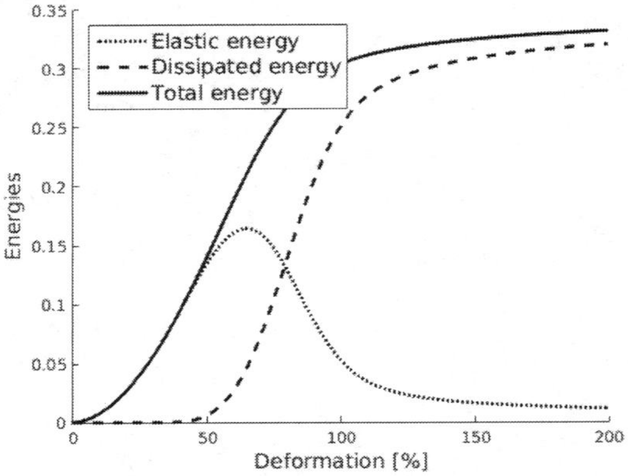

Figure 6. Evolution of the energies during the rupture of a 3D hourglass specimen under traction. The total energy is the sum of the elastic energy and the dissipated energy. Its evolution is continuous. As the sample breaks, the elastic energy tends to zero and the dissipated energy tends to a critical rupture energy.

damage progressively localizes on the constricted part of the sample, because there is a stress concentration in this zone and the critical stress is thus first attained here. The damage reaches the value 1 at the centre, and the sample is broken, as can be seen on Figure 5 (right). The damage evolution is continuous, and while the dissipated energy tends to the critical value, the elastic energy goes to zero.

5 CONCLUSIONS

The gradient damage models were applied successfully to large deformations: in a one dimensional setting, it is possible to get analytical results for the damage evolution of a 1D structure submitted to a quasi-static traction test. The numerical implementation is basically the same as for small deformation: the alternate minimization algorithm can be used, and the linear solver has to be replaced by a non linear solve to solve the non linear displacement problem. Higher dimensions numerical simulations were performed, and although the numerical parameters are sensitive, it is possible to break simple specimens at large strains with a quasi-incompressible behaviour.

One interesting perspective of this work is to determine how well the gradient damage models in large deformation can account for the process of cavitation, one of the main mechanism of damage in polymer materials.

REFERENCES

Alessi, R., J.-J. Marigo, & S. Vidoli (2014). Gradient damage models coupled with plasticity and nucleation of cohesive cracks. *Archive for Rational Mechanics and Analysis 214*(2), 575–615.

Alnæs, M., J. Blechta, J. Hake, A. Johansson, B. Kehlet, A. Logg, C. Richardson, J. Ring, M. Rognes, & G. Wells (2015). The fenics project version 1.5. *Archive of Numerical Software 3*(100).

Ball, J.M. (1976). Convexity conditions and existence theorems in nonlinear elasticity. *Archive for Rational Mechanics and Analysis 63*(4), 337–403.

Ciarlet, P.G. & G. Geymonat (1982). Sur les lois de comportement en élasticité non linéaire compressible. *Comptes Rendus de l'Acadmie des Sciences 295*, 423–426.

Francfort, G.A. & J.J. Marigo (1998). Revisiting brittle fracture as an energy minimization problem. *Journal of the Mechanics and Physics of Solids* 46(8), 1319–1342.

Giref (2016). Mef++.

Marigo, J.J., C. Maurini, & K. Pham (2016). An overview of the modelling of fracture by gradient damage models. *Meccanica 51*, 3107–3128.

Sicsic, P. & J.J. Marigo (2013). From gradient damage laws to Griffith's theory of crack propagation. *Journal of Elasticity, Springer Verlag (Germany) 113*(1), 55–74.

Constitutive Models for Rubber X – Lion & Johlitz (Eds)
 ISBN 978-1-138-03001-5

Thermomechanical characterization of the dissipation fields around microscale inclusions in elastomers

T. Glanowski
Vibracoustic, CAE and Durability Prediction, Carquefou, France

Y. Marco & V. Le Saux
Institut de Recherche Dupuy de Lôme (IRDL), FRE CNRS 3744, ENSTA Bretagne, France

B. Huneau
Institut de Recherche en Génie Civil et Mécanique (GeM), UMR CNRS 6183, Ecole Centrale de Nantes, France

C. Champy & P. Charrier
Vibracoustic, CAE and Durability Prediction, Carquefou, France

ABSTRACT: The fatigue properties of filled elastomers are strongly connected to the population of inclusions induced by their complex recipes and mixing/injection processes. The description and understanding of the basic mechanisms involved around these inclusions, depending on their nature, geometry, interface and cohesion properties are therefore of primary importance to optimize the fatigue design of industrial compounds and parts. Despite numerous studies based on SEM or tomography measurements, the understanding of the basic mechanisms at the scale of inclusions remains difficult. The objectives of this study are to take advantage of thermomechanical observations at the scale of the inclusions to characterize the early stages of the fatigue damage scenario. Several difficulties are at stake here, as high thermal and spatial resolutions are required to describe accurately the heterogeneous local fields. In order to simplify the resolution of the heat equation, thin specimens are considered here, presenting inclusions of about 500 micrometers. The experimental set-up includes an infrared camera and a digital microscope to measure both thermal and strain fields. The well resolved dissipation fields obtained are then used to characterize the heat build-up response at the inclusion's scale. The curve obtained is then challenged by comparing the response obtained for several areas on a given sample and for several samples. Finally, the heat build-up curves obtained around the inclusions and for a classical diabolo-shaped fatigue sample are compared. The very good agreement obtained validates a possible shift in scales and opens the way to the transposition and validation of macroscopic constitutive models and failure criteria at microscopic scale to identify damage scenario.

1 INTRODUCTION

The fatigue properties of elastomeric parts are strongly related to the microstructure features (fillers nature, shape, size and size dispersion) because these reinforced materials are heterogeneous. The industrial development of materials resistant to fatigue is therefore very complex because the parameters driving the microstructure are incredibly numerous, including the original ingredients and the processing parameters, from mixing to injection and curing. A clear understanding of the fatigue behavior requires both the description of the thermo-mechanical response and of the microstructure. Despite very interesting data, these approaches are usually considered separately, with accurate description of the microstructure and of the initiation sites' morphology on the one hand (Gent and Park 1984, Le Cam *et al.*, 2013, Le Saux *et al.*, 2011, Le Gorgu Jago 2012, Huneau *et al.*, 2016) and of the thermo-mechanical response on the other hand (Mars 2002, Saintier *et al.*, 2006, Le Saux *et al.*, 2010, Marco *et al.*, 2013). Some recent studies tried to relate thermal measurements and micro-tomography analysis to characterize the fatigue properties of elastomeric materials (Le Saux *et al.*, 2010, Marco *et al.*, 2013), based on an average point of view. But the links between macroscopic mechanical criteria and local failure mechanisms are still missing because the main failure mechanisms identified in the literature (Gent and Park 1984, Le Saux *et al.*, 2011, Masquelier 2014) are only described from a geometrical point of view.

The aim of this study is to define and validate tools to provide data on the fields of dissipated energy at low scales thanks to temperature measurements. An experimental set-up and a devoted protocol developed and formerly validated on structural samples at the macro-scale (Marco *et al.*, 2014, Masquelier *et al.*, 2015) is improved here and applied at the inclusions' scale to investigate an industrial compound. The correlated description of the local strains allows building the heat build-up curves at this scale, which is a fully new result. The curves generated over several areas and for several samples are compared and finally challenged on the one obtained at the macroscopic scale on a classical fatigue sample.

2 EXPERIMENTAL SETTINGS

2.1 *Materials and specimens*

The material used in this study is an industrial elastomer: a blend NR/IR reinforced with carbon blacks. Classical hourglass samples (Figure 1) are used to characterize the heat build-up curve of the material at the macroscopic scale. These samples are injected and representative of the manufacturing process used for automotive anti-vibration parts.

The description of the thermomechanical response at the inclusions' scale is carried out on very thin samples (0.2–0.3 mm of thickness) cut from the hourglass sample. The protocol used to obtain this thin film is given in Figure 2.

The geometry of the sample is centered on the studied inclusions. The dimensions of the sample and the inclusion are illustrated in Figure 3.

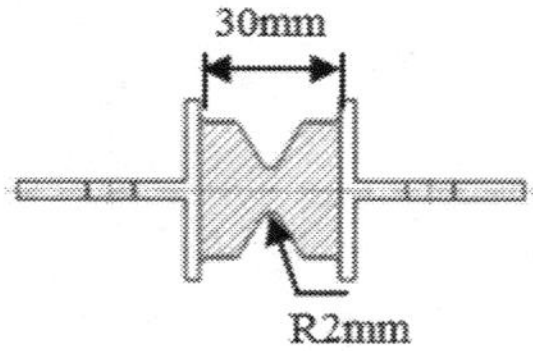

Figure 1. Geometry of the AE2 hourglass sample.

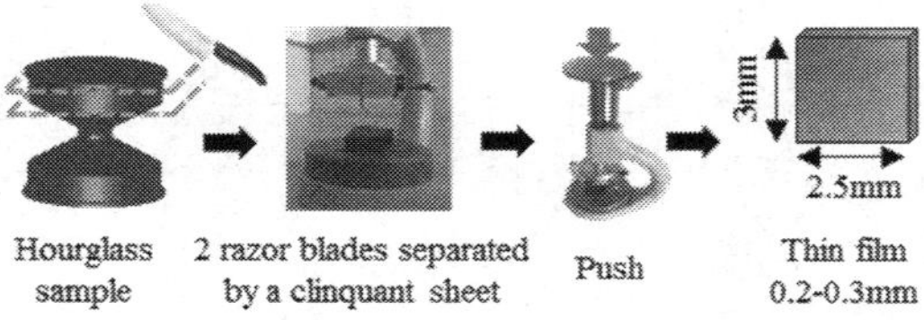

Figure 2. Protocol used to obtain the thin samples.

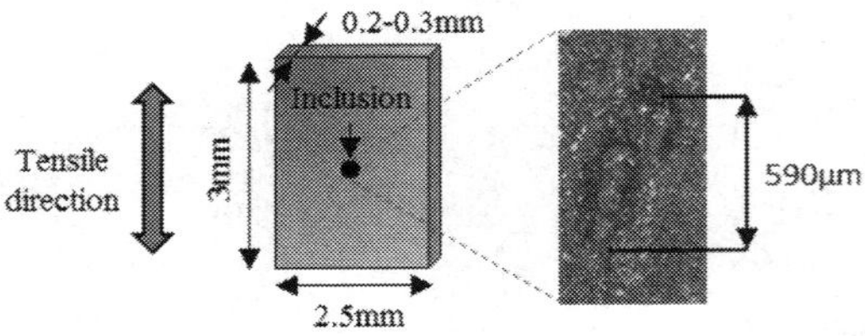

Figure 3. Global geometry of the thin samples.

The goal of using these thin samples is to reduce the analysis to a two-dimensional problem in order to facilitate the determination of the dissipation sources and the strain fields, measured on the opposite faces of the samples.

2.2 *Experimental devices*

All tests are achieved on a Bose electro-dynamical testing machine equipped with a 3.2 kN load cell and are displacement controlled. The thermal acquisition is performed thanks to a FLIR SC7600-BB infrared camera with the so-called "G3" lens with an analyzed zone of 4 mm*3 mm. The spatial resolution is 14 µm.

In order to improve the thermal resolution, a specific pixelwise calibration including housing temperature dependency is used here (Le Saux and Doudard 2017). The calibration is performed thanks to a HGH DCN1000 N4 extended black body (emissivity of 0.98 ± 0.02 and thermal stability ± 0.02°C). The digital level of each pixel of the Focal Plane Array is converted into temperature using its own function:

$$T(i,j) = \sum_{n=0}^{4} \left[\sum_{m=0}^{2} a_{n,m}(i,j).[T_{cam} - T^{0}_{cam}]^{m} \right] DL(i,j)^{n} \quad (1)$$

where T_{cam} is the actual internal temperature of the camera, T^0_{cam} is a reference internal temperature set to 30°C, $a_{n,m}$ are the coefficients of the polynome. A bad pixel detection is then performed and its temperature is replaced by the value of one of its non-defective neighbors according to a specific research algorithm. After this specific calibration, a thermal resolution of about 10 mK is obtained for differential measurements.

The other face of the sample is observed with a digital microscope (magnification lens × 1000). Beyond the observation of the damage scenario, the images recorded are used to determine the local strains by digital image correlation. The speckles are made with talcum powder sprayed on the surface.

The assumption is done here that the low thickness of the thin samples allows to achieve a match between the thermal fields measured on one side of the sample and the kinematic field measured on the other one. X-ray tomography measurements achieved on the tested films (results not shown here) indeed confirmed the geometry of the studied inclusions makes and this hypothesis fully reasonable.

2.3 *Experimental protocol*

The experimental protocol is based on the principle of heat build-up tests (see Marco *et al.*, 2016) and illustrated in Figure 5. In this protocol, the specimen is submitted to a succession of cyclic sinusoidal loading blocks of increasing displacement amplitudes. Each block is composed of an accommodation stage (10 cycles at 30 Hz) to stabilize the Mullins effect. Then the sample is kept at the maximal deformation during the cooling to return to thermal equilibrium. Finally, the thermal acquisition is performed using a rate of 200 images/s, during the loading step (50 cycles at a mechanical frequency of 30 Hz). The mechanical frequency is high in order to reach a high enough dissipated power, despite the small volume tested, and to measure a sufficient thermal signal. A special attention is paid to the experimental conditions in order to limit as much as possible the influence of the environment on the measurements (see Figure 4).

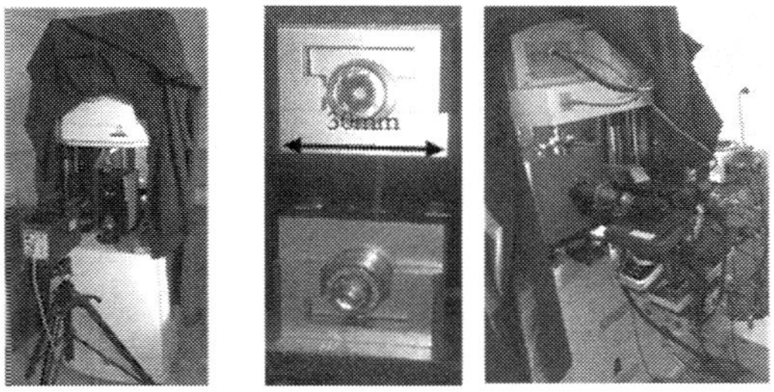

Figure 4. Experimental devices.

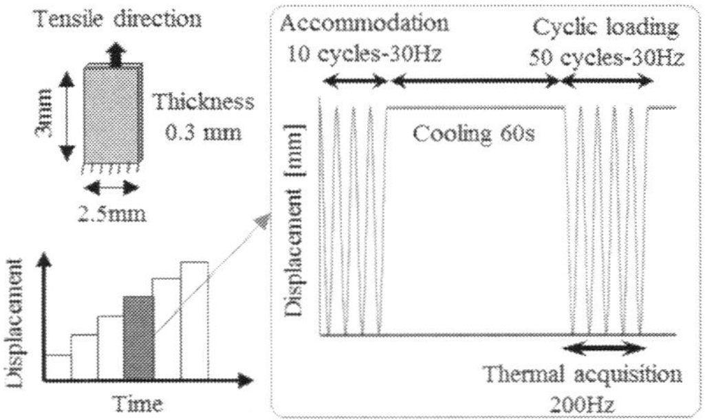

Figure 5. Mechanical protocol.

3 ANALYSIS PROTOCOL

3.1 *Thermal analysis*

3.1.1 *Evaluation of the dissipation*

The evaluation of the thermal sources from the temperature measurements requires solving the heat equation (Lemaître & Chaboche 2004).

$$\rho C\dot{T} + div\left(\vec{q}\right) = P + r + C_T + C_E \tag{2}$$

where ρ is the density, C the specific heat, T the temperature, q the thermal flux, P the intrinsic dissipation, r the contribution of external radiation, C_T the couplings of the internal variables to the temperature and C_E the thermo-elastic couplings. In our study, the temperature variations are considered to be small enough (under 4°C) to neglect the couplings of the internal variables to temperature. Equation 2 therefore becomes:

$$\rho C\dot{T} + div\left(\vec{q}\right) = P + r + C_E \tag{3}$$

In recent papers (Marco *et al.*, 2014, Masquelier *et al.*, 2015), an experimental approach has been developed and validated. Only the main ideas are recalled here.

In this protocol, the main idea is to apply an adiabatic analysis. To do so, it is necessary to reach conditions leading to neglect the diffusion term of Equation 3. Therefore, two very close instants at the beginning of the cyclic loading are considered (points A and B on Figure 6). The characteristic time to reach the thermal equilibrium of our sample is evaluated from the cooling step to about 5 s. Considering these two close cycles (cycles 2 and 5) at 30 Hz therefore leads to a ratio of 50 between the time considered to evaluate the temperature rate and the characteristic thermal time. We therefore consider that the heat diffusion is negligible between the configurations considered and Equation 3 becomes:

$$\rho C\dot{T} = P + r + C_E \tag{4}$$

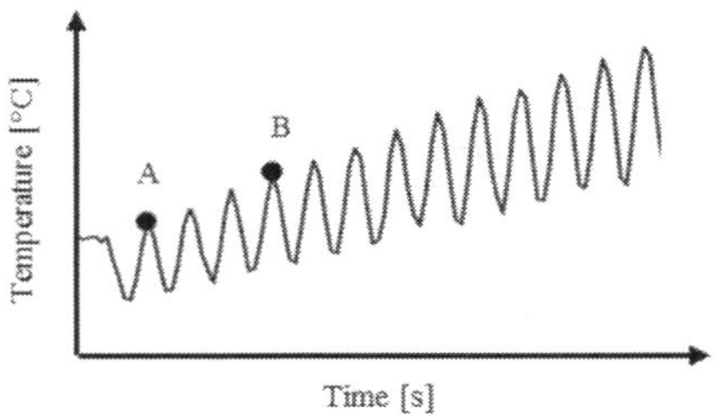

Figure 6. Protocol for dissipation mapping.

As the evaluation is done for two identical mechanical states, the contributions of elastic coupling compensate. Finally, by considering the subtraction of the temperature θ and as the radiation contribution is assumed constant for two close instants, equation 4 writes:

$$\rho C\dot{\theta} = P = f\,\Delta^* \qquad (5)$$

where f is the mechanical frequency and Δ^* the mean dissipated energy per cycle. The Equation 5 can be used to evaluate the mean temperature over a given zone but the most interesting output is to use it to evaluate directly the field of the mean dissipated energy over a cycle, by a direct subtraction of the thermal fields measured for configurations A and B. This analysis is illustrated on Figure 7 in the case of a thin sample with an inclusion at the center.

If this protocol is quite straightforward from a theoretical point of view, the difficulties come from the very high thermal and spatial resolutions needed to catch the temperature variations within microscopic distances.

3.1.2 *Thermal and energetic resolutions*

The temperature resolution of an infrared camera is given by the Noise Equivalent Temperature Difference (NETD), which quantifies its thermal sensitivity. The NETD is evaluated from the fluctuations of the measured temperatures while starring a temperature stabilized scene. The frequency distribution of the measured temperatures can be approximated by a standard normal distribution (Vollmer M. and Möllmann K. 2010):

$$f(x) = \frac{A}{\sigma\sqrt{2\pi}} e^{-\frac{(x-\mu)^2}{2\sigma^2}} \qquad (6)$$

where A is the amplitude, σ the standard deviation and μ the mean of the distribution. Using the measured temperature data, the root mean square value of the temperature fluctuations can be calculated and then used to define the NETD. Using the normal distribution, the root mean square is the standard deviation, leading to a NETD of 40 mK

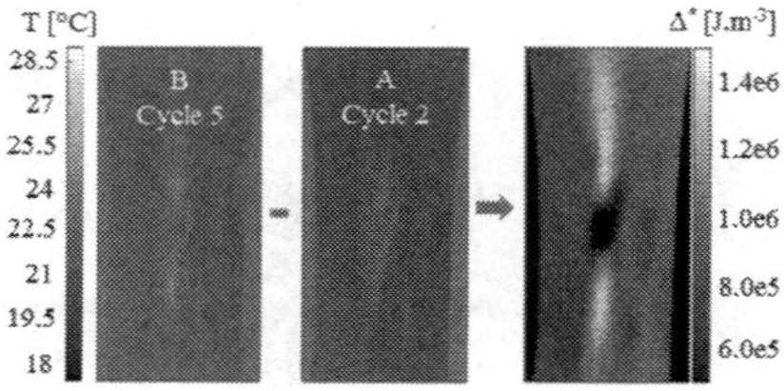

Figure 7. Dissipation mapping.

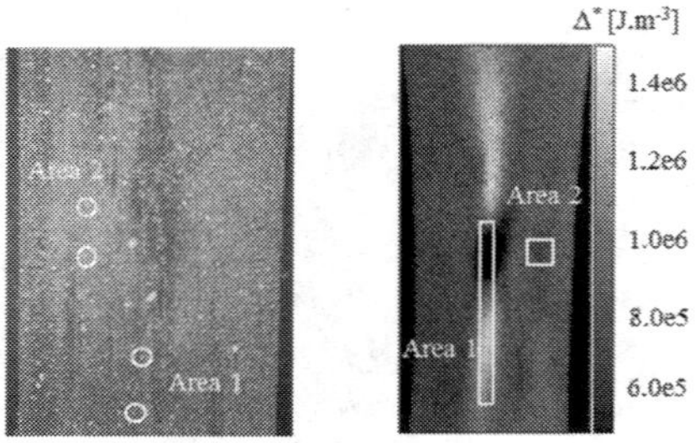

Figure 8. Measurement of local strain and of the dissipation at the pole and beside the inclusion.

in our testing conditions. Thus, the resolution of the dissipated energy is 3.10^4 J.m^{-3}.

3.1.3 *Spatial resolution*

The pixel size of the infrared image is 14 μm. The good superposition of the microscope image and the dissipated energy is illustrated by the Figure 10. We can catch also the damage evolution around the inclusion as highlighted by Figure 10.

3.2 *Evaluation of the local nominal strain*

Figure 8 illustrates the resolution obtained with the digital microscope and the used talcum speckles. The measurement of the local nominal strain is done by following two markers in the same areas as for the dissipation post treatment (see Figure 8).

4 RESULTS

4.1 *Analysis of the dissipation fields and of the failure*

Figure 9 provides an example of the obtained energetic fields (right) and of the dissipation profile (left) measured along the vertical direction (black line drawn on the dissipation field).

This profile gives a clear evaluation of the energetic resolution and of the energy gradients. A maximum dissipation at the poles of the inclusion and a low dissipation for the inclusion are observed. This is pleading for both a strong cohesion of the inclusion and a good interface connection to the surrounding matrix.

This analysis is also consistent with the crack initiation mechanism, which takes place at the pole below the inclusion, as illustrated on Figure 10.

4.2 *Heat build-up curves*

The analysis of the dissipation mapping is performed on 2 areas (see Figure 8). The first area is located at the pole of the inclusion where the crack initiates. An average is made on the width (5 pixels) of the box and then the maximal value of the profile obtained is extracted. The second area is chosen outside the near vicinity of the inclusion, where the deformations are smaller. An average value is computed over this ten per ten pixels. Figure 11 shows the evolution of the energy dissipated for these two areas for the different loading blocks, as a function of the local deformation in these areas evaluated using the talcum markers.

A unification of the heat build-up curves obtained from the two areas can be observed. This is a very interesting result as it means that the heat build-up response evaluated at the pole for a highly stretched state, leading finally to the crack initiation, and the

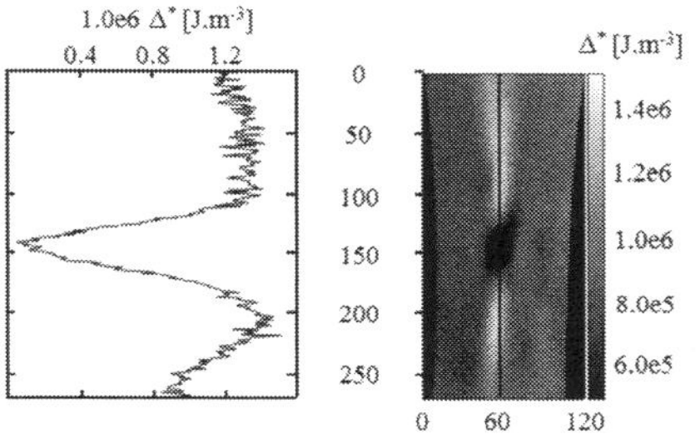

Figure 9. Example of dissipation profile.

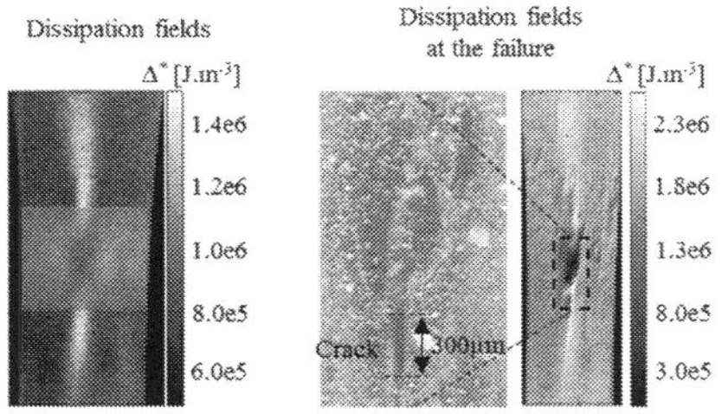

Figure 10. Comparison between dissipation fields and microscope pictures.

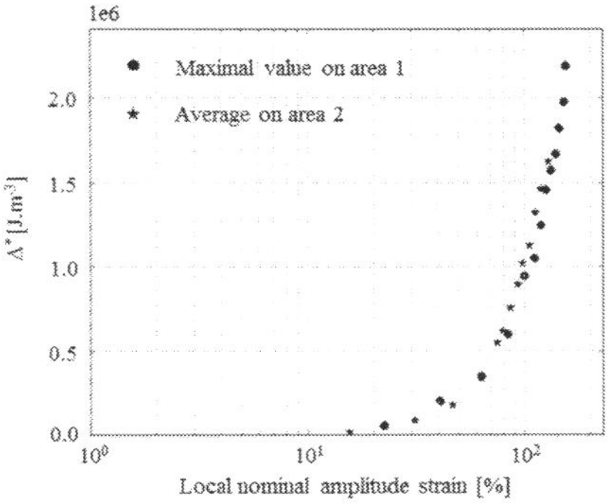

Figure 11. Heat build-up curve at the pole and beside the inclusion.

one obtained far from the inclusion are the same. This means that the material around the inclusion can be considered as homogeneous and that the heat build-up measured can be considered as intrinsic.

4.3 *Repeatability*

To check the reliability of this analysis and of the experimental protocol, another thin film with another inclusion was considered, as presented in Figure 12.

Following the same analysis than explained previously, Figure 13 shows the evolution of the maximum dissipation in area 1 and the average value of the area 2, at three different loading blocks, as a function of the local deformation evaluated using markers.

The heat build-up points obtained from this other case are very close to the first ones, both for the areas 1 and 2. One can therefore conclude a good qualitative consistency between the two experiments and for the 4 areas.

4.4 *Scale transition*

The purpose of this section is to compare the heat build-up curve measured at the microscopic and macroscopic scales. The characterization at the macroscopic scale is achieved on the hourglass sample presented previously (see Figure 1).

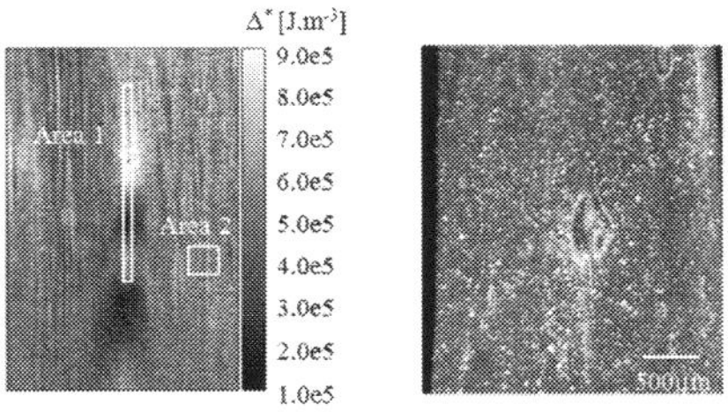

Figure 12. Dissipation mapping and microscope picture for another sample and inclusion.

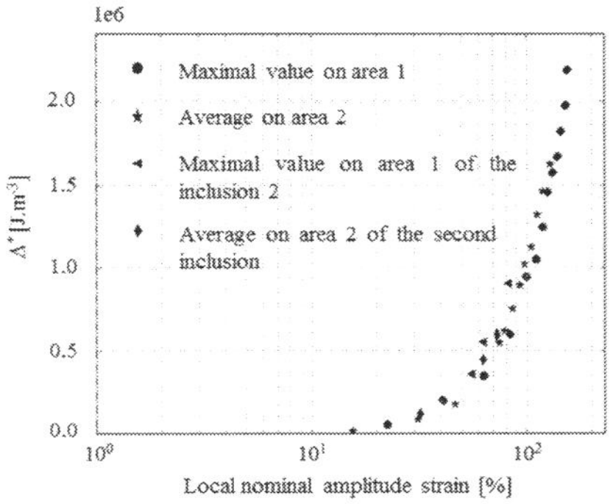

Figure 13. Dissipation of at the pole and far from the inclusion for the two samples.

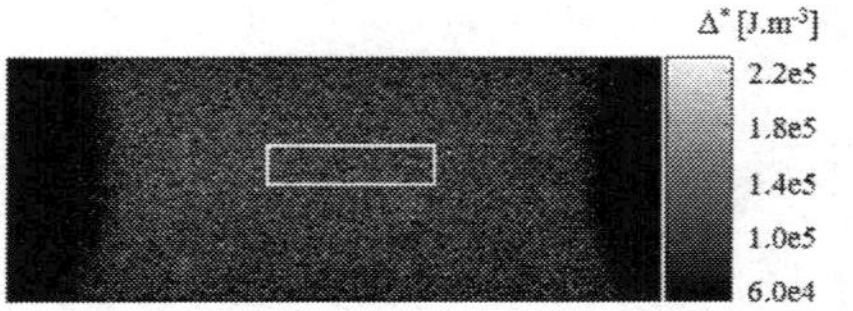

Figure 14. Post treatment area of the hourglass samples.

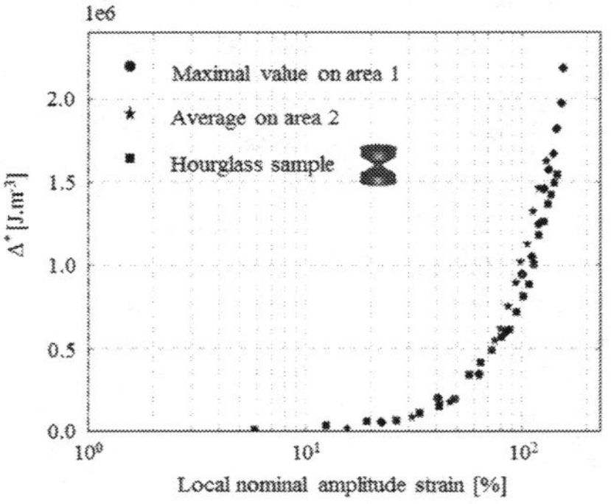

Figure 15. Comparison of the heat build-up curves at the inclusions' scale and in hourglass sample.

The mechanical protocol applied on the hourglass sample is the same than on the thin films, except for the cooling step, as the cooling time is higher for this more massive sample. The thermal acquisition is performed with the same camera but with another lens (analyzed zone of 7mm*5mm). The analyzed area is located at the mid-height of the sample like illustrated on Figure 14 and a mean evaluation is computed over this area.

The thermal analysis applied to evaluate the dissipated energy is the same than for the thin film case. Figure 15 shows the evolution of the average dissipation in the area, for the different loading blocks, as a function of the local deformation evaluated using digital image correlation.

One can observe from Figure 15 that a very good agreement is found between the heat build-up curves at the microscopic and macroscopic scales. This is an important result and opens the way to the validation of scale transition. It means that constitutive models identified at the macroscopic scale can be used to perform finite elements simulations at the inclusion's scale. It also means that the energetic failure criterion deduced at one scale can be relevant or connected to the other one, which is of great interest for the determination of the links between the failure performance and the fillers features of elastomeric materials.

5 CONCLUSION

This study allowed validating an experimental set-up to investigate the dissipation properties of elastomers at the inclusions' scale. The experimental set-up includes an infrared camera and a digital microscope to measure both thermal and strain fields. The well resolved dissipation fields obtained were used to characterize the heat build-up response at the inclusion's scale, which was never done before. The curve obtained was then challenged by comparing the response measured for several areas on a given sample and for various samples. Finally, the heat build-up curves obtained around the inclusions and for a classical diabolo-shaped fatigue sample were compared. The very good agreement between these curves validates the protocols but also a possible shift in scales and opens the way to the transposition and validation of constitutive models and failure criteria from one scale to the other one.

ACKNOWLEDGEMENT

The authors would like to thank the ANRT for its financial support and all the partners of the PROFEM 2 project, Vibracoustic, GeM, IRDL.

REFERENCES

Gent A.N. & Park B., 1984. Journal of Materials Science, 19, 1947–1956.

Huneau B., Masquelier I., Marco Y., Le Saux V., Noizet S., Schiel C. & Charrier P., 2016, Rubber Chemistry and Technology, 89, 1, 126–141.

Le Cam J.-B., Huneau B. & Verron E., 2013, International Journal of Fatigue, 52, 82–94.

Lemaître & Chaboche, 2004, Mécanique des matériaux solides, Dunod.

Le Gorgu Jago, K., 2012, Rubber Chemistry and Technology, 85, 387–407.

Le Saux V., Marco Y., Calloch S., Doudard C. & Charrier P., 2010, International Journal of Fatigue, 32, 1582–1590.

Le Saux V., Marco Y., Calloch S. & Charrier P., 2011, Polymer Engineering and Science, 51, 1253–1263.

Le Saux V., Doudard C., 2017, Infrared Physics and Technology, 80, 83–92.

Masquelier I., Marco Y., Le Saux V., Calloch S., Charrier P., Mechanics of Materials, 2015, 80, A, 113–123.

Masquelier I., 2014, PhD thesis, Université de Bretagne Occidentale.

Marco Y., Masquelier I., Le Saux V., Calloch S., Huneau B. & Charrier P., 2013, ECCMR VIII.

Marco Y., Le Saux V., Jégou L., Launay A., Serrano L., Raoult I., Calloch S., 2014, International Journal of Fatigue, 67, 142–150.

Marco Y., Masquelier I., Le Saux V., Charrier P., Rubber Chemistry and Technology, American Chemical Society, 2016.

Marco Y., Masquelier I., Le Saux V., Charrier P., 2016, Rubber Chemistry and Technology.

Mars W., Fatemi A., 2002, International Journal of Fatigue, 24, 949–961.

Saintier N., Cailletaud G. & Piques R., 2006, International Journal of Fatigue.

Vollmer M. and Möllmann K., 2010, Infrared Thermal imaging, Wiley-VCH.

Constitutive Models for Rubber X – Lion & Johlitz (Eds)
© 2017 Taylor & Francis Group, London, ISBN 978-1-138-03001-5

Influence of test specimen thickness on the fatigue crack growth of rubber

R. Stoček
PRL Polymer Research Lab., s.r.o., Zlín, Czech Republic
Centre of Polymer Systems, University Institute, Tomas Bata University in Zlín, Zlín, Czech Republic

R. Kipscholl
Coesfeld GmbH & Co. KG, Dortmund, Germany

ABSTRACT: The present paper aims to perform a complex study of the influence of test specimen thickness on the Fatigue Crack Growth (FCG) of natural and synthetic rubber to enhance the circumstantiality of the measuring methodology. In this work, the mini-Pure-Shear (mPS) test specimens (geometry ratio 1/10 "length/width") at two different thickness of 0.5 and 1.5 mm were studied using the Tear and Fatigue Analyzer. Two different rubber compounds based on NR and SBR, filled with 50 phr of carbon black and common curatives were investigated. Three double notched test specimens of each material and thickness were simultaneously tested under the Gauss pulse loading condition at varied strains. As a result, the higher slopes, represented by the material parameters, *m*, were determined for the thinner samples independent on material type. Also, significantly lower deviations of FCG values for thinner samples, independent of material was observed. Thus, it was concluded that the lower thickness of the sample represents a more critical loading case, whereby the experimental data can be determined more precisely.

1 INTRODUCTION

In the field of fracture mechanics of elastomers, the fundamental work of Rivlin & Thomas (1953) is the basis of the most theoretical and experimental works published. The theory is based on the energy balance describing the tearing energy, T, deduced as a fracture mechanics parameter for elastomeric materials. According to Eq. 1, T is the change of the deformation energy due to the energy necessary for the creation of a new crack surface.

$$T = -\frac{dW}{dA} = -\frac{dW}{B \cdot da}. \tag{1}$$

W is the recoverable elastic strain energy, where its determination is well known and based on the simple calculation of the energy at tensile loading for the unnotched specimen. For the case of plane sample with a constant thickness, B, and a straight crack front, the increment of the crack area is $dA = B.da$, where da denotes the crack length increment, whereas in the practice it means crack contour increment. Thus, it is clearly shown that the tearing energy depends highly on the complete geometry of test specimen, which are length, width and thickness.

The influence of test specimen geometry based on geometry ratio "length/width" on tearing energy with respect to observed crack length has been well described in a lot of previous works. In the fundamental work of Rivlin & Thomas (1953), they analysed varied types of specimens: however, for planar tensile test, the Single-Edge-Notched Tension (SENT) and the pure-shear (PS) specimen were used. For every specimen type, they derived specific formulas for the determination of the tearing energy, T. Yeoh (2001) firstly evaluated the tearing energy given by varied geometry ratios and the different tearing energy influenced by crack length was observed. Stoček et al. (2013) showed the influence of crack length on the tearing energy for different geometry ratios by using non-linear finite element analysis. Additionally, the effect of strain in the range up to 50% was taken into the consideration. They observed the tearing energy to be approximately linearly dependent on crack growth for short values independent on geometry ratio on condition of constant length, L_0 and varied width, Q for varied ratios. However, the tearing energy was considered to be constant for large crack length, independent on its growth, in the centre of test specimen. Based on these observations it could be concluded that the SENT is defined due to geometry ratio $L_0/Q > 1/1$ and PS due to $L_0/Q < 1/1$. Additionally, the crack growth rates in dependence on crack length at varied geometry ratios have been evaluated by Stoček et al. (2012), where it was determined that significantly higher growth rate takes place in the centre of PS specimen at large cracks in comparison to the edged region at short crack length. However, these previous observations were elaborated on the basis of quasi-static loading conditions or numerical simulations.

Stoček et al. (2013) firstly performed a complex analysis for describing the relationship between the Fatigue Crack Growth (FCG) and the tearing energy in dependence on varied geometry ratio L_0/Q for different rubber materials. The higher FCG was observed with relation to decreasing value of the geometry ratio.

However, the complex investigation of the relationship between the FCG and the tearing energy in dependence on varied geometry ratio, where the thickness would be taken in to the account, has not been performed up to date. Stadlbauer et al. in more publications (2013) studied the flat pure-shear (PS) samples with 2 mm and 4 mm thickness and additionally compared with Faint Waist Pure Shear (FWPS) samples of 2 mm thickness. The loading applied in this experimental work has to be the combination of tensile and compressional loads. They demonstrated that the FCG of the flat and waisted geometries deviated at higher tearing energies and the FWPS geometry showed higher crack growth rates than the PS geometries independent on their thickness. PS samples with 4 mm showed larger heat build-up at comparable tearing energies due to increased thickness in comparison with thinner samples.

Although the thickness of sample has a direct influence on the tearing energy (see Eq. 1), very limited investigations describing this phenomena have previously been done. Thus the aim of this work is to perform a complex study of the influence of test specimen thickness on the FCG of natural and synthetic rubber.

2 EXPERIMENTAL

2.1 *Rubber test specimens*

Rubber materials used in this study werebased on natural rubber (SMR 20 CV/BP1, Lee Rubber Co. Pte Ltd., Malaysia) and styrene butadiene rubber (Styron Sprintan® SLR 4630, Trinseo, Germany) and were filled with 50 phr of carbon black N339. Table 1 lists the formulation of rubber compounds used.

Rubber compounds were prepared by a two step mixing procedure. First, the batch was mixed in an internal mixer SYD-2L (Everplast, Taiwan) at a rotator speed of 50 rpm and at a mixing chamber temperature of 80°C. In second step, curing system consisting of accelerator and sulfur was added at the temperature of 60°C on the two-roll mill. Curing properties were determined by a moving die rheometer MDR 3000 Basic (MonTech, Germany) according to ASTM D6204 at 160°C.

Table 1. Rubber compounds formulation.

	NR	SBR
	phr	
Natural rubber	100,0	–
Styrene butadiene rubber	–	100
Carbon black	50,0	50,0
Zinc oxide	3,0	3,0
Stearic acid	1,0	1,0
IPPD[1]	1,5	1,5
CBS[2]	2,5	2,5
Sulfur	1,7	1,7

1 N-isopropy-N'-phenyl-p-phenylen-diamine.
2 n-cyklohexyl-2-benzo-thiazole-sulfenamide.

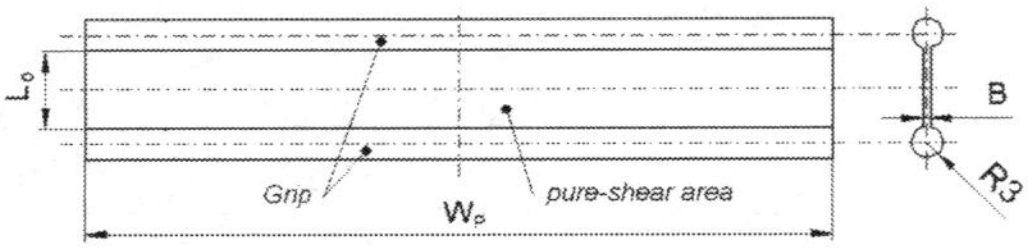

Figure 1. Test specimen geometry, where L_0 = 4 mm, W_P = 40 mm and B = 0.5 mm or 1,5 mm.

The specimens of the mini-pure-shear (mPS) geometry visualized in Figure 1, were cured in a heat press LaBEcon 300 (Fontijne Presses, Netherlands) at 160°C according to optimum curing time t90, with two varied thicknesses B = 0.5 mm and 1.5 mm. Thus, specimens represented by two geometry ratios in relation to the length, L_0: $L_0/Q/B$ = 1/10/8 and $L_0/Q/B$ = 1/10/2.67 were investigated.

2.2 *Fatigue crack growth analyses*

FCG measurements were carried out using Tear and Fatigue Analyzer (Coesfeld GmbH, Germany), which is well known and its photo is shown in the Fig. 2. The detailed description can be found e.g., in Eisele et al. (1992). However, the analyses were done on the device with implemented 3 independent loading engines, where at each engine 3 specimens were simultaneously analyzed. Each upper part of clamp attachment of test specimens was fixed to the load cell and its corresponding test specimen clamp attachment was connected to a separate computer-controlled stepping motor to ensure constant pre-stress during the whole time of testing. The crack growth of each rubber test specimen was monitored on-line through an image proc-

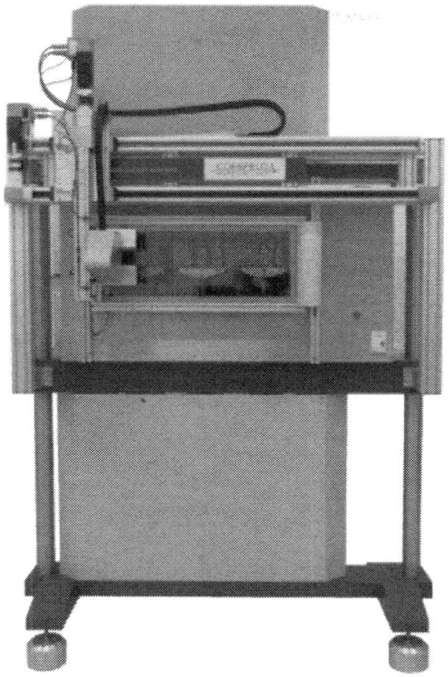

Figure 2. Photo of Tear and Fatigue Analyzer used.

ess system with Charge-Coupled Device (CCD) camera, whereby the crack contour length and data of mechanical behaviour in-situ were evaluated.

1 set of 3 samples per compound and thickness were analyzed over the complete range of applied tearing energies. The following process of the measurement was used. Firstly, the tearing energies over the broad range of loading amplitudes from 5% up to 100% were evaluated based on analyses of un-notched specimens. Secondly, each specimen was notched on both sides with the notch length $a_0 = 10$ mm in accordance to the calculation of minimal notch length in dependence on test specimen geometry ratio defined by Stoček et al. (2013). Finally, the Gauss pulse waveform of the pulse frequency 50 Hz, which corresponded with the pulse width 20 ms and loading frequency 10 Hz at the temperature 28°C was applied for the analyses over the range of tearing energies and respective loading amplitued. The R-ratio $R = 0$ was used for the controlling of pre-stress during the complete FCG analyses.

3 RESULTS AND DISCUSSION

Gent, Lindley and Thomas (1964) determined experimentally the FCG rate, *da/dn* in dependence on the tearing energy, *T*, respectively for rubber materials. Lake and Lindley (1965) derived this relationship to be mathematically represented as follow:

$$\frac{da}{dn} = b \cdot T^m, \tag{2}$$

where *b* and *m* are material constants. The plots of tearing energies, *T* vs. crack growth rates, *da/dn* for NR specimens with the thickness $B = 0.5$ mm and $B = 1.5$ mm are presented in Figure 3 in accordance to Eq. 2. Generally, the higher FCG as well as slope represented by the material parameter *m* are clearly visible for the sample of the thickness $B = 0,5$ mm. As next, it can be seen that the FCG data representing the each tearing energy have significantly lower deviation for the thinner sample.

The plots of tearing energies, *T* vs. crack growth rates, *da/dn* for specimens with the thickness $B = 0.5$ mm and $B = 1.5$ mm based on SBR are presented in Figure 5 in accordance to Eq. 2. For the material SBR more or less identical FCG were observed. However, higher slope represented by the material parameter *m* are clearly visible for the thinner sample. The lower thickness in the case of SBR material was the reason for the lower deviation of the FCG values representing each tearing energy.

Photos of the observed crack contour in NR material representing the highest applied tearing energy are shown in the Fig. 4. A crack deviation from the orthogonal direction to the main strain is visible for both of implemented cracks in the sample with the thickness $B = 1,5$ mm. This phenomenon was caused due to 3-dimensional crack growth, whereas this effect was proportionally dependent on increasing of the specimen thickness. Of course, if the crack deviates from the orthogonal direction to the main strain, the FCG decreases rapidly. However, in the case of the material SBR, crack growths were observed in the orthogonal direction to the main strain over the complete range of applied

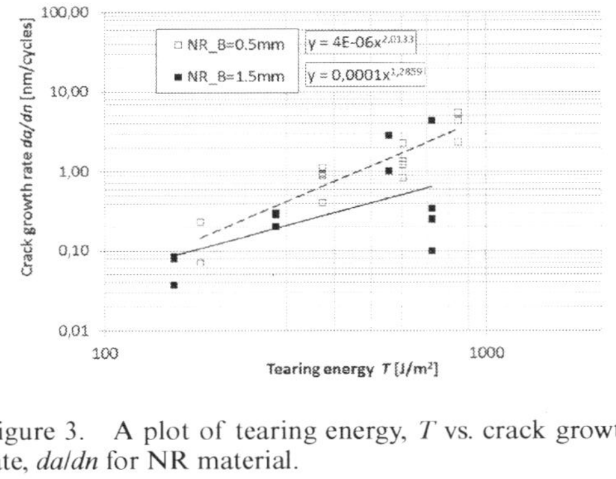

Figure 3. A plot of tearing energy, *T* vs. crack growth rate, *da/dn* for NR material.

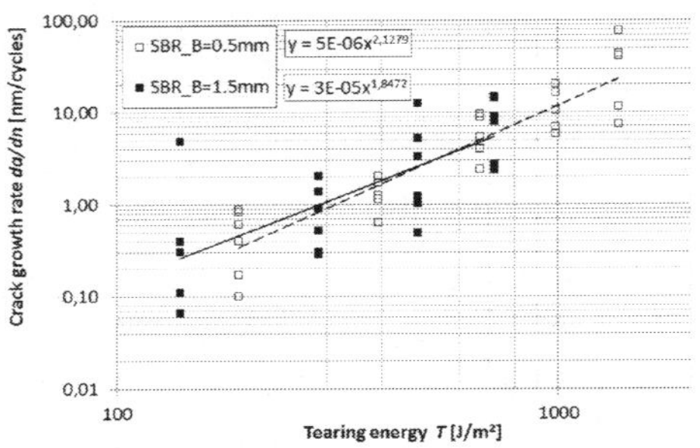

Figure 4. A plot of tearing energy, T vs. crack growth rate, da/dn for SBR material.

Figure 5. Photos of the observed crack contour in NR, left $B = 0.5$ mm, right $B = 1.5$ mm.

tearing energies. Thus, the next reason which caused the deviation was the strain induced crystallinity occurring in the NR material at larger strains.

4 CONCLUSIONS

Materials based on NR and SBR of two varied thicknesses 0.5 and 1.5 mm and respective geometry ratios 1/10/8 and 1/10/2.67 were investigated to describe the influence of the sample thickness on fatigue crack growth.

Generally, the higher FCG as well as slope represented by the material parameter m were clearly visible for the thinner NR sample, whereas for SBR more or less identical FCG was observed, independent of the thickness. However, slightly higher slope represented by material parameter m has to be determined. The FCG values representing each tearing energy independent on material showed significantly lower deviations for thinner samples.

Due to FCG analyses the deviation from the orthogonal direction to the main strain was observed for the NR sample of thickness 1.5 mm, which was predicated to strain induced crystallization.

Finally, it can be concluded that the lower thickness of the sample represents a more critical loading case, whereby the data can be determined more precisely. Thus, the sample geometry rate $L_0/Q/B$ = 1/10/8 should be preferred for FCG analyses.

ACKNOWLEDGEMENTS

This article was written with the support of Operational Program Research and Development for Innovations co-funded by the European Regional Development Fund and national budget of the Czech Republic, within the framework of the project CPS—strengthening research capacity (reg. number: CZ.1.05/2.1.00/19.0409) as well supported by the Ministry of Education, Youth and Sports of the Czech Republic—Program NPU I (LO1504).

REFERENCES

Eisele, U., Kelbch, S.A., Engels, H.-W, 1992, The Tear Analyzer—A New Tool for Quantitative Measurements of the Dynamic Crack Growth of Elastomers, Kautschuk-Gummi—Kunststoffe, 45, pp. 1064–1069.

Gent, A.N., Lindley, P.B., Thomas, A.G., 1964, Cut growth and fatigue of rubbers. I. The relationship between cut growth and fatigue. Journal of Applied Polymer Science, 455–466.

Lake, G.J., Lindley, P.B., 1965, The mechanical fatigue limit for rubber. Journal of Applied Polymer Science 9, 1233–1251.

Rivlin, R.S., Thomas, A.G., (1953), Rupture of rubber. I. Characteristic energy for tearing. Journal of Polymer Science 10, 291–318.

Stadlbauer, F., Koch, T., Planitzer, F., Fidi, W., Archodoulaki, V.-M., 2013, Setup for evaluation of fatigue crack growth in rubber: Pure shear sample geometries tested in tension-compression mode. Polymer Testing, 2013, 32, 1045–1051.

Stadlbauer, F., Koch, T., Planitzer, F., Fidi, W., Archodoulaki, V.-M., 2013, Einfluss des Rußfüllgrades von Elastomeren auf Aspekte der Ermüdung in Zug-Druck-Belastung. Kautschuk-Gummi-Kunststoffe, 66, 37–42.

Stadlbauer, F., Koch, T., Archodoulaki, V.-M., Planitzer, F., Holzner, A., 2013, Influence of Experimental Parameters on Fatigue Crack Growth and Heat Build-Up in Rubber, Materials, 6, 5502–5516.

Stocek, R., Reincke, K., Gehde, M., Grellmann, W., Heinrich, G., 2010, Einfluss der Kerbeinbringung auf die Rissausbreitung in elastomeren Werkstoffen, Kautschuk-Gummi-Kunststoffe, 63, pp. 364–370.

Stoček, R., Heinrich, G., Gehde, M, Kipscholl, R., (2012), A New Testing Concept for Determination of Dynamic Crack Propagation in Rubber Materials, Kautschuk-Gummi-Kunststoffe, 65 (2012), 49–53.

Stoček, R., Heinrich, G., Gehde, M., Kipscholl, R., 2013, Analysis of Dynamic Crack Propagation in Elastomers by Simultaneous Tensile—and Pure-Shear-Mode Testing. In: Grellmann, W., Heinrich, G., Kaliske, M., Klüppel, M., Schneider, K., Vilgis, T. (Eds.): Fracture mechanics and statistical mechanics of reinforced elastomeric blends. Springer-Verlag Berlin Heidelberg, 269–300.

Stoček, R., Horst, T., Reincke, K., 2017, Tearing energy as fracture mechanical quantity for elastomers, In: K.W. Stöckelhuber, A. Das, M. Klüppel: Designing of Elastomer Nanocomposites: From Theory to Applications. Advances in Polymer Science, Springer New York LLC, Vol. 275, 2017, pp. 361–398.

Yeoh., O.H., 2001, Analysis of deformation and fracture of 'pure shear' rubber testpiece, Plastics, Rubber and Composites 30, 391–397.

Constitutive Models for Rubber X – Lion & Johlitz (Eds)
© 2017 Taylor & Francis Group, London, ISBN 978-1-138-03001-5

Fracture analysis of a rolling tire at steady state by the phase-field method

B. Yin, M.A. Garcia & M. Kaliske
Institute for Structural Analysis, Technische Universität Dresden, Dresden, Germany

ABSTRACT: Tires are an essential part of the vehicle and therefore, the failure prediction of rubber materials due to crack initiation and propagation is of large importance for applications. A steady state rolling tire can be analysed with an Arbitrary Lagrangian-Eulerian (ALE) framework, which largely improves the computation efficiency compared to transient analysis. In terms of rubber properties, nonlinearities are involved due to friction, viscous effects and others. Therefore, in this contribution, the Arruda-Boyce model is considered for rubber modelling and a nonlinear viscous evolution law can be used for the further work. Regarding fracture, a phase-field method is proposed for crack approximation, which overcomes the limitation of numerically tracking the displacement singularity. The phase of fracture or unfracture can be indicated by a continuous scalar valued phase-field quantity. The driving force for crack propagation is derived from the elastic energy density function. Hence, the elastic energy from both the equilibrium and the non-equilibrium response is necessary to be considered for crack evolution. Moreover, the elastic energy degradation due to fracture is specifically identified, since cracks are sensitive to tensile loading. The volumetric energy is considered to contribute to fracture only if the element is undergoing tension. Representative examples are shown and corresponding discussions are presented. This evaluation ends with remarks and possible further research areas.

1 INTRODUCTION

Tires play an important role in engineering and industrial application. A tire structure contains not only several elastomers, but also reinforcement materials, such as body plies and belts. Analysis of a rolling tire may involve a series of nonlinearities due to the material properties, contact and energy dissipation, among others.

A rolling body can be modelled within the Lagrangian framework by means of the Finite Element Analysis. Some classical transient methods can be employed for the analysis of the dynamic properties of a tire, but a significant amount of computational resources is required and the simulation results are largely dependent on the time step size. However, for steady state rolling conditions, an alternative approach may be employed, which is based on the concept of an Arbitrary Lagrangian Eulerian (ALE) framework. An additional reference configuration is used to observe the tire spinning. Due to steady state rolling, the time derivative terms eventually vanish. Thus, the evolution equation of the whole system can be largely simplified in this case. For elastic material models, corresponding fundamentals and the derivation can be found in the contribution of Nackenhorst (2004). Nevertheless, for inelastic material models some further developments are proposed to solve the time derivative issue. For example, the ALE framework using streamlines is proposed by Behnke and Kaliske (2015) and Garcia et al. (2016), and it is used to investigate the viscoelastic properties of a tire at steady state rolling.

Regarding the failure of a tire, different methods can be used. In this work, the phase-field method is derived and implemented for fracture simulation. The phase-field method is a new state of the art approach to simulate fracture by means of a smeared description. A continuous scalar-valued phase-field is introduced as an additional degree of freedom to indicate if the continuum body is fractured or not. Thus, two different states are described with a smooth transition. Another parameter, the length scale, is used to regularize the width of the cracks. One of the main advantages is that the numerical tracking of discontinuities in the displacement field is not required. Therefore, it can be applied to complex geometries. However, this method requires a high resolution of the mesh, which may consume a large amount of computational resources.

A general regularization of the phase-field method is formulated by Bourdin and Chambolle (2000). In order to specify that crack evolution is only driven due to tension, some split models are adopted. Amor et al. (2009) proposed a volumetric-deviatoric split model and indicated only considering volumetric energy when the structure is undergoing tension. Miehe et al. (2010a) formulated the spectral decomposition model which separates tension and compression completely. From an algorithmic point of view, monolithic and staggered schemes are discussed as well, which are implemented by Kuhn and Müller (2011) and Miehe et al. (2010b), respectively.

The main assumption to analyse fracture properties of a spinning tire is to separate the steady state rolling and phase-field evolution processes in different frameworks, since the material flow within the ALE framework is assumed to have a constant fracture state. Based on this hypothesis, the damage state along the streamlines of a tire is fixed. Between these two processes, data has to be exchanged iteratively. First, the spinning process is in a steady state with a fixed damage state. Then, the corresponding solution data is applied as a new input to the phase-field evolution. Afterwards, an updated damage state can be obtained and will be returned back to the spinning process. These processes have to be iterated until the two equilibriums converge. In this contribution, this algorithmic scheme will be discussed and a representative example will be shown.

2 THEORETIC FUNDAMENTALS

In this section, some fundamental theoretical aspects are briefly discussed, which involve the ALE formulation, phase-field modelling, as well as a data transfer strategy.

2.1 *ALE formulation*

Within an ALE framework, an arbitrary reference configuration is introduced which is independent on the material particles and space. The kinematical relationship between each configuration is written as

$$\underline{\mathbf{F}} = \underline{\mathbf{F}}_{\varphi} \cdot \underline{\mathbf{R}}_{\mathcal{X}}, \tag{1}$$

where $\underline{\mathbf{F}}_{\varphi}$ is the deformation gradient from the reference to the current configuration and $\underline{\mathbf{R}}_{\mathcal{X}}$ describes the rigid motion from the initial to the reference configuration, respectively. The rotation tensor $\underline{\mathbf{R}}_{\mathcal{X}}$ is characterised by $det(\underline{\mathbf{R}}_{\mathcal{X}}) = 1$, and the total Jacobian determinant is given as

$$J = det(\underline{\mathbf{F}}) = det(\underline{\mathbf{F}}_{\varphi}). \tag{2}$$

In order to regularize the notation and expression, the material and reference configurations are denoted as $\mathcal{B}$ and $\mathcal{X}$, respectively. Within the ALE formulation, the material time derivative can be split into relative and convective parts, respectively. Therefore, the velocity which is derived as

$$\mathbf{v} = \left.\frac{\partial \varphi}{\partial t}\right|_{x} = \left.\frac{\partial \varphi}{\partial t}\right|_{\mathcal{X}} + \tilde{\nabla}\varphi \cdot \mathbf{w}, \tag{3}$$

is a summation of the relative velocity $\frac{\partial \varphi}{\partial t}|_{\mathcal{X}}$ and the convective velocity $\mathbf{c} = \tilde{\nabla}\varphi \cdot \mathbf{w} \cdot \tilde{\nabla}$ is a gradient operator with respect to the reference configuration. It is noteworthy that under the condition of steady state rolling, all time derivatives vanish. Therefore, in this case, $\frac{\partial \varphi}{\partial t}|_{\mathcal{X}} = 0$ is obtained. Moreover, the guiding velocity is defined as

$$\mathbf{w} = \mathbf{w}_0 + \Omega \times \mathcal{X}, \tag{4}$$

where $\mathbf{w}_0$ is the center point velocity, which is assumed to be 0 in this work. Ω is known as the angular velocity and assumed to be constant due to steady state conditions (Nackenhorst 2004).

Based on the balance law of linear momentum with respect to the reference configuration and by considering the Dirichlet boundary condition, the weak form can be formulated as

$$\int_{\mathcal{X}} \rho\dot{\mathbf{v}} \cdot \eta dV + \int_{\mathcal{X}} \underline{\mathbf{P}} \cdot \tilde{\nabla}\eta dV = \int_{\mathcal{X}} \rho\mathbf{b} \cdot \eta dV + \int_{\partial\mathcal{X}} \rho\bar{\mathbf{t}} \cdot \eta dA. \tag{5}$$

It is necessary to point out that $\underline{\mathbf{P}}$ is the first Piola-Kirchhoff stress tensor with respect to the reference configuration and yields the relation with the Cauchy stress tensor $\underline{\mathbf{P}} = J\underline{\sigma} \cdot \underline{\mathbf{F}}^{-T}$. The terms $\mathbf{b}$ and $\bar{\mathbf{t}}$ represent external volume loads and surface tractions, respectively. The first term on the left-hand side of Eq. (5) represents the inertia effects. Under the steady state rolling condition, it can be largely simplified within an ALE framework as

$$\int_{\mathcal{X}} \rho\dot{\mathbf{v}} \cdot \eta dV = -\int_{\mathcal{X}} \rho\mathbf{c} \cdot (\tilde{\nabla}\eta \cdot \mathbf{w}) dV + \int_{\partial\mathcal{X}} \rho\eta \cdot \mathbf{c}\mathbf{w} \cdot \hat{\mathbf{n}} dA, \tag{6}$$

where $\hat{\mathbf{n}}$ is the normal unit vector outwards the surface in the reference configuration. For natural boundaries, the surface traction term vanishes since $\hat{\mathbf{n}}$ is always perpendicular to the guiding velocity $\mathbf{w}$. Returning Eq. (6) to Eq. (5), the linearisation leads to a standard FEM implementation.

2.2 *Phase-field method*

The phase-field method has been developed recently for crack approximation. A continuous scalar quantity $d \in [0,1]$ is introduced as a degree of freedom to indicate if the continuum body is fractured ($d = 1$) or not ($d = 0$). The driving force of the fracture process is assumed to be the dissipated strain energy during deformation. Recalling the variational principle and regularization formulation proposed by Bourdin et al. (2008), the energy function for a quasi-static state is approximated as

$$\Psi_e \approx \int_{B} \left(g(d)\psi^{+}(\underline{\mathbf{F}}) + \psi^{-}(\underline{\mathbf{F}}) + \gamma(d, \nabla d) \right) dV. \tag{7}$$

Based on this description, the elastic energy is split into two parts and only the positive part is degradated by the function $g(d)$ as contribution to the evolution of fracture. The degradation function is defined as

$$g(d) = (1-d)^2 + k, \tag{8}$$

where the parameter k is called mobility number, which is a positive quantity and far less than 1, for example $k = 10^{-10}$. From the physical point of view, the parameter does not have any meaning. However, it can increase the numerical stability. The construction of these two energetic terms is assumed to be

$$\begin{cases} \psi^+ = f(J) \cdot \psi_{vol} + \psi_{iso}, \\ \psi^- = (1 - f(J)) \cdot \psi_{vol}, \end{cases} \tag{9}$$

where $f(J)$ is a function that controls the volumetric contribution. It equals 1 only if $J \geq 1$, otherwise, it is given as 0.

In this contribution, the constitutive characteristics of the rubbery material are described by the 8-Chain Arruda-Boyce model (1993). The volumetric and isochoric energy functions are defined as

$$\psi_{vol} = \kappa(J - ln\, J - 1) \tag{10}$$

and

$$\psi_{iso} = \mu N\left(\lambda_r \mathcal{L}^{-1}(\lambda_r) + ln \frac{\mathcal{L}^{-1}(\lambda_r)}{sinh \mathcal{L}^{-1}(\lambda_r)}\right), \tag{11}$$

respectively. Here, κ, μ and N are the material parameters.

The energetic function for a fracture surface is proposed by Miehe et al. (2010b)

$$\gamma = \frac{\mathcal{G}_c}{2l} \cdot \left(d^2 + l^2 \cdot |\nabla d|^2\right), \tag{12}$$

where $\mathcal{G}_c$ is the critical energy release rate to be determined based on experimental tests. l is a parameter defined as the internal length scale, which is used to describe the width of the cracks. For a vanishing length scale l, the crack surface consequently converges to a sharp crack. In numerical applications, the length scale is usually chosen to be twice the element's size in the potential fracture region. This not only can satisfy Γ-convergence but also gives a relatively sharp crack, shown in Fig. 1. It is noteworthy that ∇ represents the gradient with respect to the current configuration due to the finite deformation problem. Through variational principles, the minimization problem of Eq. (7) can be formulated, and the strong form of both the mechanical and the fracture responses is carried out as

$$Div\, \underline{\mathbf{P}} = 0, \tag{13}$$

which is described in the reference configuration, and

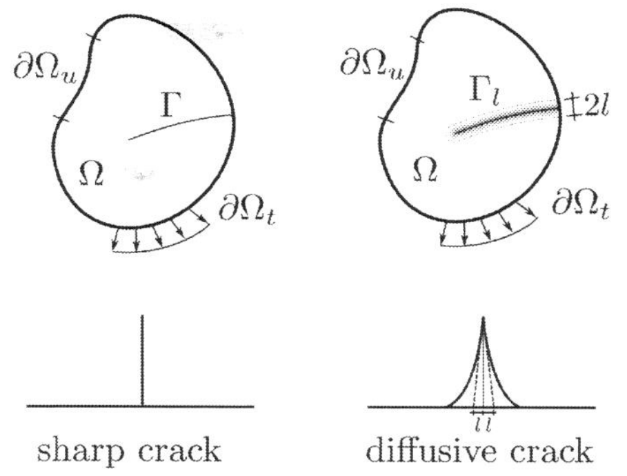

Figure 1. Sharp and diffusive crack topology.

$$g'(d) \cdot \mathcal{H} + \frac{\mathcal{G}_c}{l} \cdot d - \mathcal{G}_c l \Delta d = 0, \tag{14}$$

where $\mathcal{H}$ is defined by

$$\mathcal{H} = \max_{t \leq \tau} \psi^+ \tag{15}$$

to describe the maximum strain energy which is accounted from the initial to the current state. By this means, the initiation and propagation of cracks are numerically prescribed as irreversible processes. Considering the weak form and linearisation of Eq. (14), the FE implementation for the phase-field method can be achieved.

2.3 *Data exchange*

Considering the inelastic behaviour of a spinning tire under steady state condition, a general treatment is recommended by Le Tallec and Rahler (1994) by applying a streamline scheme for the approximation of the results. Usually, the integration points are considered to compose the streamlines, for example, see the work of Garcia et al. (2016) and Behnke and Kaliske (2015). However, in this contribution, the nodes are assumed to be the components of the streamlines. There are two important assumptions used to formulate this problem:

- the phase-field variables are fixed during the steady state rolling process,
- the phase-field evolution is driven by the decrease of fracture toughness of a tire during the steady state rolling.

For the first assumption, a constant value of phase-field is given to all nodes along a streamline. These values are measured during the phase-field evolution. The second assumption indicates that the phase-field evolves in another framework which is driven by the mechanical response of steady state rolling and a reduction of the fracture toughness parameter. A data transfer is generated between the phase-field and the mechanical loading processes.

Considering the axisymmetric model and in order to optimise computational resources, a 2D model, which is the cross-section of a tire model, is used for fracture analysis and a 3D model is proposed for the spinning simulation based on ALE formulation. The data transfer is illustrated in Fig. 2.

The mechanical analysis is started at a no fracture profile, which means that at the beginning the phase-field is given by 0 everywhere. When this analysis has been done, the mechanical solution is used as an input to the 2D fracture simulation. In this procedure, the displacement field is applied to the nodes along their streamline and kept constant. Afterwards, the phase-field equilibrium equations are solved for the cross-section of the tire. The phase-field values at this 2D model are projected to the full 3D tire model. The phase-field values are kept constant along each streamline due to the steady state rolling. Thus, a new fracture profile is prescribed for the mechanical analysis. This process will be iterated several times to reach a converged state for both frameworks.

There are two remarkable points to be noticed. The first one is that the displacement field is transferred from the 3D mechanical response to the 2D fracture simulation. In the phase-field formulation, not only energetic terms but also gradient terms with respect to the current configuration are required. In other cases, for instance, the thermomechanical analysis by Behnke and Kaliske (2015), only energetic terms are transferred since the 2D transient thermal response is analysed in the reference configuration. Another issue is related to the crack evolution in the 2D representative model. In this work, the cross-section in the contact region is considered since here the maximum strain energy is obtained. Therefore, cracks are assumed to be

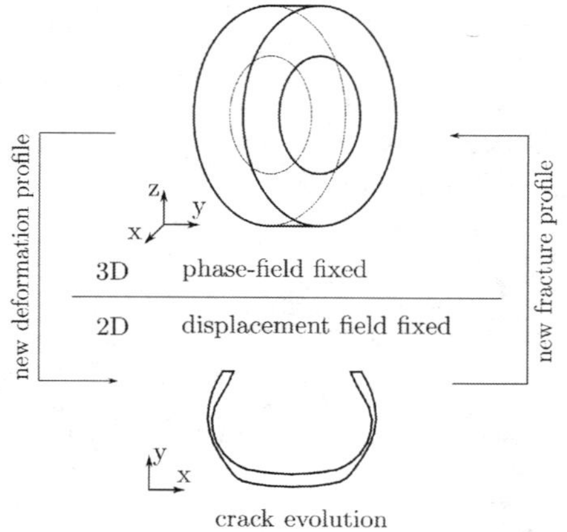

Figure 2. Data transfer between steady state rolling and fracture evolution.

initialised from the contact region and corresponding phase-field values are projected to the full 3D model along the streamlines.

3 NUMERICAL EXAMPLES

In this section, a representative example is investigated. A simplified model of a tire 175 SR14 is considered, which contains rubber, carcass, as well as steel belt reinforcement. The carcass and belts can be numerically simulated by membrane elements. But the element size is required to be much larger than the size of the rebar, see Fig. 3. In this example, fine meshes are required for the phase-field simulation. Hence, the rebar elements are approximated by standard solid elements, whose material parameters are characterised by parametric studies and a similar deformation is obtained compared to realistic results.

A realistic tire model shows unsymmetric properties due to the position and orientation of the reinforcement layers. Whereas in this example, due to the replacement of rebars by standard solid elements, a symmetrical tire model can be simplified. The model is discretized by 10540 solid elements with a refined mesh in the contact area. In the cross-section, the finer mesh is placed at the tip of the reinforcing belts, see Fig. 3, since experiments have shown that cracks are initiated in that region. The reinforcing belts are simulated by linear elastic material with material parameters as $E_c = 5 \cdot 10^9$ *MPa* and $\nu_c = 0.2$ for the carcass and $E_s = 8 \cdot 10^{10}$ *MPa* and $\nu_s = 0.2$ for the steel belts. The rubber compound is modelled by the Arruda-Boyce formulation and the parameters are $\kappa = 1.1667 \cdot 10^6$ *MPa*, $\mu = 2 \cdot 10^6$ *MPa*, as well as the segment number $N = 20$. For the 3D model, inner pressure is applied as $p = 0.2$ *MPa* at the beginning. Then, the tire undergoes contact with a rigid surface. It is noteworthy that friction is not considered yet in this example. Afterwards, the spinning velocity gradually increases up to $\Omega = 50$ *rad/s*. The density of rubber and reinforcement materials is given by $\rho_r = 1100$ kg/m^3, $\rho_c = 1500$ kg/m^3 and $\rho_s = 5900$ kg/m^3, respectively. When the inflation, contact and spinning processes are finished, the phase-field values are applied as new boundary conditions. After reaching a converged state, steady state rolling with a fixed fracture profile is achieved.

With respect to fracture evolution, the simulation is done based on the 2D model. The length scale is given by $l = 0.0017$ *m*. There are different factors which influence the crack propagation inside the rubber. In this contribution, it is assumed that crack propagation is driven by the decrease of the fracture toughness due to temperature field of the rubber during rolling of a tire. The reinforcing layer elements are not considered for damage. The fracture toughness of the rubber material is

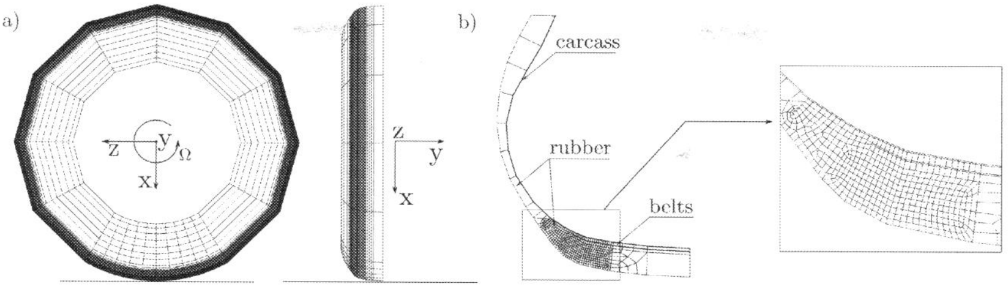

Figure 3. Discretization for tire 175SR14, a) 3D symmetric model for steady state rolling and b) 2D symmetric model for fracture.

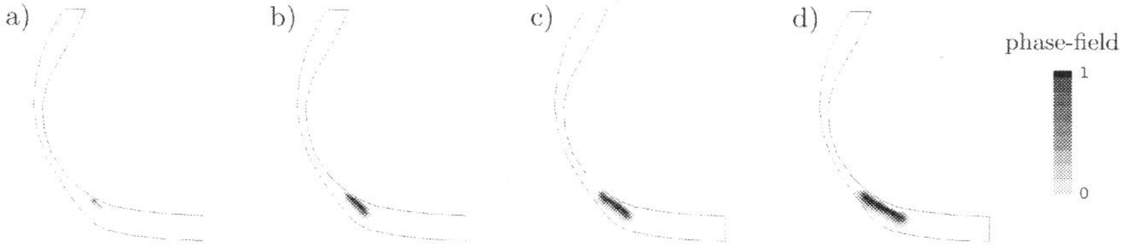

Figure 4. The evolution of fracture with different fracture toughness, a) $\mathcal{G}_c = 90\ kN/mm$, b) $\mathcal{G}_c = 80\ kN/mm$, c) $\mathcal{G}_c = 70\ kN/mm$ and d) $\mathcal{G}_c = 60\ kN/mm$.

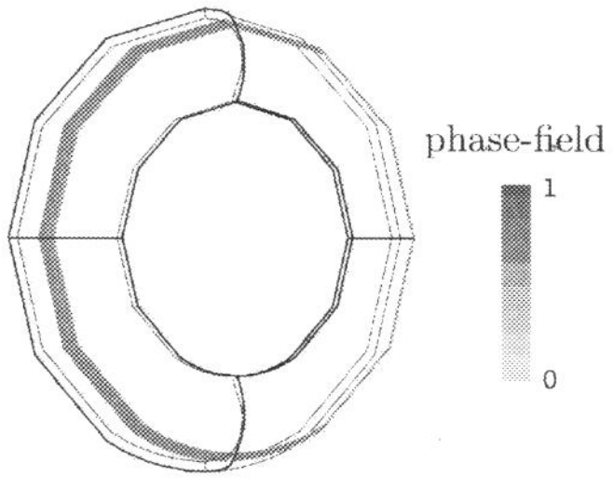

Figure 5. Steady state rolling of a 3D tire model with a fixed fracture profile $\mathcal{G}_c = 60\ kN/mm$.

initially given by $\mathcal{G}_c = 150\,kN/mm$ and, afterwards, it is gradually reduced. Another important issue is reformulating the strain energy contribution for the fracture. The phase-field is activated only if the element volume is expanded. Thus, fracture will not happen at a compressed state. Therefore, Eq. (9) is rewritten as

$$\begin{cases} \psi^+ = f(J)\cdot(\psi_{vol} + \psi_{iso}), \\ \psi^- = (1 - f(J))\cdot(\psi_{vol} + \psi_{iso}), \end{cases}$$

which may avoid damage at the tread in contact. In case of applying Eq. (9)) instead of Eq. (16), the isochoric energy at the tread region in contact will contribute to fracture, which cannot show a realistic response.

The 3D simulation indicates that the strain energy is concentrated at the tip of the reinforcement layers. Hence, the crack is initialised at that region, see Fig. 4a). Then, the fracture toughness is reduced slowly and the crack is propagated along the tensile direction, shown in Fig. 4b), c) and d). It is notable that several iterations are required between 2D and 3D simulations with respect to a certain value of the fracture toughness. For example, when $\mathcal{G}_c = 80\ kN/mm$, it demands more than 40 iterations to reach converged states between the mechanical and the phase-field equilibrium, which may consume plenty of computation resources. Therefore, the default iterations are restricted to 20 iterations. The 3D steady state rolling simulation with a constant fracture profile is presented for $\mathcal{G}_c = 60\ kN/mm$ as well in Fig. 5. It can be seen that once damage happens to the cross-section at the contact region, the whole streamline is damaged with the corresponding phase-field value.

The issue of convergence is necessary to be noticed. For the 2D simulation, both the mechanical and the phase-field equilibrium equations are solved by means of a staggered scheme. The phase-field simulation only requires 2 steps to reach a fully converged state, since it is a linear equation with respect to the phase-field d. Considering the 3D simulation, only the mechanical response is solved because the phase-field is introduced as a boundary condition to degradate the stress and material tangent tensors at each integration point. When $\mathcal{G}_c > 90\ kN/mm$, before crack initiation, it shows quadratic convergence and maximum 5 steps are required. However, while the crack starts to propagate, it requires more iteration steps but finally still shows quadratic converge. In some extreme cases, a huge reduction of the fracture toughness can result in an unconverged

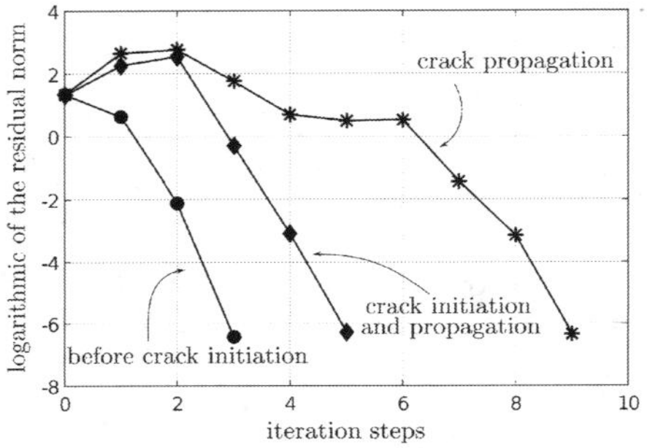

Figure 6. Iteration steps required for the 3D model by applying the phase-field at different stages.

solution. Therefore, in this example, the decrease of fracture toughness when the crack starts propagating, is controlled to be much smaller than in the case when the crack does not initiate.

4 CONCLUSIONS

First, an effective FE model of a tire is built and composite materials are considered. Rubber material is assumed to behave hyperelastically and fracture only happens inside the rubber. The rebar layers are discretized by standard solid elements.

Then, a consistent formulation for both fracture and steady state rolling is implemented. The inertia terms under steady state rolling are based on an ALE framework and fracture is simulated by means of the phase-field method. Both approaches are verified from the physical and numerical point of view.

Furthermore, the data exchange between the 2D and 3D simulations separate the ALE formulation and the phase-field method in their own frameworks. These two simulations do not influence each other directly and the inputs and outputs are applied to each other iteratively within a staggered strategy. The remarkable point is that both the steady state rolling and the fracture evolution need independent loading steps in their own simulation process. This scheme provides advantages that enhance the computation stability for each of them. Moreover, the use of streamlines satisfies the concept of steady state rolling and a constant damage along streamlines is obtained. Corresponding postprocessing has been implemented for the data transfer.

Last but not least, the capabilities of these formulations still need to be enhanced. For example, the ALE framework is only effective with a continuous tread pattern in the circumferential direction. By considering both the inelastic properties, for example, viscoelasticity and fracture evolution within an ALE framework, not only the integration point streamline approach but also the node streamlines are required for this analysis. Coupling between thermal, mechanical and fracture response can be discussed as well. Torsion failure between the rebar layers cannot be investigated at the current stage. These aspects will be included in future research.

REFERENCES

Amor, H., J.J. Marigo & C. Maurini (2009). Potential-flow instability theory and alluvial stream bed forms. *Journal of the Mechanics and Physics of Solids 57*, 1209–1229.

Arruda, E.M. & M.C. Boyce (1993). A three-dimensional constitutive model for the large stretch behavior of rubber elastic materials. *Journal of the Mechanics and Physics of Solids 41*, 389–412.

Behnke, R. & M. Kaliske. (2015). Thermo-mechanically coupled investigation of steady state rolling tires by numerical simulation and experiment. *International Journal of NonLinear Mechanics 68*, 101–131.

Bourdin, B. & A. Chambolle. (2000). Implementation of an adaptive finite-element approximation of the Mumford-Shah functional. *Numerische Mathematik 85*, 609–646.

Bourdin, B., G.A. Francfort & J.J. Marigo (2008). Regularized formulation of the variational brittle fracture with unilateral contact: Numerical experiments. *Journal of Elasticity 51*, 5–148.

Dal, H. & M. Kaliske (2009). Bergström-Boyce model for nonlinear finite rubber viscoelasticity: theoretical aspects and algorithmic treatment for the FE method. *Computational Mechanics 44*, 809–823.

Garcia, M.A., M. Kaliske, J. Wang & G. Bhashyam (2016). A consistent implementation of inelastic material models into steady state rolling. *Tire Science and Technology 44*, 174–190.

Kuhn, C. & R. Müller (2010). A continuum phase field model for fracture. *Engineering Fracture Mechanics 77*, 3625–3634.

Kuhn, C. & R.Müller (2011). A new finite element technique for a phase field model of brittle fracture. *Journal of Theoretical and Applied Mechanics 49*, 1115–1133.

Le Tallec, P. & C. Rahler (1994). Numerical models of steady rolling for non-linear viscoelastic structures in finite deformation. *International Journal for Numerical Methods in Engineering 37*, 1159–1186.

Miehe, C., F. Welschinger & M. Hofacker (2010b). Thermodynamically consistent phase field models for fracture: Variational principles and multi-field FE implementations. *International Journal for Numerical Methods in Engineering 83*, 1273–1311.

Miehe, C., M. Hofacker & F. Welschinger (2010a). A phase field model for rate-independent crack propagation: Robust algorithmic implementation based on operator splits. *Computer Methods in Applied Mechanics and Engineering 199*, 2765–2778.

Nackenhorst, U. (2004). The ALE-formulation of bodies in rolling contact: Theoretical foundations and finite element approach. *Computer Methods in Applied Mechanics and Engineering 193*, 4299–4322.

Constitutive Models for Rubber X – Lion & Johlitz (Eds)
© 2017 Taylor & Francis Group, London, ISBN 978-1-138-03001-5

Modelling and finite element analysis of cavitation and isochoric failure of hyperelastic adhesives

A. Nelson & A. Matzenmiller
Department of Mechanical Engineering, Institute of Mechanics, University of Kassel, Kassel, Germany

ABSTRACT: Finite element analysis is an established tool for the industrial design of adhesively bonded structures. Physical reliable predictions of the load capacity of bonded structures are only possible if suitable constitutive models including the failure are used for the adhesive. Test data of the adhesive to be modelled show rubber-like nonlinear elastic behaviour. Failure is experimentally found to be caused by two different mechanisms, depending on the geometrical shape of the adhesive layer and the load conditions. In order to describe the measured behaviour, a hyperelastic material model is chosen and extended with a damage model to take volumetric and isochoric failure into account. Based on test data, an identification procedure for the elastic and the failure parameters is described. To validate the model, the loading of a component-like specimen is simulated and the results are compared with the corresponding test data.

1 INTRODUCTION

The BETAFORCE 2850 of the DOW Chemical Company is a rubber-like adhesive which is used in a wide range of industrial areas, such as bonding of windshields and joining of structural components in automotive applications. This adhesive is suitable to compensate considerably different deformations of adherends due to its ability to withstand large elastic deformations. Depending on the application, failure can be caused by isochoric as well as volumetric deformations. That has to be taken into consideration properly with a constitutive model.

Developing constitutive models for the industrial application is a combined experimental, analytical and numerical task. The experimental part is regarded as the basis. In what follows, firstly, the experimental program is explained and the finite element representation of used specimens is described. Subsequently, the constitutive equations for the adhesive are shown and motivated. Then, the model parameters are identified. Finally, verification and validation analyses are carried out with the finite element code LS-DYNA using an explicit time integration scheme.

2 TESTS AND FE-MODELLING

The quasistatic behaviour of four specimens is investigate with displacement-controlled tests which finite element models are depicted in Figure 1. All adherends consist of steel and are modelled with linear elastic material. The displacement boundary conditions are imposed through either bolts or clamps, which both are modelled as rigid bodies. One of the rigid bodies is fixed and the other one is used to prescribe the boundary condition in z-direction. The discretization of the adhesive layers is as equal as possible for all specimens, i.e., in-layer element edges are 1 mm and over-thickness element size is 0.5 mm. The equal

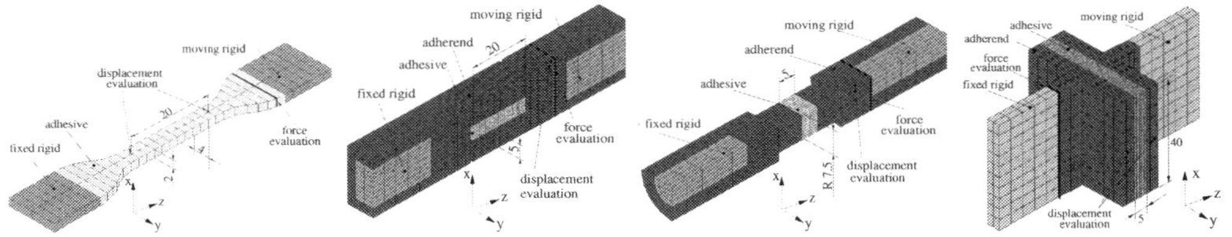

Figure 1. Discretization and dimensioning of (from left to right) the uniaxial tension specimen, the thick adherend shear specimen (half specimen is used), the buttjoint specimen (quater specimen is used) and the T-joint specimen.

discretisation is a deliberate choice to account for (damage) localisation phenomena.

2.1 *Tests with isochoric deformations*

The uniaxial tensile specimen and the thick adherend shear specimen are used for the idenfication of the elasticity constants and the failure parameters of the isochoric part of the material model. The tensile specimen is standardised in DIN 53504 (shape S) and is designed to create a homogeneous tensile stress state in the gauge. Its geometry enables a transverse contraction in the gauge and thereby reduces significantly the hydrostatic pressure so that the influence of the bulk modulus on the stress state is negligible. The deformation is described with the deformation gradient using the transverse nominal strain ε_t and the lateral nominal strain ε

$$\mathbf{F}^{\text{tensile}} = (1-\varepsilon_t)\mathbf{e}_x \otimes \mathbf{e}_x + (1-\varepsilon_t)\mathbf{e}_y \otimes \mathbf{e}_y + (1+\varepsilon)\mathbf{e}_z \otimes \mathbf{e}_z, \tag{1}$$

where $\mathbf{e}_i \otimes \mathbf{e}_j, i,j = x,y,z$ are the unity dyads with respect to the depicted coordinate systems.

The thick adherend shear specimen according to DIN EN 14869-2 is tested with two adhesive layer geometries with equal ratios of overlap length to thickness. One test is conducted for an adhesive layer with an area of 20 × 20 mm and a thickness of 5 mm and the second one with an area of 20 × 8 mm and a thickness of 2 mm. Symmetry is exploited to increase the computational efficiency. Assuming a homogeneous simple shear deformation and using κ as the tangent of the shear angle, the deformation gradient

$$\mathbf{F}^{\text{shear}} = \mathbf{e}_x \otimes \mathbf{e}_x + \mathbf{e}_y \otimes \mathbf{e}_y + \mathbf{e}_z \otimes \mathbf{e}_z + \kappa \mathbf{e}_y \otimes \mathbf{e}_z \tag{2}$$

describes the kinematics analytically.

2.2 *Tests with volumetric deformations*

The buttjoint specimen according to DIN EN 26922 is used to identify the bulk modulus and the failure parameters of the volumetric part of the damage model. The "poker-chip" shaped adhesive layer is tested with a diameter of 15 mm and thicknesses of 1 mm, 2 mm, and 5 mm. Computational efficiency is increased by exploiting symmetry and simulating one quarter of the full geometry. The deformation of the adhesive is inhomogeneous due to the transverse contraction and the geometrical constraint caused by the adherends. Thus, the stress state is a combination of an isochoric and an volumetric part, which contributions depend on the adhesive layer thickness. This is a crucial point which will be exploited for the inverse identification of the bulk modulus in section 4.1.

2.3 *Test of a component-like specimen*

The T-joint specimen is considered as a component-like specimen and therefore is used to validate the constitutive model. The cuboidal adhesive layer has an area of 40 × 40 mm and a thickness of 5 mm.

3 EXPERIMENTALLY BASED CONSTITUTIVE APPROACH

The adhesive is known to behave rubber-like so that the theory of hyperelasticity is applicable. However, there are still uncertainties regarding the amount of compressibility, i. e. the ratio of the bulk modulus to the shear modulus. Therefore, experimental and analytical studies are performed to confirm the assumption of nearly incompressibility. Based on that, a strain energy density for the hyperelastic model is chosen. Next, the failure behaviour of the adhesive is considered experimentally and numerically using test data and elastic FE-analyses. These results of the elastic analyses are exploited to develop a failure criterion.

3.1 *Consideration of the compressibility*

The compressibility behaviour is studied experimentally by measuring the lateral and the transverse displacements of the uniaxial tensile specimen under loading. Assuming isotropy, Figure 2 shows the transverse nominal strain plotted against the lateral nominal strain. With the incompressibility constraint $J = 1$, the kinematics of the

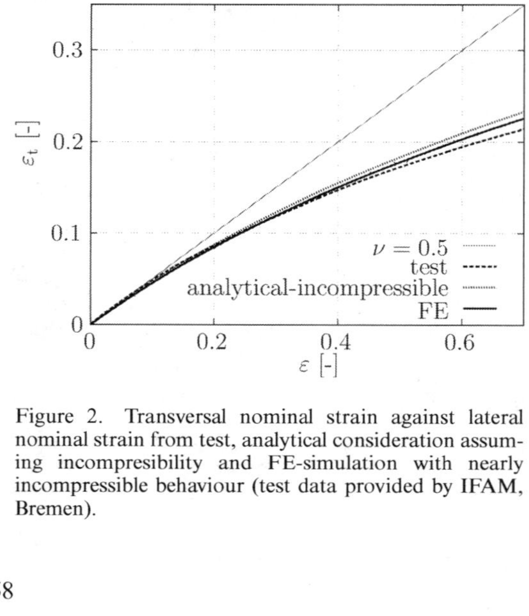

Figure 2. Transversal nominal strain against lateral nominal strain from test, analytical consideration assuming incompresibility and FE-simulation with nearly incompressible behaviour (test data provided by IFAM, Bremen).

uniaxial tensile specimen in equation (1) leads to the analytical expression

$$\varepsilon_t = 1 - \sqrt{\frac{1}{1+\varepsilon}}. \tag{3}$$

The small deviations between the analytical and test results confirm that nearly incompressibility is a reasonable assumption for the considered material.

3.2 *Hyperelastic model*

The model approach is based on the volumetric-isochoric split (Flory 1961) of the deformation gradient

$$\mathbf{F} = \bar{\mathbf{F}} \cdot \hat{\mathbf{F}}, \quad \bar{\mathbf{F}} = J^{-1/3}\mathbf{F}, \quad \hat{\mathbf{F}} = J^{1/3}\mathbf{1}, \tag{4}$$

In which $J = III_F = \det(\mathbf{F})$ is the volume ratio and $\mathbf{1}$ the second order identity tensor. The combination of equation (4) and the definition of the left CAUCHY-GREEN deformation tensor $\mathbf{B} = \mathbf{F} \cdot \mathbf{F}^{\mathrm{T}}$ defines the isochoric left CAUCHY-GREEN deformation tensor

$$\bar{\mathbf{B}} = J^{-2/3}\,\mathbf{B}, \tag{5}$$

which is used for the definition of the isochoric part of the strain energy density function. Applying the representation theorem for invariants, the strain energy density function for the nearly incompressible adhesive reads as

$$W(I_{\bar{B}}, II_{\bar{B}}, J) = W_{\mathrm{iso}}(I_{\bar{B}}, II_{\bar{B}}) + W_{\mathrm{vol}}(J), \tag{6}$$

where $I_{\bar{B}} = \mathrm{tr}(\bar{\mathbf{B}})$ and $II_{\bar{B}} = \frac{1}{2}\left(\mathrm{tr}(\bar{\mathbf{B}})^2 - \mathrm{tr}(\bar{\mathbf{B}}^2)\right)$ denote the main invariants. The MOONEY-RIVLIN model

$$W_{\mathrm{iso}}(I_{\bar{B}}, II_{\bar{B}}) = \tfrac{1}{2}c_{10}(I_{\bar{B}} - 3) + \tfrac{1}{2}c_{01}(II_{\bar{B}} - 3) \tag{7}$$

with the MOONEY-RIVLIN parameters c_{10} and c_{01} is chosen for the isochoric part. From the standpoint of industrial application, this is a reasonable compromise between accuracy and simplicity (and thus efficiency). The chosen ansatz for the volumetric part

$$W_{\mathrm{vol}}(J) = K(J - 1 - \ln J) \tag{8}$$

is adopted from (Miehe 1994) with the bulk modulus K. The effective CAUCHY stresses are calculated by the derivative of the strain energy density function with respect to the left CAUCHY-GREEN deformation tensor

$$\begin{aligned}\hat{\mathrm{T}} &= 2\frac{1}{J}\mathbf{B}\frac{dW(I_{\bar{B}}, II_{\bar{B}}, J)}{d\mathbf{B}} \\ &= \frac{1}{J}\left(\left(c_{10} + c_{01}I_{\bar{B}}\right)\bar{\mathbf{B}}^{\mathrm{D}} - c_{01}\left(\bar{\mathbf{B}}^{2}\right)^{\mathrm{D}}\right) \\ &\quad + \frac{1}{J}K(J-1)\mathbf{1},\end{aligned} \tag{9}$$

in which $\bar{\mathbf{B}}^{\mathrm{D}}$ is the deviatoric part of the left CAUCHY-GREEN deformation tensor.

3.3 *Occurence of cavitation*

Rubber-like materials can with stand high pressure under compressional load. Under dilatational load internal rupture known as cavitation occurs. Cavitation is nucleated by so called precursors in the interior of the material and forms enclosed cracks when a critical dilatational load is applied. These cracks tear open at the center, from where they expand. Thus, there is no possibility of visible detection from the outside. However, two measurable phenomena are associated with cavitation. The first is an audible cracking sounds. The second is a significant decrease of the bulk modulus, i.e., a sudden change in the load-extension response.

The experimental results of the buttjoint specimen indicate two failure mechanisms depending on the layer thickness t_a. For $t_a = 5$ mm, a crack starts to grow slowly from the outside at a relative displacement of 0.3. In contrast to that, no crack is observed at the outside if $t_a = 2$ mm and $t_a = 1$ mm. The separation of the adherends and thus the fracture of the adhesive appears suddenly. In Figure 3, results

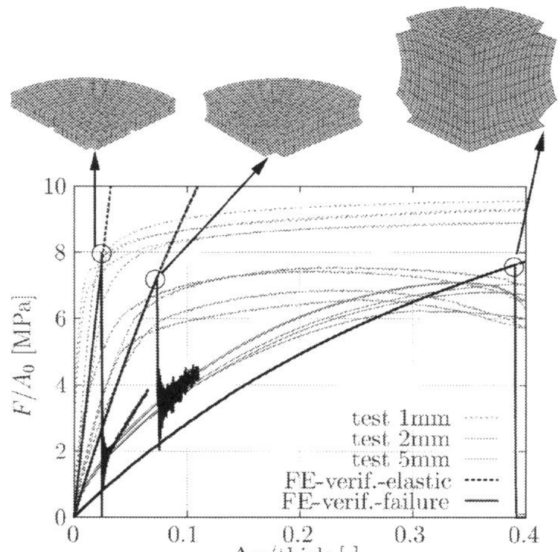

Figure 3. Verification results of the elastic range for the buttjoint specimen.

of the elastic FE-analyses are marked with dashed lines. The calculated curves are in the scatter range of the test data until about 6 MPa for $t_a=2$ mm and 7 MPa for $t_a=1$ mm. It is assumed, that the deviations are due to the occurrence of cavitation in the adhesive layers. These failure states are elaborated more concisely concerning the energy state at the corresponding critical locations. To this end, the isochoric, volumetric and total energies for 1 mm, 2 mm, and 5 mm are evaluated and plotted along the radial and axial directions in Figure 4. For $t_a=5$ mm the total energy at the outide, where failure starts, is almost equal to the isochoric energy. In both other cases, $t_a=2$ mm and $t_a=1$ mm, the total energy state at the center is dominated by the volumetric energy. Consequently, the volumetric energy is an adequate choice to model cavitation, whereas the isochoric energy is appropriate to model isochoric failure.

3.4 *Failure model*

The failure model is based on the continuum damage mechanical concept (Kachanov 1958)

$$\mathbf{T} = (1 - D)\hat{\mathbf{T}}, \tag{10}$$

with the damage variable D acting on the effective stresses according to equation (9). The chosen approach for the damage function is proposed in (Lemaitre 1985)

$$D = \begin{cases} 0, & f \le W_{\mathrm{fl}} \\ \left\langle \dfrac{f - W_{\mathrm{fl}}}{W_{\mathrm{fc}} - W_{\mathrm{fl}}} \right\rangle, & W_{\mathrm{fl}} < f < W_{\mathrm{fc}} \\ 1, & f \ge W_{\mathrm{fc}} \end{cases} \tag{11}$$

In equation (11) the argument is the equivalent energy

$$f(I_{\bar{B}}, II_{\bar{B}}, J) = W_{\mathrm{iso}}(I_{\bar{B}}, II_{\bar{B}}) + f_{\mathrm{k}} W_{\mathrm{vol}}(J) \tag{12}$$

which is driven by isochoric and volumetric energies. The model parameter f_k serves as a multiplier for the contribution of the volumetric energy to f. The mathematical form of the damage function implies the failure criterion

$$f(I_{\bar{B}}, II_{\bar{B}}, J) = W_{\mathrm{fl}} \tag{13}$$

and the fracture criterion

$$f(I_{\bar{B}}, II_{\bar{B}}, J) = W_{\mathrm{fc}}, \tag{14}$$

with W_{fl} and W_{fc} denoting the failure and the fracture energies. They both have to be determined based on test results. In the following, it will be shown that the constant approaches in equations (13) and (14) are unsuitable for the modelling of more than one loading condition. The shortcoming will be elaborated only for the failure criterion. However, the same arguments are applicable for the fracture criterion.

For the simple tension and the simple shear, the relations $II_{\bar{B}}(I_{\bar{B}})$ are derived from the corresponding deformation gradients and plotted as grey dotted curves in Figure 5, assuming isochoric deformations. Moreover, FE results of the uniaxial tensile specimen and the thick adherend shear specimen until the onset of failure (at the critical element) are plotted as thick dashed and solid lines. The FE-analyses are assumed to predict the elastic behaviour correctly, which will be addressed

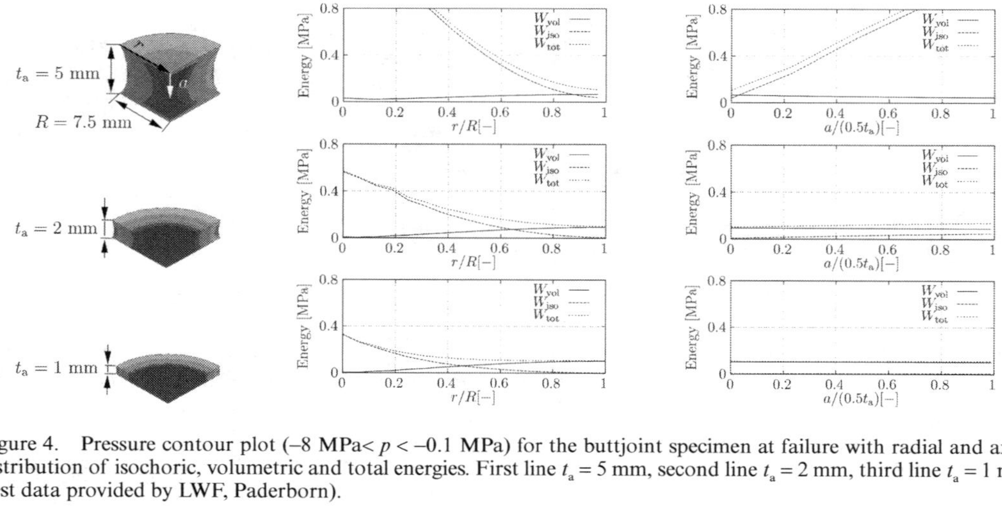

Figure 4. Pressure contour plot (–8 MPa< p < –0.1 MPa) for the buttjoint specimen at failure with radial and axial distribution of isochoric, volumetric and total energies. First line $t_a = 5$ mm, second line $t_a = 2$ mm, third line $t_a = 1$ mm (test data provided by LWF, Paderborn).

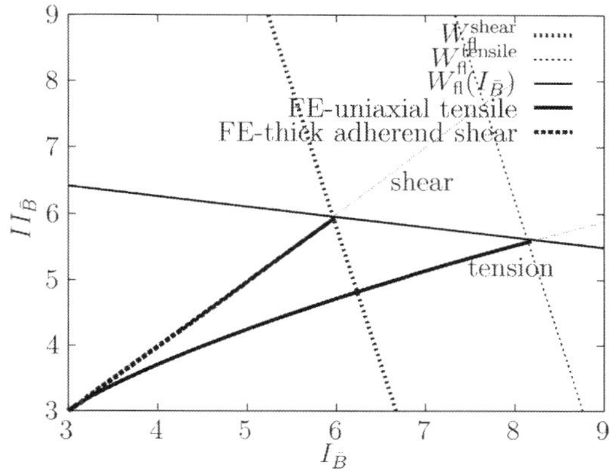

Figure 5. $II_{\bar{B}} - I_{\bar{B}}$ -space for isochoric deformations.

in section 4.2. Thus, the corresponding failure energy for tensile $W_{\rm fl}^{\rm tensile}$ MPa and for shear $W_{\rm fl}^{\rm shear}$ MPa can be extracted from the analyses. Next, the strain energy density from equation (7) and the failure criterion from (13) are inserted in (12) and the resulting expression is solved for $II_{\bar{B}}$ leading to

$$II_{\bar{B}} = \frac{2(W_{\rm fl} - f_{\rm k}W_{\rm vol}(J)) - c_{10}(I_{\bar{B}} - 3)}{c_{01}} + 3. \qquad (15)$$

Equation (15) is depicted in the $II_{\bar{B}} - (I_{\bar{B}})$ -space (see Figure 5) under the assumption of isochoric deformations as a thick dotted line for $W_{\rm fl} = W_{\rm fl}^{\rm shear}$ and as a thin dotted line for $W_{\rm fl} = W_{\rm fl}^{\rm tensile}$. The intersections of the "failure-lines" with the lines corresponding to the tensile load and the shear load show, that either the tensile load or the shear load is described adequately. A reasonable model has to meet both points. Therefore, the failure criterion $W_{\rm fl}$ and the fracture criterion $W_{\rm fc}$ are extended as

$$W_{\rm fl} = a_1 + a_2 I_{\bar{B}} \qquad (16)$$

and

$$W_{\rm fc} = b_1 + b_2 I_{\bar{B}}. \qquad (17)$$

The coefficients a_1, a_2, b_1 and b_2 are used to fit the curves in the $II_{\bar{B}} - I_{\bar{B}}$-space to test data. Practically, a_1 and a_2 in function (16) are computed via

$$\begin{bmatrix} 1 & I_{\rm fl}^{\rm tensile} \\ 1 & I_{\rm fl}^{\rm shear} \end{bmatrix} \begin{bmatrix} a_1 \\ a_2 \end{bmatrix} = \begin{bmatrix} W_{\rm fl}^{\rm tensile} \\ W_{\rm fl}^{\rm shear} \end{bmatrix} \qquad (18)$$

In which $W_{\rm fl}^{\rm tensile}$ and $W_{\rm fl}^{\rm shear}$ are the failure energies for tension and shear. $I_{\rm fl}^{\rm tensile}$ and $I_{\rm fl}^{\rm shear}$ are the corresponding first invariants. An analogue equation counts for the coefficients b_1 and b_2 by interchanging the indices $_{\rm fl}$ by $_{\rm fc}$ in equation (18). In summary, the model parameters are $W_{\rm fl}^{\rm tensile}$, $W_{\rm fl}^{\rm shear}$, $f_{\rm k}$, $W_{\rm fc}^{\rm tensile}$ and $W_{\rm fl}^{\rm shear}$, the last two of which serve for the description of post critical behaviour.

Two properties of the failure model shall be emphasised. Firstly, the choice of the approaches in equations (16) and (17) with exactly two coefficients is deliberate, given that the used test data basis consists of two tests. However, the polynomial form is not mandatory and is chosen due to its simplicity. Secondly, the failure function in equation (15) is derivable for other strain energy density functions than the MOONEY-RIVLIN model. However, the function will in general lose its linearity.

4 PARAMETER IDENTIFICATION AND VERIFICATION

The identification of the elastic model parameters and the failure parameters is performed in a consecutive order with test data of the specimens presented in sections 2.1 and 2.2.

4.1 *Identification of elastic parameters*

The three parameters c_{01}, c_{10} and K of the hyperelastic model (9) are identified in two steps using the elastic ranges of test data. The specimens of the uniaxial simple tensile test and the thick adherend shear test are suitable for the identification of the MOONEYRIVLIN parameters due to their isochoric deformations of the adhesive and consequently independence of the bulk modulus. At first, the shear modulus G is identified using the test data of the shear specimen, see Figure 6 (right). The Relation $G = c_{01} + c_{10}$ is exploited as a constraint condition for the inverse identification of the MOONEY-RIVLIN parameters c_{01} and c_{10} using the results from the uniaxial tensile test in Figure 6(center). The identified parameters are $c_{10} = 4.2$ MPa and $c_{01} = 1$ MPa. The bulk modulus K is identified inversely using the test results of the buttjoint specimen with different thicknesses of the adhesive layer. For a "poker-chip" shaped specimen the contribution of the hydrostatic stresses to the total stress state increases with increasing ratio of diameter to thickness. The critical value of the aspect ratio for a nearly incompressible material ($\nu = 0.49$) for which the stress is essentially hydrostatic is stated to be 15 (e.g. (Dorfmann, Fuller, & Ogden 2002)). The adhesive layers of the buttjoint specimen used in this work are 1 mm, 2 mm, and 5 mm which correspond to ratios of 15, 7:5 and

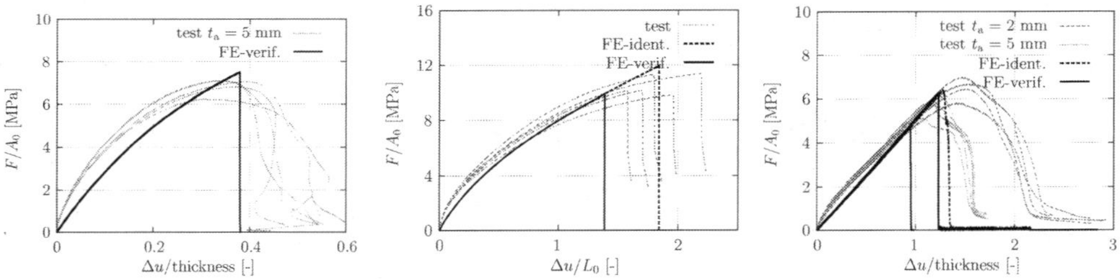

Figure 6. Verification results of buttjoint specimen, uniaxial tensile specimen and thick adherend shear specimen (test data provided by LWF, Paderborn).

3. In Figure 3 calculation results are compared with test data for the identified bulk modulus of $K = 1298$ MPa.

4.2 *Identification of failure parameters*

In this work the focus is on the computation of the failure. The post critical behaviour will be addressed in future investigations. The damage model in equation (11) reduces to a failure model by setting $W_{fc} = 1.01\ W_{fl}$. In this case the three parameters left to identify are the tensile failure energy $W_{fl}^{tensile}$, shear failure energy W_{fl}^{shear}, and the multiplier for the contribution of the volumetric energy f_k. The failure energies are identified inversely using test data of the uniaxial tensile test and the thick adherend shear specimen. To this end, FE-analyses of the uniaxial tensile specimen and the thick adherend shear specimen with 5 mm thickness are performed until failure occurs in a critical element and the invariant paths $II_{\bar{B}}(I_{\bar{B}})$ are evaluated as well as the corresponding energies $W_{fl}^{tensile} = 12.1$ MPa and $W_{fl}^{shear} = 7.7$ MPa. Evaluating equation (18) with these values, the unknown coefficients a_1 and a_2 are computed. The failure criterion is plotted in the $II_{\bar{B}} - I_{\bar{B}}$-space in Figure 5 as a thin solid line. In Figure 6 (center and right) the test data are compared to these simulation results (dashed black lines). Finally, the multiplier $f_k = 17$ is identified using the test data of the buttjoint specimen.

4.3 *Verification*

In Figure 6 the final set of parameters is verified by comparing results of verification simulations (solid, black lines) with corresponding test data of the 5 mm buttjoint test, the uniaxial tensile test and the thick adherend test. For increasing values of the multiplier f_k, the contribution of the volumetric energy to the equivalent energy f increases and therefore a premature failure can occur for isochoric deformations with very small volumetric contributions. In Figure 3 the multiplier f_k is verified quantitatively by the comparison of simulation results (dashed solid lines) with test data for the buttjoint specimen. Qualitatively, for $t_a = 2$ mm and $t_a = 1$ mm, cavitation is predicted correctly at the center of the adhesive layer. For $t_a = 5$ mm, isochoric failure starts at the outside of the adhesive layer.

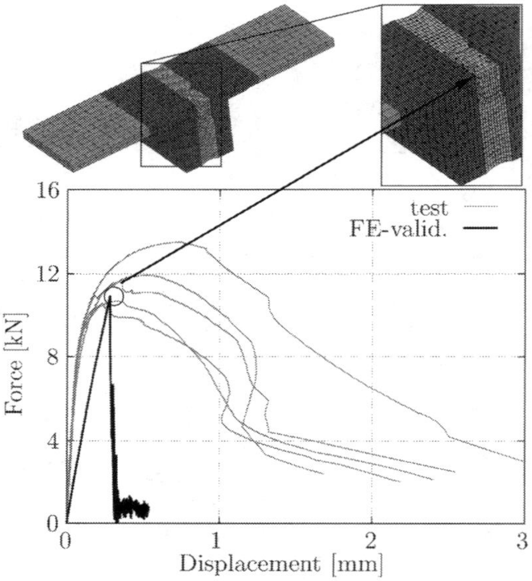

Figure 7. Validation results of the T-joint specimen (test data provided by IFAM, Bremen).

5 VALIDATION

The comparison between the numerical results and the test data from the uniaxial tensile specimens in Figure 2 can be regarded as a validation for the identified bulk modulus.

The T-joint specimen in Figure 1(right) is considered as a validation example for the failure model. In Figure 7 the measured data are compared to computed results. In the tests, no crack is detectable at the outside of the specimen, i.e. cavitation starts from the inside and leads to structural failure. Qualitatively, this observation can be confirmed numerically, where the failure starts at the center of the adhesive layer (see FE-model in Figure 7). Quantitatively, the maximum load capacity of about 11 kN is in good agreement with the test data.

6 CONCLUSIONS

For the industrially relevant rubber-like adhesive BETAFORCE 2850 of the DOW Chemical Company the hyperelastic MOONEY-RIVLIN model is extended with a damage model to account for cavitation and isochoric failure. The constitutive equations are implemented into the finite element code LS-DYNA. Using test data, the model parameters are identified and verification analyses are performed. The subsequent validation analysis of a component-like specimen confirms the ability of the model to predict the failure for industrial applications.

ACKNOWLEDGEMENTS

The IGF-project 18716 N/2 of the Forschungsvereinigung Stahlanwendung e.V. (FOSTA), Sohnstrae 65, 40237 Düsseldorf was promoted through the AiF under the program for the promotion of joint industrial (IGF) by the Federal Ministry of Economic Affairs and Energy due to a resolution of the German Bundestag. The authors want to thank the Laboratory for Materials and Joining Technology (LWF), Paderborn and Fraunhofer Institute for Manufacturing Technology and Advanced Materials (IFAM), Bremen for providing the test data.

REFERENCES

Dorfmann, A., K. Fuller, & R. Ogden (2002). Shear, compressive and dilatational response of rubberlike solids subject to cavitation damage. *INT J SOLIDS STRUCT 39*(7), 1845–1861.

Flory, P. J. (1961). Thermodynamic relations for high elastic materials. *Trans Faraday Soc 57*, 828–839.

Kachanov, L. M. (1958). Time of the rupture process under creep conditions. *Izvestia Akademii Nauk SSSR, Otdelenie Tekhnicheskich Nauk 8*, 26–31.

Lemaitre, J. (1985). A continuous damage mechanics model for ductile fracture. *J Eng Mater Techn 107*, 83–89.

Miehe, C. (1994). Aspects of the formulation and finite element implementation of large strain isotropic elasticity. *Int J Numer Meth Eng 37*, 1981–2004.

Constitutive Models for Rubber X – Lion & Johlitz (Eds)
© 2017 Taylor & Francis Group, London, ISBN 978-1-138-03001-5

The study of fatigue behavior of thermally aged rubber based on natural rubber and butadiene rubber

O. Kratina & R. Stoček
Centre of Polymer Systems, University Institute, Tomas Bata University in Zlín, Zlín, Czech Republic

B. Musil, M. Johlitz & A. Lion
Institute of Mechanics, Department of Aerospace Engineering, University of the Bundeswehr Munich, Neubiberg, Germany

ABSTRACT: This work is focused on investigation of the influence of thermal aging on fatigue behavior of carbon black filled rubber compounds which have been based on Natural Rubber (NR), Butadiene Rubber (BR) and their blend with ratio 50/50. The thermal aging was performed in thermal chamber at varied temperatures 30, 70 and 110°C for 720 hours. Firstly, the influence of thermal aging on mechanical behavior under quasi-static tensile test followed by Dynamic Mechanical Analysis (DMA) has been investigated. The fatigue behavior under sinusoidal waveform loading conditions has quantitatively been analyzed by using of dynamic testing equipment Tear and Fatigue Analyzer. The aim of this work was to investigate the influence of thermal aging on the fatigue behavior of rubber based on varied rubber types to understand the relationship between the thermal degradative processes occurred in rubber matrix under thermal aging and fatigue life. From the experimental work it was concluded based on all used testing methods, that the presence of BR rubber enhances the resistance against thermal aging and thus could be used as an efficient component reducing the aging degradation in rubber blend systems.

1 INTRODUCTION

Rubber is unique type of material due to its ability to undergo large elastic deformation. Due to this feature, rubber is used in wide range of application where rubber components are repeatedly loaded by shear or compressive stress. This category includes products such as conveyor belts, transmission belts, shock absorbers, tires, hoses etc. During application, mechanical fatigue which is represented by progressive weakening of physical properties as a result of slow crack growth occurs (Ellul, 2012). The fatigue failure of material begins by formation of cracks and following by their propagation to total rupture. The relationship between crack initiation and fatigue crack growth (FCG) in dependence of rubber type was demonstrated by Persson et al. (2005). It has been shown, that elastomers with ability to strain induced crystallization, e.g. natural rubber (NR) exhibit delayed FCG. On the other hand, non-strain crystallizing elastomers, e.g. butadiene rubber (BR), exhibit time-dependent FCG behavior. With the next increase of loading amplitude i.e. increase in tearing energy, the FCG characteristics of both elastomers analyses show difference FCG. It was observed from the measurement, that the initiation of cracks in BR starts at higher tearing energy than in NR. Whereas the FCG following crack initiation shows a counter phenomenon of more rapidly propagation of crack in BR in comparison to the rubber specimen based on NR. The results show the hypothetic relationship of crack initiation and propagation of rubber products based on nature or synthetic rubber.

Rubber products, which are loaded dynamically, heat up due to hysteresis losses. When rubber is used for a long time, it becomes thermal aging. Sufficiently elevated temperature exhibits synergistic effect on the reduction of the mechanical properties of rubber (Woo et al. 2011). Three different mechanisms of changing mechanical properties depending on temperature are visible. Moderately elevated temperature may lead to hardening of rubber due to exchange of crosslinks formation and additional crosslinking. At higher temperature, scission of crosslinks and subsequent softening of rubber occur. At very high temperatures, the molecular chains break down and charring and embrittlement of rubber are visible (Stevenson et al. 2012). Predictions and evaluations of fatigue life are key parameters for assure safety and reliability of rubber components (Frederick, 1982). In fact, the process of rubber degradation is very slow and it takes a long time to obtain rubber with degradation behaviors. Thus, accelerated aging tests are typically used to study degradation of rubbers in a comparatively short time (Jie Liu et al. 2016). Most of the efforts in this field of research were mainly concentrated to the observation of crack propagation or rupture of rubber test specimens under quasi-static loading conditions with no relationship to the fatigue behavior of the rubber matrix. Huang et al. studied the phenomenon of cyclic ageing on NR. They found that for NR, ageing at lower temperatures leads to a decrease in modulus, while at higher temperatures it leads to an increase in modulus. Bauer et al. studied the mechanical properties of skim tire based on Butadiene Rubber (BR) under oxidative ageing in the environment of 50/50 blend of N_2/O_2 at various temperatures in the range of 50°C to 70°C. They have observed a decrease in the crack resistance and an increase of the modulus with time in the whole range of temperature. Stoček et al. discussed the influence of thermal aging at 70 and 110°C of rubber compounds based on NR, SBR and their blend of 60–40 proportion on dynamic mechanical and FCG properties. They demonstrated that loss of compliance can effectively be used to detect the embrittlement or softening/hardening point of the aged rubber material. They also showed that the increase in aging temperature has deleterious effect on crack growth resistance independent on rubber materials.

In this paper, the influences of heat-aging on mechanical properties of NR and BR and their blend with ratio 50/50 were experimentally investigated. The rubber test specimens were aged in thermal chamber for 720 hours at 30, 70 and 110°C. In order to investigate heat-aging effects on mechanical properties, quasi-static tensile test, fatigue test and dynamic mechanical analysis (DMA) were performed.

2 EXPERIMENTAL PART

2.1 *Rubber test specimens*

Rubber materials used in this study have been based on natural rubber (SMR 20 CV/BP1, Lee Rubber Co. Pte Ltd., Malaysia) and butadiene rubber (Buna CB 24, Lanxess, Germany) and were filled with 50 phr of carbon black N339. Table 1 lists the formulation of rubber compounds used.

Rubber compounds were prepared by two steps mixing procedure. At first step the batch was mixed in an internal mixer SYD-2 L (Everplast, Taiwan) at rotators speed 50 rpm with temperature 80°C of mixing chamber. In second step, curing system consisting of accelerator and sulfur was added at the temperature 60°C on the two-roll mill. Curing properties were determined by moving vie

Table 1. Rubber compounds formulation.

	NR	NR/BR	BR
		phr	
Natural rubber	100,0	50,0	–
Butadiene rubber	–	50,0	100,0
Carbon black	50,0	50,0	50,0
Zinc oxide	3,0	3,0	3,0
Stearic acid	1,0	1,0	1,0
IPPD[1]	1,5	1,5	1,5
CBS[2]	2,5	2,5	2,5
Sulfur	1,7	1,7	1,7

1 N-isopropy-N'-phenyl-p-phenylen-diamine
2 n-cyklohexyl-2-benzo-thiazole-sulfenamide

rheometer MDR 3000 Basic (MonTech, Germany) according to ASTM 6204 at 160°C for 0,3 hr. and 0,50 arcs with 100 cpm frequency.

Curing of various shapes of specimens recommended for chosen characterization were performed in a heat press LaBEcon 300 (Fontijne Presses, Netherlands) at 160°C according to optimum curing time t_{90}. Prepared test specimens were additionally thermally aged at various temperatures 30, 70 and 110°C in air oven for 720 hours.

2.2 *Tensile test*

The thermal aging influence on strength in tension of rubber materials were characterized by quasi-static tensile test which was performed by machine M350–5CT (Testometric, United Kingdom). As test specimens the dumbbell shape type B according to ISO 37 was chosen, whereas the strain speed of 200 mm/min was applied. Maximal stress and elongation at break are given as results.

2.3 *Fatigue test*

Experimental analysis of mechanical fatigue was carried out using dynamic testing equipment Tear and Fatigue Analyzer (TFA) (Co. Coesfeld GmbH, Germany). The testing equipment consists of 3 individually powered electrical dynamic drives and each of them is able to analyze 3 test specimens simultaneously, thus it is possible to measure up to 9 test specimens during one analysis. Each upper clamp attachment of test specimens is fixed to the load cell and its corresponding test specimen clamp attachment is connected to a separate computer-controlled stepping motor which has ability to control amount of stress during dynamic strain controlled measurement. Fatigue test was performed using test specimens with rectangular geometry with the following dimensions: length between clamps 30 mm, width 10 mm and

Table 2. Applied testing conditions of fatigue.

Strain amplitude [%]	R ratio	Waveform	Frequency [Hz]
25	0	Sinusoidal	10

Table 3. Applied testing conditions of DMA.

Static strain [%]	Strain amplitude [%]	Waveform	Frequency [Hz]
3	0,3	Sinusoidal	0,1–10

thickness 2 mm. Used test conditions are given by Table 2. The number of cycles needed to total rupture of test specimen is given to be the resulting parameter. The level of stress required to achieve the set out displacement has also been monitored.

Materials marked by * were not measured until the rapture due to very high resistance against failure and thus highly time consuming measurement process.

2.4 *Dynamic mechanical analysis*

DMA was used to measure influence of thermal aging on viscoelastic behavior. Evolution of storage modulus and loss modulus as a function of frequency at room temperature is given as results. Characterization of viscoelastic properties in tension were performed by EPLEXOR® DMA (GABO, Germany). Test conditions are given in the Table 3.

3 RESULTS AND DISCUSSION

3.1 *Tensile test*

The results of tensile tests are shown in Figures 1–3. On the basis of these figures it is visible that mechanical properties represented by stress as well as elongation at break have a decreasing tendency with increasing aging temperature. Furthermore, it is seen that BR exhibits a lower value of stress and elongation at break in comparison with the NR as well as with the blend based on NR/BR. On the other hand, the decreasing tendency of the slopes each rubber, demonstrate higher resistance of BR against thermal aging in comparison to NR and to the blend NR/BR.

3.2 *Fatigue test*

Fatigue behaviors of analyzed materials are interpreted by course of stress depending on the number of cycles of loading. Figure 4 shows results of fatigue behavior of rubber materials aged at 30°C.

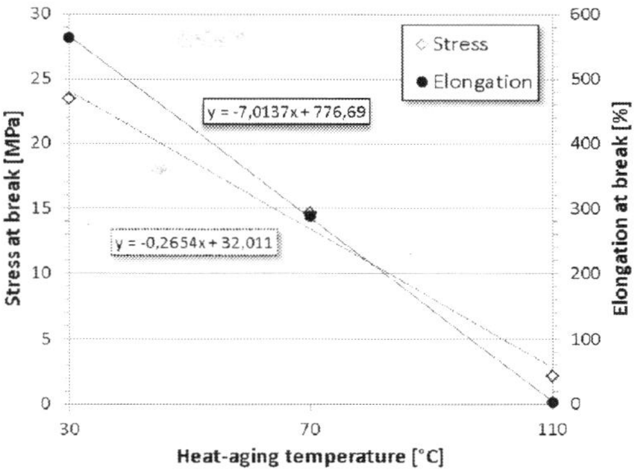

Figure 1. Stress at break at quasi-static tension in dependence on thermal aging of NR.

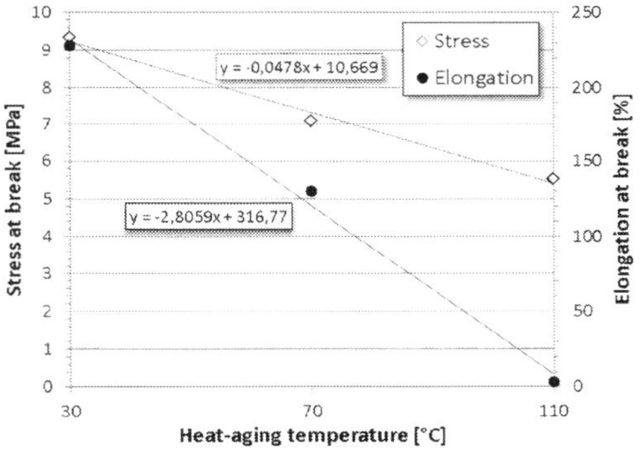

Figure 2. Stress at break at quasi-static tension in dependence on thermal aging of BR.

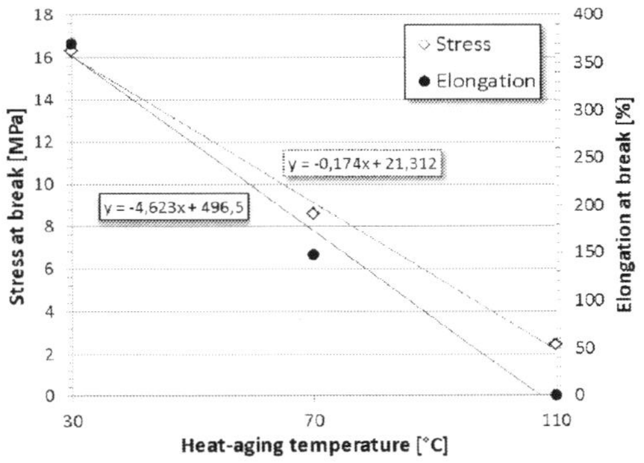

Figure 3. Stress at break at quasi-static tension in dependence on thermal aging of NR/BR.

From this graph it is visible; that rubber based on pure BR exhibits the higher fatigue resistance in comparison with NR and NR/BR. This observation is in accordance with the well-known fact that

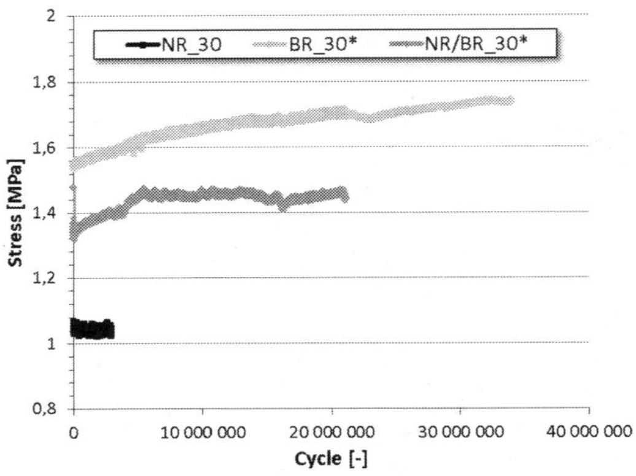

Figure 4. Fatigue behavior of all rubber materials aged at 30°C.

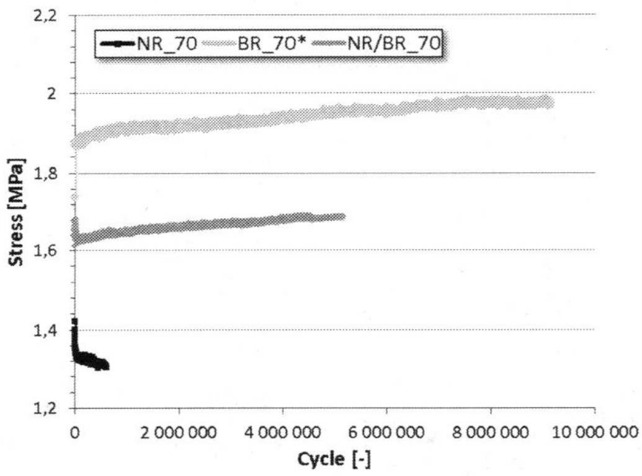

Figure 5. Fatigue behavior of all rubber materials aged at 70°C.

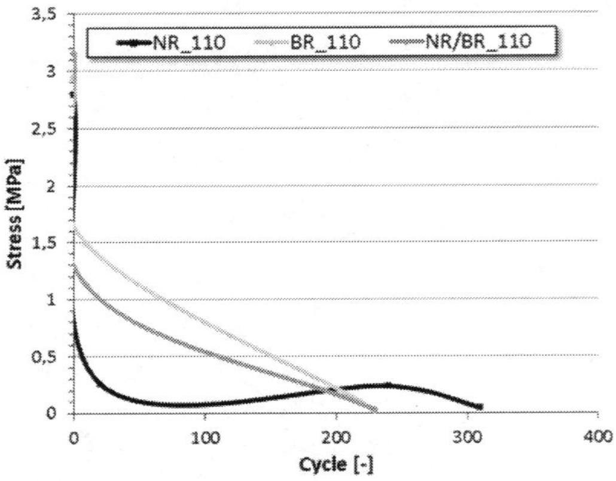

Figure 6. Fatigue behavior of all rubber materials aged at 110°C.

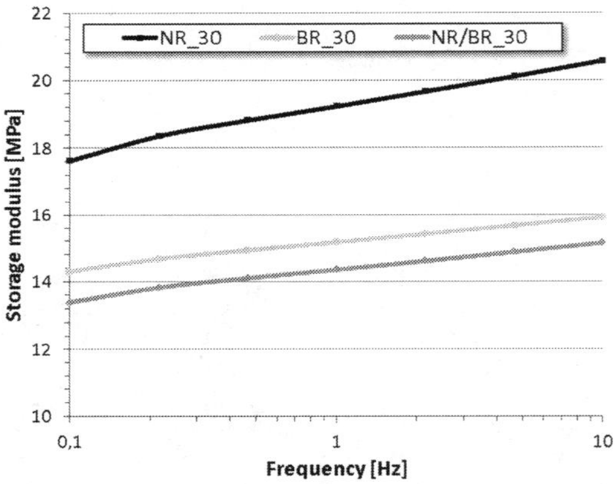

Figure 7. The dependence of storage modulus on frequency for all rubber materials aged at 30°C.

the BR exhibits high resistance against crack initiation. Furthermore, it is seen that NR exhibited decline of stress in dependence on number of cycles due to breakage and re-aggregation of damaged polymer-filler bonds of relatively soft clusters of fillers (Lorenz, 2012), whereas materials containing BR exhibited increase of stress in dependence on number of cycles at constant displacement.

In Figure 5, there is visible the course of stress during fatigue of rubbers aged at 70°C. From these fatigue curves it is evident that number of cycles necessary to rupture material is sufficiently lower than for rubbers aged at 30°C. Simultaneously, the stress necessary to reach set out strain exhibits higher value due to hardening caused by exchange of crosslinks formation and additional crosslinking.

The worst levels of mechanical properties have been observed for the rubbers aged at 110°C. Rubber matrices exhibit significant rate of degradation that is the reason for the loss of elasticity. Thus this enormous aging causes the brittleness of rubber materials. Moreover, the formation of several cracks has been visually observed on the surface of test specimens.

3.3 *Dynamic mechanical analysis*

Changes in viscoelastic behavior of analyzed rubber influenced by thermal aging were characterized using DMA. Figures 7 and 8 show the function of storage modulus of rubber in dependence on frequency at chosen temperature of thermal aging. It is seen that storage modulus increases with increasing frequency due to declining time of deformation of molecular chains. The highest values of storage modulus were exhibited for NR. This trend is most probably caused by strain induced crystal-

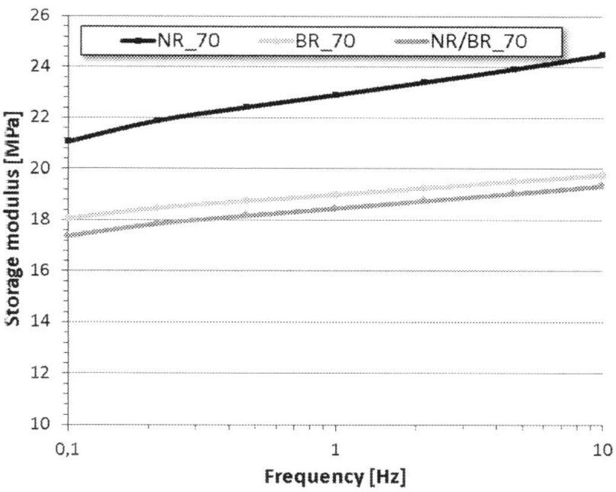

Figure 8. The dependence of storage modulus on frequency for all rubber materials aged at 70°C.

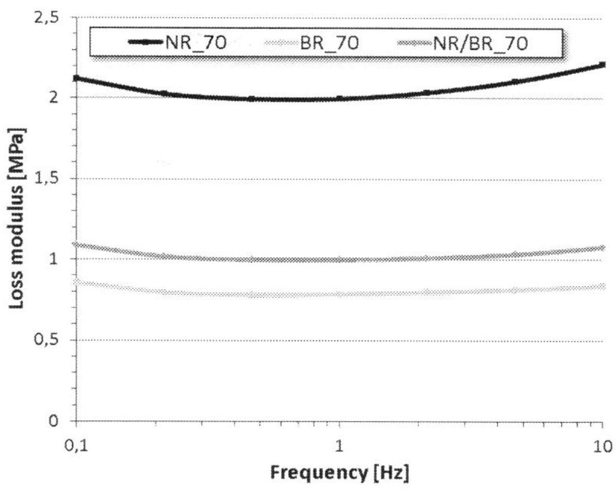

Figure 10. The dependence of storage modulus on frequency for all rubber materials aged at 70°C.

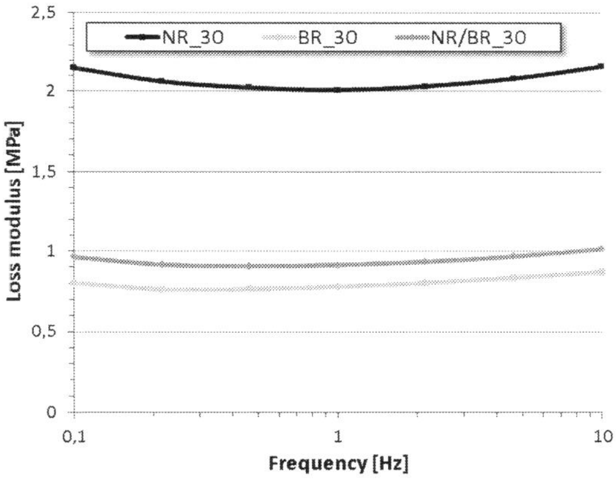

Figure 9. The dependence of loss modulus on frequency for all rubber materials aged at 30°C.

lization. Saintier et al. (2011) explained the effect of R ratio on mechanical behavior of NR based material under cyclic loading. It was argued that at zero R ratio, strain induced crystallization also is necessary to be taken in to the account, however it melts at each cycle., While at the application of the constant pre-load on the specimen, the R Ration became positive and the preventing of melting of crystalized zone become apparent. From the figures the influence if frequency in storage modulus is visible.

Rubber aged at 70°C exhibited the same trend as less aged materials but with difference that storage modulus reached higher values due to hardening causing by reorganization of chains net in rubber matrix due to thermal aging.

Figures 9 and 10 illustrate course of loss modulus representing viscous behavior of rubber in dependence on frequency and elastomer composition for aging temperature 30 and 70°C. From these results is visible, that presence of NR increases value of loss modulus. Influence of thermal aging is almost negligible.

It was not possible to perform the DMA characterization of rubber materials aged at 110°C because of their high degree of degradation. In the fact the specimens ruptured immediately during the fixing process.

4 CONCLUSION

Carbon black filled rubber compounds based on NR and BR were prepared for a purpose of description of thermal aging effect on mechanical properties. Materials which were aged for 720 hour at 30, 70 and 110°C were subjected to different types of tension loading. The quasi-static tensile test was used to describe of thermal aging effect on strength and elongation of each rubber compound. Further, the rubber materials were subjected to a fatigue tensile loading using special testing equipment Tear and Fatigue Analyzer. Finally, DMA was used to describe the influence of thermal aging on viscoelastic behavior of rubber.

The degradation of mechanical behavior in dependence on increased aging temperature of all analyzed rubbers has been observed at quasi-static tensile test. From the fatigue analyses it is evident that the thermal aging leads to significant decreasing of rubber fatigue life, which is cased due to recombination molecular network. DMA revealed that thermal aging has essential influence only on storage modulus representing elasticity of rubber.

Finally, it is evident, that the presence of BR rubber enhances the resistance against thermal aging and thus could be used as an efficient component reducing the aging degradation in rubber blend systems.

ACKNOWLEDGEMENTS

This article was written with the support of Operational Program Research and Development for Innovations co-funded by the European Regional Development Fund (ERDF) and national budget of the Czech Republic, within the framework of the project CPS—strengthening research capacity (reg. number: CZ.1.05/2.1.00/19.0409) as well supported by the Ministry of Education, Youth and Sports of the Czech Republic—Program NPU I (LO1504) and by the joint project of the Bavarian State Ministry of Education, Science and the Arts and the Ministry of Education, Youth and Sports of the Czech Republic no. 8E15B007.

REFERENCES

Bauer, D.R., Baldwin, J.M., Ellwood, K.R., 2007, Rubber aging in tires. Part 2: Accelerated oven aging tests, Polymer Degradation and Stability 92, 110–117.

Ellul, M.D., 2012, Mechanical Fatigue. In: Gent AN. Engineering with Rubber—How to Design Rubber Components. 3rd ed.: Hanser Publishers, 139–176.

Frederick, R.E., 1982, Science and technology of Rubber, Rubber Division of American Chemical Society.

Huang, D., LaCount, B.J., Castro, J.M., Ignatz-Hoover, F., 2001, Development of a service-simulating, accelerated aging test method for exterior tire rubber compounds I. Cyclic aging, Polymer Degradation and Stability 74, 353–362.

Liu, J., Li, X., Xu, L., Zhang, P., 2016, Investigation of aging behavior and mechanism of Nitrile-Butadiene Rubber (NBR) in the accelerated thermal aging environment. Polymer Testing, 54, 59–66.

Lorenz, H., Kluppel, M., 2012. Microstructure-based Modelling of Arbitrary Deformation Histories of Filler-reinforced Elastomers, Journal of Mechanics and Physics of Solids, 60, 1842.

Persson, B.N.J, et al., 2005, Crack propagation in rubber-like materials. Journal of Physics: Condensed Matter, 17, 1071–1142.

Saintier, N, Cailletaud G, Piques, R., 2011, Cyclic loadings and crystallization of natural rubber: an explanation of fatigue crack propagation reinforcement under a positive loading ratio. Materials Science and Engineering, 528, 1078–1086.

Stevenson, A., Campion R., 2012, Durabiliy. In: Gent AN. Engineering with Rubber—How to Design Rubber Components. 3rd ed.: Hanser Publishers, 139–176.

Stoček, R., Kratina, O., Ghosh, P., Maláč, J., Mukhopadhyay, R., 2017, Influence of thermal ageing process on the crack propagation of rubber used for tire application. In: W. Grellmann, B. Langer: Deformation and Fracture Behaviour of Polymer Materials. Springer, 305–316.

Woo, Ch.S., Park, H.S., 2011, Useful Lifetime Prediction of Rubber Component, Engineering Failure Analysis, 18, 1645–1651.

Constitutive Models for Rubber X – Lion & Johlitz (Eds)
© 2017 Taylor & Francis Group, London, ISBN 978-1-138-03001-5

Characterization of ageing effect on the intrinsic strength of NR, BR and NR/BR blends

R. Stoček
PRL Polymer Research Lab., s.r.o., Zlín, Czech Republic
Centre of Polymer Systems, University Institute, Tomas Bata University in Zlín, Zlín, Czech Republic

W.V. Mars
Endurica LLC, Findlay, Ohio, USA

O. Kratina, A. Machů, M. Drobilík, O. Kotula & A. Cmarová
Centre of Polymer Systems, University Institute, Tomas Bata University in Zlín, Zlín, Czech Republic

ABSTRACT: The intrinsic strengths for carbon black filled (50 phr) rubber materials based on Natural Rubber (NR), Butadiene Rubber (BR) and NR/BR blends in the volume ratio 50/50 have been evaluated for unaged material, and material aged at 50°C for 720 hours. The measurement was based on quasi-static tension loading on an edge-cracked pure shear specimen, combined with frictionless cutting via a sharp blade. The cutting force and pre-stress parameters were varied automatically using the Intrinsic Strength Analyzer (ISA©) instrument manufactured by Coesfeld. It was observed that both stiffness and the intrinsic cutting energy increase with aging. The cutting measurement provides a convenient approach for estimating fatigue threshold, a measurement with otherwise might require much longer testing periods.

1 INTRODUCTION

The strength of rubber depends on its chemical structure, as well as on the viscoelastic behavior occurring in crack near-tip fields (Lake et al. 1987, Bhowmick 1988). Due to viscoelastic behavior, the total energy required to propagate a crack in rubber is usually significantly greater than the surface energy associated with the intrinsic strength of the molecular structure. If a crack propagates in cured rubber, then all chains crossing the plane of the crack must rupture. There is a minimum energy requirement for the rupture of these chains. This minimum energy depends on the crosslink-to-crosslink length of a polymer chain, and on the weakest bond in the main polymer chain (Lake et al. 1967). It is independent of time, temperature and degree of swelling. It is therefore often called the intrinsic strength, since it reflects only the polymer chemistry and network (Gent and Mars 2013). The intrinsic strength has been associated (Lake and Lindley 1966 and Andrews 1963) with the lower fatigue crack growth limit, which is useful in design and in fatigue analysis (Mars 2007). A crack operating at an energy release rate below the intrinsic strength will operate indefinitely without growing, since there is not enough energy supplied to break polymer chains at the crack tip (Mars et al. 2016).

Rubber components are frequently subjected to thermal aging, either due to internal heat generation, or due to externally imposed thermal gradients. For instance, in a tire, as reported in Schuring et al. (1981), tread and sidewall compounds may experience temperatures of about 100°C and 60°C respectively. Several researchers e.g. Huang (2001), Celina (2000), have studied the thermal aging characteristics of rubber. Studies of aging behavior using dynamic mechanical analysis can also be found in earlier publications e.g. Payne (1962 & 1965), Lion (1999) or Baldwin et al. (2007). However, investigations on the relationship between ageing and fatigue crack growth is not common. Stoček et al. (2017) studied the influence of thermal aging at 70 and 110°C on natural rubber (NR), styrene butadiene rubber (SBR) and a 60–40 blend. He studied dynamic mechanical and fatigue crack growth properties.

The aim of this work is to determine the influence of thermal aging on the intrinsic strength of rubber for Natural Rubber, Butadiene Rubber, and a blend. The work utilizes a new commercial system, manufactured by Coesfeld GmbH, Germany, and operated with testing methodology developed by

Endurica LLC, Ohio, USA. The system uses the theoretical background and cutting method of Lake & Yeoh (1978) to estimate the intrinsic strength.

2 THEORETICAL BACKGOUND

The theory is based on the analysis of rubber cutting resistance proposed originally by Lake & Yeoh (1978). When crack tip dissipation is small, the intrinsic cutting energy *Sc* for a strained planar tension specimen undergoing a cutting process may be written as the sum of the individual energy release rates for tearing and cutting, as follows:

$$S_C = T + F, \tag{1}$$

S_c is the intrinsic cutting energy that is to be determined via measurement. T is the measured tearing energy, and F is the measured cutting energy. For an edge-cracked planar tension specimen under strain, with no blade, the tearing energy is computed as the parameter resulting from the strain energy density W and the unstrained specimen gauge height L_0.

$$T = W \cdot L_0. \tag{2}$$

During cutting by a blade, the moving force f required to maintain a constant rate of cutting does work, imparting an additional contribution to the total energy release rate driving the crack tip. We call this the cutting energy F, and its value is given by

$$F = f/t, \tag{3}$$

where the t is the thickness of the specimen.

3 MATERIAL AND EXPERIMENTAL

3.1 *Rubber composition and test specimen*

Materials in this study were based on natural rubber (SMR CV 60) and butadiene rubber (Buna CB 24, Lanxess, Germany) filled with 50 phr of carbon black N339. Composition is given in Table 1.

Rubber compounds were prepared by a 2-step process. In step 1, the batch was mixed in an internal mixer SYD-2 L (Everplast, Taiwan) at 50 rpm and temperature 80°C. In step 2, the curing system was added at a mix temperature of 60°C on a two-roll mill. Cure properties were determined via Moving Die Rheometer MDR 3000 Basic (MonTech, Germany) according to ASTM 6204 at 160°C.

The pure-shear test specimen geometry is presented in Figure 1. Specimens were cured in a heat press LaBEcon 300 (Fontijne Presses, Netherlands) at 160°C according to the optimum curing time t90. Prepared test specimens were used without ageing and additionally thermally aged at the temperature 50°C in air oven for 720 hours.

Table 1. Rubber composition.

	NR	NR/BR	BR
		phr	
Natural rubber	100,0	50,0	–
Butadiene rubber	–	50,0	100,0
Carbon black	50,0	50,0	50,0
Zinc oxide	3,0	3,0	3,0
Stearic acid	1,0	1,0	1,0
IPPD[1]	1,5	1,5	1,5
CBS[2]	2,5	2,5	2,5
Sulfur	1,7	1,7	1,7

[1]N-isopropy-N'-phenyl-p-phenylen-diamine.
[2]n-cyklohexyl-2-benzo-thiazole-sulfenamide.

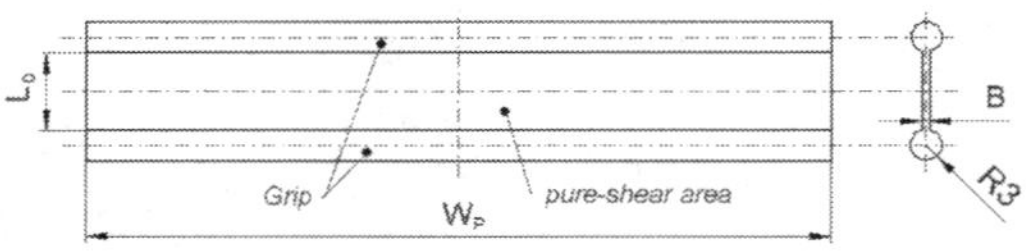

Figure 1. Test specimen geometry, where L_0 = 10 mm, W_P = 100 mm and B = 1,5 mm.

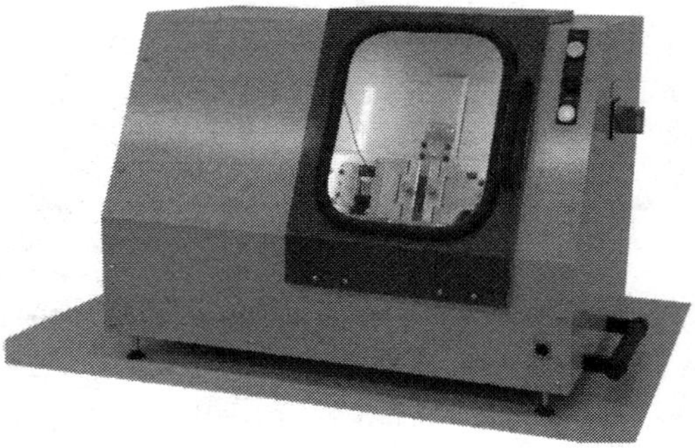

Figure 2. Intrinsic strength analyzer Test specimen geometry, where L_0 = 10 mm, W_P = 100 mm and B = 1,5 mm.

3.2 *Methodology and equipment*

The mechanical principle of the Intrinsic strength Analyzer—ISA© (Coesfeld GmbH, Germany) shown in Figure 2 used for the analyses, is illustrated in Figure 3.

The test sample is clamped into a tensile measuring station, which is instrumented to measure and control force and strain in the Y-Axis. The tensile set-up is symmetrical, meaning that the center of the

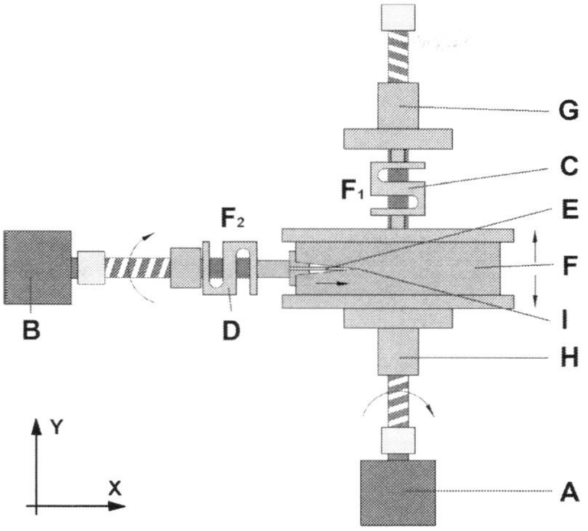

Figure 3. Measurement principle, where: A—actuator of the axis Y; B—actuator of the axis X; C—loading cell of the axis X; D—loading cell of the axis Y; E—razor blade; F—test specimen; G—upper clamping system of test specimen; H—bottom clamping system of test specimen, I—razor blade tip.

sample remains at initial position during variation of stress and/or strain. In this stable center axis a second actuator is positioned, which is also instrumented to measure and control force and/or strain (X-Axis). The tip of the x-axis is free and equipped with a cutting blade, which cuts through the sample at a rate (or series of rates) that is specified by the experimental protocol. The measurement of the intrinsic strength is performed in the pure-shear region of the test specimen determined in Stoček et al. (2013).

4 RESULTS AND DISCUSSION

A plot of F vs. T is presented in Figure 4 for the subject materials, which were not aged. In this case, a cutting energy of 464 J/m^2 is indicated. The lowest intrinsic cutting energy has clearly be observed at the rubber based on pure BR, S_{cBR} = 467,79 J/m^2, whereas the highest intrinsic cutting energy has been evaluated at the material based on NR, S_{cNR} = 943,16 J/m^2. Because of the $S_{cNR/BR}$ = 804,2 J/m^2 for the blended material, the NR seems to be a dominant material influencing the final intrinsic strength energy at the blends.

Figure 5 presents a plot of F vs. T for the subject materials, which were aged at the temperature 50°C for 720 hours. Generally, an increase of intrinsic cutting energy was caused due to thermal aging in all analyzed materials. The lowest intrinsic cutting energy again was clearly observed for the rubber based on pure BR, S_{cBR} = 1872,4 J/m^2, whereas the

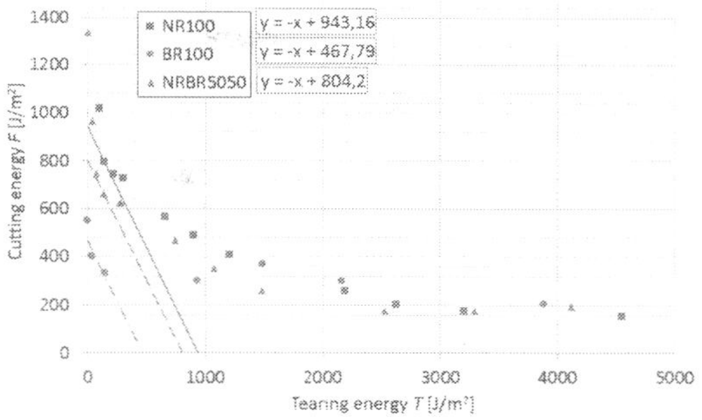

Figure 4. A plot of F vs. T for unaged materials.

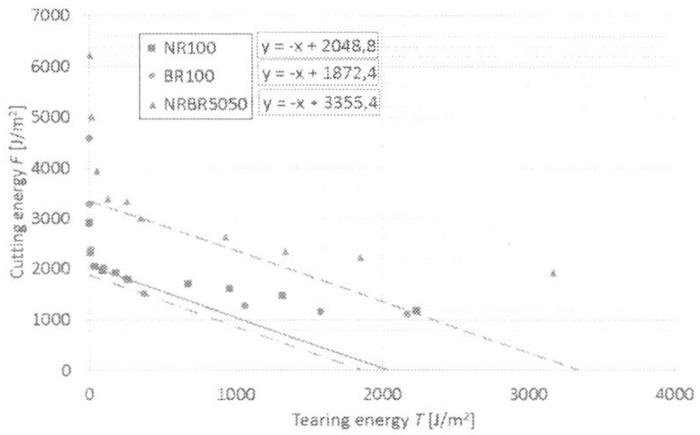

Figure 5. A plot of F vs. T of all thermally aged materials at temperature 50°C for 720 hours.

thermal aging mostly influenced the rubber blend NR/BR and thus the blend exhibits the highest intrinsic cutting energy $S_{cNR/BR}$ = 3355,4 J/m^2. The intrinsic cutting energy for the pure NR has been evaluated to be S_{cNR} = 2048,8 J/m^2, which is very close to the rubber based on BR.

Figure 6. clearly compares the values of intrinsic cutting energies of analyzed materials in dependence on aging temperature. It can be concluded that the blends based on NR/BR 50/50 exhibits the highest ability to be influenced due to thermal aging in the term of increase of intrinsic cutting energy. On the other hand the materials based on pure NR and BR show close level to each other of intrinsic cutting energy at aged state. Based on the results observed in Stoček et al. (2017), where the thermal aging has to be determined as responsible process for the enbrittlement of rubber matrix, generally the increase of intrinsic cutting energy caused due to thermal aging correspond with this previous observation.

Table 2 shows the unaged and aged stiffnesses for the subject materials. The ageing process tended to embrittle the compounds, and it was observed herein that brittle material tended to have higher intrinsic cutting energy.

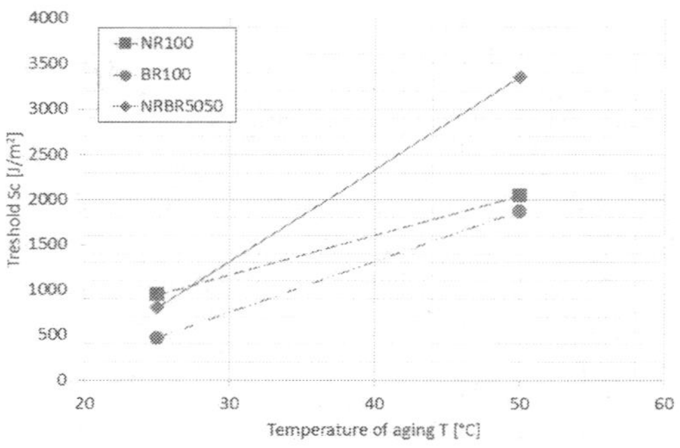

Figure 6. Intrinsic cutting energy in dependence on aging temperature.

Table 2. Young's Modulus, MPa.

	NR	NR/BR	BR
Unaged	0.426	0.521	0.669
Aged	0.615	0.779	0.940

5 CONCLUSION

The influence of thermal aging on intrinsic cutting energy of varied rubbers has been investigated by using of new commercial system, manufactured by Coesfeld GmbH, Germany and applied following methodology developed by Endurica LLC, Ohio, USA, that uses the theoretical background based on cutting method of Lake & Yeoh (1978) to estimate the intrinsic strength. For the subject materials and ageing condition, the stiffness and intrinsic cutting energy increased in correspondence with higher aging temperature.

ACKNOWLEDGEMENTS

This article was written with the support of Operational Program Research and Development for Innovations co-funded by the European Regional Development Fund and national budget of the Czech Republic, within the framework of the project CPS—strengthening research capacity (reg. number: CZ.1.05/2.1.00/19.0409) as well supported by the Ministry of Education, Youth and Sports of the Czech Republic—Program NPU I (LO1504) and by the internal grant agency of the project IGA/CPS/2017/006.

REFERENCES

Andrews, E.H., 1963, Rupture propagation in hysteresial materials: stress at a notch, Journal of the Mechanics and Physics of Solids 11, 231–242.

Bhowmick, A.K., 1988, Threshold Fracture of Elastomers, Journal of Macromolecular Science, Part C: Polymer Reviews, 28:3–4, 339–370.

Baldwin, J.M.; Bauer, D.R.; Ellwood, K.R., 2007, Rubber aging in tires. Part 1: Field results, Polymer Degradation and Stability 92, 103–109

Celina, M., Wise, J., Ottesen, D.K., Gillen, K.T., Clough, R.L., 2000, Correlation of chemical and mechanical property changes during oxidative degradation of neoprene., Polym. Degrad. Stab.,68 (2), 171–184.

Gent, A.N., & Mars, W.V., 2013, Strength of elastomers. Science and Technology of Rubber, 419–454.

Huang, D.; LaCount, B.J.; Castro, J.M.; Ignatz-Hoover, F., 2001, Development of a service-simulating, accelerated aging test method for exterior tire rubber compounds I. Cyclic aging, Polymer Degradation and Stability 74, 353–362.

Lake, G.J., & Lindley, P.B., 1966, Mechanical fatigue limit for rubber, Rubber Chemistry and Technology 39, 348–364.

Lake, G.J. & Thomas, A.G., 1967, The Strength of Highly Elastic Materials, Proc. R. Soc. Lond. A, 300, 108–119.

Lake, G.J., & Yeoh, O.H., 1978, Measurement of rubber cutting resistance in the absence of friction. International Journal of Fracture, 14(5), 509–526.

Lake, G.J. & Yeoh, O.H., 1987, Effect of Crack Tip Sharpness on the Strength of Vulcanized Rubbers, Journal of Polymer Science: Part B: Polymer Physics, Vol. 25, 1157–1190.

Lion, A., 1999, Strain-dependent dynamic properties of filled rubber: a nonlinear viscoelastic approach based on structural variables. Rubber Chemistry and Technology 72 (2), 410–429.

Mars, W.V., 2007, Fatigue life prediction for elastomeric structures. Rubber Chemistry and Technology 80, 481–503.

Mars, W.V.; Kipscholl, C.; Stocek, R., 2016, Intirnsic strength analyzer based on cutting method, Presented at the Fall 190th Technical Meeting of the Rubber Division of the American Chemical Society, Inc. Pittsburgh, PA, October 10–13.

Payne, A.R., 1962, The dynamic properties of carbon black-loaded natural rubber vulcanizates. Part I., J. Appl. Polymer Sci. 6, 57–63.

Payne, A.R., 1965, Strainwork Dependence of Filler-loaded Vulcanizates, J. Appl. Polymer Sci. 8, 2661–2686.

Schuring, D.J., Hall, G.L., 1981, Ambient Temperature Effects on Tire Rolling Loss, Rubber Chemistry and Technology, Vol. 54, No. 5, 1981, pp. 1113–1123.

Stoček, R.; Heinrich, G. Gehde, M., Kipscholl, R., 2013, Analysis of Dynamic Crack Propagation in Elastomers by Simultaneous Tensile—and Pure-Shear-Mode TestingIn: W. Grellmann et al. (Eds.): Fracture Mechanics & Statistical Mech., LNACM 70, Springer-Verlag Berlin Heidelberg, pp. 269–30.

Stoček, R.; Kratina, O.; Ghosh, P.; Maláč, J.; Mukhopadhyay, R., 2017, Influence of thermal ageing process on the crack propagation of rubber used for tire application. In: W. Grellmann, B. Langer: Deformation and Fracture Behaviour of Polymer Materials. Springer, Berlin, 305–316.

Filler reinforcement

Constitutive Models for Rubber X – Lion & Johlitz (Eds)
© 2017 Taylor & Francis Group, London, ISBN 978-1-138-03001-5

Non-entropic contribution to reinforcement in filled elastomers

P. Sotta, M. Abou Taha, A. Vieyres, R. Pérez-Aparicio & D.R. Long
Laboratoire Polymères et Matériaux Avancés, CNRS-Solvay UMR 5268, Saint-Fons, France

P.-A. Albouy
Laboratoire de Physique des Solides, CNRS-Université Paris-Sud UMR 8502, Orsay, France

C. Fayolle & A. Papon
Solvay Silica, Collonges au Mont d'Or, France

ABSTRACT: Several mechanisms have been proposed to explain the remarkable mechanical properties of elastomers filled with sub-micrometric particles. The complexity of their structure and dynamics has been evidenced through many experimental techniques. Understanding and discriminating the various reinforcement mechanisms is a key issue in the field. We shall present an approach in which the response of the elastomer matrix in a filled elastomer material is selectively observed. Different, complementary techniques are combined, namely measurements of the mechanical response in conjunction with measurements of chain segment orientation by X-ray scattering, in which the response of the elastomer matrix is selectively obtained. Crosslink densities are measured independently by multiple-quantum proton NMR. In unfilled materials, all measurements are nicely correlated, in full agreement with rubber elasticity theory. In filled materials, analyzing the deviations with respect to the behavior of the pure unfilled elastomer matrix allows discriminating chain over-stretching (entropic contribution to reinforcement) and rigid network effects (non-entropic contribution to reinforcement). The relative contributions of each mechanism are discussed, as a function of temperature, filler fraction and strain amplitude, in various elastomer materials filled with precipitated silica.

1 INTRODUCTION

In a pure, unfilled elastomer matrix, at temperatures high enough above T_g, the mechanical response and chain segment orientation are uniquely related to each other, according to rubber elasticity theory. This is the so-called stress-optical law, which writes in the case of uniaxial elongation (Treloar 1975, Doi & Edwards 1988, Erman & Mark 1997):

$$\sigma = \frac{k_B T}{b^3}\langle P_2(\cos\theta)\rangle = k_B T \nu \psi (\lambda^2 - \lambda^{-1}) \quad (1)$$

Where σ is the true stress, ν is the cross-link number density, b^3 the volume of a statistical segment and $\lambda = L/L_0$ the elongation. $\langle P_2(\cos\theta)\rangle$, the segmental orientation order parameter, is the ensemble average of the second order Legendre polynomial $(3\cos^2\theta - 1)/2$, with θ the angle of a statistical segment with respect to the stretching direction. ψ is a factor related to the model adopted to describe the response of the network. It is 1/2 for phantom network (Erman & Mark 1997).

For usual crosslink densities of the order 1 mol%, the modulus $E = k_B T \nu \psi$ is of the order 5×10^5 Pa. Equation 1 would extrapolate to a modulus of the order 10^7 Pa at an (unphysical) maximum crosslink density of about 100 mol% (one cross-link per monomer). Entropic rubber elasticity is very distinct from intermolecular forces acting in a glassy polymer, which give a shear modulus of order 10^9 Pa, that is 2–3 orders of magnitude larger at least. Equation 1 corresponds to an ideal situation in which enthalpic contributions associated with intramolecular interactions would be completely negligible (Erman & Mark 1997, Flory 1969). The nonentropic (enthalpic) contribution can be discriminated by studying the temperature variation of the modulus. This contribution is often less than 10%. In the particular case of natural rubber, it amounts to a fraction of 0.12–0.19 of the total elastic free energy (Erman & Mark 1997). This rather large contribution is basically due to the effect of chain stretching on the distribution of gauche and trans rotamers, which have different energies. Altogether, at high enough temperatures (typically $T \geq T_G + 60$), the temperature variation of the modulus follows Equation 1 within an error of the order 10%. Another typical feature of rubber elasticity is of course the ability to deform to very large amplitudes. Equation 1 eventually remains valid up to very large strain amplitudes, as illustrated in Figure 1 (from Zaghdoudi et al. (2014)).

Entropic elasticity has been quite extensively studied for decades. A new method to measure very accurately the chain segment orientation parameter $\langle P_2 \rangle$ induced upon stretching using wide angle X ray scattering has been proposed recently (Vieyres, Pérez-Aparicio, Albouy, Sanseau, Saalwächter, Long, & Sotta 2013, Albouy, Vieyres, Pérez-Aparicio, Sanséau, & Sotta 2014, Zaghdoudi, Albouy, Tourki, Vieyres, & Sotta 2014). Combining with precise measurements of the crosslink densities by NMR, it was demonstrated quantitatively that mechanical data are perfectly correlated to chain segment orientation measured by X-rays in a series of unfilled Natural Rubber (NR) materials (Vieyres, Pérez-Aparicio, Albouy, Sanseau, Saalwächter, Long, & Sotta 2013, Zaghdoudi, Albouy, Tourki, Vieyres, & Sotta 2014).

Elastomers filled with solid particles or aggregates of sub-micrometric sizes (denoted as fillers) have remarkable and complex mechanical properties which may be qualified as 'emerging' behavior, in the sense that they show qualitatively different features compared to pure elastomers (Wang 1998, Nielsen & Landel 1994, Heinrich, Klüppel, & Vilgis 2002). Their properties cannot be predicted from the mechanical response of the polymer matrix simply by combining geometrical and/or mechanical arguments like in Guth and Gold (1938), Guth and Gold (1945) or Domurath et al. (2012).

The response of the matrix in an elastomer nanocomposite may be selectively investigated by measuring the strain-induced chain segment orientation parameter $\langle P_2 \rangle$. By analyzing the deviations

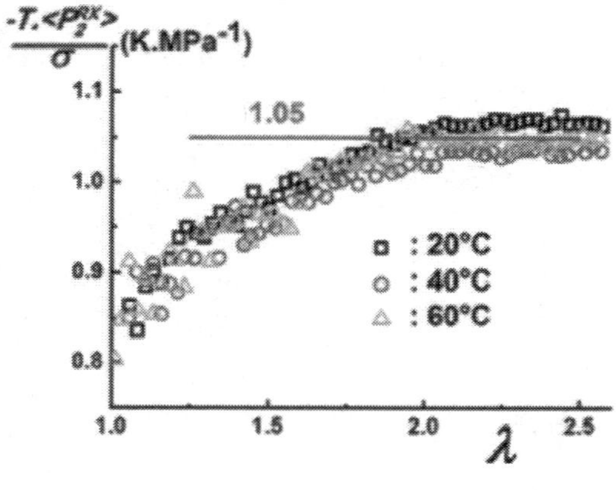

Figure 1. Ratio of the segmental orientation parameter $\langle P_2 \rangle$ over σ/T. This ratio should be constant according to Equation 1. The graph shows that it varies by about 15% over a strain range up to 150%.

with respect to the behavior of the pure unfilled elastomers, as described by Equation 1, it may give crucial pieces of information on reinforcement mechanisms. Here we propose to show and discuss measurements performed in vulcanized elastomers reinforced with nanometric silica aggregeates, in which the mechanical response exhibits drastically different features from the response of the matrix in terms of chain segment orientation, in complete violation of the generic pure elastomer behavior described by Equation 1. These strong deviations covers three distinct, though related aspects: the mechanical response deviates from the orientational response of the matrix as the strain amplitude reaches large amplitude, as temperature varies and when the filler amount varies. Also, it was shown in a series of filled Natural Rubber (NR) samples with various crosslink densities that deviations from Equation 1 occur as well when the crosslink density changes (Pérez-Aparicio, Vieyres, Albouy, Sanséau, Vanel, Long, & Sotta 2013).

Thus, we shall measure independently the various quantities in Eq. (1), i.e. the crosslink density, stress σ and segmental orientation order parameter $\langle P_2(\cos\theta)\rangle$, in elastomer samples filled with various amounts of silica. We shall examine the variation of these quantities as a function of the strain amplitude, filler fraction and temperature.

2 EXPERIMENTALS

2.1 *Materials*

The investigated elastomers are Styrene butadiene elastomers mixed and vulcanized following standard procedures, with 1.1 g of Sulfur, 2.5 g of Zinc oxide and 2 g of stearic acid per 137 g of elastomer, plus classical vulcanization accelerators. A precipitated silica (Z1165 MP from Solvay, specific surface 160 m^2/g) covalently grafted to the elastomer is used as reinforcing filler with a volume fraction varying from 0 (unfilled reference sample) up to 26.4%.

2.2 *Crosslink density*

Crosslink densities ν in the elastomer matrices were characterized by proton Multiple-Quantum (MQ) NMR (Saalwächter 2007), with a Bruker minispec mq20 spectrometer. Residual dipolar couplings between protons, partially averaged under the effect of the restriction on chain segment reorientation due to crosslinks and entanglements (Cohen-Addad 1974), are determined quantitatively. To obtain the targeted structural characterization of the crosslinking network independently of the (fast) segmental dynamics, measurements must be carried out at high temperature (70°C, i.e. $T_g + 90$ here). According to a well established procedure (Saalwächter 2007), the contribution from the sol fraction of the matrix must be substracted before computing the normalized proton Double Quantum (DQ) signal. The obtained average residual dipolar coupling D_{res} gives directly the average crosslink density v within a calibration factor: $D_{res} \propto v$. In SBR, this numerical factor is not known quantitatively, and we here shall only compare D_{res} values obtained in the various samples. D_{res} were obtained by fitting the normalized DQ signal with a function of the form

$$S_{DQ} = \frac{1}{2}\left(1 - \exp\left(-\left(D_{res}\tau_{DQ}\right)^{P_0}\right)\right) \quad (2)$$

2.3 *Mechanical characterization*

Complex dynamical oscillatory tensile (Young's) moduli G^* were measured in the linear regime (amplitude 0.089%) at a frequency 10 Hz as a function of temperature using a Metravib VA 2000 Dynamic Mechanical Analyzer (DMA). Traction curves were recorded at T = 23°C with an Instron tensile apparatus using normalized H2 test samples, at a traction speed 500 mm/min (corresponding to a strain rate $\dot{\varepsilon} = 0.33$).

2.4 *Chain segment orientation*

In a uniaxially stretched elastomer network, segments tend to orient along the stretching direction. As a result, the wide angle X-ray scattering, which comes from liquidlike short-range interferences between atoms located on neighbouring segments, becomes anisotropic. The segmental orientation parameter $\langle P_2(\cos\theta)\rangle$ which appears in Eq. 1, denoted $\langle P_2\rangle$ in the following, is evaluated from this anisotropy, with a constant proportionality factor (not calibrated here) (Mitchell 1984, Albouy, Guillier, Petermann, Vieyres, Sanseau, & Sotta 2012) $\langle P_2\rangle$ is an ensemble average over all chain segments. As stated in Eq. 1, in a pure elastomer, $\langle P_2\rangle$ is inversely proportional to the number of statistical segments between junctions, or equivalently to the crosslink density v, and proportional to the elongation parameter $\lambda^2 - \lambda^{-1}$.

A home-made uniaxial stretching device mounted on a rotating anode X-ray generator, described in details in previous papers (Albouy, Guillier, Petermann, Vieyres, Sanseau, & Sotta 2012, Vieyres, Pérez-Aparicio, Albouy, Sanseau, Saalwächter, Long, & Sotta 2013), was used. Cu Kα radiation was used (wavelength 1.542 Å). Uniaxial extension tests were performed at a constant extension rate $\dot{\lambda} = 1.66 \times 10^3 s^1$. Wide angle 2D scattering patterns were recorded as a function of λ with a Princeton indirect illumination CCD camera.

By measuring the sample width h along the test with an optical camera, the stretch ratio $\lambda = L/L_0$ was computed as $\lambda = (h_0/h)^2$, where h_0 is the width of the unstretched sample. The load F applied to the samples was simultaneously measured with a load sensor. During tensile tests, samples were placed in a temperature-regulated chamber equipped with Kapton windows, allowing incident and diffracted X-ray beams through. Filled samples were precycled three times prior to streching experiments. To analyze the variation of the scattered intensity as a function of the azimuth angle φ, a scattering range covering the amorphous halo, between roughly 7 and 17 Å^{-1}, was selected. Intensity was corrected from the response of the camera, the contributions from air scattering and from silica were subtracted. The integrated intensity was fitted with a function of the type $A + B\cos^2\varphi$, as illustrated from a representative example in Figure 2. The anisotropy parameter $\langle P_2\rangle_{RX}$ was then directly computed as (Albouy, Guillier, Petermann, Vieyres, Sanseau, & Sotta 2012, Vieyres, Pérez-Aparicio, Albouy, Sanseau, Saalwächter, Long, & Sotta 2013):

$$\langle P_2\rangle_{RX} = \frac{2B}{15A + 10B} \quad (3)$$

$\langle P_2\rangle_{RX}$ is proportional to the segmental orientation parameter $\langle P_2\rangle$ with a constant factor related to the geometry of interfering species.

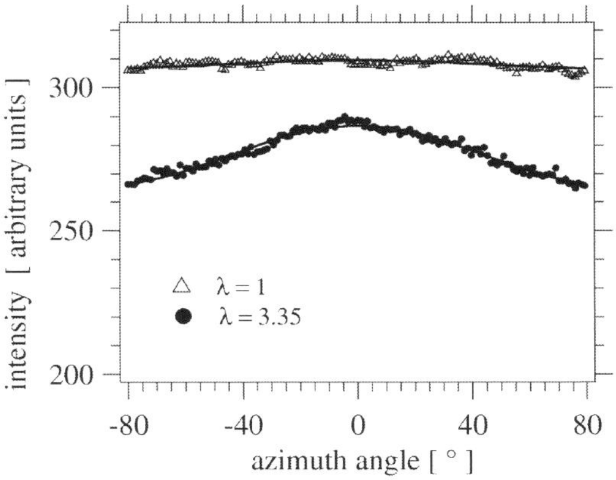

Figure 2. The intensity as a function of the azimuth angle φ in the sample with 23.3 vol% silica, in the initial, relaxed state and under stretching at $\lambda = 3.35$, together with fitting curves, showing the appearance of anisotropy. The overall change in intensity is of no relevance here. It comes from a competition between transmission and scattering power as the sample thickness changes during stretching.

3 RESULTS

3.1 *Crosslink densities characterized by NMR*

The obtained values of the NMR parameter D_{res}, which is proportional to the crosslink density, are shown in Figure 3 as a function of the silica volume fraction. A moderate, though significant decrease of the measured crosslink densiy is observed as the silica amount increases. This confirms that silica affects the crosslinking reaction to some extent. This effect has generally been attributed to partial adsorption of accelerators on the silica surface, and is coherent with the generally observed increase of the crosslinking time as the silica amount increases. This decrease of the crosslink density in the elastomer matrix in the presence of silica counterbalances to a moderate extent the reinforcement due to silica.

3.2 *Mechanical behavior*

Curves of the true stress σ as a function of the elongation parameter $\lambda^2 - \lambda^{-1}$ during tensile tests (traction curves) are shown in Figure 4 in the series of filled samples. This graph shows the strong effect of the silica amount. The strongly non linear stress-strain behavior is emphasized in Figure 5, in which the secant tensile modulus, defined here as the ratio $\sigma/\lambda^2 - \lambda^{-1}$, is plotted as a function of the elongation parameter. Thisstrongly non-linear behavior is similar to the Payne effect, usually measured in dynamical oscillatory tests performed at increasing amplitude. The non linearity is more pronounced as the silica amount increases.

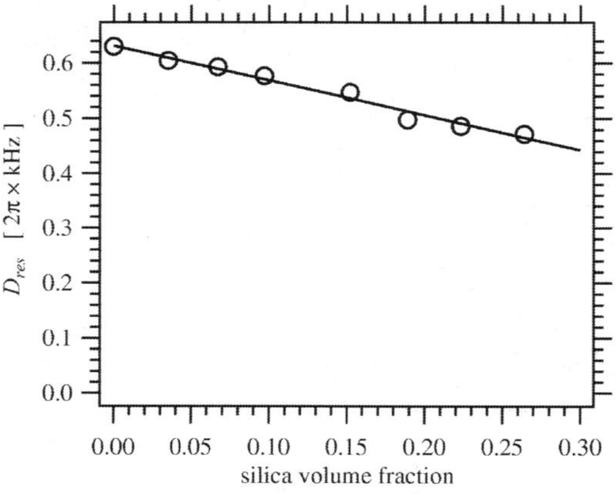

Figure 3. Fitted DQ NMR parameter D_{res}, proportional to the crosslink density, in the series of silica filled SBR samples.

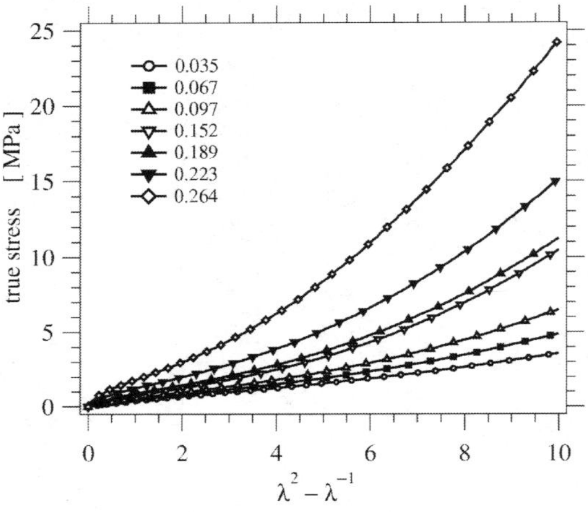

Figure 4. The true stress as a function of the elongation parameter obtained in tensile tests at temperature 23°C on the series of samples reinforced with various silica amounts. Silica volume fractions are indicated in the legend. Note the increasing non linear variation as the silica fraction increases.

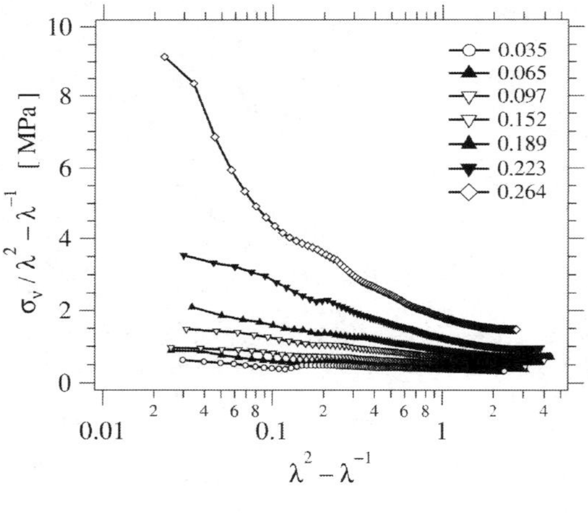

Figure 5. The secant tensile modulus $\sigma/(\lambda^2 - \lambda^{-1})$ as a function of the elongation parameter $\lambda^2 - \lambda^{-1}$ for the series of samples reinforced with various silica volume fractions (same data as in Figure 5).

3.3 *Segmental orientation in the elastomer matrix*

Curves of the induced anisotropy parameter $\langle P_2 \rangle_{RX}$ measured by X diffraction as a function of the elongation parameter $\lambda^2 - \lambda^{-1}$ measured at T = 20°C are shown in Figure 6 in the series of samples reinforced by silica. Note the strong contrast with the mechanical curves in Figure 4. Two major differences must be noticed. While in a pure elastomer, both $\langle P_2 \rangle_{RX}$ and the true stress have the same variation as a function of $\lambda^2 - \lambda^{-1}$ within

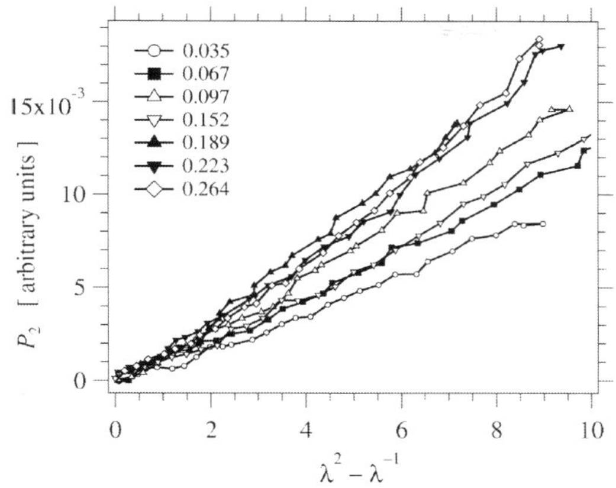

Figure 6. Anisotropy parameter $\langle P_2 \rangle_{RX}$ as a function of the elongation parameter $\lambda^2 - \lambda^{-1}$ measured at $T = 20$°C for the series of samples reinforced with various silica amounts. Silica volume fractions are indicated in the legend.

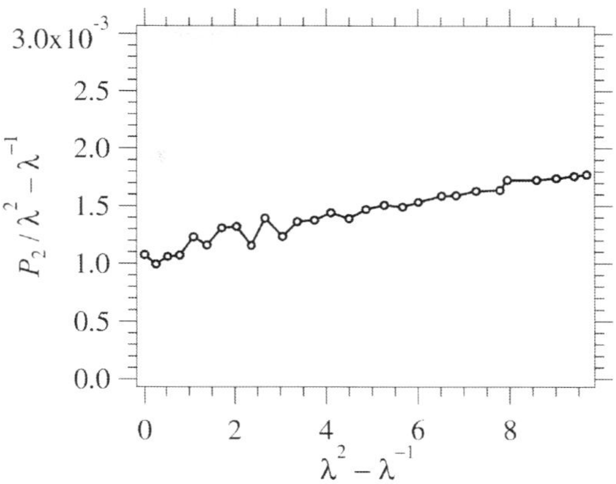

Figure 7. Representative example of the slope $\langle P_2 \rangle_{RX}/(\lambda^2 - \lambda^{-1})$ as a function of the elongation parameter $\lambda^2 - \lambda^{-1}$ (from the curve at $\Phi = 0.223$ in Figure 6). Note the drastically different behavior when compared to Figure 5.

roughly 15%, as illustrated in Figure 1 (Zaghdoudi, Albouy, Tourki, Vieyres, & Sotta 2014), the shapes of the curves are completely different in reinforced materials. This is most clearly illustrated in Figure 7. Figure 7 demonstrates that the true stress σ and the orientation parameter $\langle P_2 \rangle$ vary in completely different ways as a function of $\lambda^2 - \lambda^{-1}$, in strong contrast to what is observed in unfilled elastomers. The slope $\langle P_2 \rangle/(\lambda^2 - \lambda^{-1})$ does not show the strong decrease observed in the ratio $\sigma/(\lambda^2 - \lambda^{-1})$ as the stretch amplitude $\lambda^2 - \lambda^{-1}$ increases. This is a direct experimental proof that the stress is not only correlated to the segmental orientation in the elastomer matrix in filled elastomers.

A further proof is given by the dependence on the silica amount. As the silica fraction increases from 3.5 vol% to 26.4 vol%, the orientation parameter $\langle P_2 \rangle$ increases by a factor of order 2 at a given elongation, as shown in Figure 6. Conversely, the true stress increases by a much larger factor of order 7 (see Figure 4). This again demonstrates that the stress in reinforced materials contains some contribution not correlated to the segmental orientation in the elastomer matrix only.

Finally, the dependence on temperature is also completely different for the mechanical response and the orientational response of chain segment in the matrix. This is shown in Figure 8, in which the relative temperature variation of both the modulus and the orientational response is shown for the most reinforced sample. While the modulus decreases as temperature increases, the orientational response does not significantly vary. Thus,

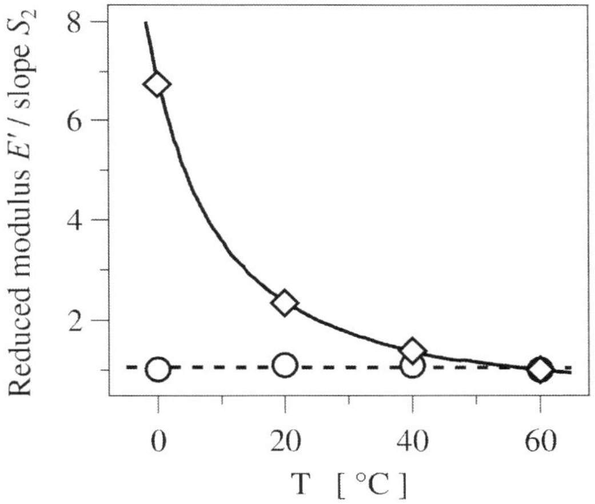

Figure 8. The dynamical oscillatory modulus measured at small amplitude (in the linear regime) and the orientational response $\langle P_2 \rangle_{RX}/(\lambda^2 - \lambda^{-1})$ as a function of temperature in the sample reinforced with 26.4 vol% silica. For comparison purpose, all values were normalized by the value at 60°C.

the orientational response of the matrix seems to follows Equation 1, which predicts no temperature dependence of the segmental orientation, while the stress behaves in a completely different way.

Figure 9 summarizes the main results. The relative variation of both the modulus and the orientational response are shown as a function of the silica volume fraction, at two temperatures. Once again, this figure illustrates the strong difference

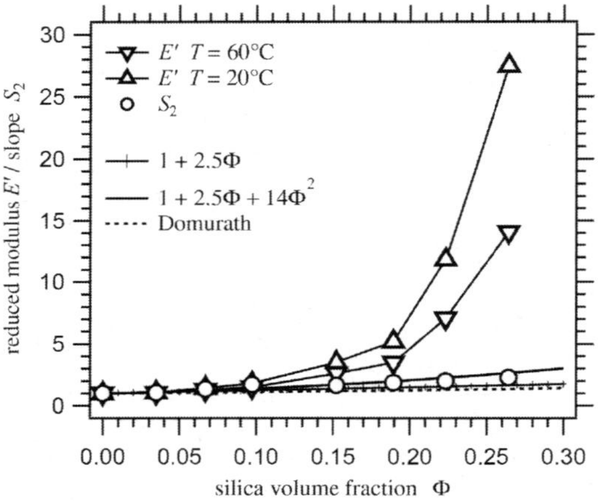

Figure 9. The dynamical oscillatory modulus measured at small amplitude (in the linear regime) and the orientational response $\langle P_2 \rangle_{RX} / (\lambda^2 - \lambda^{-1})$ as a function of the silica volume fraction, at two temperatures. For comparison purpose, all values were normalized by the value in the unfilled reference sample. The dashed curve is the Guth and Gold function $R = 1 + 2.5\Phi$, the full curve includes a quadratic term $R = 1 + 2.5\Phi + 14.1\Phi^2$, the dotted curve corresponds to the function $1/(1 - \Phi)$ proposed by Domurath et al.

in the mechanical and orientational responses of the filled materials. This demonstrates that the mechanical response is decorrelated from the orientational response of the elastomer matrix, in contrast to Equation 1.

4 CONCLUSIONS

Measuring the induced orientational parameter $\langle P_2 \rangle$ during tensile tests gives selective access to the response of the elastomer matrix in a filled material. This selective response can then be compared to the overall mechanical response of the material. The results shown in this paper provide a direct experimental proof that the mechanical response in filled elastomers does not solely originates from the elastomer matrix. The orientational response of the matrix alone generally follows the stress-optical law (no temperature dependence, linear variation with the applied strain) and seems to be reasonably well described by strain amplification models such as the one proposed by Domurath et al. (2012). Conversely, the mechanical responsestrongly deviates from this essentially geometrical/hydrodynamical description. This demonstrates the importance of rigid network effects associated to fillers in such materials. Moreover, the strong temperature dependence of the mechanical response (when compared to the matrix response) shows that these rigid network effects cannot either be described simply by the geometry of the network eventually formed by fillers in the materials. In other words, the rigidity of the network strongly depends on temperature. The distance to the glass transition of the matrix is a key parameter and shows that the modification of the dynamics in the matrix due to fillers is an essential mechanism in these systems (Berriot, Montes, Lequeux, Long, & Sotta 2002, Berriot, Montes, Lequeux, Long, & Sotta 2003, Merabia, Sotta, & Long 2008, Dequidt, Long, Sotta, & Sanséau 2012).

REFERENCES

Albouy, P.-A., G. Guillier, D. Petermann, A. Vieyres, O. Sanseau, & P. Sotta (2012). A stroboscopic x-ray apparatus for the study of the kinetics of strain-induced crystallization in natural rubber. *Polymer 53*(15), 3313–3324.

Albouy, P.-A., A. Vieyres, R. Pérez-Aparicio, O. Sanséau, & P. Sotta (2014). The impact of strain-induced crystallization on strain during mechanical cycling of cross-linked natural rubber. *Polymer 55*(15), 4022–4031.

Berriot, J., H. Montes, F. Lequeux, D. Long, & P. Sotta (2002). Evidence for the shift of the glass transition near the particles in model silica-filled elastomers. *Macromolecules 35,* 9756–9762.

Berriot, J., H. Montes, F. Lequeux, D. Long, & P. Sotta (2003). Gradient of glass transition temperature in filled elastomers. *Europhys. Lett. 64*(1), 50.

Cohen-Addad, J.P. (1974). *Journal of Chemical Physics 60,* 2440–2453.

Dequidt, A., D.R. Long, P. Sotta, & O. Sanséau (2012). Mechanical properties of thin confined polymer films close to the glass transition in the linear regime of deformation: Theory and simulations. *Eur. Phys. J. E 35*(7), 61–83.

Doi, M. & S.F. Edwards (1988). *The Theory of Polymer Dynamics.* Oxford University Press New York.

Domurath, J., M. Saphiannikova, G. Ausias, & G. Heinrich (2012). Modelling of stress and strain amplification effects in filled polymer melts. *Journal of Non-Newtonian Fluid Mechanics* 171–172, 8–16.

Erman, B. & J.E. Mark (1997). *Structures and Properties of Rubberlike Networks.* Oxford University Press New York.

Flory, P.J. (1969). *Statistical Mechanics of Chain Molecules.* New York, USA: Interscience.

Guth, E. & O. Gold (1938). *Physical Review 53,* 322.

Guth, E. & O. Gold (1945). *Journal of Applied Physics 16,* 20.

Heinrich, G., M. Klüppel, & T.A. Vilgis (2002). Reinforcement of elastomers. *Current Opinion in Solid State and Materials Science 6*(3), 195–203.

Merabia, S., P. Sotta, & D.R. Long (2008). A microscopic model for the reinforcement and the nonlinear behavior of filled elastomers and thermoplastic elastomers (payne and mullins effects). *Macromolecules 41*(21), 8252–8266.

Mitchell, G.R. (1984). A wide-angle x-ray study of the development of molecular-orientation in crosslinked natural-rubber. *Polymer 25,* 1562–1572.

Nielsen, L.E. & R.F. Landel (1994). *Mechanical Properties of Polymers and Composites.* Marcel Dekker, New York.

Pérez-Aparicio, R., A. Vieyres, P.-A. Albouy, O. Sanséau, L. Vanel, D.R. Long, & P. Sotta (2013). Reinforcement in natural rubber elastomer nanocomposites: Breakdown of entropic elasticity. *Macromolecules 46,* 8964–8972.

Saalwächter, K. (2007). Proton multiple-quantum nmr for the study of chain dynamics and structural constraints in polymeric soft materials. *Prog. Nucl. Magn. Reson. Spectrosc. 51*(1), 1–35.

Treloar, L.R.G. (1975). *The Physics of Rubber Elasticity.* Clarendon Press, Oxford, England.

Vieyres, A., R. Pérez-Aparicio, P.-A. Albouy, O. Sanseau, K. Saalwächter, D.R. Long, & P. Sotta (2013). Sulfur-cured natural rubber elastomer networks: Correlating cross-link density, chain orientation, and mechanical response by combined techniques. *Macromolecules 46,* 889–899.

Wang, M.J. (1998). Effect of polymer-filler and filler-filler interactions on dynamic properties of filled vulcanizates. *Rubber Chem. Technol. 71*(3), 520–589.

Zaghdoudi, M., P.-A. Albouy, Z. Tourki, A. Vieyres, & P. Sotta (2014). Relation between stress and segmental orientation during mechanical cycling of a natural rubber-based compound. *Journal of Polymer Science, Part B: Polymer Physics 53,* 943950.

Constitutive Models for Rubber X – Lion & Johlitz (Eds)
 ISBN 978-1-138-03001-5

A novel reinforcement structure in tire tread compounds: Organo-modified octosilicate as additive

W.R. Córdova & J.G. Meier
ITAINNOVA—Instituto Tecnológico de Aragón, Zaragoza, Spain

D. Julve, M. Martínez & J. Pérez
IQE S.A.—Industrias Químicas del Ebro S.A., C/D N°97 Polígono Industrial de Malpica, Zaragoza, Spain

ABSTRACT: Only 3phr of octosilicate (OCTO), swollen with cetyltrimethylammonium ions (CTA), used as an additive in a rubber tire tread compound (80phr of Highly Dispersible Silica, HDS) improves the end properties. The presence of the organo-modified synthetic layered silicate increases vulcanization rate, while the analogue natural silicate, montmorillonite, did not show the same effects (J. G. Meier, 2013a, 2013b, 2015). The addition of Synthetic Laminar Silicates (SLS) modifies the structure of the reinforcing filler network. As a result, the fractal exponents characterizing the reinforcement filler network were found to be different when comparing silica-only compounds to silica/SLS compounds (J. G. Meier, 2016). It points towards the formation of a genuine hybrid network, i.e. integration of layers of the SLS into the network of spheres in the silica. The formation of the hybrid filler network should depend on the concentration and on the exfoliation state of the SLS layers. The use of dimethyldioctadecylammonium ions (2HT), instead of CTA as the organic modifier, increases the interlaminar space, thus improving the exfoliation ability of the material, and furthermore, mechanical effects similar to those of OCTO-CTA can be obtained with lower OCTO-2HT concentration. Nevertheless, as a result of the change in the organic modifier and concentration levels, variations of the mixing process have also been necessary in order to obtain comparable results (W. R. Córdova, 2016). Systematic studies of our research will be expounded and discussed in this publication.

1 INTRODUCTION

The cost-effective manufacture of low rolling resistance tires is one of the most important development goals in tire industry today. In this context, a European study projects that a 3% reduction in fuel consumption with the use of low rolling resistance tires can lead to a reduction in global well-to-wheels greenhouse gas emissions by 100 million metric tons per year by 2020 (Kodjak, 2015; Pike & others, 2011). The use of swollen synthetic laminar silicate as an additive in tire tread compounds can provide significant benefits in mechanical properties, rolling resistance and ice/wet grip in comparison with normal HDS compounds (J. G. Meier, 2013b, 2015). The aforementioned papers report the results of a normal tire tread compound additivated with 3phr of organo-modified synthetic layer silicate. An improvement in the rolling resistance predictor (tan δ-value at 60°C) of ca. 35% was obtained for 3phr OCTO-CTA compounds. Furthermore, the ice/wet grip predictor (tan δ-value at ≈ −20°C) was also improved, while volume loss or abrasion was not affected. Tensile tests at room temperature also showed higher stress at break for the additivated compounds than the reference silica-only compounds. However, in the same test, at small elongations up to 100%, the stresses are lower for additivated compounds (c.f. Figure 1). Additionally, vulcanization rate was found to be accelerated in the additivated compound. This effect was considered to be related to the accelerating effect the organic surfactant can have at the typical rubber process temperatures of between

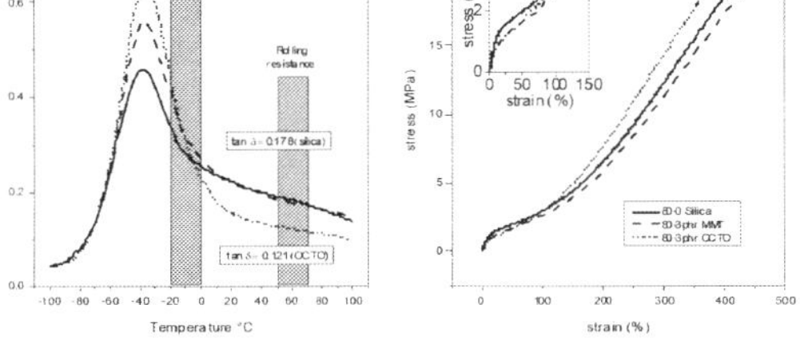

Figure 1. Caption of loss factor and tensile stress for silica reference and additivated compounds. (J. G. Meier, 2013a, 2016).

150–170°C (Galimberti Maurizio, 2009), which releases the organic modifier during the exfoliation process. However, the natural analogue reference, MMT-CTA (organo-modified montmorillonite), did not show that effect (J. G. Meier, 2013a, 2013b, 2015). The rise in vulcanization rate can reduce the cost of manufacturing low-rolling resistance tires in a production process. In addition, the accelerated vulcanization rate can be used as a quality control of successful exfoliation.

Thus, these findings combine various highly promising material properties over the entire process chain in the quest and development of low rolling resistance tires.

A qualitative theoretical model was proposed to understand these results and their relationship with the filler reinforcement structure. (J. G. Meier, 2016) The hypothesis is based on the formation of a hybrid filler network composed of silica nanoparticles and layers of silicate. The layers are capable of modifying the structure of the network, which means that connectivity between the fillers changes. As a consequence, a different scaling exponent is obtained from the additivated compounds in comparison with silica-only compounds. Furthermore, the formation of the hybrid filler network and its connectivity state should depend on the concentration of layers and hence on the exfoliation state of the SLS.

The interlaminar distance seems to play an important role in dispersing/exfoliating the synthetic layer silicate. The hydrophobization of the silicate by ion-exchange with long tallow surfactants leads to an increase of the interlayer distance, a drastic decrease in surface free energy and therefore weaker forces between the layers, thus improving exfoliation capability. (Kádár, Százdi, Fekete, & Pukánszky, 2006) Dimethyldioctadecyl ammonium ions (2HT) have two hydrogenated tails provoking a different packing of the aliphatic chains in the interlaminar space in comparison with single-tailed CTA. The substitution of CTA with 2HT as the organic modifier results in an increase in the interlaminar space, facilitating exfoliation. If exfoliation is increased, the effective concentration of exfoliated layers per mass of additive increases and comparable static and dynamic properties could be obtained at a lower overall OCTO-2HT concentration compared to OCTO-CTA.

In this paper, the addition of a few parts of OCTO-2HT to a normal tire tread compound recipe was evaluated using different compound mixing procedures.

2 EXPERIMENTAL

In this research, a basic high-loaded silica tire-tread recipe, typical of a green tire formulation, was used (Table 1). With this recipe, the effect of both the addition of a few parts of OCTO-2HT and the modification of the compounding process were evaluated at the same time. Concentration of OCTO-2HT was varied between 0 (80-0, reference compound without silicate) and 5 phr (Table 1). In regard to the compounding process, three different procedures were evaluated. These were labeled as normal (*normal*), silanization (*sila*) and modified silanization (*sila mod*). *Normal* procedure consists of a non-productive mixing step (1), where silica, silicate and silane are added; and a productive mixing step (2), where curatives are added (Table 2). Unlike the *normal* process, *sila* and *sila mod* procedures are masterbatch processes with an additional non-productive step (1.5) introduced between step 1 and the productive step. This procedure ensured that all of the silica is silanized prior to the

Table 1. Composition (phr) and sample label for *normal* compounding procedure.

Component	Concentration (phr)
S-SBR (26.3 phr TDAE)	96.3
BR	30
Stearic acid	2
ZnO	3
6PPD	1.5
OCTO-2HT	0; 1; 2; 3; 5
Silica	80
TESPT	12.8
CBS	1.5
DPG	2.0
Sulfur	1.4

Table 2. Mixing parameters and process for *normal* compounds.

Step 1: Brabender 390S (75 rpm, 50°C)	
Rubbers + stearic acid + ZnO	1 min
½ silica + ½ OCTO-2HT + TESPT	1 min open ram at 30s
½ silica + ½ OCTO-2HT + 6PPD	1 min open ram at 30s
At 145°C rotor speed is adjusted to 60 rpm	
Temperature between 145–150°C	4 min
Dump and laminate	
Step 2: Addition of curatives (After 24 h) Brabender 390S (75 rpm, 50°C)	
Mastication of masterbatch (step 1)	Until reach 90°C
CBS + DPG + Sulfur	1 min
Dump and laminate	

addition of the SLS additive. The differences between *sila* and *sila mod* compounding processes lie only in the mixing time and temperatures (lower in *sila mod*) in step 1.5, as can be seen in Tables 3 and 4, respectively. *Sila* and *sila mod* compounds were obtained twice and then unified on an open roll mill. Table 5 summarizes the compound name depending on the amount of OCTO-2HT (phr) and the mixing procedure.

Table 3. Mixing parameters and process for *sila* compounds.

Step 1.5: Brabender W50 EHT (75rpm, 90°C)	
Mastication of 80-0 *normal* (step 1) masterbatch	Until reach 130°C
Addition of OCTO-2HT	Until reach 145°C
At 145°C rotor speed is adjusted to 60rpm	
Temperature between 145–150°C	4 min
Dump and laminate	
Step 2: Addition of curatives (After 24h) Brabender W50 EHT (75rpm, 50°C)	
Mastication of masterbatch (step 1.5)	Until reach 90°C
CBS + DPG + Sulfur	1 min
Dump and laminate	

Table 4. Mixing parameters and process for *sila mod* compounds.

Step 1.5: Brabender W50 EHT (70rpm, 50°C)	
Mastication of 80-0 *normal* (step 1) masterbatch	Until reach 70°C
Addition of OCTO-2HT	Until reach 120°C
Dump and laminate	
Step 2: Addition of curatives (After 24h) Brabender W50 EHT (75rpm, 50°C)	
Mastication of masterbatch (step 1.5)	Until reach 90°C
CBS + DPG + Sulfur	1 min
Dump and laminate	

Table 5. Compound identification according to the amount of OCTO-2HT (phr) and mixing procedure.

		Mixing procedure		
		Normal	*Sila*	*Sila mod*
phr of OCTO-2HT	0	80-0 normal		80-0 sila mod
	1	80-1 normal		80-1 sila mod
	2	80-2 normal		80-2 sila mod
	3	80-3 normal	80-3 sila	80-3 sila mod
	5	80-5 normal	80-5 sila	

Cure parameters and cyclic strain sweep tests were measured using an RPA 2000 (Alpha-Technologies). All compounds and probes were vulcanized at optimal cure time (t_{90}). Uniaxial tensile testing was performed on S2 type probes obtained by die cutting from a vulcanized sheet. Hardness and abrasion tests were evaluated on cylindrical probes. Dynamic thermal mechanic behavior of the compounds was measured with a DMA$^+$ 450 (MetraviB) on cylindrical probes of 10 mm diameter and 2 mm thickness in shear mode. The measurement was performed at 1 Hz excitation frequency, 1% strain and 2°C/min as a temperature ramp.

3 RESULTS

3.1 *Compounding mixing procedures*

Torque, temperature and kneader blade speed for *normal* compounding process are shown in Figure 2.

In the *normal* mixing procedure, a high concentration of OCTO-2HT (5phr) shows an unusual behavior. From the middle of the mixing time, 240 s approximately, torque started to increase until it reached a maximum before decreasing again. In fact, mixing temperature was not easy to control even by decreasing rotor speed. This unusual behavior is also observed for *sila* mixing procedure with 5phr of OCTO-2HT (Figure 3). This behavior may be related to pre-vulcanization, as suggested by vulcametric studies (discussed below). As high doses of OCTO-2HT proved difficult to process, the 5 phr OCTO-2HT concentration was not prepared using the *sila mod* procedure. Conversely, 2 phr concentration was added to the series. The "safer" mixing parameters used in the *sila mod* procedure facilitates processability of the compounds and thus prevents pre-vulcanization. Torque, tem-

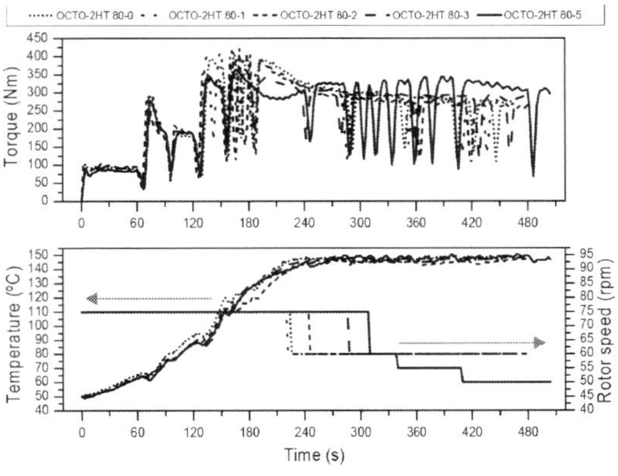

Figure 2. Caption of torque, temperature and kneader blade speed versus time for 80-0-1-2-3-5 OCTO-2HT with *normal* compounding process.

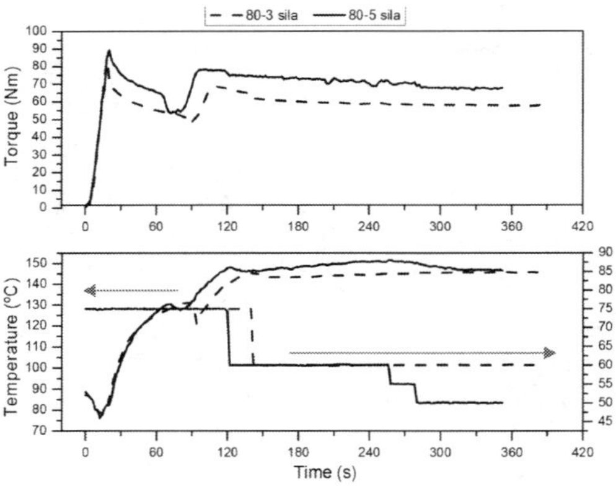

Figure 3. Caption of torque, temperature and kneader blade speed versus time for 80-3-5 OCTO-2HT with *sila* mixing procedure.

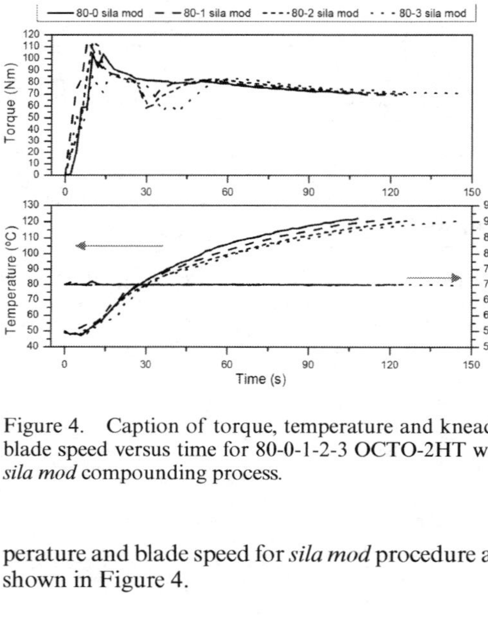

Figure 4. Caption of torque, temperature and kneader blade speed versus time for 80-0-1-2-3 OCTO-2HT with *sila mod* compounding process.

perature and blade speed for *sila mod* procedure are shown in Figure 4.

3.2 *Vulcanization (165°C, 30 min)*

Figure 5 shows S'max and S'rate, max values for all compounds extracted from the cure tests. Minimum S' modulus did not show a significant difference between all compounds. However, maximum S' modulus in the 1phr compounds are always the highest in the *normal* and *sila mod* procedures but the *sila mod* process results in a lower modulus, due to improved silanization of the silica fraction and hence reduced filler-filler interactions. This is also observed for the other concentration levels. Comparing the influence of the processing

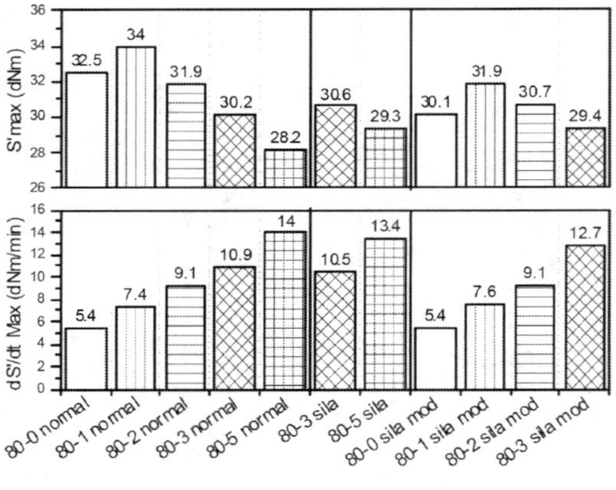

Figure 5. S'max and S'rate,max for *normal*, *sila* and *sila mod* compounds.

methods further, one observes that at the same concentrations (80-3), the *sila* process results in the same S`max and S'rate,max value as the ones in the *normal* process. Thus, changing the masterbatch process does not apparently influence curing. However, when comparing with the *sila mod* process, executing step 1.5 at 120°C instead of 145°C produced a notable decrease in S`max and an increase in S'rate,max. Therefore, the dominant factor influencing the curing behavior and the mixing state is the maximal processing temperature. This suggests issues related to pre-vulcanization of the rubber matrix during mixing.

Maximum vulcanization rate increases with OCTO-2HT concentration and also reduces induction time (onset of modulus increase), due to the increasing amount of accelerating 2HT-moieties in the compounds. Therefore, this observation indicates a successful (partial) exfoliation of OCTO-2HT. Similar to the reports by Meier *et al.* (J. G. Meier, 2013b), the analogue organo-modified natural material, montmorillonite-2HT, did not show the same tendency (data not shown).

3.3 *Uniaxial tensile tests at RT*

Stress versus strain plots from tensile tests at room temperature are given in Figure 6.

For the *normal* compounding process, one obtains qualitatively the same results as shown in Figure 1a for the 1 and 2 phr additivation levels. The tensile curves are characterized by lower stresses at small deformations and higher stresses at high deformations (except stress and strain at break for 80-2 normal). This tendency of lower stresses at high strains is even more noticeable in the samples with 3 phr and 5 phr of OCTO-2HT. The tensile curves drop clearly below the refer-

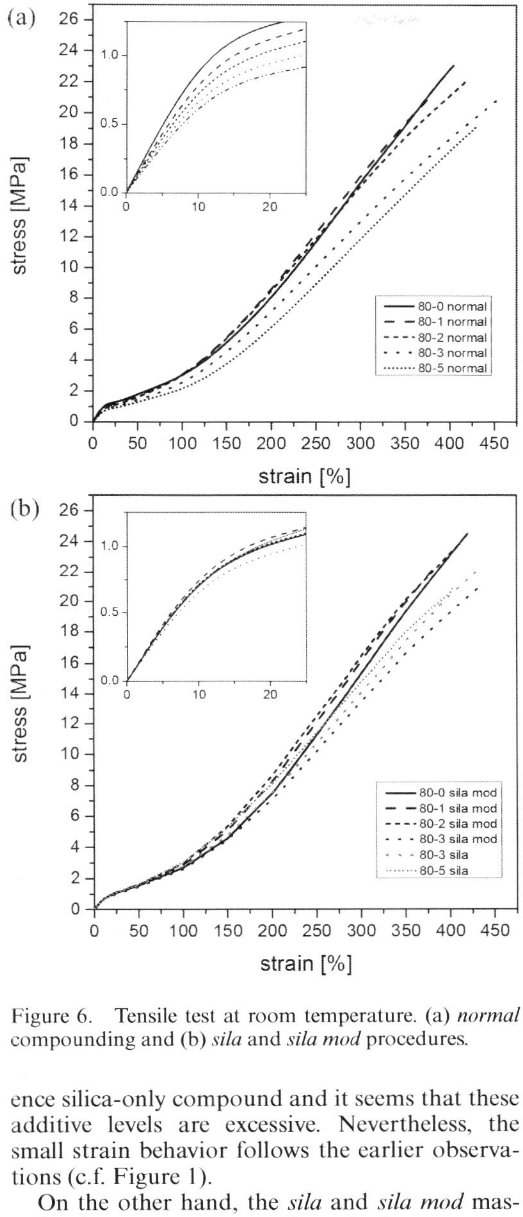

Figure 6. Tensile test at room temperature. (a) *normal* compounding and (b) *sila* and *sila mod* procedures.

ence silica-only compound and it seems that these additive levels are excessive. Nevertheless, the small strain behavior follows the earlier observations (c.f. Figure 1).

On the other hand, the *sila* and *sila mod* masterbatch processes do not show the lower stresses at small strains any more. In the case of the *sila mod* process, improved behavior at higher strains for samples 80-1 and 2 can be observed, notwithstanding sample 80-3, which shows lower stresses than the reference sample over the entire strain range. It seems almost as if the higher concentration of additive has induced a kind of transition to another state.

Sample 80-3 in the *sila* process shows higher stress values over the entire strain range like 80-3 *sila mod* but still always lower than the reference.

3.4 *Strain sweep at 60°C after vulcanization*

These results have to be contrasted with strain dependent RPA-measurements (Figure 7). Here also the masterbatch processes exhibit smaller modulus (hence stress) at low strains, recovering the behavior pattern shown in Figure 1. Moreover, the modulus systematically decreases as OCTO-2HT concentration increases.

Another important result is seen in the loss factors (tan δ) from the back strain sweep, which are shown in Figure 8. Lower loss factor readings are obtained for the *sila* and *sila mod* processes, which correspond with lower energy dissipation during deformation. In the same way, values of hysteresis are lower for *sila* and *sila mod* compounding processes (data not shown), in comparison with the *normal* compounding process.

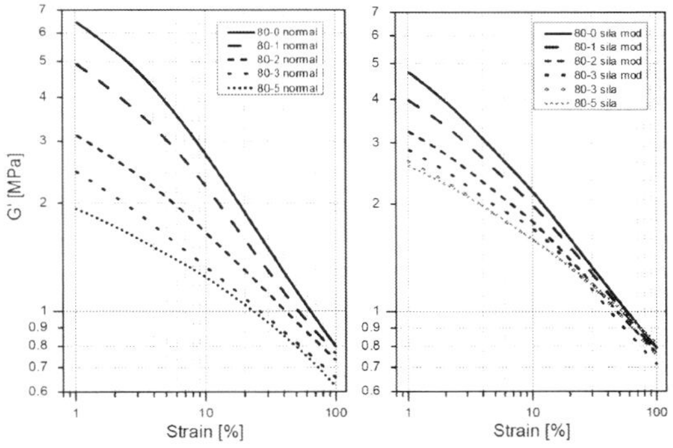

Figure 7. Caption of G' vs strain for upward strain sweep at 60°C. (Left) *normal* compounding and (Right) *sila* and *sila mod* procedures.

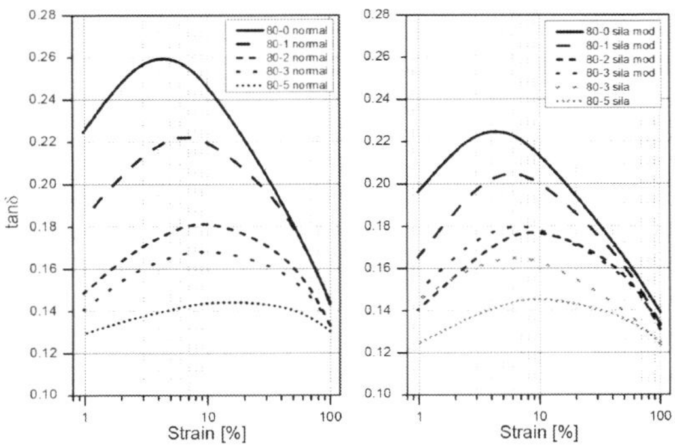

Figure 8. Caption of loss factor (tan δ) vs strain for the back strain sweep at 60°C. (Left) *normal* process and (Right) *sila* and *sila mod* processes.

The reduced Payne effect between 80-0 *normal* compound and 80-0 *sila mod* compound would seem to indicate that the additional mixing step 1.5 improves filler dispersion and reduces filler-filler interaction. This behavior is dominant in the series of *sila mod* compounds up to an additive level of 3 phr OCTO-2HT and seems to stem from the well silanized silica. The 80-3 and 5 samples of the *sila* process have in fact slightly higher small strain values than the corresponding moduli in the *normal* process samples. It seems that separate silanization can interfere with the integration of silicate layers in the silica filler network because the absolute decrease of modulus with OCTO-2HT concentration is lesser in the *sila mod* process. Such an interpretation could hint at an explanation for the tensile test results.

When combining the small strain results from RPA and the ultimate properties from tensile measurements, one can conclude that the modification of the mixing procedure results in compounds showing the behavior depicted in Figure 1 at a lower concentration of SLS-additive (1 or 2 phr).

3.5 *Hardness and abrasion tests*

Hardness and abrasion volume loss for all compounds are depicted in Figure 9.

Hardness decreases with OCTO-2HT concentration in the case of the *normal* mixing process. The reduction in hardness with increasing concentrations of SLS-additive had already been observed in previous studies. (J. G. Meier, 2013a, 2016; W. R. Córdova, 2016) On the other hand the *sila mod* mixing process gives relatively constant values of hardness with a very slight tendency to decrease with increasing concentrations of OCTO-2HT, which however, is not statistically significant.

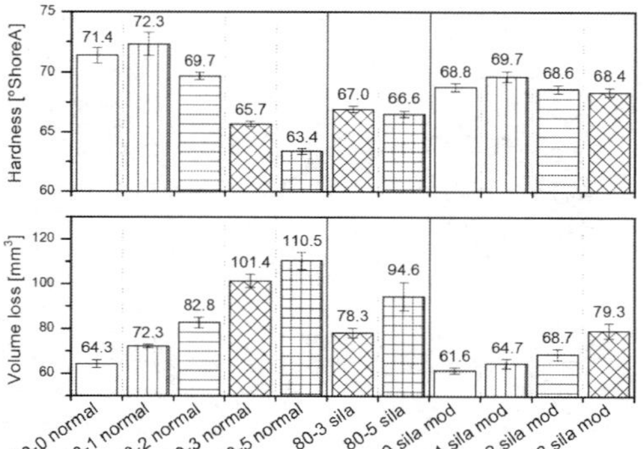

Figure 9. Caption of Hardness (up) and volume loss (down) for all compounding processes.

The *sila*-process shows slightly higher hardness values for concentration level 80-3 than the *normal* mixing process and concentration level 80-5 shows the same value, indicating no influence from the higher concentration of OCTO-2HT in this case.

Wear correlates reciprocal with hardness as typically observed.

Nevertheless, best hardness and wear results were obtained for the 1phr OCTO-2HT concentration in both masterbatch mixing procedures, showing the impact of adequate silanization of the majority fraction in the reinforcing entities.

3.6 *Dynamic Thermal Mechanical Analysis (DTMA)*

Loss factor (tan δ) for each compound was determined by dynamic thermal mechanical analysis and the results are shown in Figure 10. Similar to the RPA measurements, the additivated samples show lower tan δ values at 60°C than the non-additivated reference. Although Figure 8 shows a gradual decline of loss factor with increasing additive concentration also for the masterbatch compounds, in the DMA measurement no dependence on the OCTO-2HT concentration was resolved. All additivated samples have the same lower value than the reference.

When the OCTO-2HT is added, an increase in tan δ at -10°C is observed for all compounds. Additionally the masterbatch mixing process results in a slight increase in tan δ at -10°C and a small reduction at 60°C for the reference compound without additivation. Such an effect is normally correlated with better compatibilisation of the silica with the matrix - in this case, due to the better silanization of silica in the absence of OCTO-2HT. Nevertheless, this effect is minor in comparison to the effect caused by the addition of OCTO-2HT.

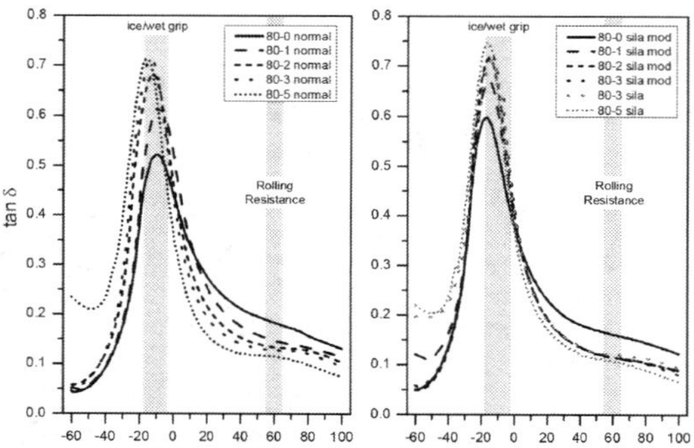

Figure 10. Caption of loss factors (tan δ) versus temperature for *normal*, *sila* and *sila mod* compounds. Gray zones indicate rolling resistance and ice/wet grip predictive parameters.

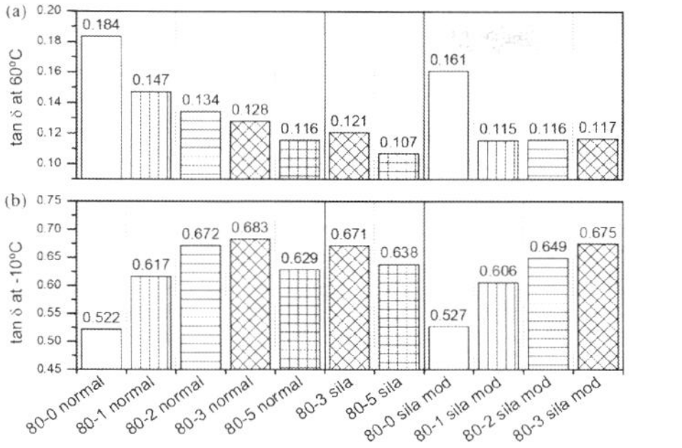

Figure 11. Caption of predictive parameters for (a) rolling resistance and (b) ice/wet grip for all evaluated compounds.

It is important to note that, with an increasing concentration of reinforcing fillers, the loss factors around the glass transition peak are normally dimished, together with a reduction in loss factors in the rubber elastic temperature regions. Here we observe an increase of the tan δ glass transition peak together with a reduction of tan δ in the rubber elastic state when the reinforcing fillers are added. This is of utmost technological interest. tan δ at 60°C is typically used as a predictor for rolling resistance performance. Improved behavior is correlated with lower values of tan δ at 60°C (lower RR) and higher values of tan δ at −10°C (better ice/wet grip). This is exactly as observed with the OCTO-2HT additivated compounds.

The predictive parameters from DTMA are shown in Figure 11.

For the *normal* process, an increase of ice/wet grip predictor of 18% and an improvement of the RR predictor of 20% are observed with the addition of 1phr of additive. The *sila mod* process improves the ice/wet predictor by 15% and the RR predictor by 29%. Such a value is obtained with the *normal* process only at a concentration of 3 phr of OCTO-2HT.

While referencing against the non-additivated sample is helpful to quantify the effect, the absolute values of tan δ are important because they quantify the dissipated energy per excitation cycle on an absolute scale, i.e. a lower tan δ value correlates with lower dissipated energy. The *sila mod* process results in a loss factor at 60°C and 1 phr of additive of 0.115 against the corresponding value of the sample obtained by the *normal* mixing process of 0.147.

4 CONCLUSIONS

The observation of a lower overall concentration of OCTO-2HT in comparison to OCTO-CTA needed to obtain similar or even better mechanical properties in the additivated silica compounds supports the view that 2HT as an organic modifier for OCTO results in better exfoliability of the octosilicate.

Changing the organic modifier required an adaptation of the mixing process. The silanization in the presence of OCTO-2HT resulted in insufficient silanization of the silica in contrast to observations with OCTO-CTA. It implies that OCTO-2HT consumes more silane than OCTO-CTA, most probably related to a higher percentage of exfoliation.

Consequently, the required concentration of additive to obtain comparable mechanical properties as in the case of OCTO-CTA additivation is also reduced from about 3 phr to about 1 phr. The same tendency is noted when changing from *normal* mixing (silanization) to the *sila mod* process (silanization of silica in the absence of OCTO-2HT). Key parameters such as small strain modulus and loss factor vary systematically with concentration of the additive, consistent with the idea of the formation of a genuine hybrid filler network, notwithstanding the possibility of some experimental inconsistency still existing.

Such dependency on the concentration of exfoliated layers opens another albeit simple-to-vary control parameter in the design of rubber compounds.

ACKNOWLEDGEMENTS

Financial support from the CDTI, Ministerio de Economía y Competitividad, Government of Spain, within the program CIEN, "NANOinTECH", grant number IDI-20141352 is gratefully acknowledged.

REFERENCES

Córdova W.R. 2016. Octosilicate modified with dimethyldioctadecylammonium (OCTO-2HT) used as additive in a tire tread compounds. Study of the concentration and silanization. *12th fall Rubber Colloquium-KHK*, Hannover (Germany).

Galimberti, Maurizio 2009. Thermal stability of ammonium salts as compatibilizers in polymer/layered silicate nanocomposites. *e-Polymers*, 056.

Kodjak, D. 2015. Policies to reduce fuel consumption, air pollution, and carbon emissions from vehicles in G20 nations. *The International Council on Clean Transportation*.

Meier, J.G. et. al. 2013a. Synthetic layered silicates as synergistic filler additive for tire tread compounds. *8th European Conference on Constitutive Models for Rubbers (ECCMR VIII)*, San Sebastián (Spain).

Meier, J.G. et. al. 2013b. Synthetic layered Silicates as Filler Additives: Synergies in Tire Tread Compounds. *KGK-Kautschuk Gummi Kunststoffe*, 46–53.
Meier, J.G. et. al. 2015. The silanization reaction of an organically modified synthetic layered silicate and its use as synergistic filler additive for tire tread compounds. *9th European Conference on Constitutive Models for Rubbers (ECCMR IX)*, Prague (Czech Republic).
Meier, J.G. et. al. 2016. On the action of synthetic layered silicates additives as filler network modifier in tire tread mixtures. *12th fall Rubber Colloquium-KHK*, Hannover (Germany).
Pike, E. 2011. Opportunities to Improve Tire Energy Efficiency. *The International Council on Clean Transportation.*
Pukánszky, B. et. al. 2006. Surface characteristics of layered silicates: Influence on the properties of clay/polymer nanocomposites. *Langmuir*, 22, 7848–7854.

Stress softening

Constitutive Models for Rubber X – Lion & Johlitz (Eds)
© 2017 Taylor & Francis Group, London, ISBN 978-1-138-03001-5

A physical interpretation for network alterations of filled elastomers under deformation: A focus on the morphology of filler–chain interactions

H. Khajehsaeid
Faculty of Engineering-Emerging Technologies, University of Tabriz, Tabriz, Iran

N. Mirzaei
Department of Chemical Engineering, Sahand University of Technology, Tabriz, Iran

ABSTRACT: Rubbers undergo a softening phenomenon called Mullins effect in first cycles of loading which is more pronounced in filled rubbers. Existence of a clear physical interpretation for this effect is of great importance in development of constitutive relations and numerical simulations. In this paper, a network alteration theory is proposed to obtain general form of evolution laws based on the physical description of the material molecular network. It is shown that the number of active monomers in material molecular network is not necessarily a decreasing function of the applied deformation but it can even increase. To find out the physical mechanism responsible for this phenomenon and also to see how it does not contravene the principle of mass conservation, the nature of broken links is investigated with focus on the role of filler particles. It is concluded that, for filled elastomers, the weak physical filler–chain interactions are the main broken links during deformation; hence, their morphology is precisely noticed and inspired by the Langmuir's theory, fractional evolution laws are formulated. Comparisons of the model results with experimental data show good correlations. It is concluded that, the developed relations well estimate the alterations of the material network and also provide a useful tool for modeling the Mullins–softening.

1 INTRODUCTION

The capability to withstand large elastic deformations under relatively small stress and also large strain recovery make elastomers extremely suitable in design of parts subjected to cyclic loadings. These materials exhibit a stress–softening in first cycles of loading called Mullins effect. Though an amount of stress–softening has also been reported for unfilled rubbers, this phenomenon is more pronounced in filled materials.

A classical approach to model this effect is the Continuum Damage Mechanics (CDM) which assigns a modified strain energy function for the softened material using a damage parameter which may correspond to any physical cause of the softening. Numerous works have been devoted to characterization and modeling of the Mullins effect by the continuum mechanics approach which make the subject well–known from the mechanical viewpoint where they mainly differ in definition of the damage criteria and/or the damage evolution laws. However, there is still no clear interpretation for its physical origin at microscopic scales which could be so advantageous for characterizing the evolution of material molecular network as well as numerical simulations (Luo et al., 2013, Khajehsaeid et al., 2013a, Khajehsaeid and Ramezani, 2014, Khajehsaeid, 2015).

An alternative approach goes through the physical theories which focus on qualitative description of microstructural alterations in material molecular network. The primary works in this context were often based on the two–phase theory (Mullins and Tobin, 1957). According to this theory, during deformation the hard phase is partly broken and transformed into soft phase which results in material softening. Also, some researches consider the chains disentanglements as the main reason of the Mullins–softening.

Macromolecular models are other alternatives for interpretation of the Mullins effect. These models are obtained by inserting the physical descriptions into equations; hence, their parameters are related to the physical properties of the materials. Network alteration theories are a category of the macromolecular models. These theories are based on the alterations induced in material molecular network during deformation.

In this paper, after briefly describing the Marckmann network alteration theory, the modification made by Chagnon et al. on this theory is thoroughly discussed. It is shown that, the results of this modification are not consistent when one uses different hyperelastic models to obtain evolutions of the network parameters. The inconsistency is more evident upon calculating the number of active monomers of the network. A general 2–step method is proposed for determining the network parameters which provides the opportunity to estimate the number of active monomers more accurately. Accordingly, it is shown that, the number of active monomers in material molecular network is not necessarily a decreasing function of the applied deformation but it can even increase. Then, the nature of broken links is investigated with focus on the role of filler particles to find out which physical mechanism is responsible for the increase of the number of active monomers and how it does not contravene the principle of mass conservation. Therefore, the morphology of the filler–chain interactions is precisely noticed and inspired by the Langmuir's theory, the adsorption and desorption of chains on fillers surface are formulated. The evolution laws for material network parameters are developed and a physically–motivated network alteration theory is proposed for the Mullins–softening of filled elastomers.

2 PHYSICAL INTERPRETATION OF NETWORK ALTERATIONS OF FILLED ELASTOMERS UNDER DEFORMATION

A recent version of the network alteration theories has been proposed by (Marckmann et al., 2002) which employs the idea of links breakage in material molecular network without emphasis on nature of the broken links. Due to this theory, during deformation some chains achieve their limiting extensibility and then break. Consequently, number of junction points between molecular chains decreases which leads to increase of average length of chains (N). On the other hand, breakage of junction points reduces the number of the chain segments per unit volume (n). Regarding this theory, number of monomer segments per unit volume (nN) must remain constant to satisfy the principle of mass conservation:

$$N = N(\lambda_{max}), \quad n = n(\lambda_{max}), \quad nN = \text{constant} \qquad (1)$$

In the present form, since this theory makes no assumption on the nature of the broken bonds it can be adapted to both filled and unfilled materials. However, it is not able to explain why the Mullins effect is more pronounced in filled rubbers compared with their unfilled counterparts. (Chagnon et al., 2006) made a minor modification on the network alteration theory of Marckmann. They assumed that, as a consequence of links breakages during first loading, some dangling chains may introduce. Therefore, the number of active monomers would decrease meanwhile maximum endured deformation by the material increases. This assumption removes the restriction (**nN = constant**) imposed by the principle of mass conservation.

Chagnon et al. proposed three modified models to predict the Mullins–softening in rubbers. They applied the modified theory to the Arruda–Boyce 8–chain, the Gent and the Hart-Smith models and proposed altered models called M2, M3 and M4 models. To adapt the modified theory to these models, the equivalency relations between AB and Gent models and also between AB and Hart–Smith models were noticed. They used the experimental data reported in (Chagnon et al., 2004) on CB-filled natural rubber to validate their theory. As expected, the number of active monomers ($\mu N = nN \times kT$) was a decreasing function of the applied deformation. The key point is that, the slope of decrease was very small meaning that only few dangling chains had been introduced. Evolutions of the network parameters are shown in Figure 1, for M2, M3 and M4 models. As seen, the numerical values of the parameters differ significantly for the altered models. This is despite the equivalency relations used in the definitions of the parameters evolutions!

The discrepancy is more clear in the case of μ. The point is that, the AB model usually underestimates the small strain part of stress–strain curve

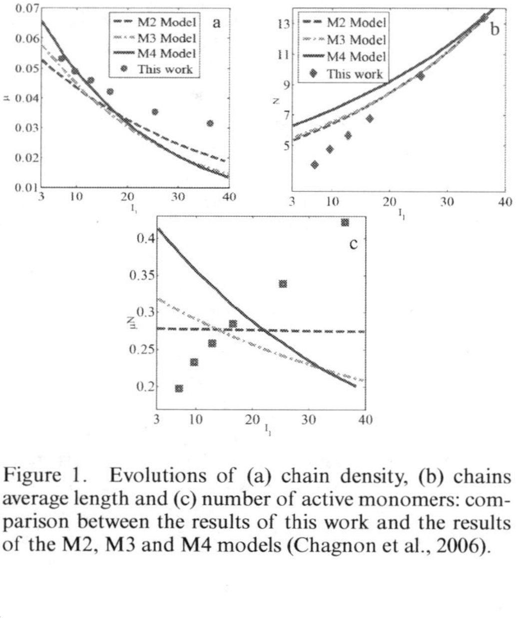

Figure 1. Evolutions of (a) chain density, (b) chains average length and (c) number of active monomers: comparison between the results of this work and the results of the M2, M3 and M4 models (Chagnon et al., 2006).

Table 1. Evolutions of network parameters and calculated number of active monomers for M2, M3 and M4 models.

Model	M2	M3	M4
Change in μ for I_1:3 $\rightarrow$ 40	−64%	−75%	−79%
Change in N for I_1:3 $\rightarrow$ 40	177%	167%	132%
Change in μN for I_1:3 $\rightarrow$ 40	−1.4%	−34.3%	−51.5%

when one attempts to obtain a good fitting over whole stress–strain regime. It is exactly the reason for M2 model to estimate lower values for μ in comparison with the latter models. More or less the Gent model is also suffered from this aspect. On the other hand, since the Hart–Smith model is driven exponentially at large stretches (rather than asymptotically), it requires a higher μ to compensate its low growth rate. Consequently as shown in Table 1, evolutions of the network parameters and especially the calculated number of active monomers obtained from M2, M3 and M4 models are totally different. Regarding the equivalency relations, one may expect to obtain–at least approximately–the same results from these models but the discrepancies arisen from the comparisons shown in Figure 1 and Table 1 motivate us to make a precise attention to the subject.

Seemingly, simultaneous determination of (n) and (N) is often problematic for accurate determination of the number of active monomers. Therefore, a new 2–step method is proposed for determination of the network parameters which provides the opportunity to estimate the number of active monomers more accurately:

1. Determination of n from small–strain regime of material stress–strain curve.
2. Determination of N from ultimately large–strain part of the stress–strain curve where the chains approach their limiting extensibility.

The hyperelastic models based on the Gaussian network theory, e.g. the neoHookean and Exp–Ln model (Khajehsaeid et al., 2013b) have reasonable accuracy at small strains while their intrinsic assumptions make them inappropriate at large strains because they do not usually take into account the limiting chain extensibility (LCE). For more accurate determination of n, it is recommended to employ such a hyperelastic model by considering only small–strain data.

On the other hand, the hyperelastic models based on non–Gaussian network theories (e.g. the AB model) basically have a crucial assumption on (LCE). Whenever the stretch of a chain approaches this limiting value, the force required for further stretching the chain increases sharply. Thus, any hyperelastic model based on (LCE) would behave asymptotically at large strains and can be employed for determination of N. Since these models often underestimate the small–strain data, in the fitting process the large–strain regime must be considered exclusively or at least given more weight.

In the following, the neoHookean and the Gent models (due to their mathematical simplicity) are respectively employed to determine $\mu(=nkT)$ (from small–strain data) and N (from large–strain data) for the experimental data reported in (Chagnon et al., 2004). The determined evolutions of μ and N are shown in Figure 1. Though trends of the results are similar to those of M2, M3 and M4 models, the numerical values are totally different especially in the case of μ. The evolution of μN is also shown in Figure 1 where one can marvelously observe that, the number of active monomers not only is not decreasing (or even constant) but also is increasing during the deformation!

The proposed procedure is also conducted on the experimental data reported by Diani et al. on CB-filled natural rubber (Diani et al., 2006). Results are shown in Figure 2 where it is observed that, the number of active monomers is increasing.

The main question which arises here is that–if the principle of mass conservation is not contravened–what is the physical mechanism responsible for increase of the number of active monomers during deformation? To this end, it is essential to investigate the nature of links which break during deformation.

2.1 *Investigation on the nature of broken links*

Based on several experiments, the Mullins–softening is more apparent in filled materials (Diaz et al., 2014, Mullins and Tobin, 1957); thus, nature of the broken links is noticed with focus on the role of filler particles and their interactions with chains.

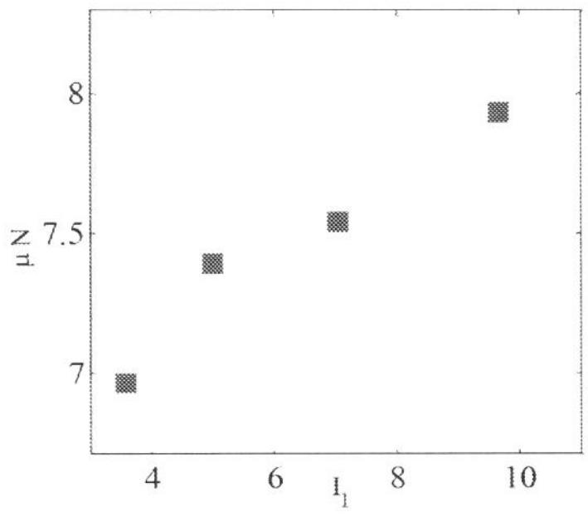

Figure 2. Evolutions of number of active monomers with respect to the applied deformation for data reported in (Diani et al., 2006).

It is known that, the nature of bonds between filler particles and molecular chains greatly depends on the type of filler, chemical composition of chains and chemical reactants on filler particles (Lorenz and Klüppel, 2012). However, these bonds are in general a combination of strong chemical (covalent type) and weak physical bonds. On the other hand, the bonds between two adjacent chains are mostly of covalent type created during the vulcanization process. Regarding the bond energy associated with physical bonds between a chain and a filler particle (7.7–11.5 kJ/mol for a single contact (Simha et al., 1953)) and the energy of covalent cross–links (351 kJ/mol for –C–C–, 285 kJ/mol for –C–S–C–, 267 kJ/mol for –C–S–S–C– and <267 kJ/mol for –C–S_x–C (Vyazovkin and Sbirrazzuoli, 2006)), one may conclude that, during deformation if some links would be supposed to break, weak physical bonds (between filler particles and chains) are more susceptible rather than covalent bonds (either between chains or between fillers and chains). Therefore, in filled materials, even if some chemical bonds are broken, their number would be negligible in comparison to the broken physical bonds. However, in the case of unfilled materials one should also notice to the broken chemical bonds. This interpretation can justify the higher Mullins–softening in filled materials compared with their unfilled counterparts.

2.2 *Morphology of the filler–chain bonds*

A filler–chain bond is made if at least one monomer segment of a chain is adsorbed on a filler surface. This bond stabilizes if its neighboring segments can also attach to the nearby free interaction positions on the surface (Maier and Goritz, 1996). The other end of the chain may be: (1) free (i.e. elastically inactive chain) or (2) adsorbed on another filler particle or (3) linked to the chains network. Each bond of types (2) and (3) increases the number of active chain segments by one where any chain of type (1) does not contribute to the material stiffness.

If there are a certain number of interaction positions on each filler particle, first chains contacting the surface will find the whole positions free. Formation of the first bond will increase the probability of adsorption of the neighboring segments. Consequently, the first chains make strong bonds (even if they are merely of physical type) with the filler particle. The later chains would find much less free positions to make an interaction. Therefore, their bonds with filler particles would be weaker (made at isolated positions) and easily removable. After a while, rates of adsorption and desorption adopt stable values and hence, a balance would be achieved. However, applying a deformation to the material may affect the balance by accelerating the breakage of weak physical filler–chain bonds.

Analogous to the Langmuir's theory–originally proposed for adsorption and desorption of gas molecules on surfaces– one can describe the aforementioned balance as:

$$\pi_a \xi^{free} = \pi_d \xi^{occ} \tag{2}$$

where π_a and π_d are the rates of chains adsorption and desorption on the filler, respectively. Also ξ^{free} and ξ^{occ} are the fractions of free and occupied interaction positions on the filler where $\xi^{free} + \xi^{occ} = 1$.

Since the number of available chains nearby a free position is constant, π_a is assumed to remain unchanged during deformation. However, the rate of desorption π_d would be affected by the deformation applied. As much as more deformation is applied more physical filler–chain bonds break. As shown in Figure 3, breakage of a filler–chain bond may result in the following phenomena:

I. Conversion of chain to a dangling one which reduces the number of (active) chains by one.
II. Activating an asleep chain which was not active so far. In this case the number of (active) chains increases by one.
III. Increase in active length of the chain. In this case average length of the chains increases.

The II and III items describe how the number of active monomers can increase during deformation without contravening the principle of mass conservation. To describe the effect of deformation, we assume the rate of desorption is a function of the applied deformation:

$$\pi_d = \pi_d^{\,0}(1 + C_1\gamma) \tag{3}$$

where C_1 is a positive constant and γ is an arbitrary measure of deformation. From (2) and (3) one obtains fraction of occupied interaction positions:

$$\xi^{occ} = \frac{\pi_a}{\pi_a + \pi_d^{\,0}(1 + C_1\gamma)} = \frac{1}{C_2 + C_3\gamma} \quad \text{where} \quad \xi_0^{\,occ} = \frac{1}{C_2} \tag{4}$$

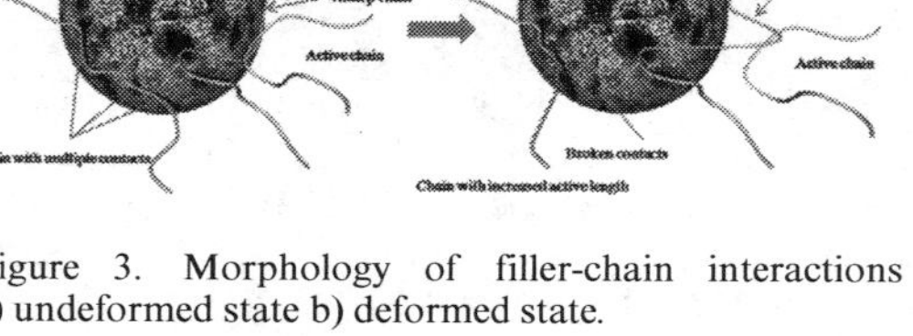

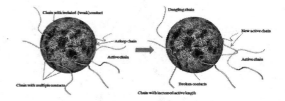

Figure 3. Morphology of filler-chain interactions a) undeformed state b) deformed state.

It means that, the number of filler–chain interactions decreases with respect to γ. For a filled network we have (Heinrich and Vilgis, 1993):

$$\mu = (1 - v_f)\mu^{unfilled} + v_f \mu^{f-ch}(\xi^{occ}) \tag{5}$$

where v_f is the filler volume fraction, $\mu^{unfilled} = n^{unfilled}kT$ corresponds to the chain density of the unfilled network and μ^{f-ch} corresponds to the filler–chain interactions. According to (I) and (II) active chains density per unit volume would depend on γ. From (4) and (5) we have:

$$\mu = (1 - v_f)\mu^{unfilled} + v_f \frac{\mu_0^{f-ch}}{1 + C_3\gamma / C_2} = \frac{1 + C_4\gamma}{C_5 + C_6\gamma} \tag{6}$$

where μ_0^{f-ch} is the initial value of μ^{f-ch} for the undeformed state and C_4, C_5 and C_6 are constant values. From the initial undeformed state and extremely high–strain state (which would not be achieved in reality because of material failure) we have the following relations for the introduced constants $\gamma = 0: \mu = 1/C_5$ and $\gamma \to \infty: \mu = C_4 / C_6$.

By experience we know that μ is a decreasing function along primary loading curves; thus one may conclude that, onset of new active chains (according to II) usually cannot dominate the onset of new dangling chains (according to I). The other consequence of (4) is that, the average length of chain segments would also depend on γ (according to III). Since N is an average quantity, for a filled network we have:

$$N = (1 - v_f)N_1 + v_f(N_2^{free} + N_2^{detached}(\xi^{occ})) \tag{7}$$

where N_1 is the average chain length of the unfiled network, N_2^{free} is the average free length of the chains which are in interaction with filler particles and $N_2^{detached}$ is the average detached length of these chains from the fillers surface. The latter term corresponds to the chain parts which were sleeping on the fillers before the deformation. Therefore, $N_2^{detached}$ is equal to the change in the average length of the sleeping parts. If a sleeping part consists of $\Upsilon(\propto \xi^{occ})$ monomer segments, we have $N_2^{detached} = \Upsilon(\xi_0^{occ}) - \Upsilon(\xi^{occ})$.

Replacing (4) in this equation we obtain $N_2^{detached}$ as a function of the applied deformation:

$$N_2^{detached} = \frac{\Gamma}{C_2} - \frac{\Gamma}{C_2 + C_3\gamma} \tag{8}$$

where Γ is a proportionality constant. Finally, replacing (8) in (7) the evolution of N is obtained:

$$N = \frac{1 + C_7\gamma}{C_8 + C_9\gamma} \tag{9}$$

Again, from the initial undeformed state and extremely high–strain state we have $\gamma = 0: N = 1/C_8$ and $\gamma \to \infty: N = C_7 / C_9$. Finally, the production of (6) and (9) will give the number of active monomer segments as a function of the applied deformation.

3 RESULTS

To validate the developed relations, evolution of material molecular network is investigated for two sets of experimental data. First we consider the experimental data reported in (Chagnon et al., 2004) on carbon black filled natural rubber. Since I_1 is related to the chains stretch, $\gamma = I_1 - 3$ is used in the relations. One may use any other measure of deformation without loss of generality (Khajehsaeid, 2016, Khajehsaeid, 2017). According to (6) and (9), the evolutions of μ and N are estimated as:

$$\mu = \frac{1 + 0.01(I_1 - 3)}{16.37 + 0.8(I_1 - 3)}, \; N = \frac{1 + 0.21(I_1 - 3)}{0.51 + 0.003(I_1 - 3)} \tag{10}$$

The model results are shown in Figure 4 where good correlations are seen between the model estimations and the experimental curves. Furthermore, the number of active monomer segments is also obtained which shows that the developed theory well describes the evolution of the number of active monomers. Moreover, the simple

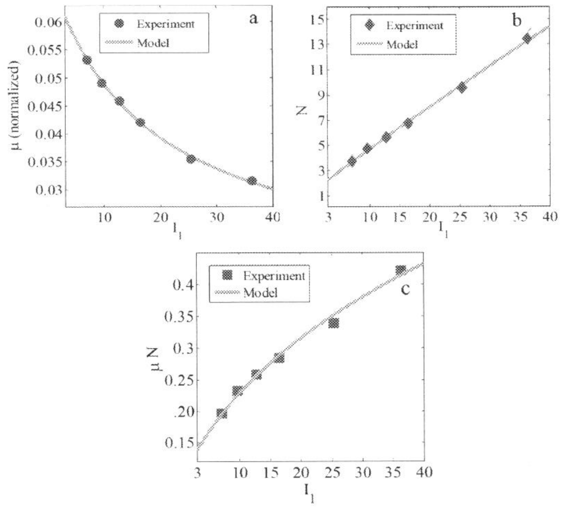

Figure 4. Evolutions of (a) chain density, (b) chains average length and (c) number of active monomers: comparison of model results with data reported in (Chagnon et al., 2004).

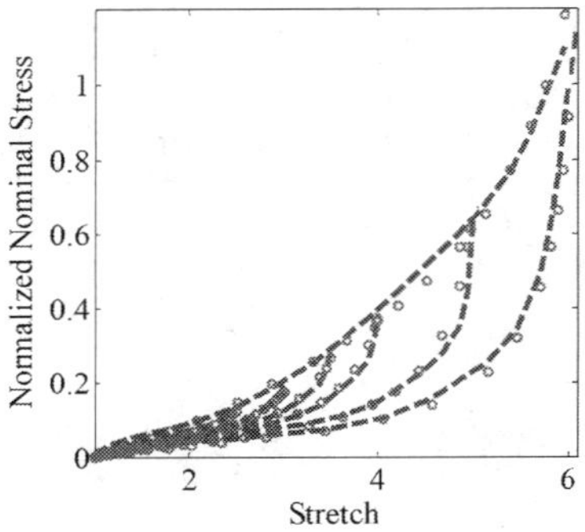

Figure 5. The model results and experimental cyclic data using Gent model. Dashed lines: experimental data from (Chagnon et al., 2004), symbols: the model results.

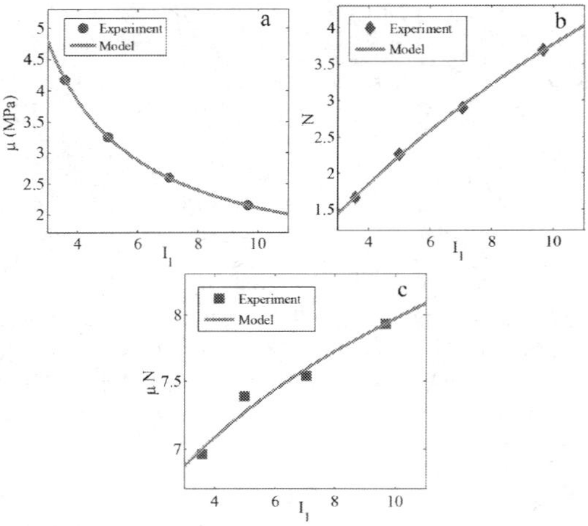

Figure 6. Evolutions of (a) chain density, (b) chains average length and (c) number of active monomers: Comparison of the model results with data reported in (Diani et al., 2006).

mathematical formulation of the evolution laws facilitates its numerical implementation (Khajehsaeid et al., 2013a).

It should be noted that, from (10) for the undeformed state we have $\mu|_{\gamma=0} = 0.061$ which is very near to the shear modulus $\mu = 0.064$ obtained from fitting the experimental loading curve by Exp–Ln model.

The developed evolution laws can be used along with any hyperelastic model defined based on the network parameters to predict the cyclic curves. In Figure 5, comparisons between the model results and experimental cyclic data are presented using Gent strain energy which implies that the architecture of the proposed theory does not depend on the explicit choice of hyperelastic models.

The experimental data reported in (Diani et al., 2006) for CB-filled NR is also considered. The evolutions laws for this material are determined as:

$$\mu = \frac{1+0.068(I_1-3)}{0.21+0.07(I_1-3)},\ N = \frac{1+0.34(I_1-3)}{0.7+0.03(I_1-3)} \quad (11)$$

Again, one can calculate the number of active monomer segments. Figure 6 shows the results of the model in comparison with those obtained from experimental curves.

4 CONCLUSIONS

Results of network alteration theories have been shown not to be consistent for evaluation of material network alterations during deformation when using different hyperelastic models. The inconsistency was more evident upon calculating the number of active monomers. Therefore, a 2–step method was proposed for determining the network parameters which proved that the number of active monomers is not necessarily a decreasing function of the applied deformation but it can even increase. Analyzing the bond energy for covalent cross–links and also for physical bonds, we concluded that, the weak physical filler–chain interactions are the main broken links during deformation. Analogous to the Langmuir's theory, the adsorption and desorption of chains on fillers surface were formulated and fractional evolution laws were developed for network parameters. It was shown that, the proposed physically–motivated theory well estimates network evolution during deformation and also provides a useful tool for modeling Mullins–softening. Architecture of the proposed theory does not depend on the explicit choice of hyperelastic models and can be applied to any model based on the network parameters.

REFERENCES

Chagnon, G., Verron, E., Gornet, L., Marckmann, G. & Charrier, P. 2004. On the relevance of Continuum Damage Mechanics as applied to the Mullins effect in elastomers. *Journal of the Mechanics and Physics of Solids,* 52, 1627–1650.

Chagnon, G., Verron, E., Marckmann, G. & Gornet, L. 2006. Development of new constitutive equations for the Mullins effect in rubber using the network alteration theory. *International Journal of Solids and Structures,* 43, 6817–6831.

Diani, J., Brieu, M. & Vacherand, J.M. 2006. A damage directional constitutive model for Mullins effect with permanent set and induced anisotropy. *European Journal of Mechanics - A/Solids,* 25, 483–496.

Diaz, R., Diani, J. & Gilormini, P. 2014. Physical interpretation of the Mullins softening in a carbon-black filled SBR. *Polymer,* 55, 4942–4947.
Heinrich, G. & Vilgis, T. 1993. Contribution of entanglements to the mechanical properties of carbon black-filled polymer networks. *Macromolecules,* 26, 1109–1119.
Khajehsaeid, H. 2015. Modeling nonlinear viscoelastic behavior of elastomers using a micromechanically motivated rate-dependent approach for relaxation times involved in integral-based models. *Constitutive Models for Rubber IX*, 165.
Khajehsaeid, H. 2016. Development of a network alteration theory for the Mullins-softening of filled elastomers based on the morphology of filler–chain interactions. *International Journal of Solids and Structures* 80, 158–167.
Khajehsaeid, H. 2017. Mullins thresholds in context of the network alteration theories. *International Journal of Mechanical Sciences* 123, 43–53.
Khajehsaeid, H., Baghani, M. & Naghdabadi, R. 2013a. Finite strain numerical analysis of elastomeric bushings under multi-axial loadings: a compressible visco-hyperelastic approach. *International Journal of Mechanics and Materials in Design* 1–15.
Khajehsaeid, H., Naghdabadi, R. & Arghavani, J. 2013b. A strain energy function for rubber-like materials. *Constitutive Models for Rubber,* 8, 205–210.
Khajehsaeid, H. & Ramezani, M.A. Year. Visco-hyperelastic modeling of automotive elastomeric bushings with emphasis on the coupling effect of axial and torsional deformations. *In:* Proceedings of the Society of Engineering Science 51st Annual Technical Meeting, October 1–3, 2014, 2014 West Lafayette, Purdue University.
Lorenz, H. & Klüppel, M. 2012. Microstructure-based modelling of arbitrary deformation histories of filler-reinforced elastomers. *Journal of the Mechanics and Physics of Solids* 60, 1842–1861.
Luo, Y., Liu, Y. & Yin, H.P. 2013. Numerical investigation of nonlinear properties of a rubber absorber in rail fastening systems. *International Journal of Mechanical Sciences,* 69, 107–113.
Maier, P. & Goritz, D. 1996. Molecular interpretation of the Payne effect. *Kautschuk Gummi Kunststoffe,* 49, 18–21.
Marckmann, G., Verron, E., Gornet, L., Chagnon, G., Charrier, P. & Fort, P. 2002. A theory of network alteration for the Mullins effect. *Journal of the Mechanics and Physics of Solids,* 50, 2011–2028.
Mullins, L. & Tobin, N.R. 1957. Theoretical Model for the Elastic Behavior of Filler-Reinforced Vulcanized Rubbers. *Rubber Chemistry and Technology* 30, 555–571.
Simha, R., Frisch, H. & Eirich, F. 1953. The adsorption of flexible macromolecules. *The Journal of Physical Chemistry,* 57, 584–589.
Vyazovkin, S. & Sbirrazzuoli, N. 2006. Isoconversional Kinetic Analysis of Thermally Stimulated Processes in Polymers. *Macromolecular Rapid Communications,* 27, 1515–1532.

Rheology and processing

Constitutive Models for Rubber X – Lion & Johlitz (Eds)
© 2017 Taylor & Francis Group, London, ISBN 978-1-138-03001-5

The evolution of viscoelastic properties of silicone rubber during cross-linking investigated by thickness-shear mode quartz resonator

A. Dalla Monta
Institute of Physics, UMR 625-CNRS, University of Rennes 1, Campus de Beaulieu, Rennes, France
ENS Rennes, SATIE—CNRS 8029, Campus de Ker Lann, Bruz, France

F. Razan
ENS Rennes, SATIE—CNRS 8029, Campus de Ker Lann, Bruz, France

J.-B. Le Cam
Institute of Physics, UMR 625-CNRS, University of Rennes 1, Campus de Beaulieu, Rennes, France
LC-DRIME, Joint Research Laboratory in Imaging, Mechanics and Elastomers, University of Rennes 1 /Cooper Standard/CNRS, Campus de Beaulieu, Rennes Cedex, France

G. Chagnon
Université Grenoble Alpes, TIMC-IMAG, Grenoble, France
CNRS, TIMC-IMAG, Grenoble, France

ABSTRACT: Characterizing the effects of cross-linking level and kinetics on the mechanical properties of rubber, especially viscoelasticity, provides information of importance to better understand and predict its final mechanical properties. Classically, the effects of cross-linking on the mechanical properties are investigated with a rheometer. Typical results give the evolution of elastic properties of rubber in the solid state with respect to time or to the cross-linking level. The frequency of the mechanical loading applied is generally a few hertz. In the case where the rubber is initially in the liquid state, such as some silicone rubbers, this type of characterization is not suitable anymore. In this study, a new characterization technique based on the Quartz Crystal Microbalance (QCM) principle has been developed in order to characterize the viscoelastic properties (elastic and viscous moduli) of a silicone rubber during cross-linking, *i.e.* from the liquid (uncross-linked) to the solid (final cross-linked) state. The device consists in a Thickness-Shear Mode (TSM) resonator generating ultrasonic waves, which provides viscoelastic properties of a material in contact with its surface from an electrical impedance analysis. In contrast to other characterization tools, it makes possible the continuous and non-destructive characterization of viscoelastic properties from a small material volume, under 1 mL. Moreover, frequencies at which these properties are characterized are of the order of magnitude of the megahertz, which provides very complementary results to classical characterization, rather in the order of the hertz.

1 INTRODUCTION

The Quartz Crystal Microbalance (QCM) is a versatile tool, first described in the founder paper work of (Sauerbrey 1959). Because it is a rather simple sensor, it has been widely used in sensing applications in various domains, allowing us to determine with precision measurands such as mass density (Stockbridge 1966b), viscosity (Kanazawa and Gordon 1985) and pressure (Stockbridge 1966a), in a continuous and non-destructive manner, with sample as small as a microliter. In mechanics, for instance, it is used for measuring complex shear modulus $\tilde{G}$ in order to characterize a polymer (Holt et al. 2006) or follow its evolution during a specific process, such as dissolution (Hinsberg et al. 1986). In biology, where its use continues to increase (Becker and Cooper 2011), the functionalization of the QCM surface with a definite substance allows to measure active species absorption or deposition, and then to recognize specific pathologies like schistosomiasis (Wang et al. 2006) or Ebola fever (Yu et al. 2006). In chemistry, with a similar method, the sensor can detect presence of harmful molecules in the air, acting as an electronic nose (Si et al. 2007).

The QCM, as shown in Figure 1, consists in a circular thin disk of piezoelectric AT-cut quartz, with metallic electrodes on both sides. The application of a voltage between them generates a shear deformation of the crystal, which can then be excited into resonance when its thickness is near an odd multiple of half the acoustic wavelength. The deposit of a material sample on the crystal changes the resonance properties, and it appears that they are fundamentally dependent on the characteristics, either mechanical or electrical, of the sample.

For all these applications, the QCM needs an electrical interface able to apply a sinusoidal voltage between its electrodes and to measure the resonance conditions. Among all the methods of read-out, the most used is the impedance analysis (Arnau 2008), which maps the electrical admittance of the QCM as a function of the frequency, giving access to a greater number of parameters than simpler methods such as the use of an oscillator circuitry. However, it requires an impedance analyzer, which is typically relatively expensive and hardly mobile. A portable and low-cost solution would therefore be particularly useful, especially in biology, where in situ assays would help patients at home and doctors on the spot, ensure the safety in the entire chain of the food industry or improve the security against biological agents (Nayak et al. 2009). A cheap, simple and robust device is also critical for the commercialization of the whole system, opening the door to the democratization of the associated analysis.

The present paper aims at validating the use of a compact and comparatively cheap network analyzer, the "miniVNA PRO", as a read-out instrument for the QCM for the characterization of the

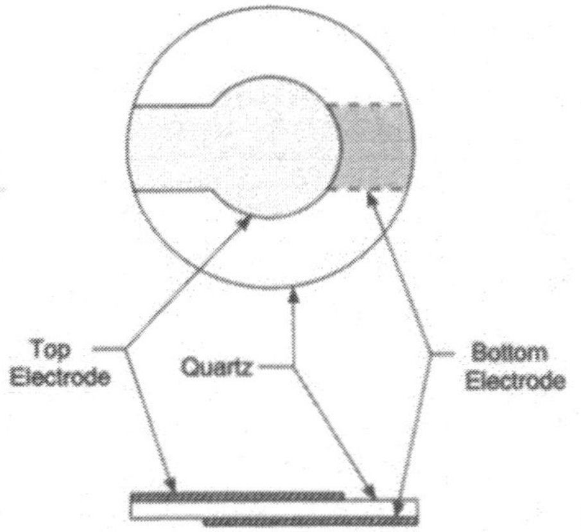

Figure 1. Top and side views of a quartz crystal microbalance (Martin et al. 1993).

cross-linking of rubbers. It is organized as follows. Section 2 describes briefly the theory governing the behavior of the QCM and the measurement principle. Section 3 presents the experimental dispositive and the protocol established in order to reduce the influence of external factors. Finally, section 4 analyses the results in light of the aforementioned objectives.

2 THEORETICAL FRAMEWORK

On a fundamental point of view, the QCM is simply a transducer, linking its load impedance with its electrical impedance. The fundamental relations governing its behavior are briefly recalled hereafter. The reader can refer to Johannsmann (2015) for further information.

2.1 *Acoustic parameterization*

In a first approximation, when its diameter is large compared to its thickness, the QCM can essentially be seen as a unidimensional device, a succession of homogeneous layers in which acoustic shear waves propagate along the z axis. Inside a layer, the amplitude $u(t,z)$ of the displacement due to the wave is described by the well-known wave equation, which writes:

$$\frac{\partial^2 u}{\partial t^2} = c^2 \frac{\partial^2 u}{\partial z^2} \tag{1}$$

This can be expressed in the frequency domain by:

$$-\omega^2 \hat{u} = \tilde{c}^2 \frac{d^2 \hat{u}}{du^2} \tag{2}$$

Here, ω is the angular frequency, $\hat{u}(z)$ is the complex amplitude of the displacement, $\tilde{c} = \sqrt{\rho / \tilde{G}}$ is the speed of sound, ρ the density and $\tilde{G}$ the shear modulus. Introducing the wavenumber $\tilde{k} = \sqrt{\omega / \tilde{c}}$, solutions of the wave equation write in the form of:

$$\hat{u}(z) = \hat{u}^+ \exp\left(+i\tilde{k}z\right) + \hat{u}^- \exp\left(-i\tilde{k}z\right) \tag{3}$$

$\hat{u}^+$ and $\hat{u}^-$ are the amplitude of a wave traveling respectively to the left and to the right. It is therefore possible to express the velocity $\hat{v}(z)$ and the shear stress $\hat{\sigma}(z)$:

$$\begin{aligned} \hat{v}(z) &= i\omega \hat{u}(z) \\ &= i\omega \hat{u}^+ \exp\left(+i\tilde{k}z\right) + i\omega \hat{u}^- \exp\left(-i\tilde{k}z\right) \end{aligned} \tag{4}$$

$$\hat{\sigma}(z) = \tilde{G}\frac{d\hat{u}}{dz}$$
$$= i\tilde{k}\tilde{G}\hat{u}^{+}\exp\left(+i\tilde{k}z\right) - i\tilde{k}\tilde{G}\hat{u}^{-}\exp\left(-i\tilde{k}z\right) \quad (5)$$
$$= i\omega\tilde{Z}\hat{u}^{+}\exp\left(+i\tilde{k}z\right) - i\omega\tilde{Z}\hat{u}^{-}\exp\left(-i\tilde{k}z\right)$$

2.2 *Mason circuit*

The acoustic shear wave is very similar to an electromagnetic wave in its behavior (Mason 1941). Establishing an analogy between electrical circuits and mechanical systems, it is therefore possible to represent a layer of the QCM as a distributed-element network. Such a representation is perfectly suitable to this device, because our goal is to link a mechanical quantity with an electrical one.

To go further, since only the quantities at the interfaces between two layers are of interest, the QCM can even be represented as a two-port network, one for each interface. This paragraph shows that the Mason circuit is a compatible representation.

In Figure 2, A is the area of the device in the plane orthogonal to the direction of propagation, and h is half the thickness of the considered layer. Using the Kirchhoff rules, it appears that forces $\hat{F}$ and velocities $\hat{v}$ are linked by the following expressions that should be validated:

$$\hat{F}_1 = iA\tilde{Z}\tan\left(\tilde{k}h\right)\hat{v}_1 - \frac{iA\tilde{Z}}{\sin\left(2\tilde{k}h\right)}\cdot\left(\hat{v}_1 + \hat{v}_2\right) \quad (6)$$

$$\hat{F}_2 = \frac{iA\tilde{Z}}{\sin\left(2\tilde{k}h\right)}\cdot\hat{v}_1 - iA\tilde{Z}\tan\left(\tilde{k}h\right)\cdot\left(\hat{v}_1 + \hat{v}_2\right) \quad (7)$$

The displacement and the shear stress can be written as:

$$\hat{u}(z) = \hat{u}_\alpha \sin\left(\tilde{k}z\right) + \hat{u}_\beta \cos\left(\tilde{k}z\right) \quad (8)$$

$$\hat{\sigma}(z) = \omega\tilde{Z}\left(\hat{u}_\alpha \cos\left(\tilde{k}z\right) - \hat{u}_\beta \sin\left(\tilde{k}z\right)\right) \quad (9)$$

Hence, being careful about the sign of the relations:

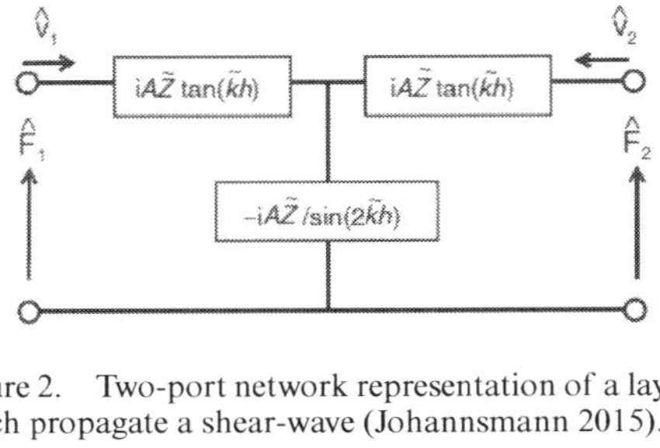

Figure 2. Two-port network representation of a layer in which propagate a shear-wave (Johannsmann 2015).

$$\begin{aligned} \hat{F}_1 &= -A\hat{\sigma}(-h) & \hat{v}_1 &= +i\omega\hat{u}(-h) \\ \hat{F}_2 &= -A\hat{\sigma}(h) & \hat{v}_2 &= -i\omega\hat{u}(h) \end{aligned} \quad (10)$$

This yields to the set of equations 6–7 after some mathematical manipulations, confirming the correctness of the circuit used. However, as such, the circuit is still incomplete: the quartz being piezoelectric, a third port electrical in nature should be added, as shown in Figure 3. Across a transformer, an electrical source connected to this port will be able to generate an acoustic wave and to be influenced by the mechanical properties of the device, with a conversion factor $\phi = A\cdot e_{26}/(2h)$, where e_{26} is the relevant component of the piezoelectric coupling tensor.

It can be shown that this three-port network satisfies the constitutive relations of piezoelectricity, namely:

$$\hat{\sigma} = \tilde{G}\frac{d\hat{u}}{dz} - \frac{e_{26}}{\tilde{\varepsilon}\varepsilon_0}\hat{D} \quad (11)$$

$$\hat{E} = -\frac{\tilde{e}_{26}}{\tilde{\varepsilon}\varepsilon_0}\frac{d\hat{u}}{dz} + \frac{1}{\tilde{\varepsilon}\varepsilon_0}\hat{D} \quad (12)$$

Finally, in practice and on one side, the QCM is in contact with the air, which has acoustic wave impedance negligible compared with the one of the quartz. Therefore $\hat{F}_1 = 0$, and the equivalent circuit is short-circuited on the left. On the other side, the QCM is in contact with the material to characterize, with acoustic impedance at the interface $\tilde{Z}_L$. Therefore $\hat{F}_2/\hat{v}_2 = A\tilde{Z}_L$, and the equivalent circuit is closed with a resistance having this value. It remains in the circuit only the electrical port, as illustrated in Figure 3, which corresponds to the fact that the QCM can only be interrogated electrically through its electrodes, and not by any acoustic or mechanical ways.

2.3 *Resonance condition*

Let us now use the Mason circuit to calculate the equivalent impedance of the circuit coming after

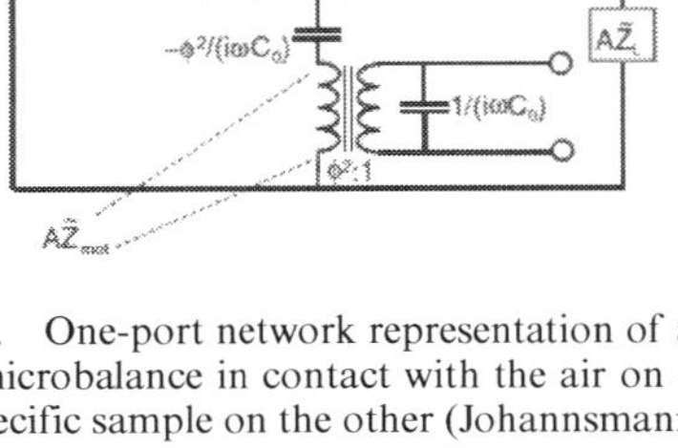

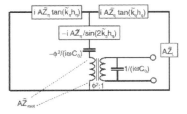

Figure 3. One-port network representation of a quartz crystal microbalance in contact with the air on one side and a specific sample on the other (Johannsmann 2015).

the transformer, the motional impedance $\tilde{Z}_{mot}$. Using the Kirchhoff rules, it follows:

$$\tilde{Z}_{mot} = -\frac{\phi^2}{i\omega C_0} - \frac{i\tilde{Z}_q}{\sin\left(2\tilde{k}_q h_q\right)} + \left(\left(i\tilde{Z}_q \tan\left(\tilde{k}_q h_q\right)\right)^{-1} + \left(i\tilde{Z}_q \tan\left(\tilde{k}_q h_q\right) + \tilde{Z}_L\right)^{-1}\right)^{-1} \tag{13}$$

This relation can be simplified by assuming that the wavenumber $\tilde{k}_q$ is close to the ideal open-circuit wavenumber $\tilde{k}_{q,OC} = n\pi / 2h_q$, solution of the equation $\tilde{Z}_{mot} = 0$ when the QCM is fully immerged in the air without influence of the piezoelectric effect. Using a Taylor expansion of the previous relation yields:

$$\tilde{Z}_{mot} \approx \frac{1}{4}\left(-\frac{4\phi^2}{i\omega C_0} + in\pi\tilde{Z}_q \frac{\tilde{\omega} - \tilde{\omega}_{OC}}{\tilde{\omega}_{OC}} + \tilde{Z}_L\right) \tag{14}$$

This relation can then be applied twice: first, in the unloaded reference state (the angular resonance frequency being called $\tilde{\omega}_{ref}$) with $\tilde{Z}_L = 0$; secondly, in charge (the angular resonance frequency being called $\tilde{\omega}_{sample}$). Calculating the frequency shift by using the resulting equations brings the so-called Gordon–Kanazawa–Mason result:

$$\frac{\Delta\tilde{f}}{f_0} = \frac{\tilde{\omega}_{ref} - \tilde{\omega}_{sample}}{\frac{\tilde{\omega}_{OC}}{n}} = \frac{i}{\pi\tilde{Z}_q}\tilde{Z}_L \tag{15}$$

This is a direct relation between frequencies measurement and the load impedance of the tested sample. To go further, it is assumed that the sample is a semi-infinite medium and equations 4 and 5 yield:

$$\frac{\hat{\sigma}(z)}{\hat{v}(z)} = \frac{i\omega\tilde{Z}_{sample}\hat{u}^{+}\exp\left(+i\tilde{k}_{sample}z\right)}{i\omega\hat{u}^{+}\exp\left(+i\tilde{k}_{sample}z\right)} = \tilde{Z}_{sample} \tag{16}$$

Hence:

$$\tilde{Z}_L = \frac{\hat{\sigma}\left(h_q\right)}{\hat{v}\left(h_q\right)} = \tilde{Z}_{sample} = \sqrt{\rho_{sample}\tilde{G}_{sample}} \tag{17}$$

Given that the density of the sample is already known, the shear modulus can be deduced. Of course, the quantities characterizing the quartz in the previous relation ($\tilde{Z}_q$ and f_0) are not known a priori. However, they can be evaluated by making an additional measurement with a sample in a well-known material, for instance water.

2.4 *Extraction of the resonance properties*

The impedance analyzer used in these experiments measures the electrical admittance $\tilde{Y}_{el,measure}$ as a function of the frequency. This paragraph explains how the complex resonance frequency can be deduced from this measurement. It can be shown that the linearization around the open-circuit frequency used previously allows getting a simpler representation of the equivalent circuit, using only standard electrical elements (resistance, inductance and conductance). That is the well-known Butterworth-Van-Dyke model, shown in Figure 4.

The Kirchhoff rules yield the electrical admittance of this circuit:

$$\tilde{Y}_{el} = i\omega C_0 + \left(i\omega\bar{L}_1 + \frac{1}{i\omega\bar{C}_1} + \bar{R}_1\right)^{-1} \tag{18}$$

It is possible to fit the measurement $\tilde{Y}_{el,measure}(\omega)$ with this theoretical $\tilde{Y}_{el}(\omega)$ and deduce the complex resonance frequency from the values of the four elements, but it is helpful to use directly the following expanded Lorentzian functions, that reduce the scatter in the fit parameters:

$$\Re e\left(\tilde{Y}_{el}\right) = G_{el,max}\Gamma\frac{f}{f_r}\left(\frac{\Gamma}{\left(f_r - f\right)^2 + \Gamma^2}\cos(\theta) + \frac{f_r - f}{\left(f_r - f\right)^2 + \Gamma^2}\sin(\theta)\right) + G_{el,off} \tag{19}$$

$$\Im m\left(\tilde{Y}_{el}\right) = G_{el,max}\Gamma\frac{f}{f_r}\left(\frac{\Gamma}{\left(f_r - f\right)^2 + \Gamma^2}\cos(\theta) - \frac{f_r - f}{\left(f_r - f\right)^2 + \Gamma^2}\sin(\theta)\right) + B_{el,off} \tag{20}$$

The complex resonance frequency is then simply deduced from these parameters:

$$\tilde{f}_r = f_r + i\Gamma \tag{21}$$

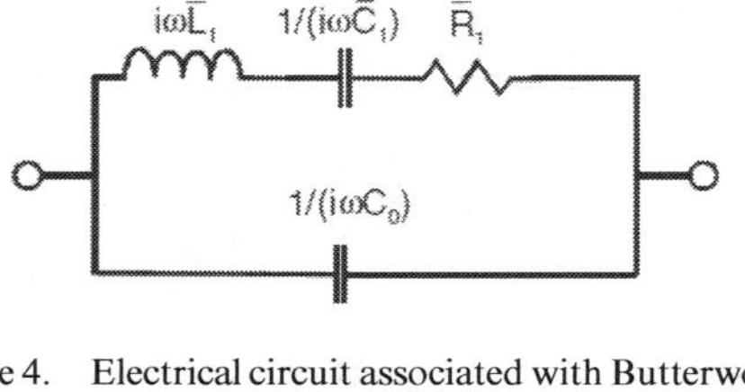

Figure 4. Electrical circuit associated with Butterworth-Van-Dyre model (Johannsmann 2015).

Figure 5. Photograph of the network analyzers used in the experiment: the mini-VNA PRO.

3 MATERIALS AND METHODS

The quartz crystal and its holder used in the experiments are the commercially available QCM200 (Stanford Research Systems, CA, USA). The crystal has a resonance frequency near 5 *Mhz* and a diameter of 2.54 *cm*. It is covered with circular electrodes of titanium and gold. It is physically maintained with one O-ring on both side and connected with the electrical interface via BNC connectors, which are then adapted to an SMA connection. The portable network analyzer is the miniVNA PRO (mini Radio Solutions, Germany), shown in Figure 5.

The QCM is placed inside a thermostatic chamber at a given controlled temperature. The network analyzer is initially calibrated by the short-open-load method. The quartz crystal is washed with acetone, rinsed with water and dry with nitrogen before being placed inside its holder. Once the resonance frequency is stabilized, the crystal is loaded with a volume of 900 μL of distilled water as the reference material, creating a layer thick enough to be considered semi-infinite. Finally, after stabilization of the resonance frequency, the water is removed and replaced by the same volume of 900 μL of PDMS rubber RTV615 (Momentive Performance Materials Inc., NY, USA) in the liquid state, *i.e.* non cross-linked. Thus, the evolution of the viscoelastic properties will be measured continuously from the liquid (uncross-linked) state to the solid (cross-linked) state.

4 RESULTS

Figure 6 gives the evolution of the viscoelastic properties in terms of shear modulus G' (a) and loss factor $\tan\delta$ (b) at different ambient temperatures, respectively **25°*C***, **50°*C*** and **80°*C***. At **25°*C***, the value of the shear modulus is stabilized from 35 hours on. Its evolution is strongly nonlinear from 4 MPa for the liquid state (only slightly cross-linked) to the stabilized value of 12 MPa for the solid state. These values are high compared to values obtained under classical mechanical spectroscopy characterization, which is consistent with the

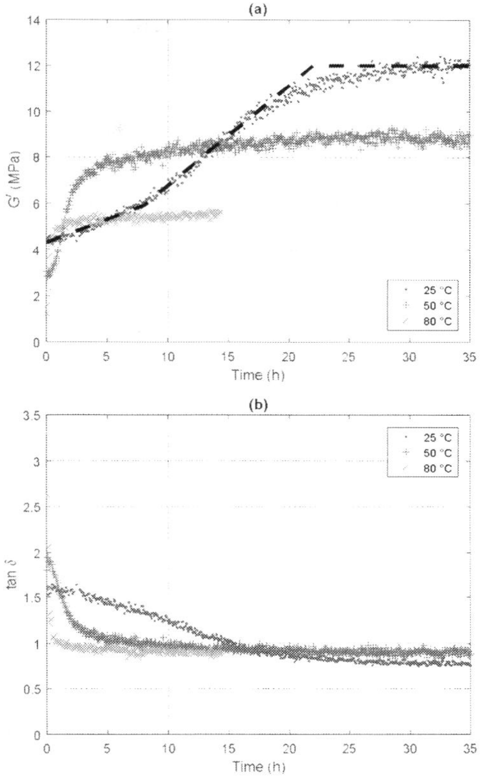

Figure 6. Evolution of the shear modulus (a) and the loss factor (b) of PDMS from the liquid to the solid state. The three dashed segments illustrate the various regimes in the cross-linking kinetics.

fact that the applied frequency is in the megahertz domain, *i.e.* much higher that classical characterization tests. Moreover, three regimes of evolution of the elastic modulus are observed, as illustrated by the dashed line in Figure 6, which can be associated with variation in the cross-linking kinetics. The loss factor decreases in a same nonlinear way from 1.6 to 0.8.

While illustrating a similar behavior, the curves obtained at **50°*C*** and **80°*C*** highlight that the higher the temperature, the faster the evolution of viscoelastic properties. Such results are no more discussed in the present paper, which only aims at presenting QCM technology as a new way of investigation of the cross-linking from liquid to solid state and at different temperatures.

5 CONCLUSION AND PERSPECTIVES

The paper aims at presenting the QCM as a relevant tool for measuring the viscoelastic properties of rubbers in the megahertz domain. The advantage of such a technique is the fact that the measurement can be carried out for liquids as well for solids, i.e. for rubbers from the uncross-linked to the cross-linked state, with a small material volume.

The kinetics of cross-linking and the nonlinear relationship between the number of cross-links and the mechanical properties can therefore be quantified and investigated through the variation in the viscoelastic response.

REFERENCES

Arnau, Antonio. 2008. "A Review of Interface Electronic Systems for AT-Cut Quartz Crystal Microbalance Applications in Liquids." *Sensors* 8 (1): 370–411.

Becker, Bernd, and Matthew A Cooper. 2011. "A Survey of the 2006–2009 Quartz Crystal Microbalance Biosensor Literature." *Journal of Molecular Recognition* 24 (5): 754–787.

Hinsberg, WD, CG Willson, and KK Kanazawa. 1986. "Measurement of Thin-Film Dissolution Kinetics Using a Quartz Crystal Microbalance." *Journal of The Electrochemical Society* 133 (7): 1448–1451.

Holt, Robert C., Gideon J. Gouws, and John. Z. Zhen. 2006. "Measurement of Polymer Shear Modulus Using Thickness Shear Acoustic Waves." *Current Applied Physics* 6 (3): 334–39. doi:10.1016/j.cap.2005.11.013.

Johannsmann, Diethelm. 2015. *The Quartz Crystal Microbalance in Soft Matter Research: Fundamentals and Modeling*. 1st ed. Soft and Biological Matter. Springer International Publishing. https://books.google.fr/books?id = IkAqBAAAQBAJ.

Kanazawa, K. Keiji, and Joseph G. Gordon. 1985. "The Oscillation Frequency of a Quartz Resonator in Contact with Liquid." *Analytica Chimica Acta* 175: 99–105. doi:10.1016/S0003-2670(00)82721-X.

Martin, Stephen J., Gregory C. Frye, Antonio J. Ricco, and Stephen D. Senturia. 1993. "Effect of Surface Roughness on the Response of Thickness-Shear Mode Resonators in Liquids." *Analytical Chemistry* 65 (20): 2910–2922. doi:10.1021/ac00068a033.

Mason, WP. 1941. "Electrical and Mechanical Analogies." *Bell System Technical Journal* 20 (4): 405–414.

Nayak, Madhura, Akhil Kotian, Sandhya Marathe, and Dipshikha Chakravortty. 2009. "Detection of Microorganisms Using Biosensors — A Smarter Way towards Detection Techniques." *Biosensors and Bioelectronics* 25 (4): 661–667.

Sauerbrey, Günter. 1959. "Verwendung von Schwingquarzen zur Wägung dünner Schichten und zur Mikrowägung." *Zeitschrift für Physik* 155 (2): 206–222.

Si, Pengchao, John Mortensen, Alexei Komolov, Jens Denborg, and Preben Juul Møller. 2007. "Polymer Coated Quartz Crystal Microbalance Sensors for Detection of Volatile Organic Compounds in Gas Mixtures." *Analytica Chimica Acta* 597 (2): 223–230.

Stockbridge, CD. 1966a. "Effects of Gas Pressure on Quartz Crystal Microbalances." *Vacuum Microbalance Techniques* 5: 147–178.

———. 1966b. "Resonance Frequency versus Mass Added to Quartz Crystals." *Vacuum Microbalance Techniques* 5: 193.

Wang, Hua, Yun Zhang, Bani Yan, Li Liu, Shiping Wang, Guoli Shen, and Ruqin Yu. 2006. "Rapid, Simple, and Sensitive Immunoagglutination Assay with SiO2 Particles and Quartz Crystal Microbalance for Quantifying Schistosoma Japonicum Antibodies." *Clinical Chemistry* 52 (11): 2065–2071.

Yu, Jae-Sung, Hua-Xin Liao, Aren E Gerdon, Brian Huffman, Richard M Scearce, Mille McAdams, S Munir Alam, et al. 2006. "Detection of Ebola Virus Envelope Using Monoclonal and Polyclonal Antibodies in ELISA, Surface Plasmon Resonance and a Quartz Crystal Microbalance Immunosensor." *Journal of Virological Methods* 137 (2): 219–228.

Special elastomers

Constitutive Models for Rubber X – Lion & Johlitz (Eds)
© 2017 Taylor & Francis Group, London, ISBN 978-1-138-03001-5

Modeling and simulation of magnetic-sensitive elastomer immersed in surrounding medium

Qimin Liu, Hua Li & K.Y. Lam
School of Mechanical and Aerospace Engineering, Nanyang Technological University, Singapore, Republic of Singapore

ABSTRACT: Magnetic-Sensitive Elastomer (MSE) is a polymer-based composite with embedded magnetically permeable particles at micro- or nano-level. Owing to its soft nature, the MSE undergoes a fast, reversible, and large deformation, which makes it potential candidate for many applications. In this work, a three-dimensional magnetic-mechanical model is developed for the MSE under magneto-mechanical coupled fields. The present model covers both the MSE and surrounding medium, where the magnetic field distribution is governed by the Maxwell's equations and the deformation by the mechanical equilibrium equation. The model is employed to investigate the responsive behavior of the magnetic-active elastomer subject to different magnetic permeability, for a deeper insight into the fundamental mechanism of the MSE.

1 INTRODUCTION

Magnetostriction is a property of magnetic material that may change its shape and dimension when subjected to an external magnetic field (Zrínyi et al., 1996). It was initially identified by James Joule in 1842 when observing the elongation of an iron rod exposed to a magnetic field (Joule, 1842), and also captured by other hard magnetic materials. Recently, magnetic-sensitive elastomer (MSE) attracts considerable attention due to the tunable elasticity and geometric shape. A typical MSE is composed of a polymer matrix filled with magnetically susceptible particles at micro or nano level. Owing to its soft nature, the MSE undergoes a large and reversible deformation that is of interest for versatile applications including soft actuators (Zrínyi et al., 1997), valves (Voltairas et al., 2003), and biomedicines (Hu et al., 2007).

In order to facilitate the applications of the MSEs, experimental and theoretical works were conducted for decades. In general, the MSEs are mainly divided into two categories, isotropic and anisotropic MSEs, by the magnetic particle distribution. It was found that the magnetostrictive strain in the isotropic MSE may be one order of magnitude larger than that in the anisotropic one (Han et al., 2015). Experiments also demonstrated that the strain is much dependent on the applied magnetic field (Ginder et al., 2002, Han et al., 2015). In addition, multiple shape reconfigurations were reported for the MSE in response to a uniform or non-uniform magnetic field including stretching, contraction, rotation, bending, and coiling (Zrínyi et al., 1997, Szabó et al., 1998, Zrinyi, 2000, Narita et al., 2003). Furthermore, the magnetic-induced instability and hysteresis were also investigated in detail (Zrínyi et al., 1996). A literature review reveals that the continuum and microscopic models are the two commonly used approaches for modeling of the MSE. Usually the linear/nonlinear elasticity was coupled with the Maxwell's equations in the continuum models (Dorfmann and Ogden, 2004b, Dorfmann and Ogden, 2004a, Raikher and Stolbov, 2005). Regarding the microscopic models, the most conventional one is based on the magnetic dipole interactions between two adjacent particles, in order to predict the change of the shear modulus (Jolly et al., 1996), the elongation and contraction in an external magnetic field (Biller et al., 2014).

Recently, the boundary value problems (BVPs) in nonlinear magneto-mechanics draw much attention from researchers, such as the shear of a slab, the elongation and torsion of a circular cylinder, and the rotation of a tube. In order to simplify the theoretical analysis and obtain the exact solution, the MSE size is generally assumed infinite one (Dorfmann and Ogden, 2005, Zubarev and Elkady, 2014, Raikher and Stolbov, 2005), where the magnetic field within the MSE is homogeneous. Accordingly, the field-induced deformation is also homogeneous. It is known that usually the finite MSE is immersed in the surrounding including a non-magnetic matrix or air (Zrínyi et al., 1997, Zrínyi et al., 1996), where the magnetic and mechanical fields are required to satisfy the boundary conditions over the interface between the MSE and its surrounding. However, the developed theoretical work either considered the MSE domain only or ignored the interface on the responsive behavior of the MSE (Nedjar, 2015, Bustamante et al., 2011), such that they may characterize the deformation inappropriately.

In order to address the limitations, a magneto-mechanical model is developed to simulate the responsive behavior of the MSE with inhomogeneous large deformation. The domain covers the MSE and the surrounding. The governing equations consist of the Maxwell's equation for the spatial distribution of the magnetic field, and the mechanical equilibrium equation for the deformation of the MSE. The model is implemented into a finite-element code using finite element software COMSOL/Multiphysics. After that, the influence of the relative magnetic permeability on the responsive behaviors of a spherical MSE is investigated, including the distribution of MSE inhomogeneous large deformation, and the magnetic field distributed in both the MSE and its surrounding.

2 MODEL DEVELOPMENT

To characterize the magnetic field, two primary magnetic vectors, the magnetic induction **b** and intensity **h** defined in the current configuration, are required. In the absence of free current and only the steady state is considered, they satisfy the following equations

$$\mathbf{curl\,h} = 0, \quad \mathrm{div}\,\mathbf{b} = 0 \tag{1}$$

where the curl and div operators are respect to **x**. According to the magnetostatics (Dorfmann and Ogden, 2003), the vectors **b** and **h** are associated with the magnetization vector **m** by the following relation

$$\mathbf{b} = \mu_0(\mathbf{h} + \mathbf{m}) \tag{2}$$

where μ_0 the magnetic permeability in a vacuum.

After a pull-back to the reference configuration, the Eulerian equation (1) as well as the relation (2) are presented in the Lagrangian forms of

$$\mathrm{Curl}\,\mathbf{H} = 0, \quad \mathrm{Div}\,\mathbf{B} = 0 \tag{3}$$

where $\mathbf{H} = \mathbf{F}^T\mathbf{h}$ and $\mathbf{B} = J\mathbf{F}^{-1}\mathbf{b}$ are the Lagrangian counterparts of the magnetic variables (Bustamante et al., 2011), **F** the deformation gradient. The Curl and Div operators are respect to **X** respectively. Across the interface between the MSE and the surrounding media, the magnetic variables are required to satisfy the jump conditions (Dorfmann and Ogden, 2005)

$$\mathbf{n}\cdot[[\mathbf{b}]] = 0, \quad \mathbf{n}\times[[\mathbf{h}]] = 0 \tag{4}$$

or

$$\mathbf{N}\cdot[[\mathbf{B}]] = 0, \quad \mathbf{N}\times[[\mathbf{H}]] = 0 \tag{5}$$

where **n** and **N** are the unit outwards normal to the domain surface S and S_0 respectively, the double square bracket denotes a quantity jump across the surface from the inside to outside of the material.

If only the steady state is considered, the mechanical equilibrium equation is written as

$$\nabla\cdot\boldsymbol{\sigma} + \mathbf{f}_b = 0 \tag{6}$$

where $\mathbf{f}_b$ is the external body force density, $\boldsymbol{\sigma}$ the total Cauchy stress tensor and also the summation of the elastic stress $\boldsymbol{\sigma}_e$ and the magnetic stress $\boldsymbol{\sigma}_m$ (Dorfmann and Ogden, 2004b). At the MSE-surrounding interface, it satisfies the mechanical boundary condition

$$\boldsymbol{\sigma}\cdot\mathbf{n} = \mathbf{t}_a + \mathbf{t}_m \tag{7}$$

where $\mathbf{t}_a$ and $\mathbf{t}_m$ are the mechanical the magnetic tractions respectively, and $\mathbf{t}_m$ is attributed to Maxwell stress $\tilde{\boldsymbol{\sigma}}_m$ acted on the outside boundary and $\mathbf{t}_m = \bar{\boldsymbol{\sigma}}_m\cdot\mathbf{n}$. The Maxwell stress tensor $\tilde{\boldsymbol{\sigma}}_m$ is represented by $\tilde{\boldsymbol{\sigma}}_m = \mathbf{bh} - (\mathbf{b}\cdot\mathbf{h})\mathbf{I}/2$, where **I** is the second-order identity. In the reference configuration, the mechanical equation (6) is rewritten as

$$\nabla_{\mathbf{X}}\cdot\mathbf{P} + \mathbf{F}_b = 0 \tag{8}$$

where **P** is the first Piola-Kirchhoff stress and associated with the Cauchy stress $\boldsymbol{\sigma}$ by $\mathbf{P} = J\boldsymbol{\sigma}\mathbf{F}^{-T}$, $\mathbf{F}_b$ the mechanical body force and $\mathbf{F}_b = J\mathbf{f}_b$.

Following Dorfmann *et al.* (Dorfmann and Ogden, 2003), the stress **P** and magnetic induction **B** associated with the deformation gradient **F** and the magnetic induction **B** are given as

$$\mathbf{P} = \frac{\partial\Phi}{\partial\mathbf{F}}, \quad \mathbf{B} = -\frac{\partial\Phi}{\partial\mathbf{H}} \tag{9}$$

where Φ is the total free energy per unit reference volume in the reference configuration.

The total free energy density Φ consists of the elastic energy of the polymer networks Φ_e, and the energy of magnetization Φ_m, namely, $\Phi = \Phi_e + \Phi_m$. The free energy is thus given by

$$\Phi = \frac{G}{2}[\mathrm{tr}(\mathbf{F}^T\mathbf{F}) - 1] - \frac{\mu_m}{2}[J\mathbf{C}^{-1} : \mathbf{HH}] \tag{10}$$

where G is the shear modulus of the MSE, μ_m denotes the magnetic permeability and $\mu_m = \mu_0\mu_r$, and $\mathbf{C}$ the right Green-Cauchy strain tensor and $\mathbf{C} = \mathbf{F}^T\mathbf{F}$.

By the energy function (10) and relation (9), if the material is incompressible, the nominal stress $\mathbf{P}$ and magnetic induction $\mathbf{B}$ are rewritten respectively as

$$\mathbf{P} = G\mathbf{F} + \mu_m\mathbf{F}^{-T}(\mathbf{HH})\mathbf{C}^{-1} - p\mathbf{F}^{-T} \tag{11}$$

$$\mathbf{B} = \mu_m\mathbf{C}^{-1}\mathbf{H} \tag{12}$$

where p is the hydrostatic pressure.

Alternatively, the constitutive equations (11) and (12) may also be expressed in terms of true quantities below

$$\boldsymbol{\sigma} = G\mathbf{FF}^T + \mu_m\mathbf{hh} - p\mathbf{I} \tag{13}$$

$$\mathbf{b} = \mu_m\mathbf{h} \tag{14}$$

By Equation (3), the magnetic intensity vector $\mathbf{H}$ is associated with the magnetic scalar potential ψ by $\mathbf{H} = -\nabla_\mathbf{X}\psi$. By the Equation (3) and the magnetic constitutive equation (12), the magnetic potential ψ in both the MSE and its surrounding satisfies the following governing equation

$$\nabla_\mathbf{X} \cdot (\mu_m\mathbf{C}^{-1}\nabla_\mathbf{X}\psi) = 0 \tag{15}$$

So far the magneto-mechanical model for the responsive behavior of the MSEs is formulated. It consists of the magnetostatic equations (15) and mechanical balance equation (8) in the reference configuration. In addition, the constitutive relations are formulated by Equations (11) and (12). The equations can be implemented directly in COMSOL Multiphysics for the following analyses.

3 RESULTS AND DISCUSSION

For further understanding of the fundamental characteristics and responsive behavior of the MSE immersed in air surrounding when subjected to a uniform magnetic field, the effect of the relative magnetic permeability μ_r on the variation of magnetic field distribution and the deformation are investigated. Here, a spherical MSE of radius $r_0 = 0.01$ m is centered in the air surrounding under a uniform magnetic field $H_0 = 120$ kA/m. The size of the surrounding is $L_R = 0.08$ m, and $L_H = 0.16$ m. The shear modulus of the MSE is $G = 1$ kPa.

Figure 1 shows the deformation behavior of the MSE subject to different relative magnetic permeability μ_r. As observed from the figure, the deformation is inhomogeneous along the z-axis direction, probably due to the different magnetic force over the MSE-surrounding interface. It is also seen that, with the increase of the relative magnetic permeability μ_r, the deformation increases, probably due to the increase of the magnetic force that acts on the MSE.

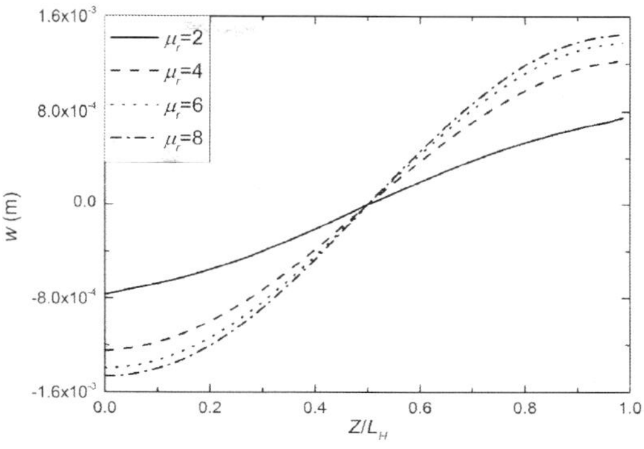

Figure 1. Influence of magnetic permeability μ_r on the variation of the axial deformation w along the $R = 0$.

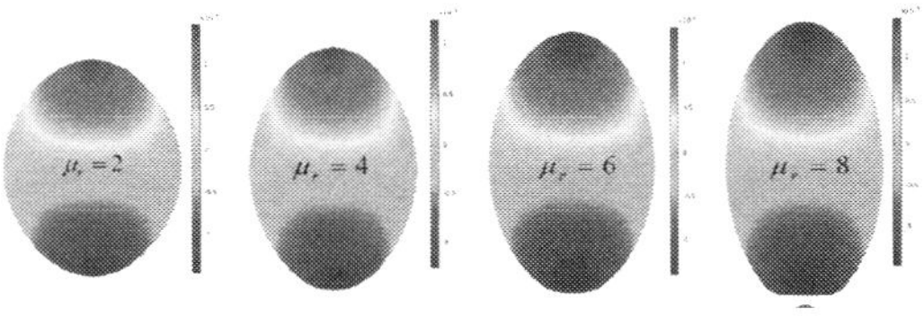

Figure 2. Influence of magnetic permeability μ_r on the deformation evolution of the MSE.

Figure 2 is plotted to demonstrate the deformation evolution of the MSE with different magnetic permeability μ_r. It is seen that the MSE elongates along the field direction and contracts in the transverse direction, which is also consistent with the experimental work qualitatively (Genoveva and Miklos, 2010).

4 CONCLUSION

The magneto-mechanical model has been formulated for analysis of the responsive behavior of the MSEs under the coupled magnetic and mechanical fields. The present model covers the MSE and its surrounding domains, in which the distribution of magnetic field is governed by the Maxwell's equation and the deformation of the MSEs by the mechanical equilibrium equation. In order to account for the large deformation of the MSE, an explicit free energy density is proposed. For deeply understanding the responsive mechanism of the

MSE, the influence of different relative magnetic permeability on the spherical MSE is conducted. It shows that the magneic permeability plays a vital role on the distribution profile of the magnetic field and deformation.

REFERENCES

Biller, A.M., Stolbov, O.V. & Raikher, Y.L. 2014. Modeling of Particle Interactions in Magnetorheological Elastomers. *Journal of Applied Physics*, 116, 114904.

Bustamante, R., Dorfmann, A. & Ogden, R.W. 2011. Numerical Solution of Finite Geometry Boundary-value Problems in Nonlinear Magnetoelasticity. *International Journal of Solids and Structures*, 48, 874–883.

Dorfmann, A. & Ogden, R.W. 2003. Magnetoelastic Modelling of Elastomers. *European Journal of Mechanics - A/Solids*, 22, 497–507.

Dorfmann, A. & Ogden, R.W. 2004a. Nonlinear magnetoelastic deformations. *Quarterly Journal of Mechanics and Applied Mathematics*, 57, 599–622.

Dorfmann, A. & Ogden, R.W. 2004b. Nonlinear Magnetoelastic Deformations of Elastomers. *Acta Mechanica*, 167, 13–28.

Dorfmann, A. & Ogden, R.W. 2005. Some Problems in Nonlinear Magnetoelasticity. *ZAMP*, 56, 718–745.

Genoveva, F. & Miklos, Z. 2010. Magnetodeformation Effects and the Swelling of Ferrogels in a Uniform Magnetic Field. *Journal of Physics: Condensed Matter*, 22, 276001.

Ginder, J.M., Clark, S.M., Schlotter, W.F. & Nichols, M.E. 2002. Magnetostrictive phenomena in magnetorheological elastomers. *International Journal of Modern Physics B*, 16, 2412–2418.

Han, Y., Mohla, A., Huang, X., Hong, W. & Faidley, L.E. 2015. Magnetostriction and Field Stiffening of Magneto-active Elastomers. *International Journal of Applied Mechanics*, 07, 1550001.

Hu, S.-H., Liu, T.-Y., Liu, D.-M. & Chen, S.-Y. 2007. Controlled Pulsatile Drug Release from a Ferrogel by a High-frequency Magnetic Field. *Macromolecules*, 40, 6786–6788.

Jolly, M.R., Carlson, J.D. & Munoz, B.C. 1996. A Model of the Behaviour of Magnetorheological Materials. *Smart Materials and Structures*, 5, 607–614.

Joule, J.P. 1842. On a new class of magnetic forces. *Annals of Electricity, Magnetism, and Chemistry*, 8, 219–224.

Narita, T., Knaebel, A., Munch, J.-P., Candau, S.J. & Zrínyi, M. 2003. Diffusing-Wave Spectroscopy Study of the Motion of Magnetic Particles in Chemically Cross-Linked Gels under External Magnetic Fields. *Macromolecules*, 36, 2985–2989.

Nedjar, B. 2015. A Theory of Finite Strain Magneto-poromechanics. *Journal of the Mechanics and Physics of Solids*, 84, 293–312.

Raikher, Y.L. & Stolbov, O.V. 2005. Deformation of an Ellipsoidal Ferrogel Sample in a Uniform Magnetic Field. *Journal of Applied Mechanics and Technical Physics*, 46, 434–443.

Szabó, D., Szeghy, G. & Zrínyi, M. 1998. Shape transition of magnetic field sensitive polymer gels. *Macromolecules*, 31, 6541–6548.

Voltairas, P.A., Fotiadis, D.I. & Massalas, C.V. 2003. Modeling the hyperelasticity of magnetic field sensitive gels. *Journal of Applied Physics*, 93, 3652–3656.

Zrinyi, M. 2000. Intelligent polymer gels controlled by magnetic fields. *Colloid and Polymer Science*, 278, 98–103.

Zrínyi, M., Barsi, L. & Büki, A. 1996. Deformation of Ferrogels Induced by Nonuniform Magnetic Fields. *The Journal of Chemical Physics*, 104, 8750–8756.

Zrínyi, M., Barsi, L. & Büki, A. 1997. Ferrogel: A New Magneto-controlled Elastic Medium. *Polymer Gels and Networks*, 5, 415–427.

Zubarev, A.Y. & Elkady, A.S. 2014. Magnetodeformation and Elastic Properties of Ferrogels and Ferroelastomers. *Physica A: Statistical Mechanics and its Applications*, 413, 400–408.

Constitutive Models for Rubber X – Lion & Johlitz (Eds)
 ISBN 978-1-138-03001-5

A simple Mullins model applied to a constitutive model for foamed rubber

M.W. Lewis
Los Alamos National Laboratory, Los Alamos, New Mexico, USA

ABSTRACT: The Mullins effect (Mullins 1948) as observed in uniaxial compression of a polydimethylsiloxane (PDMS) foam is modeled with a simplified continuum damage approach. This particular foam shows a characteristic cyclic softening behavior in which all softening is achieved in the first cycle. By this, we mean that the second cycle to the same strain level as achieved in the first cycle demonstrates the same stress-strain curve as subsequent cycles to the same strain level.

This particular behavior is well modeled with a continuum damage approach combined with the CHIPFoam model (Lewis 2016) for elastomeric foams. In this work, we demonstrate that fitting the CHIPFoam model to a softened response and developing an evolution equation for a modulus from first cycle data provides a reasonably good prediction of response seen in multiple cycles to different strain levels.

1 INTRODUCTION

The Mullins effect (Mullins 1948) is a strain-softening effect observed in elastomers. There have been several attempts to represent the Mullins effect in mechanical constitutive models for rubber (Ogden & Roxburgh 1999, Rickaby & Scott 2012, Neumann & Ihlemann 2014). The approaches can generally be classified as using two mechanisms for the softening effect: continuum damage mechanics and inelasticity. Rickaby & Scott (2012) combine the two approaches. Uniaxial compression stress-strain results on a foamed PDMS material demonstrating the Mullins effect are presented in Figure 1.

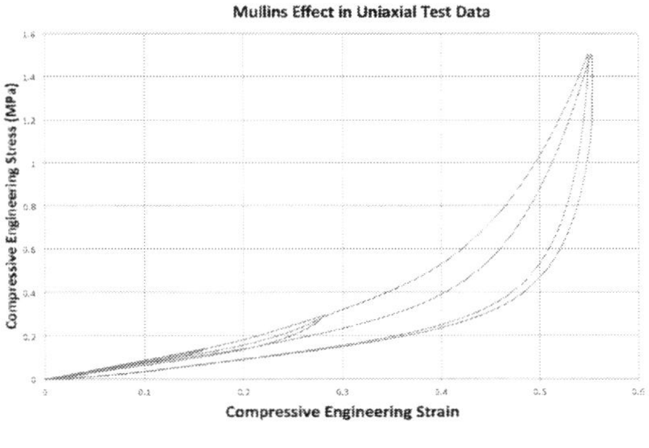

Figure 1. Uniaxial compression stress versus strain results that demonstrate the Mullins cyclic softening effect. The material tested is a foamed polydimethylsiloxane (PDMS) rubber.

In the current research, we consider the Mullins effect as observed in foamed polydimethylsiloxane (PDMS) rubber containing a filler (diatomaceous earth) that appears to be relatively inactive. The material is characterized in uniaxial compression. The Mullins effect in this material seems to consist of a softening response that leads to an observable difference between the first and second cycles to a given strain level, but subsequent cycles of loading to this strain produce stresses that are essentially identical to the second cycle response. For this reason, it seems appropriate to use a simple continuum damage response that is energy based.

In its simplest form, classic elastic continuum damage mechanics (Kachanov 1986, Chaboche 1988, Krajcinovic 1989, Addessio & Johnson 1989) is the degradation of elastic moduli as material is loaded. In the current work, we consider as an underlying mechanical constitutive model for the response of foamed rubber, the Compressible Hyperelastic Isotropic Porosity-based Foam Model (CHIPFoam) proposed by Lewis (2016). The state of damage is captured in the degradation of one modulus in the model, namely the modulus associated with the modified Danielsson function (Danielsson et al. 2004, Lewis & Rangaswamy 2015). An evolution equation for this modulus in terms of the porosity-modified first isochoric invariant is shown to reproduce compressive uniaxial test data with acceptable accuracy.

The theory underlying this approach to the Mullins effect in foamed rubber is presented in the THEORY section of this paper. Examples of uniaxial compression test data are presented in the EXPERIMENT section. Comparisons of the model to test data are presented in the NUMERICAL RESULTS section of this paper. Consideration of the advantages and limitations of this approach to modeling the Mullins effect is presented in the DISCUSSION section. The CONCLUSIONS section consists of conclusions drawn regarding this modeling approach and potential extensions to this model framework for representing other dissipative (nonconservative) effects observable in the test data.

2 THEORY

We have used the CHIPFoam strain energy function (Lewis 2016) to describe the hyperelastic response of foamed rubber. We have intentionally neglected any viscoelastic or inelastic effects in our approach so that we could focus on the Mullins effect as solely a hyperelastic damage phenomenon.

A plot of engineering stress versus engineering strain data from a repeated uniaxial compression test of a foamed PDMS material is presented in Figure 2. Compressive stress and strain are presented as positive for ease of illustration. The specimen is a circular disk with a height of approximately 10 mm and an aspect ratio (height/diameter) of approximately 0.3. The porosity of the material was estimated to be 49.5%, giving the foam a relative density or solid fraction of 0.505.

2.1 *The truncated CHIPFoam model*

Specifically, we have used a truncation of the CHIPFoam strain energy function because our test data showed no obvious compressive stress

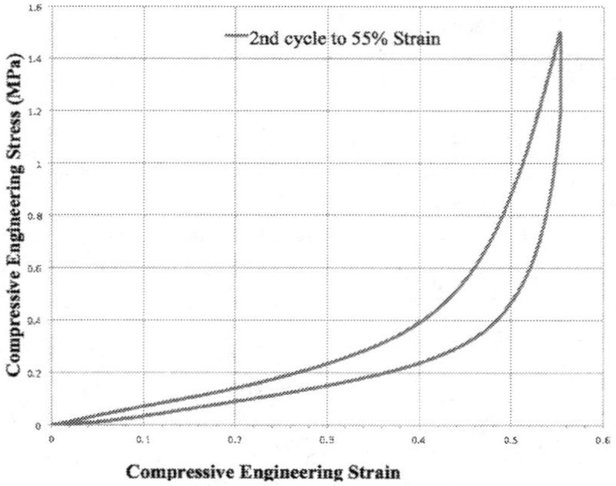

Figure 2. Compressive engineering stress versus compressive engineering strain for the second compression of a foamed PDMS sample to approximately 55%.

plateau behavior typically associated with a local buckling behavior in the foamed rubber in uniaxial compression and because the porosity of the foamed material is high enough that a percolation limit has been reached, allowing atmospheric gases to flow in and out of the pores during testing at the slow rates of this test. The nominal strain rate during the test was 0.01 s^{-1}.

The truncated CHIPFoam strain energy function, expressed as strain energy per unit initial (undeformed) volume may be written as follows:

$$W = W_{NH} + \tilde{W}_D + W_{MC}, \tag{1}$$

where the three strain energy functions that compose the right hand side of Equation 1 are as follows:

$$W_{NH} = \frac{\hat{G}}{2}\left(\bar{I}_1 - 3\right), \tag{2}$$

$$\tilde{W}_D = C_{10} J_m \left[\tilde{\bar{I}}_1 - 3\left(1-\phi_o\right)\right], \tag{3}$$

and

$$W_{MC} = K\left(1-\phi_o\right)\left[J_m \ln\left(J_m\right) - \left(J_m - 1\right)\right]. \tag{4}$$

The first of the strain energy functions, presented in Equation 2, is a simple Neo-Hookean function of the first isochoric invariant, which is defined as follows:

$$\bar{I}_1 = J^{-2/3} F_{ij} F_{ij}, \tag{5}$$

where J is the relative volume, which is the determinant of the deformation gradient tensor, F_{ij}. In Equation 2, the quantity $\hat{G}$ is a macroscale shear modulus.

The second of the strain energy functions that make up the truncated CHIPFoam model is referred to as the modified Danielsson function after Danielsson (Danielsson et al. 2004) and is presented in Equation 3. In Equation 3, C_{10} is the Neo-Hookean modulus of the parent material, J_m is the part of the relative volume that arises from the compressibility of the parent material, and the porosity-modified first isochoric invariant, $\tilde{\bar{I}}_1$, is defined as follows:

$$\tilde{\bar{I}}_1 = \bar{I}_1 f\left(\bar{J}, \phi_o\right). \tag{6}$$

The quantity ϕ_o that appears in Equations 3, 4, and 6 is the initial porosity of the specimen. In Equation 6, the function f is defined as follows:

$$f(\bar{J},\phi_o) = \frac{2\bar{J}-1}{\bar{J}^{1/3}} + (2-2\bar{J}-\phi_o)\left(\frac{\phi_o}{\bar{J}-1+\phi_o}\right)^{1/3}. \quad (7)$$

The quantity $\bar{J}$ that appears in Equations 6 and 7 is the part of the relative volume that is not a result of parent material compressibility, and is found by solving the auxiliary equation for the multiplicative split of the relative deformation as follows:

$$J = \bar{J}J_m, \quad (8)$$

where J_m is the part of the relative volume arising from matrix compressibility and is also related to $\bar{J}$ as follows:

$$J_m = exp\left[\frac{-C_{10}}{K}\left(\phi_0^{1/3}\frac{4\bar{J}-4+5\phi_o}{(\bar{J}-1+\phi_o)^{4/3}} - \frac{16\bar{J}^2-1}{3\bar{J}^{4/3}}\right)\right]. \quad (9)$$

In Equations 4 and 9, K is the parent material effective bulk modulus.

The third of the strain energy functions that make up the truncated CHIPFoam model represents energy stored in the compression of the parent material and is presented in Equation 4. All of the terms appearing in Equation 4 have been previously defined.

2.2 *A simple damage mechanics approach for the Mullins effect in the CHIPFoam model*

The simplest concepts of elastic damage mechanics involve making material elastic moduli functions of variables associated with the deformation history of a material (Kachanov 1986, Chaboche 1988, Krajcinovic 1989). We propose that a similar approach be used to model the loading behavior of foamed rubber by using the CHIPFoam model and making the material parameter C_{10} be a function of the maximum value of the porosity-modified first isochoric invariant, $\hat{\bar{I}}_1$, obtained during the deformation history of the material. To summarize this postulate mathematically, we assume a functional dependence of the modulus as follows:

$$C_{10} = C_{10}^0 g\left(\frac{\hat{\bar{I}}_1^{max}}{3(1-\phi_o)}\right), \quad (10)$$

where

$$\hat{\bar{I}}_1^{max} \equiv \max_{0\le t\le t_c} \hat{\bar{I}}_1(t), \quad (11)$$

t is the parameter associated with deformation history, typically time, and t_c is the current value of this parameter.

In order to use this method, one must have first and second cycle stress-strain data to a given strain state. The process of determining the function g of Equation 10 is then achieved by identifying the best fit of the model to second cycle data. In this process, one identifies the maximum value of $\hat{\bar{I}}_1$ seen in the deformation in the first cycle. The value of C_{10} is then systematically increased and the intersections of the resulting model stress-strain curves with the first cycle loading curved are identified. The corresponding values of $\hat{\bar{I}}_1$ and the values of C_{10} are then tabulated. These values may be treated as the values of $\hat{\bar{I}}_1^{max}$ and C_{10} from Equation 10. One can either use the table and linearly interpolate values of $C_{10}\left(\hat{\bar{I}}_1^{max}\right)$ or try to fit a functional form to the function g.

3 EXPERIMENT

Material disks of a cellular PDMS material were cored from larger parts. The nominal initial porosity of the material was 48%, but there was variation of approximately ±3% in the actual porosity of the samples. The disks were approximately 10 mm thick and 30 mm in diameter. The specimens were each weighed to determine density and approximate actual porosity.

The disks were individually compressed at 20°C in room air between platens without lubrication. The distance between the platens was continuously measured using a clip gauge. The specimens were compressed at a nominal strain rate of 0.01 s^{-1} to compressive engineering stresses of approximately 125 kPa, 275 kPa, and 1.5 MPa. After reaching these stress values, the specimens were immediately unloaded at the same strain rate, and were allowed to recover for one minute before beginning the

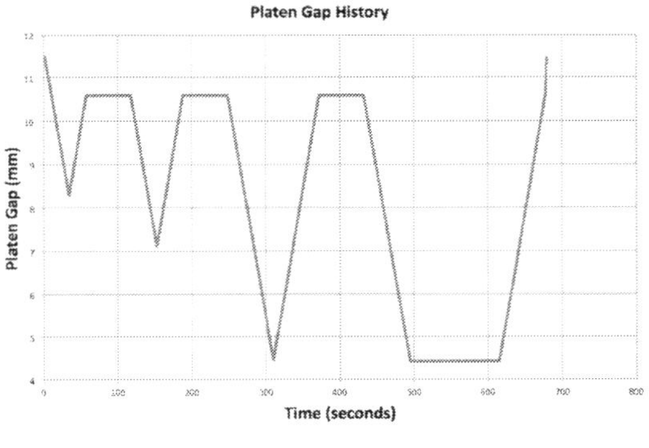

Figure 3. A time history of the gap between platens for the compression test data shown in Figure 1.

next cycle. After the recovery following the cycle to 1.5 MPa, each specimen was then compressed to a compressive engineering stress of 1.5 MPa again at a nominal strain rate of 0.01 s^{-1}. The platens were then held in place for two minutes to measure stress relaxation before an unloading at a nominal strain rate of 0.01 s^{-1}.

An example plot of the platen separation versus time is shown in Figure 3. The corresponding engineering stress vs. strain plot was presented as Figure 1.

4 NUMERICAL RESULTS

The procedure presented in the THEORY section of this paper was applied to the data shown in Figure 1. The parameters for the reduced CHIPFoam fit to the data from the second loading to approximately 1.5 MPa are presented in Table 1 and a plot showing the quality of the fit is presented in Figure 4.

The main assumption associated with this model fitting process is that the compression test is a uniaxial strain test. This is an approximation that is typically acceptable for very thin samples with no lubrication.

The next step in the fitting process for the Mullins effect model consists of identifying values of C_{10} and corresponding values of $\hat{\bar{I}}_1^{max}$. These were identified by generating stress-strain curves with the first three parameters in Table 1 fixed while C_{10} was increased and the points where the stress-strain curves intersected the virgin stress-strain curve were identified and correlated to the values of $\bar{I}_1$ that obtained there. The resulting data are presented in Table 2 and plotted in Figure 5.

An equation which provides a reasonably good fit to the data of Table 2 is as follows:

$$C_{10} = (0.0509\ \text{MPa})\left(\frac{\hat{\bar{I}}_1^{max}}{3(1-\phi_o)}\right)^{-1.545}. \tag{12}$$

This equation is included in Figure 5.

Using the approximation of Equation 12 with the reduced CHIPFoam model under the assumption of uniaxial strain, it is possible to generate the virgin compression curve to compare to test data. The comparison is presented in Figure 6 and shows reasonably good agreement for such a simple theory.

Additionally, we have calculated the third cycle response, namely the reload to approximately 28% strain at a little under 300 kPa from this fit.

Table 1. Parameters for the reduced CHIPFoam fit of Equation 1 to the second loading cycle to 1.5 MPa as shown in Figure 1.

ϕ_o	$\hat{G}$ (MPa)	K (MPa)	C_{10} (MPa)
0.495	0.35	6.6	0.0037

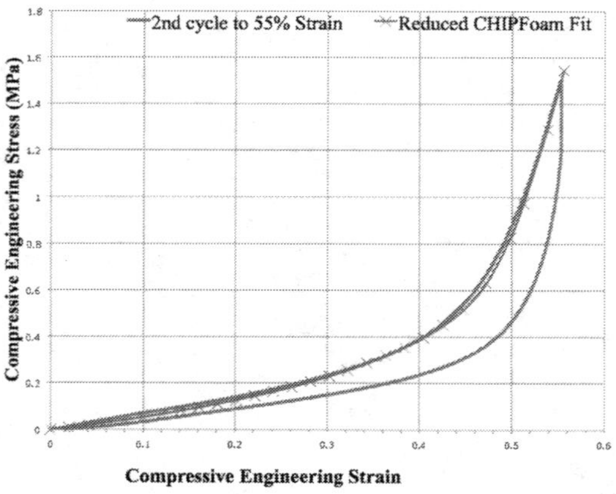

Figure 4. The second cycle stress-strain data presented in Figure 2 and the reduced CHIPFoam model fit using parameters shown in Table 1 under assumptions of uniaxial strain.

Table 2. Values of C_{10} and $\hat{\bar{I}}_1^{max}$.

$\hat{\bar{I}}_1^{max}$	C_{10} (MPa)
1.05	0.060
1.192	0.028
1.193	0.037
1.92	0.020
3.42	0.01
4.77	0.0037

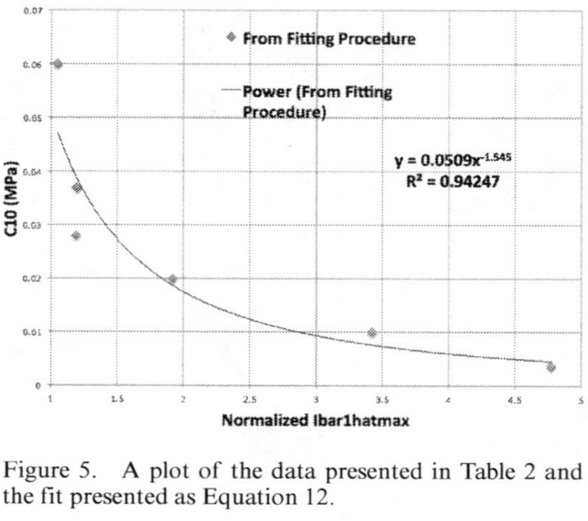

Figure 5. A plot of the data presented in Table 2 and the fit presented as Equation 12.

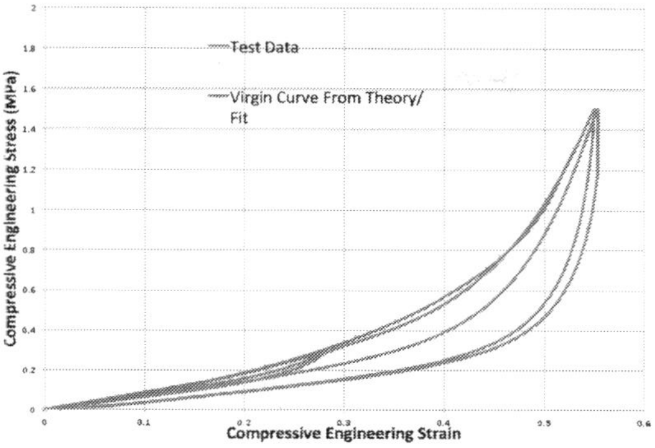

Figure 6. A comparison of the virgin uniaxial strain compression stress-strain curve to the cyclic data presented in Figure 1.

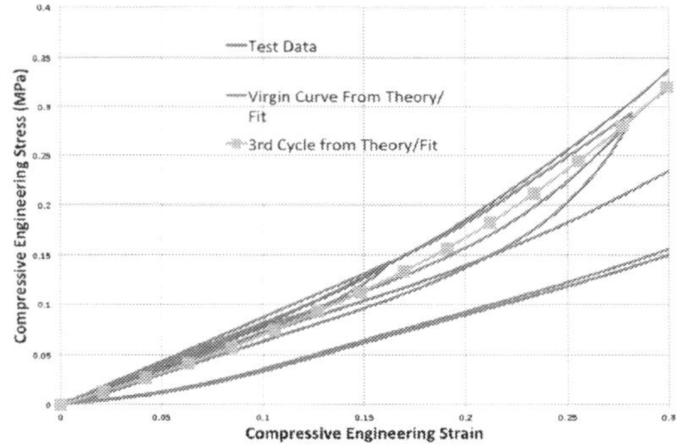

Figure 7. The calculated third cycle loading (reloading to 28% and subsequent loading) from the model fit shown with the test data and the calculated virgin curve.

The result is plotted along with the test data and the calculated virgin curve in Figure 7. The agreement with the loading curve is acceptably good.

5 DISCUSSION

We have used a simple elastic damage mechanics approach to capture the Mullins effect for filled, foamed PDMS. We used a somewhat reduced form of the CHIPFoam model to represent the non-damaging response of the foam and evolved the modulus associated with the modified Danielsson function. This evolution was based on the maximum value of the porosity-modified first isochoric invariant seen during the deformation history. Because of the micromechanical basis of the CHIPFoam model, this sort of modification preserves the porosity as a model constant and only a modulus was degraded.

More involved state variable models characterized as pseudo-elastic (Ogden & Roxburgh 1999, Neumann & Ihlemann 2014) might be necessary if softening is not completely accomplished in one cycle, that is if successive cycles after the second cycle do not reproduce the second cycle stress-strain response. Our experimental data does not show this behavior for the material in which we are most interested.

The simple model for the Mullins effect presented here is isotropic. In this model, the damage is isotropic, so loading in an orthogonal direction to the direction of initial cycling will demonstrate a softer response than the virgin response, and will not demonstrate additional softening until the porosity-modified first isochoric invariant exceeds the maximum value seen in previous loading. Work by others has indicated that the Mullins effect is in fact directional (Dianni et al. 2006, Merckel et al. 2013). We have not tried to address the anisotropic nature of the Mullins effect here as it does not appear to be important to our application of interest. If it were necessary to model cyclic softening in multiple loading directions with arbitrary strain histories in a foamed rubber, this simple model would probably not be satisfactory and additional theory development would be necessary.

The model presented here is being evaluated in the finite element analysis code ABAQUS/Standard through a user subroutine (UHYPER) with a state variable approach. At each integration point the value of $\bar{I}_1^{max}$ experienced at that integration point is stored and updated based on the deformation history. Results of this evaluation are forthcoming.

6 CONCLUSIONS

A simple approach for representing the Mullins effect in foamed rubbers has been presented. If cyclic loading response stabilizes so that third cycle response and second cycle response to the same strain state are essentially the same, this approach based on simple elastic damage mechanics provides good results. The model represents an application of the CHIPFoam (Lewis 2016) model with a damaging modulus. As this modeling approach is isotropic, it is appropriate if loading directions do not change during a deformation history. Otherwise, an anisotropic damaging theory for foamed rubber needs to be developed.

This work represents the first introduction of a dissipative mechanism into the CHIPFoam framework. Several other extensions that are desired include hysteretic modeling capability, thermal effects, and representation of material aging.

ACKNOWLEDGMENTS

The author wishes to thank Los Alamos National Laboratory (LANL) for providing funding for this work and for Carl Cady, also from LANL, for providing exceptionally good data for model development and fitting. Discussions with Bob Stevens and Partha Rangaswamy have helped in the development of the CHIPFoam model. Additionally, Dr. Rangaswamy has overseen much of the testing that has driven model development.

REFERENCES

Addessio, F.L. & Johnson, J.N. 1989, A Constitutive Model for the Dynamic Response of Brittle Materials. *J. Appl. Phys.* 67(7): 3275–3286.

Chaboche, J.-L. 1988, Continuum Damage Mechanics: Parts I and II. *J. Appl. Mech.* 55: 59–72, ASME.

Danielsson, M., Parks, D.M. & Boyce, M.C. 2004. Constitutive modeling of porous hyperelastic materials. *Mechanics of Materials* 36: 347–358.

Dianni, J., Brieu, M. & Vacherand, M. 2006. A damage directional constitutive model for Mullins effect with permanent set and induced anisotropy. *Eur. J. Mech. A Solids* 25: 483–496.

Kachanov, L.M. 1986, Introduction to Continuum Damage Mechanics, Martinus Nijhoff, Dordrecht, The Netherlands.

Krajcinovic, D. 1989, Damage Mechanics, *Mechanics of Materials* 8: 117–197.

Lewis, M. 2016. A robust, compressible, hyperelastic constitutive model for the mechanical response of foamed rubber. *Technische Mechanik* 36(1–2): 88–101.

Lewis, M.W. & Rangaswamy, P. 2015. Volume decomposition for a robust, mechanics-based, hyperelastic foam model. In Bohdana Marvalova & Iva Petrikova (eds.), *Constitutive Models for Rubber IX, Proc. of ECCMR IX, PragueCzech Republic, 1–4 September 2015*, London: Taylor & Francis Group.

Merckel, Y., Brieu, M., Diani, J. & Caillard, J. 2013. Understanding and modeling the Mullins softening from a mechanical point of view. In Nere Gil-Negrete & Asier Alonso (eds.), *Constitutive Models for Rubber VIII, Proc. of ECCMR VIII, San Sebastian, Spain, 25–28 June, 2013*, London: Taylor & Francis Group.

Mullins, L. 1948. Effect of Stretching on the Properties of Rubber. *Rubber Chemistry and Technology* 21: 281–300.

Neumann, C. & Ihlemann, J. 2014. On the thermodynamics of pseudo-elastic material models to reproduce the MULLINS effect. In arXiv:1408.0671 Condensed Matter-Materials Science, arXiv.org.

Ogden, R. & Roxburgh, D. 1999. A pseudo-elastic model for the Mullins effect in filled rubber. *Proceedings of the Royal Society of London* 455: 2861–2877.

Rickaby, S.R. & Scott, N.H. 2012. The Mullns effect. In Stephen Jerrams & Niall Murphy (eds.), *Constitutive Models for Rubber VII, Proc. of ECCMR VII, Dublin, Ireland, 20–23 September 2011*, Leiden: CRC Press/Balkema.

Constitutive Models for Rubber X – Lion & Johlitz (Eds)
© 2017 Taylor & Francis Group, London, ISBN 978-1-138-03001-5

Torsional wave propagation in tough, rubber like, doubly crosslinked hydrogel

L. Kari
KTH Royal Institute of Technology, Stockholm, Sweden

ABSTRACT: Torsional wave amplitude for tough, rubber like, doubly crosslinked hydrogel decreases rapidly with propagation distance where maximum decrease after one wave length is at frequency with maximum loss factor of shear modulus. Doubly crosslinked hydrogel is containing macromolecular polymer network with both chemical and physical crosslinks where it is possible to adaptively tune the frequency for maximum loss factor of shear modulus. This rubber like material has thus a potential to serve as vibration isolator and damper. Torsional dynamic stiffness of a cylindrical vibration isolator is modelled, displaying resonances and anti-resonances with a strong damping.

1 INTRODUCTION

Hydrogel is a rubber like material consisting of a polymer network abundantly swollen with water. See Ahmed (2015) for a review. Recently, tough hydrogels are developed, containing macromolecular polymer network with both chemical and physical crosslinks where it is possible to adaptively tune the frequency for maximum loss factor of shear modulus (e.g. Czarnecki et al., 2016; Kari, 2017; Lin et al., 2010; Long et al., 2014; Mayumi et al., 2013; Rose et al., 2013). In this paper, torsional wave propagation in a cylindrical rod made of this adaptive, rubber like material is studied and the corresponding dynamic stiffness of a finite cylindrical vibration isolator is determined.

2 METHOD

2.1 *Constitutive equations*

Hydrogel is assumed homogeneous and isotropic with a deviatoric response

$$\text{dev}\,[\boldsymbol{\sigma}] = 2\int_{-\infty}^{t} \mu(t-s)\frac{\partial \nabla \text{dev}\,[\mathbf{u}(s)]}{\partial s}\,\text{d}s, \tag{1}$$

where shear relaxation function

$$\mu(t) = \mu_\infty\left[h(t) + \Delta E_\alpha\left(-\Delta\left[\frac{t}{\nu}\right]^\alpha\right)\right], \tag{2}$$

dev is deviatoric operator, t is time, $\boldsymbol{\sigma}$ is second order stress tensor, $\mathbf{u}$ is displacement vector, ∇ is covariant derivative, $\mu_\infty = \lim_{t\to\infty}\mu(t)$ is equilibrium shear modulus, h is Heaviside step function, Δ is stress relaxation intensity, ν and $0 < \alpha < 1$ are material parameters, Mittag–Leffler function

$$E_\alpha(t) = \sum_{n=0}^{\infty}\frac{t^n}{\Gamma(1+n\alpha)} \tag{3}$$

and Γ is Gamma function. Temporal Fourier transform ($\tilde{\ }$) results in

$$\text{dev}\,[\tilde{\boldsymbol{\sigma}}] = 2\hat{\mu}\,\text{dev}\,[\nabla\tilde{\mathbf{u}}], \tag{4}$$

where shear modulus

$$\hat{\mu} = \mu_\infty\left[1 + \frac{\Delta(\text{i}\nu\omega)^\alpha}{\Delta + (\text{i}\nu\omega)^\alpha}\right], \tag{5}$$

$\omega = 2\pi f$ is angular frequency in radians per second, f is frequency in Hertz and i is imaginary unit.

2.2 *Hydrogel cylinder*

The studied semi-infinite cylindrical tough hydrogel of radius a is in Figure 1 with corresponding cylindrical co-ordinate system (r,θ,z). The cylinder is excited an angle

$$u_\theta = \Re[\hat{u}_\theta] = \Re[\phi\,\text{e}^{\text{i}\omega t}], \tag{6}$$

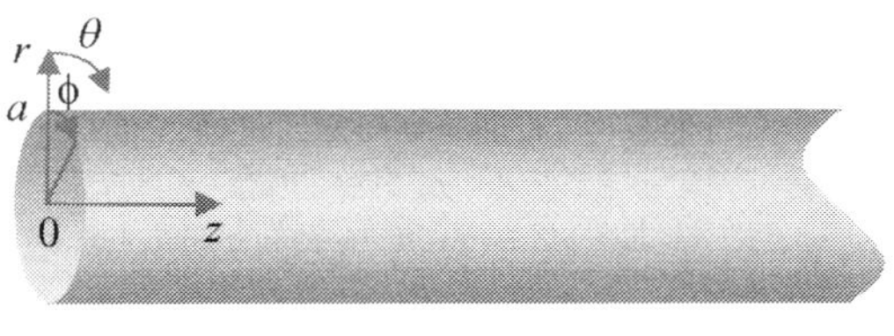

Figure 1. Semi-infinite cylindrical tough hydrogel of radius a and the corresponding cylindrical co-ordinate system (r,θ,z).

at origin $z = 0$, where $\mathcal{R}$ is real part.

2.3 *Wave propagation*

The wave equation for torsional waves in solid cylinders reads (e.g. Fredette et al, 2017)

$$c^2 \frac{\partial^2 \hat{u}_\theta}{\partial z^2} + \omega^2 \hat{u}_\theta = 0, \tag{7}$$

with a solution

$$\hat{u}_\theta = \phi \mathrm{e}^{\mathrm{i}[\omega t - kz]}, \tag{8}$$

satisfying boundary condition (6) at $z = 0$ and Sommerfeld radiation condition, where torsional wave velocity

$$c = \sqrt{\frac{\hat{\mu}}{\rho}}, \tag{9}$$

torsional wave number

$$k = \frac{\omega}{c} \tag{10}$$

and ρ is density.

2.4 *Dynamic stiffness*

A cylindrical vibration isolator of radius a and length l, made of the same hydrogel, excited at one end and blocked at the other end, is shown in Figure 2. In this case, boundary condition (6) applies, in addition to $u_\theta = 0$ at $z = l$. The solution is

$$\hat{u}_\theta = \phi \mathrm{e}^{\mathrm{i}\omega t} \frac{\mathrm{e}^{-\mathrm{i}k(z-l)} - \mathrm{e}^{+\mathrm{i}k(z-l)}}{\mathrm{e}^{+\mathrm{i}kl} - \mathrm{e}^{-\mathrm{i}kl}}, \tag{11}$$

while including propagating torsional waves in both positive and negative z-direction. The torsional moment reads

$$\hat{M}_\theta = -a^4 \hat{\mu} \frac{\pi}{2} \frac{\partial \hat{u}_\theta}{\partial z} \tag{12}$$

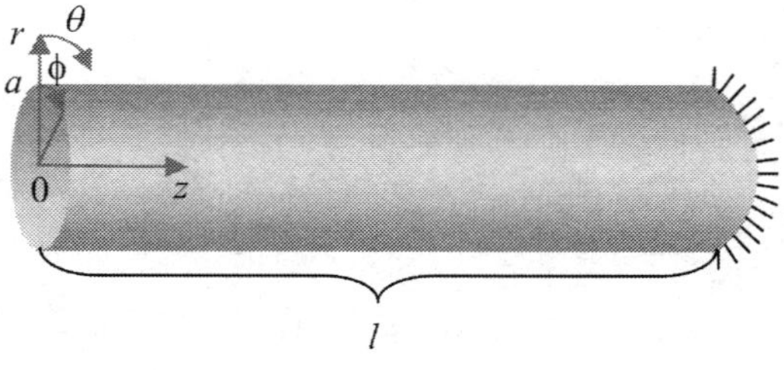

Figure 2. A cylindrical vibration isolator of radius a and length l made of tough hydrogel, excited at one end and blocked at the other end.

or

$$\hat{M}_\theta = \mathrm{i}ka^4 \hat{\mu} \phi \mathrm{e}^{\mathrm{i}\omega t} \frac{\pi}{2} \frac{\mathrm{e}^{-\mathrm{i}k(z-l)} + \mathrm{e}^{+\mathrm{i}k(z-l)}}{\mathrm{e}^{+\mathrm{i}kl} - \mathrm{e}^{-\mathrm{i}kl}}. \tag{13}$$

Torsional driving point stiffness reads

$$k_D = \mathrm{i}ka^4 \hat{\mu} \frac{\pi}{2} \frac{\mathrm{e}^{+\mathrm{i}kl} + \mathrm{e}^{-\mathrm{i}kl}}{\mathrm{e}^{+\mathrm{i}kl} - \mathrm{e}^{-\mathrm{i}kl}} \tag{14}$$

or

$$k_D = ka^4 \hat{\mu} \frac{\pi}{2} \frac{1}{\tan(kl)} \tag{15}$$

and torsional transfer stiffness reads

$$k_T = \mathrm{i}ka^4 \hat{\mu} \pi \frac{1}{\mathrm{e}^{+\mathrm{i}kl} - \mathrm{e}^{-\mathrm{i}kl}} \tag{16}$$

or

$$k_T = ka^4 \hat{\mu} \frac{\pi}{2} \frac{1}{\sin(kl)}. \tag{17}$$

3 RESULTS AND DISCUSSION

3.1 *Material*

The material is rubber like, tough hydrogel of radius $a = 50$ mm, density $\rho = 1000$ kg/m^3, equilibrium shear modulus $\mu_\infty = 200\,000$ N/m^2, stress relaxation intensity $\Delta = 4$, material parameters $\nu = 2.55 \times 10^{-3}$ s and $\alpha = 0.5$, resulting in a maximum loss factor at 200 Hz. The length of vibration isolator is $l = 50$ mm.

3.2 *Shear modulus*

The absolute value $|\hat{\mu}|$ and loss factor $\eta = \Im\hat{\mu} / \Re\hat{\mu}$ of shear modulus versus frequency are in Figure 3 and Figure 4 over the whole audible frequency

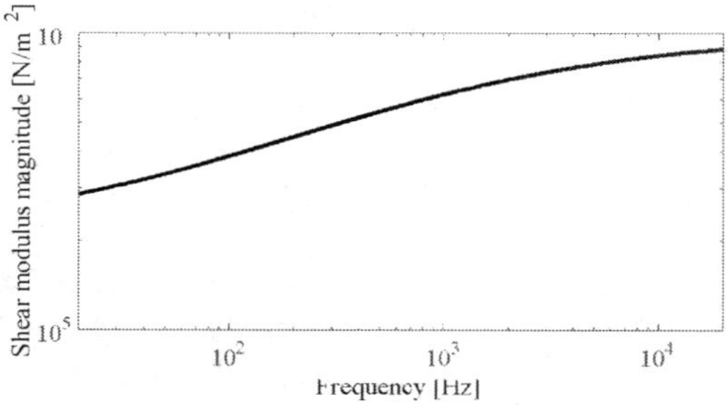

Figure 3. Absolute value of shear modulus $|\hat{\mu}|$ versus frequency.

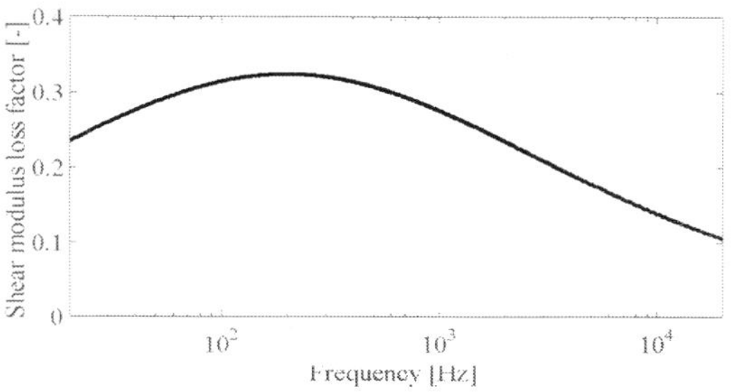

Figure 4. Loss factor of shear modulus $\eta = \Im\hat{\mu} / \Re\hat{\mu}$ versus frequency.

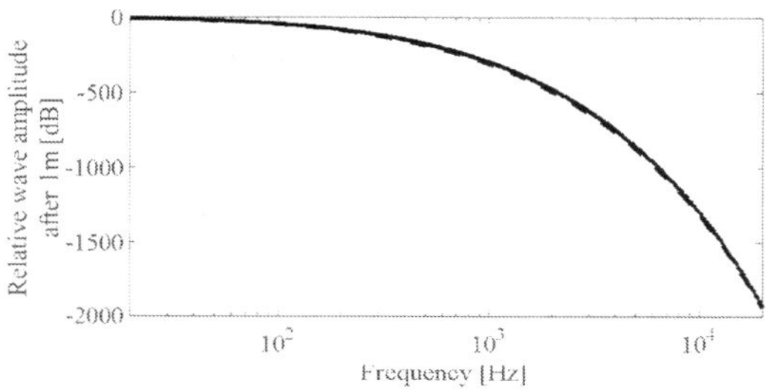

Figure 5. Magnitude ratio of torsional waves in dB after 1 m propagation versus frequency, according to Equation (19) (solid) and approximation (22) (dashed).

range, from 20 to 20 000 Hz, where is $\mathcal{I}$ imaginary part. Clearly, shear modulus magnitude increases with increasing frequency while shear modulus loss factor first increases and then decreases with increasing frequency with peak value at 200 Hz.

3.3 *Wave amplitude decrease*

The torsional wave solution (8) is possible to write as

$$\hat{u}_\theta = \phi \mathrm{e}^{+z\Im(k)}\mathrm{e}^{\mathrm{i}[\omega t - z\Re(k)]}, \tag{18}$$

where $\Im\ k < 0$. The magnitude ratio in dB after 1 meter of propagation is

$$20\log_{10}\left|\frac{\hat{u}_\theta(1\ \mathrm{m})}{\hat{u}_\theta(0\ \mathrm{m})}\right| = \Im[k]20\log_{10}[\mathrm{e}]\ 1\ \mathrm{m} \tag{19}$$

or

$$20\log_{10}\left|\frac{\hat{u}_\theta(1\ \mathrm{m})}{\hat{u}_\theta(0\ \mathrm{m})}\right| \approx 8.69\Im[k]\ 1\ \mathrm{m}, \tag{20}$$

and is shown in Figure 5 versus frequency over the whole audible frequency range, from 20 to 20 000 Hz. Approximatively,

$$20\log_{10}\left|\frac{\hat{u}_\theta(1\ \mathrm{m})}{\hat{u}_\theta(0\ \mathrm{m})}\right| \approx -10\log_{10}[\mathrm{e}]\omega\eta\sqrt{\frac{\rho}{\Re\hat{\mu}}}1\ \mathrm{m} \tag{21}$$

or

$$20\log_{10}\left|\frac{\hat{u}_\theta(1\ \mathrm{m})}{\hat{u}_\theta(0\ \mathrm{m})}\right| \approx -27.3f\eta\sqrt{\frac{\rho}{\Re\hat{\mu}}}1\ \mathrm{m}, \tag{22}$$

and is also shown in Figure 5 versus frequency. Clearly, wave amplitude after 1 m propagation decreases rapidly with increased frequency and the result from approximate expression (22) is close to that of Equation (19). The higher the frequency is, the shorter is the wave length and, thus, the more waves fit within 1 m. An alternative normalization is to calculate magnitude ratio after one wave-length, here given by $\lambda = \mathrm{R}(2\pi/k)$. Consequently, magnitude ratio in dB after a wave length of propagation is

$$20\log_{10}\left|\frac{\hat{u}_\theta(\lambda)}{\hat{u}_\theta(0\ \mathrm{m})}\right| = \Im[k]20\log_{10}[\mathrm{e}]\ \Re\left[\frac{2\pi}{k}\right] \tag{23}$$

or

$$20\log_{10}\left|\frac{\hat{u}_\theta(\lambda)}{\hat{u}_\theta(0\ \mathrm{m})}\right| \approx 54.6\ \Re\left[\frac{\Im[k]}{k}\right], \tag{24}$$

and is shown in Figure 6 versus frequency over the whole audible frequency range, from 20 to 20 000 Hz. Approximatively,

$$20\log_{10}\left|\frac{\hat{u}_\theta(\lambda)}{\hat{u}_\theta(0\ \mathrm{m})}\right| \approx -20\log_{10}[\mathrm{e}]\pi\eta \tag{25}$$

or

$$20\log_{10}\left|\frac{\hat{u}_\theta(\lambda)}{\hat{u}_\theta(0\ \mathrm{m})}\right| \approx -27.3\eta, \tag{26}$$

and is also shown in Figure 6 versus frequency. Clearly, wave amplitude ratio after one wave length propagation first decreases with increased frequency and then increases with increased frequency with minimum at 100 Hz. Consequently, the amplitude ratio displays similar behavior with frequency as the negative of loss factor. Furthermore, result from approximate expression (26) is close to that of Equation (23).

3.4 *Dynamic stiffness*

The magnitude of driving point and transfer stiffness for finite vibration isolator made of the same

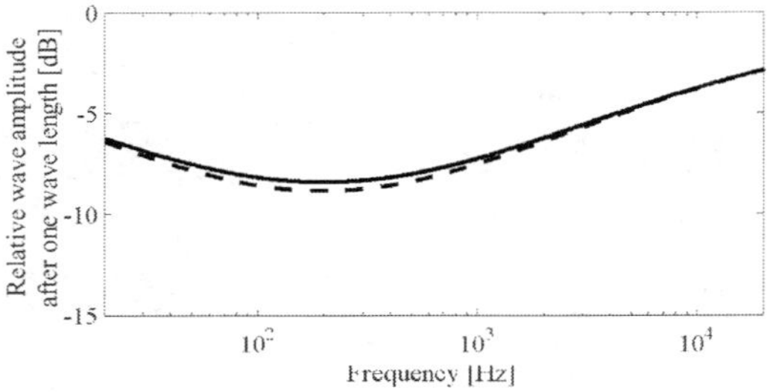

Figure 6. Magnitude ratio of torsional waves in dB after one wave length propagation versus frequency, according to Equation (23) (solid) and approximation (26) (dashed).

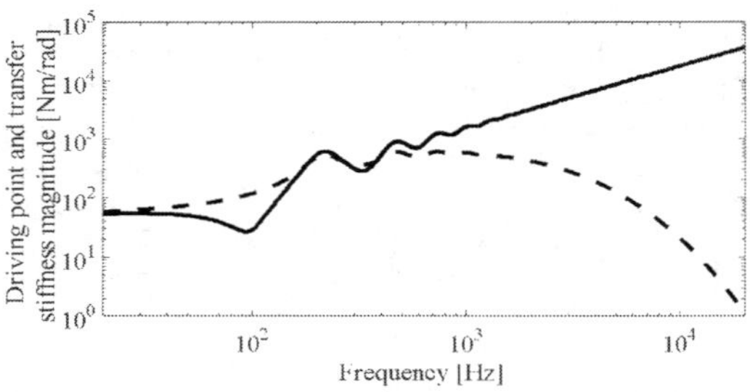

Figure 7. Magnitude of torsional driving point (solid) and transfer (dashed) dynamic stiffness magnitude versus frequency.

tough, rubber like hydrogel, versus the whole audible frequency, from 20 to 20 000 Hz, are shown in Figure 7. Clearly, vibration isolator shows a strong frequency dependence; the driving point stiffness shows consecutive resonance and anti-resonance behavior, with the first resonance (stiffness magnitude trough) at 100 Hz and first anti-resonance (stiffness magnitude peak) at 200 Hz, while transfer stiffness behavior is more nonconsecutive. At higher frequencies, the vibration isolator shows a more local behavior with transfer stiffness rapidly declining; not surprising as wave amplitude rapidly declines with distance due to strong damping. The corresponding stiffness phase $\angle k_{D,T}$ in Figure 8, where $k_{D,T} = |k_{D,T}| \exp(i \angle k_{D,T})$, confirms the resonance and anti-resonance behavior, where driving point stiffness phase is constrained within 0 and 180 while transfer stiffness phase is non-constrained and, thus, forced to be within those limits by wrapping.

3.5 *Conclusion*

Tough, rubber like doubly crosslinked hydrogel displays a strong wave amplitude decrease with

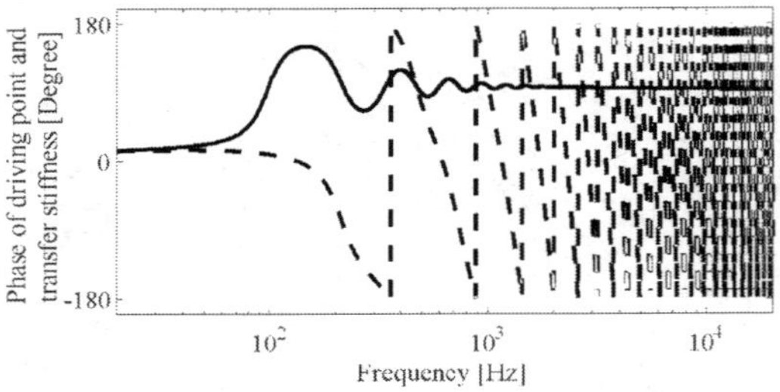

Figure 8. Wrapped torsional driving point (solid) and transfer (dashed) dynamic stiffness phase versus frequency.

propagation distance where maximum decrease after one wave length is at frequency with maximum loss factor of shear modulus. This frequency is possible to adaptively tune and, thus, making this rubber like hydrogel into an interesting material for vibration isolation and damping purposes.

REFERENCES

Ahmed, E.M. 2015. Hydrogel: Preparation, characterization, and applications: A review. *J. Adv. Res.* 6: 105–121.

Czarnecki, S., Rossow, T. & Seiffert, S. 2016. Hybrid polymer-network hydrogels with tunable mechanical response. *Polymers* 8: 82.

Fredette, L. & Singh, R. 2017. High frequency, multi-axis dynamic stiffness analysis of a fractionally damped elastomeric isolator using continuous system theory. *J. Sound Vib.* 389: 468–483.

Kari, L. 2017. Dynamic stiffness of a hybrid, tunable elastomer mount filled with doubly crosslinked hydrogel—the waveguide solution. *J. Mech. Phys. Solids* Submitted 2017.

Koeller, R.C. 1984. Applications of fractional calculus to the theory of viscoelasticity. *J. Appl. Mech.* 51: 299–307.

Lin, W.C., Fan, W., Marcellan, A., Hourdet, D. & Creton, C. 2010. Large strain and fracture properties of poly (dimethylacrylamide)/silica hybrid hydrogels. *Macromolecules* 43: 2554–2563.

Long, R., Mayumi, K., Creton, C., Narita, T. & Hui, C.Y. 2014. Hydrogel: Time dependent behavior of a dual cross-link self-healing gel: Theory and experiments. *Macromolecules* 47: 7243–7250.

Mayumi, K., Marcellan, A., Ducouret, G., Creton, C. & Narita, T. 2013. Stress-strain relationship of highly stretchable dual cross-link gels: Separability of strain and time effect. *ACS Macro Lett.* 2: 1065–1068.

Rose, S., Dizeux, A., Narita, T., Hourdet, D. & Marcellan, A. 2013. Time dependence of dissipative and recovery processes in nanohybrid hydrogels. *Macromolecules* 46: 4095–4104.

Constitutive Models for Rubber X – Lion & Johlitz (Eds)
© 2017 Taylor & Francis Group, London, ISBN 978-1-138-03001-5

Preparation of electroactive elastomers: Stress relaxation and crosslinking aspects

A. Babapour, F.A. Nobari Azar, E. Kaymazlar & M. Şen
Department of Chemistry, Polymer Chemistry Division, Hacettepe University, Beytepe Ankara, Turkey

ABSTRACT: The ultimate purpose of this research is to develop bio-inspired control systems for artificial eyeballs with short reaction time and low voltage demand using Dielectric Elastomer Actuators (DEA). To reach this goal compositions of acrylic or silicone have been prepared as the first step We have planned to develop stretchable electrode systems with high fatigue resistance using the shrinkage property of these double-networked material in the electric field and performance improvements for DEA materials after optimization of rubber formulation. As the first step of this research silicone blends with carbon black content have been prepared in internal mixer. Mechanical properties of blends such as tensile strength and elongation at break values have been obtained for all the samples. Temperature Scanning Stress Relaxation (TSSR) testing machine have been used to evaluate stress relaxation properties of the samples which are related to the network structure of the vulcanizates [1]. Crosslinking density of the vulcanizates have been calculated using Pulse-NMR technique. T2 relaxation time has been used to calculate crosslinking density of the vulcanizates according to the Hann spin-echo principle [2].

1 INTRODUCTION

Silicone rubber is a well-known dielectric elastomer, which can be used for applications such as actuation and for devices able to convert electrical energy into mechanical energy and vice versa [3]. For such applications it is necessary to have a material with low stiffness (low Young's modulus), high breakdown strength and good values for dielectric permittivity [4].

The dielectric permittivity of PDMS is very low therefore, the incorporation of inorganic fillers with high dielectric permittivity in a polymer matrix is a well-known technique to improve the dielectric constant of the material [5–6] conductive particles, such as carbon nanotubes, carbon black, titanium dioxide, barium titanate, magnesium niobate, lead magnesium niobate-lead titanate, and strontium titanate nanoparticles improving the permittivity of the dielectric elastomers.

2 EXPERIMENTAL

2.1 *Materials*

VMQ silicone rubber (Elastosil R752/50) was obtained from Wackers. Stabilizator (Elastosil AUX H3) was also obtained from Wackers. Peroxan HX-Paste 75 SI was used as cross linking agent from PERGAN GmbH.

2.2 *Methods*

In the first step of study Different compositions contain 0 phr, 3 phr and 6 phr carbon black have been prepared using ketjenblack EC-300j carbon black type. Formulations contains 100 phr silicone rubber 1 phr stabilizator agent 1.25 phr mold release agent and 0.8 phr peroxide as a crosslinking agent. All the formulations were prepared using HAKKE RHEOMIX OS LAB internal mixer. Mixing speed was 50 rpm, the mixing temperature was 50°C and the mixing time were 20 min.

All the samples have been analyzed using moving die rheometer. Mechanical properties of blends such as tensile strength and elongation at break values of all blends were determined according to ASTM D412-06ae2 standard.

Temperature Scanning Stress Relaxation machine (TSSR) has been used to evaluate stress relaxation properties of the samples which in turn are related to the network structure and crosslinking density of the vulcanizates. Crosslinking density of all the vulcanizates have been calculated using Pulse-NMR technique.

3 RESULT AND DISCUSSION

Considering mechanical test results we can see that mechanical properties increase with increase of the carbon black amount.

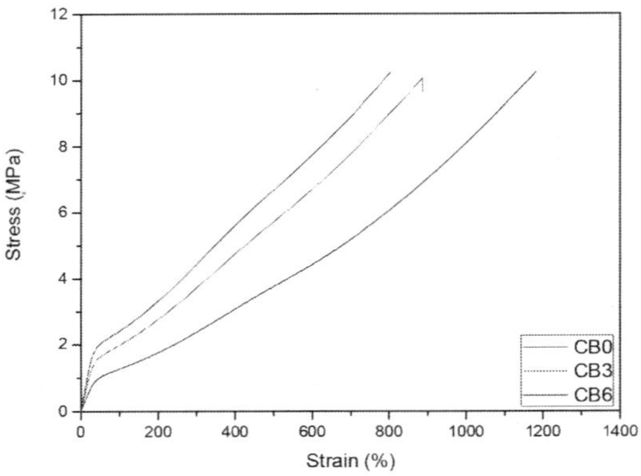

Figure 1. Stress-strain curve for elastomers with different amounts of carbon black.

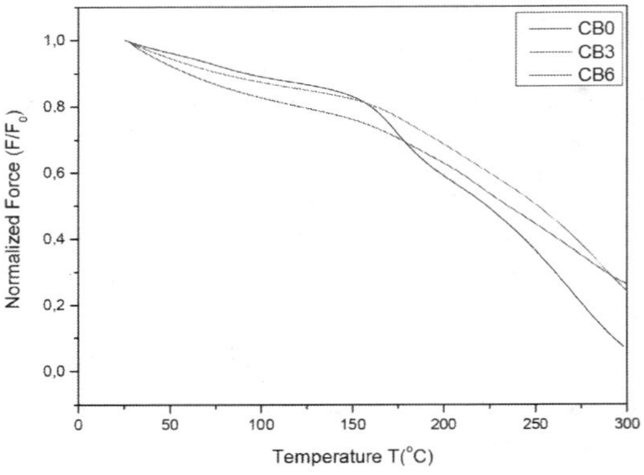

Figure 2. Normalized force-temperature curve of elastomers with different amounts of carbon black.

In the second step of the study, In order to analyze relaxation behaviors of elastomers, samples have been tested using TSSR-Meter device at temperature range of 25–300°C (Fig. 2). This method consisted of an isothermal step at temperatures about room temperature followed by an un-isothermal step with a temperature scan rate of 2 K/min. As can be seen in the figure, relaxation increase a bit with increase of the amount of carbon black at first, but at high temperatures the sample without carbon black relax faster.

Crosslink density data has been obtained using Pulse-NMR spectroscopy to calculate crosslink density. Pulse-NMR curves of samples with different amounts of carbon black have been fitted to Equation 1 using OriginLab software and motion factors (q) of elastomers have been determined

$$M(t) = A*\exp(-t/T2 - q\ M2\ t) + B*\exp(-t/T2) \quad (1)$$

Molecular weight of the segments between crosslinks (M_c) for samples have been calculated using q values and crosslink density of the vulcanizates have been calculated according to Equation 2.

$$M_c = \frac{3}{5\sqrt{q - q_0}} C_\infty \frac{M_Q}{N} \quad (2)$$

Crosslink density values indicate that crosslink density of elastomers do not change with the amount of carbon black. It can be concluded that amount of carbon black does not have any effect on network structure of elastomers but increases the mechanical properties and toughness of elastomers. Studies on the effect of Ketjenblack EC-300J carbon black on the electrical properties of silicone elastomers are underway.

REFERENCES

[1] Barbe A. & Bokamp K. & Kummerkowe C. 2005. Investigation of modified SEBS-based thermoplastic elastomers by temperature measurements scanning stress relaxation. Polym Eng Sci 45: 1498–1507.

[2] Larsen A.L. & Hansen K. & Sommer-Larsen P. & Hassage O. & Bach A. & Ndoni Jørgensen S. 2003. Elastic Properties of Nonstoichiometric Reacted PDMS. Networks *Macromolecules 36* (26): 10063–10070.

[3] Gharavi N. & Razzaghi-Kashani M. & Golshan-Ebrahimi N. 2010. Effect of organo-clay on the dielectric relaxation response of silicone rubber. Smart Mater. Struct 19: 964–1726.

[4] Enis T. & Sauers I. 2009. Industrial applications perspective of nanodielectrics. Dielectric Polymer Nanocomposites. New York: Sipringer

[5] Momen D. & Farzaneh M. 2011. Survey of micro/nano filler use to improve silicone rubber for outdoor insulators. Rev. Adv. Mater. Sci. 25: 1–13.

[6] Khastgir D. & Adachi K. 2000. Rheological and dielectric studies of aggregation of barium titanate particles suspended in polydimethylsiloxane. Polymer 41: 6403–6413.

Industrial applications

Constitutive Models for Rubber X – Lion & Johlitz (Eds)
 ISBN 978-1-138-03001-5

Modelling of the mechanical behaviour of elastomer seal at low temperature

J. Troufflard, H. Laurent & G. Rio
University Bretagne-Sud (UBS), FRE CNRS 3744, IRDL, Lorient, France

B. Omnès
Cetim, Nantes, France

ABSTRACT: The aim of the study is to describe the mechanical behaviour of a rubber seal at low temperature. The seal must ensure tightness under this extreme condition. In previous studies, presented in ECCMR VIII and ECCMR IX congress, a material database was built on a HNBR and a first numerical approach was performed. To improve the description of the O-ring behaviour at low temperature, Cetim and UBS (IRDL) use an original phenomenological model, named Hyper-Visco-Hysteresis (HVH), in association with the existing material experimental characterization database. The procedure for the identification of material parameters of HVH model and the numerical results are detailed in this paper.

1 INTRODUCTION

For many industries, such as Oil and Gas, Nuclear and Aerospace, ensure the tightness in extreme conditions, notably at low temperature, is a major issue. Recently, European Sealing Association carried out studies in order to establish a procedure to determine the low-temperature operating limits of elastomeric seals (Kerwin et al. 2016, Douglas et al. 2016). Nuclear Industry is also a major player in this field of activity, notably with studies of (Jaunich et al. 2013, Grelle et al. 2017). The authors investigate the relationship between the interfacial leak and the Compression Set (CS) with temperature.

Cetim, in collaboration with LRCCP, has performed experimental studies on rubber at low temperature (Rouillard et al. 2013) and has investigated numerical models to describe the thermo-mechanical behaviour of elastomer. One of the selected models was a hyper-viscoelastic model (Prony series) with the time-temperature equivalence (WLF), implemented in commercial code (Omnès and Heuillet 2015) and (Lejeune et al. 2015). The first numerical results highlighted some difficulties to describe the reaction force evolution of O-ring during an anisotherm cycle (Omnès and Heuillet 2015). Therefore, Cetim studied the Hyper-Visco-Hysteretic model (HVH) (Laurent et al. 2007, Vandenbroucke et al. 2010, Laurent et al. 2014) in collaboration with UBS (IRDL).

This paper shows the numerical approach carried out recently with this model and the existing experimental database. The aim was to evaluate the ability of HVH model to describe the mechanical behaviour of the rubber seal at low temperature. This model and the associated procedure for the identification of parameters are detailed here. Moreover, the confrontation between numerical and experimental results obtained on an O-ring during anisotherm cycles is also presented.

2 EXPERIMENTAL DATABASE

2.1 *Material*

The rubber compound studied was formulated by the LRCCP (French Rubber and Plastics Research and Testing Laboratory). The elastomer belongs to the hydrogenated nitrile butadiene rubber (HNBR) family. Its composition was presented in the paper (Rouillard et al. 2013).

The matrix of HNBR is a Therban A 3407 reinforced with a carbon black filler N550 (45 phr). The hardness is equal to 70 Shore A. O-rings (Inside Diameter 50.17 mm; Cross-section Diameter 5.33 mm) were moulded with this compound in order to study both tightness performances and their low temperature sensitivity. The glass transition temperature (Tg) was determined by Differential Scanning Calorimetry (D.S.C.) and the value is around –28°C (Omnès and Heuillet 2015).

2.2 *Mechanical properties with temperature*

A thermomechanical database on the HNBR was established in 2012. Different thermomechanical tests were performed such as:

- Dynamic Mechanical Analysis (D.M.A.);
- Thermal expansion test;
- Quasi-static test with a relaxation phase at different temperatures.

The D.M.A. tests were carried out on the test apparatus METRAVIB 450+ analyses, on the temperature range [–80:+150°C], to highlight the viscoelastic behaviour of the rubber. The peaks of the loss factor give the values of the Tg which depends of the frequency: $Tg_{1Hz} = -18°C$, $Tg_{10Hz} = -13°C$, $Tg_{95Hz} = -7°C$.

The thermal expansion was determined on a cubic sample (5 × 5 × 10 mm) according to ISO 11359-2. The increase or decrease of the temperature is fixed to 5°C/min. The glass transition temperature could be observed around –25°C, with the curve slope change on Figure 1. The values of the expansion coefficient above the Tg is around $175.10Tg^{-06}/°C$ and below Tg, the value is around $70.10^{-06}/°C$.

Furthermore, stress relaxation tests were performed to characterise the viscoelastic behaviour in quasi-static conditions. This test reproduces the classical use of O rings with imposed deformation by the groove.

The chosen sample is a Goodrich cylinder sample (high: 31 mm; diameter: 18 mm) as shown on the Figure 2. To measure exactly the temperature in the sample and the duration of the thermal stabilisation phase, a thermocouple was introduced in a second sample, which is close to the main sample in the climate chamber Figure 2.

After the thermal conditioning (1.5 hour), a strain was imposed on the sample to obtain a squeeze of 25% with a strain rate equal to $4.10^{-03}.s^{-1}$. The stress relaxation was measured during 3 hours followed by an unloading to null stress. The temperature range is [–34:+80°C] (Figure 3).

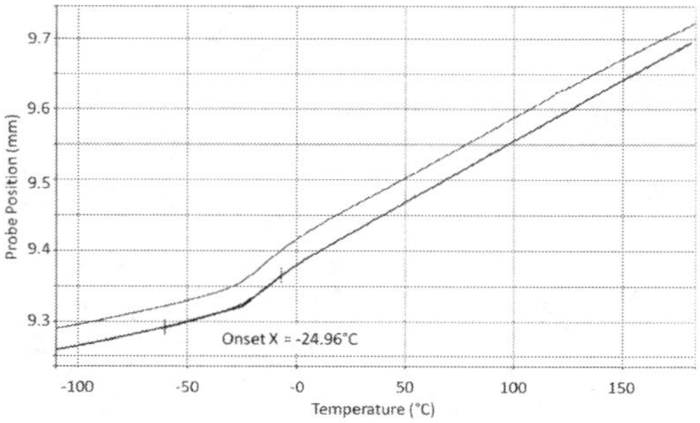

Figure 1. Evolution of the thermal expansion coefficient with temperature for HNBR; Tg observed at –25°C.

Figure 2. Stress relaxation test: climate chamber and samples used to compression test and thermal information.

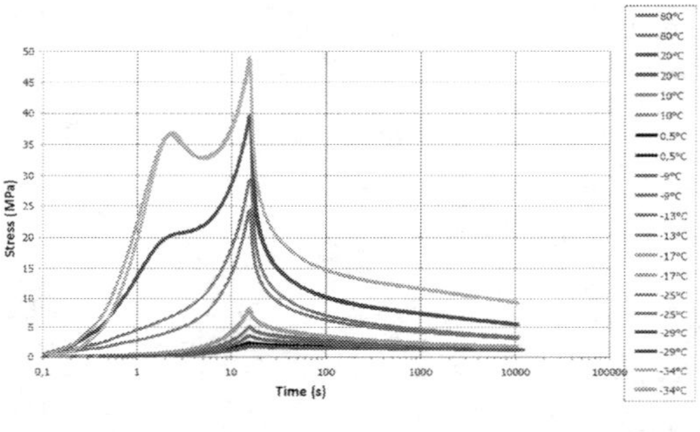

Figure 3. Relaxation test on the HNBR between [–34:+80°C].

2.3 *Leak-tightness of O-ring*

2.3.1 *Leakage detection method*

For sealing application at low temperature (close to Tg domain), only interfacial leak was studied. Indeed, this type of leak appears due to the viscoelastic evolution and expansion/retraction of the rubber in its groove. The thermomechanical evolutions induce a lack of contact pressure at seal/rigid parts interface.

Furthermore, when the temperature decreases, the permeation phenomenon impact gets lower because the gas solubility and diffusion through the rubber depend on the temperature following Arrhenius type equation.

Helium leak detectors were used in the previous study (Omnès and Heuillet 2015). But during this study a binary behaviour (leak/no leak) was observed. Moreover, when a leakage occurred at low temperature, a high quantity of helium was collected which can lead to a leak detector saturation. Therefore, a differential sensor pressure was preferred in this new study.

2.3.2 *Description of the test fixture*

The specific test apparatus used in this study (identical to previous study) is illustrated on Figure 4. The assembly consists in two plates at each end (one equipped with a force sensor) and in the middle a plate with rectangular groove.

The O-ring is mounted in the test rig without grease. The squeeze imposed on the seal is around 24%. Another plate was used to impose a squeeze around 10% to study the influence of the compression on the tightness efficiency.

The reaction force is measured with the force sensor placed over the O-ring. A chamber is created between the test seal and the secondary seal to collect the leakage. The service pressure applied in the center of the fixture is around 0.5 MPa.

2.4 *Low temperature behaviour of seal*

A new thermal profile was imposed on the test rig. The O-ring mounted in the groove was left at room temperature (20°C) for 12 hours. After this thermal conditioning period, the temperature decreases progressively by 10°C steps, down to –60°C with dwell times of four hours. When the minimal temperature was reached, the temperature increases progressively by 10°C steps, up to 20°C with twelve hours relaxation periods. The evolution of reaction force of O-ring with the new thermal profile is shown on Figure 5.

When the temperature decreases, a non-negligible loss of the reaction force appears and the value is close to zero from –20°C. Nevertheless, this lack of contact pressure is non-permanent. Indeed, a non-zero value is obtained at –10°C during the

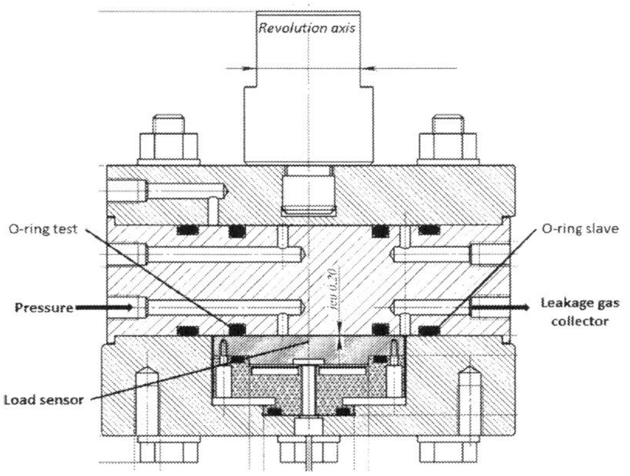

Figure 4. Test apparatus developed with a force sensor.

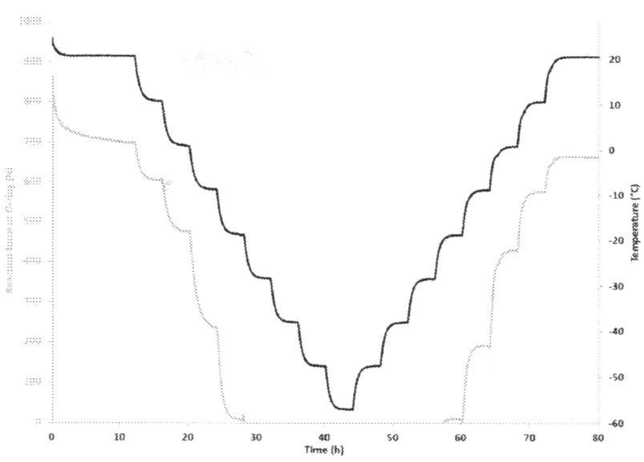

Figure 5. Evolution of the reaction force of O-ring with the thermal profile.

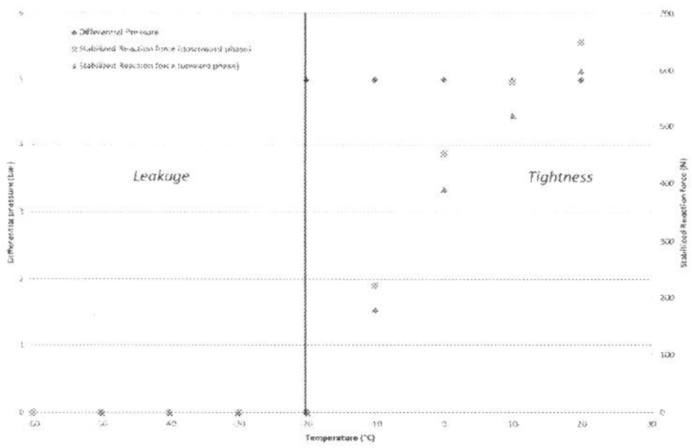

Figure 6. Evolution of the seal "leakage/tightness" boundary and the stabilized reaction force of O-ring with temperature evolution.

rise of temperature. This loss of load was observed by different authors (Laurent et al. 2014) and (Jaunich et al. 2013).

The tightness measurement (differential pressure between the chambers) is performed at the end of each holding time with a working pressure fixed at 0.5 MPa for 30 seconds. Figure 6 shows the "leakage/tightness" boundary and the stabilized reaction force with the temperature. The leakage occurs when the load is null and the value is lower during the temperature rise.

3 HYPER-VISCO-HYSTERESIS BEHAVIOUR MODEL

To simulate the behaviour of the HNBR material, an original Hyperelasto-Visco-Hysteresis (HVH) model is used. This model has been successfully applied to a fluoro-elastomer in function of temperature (Vandenbroucke et al. 2010, Laurent et al.

2011, Laurent et al. 2014). In this study, based on the previous database, the identification of material parameters of this model is presented and the previous tightness test is used to validate the response of this model.

3.1 *Model description*

The general framework is the balance of the variation of external and internal mechanical powers. The internal power is related to both stress and strain rate such as:

$$d\mathcal{P}_{ext} = d\mathcal{P}_{\text{int}} = \boldsymbol{\sigma} : \boldsymbol{D} dv \tag{1}$$

where $\mathcal{P}_{ext}$ and $\mathcal{P}_{int}$ are external and internal powers, $\boldsymbol{\sigma}$ the Cauchy stress tensor and $\boldsymbol{D}$ the strain rate tensor.

The internal power is divided into mechanical contributions of stress such as:

$$\boldsymbol{\sigma} : \boldsymbol{D} dv = \left(\boldsymbol{\sigma}_e + \boldsymbol{\sigma}_v + \boldsymbol{\sigma}_h\right) : \boldsymbol{D}\, dv \tag{2}$$

where $\boldsymbol{\sigma}_e$, $\boldsymbol{\sigma}_v$ and $\boldsymbol{\sigma}_h$ respectively denote a reversible hyperelastic part, a viscous part and a non-reversible hysteretic part (Laurent et al. 2011). This phenomenological approach is the fundamental basis for both the Hyper-Visco-Hysteresis (HVH) model setup and its strategy for material parameters identification. On one hand, these contributions can be observed and separated from macroscopic tests. On the other hand, measuring the strain rate $\boldsymbol{D}$ is achievable and the macroscopic total strain is obtained easily.

A more general level of partitioning of the stress is given by the volumetric and distortional separation. The suitable choice of the strain measures and their invariants is necessary to well establish this decoupling by mean of the spherical and deviatoric parts of the strain tensor.

The reversible part $\boldsymbol{\sigma}_e$ is defined by the Hart-Smith potential defined as:

$$\begin{aligned} W = C_1 \int_3^{J_1} \exp[C_3(J_1 - 3)^2] dJ_1 \\ + C_2 \ln\left(\frac{J_2}{3}\right) + \frac{K}{2}\ln^2(V) \end{aligned} \tag{3}$$

where J_1, J_2 are the first and second modified invariants of the right Cauchy-Green strain $\boldsymbol{B}$ and $V = v/v_0$ the relative volume. C_1, C_2 and C_3 are material coefficients and K the bulk modulus.

The generalized Maxwell model is chosen for the viscous part $\boldsymbol{\sigma}_v$. Five parallel branches are used and each one is governed by the following equation:

$$\bar{\boldsymbol{D}} = \frac{1}{2G_i}\dot{\bar{\boldsymbol{\sigma}}}_{v,i} + \frac{1}{\eta_i}\bar{\boldsymbol{\sigma}}_{v,i} \tag{4}$$

where the symbols $\bar{\ }$ and $\dot{\ }$ denote respectively the deviatoric part of the tensor and the Jauman derivative. The subscript i is the branch number. G_i and η_i are the shear modulus and the viscous coefficient of the i-th branch. These two coefficients are linked by the following relation:

$$\eta_i = 2G_i\tau_i \tag{5}$$

where τ_i is a characteristic time related to the kinetic of the stress relaxation.

The non reversible part $\boldsymbol{\sigma}_h$ is defined by the elasto-hysteresis model. This model is the integral form of the Saint-Venant rheological model, namely an infinite assembly in parallel of elastic and slip elements. Its use have an interest in modeling metal plasticity and shape memory alloys (Rio et al. 2009) but is also suitable for polymers (Bles et al. 2009). In particular, it uses a discrete storage of loading inversion points that is very efficient to capture the non-viscous hysteresis loops in cyclic loading. Using the Jauman derivative, the following constitutive equation is integrated:

$$\dot{\bar{\boldsymbol{\sigma}}}_h = 2\mu\bar{\boldsymbol{D}} + \beta\phi\Delta_R^t\bar{\boldsymbol{\sigma}}_h \tag{6}$$

The two following relations give the expressions for β and ϕ:

$$\beta = \frac{-2\mu}{(w' Q_0)^{n_p}(Q_{\Delta\bar{\sigma}_h})^{2-n_p}} \tag{7}$$

$$\phi = \Delta\bar{\boldsymbol{\sigma}}_h : \bar{\boldsymbol{D}} - \frac{(Q_{\Delta\bar{\sigma}_h})^2}{2\mu}\cdot\frac{\dot{w}'}{w'} \tag{8}$$

where β is relative to the intensity of $\Delta_R^t\bar{\boldsymbol{\sigma}}_h$ and ϕ is a non reversible intrinsic dissipated rate. w' is the Masing similarity function. μ and Q_0 are respectively the initial slope and the yield limit of the hysteresis stress-strain curve under shear loading conditions. n_p is the Prager's parameter which manages the non linear transition to reach the stress saturation Q_0.

The thermal expansion is taken into account according to the following formula:

$$\varepsilon_{th} = \alpha_T \Delta T.\boldsymbol{I} \tag{9}$$

where ε_{th} denotes the thermal strain, α_T the coefficient of thermal expansion, ΔT the temperature variation and $\boldsymbol{I}$ the second-order identity tensor.

It is important to note that the thermal dependency of the HVH model is introduced by mean of parameters evolution. The temperature field is known a priori and imposed as a nodal boundary condition in the further finite elements simulations.

3.2 *Strategy for material parameters identification*

In previous study, the material parameters of the HVH model were identified with multi-step relaxation test (Laurent et al. 2011). In the present work, the previously described HNBR database will be used. The aim is to separate the different contributions of stress from the experimental data.

The following identification steps are undergone. First, oedometric tests are directly used for bulk modulus evaluation as these tests involve a pure volume change. Secondly, the time-temperature equivalence is applied to extrapolate the isothermal relaxation tests. Using the first eight days extrapolation, the Maxwell part of the model is identified following (Laurent et al. 2011). Finally, the hyperelastic-hysteretical non viscous part is determined by the use of anisothermal relaxation tests. The force gap between two subsequent temperature levels gives an information of the non viscous modification of the material behaviour. The non homogeneous anisothermal test is approached by a simplified VER-like modelization. An inverse identification of and is performed to match the relative loss of measured contact force.

Tables 1, 2 and 3 present the identified parameters as a function of temperature, respectively for the hyperelastic, viscous and hysteretical part.

3.3 *Numerical simulations*

The functional test on an O-ring is used for the validation of the identified HVH model.

Table 1. Hyperelastic parameters.

T (°C)	−34	−25	−13	0	20
K (MPa)			2700		
C_1 (MPa)	1e-6	1e-6	1.53e-2	3.89e-2	4.98e-1
$C_2 = C_1/10$, $C_3 = C_1/100$					

Table 2. Viscous parameters (examples for first, third and fifth Maxwell branches).

T (°C)	−34	−25	−13	0	20
G_1 (MPa)	83.5	45.4	3.8	1.2	0.8
η_1 (MPa.s)	7.9e+2	4.9e+2	5.4e+1	2.0e+1	1.4e+1
τ_1 (s)	4.7	5.4	7	8.3	8.4
G_3 (MPa)	1.6	1	0.3	0.2	0.1
η_3 (MPa.s)	7.5e+3	4.6e+3	1.5e+3	8.5e+2	5.7e+2
τ_3 (s)	2367	2356	2654	2574	2602
G_5 (MPa)	2.8	0.3	0.1	0.1	0.1
η_5 (MPa.s)	1.4e+6	1.4e+5	8.2e+4	4.2e+4	3.3e+4
τ_5 (s)	2.6e+5	2.2e+5	2.8e+5	2.7e+5	2.3e+5

Table 3. Elasto-hysteresis parameters.

T (°C)	−34	−25	−13	0	20
Q_0 (MPa)	3.51	1.34	0.53	0.15	1e-6
$\mu = 4Q_0$, $n_p = 2$					

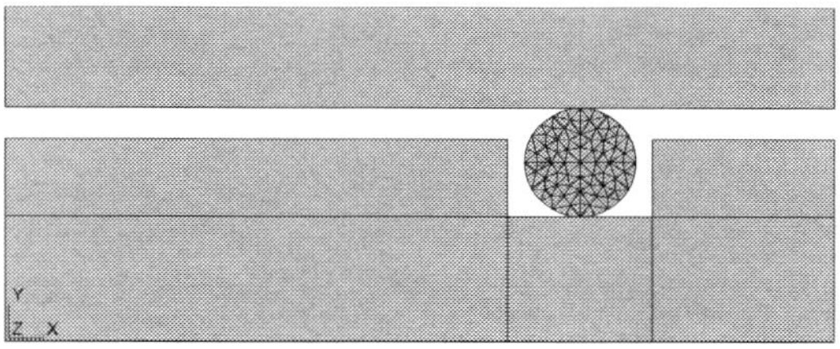

Figure 7. FE model (test rig: coarse grid of linear elements, O-ring: 281 nodes and 128 quadratic triangular elements).

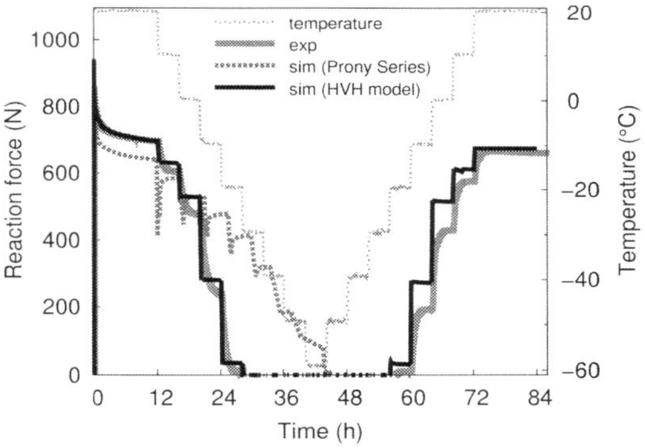

Figure 8. Evolution of the reaction force of the O-ring.

The in-house FEA software Herezh++ (Rio 2016) is used here both for Finite Element Analysis and material behaviour. It should be noted that Herezh++ can be plugged with Abaqus for the material behaviour only as a user material UMAT (Rio et al. 2008).

A 50.17 × 5.33 mm O-ring seal is modeled with axisymetric conditions. Quadratic triangular elements are used for rubber elements. A rigid test rig imposes a 24% squeeze on the rubber elements by frictionless contact conditions. The overall model is shown on Figure 7.

The numerical reaction force is compared to the experimental profile in Figure 8. The previous results with Prony series and WLF equation (Omnès and Heuillet 2015) are also given in this figure. The results with the present HVH model are in good agreement with experimental data. The model gives a good prediction of the total loose of contact pressure between −20°C and −30°C.

4 CONCLUSIONS

In this paper, the thermomechanical tests which were performed on a HNBR, were used to create an experimental database. Furthermore, functional tests were carried out on O-ring with the same compound. The aim was to describe the relationship between the thermomechanical behaviour and the tightness close to the glass transition temperature (Tg).

Based on these experimental results, material parameters of the Hyperelasto-Visco-Hysteresis model were identified according to an original strategy. This work was completed with a Finite Element Analyses to validate the model on an O-ring simulation. The evolution of the contact pressure on the seal with temperature was simulated and the numerical results were compared to the experimental values. The comparison highlights a good prediction of the drop and the lack of contact pressure during the decrease of temperature on the temperature range [–30:–20°C].

The future work between Cetim and UBS (IRDL) is to estimate the evolution of the thermomechanical behaviour of the seal with different levels of the squeeze and thermal profile.

REFERENCES

Bles, G., W. Nowacki, & A. Tourabi (2009). Experimental study of the cyclic visco-elasto-plastic behaviour of a polyamide fibre strap. *International Journal of Solids and Structures 46*(13), 2693–2705.

Douglas, A., O. Devlen, M. Mitchell, D. Edwin-Scott, & M. Neal (2016). Development of a procedure to accurately measure the low-temperature operating limits of elastomeric seals. *Sealing Technology 2016*(9), 7–12.

Grelle, T., D. Wolff, & M. Jaunich (2017). Leakage behavior of elastomer seals under dynamic unloading conditions at low temperatures. *Polymer Testing 58*, 219–226.

Jaunich, M., W. Stark, D. Wolff, & H. Völzke (2013). Investigation of elastomer seal behavior for transport and storage packages. In *Proceedings of the 17th PATRAM*, San Francisco.

Kerwin, J., O. Devlen, & D. Edwin-Scott (2016). Development of a procedure to accurately measure the low temperature operating limits of elastomeric seals. In *Fluid Sealing*, Manchester, UK.

Laurent, H., G. Rio, A. Vandenbroucke, & N. Aït Hocine (2014). Experimental and numerical study on the temperature-dependent behavior of a fluoro-elastomer. *Mechanics of Time-Dependent Materials 18*(4), 721–742.

Laurent, H., A. Vandenbroucke, S. Couëdo, G. Rio, & N. Aït Hocine (2007). An hyper-visco-hysteretic model for elastomeric behaviour under low and high temperatures: Experimental and numerical investigations. In *Constitutive Models for Rubber V*, Paris.

Laurent, H., A. Vandenbroucke, G. Rio, & N.A. Hocine (2011). A simplified methodology to identify material parameters of a hyperelasto-visco-hysteresis model: application to a fluoro-elastomer. *Modelling and Simulation in Materials Science and Engineering 19*(8), 085004.

Lejeune, J., C. Brung, N. Arnault, & M. Barry Maizeroi (2015). Simulation of an AEM rubber at the transition temperature and above for sealing applications. In *Constitutive Models for Rubber IX*, Prague.

Omnès, B. & P. Heuillet (2015). Leak tightness of elastomeric seal at low temperature: Experimental and fem-simulation. In *Constitutive Models for Rubber IX*, Prague.

Rio, G. (2016). *Herezh++, manuel d'utilisation (user manual in french), v 6.781*. university of Bretagne Sud.

Rio, G., D. Favier, & Y. Liu (2009). Elastohysteresis model implemented in the finite element sofware herezh++. In *ESOMAT 2009 The 8th European Symposium on Martensitic Transformations*, Prague.

Rio, G., H. Laurent, & G. Blès (2008). Asynchronous interface between a finite element commercial software ABAQUS and an academic research code HEREZH++. *Advances in Engineering Software 39*(12), 1010–1022.

Rouillard, F., P. Heuillet, & B. Omnès (2013). Viscoelastic characterization at low temperature on an hnbr compound for sealing applications. In *Constitutive Models for Rubber VIII*, San Sebastian.

Vandenbroucke, A., H. Laurent, N. Aït Hocine, & G. Rio (2010). A hyperelasto-visco-hysteresis model for an elastomeric behaviour: Experimental and numerical investigations. *Computational Materials Science 48*(3), 495–503.

Constitutive Models for Rubber X – Lion & Johlitz (Eds)
© 2017 Taylor & Francis Group, London, ISBN 978-1-138-03001-5

FE analysis of hybrid cord-rubber composites

H. Donner & J. Ihlemann
Professorship of Solid Mechanics, Chemnitz University of Technology, Chemnitz, Germany

ABSTRACT: Twisted polymeric cords are crucial to reinforce important industrial rubber components like tyres, driving belts, and air springs. Standard cords, consisting of multifilament yarns of only one material, are not always suitable to meet the component's demands. To this end, yarns of different materials like aramid and nylon are combined, yielding a so-called hybrid cord. In this contribution, a Finite Element (FE) model of a hybrid cord including its geometry and filament structure is presented. In addition, a shell-like Representative Volume Element (RVE) is proposed, which enables the analysis of the stresses and strains at the highly loaded parts of a hybrid cord-rubber component like the rolling lobe of an air spring. The FE models are implemented in Abaqus utilizing several of its user subroutines. Further, the particular simulations are coupled by means of MATLAB to carry out parameter identifications, parameter studies, and optimizations. To demonstrate the opportunities emerging from these models, the optimization of a hybrid cord concerning its fracture force and the optimization of a cross ply-reinforced composite concerning the maximum strain of the rubber are presented.

1 INTRODUCTION

Hybrid cords enable the combination of the favorable mechanical properties of different materials. For example, a nylon-aramid-hybrid cord shows an increased fracture strain (compared to a pure aramid cord), and an increased fracture force (compared to a pure nylon cord). Moreover, its tensile curve and thus its stiffness properties can be varied in a wide range. Figure 1 shows three experimental tensile curves of differently twisted nylon-aramid-hybrid cords. The principal twisting parameters like yarn counts[1], yarn twists[2], and the cord twist[3] are equal. Clearly, significant differences in the tensile curve, fracture strain, and fracture force can be observed. Microscopy images of the corresponding cross sections reveal that the hybrid cords' geometry and thus their filament structure differ. To distinguish between the different hybrid cords, their mass properties can be analyzed. Weighing showed different amounts of nylon and aramid within the hybrid cords.[4] As a rule, the hybrid cord's stiffness decreases and its fracture strain increases for an increasing amount of nylon. However, its fracture force is not monotonically depending on the nylon content and has an optimum whose experimental determination is quite expensive.

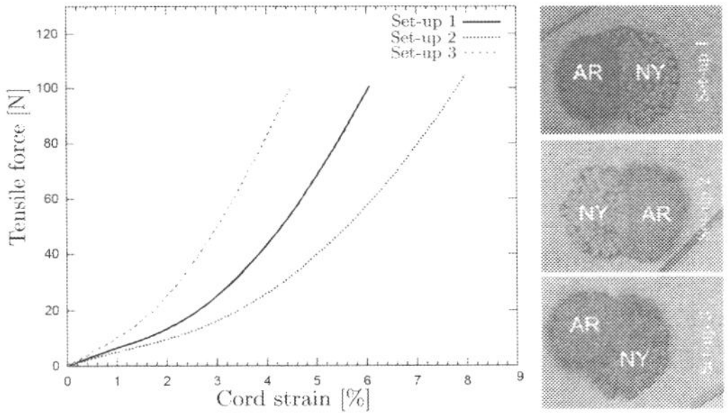

Figure 1. Experimental tensile curves of differently twisted nylon-aramid-hybrid cords and corresponding microscopy images of their cross section. The principal twisting parameters of the hybrid cords are equal.

In order to support the optimization of hybrid cords and hybrid cord-rubber composites, a FE model of a nylon-aramid-hybrid cord (Section 2) and a FE model of a corresponding cross ply-reinforced RVE (Section 3) is presented. The approaches base on the ideas presented in Donner & Ihlemann (2013). Thus, the filament structure of the hybrid cord is modeled by means of a transversely-isotropic material model whose preferred direction at each integration point is coaxially oriented to

[1] $\tilde{T}_{Ny} = 470$ **dtex**, $\tilde{T}_{Ar} = 440$ dtex, 1 **dtex** = 1 g/10,000 m
[2] $\nu_{Ny} = \nu_{Ar} = 480\ m^{-1}$
[3] $\nu_{HC} = 480\ m^{-1}$ the yarn twist and the cord twist are of opposite direction.
[4] Set-up 1: $\alpha_{Ny} = +3.8\%$ nylon, set-up 2: $\alpha_{Ny} = +1.5\%$ nylon, set-up 3: $\alpha_{Ny} = +7.2\%$ nylon. Compared with the mean yarnmass $\bar{m}_Y = m_{HC}/2$. The aramid content equals the negative nylon content $\alpha_{Ar} = -\alpha_{Ny}$.

the filament curve's tangent.[5] The geometry of the hybrid cord is obtained by the FE simulation of its twisting process. An analytical model of the hybrid cord's geometry, whose free parameters are identified by means of the results of the twisting process simulation, enables the computation of the filament structure. The geometry model is required within the composite analysis to distinguish between the different materials at each integration point (referred as gauss-point method or multiphase elements, cf. Steinkopff & Sautter 1995, Kreikemeier 2012). Consequently, a simply structured mesh is used to discretize the RVE and to reduce the computational effort, which neglects the interfaces between the yarns and the elastomer. To analyze the stresses and strains near the interface, an additional submodel of a hybrid cord surrounded by a rubber layer is utilized. The particular simulations are coupled via MATLAB to maximize the hybrid cord's fracture force and to minimize the elastomer strain within the composite (Section 4).

2 HYBRID CORD MODEL

The initial configuration of the simulation of the hybrid cord's twisting process is given by two parallel and already twisted yarns (see left side of Figure 2). The yarn twists v_Y[6] enter the simulation via the structural tensor $\underline{\underline{A}}$ of the transversely-isotropic material model which is used for both yarn materials. The Kirchhoff stress tensor $\underline{\underline{\tau}}$ of the material model reads

$$\underline{\underline{\tau}} = (1-\beta_Y)G_Y\left[\left(\bar{\lambda}^F\right)^2 - \frac{1}{\bar{\lambda}^F}\right]\underline{\underline{A}}^D + \beta_Y G_Y \underline{\underline{\bar{b}}}^D + K_Y J(J-1)\underline{\underline{I}} \quad (1)$$

with $\bar{b}$ as the unimodular part of the left Cauchy-Green tensor, J as the local volume stretch, $\underline{\underline{A}}$ as the structural tensor of the current configuration, and $\bar{\lambda}^F$ as the filament's stretch through the isochoric part of the deformation. The actual material parameters are the shear modulus G_Y[7], the bulk modulus K_Y[8], and the anisotropy ratio β_Y[9]. The superscript D denotes the deviator and $\underline{\underline{I}}$ represents the second-order identity tensor. The material model is implemented via Abaqus' UMAT.

[5] Similar approaches are Maßmann (1995) & Jacob (2005).

[6] The subscript Y is either Ny (nylon) or Ar (aramid).

[7] Identified: $G_{Ny} = 2{,}368$ MPa, $G_{Ar} = 37{,}635$ MPa

[8] Identified: $K_{Ny} = 11{,}048$ MPa, $K_{Ar} = 175{,}633$ MPa

[9] β_Y is assumed to be 0.001 for both yarn materials.

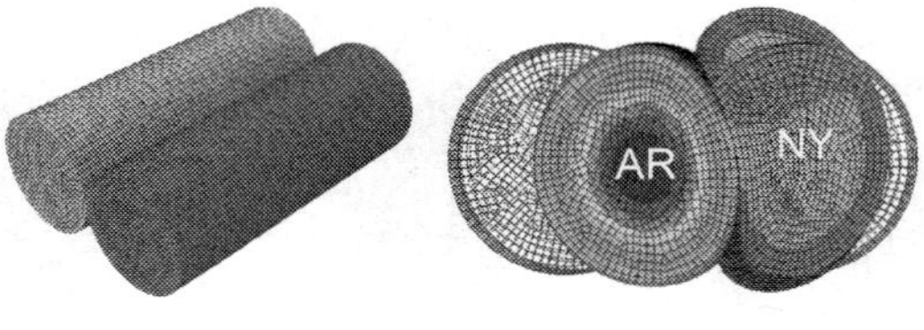

Figure 2. Initial configuration of the twisting process simulation (left) and deformed shape of the twisted hybrid cord (right). The initial configuration is depicted as grid.

Since the filaments are assumed to be initially helical with an increasing pitch for an increasing distance to the yarn center $\underline{\tilde{r}}_{YC}$, the initial tangent vector $\underline{\tilde{a}}$ at the initial position $\underline{\tilde{r}}$ reads

$$\underline{\tilde{a}}(\underline{\tilde{r}}) = \frac{2\pi v_Y\left(\Delta\tilde{y}\underline{e}_x - \Delta\tilde{x}\underline{e}_y\right) + \underline{e}_z}{\sqrt{1+4\pi^2 v_Y^2(\Delta\tilde{x}^2+\Delta\tilde{y}^2)}} \quad (2)$$

with $\Delta\underline{\tilde{r}} = \underline{\tilde{r}} - \underline{\tilde{r}}_{YC}$. The structural tensor of the current configuration is given by $\underline{\underline{A}} = \underline{\underline{F}}\cdot\underline{\underline{\tilde{A}}}\cdot\underline{\underline{F}}^T / \mathrm{tr}\,(\underline{\underline{F}}\cdot\underline{\underline{\tilde{A}}}\cdot\underline{\underline{F}}^T)$ with $\underline{\underline{\tilde{A}}} = \underline{\tilde{a}}\circ\underline{\tilde{a}}$ and $\underline{\underline{F}}$ as the deformation gradient. The initial distribution of the preferred direction is prescribed using the subroutine SDVINI. For more information on anisotropic, hyperelastic material models, the reader is referred to Holzapfel (2000) and Balzani, Neff, Schröder, & Holzapfel (2006).

From a physical point of view, within a perfectly twisted cord, two arbitrary cross sections can be transformed into each other by a rotation about the cord axis and a translation along the cord axis[10]. This equals an euclidean transformation of the cross section and enters the simulation as nonlinear constraint between the position vector of the slave nodes $\underline{r}^S$ and the position vector of the master nodes $\underline{r}^M$ of two cross sections

$$\underline{r}^S = \underline{\underline{R}}_Y\left(v_{HC}, \varepsilon_Y\right)\cdot\underline{r}^M + \left(1+\varepsilon_Y\right)\Delta\tilde{z}\underline{e}_z \quad (3)$$

with the orthogonal tensor $\underline{\underline{R}}_Y^{-1} = \underline{\underline{R}}_Y^T$

$$\underline{\underline{R}}_Y = \exp\left(\left(-2\pi v_{HC}\left(1+\varepsilon_Y\right)\Delta\tilde{z}\underline{\underline{\in}}\cdot\underline{e}_z\right)\right), \quad (4)$$

where $\in$ stands for the permutation tensor and $\Delta\tilde{z}$ for the initial distance between both cross sections. This constraint has already been used in Jiang, Yao, & Walton (1999) and Donner & Ihlemann (2014) and is implemented in Abaqus via the user subroutine MPC. Due to the constraint,

[10] The cord axis coincides with the z axis.

surface effects are completely avoided. Moreover, since only few element layers in cord direction are necessary for the yarn contact, the computational effort is significantly reduced. In contrast to v_{HC}, which is equal for both yarns, ε_{Y} controls the current distance between two cross sections and can be independently prescribed for both yarns. A change in ε_{Ny} and ε_{Ar} causes a change in the total tensile force during the twisting as well as their distribution to both yarns. Consequently, hybrid cords of different masses and yarn contents can be simulated. The deviation of the nylon yarn's mass α_{Ny} with respect to the mean yarn mass $\bar{m}_{\text{Y}}$ is $\alpha_{\text{Ny}} = (m_{\text{Ny}} - \bar{m}_{\text{Y}})/\,\bar{m}_{\text{Y}}$ and can be expressed in terms of the hybrid cord's titer T_{HC}, its nominal yarn count $\tilde{T}_{\text{Ny}}$, and ε_{Ny} as

$$\alpha_{\text{Ny}} = \left[\left(2\tilde{T}_{\text{Ny}} / (1+\varepsilon_{\text{Ny}}) - T_{\text{HC}}\right)\right] / T_{\text{HC}}. \tag{5}$$

The result of an exemplary simulation of the twisting process of a hybrid cord can be seen in Figure 2. Clearly, the cross section's shape matches with the microscopy images in Figure 1.

The simulation of the hybrid cord's twisting process contains the filament structure only implicitly. In order to compute filament curves and tangent vectors, the analytical model as proposed in Donner & Ihlemann (2013) is extended in view of hybrid cords. Thus, two convective coordinates $\bar{r}$ and $\bar{\varphi}$ are introduced, which form a cylinder coordinate system in the initial configuration with the yarn center as their origin (see Figure 3). Due to the twisting process, the cylinder coordinate system becomes a general, curvilinear coordinate system. Since the periodicity with respect to $\bar{\varphi}$ remains, the following ansatz for the position vector r_{P} of the point P is proposed

$$\Delta \underline{r}_{\text{P}} = \sum_{j=1}^{\infty} \bar{r}^{j} \left[\sum_{i=0}^{\infty} \underline{r}_{ij}^{\text{C}} \cos(i\bar{\varphi}) + \underline{r}_{ij}^{\text{S}} \sin(i\bar{\varphi}) \right] \tag{6}$$

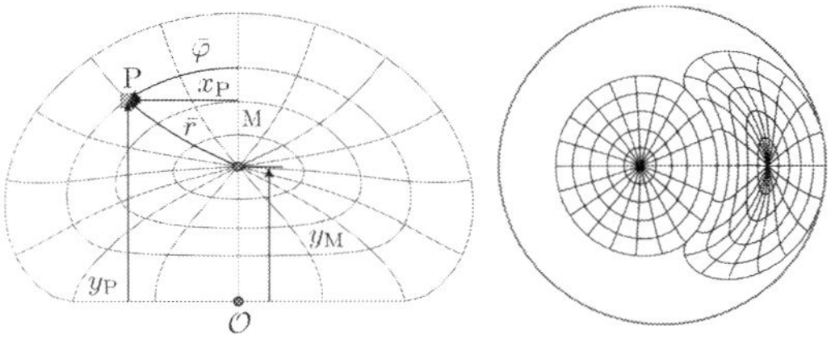

Figure 3. Convective coordinates $\bar{r}$ and $\bar{\varphi}$ of an initially cylindrical yarn cross section (left) and computed hybrid cord cross section with the hybrid cord's circumcircle (right).

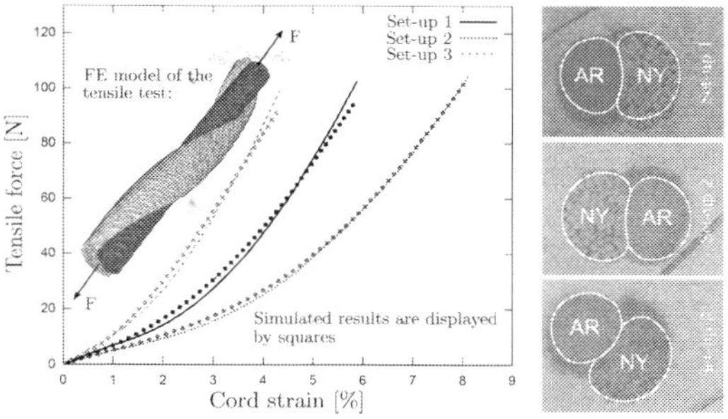

Figure 4. Comparison of simulated and experimental tensile curves (left) and superposition of computed hybrid cord cross sections with their corresponding microscopy images (right).

with $\Delta\underline{r}_{\text{P}} = \underline{r}_{\text{P}} - \underline{r}_{\text{M}}$ and $\underline{r}_{\text{M}}$ as position vector of the yarn center. For each yarn, the vector-valued coefficients $\underline{r}_{ij}^{\text{C}}$ and $\underline{r}_{ij}^{\text{S}}$ are identified by means of the nodal coordinates of the twisting process simulation solving a least-squares problem. Since equation (6) only represents a model of a former cross section, the geometry of an entire yarn including its filament structure can be obtained by the following extension

$$\underline{r}_{Y} = \exp\left(\left(-\underline{\underline{\epsilon}} \cdot \psi(\bar{z})\underline{e}_{z}\right)\right) \cdot \underline{r}_{\text{P}} + \bar{z}\underline{e}_{z} \tag{7}$$

with $\psi(\bar{z}) = 2\pi v_{\text{HC}}\bar{z}$ and $\bar{z}$ as third convective coordinate. Equation (7) represents the cross section's rotation about the cord axis and its translation along the cord axis and is equivalent to constraint (3). To compute filament curves $\underline{r}_{\text{F}}$, $\bar{\varphi}(\bar{z})$ must be set to $\bar{\varphi}(\bar{z}) = \bar{\varphi}_{\text{F}} - 2\pi v_{\text{Y}}\bar{z}/(1+\varepsilon_{\text{Y}})$.[11] Each filament is thus characterized by the unique values $\bar{r} = \bar{r}_{\text{F}}$ and $\bar{\varphi}_{\text{F}}$. The driving parameter is $\bar{z}$. Thus, the tangent vector $\underline{\tilde{a}}$, which is the necessary input for the transversely-isotropic material model (1), can be calculated by $\underline{\tilde{a}} = (\text{d}\underline{r}_{\text{F}}/\text{d}\bar{z})/\,\|\text{d}\underline{r}_{\text{F}}/\text{d}\bar{z}\|$ with $\|\underline{\tilde{a}}\| = 1$. For more information on the geometry model and its application to a cross ply-reinforced air spring, the reader is referred to Heinrich, Donner, & Ihlemann (2016).

In order to validate the cord model, Figure 4 shows the comparison between computed and experimental tensile curves (left) and the superposition of computed cross sections with their microscopy counterparts (right). The parameter setsof the simulated tensile curves only differ in the nylon content α_{Ny}, which was adapted by means of MATLAB to meet the experiments[12].

[11] $\underline{r}_{\text{F}}(\bar{z}) = \underline{r}_{\text{Y}}(\bar{r} = \bar{r}_{\text{F}}, \bar{\varphi} = \bar{\varphi}_{\text{F}} - 2\pi v_{\text{Y}}\bar{z}/(1+\varepsilon_{\text{Y}}), \bar{z})$

[12] Identified values of α_{Ny}: set-up 1: $\alpha_{\text{Ny}} = +2.4\%$, set-up 2: $\alpha_{\text{Ny}} = +1.4\%$, set-up 3: $\alpha_{\text{Ny}} = +4.2\%$

As can be seen, a variation of only one parameter leads to a remarkable agreement of the significantly different tensile curves. The deviations in the identified nylon contents and the computed cross sections result from unavoidable simplifications of the model which will be dropped in the future.

3 SHELL-LIKE RVES FOR HYBRID CORD-RUBBER COMPOSITES

In the following, a cross ply-composite reinforced by the hybrid cord of Section 2 is considered. Figure 5 shows the sketch of a cross ply-reinforced RVE[13] with its geometrical parameters thickness h, cord angle α_C, the layer positions e_i and e_o, and the pressure load p. Further, thecord spacing t_C is a necessary parameter. The RVE is discretized with a simply structured mesh of C3D8H elements, which neglects the interfaces between the elastomer and the yarns. At each integration point of the RVE, its affiliation to either the elastomer or one of the yarns is determined by means of the geometry model of Section 2. In case of a yarn-affiliated gauss-point, the structural tensor $\underline{\underline{A}}$ is computed and stored as state variable using the subroutine SDVINI. Further, the corresponding material parameters G_Y, K_Y, and β_Y are assigned. In case of an elastomer-affiliated gauss-point, the material parameters of equation (1) are set to $G_E = 1\,\text{MPa}$, $K_E = 1{,}000\,\text{MPa}$, and $\beta_E = 1$, yielding the classical Neo-Hookean model. The distribution of the materials is depicted in Figure 5 (right), in which each color represents a different material. Due to the cord twist, the composite is not exactly periodic. To reduce perturbations in the RVE's center due to surface effects, a RVE with two cords per layer is considered. The shell-like boundary conditions are applied by coupling the material line elements of the RVE's opposing cut-surfaces 1 and 2 (see Figure 6) through the following nonlinear constraint[14]

$$\underline{r}_j^{S_i} - \underline{r}^{P_i} = \Delta\underline{\underline{R}}_i \cdot \left(\underline{r}_j^{M_i} - \underline{r}^{B}\right), i = 1,2. \tag{8}$$

Equation (8) is similar to equation (3) and represents an euclidean transformation of the slave surfaces. The rotation tensor $\Delta\underline{\underline{R}}_i$ is defined as

$$\Delta\underline{\underline{R}}_i = \exp\left(\left(-\underline{\underline{\underline{\epsilon}}}\cdot\underline{\underline{\Phi}}\cdot\left(\tilde{\underline{r}}^{Pi} - \tilde{\underline{r}}^{B}\right)\right)\right), \tag{9}$$

[13] Since the hybrid cord model is not limited to two yarns, a hybrid cord made of three yarns is depicted in this case.
[14] The top and the bottom surface remain independent.

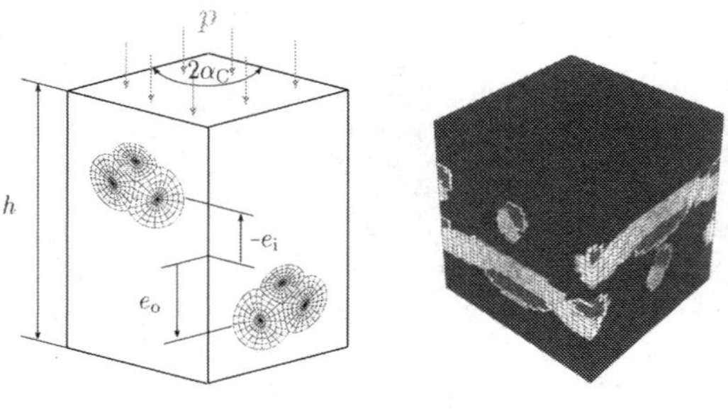

Figure 5. Sketch of an exemplary cross ply-reinforced hybrid cord-rubber composite (left) and distribution of the materials within the FE model of the shell-like RVE (right).

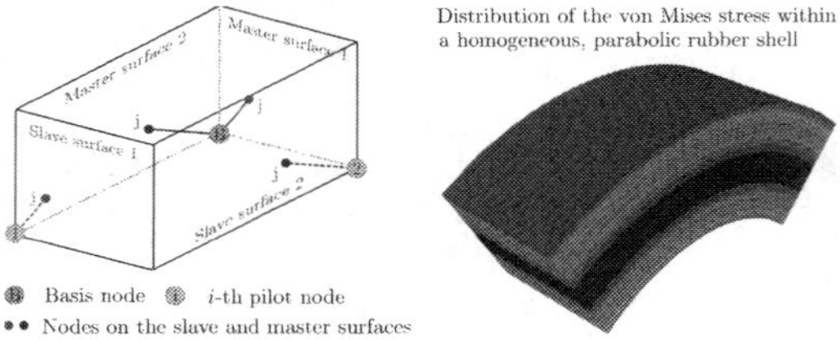

Figure 6. Material line elements of the RVE's constrained surfaces (left) and distribution of the von Mises stress within a homogeneous, parabolic rubber shell (right).

with $\underline{\underline{\Phi}}$ as rotation gradient. $\underline{r}_j^{S_i}$ and $\underline{r}_j^{M_i}$ are the current position vectors of the j-th slave and master node of the i-th surface, respectively. $\underline{r}^{P_i}$ and $\underline{r}^{B}$ are the so-called pilot node of the i-th surface and the basis node, respectively. The position vectors of the reference configuration $\tilde{\underline{r}}^{P_i}$ and $\tilde{\underline{r}}^{B}$ are connected with their counterparts of the current configuration through the membrane deformation gradient $\underline{\underline{F}}^{M}$ by

$$\underline{r}^{P_i} - \underline{r}^{B} = \underline{\underline{F}}^{M} \cdot \left(\tilde{\underline{r}}^{P_i} - \tilde{\underline{r}}^{B}\right). \tag{10}$$

Due to the constraints, the remaining degrees of freedom are $\underline{\underline{\Phi}}, \underline{\underline{F}}^{M}$, and the deformation of the master surfaces defined by $\underline{r}_j^{M_i}$. The constraints are an extension of the classical periodic boundary conditions which only enable applying membrane deformations. However, curvatures cannot be applied. Like the membrane deformation gradient $\underline{\underline{F}}^{M}$, $\underline{\underline{\Phi}}$ belongs neither to the reference nor to the current configuration. An eulerian rotation gradient $\underline{\underline{\varphi}}$ can be obtained by $\underline{\underline{\varphi}} = \underline{\underline{\Phi}} \cdot (\underline{\underline{F}}^{M})^{-1}$. It can be shown that $\underline{\underline{\varphi}}$ is connected with the classical curvature tensor $\underline{\underline{\kappa}}$ by $\underline{\underline{\kappa}} = -\underline{d} \times \underline{\underline{\varphi}}$ with $\underline{d}$ as the shell's director. The gaussian curvature K_G equals $\det[\underline{\underline{\varphi}}]$.

The constraints (8) and (10) are implemented in Abaqus via its subroutine MPC.

In the following, parabolic RVEs[15] are considered. Thus, the rolling lobe of an air spring is locally approximated by a cylinder. The distribution of the von Mises stress within a homogeneous, parabolic rubber shell is depicted on the right of Figure 6. As can be seen, the stress varies with respect to the out-of-plane coordinate but is homogeneous with respect to the in-plane coordinates. Thus, parabolic RVEs, as detail of an infinite cylinder, can be analyzed without surface effects. In case of parabolic RVEs, it can be shown that the stress resultant tensor $\underline{\underline{T}}^{\mathrm{M}}$ and the moment tensor $\underline{\underline{M}}$ are work-conjugatenote[16] to $\underline{\underline{F}}^{M}$ and $\underline{\underline{\Phi}}$, respectively. Thus, multiplying one of the tensors with the normal vector $\underline{\tilde{n}}$ of a former cross section yields the resulting force vector $\underline{t}$ or moment vector $\underline{m}$ with respect to the initial cut length[17]. If a bending about the former x axis is considered $(\Phi_{xy} \neq 0)$, the following equilibrium condition can derived by evaluating the RVE's balance of linear momentum

$$\underline{e}_y \cdot \underline{\underline{T}}^{\mathrm{M}} \Delta\tilde{x} - \underline{e}_y \cdot \underline{\underline{T}}^{\mathrm{M}} \cdot \Delta \underline{\underline{R}}_y \Delta\tilde{x} + \underline{f} = \underline{0} \tag{11}$$

with $\underline{f} = -\int^{A_p} p\underline{n}\mathrm{d}A$. Equation (11) simply states that for a curved RVE, a normal force is required to carry the pressure load p acting on its top surface (see Figure 5). This relation is well-known from the pressure vessel formulas of technical mechanics and is implemented in Abaqus via user elements and the subroutine UEL.

Figure 7 shows the max. principal strain within the RVE of the cross ply-reinforced hybrid cord-rubber composite from Figure 5. The boundary conditions were set to $p = 8$ bar, $\Phi_{xy} = 0.1\ \mathrm{mm}^{-1}$, and $F^{\mathrm{M}}_{xx} = 1.3$. T^{M}_{yy} is prescribed by Equation (11) and thus F^{M}_{yy} is determined by the equilibrium iteration. All other coordinates of $\underline{\underline{F}}_{\mathrm{M}}$ and $\underline{\underline{\Phi}}$ were set to zero. The composite's geometrical properties according to Figure 5 are $\alpha_{\mathrm{C}} = 40°$, $h = 1.5$ mm, $e_{\mathrm{i}} = -15\%$, $e_0 = 15\%$. and $t_{\mathrm{C}} = 0.61$ mm. For the choosen boundary conditions, the maximum strain occurs between the cords of the outer layer. However, the rubber's strain between the cords of the inner layer is lower. This asymmetry is caused by the curvature. In case of applying the classical periodic boundary conditions $(\underline{\underline{\Phi}} = \underline{\underline{0}})$, the strain distribution would be symmetric with respect to the middle plane. Consequently, the extension of the classical periodic boundary conditions enables a more realistic simulation of the loadings within curved structures like the rolling lobe of an air spring.

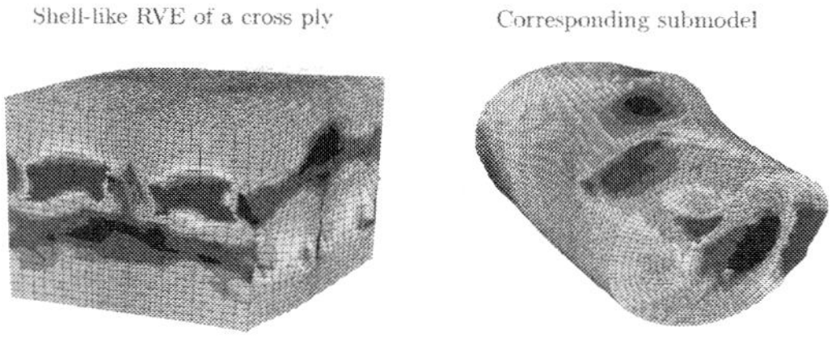

Figure 7. Max. principal strain within the cross-ply reinforced RVE (left) and max. principal strain within the corresponding submodel of a hybrid cord surrounded by a rubber layer (right).

Due to the gauss-point method, there is no reliable information about the stresses and strains near the interface between the rubber and the yarns. To overcome this deficiency, the submodel of a hybrid cord surrounded by a rubber layer is analyzed additionally. Within this model, all constituents are modeled with separate bodies which interact through contact definitions.[18] Using the Abaqus-native submodel technique, the displacements of the RVE's analysis are mapped to the boundaries of the submodel. Due to Abaqus' capabilities, only a minor effort is necessary to create the submodel. In addition to the stresses and strains within the constituents, the contact stresses can be analyzed this way. The distribution of the max. principal strain within the submodel of the cross ply-reinforced hybrid cord rubber composite is exemplarily depicted in Figure 7 (right). The combination of the shell-like RVE and the submodel enables an efficient and realistic analysis of hybrid cord-rubber composites.

4 EXEMPLARY OPTIMIZATIONS

To demonstrate the capabilities of the presented models, in a first example, a hybrid cord is optimized with respect to its fracture force by a parameter study. The design variables are the hybrid cord's nylon content α_{Ny} and the hybrid cord's twist ν_{HC}[19]. The procedure is as follows: The twisting process is simulated for each parameter set first. Subsequently, a tensile test up to a cord strain of 20% is simulated. Within the post processing, the mean filament strain of each yarn is determined by $\bar{\varepsilon}_{\mathrm{Y}} = \int^{V_{\mathrm{Y}}} \sqrt{\underline{\underline{C}} \cdot\cdot \underline{\underline{\tilde{A}}}} - 1 dV / V_{\mathrm{Y}}$ with $\underline{\underline{C}}$ as the right Cauchy-Green tensor. The hybrid cord is

[15] Parabolic RVEs are characterized by $K_G = 0$ but $\underline{\underline{\varphi}} \neq \underline{\underline{0}}$.

[16] $\delta W = \tilde{A}_s(\underline{\underline{T}}^{\mathrm{M}} \cdot\cdot \delta\underline{\underline{F}}^{\mathrm{M}} + \underline{\underline{M}} \cdot\cdot \delta\underline{\underline{\Phi}})$, $\tilde{A}_s$: initial shell area

[17] Cauchy-theorem: $\underline{t} = \underline{\tilde{n}} \cdot \underline{\underline{T}}^{\mathrm{M}}$ and $\underline{m} = \underline{\tilde{n}} \cdot \underline{\underline{M}}$.

[18] Static friction is assumed for simplicity.

[19] $\alpha_{\mathrm{Ny}} \in [-10\% + 10\%]$, and $\nu_{\mathrm{HC}} \in [240\ \mathrm{m}^{-1}, 600\ \mathrm{m}^{-1}]$

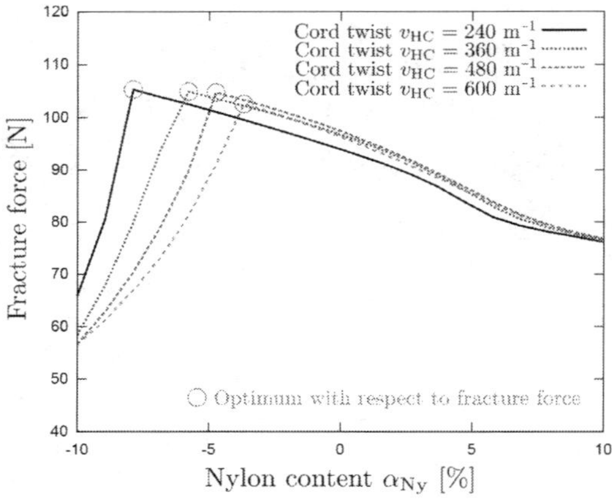

Figure 8. Fracture force of a nylon-aramid-hybrid cord versus its nylon content α_{Ny} for different cord twists v_{HC}.

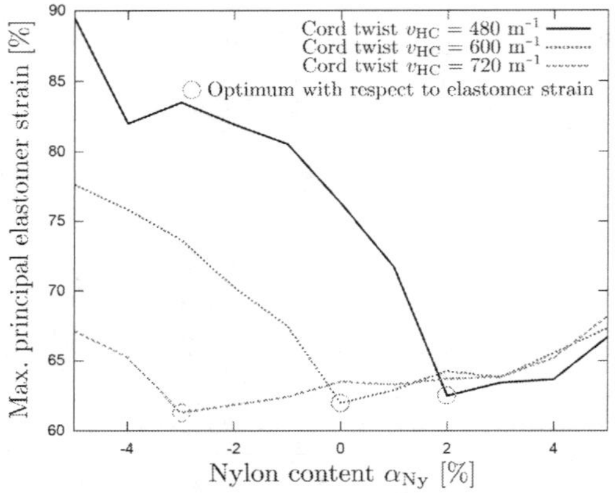

Figure 9. Maximum principal strain of the elastomer versus the nylon content α_{Ny} for different cord twists v_{HC}.

assumed to lost its resistance if one of the yarns reaches its ultimate strength. The corresponding force is seen as the hybrid cord's fracture force. The ultimate strengths $\bar{\varepsilon}_Y^{max}$ are assumed to be 17.5% (nylon) and 2.5% (aramid). The parameter study is performed using MATLAB which launches and evaluates the particular simulations and handles the data management. Figure 8 shows the fracture force versus the nylon content for different cord twists. Obviously, for each cord twist, there is an optimum with respect to the nylon content. At the optima, both yarns reach their ultimate strength simultaneously. In all other cases, either the nylon yarn or the aramid yarn reaches its ultimate strength first, whereas the other yarn remains undamaged. Consequently, a sequential yarn fracture is related with a loss of strength of the hybrid cord. Moreover, the optimum amount of nylon depends on the cord twist. For an increasing cord twist, less aramid is required. Since aramid is more expensive than nylon, an increased cord twist is beneficial concerning the material effort. On the other hand, an increasing cord twist decreases the productivity. In addition to the presented optimization, hybrid cords can be analyzed with respect to their tension-torsion coupling[20] or the robustness of the manufacturing process[21]. Thus, using the presented hybrid cord model, hybrid cords can be optimized with respect to their mechanical and economical properties.

The second example is the optimization of the cross ply-reinforced hybrid cord-rubber composite of Section 3 with respect to the maximum elastomer strain using a parameter study. The sequence of the optimization is as follows: For each parameter set, the twisting process is simulated first[22]. Subsequently, the loading of the cross ply-reinforced RVE is simulated[23]. Finally, the submodel is analyzed. MATLAB is used again to launch and evaluate the particular simulations and to handle the data management. Withinthe post processing, the maximum principal strain[24] of the elastomer is determined using the results of the RVE's and the submodel's analysis. Since the mesh is far from enabling converged results in a mathematical sense, the max. principal strain is integrated over the rubber's 5% highest-loaded area. Unfortunately, this reduces the sensitivity of the parameters. On the other hand, artificial effects are reduced and thus only physically reasonable effects are observed. The design variables are again the nylon content α_{Ny} and the hybrid cord's twist v_{HC}[25]. The results are depicted in Figure 9. Clearly, both, the nylon content and the cord twist, have a significant influence on the elastomer strain. Further, there is a twist-dependent optimum with respect to the nylon amount. The optima are characterized by a minimal radius of the cylinder enveloping the hybrid cord (see Figure 3). The minimal radius is related to a maximum thickness of the rubber between the cords of a layer. Thus, the rubber's shear deformation,

[20] This is very important in view of driving belts.

[21] Note the loss of fracture force away from the optima.

[22] To obtain converged results for the entire parameter ranges, the anisotropy ratios β_Y were reduced to 0.01.

[23] See Section 3 for the boundary conditions and the composite's geometric parameters.

[24] Note, engineering strains are considered.

[25] $\alpha_{Ny} \in [-5\% + 5\%]$, and $v_{HC} \in [480\ \text{m}^{-1}, 720\ \text{m}^{-1}]$.

which is caused by the change in cord angle α_C, decreases for an increasing thickness of the rubber. Note, the radius of the enveloping cylinder is not equal to the radius of the circle enveloping the hybrid cord's cross section, which is usually experimentally determined by microscopy images. This demonstrates the need for a FE-based optimization of hybrid cord-rubber composites.

5 SUMMARY AND CONCLUSIONS

In this paper, efficient and realistic FE models for hybrid cords and hybrid cord-rubber composites were presented. The principal idea in modeling hybrid cords is to orient the preferred direction of a transversely-isotropic material to be coaxial to the filament curve's tangent at each point. The preferred directions are obtained by a simulation of the hybrid cord's twisting process in combination with a geometry model whose free parameters are adapted by means of the results of the twisting process simulation. It was shown that the model is able to match the experiments. The basis of the composite model is the extension of the periodic boundary conditions which now enable the simulation of curved RVEs subjected to a pressure load. In order to analyze the region near the interface, a submodel of a hybrid cord surrounded by an elastomer layer is utilized. The models' capabilities were demonstrated by an optimization of a hybrid cord with respect to its fracture force and the optimization of a composite with respect to the max. elastomer strain. The results show that cord-related parameters have a significant influence on the loading of the rubber within the composite. These dependencies can be analyzed with the proposed models. Thus, the models can be used to limit the design space and to reduce the amount of expensive experimental investigations. It is important to note that in case of an experimental-based optimization of cord-related parameters, each step of the entire, time-consuming manufacturing process like twisting, dipping, and vulcanizing has to be done for each parameter set. Consequently, the FE-based optimization of hybrid cord-rubber composites accelerates the development process.

ACKNOWLEDGEMENTS

The financial support by ContiTech AG, Goodyear S.A., Mehler Engineered Products GmbH, and Vibracoustic GmbH & Co. KG is gratefully acknowledged.

REFERENCES

Balzani, D., P. Neff, J. Schröder, & G. A. Holzapfel (2006). A polyconvex framework for soft biological tissues. Adjustment to experimental data. *International Journal of Solids and Structures 43*, 6052–6070.

Donner, H. & J. Ihlemann (2013). On the efficient finite element modelling of cord-rubber composites. In N. Gil-Negrete and A. Alonso (Eds.), *Constitutive Models for Rubber VIII*, London, pp. 149–155. Taylor & Francis Group.

Donner, H. & J. Ihlemann (2014). An anisotropic large strain plasticity model for multifilament yarns. *Proc. Appl. Math. Mech.* 14, 373–374.

Heinrich, N., H. Donner, & J. Ihlemann (2016). Volumetric finite element models for textile reinforced rubber components. In *Proceedings of the 12th Fall Rubber Colloquium*, Hannover.

Holzapfel, G. A. (2000). *Nonlinear Solid Mechanics*. Chichester: John Wiley & Sons Verlag.

Jacob, H.-G. (2005). *Ein Beitrag zur Berechnung von Faserverbunden aus Elastomeren mit Polyamidfaserverstärkungen*. Habilitation, Universität Hannover, Hannover.

Jiang, W. G., M. S. Yao, & J. M.Walton (1999). A concise finite element model for simple straight wire rope strand. *International Journal of Mechanical Sciences 41*, 143–161.

Kreikemeier, J. (2012). Modelling of phase boundaries via the gauss-point method. *Technische Mechanik 32*, 658–666.

Maßmann, C. (1995). *Simulation des Verhaltens von textile Festigkeitsträgern in Cord-Gummi-Kompositen*. Dissertation, Universität Hannover, Hannover.

Steinkopff, T. & M. Sautter (1995). Simulating the elastoplastic behavior of multiphase materials by advanced finite element techniques part i: a rezoning technique and the multiphase element method. *Computational Materials Science 4*(1), 10–14.

Constitutive Models for Rubber X – Lion & Johlitz (Eds)
 ISBN 978-1-138-03001-5

Nanoparticles effects on the thermomechanical properties of a fluoroelastomer

D. Berthier
Zodiac Aerosafety Systems—CERMEL—PCM2E, France

M.P. Deffarges, F. Lacroix & S. Méo
Laboratoire LMR-CERMEL (Centre d'étude et de Recherche sur les Matériaux Elastomères), Université François Rabelais, Tours, France

B. Schmaltz, N. Berton & F. Tran Van
Laboratoire PCM2E (Physico-Chimie des Matériaux et des Electrolytes pour l'Energie), Université François Rabelais, Tours, France

Y. Tendron & E. Pestel
Zodiac Aerosafety Systems, Loches, France

ABSTRACT: Incorporation of nanoparticles is a classical way of improving polymer properties requiring low quantity of fillers. A novel approach to significantly enhance the thermal degradation temperature is the integration of a high fillers quantity. The aim of this study is to optimize the thermal properties of a fluoroelastomer (FKM) by the addition of two types of nanoparticles: Carbon nanotubes (MWNTs) and POSS (Polyhedral Oligomeric Silsesquioxanes). At the laboratory scale, the elaboration of the rubber is carried out in thin films, processed from solution and deposited with a Hand Coater. Nanofilled blends will be compared with the manufacturer's cross-linked rubber (without nanofillers).

The thermomechanical properties of the new blends have been studied by DSC (Differential Scanning Calorimetry), TGA (ThermoGravimetric Analysis), DMA (Dynamic Mechanical Analysis). For practical reasons this work presents only the results with POSS Acryloisobutyl. Simply blending these POSS-based fillers into Fluoroelastomer (FKM) had expected effect on the mechanical properties, but slight improvement on thermal resistance. POSS can provide a considerable reinforcement of thermal degradation but in this case, the quantity of fillers exceeds the limits to get the expected results.

1 INTRODUCTION

Since their first use, fluoroelastomers have become of great interest. This rubber family includes different kinds of products with specific properties, such as thermoplastic and elastomer (Ameduri et al. 2001). Their structures can be amorphous or semi-crystalline. The specificity of fluoroelastomer consists in a unique properties combination: good thermal, aging, weather and chemical resistances, low surface energy, low inflammability and low moisture absorption (Rhein 1983) (Ameduri & Boutevin 2005). These properties can be explained by a low polarizability and a strong electronegativity provided by fluor atoms because of its low molecular interactions and strong bond energy (Drobny 2007).

It should be mentioned that FFKM has better thermal and chemical resistances but poor mechanical properties (Arnold et al. 1973) (Biron 2008). The aim of this study is to optimize the thermal properties of a fluoroelastomer (FKM). To enhance thermal properties, it is important in the first step to understand the phenomenon involved during the degradation of the polymer. Degradation is caused by oxidation and thermolysis of the polymer through by free-radical reactions (Celina et al. 1998) (Banik et al. 1999). From thermal standpoint, the vinylidene fluoride used like a cross-linking bond, constitutes the lowest molecular groups because of the hydrogen groups (Sugama et al. 2015) (Mittra et al. 2004). Because chemical structure of FKM cannot be modified, another way is to integrate nanoparticles. This is the simplest mean to improve different properties including thermal properties (Koo 2006).

One of the more interesting aspects of nanocomposite is that a low nanofillers amount is enough to improve sufficiently the characteristics of the composite (Zimmermann & Schuster 2012) (Hassar 2013). For example: the tensile resistance of a natural rubber is 4.9 MPa with carbon black and 15 MPa with nanoclay, with the same composition and filler content (Arroyo et al. 2003). Indeed, when the fillers amount increases, a continuous network of fillers can be developed below the geometric percolation threshold.

In the present study, carbon nanotubes and POSSs (polyhedral oligomeric silsesquioxane) are selected. These two nanoparticles have different natures and morphologies (Fig. 1).

Two types of carbon nanotubes exist: Single-Walled NanoTube (SWNT) and Multi-Walled NanoTube (MWNT). MWNT improves the crystallinity and the properties of fluoropolymers, such as electrical response, mechanical properties, viscoelastic behavior, etc, and therefore their thermal stability (Xu et al. 2010) (Heidarian & Hassan 2014) (Heidarian et al. 2015). Recently, a novel approach to significantly enhance the thermal degradation is the integration of a high fillers quantities. It has been reported that the creation of a cellular structure leads to enhance performance of MWCNT embedded in fluorinated rubber (Endo, et al. 2008).

Polyhedral oligomeric silsesquioxanes (POSS) are organic-inorganic molecules with diameter in the range of 1–3 nm. POSS is described as a cubic cage with a silicon-to-oxygen ratio of 1:1.5, and organic peripheral groups of each of the silicon

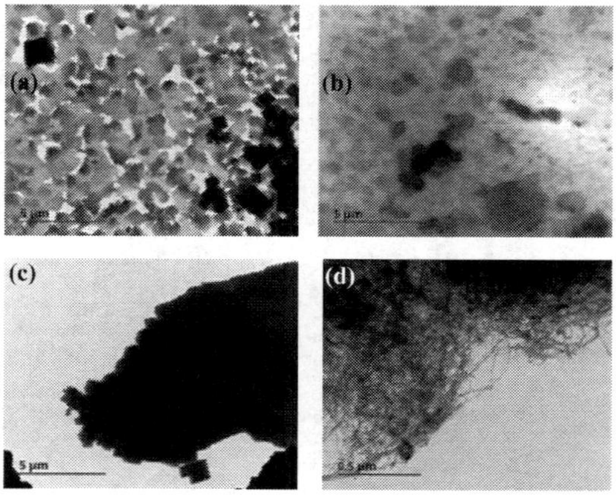

Figure 1. TEM micrographs of POSS and CNT nanoparticles. (a) POSS-A, (b) POSS-F, (c) POSS-P and (d) CNT-P.

atoms (Li et al. 2001). Currently, incorporation of POSS into polymers increases the oxidative resistance, application temperature, glass transition temperature, and storage modulus of the polymers (Sahoo & Bhowmick 2007) (Seurer & Coughlin 2008). In this case, level of fillers to obtain an efficient level of reinforcement is between 10 and 20 wt.%.

2 EXPERIMENTAL

2.1 *Materials*

The matrix is a fluororubber (terpolymer). The main ingredients used are organic peroxide, crosslinking coagent, carbon black and silica. Three types of POSS $(RSiO1.5)_n$ are employed: POSS-P Octaphenyl, POSS-F Trifluoropropyl and POSS-A Acryloisobutyl.

Two types of MWNTs are chosen: CNT-P unfunctionalized and CNT-COOH functionalized by carboxylic acid. Both have an outside diameter of 9.5 nm, a purity superior to 95%, and a length around 1.5 μm.

2.2 *Compounding procedure*

The nanocomposites were prepared by solution casting. The formulations are given in Table 1. Grey cells indicate the absence of ingredient.

The matrix is first dispersed in acetone by magnetic agitator at room temperature. Then the other additives are added. For each POSSs three filler loadings are studied: 10, 15 and 20 wt.%. Composite designation is in Table 1, following by the POSS type and the loading. Examples, F_P composite with POSS-A at 15 wt.% is called: $F_{PA\text{-}15}$.

The rubber elaboration is carried out in thin films, processing from solution and deposed with a Hand Coater. The curing and the post curing were realized in an oven.

2.3 *Characterization*

2.3.1 *Scanning Electron Microscopy (SEM)*

A Zeiss field emission model Ultra Plus Scanning Electron Microscope, operating at 5 kV, was used to examine the surface of nanocomposites.

Table 1. Composition of the composites.

Designation	F_0	F_P	F_R	F_F
FKM gum				
Crosslinking system				
Carbon black & Silica				
POSS				

This allows to determine the effect of the process on particles dispersion. The samples were cleaved with a sharp razor and coated with a Platinum film to increase electrical conductivity.

2.3.2 *Transmission Electron Microscopy (TEM)*
For characterization of nanoparticles by Transmission Electron Microscopy (JEOL 1230, 120 kV), the nanoparticles were dispersed in acetone and deposited as a drop on carbon coated copper grid and was analyzed under an accelerating electron beam.

Nanocomposites were ultramicrometered using a diamond knife at –40°C with a Lica Ultracut UCT. These ultra thin sections were transferred onto a Cu+C mesh grid.

2.3.3 *Differential Scanning Calorimetry (DSC)*
DSC was performed under a nitrogen atmosphere on a Netzch Instruments (DSC 200 F3). Samples were cooled up to –140°C and heated until 250°C with a rate of 20°C.min^{-1}. A second cooling was imposed from –80°C to 250°C at a rate of 20°C.min^{-1}. The averaged values of glass transition temperature (Tg) and enthalpy cross-linking of reaction of two tests separate were reported for each case.

2.3.4 *Thermal Gravimetric Analysis (TGA)*
TGA was performed using a Mettler–Toledo instrument (TGA-2). Samples were heated from 25°C to 600°C under nitrogen atmosphere, then from 600°C to 800°C under air, at a testing rate of 30°C.min^{-1}. The averaged values of T_{onset} (temperature for 5% weight loss), $T_{20\%}$ (Temperature for 20% weight loss) and T_{max} (temperature of pick decomposition of the derivative curve) of three tests were reported for each case.

2.3.5 *Dynamic Mechanical Analysis (DMA/TA)*
The dynamic mechanical properties of the specimens were analyzed with a Thermal Analysis Instrument, (DMA 2980 dynamic mechanical analyzer). Test specimens were taken from the post-cured thin film obtained (width 9 mm and length 20 mm) and were subjected to a tension mode. Payne effects were obtained through the strain-sweep mode from 1 μm to 4000 μm, at room temperature, at fixed frequency of 1 Hz. Temperature sweeps were conducted over a temperature range of –30°C to 250°C with a heating rate of 3°C.min^{-1} at a constant frequency of 1 Hz and with an amplitude deformation of 0.5 μm. The averaged values of glass transition temperature, tanδ at glass transition temperature and the storage modulus at 100°C and 200°C of three tests were reported for each case.

3 RESULTS AND DISCUSSION

For practical reasons only results with POSS-A (Figure 1-A) will be presented here.

3.1 *SEM/TEM*

POSSs have appearance of white powder. TEM micrographs show structure of the nanoparticles (Fig. 2). POSSs have crystal aspect.

In the matrix, the distribution of fillers can have an impact on properties of the products. The morphology of the nanocomposites was observed by TEM and SEM. SEM micrographs present the surface of F_{RA-20} (Fig. 3). POSS-A are uniformly distributed. Figure 3 a et b shows the POSS distribution at different magnifications. The TEM technique allows knowing fillers repartition. This also makes it possible to get information concerning the aggregates and their size. Figures 2 and 4 respectively present the compound with the matrix and POSS-A and the compound with the matrix, the crosslinking system and POSS-A. From the Figure 2, two types of particles are observed. To identify the difference between them, EDX (Energy Dispersive X-ray spectrometry) technique is used. Sulfur and barium chemical elements have been identified in black particles (Fig. 2 and 4-a) and thus correspond to barium sulfate. This particle is an inorganic filler used during polymerization

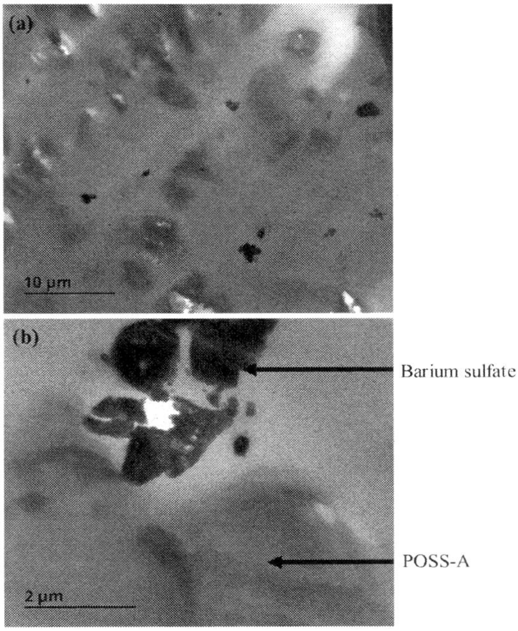

Figure 2. TEM micrograph of F_{PA-20}.

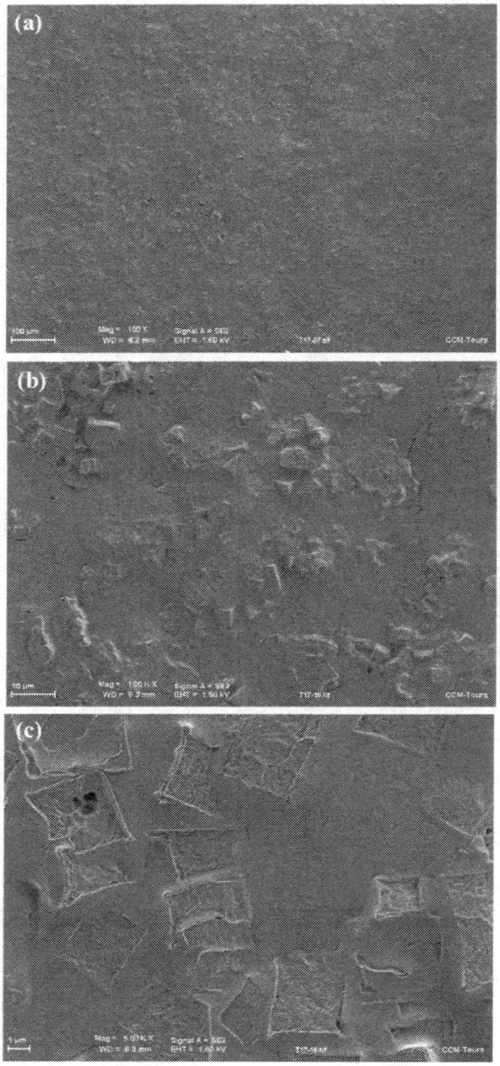

Figure 3. SEM micrograph of F_{RA-20}. a) × 100, b) × 1000 and c) × 5000.

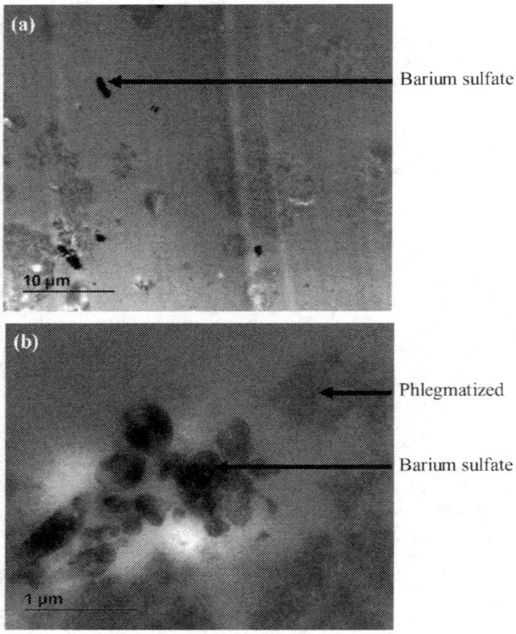

Figure 4. TEM micrograph of F_{FA-20}. (a) at 10 µm (b) at 1 µm.

Table 2. DSC results for nanocomposites F_R raw and vulcanized.

		Glass transition temperature Tg (°C)	Enthalpy of cross-linking reaction ΔH (J/g)
Raw	FRA-0	–17	25.1
	FRA-10	–21	22.6
	FRA-15	–18.5	24.9
	FRA-20	–16	25.8
Vulcanized	FRA-0	–16	
	FRA-10	–16	
	FRA-20	–16	

to avoid coalescence by agglomeration. The other particles are POSS-A. On the Figure 4-b, the phlegmatized of the crosslinking system composed by silica and calcium carbonate is shown. The large aggregates are attributed to crosslinking system dispersed in matrix and not to the POSS nanoparticles (Fig. 4).

3.2 *DSC*

DSC thermogram is a technique that allows to get the glass transition of raw and vulcanized sample. This technique realized on raw sample gives also information concerning the enthalpy of crosslinking reaction. In F_{RA} raw copolymers, the Tg and enthalpy values increase with increase in POSS loadings, with the largest jump occurring between the F_{RA-0} and F_{RA-10} samples (Table 2). The Tg values remain equivalent with all POSS loadings when samples are vulcanized. A possible explanation is that two networks are created (the first one with POSS bonds and the second one without). The corresponding Tg values could be closer due to a POSS grafting on the main chain. Nanoparticle effects in a nanocomposite are through a maximum; at 10 wt.%, the maximum is exceeded.

3.3 *TGA*

The thermal stability of the nanocomposites was also studied by thermal gravimetric analysis.

As shown in Table 3, POSS nanocomposites containing different contents of POSS show a decrease of T_{onset} for each mixture. For each case, 10 wt.% of fillers remain the best compromise to obtain good thermal behavior from T_{onset}. Concerning $T_{20\%}$, the trend is inversed (Fig. 5). An increase of fillers loading leads to an enhancement of the degradation temperature. Different phenomena might be coexisting suggesting these observations (thermal conductivity, thermal barrier, volumic ratio between rubber and fillers ...).

Table 3. TGA results for nanocomposites F_0, F_P and F_F with POSS-A.

Sample	Temperature for 5% weight loss T_{onset} (°C)	Temperature for 20% weight loss $T_{20\%}$ (°C)	Temperature of peak decomposition T_{max} (°C)
F0	452	473	492
FPA-10	351	485	498
FPA-15	335	477	498
FPA-20	326	469	498
FRA-0	458	481	497
FRA-10	449	485	501
FRA-15	430	485	502
FRA-20	419	488	504
FFA-0	457	483	501
FFA-10	453	490	505
FFA-15	439	487	503
FFA-20	430	488	503

After 500°C, the degradation mechanism is changed. An increase of $T_{20\%}$ and T_{max} are observed (Table 3). The explanation of this fact is the same given by DSC. The POSS loadings might be underestimated to drastically enhance thermal properties. Expected results could be probably obtained between 1 and 10 wt.%.

3.4 *DMA*

Filler–matrix and filler–filler interactions play an important role in the mechanical properties of filled elastomers due to the increase in cross-link density and the occlusion of a certain amount of the elastomeric matrix by filler aggregates. Formation of rigid filler network usually provides additional levels of reinforcement. Filler–matrix and filler–filler interactions are unanimously associated with the decrease of the dynamic modulus with the strain amplitude, which is known as the Payne effect.

Payne effects were observed for the two mixtures: F_R and F_F, they are respectively shown on Figure 6 and 7. Storage modulus (E') increases

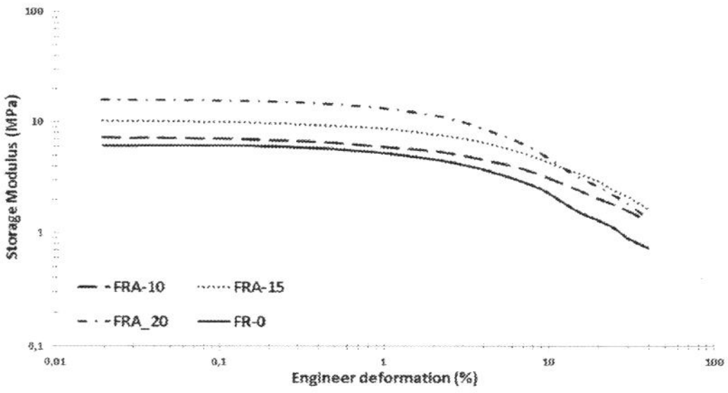

Figure 6. Payne effects for different formulations of F_{RA} nanocomposites.

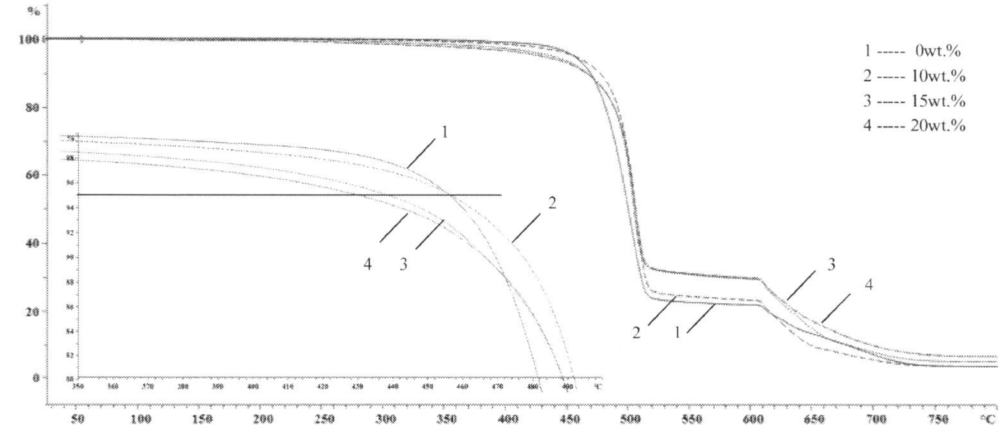

Figure 5. Temperature sweeps of F_{FA} formulation.

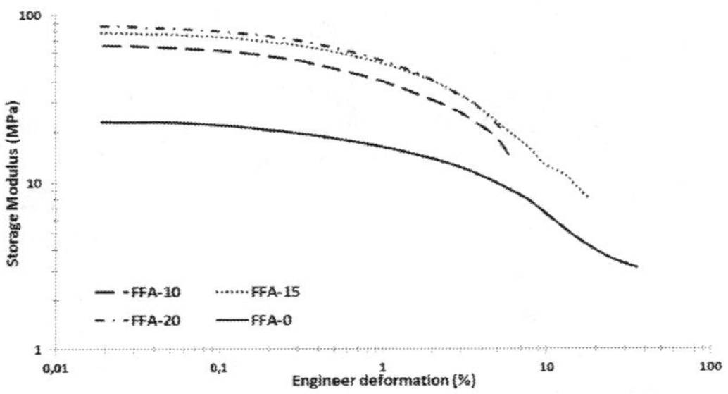

Figure 7. Payne effects for different formulations of F_{FA} nanocomposites.

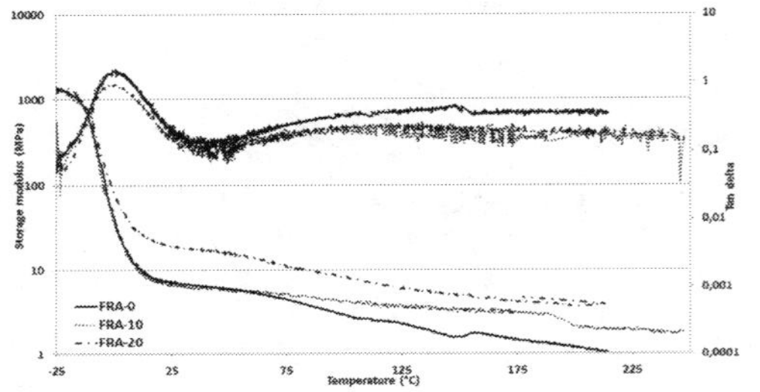

Figure 8. Temperature sweeps of F_{RA} formulation.

gradually with increasing of POSS loading. For example, the F_{PA} rubber has a E' in the linear region equal to 6.1 MPa ± 0.1, 7.1 MPa ± 0.05 and 15.7 MPa ± 0.05 for respectively 0 wt.%, 10 wt.% and 20 wt.%. On Figure 6, the inflection point with 20 wt.% occurs earlier due to the stiffening of the structure. This could be explained by more interactions between fillers and matrix. The unfilled elastomer $F_{R\text{-}0}$ does display a change in the storage modulus with strain amplitude, at least on the investigated strain range. This could be directly related to the steric hindrance of the atoms on the main chain, as expected. For mixture filled with conventional fillers, the breakdown of a filler network formed by filler-filler interactions occurs more quickly. The high fillers loading accentuated this effect. The F_{FA} rubber has E' (linear region) equal to 22.5 MPa ± 0.3, 64.5 MPa ± 1.3 and 82.5 MPa ± 1.8 for respectively 0 wt.%, 10 wt.% and 20 wt.%.

Temperature sweeps realized by DMA allow to obtain glass transition temperature and the rubbery state. Temperature sweeps have been realized on the same mixture analyzed by DSC (Fig. 8). Two main phenomena can be observed. On the first hand, a decrease of tanδ peak with increase of fillers loading and on the other hand, the evolution of storage modulus shows an improvement in temperature. For FRA-20, E' is shifted to high modulus (Fig. 8 and Table 4). POSS increases the values of E' in the rubbery plateau region. These characteristics suggest less motions of the macromolecular structure in the network. The Tg seems to be optimized with 10 wt.% of fillers. Above this amount, the presence of fillers leads to an increase of interactions with the matrix and can modify the structure (physical and chemical interactions). This weakens the thermomechanical response of the nanocomposite. This substantiates the fact that the quantity of nanoparticle effects in temperature nanocomposite improved the stiffness at

Table 4. DMA results for temperature sweeps realized on F_0, F_P and F_F with POSS-A.

	Temperature at the peak of tanδ Tg (°C)	Tanδ at Temperature of Tg (T = Tg) Tanδ	Storage modulus at 100°C E'_{100} (MPa)	Storage modulus at 200°C E'_{200} (MPa)
FR-0	−3.8 ± 0.09	1.2 ± 0.01	2.8 ± 0.73	1.1 ± 0.16
FRA-10	1.6 ± 0.55	1.5 ± 0.04	4.3 ± 0.05	2.4 ± 0.44
FRA-20	0.2 ± 0.07	1.0 ± 0.01	8.4 ± 0.68	3.9 ± 0.84

high temperature but from 10 wt.% can alter the physical-chemical behavior.

4 CONCLUSIONS

In this work, POSS-A were included into three compounds of FKM. The mixtures were observed to by SEM and by TEM. The POSS-A seems to be well dispersed in the matrix. The goal of this study is to enhance the resistance of thermal degradation and of thermo-oxidative decomposition by adding POSSs. No significant improvement of T_{onset} by TGA is observed. However the decomposition mechanism is impacted by the presence of POSS at higher temperature. The DMA results show a gradual increase of E' with loading of POSS. Temperature sweeps by DSC and DMA suggest an improvement of Tg and an impact of the quality of crosslinking. These results correlated with TGA results suggest that the POSS loading is too high to impact the temperature as expected. The POSS loading might be smaller than 10 wt.% to drastically enhance thermal properties. Further studies are under investigation on the other types of nanofillers presented. One aspect of POSS not shown in this paper is the ability of POSS to act as a crosslinking agent. Tensile tests coupled with stereo-correlation to follow the evolution of the strain field will be performed to demonstrate this capacity.

ACKNOWLEDGEMENT

The authors are pleased to acknowledge S. Georgault of the *Laboratoire de biologie cellulaire et microscopie électronique* for his help with SEM/TEM experiments and measurements. The authors also thank M. Venin of the CERMEL (*Centre d'étude et de Recherche sur les Matériaux Elastomères*) for its assistance in setting up the mechanical experiments.

REFERENCES

Améduri B. & B. Boutevin (2005). Update on fluoroelastomers: from perfluoroelastomers to fluorosilicones and fluorophosphazenes. *J. Fluor. Chem.* 126(2): 221–229.

Améduri B., B. Boutevin & G. Kostov (2001). Fluoroelastomers: synthesis, properties and applications. *Prog. Polym. Sci.* 26(1): 105–187.

Arnold R.G., A.L. Barney & D.C. Thompson (1973). Fluoroelastomers. *Rubber Chem. Technol.* 46(3): 619–652.

Arroyo M., M. Lopez-Manchado, B. Herrero (2003). Organo-montmorillonite as substitute of carbon black in natural rubber compounds. *Polymer.* 44(8): 2447–2453.

Banik I., A.K. Bhowmick, S.V. Raghavan, A.B. Majali, and V.K. Tikku (1999). Thermal degradation studies of electron beam cured terpolymeric fluorocarbon rubber. *Polym. Degrad. Stab.*63(3): 413–421.

Biron M. (2008). Elastomères Fluorocarbonés. *Tech. de l'Tingénieur.* N2820: 5.

Celina M., J. Wise, D.K. Ottesen, K.T. Gillen & R.L. Clough (1998). Oxidation profiles of thermally aged nitrile rubber. *Polymer Degrad. Stab.* 60(2–3): 493–504.

Drobny J.G. (2007). Fluoropolymers in automotive applications. *Polym. Adv. Technol.* 18: 117–121.

Endo M., T. Noguchi, M. Ito, K. Takeuchi, T. Hayashi, Y.A. Kim, T. Wanibuchi, H. Jinnai, M. Terrones & M.S. Dresselhaus (2008). Extreme-Performance Rubber Nanocomposites for Probing and Excavating Deep Oil Resources Using Multi-Walled Carbon Nanotubes. *Advanced Functional Materials.* 18: 3403–3409.

Hassar M. (2013). Influence des nanocharges de noire de carbone sur le comportement mécanique de matériaux composites: application au blindage électromagnétique. Thesis. Université de Technologie de Compiègne.

Heidarian J. & A. Hassan (2014). Microstructural and thermal properties of fluoroelastomer/carbon nanotube composites. *Comp. Part B Eng.* 58: 166–174.

Heidarian J., A. Hassan & M.M.A. Rahman (2015). Improving the thermal properties of fluoroelastomer (Viton GF-600S) using acidic surface modified carbon nanotube. *Polimeros.* 25(4): 392–401.

Koo J.H. (2006). Polymer Nanocomposites: Processing, Characterization and Applications. New York: Nanoscience.

Li G., L. Wang, H. Ni & C.U. Pittman (2001). Polyhedral Oligomeric Silsesquioxane (POSS) Polymers and Copolymers: A Review. *Journal of Inorg. Organometallic Polymer.* 11(3): 123–154.

Mitra S., A. Ghanbari-Siahkali, P. Kingshott, K. Almdal, H.K. Rehmeier, and A.G. Christensen (2004). Chemical degradation of fluoroelastomer in an alkaline environment. *Journal of Inorg. Organometallic Polymer.* 83(2): 195–206.

Rhein R.A. (1983). *Thermally Stable elastomers: A review*. California: Naval weapons center china lake.

Sahoo S. & A.K. Bhowmick (2007). Polyhedral oligomeric silsesquioxane (POSS) nanoparticles as new crosslinking agent for functionalized rubber. *Rubber Chem. Technol.* 80(5): 826–837.

Seurer B. & E.B. Coughlin (2007). Fluoroelastomer Copolymers Incorporating Polyhedral Oligomeric Silsesquioxane. *Macromol. Chem. Phys.* 209(19): 2040–2048.

Sugama T., T. Pyatina, E. Redline, J. McElhanon, and D. Blankenship (2015). Degradation of different elastomeric polymers in simulated geothermal environments at 300°C. *Polym. Degrad. Stab.* 120: 328–339.

Xu M., D.N. Futaba, T. Yamada, M. Yumura & K. Hata (2010). Nanotubes with Temperature-Invariant Viscoelasticity from −196°C to 1000°C. *Science.* 330: 1364–1368.

Zimmermann H. and R.H. Schuster (2012). Chemical degradation of fluoroelastomer in an alkaline environment. *World J. Eng.* 1369–1370.

Constitutive Models for Rubber X – Lion & Johlitz (Eds)
© 2017 Taylor & Francis Group, London, ISBN 978-1-138-03001-5

Improvement of leak tightness for swellable elastomeric seals through the shape optimization

Y. Gorash
Department of Mechanical and Aerospace Engineering, University of Strathclyde, Glasgow, UK

A. Bickley
Weir Advanced Research Centre, Weir Group, Technology and Innovation Centre, Glasgow, UK

F. Gozalo
Weir Minerals, Weir Rubber Engineering, Weir Group, Salt Lake City, UT, USA

ABSTRACT: Swellable packers have been widely employed in various oil & gas applications. Downhole conditions are difficult to reproduce using physical testing environments, but can be simulated in a virtual environment using CAE software. A better understanding of packers' mechanical behaviour in downhole conditions would provide a higher confidence and improvement in existing engineering design practices for the manufacturing of packers. The numerical simulation can be incorporated into optimisation procedures searching for an optimal shape of packers aiming to minimise the time to seal the borehole and maximise the contact pressure between the seal and borehole. Such an optimisation would facilitate the development of a packer with various designs optimised for different downhole conditions. The objective of this work is to develop a design tool integrated into Abaqus/CAE to implement parametric numerical studies using implicit and explicit FE-simulations. However, development of such a CAE plugin is associated with a number of technical challenges specific to this class of multiphysics problems, which are addressed in this research and discussed in the paper.

1 INTRODUCTION

Swellable elastomers are a special type of polymer with the ability to swell when exposed to water or oil. They are used in production of swellable elastomeric seals, a type of specifically engineered packer that activates upon contact with wellbore fluids. Such packers have been widely employed in various oil & gas applications including slimming of well design, zonal isolation, water shut-off, and multi-stage fracturing. Referring to Lou & Chester (2014), downhole service conditions are extremely challenging to reproduce using physical testing environments, but can be simulated in a virtual environment using software for multiphysics engineering analysis. A better understanding of packers' mechanical behaviour in downhole conditions in a virtual environment would provide a higher confidence and improvement in existing engineering design practices for manufacturing of packers, as demonstrated e.g. by Akhtar, Qamar, Pervez, & Al-Jahwari (2013).

The service characteristics of packers, which are generally measured from full-scale packer tests, can be predicted through numerical FE-simulations based on material data obtained from basic mechanical experiments. The experiments required for comprehensive material input would focus on evolution of hyperelastic properties (in tension, compression and shear) with the change of volume/density of specimens during swelling. The numerical simulation of packers can be incorporated into optimisation procedures finding an optimal shape of packers aiming to minimise the time to seal the borehole and maximise the contact pressure between the seal and borehole (Lou & Chester 2014). Such an optimisation procedure would facilitate the development of a packer with various designs optimised for different downhole conditions considering the borehole type, size and temperature. The objective of this research project is to develop a design tool integrated into Abaqus/CAE to implement parametric numerical studies using advanced FE-simulation to provide an improved design of packers for various downhole conditions. However, the implementation of the packer swelling simulation is associated with a number of technical challenges specific to this particular class of multiphysics problems, which are illustrated in Fig. 1 and listed below:

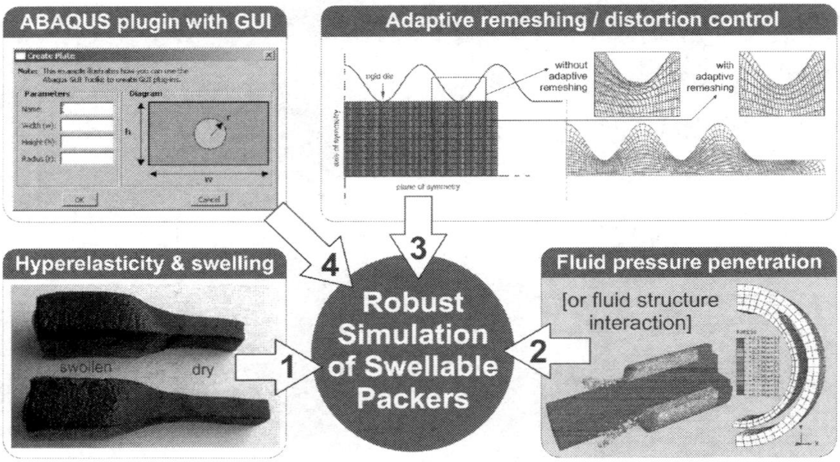

Figure 1. Diagram of technical requirements for a robust FE-simulation of swellable packers for progressive failure and leakage.

1. The key component is an advanced material model comprising both hyperelasticity and moisture swelling. It has to consider two-way interaction between mechanical response and swelling capacity. Implementation of such a material model requires programming of a Fortran subroutine for the user defined material using the Flory & Rehner Jr. (1943) theoretical background, which is presented in the first instance by Flory-Rehner equation:

$$-\left[\ln(1-\nu_2)+\nu_2+\chi_1\nu_2^2\right]=V_1 n\left(\nu_2^{\frac{1}{3}}-\frac{\nu_2}{2}\right), \quad (1)$$

where ν_2 is the volume fraction of polymer in the swollen mass, V_1 the molar volume of the solvent, n is the number of network chain segments bounded on both ends by cross-links, and χ_1 is the Flory solvent-polymer interaction term.

In polymer science Eq. (1) describes the mixing of polymer and liquid molecules as predicted by the equilibrium swelling theory of Flory & Rehner Jr. (1943). It describes the equilibrium swelling of a lightly cross-linked polymer in terms of crosslink density and the quality of the solvent. The theory considers forces arising from three sources:

- the entropy change caused by mixing of polymer and solvent;
- the entropy change caused by reduction in number of possible chain conformations via swelling;
- the heat of mixing of polymer and solvent, which may be positive, negative, or zero.

2. The moisture swelling process is not uniform and starts on the surfaces which are subject to fluid. Adsorption, which governs the progress of swelling can occur only at free surfaces. Therefore, the fluid pressure penetration needs to be incorporated into the simulation (Simulia [PPL] 2016) and directly linked to swelling. Distributed pressure penetration load allows for the simulation of fluid penetrating into the surface between two contacting bodies, penetration of fluid from multiple locations on the surface, and application of the fluid pressure normal to the surfaces. It automatically adjusts the application of a fluid pressure depending on changes of contact conditions.
3. Non-uniform swelling is associated with a localised increase of material volume. It may cause a significant distortion of FE mesh and arouse FEA convergence problems. To overcome this, there are a few options available in the setup of the FE-model (Simulia [AT] 2016) including a mesh-to-mesh solution mapping (Abaqus/Standard), adaptive remeshing (Abaqus/Explicit) and element distortion control.
4. Parametric study assumes considering a large number of different geometric configurations, looking at material properties variation and different downhole conditions. Basically this means a search for an optimal geometry through a sensitivity study, which would result in specific design recommendations for the geometry of a packer. Therefore, it would be reasonable to automate the analysis procedure through an Abaqus plug-in (Puri 2011) with a convenient graphical user interface (GUI), which provides access to the parameters of geometry, material properties and service conditions.

2 SHAPE OPTIMISATION

Over recent years, non-parametric optimisation established as standard methods to improve the overall design and reliability of structural components under extreme loading conditions. The purpose of nonparametric optimization is to give engineers a method to define a design space in regions or whole components without the process of defining the problem in design parameters (Brieger 2016). Much more freedom in terms of possible designmodification is a clear advantage of non-parametric methods (presented by topological and shape optimisation), when compared to parametric optimisation.

In relation to swellable packers, their principal dimensions (diameter and length) can be optimised parametrically, while the external contact surface can be optimised non-parametrically for better sealing capability considering its contact with the borehole. But non-parametric optimisation of structures under contact conditions is a quite specific problem because of its non-smooth character, which is not extensively studied. However, Wagner & Helfrich (2016) revealed great potential of the topology and shape optimization under contact conditions using examples from literature and industry. Various constraints, such as contact pressure, compliance and stresses can be used as objective functions for this class of problems. In addition, release directions, symmetry conditions and frozen (or fixed) regions can be considered as side constraints of the optimization process with limits in terms of weight and volume. In contrast to Wagner & Helfrich (2016) who performed optimisation with industrial codes (PERMAS and VisPER), in this work Simulia Tosca Structure is used. This is a more widely accessible software system for non-parametric structural optimisation that provides topology, sizing, shape and bead optimisation using industry standard FE-solvers (ANSYS, Abaqus, MSC Nastran).

A special usage of the shape controller algorithm in Simulia Tosca Structure is the optimization of contact zones, because small changes in the contact surface usually have a big influence on the contact pressure. Contact pressure can be either minimised as explained by Simulia [MCP] (2016), or maximised as needed in this work. Therefore, shape optimisation is used here to improve the grip of a packer with the surface of a borehole. The normal shape optimization stimulates the surface growth in contact zones, which results in a higher contact pressure and shrinkage in a lower.

For a test shape optimisation study, the trimmed version of a packer geometry from Lou & Chester (2014) was used as benchmark problem with L reduced from 16" to 2". The optimisation analysis resulted in a rippled external surface of a packer as shown in Fig. 2a with comparison to the original rectangular profile. The distribution of contact pressure became very non-uniform as shown in Fig. 2b with four maximums which are about 5 times higher than the original smooth contact pressure. The final result presented in Fig. 2 has been obtained after 30 iterations when a target convergence was achieved as illustrated in Fig. 3 comparing the evolution of objective (contact pressure) to the variation of constraint (volume), which had to remain constant. The external surface was completely free to evolve considering free movement of nodes within the packer volume with a few geometric restrictions including 1) mirror symmetry; 2) packer sides remain flat; 3) demold control considering pull direction.

An important and final part of optimisation analysis is a validation of the obtained design, which in this study is expressed in terms of comparative sealing capability. The first attempt at validation analysis was performed using quasistatic simulation in Abaqus/Standard (Gorash, Bickley, & Gozalo 2016). From obtained results, it was confirmed that the original flat surface packer can't seal effectively, because it allows the fluid to penetrate into the contact and eventually to separate the packer from the borehole, as shown in Fig. 4b. The sealing capability is better for the rippled surface packer, which doesn't let the fluid pressure penetrate at all, because of a stronger grip, as shown in Fig. 4a. This can be considered as a basic and qualitative validation of the optimisation

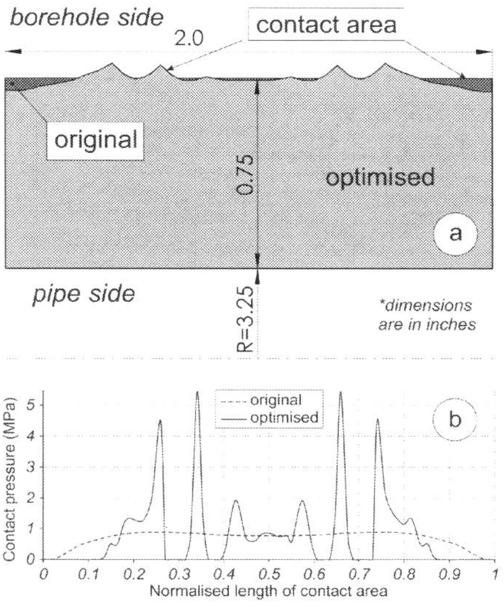

Figure 2. Packer profile shape optimisation with Tosca Structure: a) change of profile geometry, b) change of contact pressure.

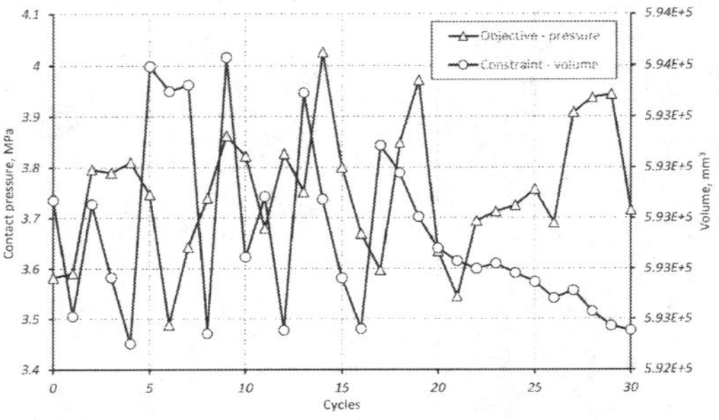

Figure 3. Convergence of the shape optimisation analysis with objective function and constraint vs time.

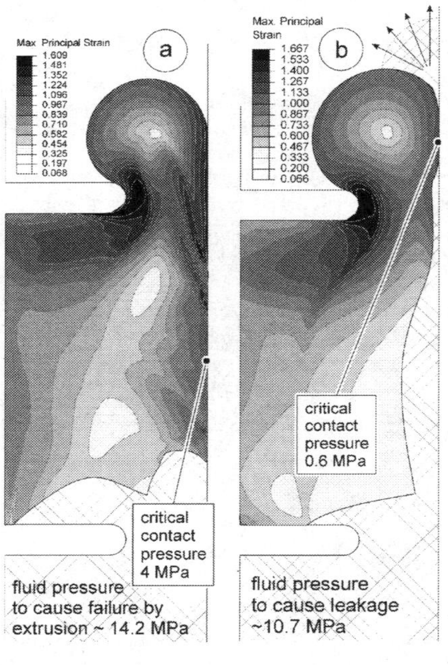

Figure 4. Validation simulation of the benchmark packer using (a) optimised and (b) original geometry with Abaqus/Standard.

result, which is sufficient to get an idea of the challenges and level of complexity associated with realistic simulation of swell packers. For a more clear, quantitative and comprehensive validation of the packer design, the simulation capabilities of Abaqus/Standard solver are insufficient.

3 VALIDATION SIMULATION

The validation simulation with Abaqus/Standard fails when trying to push the packer harder and to extrude it from the protective rings by applying excessive pressure. So there is no way to compare the maximum applicable pressure for original and optimised designs, because the maximum pressure for an optimised design is not achievable. The quasistatic analysis fails to converge because of excessive distortion of elements attributed to the seal, which stick to the ring, and borehole surfaces. Automatic adaptive remeshing is not available as a part of Abaqus/Standard functionality, therefore extrusion problems with extreme deformation can't be effectively solved using this product. But the good thing about Abaqus/Standard is the availability of pressure penetration load (PPL) explained in Simulia [PPL] (2016). This functionality replaces the computationally expensive fluid structure interaction, when the structural analysis is in focus. In application to this work, PPL allows to consider for a leakage probability, when it occurs through the contact surface without excessive extrusion.

Therefore, following the presentation of Gorash, Bickley, & Gozalo (2016) and subsequent discussion, it was decided to switch to the more advanced solver—Abaqus/Explicit, which is recognised as a more robust solver when it comes to very non-linear problems and extremely large deformations, which can be experienced by packer under very high pressure. The validation simulation using PPL for a full-size packer geometry from Lou & Chester (2014), not a benchmark, can't be completed when fluid pressure approaches a full penetration through the contact surface, because ABAQUS/Standard solver fails to converge with errors related to excessive distortion of elements caused obviously by extrusion. It would have taken a vast amount of efforts to extend the functionality of Abaqus/Standard to a suitable level to proceed with validation. The first drawback of Abaqus/Standard solver with excessive element distortion could be fixed by application of the mesh-to-mesh solution mapping (Simulia [AT] 2016), but it is not automatic. In order to automate the mesh-to-mesh solution mapping, customisation via Python script would be required. The second drawback of Abaqus/Standard solver with the fluid pressure only penetrating (available as standard feature), is that it would require a customisation via FORTRAN subroutine to enable the receding of the fluid pressure from the closed contact surfaces.

Abaqus/Explicit is a special-purpose solver that employs an explicit time integration scheme to solve highly non-linear systems with many complex contacts under transient loads, which is appropriate in many dynamic applications, such as drop tests, crushing, manufacturing processes and hydraulic fracturing. Abaqus/Explicit is more computationally expensive compared to Abaqus/Standard, but this obstacle is possible to overcome by running simulations on an HPC facility for simulations with big size models. This solver significantly expands the progressive failure analysis capabilities, and actually eliminates any limitations related to non-linearities, large deformations and transient/dynamic effects. The best prove of its efficiency is a solution of a so-called pressfit problem (Wriggers 2006), when a cylindrical rubber block compressed from the tube of bigger diameter into the tube with a smaller diameter. In previous work by Connolly, Gorash, & Bickley (2016), a simple and stable solution for such a benchmark problem using standard solvers in Ansys and Abaqus couldn't be obtained in order to develop a robust approach to simulation. It should be noted that successful simulation became possible due to modification of the friction model used in analysis from the linear Coulomb to the bi-linear Coulomb-Orowan law (Raous 1999) expressed in terms of friction force as

$$F_f = \min\left(\mu \mid F_n \mid, F_\tau\right), \tag{2}$$

where μ is a coefficient of friction, F_n is a normal force, and F_τ is a critical share force, which corresponds to a critical shear stress τ_c in the FEA setup. The Coulomb term $\mu| F_n |$ is linear and describes the partial slip. When the critical value of τ_c is reached, the total slip occurs, which plays a key role in simulation convergence, because it prevents the rubber material from sticking to the rigid walls.

Therefore, the recent work focused on a development of practical approach to simulations of packers with Abaqus/Explicit, since the setup of analyses in Standard and Explicit solvers is quite different. The biggest advantages attributed to Explicit solver are automatic adaptive remeshing (in application to large plastic deformations) or distortion control of elements (in application to large hyperelastic deformations) and stable solution of contact problems with large relative displacements.

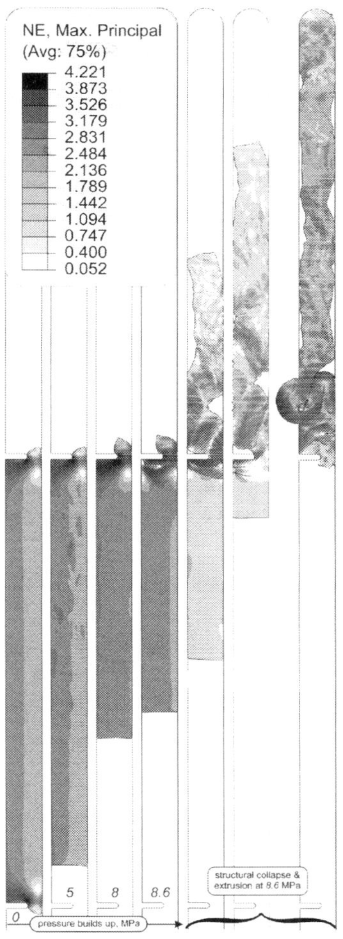

Figure 6. Validation simulation of the full-size packer failure using optimised geometry with Abaqus/Explicit.

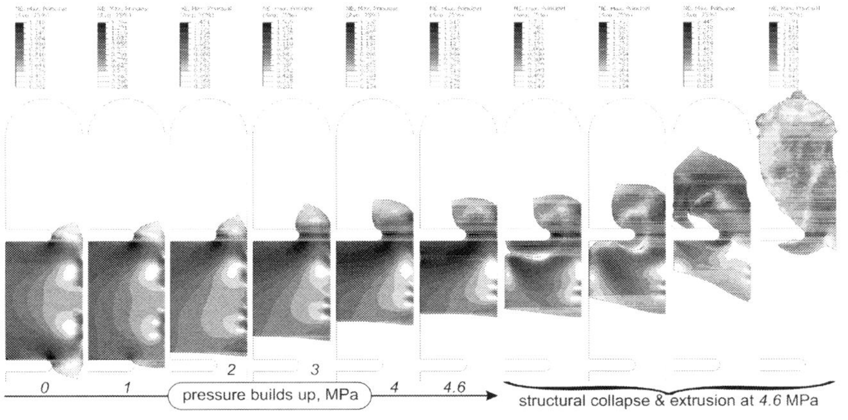

Figure 5. Validation simulation of the benchmark packer failure using optimised geometry with Abaqus/Explicit.

Considering the superior robustness of Abaqus/Explicit, it is a minor drawback that PPL functionality is unavailable for Explicit analysis. The robustness of extrusion failure simulations for packers can be demonstrated on advanced validation analysis of benchmark problem. Since PPL is unavailable, the pressure is applied to the bottom surface and ramped in the course of simulation for both benchmark packers—original (with smooth surface) and optimised (with a rippled surface). Comparison of simulation results shows that the optimised packer (see Fig. 5) can bear about 10% of more pressure compared to the packer with a smooth surface. Due to application of Coulomb-Orowan friction law (2), the whole simulation splits into two parts—slow and gradual pressure build-up and very quick failure by extrusion when reaching a critical pressure of 4.6 MPa. This can be considered as an additional validation of the optimisation result, which is extended with the comprehensive extrusion simulation, that was not possible with the Standard solver.

The validation simulation of a full-size real packer (Lou & Chester 2014) in Fig. 6 demonstrates a complete extrusion of the packer. It also shows that extrusion is not gradual, it is rather abrupt with a distinctive critical pressure (8.6 MPa) when sticking to protective rings can't stop progressive slipping any more.

4 CONCLUSIONS

It should be noted that the conducted validation simulations can lack realism, because interaction with fluid was not considered, which might result in a leakage through the contact surface and less severe extrusion. This simulation challenge can be effectively addressed by application of the Coupled Eulerian-Lagrangian (CEL) approach in Abaqus/Explicit which provides engineers with the ability to simulate a class of problems where the interaction between structures and fluids is important. This capability does not rely on the coupling of multiple software products, but instead solves the fluid-structure interaction (FSI) simultaneously within Abaqus. There is a great potential in CEL approach for packers' leakage simulation. The highest level of realism in simulation of leakage process can be achieved engaging ABAQUS/Explicit and CEL—this is where the future work will be focused.

ACKNOWLEDGEMENTS

The authors greatly appreciate the financial and material support of Weir Group PLC given within the WARC project "Design Optimisation of Swell Packers" and the University of Strathclyde for hosting during the course of this work. Results were obtained using the EPSRC funded ARCHIE-WeSt High Performance Computer (www.archie-west.ac.uk)—EPSRC grant no. EP/K000586/1.

REFERENCES

Akhtar, M., S. Z. Qamar, T. Pervez, & F. K. Al-Jahwari (2013). FEM simulation of swelling elastomer seals in downhole applications. In *Proc. ASME 2013 Int. Mechanical Engineering Congress & Exposition (15–21 Nov 2013)*, Number IMECE2013-64312, San Diego, California, USA, pp. 1–7.

Brieger, S. (2016). Non-parametric optimization. In R. Steinbuch and S. Gekeler (Eds.), *Bionic Optimization in Structural Design: Stochastically Based Methods to Improve the Performance of Parts and Assemblies*, Section 2.6, pp. 37–42. Berlin, Heidelberg: Springer.

Connolly, S., Y. Gorash, & A. Bickley (2016). A comparative study of simulated and experimental results for an extruding elastomeric component. In *23rd Int. Conf. on Fluid Sealing 2016 (2–3 Mar 2016)*, Manchester: BHR Group, pp. 31–41.

Flory, P. J. & J. Rehner Jr. (1943). Statistical mechanics of crosslinked polymer networks ii. swelling. *The Journal of Chemical Physics 11*(11), 521–526.

Gorash, Y., A. Bickley, & F. Gozalo (2016). Design optimization of swellable elastomeric seals using advanced material modelling and FEM simulations. In *Poster—Int. Conf. on Innovations in Rubber Design (7–8 Dec 2016)*, London: IOM3.

Lou, Y. & S. Chester (2014). Kinetics of swellable packers under downhole conditions. *Int. J. Appl. Mechanics 06*(06), 1450073 [18p].

Puri, G. (2011). *Python Scripts for Abaqus: Learn by Example*. USA: Kan sasana Printer.

Raous, M. (1999). Quasistatic signorini problem with Coulomb friction and coupling to adhesion. In P. Wriggers and P. Panatiotopoulos (Eds.), *New Developments in Contact Problems*, Number 388 in CISM International Centre for Mechanical Sciences, Chapter 3, pp. 101–178. Vienna: Springer-Verlag.

Simulia [AT] (2016). *ABAQUS Analysis User's Guide – 12.1.1 Adaptivity techniques* (Version 2016 ed.). Providence, RI, USA: Dassault Systèmes Simulia Corp.

Simulia [MCP] (2016). *Tosca Structure 2016 Documentation—Minimizing contact pressure* (Version 2016 ed.). Karlsruhe, Germany: Dassault Systèmes Simulia Corp.

Simulia [PPL] (2016). *ABAQUS Analysis User's Guide – 37.1.7 Pressure penetration loading* (Version 2016 ed.). Providence, RI, USA: Dassault Systèmes Simulia Corp.

Wagner, N. & R. Helfrich (2016). Topology and shape optimization of structures under contact conditions. In *Proc. 1st Euro. Conf: Simulation-Based Optimisation (12–13 Oct 2016)*, Manchester, UK, pp. 127–130.

Wriggers, P. (2006). Discretization, large deformation contact. In *Computational Contact Mechanics*, pp. 225–307. Berlin, Heidelberg: Springer.

Constitutive Models for Rubber X – Lion & Johlitz (Eds)
© 2017 Taylor & Francis Group, London, ISBN 978-1-138-03001-5

Comparison of experimental and numerical fatigue lives of rolling lobe air-springs for different diameters, inner pressures and temperatures

A. von Eitzen & U. Weltin
Institute for Reliability Engineering, Hamburg University of Technology, Hamburg, Germany

M. Flamm & T. Steinweger
Beratende Ingenieure Flamm, Buchholz, Germany

ABSTRACT: The significant functional and comfort advantages of air-spring suspension systems compared to conventional suspension systems lead to an increasing application in automotive industries. The main parts of such suspension systems are the damper and a cavity, filled with compressed air dealing as a gas spring. The cavity is surrounded by a bellow, consisting of a cord-rubber composite forming a rolling lobe during service. Due to the material heterogeneity and the large deformations of the rubber matrix between the cords, the development of a new air-spring system is still a challenging process. In the present paper numerical fatigue lives are compared to experimental fatigue lives taken from the literature. The numerical analysis is based on a global to local scale analysis as described in (von Eitzen, Weltin, Flamm, & Steinweger 2015). The fatigue life based on crack nucleation is calculated via the Wöhler concept (Flamm, Steinweger, & Weltin 2003) using different damage parameters and via the concept of cracking energy density described in (Mars 2001). S-N curves are determined using simple tension test specimens and a newly developed cord-rubber specimen. Missing parameters are fitted to experimental data for better results. The influence of the diameter of the rolling lobe and the inner pressure of the gas cavity are considered in the finite element analysis and the influence of temperature is taken into account using the Arrhenius equation. The methods and results are discussed. Furthermore, the descriptions of the cord-rubber interface and their deficiencies in analysing the deformation behaviour are mentioned.

1 INTRODUCTION

Smarter suspension systems, higher loads and comfort requirements lead to an increasing usage of bellow air-spring suspension systems in automotive applications. For such suspension systems, the bellow of the air-spring forms a rolling lobe which allows a wide range of vertical motion at low eigen frequencies. The bellow mostly consists of one or two different cord layers embedded in a rubber matrix. The cords are twisted from multifilament yarns. The gas filled cavity is gasketed by the rubber while the cords carry the loads resulting from vehicle motions.

The inhomogeneity and anisotropic properties of the bellow result in complex damage and failure processes. Due to the increasing demand, there is a considerable interest to understand the reliability and the failure mechanisms and improve existing damage calculation methods for air-springs under specific loading conditions. This is especially helpful during the development process because experimental fatigue test are very cost and time consuming.

The aim of the present paper is to compare numerically achieved fatigue lives with experimental fatigue lives taken from the literature. For numerical damage calculations it is important to analyse the local loading conditions of the cord-rubber composites, which will be the object of the first section. Afterwards two different approaches for damage calculations, namely the crack nucleation and the crack growth approach, are shortly summarized in section 3. For the crack nucleation approach, tests on simple tension test specimens (S2 specimen) and a newly developed cord-rubber composite specimen (3FP specimen) are presented to obtain predictor-fatigue curves (S-N curves). The concept of cracking energy density is applied for the crack growth approach. The experimental procedure and the results of fatigue tests on rolling lobe air-springs are described in section 4 with all relevant input parameters. To conclude, the results from the numerical and experimental fatigue lives are compared and discussed in section 5.

2 ANALYSIS OF LOADING CONDITIONS IN AIR-SPRINGS

A bellow of an air-spring consists of cords and rubber. Cords are formed by twisted yarns, twisted from filaments, resulting in a complex deformation behaviour and material orientation respectively. Due to their composition, cord-rubber composites are characterized by a high tensile stiffness and flexibility to shear loading and bending. The materials used for the cords are for example polyamide, polyester or aramid fibres and natural rubber (NR) or chloroprene (CR) or a combination for the matrix material.

Due to the nonlinear material response of the two constituents, the anisotropy, the nearly incompressible rubber matrix and the large displacements and rotations, lots of approaches have been developed to cover all these effects. To analyse the local loadings in an air-spring bellow, the consideration of both materials on a global scale will result in too many elements. In (Meschke & Helnwein 1994) a smeared material formulation based on a superposition of the stiffness of the cords or cord layers and the rubber material is used to model the material behaviour in finite element procedures (rebar-formulation). This formulation is well suited for global simulations. The local loadings in the rubber surrounding the cord are a combination of tension, shear and bending. These local loadings are achieved afterwards with a second model on the meso-scale driven by the global model as described in (Brüger, Merk, & Pelz 2006) and (von Eitzen, Weltin, Flamm, Steinweger, & Brüger 2013). In the described procedures the global model uses the rebar formulation while the local model represents both the matrix and the cord material separately and allows a detailed analysis of the local loadings of the rubber matrix between the cords.

The rubber matrix is assumed to be hyperelastic, isotropic and incompressible. A Yeoh-material is used to model the rubber matrix. The parameters are fitted for simple tension tests ($C_1 = 0.3409$ MPa, $C_2 = -0.0711$ MPa, $C_3 = 0.0224$ MPa). The cord is assumed to follow an incompressible, transversally isotropic Holzapfel-Gassner-Ogden material model (Holzapfel, Gassner, & Ogden 2000). The material parameters are fitted to tension tests on cords used in the bellows ($k_1 = 750.1$ MPa, $k_2 = 27.3[-]$.)

Figure 1 shows the maximum principal Cauchy stress σ_1 and the maximum principal stretch ratio λ_1. The values of the maximal loaded integration point in the finite element model for the air-spring bellow, described in section 4, are given. The increments represent the passing through the rolling lobe from the outer to the inner diameter

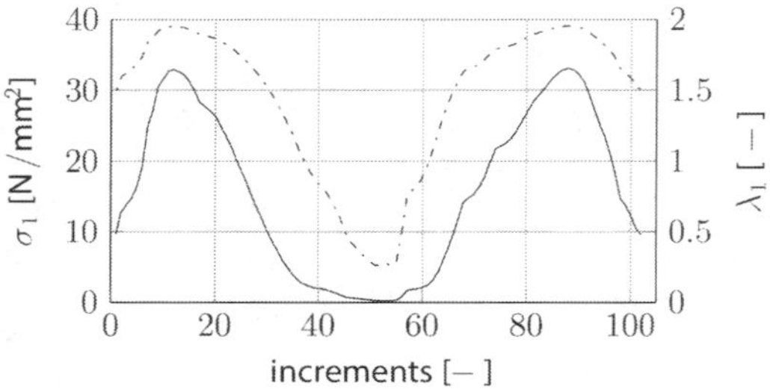

Figure 1. Local loading condition in an rolling lobe, σ_1 (—) and λ_1 (– –).

and backward respectively for compression and rebound. Increment 24 and 72 respectively represent the rolling lobe position of the integration point. The Cauchy stress reaches a maximum at increment 12 and 88 at the vicinity of the rolling lobe. At the inner diameter, σ_1 decreases to zero at increment 52. So, for a compression and rebound of the suspension system, the local model passes the rolling lobe twice. Although the loadings in a rolling lobe are non-proportional, the R-Ratio is zero. The local loadings in the air-spring bellow during a throughput of the rolling lobe are used in the following section for the numerical damage analysis.

3 NUMERICAL DAMAGE ANALYSIS

Failures in air-spring bellows are for example: crack propagation in the rubber between different cords or between the cords in the same layer (interply cracks), fibre-matrix debonding around one single cord (socketing) and at a crossing point between two cords of different layers. Failures will emerge in the area of maximum loading, the rolling lobe in the case of an air-spring. The calculation of the fatigue life can be divided into two approaches, crack nucleation and crack growth which are summarised in the following two subsections.

3.1 *Crack nucleation approach*

In the crack nucleation approach, the fatigue life can be determined from the history of quantities defined at each material point in the material. Some few applicable quantities for rubber are for example the maximum principal stretch ratio, the maximum principal Cauchy stress or the strain energy density. Fatigue tests until failure, on simple tension test specimens for example, give a relation between the cycles to failure and the critical quantity of the so called predictor. The predictor-life data follows a power law as presented in equation 1

$$N = (S_a/S_0)^k. \tag{1}$$

N is the number of cycles to failure, S_a is the predictor amplitude and S_0 and k are the empirical constants. Following a linear damage accumulation (Miner 1945), using the critical values of the specific predictor and the local distribution of the predictor from finite element analysis at each material point, the cycles to failure as well as the spatial distribution can be determined. In the following, the maximum principal Cauchy stress σ_1 is used as the predictor. As seen in Figure 1, the value of σ_1 reaches zero and a maximum for a pass of the rolling lobe. So the effect of increasing fatigue life for increasing minimum load for strain crystallizing elastomers like CR does not have to be taken into account. Otherwise, a Haigh-field has to be used for damage calculations to cover this effect.

The used specimen should be able to describe the loading conditions in the rolling lobe. For the S2 specimen this might not be the case. Therefore, a previously developed cord-rubber composite specimen is used furthermore. The specimen consists of three cords embedded in a rubber block (3FP). The central cord is cyclically loaded until failure to determine the parameters for relation 1. For more information see (von Eitzen, Weltin, Flamm, & Steinweger 2015).

Figure 2 shows the σ_1-life data for the S2 specimens and 3FP specimens. The differences in the curves are explained by the different loading conditions in the specimens. A bellow and a simple tension test specimen have approximately the same thickness of 2 mm but the loading conditions are different. A S2 specimen represents a simple tension loading case, while the 3FP specimen represents the loading of the rubber close to a cord in tension. The empirical constants for the fatigue life calculations are given in Table 1.

3.2 *Crack growth approach*

The crack growth approach first applied to rubber by Rivlin and Thomas (Rivlin & Thomas 1953)

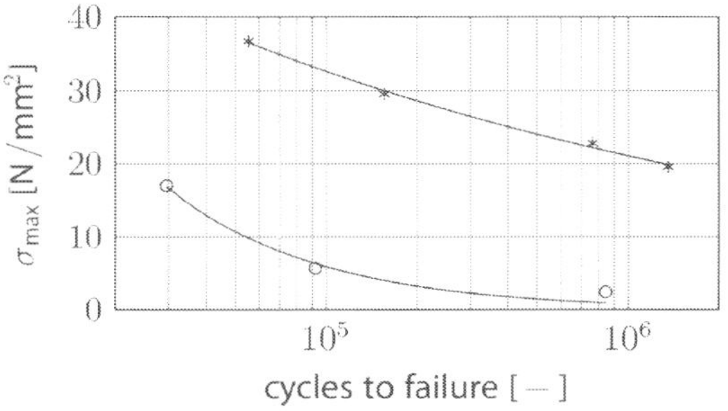

Figure 2. σ_1-life curve for S2 (°)and 3FP (*).

Table 1. Empirical constant for σ_1-life data for S2 and cord-rubber composite (3FP) specimens.

S2		3FP	
S_0 [MPa]	k [–]	S_0 [MPa]	k [–]
12570.68	–1.15	289.97	–5.27

considers the stored energy surrounding pre-existing flaws in the material. The critical energy at the flaw (tearing energy T) is used to predict the cycles to failure by integrating the crack growth rate. In a crack opening mode, the tearing energy is proportional to the size of the crack a and the strain energy density W of the far-field

$$T = 2kWa. \tag{2}$$

The parameter k depends on strain and is mostly set equal to three. For the major regime the crack growth rate follows a power law, which leads, under the assumption of small initial flaws a_0, compared to the crack length of failure, to equation 3

$$N = \frac{1}{\beta - 1}\frac{1}{B(2kW)^{\beta}}\frac{1}{a_0^{\beta-1}} \tag{3}$$

for the cycles to failure N. The associated material parameters are B and β. For multiaxial, non-proportional loadings, Mars (Mars 2001) studied the part of the strain energy which is available for crack growth. The cracking energy density W_c is the portion of energy available to be released by crack growth on a specific cracking plane

$$W_c = \int_0^{\varepsilon} \vec{r}^T \sigma d\varepsilon \vec{r}. \tag{4}$$

The cracking plane normal vector is $\vec{r}$, σ is the Cauchy stress tensor and $d\varepsilon$ is the increment in strain tensor. Using equation 3 and the maximum cracking energy density for all possible planes, the cycles to failure can be determined for each plane. An optimisation with respect to the cracking plane gives the cycles to fatigue as well as the plane orientation for the initiating crack.

Unfortunately no crack growth parameter β and B for the used mixture are given. Under the assumption of constant k and an initial flaw size regarded intrinsic to the material, equation 3 may be simplified to

$$N = DW_c^{\alpha} \tag{5}$$

with D as the material parameter. Therefore for comparison in section 5 the empirical constants D and α in equation 5 are fitted to the experimental results. The fit yields $\alpha = -2.00[-]$ and $D = 2.8810^7[MPa^{-\alpha}]$. In the following section the experimental set-up and the results of the rolling lobe fatigue test are given.

4 FATIGUE TESTS OF ROLLING LOBE AIR-SPRINGS

The experimental fatigue lives are taken from the PhD-thesis by Förster (Förster 2012). The testing procedure is described in subsection 4.1 with the specific parameters taken into account. In subsection 4.2 the transformation of the fatigue lives from testing temperature to room temperature is given.

4.1 *Testing procedure*

Cylindrical bellows with two cord layers, oriented in specific angles, are fixed on both sides in a test-rig. The fixations on both sides form a cylinder with a constant diameter. The bellow is in contact with the fixation on one side, forms the rolling lobe and is radial free to deform on the other side. So the diameter of the fixation influences the diameter of the rolling lobe. Before testing, the bellow is pressurised and the fixations are moved together in a way, that the bellow forms two rolling lobes on each side. During testing, a clamp at the middle of the bellow is fixed and the fixations on each side are coupled and cyclically moved up and down until a failure occurs in one rolling lobe. Due to the constant volume of the bellow resulting from two symmetric rolling lobes, the inner pressure is constant during one fatigue test. The failure is detected by a drop in inner pressure which represents a crack in the matrix material. For the comparison in section 5, the variation of the fixation diameter, the testing temperature and the inner pressure are considered. The average fatigue lives and the number of test results (in brackets) are given in Table 2. For diameter 90 mm and 114 mm, five test results for 1.0 MPa, three test results for 1.4 MPa and two test results for 1.5 MPa respectively are given. No results are given for 110 mm diameter for 1.4 MPa and two test results respectively for the other values.

4.2 *Influence of temperature*

The tests are performed at higher temperatures for test time reduction, while the tests on specimens described in section 3 are performed at room temperature. Thus a transformation of the calculated fatigue lives to elevated temperatures isnecessary. In (Bauman 2001) the researcher shows that, the Arrhenius approach might be applicable for

Table 2. Experimental fatigue lives for different diameters d, temperature ϑ and inner pressures p_i (Förster 2012).

$\vartheta = 60°$			
	$d_i[mm]$		
$p_i[MPa]$	90	110	114
1.0	2724978 (5)	1646821 (2)	744835 (5)
1.4	946034 (3)	(0)	395018 (3)
1.5	173808 (2)	79683 (2)	53352 (2)
$\vartheta = 100°$			
	90	110	114
1.0	1517136 (5)	1402773 (2)	558157 (5)
1.4	656456 (3)	(0)	361078 (3)
1.5	65290 (2)	69383 (2)	54943 (2)

comparing numerical to experimental results where the testing itself is conducted at different temperatures as described in equation 6.

$$N_\vartheta = N_0 exp\left[\frac{E_a}{Rg}\left(\frac{1}{\vartheta} - \frac{1}{\vartheta_0}\right)\right] \quad (6)$$

N_0 and N_ϑ are the cycles to failure at room temperature ϑ_0 and at the desired temperature ϑ in Kelvin respectively. $R_g = 8.3145\, J/molK$ represents the gas constant and E_a the activation energy accounting for different elastomers. The activation energy is determined by a fit of the data shown in Table 2 for specific diameters and inner pressures at $60°C$ and $100°C$. The fit gives an activation energy of $E_a = 13.3\, kJ/mol$ which is roughly in the range found in the literature. Equation 6 is used in the following section to transform the numerically determined cycles to failure to the testing results.

5 COMPARISON OF NUMERICAL AND EXPERIMENTAL FATIGUE LIVES

In Table 3, the results of the experimental fatigue lives are compared to the numerical results from section 3. For a better comparison, the calculated cycles to failure are given relative to the mean experimental cycles to failure (see Table 2). In the first three rows, the results for the S2 specimen are given for the different inner pressures, diameters at the bellow fixation (rolling lobe) and different temperatures respectively. In the following three rows, the results for the cord-rubber specimen are given followed by the results of the cracking energy density. The results from the S2 specimens are in a range of

Table 3. Comparison of numerical and experimental fatigue lives for rolling lobe air-springs.

		$\vartheta = 60°$			$\vartheta = 100°$		
		$d_i\,[mm]$			$d_i\,[mm]$		
	$p_i\,[MPa]$	90	110	114	90	110	114
S_2	1.0	0.005	0.004	0.007	0.005	0.003	0.006
	1.4	0.006	(–)	0.009	0.006	(–)	0.006
	1.5	0.029	0.042	0.057	0.046	0.029	0.033
$3FP$	1.0	3.140	0.244	0.245	3.368	0.171	0.196
	1.4	0.302	(–)	0.050	0.260	(–)	0.033
	1.5	0.700	0.295	0.224	1.113	0.202	0.130
W_c	1.0	1.032	0.343	1.089	1.107	0.403	0.868
	1.4	1.095	(–)	1.013	0.943	(–)	0.662
	1.5	2.125	2.771	6.167	3.379	3.182	3.578

0.003 to 0.057, meaning that during testing the air-spring lasts 17 to 333 times longer than the prediction. This reflects the inability of the S2 specimen to represent the loading conditions in the cord-rubber interface sufficiently. As mentioned before, the S2 specimen reflects the simple tension loading case. The loading of the cord-rubber interface is a combination of tension, shear and bending, so the differences are as expected. Furthermore, tension tests on dumbbell specimens are performed but show no improvements in the prediction of the fatigue life. Reasons might be problems during the manufacturing process of the specimens. The bellow of an air-spring has an approximate thickness of 2 mm which is small compared to the thickness of a dumbbell specimen of approximately 25 mm. So, the applicability of the S-N curve from the dumbbell specimens is questionable. The results are not shown in this paper.

The results from the cord-rubber specimen are mostly in a range of 0.033 to 3.368, meaning that during testing the air-spring bellow lasts 30 longer than the predicted fatigue life or the predicted fatigue life was 3.4 times longer than the tested fatigue life respectively. For an inner pressure of 1.0 MPa and a diameter of 90 mm, the predicted fatigue lives are for both temperatures higher than the tested ones. All other data predict a shorter life. The minimum fatigue lives are predicted for a diameter of 114 mm and an inner pressure of 1.4 MPa at both temperatures. Figure 3 shows the experimental S-N curves in logarithmic scale.

The data do not follow a power law relation (two slopes) with respect to the inner pressure due to the two slopes. This may be an indicator for two different failure mechanisms. In (Förster 2012) no evidence is given for two failure mechanisms, due to the fact, that the cycles to failure are detected by a decrease of inner pressure. This decrease postu-

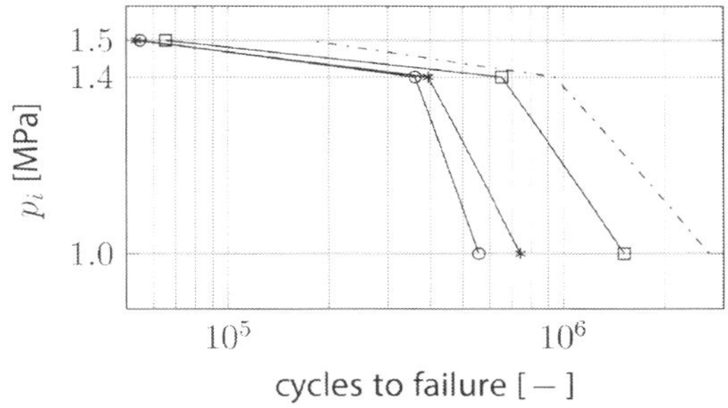

Figure 3. Experimental cycles to failure, 90 *mm*, 60°C (– –) (114 *mm*, 60°C (*) 90 *mm*, 100°C (□) and 114 *mm*, 100°C (°).

lates a crack through the bellow, but the local position of the crack initiation is not detectable. This may lead to the kinks in Figure 3. On the other hand, if there are two failure mechanisms present for the accounted parameter, the fitting of the activation energy may be invalid. This can result in longer predicted fatigue lives at a diameter of 90 mm and an inner pressure of 1.0 MPa. No further information can be given because for a new fit, it has to be determined if one failure mechanism is active at 1.4 and 1.5 MPa or at 1.0 and 1.4 MPa respectively. Compared to results of the S2 specimen, the cord-rubber specimen improves the fatigue live prediction.

Due to the fitting of the parameter D and k in case of cracking energy density, the results presented in Table 5 are sufficiently good. They are in a range of 0.403 to 6.167. For an inner pressure of 1.5 MPa the predicted fatigue life is up to six times longer. This reflects Figure 3 as well if the data for 1.0 MPa and 1.4 MPa are extrapolated. Due to the fit of the parameters α and D from equation

5 and the results of the cracking energy density, the failure mechanism may be assumed to change from 1.4 to 1.5 MPa, but this cannot be proven. The resulting orientation of the cracking plane maximising the cracking energy density coincides with cracks seen in the experiments.

6 CONCLUSION

The paper summarises numerical and experimental damage analysis of rolling lobe air-springs. First, the analysis of the local loadings in a rolling lobe is presented using a two scale finite element analysis. The used materials and constitutive equation are mentioned with references to the relevant literature. As long as the interest is on the local loadings of the rubber between the cords, a simplification of the cord to an incompressible, hyperelastic, transversely isotropic material with one orientation along the cord axis seems to be sufficient. In section 3, two numerical damage analyses are illustrated and the needed parameters are given. For the crack nucleation approach, the S-N curve of the chosen predictor for simple tension test specimens is given as well for a newly developed cord-rubber composite specimen. For the crack growth approach in combination with the cracking energy density a power law is fitted to the test-data because no crack growth parameters are given for the used mixture.

The experimental set-up is described in section 4. The test data are taken from literature (Förster 2012). Fatigue lives for different diameters of the fixation, inner pressure of the bellow and testing temperature are chosen for a comparison to numerically predicted fatigue lives in section 5.

The predicted fatigue lives related to the experimental fatigue lives are given in Table 3. It is shown that the representation of the local loadings at the cord-rubber interface is improved by the newly developed cord-rubber specimen. This results in a better numerical prediction of the fatigue lives of air-spring bellows. Due to the fitted parameters for the cracking energy density, the results are better compared to the results from the crack nucleation approach. Because of the directional characteristic of the cracking energy density, the plane of crack initiation can be determined. This will also be achieved using the crack energy density as a predictor with a corresponding S-N curve taken from the cord-rubber specimen as done for the maximum principal Cauchy stress as the predictor. In conclusion, a numerical fatigue life prediction for rolling-lobe air-springs is qualitatively possible. Improvements may be obtained with a more accurate modelling of the cord-rubber interface and a wider experimental data base referring to the temperature dependencies of the specimens and the air-spring fatigue lives.

ACKNOWLEDGEMENT

The authors would like to gratefully acknowledge the financial support of the research reported herein by the DFG under grant WE 4047/1-1/2. In addition, the authors like to thank the Vibracoustic GmbH & Co KG. for providing experimental data, specimen material and support during their manufacturing.

REFERENCES

Bauman, J.T. (2001). Fatigue life equation incorporating varying amplitude, R-ratio and temperature. In *Energy Rubber Group, Rubber Division.*

Brüger, T., J. Merk, & P. Pelz (2006). Numerische Festigkeitsauslegung von Luftfedern. In *DVM Bericht 133*, Berlin, pp. 23–36.

Flamm, M., T. Steinweger, & U.Weltin (2003). Lifetime prediction of multiaxially loaded rubber springs and bushings. In *Constitutive Models for Rubber III*, pp. 49–53.

Förster, K. (2012). *Ein Beitrag zur Untersuchung der Lebensdauerabschäzung und des Walkverhaltens von Luftfederbälgen mittels statistischer Methoden.* Ph. D. thesis, Helmut Schmidt Universität Hamburg.

Holzapfel, G.A., T.C. Gassner, & R.W. Ogden (2000). A new constitutive framework for arterial wall mechanics and a comparative study of material models. *J. Elast.* 61, 1–48.

Mars, W.V. (2001). Multiaxial fatigue crack initiation in rubber. *Tire Science and Technology 29*(3), 171–185.

Meschke, G. & P. Helnwein (1994). Large strain 3D-analysis of fibre-reinforced composites using rebar elements: hyperelastic formulations for cords. *Computational Mechanics 13*, 241–254.

Miner, M.A. (1945). Cumulative damage in fatigue. *J. Appl. Mech. 67*, 159–164.

Rivlin, R.S. & A.G. Thomas (1953). Rupture of rubber. I. characteristic energy for tearing. *J. Polym. Sci. 10*, 291–318.

von Eitzen, A., U. Weltin, M. Flamm, & T. Steinweger (2015). A new specimen for fatigue analysis of cord-rubber composites. In *Constitutive Models for Rubber IX*, pp. 373–378.

von Eitzen, A., U.Weltin, M. Flamm, T. Steinweger, & T. Brüger (2013). Modelling of cord-rubber composites of bellow airsprings. In *Constitutive Models for Rubber VIII*, pp. 631–636.

Design issues

Constitutive Models for Rubber X – Lion & Johlitz (Eds)
© 2017 Taylor & Francis Group, London, ISBN 978-1-138-03001-5

Computational material design of filled rubbers using multi-objective design exploration

M. Koishi & N. Kowatari
The Yokohama Rubber Co. Ltd., Hiratsuka, Japan

B. Figliuzzi, M. Faessel, F. Willot & D. Jeulin
MINES ParisTech, PSL Research University, CMM, Fontainebleau, France

ABSTRACT: The rubber materials of tires contain nano-fillers i.e. carbon black and silica for the improvement of tire performances. The mechanical properties of filled rubber depend on the morphology of fillers. In this work Multi-Objective Design Exploration (MODE) is applied to material design of filled rubbers to get the information between mechanical properties and morphological design variables. A multi-scale random model based on a Poisson point process is used to generate a simulation model of filled rubber, and FFT (Fast Fourier Transform) based scheme is applied to solve large-scale dynamic viscoelastic simulation. To get big data including Pareto solution for data mining, multi-objective genetic algorithm are conducted on TSUBAME, supercomputer at Tokyo Institute of Technology. Data mining is employed to highlight properties-sensitive features in the microstructure. First, the volume fraction of bound rubber plays a major role in the material design of filled rubbers. Second, the radius of aggregates contributes to mechanical properties of filled rubbers.

1 INTRODUCTION

The rubber materials of tires contain nano-fillers i.e. carbon black and silica for the improvement of tire performances e.g. rolling resistance, wear and so on. The mechanical properties of filled rubber depend on the morphology of fillers. It is said that uniformly dispersed fillers are good to reduce loss tangent, however, guideline of material design to improve loss tangent, modulus and strength at the same time is not clear.

For clarification of the mechanism and origin of mechanical properties of filled rubbers, numerical simulation of filled rubbers has been conducted in the past few decades. Recently, large-scale simulation using the model generated by 3D-TEM (transmission electron microtomography) was conducted by Akutawa et al. (2008) and Kadowaki et al. (2016) to compute effective material properties. Two-dimensional pattern reverse Monte Carlo analysis was performed to make structural model from the data obtained by time-resolved two-dimensional ultra-small angle x-ray scattering (Hagita et al. 2007 & Hagita et al. 2008). And to compute effective shear modulus, the complex multi-scale microstructure of filled rubber was generated numerically from a morphological model that was identified from statistical moments out of transmission electron microscopy images (Jean et al. 2011, for a simpler version).

The information on effective material properties and morphological design variables is helpful for the material design of filled rubbers. So, the objective of this work is to get the information using multi-objective design exploration (MODE). MODE has been already applied to tire design (Koishi et al. 2006 & Koishi et al. 2014). MODE consists of numerical simulation, multi-objective genetic algorithm and data mining technique, and is a methodology to discover an evolutional idea for making technical innovation. However, to apply MODE to material design, especially morphological design of filled rubbers, there are two big issues i.e. simulation models in which morphology of fillers can be changed parametrically and large-scale viscoelastic simulation. In this study, the issues were solved using multi-scale random model based on Poisson point process and complex FFT (fast Fourier transform) based scheme proposed by Figliuzzi et al. (2016). In addition, MODE for the morphological design of filled rubbers was conducted on TSUBAME, supercomputer at Global Scientific Information and Computing Center in Tokyo Institute of Technology.

2 SIMULATION MODELS AND LARGE-SCALE DYNAMIC VISCO-ELASTIC SIMULATION

2.1 *Morphology of filled rubbers and their feature value*

This work concerns a rubber material reinforced by carbon black and silica particles as fillers, which are embedded in a matrix of rubber and can be geometrically well-approximated by spheres. A key feature of the material is that the rubber matrix is constituted of two distinct polymers: an exclusion polymer, which cannot contain any filler, and its complementary polymer, which can contain fillers. Fillers tend to agglomerate together within the rubber matrix. To analyze the rubbers microstructure, we dispose of a set of 50 TEM micrographs of size 1024 by 1024 pixels with resolution 2.13 nm per pixel. The slices of material probed by the microscope have a thickness around 40 nm. The first task is to segment these images, in order to identify the spatial distribution of the fillers as shown in Figure 1.

An efficient way to keep track of the information embedded in the segmented images is to rely on a morphological characterization of the material. In this work, the covariance and the granulometry of fillers are used as feature value of morphology. The covariance and the cumulative granulometry curve of silica fillers are shown in Figure 2. Using the covariance and the granulometry curves, we can select parameters for the identification of multi-scale random models as explained in the next section. This is important way to merge the information between real material and simulation model at a material design stage.

2.2 *Multi-scale random model*

A multi-scale random model is used to describe the microstructure (Figliuzzi et al. 2016) as shown in Figure 3. The first scale corresponds to the aggregates, while the second one describes more

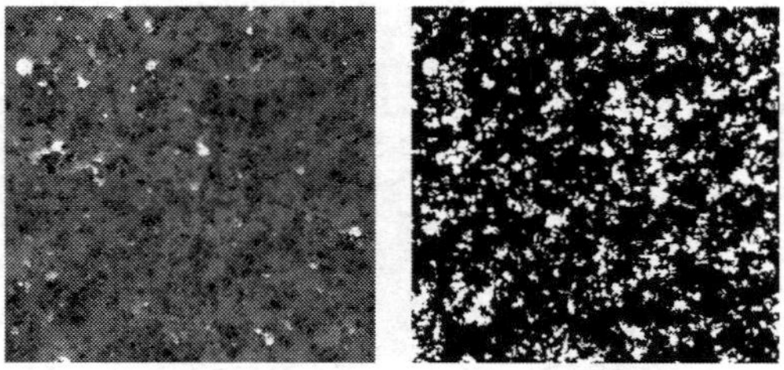

(a) (b)

Figure 1. (a) Original TEM image of silica filled rubber, (b) segmented binary image showing the filler in white.

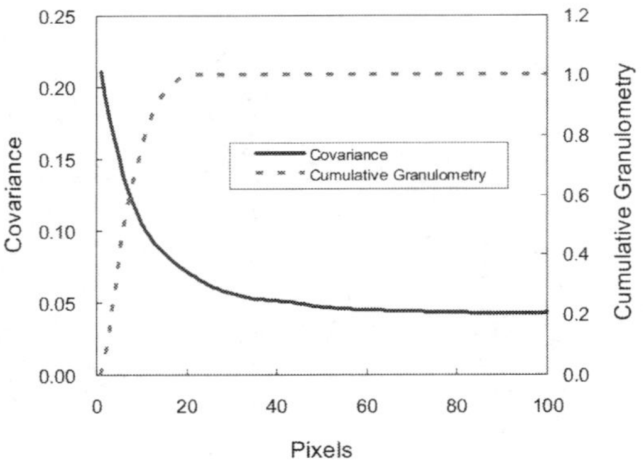

Figure 2. Covariance (solid line) and cumulative granulometry (broken line) computed from TEM image shown in Figure 1.

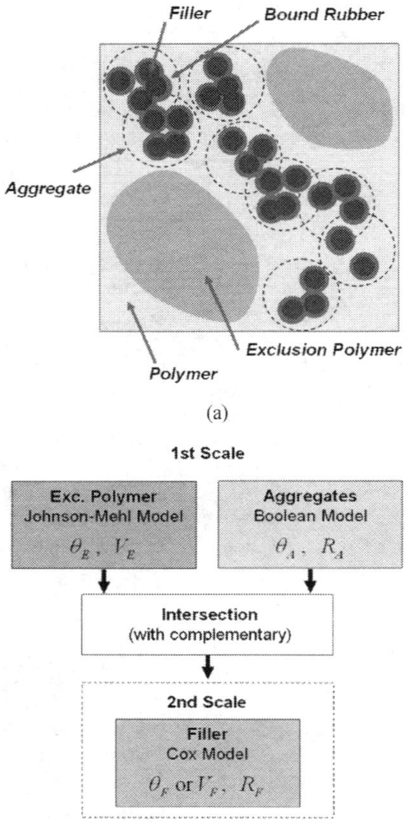

Figure 3. (a) Schematic figure of filled rubber constituted of two distinct polymers and (b) procedure of a multi-scale random model.

specifically the single particles inside the aggregates. In addition, several alternatives can be considered to locate the aggregates outside of the exclusion polymer. Bound rubber in Figure 3 is the interface layer between polymer and filler. Morphological parameters of a multi-scale random model are followings;

- Poisson point intensity and volume fraction of exclusion polymer,
- Poisson point intensity and radius of aggregate,
- Poisson point intensity or volume fraction and radius of filler,
- Overlapping distance of fillers,
- Thickness of bound rubber.

These parameters can be used as design variables for parametric study, optimization and MODE. For example, cross sections of six 3D simulation models generated by multi-scale random modeling are shown in Figure 4. These models are generated with different Poisson point intensity and radius of aggregate as design parameters, however, the volume fraction and radius of filler remain constant.

2.3 *Computation of the effective dynamic viscoelastic properties using FFT-based scheme*

According to our research, RVE (representative volume element) size of multi-scale random model should be 20 times larger than the radius of filler. And for sufficient resolution of bound rubber, the element size of the bound rubber should be smaller than its thickness. Moreover more than thousand computations should be done to get a Pareto solution using multi-objective genetic algorithm. Therefore, a fast simulation scheme is needed for large-scale dynamic viscoelastic problems. Though the homogenization procedure based on FEM for dynamic viscoelasticity is proposed by Koishi et al. (1997), a FFT-based scheme (Figliuzzi et al. 2016 & Willot 2015) is used in this study.

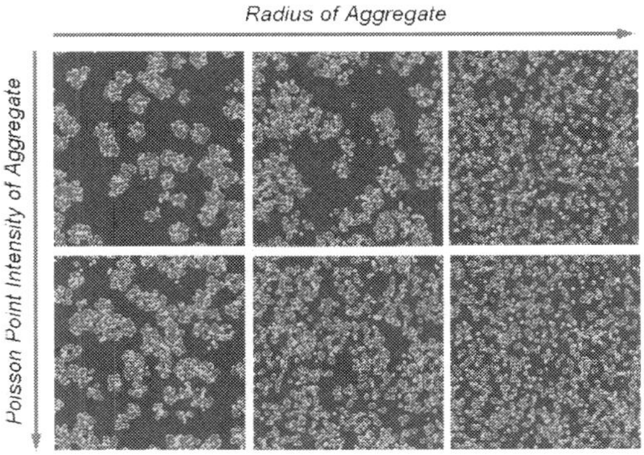

Figure 4. Cross section of six 3D simulation models generated with different Poisson point intensity and radius of aggregate, domain size: 1,000 nm × 1,000 nm × 1,000 nm, the volume fraction of filler: 15%, the radius of filler: 10 nm, the thickness of bound rubber: 5 nm.

Here, we demonstrate the efficiency of FFT-based scheme as a solver of large-scale problems. A single spherical filler embedded in rubber matrix shown in Figure 5 is considered for a benchmark test. Simulation models are discretized in uniform voxel mesh. The largest model consists of over 1 billion elements (1024 × 1024 × 1024 elements). Figure 5 and Figure 6 show computation time and required memory of the FFT-based scheme against the number of elements, respectively. Those of FEM are also shown in the same figures. Both of FFT-based scheme and FEM were run on 8 CPUs. The computational time and required memory size

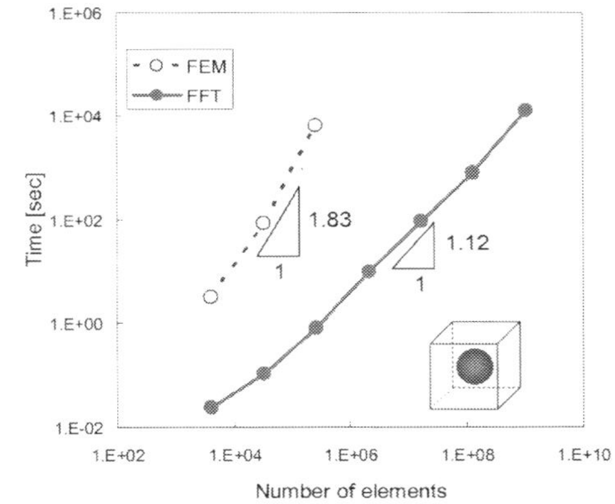

Figure 5. Computation time of the FFT-based scheme and FEM against the number of elements.

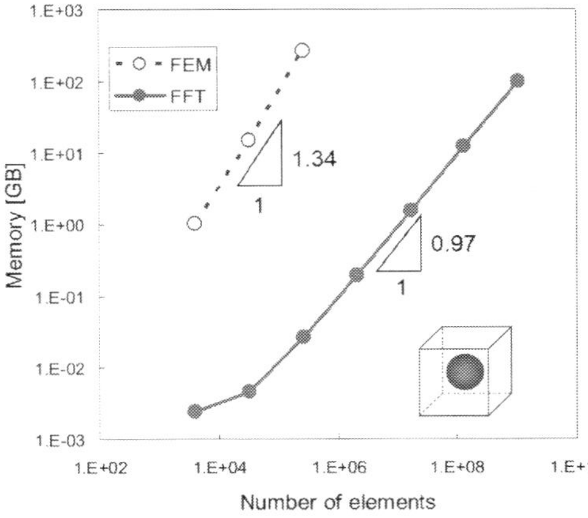

Figure 6. Required memory size of the FFT-based scheme and FEM against the number of elements.

of the FFT-based scheme are almost proportional to the number of elements. The figures show that as the number of elements increases, the FFT-based scheme is much faster than FEM.

3 MULTI-OBJECTIVE DESIGN EXPLORATION FOR MORPHOLOGICAL DESIGN OF FILLED RUBBERS

3.1 *Multi-Objective Design Exploration (MODE)*

The decision making in the real world depends on multiple criteria. Multiple criteria decision making is indispensable even in the design process of rubber materials of tires. So the material design of rubber is so-called multi-objective design optimization. The multi-objective design optimization is to find design variables which minimize or maximize each objective function i.e. mechanical properties of filled rubbers. Since some of objective functions have trade-off, a set of optimal solutions called Pareto solution is found out in the multi-objective optimization. We can obtain design information by clarifying the relationship between objective functions and design variables in design space including Pareto solutions. The information and the evolutional idea inspired by the information are helpful for the multi-criteria decision making at material design stage.

MODE is a methodology to get helpful information between objective functions and design parameters using multi-objective optimization and data mining. From the information, we can find out an evolutional idea for making new productions. Figure 7 shows procedure of MODE. Here the procedure of MODE consists of FFT-based scheme to simulate properties of filled rubber generated by multi-scale random model, evolutional computation e.g. multi-objective genetic algorithm to get desirable solutions including Pareto solution, and data mining using self-organizing map (SOM) to find out morphological design information of filled rubbers as shown in Figure 8. Since the viscoelastic simulation of each individual can be conducted independently, parallel computing is effective in this study.

3.2 *Problem set-up*

The simulation model of the filled rubber which consists of three domain i.e. polymer, filler and bound rubber is considered here. Modulus of polymer, filler and bound rubber are 2 MPa, 2000 MPa

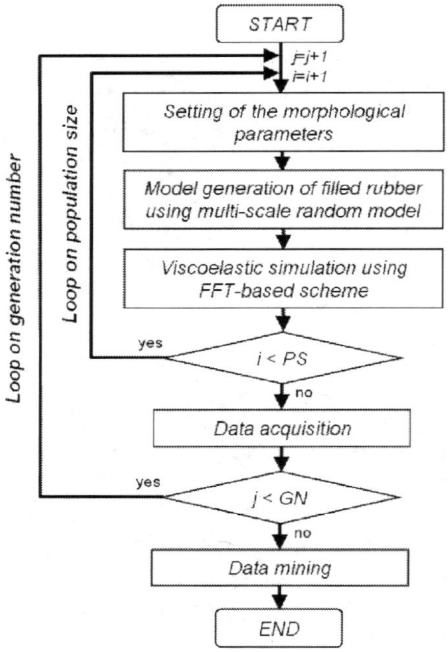

Figure 8. MODE for morphological design of filled rubbers. PS and GN stand for population size and generation number in multi-objective genetic algorithm.

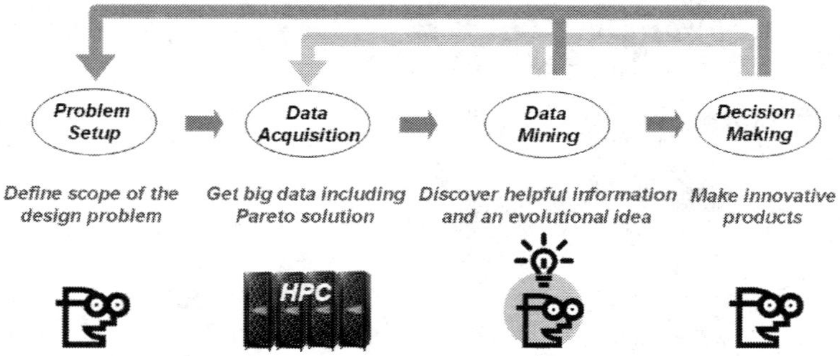

Figure 7. Procedure of multi-objective design exploration (MODE) to discover helpful information and an evolutional idea for making new productions. HPC stands for high performance computing.

and 20 MPa, respectively. And loss tangent of polymer, filler and bound rubber are 0.05, 0.0 and 0.1, respectively. The radius of filler is 10 nm and the size of RVE is 500 nm^3. Figure 9 shows a simulation model which consists of 134 million elements (512 × 512 × 512 elements) to resolve the bound rubber of the thickness of 5 nm.

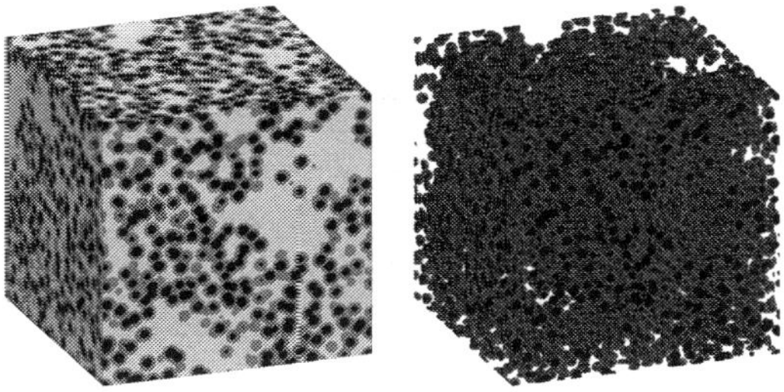

Figure 9. (a) 3D simulation model of filled rubber generated by multi-scale random model and (b) fillers embedded in the model. The number of elements is 134 million (512 × 512 × 512).

The design variables are the following parameters that can change the morphology of filler;

- Poisson point intensity of aggregate,
- Radius of aggregate,
- Volume fraction of filler,
- Overlapping distance of fillers.

The objective functions are following;

- Effective loss tangent (to be minimized),
- Effective modulus (to be maximized),
- Average Mises stress in polymer domain (to be minimized),
- Standard deviation of Mises stress in polymer domain (to be minimized).

The last two objective functions are as indicators of resistance to wear. A multi-objective genetic algorithm, NSGA-II (Deb et al. 2000) is used to get data set including Pareto solution. The population size is 100, and the generation number is 40. Therefore, the computation of 4,000 large-scale models which consist of 134 million elements is carried out in this study.

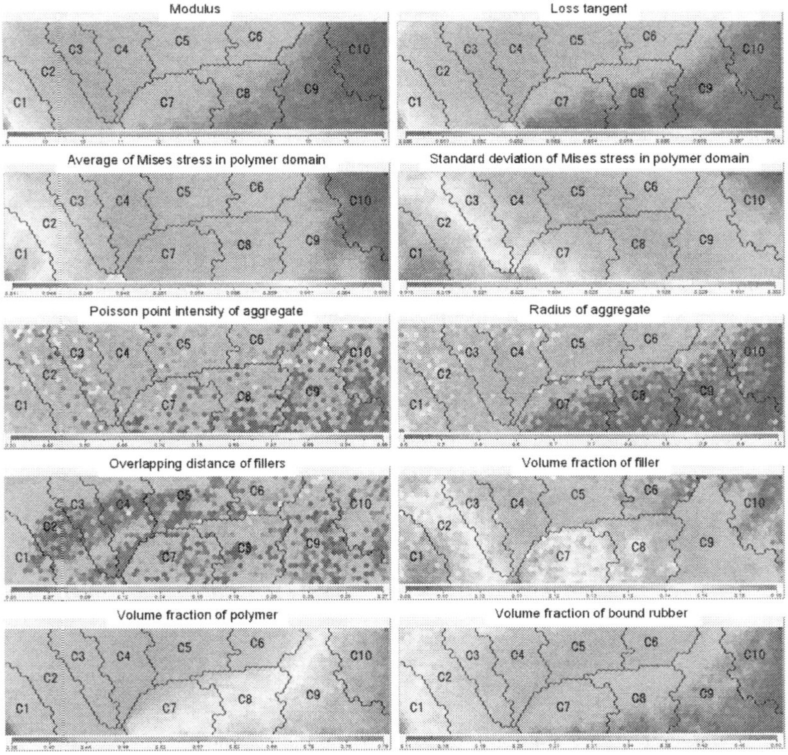

Figure 10. Contour plot of four objective functions, four design variables and two additional variables on the self-organizing map constructed using four objective functions.

3.3 *Data mining using SOM (Self-Organizing Map)*

A sophisticated visualization technique is required to show high-dimensional objective functions and design parameters simultaneously. To visualize high-dimensional data, self-organizing map (SOM) proposed by Kohonen (1995) is employed in this work. SOM is one of the neural network model based on unsupervised and competitive learning. It provides a mapping with preserving topology from the high-dimensional space to two-dimensional plane, so called map. Although SOM does not remain the direction and distance in the original high-dimensional space, nearby points in the high-dimensional space are mapped to the nearby points on SOM. Roughly speaking, a relation between high-dimensional data and SOM is similar to the relation between the earth and a world map. Nearby countries on the earth are mapped to nearby positions on a world map. SOM is useful not only for visualization of high-dimensional data but also for the cluster analysis for design problems in industry (Koishi et al. 2006 & Koishi et al. 2014).

Using NSGA-II, ten-dimensional data of about 4,000 models were obtained for data mining. Each data contains four objective functions and four design variables and two additional variables, volume fraction of polymer and that of bound rubber. Figure 10 shows the contour plot of four objective functions, four design variables and two additional variables on the same SOM generated using four-dimensional objective functions space.

Each map shown in Figure 10 is divided into ten clusters, C1-C10. The relation between each objective function can be understood visually by comparing upper four maps in this figure. It is shown that modulus (to be maximized) has a tradeoff with loss tangent (to be minimized). Cluster C10 is the best design domain on minimization of loss tangent. The desirable cluster to maximize modulus is C1. Furthermore, the figures show the correlation between average and standard deviation of Mises stress. According to visual data mining using Figure 10, we obtain the following results:

- The volume fraction of bound rubber, more than that of filler, greatly contributes to mechanical properties of filled rubbers,
- The radius of aggregate contributes to mechanical properties of filled rubbers.

4 CONCLUSIONS

For the morphological material design of filled rubbers, multi-objective design exploration based on multi-scale random model, FFT-based scheme, NSGA-II and self-organizing map was proposed in this work. The parametric models of morphology are generated using multi-scale random model. The FFT-based scheme has big advantage against FEM for large-scale viscoelastic simulation from the view points of time and memory size. For the MODE of filled rubbers, the computations of 4,000 large-scale models which consist of 134 million elements were carried out on TSUBAME, super computer at Tokyo Institute of Technology. As a result of data mining, the following conclusions are drawn. The volume fraction of bound rubber plays a major role in the material design of filled rubbers. Furthermore, the radius of aggregates contributes to mechanical properties of filled rubbers.

MODE for filled rubbers in which the rubber matrix is constituted of two distinct polymers will be studied in a next step.

ACKNOWLEDGMENT

This research used TSUBAME, supercomputer at Global Scientific Information and Computing Center in Tokyo Institute of Technology though the HPCI System Research Project (Project ID: hp160039).

REFERENCES

Akutagawa, K. & Yamaguchi, K. & Yamamoto, A. & Heguri, H. & Jinnai, H. & Shinbori, Y., 2008, Mesoscopic Mechanical Analysis of Filled Elastomer with 3D-Finite Element Analysis and Transmission Electron Microtomography, *Rubber Chemistry and Technology*, 81, 2, 182–189.

Deb, K. & Agrawal, S. & Pratab, A. & Meyarivan, T., 2000, A Fast Elitist Non-Dominated Sorting Genetic Algorithm for Multi-Objective Optimization: NSGA-II, *KanGAL report* 200001.

Figliuzzi, B. & Jeulin, D. & Faessel, M. & Willot, F. & Koishi, M. & Kowatari, N., 2016, Modeling the Microstructure and the Viscoelastic Behaviour of Carbon Black Filled Rubber Materials from 3D Simulations, *Technische Mechanik*, 36, 1–2, 32–56.

Hagita, K. & Arai, T. & Kishimoto, H. & Umesaki, N. & Shinohara, Y. & Amemiya, Y., 2007, Two-Dimensional Pattern Reverse Monte Carlo Method for Modelling the Structures of Nano-Particles in Uniaxial Elongated Rubbers, *Journal of Physics: Condensed Matter*, 19, 33, 335317.

Hagita, K. & Arai, T. & Kishimoto, H. & Umesaki, N. & Suno, H. & Shinohara, Y. & Amemiya, Y., 2008, Structural Change of Silica Particles in Elongated Rubber by Two-Dimensional Small-Angle X-Ray Scattering and Extended Reverse Monte Carlo Analysis, *Rheologica Acta*, 47, 537–541.

Jean, A. & Willot, F. & Cantournet, S. & Forest, S. & Jeulin, D., 2011, Large-Scale Computations of

Effective Elastic Properties of Rubber with Carbon Black Fillers, *International Journal for Multiscale Computational Engineering*, 9, 3, 271–303.

Kadowaki, H. & Hashimoto, G. & Okuda, H. & Higuchi, T. & Jinnai, H. & Seta, E. & Saguchi, T., 2016, Evaluation of the Appropriate Size of the Finite Element Representative Volume for Filled Rubber Composite Analysis, *Mechanical Engineering Journal*, 3, 5, 16-00372.

Kohonen, T., 1995, *Self-Organizing Map*, Springer.

Koishi, M. & Shiratori, M. & Miyoshi, T. & Kabe, K., 1997, Homogenization Method for Dynamic Viscoelastic Analysis of Composite Materials, *JSME International Journal, Series A*, 40, 3, 306–312.

Koishi, M. & Shida, Z., 2006, Multi-Objective Design Problem of Tire Wear and Visualization of Its Pareto Solutions, *Tire Science and Technology*, 34, 3, 170–194.

Koishi, M. & Miyajima, H. & Kowatari, N., 2014, Conceptual Design of Tires using Multi-Objective Design Exploration, *Proceedings of the Jointly Organized 11th World Congress on Computational Mechanics—WCCM XI, 5th European Congress on Computational Mechanics—ECCM V and 6th European Congress on Computational Fluid Dynamics—ECFD VI*, Barcelona, Spain, July 20–25, 3180–3189.

Willot, F., 2015. Fourier-based schemes for computing the mechanical response of composites with accurate local fields, *Comptes Rendus Mecanique*, 343, 3, 232–245.

Modelling of viscoelastic and hyperelastic behaviour

Constitutive Models for Rubber X – Lion & Johlitz (Eds)
© 2017 Taylor & Francis Group, London, ISBN 978-1-138-03001-5

Isolation and damping properties of rubber-buffers

D. Willenborg & M. Kröger
Institute for Machine Elements, Design and Manufacturing, University of Technology Bergakademie Freiberg, Germany

ABSTRACT: Rubber-buffers can be analysed concerning different aspects. Static and acoustic properties as well as durability are essential topics. In this case, acoustic issues will be discussed. Rubber-buffers have a simple geometry compared to suspension mounts. Rubber-buffers are analysed experimentally and by simulations. The simple geometry makes the understanding of general aspects concerning damping and isolation simpler. A harmonic force excites cylindrical and waisted rubber-buffers and accelerations are measured on a rod-structure in the front and in the back of the rubber-buffer to calculate damping and isolation properties. By this means, a frequency response analysis is done. A data analysis program calculates mechanical impedance, the amplification and transmission loss of different buffer types. An analytical model displays the frequency response with good accordance to the experiment. Rubber-buffers of cylindrical and waisted geometry are compared concerning their damping and isolation properties in dependence of the frequency. The rubber-buffer type with the best isolation and damping properties is pointed out.

1 INTRODUCTION

In many technical applications parts made of rubber damp and isolate vibrations or ensure an elastic behavior of assemblies. Different excitations are possible for the investigations of the dynamic properties of rubber-parts. Relevant for technical applications can be transient, random or harmonic excitations. In this case, a harmonic load excites a rubber-buffer in a test rig, which enables to quantify isolation and damping properties. A frequency-response analysis concerning the transmission loss and damping properties shows the characteristic properties of different buffer-types.

Several other investigators, for example Cramer, has investigated dynamic properties of rubber-parts. Cramer excites rubber-rods by a harmonic load and measures the complex modulus in dependence the frequency and temperature.

Lee et al. measure and compute the transmission behavior of rubber-bushings in marine diesel engines. A 6-degrees-of-freedom-oscillator simulates the transmission behavior and specific velocities of the bushings.

Retka describes the dynamic behavior of prestressed elastomer parts by constitutive equations and implements a material model, which is able to consider quasi-static prestress-effects.

Boulanger et al. calculate wave propagation properties by hyperelastic material models by Finite Elements.

Lohse investigates the interactions in suspension mounts and the influence on the dynamic behavior of the front wheel suspension.

Willenborg et al. excite a rubber-buffer-rod-structure by a pulse-shaped force and determine different dynamic properties.

No mentioned author has identified damping and isolation properties based on acceleration measurements. The advance of this paper is the comparison of different geometries and hardnesses regarding damping and isolation properties.

One another very important focus of this paper is the analytical description of the dynamic behavior with an efficient simulation-model. A 2-degrees-of-freedom-oscillator models the dynamic behavior of the test device with the rubber-buffers.

The paper is structured as follows. The second section describes the test rig together with the different specimens. The third section shows the measurement and data analysis procedure including the experimental results. Section four consists of the modeling approach resulting in a comparison between experimental and simulative data. Section five discusses the results and points out model limitations.

2 DIFFERENT SPECIMENS AND TEST RIG

2.1 *Different specimens*

A rubber-buffer-test-rig is used to investigate the damping and isolation properties of waisted and cylindrical rubber-buffers based on the signals of the acceleration sensors, see Figure 1. The cylindrical and waisted rubber-buffers are available in two different Shore-A-hardnesses. The soft rubber-buffers have a Shore-A-hardness of 38 and the hard rubber-buffers have Shore-A-hardness of 68. Summarized four different types of rubber-buffer types are available. All rubber-buffers are made of carbon-black-filled natural rubber.

In general, the rubber-buffers are made of two different parts. There are always the two connecting plates, which are vulcanized with the rubber-part of the buffer.

The investigation of six pieces of every buffer type quantifies the scattering of the damping and isolation properties concerning hardness deviations.

2.2 *Test rig*

The Institute of Machine Elements, Design and Manufacturing of the University of Technology Bergakademie Freiberg has built a rubber-buffer-test-rig, which enables to measure and quantify the damping and isolation properties based on acceleration measurements, see Figure 2. The excitation is a harmonic force, whose

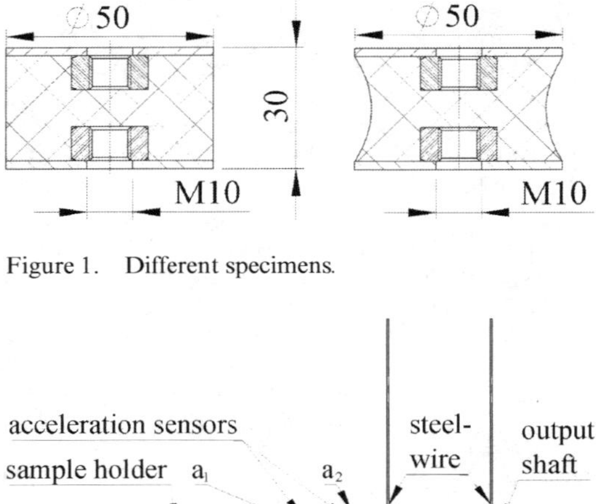

Figure 1. Different specimens.

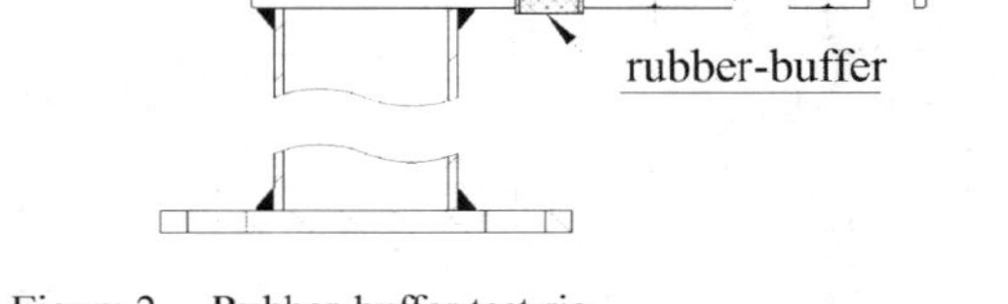

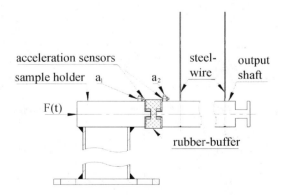

Figure 2. Rubber-buffer test rig.

frequency is varied from 40 Hz to 1000 Hz for the soft waisted rubber-buffer and 55 Hz to 1000 Hz for all other rubber-buffers. Two acceleration sensors placed in the front and in the back of the rubber-buffer to measure the system response.

The test rig consists of five parts. The first part is the sample holder, which has an external thread to mount the rubber-buffer. It is composed of three parts. There is one base plate, that is mounted to a test rig fundament. Overall, the sample holder is a welded assembly. One special aspect is the spot-welded connection between the input shaft and the rectangular steel-profile, which minimizes wave-reflections. As a result, the sample holder is much stiffer than the samples. This is important for displacement-controlled experiments, because the same displacement excitation amplitude leads to the same force response. The other way round, the same force amplitude should lead to the same acceleration response a_1. As a whole, the excitation is nearly independent from the sample stiffness. The second part is the output shaft with an external thread. The rubber-buffer is the third part, which can be mounted to the test rig by the two mentioned external threads. Rubber-buffers with an internal thread ensure a full-surface contact, which is very important for wave-propagation. A steel-wire mounts the output shaft to the surroundings and is the fourth part of the test rig. It is necessary to decouple the suspension of the steel-wire from the fundament. For that reason, the steel-wire-suspension has an elastic mounting, realised by elastomer plates. The acceleration sensors, part five, measure the system response.

3 MEASUREMENT DATA AND EXPERIMENTAL RESULTS

3.1 *Measurement data*

The measurement chain consists of three measurement: One force sensor and two acceleration sensors. All sensors are piezoelectric sensors. Furthermore, the sensors are connecting to an amplifier, whose signals are converted and analyzed by an AD-card. A measurement program records and analyzes the data with a sampling rate of 80 kHz.

As a first step, the relevant characteristic values are defined.

The dynamic behavior of a system can be defined by the amplification, which characterizes the transfer behavior in the frequency domain:

$$V(f) = \frac{a_{2\max}(f)}{a_{1\max}(f)} \tag{1}$$

where V is the amplification, a_{2max} the acceleration at the measurement point behind the rubber-buffer,

a_{1max} the acceleration signal in the front of the rubber-buffer and f the excitation frequency.

One simple way to define the isolation properties of a dynamic system is the transmission loss:

$$D_e(f) = \frac{v_{1\max}(f)}{v_{2\max}(f)} = \frac{a_{1\max}(f)}{a_{2\max}(f)} \tag{2}$$

where D_e is the transmission loss, v_{1max} the velocity at the measurement point one and v_{2max} the velocity at the measurement point two. For harmonic excitation the isolation D_e is the inverse of the amplification V.

Another characteristic acoustic value is the mechanical impedance, which is a measure for the resistance of an elastic body against mechanical load:

$$z(f) = \frac{F_{\max}(f)}{v_{2\max}(f)} \tag{3}$$

where F_{max} is the maximal value of the excitation force in dependence of the frequency.

After filtering and removing zero-offset the frequency response based on the acceleration signal can be presented. The influence of the Shore-hardness A tolerance is analyzed for soft cylindrical rubber-buffers.

Figure 3 shows the Amplification in dependence of frequency of the soft cylindrical rubber-buffer. Specimen 3 causes a little bit smaller eigenfrequency than the other specimens. All other soft cylindrical rubber-buffers despite number 3 show nearly the same eigenfrequency. Some differences concerning the damping properties are observable.

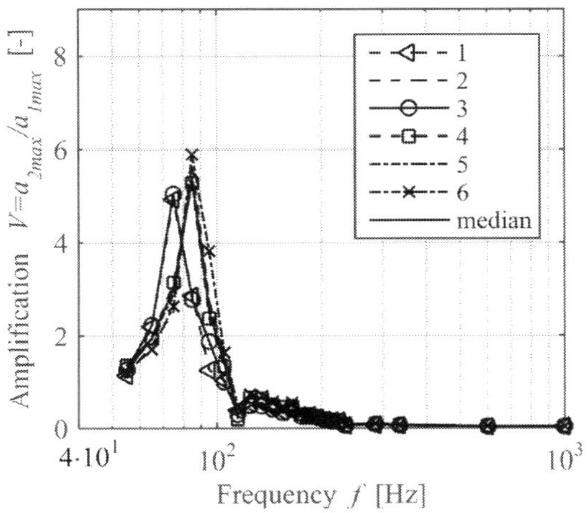

Figure 3. Amplification of the soft cylindrical rubber-buffers.

Specimen number 6 has the smallest damping properties. A possible reason for all differences is the material tolerances, which are given by +/– 5 Shore-hardness A.

All other specimens show nearly the same dynamical properties, especially concerning excitation frequencies larger than 100 Hz.

Figure 4 shows the transmission loss of the soft cyldrical rubber-buffers in dependence of frequency. Specimen 3 has the largest transmission loss up to $D_e \approx 37$. Specimen 5 reaches the smallest transmission loss up to $D_e \approx 30$. The deviation of the transmission loss caused by the tolerances is 19%.

Figure 5 presents the mechanical impedance of the soft cylindrical rubber-buffers in dependence

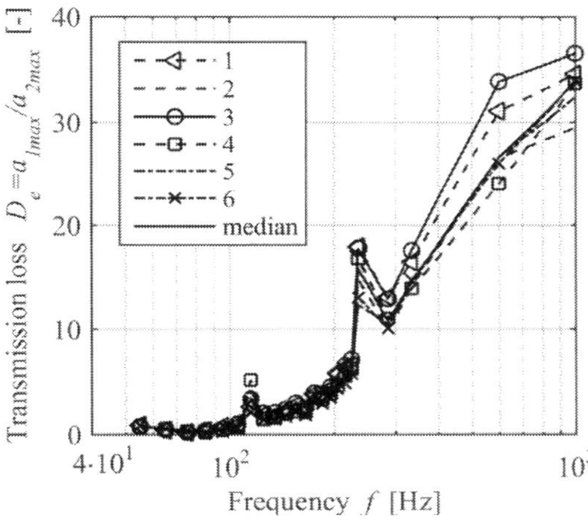

Figure 4. Transmission loss of the soft cyl. rubber-buffers.

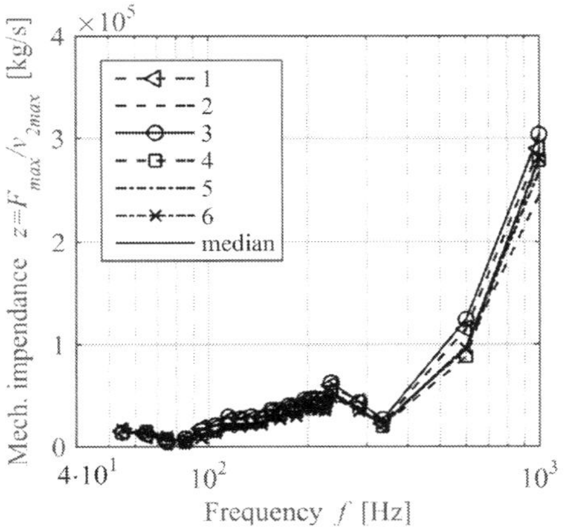

Figure 5. Mechanical impedance of the soft cyl. rubber-buffers.

of frequency. The differences of the six buffers are small. Specimen 3 has the largest mechanical impedance at 1000 Hz.

For the mounting situation in the rubber-buffer test rig a large transmission loss accompanies a large mechanical impedance as well as a small transmission loss is combined with a small mechanical impedance. The deviations of the mechanical impedances and the transmission losses are increasing with frequency, whereby the amplification shows the largest deviations concerning the resonance frequency.

For further considerations, the median values of transmission loss, amplification and mechanical impedance for every rubber-buffer type is used. The median values are considered instead of the mean values, because the median values are more robust against outliers than mean values for small numbers of samples.

3.2 *Experimental results*

The waisted hard rubber-buffers have manufacturing defects, which lead to unusable measurement data. The front surfaces have an angular deviation regarding the inner thread, that affects the wave propagation from the sample holder to the rubber-buffer. Measured dynamic properties are too large.

The comparison of the median values of the amplification in dependence of frequency is presented in Figure 6. The smallest eigenfrequency can be observed in the case of the waisted soft rubber-buffer. The hard cylindrical rubber-buffers together with the soft waisted rubber-buffers show nearly the same maximum amplification value of $V \approx 6.5$.

It can be observed, that the eigenfrequency of the hard cylindrical rubber-buffer is the largest, what is a result of the largest specimen stiffness.

The cylindrical and waisted soft rubber-buffers have the largest transmission loss with a maximum value of nearly $D_e \approx 34$, see Figure 7. Different is the frequency, at which the specimens achieve the maximal transmission loss value. The waisted soft rubber-buffers reach the maximal transmission loss value at a lower frequency than the cylindrical soft rubber-buffers.

Hard cylindrical rubber-buffers show only poor isolation properties with maximal transmission loss $D_e \approx 20$.

Over the whole investigated frequency band, the transmission loss of the waisted soft rubber-buffers rise most of all specimens. Hard cylindrical rubber-buffers show the smallest rise over the frequency concerning the transmission loss.

The mechanical impedance of the soft rubber-buffers has the largest values of the investigated specimens, see Figure 8. The sample holder is much stiffer than all of the rubber-buffers. Therefore, the same excitation force causes almost the same velocity at measurement point one.

If the stiffness of the sample holder is smaller or equal to the stiffness of the rubber-buffers, there would be a different result concerning the mechanical impedances. The velocity at measurement point one at a defined excitation force maximum would be dependent on the specimen stiffness.

For characterization purpose the same excitation force value should cause the same velocity at measurement point one for each rubber-buffer type. As a result, the mechanical impedances shown in Figure 8 are in general characteristic

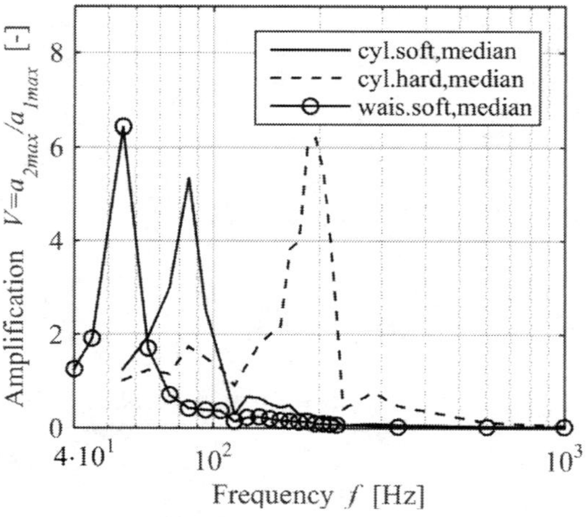

Figure 6. Comparison of the amplification of different specimens.

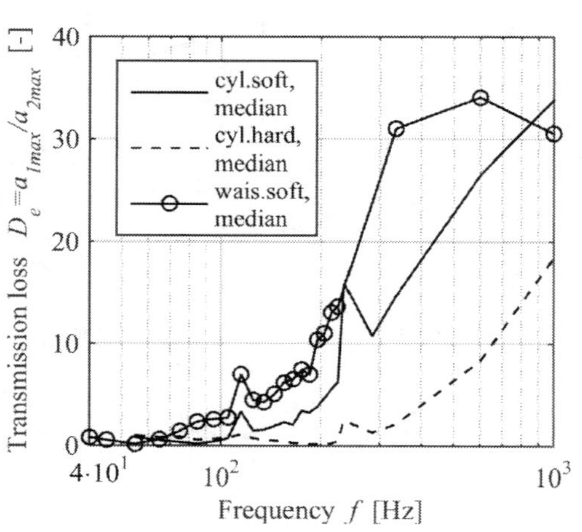

Figure 7. Comparison of the transmission loss of different specimens.

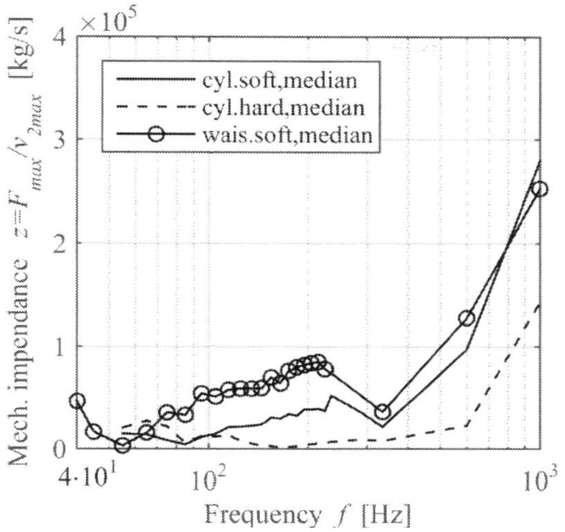

Figure 8. Comparison of the mechanical impedance of different specimens.

values of the assembly consisting of the sample holder and the rubber-buffer. For this reason, the damping and isolation influence of the rubber-buffers in the shown mounting situation can be determined clearer than in the case of a softer mounting situation. In a softer mounting situation (stiffness of the sample holder smaller or equal to the stiffness of the rubber-buffer) the rubber-buffer stiffness influences the velocity at measurement point one.

Similarities between Figure 7 and Figure 8 are the rising behavior and the maximum values. Hard cylindrical rubber-buffers show a small mechanical impedance and soft rubber-buffers show a large mechanical impedance. Over the whole frequency range waisted soft rubber-buffers show the largest mechanical impedance. For the mounting situation in the rubber-buffer test rig, it can be concluded, that large isolation values accompany large mechanical impedances.

4 MODELING APPROACH

4.1 *Model description*

The goal of the modeling approach is to compute the amplification of different rubber-buffers in dependence of frequency. Requirement is a small computation time. Therefore, a 2-degrees-of-freedom-oscillator models the transfer behavior of two selected rubber-buffers.

Equation (4) defines the discrete mechanical model, see Figure 9. The model has two degrees of freedom, because it consists of two masses with

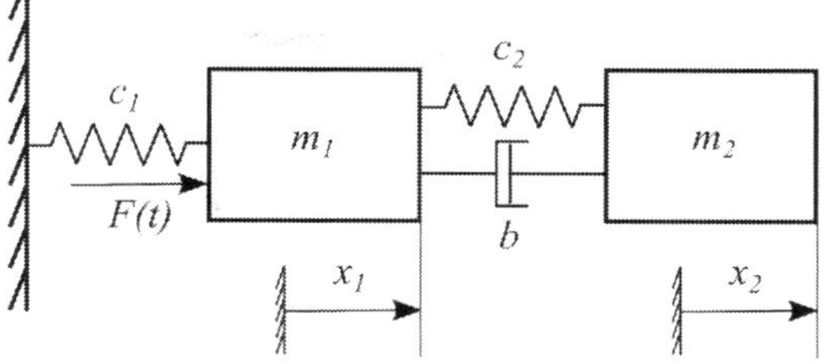

Figure 9. 2-degrees-of-freedom-oscillator.

one degree of freedom each and is connected to the surroundings by spring c_1. The first mass m_1 is connected to the surroundings by spring c_1. The coupling of m_1 and m_2 is realized by a Kelvin-Voigt-Model, which represents the mechanical properties of the rubber-buffers. The Kelvin-Voigt-Model consists of a spring c_2 and damper b. m_1 represents the mass of the sample holder with half of the mass of the rubber-buffer. c_1 reflects the stiffness of the sample holder to the shaker. m_2 is the mass of the output shaft together with half the mass of the rubber-buffer.

The differential equation system is as follows:

$$\begin{pmatrix} m_1 & 0 \\ 0 & m_2 \end{pmatrix}\begin{pmatrix} \ddot{x}_1 \\ \ddot{x}_2 \end{pmatrix} + \begin{pmatrix} b & -b \\ -b & b \end{pmatrix}\begin{pmatrix} \dot{x}_1 \\ \dot{x}_2 \end{pmatrix} + \begin{pmatrix} c_1 + c_2 & -c_2 \\ -c_2 & c_2 \end{pmatrix}\begin{pmatrix} x_1 \\ x_2 \end{pmatrix} = \begin{pmatrix} F(t) \\ 0 \end{pmatrix} \quad (4)$$

Solving the eigenvalue problem, the eigenfrequencies can be computed as follows:

$$\xi = \frac{m_1 c_2 + m_2 c_1 + m_2 c_2}{2 m_1 m_2} \quad (5)$$

$$f^2_{1/2} = \frac{1}{4\pi^2}\left(\xi \pm \sqrt{\xi^2 - \frac{c_1 c_2}{m_1 m_2}} \right) \quad (6)$$

The system parameters m_1 and m_2 are measured. The spring constant c_1 is determined by Finite-element-analysis. Spring constant c_2 is identified using equation (6). The damper constant b is determined numerically.

4.2 *Comparison of simulation and experiment*

Both simulations are carried out for a soft cylindrical rubber-buffer and a soft waisted rubber-buffer.

Figure 10 shows a good accordance between simulation and experiment of the soft cylindrical rubber-buffer. From 55 Hz to the resonance frequency the largest deviation is observable.

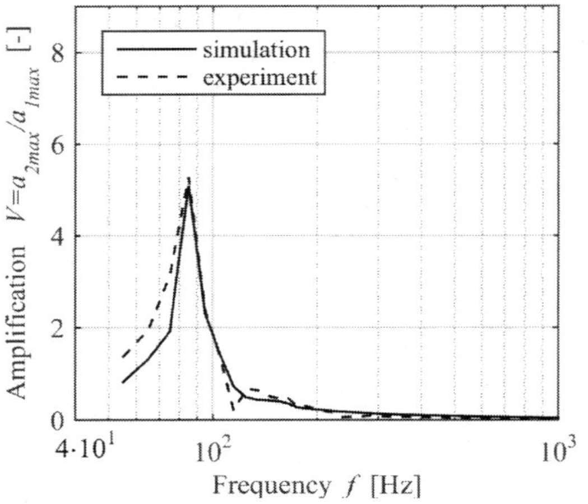

Figure 10. Soft cylindrical rubber-buffer (buffer 4).

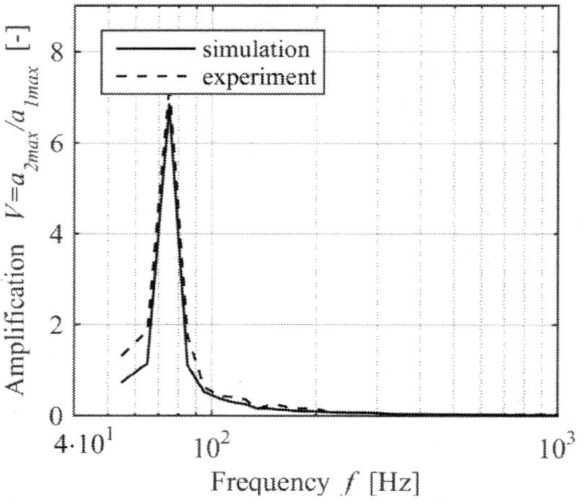

Figure 11. Soft waisted rubber-buffer (buffer 4).

Figure 11 presents the comparison of the simulation and experiments of the soft waisted rubber-buffer. The accordance of this simulation concerning the experiment is better than in the case of soft cylindrical rubber-buffer. The deviation is large from 55 Hz to 65 Hz. The rest of the frequency range can be computed with a good accordance.

Comparing simulation and experiment for both specimen types, it can be summarized, that the simple 2-degrees-of-freedom-oscillator is able to compute the amplification in dependence of frequency with a good accordance between simulation and experiment.

5 CONCLUSION AND OUTLOOK

The rubber-buffer-test-rig is able to investigate the dynamic properties of different rubber-buffer-types with an internal thread.

The influence of the material-hardness-tolerance on the dynamic properties is clarified and the rubber-buffer with the best isolation and damping properties can be pointed out.

Furthermore, a mechanical model is able to compute the amplification in dependence of excitation frequency with good accordance between experiment and simulation.

Future investigations will focus on different modeling approaches, e.g. Finite-elements and continuous models. Other excitations like impulses will be investigated and modeled with regard to the dynamic properties.

REFERENCES

Boulanger, P.; Hayes M. 1997 Wave propagation in sheared rubber, Acta Mechanica 122, pp.75–87.

Cramer, W. 1957 Propagation of Stress Waves in Rubber Rods, Journal of Polymer Science, Vol. 26, pp.57–65.

Lee, D.C.; Brennan, M.J.; Mace, B.R. 2004 Dynamic Behaviour and Transmission Characteristics of Structure-Borne Noise of Marine Diesel Engine Generator with Resilient Rubber Mounts and Elastic Foundation, ISVR Technical Memorandum No 943.

Lohse, C. 2016 Über Wechselwirkungen in Elastomerlagern und deren Einfluss auf die Elastokinematik einer Vorderradaufhängung, Dissertation, Technische Universität Bergakademie Freiberg.

Retka, J. 2012 Vibroakustisches Verhalten von viskoelastischen Strukturen unter finiter Vordeformation, Dissertation, Universität der Bundeswehr München.

Willenborg, Kröger 2016 Wave propagation and damping in rubber—steel-interfaces of suspensions, 12th Fall Rubber Colloquium, Hannover.

Constitutive Models for Rubber X – Lion & Johlitz (Eds)
© 2017 Taylor & Francis Group, London, ISBN 978-1-138-03001-5

Constitutive modelling of nonlinear viscoelastic behaviour for Poly (L-Lactic Acid) above glass transition

H.D. Wei, G.H. Menary, F. Buchanan & S.Y. Yan
Queen's University Belfast, Belfast, UK

ABSTRACT: Poly (L-Lactic Acid) (PLLA) has excellent bioresorbable properties and applied in manufacturing Bioresorbable Vascular Scaffold (BVS). In order to improve the mechanical performance of PLLA BVS, biaxial orientation via stretch blow moulding process has been introduced to present products with thinner but stronger struts. In order to better understand this deformation process above the glass transition, the mechanical behaviour of amorphous PLLA materials was obtained by biaxial stretch testing under displacement controlled conditions. The Glass-Rubber (GR) constitutive model was adopted to simulate its nonlinear viscoelastic behaviour and showed its ability for the simulation of blow forming process.

1 INTRODUCTION

Poly (L-lactic acid) (PLLA) is a kind of bioresorbable polymer that has been widely applied in manufacturing medical implants. For its excellent degradable properties, bioresorbable vascular scaffolds (BVS) based on PLLA materials have shown many advantages over the traditional metallic alloy scaffolds (Garcia-Garcia et al., 2014). However, due to the weak mechanical properties of polymers, thicker struts than the similar metallic materials for PLLA BVS are always required to provide the comparable mechanical performance and it significantly limits its applications (Ang et al., 2017).

The morphology of many polymers can be changed by introducing biaxial stretch at temperatures above glass transition and with a subsequent quenching process. For polylactide (PLA) materials, the mechanical properties can be enhanced a lot by this physical stretching process due to the effect from orientation of molecular chains and strain induced crystallization (Chapleau 2007). Benefited from this characteristics, stretch blow moulding method has been introduced during the manufacturing process of polymer BVS (Glauser & Gueriguian 2013). In this process, an extruded tube is placed in a closed mold and temperature is increased to above the glass transition. Axial stretch is conducted by drawing one or two ends of the tube and circumferential stretch is performed by pressurization with air internally.

In order to get the qualified polymer products in blow moulding process, suitable processing parameters, like temperature, stretching speed, internal pressure etc. need to be determined. A lot of trial-and-error tests are always required to achieve that, which is consumable on time and cost. This traditional method can be significantly improved by understanding the mechanical behaviour of polymer materials and simulating its deformation process by a reliable constitutive model. In stretch blow moulding method process for PET material, it shows great success in predicting the shape evolution of bottle products by using the glass-rubber (GR) model (Yang et al., 2004, Menary et al. 2010a). Similar mechanical response at lower strain rate has been found between PLLA and PET at temperatures above glass transition (Ou & Cakmak 2008). In order to achieve this for PLLA materials, its mechanical behaviour above glass transition will be studied by a biaxial stretch testing method. More concentration will be paid on more broad temperature range and higher strain rate that may happen in the stretch blow moulding process. Then the GR model is adopted and calibrated to describe its mechanical behaviour under biaxial stretch.

2 BIAXIAL STRETCH TESTING

2.1 *Material preparation and testing set-up*

PLLA pellets material was purchased from Corbion PURAC LX175L. The pellets were extruded into sheets with thickness of 0.5 mm. The temperature window for the biaxial stretch is set ranging from 70°C to 100°C. Tests were conducted by an in-house developed biaxial stretch testing machine (Menary et al. 2012b), which is shown in Figure 1. PLLA sheet samples with dimension of 75 mm × 75 mm

Figure 1. Biaxial stretch testing machine in QUB.

were fixed on the machine by four groups of grips. The samples were heated to the desired processing temperature by measuring temperatures near the sheet surfaces with two thermocouples and controlling the power ate of two heaters. Displacement of the sheet in two directions can be obtained by actuating motions of grips in two perpendicular axes. With different setting on the motions, two deformation modes, the equal biaxial stretch (EB) and constant-width stretch (CW) can be achieved. The nominal strain rate under different stretching speed resulted from the motors were set at 1, 4 and 16 s^{-1}. The testing conditions covered the critical cases that might happen in the stretch blow moulding process. Displacement and force data were recorded by the computer and used to calculate the stress-strain relations of the materials in various deformation process.

2.2 *Mechanical properties*

The mechanical properties of PLLA materials above glass transition from influence of temperature, strain rate and deformation mode are shown in Figures 2–4. In Figure 2, under EB deformation at strain rate of 1 s^{-1}, the mechanical behaviour has strong temperature dependence. Three different stages of mechanical response can be found. In the first stage before yield, an initial stiff response happens. Then stress flows after yielding point with very smooth increase with the strain. Finally, after a critical strain point, a dramatic strain hardening behaviour is encountered. There is more obvious initial stiff response and earlier onset of strain hardening at lower temperature conditions. Oppositely, the flowing stage covers bigger strain range at high temperatures. The strain rate dependence is shown in Figure 3 at temperatures of 70°C and 80°C. Under higher strain rate, stiffer material response is indicated at same strain. Mechanical behaviour at 1 s^{-1} at 70°C and 80°C for two deformation modes is shown in Figure 4. It shows that the initial stress-strain response under two modes

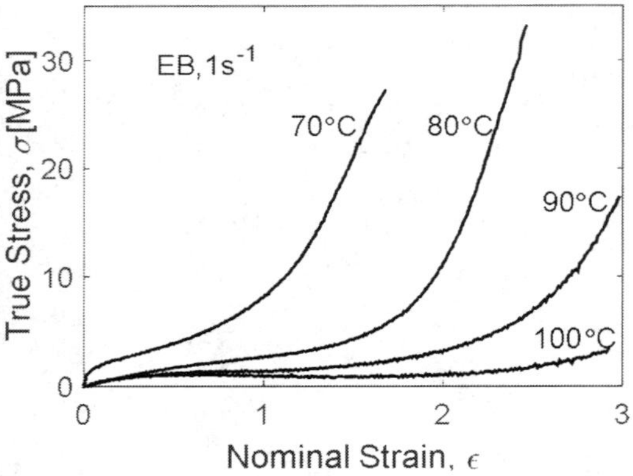

Figure 2. Influence from temperature under EB deformation.

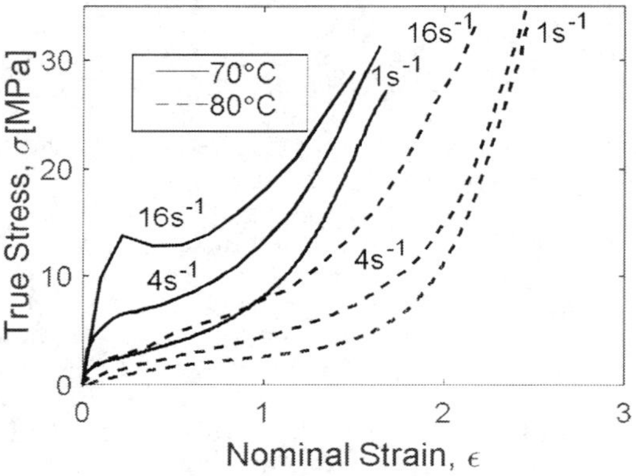

Figure 3. Influence from strain rate under EB deformation.

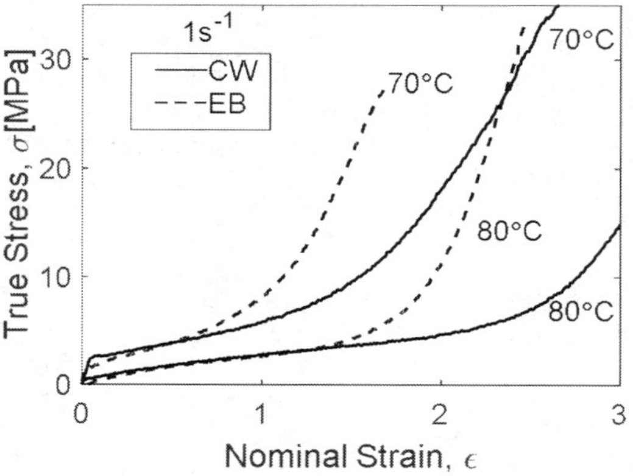

Figure 4. Influence from deformation mode under strain rate of 1 s^{-1}.

is similar at low strain. But as the strain increases, stress is much higher for the EB deformation than CW deformation at the same strain.

From those above findings, it can be summarized that there is nonlinear viscoelastic mechanical behaviour of PLLA at temperature above its glass transition, influenced by the temperature, strain rate and deformation mode. The similar mechanical response has been found in other amorphous thermoplastics, like PET. The constitutive model shall incorporate theses effects from temperature, strain rate and deformation mode. The glass-rubber (GR) model is one kind of model that use this description and successfully applied on modelling the mechanical behaivour of PET above glass transition (Buckley & Jones 1995a) and will be used to describe the mechanical properties of PLLA materials.

3 CONTITUTIVE MODEL

3.1 *Glass-rubber model*

According to the mechanical characteristics of PLLA material, it is natural to divide the stress into two parts: the initial stiff response arising from bond-stretching of molecules and hyper-elasticity from conformational stretch of molecular chains. This mechanism has been adopted by glass-rubber (GR) model to describe the viscoelastic behaviour. The total stress tensor $\boldsymbol{\sigma}$ consists of a bond-stretching stress $\boldsymbol{\sigma}^b$ and a conformational stress $\boldsymbol{\sigma}^c$ by Equation 1.

$$\boldsymbol{\sigma} = \boldsymbol{\sigma}^b + \boldsymbol{\sigma}^c \tag{1}$$

In the principal directions the stress balance for the total stress can be expressed as Equation 2 and 3 in three axes.

$$\sigma_i = s_i^b + K^b \sum_{j=1}^{3} e_j + \sigma_i^c \left(i = 1, 2, 3\right) \tag{2}$$

$$s_i^b = \sigma_i^b - \frac{1}{3}\sum_{i=1}^{3} \sigma_i^b \tag{3}$$

where, s_i^b = deviatoric stress; K^b = bulk modulus; and e_i is principal natural strain.

The bond-stretching part has Hooke elasticity and non-Newtonian viscosity that can be express as Equation 4.

$$2G^b \frac{de_i}{dt} = \frac{ds_i^b}{dt} + \frac{s_i^b}{\tau}\left(i = 1, 2, 3\right) \tag{4}$$

where G^b = shear modulus; and τ = relaxation time. The relaxation time is defined as $\tau = \mu/2G^b$, where μ = shear viscosity.

The nonlinear effect for the viscous part is incorporated into relaxation time τ by changing the initial relaxation time τ^*_0 in Equation 5, where the effects of evolution from stress ($a_{\sigma,j}$), structure ($a_{s,j}$) and temperature (a_T) are considered.

$$\tau = a_{\sigma,j} a_{s,j} a_T \tau_0^* \left(j = 1, 2, 3\right) \tag{5}$$

The conformational part has hyper-elasticity and non-Newtonian viscosity that can be expressed as Equation 6.

$$\frac{de_i}{dt} = \frac{1}{\lambda_i^n} \sum_{j=1}^{3} C_{ij}^c \frac{d\sigma_j^c}{dt} + \frac{s_i^c}{\gamma}\left(i = 1, 2, 3\right) \tag{6}$$

where λ_i^n = invariant of the network stretch; γ = conformational viscosity; and C_{ij}^c = tangent conformational compliance matrix. This matrix can be obtained from the Edwards-Vilgis model by Equation 7 and 8, where contribution from slip-links of hyper-elasticity is considered.

$$A^c = \frac{N_s k_b T}{2}\left[\frac{(1+\eta)(1-\alpha^2)}{1-\alpha^2 \sum_{i=1}^{3} \lambda_i^2} \sum_{i=1}^{3} \frac{\lambda_i^2}{1+\eta\lambda_i^2} + \sum_{i=1}^{3} \ln\left(1+\eta\lambda_i^2\right) + \ln\left(1-\alpha^2 \sum_{i=1}^{3} \lambda_i^2\right)\right] \tag{7}$$

$$\sigma_j^c = \frac{1}{\det \Lambda} \frac{\partial A^c}{\partial \ln \lambda_i} \tag{8}$$

where A^c = strain energy function; N_s = number density of slip-links; k_b = Boltzmann constant; η = freedom of movement of slip-links; α = the degree of inextensibility of chains; Λ = the volume change during the deformation process; and λ_i = the eigenvalues of left stretch tensor. More details on the GR model can be found in the references (Buckley & Jones 1995a, Buckley et al. 1996b, Adams et al. 2000).

3.2 *Numerical solution calibration*

A homogeneous deformation along the PLLA sheet samples can be achieved in the biaxial stretch testing process. This can be validated by an after-stretch PLLA film, which is shown in Figure 5. The incompressibility of PLLA materials during stretch above glass transition can be assumed. Deformation gradient F_{EB} and F_{CW} under EB and CW stretch with reference to the initial coordinate system can be expressed with the variable of stretch ratios λ in the stretching direction by Equation 9 and 10. For the biaxial stretch with constant speed, the stretch ratio at time t can be expressed

Figure 5. After-stretch PLLA film.

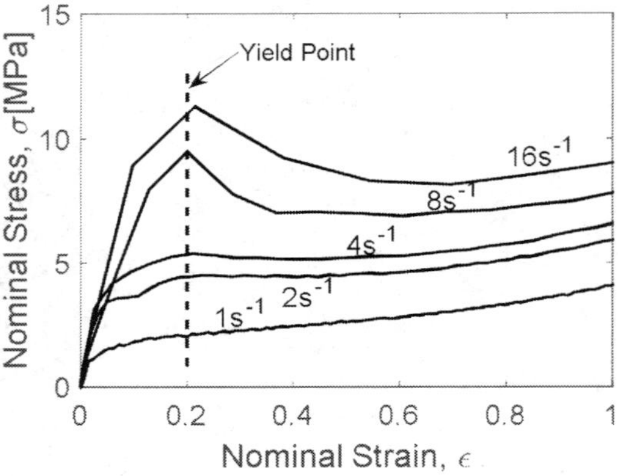

Figure 6. Yielding behaviour of EB deformation at 70°C.

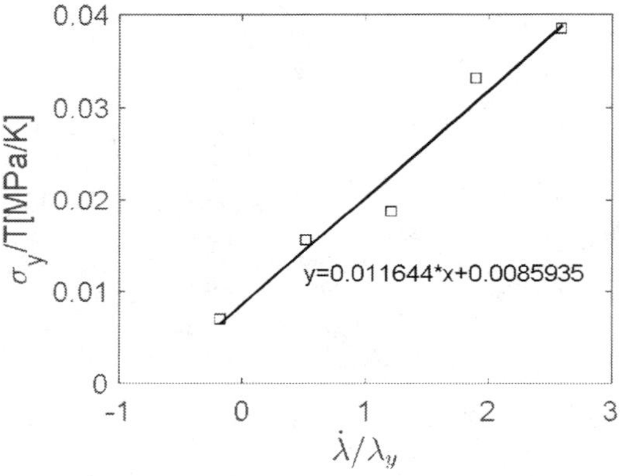

Figure 7. Relationship between yielding stress and strain rate.

as $\lambda(t) = 1 + Kt$ and K is the nominal strain rate. An explicit integration method can be adopted to obtain the stress evolution under different strain rates by solving the above equations (Li & Buckley 2009).

$$F_{EB} = \begin{bmatrix} \lambda & & \\ & \lambda & \\ & & \frac{1}{\lambda^2} \end{bmatrix} \tag{9}$$

$$F_{CW} = \begin{bmatrix} \lambda & & \\ & 1 & \\ & & \frac{1}{\lambda} \end{bmatrix} \tag{10}$$

4 MODEL CALIBRATION

4.1 *Bond-stretching stress*

As there is significant yielding behaviour at lower temperatures near the glass transition, the nominal stress-strain curve at strain rate from 1 s^{-1} to 16 s^{-1} for EB deformation at 70°C is reached and shown in Figure 6. By extracting the yield stress, a linear fitting between the ratio of yield stress to temperature and the ratio of extension rate to yield stretch can be found in Figure 7. This is the Erying process that is used to determine the strain-rate dependence of bond-stretching stress at different temperature conditions. Another assumption is made that the full relaxation of bond-stretching stress happens at the highest temperature in the flowing region. The total stress at the rubbery plateau at higher temperatures was considered as the contribution from conformational part. As the conformational stress from hyperelastic response is mainly depended on the stretch ratios, the bond-stretching stress at temperatures at lower temperatures can be obtained at the same stretch ratio to get the viscosity and its dependence by subtracting the conformational contribution.

4.2 *Conformational stress*

At lowest temperature, the conformational slippage is minimized and the conformational stress can be obtained after getting the difference between total stress and corresponding bond-stretching stress. Again, there is primary dependence on effective stretch for the EV model. The material parameters for this model can be calibrated based on the

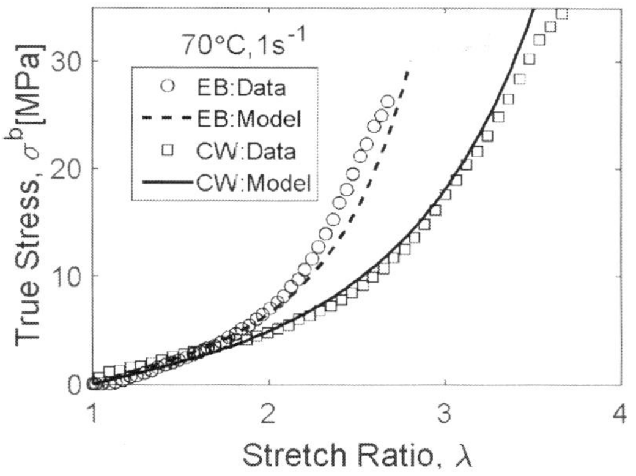

Figure 8. Conformational stress-stretch ratio relations at 70°C.

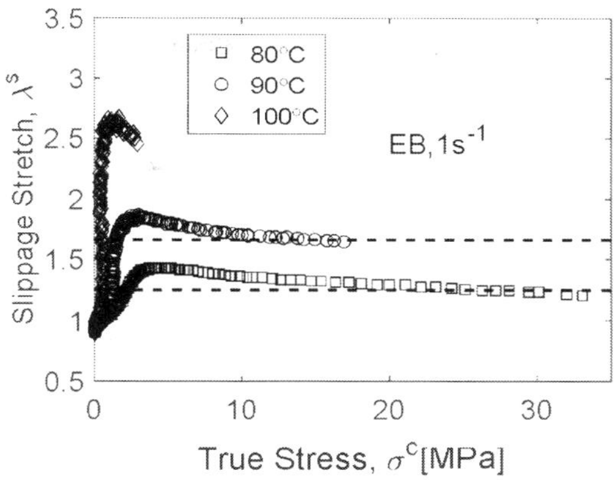

Figure 9. Slippage stretch at different temperatures.

conformational response at temperature of 70°C by least-square fitting method, which is shown in Figure 8. It indicates that both the conformation stress-stretch relations under EB and CW deformation can be fitted very well by the model. But as the temperature dependence for the EV model is very small, it is able make conclusions that the model cannot give reasonable mechanical response at other temperatures as there are very different mechanical response from the experiment.

4.3 *Conformational slippage*

The temperature and strain rate dependence of conformational stress is incorporated by considering the viscous effects from the slippage stretch. Here the EV model is fitted at 70°C so the slippage effect at this temperature is neglected. Under the same conformational stress level, by comparing the stretch at 70°C and stretch at other temperature conditions at EB deformation modes, the network stretch λ^n that contribute to the stress response and slippage stretch λ^s from viscosity can be obtained by the formation $\lambda = \lambda^n\lambda^s$. By this way, the slippage stretch at strain rate of 1 s^{-1} at higher temperatures than 70°C is shown in Figure 9. It shows that at the same stress value, slippage stretch increased with temperature. An arresting effect of slippage (dash line) after certain stress was reached. This is corresponding to the strain hardening stage of the conformational response. Therefore the relationship between values of arrested critical slippage stretch and temperatures can be built. A mathematical formulation can be introduced to show this effect by letting the slippage viscosity go infinity as the maximum slippage stretch reach the critical value. Then the slippage viscosity can be obtained by a trial-and-error process in the modified GR model.

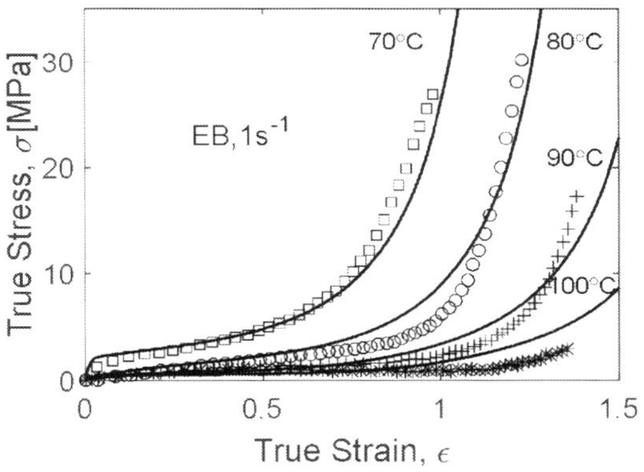

Figure 10. Results of EB deformation at 1 s^{-1}.

5 MODELLING RESULTS

The calibrated model was used to describe the constitutive behaviour in the biaxial stretch testing. The comparison between results from modelling (lines) and testing data (symbols with the same colours) is shown in Figures 10–12. A good agreement with biaxial stretch testing from numerical modelling can be found for all the conditions. Both the temperature and strain rate dependence can be well captured by the GR model indicated in Figures 10 and 12. It also shows excellent suitability in EB and CW deformation from Figure 11. At temperature of 70°C at 16 s^{-1} under EB deformation, the strain softening behaviour in the experimental study is not indicated by the modelling results because the current structural temperature for bond-stretching stress is assumed in an equilibrium state. It can be revised and incorporated into the model by an evolutional

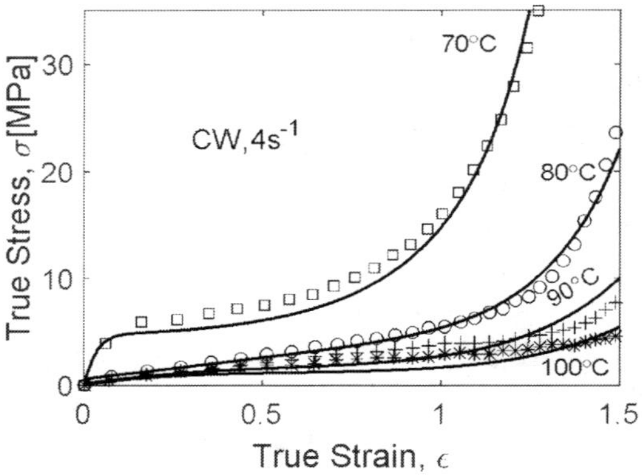

Figure 11. Results of CW deformation at 4 s^{-1}.

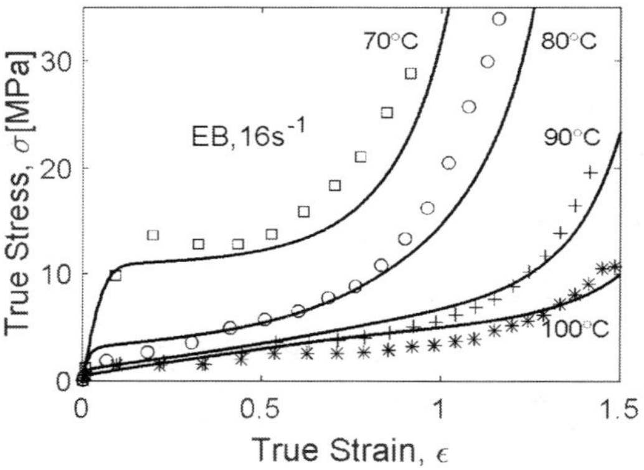

Figure 12. Results of EB deformation at 16 s^{-1}.

structural fictive temperature to introduce this effect (Buckley & Lew 2011c, Figiel et al., 2010).

6 CONCLUSIONS

The viscoelastic mechanical behaviour of amorphous PLLA materials under biaxial stretch was investigated at temperatures above glass transition. Strong temperature, strain rate and deformation mode dependence can be found. The GR model was adopted to simulate the constitutive behaviour at the different process conditions. A calibration process is introduced based on the features of testing data. A good agreement was found between the results from test and simulation by charactering the material parameters in the model. It can be concluded that the GR model can be used to describe the deformation properties for PLLA materials at temperature above glass transition and has great potential for application in stretch blow moulding process for the manufacture of PLLA BVS.

REFERENCES

Adams, A.M., Buckley, C.P., Jones, D.P., 2000. *Biaxial hot drawing of poly (ethylene terephthalate): measurements and modelling of strain-stiffening*. Polymer 41, 771–786.

Ang, H.Y., Bulluck, H., Wong, P., Venkatraman, S.S., Huang, Y., Foin, N., 2017. *Bioresorbable stents: Current and upcoming bioresorbable technologies*. Int. J. Cardiol. 228, 931–939.

Buckley, C.P., Jones, D.C., 1995. *Glass-rubber constitutive model for amorphous polymers near the glass transition*. Polymer 36, 3301–3312.

Buckley, C.P., Jones, D.C., Jones, D.P., 1996. *Hot-drawing of poly (ethylene terephthalate) under biaxial stress: application of a three-dimensional glass—rubber constitutive model*. Polymer 37, 2403–2414.

Buckley, C.P., Lew, C.Y., 2011. *New progress in the modelling of stretching flows in a stress-crystallising polymer*. AIP Conference Proceedings 1353, 720–725.

Chapleau, N., 2007. *Biaxial Orientation of Polylactide/Thermoplastic Starch Blends*. IPP XXII, 402.

Figiel, Ł., Dunne, F.P.E., Buckley, C.P., 2010. *Computational modelling of large deformations in layered-silicate/PET nanocomposites near the glass transition*. Modelling and Simulation in Materials Science and Engineering 18, 1–27.

Garcia-Garcia, H.M., Serruys, P.W., Campos, C.M., Muramatsu, T., Nakatani, S., Zhang, Y., Onuma, Y., Stone, G.W., 2014. *Assessing Bioresorbable Coronary Devices: Methods and Parameters*. JACC: Cardiovascular Imaging 7, 1130–1148.

Glauser, T., Gueriguian, V., 2013. *Controlling crystalline morphology of a bioabsorbable stent*. 12/559, 400.

Li, H.X., Buckley, C.P., 2009. *Evolution of strain localization in glassy polymers: A numerical study*. Int. J. Solids Structures 46, 1607–1623.

Menary, G.H., Tan, C.W., Armstrong, C.G., Salomeia, Y., Picard, M., Billon, N., Harkin-Jones, E.M.A., 2010. *Validating injection stretch-blow molding simulation through free blow trials*. Polymer Engineering & Science 50, 1047–1057.

Menary, G.H., Tan, C.W., Harkin-Jones, E.M.A., Armstrong, C.G., Martin, P.J., 2012. *Biaxial deformation and experimental study of PET at conditions applicable to stretch blow molding*. Polymer Engineering & Science 52, 671–688.

Ou, X., Cakmak, M., 2008. *Influence of biaxial stretching mode on the crystalline texture in polylactic acid films*. Polymer 49, 5344–5352.

Yang, Z.J., Harkin-Jones, E., Menary, G.H., Armstrong, C.G., 2004. *Coupled temperature–displacement modelling of injection stretch-blow moulding of PET bottles using Buckley model*. J. Mater. Process. Technol. 153–154, 20–27.

Constitutive Models for Rubber X – Lion & Johlitz (Eds)
© 2017 Taylor & Francis Group, London, ISBN 978-1-138-03001-5

Thermo-mechanical properties of strain-crystallizing elastomer nanocomposites

J. Plagge, T. Spratte, M. Wunde & M. Klüppel
Deutsches Institut für Kautschuktechnologie e.V., Hannover, Germany

ABSTRACT: Strain-Induced Crystallization (SIC) of unfilled and filled elastomers is investigated by mechanical and thermal analysis during stretching and retraction. The expected heating of the sample due to mechanical work done on the samples is compared to on-line measurements of the surface temperature by IR camera. SIC is quantified by taking into account that crystallization is an exothermal process leading to an additional heating besides entropy caused reversible heating and dissipative heat losses. From the measured excess temperature and the known crystallization enthalpy the degree of crystallinity in the stretched rubber sample is estimated. It is found that the crystallinity of unfilled and filled rubbers shows a pronounced hysteresis, which correlates with the mechanical hysteresis. By comparing carbon black— with silica/silane-filled NR, the influence of filler type on SIC and reinforcing properties is analyzed. A pronounced SIC is also found for carbon black filled EPDM, which can be related to the crystallization of ethylene sequences. For carbon black filled SBR composites no SIC is detected in the range of experimental errors indicating that SBR is not able to crystallize under strain. The characterization of SIC by temperature measurements is a promising technique to investigate synergetic interactions between SIC and filler reinforcement on a broader experimental scale. This will probably deliver a better understanding of the worse fatigue and wear properties of silica/silane filled NR composites in comparison to carbon black filled systems.

1 INTRODUCTION

Strain-induced crystallization (SIC) plays a major role in tire industry, since it leads to a self-reinforcement of the rubber in highly deformed regions and therefore impedes the formation of cracks, which could be caused by sharp objects. Since the development of green tire technology there is great interest in combining the unmatched wear properties of natural rubber (NR) with the energy-saving potential of replacing carbon black with silica. Unfortunately even silane coupled silica NR compounds perform far worse than comparable carbon black filled compounds in terms of wear. The reason for this is still unknown, but could be a disturbed SIC or the interaction of silane with natural ingredients (proteins, phospholipids) of the NR.

This question of high technological relevance will be addressed in the present paper. Thereby, we will refer to a new evaluation procedure for SIC based on temperature measurements at stretched samples (Spratte et al. 2017). A number of measurement techniques, such as differential scanning calorimetry (DSC), nuclear magnetic resonance spectroscopy (NMR) or X-ray diffraction (XRD) can be used for the characterization of SIC and have been analysed extensively by several authors, e.g. (Rault et al. 2006, Brünning et al. 2013).

2 EXPERIMENTAL

Three samples of natural rubber (Pale Crepe) have been prepared in an internal mixer containing different fillers (i) unfilled; (ii) 50 phr carbon black N330; (iii) 50 phr silica U7000 and 4.17 phr silane Si69. For vulcanization accelerator (1.5 phr CBS) and cross-linker (1.5 phr sulfur) was used. For comparison, an amorphous ethylene propylene diene rubber (EPDM, Keltan 4450) filled with 50 phr N339 and a styrene butadiene rubber (SBR, Buna VSL 4526) filled with 50 phr N330 were prepared. The SBR was vulcanized with 1.2 phr sulfur and 2.0 phr CBS, for EPDM 1.8 phr sulfur, 2.4 phr CBS and 1.5 phr DPG was used. Additionally, all samples contain activator (3 phr ZnO and 1 phr stearic acid) and antioxidant (1 phr IPPD).

Vulcanization curves have been recorded for each batch and the cross-linking was done by compression molding at 160°C. A sketch of the shape of the tensile test samples can be seen in Fig. 1. The dimensions are 2.5 cm width, 6.5 cm length and 0.15 cm thickness.

The heat capacities and densities of all materials have been determined by differential scanning calorimetry (DSC) and a pycnometer. The values are listed in Table 1.

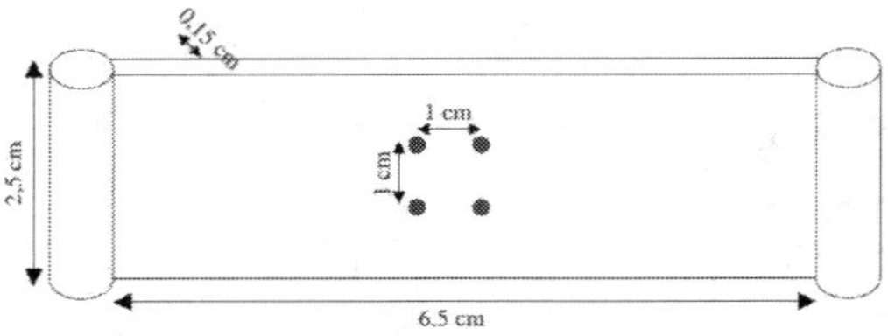

Figure 1. Geometric shape of tensile test samples.

Table 1. Heat capacities and densities of all samples.

Sample	c_p [J/gK]	ρ [g/cm³]
NR	1.78	0.93
NR/N330	1.52	1.11
NR/U7000+Si69	1.57	1.14
SBR/N330	1.49	1.09
EPDM/N339	1.52	1.06

A mechanical analysis of the different samples has been performed by standard tensile test (Zwick UPM-03). For this purpose 3 tensile specimens per batch have been extended until fracture and the stress-strain curves have been recorded. The strain was measured via an optical system consisting of two cameras tracking fixed light reflection points on the samples.

The influence of filler content on the crystallization process has been investigated by periodic stretching of all samples over four cycles from the relaxed state to 350% strained for filled and 600% strained for unfilled rubber respectively and relaxation back to 0% strain. The strain rate was set constant to 2000 mm/min. A thermal imaging camera (Vario Therm Jenoptic) was used to record the behaviour of surface temperature in dependency of the time. The camera was positioned in a distance of 1 m from the sample and operated with an emission factor of $\varepsilon = 1$. Four light reflection points were glued on the samples in a square of 1 cm × 1 cm to indicate a reference area. Consequently the average surface temperature of this area was read out with IBRIS PLUS software and plotted against time.

3 RESULTS AND DISCUSSION

In the following we will focus on the thermo-mechanical response of strain-crystallizing elastomers like NR. It consists of 99.9% cis 1,4-polyisoprene, which is a stereoregular polymer and thus can crystallize partially (Abts 2007). If an NR sample is stretched, the crystallization ability is enhanced since polymer chains align in direction of stretching. In this more ordered state entropy is decreasing and the free energy increases. It can be reduced again by the formation of micro-crystallites (Ozbas et al. 2012), which is an exothermal process. The crystallites have a fibrillar or lamellar shape (Yeh & Hong 1997) and act as physical cross-links, leading to a self-reinforcement of the rubber, which prevents crack propagation. The release of crystallization energy ΔU_{cryst} implies an increase of sample temperature ΔT_{cryst}, which can be expressed in terms of density and heat capacity of the sample:

$$\Delta U_{\text{cryst}} = \rho c_p \Delta T_{\text{cryst}}$$

The same amount of heat is consumed during melting of the crystallites under relaxation. Note that in the presence of fillers, the crystallization might be amplified, since regions of polymer between non-deformable filler particles can be exposed to higher local strains than the external strain.

Our theoretical approach is based on the assumption of pure entropy elasticity of rubbers, which describes the appearance of an entropic restoring force upon deformation. In the deformed state the polymer chains are aligned in direction of the end-to-end vector, which means there are less possible conformations than in the relaxed state, where the polymer chains form coils (Mark 2003). Moreover, for an adiabatic process, where the produced heat cannot flow into the surrounding, all work done on the sample immediately increases the temperature of the sample. This is due to the purely kinetic character of the internal energy of rubbers, while e.g. in steel mainly potential energy is stored upon stretching. Considering the first law of thermodynamics, work and internal energy can be set equal in the case of an adiabatically stretched rubber since $dQ = 0$:

$$\mathrm{d}U = c_p \rho \,\mathrm{d}T = \mathrm{d}W + \mathrm{d}Q = \mathrm{d}W \tag{1}$$

This means that all mechanical work done on a rubber sample will be stored in the internal degrees of freedom of the rubber and thus increase the temperature of the sample.

The recorded stress-strain-curves of the pure, carbon black and silica filled NR sample under repeated loading at room temperature are shown in Fig 2. Both curves show a pronounced hysteresis that is possibly related to a mix of strain-induced crystallization and filler networking effects. The first cycle shows the strongest hysteresis compared to the following ones, which is typical for the filled rubbers and can be related to stress softening as explained by the Mullins-effect. It can be explained on microscopic level by breakage of the primary filler network during the first cycle, which results in a significantly larger release of dissipated energy than in the later cycles (Klüppel 2003, Vilgis et al. 2009).

Integration of the measured stress-strain-curves delivers the energy density, whereby the area in the hysteresis loops is a measure of energy dissipation due to internal friction of the rubber (as justified by the first law of thermodynamic). Beside the dissipated heat also reversibly stored entropic energy must be considered, which can be calculated by integrating over the up-cycles. Part of it is gained back during relaxation of the samples, given by the integral over the down-cycles. The mechanic work, representing the sum of stored and dissipated energy, can be calculated by integration of the stress-strain-curves with respect to measurement time:

$$W = \int_0^t \sigma(t') \frac{d\varepsilon(t')}{dt'} dt' \quad (2)$$

Taking into account that heat is related to an increase in temperature via heat capacity and density of the material, one can calculate the temperature resulting from mechanical work T_{mech} from stress strain data using Eq. (1) and Eq. (2):

$$T_{mech}(t) = T_0 + \frac{1}{c_p \rho} \int_0^t \sigma(t') \frac{d\varepsilon(t')}{dt'} dt' \quad (3)$$

Note that T_{mech} as given by Eq. (3) contains both the irreversible dissipative heat and the reversible heat arising from entropic changes during stretching of the rubber.

In Fig. 3 the temperature calculated with Eq. (3) is compared to the temperature measured by IR for various samples, which both oscillate in time. The maxima are reached in the stretched state. The first peak, which represents the first strain cycle, shows the highest rise in temperature due to the large hysteresis of this cycle (Fig. 2). However, the mechanically calculated temperature shows a systematic deviation from the measured temperature since it increases with each cycle while the measured temperature decreases. This can be related to heat losses into the environment and the contribution of SIC, which are not considered, so far. The higher the temperature difference between sample and environment the higher the heat losses, which are caused by thermal radiation and convection.

The heat transfer between rubber and environment is, to first order, proportional to the difference between surface and environmental temperature $\Delta T = T_{IR} - T_0$. In addition, the heat transfer scales with the surface area S of the sample, which increases during uniaxial straining with the square root of the stretching ratio $\lambda = 1 + \varepsilon$:

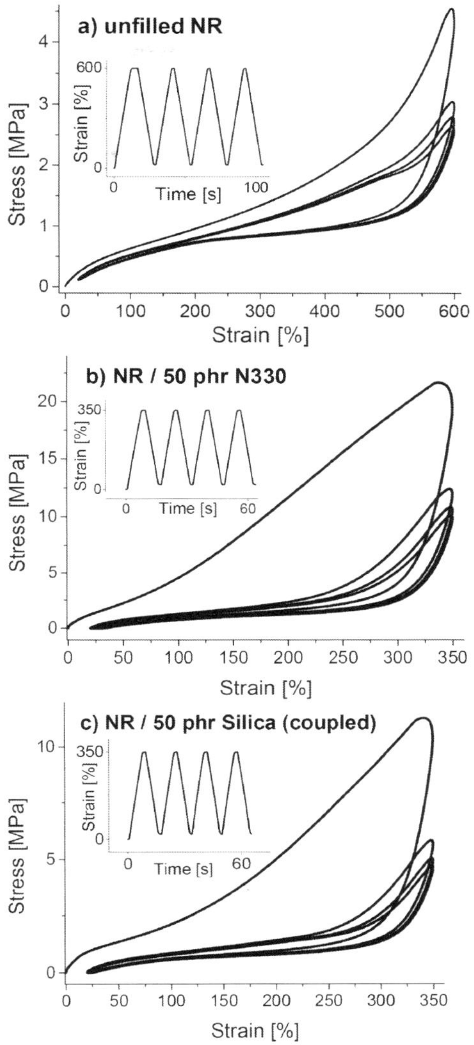

Figure 2. Stress-strain response of (a) unfilled NR, (b) 50 phr carbon black filled NR and (c) 50 phr silica/silane filled NR under cyclic loading as indicated in the inset.

$$S \propto \lambda \frac{1}{\sqrt{\lambda}} = \sqrt{\lambda}$$

Here, the front and back side of the flat sample is considered, only (Fig. 1), and constant volume of the sample during stretching is assumed, implying that the lateral size of the samples decreases $\alpha 1/\sqrt{\lambda}$. Then, the loss of temperature can be calculated as

$$T_{loss} = a \int_0^t \Delta T(t') \sqrt{\lambda(t')} dt' + b \quad (4)$$

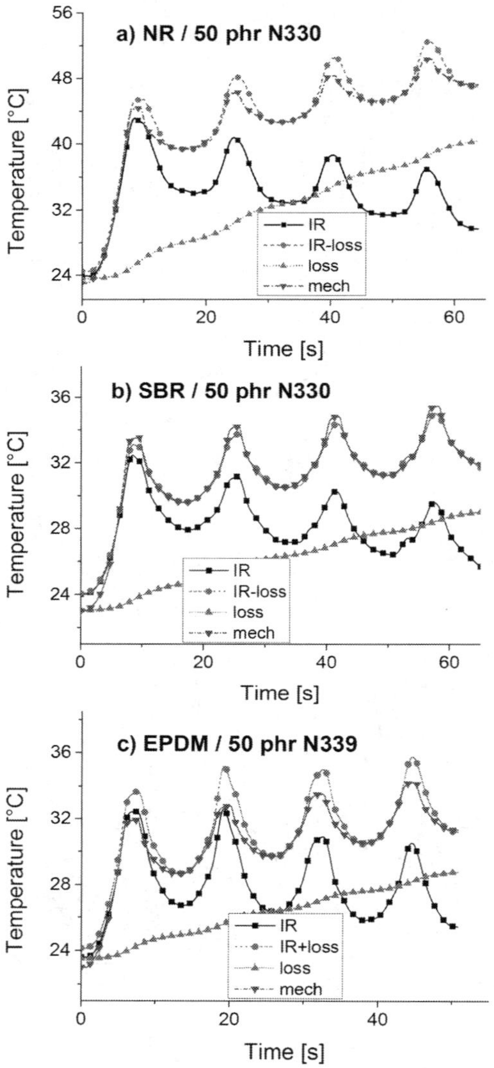

Figure 3. Measured and calculated temperatures obtained for (a) carbon black filled NR, (b) SBR and (c) EPDM; Temperatures are indicated as follows: (IR) measured by infrared camera, (mech) calculated with Eq. (3) from mechanical data given in Fig. 2, (loss) loss temperature calculated according to Eq. (4) and fitted such that the sum of IR and loss temperature (IR+loss) equals the mechanical temperature (mech) at the minima.

where a and b are a proportionality factors collecting heat-transfer constants and higher order terms, respectively.

The measured temperature can be split up into the mechanical contribution, the lost temperature and the contribution of SIC

$$T_{\text{IR}} = T_{\text{mech}} + T_{\text{cryst}} - T_{\text{loss}} \tag{5}$$

For amorphous rubber, no crystallites are present at zero elongation implying that T_{cryst} is zero at these points. So the equation can be rewritten for zero elongation:

$$T_{\text{mech}}(\varepsilon = 0) = T_{\text{IR}}(\varepsilon = 0) + T_{\text{loss}}(\varepsilon = 0)$$

The free parameters a and b are fitted such, that this equation is fulfilled.

This is visualized in Fig. 3, where the loss temperature is shown as step-like increasing line with higher slope in the stretching ranges. The results show that for the carbon black filled SBR the calculated temperature T_{mech} agrees fairly well with the corrected temperature $T_{\text{IR}} + T_{\text{loss}}$, indicating that $T_{\text{cryst}} = 0$, i.e. no crystallization takes place. In contrast, the calculated temperature T_{mech} for the carbon black filled NR and EPDM is higher than the corrected temperature $T_{\text{IR}} + T_{\text{loss}}$. The difference must be related to the heat arising from strain induced crystallization (Eq. (5)). The crystallinity K is evaluated from T_{cryst} by:

$$K = \frac{\rho c_p T_{\text{cryst}}}{(1 - \Phi_{\text{filler}})\Delta H_{\text{cryst}}} \cdot 100\%$$

Here, ΔH_{cryst} is the crystallization enthalpy of NR taken as $\Delta H_{\text{cryst}} = 61$ J/cm³ (Kim & Mandelkern 1972), ρ and c_p denote density and heat capacity of the material (Table 1) and Φ_{filler} is the filler volume fraction.

Fig. 4 shows the degree of crystallization plotted against strain for the second, third and fourth cycle. It is seen that the crystallinity shows a hysteresis similar to the mechanical data, i.e. the way to maximum elongation differs from the way back to the relaxed state. This means that the melting of crystallites happens delayed compared to the formation of crystallites. This could be related to metastable states during formation of the crystallites with lower heat of fusion in comparison to the melting enthalpy (Strobl 2007).

Unfilled and carbon black filled NR show a comparable high degree of crystallization at 600% and 350%, respectively. This indicates that the polymer chains between the non-deformable carbon black particles are exposed to locally higher strain than the external strain (strain amplification). The maximum crystallinity of the silica/silane filled sample is about 30% lower than for the carbon black filled NR. A similar reduction of about 40% is observed for the stress level of the second, third and fourth cycle (Figs. 2b and 2c). This indicates that at the same filler loading the strain amplification factor is lower for the silica/silane filled system, which can

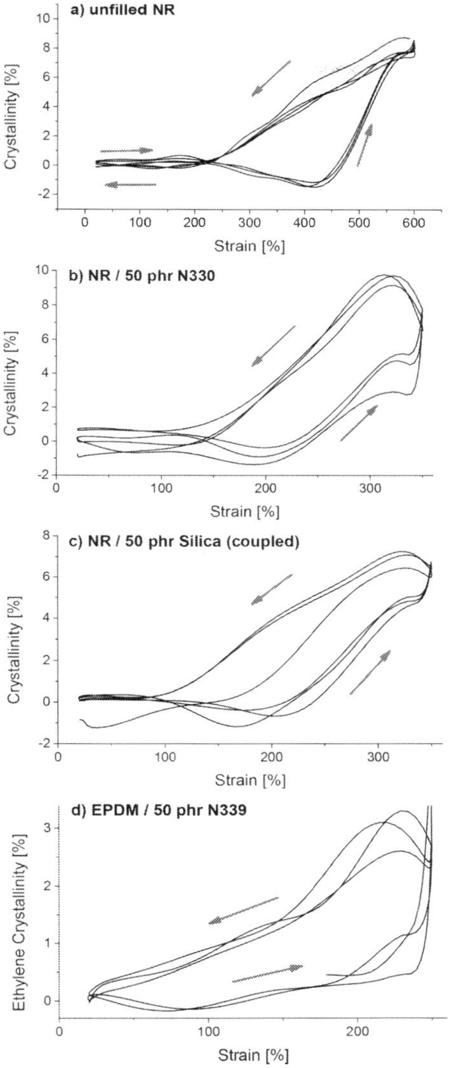

Figure 4. Degree of crystallization in dependence of strain for the second, third and fourth cycle: a) unfilled NR, b) carbon black filled NR, c) silica/silane filled NR and d) carbon black filled EPDM.

be related to less stable silica clusters compared to carbon black. This is a characteristic property of silica/silane filler systems and can be quantified by fitting the stress-strain cycles with the Dynamic Flocculation Model (comp. e.g. Fig. 10.18 in Vilgis et al. 2009). In addition, the onset of crystallization seems to be roughly the same for both filled NR samples and is located at around 200% elongation. Carbon black filled EPDM shows a weak crystallization of about 3% (with respect to the ethylene content) at 250% strain, which can be related to the stereoregular ethylene sequences of this rubber. The crystallinity was estimated by taking 52 wt.% ethylene units with crystallization enthalpy $\Delta H_{cryst} = 280$ J/cm^3 (Patki et al. 2007).

The crystallization cycles of Fig. 4 show some peculiarities that should be mentioned. In particular, the pronounced increase in crystallinity for the carbon-black and silica/silane filled samples at maximum strain may be explained by fast, filler-induced relaxation processes at the upper turning point of each cycle. At these points the testing machine has to slow down to allow the change of direction, which yields a strong decrease in stress due to relaxation at almost constant strain. It is clear that this relaxation process dissipates energy, which is not captured by the work integral Eq. (2). Hence calculated temperature stays below reality at the turning points, generating an overestimation of crystallinity. The "negative crystallinity" found for the unfilled NR between 200% and 450% strain and for the filled NR samples between 100% and 200% strain could be an effect of non-entropic elastic contributions, e.g. stretching or rotation of filler-filler bonds or long time relaxation effects.

Apart from these flaws the crystallinity measurements for the NR samples are in good quantitative agreement with data obtained by WAXS measurements of Brünning et al. (2013) but are systematically lower if compared to WAXS data of Rault et al. (2006). We note that that an evaluation of absolute values for the crystallinity is generally difficult, partly due to missing standards for the evaluation of WAXS patterns, or partly due to possible errors in the crystallization enthalpy, obtained e.g. from DSC measurements, since this requires already the degree of crystallization of the semi-crystalline samples.

4 CONCLUSIONS

It has been shown that measurements of surface temperature together with mechanical hysteresis obtained from cyclic tensile tests of unfilled and filled rubbers can be used to evaluate the degree of strain-induced crystallinity. The dissipated energy can be calculated by integration of the stress-strain-curves and set equal to the dissipated heat which is consequently converted into an increase of temperature. Considering a correction of the measured temperature due to heat losses, the difference in calculated and corrected temperature can be converted into a degree of crystallinity. A drawback of this method is the appearance of a negative degree of crystallinity for low strains, which could be related to energetic elastic contributions not captured by our model.

The results obtained from this kind of experiments show that SIC shows a pronounced hysteresis similar to the mechanical hysteresis. Furthermore, silica/silane filled NR shows less crystallization compared to carbon black filled NR indicating that the strain amplification factor is lower. Nevertheless, the onset of crystallization is roughly the same for both samples. For carbon black filled SBR no SIC is detected, while the EPDM shows a weak SIC due to the stereoregular ethylene sequences of this rubber.

REFERENCES

Abts, G. 2007. *Einführung in die Kautschuktechnologie.* Carl Hanser Verlag, München.

Brünning, K., Schneider, K. & Heinrich, G., In-situ Characterization of Rubber during Deformation and Fracture, chap. 2 in Grellmann, W. et al. (eds.) 2013. *Fracture Mechanics and Statistical Mechanics of Reinforced Elastomeric Blends.* LNACM 70, Springer-Verlag Berlin, Heidelberg.

Kim, H. G. & Mandelkern, L. 1972. Multiple Melting Transitions in Natural Rubber. *J. Polym. Sci., Part A2: Polym. Phys.* 10(6):1125.

Klüppel M. 2003. The Role of Disorder in Filler Reinforcement of Elastomers on Various Length Scales. *Adv. Polym. Sci.* 164:1.

Mark, J., et al. (eds.) 2003. *Physical Properties of Polymers.* Cambridge University Press, New York.

Ozbas, B., Toki, S., Hsiao, B.S., Chu, B, Register, R.A., Aksay, I.A. & Adamson, D.H. 2012. Strain-induced crystallization and mechanical properties of functionalized graphene sheet-filled natural rubber. J. Polym. Sci. B Polym. Phys. 50 (10):718; DOI: 10.1002/polb.23060.

Patki, R., Mezghani, K. & Philips, P.J. Krystallization Kinetics of Polymers, chap. 39 in Mark, J.E. (ed.) 2007. *Physical Properties of Polymers Handbook*, Springer Science +Busyness Media, LLC.

Rault, J., Marchal, J., Judeinstein, P. & Albouy, P.A. 2006. Stress-Induced Crystallization and Reinforcement in Filled Natural Rubbers: 2 H NMR Study. *Macromolecule.* 39 (24):8356; DOI: 10.1021/ma0608424.

Spratte, T., Plagge, J., Wunde, M. & Klüppel, M. 2017. Investigation of Strain Induced Crystallization of Carbon Black and Silica Filled Natural Rubber Composites based on Mechanical- and Temperature Measurements, *Polymer* 115:12; DOI: 10.1016/j.polymer.2017.03.019.

Strobl, G. 2007. *The Physics of Polymers*, Springer, Berlin, Heidelberg, N. Y.

Vilgis, T., Heinrich, G. & Klüppel, M. 2009. *Reinforcement of Polymer Nano-Composites: Theory, Experiments and Applications.* Cambridge University Press, Cambridge, N. Y.

Yeh, G. S. Y. & Hong, K. Z. 1979. *Strain-induced crystallization, Part III: Theory.* Polym. Eng. Sci. 19 (6):395; DOI: 10.1002/pen.760190605.

Constitutive Models for Rubber X – Lion & Johlitz (Eds)
© 2017 Taylor & Francis Group, London, ISBN 978-1-138-03001-5

Vibration isolators with stiffness nonlinearity using Maxwell-Voigt models

S. Kaul
Western Carolina University, Cullowhee, NC, USA

S. Karimi & M. Shabanisamghabady
Clemson University, Clemson, SC, USA

ABSTRACT: Maxwell elements have been commonly used for modelling elastomeric vibration isolators. Their most common form of usage is in Maxwell-Voigt (MV) models with one Maxwell element or in Maxwell-Maxwell-Voigt (MMV) models with two Maxwell elements. These models have been found to represent a vibration isolation system well in time domain and frequency domain. This paper examines the influence of nonlinear stiffness in MV and MMV models. The main aim of introducing stiffness nonlinearity is to overcome the trade-off between mitigation of resonance peaks and the reduction of transmissibility at relatively higher frequencies. The investigation of nonlinear vibration isolators is in response to an increasing demand in multiple applications where it is important to meet conflicting design requirements with several constraints while providing isolation over a large range of excitation frequencies. It is observed that one of the two models is reasonably successful in overcoming some design trade-offs for vibration isolation.

1 INTRODUCTION

The use of passive vibration isolators is common in automotive, aerospace, railroad and structural applications (De Silva, 2005; Ibrahim, 2008). Due to the inherent trade-off between reducing transmissibility at high frequencies and mitigating resonance peaks, there has been a widespread interest in investigating the design of nonlinear vibration isolators that may be able to overcome the constraints that are typically associated with linear isolators. Design features that have been investigated in the existing literature to overcome the trade-off and constraints include introduction of specific damping nonlinearities, torsion-crank linkages, etc. (Ibrahim, 2008).

Maxwell elements have been found to be suitable for modelling linear characteristics of passive elastomeric isolators, this includes the frequency response as well as time domain characteristics (Zhang and Richards, 2007; Kaul, 2015). One or two Maxwell elements are most commonly used in isolator models, these are referred to as Maxwell-Voigt (MV) or Maxwell-Maxwell-Voigt (MMV) models respectively. The use of a Maxwell element introduces an additional displacement node that in turn introduces an internal state. The introduction of this internal state has been found to significantly influence the system response (Rubin, 2013). Some generalized Maxwell models with multiple Maxwell elements have also been investigated in the literature in order to improve model characterization (Renaud et al., 2011; Lu et al., 2012). An alternative form of the MV model has also been investigated in the literature, called as the Zener model (Brennan et al., 2008), some possible advantages of this model are pointed out for harmonic excitation.

The use of auxiliary horizontal stiffness and cubic damping has been investigated in the existing literature with positive results. It has been reported that some of the conflicting design requirements can be satisfied with a specific introduction of nonlinearities (Ho et al., 2014). Other nonlinear models have also been investigated with such features as auxiliary horizontal damping (Tang and Brennan, 2012), or models with fractional derivatives (Sjoberg and Kari, 2002). The influence of horizontal damping has been found to be similar to cubic damping for the purposes of force transmissibility (Tang and Brennan, 2012), and fractional derivatives have been reported to be successful in representing some nonlinear attributes of elastomeric isolation systems (Sjoberg and Kari, 2002). Cubic damping has also been investigated by using alternative techniques (Peng et al., 2012). Other forms of nonlinearities such as piecewise behavior have also been investigated (Narimani et al., 2004; Kaul, 2012) in order to balance the need for vibration isolation with the need for overall motion control of an isolated system. This is particularly important in applications where the space envelope around the isolated system is limited due to design constraints and packaging requirements.

Overall, it can be stated that the investigation of nonlinear vibration isolators is in response to an ever increasing demand in multiple applications where it is important to meet conflicting design requirements with several constraints while providing isolation over a large range of excitation frequencies (Yarmohamadi and Berbyuk, 2011). This paper discusses the results from two design configurations of nonlinear vibration isolators that incorporate the commonly-used Maxwell elements in conjunction with stiffness nonlinearities.

2 MODELS

This section presents the governing models for vibration isolators that incorporate stiffness nonlinearities while using the MV and MMV models. The frequency response and the transmitted force for each model are derived from the equations of motion and the viscoelastic stress-strain relationship.

The governing behavior of the MV model, shown in Figure 1, can be represented by the viscoelastic stress-strain model for an elastomeric isolator that is under shear as follows:

$$\tau + D_1 \frac{d\tau}{dt} = G_0\gamma + G_1 \frac{d\gamma}{dt} + G_2 \frac{d^2\gamma}{dt^2} \tag{1}$$

In Equation (1), τ is the shear stress and γ is the shear strain, and D_1, G_0, G_1, G_2 are positive coefficients. For the MV model shown in Figure 1, the corresponding complex frequency response can be derived from the stress-strain model in Equation (1) as:

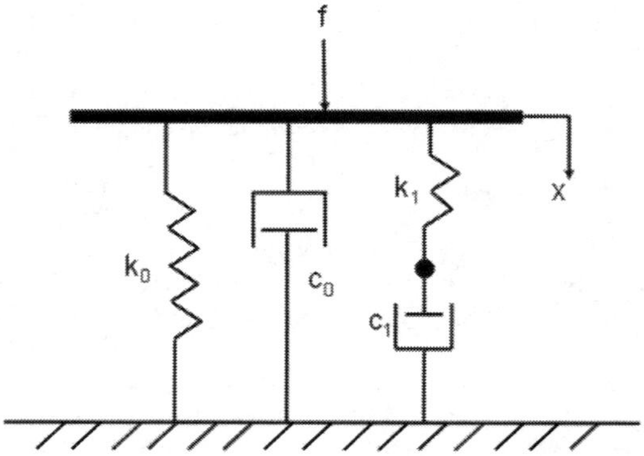

Figure 1. MV Model.

$$\frac{X(\omega)}{F(\omega)} = \frac{h}{2A} \frac{1 + i\omega D_1}{G_0 - \omega^2 G_2 + i\omega G_1} \tag{2}$$

In Equation (2), h and A are the thickness and cross-sectional area of the elastomer for a sandwich isolator design where the elastomer is in shear, and $\tau = \frac{f}{2A}$, $\gamma = \frac{x}{h}$, where f is the external force and x is the displacement. The coefficients for the MV model of such an elastomeric isolator are derived in terms of the stiffness and damping constants, shown in the model in Figure 1, as follows:

$$\begin{aligned} D_1 &= \frac{c_1}{k_1} \\ G_0 &= k_0 \frac{h}{2A} \\ G_1 &= \frac{k_0 c_1 + k_1 c_0 + k_1 c_1}{k_1} \frac{h}{2A} \\ G_2 &= \frac{c_0 c_1}{k_1} \frac{h}{2A} \end{aligned} \tag{3}$$

In Equation (3), k_0, k_1, c_o and c_1 are the stiffness and damping constants, as seen in Figure 1 and are directly related to the coefficients from the stress-strain behaviour. For a MV element supporting a lumped mass (m), the frequency response in Equation (2) can be modified as:

$$\frac{X(\omega)}{F(\omega)} = \frac{1}{-m\omega^2 + i\omega c_0 + k_0 + k_1 - \frac{k_1^2}{k_1 + i\omega c_1}} \tag{4}$$

Equation (4) is identical to Equation (2) for $m = 0$ after substituting the coefficients from Equation (3). It can be seen that the frequency response in Equation (4) can be directly derived from the governing equations of motion of the MV element supporting a one degree-of-freedom lumped mass. For the MV model, the output force transmitted to the base can be expressed as follows:

$$\frac{F_{out}(\omega)}{X(\omega)} = k_0 + i\omega c_0 + \frac{i\omega c_1 k_1}{k_1 + i\omega c_1} \tag{5}$$

In Equation (5), F_{out} is the force transmitted through the vibration isolator to the base or ground, for the model shown in Figure 1. It may be noted that zero initial conditions have been assumed for all the derivations.

The governing behaviour of the MMV model, shown in Figure 2, can be expressed in the form of the stress-strain model as:

$$\tau + D_1 \frac{d\tau}{dt} + D_2 \frac{d^2\tau}{dt^2} = G_0\gamma + G_1 \frac{d\gamma}{dt} + G_2 \frac{d^2\gamma}{dt^2} + G_3 \frac{d^3\gamma}{dt^3} \tag{6}$$

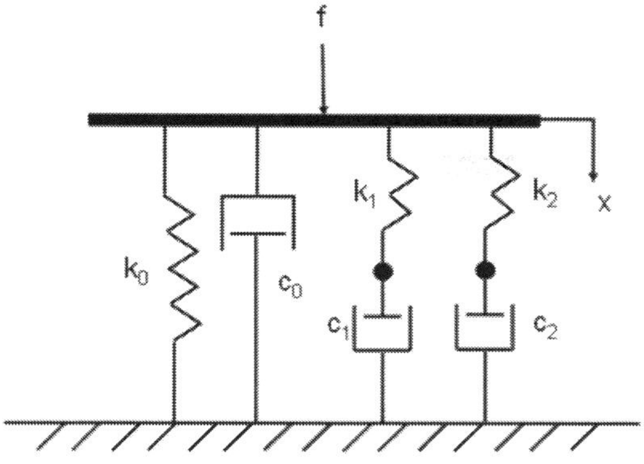

Figure 2. MMV Model.

In Equation (6), D_1, D_2, G_0, G_1, G_2, G_3 are positive coefficients associated with the MMV model.

For the MMV model shown in Figure 2, the complex frequency response can be derived from the viscoelastic stress-strain model in Equation (6) as follows:

$$\frac{X(\omega)}{F(\omega)} = \frac{h}{2A} \frac{1 - \omega^2 D_2 + i\omega D_1}{G_0 - \omega^2 G_2 + i\omega G_1 - i\omega^3 G_3} \tag{7}$$

In Equation (7), h and A are the thickness and cross-sectional area of the elastomer and the coefficients are analogous to the MV model. The coefficients for the MMV model of such an elastomeric isolator are derived and listed below:

$$D_1 = \frac{c_2}{k_2} + \frac{c_1}{k_1}$$

$$D_2 = \frac{c_1 c_2}{k_1 k_2}$$

$$G_0 = k_0 \frac{h}{2A}$$

$$G_1 = \frac{k_0 k_2 c_1 + k_1 k_2 c_0 + k_1 k_2 c_1 + k_0 k_1 c_2 + k_1 k_2 c_2}{k_1 k_2} \frac{h}{2A}$$

$$G_2 = \frac{k_2 c_0 c_1 + k_0 c_1 c_2 + k_1 c_0 c_2 + k_1 c_1 c_2 + k_2 c_1 c_2}{k_1 k_2} \frac{h}{2A}$$

$$G_3 = \frac{c_0 c_1 c_2}{k_1 k_2} \frac{h}{2A} \tag{8}$$

In Equation (8), k_0, k_1, k_2, c_o, c_1 and c_2 are the stiffness and damping constants, as seen in Figure 2. Substituting the coefficients from Equation (8) in Equation (7) yields the alternate form of the complex response for the MMV model:

$$\frac{X(\omega)}{F(\omega)} = \frac{1}{-m\omega^2 + i\omega c_0 + k_0 + k_1 + k_2 - \frac{k_1^2}{k_1 + i\omega c_1} - \frac{k_2^2}{k_2 + i\omega c_2}} \tag{9}$$

Equation (9) is derived directly from the governing equation of the motion of a single degree-of-freedom system supported by a MMV element. The force transmitted to the base is used to compute force transmissibility, and is derived for the MMV model as:

$$\frac{F_{out}(\omega)}{X(\omega)} = k_0 + i\omega c_0 + \frac{i\omega c_1 k_1}{k_1 + i\omega c_1} + \frac{i\omega c_2 k_2}{k_2 + i\omega c_2} \tag{10}$$

In Equation (10), the ratio between the transmitted force, F_{out}, and the displacement is derived from the governing equations of motion of the MMV model.

The MV and MMV models discussed thus far do not exhibit any stiffness or damping nonlinearity. The MV model with horizontal stiffness is shown in Figure 3. The equations of motion (EOM) for a single degree-of-freedom system (with mass m) supported by the MV element shown in Figure 3 are derived with the assumption of small displacements as follows:

$$m\ddot{x} + k_0 x + c_0 \dot{x} + k_1(x - x_1) + 2k_h\left(1 - \frac{l_0}{\sqrt{l^2 + x^2}}\right)x = f \tag{11}$$

$$k_1(x - x_1) = c_1 \dot{x}_1 \tag{12}$$

In Equation (11) and Equation (12), l_0 is the free length of the horizontal spring (with a stiffness constant of k_h), l is the compressed length of the spring in its horizontal position, and x_1 is the displacement at the node of the Maxwell element. The EOM for stiffness nonlinearity are derived by

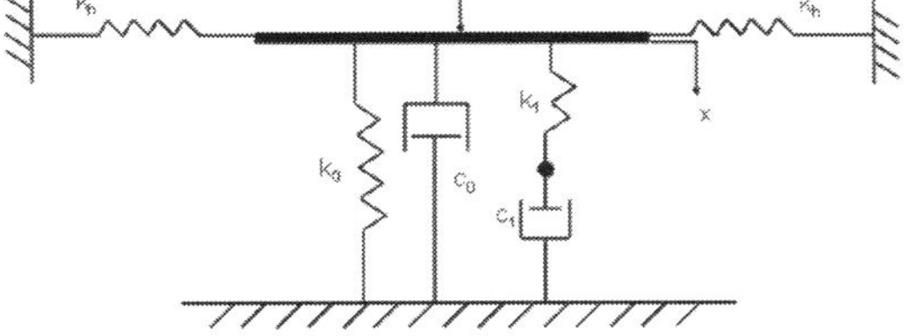

Figure 3. MV Model with stiffness nonlinearity.

assuming small displacements of the isolated mass. The output force transmitted to the base for the MV model with the horizontal spring is:

$$F_{out} = k_0 x + c_0 \dot{x} + k_1 (x - x_1) + 2k_h \left(1 - \frac{l_0}{\sqrt{l^2 + x^2}}\right) x \tag{13}$$

The MMV model with an additional horizontal stiffness is shown in Figure 4. This model is a slight modification of the model in Figure 3, simply adding a Maxwell element to the MV model while retaining the horizontal stiffness. Assuming small displacements, the EOM for the model in Figure 4 supporting a single degree-of-freedom lumped mass are as follows:

$$m\ddot{x} + k_0 x + c_0 \dot{x} + k_1 (x - x_1) + k_2 (x - x_2) + 2k_h \left(1 - \frac{l_0}{\sqrt{l^2 + x^2}}\right) x = f \tag{14}$$

$$k_1 (x - x_1) = c_1 \dot{x}_1 \tag{15}$$

$$k_2 (x - x_2) = c_2 \dot{x}_2 \tag{16}$$

In Equations (14), (15) and (16), l is the pre-compressed length of the horizontal spring, and x_1 and x_2 are the displacements at the nodes associated with the two Maxwell elements. The output force transmitted to the base through the isolator can be calculated as:

$$\begin{aligned} F_{out} &= k_0 x + c_0 \dot{x} + c_1 \dot{x}_1 + c_2 \dot{x}_2 + 2k_h \left(1 - \frac{l_0}{\sqrt{l^2 + x^2}}\right) x \\ &= k_0 x + c_0 \dot{x} + k_1 (x - x_1) + k_2 (x - x_2) \\ &\quad + 2k_h \left(1 - \frac{l_0}{\sqrt{l^2 + x^2}}\right) x \end{aligned} \tag{17}$$

The experimental setup used to characterise the models discussed in this section is briefly presented in Section 3.

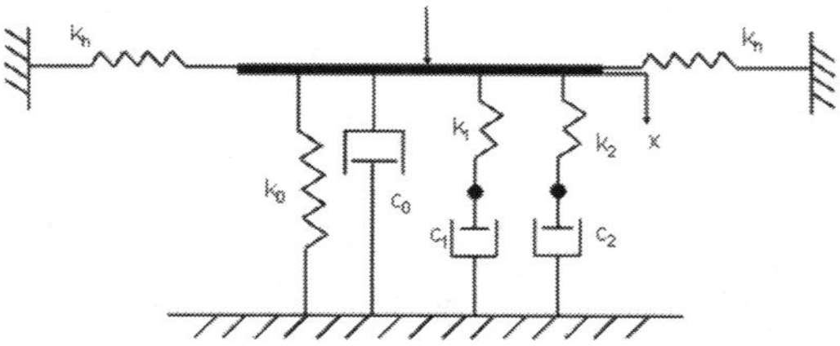

Figure 4. MMV Model with stiffness nonlinearity.

3 EXPERIMENTAL SETUP

The experimental setup used to characterise the elastomeric isolators discussed in Section 2 is briefly presented in this section. The variables associated with the MV and MMV models are identified by using a commercially available vibration isolator with a double shear plate design and the elastomer under shear. The variables are determined by solving an optimisation problem that minimizes the difference between the measured transmitted force and the predicted transmitted force. The optimisation toolbox in MATLAB® (MATLAB User Guide, 2013) is used to solve an unconstrained optimisation problem, and multiple runs of the optimisation problem are performed with varying starting guesses. It may be noted that the time history of the measured response has been used for optimization, while some studies in the literature have instead used the frequency response to identify the variables of the elastomeric isolator (Zhang and Richards, 2007).

The isolator used for testing was pre-loaded radially (in shear) with a force of 460 N. An axial pre-load corresponding to the assembly load was also used. The pre-loading condition represents the state of static equilibrium. The isolator had a circular cross-section with the elastomer (neoprene) under shear. A servo-hydraulic actuator was used to excite the isolator under displacement control. The reaction force that was transmitted to the frame was measured by a load cell at a sampling frequency of 500 Hz. The measured transmitted force was used in the optimisation problem. Data collection was initiated 10 minutes after the start of testing in order to allow the elastomeric material to settle and reach isothermal conditions.

The variables computed for the MV model shown in Figure 1 are as follows: k_0 = 251.26 N/mm, c_0 = 3.23 N-s/mm, k_1 = 237.38 N/mm, c_1 = 121.921 N-s/mm. The variables computed for the MMV model shown in Figure 2 are as follows: k_0 = 251.26 N/mm, c_0 = 3.23 N-s/mm, k_1 = 237.38 N/mm, c_1 = 121.92 N-s/mm, k_2 = 180.59 N/mm, c_2 = 1.89 N-s/mm. The variables associated with the nonlinear models in Figure 3 and Figure 4 were not calculated directly. Instead, the influence of variables such as horizontal stiffness is analyzed individually, these results will be discussed in Section 4. For the isolator used in this study, the displacement is limited between 4.5 mm and −5.25 mm about the position of static equilibrium to prevent the engagement of the displacement-limiting mechanism. The thickness (h) of the isolator used in this study is 24 mm and the cross-sectional area (A) is 3269.1 mm^2. Simulation results for the models with stiffness nonlinearity presented in Section 2 are presented in Section 4 by using the variables identified in this section.

4 RESULTS

The models presented in Section 2 are used in this section to simulate the response of the models. The MV model with stiffness nonlinearity is analysed first. For this model, two parameters associated with the horizontal spring are investigated—horizontal stiffness and the length of the horizontal spring. The results from investigating these two parameters are shown in Figure 5 and Figure 6. A mass of 125 kg has been used for all the simulations along with the variables identified in Section 3.

For low pre-compression of the horizontal spring, horizontal stiffness does not exhibit any influence on frequency response—either at the resonant peak or at the higher frequencies. This behaviour is exhibited regardless of the magnitude of horizontal stiffness, as seen in Figure 6. For relatively higher pre-compression of the horizontal spring (around 50%), as the horizontal stiffness increases, the resonant peak increases but the peak frequency reduces. The response is not seen to differ much after the peak frequency. This behavior can be seen from Figure 5.

The trends of increasing pre-compression remain the same regardless of the horizontal stiffness, although with slightly varying magnitudes. Increasing pre-compression results in an increase in the resonant peak of the response and a slight shift (reduction) in the peak frequency. This can be seen from the response in Figure 7 and Figure 8.

In all the simulation results discussed thus far, it can be seen that the two parameters of the horizontal spring can be used to directly shift the peak frequency and mitigate the peak response without affecting the transmissibility at relatively higher frequencies. This is an advantage over a linear isolator where a decrease in the resonant peak results in a corresponding increase in transmissibility at higher frequencies. The attribute exhibited by the

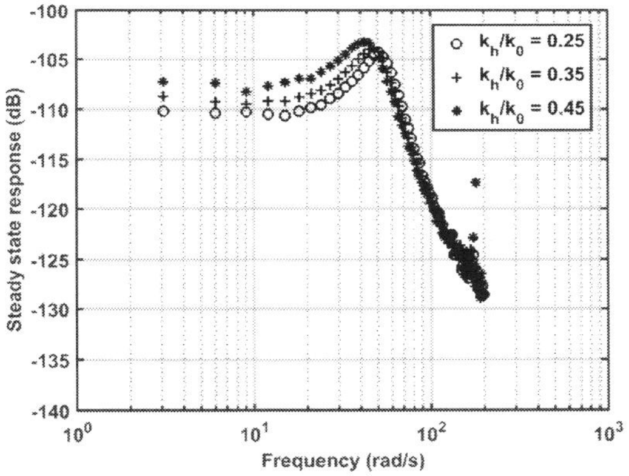

Figure 5. Frequency response – $l = 0.5\ l_0$ – varying horizontal stiffness—MV model.

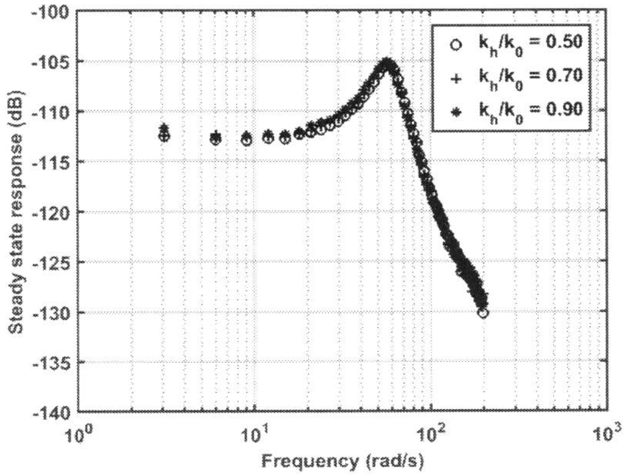

Figure 6. Frequency response – $l = 0.9\ l_0$ – varying horizontal stiffness—MV model.

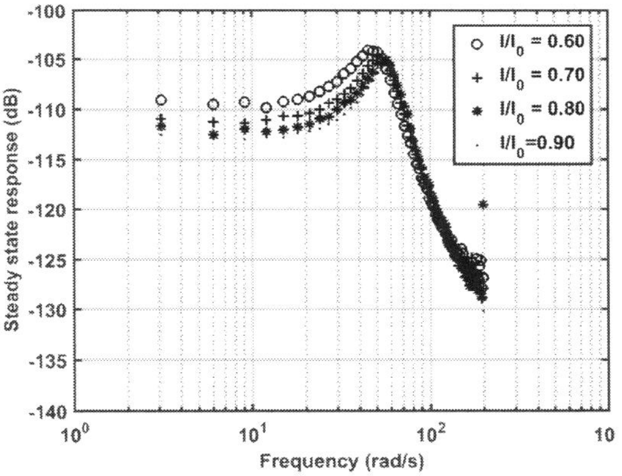

Figure 7. Frequency response – $k_h = 0.5\ k_0$ – varying pre-compression of horizontal spring—MV model.

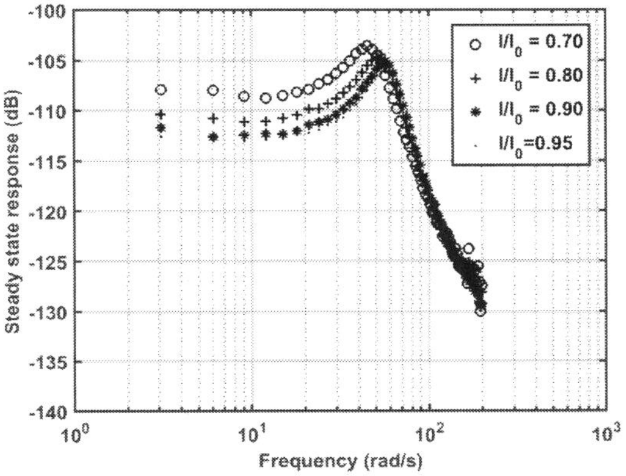

Figure 8. Frequency response – $k_h = 0.9\ k_0$ – varying pre-compression of horizontal spring—MV model.

nonlinear isolator with a horizontal stiffness provides design flexibility that can be used to control the characteristics of an isolation system without the trade-off associated with a linear isolator.

The simulation results of the MMV model with stiffness nonlinearity show similar trends to the MV model. However, there are some distinctions between the two models. The results in Figure 9 and Figure 10 indicate that a significant pre-compression of the horizontal spring is necessary in order for the horizontal stiffness to have an influence on the frequency response.

The direct influence of varying pre-compression of the horizontal spring on the frequency response of the MMV model is demonstrated in Figure 11.

Overall, the MMV model exhibits a lower amplitude of the frequency response for all levels of horizontal stiffness and pre-compression. However, the influence of the stiffness nonlinearity is somewhat limited in the MMV model. This can be observed by directly comparing the response from Figure 5 (MV model) to Figure 9 (MMV model), and also comparing the response from Figure 8 (MV model) to Figure 11 (MMV model). There are no apparent advantages to using the MMV model for the design with stiffness nonlinearity that is incorporated in the form of a horizontal stiffness.

The force transmitted to the frame for the MV model is evaluated in Figure 12 and Figure 13 for a relatively low pre-compression of the horizontal spring (30%).

It may be noted that Figure 13 is just a zoomed in plot for the transmitted force in Figure 12. Increasing stiffness of the horizontal spring is seen to result in a slight reduction in the peak transmitted force as well as a limited reduction in the transmitted force at all higher frequencies.

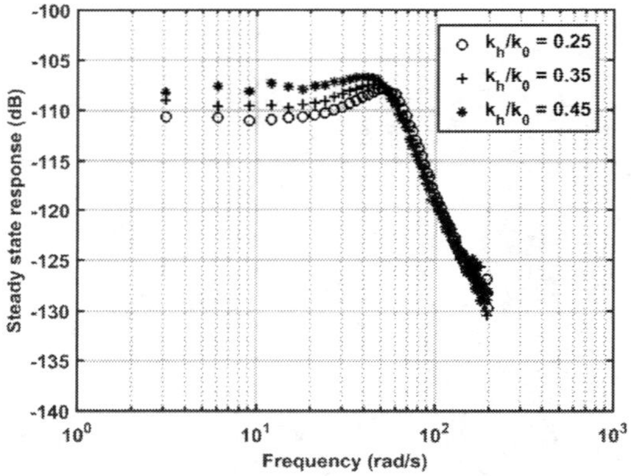

Figure 9. Frequency response – $l = 0.5\,l_0$ – varying horizontal stiffness—MMV model.

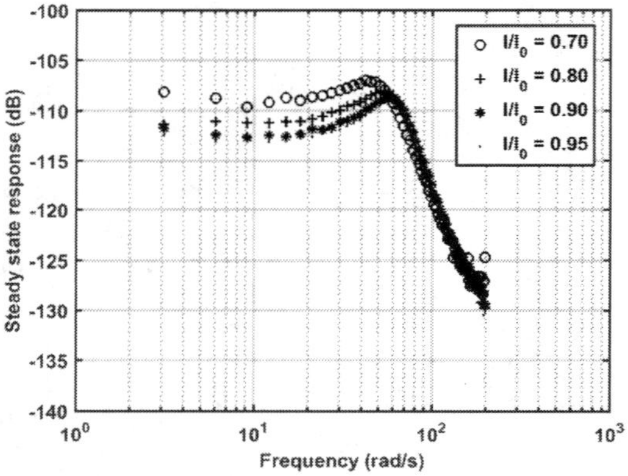

Figure 11. Frequency response – $k_h = 0.9\,k_0$ – varying pre-compression of horizontal spring—MMV model.

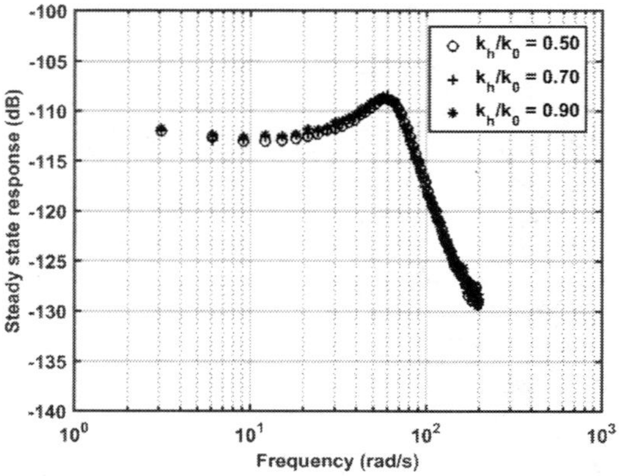

Figure 10. Frequency response – $l = 0.9\,l_0$ – varying horizontal stiffness—MMV model.

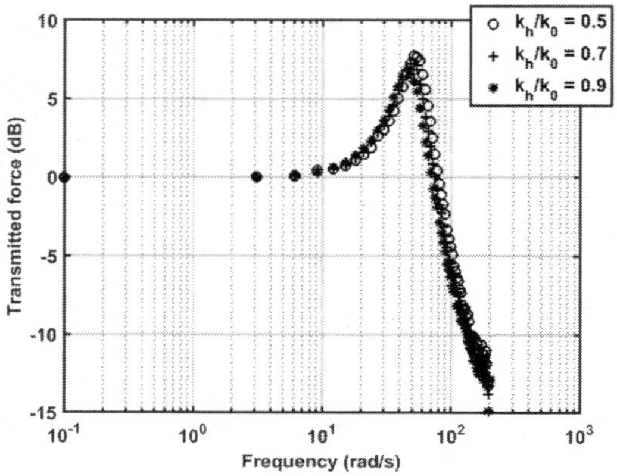

Figure 12. Transmitted force – $l = 0.7\,l_0$ – varying horizontal stiffness—MV model.

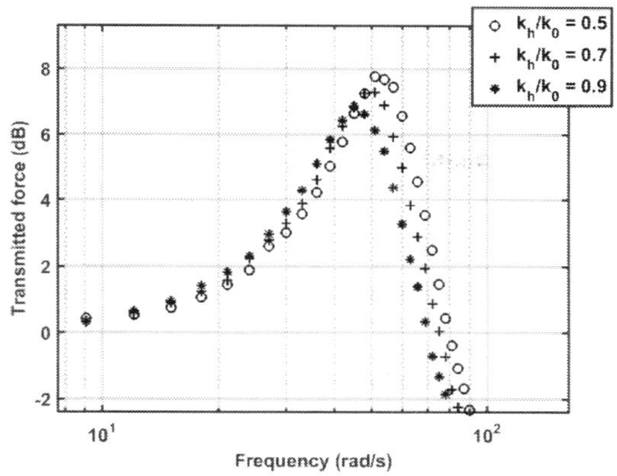

Figure 13. Transmitted force – $l = 0.7\, l_0$ – varying horizontal stiffness—MV model (zoomed in).

5 CONCLUSIONS

This paper presents results from incorporating stiffness nonlinearity in the MV and MMV models. Simulation results indicate that the incorporation of stiffness nonlinearity can be useful to overcome the trade-off between resonant peaks and transmissibility at higher frequencies. The proposed design also allows the resonant frequency to be shifted (up to 8 Hz in the simulations for this study) significantly. This can provide additional flexibility for the design of the isolation system. However, there are no specific advantages with the use of the MMV model. It is observed that the pre-compression of the horizontal spring needs to be reasonably high (at least 40% in the simulations for this study) in order to have a significant effect on the frequency response. The effect of the parameters associated with the horizontal spring is somewhat similar to the effect of the damping ratio in a linear isolator. However, the use of a horizontal spring does not involve the trade-off between the resonant peak and high frequency transmissibility, as typically seen in a linear vibration isolator. The advantages of the horizontal spring could be further reinforced by incorporating stiffness and damping nonlinearities in the isolator design.

Future work will investigate the effect of coupled stiffness and damping nonlinearities on MV and MMV models in conjunction with the horizontal stiffness investigated in this study. Parameters associated with the stiffness nonlinearity will be investigated in the future work to optimize the design of the isolator for specific objectives and design constraints. The main aim of investigating such a design is to attempt to overcome the restrictive trade-offs associated with a linear vibration isolator.

REFERENCES

Brennan, M.J., Carrella, A., Waters, T.P. & Lopes, V. 2008. On the dynamic behaviour of a mass supported by a parallel combination of a spring and an elastically connected damper. *Journal of Sound and Vibration* 309: 823–837.

De Silva, C.W. 2005. *Vibration and Shock Handbook*, Taylor and Francis, Boca Raton, FL.

Ho, C., Lang, Z. & Billings, S.A. 2014. Design of vibration isolators by exploiting the beneficial effects of stiffness and damping nonlinearities. *Journal of Sound and Vibration* 333: 2489–2504.

Ibrahim, R.A. 2008. Recent Advances in Nonlinear Passive Vibration Isolators. *Journal of Sound and Vibration* 314: 371–452.

Kaul, S. 2012. Dynamic Modeling and Analysis of Mechanical Snubbing. *Journal of Vibration and Acoustics* 134: 021020.

Kaul, S. 2015. Maxwell–Voigt and Maxwell Ladder Models for Multi-Degree-of-Freedom Elastomeric Isolation Systems. *Journal of Vibration and Acoustics* 137: 021021.

Lu, L., Lin, G. & Shih, M. 2012. An experimental study on a generalized Maxwell model for nonlinear viscoelastic dampers used in seismic isolation. *Engineering Structures* 34: 111–123.

MathWorks. 2013. *MATLAB User Guide*, MathWorks, Natick, MA.

Narimani, A., Golnaraghi, M.F. & Jazar, G.N. 2004. Frequency Response of a Piecewise Linear Vibration Isolator. *Journal of Vibration and Control* 10: 1775–1794.

Peng, Z.K., Meng, G., Lang, Z.Q., Zhang, W.M. & Chu, F.L. 2012. Study of the effects of cubic nonlinear damping on vibration isolators using Harmonic Balance Method. *International Journal of Non-Linear Mechanics* 47: 1073–1080.

Renaud, F., Dion, J., Chevallier, G., Tawfiq, I. & Lemaire, R. 2011. A new identification method of viscoelastic behavior: Application to the generalized Maxwell model. *Mechanical Systems and Signal Processing* 25: 991–1010.

Rubin, M.B. 2013. Influence of the Internal State of a Maxwell Damper on Free Critically Damped Vibrations. *Journal of Vibration and Acoustics* 135: 064503.

Sjoberg, M. & Kari, L. 2002. Nonlinear behavior of a rubber isolator system using fractional derivatives. *Vehicle System Dynamics* 37: 217–236.

Tang, B. & Brennan, M.J. 2012. A comparison of the effects of nonlinear damping on the free vibration of a single-degree-of-freedom system. *Journal of Vibration and Acoustics* 134: 024501.

Yarmohamadi, H. & Berbyuk, V. 2011. Computational model of conventional engine mounts for commercial vehicles: validation and application. *Vehicle System Dynamics* 49: 761–787.

Zhang, J. & Richards, C.M. 2007. Parameter identification of analytical and experimental rubber isolators represented by Maxwell models. *Mechanical Systems and Signal Processing* 21: 2814–2832.

Constitutive Models for Rubber X – Lion & Johlitz (Eds)
© 2017 Taylor & Francis Group, London, ISBN 978-1-138-03001-5

Calibration of advanced material models for elastomers

T. Dalrymple
DS SIMULIA Corp., Johnston, USA

A. Pürgstaller
Universität für Bodenkultur Wien, Wien, Austria

ABSTRACT: It is common practice in the application of Finite Element Analysis (FEA) to model elastomers with hyperelasticity, capturing the nonlinear elasticity at large strains. In the application area of sealing, FEA practitioners have successfully used hyperelasticity and linear viscoelasticity for over two decades now, Gent (1992). Commercial FEA software packages offer more advanced capabilities, such as nonlinear viscoelasticity, stiffness damage (Mullins effect) and plasticity, yet it is rare to see these advanced capabilities used in practice, Volgers (2016). We believe this is largely due to the lack of calibration strategies and tools, and perhaps the lack of good examples of these advanced modeling features used to capture elastomer behaviors with higher fidelity. This paper speaks to the calibration strategies and methodologies required to construct advanced material models. Our goal is to raise the bar in terms of common practice of modeling elastomer behavior, hopefully to include these advanced features more routinely in elastomer applications of FEA.

1 INTRODUCTION

The educational background of today's FEA practitioners often begins with one or more courses at the BS, or MS university level. But these classes cover the basics of the finite element method and do not delve into aspects of material modeling. Much of the educational information for material modeling is provided by the commercial software providers, or other niche sources. Most of the commercial FEA providers have offered hyperelasticity models for modeling elastomers for many years, offered education on their use, and perhaps offered calibration of the material model parameters within their tools. The education provided often describes the actual material behavior, compared to the behavior that can be described by the FEA tools. Figure 1 shows a typical elastomer stress-strain curve when loaded cyclically to three strain levels. The uniaxial specimen is first loaded to 10% strain for 10 cycles, then loaded to 30% strain for 10 cycles, then loaded to 100% strain for 10 cycles. The nature of this test may cause viscous effects to appear as if there is permanent set in the specimen. Before the advent of Mullins (1969) damage effect capability in commercial FEA software, the FEA practitioner was taught that the stiffness is a function of the previous strain level; be careful to pre-condition the specimen to a strain appropriate for the application. Thus, a "first-use" application, such as assembly in the factory, needs no pre-conditioning (scragging) of the specimen. While

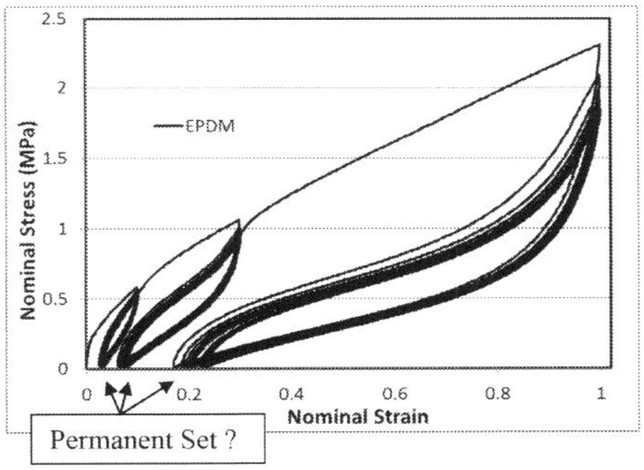

Figure 1. Typical uniaxial response to cyclic load.

this is commonly covered in FEA education on modeling of elastomers, it is still frequently the source of discrepancy, or lack of correlation, in component simulations.

As material models in commercial software have become more advanced, the education and calibration tools have lagged behind. As an example, the Mullins effect (stiffness damage) has been available for over a decade in the commercial software Abaqus®, see Bose, (2003), and Paige (2004), yet no tools are provided for the calibration of parameters.

In today's FEA tools, there are typically capabilities to model nonlinear elasticity, linear or nonlinear viscoelasticity, plasticity, and Mullins effect. While it is debatable whether plasticity is

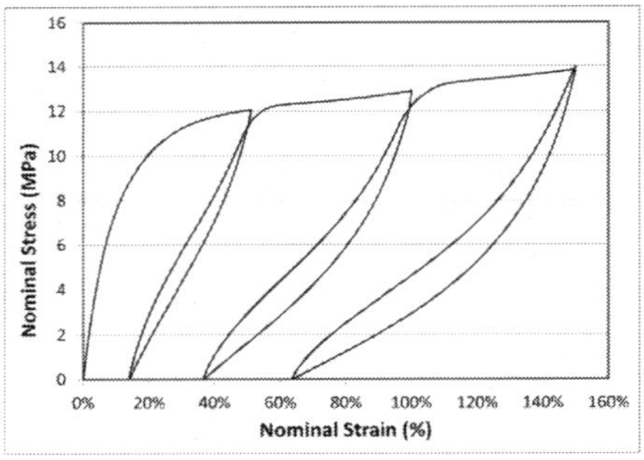

Figure 2. Cyclic tensile test data of DuPont Hytrel® 50 shore D, Volgers (2016).

truly needed for typical elastomers, materials such as thermoplastic elastomers (TPEs) exhibit a clear permanent set, as shown in Figure 2.

The paper by Volgers (2016) shows a nice example of modeling an automotive jounce bumper, by representing the Hytrel® material as hyperelastic, with Mullins effect and plasticity (viscoelasticity effects were assumed to be of little importance to the application). Interestingly, Govindarajan (2007) used the same material model of hyperelastic, Mullins effect and plasticity to model an EPDM rubber, similar to Figure 1. As mentioned earlier, it is debatable whether permanent set really occurs, or whether the cyclic testing under prolonged positive load leads to the appearance of permanent set (the strain would have returned to zero if given sufficient time at zero stress). In any case, it is often the application that dictates whether capturing the material's viscous behavior is important.

2 APPLICATION

Post-installed anchors in concrete find widespread applications in fastening non-structural components (NSC) to structures and are commonly used for structures located in seismic regions. Over the last decades, special effort was put in various research projects to improve the understanding of fastener behavior under seismic conditions. Current design guidelines, Eurocode (2012) do not allow to for dissipation mechanisms and to the best knowledge of the authors no such dissipation devices are currently on the market. Nonstructural components (NSC), often termed secondary systems, are those elements attached to floors, roof and walls of a building, that do not contribute to the primary structural systems load bearing task and are subject to the dynamic environment within the building. They constitute about 60 to 70% of the total construction costs in typical buildings, Taghav (2003), and have resulted in major economic losses in past earthquakes. One of the primary goals in Pürgstaller (2017) was to develop, optimize and investigate a supplemental viscoelastic damping device for traditional post-installed anchors in concrete. That damping device has the ability to enhance the seismic performance of traditional anchors and at the same time reduce the damage to the attached NSC. The damping is provided by a supplemental viscoelastic damping device. Since the damping is of keen interest in this application, we must represent the viscoelastic nature of the rubber material. Complex three-dimensional FE models with special focus on a prototype and the rubber damping material were investigated. For the current paper, we examine the cyclic load experimental data for the rubber damping material, and comment on the calibration process to derive a hyperelastic, viscoelastic, Mullins effect material model in the Abaqus finite element software. We know from past experience that for many traditional rubber materials a linear viscoelastic (Prony series) representation matches the actual behavior very well. A representative sample of the cyclic test data is shown in Figure 3 and Figure 4. Figure 3 shows the stress-strain response to the cyclic load history, and Figure 4 is used to more clearly show the load history itself. The tensile test specimen is loaded for 5 cycles to a maximum strain of 10%, followed by 5 cycles at

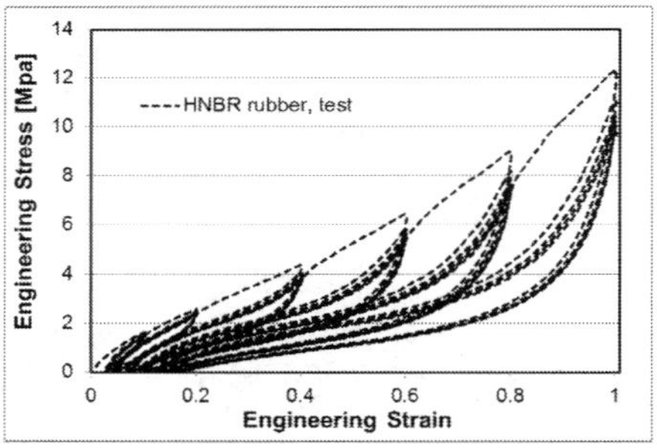

Figure 3. HNBR uniaxial response to cyclic load.

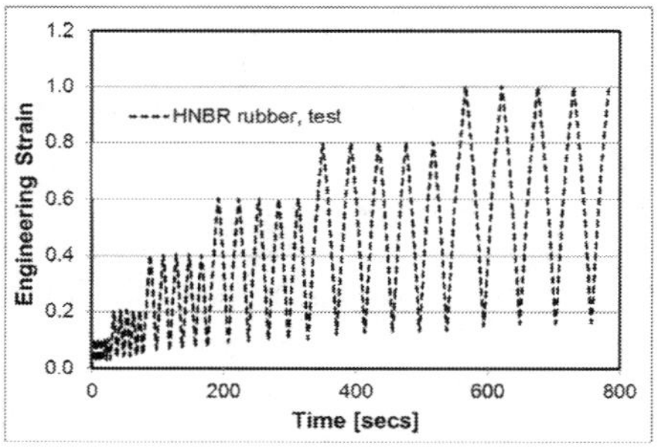

Figure 4. HNBR uniaxial cyclic load vs. time.

20% strain, etc. After each tensile cycle the specimen is unloaded to zero stress—we do not want negative stress so as not to buckle the specimen. The strains at unload do not return to zero, and we hypothesize that this is due to viscoelastic effects.

3 CALIBRATION

3.1 *Calibration tools*

One of the biggest challenges for industrial users of commercial FEA tools is the lack of calibration software. Most commercial packages offer some calibration, but are typically geared towards the basics, such as a simple hyperelastic material. For more advanced material models, including Mullins effect, plasticity, and/or viscoelasticity, academics and researchers will often write their own tools. In the last few years there have also been many papers published about using general purpose optimization packages, coupled to running FEA solvers for calibration, for example, Swaminathan (2011). Industrial FEA practitioners typically do not have the time resources to do this, and must rely on available tools. Two commercially available tools are worth mentioning here. The first is software called Hyperfit®, (2017) from Pavel Skacel, Brno University of Technology, in the Czech Republic. As the name suggests this tool is focused on hyperelasticity, and can calibrate hyperelastic-viscoelastic models as well as hyperelastic with Mullins. One cannot add plasticity to these attributes. The second is MCalibration® (2017), from Veryst Corporation. MCalibration is a very substantial tool, with calibration capabilities that support many major commercial FEA packages. It supports calibration of about 20 Abaqus material models by way of equations—or its own ODE solver. Hyperelastic+Mullins and hyperelastic+Prony viscoelasticity are supported via equations. It also supports the calibration of any Abaqus material model by running the Abaqus solver to determine the stress from a known test time-strain history. It automatically runs unit-cube analyses for each piece of test data and post-processes to compare the resulting stress-strain response to the test data. Hyperelastic+Mullins+Prony series viscoelasticity is not supported by equation and must be solved by running unit-cube models in the Abaqus solver.

3.2 *Calibration strategies*

What we mean by calibration strategies is using and following a methodology, or recipe, to improve the calibration process. By developing a strategy or recipe we hope to develop a repeatable process leading to consistently good material model results. In this section we will make some general remarks, then later in the calibration results section we will make some more specific comments on strategies used in this calibration of HNBR.

3.2.1 *Gathering test data*

The first and foremost concern as we gather test data is to keep the intended application in mind. The scope of the testing needs to span the scope of the application. By scope, we mean the range of strain, the range of strain-rate, the span of time, the span of temperature, etc. It is dangerous to calibrate a material model over a certain scope, then use that material model in a simulation that goes outside the original scope; this leads to an extrapolation that may be entirely invalid. In keeping with this thought, for any material model, it is important to document the scope of its intended application. This idea may seem obvious, but many FEA practitioners ask about the suitability of using DMA test data to calibrate an elastomer. FEA practitioners are often unaware that DMA testing is performed at very small strains, often some fraction of 1%. If we need an elastomer material model for a simulation application out to 50% strain, then we need to gather test data for calibration that spans this strain range.

When calibrating elastomers, we know that there are multiple mechanical behaviors in play—nonlinear elasticity, Mullins damage and time-dependence. As we move to thermoplastic elastomers and perhaps polymers in general, we see the above behaviors plus plasticity, or permanent set. As we gather test data for our material, we would perform tests that help us separate or delineate these behaviors. For instance, if we believe that plasticity is significant, we would like to perform a test like the 'strain recovery' test (aka load-unload-recover), Brusselle-Dupend (2001, 2003) and Quinson (1996, 1997). This test is very useful because viscous effects are often mistaken for plasticity. What these authors have shown is that the yield point for polymers is often higher than expected, since viscous effects at lower strains lead to nonlinearity of the stress-strain relation, which is often mistaken for plasticity. Once we have determined the approximate yield point, we can then perform stress relaxation or creep experiments below yield in an effort to test these behaviors in a separate manner.

Because we expect time-dependence, time should be recorded in every test that is performed—even for simple uniaxial pull tests. Another aspect that is often over-looked is to gather unloading information. For any test that does not go to failure, record the process of unloading (with perhaps an unloaded 'recover' time at the end of the test).

Besides performing tests that help us separate the various behaviors, we recognize that obtaining information from many different tests is very helpful. We have already shown test data from a cyclic test at progressively higher strains. This test starts with a virgin specimen and helps us understand the Mullins damage effect. The cyclic nature of the test also gives us some viscous information in the hysteresis loops generated. It is helpful to have test data from a family of stress relaxation (or creep) tests performed at various load levels. The different load levels help us determine if the material behaves in a linearly viscous or nonlinearly viscous manner. Another helpful test to perform is a family of strain-rate tests, with a span of strain-rates chosen from the intended application of the material.

3.2.2 *Test data adjustments/corrections*

It is quite common to find inconsistencies or inaccuracies in test data. The most common is a zero-shift in either the initial strain or initial stress in the dataset. This arises due to the process and order of gripping the specimen versus zeroing the load cell and the strain measurement device. It is very useful to overlay plots of the strain-time and the stress-time signals to make zero-shift corrections. Discussions with the test lab can often help determine which signal needs correction. Because our material model is going to start at time = strain-stress = 0, it is import!nt that our test data begin at this point as well.

Another common inconsistency is between the various test datasets that you might have. It is good practice to plot all your data on one plot and look for consistency among the various data. You might, for instance, overlay a family of stress relaxation testing with a family of strain-rate testing. The loading rates of the stress relaxation tests should make sense when compared to the loading rates and response of the strain-rate testing. Inconsistencies between various pieces of your test data can lead to 'tug-of-war' in the general optimization process and lead to slow convergence or no convergence of the process.

3.2.3 *Using test data in calibration*

When calibrating simple material models, like a Yeoh hyperelastic model, it is possible to use the linear least squares approach to calibration. As the material model chosen becomes more complex, and nonlinear in its parameters, many times a general optimization framework is used to perform the calibration. For advanced material models we also need to recognize that the design space may have local minima, and in general we may have non-uniqueness issues. When using a general optimization framework for calibration the initial values of the design variables (the model parameters) can be very important to a successful process. For complex material models with a large number of material parameters, knowing where to start can be the biggest challenge. For these cases we advocate a multi-step process: use a portion of the test data to get a good estimate of a few of the parameters, then use other data (or more data) to determine a broader set of the overall parameters. For example, when our target is a hyperelastic+Prony series model, use just the test data from a high-rate pull test to determine the instantaneous hyperelastic parameters (with other deformation modes as needed). Then use all the data together to fit all of the material model parameters. In our next section on calibration results we will show another example of this multi-step process for a hyperelastic+Mullins +Prony series material model. When plasticity is added, this multi-step process becomes even more valuable.

3.3 *Calibration results for HNBR*

The test data for our HNBR material has already been shown in Figures 3 and 4. In this case, this was our only test data and we did not have a family of stress relaxation testing or a family of strain-rate testing. As was previously discussed, we have seen examples of using a hyperelastic+Mullins+plasticity model to represent similar test data. However, we do not believe our HNBR material undergoes plasticity, and the attribute labeled "Permament Set?" in Figure 1 is really a viscoelastic effect. Also, for our application to NSC anchoring systems with damping, it is crucial to capture the viscous behavior of the material. Our target material model is a hyperelastic+Mullins+Prony series model. The test data was pretty clean, but we did adjust the data slightly to make it begin at time = strain = stress = 0. Because we only had uniaxial deformation data, we restricted our choice of hyperelastic models to those based on the first strain invariant, choosing the Yeoh model.

Following our multi-step calibration process, we simplified the test data somewhat (eliminated many of the 5 cycles), and fit just the Yeoh hyperelastic portion of the model. The parameters used and the stress-strain response are shown in Figure 5. We are not trying to get a highly accurate material model, just trying to establish a good first guess at parameters. While there are 3 Yeoh parameters (we are assuming the material is incompressible), we use a long standing rule of thumb that the 2nd parameter ought to be negative and about 10% of the 1st parameter. The 3rd parameter is about the same magnitude as the 2nd parameter, but positive. This 'calibration' was just typing in a few successive guesses of C10, setting C20 and C30 by the rule of thumb and looking at the calculated stress-strain response. This took only a few minutes using the MCalibration tool.

With good estimates of the Yeoh parameters established, we use the same simplified data and promote the material model to a Yeoh+Mullins model. We also have some experience that reasonable

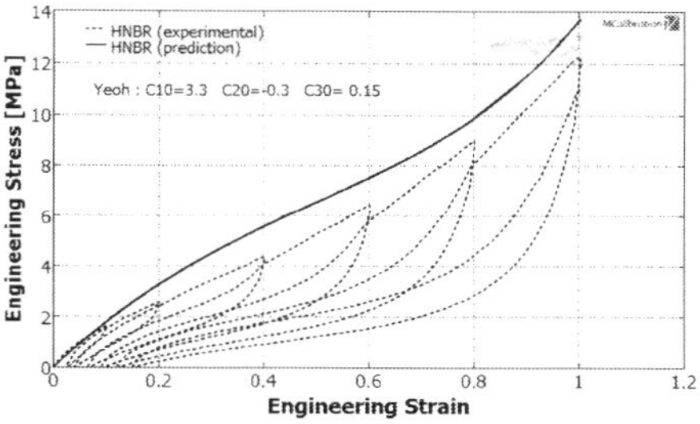

Figure 5. Yeoh hyperelastic response for HNBR.

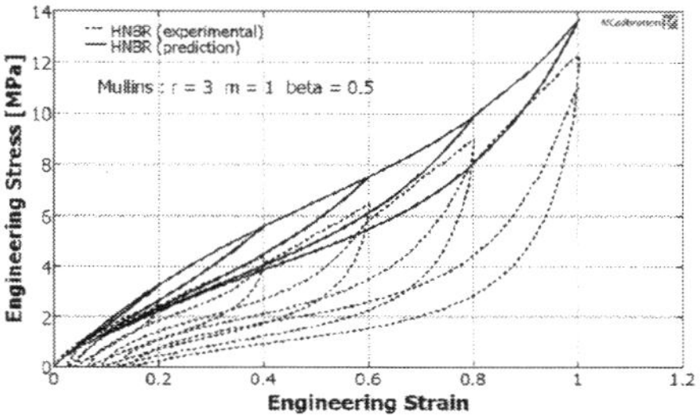

Figure 6. Yeoh+Mullins response for HNBR.

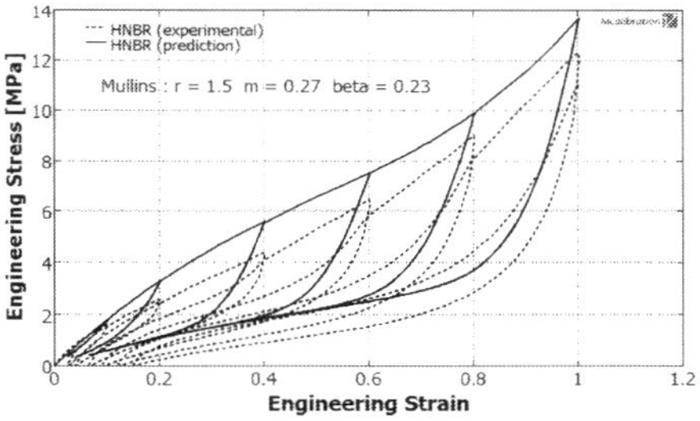

Figure 7. Yeoh+Mullins response after calibration.

Mullins parameters for natural rubber are r = 3.0, m = 1.0 and beta = 0.5. The response using these natural rubber values is shown in Figure 6. A few minutes of calibration gives us a better matching re3ponse as shown in Figure 7. There are just 3 design variables, the Mullins parameters, during this phase we have held the Yeoh parameters constant.

For the last phase of this work, we promote the material model to Yeoh+Mullins+Prony series viscoelasticity and we revert to the full test data. This material model requires the use of FEA solutions to populate the predicted stress values, so the calibration takes longer. We again use a rule of thumb to determine the Prony time constants, tau's. We look at the time span of the test data and define one tau per decade of time. From Figure 4 we see that the test data spans about 800 seconds, with the $\Delta t \sim 0.02$. Based on this, we chose to use 4 tau values of 0.1, 1, 10 and 100. From the Prony series equations we also know that response is not very

Table 1. Summary of parameters, 2nd estimate.

Yeoh	C10 = 6.0	C20 = −0.6	C30 = 0.35
Mullins	r = 1.5	m = 0.27	beta = 0.2
Prony	G	Tau	
	0.25	0.1	
	0.20	1.0	
	0.10	10.0	
	0.05	100.0	

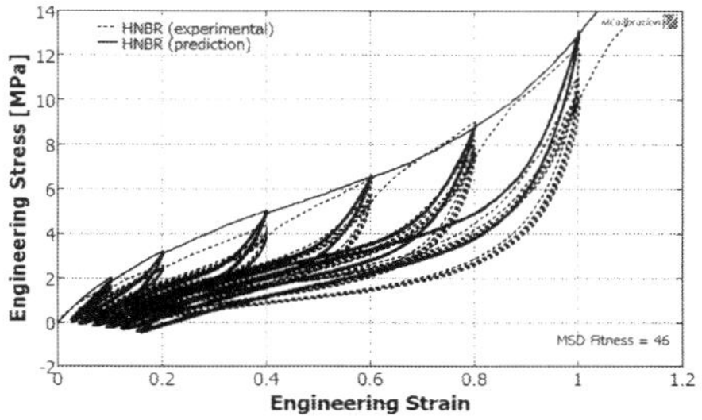

Figure 8. Yeoh+Mullins+Prony response for HNBR.

sensitive to even large changes in tau. For this work we will leave the tau's fixed, thus we have a total of 10 design variables. The Prony G terms are initially set to 0.1. After a one-time run to see the response (not shown) it is clear from the small hysteresis loops that the Prony G parameters need to be increased. From prior experience with stress relaxation calibration we also know to bias the G's towards early time. The 2nd estimate of the material model parameters is shown in Table 1.

Using the parameters from Table 1, the calculated response is shown in Figure 8. This response was generated by running a one-element FEA analysis. Care should be taken to understand the runtime, since further calibration with optimization methods will run this model hundreds of times. This one-element model initially took over 40 minutes to run. Some investigation pointed to the time integration controls. In Abaqus, changing the CETOL value from its default of 1e-3 to 1-e2 shortened the runtime to about 4 minutes. This material model response is a good starting point for further optimization. We have developed the above step-by-step strategy in order to get an initial set of model parameters that will lead to successful optimization.

Our final calibration run, using automated optimization, ran for about 24 hours on a Win7 laptop. The resulting stress-strain response is shown in Figure 9. This result matches the viscous hysteresis loops very well, and it also demonstrates that what one might have thought of as permanent set (Figure 1), is really a viscous effect.

The final set of parameters is shown in Table 2.

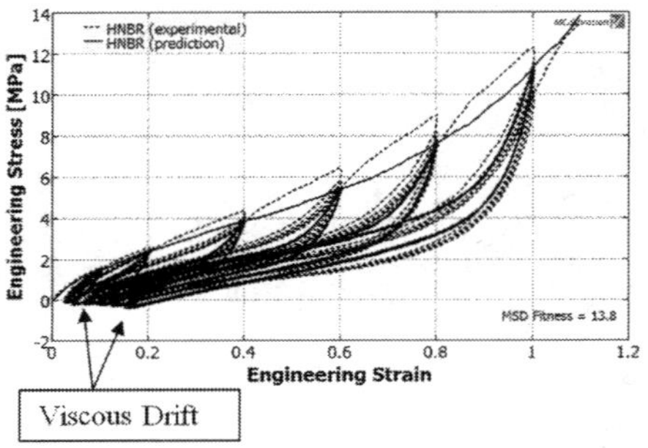

Figure 9. Yeoh+Mullins+Prony final response.

Table 2. Summary of final parameters.

Yeoh	C10 = 6.0	C20 = 0.0	C30 = 0.3
Mullins	r = 1.43	m = 0.290	beta = 0.212
Prony	G	Tau	
	0.294	0.1	
	0.352	1.0	
	0.036	10.0	
	0.034	100.0	

4 CONCLUSIONS

We have made some general comments on tools and strategies for calibration of advanced material models for elastomers. While hyperelasticity by itself and in conjunction with Prony series linear viscoelasticity is fairly common in FEA practice, it is surprising that the use of Mullins effect to capture the stiffness damage of elastomers is still rather uncommon—especially since this capability was added to commercial FEA software about a decade ago. The authors believe this is largely due to the lack of calibration tools and strategies for their successful use. We have shown a successful calibration of HNBR rubber using a Yeoh+Mullins+Prony series material model and demonstrated what is hopefully a repeatable calibration process, or recipe.

ACKNOWLEDGEMENTS

The authors want to thank Kurt Miller of Axel Products Test Lab for many years of fruitful discussions on the testing of elastomer and polymers.

REFERENCES

Bose, K. Hurtado, J. Snyman, M. Mars, W.V. & Chen, J.Q. 2003. Modeling of stress softening in filled elastomers, *Proceedings of the 3rd ECCMR*, Busfield, J.J.C. & Muhr, A.H. (eds), Swets and Zeitlinger, Netherlands 223–230.

Brusselle-Dupend, N., Lai, D., Feaugas, X., Guigon, M., Clave, M., 2001, Mechanical behavior of a semicrystalline polymer before necking. Part I: Characterization of Uniaxial Behavior, Polym. Eng. Sci. 41(1), p. 66–76.

Brusselle-Dupend, N., Lai, D., Feaugas, X., Guigon, M., Clave, M., 2003, Mechanical behavior of a semicrystalline polymer before necking. Part II: Modeling of uniaxial behavior, Polym. Eng. Sci. 43(2), p. 501–518.

EN1992-4, 2015, *Eurocode 2–4:* Design of Fastenings for Use in Concrete, Draft.

Gent, A.N., 1992, *Engineering with rubber: How to design rubber components*, New York, Hanser.

Govindarajan, S.M., J.A. Hurtado, and W.V. Mars. 2008. Simulation of Mullins effect and permanent set in filled elastomers using multiplicative decomposition. *Constitutive Models for Rubber*, Vol. 5, Balkema.

Hyperfit, 2017, A fitting utility for parameter identification, http://www.hyperfit.wz.cz/

Mars, W.V. 2004. Evaluation of a pseudo-elastic model for the Mullins effect. *Tire Science and Technology* 32: 120–145.

Mullins, L. 1969. Softening of rubber by deformation. *Rubber Chemistry and Technology* 42: 339–362.

Paige, R.E., & Mars, W.V., 2004. Implications of the Mullins Effect on the Stiffness of a Pre-loaded Rubber Component. In *ABAQUS Users' Conference* (pp. 1–15).

Pürgstaller, A., 2017, *Seismic performance of post-installed fasteners in concrete with supplemental damping device at Structure-Fastener-Nonstructural-(SFN)-level.*, Dissertation, University of Applied Sciences, BOKU, Vienna.

Quinson R., Perez J., 1996, Components of non-elastic deformation in amorphous glassy polymers, Journal of Material Science 31 p. 4387–4394.

Quinson R., Perez J., 1997, Yield criteria for amorphous glassy polymers, Journal of Material Science, 32 p. 1371–1379.

Swaminathan. S, Hill, L., Urankar, S., 2011, Calibrating Material Constants from Experimental Data for Lead-Free Solder Materials using a Parametric Optimization Approach, NAFEMS World Congress, Boston, USA.

Taghavi, S. & Miranda, E., 2003, Response assessment of nonstructural building elements. PEER report 2003/05. Pacific Earthquake Engineering Research Center, College of Engineering, 2003, p 84.

Veryst 2017, MCalibration software from Veryst Engineering, Needham, MA. http://www.veryst.com

Volgers, P., 2016, Modelling of Hytrel® Thermoplastic Elastomer Material for High-Strain Cyclic Loading, In *SIMULIA Community Conference* (pp. 1–15).

Constitutive Models for Rubber X – Lion & Johlitz (Eds)
© 2017 Taylor & Francis Group, London, ISBN 978-1-138-03001-5

Influence of nonlinear viscoelasticity for steady state rolling

M.A. Garcia & M. Kaliske
Institute for Structural Analysis, Technische Universität Dresden, Dresden, Germany

ABSTRACT: Tires and rolling rubber wheels are plenty used in many applications and their analysis is often simplified as steady state motion, in which an Arbitrary Lagrangian-Eulerian (ALE) formulation provides an efficient framework for the numerical analysis using the Finite Element Method. While many nonlinearities due to the complex geometry, large deformations, and material properties must be taken into account, the proper modeling of rubber-like materials represents the most challenging task. Withinthe ALE-formulation, hyperelastic materials can be used, as the reference frame does not coincide with the material, and is not fixed in space. Nevertheless, for the consideration of inelastic materials, the internal variables have to be taken into account by means of streamlines.

In this contribution, the inelastic properties of rolling wheels are computed using a nonlinear viscoelastic formulation, based on the Bergström-Boyce model. The implementation takes into account the contribution of all points along a streamline. Therefore, the elastic and viscous contributions are obtained in a single step and no advection algorithm is required. Additionally, the energy dissipation is computed directly and the rolling resistance can be estimated. Numerical examples show the capabilities of this formulation. A discussion on the results obtained, important remarks and an outlook to further research close this presentation.

1 INTRODUCTION

The use of different types of rubber compounds in tire industry makes the numerical simulation of rolling tires a challenging task. Large nonlinearities inherent to the material properties, complex geometry and large deformations must be taken into account.

The Finite Element Method (FEM) is commonly used for the analysis of complex materials. Due to the dynamic nature of the rolling phenomenon, a standard transient approach is often used for the numerical simulation of tires. However, this requires a large amount of computational resources.

Nowadays, the use of an Arbitrary Lagrangian-Eulerian (ALE) formulation is widely used for the rolling analysis, as the process is decomposed into rigid body rotation (Eulerian) followed by the deformation (Lagrangian). This framework is particularly efficient for steady state conditions, leading to a formulation completely independent of time (Nackenhorst 2004). However, the consideration of path dependent material models becomes complex.

Early works related to the analysis of rolling bodies focused on the transient analysis of rolling cylinders (Oden & Lin 1986, Padovan 1987), where viscoelastic properties where already playing an important role. Extensive analyses of the treatment of internal variables (Benson 1992) distinguish between two main schemes: *split* and *unsplit* methods. While the latter approach has been frequently applied in rolling tires (Ziefle & Nackenhorst 2008, Suwannachit & Nackenhorst 2013), a convective step is required in order to transport the internal variables through the mesh.

The unsplit approach leads to a consistent linearization by means of additional coupling terms. Whether the linearization is made for only a previous point (Wollny & Kaliske 2013, Behnke & Kaliske 2015) or considers all contributions in a streamline (Nasdala, Kaliske, Becker, & Rothert 1998, Garcia, Kaliske, Wang, & Bhashyam 2015), the computation of the inelastic effects in only one step represents an important advantage.

In this contribution, an unsplit treatment of the internal variables is implemented in order to compute the inelastic response of rubber-like materials at steady state rolling. Two material models for finite deformations are compared: linear and nonlinear viscoelasticity.

2 ALE FRAMEWORK

Within the Arbitrary Lagrangian Eulerian (ALE) framework, the reference configuration is not attached to the material, nor fixed in space. Originally used for fluid dynamics, this formulation allows an efficient description of the rolling process at steady state. The mapping of a material point $\mathbf{X}$ from the initial to the current configuration

$$\mathbf{x} = \Phi(\mathbf{X}, t) = \hat{\Phi}(\chi(\mathbf{X}, t)) \quad (1)$$

is decomposed into rigid body rotation $\chi(\mathbf{X}, t)$ and the deformation $\hat{\Phi}(\chi, t)$, respectively. Therefore, the deformation gradient $\underline{\mathbf{F}}$ is also split into rotation $\underline{\mathbf{F}}_\chi$ and deformation $\underline{\mathbf{F}}_\varphi$, thus

$$\underline{\mathbf{F}} = \underline{\mathbf{F}}_\varphi \underline{\mathbf{F}}_\chi. \quad (2)$$

The volume in reference and current configuration $dv = JdV$ is related by the Jacobian $J = \det \underline{\mathbf{F}}_\varphi$. Notice that for the rigid body rotation, $\det \underline{\mathbf{F}}_\chi = 1$ holds. In the following, the subindices φ and χ are dropped in order to simplify the notation as all quantities are related to the reference configuration unless specified otherwise.

For the consideration of the inertia effects, the vector Ω is composed by the angular velocities

$$\Omega^T = [\Omega_x, \Omega_y, \Omega_z], \quad (3)$$

and the guiding velocity of the material $\mathbf{w} = \mathbf{\Omega} \times \chi$ defines the spatial velocity. Furthermore, the material velocity is defined within this framework as

$$\mathbf{v}(\varphi, t) = \left.\frac{\partial \varphi}{\partial t}\right|_\chi + \mathbf{Grad}\varphi \cdot \mathbf{w}, \quad (4)$$

thus, the relative velocity of the mesh $\left.\frac{\partial \varphi}{\partial t}\right|_\chi$ and the convective velocity of the material $\mathbf{c} = \mathbf{Grad}\varphi \cdot \mathbf{w}$ are taken into account independently of each other (Donea, Huerta, Ponthot, & Rodriguez-Ferran 2004).

The strong form of the balance of linear momentum

$$\rho \dot{\mathbf{v}} = \rho \mathbf{b} + \mathrm{Div}\, \underline{\mathbf{P}}, \quad (5)$$

takes into account the inertia effects, with the density of the material given by ρ. The body forces are denoted by $\mathbf{b}$ and defined per unit spatial mass. The first Piola-Kirchhoff stress tensor is given by $\underline{\mathbf{P}}$, is defined in the reference configuration as $\underline{\mathbf{P}} = J \underline{\sigma} \cdot \underline{\mathbf{F}}^{-T}$ and with the Cauchy stress $\underline{\sigma}$.

Within this framework, the weak form of the equation of motion

$$\int_\chi \rho \dot{\mathbf{v}} \cdot \eta dV + \int_\chi \underline{\mathbf{P}} : \mathrm{Grad}\, \eta dV = \int_\chi \rho \mathbf{b} \cdot \eta dV + \int_{\partial\chi} \bar{\mathbf{t}} \cdot \eta dA, \quad (6)$$

is obtained by applying Dirichlet and Neumann boundary conditions. The virtual displacements are denoted by η. On the left-hand side, the inertial and the internal forces are described, while the external forces and surface traction $\bar{\mathbf{t}}$ are given on the right-hand side.

The inertial effects can be expressed as

$$\int_{\chi(\mathcal{B})} \rho \dot{\mathbf{v}} \cdot \eta dV = -\int_{\chi(\mathcal{B})} \rho \mathbf{c} \cdot (\mathrm{Grad}\, \eta \cdot \mathbf{w}) dV + \int_{\partial\chi(\mathcal{B})} \rho \eta \cdot \mathbf{c} \mathbf{w} \cdot \hat{\mathbf{n}} dA, \quad (7)$$

in order to avoid the computation of second order gradients (Nackenhorst 2004). It can be noticed that a second term in the right-hand side appears, which describes the material flux over the boundary. This term can be neglected as the outward normal vector $\hat{\mathbf{n}}$ is perpendicular to the velocity $\mathbf{w}$ on the boundary $\partial\chi(\mathcal{B})$. However, numerical fluctuations have been reported in finite element meshes which are not regular in circumferential direction (Suwannachit & Nackenhorst 2013).

In Eq. (7), all time derivatives have vanished due to the steady state assumption and this formulation is now independent of time. As only spatial quantities are required, the implementation for linear elastic and hyperelastic materials is done in a straightforward manner. However, for history-dependent materials, a time measure is required in order to account for the viscoelastic stresses in Eq. (6).

From Eq. (3), a constant velocity Ω_y is assumed and a particle can be traced from a position φ_i to a position φ_s in a so-called streamline for a time $\Delta t_{is} = \Delta\varphi_{is} / \Omega_y$ and knowing the angle separating both points, as seen in Fig. 1. The creation of the streamlines is only possible for axisymmetric models, as all the integration points share the same position in radial direction in the reference configuration.

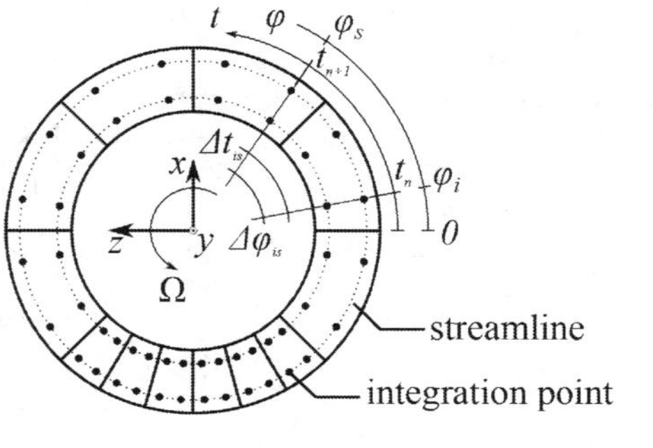

Figure 1. Spatial and time measures along streamlines.

3 VISCOELASTIC MODELS

The inelastic properties of rubber-like materials are described using a viscoelastic model and treating the internal variables in an unsplit scheme. Nasdala et al. (Nasdala, Kaliske, Becker, & Rothert 1998) presented a framework for steady state rolling using viscoelastic materials, taking into account the contribution of all elements connected in circumferential direction and a set of coupling terms obtained due to the consistent linearization. However, Fourier transformations were required.

By using streamlines, Wollny and Kaliske (2013) implemented an ALE formulation for inelastic materials, which considers the influence of a point φ_{i-1}. As this implementation was focused on pavement structures, their straight streamlines are not closed and no equilibrium along them was required. Using a similar scheme, Behnke and Kaliske (2015) show good results for the thermo-mechanically coupling of a linear viscoelastic model for rolling tires. However, a sub-quadratic convergence rate was achieved.

In this contribution, two viscoelastic material models are presented and compared. While the ground response of the material is given by hyperelastic springs, the inelastic behavior is described by an arbitrary number of Maxwell branches.

The strain energy function for a unit reference volume

$$\psi = \psi^e(\underline{\mathbf{C}}) + {}^{-}\bar{\psi}^i(\underline{\mathbf{C}}) \tag{8}$$

can be derived as the sum of the deformation energy associated with the ground state response and the inelastic branch. The elastic response ψ^e depends on the right Cauchy-Green tensor $\underline{\mathbf{C}}$ and it can be further split into volumetric and isochoric parts

$$\psi^e(\underline{\mathbf{C}}) = U(J) + \gamma_\infty \bar{\psi}^e(\bar{\underline{\mathbf{C}}}), \tag{9}$$

being the unimodular right and left Cauchy-Green tensors denoted by

$$\bar{\underline{\mathbf{C}}} := \bar{\underline{\mathbf{F}}}^T \bar{\underline{\mathbf{F}}} \quad \text{and} \quad \bar{\underline{\mathbf{b}}} := \bar{\underline{\mathbf{F}}}\bar{\underline{\mathbf{F}}}, \tag{10}$$

respectively, with $\bar{\underline{\mathbf{F}}} = J^{-1/3}\underline{\mathbf{F}}$.

The volumetric part depends on the determinant of the deformation gradient $J = \det \underline{\mathbf{F}}$, which describes the volume change, and is defined as

$$U(J) = \kappa (J-1)^2, \tag{11}$$

with the bulk modulus given by κ large enough in order to enforce incompressibility. For the isochoric part of the ground response of the material, the free energy function is based on the Yeoh hyperelastic model

$$\begin{aligned}\bar{\psi}^e(\bar{\underline{\mathbf{C}}}) = {} & C_1(\bar{I}_1 - 3) + C_2(\bar{I}_1 - 3)^2 \\ & + C_3(\bar{I}_1 - 3)^3,\end{aligned} \tag{12}$$

and the model parameters C_1, C_2, and C_3. The first invariant of the right Cauchy-Green tensor is given by $\bar{I}_1 = \operatorname{tr} \bar{\underline{\mathbf{C}}}$.

The inequilibrium response $\bar{\psi}^i(\bar{\underline{\mathbf{C}}}^e)$ depends on the viscoelastic model chosen and it is analyzed in detail in the next sections.

Furthermore, the Kirchhoff stress can be also additively decomposed into volumetric and deviatoric contributions

$$\underline{\boldsymbol{\tau}} = \underline{\boldsymbol{\tau}}_{vol} + \underline{\boldsymbol{\tau}}_{iso}, \tag{13}$$

where

$$\underline{\boldsymbol{\tau}}_{vol} = JU'(J) \quad \text{and} \quad \underline{\boldsymbol{\tau}}_{iso} = \underline{\underline{\mathbb{P}}} : \bar{\underline{\boldsymbol{\tau}}}, \tag{14}$$

using $\underline{\underline{\mathbb{P}}} := \underline{\underline{\mathbb{I}}} - \frac{1}{3}\underline{\mathbf{1}} \otimes \underline{\mathbf{1}}$ as the fourth order deviatoric projection tensor and a further split into

$$\begin{aligned}&\bar{\underline{\boldsymbol{\tau}}} = \bar{\underline{\boldsymbol{\tau}}}^e + \bar{\underline{\boldsymbol{\tau}}}^i, \text{ with} \\ &\bar{\underline{\boldsymbol{\tau}}}^e = 2\partial_{\bar{\underline{\mathbf{C}}}}\bar{\psi}^e(\bar{\underline{\mathbf{C}}}) \text{ and } \bar{\underline{\boldsymbol{\tau}}}^i = 2\partial_{\bar{\underline{\mathbf{C}}}}\bar{\psi}^i(\bar{\underline{\mathbf{C}}}).\end{aligned} \tag{15}$$

3.1 *Linear viscoelastic model*

The linear viscoelastic model, with a consistent linearization, is based on the well established formulation for finite deformations and large strains (Kaliske & Rothert 1997, Simo & Hughes 1998). This model makes use of a stress-type internal variables $\underline{\mathbf{H}}_k$ and the Maxwell elements are described by the linear rate constitutive equation

$$\dot{\underline{\mathbf{H}}}_k + \frac{1}{\tau_k}\underline{\mathbf{H}}_k = \gamma_k \underline{\mathbf{S}}^e_{iso}. \tag{16}$$

The viscosity in a branch k is defined by the relaxation time τ_k, while the parameter γ_k denotes the contribution of this rheological element to the total stiffness of the material. The isochoric part $\underline{\mathbf{S}}^e_{iso}$ of the second Piola-Kirchhoff stress tensor $\underline{\mathbf{S}}^e$ is obtained from

$$\underline{\mathbf{S}}^e_{iso} = \mathbf{Dev}\,\underline{\mathbf{S}}^e = J^{-\frac{2}{3}}\mathrm{Dev}\left[2\partial_{\bar{\underline{\mathbf{C}}}}\bar{\psi}^e(\bar{\underline{\mathbf{C}}})\right]. \quad 17)$$

Therefore, within a steady state rolling approach and by considering discrete time steps along the circumference (Garcia, Kaliske, Wang, & Bhashyam 2015), the viscoelastic stress $\underline{\mathbf{H}}_{k,s}$ at a point $\varphi = s$ is the sum of all the points $N\oplus$ in the streamline

$$\underline{\mathbf{H}}_{k,s} = \alpha_k \sum_{i=1}^{N\oplus} \bar{h}_{s,i,k} \underline{\mathbf{T}}^{si^T} \underline{\mathbf{S}}^e_{iso,i} \underline{\mathbf{T}}^{si}, \tag{18}$$

where the stress $\underline{\mathbf{S}}^e_{iso,i}$ at all points i must be known a priori. Transformation matrices $\underline{\mathbf{T}}^{si}$ between points i and s are required. The factor α_k relates the decay functions

$$\alpha_k = \frac{\exp(T/\tau_k)}{\exp(T/\tau_k) - 1}, \tag{19}$$

for an infinite number of full revolutions $T = 2\pi/\Omega$ along the streamline. Additionally, the difference between the contributions $\bar{h}_{s,i,k} = h_{s,i,k} - h_{s,i+1,k}$ at points i and $i + 1$ has to be computed, with

$$h_{s,i,k} = \frac{\tau_k}{\Delta t_i} \exp\left(\frac{-\Delta t_{is}}{\tau_k}\right)\left[1 - \exp\left(\frac{-\Delta t_i}{\tau_k}\right)\right]. \tag{20}$$

Notice that the time measure Δt_i is defined as the time between a point i and $i - 1$, while Δt_{is} is measured between i and s.

The proposed form of the stored energy yields an additive decomposition of the stress into volumetric and isochoric parts, therefore the total stress yields

$$\underline{\mathbf{S}} = \underline{\mathbf{S}}_{vol} + \gamma_\infty \underline{\mathbf{S}}^e_{iso} + \sum_{k=1}^{m} \gamma_k \underline{\mathbf{H}}_k \tag{21}$$

and, while the linearization of the first two terms on the right-hand side is used as standard Lagrangian formulation (Simo & Hughes 1998), the linearization of the last term leads to

$$D_{\Delta\varphi} \underline{\mathbf{S}}^v(\varphi, \eta) = \int_\oplus \bar{h}^\oplus D_{\Delta u}(\underline{\mathbf{T}}^{\oplus^T} \underline{\mathbf{S}}^\oplus_{iso} \underline{\mathbf{T}}^\oplus)(\mathbf{u}) d\oplus, \tag{22}$$

including continuous expressions for Eq. (18) along the streamline $\oplus$.

Therefore, additional coupling terms

$$\underline{\mathbf{K}}^{ef} = \int_{x(\mathcal{B})} \int_\oplus \left[\frac{1}{J} h^\oplus \mathbf{B}^T (\underline{\underline{\mathbb{C}}}^\oplus_{iso}) \mathbf{B}^\oplus \right] d\oplus\, dv \tag{23}$$

appear in the stiffness matrix for the Finite Element implementation, where the 4th order tangent moduli

$$\begin{aligned} \underline{\underline{\mathbb{C}}}_{iso} = & \frac{2}{3} \mathrm{tr}(\underline{\bar{\boldsymbol{\tau}}}) \underline{\underline{\mathbb{P}}} - \frac{2}{3}(\underline{\boldsymbol{\tau}}_{iso} \otimes \underline{\mathbf{1}} + \underline{\mathbf{1}} \otimes \underline{\boldsymbol{\tau}}_{iso}) \\ & + \underline{\underline{\mathbb{P}}} : \underline{\underline{\bar{\mathbb{C}}}} : \underline{\underline{\mathbb{P}}} \end{aligned} \tag{24}$$

must be computed for all points in the streamline. The fictitious tangent and Kirchhoff stress tensor are given in the current configuration by

$$\underline{\underline{\bar{\mathbb{C}}}} = 4\underline{\bar{\mathbf{b}}} \partial^2_{\underline{\bar{\mathbf{b}}}\underline{\bar{\mathbf{b}}}} \bar{\psi}^e(\underline{\bar{\mathbf{b}}}) \underline{\bar{\mathbf{b}}} \quad \text{and} \quad \underline{\bar{\boldsymbol{\tau}}} = 2\partial_{\underline{\bar{\mathbf{b}}}} \bar{\psi}^e(\underline{\bar{\mathbf{b}}}) \underline{\bar{\mathbf{b}}}, \tag{25}$$

respectively.

This consistent linearization and additional coupling terms lead to a quadratic rate of convergence, computing the numerical solution in only one load step and no advection step is required. Nevertheless, the system becomes unsymmetric and the sparseness is affected, which must be taken into account while using a numerical solver.

3.2 *Non-linear viscoelastic model*

The numerical formulation for path independent materials with inelastic characteristics presented by Wollny and Kaliske (2013), combines split and unsplit techniques for the transportation of internal variables along straight and open streamlines within an ALE framework. As the implementation is focused on pavement structures, a direct application to the field of rolling bodies is not possible. Nevertheless, in this contribution, an alternative is presented as the nonlinear viscous effects, based on the Bergström-Boyce model, along the streamline are iterated until a balance is achieved. To obtain an analytical solution involving all points in the streamline is complex, due to the local iterations that this particular model requires.

The unimodular part of the deformation gradient

$$\underline{\bar{\mathbf{F}}} = \underline{\mathbf{F}}^e \underline{\mathbf{F}}^i \tag{26}$$

is further split into elastic and inelastic contri butions.

The inequilibrium response for this material model is given by the Yeoh model as

$$\begin{aligned} \bar{\psi}^i(\underline{\bar{\mathbf{C}}}^e) = & C_1(\bar{I}_1 - 3) + C_2(\bar{I}_1 - 3)^2 \\ & + C_3(\bar{I}_1 - 3)^3, \end{aligned} \tag{27}$$

and the evolution law (Bergström & Boyce 2000) for the inelastic rate of deformation tensor

$$\underline{\tilde{\mathbf{d}}}^i = \dot{\gamma} \underline{\mathbf{N}}, \quad \text{with} \quad \underline{\mathbf{N}} = \frac{\underline{\boldsymbol{\tau}}^i_{iso}}{\left\|\underline{\boldsymbol{\tau}}^i_{iso}\right\|} \tag{28}$$

and considering the isochoric part of the inelastic Kirchhoff tensor $\underline{\boldsymbol{\tau}}^i_{iso} = \underline{\underline{\mathbb{P}}} : \underline{\bar{\boldsymbol{\tau}}}^i$. The effective creep rate is denoted by $\dot{\gamma}$ as

$$\dot{\gamma} = \dot{\gamma}_0 (\lambda^e)^c \left(\frac{\tau_v}{\hat{\tau}} \right)^m, \quad \text{where} \quad \tau_v = \frac{\| \boldsymbol{\tau}^i_{iso} \|}{\sqrt{2}}. \tag{29}$$

The reference creep rate is defined by $\dot{\gamma}_0$, the stretch in the hyperelastic spring is given by λ^e and the parameter $\hat{\tau}$ is introduced for dimensional purposes, with the same units as $\boldsymbol{\tau}_v$. The coefficient c controls the kinetics of the material relaxation. The parameters $\dot{\gamma}_0 > 0$, $c > 0$ and $m > 0$ represent the material parameters for each viscoelastic branch, with $c = 0$ and $m = 1$ leading to the deviatoric part of the model proposed by Reese and Govindjee (1998).

The computation of the elastic strains in the inelastic branch depends on the treatment of the evolution law and its implementation into FEM requires a local iterative procedure (Dal & Kaliske 2009). Therefore, obtaining an analytical expression for the viscous effects in an arbitrary point in relation to all other points in a streamline becomes a complex task. For this model, the linearization is done taking into account the contribution from a point i and $i - 1$.

An additional iterative process is applied for an appropriate initialization of the internal variables at the beginning of the streamline. Storing the displacement fields for all elements connected in a circumferential ring, it is possible to make a loop along the points in a streamline and obtain an initial elastic strain $\underline{\mathbf{b}}^e$ in the viscoelastic branch at the point $i = 1$. Notice that, in a similar way, an infinite number of iterations along the streamline for the linear viscoelastic model leads to Eq. (18).

Due to the lack of additional coupling terms, unlike the linear model, sub-quadratic rate of convergence is achieved.

Some of the advantages of using a nonlinear viscoelastic model are related to a better description of important phenomena like amplitude dependency, preload dependent behavior and stress softening, among others. A comparison between linear and nonlinear viscoelastic models, especially highlighting the temperature dependencies is presented by Berger et al. (Berger, Behnke, & Kaliske 2016), but only using a linearization for points i and $i - 1$ in both models.

4 NUMERICAL EXAMPLES

For the following examples, a tire model with simplified geometry is presented. The cross-section is discretized into 40 elements. The boundary conditions includes a constraint of some nodes, as seen in Fig. 2. The three-dimensional FEM model consists of 22 circumferential sections and a total of 880 solid elements, with a refined mesh near the contact region.

The ground response of the rubber compounds is defined by the Yeoh hyperelastic model, see Table 1, with density $\rho_{rub} = 1100$ kg/m³. The model considers additional anisotropic linear elastic reinforcement layers, with $E = 1.72 \times 10^5$ MPa and $\nu = 0.3$, describing the body-ply and two belts and with densities $\rho_{ply} = 5900$ kg/m³ and $\rho_{belt} = 1500$ kg/m³, respectively. Internal pressure $P = 0.2$ MPa is applied during a first load step. In a second load step, the tire is brought into contact with a frictionless rigid flat surface. A final load step is defined by a variable angular velocity Ω around the y-direction.

The viscous behavior is defined by Prony series, with coefficients given by the first 3 columns in Table 2. For the nonlinear model, the same branches are applied considering the parameter c as a constant value $c = 0$ or as a variable, as seen

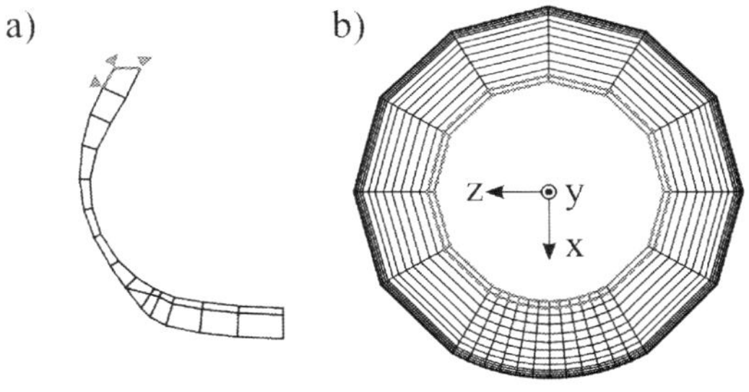

Figure 2. Tire model with simplified geometry, a) 2D cross-section and b) 3D-model.

Table 1. Material properties for tire model.

Parameter	Value	Units
C_1	3.095	MPa
C_2	–0.592	MPa
C_3	1.066	MPa
κ	35.515	MPa

Table 2. Parameters for viscoelastic material.

branch	γ_k	τ_k	C
∞	0.0939	–	–
1	0.0864	7.3702	0
2	0.0692	0.6859	0
3	0.0859	0.1409	300
4	0.0746	0.0217	500
5	0.1499	0.0045	4000
6	0.1952	0.0005	4000
7	0.3346	0.00015	4000
8	2.0892	0.000012	4000

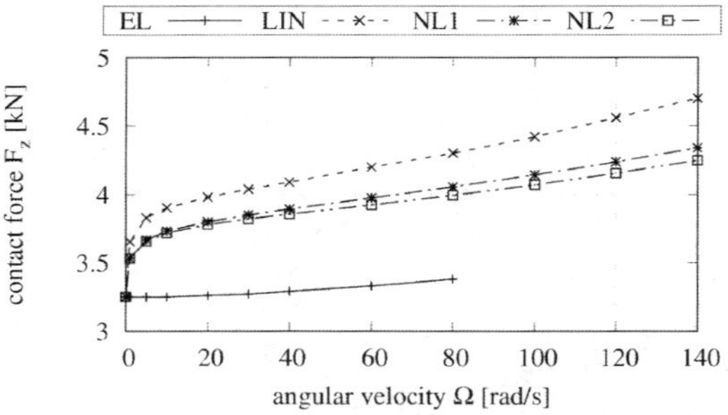

Figure 3. Contact force at different angular velocities.

in Table 2. In both cases, the coefficient $m = 1$ has been used.

The contact forces are shown in Fig. 3 for the long-term elastic response (EL), linear viscoelastic model (LIN) and nonlinear viscoelastic model with $c = 0$ (NL1) and c variable (NL2). It can be seen that the reaction force for the pure elastic material increases smoothly with the angular velocity due to the inertia effects, but it is not possible to obtain a converged solution for angular velocities $\Omega > 90$ rad/s.

An averaged difference of 6% arises between the forces due to linear and nonlinear viscoelastic approaches with the increase of velocity. This is understood as one of the main differences between obtaining the viscous contributions in a consistent manner (linear) or through the partial linearization made for the nonlinear model.

The viscoelastic models show an important increase of the contact force due to the relaxation processes happening in the inelastic branches, which are active at different frequencies. It can also be noticed that the consideration of coefficients $c = 0$ leads to a linear viscoelastic behavior, with the curve NL1 being closer to the one defined by the linear model. Different coefficients $c > 0$ and $m \neq 1$ provide additional flexibility for a fitting procedure, as the behavior of the material can be adjusted by keeping constant the rest of the parameters of the viscoelastic branches.

The consideration of inelastic properties allows the computation of energy dissipated during the rolling process. Fig. 4 shows the averaged dissipation along the streamlines for the tire rolling at 80 rad/s. This dissipation is shown in the cross-section for the linear and nonlinear viscoelastic models, where it is visible the higher dissipation located at the end of the belts and also in the tread. A maximum dissipated energy of 2.996 J/mm^2 is obtained for the linear model, Fig. 4a), while the nonlinear model shows a maximum value of 2.451 J/mm^2, Fig. 4b).

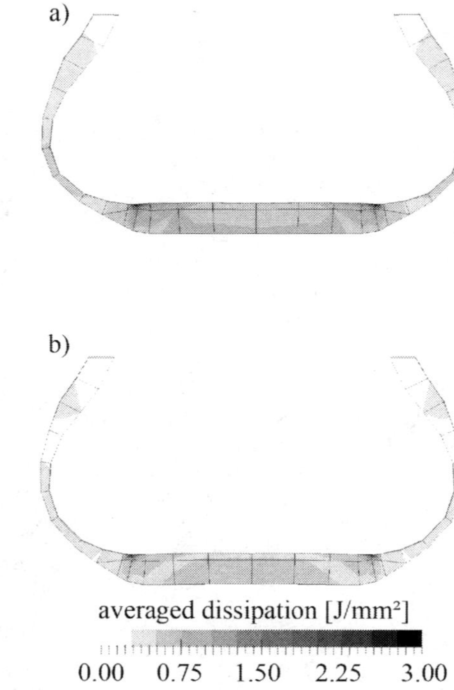

Figure 4. Averaged dissipation in the cross-section for 80 rad/s in a) linear and b) nonlinear (NL2) viscoelastic models.

5 CONCLUSIONS

In this work, the inelastic behavior of rubber materials is described by two different approaches: linear and nonlinear viscoelasticity, implemented for large strains into the in-house FEM code.

The consistent implementation of linear viscoelasticity allows the computation of the solution in a single step, with a quadratic rate of convergence. The main disadvantage is the need of special solution techniques for the nonsparse and nonsymmetric system of equations. The nonlinear viscoelastic model requires additional local iterations, but the overall system remains sparse. In both cases, using a finite number of inelastic branches, it is possible to model rubber-like behavior, especially at small strains. Rolling resistance can be estimated from the energy dissipated in the material.

By adjusting only a few parameters, the nonlinear viscoelastic model can be set to describe the inelastic behavior for small and large strains, with the possibility of reducing the number of branches required, which proves the versatility for describing the behavior of rubber-like materials in a more accurate manner. Moreover, the dissipative characteristics of the material can be controlled by the parameters c and m. However, the linearization employed shows a sub-quadratic rate of convergence due to the lack of additional coupling terms. The computation of these coupling terms is still a complex task.

The examples here presented are carried out for isothermal conditions and further comparison between these two approaches should be done using a thermo-mechanical coupling, where the dissipated energy in the contact area due to frictional forces can be taken into account. Within this framework, additional damage variables can be included. The analysis and validation of the results with experimental data is part of the outlook.

REFERENCES

Behnke, R. & M. Kaliske (2015). Thermo-mechanically coupled investigation of steady state rolling tires by numerical simulation and experiment. *International Journal of Non-Linear Mechanics 68*, 101–131.

Benson, D. J. (1992). Computational methods in Lagrangian and Eulerian hydrocodes. *Computer Methods in Applied Mechanics and Engineering 99*, 235–394.

Berger, T., R. Behnke, & M. Kaliske (2016). Viscoelastic linear and nonlinear analysis of steady state rolling rubber wheels: a comparison. *Rubber Chemistry and Technology 89*, 499–525.

Bergström, J. & M. Boyce (2000). Large strain time-dependent behavior of filled elastomers. *Mechanics of Materials 32*, 627–644.

Dal, H. & M. Kaliske (2009). Bergström-Boyce model for nonlinear finite rubber viscoelasticity: theoretical aspects and algorithmic treatment for the FE method. *Computational Mechanics 44*, 809–823.

Donea, J., A. Huerta, J.-P. Ponthot, & A. Rodriguez-Ferran (2004). Arbitrary Lagrangian-Eulerian Methods. In *Encyclopedia of Computational Mechanics*. JohnWiley & Sons, Ltd.

Garcia, M.A., M. Kaliske, J. Wang, & G. Bhashyam (2015). A Consistent Implementation of Inelastic Material Models into Steady State Rolling. *Tire Science and Technology 44*, 174–190.

Kaliske, M. & H. Rothert (1997). Formulation and implementation of three-dimensional viscoelasticity at small and finite strains. *Computational Mechanics 19*, 228–239.

Nackenhorst, U. (2004). The ALE-formulation of bodies in rolling contact: Theoretical foundations and finite element approach. *Computer Methods in Applied Mechanics and Engineering 193*, 4299–4322.

Nasdala, L., M. Kaliske, A. Becker, & H. Rothert (1998). An efficient viscoelastic formulation for steady-state rolling structures. *Computational Mechanics 22*, 395–403.

Oden, J. & T. Lin (1986). On the general rolling contact problem for finite deformations of a viscoelastic cylinder. *Computer Methods in Applied Mechanics and Engineering 57*, 297–367.

Padovan, J. (1987). Finite element analysis of steady and transiently moving/rolling nonlinear viscoelastic structure-I. Theory. *Computers and Structures 27*, 249–257.

Reese, S. & S. Govindjee (1998). A theory of finite viscoelasticity and numerical aspects. *International Journal of Solids and Structures 35*, 3455–3482.

Simo, J. & T. Hughes (1998). *Computational Inelasticity*, Volume 7 of *Interdisciplinary Applied Mathematics*. Springer.

Suwannachit, A. & U. Nackenhorst (2013). A Novel Approach for Thermomechanical Analysis of Stationary Rolling Tires within an ALEKinematic Framework. *Tire Science and Technology 41*, 174–195.

Wollny, I. & M. Kaliske (2013). Numerical simulation of pavement structures with inelastic material behaviour under rolling tyres based on an arbitrary Lagrangian Eulerian (ALE) formulation. *Road Materials and Pavement Design 14*, 71–89.

Ziefle, M. & U. Nackenhorst (2008). An Internal Variable Update Procedure for the Treatment of Inelastic Material Behavior within an ALE-Description of Rolling Contact. *Applied Mechanics and Materials 9*, 157–171.

Constitutive Models for Rubber X – Lion & Johlitz (Eds)
© 2017 Taylor & Francis Group, London, ISBN 978-1-138-03001-5

Comparison of the implicit and explicit finite element methods in quasi-static analyses of rubber-like materials

V. Yurdabak & Ş. Özüpek
Department of Mechanical Engineering, Boğaziçi University, Istanbul, Turkey

ABSTRACT: The accuracy and efficiency of the explicit integration in the quasi-static analysis of rubber is investigated. In particular, boundary value problems with varying degree of confinement are analyzed with both implicit and explicit time integration. As an example of a lightly constrained problem uniaxial stretch of a rubber plate with a central hole is considered. As an example of a highly constrained problem the hole is replaced by a crack. For all problems, the rubber density and the loading velocity are varied in order to determine the transition from quasi-static to dynamic state. Dependence of the predictions on the compressibility of rubber and the magnitude of the applied load are evaluated.

The conclusions may be used as a guidance in quasi-static analysis and fatigue life prediction of real applications such as seismic isolators, bearings and engine mounts.

1 INTRODUCTION

In static finite element analysis of structures made of rubber-like materials convergence problems may occur due to contacts, large deformation, complex loading conditions, incompressibility and nonlinear material properties. In order to overcome convergence difficulties, explicit time integration algorithms which are typically employed in dynamic analyses, may be used to provide quasistatic solutions. In the literature, such studies may be found for non-rubbery materials. Hu, Wagoner and Daehn simulated (1994) a uniaxial tension test with both implicit and explicit time integration. Limiting velocity values for a valid quasistatic analysis were found to depend on the plasticity of the material. Rust and Schweizerhof (2003) investigated thin-walled structures. The study concluded that the explicit time integration could be employed for the quasi-static limit load analysis of structures comprised of thin-walls. B. Egan et. al. (2014) studied the bearing failure of countersunk composite joints, and concluded that explicit time integration algorithms were more robust than the implicit ones due to the lack of convergence difficulties.

In the literature, explicit methods are not extensively employed in quasi-static analysis of rubber-like materials. Therefore, the main objective of this work is to systematically investigate the explicit time integration for the quasi-static solution of boundary value problems with varying degree of confinement. The effects of compressibility, loading velocity and material density on the explicit quasi-static solution are studied. The results of static analysis with implicit integration are used as reference values to evaluate the accuracy of the explicit method. All analyses are done in ABAQUS (2013).

The remainder of the paper is organized as follows: Section 2 provides a background for the implicit and explicit solution methods. Analyses of a plate with a circular hole subjected to uniaxial loading and a plate with a central crack subjected to uniaxial loading are given in Section 3. Conclusions are presented in Section 4.

2 COMPARISON OF IMPLICIT AND EXPLICIT TIME INTEGRATION

Static finite element analysis with implicit time integration seeks solution by following the equilibrium path of a given boundary value problem. In a nonlinear analysis, the load is applied incrementally and each increment requires several iterations for the solution to converge. For the iterations, most commonly Newton-Rapson method is used. Due to the equilibrium check, analysis with implicit time integration is accurate and can use bigger time steps to achieve convergence. On the other hand, due to formation of the stiffness matrix and its inversion for each iteration the computational cost is considerable. The presence of contact, large deformation, complex loading conditions, incompressibility and nonlinear material properties not only increases the cost, but also enforces smaller time increments. Since implicit time integration is unconditionally stable, both compressible and incompressible materials can be analyzed with it.

For the later, typically mixed finite element formulation is used. A representative flowchart of a static finite element analysis with implicit time integration is presented in Figure 1 (2013).

Analysis with explicit time integration does not enforce equilibrium, therefore special attention is needed when solving quasi-static problems. Since stiffness matrix is not formed and only the diagonal mass matrix is inverted, the computational cost is reduced. On the other hand the stability is conditional due to the requirement on a time step Δt given as:

$$\Delta t < \alpha \frac{L_{eff}}{c} \quad c = \sqrt{\frac{K+\mu/3}{\rho}}$$

where α is a correction factor, L_{eff} is a characteristic element dimension and c is the wave speed of the material calculated in terms of density ρ, bulk and shear moduli, K and μ. As compressibility increases, time step decreases, so that for an incompressible material it becomes zero. Therefore, for the explicit lgorithm to work efficiently some compressibility must be provided. In ABAQUS, the maximum value for compressibility is suggested as 100. A representative flowchart of an analysis with explicit time integration is given in Figure 2 (2013).

1 Initialization: σ^0; $\mathbf{d}^0 = 0$, n=0, t=0,$\mathbf{d}_{new} = \mathbf{d}^0$
2 Iteration for the (n+1)th increment
1 ForceVector $\rightarrow$ $\mathbf{f}(\mathbf{d}_{new}, t^{n+1})$; $\mathbf{r} = \mathbf{f}(\mathbf{d}, t^{n+1})$
2 Compute the Jacobian $\mathbf{A}(\mathbf{d}_{new})$
3 Solve $\mathbf{\Delta d} = - \mathbf{A}^{-1}\, \mathbf{r}$
4 $\mathbf{d}_{new} = \mathbf{d}_{new} + \mathbf{\Delta d}$
5 **if** The error criterion is not satisfied $\rightarrow$
Return to the beginning of the loop
3 Updates: $\mathbf{d}^{n+1} = \mathbf{d}_{new}$, n+1 $\rightarrow$ n, t $\rightarrow$ t + Δt
4 **if** Analysis is not complete $\rightarrow$ Step **2**
5 Output

Figure 1. Implicit solution of equilibrium equation. $\mathbf{f}_{int} = \mathbf{f}_{ext}$.

1 Initialization: $\mathbf{v}^0$ and σ^0; $\mathbf{d}^0 = 0$, n=0, t=0, $\mathbf{M}$
2 Calculation of Forces: $(\mathbf{f}_{ext} - \mathbf{f}_{int})$
3 Accelerations: $\mathbf{a}^n = \mathbf{M}^{-1}(\mathbf{f}_{ext} - \mathbf{f}_{int})$
4 First Velocity and Displacement Update for t = n + $\frac{1}{2}$
5 Update Displacements $\mathbf{d}^{n+1}$
6 Calculation of Forces $(\mathbf{f}_{ext} - \mathbf{f}_{int})$ for t = n+1
7 Calculation of Displacements $\mathbf{a}^{n+1}=\mathbf{M}^{-1}(\mathbf{f}_{ext} - \mathbf{f}_{int})$
8 Second Velocity Update for t = n + 1
9 Energy Balance Control at t = n + 1
10 **if** Analysis is not complete $\rightarrow$ Step **4** and n = n + 1
11 Output

Figure 2. Explicit solution of dynamic equation $\mathbf{Ma} + \mathbf{f}_{int} = \mathbf{f}_{ext}$.

3 BOUNDARY VALUE PROBLEMS

Two boundary value problems concerned with a uniaxially loaded plate were analyzed. In the first problem the plate had a central circular hole, in the second problem it had a central crack.

Material model

For all analyses, the material was represented as hyperelastic with the Yeoh strain energy form.

Compressibility was defined as the ratio of the initial bulk modulus to the initial shear modulus. In this work $K_0/\mu_0 = 10$, 50, 100 were used in analyses with explicit integration. For static analysis with implicit integration incompressible case was also considered.

Load

For all quasi-static analyses with explicit time integration the loads were applied with a "smoothstep" function with respect to time in order to avoid oscillations in the response (2003). In a "smoothstep" shown in Figure 3, the load has a quadratic time variation with an increasing rate from zero to half of the load and a decreasing rate from half of the load to full load. As a result the internal energy and nodal displacements also have a "smoothstep" history.

Parameters

The effects of the following parameters on the results of a quasi-static analysis with explicit time integration were investigated:

- Loading velocity
- Rubber density
- Rubber compressibility
- Applied load

Since when using explicit integration it is not possible to model the rubber as incompressible or highly compressible, results of static analyses with implicit time integration were evaluated for the sensitivity of results with respect to the following parameters:

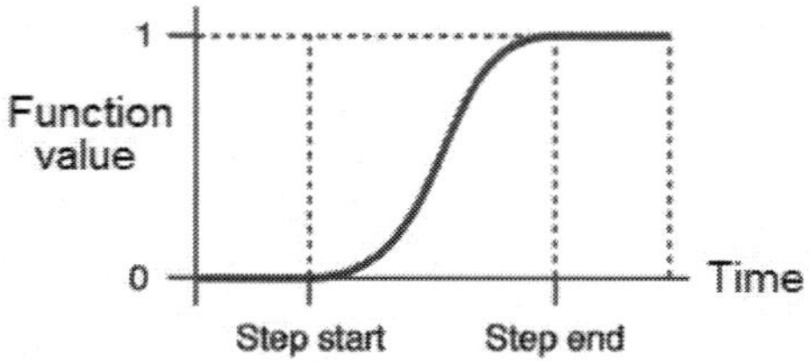

Figure 3. The "smoothstep" function.

- Rubber compressibility
- Applied load

3.1 *BVP1: Plate with a Circular Hole*

A square plate with a circular hole was uniaxially stretched. Plane stress was assumed and the load was applied as a displacement in the amounts of 25 mm, 100 mm and 150 mm. The mesh shown in Figure 4 consisted of 4-node bilinear reduced integration elements, the only quadrilateral element type supported by ABAQUS/Explicit.

3.1.1 *Quasi-static analyses with explicit time integration*

Effect of parameters listed in Section 3, on the maximum von Mises stress and total load were investigated.

Effect of velocity

The quasi-static analysis results obtained from the explicit time integration should have negligible inertia effects. The quantity that can be used to measure this effect is the ratio of the kinetic energy (KE) to the total internal energy (IE). The maximum KE/IE ratio is required to remain below some allowable value. In addition, the histories of the kinetic energy, nodal displacements and internal energy must be checked.

The results with 1 g/cm^3 density and various velocities are presented in Figure 5. The predictions were normalized by those of the static analyses with implicit integration. The corresponding KE/IE ratios are presented in Table 1 where values that are not acceptable for quasi-static analysis are shown in italic. It was determined that the analysis becomes dynamic when the KE/IE ratio is more than 10%. Also when KE/IE ratio is below 10^{-4}%, the results start to

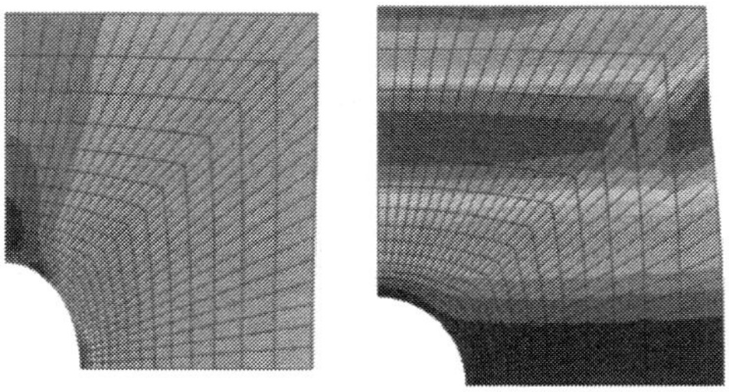

Figure 4. Quasi-static and dynamic (right) solutions of BVP1 for $K_0/\mu_0 = 10$ and 25 mm displacement.

Table 1. BVP1: KE/IE (%) for density of 1 g/cm^3.

Displacement (mm)	Velocity (m/s)	K_0/μ_0 10	50	100
25	0.01	$3.04\ 10^{-4}$	$3.01\ 10^{-4}$	$3.00\ 10^{-4}$
25	0.1	0.03	0.03	0.03
25	1	3.75	3.69	3.7
25	10	*99*	*100*	*101*
100	0.01	$2.05\ 10^{-5}$	$1.96\ 10^{-5}$	$1.97\ 10^{-5}$
100	0.1	$2.05\ 10^{-3}$	$1.99\ 10^{-3}$	$1.98\ 10^{-5}$
100	1	0.2	0.2	0.2
100	10	*33*	*32*	*28*
150	0.01	$8.66\ 10^{-6}$	$8.30\ 10^{-6}$	$8.11\ 10^{-6}$
150	0.1	$9.45\ 10^{-4}$	$8.18\ 10^{-4}$	$8.11\ 10^{-4}$
150	1	0.09	0.08	0.08
150	10	*12*	*11*	*11*

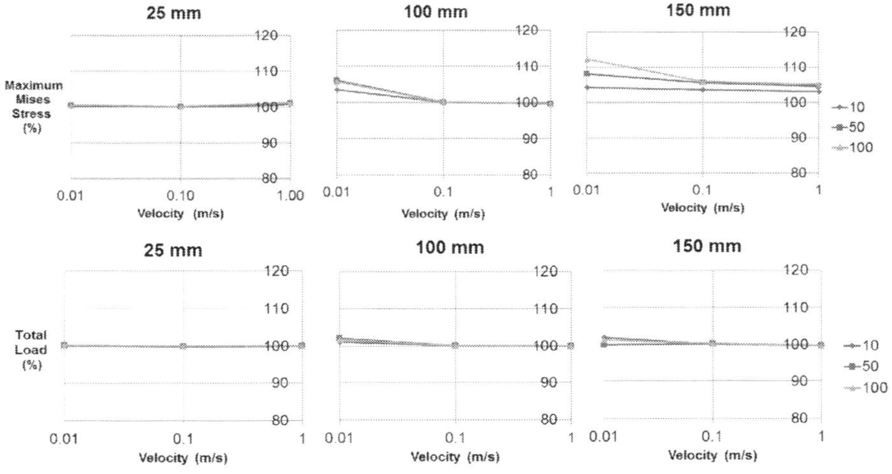

Figure 5. BVP1: Maximum von Mises stress (above) and total load for density of 1 g/cm^3 at different compressibilities and applied displacements.

deviate from the static analysis due to the fluctuations of the kinetic energy created by numerical errors.

To demonstrate the effect of the velocity on the results, von Mises stress distributions of quasi-static and dynamic solutions are compared in Figure 4. It is observed that even if the difference in maximum von Mises stress is insignificant, the stress distribution is completely different. Therefore, both the maximum value and the distribution of von Mises stress need to be considered in order to verify quasi-static analyses.

Effect of density
As discussed in Section 2 the increase in material density decreases the wave speed through the material. Since, this may yield a significant reduction in solution time, the quasi-static analyses with a density of 1000 g/cm^3 were completed and compared to those with a density of 1 g/cm^3. The allowable limits for KE/SE were not affected by the increase in density, however the permissible velocity range became narrower.

Effect of applied displacement
It is observed in Figure 5 that the total load is not significantly affected by the amount of the displacement while the accuracy of maximum von Mises decreases for large values of applied displacement.

Effect of compressibility
According to Figure 5 the accuracy of explicit solution for total load and maximum von Mises stress is not significantly affected by the compressibility of the rubber.

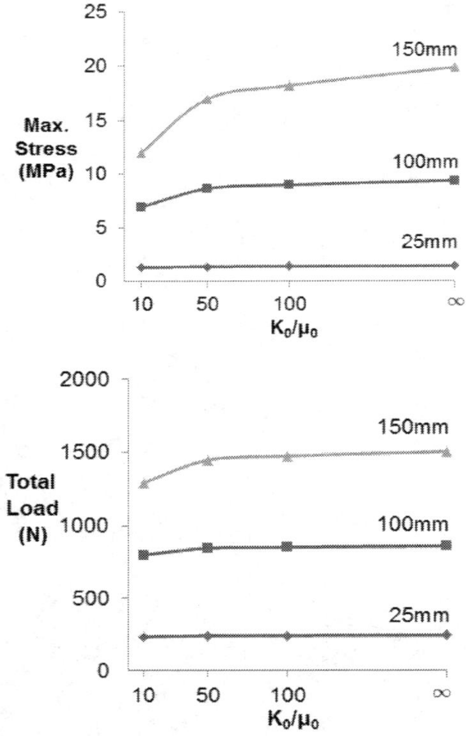

Figure 6. BVP1: Maximum von Mises stress (above) and total load with implicit time integration at different compressibilities and applied displacements.

3.1.2 *Static analyses with implicit time integration*
Implicit solution for the maximum von Mises stress and total load are shown in Figure 6. It is observed that for small values of applied displacement, both quantities remain the same for all K_0/μ_0 values. As the applied load increases, maximum von Mises stress shows increasing dependence on compressibility. The total load, on the other hand, is less sensitive to the level of compressibility. That is, if the total load is a concern the material may be assumed as incompressible for $K_0/\mu_0 > 50$ even when the applied displacement is large.

3.1.3 *Conclusions for BVP1*
Based on the results presented, the following were concluded:

- For proper quasi-static analysis with low degree of confinement, the ratio of kinetic energy to internal energy of the system should remain between 0.0001% and 10%.
- If the ratio of kinetic energy to internal energy of the system is less than 10%, the difference between static and quasi-static analysis in terms of total load is less than 3%.
- When the loading is prescribed via the smooth-step function oscillations in the system during the explicit solution are avoided.
- In order to decrease the solution time of a quasistatic analysis, density may be increased. The disadvantage is that this limits the rate at which the system may be loaded.
- For small to moderate values of applied displacement static and quasi-static analyses predict the same total load and maximum von Mises stress.
- For a large applied displacement static and quasistatic analyses predictions are the same for the total load, however differ for the maximum von Mises stress. The difference is sensitive to material compressibility.

3.2 *BVP2: Plate with a crack*

A square plate with a central crack was uniaxially stretched. Plane strain was assumed and the load was applied as a displacement in the amounts of

50 mm, 200 mm and 300 mm. The mesh consisted of 4- node bilinear reduced integration elements. The problem may be considered as highly constrained due to the presence of the crack and the condition of plane strain.

3.2.1 *Quasi-static analyses with explicit time integration*

Effect of various parameters on the crack tip displacement and total load were investigated.

Effect of velocity
The crack tip displacement predictions for 1 g/cm³ density and various velocities are presented in Figure 7. The results were normalized by those of the static analyses with implicit integration. The corresponding KE/IE ratios are presented in Table 2 where values that are not acceptable for quasi-static analysis are shown in italic. It was determined that the analysis becomes dynamic when the KE/IE ratio is more than 1%, a value lower than that for BVP1. The conclusions regarding the total load were similar to those of BVP1.

Effect of density
The quasi-static analyses with a density of 1000 g/cm³ were completed. The conclusions were found to be similar to those to those with a density of 1 g/cm³, that is the allowable limits for KE/SE did not show dependence on the density, however the permissible maximum velocity decreased.

Effect of applied displacement
The total load results (not shown) agree with the corresponding static analysis results. The accuracy of the crack tip displacement, on the other hand, decreases as the applied load increases as observed in Figure 7.

Effect of compressibility
The accuracy of total load and crack tip displacement were not significantly affected by rubber compressibility for low values of applied load. For greater loads, the accuracy decreases as compressibility increases.

3.2.2 *Static analyses with implicit time integration*

Implicit solution for the crack tip displacement and total load are shown in Figure 8. For low values of applied displacement, both quantities do not change with compressibility level. As the displacement increases, dependence on compressibility becomes significant for both quantities.As a consequence the use of lower than that of the actual material response would give inaccurate results.

3.2.3 *Conclusions for BVP2*

Based on the results presented, the following were concluded:

- For proper quasi-static analyses with high degree of confinement, the ratio of kinetic energy to internal energy of the system should remain between 0.001% and 1%.
- If the ratio of kinetic energy to internal energy of the system is less than 1%, the difference between static and quasi-static analysis in terms of total load is less than 2%.
- When the loading is prescribed via the smooth-step function oscillations in the system during the explicit solution are avoided.

Table 2. BVP2: KE/IE (%) for density of 1 g/cm³.

Displacement (mm)	Velocity (m/s)	K_0/μ_0 10	50	100
50	0.01	$6.72\ 10^{-5}$	$6.86\ 10^{-5}$	$6.83\ 10^{-5}$
50	0.1	$6.72\ 10^{-3}$	$6.84\ 10^{-3}$	$6.86\ 10^{-3}$
50	1	0.68	0.68	0.68
50	10	*65*	*73*	*63*
200	0.01	$3.70\ 10^{-6}$	$3.19\ 10^{-6}$	$3.22\ 10^{-6}$
200	0.1	$3.71\ 10^{-4}$	$3.26\ 10^{-4}$	$3.16\ 10^{-4}$
200	1	0.03	0.03	0.03
200	10	*4.24*	*3.63*	*3.50*
300	0.1	$1.24\ 10^{-4}$	$9.11\ 10^{-5}$	$8.37\ 10^{-5}$
300	1	$1.23\ 10^{-2}$	$9.20\ 10^{-3}$	$8.49\ 10^{-3}$
300	10	1.23	0.92	0.85
300	100	*65*	*73*	*60*

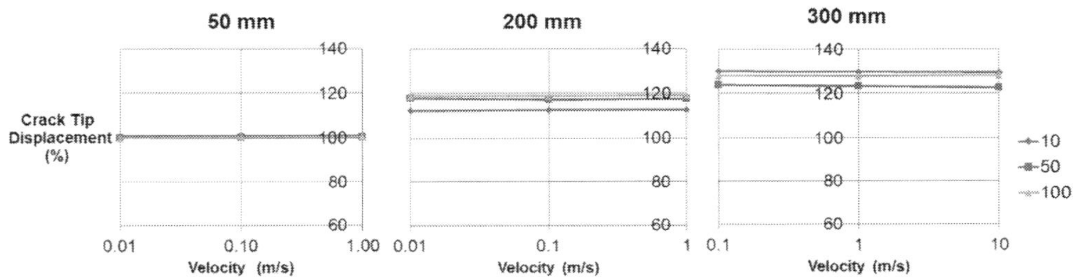

Figure 7. BVP2: Crack tip displacement for density of 1 g/cm³ at different compressibilities and applied displacements.

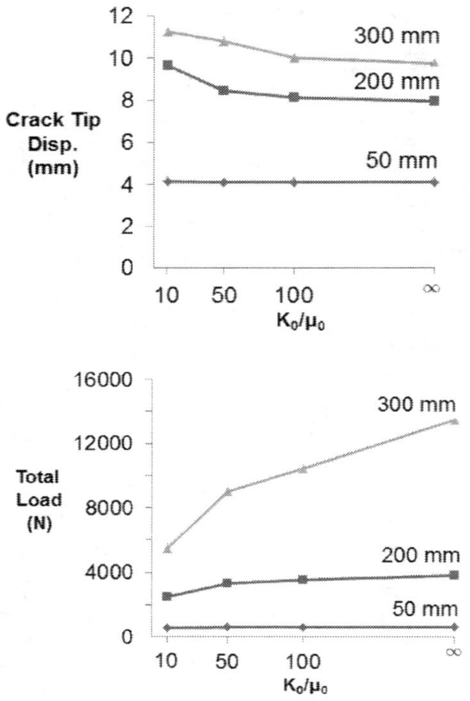

Figure 8. BVP3: Crack tip displacement and total load with implicit time integration at different compressibilities and applied displacements.

- In order to decrease the solution time of a quasistatic analysis, density may be increased. The disadvantage is that this limits the rate at which the system may be loaded.
- For a given compressibility, the difference between static and quasi-static analysis in terms of crack tip displacement increases with the increase in applied load.
- For a low value of applied displacement static and quasi-static analyses predict the same total load and crack tip displacement.
- For moderate or large applied displacements static and quasi-static analyses predictions are the same for the total load, however differ for the crack tip displacement. The difference is sensitive to material compressibility.

4 CONCLUSIONS

The objective of this work was to investigate the effects of rubber compressibility, rubber density, magnitude and rate of the applied load, and degree of confinement on the quasi-static analysis of rubber-like materials.

Two boundary value problems were analyzed with implicit and explicit integration techniques. The accuracy of the explicit analysis was evaluated against the results of static analysis with implicit integration. It was determined that loading velocity range for a valid quasi-static solution varies with the degree of confinement. The accuracy of global values such total load were not significantly affected by compressibility, applied displacement and rubber density. The accuracy of local quantities such as maximum Mises stress at a stress concentration or crack tip displacement have strong dependence on compressibility, applied displacement and rubber density. It was further concluded that for low degree of confinement and moderate loads quasistatic solution of problems is not significantly affected by compressibility. For highly constrained problems, on the other hand, reducing compressibility for solvability with explicit technique will result in significant difference in the predictions as compared to the solution for a highly compressible material behavior.

The results of the work may be used as a guidance in quasi-static analysis of problems with highly nonlinear material behavior, complex contact and loading conditions. Future work will focus on the use of explicit integration techniques in fracture mechanics and fatigue life prediction of elastomers.

REFERENCES

Belytschko, T., W. Liu, B. Moran, & K. Elkhodary (2013). *Nonlinear Finite Elements for Continua and Structures.* Northwestern University, Chicago, Illinois: John Wiley & Sons, Inc.

Egan, B., K.M.A. McCarthy, R.M. Frizzel, P.J. Gray, & C.T. McCarthy (2014). Modeling bearing failure in countersunk composite joints under quasi-static loading using 3d explicit finite element analysis. *Composite Structures 108,* 963–977.

Hu, X., R.H. Wagoner, & G.S. Daehn (1994). Comparison of implicit and explicit finite element methods in the quasistatic simulation of uniaxial tension. *Communications in Numerical Methods in Engineering 10,* 993–1003.

Rust, W. & K. Schweizerhof (2003). Finite element limit load analysis of thin-walled structures by ansys (implicit), ls-dyna (explicit) and in combination. *Thin-Walled Structures 41,* 227–244.

SIMULIA (2013). *ABAQUS 6.13 Documentation.* Dassault Systems Simulia Corp.

Constitutive Models for Rubber X – Lion & Johlitz (Eds)
© 2017 Taylor & Francis Group, London, ISBN 978-1-138-03001-5

Micro-mechanical modeling of visco-elastic behavior of elastomers with respect to time-dependent response of single polymer chains

L. Khalili, V. Morovati & R. Dargazany
Department of Civil and Environmental Engineering, Michigan State University, USA

J. Lin
Department of Material Science and Engineering, Massachusetts Institute of Technology, USA

ABSTRACT: We developed a micro-mechanically based model of nonlinear behavior of elastomers under different strain-rates. The model is based on the concept of local relaxation of polymer chains in a phantom polymer network. The concept has been further used to describe the decay of stiffness of the polymer chains and eventually creep. Accordingly, the time-dependent damage in one polymer chain is derived with respect to the time and deformation history. The concept is based on experimental data that suggests the time dependent behavior of network is mostly resulted due to the sliding or partial breakage of physical links over time, which results in different distribution of chain lengths over time. The proposed model includes a few number of physically motivated material constants and demonstrates good agreement with experimental data.

1 INTRODUCTION

Elastomers show an excellent range of desirable characteristics that makes them ideal material for many sensitive applications. However, they exhibit a complicated response featuring nonlinear elasticity with time-dependent and time-independent inelastic effects when during large deformations. With significant increase in the uses of elastomers in the last decades, a detailed understanding of their behavior becomes vital for design, maintenance and improvement of them. Early studies on elastomers reveals the visco-elastic nature and their fast stress relaxation rate (Bergstrm & Boyce 1998).

Micromechnical modeling of the viscoelastic behavior of elastomers remains a challenge due to the extreme nonlinear nature of their response. While classical linear viscoelastic models, such as the Voigt, Maxwell, and Zener models, have been often hired in simulations, they were used mainly due to their simple representation of the matrix with spring and dashpots. However, linear viscoelastic models have significant limitations in large deformations since their relaxation modulus (or creep compliance) is independent of strain magnitude. In nonlinear viscoelastic models, the decrease of the relaxation modulus at higher strains is considered as well as shear thinning, thermal softening at higher temperatures and nonlinear dependence on deformation and loading rates. Many nonlinear theory of viscoelasticity have been proposed in which either the stress depends on the entire deformation history; or alternatively the strain depends on the entire stress history (Khan, Lopez-Pamies, & Kazmi 2006).

While considerable progress has been made in developing empirical and phenomenological models for the finite strain for specific range of strain rates and temperatures, much less progress has been made on development of physically motivated models (Dal & Kaliske 2009, Reese 2003), and even less on physically-based models that are valid on a wide range of strain rates.

Most micro-mechanically motivated models are extremely difficult to use due to their complicated formulations and oversimplifying assumptions of the network. The main objective of this work is to propose a two-scale micromechanical model to predict the visco-elastic behavior of elastomers in long deformation ranges. In micro-scale, the strain energy of a single polymer chain is defined with respect to local relaxation of the chain and in meso-scale is implemented into network evolution model (Dargazany, Khiêm, Poshtan, & Itskov 2014, Dargazany, Khiêm, & Itskov 2015) to provide a continuum-scale response. The model predictions are verified against the experimental results of various relaxation tests on the Adiaprene-L100 (Khan, Lopez-Pamies, & Kazmi 2006) at increasing stretch amplitudes.

2 SINGLE CHAIN MECHANICS

The purely entropic non-Gaussian strain energy of a polymer chain with n segments, each with length l, and end to end distance r, is given by

$$\psi = nKT\left(\frac{r}{nl}\beta + ln\frac{\beta}{\sinh\beta}\right), \qquad (1)$$

where T is the absolute temperature, K the Boltzmann's constant, $\beta = \mathcal{L}^{-1}(\frac{r}{nl})$ and $\mathcal{L}^{-1}$ denotes the inverse Langevin function (Dargazany, Hörnes, & Itskov 2013). Let us further use the relative length $\bar{r} = \frac{r}{n}$. Accordingly, the strain energy of a polymer chain can be mainly described with respect to its extensibility ratio $S_l = \frac{\bar{r}}{n} \in [0,1]$. Recent experimental data suggests that the isothermal stiffness of polymer chains not only depends on extensibility ratio but also on the time (Ganss, Satapathy, Thunga, Weidisch, Pötschke, & Janke 2007). Hence, the time-dependent phenomena such as creep and recovery which are particularly important in the macro-scale response of the network have induced mainly by changes in the mechanics of polymer chains at micro-scale (Yang, Zhang, Friedrich, & Schlarb 2007). Here our goal is to understand and formulate the creep and relaxation behavior of a polymer chain.

2.1 *CREEP*

Polymers, when subjected to deformation, undergo micro-structural changes that is induced from several deformation mechanisms. In case of creep where a constant load is applied on a sample for a long time the constitutive response of the polymer is as follows; First, polymers show an instantaneous deformation which is associated to the rapid stretch of the van der Waals' bonds and dislocation of the valence bonds. Then, stiffness of the polymer chains gradually decays due to partial release of entanglement which will be discussed later. To counteract the reduced stiffness, the polymers are pushed to jump to a new conformation with least potential energy (see Fig. 1) and this process reaches a balance state until the convoluted polymer chain reaches its final conformation where it becomes perfectly linear. In the latter case,, the force applied on the chain, similar to its entropy goes to infinity. While entanglements and physical bonds prevent the fluctuations of the polymer chains by increasing their entropy, they dissolve or gradually slide over time which results in partial release of chains. The release generally results in higher length of the chain, and also loss of entropy which can be translated into the mechanical decay of the polymer chains (Fig. 1c).

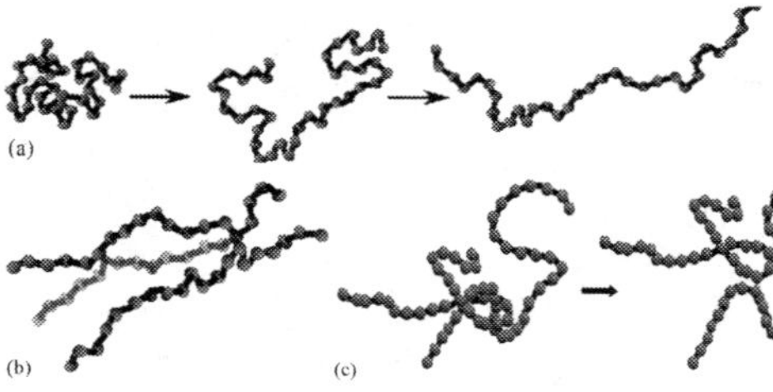

Figure 1. a) Conformational changes during tension, b) sliding of entanglement and c) partial release.

When a chain of length $\bar{r}_0$ is subjected a step constant stress, it illustrates a time-independent mechanical deformation to $\bar{r}_1 = \lambda_1 \bar{r}_0$ which is followed by a time-dependent increase in strain known as viscoelastic creep $\bar{r}_1^* = \lambda_1^* \bar{r}_1$ (See Fig. 2). One can define $\lambda_1 = \lambda(t_1)$ and λ_1^* as the mechanical and the creep stretches, respectively. To this end, for a sample subjected to mechanical stretch λ at time t_1, as time passes, it experiences also a creep stretch of $\lambda^*(t - t_1)$ which is 1 for $t < t_1$. While in linear visco-elasticity, the creep stretch λ^* is independent of the stretch values and only depends on time, in nonlinear visco-elasticity, λ^* depends on the both time and mechanical stretch as given below (see Fig. 1b)

$$\begin{aligned} &\lambda^*(t - t_i) = \lambda_i^* && \text{linear visco-elasticity} \\ &\lambda_j^*(t - t_i) = \lambda_i^*\big|_{\lambda=\lambda_j} && \text{nonlinear visco-elasticity} \end{aligned} \qquad (2)$$

Accordingly in nonlinear cases, the actual stretch of the chain can be given by $\hat{\lambda}(t)$ which is

$$\bar{r}_1^* = \hat{\lambda}(t_1)\bar{r}_0 = \lambda_1 \lambda_1^*(t - t_1)\bar{r}_0. \qquad (3)$$

Considering a chain subjected to a stepwise deformation history composed of three stretch

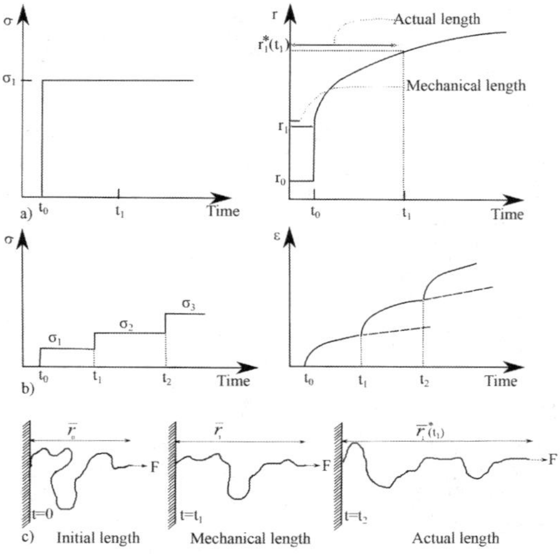

Figure 2. Creep in elastomers; a) the stress and the strain curves with respect to the time. b) Principle of superposition to describe the actual stretch with respect to the mechanical stretch history, and (c) Changes of the end-to-end length of a singlepolymer chain during mechanical and creep deformations.

jump discontinuities λ_0, λ_1, and λ_2, the actual stretch at time $\hat{\lambda}(t)$ can be calculated by superposing step functions (see Fig. 2b) as

$$\begin{aligned}\hat{\lambda}(t) &= \lambda_0\left(\lambda_0^*(t-t_0) - \lambda_0^*(t-t_1)\right) \\ &+\lambda_1\left(\lambda_1^*(t-t_1) - \lambda_1^*(t-t_2)\right) + \lambda_2\lambda_1^*(t-t_2).\end{aligned} \tag{4}$$

The equation describes that each stress jump σ_i induces an instantaneous mechanical stretch jump λ_i which is followed by immediate initiation of the creep stretch $\lambda_i^*(t-t_i)$. For closer approximation of actual stretch of a polymer chain subjected to a varying continuous deformation history with several jumps, the history is decomposed into a number of fine stretch increments treated as stretch jumps. In standard linear solids, the Boltzmann superposition principle (BSP) is often used for calculation of the actual deformation resulting from infinite steps in Eq. 4. However, this principle is not applicable to elastomers because of dependence of relaxation times on stretch rates (Kolarik & Pegoretti 2008). Here, we propose an incremental multiplicative superposition of stretches (see Fig. 3) which allows rewriting Eq. 4 as

$$\begin{aligned}\hat{\lambda}(t) &= \lambda_0\left(\lambda_0^*(t-t_0) - \lambda_0^*(t-t_1)\right) \\ &+\lambda_1\left(\lambda_1^*(t-t_1) - \lambda_1^*(t-t_2)\right) \\ &+\lambda_{n-1}\left(\lambda_{n-1}^*(t-t_{n-1}) - \lambda_{n-1}^*(t-t_n)\right) + \lambda_n\lambda_n^*(t-t_n) \\ &= \sum_{i=0}^{n-1}\lambda_i\left[\lambda_i^*(t-t_i) - \lambda_i^*(t-t_{i+1})\right] + \lambda_n\lambda_n^*(0) \\ &= \lambda(t) + \int_{T=0}^{t}\lambda(T)d\lambda_i^*(t-T)\end{aligned} \tag{5}$$

where

$$\begin{aligned}d\lambda_T^*(t-T) &= \frac{d\lambda_T^*(t-T)}{dT}dT \\ &= \left[\frac{d\lambda_T^*(t-T)}{d\lambda(T)}\frac{d\lambda(T)}{dT} + \frac{d\lambda(t-T)}{d(t-T)}\frac{d(t-T)}{dT}\right]dT \\ &= \left[\frac{d\lambda_T^*(t-T)}{d\lambda}\dot{\lambda}(T) - \dot{\lambda}_T^*(t-T)\right]dT\end{aligned} \tag{6}$$

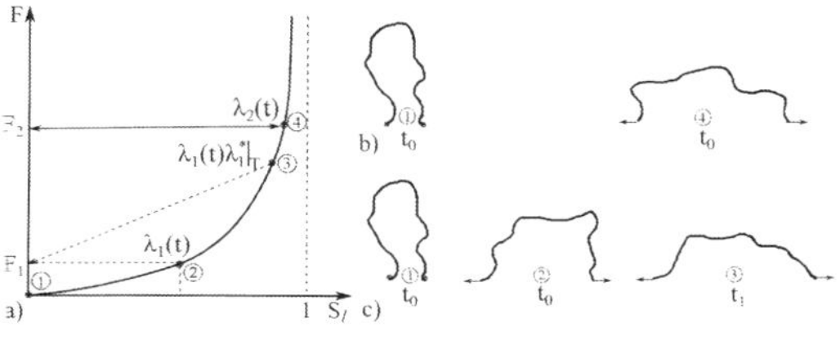

Figure 3. a) Force developed by chains at different S_l, b) instantaneous response of a polymer chain with the stretch of $\lambda_2(t_2)$, c) instantaneous and time dependent response of a single polymer chain with the stretch of $\lambda_1(t_0)$ and $\lambda_1(t_0)\lambda_1^*(t_1)$.

where superscript $\dot{}$ represents time derivative of a function. Thus, one can rewrite Eq. 5 as

$$\begin{aligned}\hat{\lambda} &= \lambda(t) \\ &+ \int_{T=0}^{t}\lambda(T)\left[\frac{d\lambda_T^*(t-T)}{d\lambda}\dot{\lambda}(T) - \dot{\lambda}_T^*(t-T)\right]dT\end{aligned} \tag{7}$$

In creep, due to sliding, partial release and stiffness decay, the end-to-end distance of the chain, $\lambda\bar{r}_0$, increases by the creep stretch over time while the mechanical stretch λ is kept constant. In this condition, the strain energy of the chain and consequently its force remain constant despite the fact that partial sliding gradually increases the length of the chain. Therefore, the strain energy and the force are only functions of the mechanical length $\bar{r} = \lambda\bar{r}_0$ and not the actual stretch. Thus, one has

$$\begin{aligned}F(n,\bar{r}_1) &= KT\mathcal{L}^{-1}\left(\frac{\bar{r}_1}{n}\right) = KT\mathcal{L}^{-1}\left(\frac{\bar{r}_1^*}{n\lambda_1^*|_{\lambda_1}}\right), \\ \psi(n,\bar{r}_1) &= \psi\left(n, \frac{\bar{r}_1^*}{\lambda_1^*|_{\lambda_1}}\right).\end{aligned} \tag{8}$$

Theoretically during creep, a chain can relax as long as its extensibility limit, $S_l \leq 1$, allows. In practice, however, the extensibility limit is correlated with singularity of force. Thus the mechanical stiffness of chain does not allow extension beyond certain limit. Let us define the maximum possible creep stretch of the chain and its corresponding end-to-end distance by λ_{max}^* and $\bar{r}_{max}$, respectively.

Now, by comparing the resistive force of that chain, $F(n,\bar{r}_1)$, against the resistive force of that chain when its current length $\bar{r}_{max}$ is achieved only by mechanical stretch $F(n,\bar{r}_{max})$, one can write

$$\begin{aligned}&\frac{F(n,\bar{r}_1)}{F(n,\bar{r}_{max})} = \gamma, \quad \text{where} \quad \bar{r}_{max} = \lambda_{max}^*\bar{r}_1, \\ &\frac{\mathcal{L}^{-1}(S_l)}{\mathcal{L}^{-1}(\lambda_{max}^*S_l)} = \gamma \Rightarrow \lambda_{\max}^*(\lambda) = \left\{x \middle| \frac{1-x^3S_l^3}{1-S_l^3} = x\gamma\right\}\end{aligned} \tag{9}$$

where $S_l = \frac{\lambda\bar{r}_0}{n}$. Here, $(1-\gamma)$ (see Fig. 3) represents the ratio of the resistive force lost due to the creep or structural damage within a chain, and will be considered as a material parameter. Associating the creep stretch $\lambda_T^*|_\lambda$ with a decay function and considering its boundary values, is sufficient to derive it as

$$\begin{aligned}&\lambda_T^*|_\lambda = \begin{cases}1 & t \leq T \\ \lambda_{\max}^*(\lambda) & t = \infty\end{cases} \\ &\Rightarrow \lambda_T^*|_\lambda = 1 + (\lambda_{\max}^*(\lambda) - 1)(1 - K(t)).\end{aligned} \tag{10}$$

Here, $K(t)$ represents degradation in mechanical response of a polymer chain over time in which

$K(0) = 1$ and $K(\infty) = 0$. The equation that describes exponential decay is

$$\frac{dK(t)}{dt} = -\alpha K(t) \Rightarrow K(t) = e^{-\alpha t}, \tag{11}$$

where the notation α is the decay constant and will be considered as a material parameter.

2.2 *Relaxation*

The proposed mechanism can also describe other nonlinear visco-elastic effects in elastomers such as the relaxation or shear thickening. In relaxation, the mechanism is similar to that of creep. A polymer chain of initial length $\bar{r}_0$ will be stretched to actual length $\bar{r}*_1 = \hat{\lambda} r_0$ and then stretch is fixed and the changes in the stress are measured. Accordingly, considering that, the actual length will be kept constant during relaxation, in view of Eq. 3, one has

$$\hat{\lambda} = \lambda(T)\lambda_T^*|_\lambda = C \quad \Rightarrow \quad \lambda(T) = \frac{C}{\lambda_T^*|_\lambda}, \tag{12}$$

where C is the relaxation stretch. Similarly, from Eq. 8 one can calculate the relaxation force as

$$\frac{F(n,\bar{r}_1)}{KT} = \mathcal{L}^{-1}\left(\frac{\lambda \bar{r}_0}{n}\right) = \mathcal{L}^{-1}\left(\frac{\bar{r}_0}{n}\frac{C}{\lambda_T^*|_\lambda}\right) \tag{13}$$

In Fig. 4, the model predictions on nonlinear relaxation and creep behavior for a polymer chain is depicted. Since nonlinear visco-elasticity is important in determining material behavior at moderate or higher stress levels where materials behave hyperelastic in quasi-static deformation. The proposed model can be coupled to existing hyperelastic models to account for visco-elasticity in addition to hyperelasticity. Interested readers can visit the review of hyperelastic models by (Marckmann & Verron 2006).

2.3 *Viscoelasticity in large deformations*

Classical linear theory of viscoelasticity describes fairly well the characteristic behavior of elastomers such as creep, relaxation, and strain rate dependence in a qualitative manner. The quantitative description, however, is generally limited to very narrow loading rate and temperature regimes. In order to approximate an actual transient stress relaxation and creep behavior, linear viscoelastic differential constitutive relations necessitate a large number of material parameters (Christensen 1982). In this section, the nonlinear viscoelastic constitutive law of a single polymer chain which was derived in the previous section will be implemented into a network evolution model to provide micro-macro scale transition (Dargazany, Khiêm, & Itskov 2015). Elastomers show limited compressibility and are often considered as nearly incompressible materials. Accordingly, their strain energy per unit reference volume ψ can be given as a decoupled summation of an isochoric $\Psi^{iso}(\bar{\mathbf{C}})$ and a volumetric part $U(J)$ as

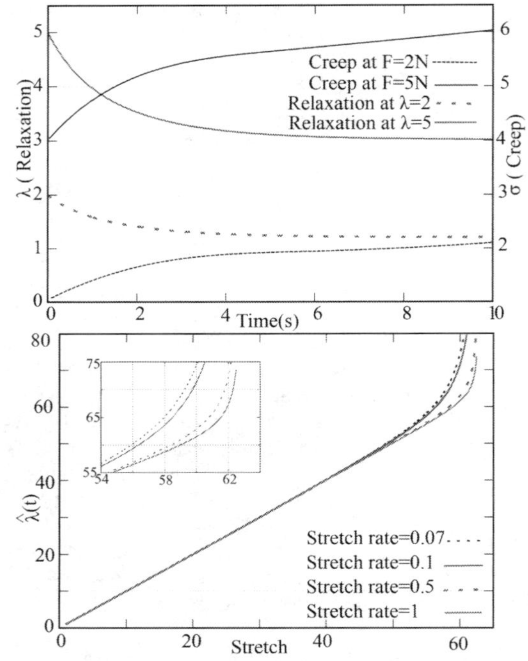

Figure 4. a) Constitutive behavior of a chain ($\bar{r}_0 = 1, n = 10$) in creep due to forces $F = 2N$ and $5N$, and in relaxation due to stretches $\lambda = 2$ and $F = 5$ b) actual stretch versus stretch graph showing the nonlinear visco-elasticity of a polymer chain (encapsulated shows the graphs in larger scale).

$$\Psi(\mathbf{C}) = \Psi^{iso}(\bar{\mathbf{C}}) + U(J) \tag{14}$$

where $\mathbf{C}$ denotes the right Cauchy-Green tensor, $J^2 = \det(C)$, and $\bar{\mathbf{C}} = J^{-\frac{2}{3}}\mathbf{C}$. For the nearly incompressible materials, the volumetric energy $U(J)$ can be imported as a penalty function

$$U(J) = \frac{k}{2}(J-1) \tag{15}$$

to enforce the incompressiblity condition (Göktepe & Miehe 2005). Here, the penalty parameter k is a positive material constant. In elastomers, the hyperelastic and viscous responses can be exclusively associated to the isochoric part of the deformation $\bar{\mathbf{C}}$. The hyperelastic behavior yields the reversible elastic processes in equilibrium state, while viscous behavior results from the irreversible processes such as relaxation or creep effects which belong to the realm of non-equilibrium thermodynamics. Here, the material matrix is decomposed into two networks, Hyperelastic (H), and Permanent damage (P) network. Each network is a 3D composition of many 1D subnetworks that are distributed in all spatial directions. Subnetworks experience different deformation and damage history based on their directions. Accordingly, the generalized strain energy function of the rubber matrix, Ψ_M, can be represented as the sum of the energies of all networks by

$$\Psi_M = \Psi_H + \Psi_P, \tag{16}$$

where $\Psi_{\bullet}$ is the strain energy of a system. The subscripts $\bullet_M$, $\bullet_H$, and $\bullet_P$ represent the parameters related to the matrix 'M', hyperelastic network 'H', and Permanent damage network 'P', respectively (Dargazany, Khiêm, & Itskov 2015).

Subnetworks are subjected to different uniaxial deformations and damage histories based on their directions. Integrating a subnetwork in all directions, the consequent network is a representation of that model in a 3D configuration. Here, by assuming the isotropic spatial distribution of polymer chains in all directions, the macroscopic energy of each network, $\psi_{\bullet}$ can be written

$$\Psi_{\bullet} = \frac{1}{A_s}\int_S \overset{d}{W}_{\bullet}\, \overset{d}{du}, \tag{17}$$

where A_s represents the surface area of a micro-sphere S, and $\overset{d}{u}$ the unit area of the surface with the normal direction d. The parameter $\overset{d}{W}_{\bullet}$ represents the energy of the subnetwork in direction d. Accordingly, the macroscopic energy of the matrix, ψ_M, equals to

$$\Psi_M = \frac{1}{A_s}\int_S \overset{d}{W}_M\, \overset{d}{du}, \quad \text{where} \quad \overset{d}{W}_M = \overset{d}{W}_H + \overset{d}{W}_P. \tag{18}$$

The integration of the macroscopic energy of the three-dimensional matrix can be carried out numerically as follows

$$\Psi_M \cong \sum_{i=1}^{k} \overset{d_i}{W}_M\, w_i, \tag{19}$$

where w_i are weight factors corresponding to the collocation directions $d_i(i=1,2,\ldots,k)$. A set of k = 45 integration points on the half sphere was found to yield the best optimization between computational expenses and the resulted error due to the induced anisotropy (Ehret, Itskov, & Schmid 2010).

H Netowrk: A sample model for H network can be developed by assuming a subnetwork of N_H chains with identical number of segments n_H. Then, the energy of the subnetwork yields

$$W_H = N_H \psi(n_H, \lambda \bar{r}_0). \tag{20}$$

where $\bar{r}_0$ is the initial chain end-to-end distance. Such a simple module is the basis of many current hyperelastic models such as the full-network model by Wu and van der Giessen (Wu & Giessen 1993).

P Network: Here, the subnetwork is assumed to be made by different lengths of the polymer chains. $\tilde{N}(n,\bar{r})$ introduces the number of chains with the number of segments (relative length) n and the relative end-to-end distance $\bar{r}$. Since the filler aggregates are significantly stiffer than the polymer chains (Dargazany, Chen, Lin, Azad, & Alexander-Katz 2017, Dargazany, Lin, Khalili, Itskov, Chen, & Alexander-Katz 2016), the strain energy of the corresponding polymer-filler subnetwork is expressed by $\tilde{N}(n,\bar{r})\psi(n,\bar{r})$ Where ψ is given by Eq.16. Further, by integrating over the whole set of relative chain lengths n available in the subnetwork d, the strain energy of the subnetwork in direction d is formulated as

$$\overset{d}{W}_P = \int_{D_A} \tilde{N}(n,\bar{r})\psi(n,\bar{r})dn. \tag{21}$$

where D_A is the range of average chain lengths. The network evolution describes the competition of two simultaneous processes namely aggregate-polymer debonding and network rearrangement. (interested readers see (Dargazany, Khiêm, & Itskov 2015)).

The accuracy of the proposed model is evaluated against experimental tests of Adiprene-L100, a polyether urethane-based rubber, under various uniaxial relaxation tests provided by (Khan, Lopez-Pamies, & Kazmi 2006). Adiprene-L100 is almost totally viscoelastic material, with very little visco-plastic deformation. Thus, it is a suitable candidate for validation of the proposed viscoelastic model. Several relaxation tests were preformed by subjecting the sample to compressive load

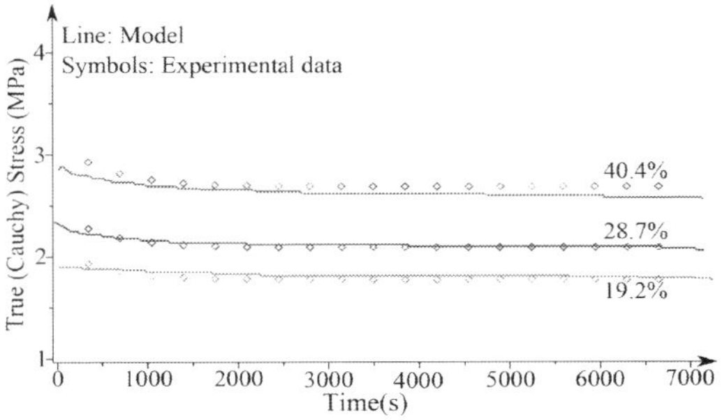

Figure 5. Validation of model predictions against experimental relaxation tests on Adiprene-L100 at different strain.

cycles with increasing amplitude. The material parameters γ, α, N_P, n_p and $\bar{r}_0$ were determined from the stress relaxation tests at small and large strain levels, i.e. 19.2, 28.7, and 40.4%. Fig. 5 shows the model predictions along with experimental data of the stress relaxation tests on other strain levels, where the results show excellent agreement (Fig. 5). The values of the fitting parameters in this model are $\alpha = 0.19e^{-2}$, $n = 1.85$, $\gamma = 0.85$ and $N = 0.78$.

3 CONCLUSION

A micro-mechanical based viscoelasticity model for large deformation of elastomers is developed. The model is based on a novel approach which represents visco-elasticity of the polymer matrix as an accumulative effect associated with the visco-elasticity of single polymer chains at micro-scale. The micro-scale description of visco-elasticity allows the model to be incorporated into many current macro-scale micro-mechanical models of quasi-static deformation and enables them to consider time-dependent changes of the material. The model has an extra advantage of few material parameters. We have shown that with a set of properly fitted material parameters, the developed model can predict, with reasonable accuracy, the experimental tests on relaxation tests not only for large deformations, but also for small deformation ranges. We believe that such micro-mechanical models are strongly advantageous in identifying the contribution of the elements of the micro-structure and the intrinsic processes responsible for different aspects of the rubber response at different conditions.

ACKNOWLEDGEMENT

This research was supported by the USDOT Tier 1 UTC Center for Highway Pavement Preservation (CHPP) at Michigan State University.

REFERENCES

Bergstrm, J. & M. Boyce (1998). Constitutive modeling of the large strain time-dependent behavior of elastomers. *Journal of the Mechanics and Physics of Solids 46* (5), 931–954. cited By (since 1996)358.

Christensen, R. M. (1982). Theory of viscoelasticity [m].

Dal, H. & M. Kaliske (2009). Bergström–boyce model for nonlinear finite rubber viscoelasticity: theoretical aspects and algorithmic treatment for the fe method. *Computational Mechanics 44* (6), 809–823.

Dargazany, R., H. Chen, J. Lin, A. I. Azad, & A. Alexander-Katz (2017). On the validity of representation of the inter-particle forces of a polymer-colloid cluster by linear springs. *Polymer 109*, 266–277.

Dargazany, R., J. Lin, L. Khalili, M. Itskov, H. Chen, & A. Alexander-Katz (2016). Micromechanical model for isolated polymer-colloid clusters under tension. *Physical Review E 94* (4), 042501.

Dargazany, R., K. Hörnes, & M. Itskov (2013). A simple algorithm for the fast calculation of higher order derivatives of the inverse function. *Applied Mathematics and Computation 221*, 833–838.

Dargazany, R., V. N. Khiêm, & M. Itskov (2015). A generalized network decomposition model for the quasi-static inelastic behavior of filled elastomers. *International Journal of Plasticity 63*, 94.

Dargazany, R., V. N. Khiêm, E. A. Poshtan, & M. Itskov (2014). Constitutive modeling of strain-induced crystallization in filled rubbers. *Physical Review E 89* (2), 022604.

Ehret, A., M. Itskov, & H. Schmid (2010). Numerical integration on the sphere and its effect on the material symmetry of constitutive equations—a comparative study. *International Journal for Numerical Methods in Engineering 81*, 189.

Ganss, M., B. K. Satapathy, M. Thunga, R. Weidisch, P. Pötschke, & A. Janke (2007). Temperature dependence of creep behavior of pp–mwnt nanocomposites. *Macromolecular rapid communications 28* (16), 1624–1633.

Göktepe, S. & C. Miehe (2005). A micro-macro approach to rubber-like materials. Part III: The micro-sphere model of anisotropic Mullins-type damage. *Journal of the Mechanics and Physics of Solids. 53*, 2259.

Khan, A. S., O. Lopez-Pamies, & R. Kazmi (2006). Thermo-mechanical large deformation response and constitutive modeling of viscoelastic polymers over a wide range of strain rates and temperatures. *International Journal of Plasticity 24* (4), 581–601.

Kolarik, J. & A. Pegoretti (2008). Proposal of the boltzmann-like superposition principle for nonlinear tensile creep of thermoplastics. *Polymer Testing 21* (5), 596–606.

Marckmann, G. & E. Verron (2006). Comparison of hyperelastic models for rubber-like materials. *Rubber Chemistry and Technology 79*, 835.

Reese, S. (2003). A micromechanically motivated material model for the thermo-viscoelastic material behaviour of rubber-like polymers. *International Journal of Plasticity 19* (7), 909–940.

Wu, P. & E. v. d. Giessen (1993). On improved network models for rubber elasticity and their applications to orientation hardening in glassy polymers. *Journal of the Mechanics and Physics of Solids. 41*, 427.

Yang, J., Z. Zhang, K. Friedrich, & A. K. Schlarb (2007). Creep resistant polymer nanocomposites reinforced with multiwalled carbon nanotubes. *Macromolecular rapid communications 28* (8), 955–961.

Constitutive Models for Rubber X – Lion & Johlitz (Eds)
© 2017 Taylor & Francis Group, London, ISBN 978-1-138-03001-5

A framework for analyzing hyper-viscoelastic polymers

A.R. Trivedi & C.R. Siviour
Solid Mechanics and Materials Engineering Group, Department of Engineering Science, University of Oxford, Oxford, UK

ABSTRACT: Hyper-viscoelastic polymers have multiple areas of application including aerospace, biomedicine, and automotive. Their mechanical responses are therefore extremely important to understand, particularly because they exhibit strong rate and temperature dependence, including a low temperature brittle transition. Relationships between the response at various strain rates and temperatures are investigated and a framework developed to predict large strain response at rates of c. 1000 s^{-1} and above where experiments are unfeasible. A master curve of the storage modulus's rate dependence at a reference temperature is constructed using a DMA test of the polymer. A frequency sweep spanning two decades and a temperature range from pre-glass transition to pre-melt is used. A fractional derivative model is fitted to the experimental data, and this model's parameters are used to derive stress-strain relationships at a desired strain rate.

1 INTRODUCTION

The overall research goal is to be able to predict the mechanical response of polymers and their composites at high strain rates purely by knowing their composition. And vice versa, this would also allow a material to be designed to perform to a certain set of mechanical requirements or constraints. In order to obtain the mechanical properties of polymers at high rates, there are three potential options that were investigated in the current research.

Most simply, to obtain the material response at high strain rates, one could test the polymer at high rates. This would be ideal, however it is not as simple as it initially sounds. Due to the low stiffness of the material, the ensuing low wave speed means that stress equilibrium is not easily reached, and under dynamic loading, the forces measured at the ends of the specimen may not be representative of the material response. This can also be understood as structural vibrations masking the material response as the resonant frequency of the sample is similar to the frequencies in the loading pulse. A second problem, encountered in the split-Hopkinson pressure bar (SHPB), the standard apparatus for material characterization at these rates, is that the forces supported by soft specimens are low, making measurements difficult. Some effort has been made in trying to match bar and specimen impedances to allow the force transmitted to be maximized and this issue to be overcome (Chen 1999, Siviour 2008). A good description of high strain rate testing of low impedance materials is given in Gray & Blumenthal (2000).

The second option allows for a qualitative understanding of the temperature-rate interdependence through a one-to-one equivalence between a characteristic stress at a particular rate and the same characteristic stress at a lower rate, but different temperature. This is based on the classic result of the time-temperature superposition (TTS) principle in which one can quantify changes in rate to changes in temperature (Siviour et al. 2005, Roland 2011, Kendall & Siviour 2014). The issue with this method is that it is only useful for obtaining a descriptive relationship between a particular value of stress and the temperature at which one must conduct the test to obtain similar mechanical properties at a lower strain rate. It is not guaranteeing the full stress-strain relationship would be possible to obtain by this equivalence, and nor is it allowing us to predict the mechanical response since it refers to experimental data points that must have been collected prior to conducting this time-temperature equivalence.

The third option is where the majority of the work done in the current research is concentrated. It relies on a dynamic mechanical and thermal analysis (DMA) test and the ability to use TTS to obtain master curves and shift factors detailing the material's rate and temperature dependences. By fitting a suitable model to the experimental data from the DMA test, it is possible to obtain material parameters for a constitutive model that would allow for the prediction of the stress-strain response at any appropriate strain rate.

2 EXPERIMENTAL OVERVIEW

2.1 *Material*

The polymer material used for the current research as presented in this paper was a commercial black neoprene rubber supplied in the form of a 5 mm sheet by Brammer, UK.

2.2 *Dynamic mechanical and thermal analysis*

For the DMA tests, rectangular samples of the material were used with dimensions 50 × 10 × 5 mm. The tests were conducted on a TA Instruments Q800 and all tests were performed in the dual cantilever configuration. An isothermal frequency sweep in 2°C increments was used with values of 0.5, 2, 5, and 10 Hz and a temperature range from −50°C to 80°C.

2.3 *Compression tests*

From the neoprene rubber sheet, cylindrical samples with a diameter of 5 mm were cut out for use in the compression tests. Compression tests were conducted at a variety of strain rates from 10^{-3} to 2100 s^{-1} at a temperature of 25°C, and at a variety of temperatures from −100°C to 80°C at a rate of 10^{-2} s^{-1}.

2.3.1 *Low strain rate compression tests*

A commercially available Instron 5980 electromechanical static testing machine was used for tests at strain rates of 10^{-3}, 10^{-2}, and 10^{-1} s^{-1}. A 5 kN load cell was used for measurements and the machine crosshead displacement to obtain strain. Preliminary experiments showed that at room temperature there was minimal machine compliance, so the cross-head displacement gave an accurate measure of the movement of the loading platens. Constant time strain rate control was used in all experiments.

When testing at a rate of 10^{-2} s^{-1} at sub-ambient temperatures, an environmental chamber was used with a liquid nitrogen feed. Temperature was controlled with the default thermostat which was accurate to ±1°C. The environmental chamber's feedback control thermocouple is situated at the back of the chamber away from the specimen, and so a second thermocouple was inserted into the loading platen itself to measure the temperature as close to the specimen as possible. Only when this thermocouple reading matched the chamber's thermocouple, was the test conducted. The same chamber and temperature measurements were used for supra-ambient temperature tests.

At lower temperatures, due to higher loads experienced, there was a need to increase the load cell to 50 kN and as the higher loads could lead to

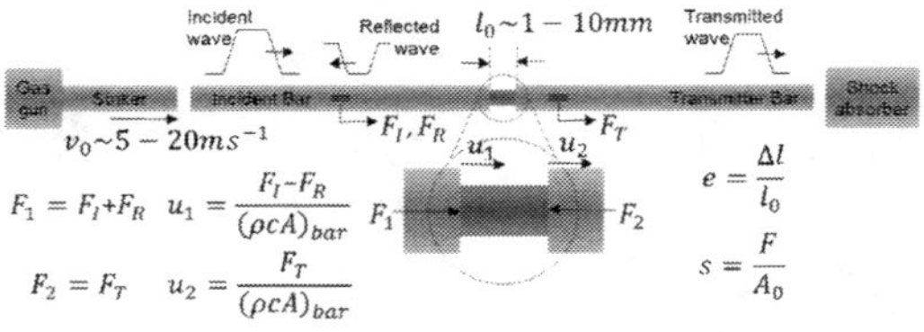

Figure 1. Schematic of the SHPB set-up used for the high strain rate tests.

rig compliance, an extensometer with a sufficiently large gauge length of 12 mm was used to obtain local strain measurements.

2.3.2 *High strain rate compression tests*

For compression tests at a strain rate of the order 10^3 s^{-1}, an in-house SHPB was used. Details of the analysis procedure can be found in Gray & Blumenthal (2000).

Figure 1 shows a schematic of the SHPB set-up. Ti-6 Al-4V alloy bars with a 12.7 mm diameter were used. Using compressed gas to fire the striker bar, speeds of up to around 20 m s^{-1} were achievable giving the possibility of testing at rates of order 10^3 s^{-1}.

3 RESULTS AND ANALYSIS

3.1 *Varying rate tests*

The results of the low strain rate tests on the rubber sample are presented in Figure 2. It is evident from the results of the low strain rate tests that when the strain rate increases, the stress experienced at any level of strain is higher. If we consider a characteristic stress at a strain value of 0.1, then the stress is seen to increase from 0.79 MPa at a rate of 10^{-3} s^{-1} to 1.00 MPa at a rate of 10^{-1} s^{-1}. This represents on average a 13% increase in characteristic stress per decade at these low strain rates.

Results of the high strain rate tests are presented in Figure 3. If we again consider the characteristic stress at a strain value of 0.1, we find that average at a strain rate order of 10^{-3} s^{-1} is 6.5 MPa. This is a significant increase from the lower orders of strain rate. The oscillations on the stress-strain curves are a result of structural vibrations in the specimen. These would be more significant for lower modulus specimens or higher strain rates.

The overall behavior can be characterized by the graph in Figure 4. Although further data are required to fully define the rate dependence, from previous studies we expect an approximately bilinear relationship between stress and strain rate, with a stronger dependence in the high rate experiments. The effect of glass transition leading

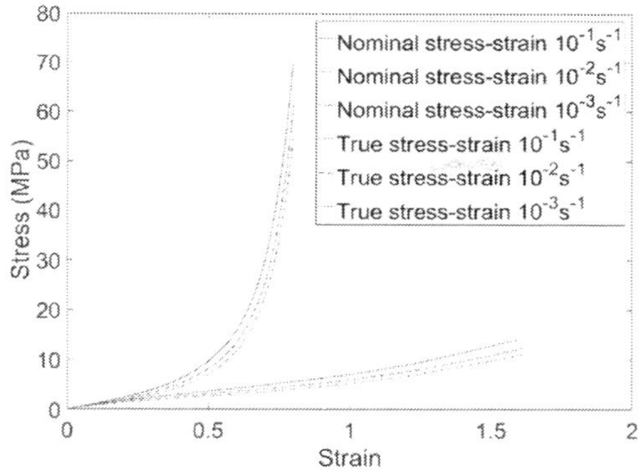

Figure 2. Comparison of stress-strain relationships for the rubber sample averaged over a series of three tests conducted at a variety of low strain rates.

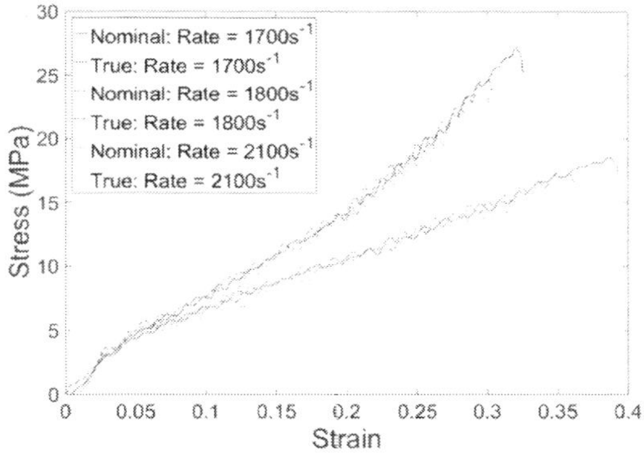

Figure 3. Comparison of stress-strain relationships for the rubber sample over a series of three tests conducted at a variety of high strain rates.

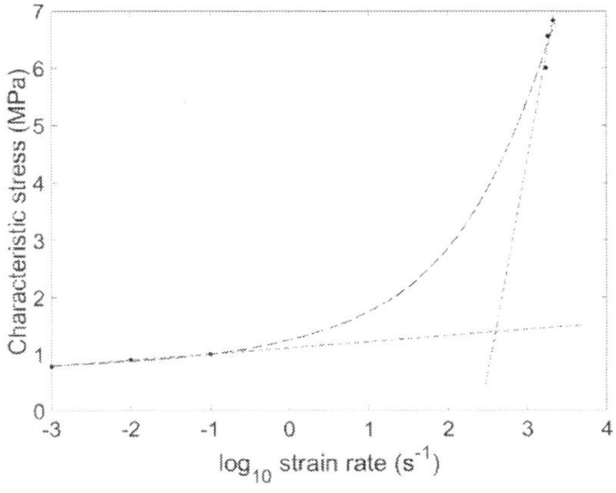

Figure 4. Behavior of characteristic stress as a function of the strain rate.

to the rate dependence of the mechanical properties of the rubber is clear to see, and will now be demonstrated by comparison to tests at different temperatures.

3.2 *Varying temperature tests*

The results of the compression tests conducted at a variety of temperatures are presented in Figure 5.

From these uniaxial tests, it is clear to see that there is a remarkable distinction in mechanical properties at lower temperatures compared to higher temperatures. The transition occurs just below –40°C. Above this temperature, the response of the material is rubbery. Below this temperature, the response is glassy and shows brittle fracture characteristics. In fact, this was noticeable from physical observations of the rubber post testing.

Figure 6 shows a similar dependence on temperature as Figure 4 showed with the strain rate.

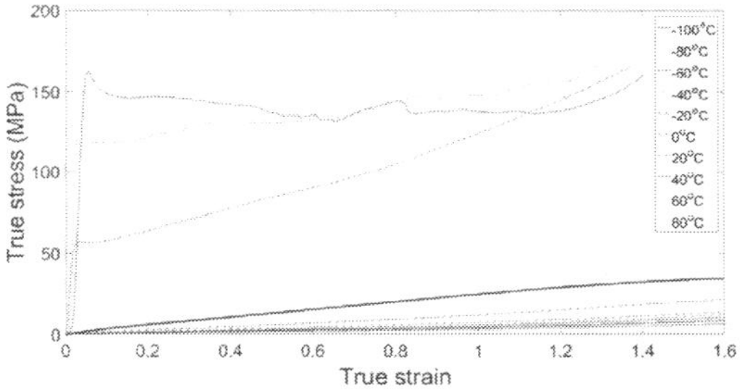

Figure 5. Comparison of stress-strain relationships for the rubber sample averaged over a series of three tests conducted at a variety of temperatures at a strain rate of 10^{-2} s^{-1}.

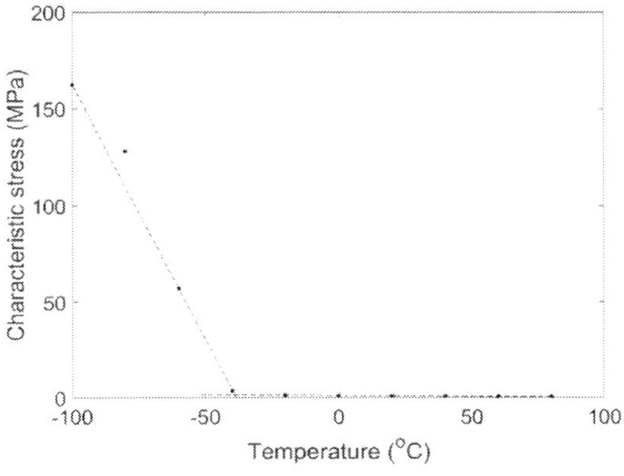

Figure 6. Behavior of characteristic stress as a function of the temperature.

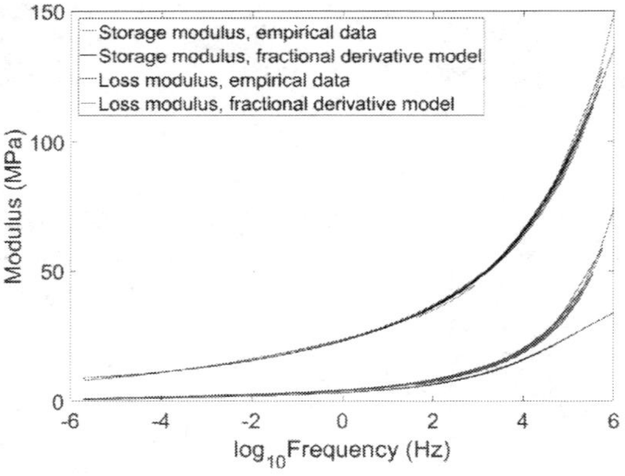

Figure 7. Master curve for averaged DMA data at a reference temperature of 25°C.

3.3 *Temperature-rate equivalence*

Qualitatively, it is already possible to note using Figures 4, 6 that high stresses occur at both high strain rates and low temperatures. So it does not seem too ridiculous to postulate that there may be some sort of equivalence in material response between high rate compression tests and low temperature compression tests conducted at a lower strain rate.

For example, if we consider where we want to know the likely stress-strain relationship of this rubber at a strain rate of 2000 s^{-1}, using Figure 4, we note that the characteristic stress at this strain rate is around 6.5 MPa. This same value of characteristic stress is present in Figure 6 at a temperature of –45°C. This means we can obtain a representative stress-strain relationship for the rubber by using this equivalence and testing it under compression at –45°C and a strain rate of 10^{-2} s^{-1}.

3.4 *DMA tests*

A DMA test was conducted as explained in § 2.2, and TTS was performed with the data by shifting data at various temperature steps parallel to the logarithmic frequency axis by a certain shift factor to create a master curve from this series of overlapping data.

Once this has been completed with all the temperature steps, the master curve as shown in Figure 7 is obtained. Further details of how to construct master curves can be found in literature (Morrison 2001, Tobolsky 1960).

4 MODELLING AND SIMULATIONS

4.1 *Hyperelastic constitutive model*

To model the hyperelastic behavior of the rubber, an Ogden model (1972, 1984) was used as it has a well-established base in literature (Shergold et al. 2006, Yoon 2016, Kossa & Berezvai 2016). It is based on isotropic, isothermal, and incompressible assumptions. For the case of the rubber in this paper, these assumptions are valid.

$$\phi = \frac{2\mu}{\alpha^2}\left(\lambda_1^\alpha + \lambda_2^\alpha + \lambda_3^\alpha - 3\right) \quad (1)$$

$$\sigma_3 = \frac{2\mu}{\alpha}\left[\lambda_3^{\alpha-1} - \lambda_3^{-\left(1+\alpha/2\right)}\right] \quad (2)$$

The Ogden model in Equation 1 is based on ϕ, a strain energy density per unit volume; μ = representative shear modulus; α = strain hardening exponent; and λ_i = principle stretch ratios for the three Cartesian directions. Based on the special case of the Ogden where the material is incompressible and exposed to uniaxial stress loading, it is possible to obtain the stress relationship as shown in Equation 2.

In Figure 8a, the Ogden model applied using a rate independent strain hardening exponent and a rate dependent shear modulus term shows an excellent fit to the experimental data. In Figure 8b, the fit

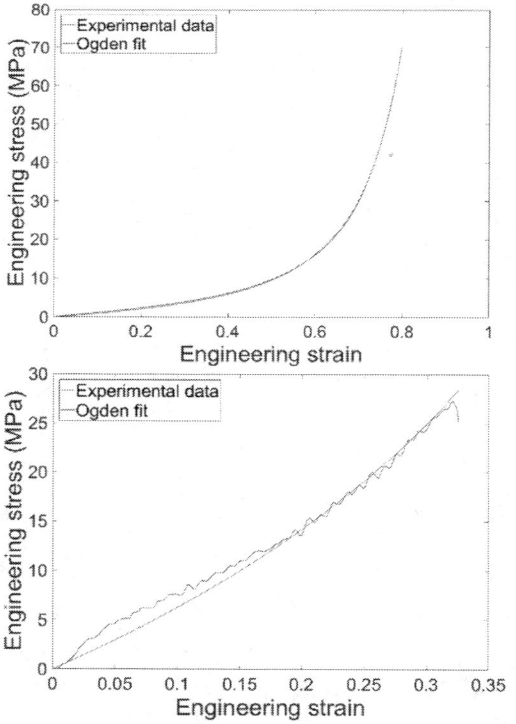

Figure 8. Ogden model fit for samples tested at a rate of 10^{-1} s^{-1} (a, top) and one tested at a rate of 2100 s^{-1} (b, bottom).

is still very good but deviates from the experiment at low levels of strain. This is due to the added rate dependence of the viscoelastic component to the stress at the higher rate.

4.2 *Viscoelastic constitutive model*

To capture the low strain viscosity that was lacking in the hyperelastic model in Figure 8b, the next step of the modelling process was to decompose the stresses into their hyperelastic and viscoelastic components. The hyperelastic element was the same as the Ogden model in §4.1, but a Standard Linear Solid (SLS) model was used for the viscoelastic component. A diagrammatic representation of which can be found in Figure 10a. This decomposition is seen in recent literature like Bai (2016) and is well suited to model the rubber here.

Figure 9 shows the result of the superposition of the hyperelastic stress and viscoelastic stress components to form the SLS curve which clearly follows the experimental data much better than the simple Ogden model.

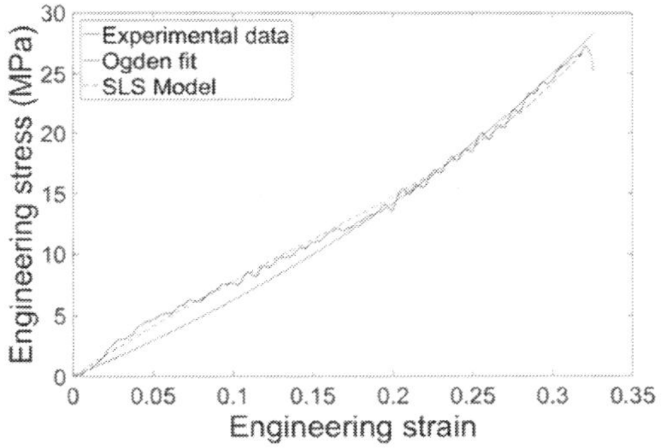

Figure 9. Comparison of the fits to the experimental data using either a hyperelastic Ogden model or an SLS model.

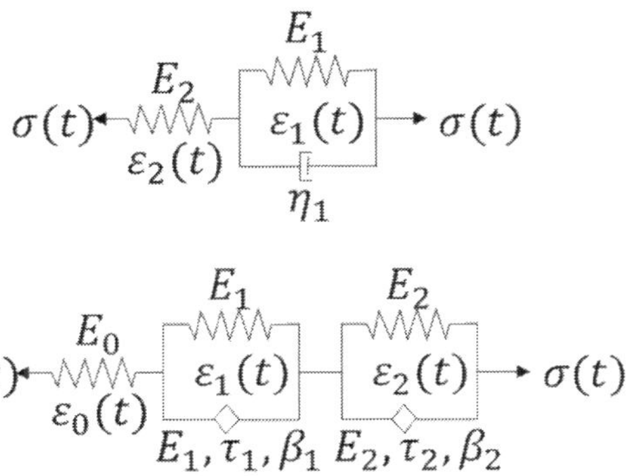

Figure 10. Diagrammatic representation of the SLS model (a, top) and the fractional SLS model (b, bottom).

4.3 *Fractional derivative constitutive model*

Although it has been possible to use constitutive models to provide excellent fits to the experimental data, the aim of the current research is to be able to predict the high strain rate response of the rubber doing a minimum number of experimental tests, especially those conducted at an unfeasible higher strain rate. In order to achieve this, a fractional derivative constitutive model is fitted directly to the DMA results; the results can be seen again in Figure 7.

In modelling the storage modulus, the fractional SLS model shown in Figure 10b provides an excellent fit. There is also a decent fit at lower frequencies for the loss modulus. The storage and loss moduli are obtained by solving the differential equation for the fractional SLS model to obtain a complex modulus as shown in Equation 3 and then taking either the real part (storage modulus) or the imaginary part (loss modulus).

In order to obtain a stress-strain relationship, the modulus in the time domain needs to be found first. To convert the complex modulus into time domain is quite cumbersome when the values of β are not equal and take the value of one (reduces the problem to that of the SLS leading to an exponential time domain term) or a half (leads to an error function term in time domain as seen in Koeller (2007)). Hence, an approximate interconversion method is used with Equations 4–6, which has been verified with examples (Schapery & Park 1999).

Once the modulus in time domain is found, a standard hereditary integral operation using Equation 7 can be performed of which examples can be found in Park (2001) and Bai (2016). Since the majority of points from the results of the TTS are at low values of time, it makes sense to change variables such that the integral can be performed using ln t instead as shown in Equation 8. To counter the issue with the limit of the integral starting from ln 0, an approximate initial value of stress $\sigma_0 = E_0\dot{\varepsilon}e^{du}$ is used instead, where E_0 is the instantaneous modulus, $\dot{\varepsilon}$ the constant strain rate, and du the first increment in ln-time.

The viscoelastic stress obtained in this manner can then be superposed with the hyperelastic component from the Ogden model described in § 4.1 with the rate-dependent terms replaced with independent ones. The results of this superposition can be seen in Figure 11.

$$E^* = E_0 + \frac{E_1(i\omega)^{\beta 1}}{(i\omega)^{\beta 1} + 1/\tau_1^{\beta 1}} + \frac{E_2(i\omega)^{\beta 2}}{(i\omega)^{\beta 2} + 1/\tau_2^{\beta 2}} \tag{3}$$

$$E(t) \cong \frac{1}{\lambda}\tilde{E}(s)\Big|_{s=1/t} \tag{4}$$

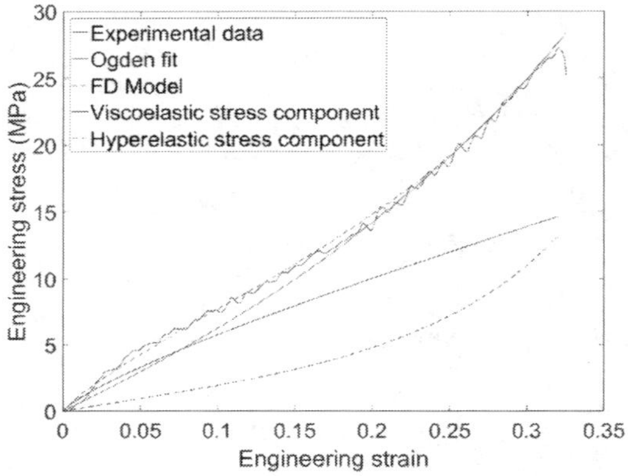

Figure 11. Comparison of the fits to the experimental data using the hyperelastic Ogden model as seen before, and the result of the fractional derivative model, along with its components.

$$\tilde{\lambda} = \Gamma(1-n) \tag{5}$$

$$n = \frac{d \log \tilde{E}(s)}{d \log s} \tag{6}$$

$$\sigma = \int_0^t E(t-\xi) \frac{d\varepsilon}{d\xi} d\xi \tag{7}$$

$$\sigma = \dot{\varepsilon} \int_{\ln 0}^{\ln t} E(t-e^u) e^u du \tag{8}$$

5 CONCLUSIONS

This paper presents what has been possible so far in providing a framework for better understanding mechanical properties of polymers at high strain rates. This has been done mainly by fitting a fractional constitutive model to DMA data, and then using that model to verify the stress-strain relationship at a strain rate of $2100 s^{-1}$. However, there are some challenges which allows for further research to be conducted in this area.

The predictive intention of the framework needs to be fully verified with the investigation of previously unexplored strain rates; finding the stress-strain relationship from the model before doing the actual experiment to see whether it is indeed correct as expected. An FEA model will also be produced which will be able to simulate the experimental behaviour and can be used to verify the predictive framework.

Overcoming these obstacles will allow for a truly cohesive framework to be developed to allow for the prediction of high strain rate properties of hyper-viscoelastic polymers and their composites.

ACKNOWLEDGEMENTS

This material is based upon work supported by the Air Force Office of Scientific Research, Air Force Materiel Command, USAF under Award No. FA9550-15-1-0448. Any opinions, findings, and conclusions or recommendations expressed in this publication are those of the author(s) and do not necessarily reflect the views of the Air Force Office of Scientific Research, Air Force Materiel Command, USAF.

This author would also like to thank Richard Duffin of the Engineering Science Solid Mechanics workshop for his technical assistance, and Nick Hawkins of the Oxford Silk Group in the Zoology Department, University of Oxford for his ongoing support with the TA Q800 DMA apparatus.

REFERENCES

Bai, Y. et al. 2016. A Hyper-Viscoelastic Constitutive Model for Polyurea under Uniaxial Compressive Loading. *Polymers*, 8(4): 133.

Chen, W., Zhang, B. & Forrestal, M.J., 1999. A Split Hopkinson Bar Technique for Low Impedance Materials. *Experimental Mechanics*, 39(2): 81–85.

Gray, G.T.III. & Blumenthal, W.R., 2000. *Split-Hopkinson Pressure Bar Testing of Soft Materials.* ASM Handbook Mechanical Testing and Evaluation. Vol. 8. ASM International. pp. 488–496.

Jeong, H.T. et al. 2004. The mechanical relaxations of a Mm(55)Al(25)Ni(10)Cu(10) amorphous alloy studied by dynamic mechanical analysis. *Materials Science and Engineering A – Structural Materials Properties Microstructure and Processing*, 385(1–2): 182–186.

Kendall, M.J. & Siviour, C.R. 2014. Rate dependence of poly(vinyl chloride), the effects of plasticizer and time-temperature superposition. *Proceedings of the Royal Society A: Mathematical, Physical and Engineering Sciences*, 470(2167): 20140012–20140012.

Koeller, R.C., 2007. Toward an equation of state for solid materials with memory by use of the half-order derivative. *Acta Mechanica*, 191(3–4): 125–133.

Kossa, A. & Berezvai, S. 2016. Novel strategy for the hyperelastic parameter fitting procedure of polymer foam materials. *Polymer Testing*, 53: 149–155.

Morrison, F.A. 2001. *Understanding Rheology.* 1st ed. Oxford: Oxford University Press.

Ogden, R.W. 1972. Large deformation isotropic elasticity - On the correlation of theory and experiment for incompressible rubber like solids. *Proceedings of the Royal Society A: Mathematical, Physical and Engineering Sciences*, 326(1567): 565–584.

Ogden. R.W. 1984 *Non-linear Elastic Deformations.* 1st ed. Chichester: Wiley.

Park, S.W. 2001. Analytical modeling of viscoelastic dampers for structural and vibration control. *International Journal of Solids and Structures*, 38(44–45): 8065–8092.

Roland, C.M. 2011. *Viscoelastic Behavior Of Rubbery Materials*. 1st ed. Oxford: Oxford University Press.

Schapery, R.A. & Park, S.W. 1999. Methods of interconversion between linear viscoelastic material functions. Part II—an approximate analytical method. *International Journal of Solids and Structures*, 36(11): 1677–1699.

Shergold, O.A., Fleck, N.A. & Radford, D. 2006. The uniaxial stress versus strain response of pig skin and silicone rubber at low and high strain rates. *International Journal of Impact Engineering*, 32(9): 1384–1402.

Siviour, C.R. et al., 2005. The high strain rate compressive behaviour of polycarbonate and polyvinylidene difluoride. *Polymer*, 46(26): 12546–12555.

Siviour, C.R. et al., 2008. High strain rate properties of a polymer-bonded sugar: their dependence on applied and internal constraints. *Proceedings of the Royal Society A: Mathematical, Physical and Engineering Sciences*, 464(2093): 1229–1255.

Tobolsky, A.V. 1960. *Properties and Structures of Polymers.* New York: Wiley.

Williams, M.L., Landel, R.F. & Ferry, J.D., 1955. The Temperature Dependence of Relaxation Mechanisms in Amorphous Polymers and Other Glass-forming Liquids. *Journal of American chemical society*, 77(12): 3701–3707.

Yoon, S.h. Winters, M. & Siviour, C.R. 2016. High Strain-Rate Tensile Characterization of EPDM Rubber Using Non-equilibrium Loading and the Virtual Fields Method. *Experimental Mechanics*, 56(1): 25–35.

Constitutive Models for Rubber X – Lion & Johlitz (Eds)
© 2017 Taylor & Francis Group, London, ISBN 978-1-138-03001-5

On the influence of swelling on the viscoelastic material behaviour of natural rubber

F. Neff, A. Lion & M. Johlitz
Department of Aerospace Engineering, Institute of Mechanics, University of the Bundeswehr Munich, Neubiberg, Germany

ABSTRACT: Liquid uptake of polymers, especially of elastomers, is associated with swelling. The swelling process is induced by diffusion and causes stresses, which interact with the diffusion again. Thus, swelling is a coupled phenomenon of mechanical behaviour and mass transport. The material behaviour of a swelling elastomer is investigated by experiments referring to the equilibrium state of swelling, when the liquid concentration in the elastomer reaches a saturation level. Following established viscoelastic material models for large deformations, a split into equilibrium and viscose part is common. Both parts are influenced by the resence of fluid. This influence is the focus of this paper by using a typical experimental set up to determine material parameters. Furthermore, experiments show that swelling depends on the externally applied stress. This effect leads to the discussion if the presented well-established experimental set up is usable to determine viscoelastic material parameters. In this context it is pointed out, that the usability of the experiments depends on the time scale of the adoption.

1 INTRODUCTION

Swelling behaviour of polymers has been investigated by several authors for almost one century. The basic equations of diffusion go back to investigations of Fick (Fick 1855) in the mid 19th century. The mathematics of the diffusion transport are discussed by Crank (Crank 1956), who gives a lot of approaches to solve the diffusion equations with several initial and boundary conditions. To determine the chemical potential of polymers, fundamental works are published by Flory (Flory 1942), (Flory 1961) and Huggins (Huggins 1942), which give the opportunity to calculate liquid uptake and dilution of polymers with a statistical thermodynamics approach. Treloar (Treloar 1975) uses this formulation and extends it to cross-linked materials like rubber and carried out, that the swelling process is coupled with the mechanical loads.

Fundamental works of isotropic, hyperelastic and incompressible material behaviour are provided by Treloar, Rivlin and Mooney. Treloar proposed a strain-energy function for nonlinear incompressible materials derived from the network theory with Gaussian statistics (Treloar 1944). This strain-energy function is called neo-Hookean form. Furthermore, Weitsmann (Weitsmann 2012) proposes models to describe hyperelastic and linear-viscoelastic materials under swelling influences. Other models were founded by e.g. (Weitsman 1987), (Derrien & Gilormini 2006), (Derrien & Gilormini 2009). In those models the presence of fluid has no influence on the mechanical properties. Chester (Chester & Anand 2010) publishes a model for large deformations under swelling using the theory of Flory-Huggins. This work proposes a similar approach to model large deformations of swelling elastomers with an employed hyperelastic material behaviour as well. Additionally, the material parameters of the elastomer depend on the current swelling state. In addition to this model, the viscous part of the material behaviour should be taken under consideration. Thanks to the multiplicative split of the deformation gradient, which was indirectly proposed by Flory (Flory 1961) and formulated by Lubliner (Lubliner 1985), the constitutive model of viscoelastic material behaviour under large deformation can be seen as the sum of an equilibrium stress and a non-equilibrium part, the so called overstress. Thus, under large deformations the rheological model used to describe linear viscoelastic behaviour (Ferry 1980) could be taken into account, although its elements are not linear anymore. Therefore, equations for the stress-strain relation cannot be formulated directly out of the rheological model like in the linear case. For derivation of a continuum mechanic model for viscoelastic material behaviour see e.g. (Lion 1998), (Johlitz, Dippel, & Lion 2015).

Experiments distinguish between equilibrium stress and non-equilibrium stress. Taking the equilibrium part under observation, a multistep relaxation test or a multi hysteresis test is established (Diercks 2015). A standard strain-controlled relaxation test is usually conducted to determine the non-equilibrium part. Dynamic mechanical analysis (DMA) are established especially for linear viscoelastic materials. Due to the maximum amplitude of the accessible DMA machines large deformation cannot be set up with this method for the samples in this research.

2 EXPERIMENTS

The carbon filled natural rubber (NR) used in this research is provided by BOGE Rubber & Plastics, see Table 1. The alkane n-octane is chosen to observe the swelling effect. This composition was found by gravimetric, calorimetric and mechanical pretests, because it shows the required reversible material behaviour with respect to the swelling process at isothermal conditions. All experiments are executed at constant room-temperature. The swelling state is characterised by the mass proportion $\rho_{\text{rel}} = \frac{m_{\text{octan}}}{m_{\text{NR}}}$, which defines the octane concentration in the NR. After a storage time of more than 24 hours for the NR in pure octane, an equilibrium mass proportion is reached.

To reduce the influence of strain depended softening (Mullins effect) (Mullins 1969), all used samples are preconditioned by applying 10 of cycles of up to 25% additional to the maximum strain applied in following the experiment. After the preconditioning, the specimens need at least one week to recover.

The boundary concentration has to be changed for testing the mechanical properties of different fluid contents. Therefore, octane is mixed with ethanol, as ethanol shows no swelling effect with the used NR. Figure 1 shows the equilibrium fluid content with different boundary concentrations. Storing the samples in pure octane leads to an equilibrium mass proportion of $\rho_{\text{rel}}^{(3)} = 0{,}85$. Thinning down the pure octane to 50 Vol% with ethanol at room-temperature yields to $\rho_{\text{rel}}^{(2)} = 0{,}49$. Besides

Table 1. Composition of the used carbon filled natural rubber.

Description	Content [phr]
NR SVR CV60	100.00
Carbon black N-772	50.00
ZnO	5.00
Stearic Acid	2.00
Sulphur	1.50
CBS	2.00
TBzTD	0.20

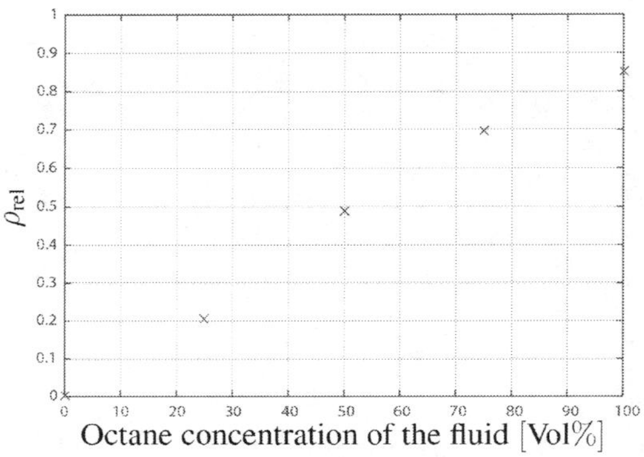

Figure 1. Equilibrium mass proportion for NR samples stored in fluid with different octane concentration.

these two swollen states the unswollen state is taken into account $\rho_{\text{rel}}^{(1)} = 0$. The mechanical testing is executed with those three fluid contents.

All experiments are uniaxial tension or uniaxial pressure tests and the applied stress refers to the cross-section area of the unloaded sample (1st Piola-Kirchhoff-stress). With an increase of the fluid content, the cross-section area of the samples increases as well. Therefore, the stress has to be calculated with respect to this change. For the tension tests S2 tension samples (DIN 53504, type 2)[1] are used. For the pressure tests, cylindrical samples with a radius and height of 6,25 mm are used. Before the mechanical tests are done at different fluid contents, the samples are stored in the fluid to reach equilibrium. All mechanical tests are performed in the fluid to provide a loss of fluid due to shrinking during the experiment.

2.1 *Equilibrium stress*

2.1.1 *Multistep relaxation*

First, the multistep relaxation test is performed to investigate the equilibrium stress. This multistep relaxation test is carried out with four steps and an increase of 30% strain at each step. The holding time per step is chosen to 1,5h, see Figure 2. The mean value of the stress at the end of two equivalent steps (increasing and decreasing path) approximates the equilibrium stress. The accuracy of this method depends directly on the holding time at each step. As longer this time the more detailed the equilibrium value is approximated. But the research time increases dramatically. Therefore, a

[1]DIN 53504 is a German standard titled: 'Testing of rubber—determination of tensile strength at break, tensile stress at yield, elongation at break and stress values in a tensile test'. This standard is set by the German Institute of Standardization.

compromise has to be made between accuracy and experimental duration. The multistep relaxation tests are executed for pressure with an increasing 7.5% strain per step, 2 h holding time and six steps of increasing pressure.

Those tests are executed for the three different equilibrium fluid contents. The calculated equilibrium stress over strain is depicted Figure 3.

2.1.2 *Multi hysteresis test*

Another method to determine the equilibrium stress is the multi hysteresis test. Strain cycles are performed under a low strain rate of $1\frac{\%}{s}$. After a certain number of cycles the strain increases to the next strain step. Five cycles are performed, because the material showed a static hysteresis curve in pretests after three or four cycles regarding the above mentioned strain rate. The fifth cycle is used to determine the equilibrium stress. The diagram of stress over strain denotes the centre of the area embedded by the hysteresis curve the equilibrium stress at the related strain, see Figure 4. For further discussion and details about this experimental set up see (Diercks 2015). Under tension ten hysteresis curves are recorded with an increase of 10% strain for each step. Under pressure eight curves are recorded with an increase of 5% strain for each step. Calculating the equilibrium stresses depending on the strain yields Figure 5.

2.2 *Overstress*

Taking the overstress due to viscous material response under observation, it is common to use relaxation tests in the time domain. Therefore, the sample is deformed with a constant strain. This deformation is brought to the sample with the maximal strain rate of the test machine. The relaxation of the stress over time is recorded as soon as the maximum strain is reached. Out of conception of the model for viscoelastic material behaviour this stress converges the equilibrium stress for infinite time ranges. Hence, as the experimental time is finite, a compromise was made between accuracy and experimental duration.

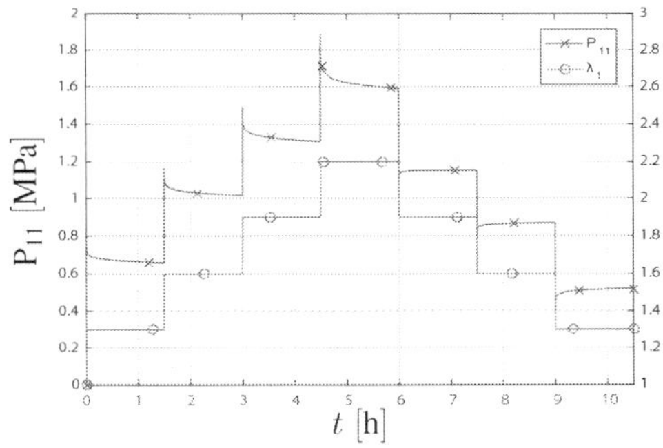

Figure 2. Strain controlled multistep relaxation test.

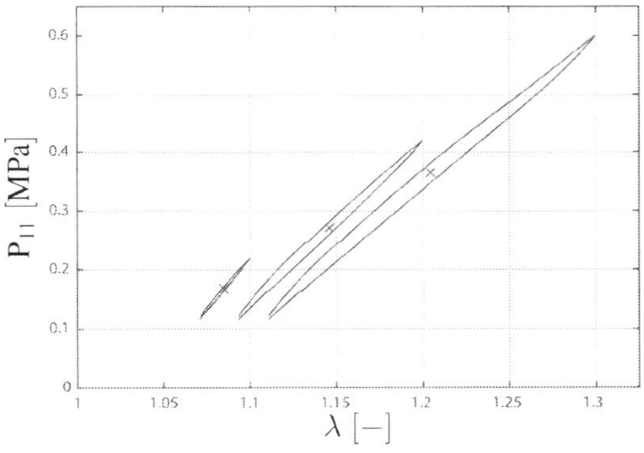

Figure 4. Equilibrium stress (x) out of the cycles of a multi hysteresis test.

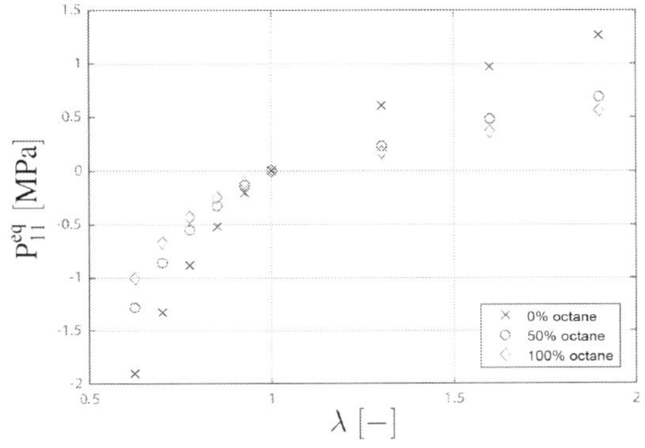

Figure 3. Equilibrium stress over strain out of multistep relaxation tests for different fluid contents.

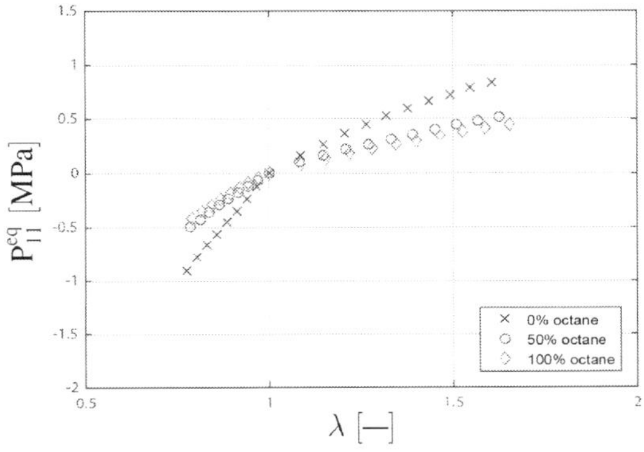

Figure 5. Equilibrium stress over strain out of multi hysteresis tests for different fluid contents.

3 RESULTS AND DISCUSSION

3.1 *Equilibrium stress: Comparison between methods of determination*

The duration of the experiments differs significantly. The multi hysteresis tests exceeded for tension take around 25 min and the equivalent multistep relaxation tests take 10,5h. On the one hand, with a view to a fast testing process, the multi hysteresis test is advantageously. On the other hand, the first step out of the multistep relaxation test can be used to determine the viscous behaviour. Therefore, it is necessary to execute a second experiment when using multi hysteresis test.

First, the results of unswollen elastomers depict in Figure 6 are under consideration. Both experiments lead to good accordance for the equilibrium stress under pressure. With a tension strain, the equilibrium stress is determined higher by the multistep relaxation test. Comparing the relaxation under tension of one step in the increasing path with the equivalent step in the decreasing path, shows that the relaxation in the decreasing path is faster. Thus, the equilibrium stress for tension calculated out of the multistep relaxation test is slightly overestimated by building the mean value.

Second, the results of the swollen elastomer in pure octane (Figure 7) are under consideration. Comparing multi hysteresis and multistep relaxation again, yields to the observation that the estimated equilibrium stress under tension with the multistep relaxation test is now lower than the estimated equilibrium stress with the multi hysteresis test. This phenomenon could be explained the dependence of the equilibrium mass proportion on the current loading, as established by (Treloar 1975). A tension load increases the ability of the elastomer to absorb fluid, which leads to an increase in volume of the sample. This additional volumetric strain reduces the stress in the multistep relaxation test. Therefore, the tension stress relaxes due to viscous effects and the additional absorption of fluid with a fixed strain step. Those effects cannot be separated with this experimental set up. It is shown that the absorption behaves inert to the deformation rate. This leads to the conclusion, that as longer the holding time as higher the influence of the additional absorption. But this influences the multi hysteresis test in a minor way.

Besides the comparison of both experiments, there is a significant influence of the fluid content in the elastomer to its stiffness, see Figure 3 and 5. With increasing fluid contents the stiffness is decreasing nonlinear. Thus, the absorbed octane behaves like a plasticiser, which is observed for natural rubber (Domininghaus 2013). While modelling the equilibrium stress of the elastomer, this has to be taken into account and that leads to a nonlinear dependence of the material parameters, e.g. shear modulus.

3.2 *Overstress: Relaxation under different fluid contents*

Relaxation tests at 20% pressure strain are performed to observe the time-dependent overstress. It is common to illustrate the stress response over time in a semi logarithmic scale. For a four-hour holding time, the results are printed in Figure 8. For the elastomer without fluid the relaxation curve in the semi logarithmic scale shows an almost linear development over time. At the beginning, the gradient of the relaxation curves with absorbed fluid is smaller in comparison to the curve of the relaxation without fluid. With a view to higher decades, the gradient of the relaxation curves with fluid increases. The determined equilibrium stress is compared with the measured 1st Piola-Kirchhoff-stress after the four hours holding, see Table 2. Without fluid, the stress after four hours is higher than the calculated

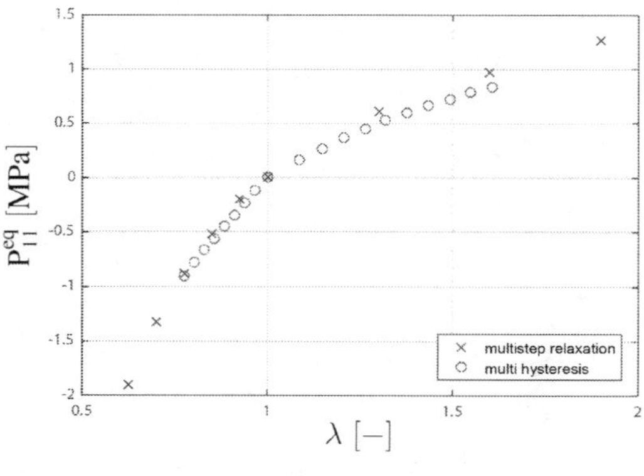

Figure 6. Equilibrium stress for fluid content $\rho_{rel} = 0$.

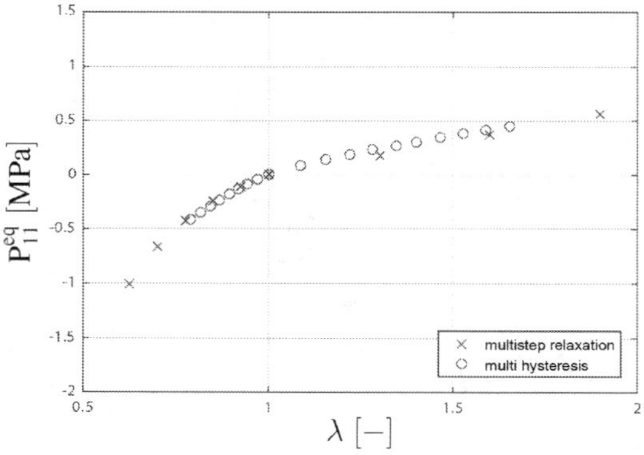

Figure 7. Equilibrium stress for fluid content $\rho_{rel} = 0.85$.

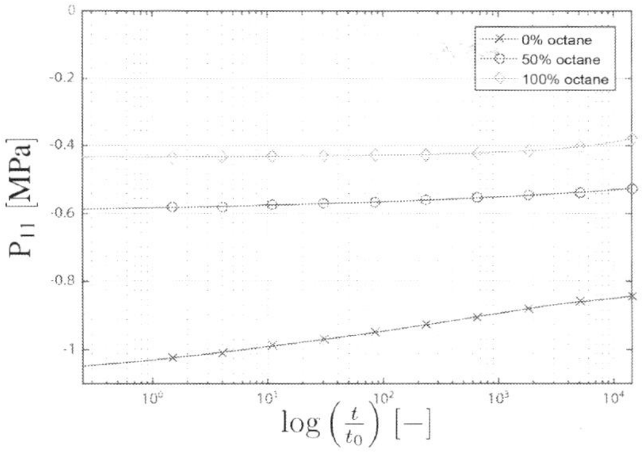

Figure 8. Relaxation of the overstress for different fluid contents ($t_0 = 1[s]$).

Table 2. 1st Piola-Kirchhoff-stress for 20% pressure strain.

ρ_{rel}	P_{11}^{equ} [MPa]	$P_{11}(4\text{h})$ [MPa]	$m_1 - m_0$ [g]
0	−0.78	−0.84	0
0.49	−0.47	−0.53	0.008
0.85	−0.39	−0.38	0.031

equilibrium stress. This behaviour is predicted by the model, because the material cannot relax completely in the observed time range. The sample stored in pure octane shows a lower stress response after four hours of loading than the calculated equilibrium stress. At a first view this contradicts with the model. At a second view the sample weight before m_0 and directly after m_1 the loading has to be taken into account additionally. The swollen samples lose weight during the experiment, so the swelling equilibrium is load dependent. The loss in weight causes a shrink of the sample. Thus, the effective mechanical strain is reduced. The relaxation experiment shows superposed effects: the relaxation of the overstress and the shrinking of the sample due to the applied stress. For high fluid contents almost no overstress relaxation appears. This is consistent with the statement (Domininghaus 2013) that plasticiser reduce the viscosity of the natural rubber.

4 CONCLUSIONS

Fluid uptake due to swelling influences the mechanical properties significantly. With respect to the stiffness of the elastomer samples, absorbed fluid reacts like a plasticiser and soften the material. Mechanical loads influence the liquid uptake of the elastomer and the equilibrium saturation level. That leads to a both-sided coupling between mechanical and swelling behaviour. Especially for long holding times at a constant strain, the stress signal is tampered by load-induced swelling or shrinking processes. This has to be taken into account while choosing suitable experiments to determine material parameters for modelling. Therefore, the multi hysteresis test is be preferred to identify the parameters for the equilibrium stress.

It is shown that only the results of relaxation tests are not usable to adapt the viscous part of model. Additionally, the equilibrium fluid content dependent on the mechanical load is needed to estimate the viscous behaviour. This dependence is investigated by measuring the change in weight of a swollen sample due to applied loads. Another option would be to use dynamic mechanical analysis (DMA), (Engelhard 2014). In the context of large deformations and the circumstance that the experiment has to be done in the liquid, the DMA is not appropriate. The increase of the fluid content decreases the viscosity. With respect to the used time scale it leads to the assumption that for high fluid contents the viscosity is small enough and thus the stress response is not time dependent anymore.

ACKNOWLEDGEMENTS

Parts of this research were supported by the company BOGE Rubber & Plastics, which provided the natural rubber. The financial support of the German science foundation (DFG) under the grant: JO818/3-1 is greatfully acknowledged.

REFERENCES

Chester, S.A. & L. Anand (2010). A coupled theory of fluid permeation and large deformations for elastomeric materials. *Journal of the Mechanics and Physics of Solids 58*(11), 1879–1906.

Crank, J. (1956). *The Mathematics of Diffusion*. Clarendon Press.

Derrien, K. & P. Gilormini (2006). Interaction between stress and diffusion in polymers. In *Defect and Diffusion Forum*, Volume 258, pp. 447–452. Trans Tech Publ.

Derrien, K. & P. Gilormini (2009). The effect of moistureinduced swelling on the absorption capacity of transversely isotropic elastic polymer–matrix composites. *International Journal of Solids and Structures 46*(6), 1547–1553.

Diercks, N. (2015). *The dynamic behaviour of rubber under consideration of the Mullins and the Payne effect.* Ph. D. thesis, Universitt der Bundeswehr Mnchen.

Domininghaus, H. (2013). *Kunststoffe: Eigenschaften und Anwendungen.* Springer-Verlag.

Engelhard, M. (2014). *Thermomechanische Beschreibung der Fluiddiffusion in Polymeren*. Ph. D. thesis, Universitt der Bundeswehr Mnchen.

Ferry, J.D. (1980). *Viscoelastic properties of polymers*. John Wiley & Sons.

Fick, A. (1855). ber diffusion. *Annalen der Physik 170*(1), 59–86.

Flory, P. (1961). Thermodynamic relations for high elastic materials. *Transactions of the Faraday Society 57*, 829–838.

Flory, P.J. (1942). Thermodynamics of high polymer solutions. *The Journal of chemical physics 10*(1), 51–61.

Huggins, M.L. (1942). Theory of solutions of high polymers 1. *Journal of the American Chemical Society 64*(7), 1712–1719.

Johlitz, M., B. Dippel, & A. Lion (2015). Dissipative heating of elastomers: a new modelling approach based on finite and coupled thermomechanics. *Continuum Mechanics and Thermodynamics*, 1–15.

Lion, A. (1998). Thixotropic behaviour of rubber under dynamic loading histories: experiments and theory. *Journal of the Mechanics and Physics of Solids 46*(5), 895–930.

Lubliner, J. (1985). A model of rubber viscoelasticity. *Mechanics Research Communications 12*(2), 93–99.

Mullins, L. (1969). Softening of rubber by deformation. *Rubber chemistry and technology 42*(1), 339–362.

Treloar, L. (1944). The elasticity of a network of longchain molecules. ii. *Rubber Chemistry and Technology 17*(2), 296–302.

Treloar, L. (1975). *The Physics of Rubber Elasticity*. Clarendon Press.

Weitsman, Y. (1987). Stress assisted diffusion in elastic and viscoelastic materials. *Journal of the Mechanics and Physics of Solids 35*(1), 73–93.

Weitsmann, Y. (2012). *Fluid Effects in Polymers and Polymeric Composits*. Springer.

Micro-structural theories of rubber

Constitutive Models for Rubber X – Lion & Johlitz (Eds)
© 2017 Taylor & Francis Group, London, ISBN 978-1-138-03001-5

Electroelasticity of dielectric elastomers based on molecular chain statistics

M. Itskov, V.N. Khiêm & S. Waluyo
Department of Continuum Mechanics, RWTH Aachen University, Aachen, Germany

ABSTRACT: Mechanical response of dielectric elastomers can be influenced or even controlled by an imposed electric filed. It can, for example, cause mechanical stress or strain without any applied load. The latter phenomenon is referred to as electrostriction. There are many phenomenological hyperelastic models describing this electro-active response of dielectric elastomers. There, coupled electro-elastic terms are of special importance. So far, these terms have not got any physical reasoning. In this contribution, we proposed an electro-mechanical constitutive model based on molecular chain statistics. The model considers polarization of single polymer chain segments and takes into account their directional distribution. The latter one results from the non-Gaussian chain statistics taking finite extensibility of polymer chains into account. The so resulting electric potential of a single polymer chain is further generalized to the network potential serving as a basis for the prediction of the above mentioned electro-active response. The model includes a few number of physically interpretable material constants and demonstrates good agreement with experimental data.

1 INTRODUCTION

Dielectric elastomers are often applied in elastomeric actuators which can produce considerable displacements under an external electric field (see, e.g., (BarCohen 2004)). To this end, usually a film of such a dielectric elastomer is coated on both sides with a compliant electrode material. Subject to a voltage these electrodes generate an electric field in the thickness direction of the film. The electric field, in turn, causes deformation of the elastomer without any mechanical load. This electrostriction results non only from the charges of the electrodes but also from the own polarization of the elastomer in the external electric field.

The response of dielectric elastomers to the external electric field has so far been described mostly by phenomenological models. They are usually based on elctroelastic invariants (see, e.g., (Dorfmann & Ogden 2005)). In the most simple case, the electric potential is represented by a quadratic function of an electric variable such as electric field, displacement or polarization (Zhao, Hong, & Suo 2007, Zhao & Wang 2014, Castañeda & Siboni 2012, Aboudi 2015, Miehe, Vallicotti, & Teichtmeister 2016). This is also the case for the so-called neo-Hookean electroelastic material (Vu, Steinmann, & Possart 2007, Bustamante & Merodio 2011, Cesana & DeSimone 2011, Rudykh & deBotton 2011, Thylander, Menzel, & Ristinmaa 2012, Rudykh, Bhattacharya, & deBotton 2013, Dorfmann & Ogden 2014a). More elaborated electroelastic strain energies are based on the Brillouin function (Li, Chen, Qiang, & Zhou 2012) or three elastic and three electroelastic invariants (Vertechy, Berselli, Castelli, & Bergamasco 2013). In this context, polyconvex electroelastic strain energies can also be mentioned (Itskov & Khiêm 2016).

Very recently some micromechanically motivated electroelastic strain energies have also been proposed (Cohen & deBotton 2014, Cohen & deBotton 2015a, Cohen & deBotton 2015b, Cohen, Dayal, & deBotton 2016). They are based on the enthalpy of long polymer chains and Gaussian chain statistics. Another important feature of the model is an orientational distribution of polymer chain segments around the direction of the applied electric field. This is, however, not in accord with non-Gaussian chain statistics which predicts an orientational distribution of polymer chain segments around the end-to-end direction of the polymer chain. The later concept is plausible because this distribution should exist also without any electric field.

In the present contribution we propose a model of electroelasticity for dielectric elastomers. It is based on the polarization of chain segments in an applied external electric field. The resulting electric potential energy depends on the angle between the direction of the monomer and the end-to-end direction of a polymer chain which, in turn, is described by molecular chain statistics in the non-Gaussian setting taking finite extensibility of polymer chains into account. By this means, the electric potential energy of the polymer chain is expressed. The so resulting one-dimensional model is generalized to three dimensions by means of the directional averaging based on the numerical integration over the unit sphere.

2 DEFORMATION DEPENDENT DIRECTIONAL DISTRIBUTION OF POLYMER CHAIN SEGMENTS

In the current configuration of the body we consider a polymer chain with N segments each of the length l. The end-to-end distance of the chain in this state can be expressed by $r = \lambda\sqrt{N}l$, where λ denotes the stretch in the end-to-end direction. Here, we assume a distribution $\rho(\theta)$ of an angle θ between a chain segment and the end-to-end direction of the chain. For a sufficiently large number of segments N we can thus write

$$N\int_0^{2\pi}\int_0^{\pi}\rho(\theta)l\cos\theta\sin\theta d\theta d\phi = \lambda\sqrt{N}l, \tag{1}$$

where ϕ denotes an angle specifying the position of the chain segment in the plane orthogonal to the end-to-end direction. Thus,

$$2\pi\int_0^{\pi}\rho(\theta)\cos\theta\sin\theta d\theta = \lambda_r, \tag{2}$$

where

$$\lambda_r = \frac{\lambda}{\sqrt{N}} \tag{3}$$

is referred to as the relative chain stretch (Beatty 2003). In the following we apply the von Mises-Fischer distribution with the following directional probability density function

$$\rho(\theta) = \frac{\kappa}{4\pi\sinh\kappa}\exp(\kappa\cos\theta), \tag{4}$$

where κ denotes the concentration parameter. This distribution results from the non-Gaussian chain statistics by Kuhn and Grün (Kuhn & Grün 1942). Inserting (4) into (2) we obtain

$$\mathcal{L}(\kappa) = \lambda_r, \quad \kappa = \mathcal{L}^{-1}(\lambda_r), \tag{5}$$

where $\mathcal{L}^{-1}$ denotes the inverse of the Langevin function defined by

$$\mathcal{L}(x) = \coth(x) - \frac{1}{x}. \tag{6}$$

Thus, the concentration parameter of the directional distribution function of chain segments (4) depends on the chain stretch.

According to the non-Gaussian statistical theory of rubber elasticity (Kuhn & Grün 1942) the force developed by the polymer chain is given by

$$f(\lambda) = \frac{k_B T}{l}\mathcal{L}^{-1}(\lambda_r), \tag{7}$$

where k_B denotes Boltzmann's constant and T is the absolute temperature. By (5) we thus obtain

$$f(\lambda) = \frac{k_B T\kappa}{l}. \tag{8}$$

In the special case $\kappa \to 0$ and $f \to 0$ the ends of the chain coincide, its end-to-end vector degrades to a point. The directional distribution of chain segments becomes isotropic. In the opposite case, when $\kappa \to \infty$ and $f \to \infty$ the chain is in the fully stretched straight state, where the end-to-end distance coincides with contour length. All the chain segments are almost unidirectional.

3 POLARIZATION AND ELECTRIC POTENTIAL OF POLYMER CHAINS AND THE NETWORK

Now we consider the polymer chain in an external electric field $\boldsymbol{E} = E\boldsymbol{e}$, where $E = \|\boldsymbol{E}\|$ represents its magnitude and $\boldsymbol{e}$ denotes a unit vector in the direction of the field. The electric field is assumed to be constant over the contour of the polymer chain and very large in comparison to the own electric fields of the molecule.

The electric field causes displacements of electrons with respect to positive nucleus of atoms and distorts the arrangement of atomic nuclei in the polymer molecule (Blythe & Bloor 2008), which further leads to a polarization of the chain monomers along and perpendicular to their directions (Stockmayer 1967, Blythe & Bloor 2008).

Now, we consider a monomer of the polymer chain with the direction specified by a unit vector $\boldsymbol{u}$. The latter one can expressed by means of the (spherical) angles θ and ϕ introduced before as follows

$$\boldsymbol{u} = \cos\theta\,\boldsymbol{v}_0 + \sin\theta\cos\phi\,\boldsymbol{v}_1 + \sin\theta\sin\phi\,\boldsymbol{v}_2, \tag{9}$$

where the vectors $\boldsymbol{v}_i\,(i = 0,1,2)$ represent an orthonormal basis in three-dimensional Euclidean space, while $\boldsymbol{v}_0$ specifies the end-to-end direction of the polymer chain.

Assuming the linear induced polarization of the monomer its dipole can thus be represented by (see Fig. 1) (Cohen 2003, Blythe & Bloor 2008, Cohen & deBotton 2015b)

$$\boldsymbol{p} = \mu_1\boldsymbol{u}\otimes\boldsymbol{u}\boldsymbol{E} + \mu_2(\mathbf{I} - \boldsymbol{u}\otimes\boldsymbol{u})\boldsymbol{E}, \tag{10}$$

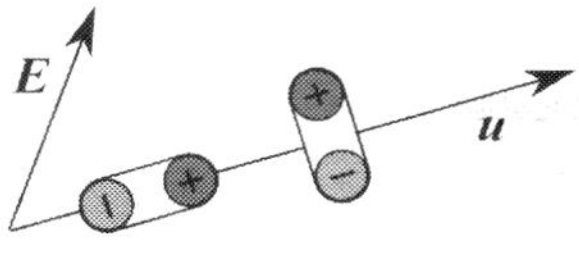

Figure 1. Polarization of a polymer chain segment in an external electric field $\boldsymbol{E}$.

where the material constants μ_1 and μ_2 represent the polarizability of the monomer in its axial and transverse direction, respectively, while $\mathbf{I}$ denotes the identity tensor. The force applied to the dipole is $\boldsymbol{F} = \boldsymbol{p} \cdot \text{grad}\, \boldsymbol{E} = \boldsymbol{0}$ (see, e.g. (Kovetz 2000)) due to the above assumption of the constant electric field. The moment on the dipole in the electric field can further be expressed by virtue of (10) as

$$\boldsymbol{m} = \boldsymbol{p} \times \boldsymbol{E} = \mu(\boldsymbol{u} \cdot \boldsymbol{E})\boldsymbol{u} \times \boldsymbol{E}, \tag{11}$$

where $\mu = \mu_1 - \mu_2$. The potential energy of the dipole in the electric field $\boldsymbol{E}$ can be obtained as the work done by the moment (11) on the rotation angle $\hat{\alpha}$ above the axis of $\boldsymbol{m}$

$$w_e(\boldsymbol{u}) = \int_0^{\alpha} \|\boldsymbol{m}\| d\hat{\alpha}, \tag{12}$$

where α is the current angle between $\boldsymbol{u}$ and $\boldsymbol{E}$ such that $\cos\alpha = \boldsymbol{u} \cdot \boldsymbol{e}$. By (11) we thus can write

$$w_e(\boldsymbol{u}) = \frac{1}{2}\mu E^2(1 - \cos^2\alpha) \tag{13}$$

and consequently

$$w_e(\boldsymbol{u}) = \frac{1}{2}\mu\left[\boldsymbol{E} \cdot \boldsymbol{E} - (\boldsymbol{u} \cdot \boldsymbol{E})^2\right]. \tag{14}$$

The dipole vector resulting from this potential energy can be expressed by

$$\bar{\boldsymbol{p}} = -\frac{\partial w_e}{\partial \boldsymbol{E}} = \mu\left[(\boldsymbol{u} \cdot \boldsymbol{E})\boldsymbol{u} - \boldsymbol{E}\right] \tag{15}$$

and coincides with (10) up to the part work-conjugate to $\boldsymbol{E}$.

The total electric potential of the polymer chain ψ_e can now be obtained by averaging (14) with the probability distribution function (4). To this end, we first express in (14) the directional cosine of the angle between the electric field vector and the monomer direction as (see also Fig. 2)

$$\begin{aligned}\cos\alpha &= \boldsymbol{e} \cdot \boldsymbol{u} \\ &= e_0 \cos\theta + e_1 \sin\theta\cos\phi + e_2 \sin\theta\sin\phi,\end{aligned} \tag{16}$$

where $e_i = \boldsymbol{e} \cdot \boldsymbol{v}_i \, (i = 0, 1, 2)$. In particular, $e_0^2 = 1 - e_1^2 - e_2^2 = \cos^2\gamma$, where γ denotes the angle between the electric field vector and the chain end-to-end direction. By virtue of (4) we thus obtain

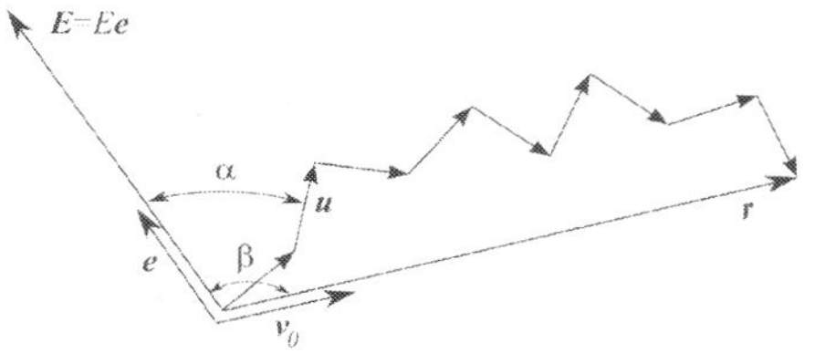

Figure 2. A polymer chain in an external electric field $\boldsymbol{E}$.

$$\begin{aligned}\psi_e(\boldsymbol{v}_0) &= N\int_0^{2\pi}\int_0^{\pi} \rho(\theta) w_e \sin\theta d\theta d\phi \\ &= \frac{\kappa\mu E^2 N}{4\sinh\kappa}\int_0^{\pi} e^{\kappa\cos\theta}\left[1 - \frac{1}{2}(e_1^2 + e_2^2)\sin^2\theta \right. \\ &\left. - e_0\cos^2\theta\right]\sin\theta d\theta,\end{aligned} \tag{17}$$

which by (5) finally yields

$$\begin{aligned}\psi_e(\boldsymbol{v}_0) = \frac{1}{2}\mu N E^2 &\left[1 - e_0^2 \right. \\ &\left. + \frac{\lambda_r}{\mathcal{L}^{-1}(\lambda_r)}(3e_0^2 - 1)\right].\end{aligned} \tag{18}$$

Taking into account that $\lim_{\kappa\to 0}\frac{\mathcal{L}(\kappa)}{\kappa} = \frac{1}{3}$ the potential (18) becomes independent of $\boldsymbol{v}_0$ in the case of isotropically distributed chain segments with $\kappa \to 0$. This result is plausible because in this case the end-to-end vector of the chain degrades to a point and its direction is thus undefined.

The electric potential of the whole polymer network can be obtained by averaging (18) over all spatial directions and taking into account the number of polymer chains n per unit reference volume as

$$\begin{aligned}\Psi_e &= n\langle\psi_e\rangle \\ &= \frac{1}{2}c_e E^2\left\langle 1 - e_0^2 + \frac{\lambda_r}{\mathcal{L}^{-1}(\lambda_r)}(3e_0^2 - 1)\right\rangle,\end{aligned} \tag{19}$$

where $c_e = \mu n N$. Assuming an initially isotropic distribution of polymer chains in the reference configuration the averaging operator in (19) can be represented by

$$\langle g(\Theta, \Phi)\rangle = \frac{1}{4\pi}\int_0^{2\pi}\int_0^{\pi} g(\Theta, \Phi)\sin\Theta d\Theta d\Phi, \tag{20}$$

where g is a function of spherical angles Θ and Φ of the chain end-to-end direction with respect to a fixed reference frame.

The above discussed special case of isotropically distributed chain segments with $\kappa \to 0$ can be obtained by linearizing the inverse Langevin function as

$$\mathcal{L}^{-1}(y) = 3y + \frac{9}{5}y^3 + \frac{297}{175}y^5 + \frac{1539}{875}y^7 \ldots \tag{21}$$

In this case, the non-Gaussian chain statistics reduces to the Gaussian one while the electric potential (19) simplifies to

$$\Psi_e^L = \frac{1}{3}c_e E^2. \tag{22}$$

Accordingly, it no more depends on deformation and cannot thus describe the deformation dependent polarization of dielectric elastomers.

The polarization density of the network resulting from the electric potential (19) can be expressed by

$$\boldsymbol{P} = -\frac{1}{J}\frac{\partial \Psi_e}{\partial \boldsymbol{E}} = \frac{c_e}{J}\left[\left\langle \frac{\lambda_r}{\mathcal{L}^{-1}(\lambda_r)} - 1 \right\rangle \boldsymbol{E} + \left\langle \left(1 - \frac{3\lambda_r}{\mathcal{L}^{-1}(\lambda_r)}\right)(\boldsymbol{E}\cdot\boldsymbol{v}_0)\boldsymbol{v}_0 \right\rangle\right], \tag{23}$$

where $J = \det \mathbf{F}$ represents the relative volume change expressed in terms of the deformation gradient $\mathbf{F}$.

4 ELECTRO-MECHANICAL STRESS

The additional Cauchy stress induced by the network of polymer chain under the electric field $\boldsymbol{E}$ can be expressed by

$$\boldsymbol{\sigma}_e = J^{-1}\frac{\partial \Psi_e}{\partial \boldsymbol{F}}\boldsymbol{F}^T + \tau_e, \tag{24}$$

where τ_e represents the electrostatic Maxwell stress. The latter one is expressed in a dielectric material by (see, e.g., (Dorfmann & Ogden 2014b))

$$\tau_e = \boldsymbol{E}\otimes\boldsymbol{D} - \frac{1}{2}\epsilon_0 E^2 \boldsymbol{I}, \tag{25}$$

where ϵ_0 denotes the electric permittivity of vacuum and

$$\boldsymbol{D} = \boldsymbol{P} + \epsilon_0 \boldsymbol{E} \tag{26}$$

represents the electric displacement.

Inserting (26) and (23) into (25) we obtain

$$\tau_e = \left[\frac{c_e}{J}\left\langle \frac{\lambda_r}{\mathcal{L}^{-1}(\lambda_r)} - 1 \right\rangle + \epsilon_0\right]\boldsymbol{E}\otimes\boldsymbol{E} + \frac{c_e}{J}\boldsymbol{E}\otimes\left\langle \left(1 - \frac{3\lambda_r}{\mathcal{L}^{-1}(\lambda_r)}\right)(\boldsymbol{E}\cdot\boldsymbol{v}_0)\boldsymbol{v}_0 \right\rangle - \frac{1}{2}\epsilon_0 E^2 \boldsymbol{I}. \tag{27}$$

In order to express the first stress component on the right hand side of (24) we insert there (19) and take into account that

$$e_0 = \lambda^{-1} e \cdot (\mathbf{F} V_0) \tag{28}$$

and consequently

$$\frac{\partial e_0}{\partial \mathbf{F}} = \lambda^{-1} e \otimes \boldsymbol{V}_0 + \lambda^{-2}\boldsymbol{v}_0 \otimes \boldsymbol{V}_0,$$
$$\frac{\partial e_0^2}{\partial \mathbf{F}}\mathbf{F}^T = 2e_0\left(e\otimes\boldsymbol{v}_0 + \lambda^{-1}\boldsymbol{v}_0\otimes\boldsymbol{v}_0\right), \tag{29}$$

where the unit vector $\boldsymbol{V}_0 = \lambda \mathbf{F}^{-1}\boldsymbol{v}_0$ represents the counterpart of $\boldsymbol{v}_0$ in the reference configuration. Thus,

$$\frac{1}{J}\frac{\partial \Psi_e}{\partial \mathbf{F}}\mathbf{F}^T = \frac{c_e}{2J}E^2\left\langle (3e_0^2 - 1)\frac{\partial}{\partial \mathbf{F}}\frac{\lambda_r}{\mathcal{L}^{-1}(\lambda_r)}\mathbf{F}^T + \frac{2e_0}{\lambda}\left(\frac{3\lambda_r}{\mathcal{L}^{-1}(\lambda_r)} - 1\right)\boldsymbol{v}_0\otimes\boldsymbol{v}_0 \right\rangle + \frac{c_e}{2J}\boldsymbol{E}\otimes\left\langle \left(\frac{3\lambda_r}{\mathcal{L}^{-1}(\lambda_r)} - 1\right)(\boldsymbol{E}\cdot\boldsymbol{v}_0)\boldsymbol{v}_0 \right\rangle.$$

The Cauchy stress (24) induced by the electric field takes finally the form

$$\sigma_e = \frac{c_e}{2J}E^2\left\langle (3e_0^2 - 1)\frac{\partial}{\partial \mathbf{F}}\frac{\lambda_r}{\mathcal{L}^{-1}(\lambda_r)}\mathbf{F}^T + \frac{2e_0}{\lambda}\left(\frac{3\lambda_r}{\mathcal{L}^{-1}(\lambda_r)} - 1\right)\boldsymbol{v}_0\otimes\boldsymbol{v}_0 \right\rangle - \frac{1}{2}\epsilon_0 E^2 \mathbf{I} + \left[\frac{c_e}{J}\left\langle \frac{\lambda_r}{\mathcal{L}^{-1}(\lambda_r)} - 1 \right\rangle + \epsilon_0\right]\boldsymbol{E}\otimes\boldsymbol{E} \tag{30}$$

and is thus symmetric. Indeed, the derivative in the first term of this expression can be represented by

$$\frac{\partial}{\partial \mathbf{F}}\frac{\lambda_r}{\mathcal{L}^{-1}(\lambda_r)}\mathbf{F}^T = \lambda_r \frac{d}{d\lambda_r}\left(\frac{\lambda_r}{\mathcal{L}^{-1}(\lambda_r)}\right)\boldsymbol{v}_0\otimes\boldsymbol{v}_0 = \frac{1}{\sqrt{N}}\mathbf{F}\left[\frac{1}{\lambda}\frac{d}{d\lambda_r}\left(\frac{\lambda_r}{\mathcal{L}^{-1}(\lambda_r)}\right)\boldsymbol{V}_0\otimes\boldsymbol{V}_0\right]\mathbf{F}^T, \tag{31}$$

where

$$\frac{d}{dy}\left(\frac{y}{\mathcal{L}^{-1}(y)}\right) = \frac{1}{\mathcal{L}^{-1}(y)}\left[1 - \frac{y}{\mathcal{L}^{-1}(y)(1-y^2) - 2y}\right]. \tag{32}$$

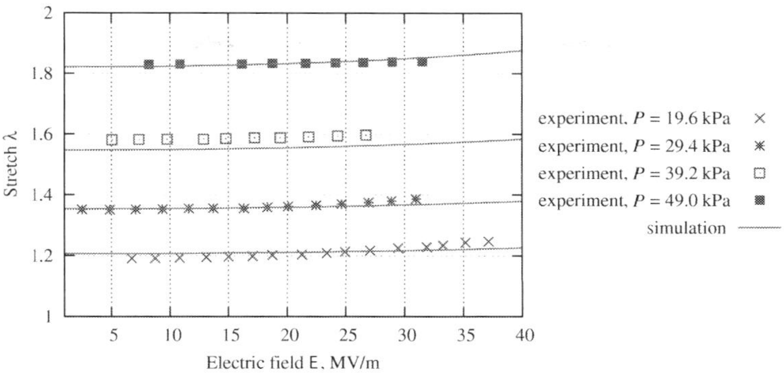

Figure 3. Model predictions versus experiential data by Carpi et al. (2003) from the dielectric planar actuator.

Within a network model the mechanical stress contribution can be obtained by using the entropic energy of the polymer chain. It results from (7) as

$$\begin{aligned}\psi_M &= \sqrt{N}l\int f(\lambda)d\lambda \\ &= k_B TN\left[\lambda_r \mathcal{L}^{-1}(\lambda_r) + \ln\frac{\mathcal{L}^{-1}(\lambda_r)}{\sinh \mathcal{L}^{-1}(\lambda_r)}\right] + C,\end{aligned} \tag{33}$$

where C denotes an integration constant. After averaging we further obtain

$$\begin{aligned}\Psi_m &= n\langle\psi_M\rangle \\ &= c_m\left\langle \lambda_r \mathcal{L}^{-1}(\lambda_r) + \ln\frac{\mathcal{L}^{-1}(\lambda_r)}{\sinh \mathcal{L}^{-1}(\lambda_r)}\right\rangle,\end{aligned} \tag{34}$$

where $c_m = k_B TnN$. The resulting Cauchy stress takes thus the form

$$\begin{aligned}\sigma_m &= \frac{c_m}{J}\frac{\partial}{\partial \mathbf{F}}\left\langle \lambda_r \mathcal{L}^{-1}(\lambda_r) + \ln\frac{\mathcal{L}^{-1}(\lambda_r)}{\sinh \mathcal{L}^{-1}(\lambda_r)}\right\rangle \mathbf{F}^T \\ &= \frac{c_m}{J\sqrt{N}}\mathbf{F}\left\langle \frac{\mathcal{L}^{-1}(\lambda_r)}{\lambda^2}\boldsymbol{V}_0 \otimes \boldsymbol{V}_0\right\rangle \mathbf{F}^T.\end{aligned} \tag{35}$$

The total Cauchy stress results as a sum of the electric (30) and the mechanical contribution (35). In view of (31) it can be written by

$$\begin{aligned}\sigma = \sigma_e + \sigma_m &= -\frac{1}{2}\epsilon_0 E^2 \mathbf{I} \\ &+ \left[\frac{c_e}{J}\left\langle \frac{\lambda_r}{\mathcal{L}^{-1}(\lambda_r)} - 1\right\rangle + \epsilon_0\right]\boldsymbol{E}\otimes\boldsymbol{E} \\ &+ \mathbf{F}\left\langle\left\{\frac{c_e E^2}{J\lambda}\left[e_0\left(\frac{3\lambda_r}{\mathcal{L}^{-1}(\lambda_r)} - 1\right)\right.\right.\right. \\ &\left.+ \frac{3e_0^2 - 1}{2\sqrt{N}}\frac{d}{d\lambda_r}\left(\frac{\lambda_r}{\mathcal{L}^{-1}(\lambda_r)}\right)\right] \\ &\left.\left.+ \frac{c_m}{J\sqrt{N}}\frac{\mathcal{L}^{-1}(\lambda_r)}{\lambda^2}\right\}\boldsymbol{V}_0\otimes\boldsymbol{V}_0\right\rangle \mathbf{F}^T.\end{aligned} \tag{36}$$

5 NUMERICAL RESULTS

For the averaging in (20), (34) and (36) we used numerical integration over the unit sphere (microsphere model) (Bažant & Oh 1986). A numerical integration scheme with 45 integration points on a hemisphere (Heo & Xu 2001) has been applied as it provides a good compromise between numerical errors and artificially induced anisotropy on the one hand and numerical efforts on the other hand (Ehret, Itskov, & Schmid 2010).

For the validation of the electroelastic model presented above we compared it predictions to the experimental data from a dielectric planar actuator (Carpi, Chiarelli, Mazzoldi, & Rossi 2003). In these experiments a 50 μm-thick film of an acrylic elastomer coated with a compliant electrode was subject to pretension in the longitudinal direction and the electric field along the thickness direction. The electric field leads to a change of the stretch in the longitudinal direction. The electroelastic response depends, however, on the material of the electrode. For the verification of the proposed model we used experimental data obtained with graphite powder coating. In Fig. 3 the stretch of the film in the longitudinal direction is plotted versus the electric field in the thickness direction for various values of the prestress P. One can observe good agreement between the experimental data and model predictions obtained with the following material constants $c_e = 8.45\cdot 10^{-12}\,F/m$, $c_m = 1.2$ MPa and $N = 25.3$.

6 CONCLUSION

A model of electroelasticity for dielectric elastomers has been presented. The model is based on the polarization of polymer chain segments in an applied external electric field and the directional distribution of these segments according to non-Gaussian chain statistics. Model predictions demonstrate good agreement with experimental data with respect to electrostriction resulting from the electric field applied to an acrylic film in the thickness direction. A simplified model resulting in the special case of Gaussian statistics cannot describe deformation dependent polarization of dielectric elastomers.

REFERENCES

Aboudi, J. (2015). Micro-electromechanics of soft dielectric matrix composites. *International Journal of Solids and Structures 64*, 30–41.

Bar-Cohen, Y. (Ed.) (2004). *Electroactive Polymer (EAP) Actuators as Artificial Muscles: Reality, Potential, and Challenges* (2nd ed.). Spie Press, Bellingham, Washington USA.

Bažant, Z. & B. Oh (1986). Efficient numerical integration on the surface of a sphere. *Z. Angew. Math. Mech. 66*, 37.

Beatty, M.F. (2003). An average-stretch full-network model for rubber elasticity. *Journal of Elasticity 70*(1), 65–86.

Blythe, T. & D. Bloor (2008). *Electrical Properties of Polymers* (2nd ed.). Cambridge University Press.

Bustamante, R. & J. Merodio (2011). Constitutive structure in coupled non-linear electro-elasticity: Invariant descriptions and constitutive restrictions. *International Jounal of Nonlinear Mechanics 46*, 1315–1323.

Carpi, F., P. Chiarelli, A. Mazzoldi, & D.D. Rossi (2003). Electromechanical characterisation of dielectric elastomer planar actuators: comparative evaluation of different electrode materials and different counterloads. *Sensors and Actuators A: Physical 107*(1), 85–95.

Castañeda, P.P. & M. Siboni (2012). A finite-strain constitutive theory for electro-active polymer composites via homogenization. *International Journal of Non-Linear Mechanics 47*(2), 293–306.

Cesana, P. & A. DeSimone (2011). Quasiconvex envelopes of energies for nematic elastomers in the small strain regime and applications. *Journal of the Mechanics and Physics of Solids 59*(4), 787–803.

Cohen, A.E. (2003, Dec). Force-extension curve of a polymer in a high-frequency electric field. *Phys. Rev. Lett. 91*, 235506.

Cohen, N. & G. deBotton (2014). Multiscale analysis of the electromechanical coupling in dielectric elastomers. *European Journal of Mechanics - A/Solids 48*, 48–59. Frontiers in Finite-Deformation Electromechanics.

Cohen, N. & G. deBotton (2015a). The electromechanical response of polymer networks with long-chain molecules. *Mathematics and Mechanics of Solids 20*(6), 721–728.

Cohen, N. & G. deBotton (2015b). The electromechanical response of polymer networks with long-chain molecules. *Mathematics and Mechanics of Solids 20*(6), 721–728.

Cohen, N., K. Dayal, & G. deBotton (2016). Electroelasticity of polymer networks. *Journal of the Mechanics and Physics of Solids 92*, 105–126.

Dorfmann, A. & R.W. Ogden (2005). Nonlinear electroelastic deformations. *Journal of Elasticity 82*, 99–127.

Dorfmann, L. & R.W. Ogden (2014a). Instabilities of an electroelastic plate. *International Journal of Engineering Science 77*, 79–101.

Dorfmann, L. & R.W. Ogden (2014b). *Nonlinear Theory of Electroelastic and Magnetoelastic Interactions*. Springer.

Ehret, A.E., M. Itskov, & H. Schmid (2010). Numerical integration on the sphere and its effect on the material symmetry of constitutive equationsa comparative study. *International Journal for Numerical Methods in Engineering 81*(2), 189–206.

Heo, S. & Y. Xu (2001). Constructing fully symmetric cubature formulae for the sphere. *Mathematics of Computation 70*(233), 269–279.

Itskov, M. & V.N. Khiêm (2016). A polyconvex anisotropic free energy function for electro- and magneto-rheological elastomers. *Mathematics and Mechanics of Solids 21*(9), 1126–1137.

Kovetz, A. (2000). *Electromagnetic theory*. Oxford University Press.

Kuhn, W. & F. Grün (1942). Beziehungen zwischen elastischen Konstanten und Dehnungsdoppelbrechung hochelastischer Stoffe. *Kolloid-Zeitschrift 101*, 248–271.

Li, B., H. Chen, J. Qiang, & J. Zhou (2012). A model for conditional polarization of the actuation enhancement of a dielectric elastomer. *Soft Matter 8*, 311–317.

Miehe, C., D. Vallicotti, & S. Teichtmeister (2016). Homogenization and multiscale stability analysis in finite magnetoelectro-elasticity. application to soft matter ee, me and mee composites. *Computer Methods in Applied Mechanics and Engineering 300*, 294–346.

Rudykh, S. & G. deBotton (2011). Stability of anisotropic electroactive polymers with application to layered media. *Zeitschrift für angewandte Mathematik und Physik 62*(6), 1131–1142.

Rudykh, S., K. Bhattacharya, & G. deBotton (2013). Multiscale instabilities in soft heterogeneous dielectric elastomers. *Proceedings of the Royal Society of London A: Mathematical, Physical and Engineering Sciences 470*(2162).

Stockmayer, W.H. (1967). Dielectric dispersion in solutions of flexible polymers. *Pure Appl. Chem. 15*(3–4), 539–554.

Thylander, S., A. Menzel, & M. Ristinmaa (2012). An electromechanically coupled micro-sphere framework: application to the finite element analysis of electrostrictive polymers. *Smart Materials and Structures 21*(9), 094008.

Vertechy, R., G. Berselli, V.P. Castelli, & M. Bergamasco (2013). Continuum thermo-electro-mechanical model for electrostrictive elastomers. *Journal of Intelligent Material Systems and Structures 24*(6), 761–778.

Vu, D.K., P. Steinmann, & G. Possart (2007). Numerical modeling of non-linear electroelasticity. *International Journal for Numerical Methods in Engineering 70(6)*, 685704.

Zhao, X. & Q. Wang (2014). Harnessing large deformation and instabilities of soft dielectrics: Theory, experiment, and application. *Applied Physics Reviews 1*(2), 021304.

Zhao, X., W. Hong, & Z. Suo (2007, Oct). Electromechanical hysteresis and coexistent states in dielectric elastomers. *Phys. Rev. B 76*, 134113.

Constitutive Models for Rubber X – Lion & Johlitz (Eds)
© 2017 Taylor & Francis Group, London, ISBN 978-1-138-03001-5

Analytical network averaging: A general concept for material modeling of elastomers

V.N. Khiêm & M. Itskov
Department of Continuum Mechanics, RWTH Aachen University, Germany

ABSTRACT: In this contribution, an attempt to bridge the gap between macroscopic and microscopic response of elastomers is made on the basis of statistical mechanics. The transition between macroscopic and mesoscopic deformation fields is carried out by means of an analytical network averaging concept. Accordingly, we assume the existence of a directional distribution function of polymer chains in the polymer network. The mean field mesoscopic deformations on the subnetwork level are computed by averagingthe macroscopic deformations over the unit sphere. The directional distribution of polymer chains introduces non-affine deformations, stress softening or strain hardening mechanisms into the model. Furthermore, the meso-micro bridging between the subnetwork and the single chain is done by the mesostretch amplification which is essential to capture the limited extension of the polymer network. In contrast to previous works, within the presented model the network averaging can be derived analytically. Furthermore, the free energy of polymer chains is developed from a closed-form of Rayleigh's exact non-Gaussian distribution function based on quantum mechanics. Thus, the inverse Langevin function is entirely bypassed. Evolutions of the Mullins effect, hysteresis and crystallinity are also analytically evaluated. The model includes a few physically motivated material parameters and demonstrates good agreement with multi-dimensional experimental data of different elastomers.

1 INTRODUCTION

The term elastomer (or rubber-like material) generally refers to materials which have chemical and physical properties similar to vulcanized natural rubbers, including synthetic rubbers, biopolymers and bionanocomposites. They are characterized by the ability to undergo large deformations and return to the original shape with negligible permanent strain. The exceptional properties of rubber-like materials are attributed to their molecular structure. In the undeformed configuration, long polymer molecules are in highly disordered states. Upon application of external load, the polymer chains are straightened in the direction of stretch by rotations between their segments. Since this deformation is only due to uncoiling of the polymer chains, they can be elongated significantly without break. The cross-links between polymer chains act as memory anchors and permit them to return to their initial configurations after release of the load. Today, rubber-like materials can be found in traditional engineering products such as carcass of automotive tires, suspension mounts, building and bridge bearings, elastomericflex-elements as well as in modern technology like wearable electronics, artificial muscle and soft robots.

Despite diverse applications, physics of rubber-like materials has not been fully understood. Their mechanical properties differ significantly when varying the structural parameters, such as the kind of vulcanizing agents, cross-link density, filler volume fraction, degree of dispersion, particle size, aspect ratio and chemical nature of filler. Furthermore, imperfections in the polymer network may lead to discrepancy between theoretical prediction and experimental value (Hild 1998). Physically-based constitutive theories with explicit relations between the molecular parameters and macroscopic mechanical response of elastomers are therefore desired to improve the construction of rubber components.

Nevertheless, from the continuum mechanical point of view, it is challenging to develop a material model that can provide reliable predictions in all admissible deformations of elastomers using a single set of material parameters. Indeed, as reported in (Marckmann & Verron 2006), only three models among 20 examined hyperelastic constitutive models, namely the extended-tube model (Kaliske & Heinrich 1999), the Shariff model (Shariff 2000) and the tube-like microsphere model (Miehe, Göktepe, & Lulei 2004) are able to capture stress-strain relations for vulcanized rubber in multiple states of deformation.

In comparison to phenomenological models, the tube-like microsphere model (Miehe, Göktepe, & Lulei 2004) is physically motivated and includes a first order approximation non-Gaussian phantom network and a topological constraint. Thus, its parameters can be defined by studying the microstructure of the material and could be physically measured. In microsphere type models, polymer chains are assumed to be continuously distributed in all spatial directions. The network strain energy function is calculated by integration of the one dimensional free energy of polymer chains over the unit sphere (Wu & van der Giessen 1993). Therefore, microsphere type models require, in general, numerical cubature. However, large errors in the stress-strain relation of microsphere models due to the numerical integration over the unit sphere have recently been reported (Verron 2015, Khiêm & Itskov 2016).

The mechanical effects in polymeric materials such as rubber elasticity, stress softening and strain hardening have mostly been studied and simulated separately of each other, rather than from a unified perspective. For example, filled and unfilled rubbers are frequently treated as different materials, described by different constitutive theories. To the best of our knowledge, there exists no physically based constitutive model that is adaptable to different types of elastomers and at the same time demonstrates satisfactory agreement with multidimensional experimental data for these materials.

In the current work, we present a general physically-based constitutive theory for various types of rubber-like materials under mutltiaxial states of deformation. The theory is developed on the basis of two novel ingredients: a closed form of the Rayleigh exact non-Gaussian distribution function (Section 2) and a generalized analytical network-averaging concept (Section 3). By means of the former, the inverse Langevin function is avoided. The latter assumes the existence of a directional distribution function of polymer chains in the rubber network. Due to the particular form of the distribution function, averaged deformation measures can be computed analytically. Variations of the distribution function of polymer chains with deformation elucidate the elastic and inelastic effects in rubber-like materials (Fig. 1). Accordingly, the rubber elasticity is based on the isotropic chain distribution. The anisotropic Mullins effect and hysteresis in filled elastomers are captured by the deformation-induced contraction of the distribution surface. The network hardening during strain-induced crystallization of natural rubbers results from the deformation-induced extension of the distribution surface.

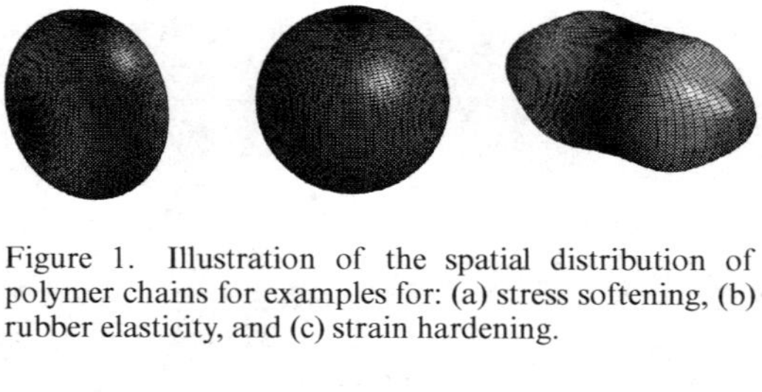

Figure 1. Illustration of the spatial distribution of polymer chains for examples for: (a) stress softening, (b) rubber elasticity, and (c) strain hardening.

2 MOLECULAR STATISTICAL FOUNDATION

The rubber network is considered as a composition of several one-dimensional subnetworks dispersed in different spatial directions n. The molecular conformation in each subnetwork is restricted by neighbouring network chains. Accordingly, each polymer molecule in a directional subnetwork is considered as a freely jointed chain restricted within a tube-like zone. The conformational probability density function of a polymer chain with n segments and normalized end-to-end distance $r = \lambda^n R$ lying within a tube of length r and diameter d can be considered as a joint probability of a random walk chain (Khiêm & Itskov 2016)

$$P_c(r) = A\left[\frac{n}{\pi r}\sin\left(\frac{\pi r}{n}\right)\right]^{\kappa n}, \tag{1}$$

and a topological constraint

$$P_t(d) = \exp\left[-a\frac{\pi^2 R}{3d^2}\right]. \tag{2}$$

Therein A is a mode normalization, λ^n denotes the stretch in the direction n, R is the nomalized

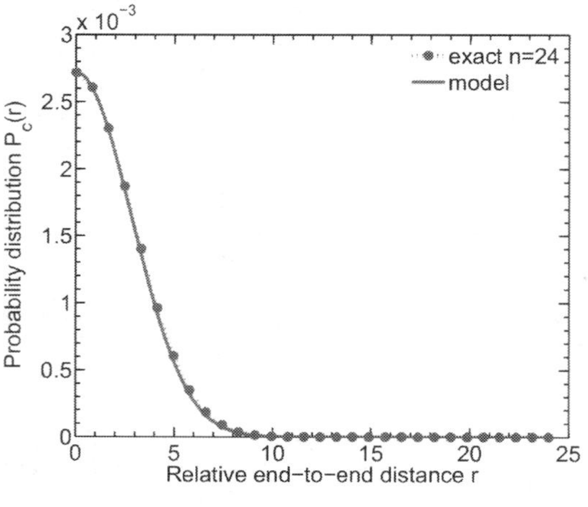

Figure 2. Validation of the closed-form (1) with the chain length $n = 24$. The limit of chain extensibility can be seen here: the likelihood of a polymer chain reaching its contour length ($r = n$) is zero.

end-to-end distance of polymer chains in the reference configuration, and a is a constraint normalization.The constant $\kappa = \frac{9}{\pi^2}$ arises from the spring constant of the linear coarse-grained model. The probability distribution function (1) is a quantum mechanically based closed-form replacement of the Rayleigh exact distribution function for the non-Gaussian chain. As seen in Fig. 2, (1) demonstrates very good agreement with the exact non-Gaussian distribution.

3 TWO-STEP SCALE BRIDGING

In this section we present a link between macroscale and microscale via a two-step transition. First, the scale bridging between macroscopic deformation and mesoscopic deformation (at the sub-network level) is made by the anistropic network averaging. Afterwards, the microscopic stretch (of a single chain) is related to the mesoscopic stretch by the stretch amplification.

3.1. *Analytical network-averaging*

According to the analytical network-averaging concept (Khiêm & Itskov 2016), each directional chain in the network is represented by a mean chain confined in a mean tube undergoing an average-tube stretch and an average-tube contraction.

The spatial distribution function of polymer chains in arbitrary direction n (see Fig. 1) can be developed into a series of probability density functions ρ_i related to a set of predefined vectors $\boldsymbol{E}_i$ ($i = 1, 2, \ldots, m$) as follows

$$\rho(\boldsymbol{n}) = \sum_{i=1}^{m} \frac{1}{m} \cdot \tilde{\rho}_i(\boldsymbol{n}) = \sum_{i=1}^{m} \frac{1}{m} \cdot \rho_i(\theta_i, \varsigma_i, \boldsymbol{E}_i). \tag{3}$$

Therein, m is the number of equidistant collocation points on a half-sphere specifying the deformation-induced anisotropy. $\boldsymbol{E}_i$ ($i = 1, 2, \ldots, m$) are the corresponding unit vectors in the radial directions. Each constituent $\rho_i(\theta_i, \varsigma_i, \boldsymbol{E}_i)$ of the probability density (3) is defined on the basis of an even von Mises-Arnold-Fisher distribution function (Fisher 1953)

$$\rho_i(\theta_i, \varsigma_i, \boldsymbol{E}_i) = \frac{\varsigma_i}{4\pi \sinh(\varsigma_i)} \cosh(\varsigma_i \cdot \cos\theta_i), \tag{4}$$

where θ_i indicates the angle between the directional unit vector $\boldsymbol{n}$ and the vector $\boldsymbol{E}_i$, ς_i, denotes the concentration parameter.

The most general formulation of the concentration parameters can be given at time t by (for the sake of brevity, explicit reference to the material point $\boldsymbol{X}$ is omitted in the response functional)

$$\varsigma_i(t) = \underset{\tau \in (-\infty, t]}{\mathcal{M}} \left[\mathbf{F}(\tau), \Xi \right], \tag{5}$$

where $\mathbf{F}(\tau)$ denotes the deformation gradient at time τ in the past and Ξ is an internal variable.

The mesoscopic stretch is evaluated as the root mean square of the macroscopic stretch over the unit sphere by taking the directional chain distribution into account (Khiêm & Itskov 2017)

$$\bar{\Lambda} = \sqrt{\sum_{i=1}^{m} \frac{1}{m} \cdot \left((1 - w_i) \frac{I_1}{3} + w_i \Lambda_i^2 \right)}, \tag{6}$$

where θ_i and φ_i are spherical coordinates of the vector n with respect to orthonormal vectors based on the collocation directions $\boldsymbol{E}_i$. I_1 denotes the first principal invariant of the right Cauchy-Green tensor $\mathbf{C} = \mathbf{F}^{\mathrm{T}}\mathbf{F}$, and $\Lambda_i^2 = \mathbf{C} : \boldsymbol{E}_i \otimes \boldsymbol{E}_i$ is the square of the macro-stretch in direction $\boldsymbol{E}_i$. The fraction of chains w_i aligned in each direction i ($i = 1, 2, \ldots m$) is expressed by

$$w_i = \frac{\varsigma_i^2 - 3\coth(\varsigma_i)\varsigma_i + 3}{\varsigma_i^2}. \tag{7}$$

In the next step, the mesoscopic tube contraction is evaluated as the root mean square of the macroscopic tube contraction over the unit sphere. In the same manner as (6), the mesoscopic tube contraction is derived as

$$\bar{\Upsilon} = \sqrt{\sum_{i=1}^{m} \frac{1}{m} \cdot \left((1 - w_i) \frac{I_2}{3} + w_i \Upsilon_i^2 \right)}, \tag{8}$$

where I_2 denotes the second principal invariant of the right Cauchy-Green tensor $\mathbf{C}$ and Υ_i is the macro tube contraction in direction i. It is expressed as $\Upsilon_i^2 = \mathbf{cofC} : E_i \otimes E_i$, where $\mathrm{cof}\mathbf{C} = \mathbf{C}^{-\mathrm{T}} det\mathbf{C}$. In contrast to the previous works (see e.g. (Miehe, Göktepe, & Lulei 2004)), the mesoscopic deformation measures are governed here not only by the macroscopic deformation gradient, but also by the history of deformation. Thus, this approach allows us to include both non-affine deformation and intrinsic microstructural evolutions into the model.

3.2 *Mesostretch amplification*

By comparison of different physically-based constitutive models with molecular dynamics simulation of several polymers, it was demonstrated (Khiêm & Itskov 2017) that an additional transition between the mesoscopic and microscopic scales is essential to capture the failure stretch of the polymer network. Accordingly, the micro-stretch is expressed by

$$\bar{\lambda} = \bar{\Lambda}^{q}, \tag{9}$$

where q is the stretch amplification exponent related to the degree of inhomogeneity of the rubber network.

Replacing in (3) the normalized end-to-end distance by the root mean square end-to-end distance of the freely jointed chain $R = \sqrt{n}$, we obtain

$$\Psi_c\left(n,\bar{\lambda}\right) = \mu_c n \ln \frac{\pi\bar{\lambda}}{\sqrt{n}\sin\left(\frac{\pi}{\sqrt{n}}\bar{\lambda}\right)}, \tag{10}$$

where $\mu_c = N_c k_B T \kappa$ is the effective shear modulus.

We assume that the average micro tube contraction and the average macro tube contraction are identical $\bar{\upsilon} = \bar{\Upsilon}$. In view of (2), the network strain energy due to the topological constraint can be expressed as

$$\Psi_t = \mu_t \bar{\Upsilon}, \tag{11}$$

where $\mu_t = N_c k_B T \omega$ is the topological shear modulus, and ω is the tube geometrical parameter.

Therefore, the strain energy density of the whole rubber network is given in view of (10) and (11) by

$$\Psi = \mu_c n \ln \frac{\pi\bar{\Lambda}^{q}}{\sqrt{n}\sin\left(\frac{\pi}{\sqrt{n}}\bar{\Lambda}^{q}\right)} + \mu_t \bar{\Upsilon}. \tag{12}$$

4 MECHANISMS OF INELASTIC EFFECTS

4.1 *Stress softening*

Recent experimental results (Diaz, Diani, & Gilormini 2014, Huneau, Masquelier, Marco, Le Saux, Noizet, Schiel, & Charrier 2016) confirm that the stress softening is not due to the deformation of carbon black agglomerates. Thus, filler aggregates are regarded in this work as rigid inclusions. The Mullins effect can be explainedby permanent debonding of polymer chains at the irreversible adsorption area on the aggregate surface (Bueche 1960).

Each directional damage subnetwork of polymer chains is considered as a series of M_c polymer chains connecting two adjacent aggregates. Let M_c^{max} be the maximal number of chains in a subnetwork. The probability $P_D(l_c)$ of existence of a subnetwork of polymer chains with the contour length l_c connecting two aggregates can be obtained as (Khiêm & Itskov 2017)

$$P_D\left(l_c\right) = \frac{\alpha \sin^2\left(\frac{\pi l_c}{M_c^{max} n}\right)\exp\left[\frac{\alpha(l_c-1)}{2}\right]}{\exp\left[\frac{\alpha M_c^{max} n}{4\pi}\left(\sin\frac{2\pi l_c}{M_c^{max} n} - \sin\frac{2\pi}{M_c^{max} n}\right)\right]}, \tag{13}$$

where α is the average functionality of the active adsorption site on the aggregate surface (i.e. the number of polymer chains that can connect to the active adsorption site).

Let L be the average referential end-to-end distance of the subnetworks of polymer chains. During deformation, subnetworks oriented in the principal direction i with the contour length l_c shorter than $\Lambda_i^{max} L$ will break and no more contribute to the network entropy, where Λ_i^{max} is the maximal stretch previously reached in direction i. This process continues with other subnetworks with longer contour lengths and leads to significant stress softening. Thus, the domain of survival chains can be given by $D_a = \left\{ l_c \middle| \Lambda_i^{max} L \le l_c < M_c^{max} n \right\}$. According to experimental observations, filler aggregates also rearrange in the stretching direction (Oberdisse, Pyckhout-Hintzen, & Straube 2009). Therefore the fraction of available chains in one direction can be calculated as

$$w_i = w_{0D} \exp\left(\frac{\bar{\alpha}}{4\pi}\left[n \sin\frac{2\pi\Lambda_i^{max}\bar{L}}{n} - 2\pi\Lambda_i^{max}\bar{L}\right]\right), \tag{14}$$

where $\bar{\alpha} = \alpha M_c^{max}$ and the normalized average end-to-end distance of the subnetwork $\bar{L} = \frac{L}{M_c^{max}}$ are considered as material constants.

By considering the volume fraction of fillers as the fraction of polymer chains connected to filler particles, its initial value w_{0D} can be given by

$$w_{0D} = w_f^{\frac{1}{3}}, \tag{15}$$

where w_f is the volume fraction of fillers.

4.2 *Strain induced crystallization*

In this work, the crystallite is assumed to be bundle-like with a cylindrical geometry (cf. (Candau, Laghmach, Chazeau, Chenal, Gauthier, Biben, & Munch) 2014, Liu, Tian, Huang, Cui, Wang, Hu, Yang, Li, & Li 2014, Gros, Tosaka, Huneau, Verron, Poompradub, & Senoo 2015)). Each crystallite is built by K neighbouring polymer chains. The change in free energy of a subnetwork due to the nucleationof a crystallite is given by

$$\Delta G = \gamma_t 2Ks + \gamma_s 2\sqrt{\pi K\, sl} - K\, sl\Delta\Psi_i, \tag{16}$$

where γ_t is the surface tension at the top surfaces of the crystallite. γ_s is the surface tension at the side of the crystallite. $\Delta\psi_i$ is the difference in bulk free energies between the semicrystalline andthe amorphous subnetworks. s is the cross-section area of a single polymer chain, and l is the length of the crystallite.

Thus, according to the Lauritzen-Hoffman nucleation theory (Hoffman & Lauritzen 1961), the number of molecules K^b in the critical nucleus size can be obtained as

$$K^b = \frac{4\pi\gamma_s^2}{s\Delta\Psi_i^2}. \tag{17}$$

Crystallites with the number of molecules larger than this critical value are likely to grow. The free energy barrier is given in view of (16) by

$$\Delta G^b = \frac{8\pi\gamma_t\gamma_s^2}{\Delta\Psi_i^2}. \tag{18}$$

Therefore, the fraction of chains aligned in each direction can be calculated as

$$w_i = w_C \exp\left(-\frac{8\pi\gamma_t\gamma_s^2}{k_B T\Delta\Psi_i^2}\right), \tag{19}$$

where w_C denotes the limit of chain alignment (Toki, Sics, Ran, Liu, Hsiao, Murakami, Senoo, & Kohjiya 2002).

5 RESULTS

5.1 *Rubber elasticity*

Since the elastomers considered in this subsection are unfilled and the strain-induced crystallization is not taken into account, $w_i = 0\ \forall \mathbf{i}$. In view of (7), $\varsigma_i = 0\ \forall i$, so that the distribution function of polymer chains (3) becomes isotropic (see Fig. 1). Thus, $\bar{\Lambda} = \sqrt{\frac{I_1}{3}}$ and $\bar{\Upsilon} = \sqrt{\frac{I_2}{3}}$, (12) reproduces the constitutive model proposed by (Khiêm & Itskov 2016) for rubber elasticity.

The model is first fitted to experimental data by Treloar on uniaxial tension, pure shear and equibiaxial tension of vulcanized rubbers containing 8 phr sulfur (Treloar 1944). The same set of material constants is also used to predict the stress-strain relation in a series of biaxial tensions by (Kawabata, Matsuda, Tei, & Hawai 1981). This benchmark is very tough because there are only three models known in literature able to capture both (Kawabata, Matsuda, Tei, & Hawai 1981) and (Treloar 1944) data using the same set of material constants (Marckmann & Verron 2006). As seen in Fig. 2, the proposed model demonstrates very good agreement with both sets of experimental data.

5.2 *Stress softening*

The model is further compared to experimental data of filled silicone rubber undergoing uniaxial tension, pure shear and equibiaxial tension (Machado, Chagnon, & Favier 2010). Despite the low number of fitting parameters, good agreement between model predictions and the experimental data with respect to both the Mullins effect and hysteresis is found (Fig. 4).

5.3 *Strain-induced crystallization*

Finally, predictive capability of the model in the case of strain-induced crystallization in natural rubber is demonstrated by a comparison with

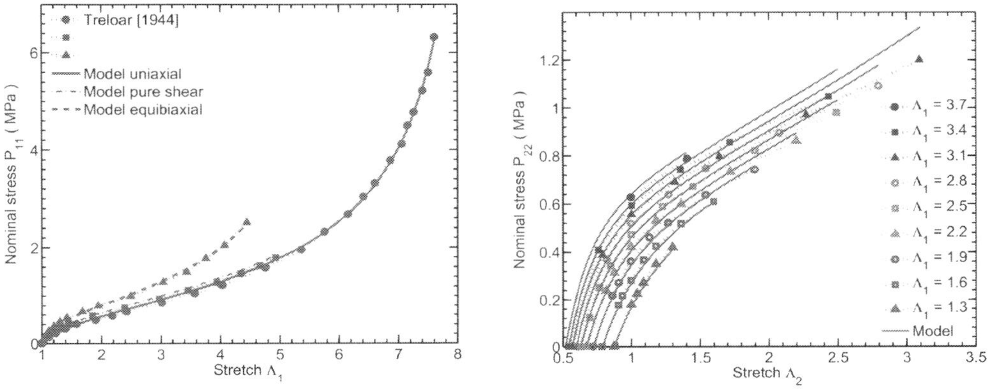

Figure 3. Prediction of the proposed model in comparison with the multi-dimensional experimental data of unfilled rubber: in uniaxial tension, pure shear and equibiaxial tension (left), and in a series of biaxial tensions by Kawabata et al. 1981 (right).

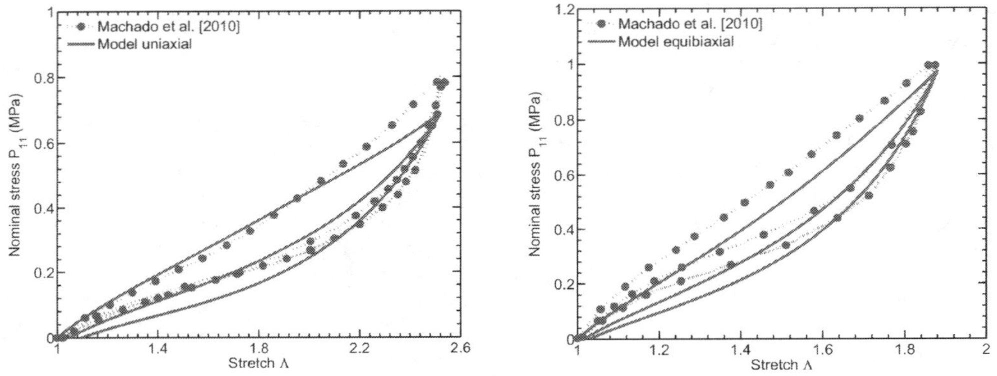

Figure 4. Prediction of the proposed model in comparison to the experimental data by Machado et al. 2012: uniaxial tension (left), and equibiaxial tension (right).

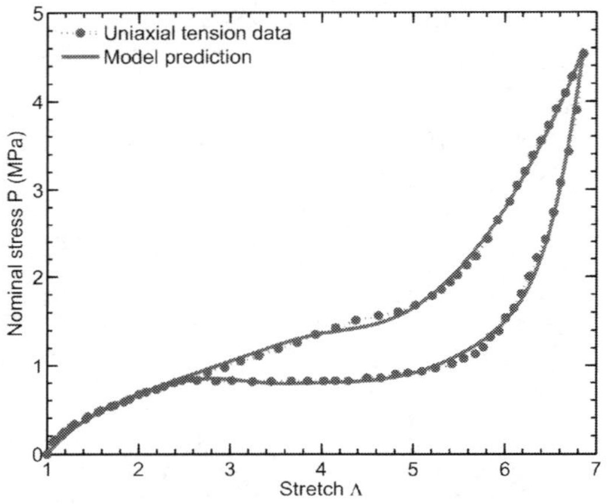

Figure 5. Prediction of the proposed model in the case of straininduced crystallization in comparison to the uniaxial tension of natural rubber by Albouy et al. 2005.

experimental data by (Albouy, Marchal, & Rault 2005). One observes very good agreement between calculated and measured values (Fig. 5).

6 CONCLUSIONS

In the current work, we presented a general physically-based constitutive law (12) for rubber-like materials. The model is developed on the basis of the analytical network averaging concept by considering deformation history dependent distribution of polymer chains. Accordingly, the spatial arrangement of polymer chains is driven by microstructural evolutions, and consequently alters the mean field deformation measures (mesoscopic stretch and tube contraction). From the computational point of view, this unified approach is of highly advantageous as only one finite element code is required for predicting a wide spectrum of various elastic and inelastic phenomena in elastomers.

REFERENCES

Albouy, P.-A., J. Marchal, & J. Rault (2005, jul). Chain orientation in natural rubber, Part I: the inverse yielding effect. *The European Physical Journal E 17*(3), 247–59.

Bueche, F. (1960). Molecular basis for the Mullins effect. *Journal of Applied Polymer Science 4*, 107–114.

Candau, N., R. Laghmach, L. Chazeau, J.-M. Chenal, C. Gauthier, T. Biben, & E. Munch (2014). Strain-Induced Crystallization of Natural Rubber and Cross-Link Densities Heterogeneities. *Macromolecules 47*(16), 5815–5824.

Diaz, R., J. Diani, & P. Gilormini (2014). Physical interpretation of the Mullins softening in a carbon-black filled SBR. *Polymer 55*(19), 4942–4947.

Fisher, R. (1953). Dispersion on a sphere. *Proceedings of the Royal Society of London. Series A, Mathematical and Physical Sciences 217*(1130), 295–305.

Gros, A., M. Tosaka, B. Huneau, E. Verron, S. Poompradub, & K. Senoo (2015). Dominating Factor of Strain-induced Crystallization in Natural Rubber. *Polymer 76*, 230–236.

Hild, G. (1998). Model networks based on 'endlinking' processes: synthesis, structure and properties. *Progress in Polymer Science 23*(6), 1019–1149.

Hoffman, J.D. & J. Lauritzen (1961). Crystallization of Bulk Polymers With Chain Folding: Theory of Growth of Lamellar Spherulites. *Journal of Research of the National Bureau of Standards-A. Physics and Chemistry 65A*(4), 297–336.

Huneau, B., I. Masquelier, Y. Marco, V. Le Saux, S. Noizet, C. Schiel, & P. Charrier (2016). Fatigue Crack Initiation in a Carbon BlackFilled Natural Rubber. *Rubber Chemistry and Technology 89*(1), 126–141.

Kaliske, M. & G. Heinrich (1999). An extended tube-model for rubber elasticity: Statistical-mechanical theory and finite element implementation. *Rubber Chemistry and Technology 72*(4), 602–632.
Kawabata, S., M. Matsuda, K. Tei, & H. Hawai (1981). Experimental survey of the strain energy density function of isoprene rubber vulcanizate. *Macromolecules 14*, 154–162.
Khiêm, V.N. & M. Itskov (2016). Analytical network-averaging of the tube model: Rubber elasticity. *Journal of the Mechanics and Physics of Solids 95*, 254–269.
Khiêm, V.N. & M. Itskov (2017). An averaging based tube model for deformation induced anisotropic stress softening of filled elastomers. *International Journal of Plasticity 90*, 96–115.
Liu, D., N. Tian, N. Huang, K. Cui, Z. Wang, T. Hu, H. Yang, X. Li, & L. Li (2014). Extension-induced nucleation under near-equilibrium conditions: The mechanism on the transition from point nucleus to shish. *Macromolecules 47*(19), 6813–6823.
Machado, G., G. Chagnon, & D. Favier (2010). Analysis of the isotropic models of the Mullins effect based on filled silicone rubber experimental results. *Mechanics of Materials 42*(9), 841–851.
Marckmann, G. & E. Verron (2006). Comparison of hyperelastic models for rubber-like materials. *Rubber Chemistry and Technology 79*, 835–858.
Miehe, C., S. Göktepe, & F. Lulei (2004). A micro-macro approach to rubber-like materials—Part I: the non-affine microsphere model of rubber elasticity. *Journal of the Mechanics and Physics of Solids. 52*(11), 2617–2660.
Oberdisse, J., W. Pyckhout-Hintzen, & E. Straube (2009). Structure Determination of Polymer Nanocomposites by Small Angle Scattering. In S.V. Thomas, S.; Zaikov, G.; Valsaraj (Ed.), *Recent Advances in Polymer Nanocomposites*, pp. 397. Brill NV: Leiden, The Netherlands.
Shariff, M.H.B.M. (2000). Strain energy function for filled and unfilled rubberlike material. *Rubber Chemistry and Technology 73*(1), 1–18.
Toki, S., I. Sics, S. Ran, L. Liu, B.S. Hsiao, S. Murakami, K. Senoo, & S. Kohjiya (2002). New insights into structural development in natural rubber during uniaxial deformation by in situ synchrotron X-ray diffraction. *Macromolecules 35*(17), 6578–6584.
Treloar, L.R.G. (1944). Stress-strain data for vulcanised rubber under various types of deformation. *Transactions of the Faraday Society 40*, 59–70.
Verron, E. (2015). Questioning numerical integration methods for microsphere (and microplane) constitutive equations. *Mechanics of Materials 89*, 216–228.
Wu, P. & E. van der Giessen (1993). On improved network models for rubber elasticity and their applications to orientation hardening in glassy polymers. *Journal of the Mechanics and Physics of Solids 41*(3), 427–456.

Constitutive Models for Rubber X – Lion & Johlitz (Eds)
© 2017 Taylor & Francis Group, London, ISBN 978-1-138-03001-5

A hyperelastic physically based model for filled elastomers including continuous damage effects and viscoelasticity

J. Plagge & M. Klüppel
Deutsches Institut für Kautschuktechnologie e. V., Hannover, Germany

ABSTRACT: A novel physically based material model is presented that describes the complex stress-strain behavior of filled rubbers under arbitrary deformation histories in a constitutive manner. The polymer response is considered by the extended non-affine tube model. Stress softening is taken into account via the breakdown of highly stressed polymer-filler domains under load and self-homogenization of the medium. Simulations of the stress-strain response are in good agreement with experiments for different deformations modes, deformation speeds and temperatures. The model is extended to continuous damage and viscoelastic effects by relaxing structure-defining parameters in an appropriate way. This is based on relaxation measurements of differently filled EPDM showing logarithmic relaxation, which can be approximated as a slow powerlaw. It is shown, that powerlaw relaxation can be generated by choosing relaxation times in a specific way. The methodology presented is easily transferable to similar models. The resulting model curves closely resemble true rubber behavior in terms of hysteresis, relaxation characteristics and stress softening.

1 INTRODUCTION

An unfilled elastomer is almost ideally hyperelastic, e.g. exhibiting small hysteresis and weak rate dependency, but exhibits poor wear- and fatigue properties. The incorporation of fillers greatly toughens and stiffens the polymer, but also creates memory-like material properties as increased hysteresis and stress softening. For small dynamic deformations the latter effect is known as Payne-Effect and is attributed to the irreversible breakdown of a stiff, percolated filler network (Rendek and Lion 2010). A similar phenomenon can be observed for larger deformations, which was investigated by (Mullins and Tobin 1965): The greater the deformation the material was subject to, the softer the materials response. The change in the response is not immediate, but requires several cycles to reach a steady state. The explanation for this effect is the complex interplay and structural rearrangement of filler particles and the polymer network, as reviewed in (Vilgis, Heinrich, and Klüppel 2009). In particular, a breakdown or slippage and disentanglement of adsorbed polymer chains at the filler surface (Hamed and Hatfield 1989) or a rearrangement of network junctions in filled systems has been proposed. A constitutive model of stress-induced desorption of chains from the filler surface has been derived by Govindjee and Simo based on a statistical mechanics approach (Govindjee and Simo 1992). This idea was recently extended to include deformation of filler aggregates as well, allowing the calculation of hysteresis (Dargazany and Itskov 2013). A more general yet abstract approach was chosen by (Wulf and Ihlemann 2015), which did computer simulations of a minimal rubber material on a molecular scale. In their work a small collection of generalized interactions was sufficient to generate typical rubber stress-strain data. The so called Dynamic-Flocculation-Model (DFM) derived by Klüppel et al. (Klüppel and Schramm 2000, Klüppel 2003, Klüppel, Meier, and Dämgen 2005) states that the underlying mechanism of stress softening is hydrodynamic reinforcement by stiff filler clusters, which immobilize a certain amount of polymer. By increasing the load, more and more filler clusters are broken, decreasing the geometric constraints imposed on the polymer matrix. In addition, the broken clusters are dynamically re-aggregating and breaking during repeated loading up to a certain strain level, thereby causing hysteresis which is a further typical effect found in experiments. The DFM was shown to be in good agreement with experimental data for various elastomer nanocomposites with carbon black or silica/silane filler systems at different deformation modes (Lorenz, Meier, and Klüppel 2010, Lorenz and Klüppel 2012). In addition to these physical approaches, there exists a wealth of phenomenological models. For example, (Amin, Lion, Sekita, and Okui 2006) modeled the material by decomposing the deformation gradient into elastic and inelastic parts using the concept of the Zener model.

Recently, a new physically based model was presented, which includes stress-softening as well as temperature and rate dependent hysteresis (Plagge and Klüppel 2016). Additionally, it successfully describes several deformation modes and provides a free energy density, allowing easy calculations. All parameters are physically reasonable and have been proven to behave as expected by extensive fits to differently filled and differently crosslinked compounds. Even though the model is rate dependent, a true time dependency (e. g. stress relaxation at static strain) is still missing. Filled elastomers have been found to show power law or even sub-power law stress relaxation on timescales ranging from seconds to many hours (Ehrburger-Dolle, Morfin, Bley, Livet, Heinrich, Richter, Piché, and Sutton 2012). To the best of our knowledge no model is capable of reproducing long-time relaxation behavior, which may be responsible for continuous damage as well as large strain hysteresis behavior. This problem shall be addressed in the present paper.

2 THEORY AND DISCUSSION

2.1 *A constitutive model for filled elastomers*

The following section is a strong condensation of the theory described in (Plagge and Klüppel 2016). The free energy density of the model is based on the non-affine tube model (Heinrich, Straube, and Helmis 1988).

$$W(\bar{I}_1,\bar{I}^*)=\frac{G_c}{2}\left[\frac{\left(1-\frac{1}{n}\right)\bar{I}_1}{1-\frac{1}{n}\bar{I}_1}+\log\left(1-\frac{\bar{I}_1}{n}\right)\right]+2G_e\bar{I}^* \quad (1)$$

with $\bar{I}_1=\lambda_1^2+\lambda_2^2+\lambda_3^2-3$ and $\bar{I}^*=\lambda_1^{-1}+\lambda_2^{-1}+\lambda_3^{-1}-3$ being the first and a generalized invariant, respectively. The first invariant is a direct measure for the square of the absolute deformation of the sample. For the generalized invariant the same holds for small deformations. Thus, it can be shown, that internal strains can be amplified by replacing

$$W(\bar{I}_1,\bar{I}^*) \quad \rightarrow \quad W(X\bar{I}_1,X\bar{I}^*) \quad (2)$$

with $X>1$ being the strain amplification factor. We assume, that the material consists of differently amplified domains, maybe due to inhomogeneous crosslinking or filler dispersion. The amplification factor distribution is thus modeled as a decreasing power-law between $[X_{min},X_{max}]$, which is always normalized to 1:

$$P_X(X,X_{max})=X^{-\chi}\cdot\frac{\chi-1}{X_{min}^{1-\chi}-X_{max}^{1-\chi}} \quad (3)$$

where $\chi>0$ is the exponent determining the broadness of the distribution. The normalization guarantees, that there is always the same amount of elastically active rubber. This is the consequence of a simple physical idea: Once we increase maximum strain, highly amplified domains break down (X_{max} decreases) and rearrange less amplified (normalization increases height of distribution). This is visualized in Fig. 1. The minimum amplification factor is chosen to be $X_{min}=1$, corresponding to non-amplified domains. The maximum amplification factor decreases with maximum strain and is defined by

$$X_{max}=\mathbf{max}\left(X_{min},\frac{n}{\bar{I}_{1,max}+\gamma}\right) \quad (4)$$

where $\bar{I}_{1,max}$ is the all time maximum of the first invariant and γ is a parameter determining the maximum amplification factor at zero strain. Additionally, it guarantees, that the energy density isn't diverging. This would happen at $X_{div}=n/\bar{I}_{1,max}$, see Eq. (2) and (1). Altogether, the energy density of the inhomogeneously amplified material is calculated as a superposition of amplified energy densities given by Eq. (2) weighted by the distribution function P_X (Eq. (3)):

$$W_X(\bar{I}_1,\bar{I}^*)=\int_{X_{min}}^{X_{max}}\mathrm{d}X\,P_X(X,X_{max})W(X\bar{I}_1,X\bar{I}^*) \quad (5)$$

where stress softening is implemented via the change of X_{max} according to Eq. (4). The integral can be carried out analytically. Stress is easily calculated from Eq. (5) by standard continuum mechanical methods, e. g. (Holzapfel 2000).

In the original theory the hysteresis component was calculated by assuming structures to break

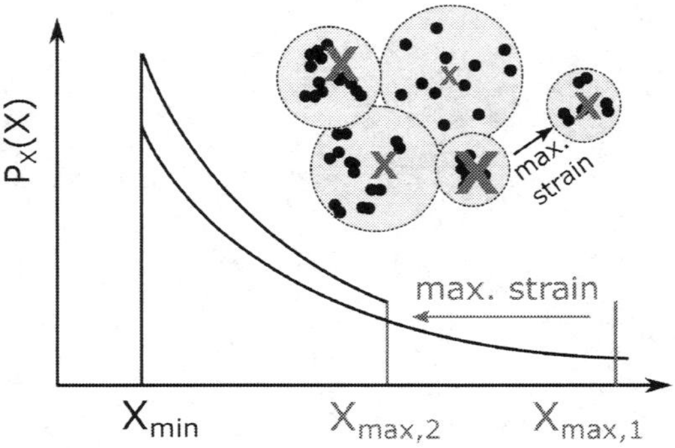

Figure 1. Powerlaw amplification factor distribution. The distribution is normalized to 1, such that a decrease of X_{max} yields an increase of occurrence of less amplified domains.

and rebind at a certain stress, defined by the stress required to pass a potential barrier, as visualized in Fig. (2):

$$f_{\mathrm{hys}} = f_{\mathrm{el}}(\varepsilon) - \sum_j \theta(f_{\mathrm{el}} - f_{\mathrm{el},j}) \left[f_{\mathrm{el},j} - f_{\mathrm{el},j-1} \right] \quad (6)$$

where the sum runs over all rebinding events j and $\theta(x)$ represents the Heaviside step function. Doing the transition from forces to stress and averaging over structure, we end up with an integral equation

$$\sigma_{\mathrm{hys}} = \int_{-\infty}^{\varsigma} \mathrm{d}\varsigma' \frac{1}{1+\frac{\varsigma-\varsigma'}{\sigma_c}} \frac{\mathrm{d}\sigma_{\mathrm{el}}(\varsigma')}{\mathrm{d}\varsigma'}; \quad \varsigma = \int_{-\infty}^{t} \mathrm{d}t' \left| \dot{P}_{\mathrm{el}}(t') \right| \quad (7)$$

where σ_{hys} denotes the hysteresis component of the first Piola Kirchhoff stress, σ_{el} is the elastic first Piola Kirchhoff stress component calculated from Eq. (5) and ς is the intrinsic time, which is running when elastic stress changes. Note that Eq. (7) is an convolution and can be efficiently evaluated numerically. The critical stress σ_c defines the average stress required to pass a potential barrier, as can be seen in the inset of Fig. 2. It has been made temperature- and rate dependent by using Kramers escape rate (Hänggi, Talkner, and Borkovec 1990):

$$r \pm = r_0 \mathrm{e}^{-\frac{E_b \mp \sigma_c V}{k_B T}} \quad (8)$$

with hopping rate in force (+) or opposite to force (–) direction, energy barrier E_b, zero-potential hopping rate r_0 and volume of hysteresis generating structures V. The net hopping rate is $r = r_+ - r_-$, such that

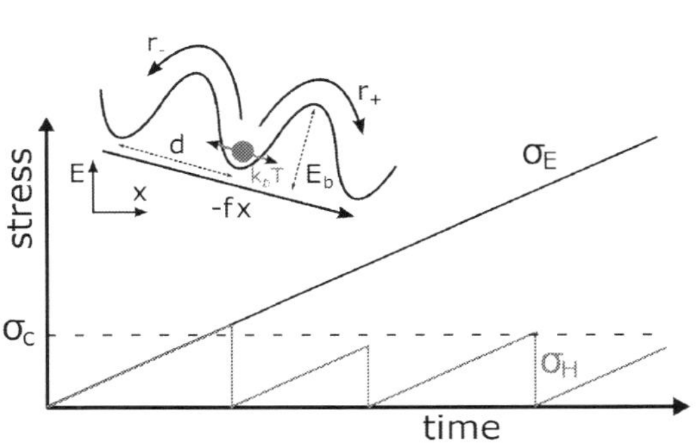

Figure 2. Visualization of the breaking and rebinding process responsible for hysteresis. At the critical stress σ_c structures break and rebind, resulting in a retardation of stress. Inset: The same process in terms of diffusion in a periodic potential. The potential is tilted by a force $f \sim \sigma_H$ until the structure breaks (crosses the barrier).

$$r = 2 r_0 \mathrm{e}^{-\frac{E_b}{k_B T}} \sinh\left(\frac{\sigma_c V}{k_B T} \right) \quad (9)$$

When doing an experiment at given strain rate r has to follow the pace of the machine R and is assumed to be proportional to it $r = aR$. Consequently, the force (or stress) acting on the structures σ_c has to change, e. g. a faster deformation requires a larger force to break the structures fast enough. Solving Eq. (9) for σ_c gives

$$\sigma_c(R,T) \approx \frac{1}{V} \left(E_b + k_B T \log\left(a R_d / r_0 \right) \right) \quad (10)$$

The logarithm already indicates, that hysteresis is only weakly dependent on strain rate. Eq. (10) can be fitted to $\sigma_c(R,T)$ obtained from experiments carried out at different rates and temperatures to determine E_b, V and a/r_0. For example, for sulfur cured 50 phr N339 filled EPDM this yields $V = 375\ \mathrm{nm}^3$, $E_b = 104$ kJ/mol and $\log(r_0/a) = 30$. An example is shown in Fig. 4, where both model curves were calculated using Eq. (10) and the parameters given above. Take into account, that $V^{1/3} \approx 7.2$ nm as well as E_b are in the range of polymer-filler interaction. For details see (Plagge and Klüppel 2016).

In scalar notation, total stress σ is then calculated as

$$\sigma = (1-\phi)\sigma_{\mathrm{el}} + \phi \sigma_{\mathrm{hys}} \quad (11)$$

with ϕ being a factor balancing elastic and inelastic forces. It has been shown to be proportional to filler volume fraction. An example of a model fit to uniaxially and biaxially deformend 50 phr carbon black filled (N339) EPDM is shown in Fig. 3. All fitting parameters are physically reasonable. This is discussed more extensively in (Plagge and Klüppel 2016). The model has been shown to correctly describe the rate- and temperature dependence of various compounds, but requires a constant deformation rate R as used in Eq. (10). Here, we want to show an alternative approach, which is able to describe relaxation, continuous damage and hysteresis.

2.2 *Stress relaxation of filled elastomers*

Large-strain deformations are almost always dominated by long time relaxation. For this purpose we investigated stress relaxation at large strain for an amorphous EPDM (Keltan 4450) filled with 20 phr, 40 phr and 60 phr N339 carbon black and cured with 2.40 phr CBS, 1.5 phr DPG and 1.8 phr Sulfur. Pure shear and S3 A samples were stretched using

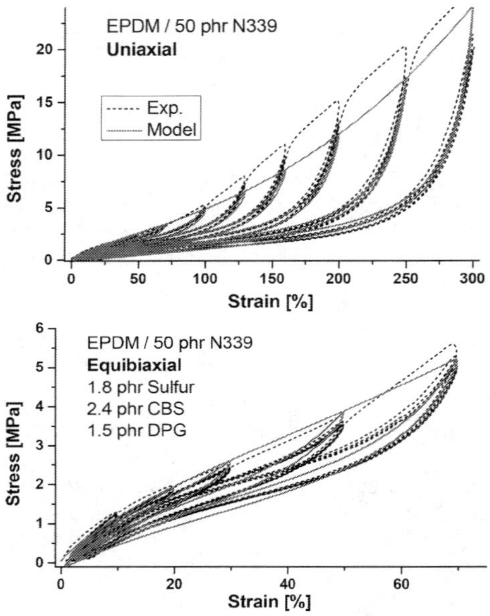

Figure 3. Example of a model fit to 50 phr N339 filled EPDM. Uniaxial (left), biaxial (right) and pure shear (not shown) deformation were fitted using one common set of fit parameters.

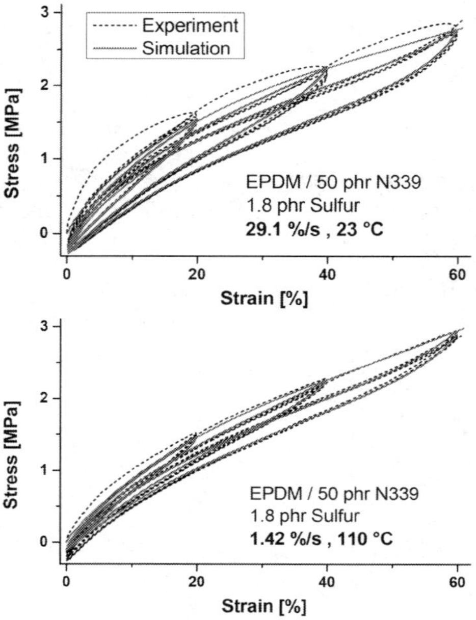

Figure 4. Example of the model with all parameters fixed except of σ_c and G_c, which was calculated in terms of rate and temperature according to Eq. (10). G_c is weakly dependent on temperature and was calculated using a linear fit.

a Zwick universal testing machine and relaxed at constant strain, controlled by an optical tracking system. To determine the relaxation characteristics, stress data was smoothed using a Savitzky-Golay filter and differentiated with respect to time to get rid of a constant offset. The results are shown in Fig. (5). At 23 °C a fit to $-\mathrm{d}\sigma/\mathrm{d}t \sim t^{-(\alpha+1)}$ for $t > 100$ s yields results ranging between $\alpha = 0.05$ for 20 phr N339 and $\alpha = 0.00$ for 60 phr N339, slightly depending on fitting range. At elevated temperature the exponent is slightly higher with $\alpha = 0.08$ for 60 phr at 80°C. All exponents are close to 0 and indicate, that stress relaxation of carbon black filled samples is logarithmic, theoretically never approaching a plateau. This has been found by different authors, too (Ehrburger-Dolle, Morfin, Bley, Livet, Heinrich, Richter, Piché, and Sutton 2012). They also stated, that non-coupled silica relaxes significantly faster according to $\alpha = 0.17$. To conclude, most compounds can be described by a slowly relaxing powerlaw with $\alpha < 0.20$.

For easy implementation a powerlaw representation in terms of exponentials isrequired. This is obtained by looking at

$$\sum_{n=1}^{\infty} \mathrm{e}^{-\frac{tn^{1/\alpha}}{\tau_{\max}}} \approx \int_{1}^{\infty} \mathrm{d}n\, \mathrm{e}^{-\frac{tn^{1/\alpha}}{\tau_{\max}}} = \left(\frac{\tau_{\max}}{t}\right)^{\alpha} \int_{t/\tau_{\max}}^{\infty} \mathrm{d}z\, \mathrm{e}^{-z} z^{\alpha-1}$$

the last integral is constant for $t \ll \tau_{\max}$. So powerlaw relaxation $\sim t^{-\alpha}$ can be generated by choosing relaxation times according to

$$\tau_n = \tau_{\max} n^{-1/\alpha} \quad \text{with} \quad n = 1 \ldots N \tag{12}$$

where N defines the smallest relaxation time. Together $\tau_{\min}$ and $\tau_{\max}$ set the interval where

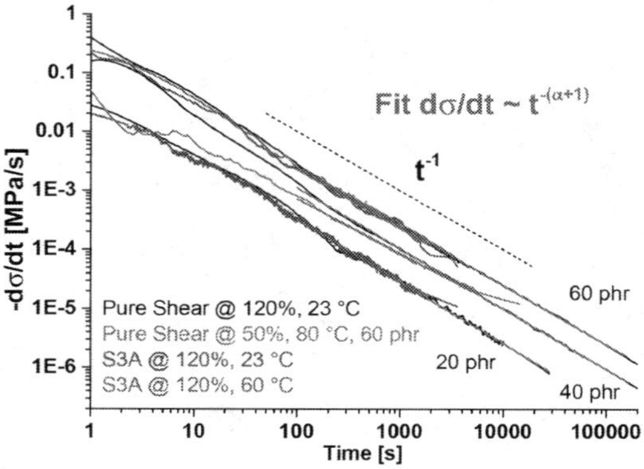

Figure 5. Powerlaw decay of stress for EPDM filled with 20 phr, 40 phr and 60 phr N339 carbon black. Irrespective of strain level, sample geometry and temperature a decay close to $\mathrm{d}\sigma/\mathrm{d}t \sim -t^{-1}$ or $\sigma \sim -\log(t)$ is observed. Color online.

relaxation ~$t^{-\alpha}$. Outside these bounds relaxation is exponential.

For continuous damage effects the upper amplification factor X_{max} defined by Eq. (4) has to be relaxed. Even though stress is nonlinear in X_{max} (near stress upturn) a powerlaw decay in X_{max} should generate similar relaxation characteristics. As described in section 2.1 X_{max} was directly determined by the all time maximum of the first invariant $\bar{I}_{1,max}$, see Eq. (4). The model diverges at $X_{max}\bar{I}_1 = n$ as can be seen from Eq. (1). Thus, when relaxing X_{max} we have to ensure, that X_{max} isn't violating this condition, which could occur by faster deformation than relaxation. A physically reasonable solution of this problem is a load dependent relaxation time spectrum, e. g. by $\tau_n = \tau_{n,0}\, e^{-\sigma V/k_{BT}}$ with V being a certain volume of amplifying structures, σ being a stress measure and $\tau_{n,0}$ chosen such, that Eq. (12) is fulfilled. The exponential will decrease relaxation time when a critical stress is surpassed. Numerically, this may be problematic, because time steps could be too large for the exponential to react.

A more simple yet phenomenological approach is to define an upper and lower bound for $X^{\pm}_{max}$ by

$$X^{\pm}_{max} = \max\left(X_{min}, \frac{n}{\bar{I}_1 + \gamma^{\pm}} \right) \tag{13}$$

this definition is the same as Eq. (4), except of $\gamma \rightarrow \gamma^{\pm}$. X^{-}_{max} shall now define the equilibrium state, thus the original model can be regarded as the stationary solution of the mechanism described here. X_{max} has to be written as a superposition of exponentially relaxing amplification factors. The evolution law is then given by

$$\dot{X}_{n,max} = \min\left(-\frac{1}{\tau_{n,X}}\left(X_{n,max} - X^{-}_{max} \right), 0 \right) \tag{14}$$

$$X_{n,max} = \min\left(X^{+}_{max}, X_{n,max} \right) \tag{15}$$

$$X_{max} = \frac{1}{N_X} \sum_{n=1}^{N_X} X_{n,max} \tag{16}$$

where $\tau_{n,X}$ are chosen according to Eq. (12). In detail, Eq. (14) guarantees decreasing amplification, Eq. (15) ensures, that the model is not diverging and Eq. (16) is an average over all exponentials, to be plugged into Eq. (5). An example for uniaxial loading is shown in Fig. 6, where the maximum and minimum stress bounds defined by X^{+}_{max} and X^{-}_{max} are shown. Hysteresis can be introduced in a similar way by directly relaxing stress according to Eq. (12), probably with N and τ_{max} different from

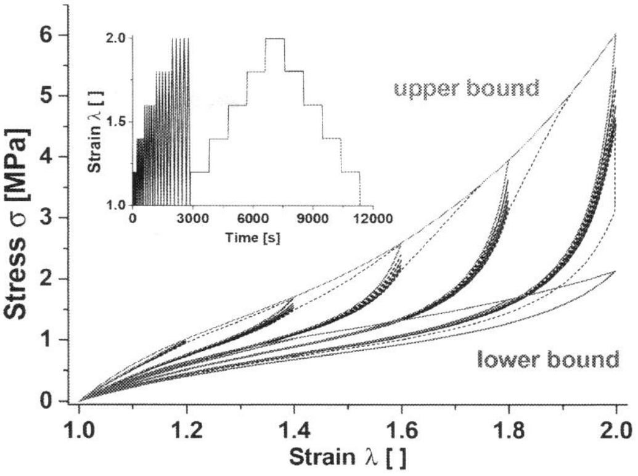

Figure 6. Model example including powerlaw ($\alpha = 0.15$) continuous damage effects. Upper and lower curves show the limiting stresses defined by X^{+}_{max} and X^{-}_{max}, respectively.

those used for amplification factor relaxation. The evolution is given by

$$\dot{\sigma}_{n,hys} = -\frac{1}{\tau_{n,hys}} \sigma_{n,hys} + \dot{\sigma}_{el} \tag{17}$$

$$\sigma_{hys} = \frac{1}{N_{hys}} \sum_{n=1}^{N_{hys}} \sigma_{n,hys} \tag{18}$$

The total stress is then calculated using Eq. (11). An example is shown in Fig. 7 where a curve is generated from an arbitrary set of fitting parameters for different deformation speeds using the same strain protocol as in Fig. 6. Except of the extreme stress recovery at $\lambda = 1.8$, curves for speed 0.1 and 1.0 have a realistic shape, showing all effects characteristic for filled elastomers. The fast sample with speed 10.0 is exemplary for a problem always occurring when superimposing viscous effects onto a hyperelastic material model with an upturn, or strain stiffening: The first cycle of each strain level has a lower maximum stress than subsequent cycles, which is untypical, because usually maximum stress is continuously decreasing. This can be explained by the smaller relaxation times, which act purely dissipative in the first cycle, where stress increases constantly at a rather slow rate. In subsequent cycles, where stress increases sharply near the end of the cycle, the same elements react elastically and contribute to stress. The already mentioned pronounced stress recovery at $\lambda = 1.8$ has a similar origin. During the relaxation step at $\lambda = 2.0$ almost all viscoelastic elements (Eq. (17)) are fully relaxed. In the following stress decreases, again, very sharply due to the S-shape of the curve, storing much energy in the viscoelastic

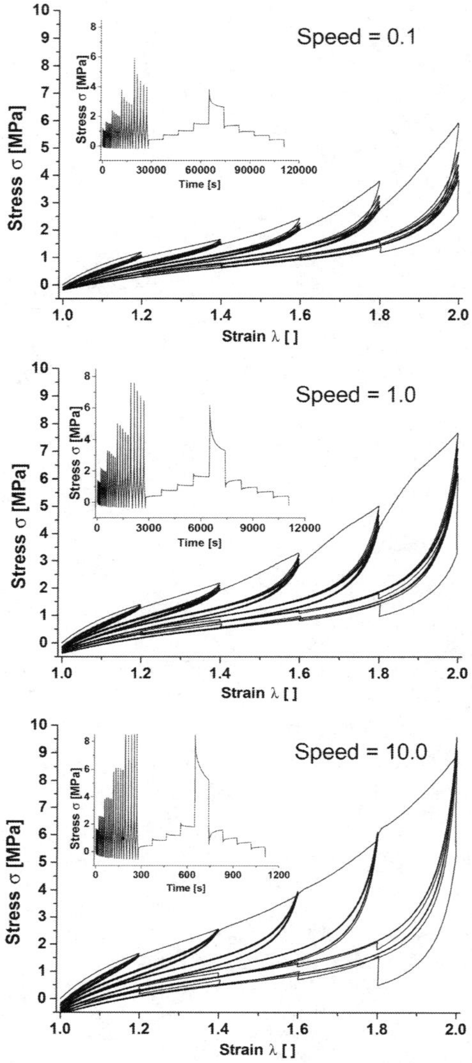

Figure 7. Model example including powerlaw ($\alpha = 0.15$) continuous damage effects and hysteresis made from arbitrary parameters. Top and bottom curves are made with 10 times lower and higher deformation speed, respectively.

elements. In the relaxation step at $\lambda = 1.8$ these elements relax, seemingly generating an overestimation of stress recovery. The conclusion is, that for models containing stress softening it is not sufficient to introduce viscoelasticity by convolving stress with viscoelastic elements as done here or in the widely used Prony-Series approach. Doing it will always open the possibility of subsequent cycles exhibiting larger stresses than the first one. Instead, viscoelasticity has to be derived from the same mechanism as stress softening. This could be done in our model by allowing the amplification to partly recover.

3 CONCLUSION

It has been shown, that stress of filled elastomers relaxes according to a slow powerlaw or logarithmically, rather insensitive to filler loading. On the basis of these findings a simple methodology to generate powerlaw relaxation from simple exponential kernels was derived and implemented into an existing model to include continuous damage and viscoelasticity. The resulting model seems to describe rubber stress-strain characteristics rather good, but exhibits some flaws regarding physical consistency. The reason for these flaws is universal and probably inherent in many other models. In future works we will try to avoid these problems by describing both continuous damage and hysteresis in terms of an universal mechanism.

ACKNOWLEDGEMENT

We thank the Arbeitsgemeinschaft industrieller Forschungsvereinigungen (AiF) grant 18079 BG for financial support.

REFERENCES

Amin, A., A. Lion, S. Sekita, and Y. Okui (2006). Nonlinear dependence of viscosity in modeling the rate-dependent response of natural and high damping rubbers in compression and shear: Experimental identification and numerical verification. *International Journal of Plasticity 22*(9), 1610–1657.

Dargazany, R. and M. Itskov (2013). Constitutive modeling of the mullins effect and cyclic stress softening in filled elastomers. *Physical Review E 88*(1), 012602.

Ehrburger-Dolle, F., I. Morfin, F. Bley, F. Livet, G. Heinrich, S. Richter, L. Piché, and M. Sutton (2012). XPCS investigation of the dynamics of filler particles in stretched filled elastomers. *Macromolecules 45*(21), 8691–8701.

Govindjee, S. and J. Simo (1992). Transition from micro-mechanics to computationally efficient phenomenology: carbon black filled rubbers incorporating mullins' effect. *Journal of the Mechanics and Physics of Solids 40*(1), 213–233.

Hamed, G.R. and S. Hatfield (1989). On the role of bound rubber in carbon-black reinforcement. *Rubber Chemistry and Technology 62*(1), 143–156.

Hänggi, P., P. Talkner, and M. Borkovec (1990). Reactionrate theory: fifty years after kramers. *Reviews of modern physics 62*(2), 251.

Heinrich, G., E. Straube, and G. Helmis (1988). Rubber elasticity of polymer networks: Theories. In *Polymer Physics*, pp. 33–87. Springer.

Holzapfel, G.A. (2000). Nonlinear solid mechanics. *24*.
Klüppel, M. (2003). The role of disorder in filler reinforcement of elastomers on various length scales. In *Filler-Reinforced Elastomers/Scanning Force Microscopy*, pp. 1–86. Springer.
Klüppel, M., J. Meier, and M. Dämgen (2005). Modeling of stress softening and filler induced hysteresis of elastomer materials. In *Constitutive Models for Rubber*, Volume 4, pp. 171. Balkema.
Klüppel, M. and J. Schramm (2000). A generalized tube model of rubber elasticity and stress softening of filler reinforced elastomer systems. *Macromolecular Theory and Simulations 9*(9), 742–754.
Lorenz, H. and M. Klüppel (2012). Microstructure-based modeling of arbitrary deformation histories of filler-reinforced elastomers. *Journal of the Mechanics and Physics of Solids 60*(11), 1842–1861.
Lorenz, H., J. Meier, and M. Klüppel (2010). Micromechanics of internal friction of filler reinforced elastomers. *LNACM* (51).
Mullins, L. and N. Tobin (1965). Stress softening in rubber vulcanizates. part i. use of a strain amplification factor to describe the elastic behavior of filler-reinforced vulcanized rubber. *Journal of Applied Polymer Science 9*(9), 2993–3009.
Plagge, J. and M. Klüppel (2016). A physically based model of stress softening and hysteresis of filled rubber including rateand temperature dependency. *International Journal of Plasticity*.
Rendek, M. and A. Lion (2010). Amplitude dependence of fillerreinforced rubber: Experiments, constitutive modelling and fem–implementation. *International Journal of Solids and Structures 47*(21), 2918–2936.
Vilgis, T.A., G. Heinrich, and M. Klüppel (2009). *Reinforcement of polymer nano-composites: theory, experiments and applications*. Cambridge University Press.
Wulf, H. and J. Ihlemann (2015). Simulation of mullins effect and relaxation due to self-organization processes in filled rubber. *Constitutive Models for Rubber IX*, 305.

Constitutive Models for Rubber X – Lion & Johlitz (Eds)
© 2017 Taylor & Francis Group, London, ISBN 978-1-138-03001-5

A micro-mechanical model based on the hydrodynamic strain amplification in filled elastomers

E. Darabi & M. Itskov
Department of Continuum Mechanics, RWTH Aachen University, Aachen, Germany

M. Klüppel
Deutsches Institut für Kautschuktechnologie e.V. (DIK), Hannover, Germany

ABSTRACT: A constitutive model for filled elastomers is developed by combining the framework of the Dynamic Flocculation Model (DFM) (Klüppel 2003) and the continuum damage model (Govindjee & Simo 1991). The model extends the previously proposed micro-mechanical formulation describing both the polymerfiller network damage and the induced filler breakage (Darabi, Itskov, & Klüppel 2016). Deformation induces damage both in the network rubbery matrix and inside the filler aggregates. This leads to the evolution of the probability density function of the number of segments and the filler size, which, in turn, causes the stress softening and the Mullins effect. These effects result in the hydrodynamic strain amplification being the major topic of this work. The model is capable of describing the deformation induced anisotropy as well as permanent set and includes a few number of physically motivated material constants characterizing the average filler cluster dimension, filler-filler and filler-matrix interaction properties.

1 INTRODUCTION

Elastomers are usually reinforced by stiff fillers e.g. carbon black (CB) and silica to enhance the mechanical and thermal properties. Addition of the fillers alters the nonlinear characteristics of the composite system and induces inelastic mechanical effects. Modeling such a sophisticated nonlinear behavior of elastomers under large cyclic deformations requires comprehensive information about the polymer chains, filler clusters and the interaction between them. Fillers increase the stiffness of the composite system as a reinforcement agent with higher stiffness modulus or provide additional cross-linking sites for polymer chains at the filler-matrix interface (Itskov & Darabi 2016). The latter causes multiple constraints on the longer polymer chains, which makes the matrix stiffer than the unfilled polymer. The Mullins effect (Mullins 1949) is mostly attributed to the debonding of the polymer chains from the filler surface and also to the breakage of inner particle-particle bonds inside the fillers which can influence the aggregate structure. To the best of our knowledge, a simultaneous failure mechanism both in the rubber network and inside filler network has so far not been applied to rubber elasticity and softening models. It appears to be very advantageous within a non-affine micro-sphere model. Several non-affine models have so far been proposed based on the chains isotropically distributed in space. The model of Govindjee and Simo (Govindjee & Simo 1991) was based on the isotropic three chain model of James and Guth (James & Guth 1943) and decomposed the rubber matrix into pure rubber and polymer-filler networks in which damage occurred only in the polymer-filler network. The eight-chain model by Arruda and Boyce (Arruda & Boyce 1993) considered the stretch along the diagonal of a unit cubic cell representing the macromolecular network structure of rubber and applied the non-Gaussian statistics for single chains. Later, this model was adopted for other materials as well (Kuhl, Garikipati, Arruda, & Grosh 2005, Palmer & Boyce 2008). Miehe et al. (Miehe, Göktepe, & Lulei 2004) extended the model to a non-affine micro-sphere model with 21 material directions by considering a specific relationship between the microscopic and macroscopic deformation. Dargazany and Itskov (Dargazany & Itskov 2009) proposed a purely micro-mechanical model describing the anisotropic Mullins effect and permanent set. The Dynamic Flocculation Model (DFM) was based on a physically motivated framework and considered filler-filler interactions by breakage and re-aggregation of filler clusters during subsequent deformation cycles. In this context, the microstructure of filler clusters plays the dominant role in the mechanical reinforcement and the cyclic loading characterisation. In the present work, a failure mechanism based on a combination of the DFM framework and the continuum damage model (Govindjee & Simo 1991) is proposed.

2 NETWORK DECOMPOSITION

We decompose the free energy density of the rubber composite as follows:

$$\Psi = (1-\varphi_{eff})[\Psi_{cc} + \Psi_{pp}] + \varphi_{eff}\Psi_A, \quad (1)$$

where ψ_{cc} and ψ_{pp} denote the free energy density of the pure rubber crosslink-crosslink network and a particle-particle rubber network, respectively (Govindjee & Simo 1991). ψ_A corresponds to the strain energy stored in the fractions of substantially deformed fragile filler clusters. φ_{eff} is the total effective filler volume fraction which takes into account the internal irregular structure of filler particles. For spherical filler particles, φ_{eff} is equal to the filler volume fraction φ, while for structured particles $\varphi_{eff} > \varphi$. Note that the decomposition in (1) implies that there is no interaction between CC, PP networks and aggregates.

The free energy of the Freely Jointed Chain (FJC) can be obtained on the basis of random walk theory (Flory, 1989; Treloar, 1975). Accordingly, the non-Gaussian strain energy of the FJC can be written by

$$\psi(n,\lambda) = nKT\left(\frac{\lambda}{\sqrt{n}}\beta + \ln\frac{\beta}{\sinh\beta}\right), \quad (2)$$

where n is the number of chain segments and λ is the stretch in the end-to-end direction. $\beta = \mathcal{L}^{-1}(\frac{\lambda}{\sqrt{n}})$, where $\mathcal{L}^{-1}$ denotes the inverse Langevin function. T stands for the absolute temperature and K is Boltzmann's constant. A good approximation of the inverse Langevin function can be achieved, for example, by the following expression (Darabi & Itskov 2015)

$$\mathcal{L}^{-1}(x) \approx x\frac{x^2-3x+3}{1-x}. \quad (3)$$

Assuming an isotropic spatial distribution of polymer chains and aggregates, the macroscopic energy of a three-dimensional network can be calculated by the integral over the unit sphere as

$$\Psi_k = \frac{1}{4\pi}\int_S \psi_k du, \quad k = cc, pp, A. \quad (4)$$

The integration is then carried out numerically as

$$\Psi_k \approx \sum_{i=1}^{m} w_i \overset{d_i}{\psi_k}, \quad k = cc, pp, A, \quad (5)$$

where w_i are weight factors corresponding to the collocation directions $d_i (i = 1,2,...,m)$. The numerical scheme with 45 integration points by Heo and Xu (Heo & Xu 2000) is used.

The motion of polymer chains inside the CC network is considered to be affine. This is due to the fact that filler clusters are not present in the network. The entropic energy of the CC network consisting of N_c chains with n_c segments each is defined in direction d_i as

$$\overset{d_i}{\psi}_{cc} = N_c\psi\left(n_c, \overset{d_i}{\lambda}\right), \quad (6)$$

where $\overset{d_i}{\lambda}$ is the micro-stretch in this direction.

The chains in the PP network are assumed to have different number of segments. The probability of finding a chain in the PP network consisting of n segments each of the length l and end-to-end distance of R, whose intermediate segments have no attachment to the filler surface, is expressed as

$$\begin{aligned} P(n,R) &= \kappa\sqrt{\frac{Al^2}{\pi R^2}}\exp(B), \quad A = \frac{3R^2}{2nl^2}, \\ B &= -A - \kappa\sqrt{\frac{6}{\pi}}\Big[\sqrt{n}\exp(-A) - \exp(-An) \\ &\quad + \sqrt{\frac{6\pi R^2}{l^2}}\left(erf\left(\sqrt{A}\right) - erf\left(\sqrt{An}\right)\right)\Big], \end{aligned} \quad (7)$$

where κ denotes the average area of active adsorption sites available for one bond.

The entropic energy of the PP network consisting of N_p chains in direction d_i can thus be expressed by

$$\overset{d_i}{\psi}_{pp} = N_p \int_{D\left(\overset{d_i}{\lambda}\right)} P(n,R)\psi\left(n, \overset{d_i}{\lambda}\right)dn, \quad (8)$$

where $D(\overset{d_i}{\lambda})$ denotes the set of available chain lengths n in the direction d_i. The deformation in the PP network is non-affine and related to the macro deformation through the strain amplification factor which will be discussed in details in the next section.

In the following, we assume linear-elastic behavior of clusters. The contribution of stiff clusters to the total stored energy is thus neglected. Accordingly, the strain energy density of the soft filler clusters is written by

$$\overset{d_i}{\psi}_A = \frac{1}{2}\int_{D_f\left(\overset{d_i}{\lambda}_A\right)} \overset{d_i}{G}_A\left(\overset{d_i}{\lambda}_A - 1\right)^2 \Phi\left(\overset{d_i}{x}\right)d\overset{d_i}{x}, \quad (9)$$

where $\overset{d_i}{x} = \frac{\overset{d_i}{\xi}}{d}$ is the normalized size of the cluster and d is the size of the particles. $\overset{d_i}{\lambda}_A$ denotes the stretch and $\overset{d_i}{G_A}$ is the elastic modulus of the soft filler clusters in the direction specified by a unit vector d_i. The integration is performed over the interval of available re-aggregated clusters $D_f(\overset{d_i}{\lambda_A})$ in the direction d_i. The critical normalized size of currently breaking soft clusters can be derived from the matrix stress acting on the clusters as

$$\overset{d_i}{x_c} = \frac{Q_d\, \varepsilon_{b,d}}{d^3\, P(\overset{d_i}{\lambda})} = \frac{S_d}{P(\overset{d_i}{\lambda})}, \qquad P(\overset{d_i}{\lambda}) = \frac{\partial \overset{d_i}{\psi}}{\partial \overset{d_i}{\lambda}}, \tag{10}$$

where $S_d = \frac{Q_d\, \varepsilon_{b,d}}{d^3}$ is the tensile strength of damaged bonds expressed in terms of their failure strain $\varepsilon_{b,d}$ and the elastic modulus Q_d / d^3. The strain of the filler clusters is computed from the rubber stress $P(\overset{d_i}{\lambda})$.

3 STRAIN AMPLIFICATION

The fractal filler clusters exhibit a particular size distribution which can be obtained using different types of microscopies. The normalized cluster size distribution $\Phi(\overset{d_i}{x})$ in a direction d_i in (9) is given by

$$\Phi(\overset{d_i}{x}) = \frac{4\overset{d_i}{x}}{\left\langle \overset{d_i}{x} \right\rangle^2} \exp\left(-\frac{2\overset{d_i}{x}}{\left\langle \overset{d_i}{x} \right\rangle} \right), \tag{11}$$

where $\langle \cdot \rangle$ represents the mean function.

Overstraining of the rubber matrix due to more rigid clusters can be quantified by a strain amplification factor $\overset{d_i}{X}$. This relates the external stretch $\overset{d_i}{\tilde{\lambda}}$ of the sample to the internal stretch $\overset{d_i}{\lambda}$ of the rubber matrix as

$$\overset{d_i}{\lambda} = 1 + \overset{d_i}{X}\left(\overset{d_i}{\tilde{\lambda}} - 1 \right). \tag{12}$$

The strain amplification factor $\overset{d_i}{X}$ is assumed to depend only on the maximal stretch $\overset{d_i}{\lambda}_{max}$ previously reached in the loading history of this direction d_i. Klüppel (Klüppel 2003) extracted a formula for the strain amplification factor based on the overlapping fractal cluster theory of Huber and Vilgis (Huber & Vilgis 1999) as follows

$$\overset{d_i}{X}_{max} = 1 + c\,\varphi_{eff}^{\frac{2}{3-d_f}} \left\{ \int_0^{\overset{d_i}{x}_{min}} \left(\overset{d_i}{x} \right)^{d_w - d_f} \Phi\left(\overset{d_i}{x} \right) d\overset{d_i}{x} + \int_{\overset{d_i}{x}_{min}}^{\infty} \Phi\left(\overset{d_i}{x} \right) d\overset{d_i}{x} \right\}, \tag{13}$$

where c is the Einstein coefficient for spherical inclusions, d_f is the fractal dimension and d_w is the anomalous diffusion exponent of fractal clusters. $\overset{d_i}{x}_{min}$ is the normalized maximum size of clusters surviving under external strains. An analytical solution of (13) can be expressed as

$$\begin{aligned} \overset{d_i}{X} &= 1 + c\,\varphi_{eff} \\ &\left\{ \left(\frac{\overset{d_i}{x}_{min}}{\alpha} \right)^{\frac{\omega}{2}} e^{\frac{-\alpha \overset{d_i}{x}_{min}}{2}} M\left(\frac{\omega}{2}, \frac{\omega+1}{2}, \omega \overset{d_i}{x}_{min} \right) \right. \\ &\left. - e^{-\alpha \overset{d_i}{x}_{min}} \left(\alpha \overset{d_i}{x}_{min}^{\;\omega+1} - \alpha \overset{d_i}{x}_{min} - 1 \right) \right\}, \\ &\omega = d_w - d_f,\; \alpha = \frac{2}{\left\langle \overset{d_i}{x} \right\rangle}, \end{aligned} \tag{14}$$

where $M(a,b,z)$ is the Whittaker function, defined as

$$M(a,b,z) = \frac{\Gamma(b)}{\Gamma(a)\Gamma(b-a)} \int_0^1 e^{zu} u^{a-1} (1-u)^{b-a-1}\, du \tag{15}$$

and $\Gamma(t)$ is the Gamma function written by

$$\Gamma(t) = \int_0^{\infty} x^{t-1} e^{-x} dx. \tag{16}$$

4 A SINGLE CLUSTER AS A FLEXIBLE CHAIN

A single cluster subject to longitudinal, bending and twisting deformations can be modeled by a series of two molecular springs (see Fig. 1):

- a soft spring with the stiffness constant $K_s \sim \overline{G}$ representing the bending-twisting mode
- a stiff spring with the stiffness constant $K_b \sim Q$ representing the tension mode

The stiff spring governs the fracture behavior of the system because it takes the longitudinal deformation and hence the spatial separation of filler-filler bonds into account. In larger clusters, the spring constant of stiff fillers is much higher than

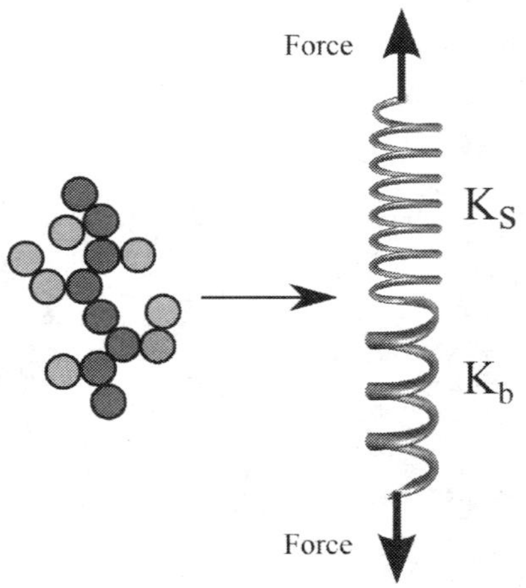

Figure 1. A model of a filler cluster as a series of two molecular springs.

the one of soft fillers $(K_b \gg K_s)$. Therefore, the deformation of the stiff spring can be neglected, and one can conclude that only the soft spring impacts the elasticity of the whole system. The elastic modulus of large clusters is calculated as

$$\overset{d_i}{G}_A \equiv \frac{\overset{d_i}{K}_S}{\overset{d_i}{\xi}} = \frac{\eta\bar{G}}{d^3}\left(\frac{d}{\overset{d_i}{\xi}}\right)^{3+d_{f,B}}, \tag{17}$$

where $\bar{G}$ is an averaged elastic constant, η is a geometrical factor of order one and $d_{f,B}$ is the backbone fractal dimension of the filler clusters. By assuming cluster-cluster aggregation the value of the latter one can be set to $d_{f,B} = 1.3$ (Meakin 1987).

The failure strain of a filler cluster can be evaluated from the stress equilibrium between the soft and stiff springs as

$$\overset{d_i}{\varepsilon}_F = \left(1 + \frac{K_b}{\overset{d_i}{K}_S}\right)\varepsilon_b. \tag{18}$$

In the case of large clusters $K_b \gg K_S$ and in view of (17)

$$\overset{d_i}{\varepsilon}_F \approx \frac{Q\varepsilon_b}{\bar{G}\eta}\left(\frac{\overset{d_i}{\xi}}{d}\right)^{2+d_{f,B}}, \tag{19}$$

where Q is the elastic tension constant of the Kantor-Webman model and $K_b = Q/d^2$. ε_b is the breakage strain of filler-filler bonds. From (19) one can realize that the fracture strain of the filler clusters has a direct relationship to the size of the clusters according to a power law.

From the stress equilibrium between the rubber matrix and clusters which holds in every direction d_i, the strain in a filler cluster is derived as

$$\overset{d_i}{\varepsilon}_A = \frac{P\left(\overset{d_i}{\lambda}\right)}{\overset{d_i}{G}_A} = \frac{d^3}{\eta\bar{G}}\left(\frac{\overset{d_i}{\xi}}{d}\right)^{3+d_{f,B}}. \tag{20}$$

Thus, with increasing size $\overset{d_i}{\xi}$, the clusters are more prone to break.

5 MULTIPLE FRACTIONS OF CLUSTERS

With increasing strain, filler clusters of the maximal size begin to break. This process continues as long as all the clusters larger than some minimal value are broken. The minimal size of clusters surviving at an external strain is written as

$$\overset{d_i}{x}_{\min} = \frac{Q_v\varepsilon_{v,b}}{d^2 P\left(\overset{d_i}{\lambda}_{\max}\right)} = \frac{S_v}{P\left(\overset{d_i}{\lambda}_{\max}\right)}, \tag{21}$$

where Q_v is the elastic tension constant of virgin filler-filler bonds. The breakage strain of the virgin filler-filler bond is denoted by $\varepsilon_{v,b}$. S_v is the tensile strength of virgin filler-filler bonds. It governs the strain amplification factor for the first loading cycle and is considered as a material constant.

In the previous models, the normalized cluster size distribution $\Phi(\overset{d_i}{x})$ in the spatial direction d_i was independent of the strain. Breakage of the filler clusters into two or more parts does not necessarily result in the complete loss of their role in the total energy of the filler. Rather, it leads to the evolution of $\Phi(\overset{d_i}{x})$ due to increase in the number of smaller clusters as shown in Fig. 2. However, the total number of the fillers remains constant. Since the integration will be carried out only over the set of available clusters during loading

$$\int_{D(\overset{d_i}{\lambda}_{\max})} \Phi\left(\overset{d_i}{x}\right) d\overset{d_i}{x} < 1. \tag{22}$$

The earlier assumption of the constant number of fillers leads to the condition

$$\int\limits_{D(\overset{d_i}{\lambda}_{\max}=1)} \Phi\left(\overset{d_i}{x}\right)\overset{d_i}{x}\, d\overset{d_i}{x} = \int\limits_{D(\overset{d_i}{\lambda}_{\max})} f\left(\overset{d_i}{\lambda}_{\max}\right)\Phi\left(\overset{d_i}{x}\right)\overset{d_i}{x}\, d\overset{d_i}{x}. \tag{23}$$

Thus, one can express the normalization function as

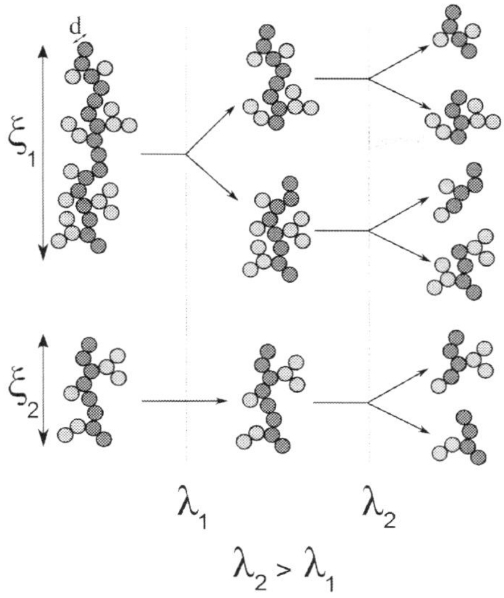

Figure 2. Illustration of the filler fraction concept at two stretch levels.

$$f(\overset{d_i}{\lambda}_{\max}) = \frac{\displaystyle\int_{D(\overset{d_i}{\lambda}_{\max}=1)} \overset{d_i}{\Phi}\,\overset{d_i}{x}\,d\overset{d_i}{x}}{\displaystyle\int_{D(\overset{d_i}{\lambda}_{\max})} \overset{d_i}{\Phi}\left(\overset{d_i}{x}\right) d\overset{d_i}{x}}. \tag{24}$$

The probability amplification factor has the analytical form of

$$f(\overset{d_i}{\lambda}_{\max}) = \frac{2}{2-\left(\alpha^2\,\overset{d_i}{x}{}^{2}_{\max} + 2\alpha\,\overset{d_i}{x}_{\min} + 2\right)e^{-\alpha\,\overset{d_i}{x}_{\min}}}. \tag{25}$$

The amplification of the probability distribution caused by the function $f(\overset{d_i}{\lambda}_{\max})$ is illustrated in Fig. 3.

Under unloading, re-aggregation of the filler particles re-creates clusters of the original shape. The previously damaged filler-filler bonds, which are newly formed again, differ from the virgin ones, annealed by the heat treatment during vulcanization. Stress softening of the rubber composite is attributed to this mechanism of the subsequent filler fraction. This implies a boundary size x_{min} which separates the unbroken virgin hard clusters from the damaged soft clusters (see Fig. 4). By increasing the strain of the already pre-strained material, the re-aggregated clusters with soft filler-filler bonds are broken sooner than before and dissipate a certain amount of energy previously

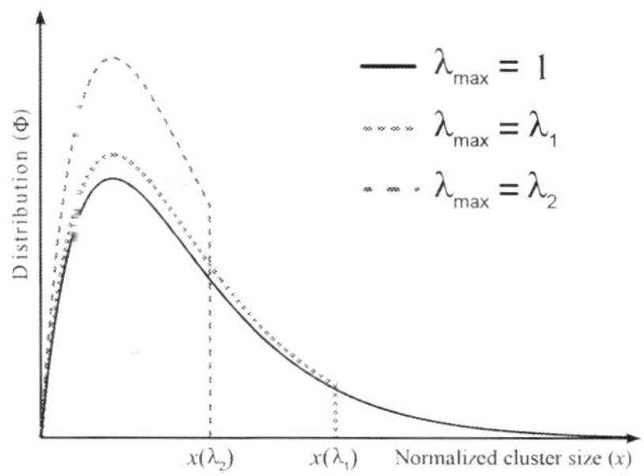

Figure 3. Effect of the normalization factor on the cluster size probability distribution by increasing stretch $\lambda_2 > \lambda_1 > 1$.

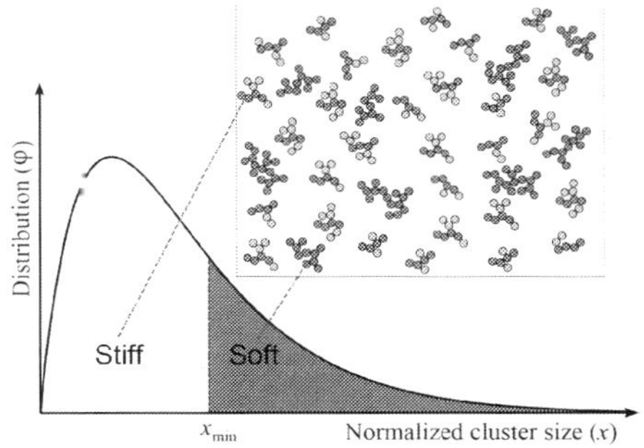

Figure 4. Decomposition of filler clusters into soft and stiff ones

stored in the clusters. Shift of the minimum boundary size x_{min} with increasing pre-strain leads to hysteresis. Filler-induced hysteresis originates from this cyclic breakdown and re-aggregation of the soft fillers. The decomposition into hard and soft filler cluster units is illustrated in Fig. 4.

6 RESULTS AND DISCUSSION

6.1 *Model prediction*

Taking into account the incompressibility condition $\det\mathbf{F} = 1$, the constitutive equation for the first-Piola Kirchhoff stress tensor $\mathbf{P}$ can be written by

$$\begin{aligned}\mathbf{P} &= \frac{\partial\Psi}{\partial\mathbf{F}} - \mathrm{p}\mathbf{F}^{-T} \\ &= (1-\varphi_{\text{eff}})\left[\frac{\partial\Psi_{\text{cc}}}{\partial\mathbf{F}} + \frac{\partial\Psi_{\text{pp}}}{\partial\mathbf{F}}\right] + \varphi_{\text{eff}}\frac{\partial\Psi_{\text{A}}}{\partial\mathbf{F}} - \mathrm{p}\mathbf{F}^{-T},\end{aligned} \tag{26}$$

Table 1. Material parameters of the multiple filler fraction model for CR with 50 phr CB.

Sample	n_c	$\langle x_0 \rangle$	$N_c KT(MPa)$	$N_p KT(MPa)$	$S_v(MPa)$	$S_d(MPa)$	ϕ_{eff}	κ	$R(nm)$
CR/CB	140	15.69	0.5	0.685	4.15	3.9	0.35	1.15	7.8

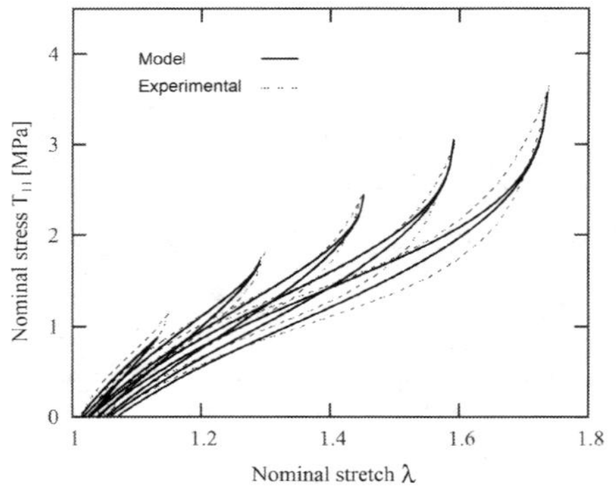

Figure 5. Uniaxial cyclic tension tests: model predictions versus experimental data for CR with 50 phr CB (Dargazany & Itskov 2009).

where **F** denotes the deformation gradient and p stands for an arbitrary scalar parameter which can be defined according to a particular boundary value problem.

In the following we compare model predictions with experimental data from quasi-static uniaxial tension tests of specimens made of polychloroprene rubber (CR) reinforced with 50 phr carbon black (Dargazany & Itskov 2009). The values of the material constants obtained by fitting to experimental data are listed in Table 1. The cross-link modulus $N_p KT$ is larger than that of the unfilled rubber $N_c KT$ because fillers with an active filler surface could act as crosslinks. The mean cluster size $\langle x_0 \rangle$ depends on the clustering adhesion coefficient and can be extracted from the statistical micrography analysis. For typical carbon black and silica clusters it is in the range between $5 \le \langle x_0 \rangle \le 18$. Taking into account the bound rubber around the filler clusters, the effective filler volume fraction lies in the range of $0.32 < \varphi_{eff} < 0.35$, which is somewhat higher than the filler volume fraction expected. The tensile strength of virgin filler-filler bonds $S_v = 4.15\ MPa$ is somehow higher than the tensile strength of damaged filler-iller bonds $S_d = 3.90\ MPa$. Parameter κ is related to the filler characteristics and concentration and directly affects the stress softening and hysteresis. R is the end-to-end distance of the polymer chains and varies with the type of polymer matrix and also with the filler volume fraction. The model predictions are verified against the stress-stretch diagrams (adopted from the above mentioned paper) and are plotted in Fig. 5. In this plot, one can observe relatively good agreement. This numerical example illustrates the ability of the proposed constitutive model to describe the Mullins effect, permanent set and hysteresis which are very pronounced in filled rubbers.

7 CONCLUSION

In this paper, a micro-mechanical model of filled elastomers has been proposed. The model is based on both the DFM framework, which captures stress-softening and filler-induced hysteresis of reinforced elastomers, and on the continuum damage concept which deals with the damage mechanism inside the rubber network. The model relies on the hydrodynamic reinforcement of rubber due to strain amplification by stiff filler clusters. Damage in different directions is governed by the evolution concept for both the fillers and the rubber network. It considers the changes in the size distribution of the clusters which then affects the strain amplification factor as well as the change in the size distribution of the chains. The latter one causes the permanent set and Mullins effect. The strain amplification factor is also derived analytically. The DFM framework is enriched by the multiple filler fraction theory, which grants a new insight into the subsequent damage mechanism. This model has a small number of physically motivated material parameters and demonstrates relatively good agreement with experimental data as well.

REFERENCES

Arruda, E. M. & M. C. Boyce (1993). A three-dimensional constitutive model for the large stretch behavior of rubber elastic materials. *Journal of the Mechanics and Physics of Solids 41*(2), 389–412.

Darabi, E. & M. Itskov (2015). A simple and accurate approximation of the inverse langevin function. *Rheologica Acta 54*(5), 455–459.

Darabi, E., M. Itskov, & M. Klüppel (2016). Modelling of hydrodynamic strain amplification in filled elastomers. *PAMM 16*(1), 315–316.

Dargazany, R. & M. Itskov (2009). A network evolution model for the anisotropic Mullins effect in carbon black filled rubbers. *International Journal of Solids and Structures 46*, 2967.

Govindjee, S. & J. Simo (1991). A micro-mechanically based continuum damage model for carbon black-

filled rubbers incorporating mullins' effect. *Journal of the Mechanics and Physics of Solids 39*(1), 87–112.
Heo, S. & Y. Xu (2000). Constructing fully symmetric cubature formulae for the sphere. *Mathematics of Computation 70*, 269.
Huber, G. & T.A. Vilgis (1999). Universal properties of filled rubbers: Mechanisms for reinforcement on different length scales. *Kautschuk und Gummi, Kunststoffe 52*(2), 102–107.
Itskov, M. & E. Darabi (2016). Constitutive modeling of carbon nanotube rubber composites on the basis of chain length statistics. *Composites Part B: Engineering 90*, 69–75.
James, H. & E. Guth (1943). Theory of the elastic properties of rubbers. *Journal of Chemical Physics 11*, 455.
Klüppel, M. (2003). The role of disorder in filler reinforcement of elastomers on various length scales. *Advances in Polymer Science 164*, 1–86.
Kuhl, E., K. Garikipati, E.M. Arruda, & K. Grosh (2005). Remodeling of biological tissue: mechanically induced reorientation of a transversely isotropic chain network. *Journal of the Mechanics and Physics of Solids 53*(7), 1552–1573.
Meakin, P. (1987). Fractal aggregates. *Advances in Colloid and Interface science 28*, 249–331.
Miehe, C., S. Göktepe, & F. Lulei (2004). A micro-macro approach to rubber-like materials part i: the non-affine microsphere model of rubber elasticity. *Journal of the Mechanics and Physics of Solids 52*(11), 2617–2660.
Mullins, L. (1949). Permanent set in vulcanized rubber. *Rubber Chemistry and Technology 22*(4), 1036–1044.
Palmer, J.S. & M.C. Boyce (2008). Constitutive modeling of the stress–strain behavior of f-actin filament networks. *Acta Biomaterialia 4*(3), 597–612.

Constitutive Models for Rubber X – Lion & Johlitz (Eds)
© 2017 Taylor & Francis Group, London, ISBN 978-1-138-03001-5

Effect of microscopic structure on mechanical characteristics of foam rubber

A. Matsuda
Faculty of Engineering, Information and Systems, University of Tsukuba, Ibaraki, Japan

S. Oketani, Y. Kimura & A. Nomoto
Graduate School of Systems and Information Engineering, University of Tsukuba, Ibaraki, Japan

ABSTRACT: In this study, the effect of microscopic structures of foam rubber were investigated numerically. A 2-dimensional finite element analysis code and a 2-dimentional homogenization finite element analysis code of the hyperelastic material were developed, respectively. The incompressible hyperelasticity in the total Lagrangian finite strain framework was applied to both of general FEM and homogenization FEM codes.

The reason of which foam rubber were used widely, is that it has some advantage properties. But effect of microscopic structure on macroscopic stress-strain relationships is not evaluated enough in the point of engineering material design. To clear this engineering issue, compression loading tests of rubber specimens which have periodic geometrical holes were conducted to evaluate the effect of periodical structure on mechanical characteristics. Relative density of natural rubber specimens were 42%, 56%, 72%, 80% and 86%, respectively. The biaxial tensile tests of rubber sheets which were made from same natural rubber used in compression loading tests were conducted to identify material coefficients of the elastic potential function of the hyperelastic matrix. The stress-strain relationships given by compression loading tests of the natural rubber specimens were simulated by the finite element analysis programs with the general boundary condition and the periodic boundary condition to investigate the applicability of homogenized method to foam rubber.

In addition, effects of number of unit cells on the macroscopic stress-strain relationships of foam rubber were investigated numerically.

1 INTRODUCTION

Foam rubbers such as polyurethane foam are used widely in transport equipment, sports equipment, living ware and so on. The advantage reasons of which the foam rubber are used, are large deformable performance, soft stiffness, durability for cyclic deformations and manufacturing formability. The form rubber include numerous microscopic cavities inside. So, the mechanical characteristics of the foam rubber are strongly depend on mechanical properties of rubber matrix and its microstructure.

The relative density is important material parameter which defined mechanical calculated by the volume and density of foam rubber. The relative density which is calculated by the ratio of rubber matrix in foam rubber are used as parameter which present material characteristics. Material design using numerical simulation considering the microstructure would be required to improve quality, size and productivity of foam rubber products.

In this study, finite element analysis codes with periodical and normal boundary conditions were performed. The microstructures of the foam rubber were assumed to have periodic structure with numerical spherical cavities. The rubber matrix was assumed to be represented by incompressible hyperelastic material.

Compression tests of natural rubber specimens with periodical holes were carried out to investigate effects of periodical structure on the mechanical characteristics. The material parameters of the natural rubber were given from the biaxial tensile test results of rubber sheet specimens of which material was same as compression test specimens.

By comparing the compression test and the numerical simulation, applicability of homogenization analysis to foam rubber were investigated.

In addition, effects of number of unit cells on the macroscopic stress-strain relationships of foam rubber were investigated numerically.

2 HOMOGENIZATION METHOD

In this study, a foam rubber was defined as a material which has a periodic microscopic structure. The formulation for the large deformation problem of porous polymers was described using $\boldsymbol{X}$ and $\boldsymbol{Y}$-coordinates. The macroscopic behavior is described with the $\boldsymbol{X}$-coordinate, and the microscopic behavior is described with the $\boldsymbol{Y}$-coordinate. The two coordinates are related using a scale ratio ε as follows:

$$\boldsymbol{Y} = \boldsymbol{X}/\varepsilon \tag{1}$$

When the scale of the microscopic structure is much smaller than the scale of the whole structure, the scale ratio ε is a very small. The macroscopic characteristics, such as stiffness, stress and strain, are calculated from the volume average of microscopic characteristics using the homogenization theory. In Figure 1, the unit cell for homogenization analysis are shown.

Here, the total deformation of microscopic structure is divided into macroscopic deformation $\boldsymbol{Y}$ and microscopic periodical deformation $\boldsymbol{w}$ as follows:

$$\boldsymbol{y} = \tilde{\boldsymbol{F}}\boldsymbol{Y} + \boldsymbol{w} \tag{2}$$

$\tilde{\boldsymbol{F}}$ is the macroscopic deformation gradient tensor. The gradient tensor $\tilde{\boldsymbol{F}}$ is applied to the numerical simulation as the deformation condition. The microscopic deformation gradient tensor is calculated from Equation 2 as follows:

$$\boldsymbol{F} = \nabla \boldsymbol{y} = \tilde{\boldsymbol{F}} + \frac{\partial \boldsymbol{w}}{\partial \boldsymbol{Y}} \tag{3}$$

The rate of displacement is also described by

$$\boldsymbol{v} = \dot{\boldsymbol{y}} = \frac{\partial \dot{\boldsymbol{x}}}{\partial \boldsymbol{X}}\boldsymbol{Y} + \dot{\boldsymbol{w}} \tag{4}$$

In this study, rubber matrix was assumed to have incompressible hyperelasticity. The second Piola-Kirchhoff stress $\boldsymbol{S}$ is given by the partial differentiation of the strain energy function W with respect to the right Cauchy-Green deformation tensor $\boldsymbol{C}$ as follows:

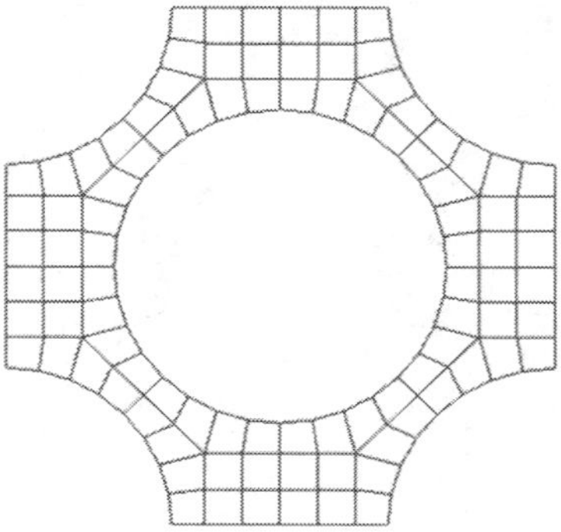

Figure 1. Unit cell for homogenization analysis.

$$\boldsymbol{S} = 2\frac{\partial W(\bar{\boldsymbol{C}})}{\partial \boldsymbol{C}} \tag{5}$$

Here, $\boldsymbol{C}$ is calculated as $\boldsymbol{C} = \boldsymbol{F}^T\boldsymbol{F}$ and $\boldsymbol{F}$ is the deformation gradient tensor. The following strain energy function of hyperelasticity was applied to the rubber matrix in this study,

$$\begin{aligned} W(\bar{\boldsymbol{C}}, p) = {} & \boldsymbol{C}_{11}(\bar{I}_1 - 3) + \frac{\boldsymbol{C}_{12}}{2}(\bar{I}_1 - 3)^2 \\ & + \frac{\boldsymbol{C}_{13}}{3}(\bar{I}_1 - 3)^3 + \frac{\boldsymbol{C}_{14}}{\boldsymbol{C}_{15}}\exp\{\boldsymbol{C}_{15}(\bar{I}_1 - 3)\} \\ & + \boldsymbol{C}_{21}(\bar{I}_2 - 3) + \frac{\boldsymbol{C}_{22}}{2}(\bar{I}_2 - 3)^2 \\ & + \frac{\boldsymbol{C}_{23}}{3}(\bar{I}_2 - 3)^3 + \frac{\boldsymbol{C}_{24}}{\boldsymbol{C}_{25}}\exp\{\boldsymbol{C}_{25}(\bar{I}_2 - 3)\} \\ & + p(J - 1) \end{aligned} \tag{6}$$

where, C_{11}, C_{12}, C_{13}, C_{14}, C_{15}, C_{21}, C_{22}, C_{23}, C_{24}, C_{25} are material parameters, and $\bar{I}_1, \bar{I}_2$ are the first and second principal invariants of the deformation tensor $\bar{\boldsymbol{C}}(= \boldsymbol{J}^{-2/3}\boldsymbol{F}^T\boldsymbol{F})$. p is the hydrostatic pressure, and J is the determinant of deformation tensor $\boldsymbol{F}$.

The material parameters of Equation 7 are identified by the biaxial tensile test of rubber sheet using biaxial tensile machine. Rubber sheet is made of natural rubber same as the rubber specimen. The sheet shaped is 50 mm in both length and width, and 1 mm in thick. Boundary condition was applied uniaxial tensile and uniaxial fixed and strain rate was set to 0.1% s^{-1} to reduce the increased viscosity associated with deformation rate.

The deformation gradient tensor $\boldsymbol{F}$ for biaxial deformation is shown by using stretch λ in tensile direction as follows:

$$\boldsymbol{F} = \begin{bmatrix} \lambda & 0 & 0 \\ 0 & 1 & 0 \\ 0 & 0 & 1/\lambda \end{bmatrix} \tag{7}$$

Here, the partial differentiation of the strain energy function W with respect to the first and the second invariants $\bar{I}_1$ and $\bar{I}_2$ using the nominal stress P_{11}, P_{22} given by biaxial tensile tests as follows:

$$\frac{\partial W(\bar{\boldsymbol{C}})}{\partial \bar{I}_1} = \frac{1}{2(\lambda^2 - 1)}\left(\frac{\lambda^3 P_{11}}{\lambda^2 - \lambda^{-2}} - \frac{P_{22}}{1 - \lambda^{-2}}\right) \tag{8}$$

$$\frac{\partial W(\bar{\boldsymbol{C}})}{\partial \bar{I}_2} = -\frac{1}{2(\lambda^2 - 1)}\left(\frac{\lambda P_{11}}{\lambda^2 - \lambda^{-2}} - \frac{P_{22}}{1 - \lambda^{-2}}\right) \tag{9}$$

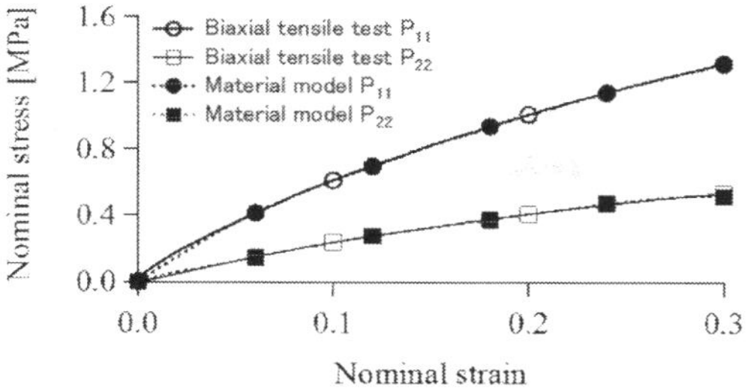

Figure 2. The comparison of biaxial tensile test and material model.

The material parameters of Equation 8 were identified by Equations 10 and 11. Material parameters of the natural rubber used in this study were given as $C_{11} = 1.79$, $C_{12} = -6.61$, $C_{13} = 13.5$, $C_{14} = 4.41$, $C_{15} = -83.4$, $C_{21} = -0.92$, $C_{22} = 6.07$, $C_{23} = -12.7$, $C_{24} = -4.32$ and $C_{25} = -85.9$, respectively.

The comparison of biaxial tensile test and material model was shown in Figure 2. Theoretical calculation of hyperelasticity showed enough agreement with the biaxial tensile test result.

3 COMPRESSION TEST OF RUBBER SPECIMENS

Compression tests of the rubber specimens were conducted to verify the applicability of simulation codes. All rubber specimens are made of natural rubber, 50 mm cubic dimensions on each side and have periodic holes as shown in Figure 3. The diameter of all hole was 6 mm. Relative density of the rubber specimens ρ is calculated by following relation,

$$\rho = \frac{\rho^*}{\rho_S} \tag{10}$$

Here, ρ_S and ρ^* are the density of rubber matrix and form rubber, respectively. The relative density of rubber specimens are 0.42, 0.56, 0.72, 0.80 and 0.86, respectively. Compression tests of rubber specimen were conducted using uniaxial loading machine (AG-20kNXplus, Shimadzu Corporation). Rubber specimens were compressed up to nominal strain 0.3. Strain rate of all loading tests were 0.1% s^{-1}.

In addition, cross markers were placed around the all holes of rubber specimen, and center holes of the rubber specimens were defined as the unit cell. The state of compression test was recorded by the digital high-speed camera, the strain of the unit cells was calculated by video analysis software TEMA 3D by Image systems motion analysis.

The stress-strain relationships of unit cell given by compression test was shown in Figure 4. They showed increase in stiffness as relative density increases. Deformation of rubber specimen of which the relative density ρ is 0.56 and 0.42, are shown in Figure 5. Buckling of the rubber matrix

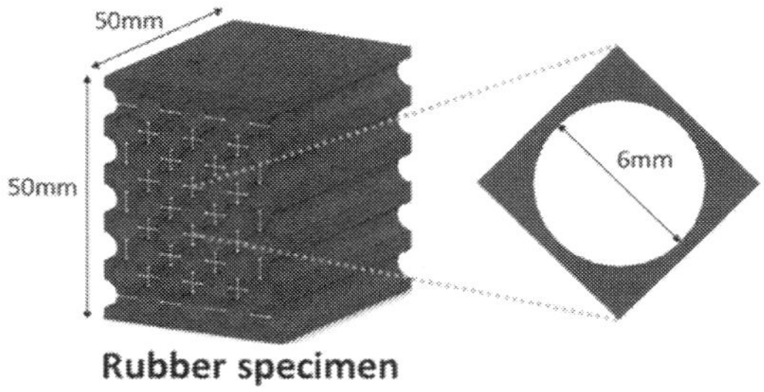

Figure 3. Rubber specimen and diameter of holes.

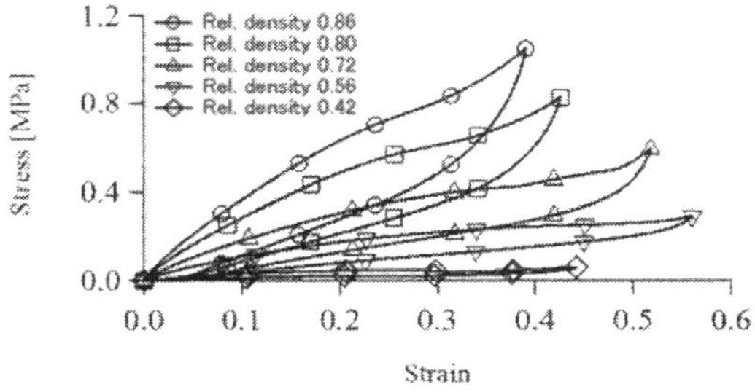

Figure 4. Compression test result of rubber specimens.

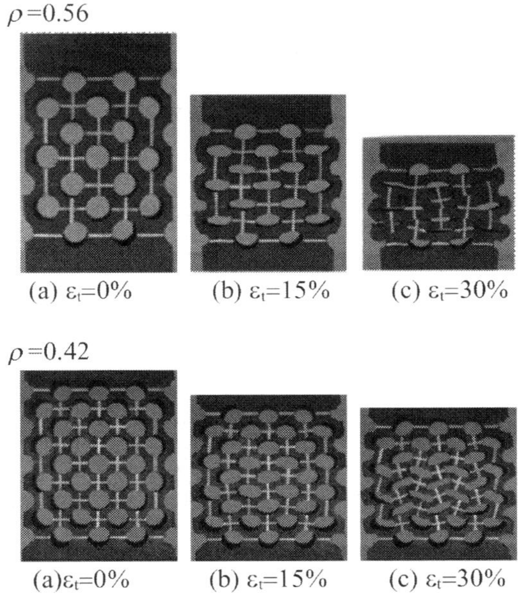

Figure 5. Deformation of rubber specimens.

was observed in the loading test of rubber specimen of which the relative density ρ is 0.42.

Comparison of numerical simulation and compression test was conducted to verify the applicability of analysis code. Numerical simulations with periodic and general boundary conditions were conducted. The macroscopic deformation and periodical boundary condition were applied to the unit cell in homogenization analysis. Mean stress distributions of homogenization analysis are shown in Figure 6. In Figure 7, mean stress distribution of general FEM analysis are shown. In general FEM analysis, n was defined as the number of unit cells arranged in the vertical direction. Total number of unit cells in general FEM analysis is the square of n.

The comparison of homogenization analysis and compression test was shown in Figure 8. The stress-strain relationships of numerical simulation and loading tests of which the relative density ρ is 0.42 are in better agreement than the results of which relative density ρ is 0.56. For this reason, unit cells included in rubber specimen of which relative density ρ is 0.56 was smaller and boundary conditions on both sides influence analysis results.

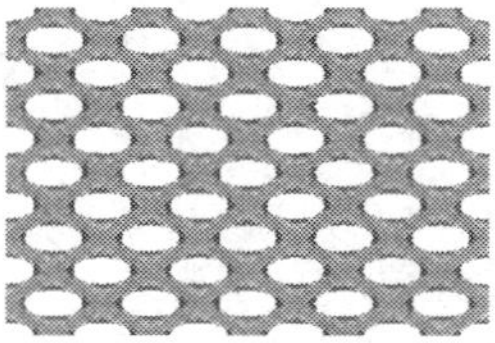

(a) ρ=0.56

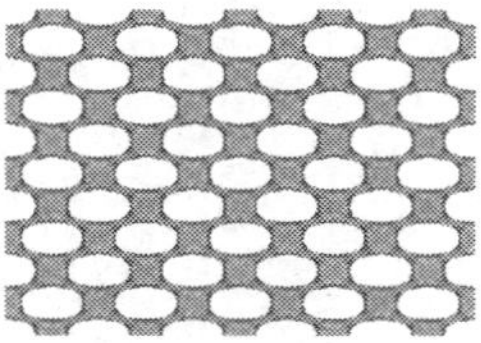

(b) ρ=0.42

Figure 6. Mean stress distribution of homogenization FEM analysis (n = 5; 25 unit cells).

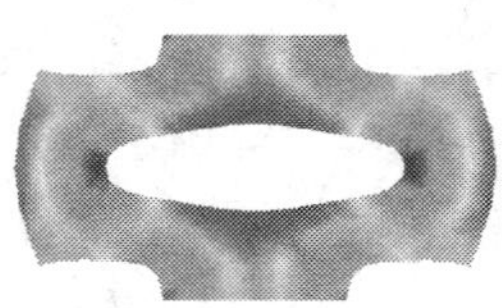

(a) n=1(ρ= 0.56)

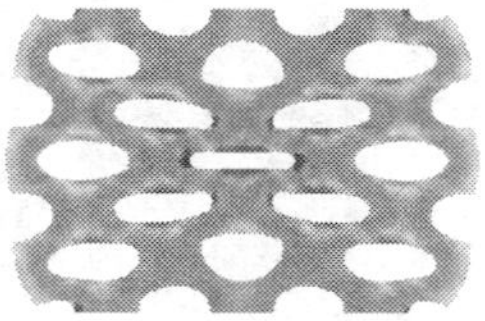

(b) n=3(ρ= 0.56)

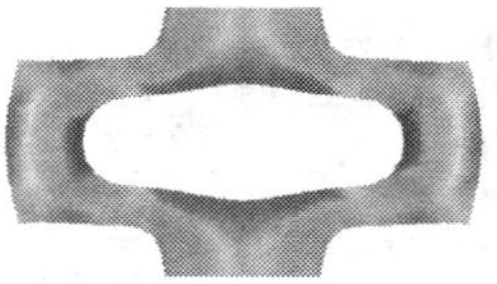

(a) n=1(ρ= 0.42)

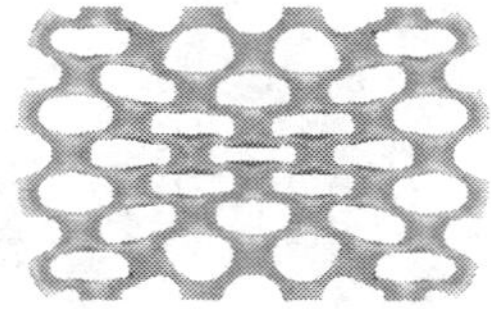

(b) n=4(ρ= 0.42)

Figure 7. Mean stress distribution of general FEM analysis.

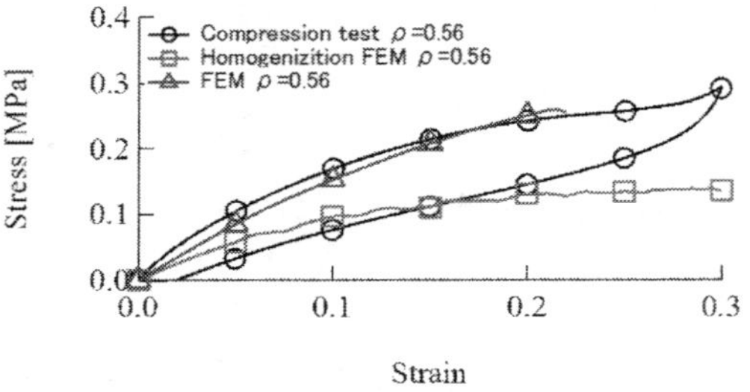

(a) Relative density ρ=0.56(n=3)

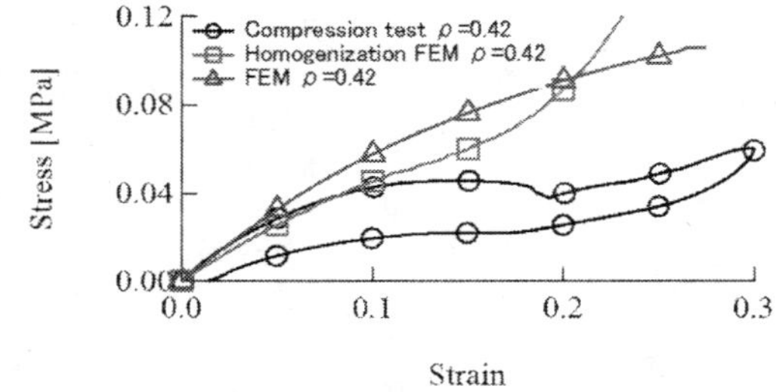

(b) Relative density ρ=0.42(n=5)

Figure 8. Comparison of compression tests results and numerical simulation.

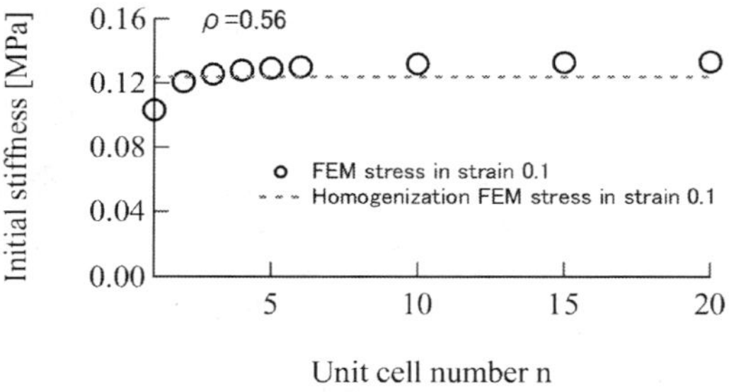

(a) Relative density ρ=0.56

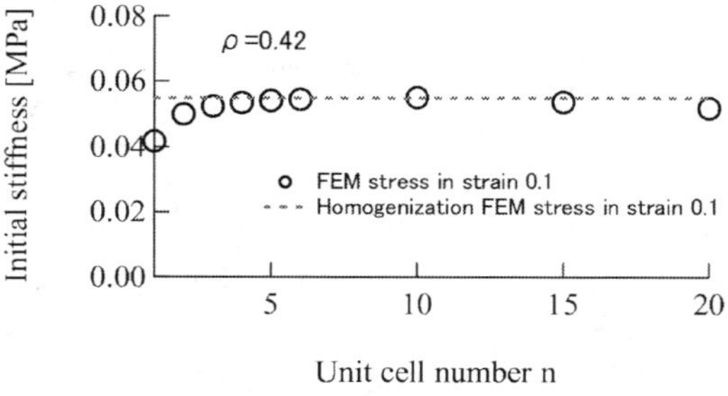

(b) Relative density ρ=0.42

Figure 9. Effect of number of unit cell on initial stiffness given by numerical calculation.

In Figure 9, effects of number of unit cells on the stress-strain relationships were shown. By increasing the number of unit cells, the result of the FEM gradually approaches the simulated result of homogenization analysis. From these results, it is necessary to prepare specimens whose unit cell number is 4 or more in order to investigate the effect of microstructure.

4 CONCLUSION

In this study, an evaluation method of foamed rubber considering the mechanical characteristics of rubber matrix and its microstructure was shown. In order to develop the evaluation method, the microstructure of the foamed rubber is assumed to the periodic structure with holes, and 2-dimentional FEM analysis code for hyperelastic material was developed. The material parameters of rubber matrix were identified from the biaxial tensile test results. Numerical analysis using the developed code was conducted under the condition that relative density of FEM model was equivalent to specimen.

Compression test of the rubber specimen reproducing FEM model was conducted to verify the applicability of analysis code.

Compared to analysis and compression test, analysis results whose relative density is 0.42 showed good agreement with the compression test results in the low strain region.

REFERENCES

Demiray, S., W. Becker and J. Hohe. Analysis of two-and three-dimensional hyperelastic model foams under complex loading conditions. *Mechanics of Materials* 38; 2006. p. 985–1000.

Kenjiro Terada, Muneo Hori, Takashi Kyoya and Noboru Kikuchi. Simulation of the multi-scale convergence incomputational homogenization approaches. *International Journal of Solids and Structure*, Vol. 37; 2000. p. 2285–2311.

Lorna J. Gibson and Michael F. Ashby. Cellular solids—structure and properties—Second edition. Cambridge University Press; 1997.

Mesut Kırca, Ayşenur Gülb, Ekrem Ekincic, Ferhat Yardımc, Ata Mugana. Computational modeling of micro-cellular carbon foams. *Finite Elements in Analysis and design*, Vol. 44; 2007. p. 45–52.

Shinoda Y., Matsuda A., Homogenization analysis of porous polymer considering microscopic structure. *6th Asia-Pacific Congress on Sports Technology* 60; 2013. p. 343–348.

Constitutive Models for Rubber X – Lion & Johlitz (Eds)
© 2017 Taylor & Francis Group, London, ISBN 978-1-138-03001-5

Three-dimensional homogenization finite element analysis of open cell polyurethane foam

S. Oketani
Graduate School of Systems and Information Engineering, University of Tsukuba, Ibaraki, Japan

A. Matsuda
University of Tsukuba, Ibaraki, Japan

A. Nomoto & Y. Kimura
Graduate School of Systems and Information Engineering, University of Tsukuba, Ibaraki, Japan

ABSTRACT: In this study, three-dimensional homogeneous finite element analysis of truncated octahedral unit cells consisting of hyperelastic beams were conducted to predict macroscopic mechanical characteristics of polyurethane foam while considering its microscopic structure. Homogenization theory was applied to original finite element analysis code, and unit cells for simulation were assumed to be subject to periodic boundary conditions. The relative density of homogenization analysis was adjusted based on the width of the beams in the unit cell model. The polyurethane matrix was assumed to be represented by incompressible hyperelasticity. The Mooney-Rivlin model and incompressible condition were applied to this hyperelasticity. The material parameters of the Mooney-Rivlin model for the polyurethane matrix were identified with the tensile loading test results. Furthermore, compression tests on polyurethane foam with various relative densities were conducted to verify the applicability of the analysis. The analysis results had good agreement with the compression test results.

1 INTRODUCTION

Polyurethane foam is made with a foam molding process using some form of polyurethane polymer. The microstructure of polyurethane foam is complex, and consists of beams and walls of polymer resin. The beams and the walls bend when polyurethane foam is compressed or stretched. Thus, polyurethane foam has good shock-absorbing characteristics in addition to having good formability and being lightweight. For that reason, polyurethane foam is widely used, for example for car seats, shoe soles and packing materials.

Now, prediction method of the mechanical characteristics of foam material is useful for the suitable design of polyurethane foam. However, such prediction is difficult because the mechanical characteristics of foam material is complicated and it vary with the relative densities. The relative density represents the ratio of the matrix in the foam. The material mechanical characteristics of polyurethane foam are evaluated through material testing of prototypes under an assumption that these materials are simple isotropic materials. Thus, an evaluation method using a numerical analysis that considers the microstructure would be a useful method to improve quality, performance and productivity.

In this study, three-dimensional homogeneous finite element analysis of truncated octahedral unit cells consisting of hyperelastic beams were conducted to predict the macroscopic mechanical characteristics of polyurethane foam while considering its microscopic structure. Previous research has found that the mechanical characteristics of polyurethane foam are determined by the characteristics of the polyurethane matrix and its microstructure. Here, to acquire the characteristics of the polyurethane matrix, tensile tests of the solid matrix specimens were conducted. The relative density of these solid specimens were 100%. In addition, the microstructure of the polyurethane foam was reproduced using a three-dimensional truncated octahedral unit cell which consisted of the beams shown in Figure 1. Homogenization theory was applied to the finite element analysis code, and the periodic boundary condition was set to boundary surfaces of the unit cell. By comparing the results of the compression tests and three-dimensional homogenization FEM analysis, the mechanical characteristics of polyurethane foam were evaluated while considering its micro structure.

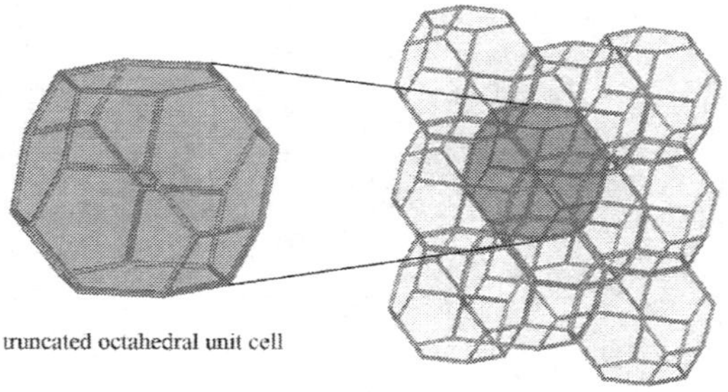

Figure 1. FEM model based on a truncated octahedron.

2 NUMERICAL ANALYSIS

2.1 *Formulation of homogenization method*

In this study, the microstructure of polyurethane foam was assumed to have a periodic structure for analysis. The formulation of homogenization analysis for large deformation is described with two different coordinates, $\boldsymbol{X}$ and $\boldsymbol{Y}$. The macroscopic behavior is defined in the $\boldsymbol{X}$-coordinate, and the microscopic behavior was defined in the $\boldsymbol{Y}$-coordinate. The two coordinates are related by the scale ratio ε as follows:

$$\boldsymbol{Y} = \boldsymbol{X} / \varepsilon. \tag{1}$$

When the scale of the microstructure was much smaller than the scale of the whole structure, the scale ratio ε had a very small value. Therefore, macroscopic characteristics, such as stiffness, stress and strain were calculated from the average of microscopic characteristics using homogenization theory. The entire microscopic structure deformed uniformly, and microscopic periodicity was kept under finite deformation. The total displacement $\boldsymbol{y}$ of the microscopic structure was divided into macroscopic displacement $\boldsymbol{Y}$ and microscopic periodical deformation $\boldsymbol{w}$ as follows:

$$\boldsymbol{y} = \tilde{\boldsymbol{F}}\boldsymbol{Y} + \boldsymbol{w}, \tag{2}$$

where $\tilde{\boldsymbol{F}}$ is the macroscopic deformation gradient tensor. The deformation gradient tensor $\tilde{\boldsymbol{F}}$ was given as the boundary condition of numerical simulation. The microscopic deformation gradient tensor $\boldsymbol{F}$ was calculated from Equation 2 as follows:

$$\boldsymbol{F} = \nabla \boldsymbol{y} = \tilde{\boldsymbol{F}} + \frac{\partial \boldsymbol{w}}{\partial \boldsymbol{Y}}. \tag{3}$$

The macroscopic deformation gradient tensor $\tilde{\boldsymbol{F}}$ was supposed to be constant and calculated as the average volume integration of the microscopic deformation gradient $\boldsymbol{F}$, so $\tilde{\boldsymbol{F}}$ was calculated as follows:

$$\tilde{\boldsymbol{F}} = \frac{1}{V}\int_{Y_0} \boldsymbol{F} dY. \tag{4}$$

Also, the average volume integration of microscopic deformation gradient $\boldsymbol{F}$ was calculated using Equation 3 as follows:

$$\frac{1}{V}\int_{Y_0} \boldsymbol{F} dY = \frac{1}{V}\int_{Y_0}\left(\tilde{\boldsymbol{F}} + \frac{\partial \boldsymbol{w}}{\partial \boldsymbol{Y}}\right) dY = \tilde{\boldsymbol{F}} + \frac{1}{V}\int_{Y_0}\frac{\partial \boldsymbol{w}}{\partial \boldsymbol{Y}} dY. \tag{5}$$

From Equations 4 and 5, it is necessary to satisfy the following equation. By giving $\boldsymbol{w}$ a periodic boundary condition, the following equation was satisfied automatically,

$$\frac{1}{V}\int_{Y_0}\left(\frac{\partial \boldsymbol{w}}{\partial \boldsymbol{Y}}\right) dY = 0. \tag{6}$$

To describe formulation of the homogenization analysis applied to the large deformation problem of polyurethane foam, the following principle of virtual work in Total Lagrange form was introduced:

$$\int (\delta\psi) dV = \int \mathbf{t} \cdot \delta\mathbf{u} dS + \int \rho_0 \mathbf{g} \cdot \delta\mathbf{u} dV, \tag{7}$$

where $\delta\Psi$ is the variation of strain energy function, dV is the volume element before deformation, dS is the area element after deformation, $\boldsymbol{g}$ is the body force, $\boldsymbol{t}$ is traction on the boundary, ρ_0 is the density of the body and $\delta\boldsymbol{u}$ is the variation of displacement vector u.

In order to produce an accurate evaluation in consideration of the incompressibility of the polymer material, a mixing method with unknown pressure and displacement was applied using unspecified constants p of Lagrange. Therefore, the strain energy function of nonlinear hyperelasticity was defined as follows:

$$\psi(\mathbf{C}, p) = \bar{W}(\bar{\mathbf{C}}) + p \cdot f(J) - U_c(p), \tag{8}$$

where $\bar{W}$ is the elastic potential function, $\boldsymbol{C}$ is the right Cauchy-Green deformation tensor, $\bar{\boldsymbol{C}}$ is the modified right Cauchy-Green deformation tensor removing the volume component, $f(J)$ is the equation for the volume constant condition and $U_c(p)$ is the complementary strain energy expressing the relation between volumetric strain and pressure. $\boldsymbol{C}$, J and $f(J)$ are expressed as follows:

$$\boldsymbol{C} = \boldsymbol{F}^T \boldsymbol{F}, \tag{9}$$

$$J = \det(\boldsymbol{F}) \text{ and} \tag{10}$$

$$f(J) = J - 1, \tag{11}$$

where J is the determinant of the deformation gradient tensor and represents volumetric deformation. Additionally, the complementary strain energy $U_c(p)$ is expressed using the bulk modulus k as follows:

$$U_c(p) = \frac{p^2}{2k}. \tag{12}$$

The elastic potential $\bar{W}$ is expressed using the modified first principal invariant $\bar{I}_1$ and the modified second principal invariant $\bar{I}_2$ in $\bar{C}$ as follows:

$$\bar{W}(\bar{C}) = \bar{W}(\bar{I}_1, \bar{I}_2) \tag{13}$$

Here, the corrected deformation gradient tensor $\bar{F}$, which does not contain volume components, $\bar{C}, \bar{I}_1$ and $\bar{I}_2$ are expressed as follows:

$$\bar{F} = J^{-1/3} \cdot F, \tag{14}$$

$$\bar{C} = \bar{F}^T \bar{F} = J^{-2/3} \cdot F^T F = J^{-2/3} \cdot C, \tag{15}$$

$$\bar{I}_1 = \bar{C} : I = J^{-2/3} \cdot C : I = J^{-2/3} \cdot I_1 \text{ and} \tag{16}$$

$$\bar{I}_2 = \frac{1}{2}\left(\bar{I}_1^2 - \bar{C}^2 : I\right) = J^{-4/3} \cdot I_2. \tag{17}$$

2.2 *Boundary condition for simulation*

The periodic boundary condition was set on the boundary surface of the FEM model.

The macroscopic deformation gradient tensor $\tilde{F}$ was given by multiplying the deformation gradient tensor $\tilde{F}_0$ by the rotation tensor R as follows:

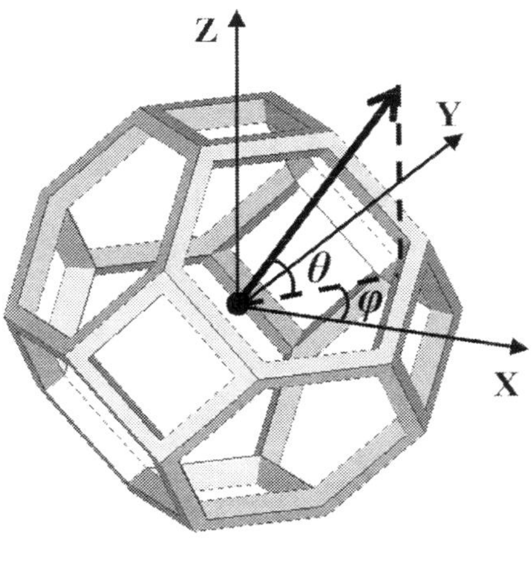

Figure 2. Definition of load direction for analysis, φ is the rotation angle around the Z-axis and θ is the rotation angle around the Y-axis.

$$\tilde{F} = R \cdot \tilde{F}_0 \cdot R^T, \tag{18}$$

$$\tilde{F}_0 = \begin{bmatrix} F_1 & 0 & 0 \\ 0 & F_2 & 0 \\ 0 & 0 & F_3 \end{bmatrix} \text{ and} \tag{19}$$

$$R = \begin{bmatrix} \cos\theta & 0 & \sin\theta \\ 0 & 1 & 0 \\ -\sin\theta & 0 & \cos\theta \end{bmatrix} \cdot \begin{bmatrix} \cos\varphi & -\sin\varphi & 0 \\ \sin\varphi & \cos\varphi & 0 \\ 0 & 0 & 1 \end{bmatrix}, \tag{20}$$

where φ is the rotation angle around the Z-axis and θ is the rotation angle around the Y-axis, as shown in Figure 2. In this study, $\theta = 0$ and $\varphi = 0$.

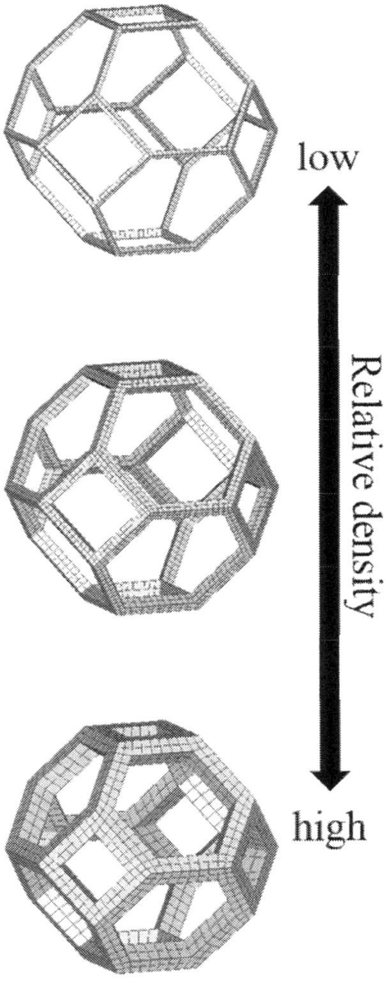

Figure 3. Three-dimensional models based on a truncated octahedron. As the beam thickens, the relative density increases.

In addition, F_1 is the load direction and F_2 and F_3 are given conditions in which the stress in each direction was 0, in order to mimic the deformation produced in the uniaxial load test. For the numerical analysis code, the penalty method was used as follows:

$$F_2^{n+1} = F_2^n - pe \cdot T_2^* \text{ and} \tag{21}$$

$$F_3^{n+1} = F_3^n - pe \cdot T_3^*, \tag{22}$$

where T_1^*, T_2^* and T_3^* are eigenvalues of the Cauchy stress tensor. T_1^* is the principal stress in the same direction of the load deformation and, T_2^* and T_3^* are the transverse Cauchy stresses.

2.3 *FEM models of unit cells*

For this study, Three original FEM unit-cell models for polyurethane foam were constructed using FEMAP, which Siemens PLM Software developed. As shown in Figure 3, they are three-dimensional models based on a truncated octahedron. Relative densities for the homogenization analysis were adjusted based on the width of the beams in unit cell. Uniform macroscopic deformation was given as analysis condition and periodic microscopic displacement was calculated by homogenization FEM analysis.

3 ACQUISITON OF MATERIAL CHARACTERISTICS OF POLYURETHANE MATRIX

In order to identify the material parameters of polyurethane matrix, tensile loading tests on 3-solid specimens were conducted. The solid specimens are specimens filled with the polyurethane foam matrix, as shown in Figure 4. Tensile tests were conducted using three solid specimens and a uniaxial machine (Autograph AG-20kNXplus, Shimadzu Corporation), strain rate of all tests was set to 0.1%/s.

The stress-strain relationships given by the tensile loading tests are shown in Figure 5. The three specimens had similar results. The results did not exhibit the complicated mechanical characteristics peculiar to polyurethane foam.

In this study, the mechanical behavior of the polymer matrix was assumed to exhibit incompressible hyperelasticity. Thus, the second Piola-Kirchhoff stress tensor $\boldsymbol{S}$ of hyperelastic material is given by the partial differentiation of the strain energy function W with respect to the right Cauchy-Green deformation tensor $\boldsymbol{C}$ as follows:

$$\boldsymbol{S} = 2\frac{\partial W(\bar{\boldsymbol{C}})}{\partial \boldsymbol{C}}. \tag{23}$$

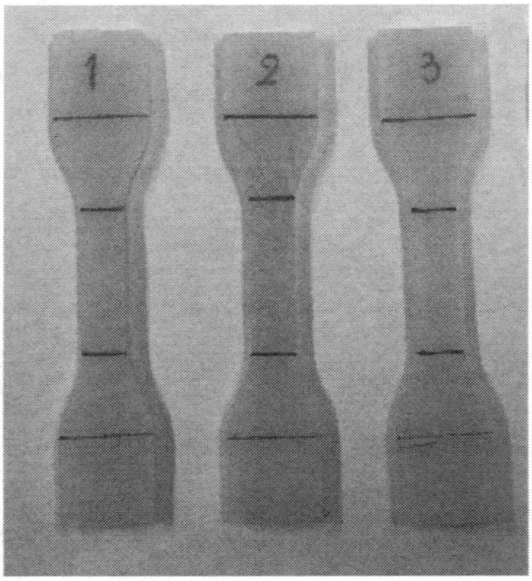

Figure 4. Specimens filled with polyurethane matrix.

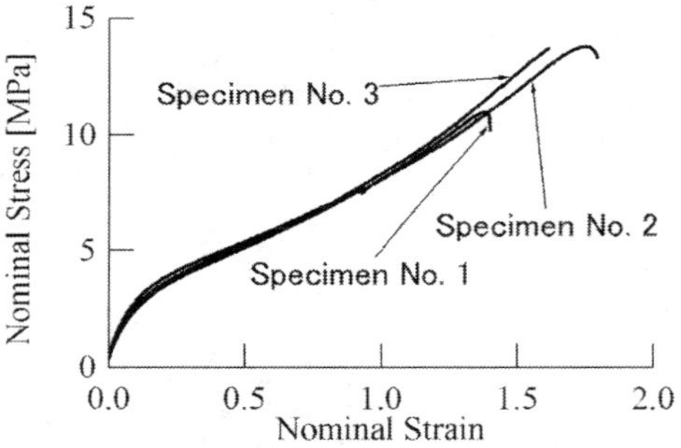

Figure 5. Stress-strain relationships from tensile tests conducted using solid specimens.

The following equation of elastic energy function was applied to represent stress-strain relationships of the polymer matrix as follows:

$$\begin{aligned} W(\bar{\boldsymbol{C}},p) &= \bar{W}(\bar{I}_1,\bar{I}_2) + p\cdot(J-1) \\ &= C_{11}(\bar{I}_1-3) + \frac{C_{12}}{2}(\bar{I}_1-3)^2 \\ &+\frac{C_{13}}{3}(\bar{I}_1-3)^3 + \frac{C_{14}}{C_{15}}\exp\{C_{15}(\bar{I}_1-3)\} \\ &+C_{21}(\bar{I}_2-3) + \frac{C_{22}}{2}(\bar{I}_2-3)^2 \\ &+\frac{C_{23}}{3}(\bar{I}_2-3)^3 + \frac{C_{24}}{C_{25}}\exp\{C_{25}(\bar{I}_1-3)\} \\ &+p\cdot(J-1), \end{aligned} \tag{24}$$

where, C_{11}, C_{12}, C_{13}, C_{14}, C_{15}, C_{21}, C_{22}, C_{23}, C_{24} and C_{25} are material parameters. The material parameters were identified by curve fitting with the tensile test results for solid specimens.

4 COMPRESSION TESTS AND NUMERICAL SIMULATION

4.1 *Compression tests of polyurethane foam*

The relative densities of compression specimens were 3.73%, 4.82% and 5.94%, respectively. The specimens were compressed up to a nominal strain 0.6, and unloaded. The strain rate of all compression tests was set to 0.1%/s.

The stress-strain relationships given by compression tests are shown in Figure 6. They reveal an increase in stiffness as relative density increases. All specimens showed linear-elastic areas, plateau

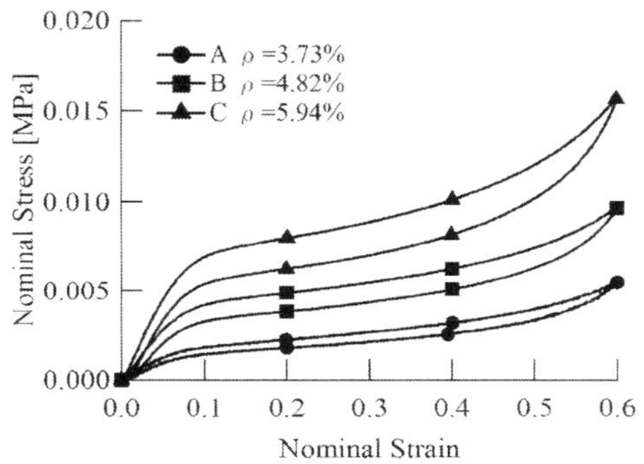

Figure 6. Stress-strain relationships from compression tests.

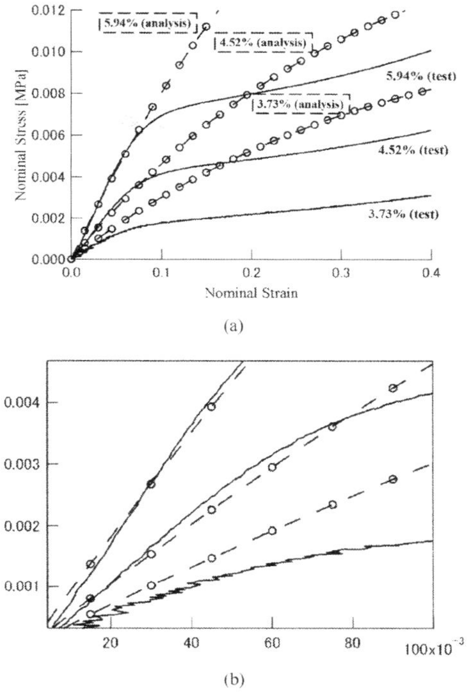

Figure 7. (a) Stress-strain relationships of compression analysis and compression tests, (b) Enlarged view of low strain region.

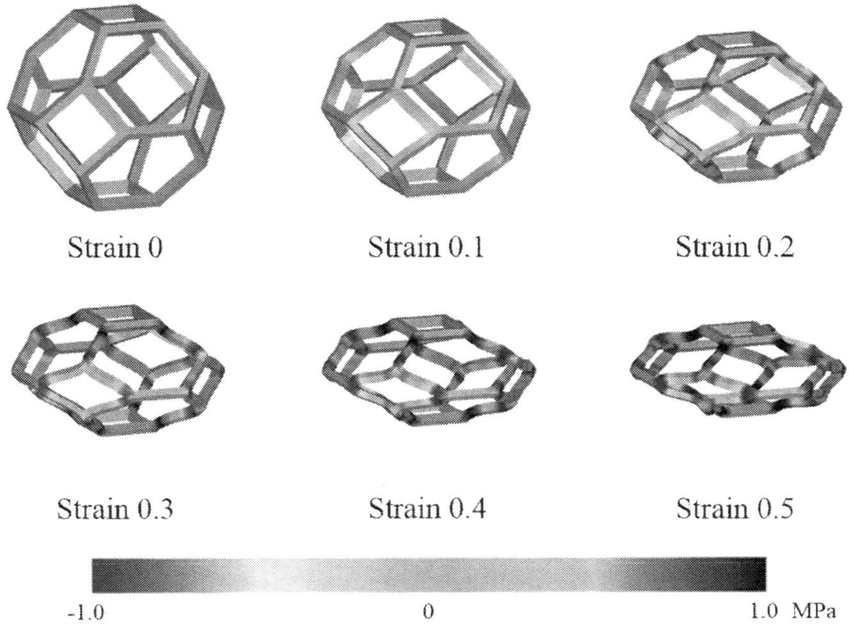

Figure 8. Stress distribution chart for a unit cell with a relative density of 5.94%. The deeper the color, the higher the stress.

areas and high-density areas. It is a characteristic peculiar to polyurethane foam that the stress-strain of the three specimens were similar.

4.2 *Comparison analysis and test*

Comparison of homogenization analysis and compression tests were conducted to verify the applicability of the analysis. The relative densities of the homogenization analysis were adjusted to the relative densities of the compression specimens: 3.73%, 4.82% and 5.94%. The comparison of the stress-strain relationships is shown in Figure 7, and diagrams of the compression deformation of the 5.94%-density unit cell are shown in Figure 8.

As can be seen in Figure 7, the analysis results had good agreement with the compression test results to a strain of 0.1. After a strain of 0.1, while the stiffness revealed in test results suddenly decline, the stiffness of analysis results did not change. This is due to the macroscopic concentration of deformation. The stiffness of polyurethane foam suddenly declines due to the propagating macroscopic concentration of deformation. In homogenization analysis, macroscopic concentration of deformation did not occur because the analysis model had a periodic structure. Therefore, the stiffness value revealed in the homogenization analysis did not decline.

5 CONCLUSION

An evaluation method for polyurethane foam that considers its microstructure was introduced in this study. In order to reproduce the foam's microstructure, three-dimensional homogeneous FEM analysis of truncated octahedral unit cells consisting of beams were conducted. The microstructure of polyurethane foam was assumed to have the periodic structure, and the three-dimensional FEM analysis were developed based on the homogenization theory. Also, in order to acquire the characteristics of the polyurethane matrix, tensile tests on solid specimens were conducted. The polyurethane matrix was assumed to be represented by incompressible hyperelasticity.

A comparison of analysis and compression test results, revealed good agreement in the small strain region (<0.1). In analysis, as relative density increased, the stiffness increased, as in compression tests. Representation of plateau behavior of foam material by homogenization analysis would be investigated as future works.

REFERENCES

Shimazu, R., et al. 2015. Three-dimensional homogenization FEM analysis of hyperelastic low density polymer foams. *European Conference on Constitutive Models for Rubbers*: 675–681.

Shinoda, Y. & Matsuda, A. 2013. Homogenization analysis of porous polymer considering microscopic structure. *6th Asia-pacific Congress on Sports Technology* 60: 343–348.

Terada, K., et al. 2000. Simulation of the multi-scale convergence in computational homogenization approaches. *International Journal of Solids and Structures* 37: 2285–2311.

Constitutive Models for Rubber X – Lion & Johlitz (Eds)
© 2017 Taylor & Francis Group, London, ISBN 978-1-138-03001-5

Derivation of full-network models with chain length distribution

Erwan Verron & Alice Gros
Institut de Recherche en Génie Civil et Mécanique, UMR CNRS 6183, École Centrale de Nantes, France

ABSTRACT: The majority of network models for rubber-like materials are derived for homogeneous materials: all chains have the same length. So, these models are not defined to predict the mechanical response of materials composed by chains of different lengths. Recently, some specific constitutive equations for inhomogeneous rubber-like materials have been proposed: most of them consist in assemblies of homogeneous networks with different chain lengths assuming equal stretch of chains. Only few papers adopt an equal stress assumption. Here, we propose a full-network model for materials with chain length distribution. In each direction of the unit sphere, chains of different lengths are supposed to bear the same force. More precisely, the affine deformation of the unit sphere containing non-Gaussian freely jointed chains is considered. It is demonstrated that: (i) the equal force assumption leads to the equality of relative stretches in chains of different length, and (ii) the response of an inhomogeneous network is similar to the one of a homogeneous network, characteristics of which being defined from the chain length distribution.

1 INTRODUCTION

Most of constitutive equations for rubber-like materials are devoted to homogeneous materials: a unique population of chains is considered, i.e. all chains admit the same length. For network constitutive equations (James & Guth 1943, Treloar 1954, Treloar & Riding 1979, Arruda & Boyce 1993, Wu & van der Giessen 1993, among others) founded on the integration of individual chains response into a threedimensional network, all chains have the same number of Kuhn segments. Obviously, such approach is not appropriate for materials with a chain length distribution such as double network of natural rubber (Mott & Roland 2000), but also gels and IPNs (Gong 2010, Dragan 2014). In the last few years, authors have proposed specific constitutive equations: in most cases, homogeneous networks with different chain lengths are mixed and an equal strain assumption is adopted (Suo & Zhu 2009, Goulbourne 2011, Zhao 2012). On the contrary, few authors adopt an equal stress assumption. von Lockette, Arruda, & Wang (2002) derive both mechanical and optical constitutive equations for bimodal elastomer networks; they assumed that chains of different lengths bear the same force invoking cross-linking simulations (von Lockette & Arruda 1999). Okumura (2004) investigates the toughness of materials composed of a hard and a soft elastic networks; for small strain, results obtained with both equal strain and equal stress assumptions are compared. From a general point of view, we believe that difficulties induced in models derivation, much more than physical reasons, prevent authors from adopting such equal force (or equal stress) assumption.

Here, we propose a derivation of full-network models for rubber-like materials with chain length distribution. The full-network framework is adopted and each direction of the unit sphere affinely deforms. Moreover, the one-dimensional response in a given direction is derived considering a series arrangement of non-Gaussian freely jointed chains which respect the whole chain length distribution and bear the same force. First, the representative chain of the inhomogeneous network is defined and its force vs. stretch response is derived. Next, the assembly in the unit sphere is proposed. Finally, generality of the theory is briefly discussed in the conclusion.

2 A REPRESENTATIVE INHOMOGENEOUS POLYMER CHAIN

2.1 *Definition of the network*

We consider an elastic, isotropic and incompressible polymer network. It is inhomogeneous, i.e. it admits a chain length distribution, and then it is made of homogeneous sub-networks defined by their number of Kuhn segments per chain N. The corresponding sub-network of N-mers is defined by its number and volume fractions in the whole network, $P(N)$ and $V(N)$ respectively. n_0 is the number density of chains in the whole network and $n(N)$ is the number density of N-mers in the pure N-mers network. Then, the number density of the N-mers in the whole network can be defined as

$$\phi(N) = n_0\, P(N) = n(N)V(N), \tag{1}$$

with

$$\sum_{N=1}^{\infty} \phi(N) = n_0. \tag{2}$$

2.2 *Equal force vs. equal stretch*

An equal strain assumption can be schematized by a parallel arrangement of chains with different number of Kuhn segments, as shown in Figure 1(a). This assumption leads to difficulties illustrated in Fig. 1(b). Consider non-Gaussian freely jointed chains of N Kuhn segments; their force f_N vs. length l_N equation is

$$f_N = \frac{kT}{b}\mathcal{L}^{-1}\left(\frac{l_N}{Nb}\right) \quad \text{for } l_N \in [0, Nb[, \tag{3}$$

where k is the Boltzmann constant, T is the temperature, b the Kuhn length, and $\mathcal{L}$ stands for the Langevin function defined as $\mathcal{L}(x) = \coth x - 1/x$. For different N, corresponding curves are shown in Fig. 1(b). Adopting the equal length assumption consists in drawing a vertical line, and the total force that exerts on the network is the sum of forces f_N given by the intersections of the line with $f_N \times (b/kT)$ vs. l_N/b curves. Nevertheless, both very large values of force and undefined force values can be encountered. Practically, this difficulty is overcome by considering strain energy functions without asymptote, or damage to render such chains inactive (Zhao 2012).

Another way to overcome the above-mentioned limitations consists in adopting the equal force assumption. Chains of all sub-networks bear the same force, and it can be schematized by a series arrangement of chains as depicted in Figure 2(a). Then, for a given N the reduced length l_N/b is obtained by inverting Eq. (3) which corresponds to the intersection of the horizontal line with chain force vs. length curves as illustrated in Fig. 2(b).

2.3 *Force vs. stretch equation*

Let define the representative inhomogeneous chain: it is the one-dimensional chain made of the same distribution of N-mers than the whole network, i.e. $P(N)$ N-mers for $N = 1,\ldots,\infty$. The force vs. stretch equation of this representative chain has to be carefully derived.

First, we consider the three different states shown in Figure 3.

(a) Firstly, chains are unstretched and not connected one to the other. For a given N, the length, stretch ratio, and force that exerts on it are

$$\tilde{l}_N^0 = \sqrt{N}b, \tag{4}$$

$$\tilde{\lambda}_N^0 = 1, \tag{5}$$

and

$$\tilde{f}_N^0 = \frac{kT}{b}L^{-1}\left(\frac{1}{\sqrt{N}}\right). \tag{6}$$

As forces $\tilde{f}_N^0$ differ, this assembly does not respect the equal force assumption.

(b) Secondly, chains are connected to respect the equal force assumption, and the total length is assumed to be the same than in state (a). For a given N, the stretch ratio is

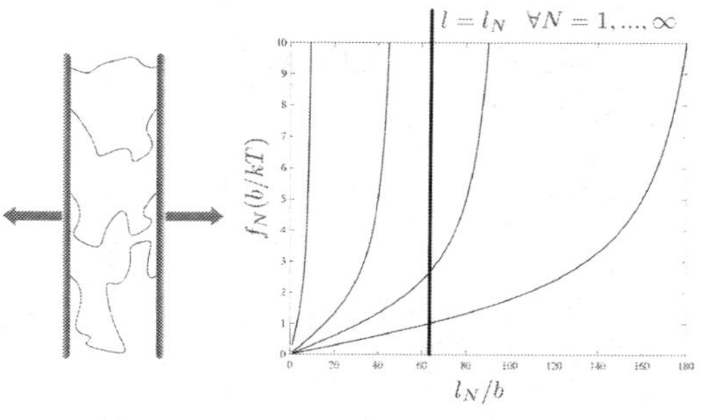

(a) (b)

Figure 1. Equal length assumption. (a) Parallel arrangement of chains, (b) force vs. length.

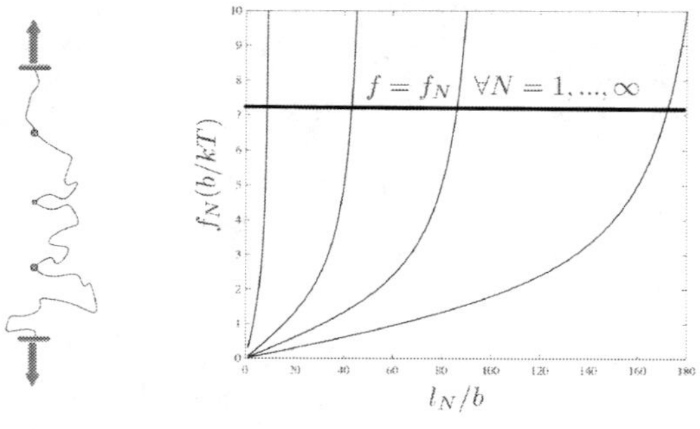

(a) (b)

Figure 2. Equal force assumption. (a) Series arrangement of chains, (b) force vs. length.

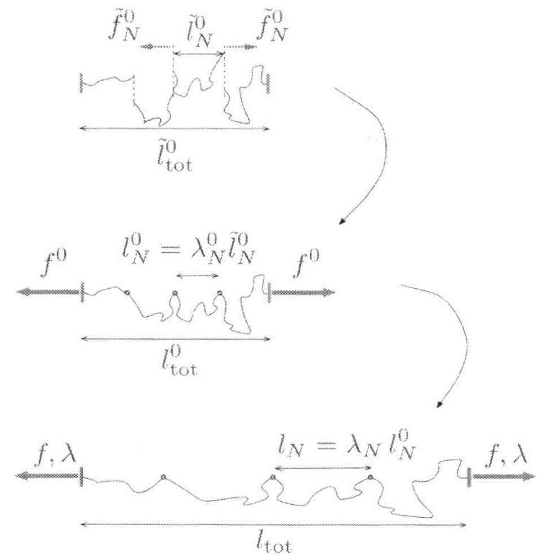

Figure 3. Stretched representative chain. (a) Before assembly, (b) reference state, (c) stretched state.

$$l_N^0 = \lambda_N^0 \tilde{l}_N^0 = \lambda_N^0 \sqrt{N} b \tag{7}$$

and the force f_N^0 which exerts on it is equal to the one which exerts on the whole assembly, i.e. f^0:

$$f_N^0 = \frac{kT}{b} \mathcal{L}^{-1}\left(\frac{\lambda_N^0}{\sqrt{N}}\right) = f^0 \quad \forall N. \tag{8}$$

(c) Thirdly, the assembled chain of state (b) is stretched, and the force exerted on it is denoted f. For a given N, the length l_N is

$$l_N = \lambda_N l_N^0 = \lambda_N \sqrt{N} b, \tag{9}$$

and the corresponding force f_N is

$$f_N = \frac{kT}{b} \mathcal{L}^{-1}\left(\frac{\lambda_N}{\sqrt{N}}\right) = f \quad \forall N. \tag{10}$$

All quantities are now defined. In the following, the force vs. stretch equation for the representative chain is now derived in two steps: the definition of the reference state (b), and the stretching of the representative chain.

Reference state. The reference state (b) is defined by the stretch ratios of all chains λ_N^0 $\forall N$. As shown in Fig. 3 (from (a) to (b)), the length of the representative chain does not change

$$l_{tot}^0 = \tilde{l}_{tot}^0, \tag{11}$$

where l_{tot}^0 and $\tilde{l}_{tot}^0$ are lengths of the representative chain in (b) and (a) states, respectively. Eq. $(7)_2$ gives

$$\lambda_N^0 = \sqrt{N} \mathcal{L}\left(\frac{f^0 b}{kT}\right), \tag{12}$$

and as the representative chain has the same length distribution than the whole network

$$\tilde{l}_{tot}^0 = \sum_{N=1}^{\infty} P(N) \tilde{l}_N^0 = \sum_{N=1}^{\infty} P(N) \sqrt{N} b. \tag{13}$$

Then,

$$\begin{aligned} l_{tot}^0 &= \sum_{N=1}^{\infty} P(N) l_N^0 = \sum_{N=1}^{\infty} P(N) \lambda_N^0 \sqrt{N} b \\ &= \sum_{N=1}^{\infty} P(N) N b \mathcal{L}\left(\frac{f^0 b}{kT}\right). \end{aligned} \tag{14}$$

Previous equations give

$$\mathcal{L}\left(\frac{f^0 b}{kT}\right) = \frac{\sum_{N=1}^{\infty} P(N)\sqrt{N}}{\sum_{N=1}^{\infty} P(N) N}. \tag{15}$$

and introducing the following notation

$$\sqrt{\mathcal{N}} = \frac{\sum_{N=1}^{\infty} P(N) N}{\sum_{N=1}^{\infty} P(N)\sqrt{N}}, \tag{16}$$

Eq. (12) leads to

$$\lambda_N^0 = \sqrt{\frac{N}{\mathcal{N}}}. \tag{17}$$

Remark 1. *Eq. (16) can be rewritten as follows*

$$\sqrt{\mathcal{N}} = \sum_{N=1}^{\infty} \varphi(N) \sqrt{N} \tag{18}$$

with

$$\varphi(N) = \frac{P(N)\sqrt{N}}{\sum_{N=1}^{\infty} P(N)\sqrt{N}}. \tag{19}$$

$\varphi(N)$ represents the length fraction of chains made of N Kuhn segments in the representative chain, and then $\mathcal{N}$ is the mean number of segments in the representative chain.

Stretching of the representative chain. As shown in Fig. 3 (from. (a) to (c)), the stretch ratio of the representative chain is given by

$$\lambda = \frac{l_{\text{tot}}}{l_{\text{tot}}^0} \tag{20}$$

in which

$$l_{\text{tot}} = \sum_{N=1}^{\infty} P(N) l_N = \sum_{N=1}^{\infty} P(N) N b \mathcal{L}\left(\frac{fb}{kT}\right). \tag{21}$$

Recalling that $l_{\text{tot}}^0 = \tilde{l}_{\text{tot}}^0 = \sum_{N=1}^{\infty} P(N)\sqrt{N} b$, Eq. (20) leads to

$$\lambda = \sqrt{\mathcal{N}} \mathcal{L}\left(\frac{fb}{kT}\right), \tag{22}$$

and finally the force vs. stretch response of the representative chain is

$$f = \frac{kT}{b} \mathcal{L}^{-1}\left(\frac{\lambda}{\sqrt{\mathcal{N}}}\right). \tag{23}$$

Remark 2. *Eqs (10) and (23) lead to*

$$\frac{\lambda_N}{\sqrt{N}} = \frac{\lambda}{\sqrt{\mathcal{N}}} \quad \forall N; \tag{24}$$

with the help of Eq. (17), it can be written as

$$\frac{\lambda_N}{\lambda_N^0} = \lambda \quad \forall N. \tag{25}$$

In this equation, the left-hand side term is the relative stretch ratio in (c) with respect to the reference configuration (b). This equation demonstrates that the equal force assumption is also an equal stretch assumption.

3 THREE-DIMENSIONAL CONSTITUTIVE MODEL

3.1 *Derivation of the full-network model*

We restrict our model to hyperelasticity: it exists a strain energy density $W(\boldsymbol{F})$ per unit of undeformed volume that depends on the deformation gradient tensor $\boldsymbol{F}$. Then, the Cauchy stress tensor is given by

$$\sigma = -p\boldsymbol{I} + \frac{\partial W}{\partial \boldsymbol{F}} \boldsymbol{F}^T, \tag{26}$$

where p is the hydrostatic pressure due to incompressibility, $\boldsymbol{I}$ is the 3×3 identity tensor and $\cdot^T$ stands for transposition.

Following works of Treloar & Riding (1979), and Wu & van der Giessen (1993) on full-network models, we can derive a constitutive equation by integrating the force vs. stretch of the representative chain over the unit sphere. Denoting $w(\lambda)$ the strain energy of the representative chain, the strain energy density of the whole network is

$$W(\boldsymbol{F}) = \frac{1}{|S|} \iint_S n(\boldsymbol{u}) w(\lambda(\boldsymbol{u})) d^2\Omega(\boldsymbol{u}), \tag{27}$$

where S is the unit sphere ($|S|$ is its surface), $\boldsymbol{n}(\boldsymbol{u})$ is the number density of chains in direction $\boldsymbol{u}$ (a radial direction in the undeformed state), $\lambda(\boldsymbol{u})$ is the stretch ratio of the chainin this direction, and $d^2\Omega(\boldsymbol{u})$ the corresponding solid angle. As mentioned above, the affine deformation of the sphere is adopted, i.e.

$$\lambda(\boldsymbol{u}) = \|\boldsymbol{F}\boldsymbol{u}\|, \tag{28}$$

and in the special case of isotropy,

$$n(\boldsymbol{u}) = n_0 \forall \boldsymbol{u}. \tag{29}$$

In Eq. (26), the derivative of W with respect to $\boldsymbol{F}$ is:

$$\frac{\partial W}{\partial F} = F \frac{1}{|S|} \iint_S n_0 w'(\lambda(\boldsymbol{u})) \times \frac{1}{\lambda(\boldsymbol{u})} \boldsymbol{u} \otimes \boldsymbol{u} \, d^2\Omega(\boldsymbol{u}), \tag{30}$$

where (Treloar 1975)

$$w'(\lambda) = \frac{\partial w}{\partial \|\boldsymbol{r}\|} \frac{\partial \|\boldsymbol{r}\|}{\partial \lambda} = f \frac{\partial \|\boldsymbol{r}\|}{\partial \lambda}. \tag{31}$$

In this equation, the force f is given by Eq. (23), and the norm of the end-to-end vector $\|\boldsymbol{r}\|$ is simply equal to l_{tot}. Thus, considering Eqs (21–22), it leads to

$$w'(\lambda) = \frac{\sum_{N=1}^{\infty} P(N) N}{\mathcal{N}} kT \sqrt{\mathcal{N}} \mathcal{L}^{-1}\left(\frac{\lambda}{\sqrt{\mathcal{N}}}\right), \tag{32}$$

and finally

$$\boldsymbol{\sigma} = -p\boldsymbol{I} + \boldsymbol{F} \frac{n_0 \sum_{N=1}^{\infty} P(N) N}{\mathcal{N}} kT \times \frac{1}{|\mathcal{S}|} \iint_{\mathcal{S}} \frac{\sqrt{\mathcal{N}}}{\lambda(\boldsymbol{u})} \mathcal{L}^{-1}\left(\frac{\lambda(\mathbf{u})}{\sqrt{\mathcal{N}}}\right) \boldsymbol{u} \otimes \boldsymbol{u} \, d^2\Omega(\boldsymbol{u}) \boldsymbol{F}^T. \tag{33}$$

The last equation is the stress-strain relationship of a standard full-network of non-Gaussian freely jointed chains defined by

- its number of segments per representative chain N,
- its chain density: the number of Kuhn segments per unit of volume, i.e. $n_0 \sum_{N=1}^{\infty} P(N)N$, divided by the number of segments per representative chain, $\mathcal{N}$.

Remark 3. *The previous model could have been derived for Gaussian chains. The corresponding result is easily recovered if one considers that*

$$\mathcal{L}^{-1}\left(\frac{\lambda}{\sqrt{\mathcal{N}}}\right) \approx 3\frac{\lambda}{\sqrt{\mathcal{N}}}, \tag{34}$$

and

$$\frac{1}{|S|}\iint_S \boldsymbol{u} \otimes \boldsymbol{u}\, d^2\Omega(\boldsymbol{u}) = \frac{\boldsymbol{I}}{3}. \tag{35}$$

Then, Eq. (33) simplifies to

$$\sigma = -p\boldsymbol{I} + \frac{n_0 \sum_{N=1}^{\infty} P(N)N}{\mathcal{N}} kT\mathbf{B} \tag{36}$$

which is a neo-Hookean like constitutive equation (Treloar 1943).

3.2 *Numerical integration*

The previous constitutive equation Eq. (33) cannot be integrated over the unit sphere analytically and necessitates the use of a numerical integration scheme. The reader can refer to the recent works of Verron (2015) and Itskov (2016) who address this issue.

4 CONCLUSION

A new approach for constitutive equations of rubber-like materials with arbitrary chain length distribution has been proposed in this paper. It is based on the equal force assumption which states that in a given direction of space all chains bear the sameforce whatever their length. Schematically, it corresponds to a series arrangement of the chains. The complete derivation has been proposed for full-network models of non-Gaussian freely jointed chains with different numbers of Kuhn segments. It is to note that even if the deformation of the unit sphere is assumed affine, i.e. the representative chains follow the continuum deformation, each individual chain does not fulfill it. Thus, the assembled model is non-affine.

The two major results of this paper are:

- the homogeneized constitutive equation reduces to a standard full-network model, its stiffness and chain length being defined by the chain length distribution;
- if the reference state is well-defined, the equal force assumption is equivalent to an equal relative stretch assumption.

REFERENCES

Arruda, E. M. & M. C. Boyce (1993). A three dimensional constitutive model for the large stretch behavior of rubber elastic materials. *J. Mech. Phys. Solids 41*(2), 389–412.

Diani, J., M. Brieu, & J.-M. Vacherand (2006). A damage directional constitutive model for mullins effect with permanent set and induced anisotropy. *Eur. J. Mech. A/Solids 25*, 483–496.

Dragan, E. S. (2014). Design and applications of interpenetrating polymer network hydrogels. A review. *Chem. Engng. J. 243*, 572–590.

Flory, P. J. & J. Rehner (1943). Statistical mechanics of crosslinked polymer networks. I. Rubberlike elasticity. *J. Chem. Phys. 11*, 512–520.

Gong, J. P. (2010). Why are double network hydrogels so tough? *Soft Matter 6*, 2583–2590.

Goulbourne, N. C. (2011). A constitutive model of polyacrylate interpenetrating polymer networks for dielectric elastomers. *Int. J. Solids Struct. 48*(7–8), 1085–1091.

Itskov, M. (2016). On the accuracy of numerical integration over the unit sphere applied to full network models. *Comput. Mech. 57*(5), 859–865.

James, H. M. & E. Guth (1943). Theory of the elastic properties of rubber. *J. Chem. Phys. 11*, 455–481.

Kuhn, W. & F. Grün (1942). Beziehungen zwichen elastischen Konstanten und Dehnungs doppelbrechunghoch elastischer Stoffe. *Kolloideitschrift 101*, 248–271.

Marckmann, G., E. Verron, L. Gornet, G. Chagnon, P. Charrier, & P. Fort (2002). A theory of network alteration for the Mullins effect. *J. Mech. Phys. Solids 50*, 2011–2028.

Miehe, C., S. Göktepe, & F. Lulei (2004). A micro-macro approach to rubber-like materials - Part I: the non-affine microsphere model of rubber elasticity. *J. Mech. Phys. Solids 52*, 2617–2660.

Mott, P. H. & C. M. Roland (2000). Mechanical and optical behavior of double network rubbers. *Macromolecules 33*, 4132–4137.

Okumura, K. (2004). Toughness of double elastic networks. *Europhys. Lett. 67*(3), 470–476.

Sloan, I. H. & R. S. Womersley (2004). Extremal systems of points and numerical integration on the sphere. *Adv. Comput. Math. 21*, 107–125.

Suo, Z. & J. Zhu (2009). Dielectric elastomers of interpenetrating networks. *Appl. Phys. Lett. 95*, 232909-1-232909-3.

Treloar, L. R. G. & G. Riding (1979). A non-Gaussian theory for rubber in biaxial strain. I. Mechanical properties. *Proc. R. Soc. Lond. A 369*, 261–280.

Treloar, L. R. G. (1943). The elasticity of a network of long chain molecules (I and II). *Trans. Faraday Soc. 39*, 36–64; 241–246.

Treloar, L. R. G. (1954). The photoelastic properties of shortchain molecular networks. *Trans. Faraday Soc. 50*, 881.
Treloar, L. R. G. (1975). *The Physics of Rubber Elasticity*. Oxford: Oxford University Press.
Verron, E. (2015). Questioning numerical integration methods for microsphere (and microplane) constitutive equations. *Mech. Mater. 89*, 216–228.
von Lockette, P. R. & E. M. Arruda (1999). Topological studies of bimodal networks. *Macromolecules 32*(6), 1990–1999.
von Lockette, P. R., E. M. Arruda, & Y. Wang (2002). Mesoscale modeling of bimodal elastomer networks: constitutive and optical theories and results. *Macromolecules 35*, 7100–7109.
Wu, P. D. & E. van der Giessen (1993). On improved network models for rubber elasticity and their applications to orientation hardening in glassy polymers. *J. Mech. Phys. Solids 41*(3), 427–456.
Zhao, X. (2012). A theory of large deformation and damage of interpenetrating polymer networks. *J. Mech. Phys. Solids 60*, 319–332.

Constitutive Models for Rubber X – Lion & Johlitz (Eds)
© 2017 Taylor & Francis Group, London, ISBN 978-1-138-03001-5

Nano-mechanical modeling for rubbery materials

Keizo Akutagawa
Bridgestone Corporation, 3-1-1 Ogawahigashi-cho, Kodaira-shi, Tokyo, Japan

ABSTRACT: In the field of rubber science and technology the digital revolution is under way and integrates technologies across the information technologies coupled with advances in characterization and experiment of materials. In this situation, accelerating the integration will be crucial to achieving competitiveness. A good simulation structural model should capture the actual structure at the moment as a function of time and space. But it is currently difficult to visualize the real motion of a rubber chain and its cross-linked network experimentally. Hence, a mathematical model based on fundamental physical principles such as thermodynamics is more desirable than a model based on virtual simulations. But no one has yet successfully derived such a model for rubber from the physical principles governing the molecular response of rubbers. The present work shows a nano-mechanical model of rubbery materials for computational simulation, which is based on the non-equilibrium thermodynamics.

Hence, a constitutive equation, which can cover the stress-strain behaviours up to higher strain region as a function of temperature, has been investigated. Since the stress-strain behaviour shows a transition to the glass-hard state at lower temperature, where the contribution of the internal energy is increased, a constitutive equation was derived from the statistical thermodynamics with Hamiltonian equations considering the change in internal energy.

1 INTRODUCTION

Since the late 1960's, tire industry began a research on application of the finite element method to tire development, and it has been developed as a method of analysis from a relatively early stage of the finite element method history. The reason is that the tire has a complicated structure made of many parts of rubber and steel cord or organic fiber and it was difficult to observe the mechanical behaviors at contact patch and inside tire experimentally due to its unique cross-sectional shape and layered structures. In recent years, it has become possible to predict the tire physical properties quantitatively by the benefit from the evolution of analytical methods and computer technology. In the early days of the finite element method, it was often used for interpretation of experimental results, but in recent years it has been used as a optimization tool for product design. For the future development of finite element method, a new approach to predict the mechanical behaviors of rubber materials in nano-scale has started, which is a more complicated and require larger calculation resources to represent the microstructure and micromechanics of materials.

Figure 1 shows a scale chart for the size of the model and the corresponding computational analysis methods [1]. Currently, the capacity of cutting-edge computers is growing very rapidly and

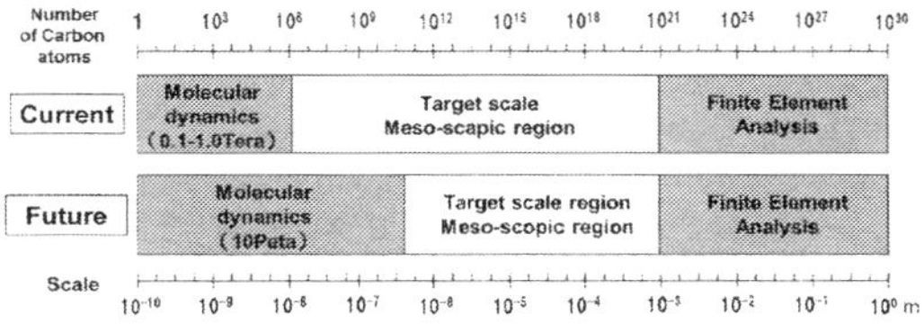

Figure 1. The scale range of the model to be calculated and the domain of the corresponding computational analysis method [1].

reached to 500 tera times (5×10^{14}) per second of four arithmetic operation of floating-point. Further development is undergoing and is expected to reach 10 peta times (10^{16}). If it is assumed to perform a model calculation with 10^6 elements on an average class computer of 100 giga (10^{11}) computer, it is expected that model calculation with 10^{11} elements will be possible in near future. But if a full carbon atom model as an exteram case is assumed to apply to calculation of 10 kg commercial goods (10^{27} elements), we have to watch for the further evolution of semiconductor more than a century according to the Moore's law. Currently, the finite element analysis is mainly utilized aiming at the larger model size with a mesh of at least a millimeter scale, but the quantum mechanics and molecular dynamics are targeting at nano-scale models. The computational scientific methods in even smaller scale regions are being developed expanding to the meso-scale region by introducing coarse-graining method and in larger scale region are developed by introducing the nano-scale finite element analysis. A lot of attempts are carried out [2–5], but a wide range of blank area on meso-scale region still exists. It is expected that in the future overlapping of the technologies in the meso-scale region will bring us to the realization of multi-scale simulation covering a wide range from nanoscale to tire size macro scale. The multi-scale modeling is one of ideas to make the materials developments faster and at lower cost than is possible today. It will bridge the material design in the molecular scale and the product design in the real scale by means of continuum methods as advanced multi-scale finite element method.

The present work shows a mathematical molecular model of rubbery materials which was proposed for the calculation of finite element analysis in nano-scale. It is expected to cover the blank at meso-scale regon with minimum calculation time. Once it was realized, it will be able to respond quickly to various demands from the market.

2 THEORY

A brief review is given here of the non-equilibrium thermodynamics derived from the analysis of phenomenological data on heat build-up experiments [6]. Temperature change on stretching and unstretching process can be reduced into two processes, reversible process and irreversible process. The reversible process can be described as the configurational entropic process of rubber elasticity. The irreversible process is partly associated with both of heat capacity, C, and glass transition temperature, Tg. The temperature change on stretching can be represented by

$$\Delta T = \frac{1}{C}\int_0^{l_x} f \cdot dl - \frac{1}{C}\cdot\left\{p\left(V - V_0\right) + A_f\left(x - x_0\right)\right\} \quad (1)$$

On the bases of this analysis the thermodynamic relations was derived as;

$$f \cdot dl = p \cdot dV - A_f \cdot dx - T \cdot dS \quad (2)$$

where ΔT is temperature change during deformation, f is the tensile force at tensile length of l, C is the heat capacity, p is the pressure, dV is the volume change, A_f is the affinity force, x is the internally controllable variable as a function of time, dS is the configurational entropy change, T is an absolute temperature of the system. The important feature of the non-equilibrium thermodynamic model here is an introduction of $A_f \cdot dx$, which represents weak interactions between molecules. The resultant constitutive equation can be given by

$$\begin{aligned} f \cdot dl = {} & B\left(I_3^{\frac{1}{2}} - 1\right)^2 - \frac{1}{\beta'}\left\{\ln\left[1 + e^{\beta'\cdot\kappa}\cosh\left(2\beta'\left(I_2' - 3\right)\right)\right]\right\} \\ & + \frac{e^{\beta'\cdot\kappa}\left\{\kappa\cosh\left(2\beta'\left(I_1' - 3\right)\right) + 2\left(I_1' - 3\right)\cdot\sinh\left(2\beta'\left(I_1' - 3\right)\right)\right\}}{e^{\beta'\cdot\kappa}\cosh\left(2\beta'\left(I_1' - 3\right)\right) + 1} \\ & - \frac{N}{\beta}\cdot\left\{\frac{1}{2}I_1 + \frac{3}{100n}\left(3I_1^2 - 4I_2\right) + \frac{99}{12250n}\left(5I_1^3 - 12I_1I_2\right)\right\} \end{aligned} \quad (3)$$

where κ is the attractive interaction energy between molecules, B is bulk modulus, β' is $x/k\Delta T$ and n is the number of statistical links between crosslinks. The detail of the equation is shown elsewhere [6–9].

3 EXPERIMENT

The rubber used for this work is SBR1500 to avoid the heat build-up from the strain-induced crystallization during the deformation. The SBR was unfilled and cross-linked using a sulfur curing system with conventional sulfur cross-links. Following mixing, the rubbers were pressed into sheets of nominal thickness of 2 mm with a pressure of 20 MPa at the temperatures of 145°C. The curing time was 33 minutes. The mechanical property was measured using a 50 N load cell and the extension ratio was calculated from the cross-head displacement. Most of the tests were carried out at a cross-head speed of 26.4 mm/s corresponding to a finite strain rate of 3.3×10^{-1}/s with triangle deformation mode.

The temperature measurements under strain were carried out at 25°C using a conventional tensile testing machine equipped with a high resolution thermography (FSV-7000E, Apiste Corporation). The temperature was captured with flame speed of 1/60 second. Rubber strip samples of dimensions 50 × 6 × 2 mm were used.

4 RESULTS AND DISCUSSION

The verification of equation (3) was carried out comparing the calculation data with the experimental data on stress-strain relations as shown in Figure 2. The stress-strain loops at both initial and second stages were measured. The number indicates the sequence of the stretching and unstretching process, ① stretching process at 1st loop, ② un-stretching process at 1st loop and ③ stretching process at 2nd loop. The calculation was carried out by fitting the parameters to the stretching

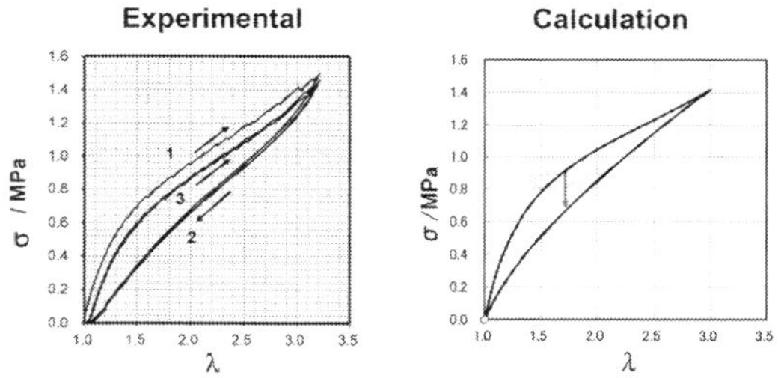

Figure 2. Comparison between experimental and calculation for stress strain relation of SBR with non-equilibrium thermodynamic equation.

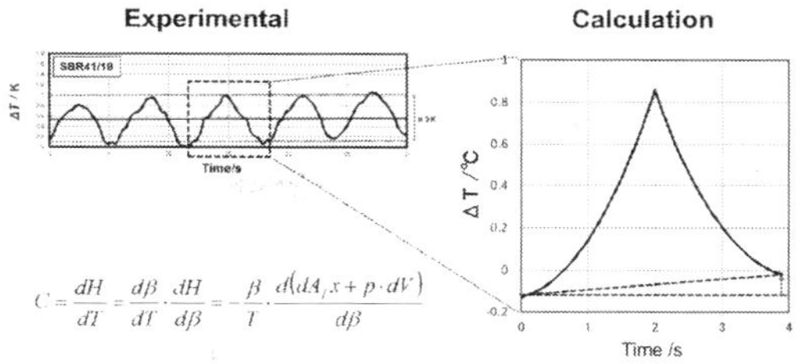

Figure 3. Comparison between experimental and calculation for heat build-up of SBR with non-equilibrium thermodynamic equation.

and un-stretching process after 2nd loop. It shows a quite good agreement between calculations and experimental results. From this results it is assumed that the mechanical behaviors calculated in nano-scale are able to represent those appeared in the actual behaviors. The stress relaxation at stretching process can be explained by the breakdown of weak interaction between molecules represented by the term of $A_f \cdot dx$,

On the other hand the temperature change, ΔT, on stretching and un-stretching process was estimated from equation (1) together with equation (3). The heat capacity, C, was derived from the equation superimposed in Figure 3. The work done on the system was calculated from the predicted stress-strain curve as shown in Figure 2. The rubber was stretched to 200% and un-stretched linearly. The 3rd loop of the experiment was picked up and compared with the calculation. It is a good agreement between calculations and experimental results, especially in the magnitude of temperature change.

The nano-mechanical model is successful in describing the stress response and temperature change in stretching and un-stretching process. From this model it was concluded that the heat build-up process can be thermodynamically explained by the following steps;

1. The temperature change on stretching and un-stretching process can be reduced into two processes, reversible process and irreversible process. The magnitude of the effects is scaled by heat capacity.
2. The reversible process can be associated with the entropic process of rubber elasticity. But in the irreversible process the work is leaked as heat to the external system or the fluctuation of the molecular motion. The irreversible process can be associated with the breakdown of weak interactions between molecules as energetic process.
3. In un-stretching process the reversible part of the temperature decreased with decreasing entropic force and some of weak interactions are recovered, but the irreversible temperature is never recovered and is observed as base temperature increases.

5 CONCLUSION

The proposed nano-mechanical model based on the non-equilibrium thermodynamics works well on describing the mechanical behaviors obtained experimentally. From the analysis of this model it was shown that the entropy is of course a major contribution to the rubber elasticity, but the energy term is also an important parameter for describing the mechanical properties such as heat build-up, dumping, friction and time-dependent behaviors.

REFERENCES

[1] Akutagawa, K. (2009) Nanoscopic Mechanical Simulation of Elastomers with Finite Element Analysis. Nippon Gomu Kyokaishi, 82(5), 277–232.

[2] Li, Y., Tang, S., Abberton, B.C., Kröger, M., Burkhart, C., Jiang, B., … Liu, W.K. (2012). A predictive multiscale computational framework for viscoelastic properties of linear polymers. Polymer, 53(25), 5935–5952.

[3] Hagita, K., Morita, H., Doi, M., & Takano, H. (2016). Coarse-Grained Molecular Dynamics Simulation of Filled Polymer Nanocomposites under Uniaxial Elongation. Macromolecules, 49(5), 1972–1983.

[4] Vineet Jha, Amir A., Hon, Alan G., & Thomas, J.J.C.B. (2008). Modeling of the effect of rigid fillers on the stiffness of rubbers. Journal of Applied Polymer Science, 15, 2572–2577.

[5] Akutagawa, K., Yamaguchi, K., Yamamoto, A., Heouri, H., Jinnai, H., & Shinbori, Y. (2008). Mesoscopic mechanical analysis of filled elastomer with 3D-finite element analysis and transmission electron microtomography. Rubber Chemistry and Technology, 81(2).

[6] Akutagawa, K., Hamatani, S., & Nashi, T. (2015). The new interpretation for the heat build-up phenomena

of rubbery materials during deformation. Polymer (United Kingdom), 66, 201–209.

[7] Keizo Akutagawa. (2009). Mesoscopic 3D_Finite Element Analysis of Filled Elastomer. In S. Gert Heinrich, Michael Kaliske, Alexander Lion, Reese (Ed.), Constitutive Models for Rubber VI—Proceedings of the 6th European Conference for Constitutive Models for Rubber, ECCMR 2009 404–411. London: CRC press.

[8] Akutagawa, K., Hamatani, S., & Kadowaki, H. (2017). Nanomechanics on non-equilibrium thermodynamics for mechanical response of rubbery materials. in print.

[9] Bridgestone Corporation/Akutagawa, K. (USA), Method for predicting elastic response performance of rubber product, Method for design, and device for predicting elastic response performance. PCT application WO 2012/046739 A1.

Constitutive Models for Rubber X – Lion & Johlitz (Eds)
© 2017 Taylor & Francis Group, London, ISBN 978-1-138-03001-5

Evaluation of rheological parameters for injection molding simulations

J. Meier
Deutsches Institut für Kautschuktechnologie e.V., Hannover, Germany

W. Villa-Ramirez & F. Hüls
Henniges Automotive GmbH & Co KG, Rehburg-Loccum, Germany

ABSTRACT: Measurement and evaluation of rheological parameters for simulation of injection molding processes are discussed. Prediction limits especially for minimal necessary machine pressure as well as temperature development of mixtures are shown to origin in thermo-rheological effects not fully taken into account via measurement interpretation. A concept for consistent evaluation of rheological parameters by simulation of high pressure capillary measurement is proposed and resulting corrections of flow curves, as given by shear rate dependency of viscosity are estimated.

1 INTRODUCTION

Nowadays, in the design phase of the production of elastomer articles mold filling simulation has proven useful for process balancing and processing verification.

As an example Figure 1 displays a filling study for an exhaust hanger, comparing simulation with partly-filled articles in accordance.

Since this method provides forecasts on real processing behavior, it reduces the necessity for trial and correction steps. Although proven as useful, sometimes such simulations show pronounced deviation from production findings especially considering pressure and heat build-up.

Therefore a more profound rheological consideration and interpretation of rheological data is required.

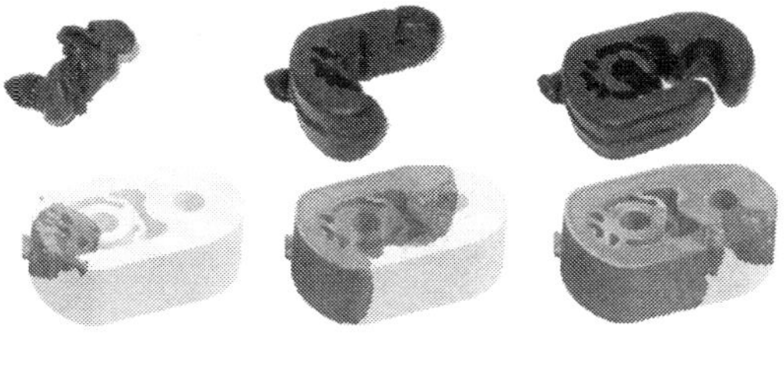

Figure 1. Filling study for exhaust hanger – reality compared to simulation; filling degree from left to right: 10%, 40%, 80%.

2 EXPERIMENTAL

Rheological flow curves were acquired by high pressure capillary viscometer (HPC) Göttfert Rheograph 6000, covering shear rates from about 30 to 1000 s^{-1} at three temperatures: 80, 100, 120°C. Figure 2 shows a sketch of the measurement setup. Chemical reaction was examined with rheometer Alpha Technologies MDR 2000 E at three temperatures: 135, 155, 175°C.

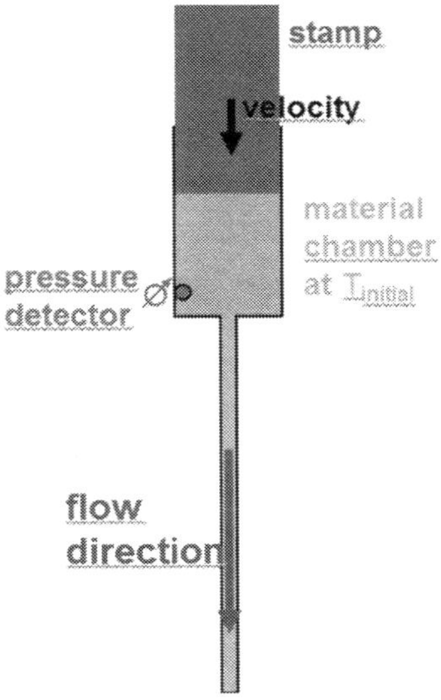

Figure 2. High pressure capillary viscometer; measurement setup.

Heat capacity was taken at an uncured mixture with differential scanning calorimeter TA-Instruments DSC 2920CE. Finally, heat conductivity was characterized at three temperatures.

3 THERMO-RHEOLOGICAL CONCEPTS

For simulation of the shear response of rubber mixtures, several thermo-rheological phenomena must be considered. At first, force equilibrium can be stated by the Navier-Stokes equation given in the form

$$\rho\frac{D_v}{D_t} = -\nabla\rho + (\lambda+\mu)\nabla(\nabla\cdot v) + \eta\Delta v + f \tag{1}$$

where the fluid's properties are denoted by density ρ, velocity v, pressure p, and dynamic viscosity μ (Sigmasoft 2014). Dynamic viscosity can be taken into account with Carreau-like approach extended by WLF-conform temperature dependency

$$\eta(T,\dot{\gamma}) = \frac{p_1 a_T}{\left(1+p_2 a_T \dot{\gamma}\right)^{p_3}} \tag{2}$$

using $\dot{\gamma}$ [1/s] as shear rate, p_1 as zero shear viscosity, p_2 as reciprocal transition shear rate, p_3 as exponent parameter, a_T as WLF temperature shift [10], and temperature T.

Since rubber mixtures are processed closed to temperatures relevant for vulcanization, the consideration of reaction kinetics at least as processibility indicator is crucial, for example with an WLF-conform of Carreau-like approach

$$\dot{c}(t) = (K_1 + K_2 c^m)(1-c)^n \tag{3}$$

for both meta-parameters K_i the activation relation is as given by:

$$K_i(T) = A_i e^{\frac{-E_i}{R_G T}} \tag{4}$$

with curing degree c, Arrhenius reaction rates K1, K2, relaxation time coefficient log10(Ai), reaction orders n and m, activation energy Ei, and gas constant RG.

4 RESULTS AND DISCUSSION

Relevant simulations rely on rheological data from appropriate measurements, characterizing the material in a manner close to processing conditions (Vlachopoulos & Strutt). For the interpretation of the measurements, it is worth to be aware of method limits intrinsic to rheological characterization instruments.

With high viscosity rubber mixtures, shear induced heating occurs at higher shear rates about thousand per second. Hence, in high pressure capillary viscometry (HPC), due to the accompanying viscosity reduction, this effect gives rise to pronounced underestimation of viscosities.

Simulation results for the local shear rates and temperatures along the channel are displayed in Figures 3 and 4, respectively.

In the high shear rate regime at the vicinity of the walls pronounced heat build up is found all along the channel, giving rise to temperature increase of about 20°C at the outlet.

This anisotropic temperature field differs strongly from correction concepts assuming homogenously distributed heat build up. Since mold filling simulations rely strongly on rheological data, this heating might be a main reason for the sometimes unsatisfying capability of simulation software in reproducing the high temperature steps mixtures undergo by being squeezed through finger gates in mold processes. Additionally, the well known phenomena of thixotropy and demixing contribute to the establishment of the shear field, where the former results in a transient shear stress to shear rate answer and the latter leads to wall slippage which has also been found to depend on the channel material and roughness (Barnes 1997 & Klie et al. 2015). Both effects can be considered as depending on molecular weight distribution of the basic polymer, which dominates the relaxation times (Sgodzaj 2002 & Meier et al. 2003). Moreover in HPC changes in channel diameter have been found to cause high additional pressure contributions, attributed to elasticity and the so-called extensional viscosity (Petrie 2006 & Aho, 2011).

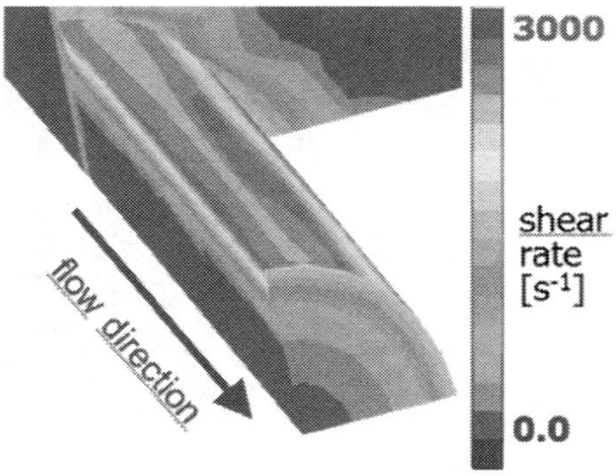

Figure 3. Simulation result for shear rate distribution at flow rate 0.1 cm³/s.

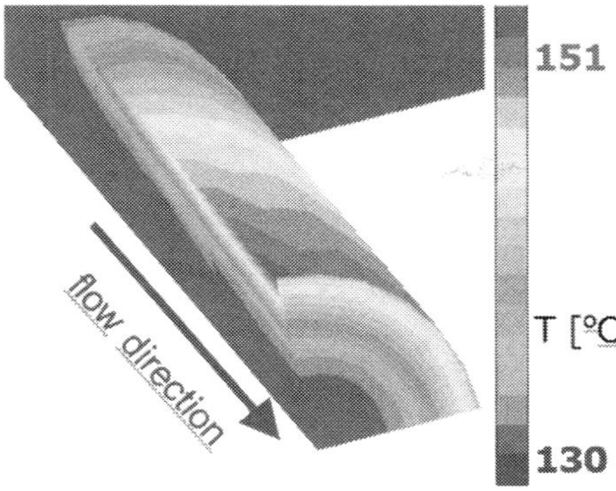

Figure 4. Simulation result for temperature distribution at flow rate 0.1 cm^3/s.

5 CONCLUSION

Thus, discussing rheological data assessment and further interpretation towards shear-viscosity flow curves together with simulative application studies might be a useful manner towards separation of several effects and acquirement of reliable rheological parameters. Here, a concept is suggested to correct the heat-falsified viscosity data by means of iterative data fitting and simulation of the original measurement (Meier et al. 2012). By comparative mold filling simulations it can be shown how this correction of viscometer data might explain typical predictive errors in rubber heating and machine injection pressure. Actually, the authors are investigating a parameter optimization method involving iterative simulations of the rheological measurement itself with the aim of matching directly the observables pressure and heat build up.

REFERENCES

Aho, J. 2011. *Rheological Characterization of Polymer Melts in Shear and Extension: Measurement Reliability and Data for Practical Processing*. Tampere: Tampere University of Technology.

Barnes, H.A. 1997. Thixotropy – a review, *J. Non-Newtonian Fluid Mech.* 70: 1–33.

Klie, B., Haberstroh, E., Giese, U., Brockhaus, S. & Schöppner, V. 2015. Influence of the flow Channel Coating of the high Pressure capillary Viscometer on the Formation of Wall slip Effects in the Case of Rubber Compounds. *Kautsch. Gummi Kunstst.* 7/8: 46–58.

Meier, J., Klüppel, M., Schuster, R.H. & Giebeler, E. 2003. Einfluß von Konstitution und Molekulargewicht auf die lineare Viskoelastizität von BR- und SBR-Schmelzen. *Kautsch. Gummi Kunstst.* 56: 427–432.

Meier, J., Hüls, F. & Villa-Ramirez, W. 2012. *Sigmasoft Version 5.0 in the Practise - Evaluation in Elastomeric Applications*. Bensheim: SIGMASOFT International User Meeting 27. + 28. September 2012.

Petrie, C.J.S. 2006. Extensional viscosity: A critical discussion. *J. Non-Newtonian Fluid Mech.* 137: 15–23.

Sgodzaj, U. 2002. *Einfluß der Polymer-Struktur auf die Rheologie und das Dispersionsverhalten von BR- und SBR-Systemen*. Clausthal: Technische Universität Clausthal.

Sigmasoft Eng. 2014. *Sigmasoft Version V5.0 – Handbook*, Aachen: Sigmasoft Engineering.

Vlachopoulos, J. & Strutt, D. *The Role of Rheology in Polymer Extrusion*, e-book, Hamilton, Ontario, Canada: Department of Chemical Engineering, McMaster University and Polydynamics Inc.,

Williams, M.L., Landel, R.F. & Ferry, J.D. 1955. The Temperature Dependence of Relaxation Mechanisms in Amorphous Polymers and Other Glass-forming Liquids. *J. Am. Chem. Soc.* 77 (14): 3701–3707.

Constitutive Models for Rubber X – Lion & Johlitz (Eds)
© 2017 Taylor & Francis Group, London, ISBN 978-1-138-03001-5

Statistical investigation of self-organization processes in filled rubber

H. Wulf & J. Ihlemann
Department of Solid Mechanics, Chemnitz University of Technology, Chemnitz, Germany

ABSTRACT: A typical property of filled rubber material is that prestraining leads to a reduction of stresses at strain levels smaller than the maximum strain in the loading history, known as Mullins effect. This induced softening is related to the direction of the prestrain and therefore results in a material anisotropy. The micro-structural mechanisms responsible for this effect are still unclear and subject of discussion. In this work, an explanation based on a self-organization process of weak physicall inks is proposed. The core idea states, that a pattern consisting of low and high linkage density areas arises and evolves with the deformation. The Mullins effect is attributed to an adaption of the pattern to previous deformation states. A simulation model is employed to validate the theory. Hereby, an extreme abstraction of the molecular structure of the material is used to achieve a very simple model structure, which can exhibit inelastic effects only due re-organization of the physical links. The model response is compared with a measurement result and reveals a striking resemblance with typical rubber behavior. Statistical investigations underline that the model is indeed based on a self-organization process and yield insights concerning typical model properties.

1 INTRODUCTION

The molecular structure of rubber is quite well known: The bulk consists of polymer chains, which exists in a highly curled and entangled state. They are connected into a flexible network by curing (classically with sulfur) which established crosslinks between the polymer chains. An industrial rubber contains a multitude of additives, but the most important one is filler, the most common being fine milled elemental carbon called carbon black. The filler has a very active surface and is structured in aggregates and agglomerates.

A variety of different explanations how rubber properties, and specifically the Mullins effect, arise from this molecular structure have been proposed. One of the oldest concepts, originating from Mullins himself (Mullins 1948), is the gradual destruction of a stiffening structure. Mullins assumed that the filler structure is this stiffening structure. A recent model based on this idea is the dynamic flocculation model (DFM) by Klüppel, Meier, & Dämgen (2005). However, it should be noted, that the discovery of unfilled rubber also being capable to exhibit a Mullins effect led to Mullins rejecting his own previous theory (Harwood, Mullins, & Payne 1966) (Mullins 1969). Moreover, various recent measurements suggest that the filler structure is rather responsible for the Payne effect and therefore destroyed by a few percent of deformation already (see e.g. Wollscheid & Lion (2013)).

Another classical theory according to Bueche is that debonding of chains from fillers causes the Mullins effect (Bueche 1960). The memory effect emerges from the fact that under increasing deformation short connections between fillers will debond first. A recent material law based on this concept was developed by Dargazany, Khiêm, & Itskov (2014). Contradicting observations where provided by Hamed & Hatfield, who showed that washing out free chains changes the material properties considerably (Hamed & Hatfield 1989). According to Bueche, such chains should be irrelevant.

After dismissing the concept of the breakdown of the filler structure, Mullins supposed that a change of the network structure under deformation is the crucial process (Mullins 1969). This term describes a slip and disentanglement of the rubber molecules and a relative movement of the crosslinking sites. Recently, strain-induced crystallization (which can be viewed as a special case of change in the network structure) has been subject of intense investigation yielding promising material models (e.g. Khiêm, Dargazany, & Itskov (2013)). A shortcoming of such approaches is always that the relevance of the fillers is neglected or reduced to very simple concepts.

Of course these examples do not aim to provide a complete overview over the current state of the research. For a more in-depth review refer for example to Diani, Fayolle, & Gilormini (2009). Nevertheless, they demonstrate that there is no single, flawless explanation for the Mullins effect. This being said, it should be noted that all said theories have been used to derive micro-mechanical material models, which reproduce rubber behavior and specifically the Mullins effect well. This suggests that each of them has some claim to describe a process which is relevant for the material behavior.

2 THEORY OF SELF-ORGANIZING LINKAGE PATTERNS

In this work, a different micro-mechanical explanation for the Mullins effect is proposed. Instead of concentrating on the micro-mechanical constituents, it considers the bond types occurring between them.

The polymer chains as well as the crosslinks consist of atomic bonds, here also called chemical linkages. These are strong links with a high bonding energy. It is usually assumed, that they do not break under mechanical loading, unless permanent damage is done to the material. Hence the network structure is static concerning its topology. Of course, given the large distance between the entanglement points, it is still very flexible and capable of a variety of internal movements. Furthermore, weak physical interaction (Van-der-Waals forces) exist between polymer-polymer, polymer-filler and filler-filler. Especially the filler provides a large active surface as attachment point for physical links. These links are significantly weaker than the chemical links. Moreover, they can be easily established and disrupted without permanent damage.

The core idea of the theory developed by Ihlemann (2003) states that a pattern in the density of physical links emerges due to self-organization. Therefore, the concept is called Self-Organizing Linkage Patterns (SOLP). The reason is a positive feedback loop: At external loading a non-affine deformation with locally very different movement directions of the chains will occur. This leads to disruption of many physical links and establishment of new ones, where potential bond partners move into vicinity. However, areas with an above average linkage density will rather resist the disruptive process and have better chances to establish new links with free chains coming along. Areas with low linkage density experience above average deformation and rather loose further links. Even starting from small initial fluctuation in linkage density, these processes lead to a pronounced differentiation into a pattern with different linkage density areas. It is self-organized, as it establishes without any external influence directly promoting it. The positive feedback is restricted by the limited extensibility of the polymer chains. As soon as a large number of highly stretched chains exists between two areas of high linkage density, large forces will be transmitted between these areas at further loading. This leads to disruption of at least one of these areas.

As a result, the pattern has a characteristic size scale, which is very typical for self-organized structures (see Haken (1977)). The exact magnitude of this meso-scale is unknown. An indicator may be obtained by considering strain-induced crystallization. By recognizing the crystallites as the arrangement of the chains allowing the maximum linkage density, this can be seen as the only directly observable corner case of the self-organization. The average size of the crystallites was determined by Poompradub, Tosaka, Kohjiya, Ikeda, Toki, Sics, & Hsiao (2005) to be remarkably stable under different conditions at 20 nm. This is most probably a lower bound for the characteristic size scale.

The linkage pattern is permanently reorganized during the deformation process. Hence, the pattern existing at a certain deformation is the result of an evolution over the complete deformation history up to the current state. Therefore, it provides the memory of the material. As previously explained, structures which exhibit little resistance to the current deformation are less likely dissolved. Hence, the pattern will evolve toward structures which exhibit reduced resistance to previously applied strain states, leading to the Mullins effect. As this process naturally incorporates the direction of strain, the induced anisotropy is explained as well.

Also note that this theory does not contradict the previously mentioned concepts, but can be seen as an abstraction and generalization. For example the attachment of polymer chains to filler can be seen as an establishment of physical links and the detachment or the break of a filler cluster can be seen as a locally concentrated destruction of links. However, there is also a central difference: All other mentioned concepts try to derive properties of the macro-scale based on statistical values concerning the micro-scale. The derivations always rely on the assumption that the values of the considered quantity are spatially statistically independent. In presence of a meso-scale pattern, this assumption does not hold. Hence, all statistical investigations must be carried out on the meso-scale (see also Ebeling, Freund, & Schweitzer (1998)).

3 SIMULATION MODEL

As an experimental verification of the theory is difficult, a simulation program is used. Its purpose is to demonstrate that the concept is capable of explaining a large variety of typical rubber properties. As explained in the last section, it must operate on the meso-scale. Very popular tools for such simulations are atomistic or molecular dynamic simulations. However, they are incapable of reaching sufficient magnitudes of simulated volume and simulated time in acceptable computing times. Hence, aggressive simplifications and abstractions are required. The central idea is to model characteristic properties of the different molecular constituents instead of specific polymers, crosslinks and fillers. Even though some model elements resemble some single molecular entities, they should be regarded as meso-scale ensembles of molecular components. Furthermore, this way the specification of an exact size scale can be avoided and instead it can be assumed that the meso-scale model elements contain enough molecular components to arrive at the characteristic size scale. Moreover, all model elements are chosen to be as simple as possible. This allows to attribute any complex properties of the model to self-organization.

The first model element is called strings and represents the high flexibility and low bending stiffness of the polymer chains. These strings consist of a certain number of equal length segments connected at a random angle. The position of the nodes at the connection points are the degrees of freedom. Each segment is a beam element. To model the limited extensibility of the chains, a progressive elastic characteristic is used.

Next, the model is connected into a flexible network structure by inserting long range connections between the strings. These are beam elements again, with a high stiffness in comparison to all other model elements. Although this model element will be called crosslinks, it should be emphasized, that these first two steps have the core purpose of generating a structure with fixed topology, but diverse non-affine local movements under deformation.

Naturally, the most important model element is representing the properties of the physical links. It is dynamically established and dissolved to allow for the self-organization. First, a certain percentage of the nodes is marked as active nodes to model the inhomogeneous distribution of potential bond partners. Only these active nodes may establish physical links. Next, a critical distance is defined. As soon as the distance of two nodes reduces below the critical length, a spring is inserted between these nodes. When a link is stretched above the critical length, it is dissolved. These links have a linear characteristic with an unstretched length of zero. In other words, they always exert a pull force proportional to the distance. To ensure a correct tracking of the self-organization process, the insertion and removal rules are frequently checked during the simulation process and the current time step is repeated, if changes occurred.

It should be noted that the filler is not explicitly modeled. Instead, its property of providing active surfaces is represented by the active nodes. Hence, the segments defined in the model can be understood as some abstract meso-scale building blocks, which will usually contain a mix of polymer and filler.

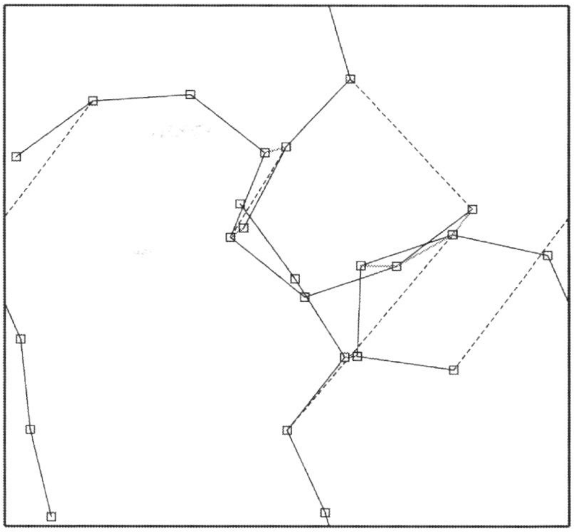

Figure 1. Minimal simulation system consisting of three strings with 8 segments each (black). Crosslinks (black dashed) are added to obtain a mesh structure. Physical bonds (gray) are inserted if the distance between active nodes drops below a critical distance. Cyclic boundary conditions are applied at top/bottom and left/right borders.

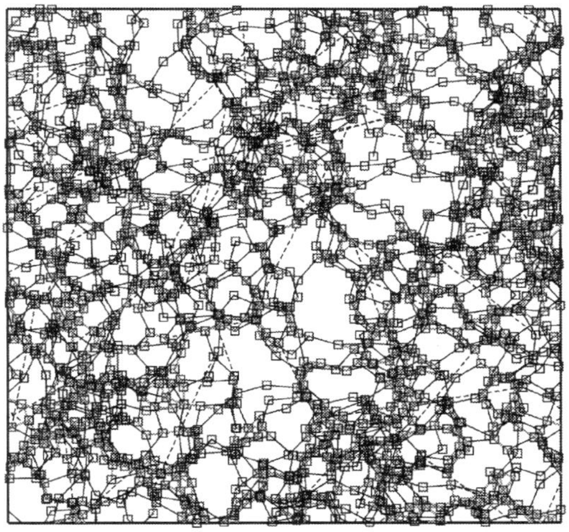

Figure 2. Regular system with 40 strings consisting of 40 segments each. Differentiation into regions of high and low linkage density is already visible and gets more pronounced after loading.

At the border of the model, cyclic boundary conditions are applied. Any model elements may cross the top/bottom or left/right border and re-enter at the other side. This avoids boundary effects and allows a smaller model size.

4 RESULTS

In order to assess the properties of the model, the simulated stress-strain curves are compared to measurement results. An experimental setup, which is capable of revealing the Mullins effect including the induced anisotropy was proposed by Besdo, Ihlemann, Kingston, & Muhr (2003). The prescribed loading schedule is shown in Figure 3. It starts as a one-sided shear with multiple load levels and cycles per load level. Subsequently, the loading is extended towards a two-sided shear in multiple steps until arriving at a completely symmetric shear. Note that the maximum shear intensity is already reached during the one-sided shear. However, as the principal loading directions are different for left- and right-sided shear, during the two-sided shear new directions are subject to the maximum deformation. The measured stress response is depicted in Figure 4. A secondary softening during the phase of two-sided shear is clearly visible. This demonstrates that the softening is indeed sensitive to the loading directions.

Due to the sudden insertion and removal of springs in the model, the stress response is discontinuous. Hence, the average of 200 model systems is computed to obtain smooth results. These systems share the same model parameters, but they are generated with different random numbers. This also helps to reduce the influence of the randomized generation.

The loading schedule used for the simulation is almost the same as shown in Figure 3 except for two changes: The first set of cycles with the smallest amplitude was omitted and the number of repetitions per cycle was reduced to three in order to save some computation time. The results of the simulation are shown in Figure 5. It can be seen that the model successfully qualitatively reproduces various typical rubber properties, especially the Mullins effect with induced anisotropy. This is remarkable, as all the single model elements are purely elastic. They only inelastic model aspect is the rule for inserting and removing physical links. Apparently, this is sufficient to generate such a complex behavior. Considering the relatively simple structure of the model, this result is a strong indicator that the most relevant processes in rubber micro structure have been successfully captured.

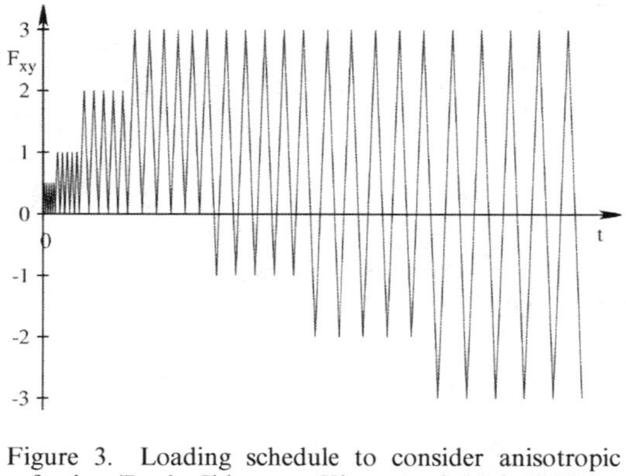

Figure 3. Loading schedule to consider anisotropic softening (Besdo, Ihlemann, Kingston, & Muhr (2003).

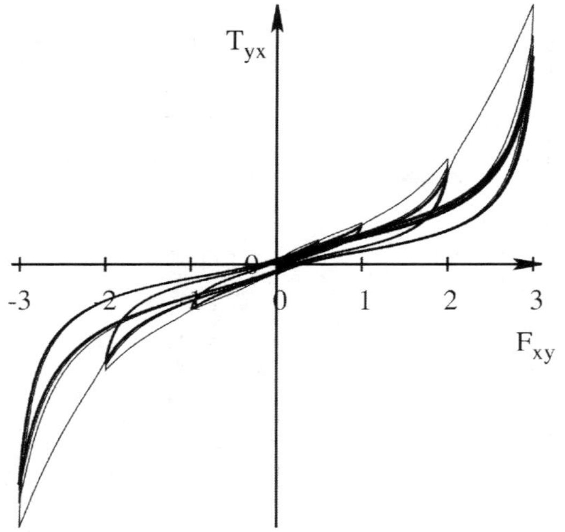

Figure 4. Measured results for the experimental scheme from Figure 3. In the third quadrant, a pronounced secondary softening during two-sided shear can be observed (Besdo, Ihlemann, Kingston, & Muhr (2003).

Naturally, one could suspect that the typical softening is just based on a reduction of the overall number of links. In Figure 6 the overall number of physical links during the first part of the loading schedule is shown. After a short initialization at the start, the fluctuations are small and certainly insufficient to explain the softening. Hence, the typical model properties arise from a rearrangement of the physical links instead of a mere change in number.

By computing spatial correlations, it can be shown that the links organize into areas of different linkage density. Specifically, the correlation between the number of links attached to two nodes as a function of the distance between the nodes was computed. Figure 7 shows the resulting correlation functions. They show that the link numbers of the nodes are highly correlated (up to a value of 0.8)

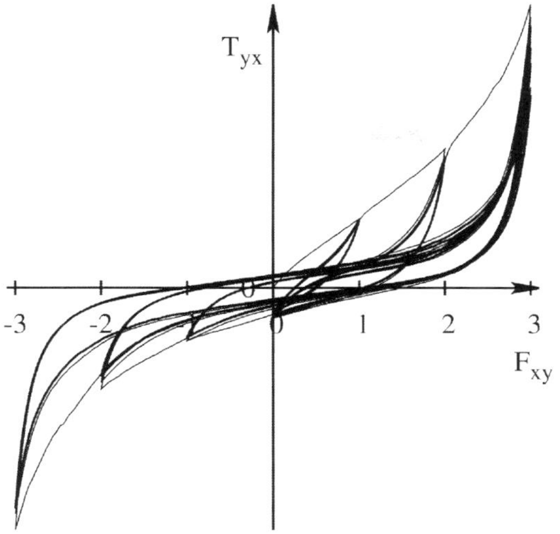

Figure 5. Simulated stress response according to the Muhr experiment (Besdo, Ihlemann, Kingston, & Muhr 2003). The softening in the third quadrant reveals the strain-induced anisotropy.

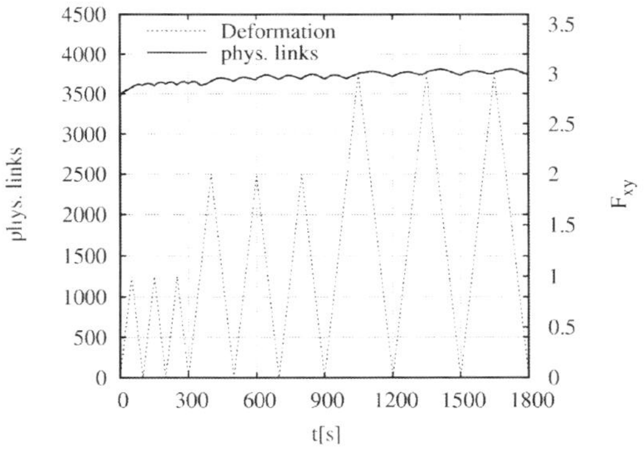

Figure 6. Overall number of links during the one-sided shear part of the simulation. After an initialization phase, little changes occur.

at small distances. This means that highly linked nodes are usually close to other highly linked nodes, forming an area of high linkage density. At a distance above five times the critical length, the correlation vanishes, indicating a limited size of these areas. When expecting a pattern with a single typical length scale, the so-called correlation length, an exponential function for this spatial correlation function, is expected. In fact, the results meet this expectation quite well except for two deviations: First, when exceeding the critical length, the correlation function drops remarkably, which is of course due to the sudden disconnection of two nodes at this point. Second, nodes which are on the same segment have an elevated correlation. This leads to a kink when the distance equals the segment length, which was about 1.4 times the critical length in this model. In Figure 7, the correlation functions for the pattern after completing the first, second and third load level are shown, but there is little difference. This demonstrates that the typical size scale of the pattern is invariant with respect to the loading history.

Apparently, the model obtains its properties just by changing the position or perhaps the internal structure of the high and low linkage density areas. A good metric to assess the change is the number of changed links during a time step. Because the overall number of links stays approximately constant, there is no big difference in considering inserted or removed links. In Figure 8 the number of changed links for each time step during the

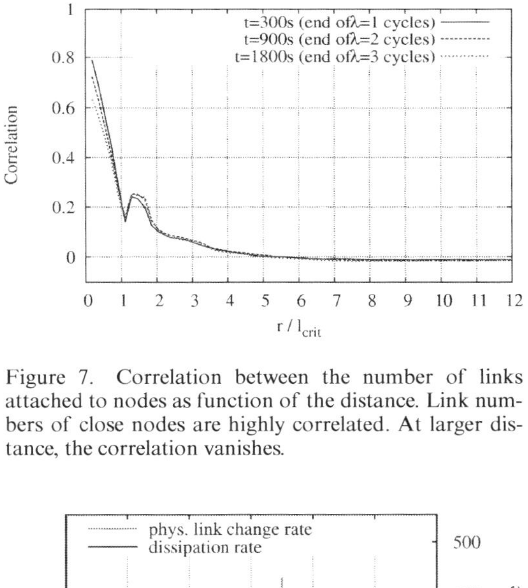

Figure 7. Correlation between the number of links attached to nodes as function of the distance. Link numbers of close nodes are highly correlated. At larger distance, the correlation vanishes.

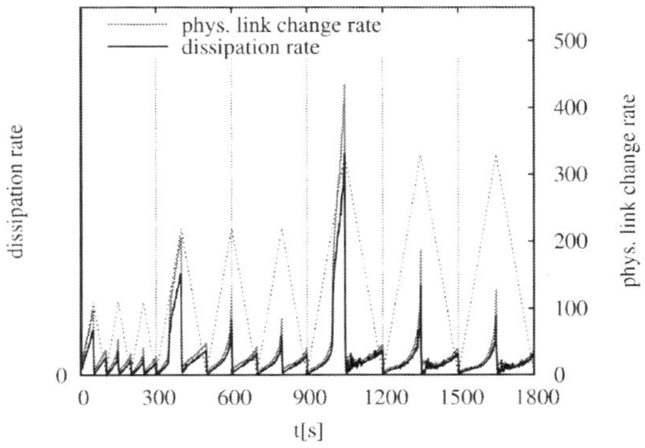

Figure 8. The rate of changed links and the dissipation rate. The two quantities are highly correlated and show clear peaks when approaching new strain maxima.

simulation is depicted. There are remarkable peaks indicating extensive reorganization whenever a new maximum in strain is approached. Furthermore, there are smaller peaks when the maximum strain is repeated. This indicates that softening after load repetition occurs due to a further adaption and optimization of the linkage pattern. The amount of energy dissipated during the loading process can be easily calculated by subtracting energy stored in the model elements from the external work done. In Figure 8 the rate of dissipation is plotted as well. There is a very high correlation between the dissipation rate and the number of changed links. It seems that the dissipation is driven by the self-organization process. This makes sense, as self-organizing systems always need to export entropy, usually by dissipation of heat, to allow the emergence of structures (see e.g. (Nicolis & Prigogine 1977), (Ebeling, Freund, & Schweitzer 1998)).

The rate of changed links is sufficient to replace the complete set of links during each loading cycle. However, a closer investigation reveals that many links are dissolved shortly after being created, but a certain share of links persists over a longer duration. The latter represent the memory of the system. To visualize this memory, a statistic of the age of the links that were present at the end of the last one-sided shear cycle is shown in Figure 9. Specifically, the figure shows all links existing at the end, but the bars indicate the time step of their creation. It can be seen, that a significant share of links (about 15%) was created when approaching the maximum strain the first time and persisted since then. Most probably, they are responsible for the Mullins effect. However, further stable structures are created during the subsequent load repetitions. Apparently, the structure gets more pronounced and fully established by repeated loading. Another

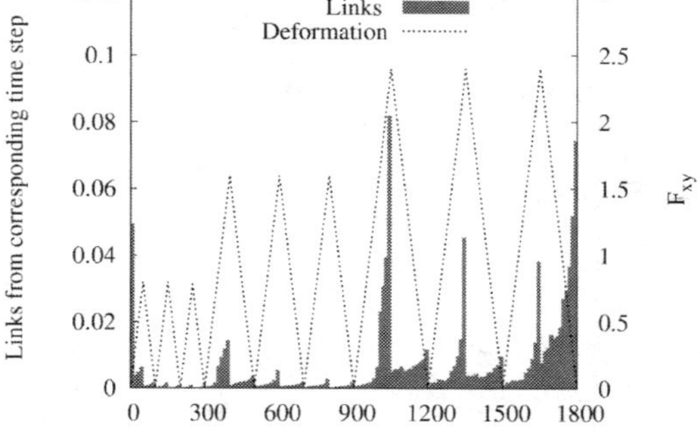

Figure 9. Time step of the creation of the links existing at the end of the depicted loading cycle. A significant share of the links was created near the maximum strain and persisted since then.

good share of links was created close to the end of the simulation. Most of these links are volatile links, that get destroyed quickly by continued loading. These are responsible for the hysteresis. Finally, some memory persists that is related to the initial state and with the previous load maxima. This indicates that the process of the material's internal memory evolution is way more complex than considered by most theories.

5 CONCLUSIONS

A theory explaining the behavior of rubber by a self-organization process on basis of weak physical interactions was presented. The core idea states that a permanently reorganized pattern of different linkage densities is responsible for the typical properties. In order to demonstrate the capabilities of the concept, a simulation program based on the theory was developed. The program uses an abstract model of the molecular structure of rubber. The stress-strain curves generated with the program are qualitatively similar to rubber behavior concerning several aspects, including the Mullins effect and the induced anisotropy. This is especially surprising, as these properties were not enforced by the model design but emerged from its internal dynamics. It can be shown that the model behavior is driven by a self-organization process. Typical properties like the existence of a characteristic pattern size and typical dissipative behavior were observed. The combination of these results provides strong evidence, that self-organization is relevant to real rubber material. Following this assumption, the model can be used to obtain knowledge about how the internal memory of the material evolves during loading.

ACKNOWLEDGEMENT

The authors gratefully acknowledge financial support from Freudenberg Technology Innovation and Vibracoustic.

REFERENCES

Besdo, D., J. Ihlemann, J. Kingston, & A. Muhr (2003). Modelling inelastic stress-strain phenomena and a scheme for efficient experimental characterization. In J. Busfield and A. Muhr (Eds.), *Constitutive models for rubber III*, pp. 309–317. Swets & Zeitlinger, Lisse.

Bueche, F. (1960). Molecular basis for the mullins effect. *J. Appl. Polymer Sci. 4*, 107–114.

Dargazany, R., V.N. Khiêm, & M. Itskov (2014). A generalized network decomposition model for the

quasistatic inelastic behavior of filled elastomers. *International Journal of Plasticity 63*, 94–109.
Diani, J., B. Fayolle, & P. Gilormini (2009). A review on the mullins effect. *European Polymer Journal 45*, 601–612.
Ebeling, W., J. Freund, & F. Schweitzer (1998). Komplexe strukturen. *Entropie und Information, Stuttgart: Teubner*.
Haken, H. (1977). *Synergetics. An Introduction*. Berlin: Springer-Verlag.
Hamed, G. & S. Hatfield (1989). On the role of bound rubber in carbon-black reinforcement. *Rubber chemistry and technology 62*(1), 143–156.
Harwood, J., L. Mullins, & A. Payne (1966). Stress softening in natural rubber vulcanizates. part ii. stress softening effects in pure gum and filler loaded rubbers. *Rubber Chemistry and Technology 39*(4), 814–822.
Ihlemann, J. (2003). *Kontinuumsmechanische Nachbildung hochbelasteter technischer Gummiwerkstoffe*. Düsseldorf: VDI-Verlag.
Khiêm, V., R. Dargazany, & M. Itskov (2013). Micromechanical model of strain-induced crystallization for filled natural rubbers. In *Constitutive Models for Rubber VIII*, Volume 8, pp. 253–258. Balkema.
Klüppel, M., J. Meier, & M. Dämgen (2005). Modelling of stress softening and filler induced hysteresis of elastomer materials. In K. Austrell (Ed.), *Constitutive models for rubber IV*, pp. 171–177. Taylor & Francis, London.
Mullins, L. (1948). Effect of stretching on the properties of rubber. *Rubber Chemistry and Technology 21*(2), 281–300.
Mullins, L. (1969). Softening of rubber by deformation. *Rubber chemistry and technology 42*(1), 339–362.
Nicolis, G. & I. Prigogine (1977). *Self-organization in nonequilibrium systems*, Volume 191977. Wiley, New York.
Poompradub, S., M. Tosaka, S. Kohjiya, Y. Ikeda, S. Toki, I. Sics, & B. Hsiao (2005). Mechanism of straininduced crystallization in filled and unfilled natural rubber vulcanizates. *Journal of applied physics 97*(10), 103529.
Wollscheid, D. & A. Lion (2013). Predeformation and frequency-dependent material behaviour of fillerreinforced rubber: Experiments, constitutive modeling and parameter identification. *International Journal of Solids and Structures 50*(9), 1217–1225.

Tyres and friction

Constitutive Models for Rubber X – Lion & Johlitz (Eds)
© 2017 Taylor & Francis Group, London, ISBN 978-1-138-03001-5

Prediction of energy release rate in opening mode of fracture mechanics for filled and unfilled elastomers

M. El Yaagoubi, J. Meier, T. Alshuth & U. Giese
German Institute for Rubber Technology, Hannover, Germany

D. Juhre
Otto-von-Guericke University Magdeburg, Magdeburg, Germany

ABSTRACT: In this work, the stretch intensity factor for filled and unfilled elastomers for different mixtures is introduced. This new stretch intensity factor allows predicting the analytically evaluated energy release rate for a cracked sample under uniaxial tension. Thus, the opening mode from the fracture mechanics (Modus *I*) was investigated. Appropriate continuum mechanical derivations are based on a non-linear hyperelastic material behavior. The energy release rate is evaluated through a closed path integral very near to the crack tip whereby the integrand includes asymptotic solution of strain, stress and energy density using the Ogden model. The decisive advantage of this method is to predict well the critical tearing energy values by the crack growth using the analytical energy release rate term. In this work the Mullins effect is not considered, because the cracked samples are tested without any preconditioning.

1 INTRODUCTION

Elastomers are very fundamental materials in human daily life due to theirs specific properties. Elastomer components are used in different application fields like automotive, sealing, industry or medicine. In elastomer components, micro-cracks can accrue through production reasons or friction. Such defects affect the strength of the component and lead mostly to early failures. These micro-cracks grow up under certain conditions and lead to total failure of the components. If the critical value of the specific surface energy is not exceeded, the micro-cracks are not able to grow. For the determination of the resulting critical energy flux values at the crack tip, two fracture mechanics criteria are applied: the energy release rate G [1] and the tearing energy T [2]. The energy release rate is evaluated analytically through a path integral around the crack, in which stresses, strains and energy density are locally determined. The tearing energy is determined from experimental outputs of the whole sample. By the derivation of the energy release rate, a non-dissipative hyperelastic material behaviour is assumed. The variations of the analytical energy release rate are based on the material model Ogden of order 2 [3]. The evaluated path integral of the energy release rate provides an expression, which depends on the material parameters of the Ogden model and an unknown parameter. Aim of this work is the determination of this parameter for the mode I. From similar studies on mode *III* (anti-plane shear state), the unknown constants for filled and unfilled compounds were determined [5, 6]. The decisive advantage of the analytically determined energy release rate is to predict the critical energy flux values with high precision. The energetic consideration of a cracked sample is divided in 2 main regions (see Figure 1). The dark grey region is well known as the dissipative zone and the light grey

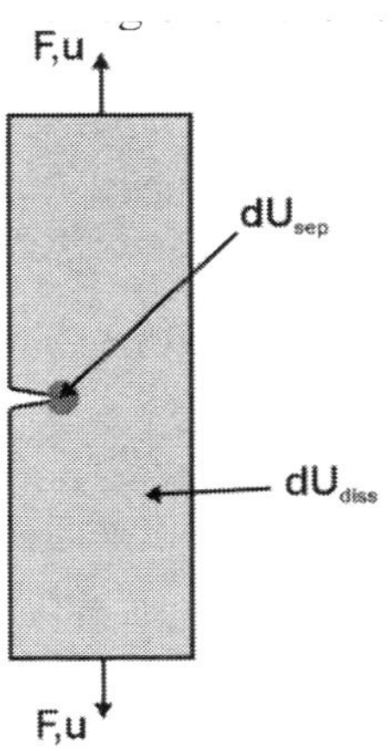

Figure 1. Illustration of dissipation and separation energies [4].

is the rest of the sample. In the dissipative zone prevail inelastic effects, which appear in form of stress-softening or energy dissipation. Such effects remain also in the rest of the sample but they are very weak. The resulting energy by crack propagation is only a part of the whole dissipated energy, since the inelastic effects in the two regions have to be considered.

The energy balance during an infinitesimal crack growth for a material with dissipative behaviour can be expressed as follows [6]:

$$\frac{dU_{Ext}}{dA}-\frac{dU_{Int}}{dA}=\frac{dU_{Sep}}{dA}+\frac{dU_{Diss}}{dA} \tag{1}$$

where U_{Ext} is the external work and U_{Int} is the internal recoverable elastic strain energy. U_{Sep} is the well-known specific surface potential, which is the necessary work for creating a new crack surface. U_{Diss} is the dissipated energy in the whole sample. During the crack propagation, the change of U_{Ext} is equal to zero since no displacement variation is applied [6]. We assume a non-linear hyperelastic material behaviour, so equation (1) is abbreviated to:

$$-\frac{dU_{Int}}{dA}=\frac{dU_{Sep}}{dA} \tag{2}$$

Equation (2) reflects the well-known Griffith fracture criterion from the linear fracture mechanics for pure elastic materials [1]. We assume that the two proposed fracture criteria give the same critical energy flux value.

$$T=G=-\frac{dU_{Int}}{dA}=-\frac{dU_{Int}}{tda} \tag{3}$$

where t is the thickness of the specimens and a is the crack length.

2 THEORY

2.1 *Energy release rate*

Stumpf and Le formulated the variational principle of total energy in a cracked hyperelastic body with the consideration of finite deformation [7]. From the variational principle formulation they derive the statical equations and the boundary conditions near the crack tip. The obtained expression gives the resulting energy flux at the crack tip and corresponds to the J-integral introduced by J. Rice.

Le and Stumpf determine the singular field for a cracked sample under mode *I* of fracture mechanics with consideration of plane deformation near the crack [8]. By the derivations, a hyperelastic behaviour and finite strain deformation are assumed. This calculation is done for the symmetrically opening crack mode, in which the movement of the crack flanks is orthogonal on the crack front (Figure 2). For the determination of singular field the Ogden model is used. The analytical solutions of strain, energy density and stress are described in detail in [8]. The energy release rate is determined using a path integral near the crack tip as the known J-integral. The energy release rate expression G_I for a straight crack extension is defined as follows [8]:

$$G_I=\mu\frac{\pi}{2}\left(\frac{2}{\alpha^3}\right)^{\frac{\alpha}{2}-1}(\alpha-1)^{\alpha-1}B_I^{\alpha} \tag{4}$$

where μ and α are the dominated part of material parameters of Ogden model and B_I is a new defined parameter. G_I depends on the unknown parameter B_I and the stiffness of the material (μ, α). Similar observation was found in the linear fracture mechanics, in which the energy release rate depends on the elasticity modulus and Poisson's ratio [9].

Similar approach for the mode *III* (anti-plane state) has been performed in [4, 5]. The continuum mechanical derivations of the analytical energy release rate are described in detail in [4]. In these both studies, filled and unfilled elastomers using the trouser sample were investigated. The prediction of the energy release rate with using the new defined stretch intensity factor gives very good results.

2.2 *Tearing energy*

The tearing energy T is a widely used fracture mechanics criterion for elastomers. The first considerations were done for pure elastic and brittle

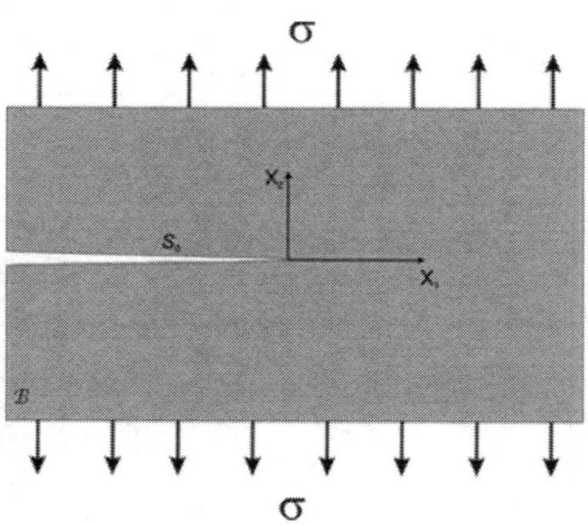

Figure 2. Mode *I* of fracture Mechanics (Opening Mode).

materials by Griffith [1]. This concept was transferred for nonlinear material behaviour by Thomas and Rivlin [2]. The tearing energy is defined as follows:

$$T = -\frac{\partial U_{Int}}{\partial A} = -\frac{\partial U_{Int}}{t\partial a} \tag{5}$$

Thomas and Rivlin derived a tearing energy expression for different load cases using the equation (5). A single edge notched tension sample (SENT-sample) fulfils the load conditions of the mode I of the fracture mechanics (opening mode). Here, the tearing energy for SENT-sample is defined as follows [10]:

$$T = \frac{2\pi}{\sqrt{\lambda}} W_{Tot} a \tag{6}$$

where λ is the stretch of the whole sample, W_{Tot} is the total energy density of the sample and a is the actual crack length. The feature of tearing energy expression for mode I in equation (6) is the linear dependency on the crack length.

3 EXPERIMENTS

3.1 *Material characterization*

Two compounds are prepared. The used polymers are natural rubber (NR) and ethylene propylene diene monomer (EPDM). The compounds are filled with 50 phr of carbon black N347. For the analytical prediction of energy release rate the Ogden parameters for both materials are needed. Therefore uniaxial tensile strength test are performed at room temperature with a cross-head speed of 200 mm/min until break.

3.2 *Fracture mechanics*

The single edge notched tension sample is notched centrally at one side of the sample. To examine the crack length influence on the evaluation of the energy release rate, different crack lengths are cutted into the samples. The samples have a width of 20 mm and a height of 110 mm. Only 75 mm represents the effective loaded length. For each sample, the actual thickness has been considered, which is about 2 mm. The measurements were performed with a tensile tester (Zwick/Roell machine) at room temperature (22°C). The preload is 1 N. The samples are uniaxial stretched with a cross-head speed of 100 mm/min.

4 RESULTS

Material model fitting is done on tensile strength measurement results for both materials compounds using Ogden model. This procedure was choosen, because the experimental investigations were done without any preconditioning. In addition high values of stress and strain are expected at the crack tip. The fitting until large strain values permits to approximate the singularities of both stress and strain at the crack tip. In Figure 3, the measured uniaxial tensile strength curves and the fitting curves are displayed.

The measured tensile strength curves are approached well with the fitting curves of the Ogden model. Although the SENT-sample is under pure uniaxial tension, there is also another deformation like necking due to the compression and shearing near the crack tip. The stability of the fitting in other modes like pure shear and biaxial tension is checked. The fit parameters for both compounds are listed in Table 1.

The tearing energy for mode I is evaluated according to equation (6) (see Table 2). The equation shows that the tearing energy depends on the crack geometry, energy density and the stretch of the cracked sample. The tearing energy is evaluated at the point, in which the crack starts to grow so that the critical tearing energy is determined. The time between the starting point of crack growth and the total

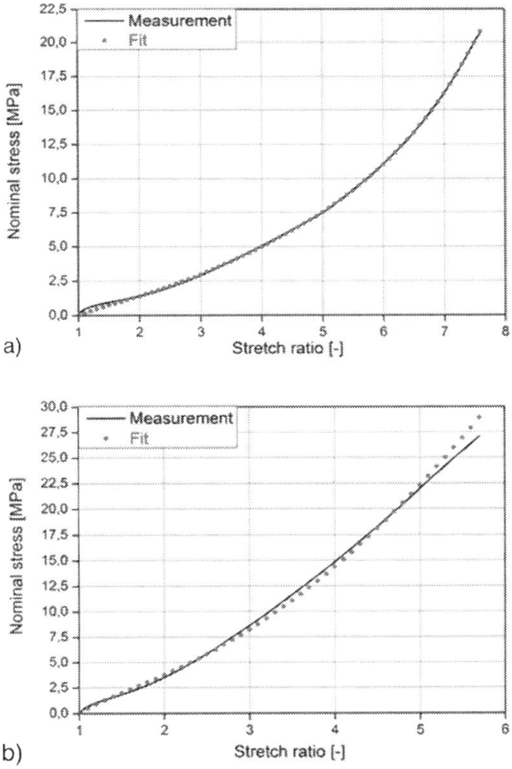

Figure 3. Uniaxial tensile strength curves (symbol) and fit (lines) with Ogden for: a) EPDM and b) NR.

Table 1. Identified fitting parameters of Ogden model.

	μ_1 (MPa)	α_1 (–)	μ_2 (MPa)	α_2 (–)
NR	1.2552	3.0326	0.5039	0.0000362
EPDM	0.6159	2.7190	0.000019	7.9225

Table 2. Calculated tearing energy from the measurements.

	W_{Tot} (N/mm^2)	λ (–)	a (mm)	T (N/mm)
NR-1	0.93	1.68	7.0	31.82
NR-2	0.71	1.57	7.5	26.66
NR-3	0.68	1.57	8.0	27.19
NR-4	0.63	1.54	8.0	25.71
NR-5	0.61	1.62	8.5	25.79
NR-6	0.43	1.43	9.00	20.30
NR-7	0.36	1.39	10.0	19.12
NR-8	0.26	1.30	11.0	15.80
NR-9	0.18	1.26	13.0	13.28
NR-10	0.14	1.22	15.0	12.66
EPDM-1	0.46	1.72	6.00	13.44
EPDM-2	0.39	1.62	6.50	12.73
EPDM-3	0.38	1.62	6.75	12.82
EPDM-4	0.43	1.70	7.0	14.47
EPDM-5	0.37	1.64	7.0	12.98
EPDM-6	0.33	1.58	7.5	12.55
EPDM-7	0.28	1.53	8.0	11.73
EPDM-8	0.18	1.46	10.0	9.45
EPDM-9	0.14	1.41	11.0	8.68
EPDM-10	0.13	1.41	13.0	8.99

failure point for EPDM is shorter compared to NR. A reasonable explanation of this phenomenon is that the strain induced crystallization at the crack tip slows down crack propagation.

The analytically energy release rate in equation (4) is solvable definitely determinable only if the parameter B_I is known. Here a new constant is designated as stretch intensity factor because it depends only on the stretch of the sample. Similar results and remarks are done for the Mode *III* of fracture mechanics for filled [4] and unfilled elastomers [5]. This newly introduced parameter takes for the used mixtures for the mode *I* the following expression:

$$B_I = 1.5\lambda \tag{7}$$

The experimental tearing energy values are compared with the analytical energy release rate (Table 3).

The table show clearly that the prediction of the energy release rate using the equations (4) and (7) provide very good approximation to the experimentally determined tearing energy. The average relative error of ten samples for each mixture is lower than 10%.

Table 3. Comparison between T and G for NR and EPDM mixture.

	T (N/mm)	G (N/mm)	Relative error $\left\|\frac{G-T}{T}\right\|$ (%)
NR-1	31.82	35.30	10.92
NR-2	26.66	29.24	9.67
NR-3	27.19	29.21	7.416
NR-4	25.71	27.52	7.02
NR-5	25.79	31.56	22.36
NR-6	20.30	21.71	6.95
NR-7	19.12	20.11	5.20
NR-8	15.80	16.35	3.48
NR-9	13.28	14.74	10.98
NR-10	12.66	13.56	7.13
EPDM-1	13.44	14.10	4.88
EPDM-2	12.73	12.13	4.68
EPDM-3	12.82	11.97	6.60
EPDM-4	14.47	13.78	4.79
EPDM-5	12.98	12.44	4.16
EPDM-6	12.55	11.29	9.99
EPDM-7	11.73	10.40	11.32
EPDM-8	9.45	9.05	4.22
EPDM-9	8.68	8.28	4.57
EPDM-10	8.99	8.29	7.71

5 CONCLUSIONS

Prediction of tearing energy for elastomers using the mode *I* of fracture mechanics is the principal objective of this investigation. This prediction must be performed with a little effort. The evaluated energy release rate using a path integral around the crack yields to an analytical equation, which depends on the material parameters and the unknown parameter B_I. The determined parameter depends on the stretch of the whole sample. The same procedure has been done for mode *III*. Similar results were achieved for filled and unfilled mixtures under Mode *III* [4] and [5].

The fracture mechanics characterisation of elastomers is done by using SENT-sample. This sample is stretched uniaxially, so that the deformation states fulfil the conditions of mode *I* of fracture mechanics. The critical tearing energy was evaluated at the point, at which the crack starts to grow.

The comparison between the two used fracture criteria for the two filled mixtures shows that the energy release rate equation predicts the measured tearing energy with high precision. The main benefit of the analytical evaluated energy release rate is to predict the resulting critical energy flux without performing any fracture mechanics measurements. If the preconditioning is also considered, the parameters of Ogden model have to be originated from multi-hysteresis fitting.

REFERENCES

[1] A.A. Griffith, The phenomena of rupture and flow in solids. Phil. Trans. Roy. Soc. London, Series A, Vol. 221 (1920), 163–198.

[2] R.S. Rivlin, A.G. Thomas, Rapture of Rubber. I. Characteristic Energy for Tearing, J. Polym. Sci. 10 (1953), 291–318.

[3] ABAQUS/Standard User's and Theory Manuals 5.8, Hibbit, Karlsson & Soreson Inc. USA (1998).

[4] M. El Yaagoubi, D. Juhre, J. Meier, T. Alshuth, U. Giese, Prediction of Tearing Energy in Mode III for Filled Elastomers, Theoretical and Applied Fracture Mechanics 88 (2017), 31–38. http://dx.doi.org/10.1016/j.tafmec.2016.11.006.

[5] M. El Yaagoubi, D. Juhre, J. Meier, K.C. Le, T. Alshuth, U. Giese, Prediction of Energy Release Rate in Anti-Plane Shear State (Mode III) for Unfilled Elastomers, Kautschuk Gummi Kunstoffe (submitted).

[6] T. Horst, G. Heinrich, Linking Mesoscopic and Macroscopic Aspects of Crack Propagation in Elastomers, Fracture Mechanics and Statistical Mechanics of Reinforced Elastomeric Blends, Volume 70 (2013), 127–165.

[7] H. Stumpf, K.C. Le, Variational principles of nonlinear fracture mechanics, Acta Mechanica, Volume 83 (1990), 25–37.

[8] K.C. Le, H. Stumpf, The singular elastostatic field due to a crack in rubberlike materials, Journal of Elasticity, Volume 32 (1993), 183–222.

[9] D. Groß, T. Seelig, Bruchmechanik mit Einführung in die Mikromechanik, Springer (2011), 100.

[10] W. Grellmann, K. Reincke, Technical Material Diagnostics-Fracture Mechanics of Filled Elastomer Blend, Fracture Mechanics and Statistical Mechanics of Reinforced Elastomeric Blends, Volume 70 (2013), 227–268.

Constitutive Models for Rubber X – Lion & Johlitz (Eds)

Steady state and sequentially coupled thermo-mechanical simulation of rolling tires

T. Berger & M. Kaliske
Institute for Structural Analysis, Technische Universität Dresden, Dresden, Germany

ABSTRACT: For a comprehensive simulation of the behaviour of a rolling tire, the coupling effects between mechanical and thermal field should be considered. Two different ways to model the coupling between these fields will be presented. First, a simulation is chosen, where the steady state of the mechanical and thermal part are computed directly. Therefore, the mechanical part is modeled by a consistent implementation of a finite linear viscoelastic material model, monolithicly coupled with the thermal linear heat equation. The main disadvantage is, that the Jacobian matrix is more complex and only the steady state can be computed. To overcome these difficulties, a sequentially coupled algorithm is presented. The rolling of the tire is computed separately and isothermally by a steady state formulation in 3D. However, to model the evolution of the temperature inside the tire, the thermal equilibrium is computed in the cross-section over time in 2D. In this approach, only in the thermal computation a time step is performed and the updated temperature distribution is then provided for a new mechanical simulation.

In this contribution, the two procedures will be presented and compared to each other for a rolling wheel. For comparison of the two approaches, the stress distribution and temperature evolution are studied and compared at steady state conditions.

1 INTRODUCTION

The simulation of rolling rubber tires is a challenging and complex task. There are various loading types active in such a model and also inertia forces will appear. The tire will come into contact with a rigid surface, for example the road or a test drum. Thus, the material points will undergo cyclic deformations, that will cause the rubber material to dissipate energy due to the time dependent properties. This viscoelastic behaviour can be modelled using a linear approach with a finite number of Maxwell branches. Recently, a nonlinear model proposed by Bergström and Boyce (Dal & Kaliske 2009) was implemented into the steady state rolling simulation framework. In (Berger, Behnke, & Kaliske 2016) both models were compared to each other and showed for the small strain regime the same results, but the different relaxation kinematics could be seen at large strains. For simplicity, only a linear viscoelastic model will be used.

The energy dissipated due to the rubber material will cause the tire to increase its temperature. Consequently, the heating will change the properties of the temperature dependent rubber material. Therefore, it is mandatory to take these coupling effects into account. In this contribution, a numerical efficient way is presented to obtain the steady state solution for a rolling tire under free rolling conditions. The steady state rolling implementation is extended with a thermal computation, so the balance of momentum and the linear heat equation can be solved simultaneously. For comparison, a sequentially coupled algorithm (Behnke & Kaliske 2015) is shown. Both simulation approaches will be compared to each other for a steady state rolling Grosch wheel.

2 STEADY STATE ROLLING

A rolling tire at a constant velocity will reach a steady state after a certain time. In that case, the tire will show no changes for an observer travelling with the same velocity. Modelling this in a transient way would be a computationally expensive task. Therefore, the steady state rolling formulation of an Arbitrary Langrangian Eulerian framework is chosen (Nackenhorst 2004). The mapping $\mathbf{x} = \hat{\varphi}(\mathbf{X},t)$ describes the different states of a particle at time t. The ALE formulation is based on the assumption of a decomposition in a purely rigid body motion $\mathbf{X}^{*} = \chi(\mathbf{X},t)$ and a deformation $\mathbf{x} = \varphi(\mathbf{X}^{*},t)$, where

$$\hat{\phi} = \phi \circ \chi \tag{1}$$

holds. The rigid body motion χ is described in an Eulerian and the deformation ϕ in a Lagrangian manner. For the special case of steady state rolling of the deformed body, the deformation mapping is time independent and it holds

$$\varphi(\mathbf{X}^*,t) = \varphi(\mathbf{X}^*,t+\Delta t). \tag{2}$$

However, single material points will still undergo a change of their position and deformation state, but the image of the deformation state of the whole tire will not be changed. To track the deformation history of every particle in the cross-section, streamlines are introduced. These are built by every integration point of the two dimensional cross-section, revolved around the rotational axis. These streamlines are formed in the reference configuration, thus, the predecessor and the successor will stay unaltered, but can undergo deformations.

For the implementation of the viscoelastic material model, a consistent linearization of the equilibrium equation is chosen. This is taken from the work of (Garcia, Kaliske, Wang, & Bhashyam 2016). With this formulation, the history of the whole streamline is taken into account. Therefore, an additional loop over the streamline has to be done. This will lead to an higher amount of nonzero elements in the general stiffness matrix but also to quadratic convergence in the isothermal state.

This leads for the second Piola-Kirchhoff stress tensor in the steady state rolling conditions to

$$\mathbf{S}_s = \mathbf{S}^{\infty}_{\mathrm{vol},s} + \gamma_{\infty}\mathbf{S}^{\circ}_{\mathrm{iso},s} + \sum_{k=1}^{m} \gamma_k \mathbf{H}_{k,s}, \tag{3}$$

with the equation for the viscoelastic stresses for one streamline after an infinite number of iterations

$$\mathbf{H}_{k,s} = \alpha_k \sum_{i=1}^{N^{\otimes}} h_{si,k}\xi_i,k \left[\mathbf{T}^{si^T}\mathbf{S}^{\circ}_{\mathrm{iso},i}\mathbf{T}^{si} - \mathbf{T}^{s,i-1^T}\mathbf{S}^{\circ}_{\mathrm{iso},i-1}\mathbf{T}^{s,i-1}\right]. \tag{4}$$

It is worth mentioning, that for the stresses at the first integration point, the stress at the end of the streamline $\mathbf{S}^{\circ}_{\mathrm{iso},0} = \mathbf{S}^{\circ}_{\mathrm{iso},N^{\otimes}}$ is needed. The rotation tensor

$$\mathbf{T}^{si} = \begin{bmatrix} \cos\Delta\varphi_{is} & 0 & \sin\Delta\varphi_{is} \\ 0 & 1 & 0 \\ -\sin\Delta\varphi_{is} & 0 & \cos\Delta\varphi_{is} \end{bmatrix} \tag{5}$$

is used to project the deformations to the current point φ_s. The factor α_k relates the sum to infinity and $h_{is,k}$ respects the position of every point φ_i on the streamline. While ξ_k represents the viscoelastic material properties. For more information on the factors and on the consistent linearization, the reader is refered to (Garcia, Kaliske, Wang, & Bhashyam 2016).

3 THERMO-MECHANICAL COUPLING

There are many different ways to model the coupling of the mechanical and thermal field of a rolling rubber wheel. In this section, the well studied sequentially coupled algorithm proposed by Behnke (Behnke & Kaliske 2015) is shown. The second approach is an instantaneously coupled algorithm, where the mechanical and thermal fields are solved simultaneously.

3.1 *Sequentially coupled*

One approach to couple the mechanical and the thermal fields is to use a staggered algorithm. Now, the coupled problem is split into two subproblems, which are computed separately. This can be done due to the different time scales of the two fields. A tire needs more time to heat up than required for one full revolution. Therefore, it can be assumed that during one rotation the material particle will have a constant temperature. For a numerically efficient procedure, the thermal equations can now be solved only in the cross-section of the tire. Instead of one system of equations, where both fields are solved instantaneously, there are now two smaller system of equations. One for the mechanical part

$$\mathbf{K}_{uu}\Delta\mathbf{u}^e = \Delta\mathbf{f}^e, \tag{6}$$

and one for the thermal part,

$$\mathbf{K}_{\theta\theta}\Delta\theta^e = \Delta\mathbf{q}^e. \tag{7}$$

The dissipated energy stems only from the viscoelastic material properties and the temperature increase is only due to this dissipated energy. The mechanical field is solved isothermally, but to take into account temperature effects of the material, the parameters of the free energy function are changing with varying temperature. Because of the isothermal computation, no derivative with respect to the temperature has to be computed. Therefore, the material parameters can be given at a specific temperature and interpolated in between. Thus, only a few measurements and fitting are needed to describe a proper temperature dependent material.

It is assumed that the volumetric part is purely elastic and the viscoelastic properties only affect the isochoric configuration. Therefore, a split of the deformation into an isochoric $\overline{\mathbf{F}}$ and a volumetric part $\mathbf{F}_{\mathrm{vol}}$ has to be done,

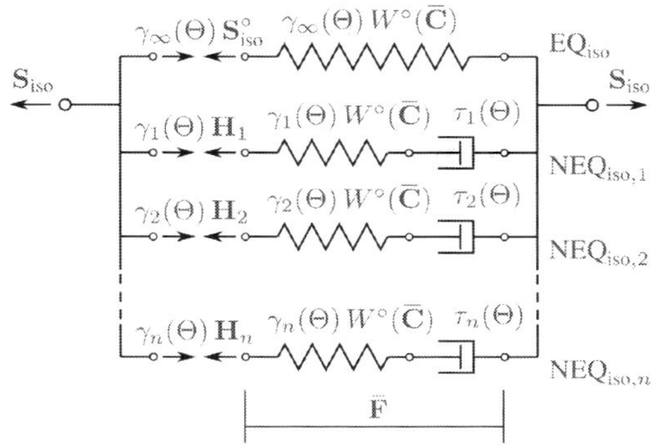

Figure 1. Material rheology with stress type internal variables.

$$\mathbf{F} = \bar{\mathbf{F}}\mathbf{F}_{\mathrm{vol}},\ \mathbf{F}_{\mathrm{vol}} = J^{\frac{1}{3}}\mathbf{1},\ \ \bar{\mathbf{F}} = J^{-\frac{1}{3}}\mathbf{F}. \tag{8}$$

To model a certain frequency range, a generalized Maxwell element is chosen for the isochoric part. The rheology model Fig. 1 with a finite number of Maxwell branches is used to represent the viscoelastic properties. The free energy function is divided into an equilibrium

$$\Psi_{\mathrm{EQ}}(\mathbf{C},\theta) = \tfrac{\kappa}{2}(J-1) + \gamma^{\infty} W^\circ(\mathbf{C},\theta) \tag{9}$$

and a non-equilibrium part with stress type internal variables

$$\Psi_{\mathrm{NEQ}}(\mathbf{C},\mathbf{H},\theta) = \sum_{k=1}^{m} \gamma_k(\theta)\mathbf{H}_k : \tfrac{1}{2}\mathbf{C} + \Xi\left(\sum_{k=1}^{m} \gamma_k(\theta)\mathbf{H}_k\right). \tag{10}$$

The equilibrium part represents the purely elastic part of the Prony series and the non-equilibrium part the m Maxwell elements. The equilibrium condition leads to the following linear rate equation

$$\gamma_k(\theta)\dot{\mathbf{H}}_k + \frac{1}{\tau_k(\theta)}\gamma_k(\theta)\mathbf{H}_k = \gamma_k(\theta)\dot{\mathbf{S}}^{\circ}_{\mathrm{iso}}. \tag{11}$$

The solution of Eq. (12) can be found using Eq. (5). For both parts, the stress response is modelled with the same type of free energy function, namely the Yeoh model (Yeoh 1990). This is a stable formulation with temperature dependent coefficients $C_1(\theta), C_2(\theta)$ and $C_3(\theta)$ and which is only depending on the first Invariant $I_1 = \mathrm{tr}\,\mathbf{C} = \mathrm{tr}\,\mathbf{b}$,

$$W^\circ(\mathbf{C},\theta) = C_1(\theta)[I_1 - 3] + C_2(\theta)[I_1 - 3]^2 + C_3(\theta)[I_1 - 3]^3. \tag{12}$$

During the isothermal mechanical analysis, the energy dissipated at each material point is computed within the steady state formulation with the fourth order viscosity tensor $\underline{\underline{\eta}}_k$ of each dashpot k,

$$D_k = \gamma_k H_k : \frac{1}{2}\dot{\bar{C}}_{i,k} = \gamma_k H_k : \underline{\underline{\eta}}_k^{-1} : \gamma_k H_k. \tag{13}$$

The values D_k of every streamline are then averaged for one cycle with the duration T

$$\bar{w}_{\mathrm{diss}} = \frac{1}{T}\int_t^{t+T} \sum_{k=1}^{m} D_k\,dt. \tag{14}$$

This value will then be the input variable for the thermal analysis in the two dimensional cross-section model. In the thermal module, the linear heat equation will be solved in the reference configuration,

$$-\nabla_{\mathbf{X}^*} \cdot \mathbf{Q} + \bar{w}_{\mathrm{diss}} - c_v\dot{\theta} = 0. \tag{15}$$

The thermal boundary conditions are applied with surface heat outflux vector,

$$\bar{\mathbf{Q}}_N = h(\theta - \theta_a)\mathbf{N}, \tag{16}$$

with the outward unit vector $\mathbf{N}$ and a constant thermal heat conduction coefficient h.

Note, that Eq. (17) is treated in a transient manner, different to the steady state formulation in the mechanical module, where the specific heat and the time derivative of the temperature are taken into account. Thus, one steady state mechanical simulation represents a certain time step, that is elapsed in the thermal simulation. This allows to show the evolution of the temperature and dissipation over time, which also gives information on how fast the tire will heat up. With that equations, the temperature distribution in the cross-section can be obtained, and will subsequently be transferred to the mechanical computation. The new temperature will lead to different material properties and, consequently, a new dissipated energy rate is computed. An overview on the simulation algorithm is given in Fig. 2. The staggered solution algorithm will be executed until a steady state is reached or a defined total time is elapsed. For more information on the sequentially coupled algorithm and the implementation, the reader is referred to (Behnke & Kaliske 2015).

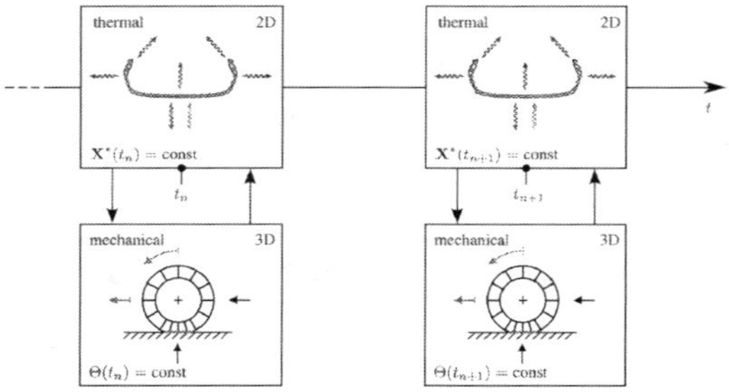

Figure 2. Simulation overview on the sequentially coupled algorithm.

3.2 *Instantaneously coupling*

In this contribution, a new approach for the coupling is shown. Differently to the first approach, both fields will be solved instantaneously. Therefore, the tangent matrix will be of higher order, but only one step has to be executed, instead of several iterations. In contrast to Eq. (7) and Eq. (8), the fully coupled system of equations is introduced,

$$\begin{bmatrix} \mathbf{K}_{uu} & \mathbf{K}_{u\Theta} \\ \mathbf{K}_{\Theta u} & \mathbf{K}_{\Theta\Theta} \end{bmatrix} \begin{bmatrix} \Delta \mathbf{u}^e \\ \Delta \theta^e \end{bmatrix} = \begin{bmatrix} \Delta \mathbf{f}^e \\ \Delta \mathbf{q}^e \end{bmatrix}. \tag{17}$$

Notice, that additional coupling terms $\mathbf{K}_{u\Theta}$ and $\mathbf{K}_{\Theta u}$ are active in the system, which leads to more non-zero entries in the tangent matrix. The temperature dependency of the material is now modelled with the help of temperature scaling functions f_{EQ} and f_{NEQ}, instead of different sets of parameters at each temperature. The free energy function is now

$$\Psi_{\text{EQ}}(\mathbf{C},\theta) = f_{\text{EQ}}(\theta)\left(\tfrac{\kappa}{2}(J-1) + \gamma_\infty W^\circ(\mathbf{C})\right) \tag{18}$$

for the equilibrium and

$$\Psi_{\text{NEQ}}(\mathbf{C},\theta) = f_{NEQ}(\theta) \left[\sum_{k=1}^{m} \gamma_k \mathbf{H}_k : \frac{1}{2}\mathbf{C} + \Xi\left(\sum_{k=1}^{m} \gamma_k \mathbf{H}_k\right)\right] \tag{19}$$

for the non-equilibrium response. The used temperature scaling function is adapted from (Reese & Gonvindjee 1998),

$$f_{EQ}(\theta) = \frac{\theta}{\theta_0} - \frac{\theta[\tanh(b(\theta-\theta_0))]}{\theta_0 + a}. \tag{20}$$

This function represents best the transformation from glassy to rubbery state, and also entropy elasticity can be modelled in a sufficient way. The parameter θ_0 can be interpreted in this regard as the glass transition temperature. The parameters $a = 70$ and $b = 0.2$ are used to change the shape of the scaling function and are taken from (Behnke, Kaliske, & Klüppel 2016). Under these considerations, the second Piola-Kirchhoff stress at steady state rolling is computed by

$$\mathbf{S}_s = f_{\text{EQ}}(\theta)\left(\mathbf{S}^\infty_{\text{vol},s} + \gamma_\infty \mathbf{S}_{\text{iso},s}\right) + f_{\text{NEQ}}(\theta)\left(\sum_{k=1}^{m} \gamma_k \mathbf{H}_{k,s}\right). \tag{21}$$

After assembling the mechanical contribution to the stiffness matrix K_{uu} and residual $\mathbf{f}^e$, the next step is to build the thermal parts. Therefore, the computed dissipation will be averaged over every streamline and will be the input value $\bar{w}_{\text{diss}}$, to ensure that its heating is equal in the circumferential direction. Different to the previous approach, the linear heat equation, will now be solved in a steady state manner, because the rolling motion and the heating are not uncoupled any more and only the steady state will be obtained. Therefore, all time-dependent values will vanish $\dot{\theta} = 0$ and the linear heat equation, Eq. (17), will simplify to

$$-\nabla^*_{\mathbf{X}} \cdot \mathbf{Q} + f_{\text{NEQ}}(\theta)\bar{w}_{\text{diss}} = 0. \tag{22}$$

Notice, that due to the constant temperature in circumferential direction the same value of the temperature scaling function f_{NEQ} can be used in one streamline.

The derivation of the additional coupling terms is now presented. Because of the temperature scaling function f_{EQ}, the volumetric and isochoric material tangent can be found,

$$\underline{\underline{\mathbb{C}}}_{\text{vol}} = f_{\text{EQ}}(\theta)\, 2\tfrac{\partial \mathbf{S}_{\text{vol}}}{\partial \mathbf{C}}, \quad \underline{\underline{\mathbb{C}}}_{\text{iso}} = f_{\text{EQ}}(\theta)\, 2\tfrac{\partial \mathbf{S}_{iso}}{\partial \mathbf{C}}. \tag{23}$$

For the derivation of the isothermal material tangents, the reader is referred to (Miehe 1995). The material coupling term $\mathbf{K}_{u\theta}$ is derived by Eq. (23) with respect to the temperature,

$$\frac{\partial \mathbf{S}_s}{\partial \theta} = \frac{\partial f_{\text{EQ}}(\theta)}{\partial \theta}\left(\mathbf{S}^\infty_{\text{vol},s} + \gamma_\infty \mathbf{S}_{\text{iso}}\right). \tag{24}$$

with Eq. (20), the derivative is obtained

$$\frac{\partial f_{\text{EQ}}(\theta)}{\partial \theta} = \frac{1}{\theta} - \frac{\tanh(b(\theta-\theta_0))^3}{\theta_0 + a} - \frac{3b\theta}{\theta_0 + a}\frac{\tanh(b(\theta-\theta_0))^2}{\cosh(b(\theta-\theta_0))^2}. \tag{25}$$

In the current state, only the equilibrium part of the material can be modelled temperature dependent. This leads to the assumption $f_{\mathrm{NEQ}} = 1$. Therefore, the derivative and the additional coupling terms from the viscoelastic material vanish

$$\frac{\partial f_{\mathrm{NEQ}}(\theta)}{\partial \theta} = 0. \tag{26}$$

For a consistent derivation and a better convergence of the solution, these parts should be derived in later studies.

4 EXAMPLE

For comparison of the two approaches, a rolling Grosch wheel is chosen. The rotational velocity is set to $\Omega = 80 s^{-1}$ and frictional contact is neglected. For modelling the contact with the road, the displacement of some of the nodes are prescribed.

The same material parameters are used and shown in Tables 1 and 2. The temperature dependency is modelled with the scaling factor f_{EQ} from Eq. (22). For the sequentially coupled simulation, 8 temperature sets are created, where the instantaneous hyperelastic factor γ_∞ is multiplied with the temperature function. The viscoelastic parameters are assumed to not change with the temperature $f_{\mathrm{NEQ}} = 0$, for later studies, this has to be implemented to model the different relaxation properties with varying temperatures. The proposed material parameters and temperature scaling function are fictitious and an academic example. To obtain the viscoelastic scaling factors and relaxation times from experiments, different techniques, for example shown in (Behnke & Kaliske 2015), are available. However, the identification of the temperature scaling function and its parameters is a complex and challenging task, which has be to studied further.

Table 1. Hyperelastic material parameter.

Parameter		
C_1	N/mm^2	22.64
C_2	N/mm^2	−8.50
C_3	N/mm^2	15.50
κ	N/mm^2	103.51

Table 2. Viscoelastic material parameter.

Parameter		
γ_∞	–	1.0
γ	–	0.2
τ_1	s^{-1}	0.1
γ_2	–	0.2
τ_2	s^{-1}	0.01

4.1 *Results*

The flow of the material through the fixed mesh with the prescribed deformation will cause every material point to undergo a cyclic excitation. Due to the viscoelastic properties, energy will be dissipated. In Fig. 3, the dissipated energy rate for both approaches are shown. They are plotted in the cross-section and are averaged using Eq. (16). It can be seen that the distribution and the values are nearly the same for the sequentially coupled and instantaneously coupled model. As it is the main reason for the heating of the material, both simulation techniques will compute nearly the same temperature distribution for the steady state rolling case, see Fig. 5. In the center of the Grosch

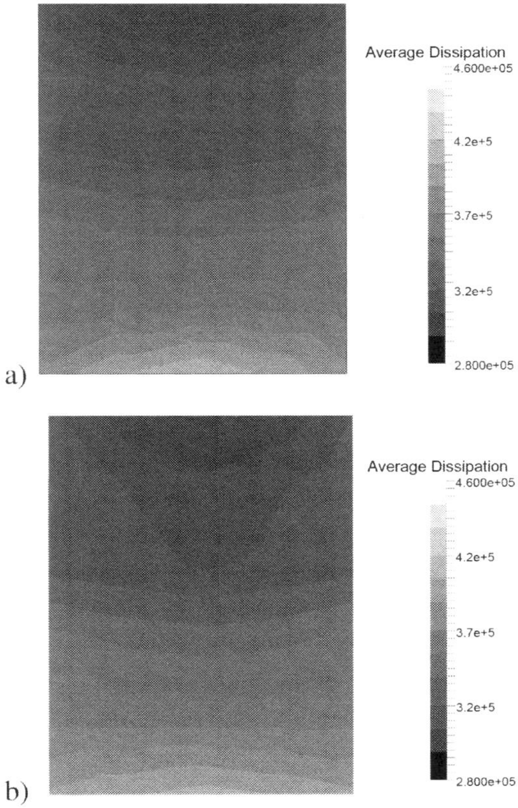

Figure 3. Average energy dissipated in the cross-section for a) sequentially coupled and b) instantaneously coupled model.

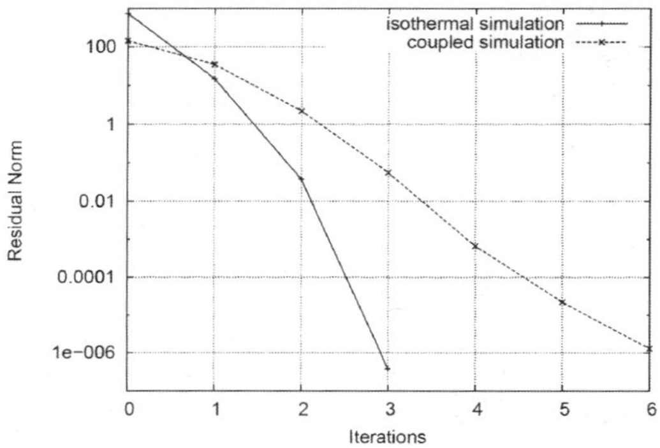

Figure 4. Residual norm of the isothermal and coupled steady state rolling simulation.

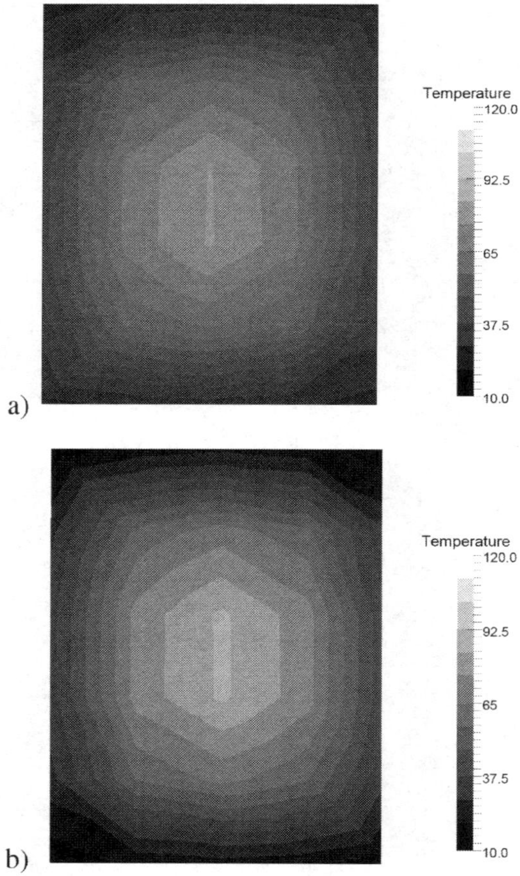

Figure 5. Temperature distribution in the wheel cross-section for a) sequentially coupled and b) instantaneously coupled model.

Table 3. Thermal material parameter.

Parameter		
k	$Wm^{-1}K^{-1}$	0.13
h_{side}	$Wm^{-2}K^{-1}$	50.0
h_{tread}	$Wm^{-2}K^{-1}$	100
h_{rim}	$Wm^{-2}K^{-1}$	100

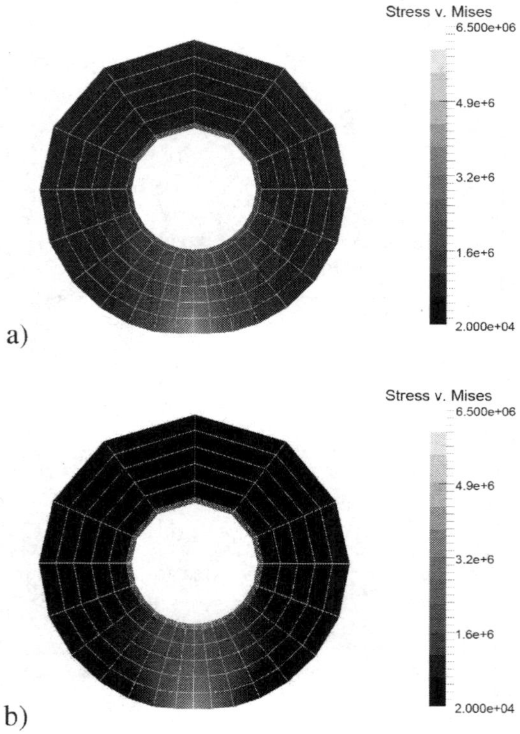

Figure 6. Von Mises stress distribution for a) sequentially coupled and b) instantaneously coupled steady state rolling Grosch wheel.

wheel, the material will have the highest temperature, because of the heat outflow to the environment close to the surface. The stress distribution in the steady state case shows also similar results, see Fig. 6.

4.2 *Convergence*

In the isothermal case, the convergence is perfectly quadratic, due to the consistent derivation of the stress with respect to all elements in circumferential direction. In the instantaneously coupled algorithm, the change of the dissipated energy is causing different power terms. In the beginning, the convergence ratio is quadratic, but then becomes suddenly linear. There are still some linearization terms missing, which should be studied further. The residual norms for the isothermal and thermal coupled approach are plotted in Fig. 4. However, with the achieved convergence ratio, the steady state of the rolling tire can be found much quicker than with the sequentially coupled method, because the system of equations has to be solved only once. If the evolution of the temperature distribution inside the tire over time is of interest, then the sequentially coupled algorithm has to be chosen.

5 CONCLUSION

Two different approaches for the thermo-mechanical coupling of rolling tires are presented and compared. A numerically stable approach is achieved with a sequentially coupled algorithm, which can handle parameter sets for different temperatures. It shows a quadratic convergence behaviour and a more complex material model can be implemented. Due to the transient computation of the thermal field, an evolution of the temperature inside the cross-section can be simulated. The main disadvantage is that a large number of iteration steps has to be performed until a steady state is reached. The second approach is a monolithic coupling of mechanical and thermal field instantaneously under steady state rolling conditions. By omitting the evolution process, the steady state solution can be found with a smaller number of iterations. Although the system of equations is more complex, the overall solution can be found quicker, because avoiding the transient computation. Additional research should be done in the consistent derivation of the nonequilibrium parts to take into account a temperature dependency of the viscoelastic properties. This will result in a fully coupled instantaneous formulation with a quadratic convergence rate.

REFERENCES

Behnke, R. & M. Kaliske (2015). Thermo-mechanically coupled investigation of steady state rolling tires by numerical simulation and experiment. *International Journal of Non-Linear Mechanics 68*, 101–131.

Behnke, R., M. Kaliske, & M. Klüppel (2016). Thermo-mechanical analysis of cyclically loaded particle-reinforced elastomer components: experiment and finite element simulation. *Rubber Chemistry and Technology 89*, 154–176.

Berger, T., R. Behnke, & M. Kaliske (2016). Viscoelastic linear and nonlinear analysis of steady state rolling rubber wheels: a comparison. *Rubber Chemistry and Technology 89*, 499–525.

Dal, H. & M. Kaliske (2009). Bergstrm-boyce model for nonlinear finite rubber viscoelasticity: Theoretical aspects and algorithmic treatment for the fe method. *Computational Mechanics 44*, 809–823.

Garcia, M., M. Kaliske, J. Wang, & G. Bhashyam (2016). A consistent implementation of inelastic material models into steady state rolling. *Tire Science and Technology 44*, 174–190.

Miehe, C. (1995). Entropic thermoelasticity at finite strains. aspects of the formulation and numerical implementation. *Computer Methods in Applied Mechanics and Engineering 120*, 243–269.

Nackenhorst, U. (2004). The ale-formulation of bodies in rolling contact: theoretical foundations and finite element approach. *Computer Methods in Applied Mechanics and Engineering 193*, 4299–4322.

Reese, S. & S. Gonvindjee (1998). Theoretical and numerical aspects in the thermo-viscoelastic material behaviour of rubber-like polymers. *Mechanics of Time-Dependent Materials 1*, 357–396.

Yeoh, O. (1990). Characterization of elastic properties of carbon-black-filled rubber vulcanizates. *Rubber Chemistry and Technology 63*, 792–805.

© 2017 Taylor & Francis Group, London, ISBN 978-1-138-03001-5

Author index